# About Pearson

*Pearson is the world's learning company, with presence across 70 countries worldwide. Our unique insights and world-class expertise comes from a long history of working closely with renowned teachers, authors and thought leaders, as a result of which, we have emerged as the preferred choice for millions of teachers and learners across the world.*

*We believe learning opens up opportunities, creates fulfilling careers and hence better lives. We hence collaborate with the best of minds to deliver you class-leading products, spread across the Higher Education and K12 spectrum.*

*Superior learning experience and improved outcomes are at the heart of everything we do. This product is the result of one such effort.*

*Your feedback plays a critical role in the evolution of our products and you can contact us –* reachus@pearson.com. *We look forward to it.*

# Problems and Solutions in Physical Chemistry

## for JEE (Main and Advanced)

# Problems and Solutions in Physical Chemistry

## for JEE (Main and Advanced)

*Second Edition*

Neeraj Kumar

Pearson

*Development Editor*: Sonal Jain
*Production Editor*: Vipin Kumar



ISBN 978-93-534-3945-3

First Impression, 2020
Eighteenth Impression, 2025
**Nineteenth Impression, 2026**

Published by Pearson India Education Services Pvt. Ltd, CIN: U72200TN2005PTC057128.

Head Office: 1st Floor, Berger Tower, Plot No. C-001A/2, Sector 16B, Noida - 201 301, Uttar Pradesh, India.
Registered Office: 6th Floor, Tower A, Unit A and B International Tech Park, CapitaLand Chennai, 200 Feet Radial Road, Zamin Pallavaram, Old Pallavaram
Chennai – 600117, Tamil Nadu, India
Website: in.pearson.com, Email: companysecretary.india@pearson.com

*Compositor*: SRS Global, Puducherry.
*Printed in India at Saurabh Printers Pvt. Ltd.*

# Contents

*Preface* *ix*
*Acknowledgement* *x*
*About the Author* *x*

**Chapter 1** Mole Concept 1.1–1.64

**Chapter 2** Equivalent Concept 2.1–2.48

**Chapter 3** Gaseous State 3.1–3.62

**Chapter 4** Thermodynamics 4.1–4.52

**Chapter 5** Thermochemistry 5.1–5.48

**Chapter 6** Chemical Equilibrium 6.1–6.52

**Chapter 7** Ionic Equilibrium 7.1–7.56

**Chapter 8** Electrochemistry 8.1–8.54

**Chapter 9** Solid State 9.1–9.40

**Chapter 10** Liquid Solution 10.1–10.40

**Chapter 11** Chemical Kinetics 11.1–11.62

**Chapter 12** Surface Chemistry 12.1–12.28

**Chapter 13** Atomic Structure 13.1–13.56

**Chapter 14** Nuclear Chemistry 14.1–14.24

# Preface

*Problems and Solutions in Physical Chemistry for JEE (Main and Advanced), 2nd edition* has been conceived to meet the specific requirements of the students preparing for engineering entrance examination. The best way to ensure that students understand the concepts of physical chemistry is to solve as many problems on each concept. It aims to impart the wide range of questions with exhaustive solutions from different concepts to practice while preparing for examination. Students should attempt range of questions, rather than spending too much time with the same problems again and again. Students should also ensure to read each problem carefully, since a small variation in the problem can make huge difference in its solution.

This enhanced second edition written in lucid and structured manner to help students prepare and practice the ample number of problems in minimum possible time and develop effective techniques to attempt the examination. Questions are arranged into two sections—JEE Main and JEE Advanced followed by topic-wise practice exercises. These include – single and multiple correct choice, comprehension-based, assertion-reason, matrix match and integer answer type. Exhaustive solutions to these practice questions both (Main and Advanced) incorporated to achieve target with good score.

Practicing these questions, aspirants will come to know about the pattern and toughness of the examination in both Main and Advanced. The soul of the title is to make the aspirants competent enough to crack the uncertainty of success in the entrance examination.

In the last, I wish to emphasize that though I made every sincere effort to make this book error-free, but I do not claim to be infallible. I hope the students will be greatly benefited with this edition. Suggestions are welcome.

# Acknowledgement

I am grateful to Pearson Education for keeping faith in me. I would like to thank Jitendra Kumar, from the editorial division, for regularly motivating me. Without his kind support, bringing out this book would not have been possible.

I would like to thank all of my friends—Akhilesh Kanthed, Amit Kumar Gupta, Ankit Sharma, Deepak Arora, Dilip Sharma, Faheem Naik, Gautam Bothra, Girish Kumar Gupta, Jitendra Hirwani, Krishnavtar Gaur, Manoj Agarwal, Mithilesh Thakur, P. C. Reddy, Puneet Hakoo, Rahul Kaushik, Rahul Mishra, Rajesh Kumar, Samvit Bissa, Vijay Jain, Vijay Kumar Tripathy, Vijit Jain—for their constructive criticism and suggestions that added value to this book.

I appreciate Vikas Bansal, Amit Jain, Rohit Suman for their support in various developmental stages of this book.

My special thanks goes to my wife Sweta, and daughter Shrija, for their encouragement, patience, sacrifice and constant support.

A very special thanks to my son Mihir Suman (Mihu) who delighted our life for a very small time (only three months) but I have written the solutions of 3–4 chapters during hospitalisation. It's great memory for me.

I am also thankful to all of my students, who made me, who am I now.

As an author, I have tried to touch the goodness of my father, Suresh Prasad. Without his blessings and motivation, it would not have been possible.

I heartily welcome suggestions from the readers for further improvement of this book. Please feel free to contact me at neerajnca@gmail.com

# About the Author

**Neeraj Kumar** is a civil engineer from Bihar College of Engineering, Patna (now NIT, Patna). He has been teaching IIT-JEE aspirants since 1992. He established Neeru's Chemistry Arena at Musallahpur Hat, Patna. He worked for Career Point, Kota (2002–04), Sri Chaitanya IIT Academy, Kota (2004–07), AMBITION (own institute at Kota) (2007–09) and Bansal Classes, Kota. Presently, he is working as Senior Lecturer in the department of Physical Chemistry (IIT Division), Allen Career Institute, Kota.

The author strongly believes that cracking any competitive examination demands theoretical knowledge. Even the current trends of IIT-JEE support the theoretical and analytical approach.

Neeraj Kumar has written a good number of books for IIT-JEE as well as for schools. He has co-authored several books and contributed papers in reputed educational magazines such as *Chemistry Patron and Junior Science Refresher.*

CHAPTER 1

# Mole Concept

## EXERCISE I (JEE MAIN)

### Laws of Chemical Combinations

1. A quantity of 10 g of a hydrocarbon exactly requires 40 g oxygen for complete combustion. The products formed are $CO_2$ and water. When $CO_2$ gas formed is absorbed completely in lime water, the mass of solution increases by 27.5 g. What is the mass of water formed simultaneously in the combustion?

   (a) 22.5 g (b) 27.5 g
   (c) 50 g (d) 10 g

2. Zinc ore (zinc sulphide) is treated with sulphuric acid, leaving a solution with some undissolved bits of material and releasing hydrogen sulphide gas. If 10.8 g of zinc ore is treated with 50.0 ml of sulphuric acid (density 1.2 g/ml), 65.2 g of solution and undissolved material remains. In addition, hydrogen sulphide (density 1.4 g/L) is evolved. What is the volume (in litres) of this gas?

   (a) 4.0 (b) 5.6
   (c) 7.84 (d) 4.4

3. When a mixture of aluminium powder and iron (III) oxide is ignited, it produces molten iron and aluminium oxide. In an experiment, 5.4 g of aluminium was mixed with 18.5 g of iron (III) oxide. At the end of the reaction, the mixture contained 11.2 g of iron, 10.2 g of aluminium oxide, and an undetermined amount of unreacted iron (III) oxide. No aluminium was left. What is the mass of the iron (III) oxide left?

   (a) 2.5 g (b) 7.3 g
   (c) 8.3 g (d) 2.9 g

4. Some bottles of colourless liquids were being labelled when the technicians accidentally mixed them up and lost track of their contents. A 15.0 ml sample withdrawn from one bottle weighed 22.3 g. The technicians knew that the liquid was either acetone, benzene, chloroform or carbon tetrachloride (which have densities of 0.792 g/$cm^3$, 0.899 g/$cm^3$, 1.489 g/$cm^3$, and 1.595 g/$cm^3$, respectively). What was the identity of the liquid?

   (a) Carbon tetrachloride
   (b) Acetone
   (c) Chloroform
   (d) Benzene

5. A sample of an ethanol–water solution has a volume of 55.0 $cm^3$ and a mass of 50.0 g. What is the percentage of ethanol (by mass) in the solution? Assume that there is no change in volume when the pure compounds are mixed. The density of ethanol is 0.80 g/$cm^3$ and that of water is 1.00 g/$cm^3$.

   (a) 20% (b) 40%
   (c) 60% (d) 45.45%

6. In a textile mill, a double-effect evaporator system concentrates weak liquor containing 4% (by mass) caustic soda to produce a lye containing 25% solids (by mass). What is the weight of water evaporated per 100 g feed in the evaporator?

   (a) 125.0 g (b) 50.0 g
   (c) 84.0 g (d) 16.0 g

7. At 373 K and 1 atm, if the density of liquid water is 1.0 g/ml and that of water vapour is 0.0006 g/ml, then the volume occupied by water molecules in 1 litre of steam at that temperature is

(a) 6 ml (b) 60 ml
(c) 0.6 ml (d) 0.06 ml

8. A person needs an average of 2.0 mg of riboflavin (vitamin $B_2$) per day. How many grams of butter should be taken by the person per day if it is the only source of riboflavin? Butter contains 5.5 μg riboflavin per g.

(a) 363.6 g (b) 2.75 mg
(c) 11 g (d) 19.8 g

9. Law of multiple proportions is not applicable for the oxide(s) of

(a) carbon (b) iron
(c) nitrogen (d) aluminium

10. Two elements A and B combine to form compound X and Y. For the fixed mass of A, masses of B combined for the compounds X and Y are in 3:7 ratio. If in compound X, 4 g of A combines with 12 g B, then in compound Y, 8 g of A will combine with ...... g of B.

(a) 24 (b) 56
(c) 28 (d) 8

## Atomic Mass

11. The mass of $3.2 \times 10^5$ atoms of an element is $8.0 \times 10^{-18}$ g. The atomic mass of the element is about ($N_A = 6 \times 10^{23}$)

(a) $2.5 \times 10^{-22}$ (b) 15
(c) $8.0 \times 10^{-18}$ (d) 30

12. A graph is plotted for an element by putting its mass on *X*-axis and the corresponding number of atoms on *Y*-axis. What is the atomic mass of the element for which the graph is plotted? ($N_A = 6.0 \times 10^{23}$)

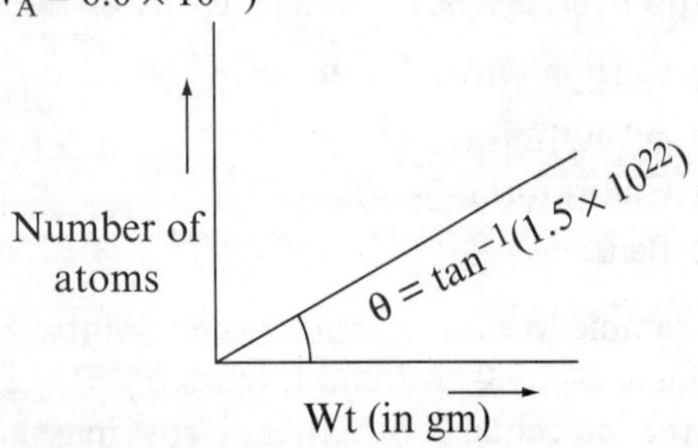


(a) 80 (b) 40
(c) 0.025 (d) 20

13. If '**NEERAJ KUMAR**' is written by a graphite pencil, it weighs $3.0 \times 10^{-10}$ g. How many carbon atoms are present in it? ($N_A = 6 \times 10^{23}$)

(a) $1.5 \times 10^{13}$ (b) $5 \times 10^{12}$
(c) $2 \times 10^{33}$ (d) $1.5 \times 10^{10}$

14. The atomic masses of two elements P and Q are 20 and 40, respectively. If '*a*' g of P contains '*b*' atoms, then how many atoms are present in '2*a*' g of Q?

(a) *a* (b) *b*
(c) 2*a* (d) 2*b*

15. The molecular formula of a compound is $X_4O_9$. If the compound contains 40% X by mass, then what is the atomic mass of X?

(a) 24 (b) 12
(c) 26 (d) 13

16. A quantity of 1 g of metallic carbonate $XCO_3$ is completely converted into a chloride $XCl_2$ weighing 1.11 g. The atomic mass of the element 'X' is

(a) 10 (b) 20
(c) 30 (d) 40

17. An element X has three isotopes $X^{20}$, $X^{21}$ and $X^{22}$. The percentage abundance of $X^{20}$ is 90% and average atomic mass of the element is 20.18. The percentage abundance of $X^{21}$ should be

(a) 2% (b) 8%
(c) 10% (d) 0%

18. A sample of hydrogen gas is collected and it is observed that it contains only hydrogen and deuterium atoms in the atomic ratio 6000 : 1. The number of neutrons in 3.0 g of such a sample should be nearly

(a) 0.0005 (b) $3.01 \times 10^{20}$
(c) $1.80 \times 10^{24}$ (d) 1.0

19. If isotopic distribution of $C^{12}$ and $C^{14}$ is 98.0% and 2.0%, respectively, then the number of $C^{14}$ atoms in 12 g of carbon is

(a) $1.032 \times 10^{22}$ (b) $1.20 \times 10^{22}$
(c) $5.88 \times 10^{23}$ (d) $6.02 \times 10^{23}$

20. The fractional abundance of $Cl^{35}$ in a sample of chlorine containing only $Cl^{35}$ (atomic weight = 34.9) and $Cl^{37}$ (atomic weight = 36.9) isotopes, is 0.6. The average mass number of chlorine is

(a) 35.7 (b) 35.8
(c) 18.8 (d) 35.77

## Molecular Mass

21. Twenty molecules of $SO_3$ will weigh as much as ..... molecules of oxygen.

(a) 100 (b) 50
(c) 15 (d) 8

22. The mass of $CO_2$ that must be mixed with 20 g of oxygen such that 27 ml of a sample of the resulting mixture would contain equal number of molecules of each gas, is

(a) 13.75 g (b) 27.50 g
(c) 41.25 g (d) 55 g

23. A mixture of $2 \times 10^{21}$ molecules of P and $3 \times 10^{21}$ molecules of Q weighs 0.60 g. If the molecular mass of P is 45, then the molecular mass of Q will be ($N_A = 6 \times 10^{23}$)

(a) 45 (b) 180
(c) 90 (d) 270

24. The shape of tobacco mosaic virus (TMV) is cylindrical, having length 3000 Å and diameter 170 Å. If the specific volume of virus is 12.5 ml/g, then the molecular mass of TMV is ($N_A = 6 \times 10^{23}$)

(a) 3.28 (b) $5.44 \times 10^{-24}$
(c) $5.44 \times 10^{-18}$ (d) $3.27 \times 10^{6}$

25. The density of a DNA sample is 1.1g/ml and its molar mass determined by cryoscopic method was found to be $6 \times 10^8$ g/mole. What is the volume occupied by one DNA molecule? ($N_A = 6 \times 10^{23}$)

(a) $5.45 \times 10^8$ ml (b) $1.83 \times 10^{-9}$ ml
(c) $9.09 \times 10^{-16}$ ml (d) $1.09 \times 10^{-13}$ ml

26. How many atoms do mercury vapour molecules consist of if the density of mercury vapour relative to air is 6.92? The average mass of air is 29 g per mole (Hg = 200).

(a) 1 (b) 2
(c) 4 (d) Infinite

27. Vapour density of a volatile substance is 1.2 ($C_2H_6$ = 1). Its molecular mass would be

(a) 1.2 (b) 2.4
(c) 36 (d) 72

28. A compound contains 7 carbon atoms, 2 oxygen atoms and $9.96 \times 10^{-24}$ g of other elements. The molecular mass of compound is ($N_A = 6 \times 10^{23}$)

(a) 122 (b) 116
(c) 148 (d) 154

29. If the mass of neutron is doubled and that of proton is halved, the molecular mass of $H_2O$ containing only $H^1$ and $O^{16}$ atoms will

(a) increase by about 25%.
(b) decrease by about 25%.
(c) increase by about 16.67%.
(d) decrease by about 16.67%.

30. Out of 1.0 g dioxygen, 1.0 g atomic oxygen and 1.0 g ozone, the maximum number of oxygen atoms are contained in

(a) 1.0 g of atomic oxygen.
(b) 1.0 g of ozone.
(c) 1.0 g of oxygen gas.
(d) All contain the same number of atoms.

31. Total number of electrons present in 4.4 g oxalate ion ($C_2O_4^{2-}$) is

(a) $0.05N_A$ (b) $2.3N_A$
(c) $2.2N_A$ (d) $2.1N_A$

32. Total number of valence electrons present in 6.4 g peroxides ion ($O_2^{2-}$) is

(a) $0.2N_A$ (b) $3.2N_A$
(c) $3.6N_A$ (d) $2.8N_A$

33. The number of $F^-$ ions in 4.2 g $AlF_3$ is (Al = 27, F = 19)

(a) 0.05 (b) $9.03 \times 10^{22}$
(c) $3.01 \times 10^{22}$ (d) 0.15

34. A quantity of 13.5 g of aluminium when changes to $Al^{3+}$ ion in solution will lose (Al = 27)

(a) $18.0 \times 10^{23}$ electrons
(b) $6.02 \times 10^{23}$ electrons
(c) $3.01 \times 10^{23}$ electrons
(d) $9.03 \times 10^{23}$ electrons

35. If an iodized salt contains 1% of KI and a person takes 2 g of the salt every day, the iodine ions going into his body everyday would be approximately (K = 39, I = 127)

(a) $7.2 \times 10^{21}$ (b) $7.2 \times 10^{19}$
(c) $3.6 \times 10^{21}$ (d) $9.5 \times 10^{19}$

## Calculation of Mole

36. Dopamine is a neurotransmitter, a molecule that serves to transmit message in the brain. The chemical formula of dopamine is $C_8H_{11}O_2N$. How many moles are there in 1 g of dopamine?

(a) 0.00654 (b) 153
(c) 0.0654 (d) None of these

37. Ethanol is a substance, which is commonly called alcohol. The density of liquid alcohol is 0.8 g/ml at 293 K. If 1.2 moles of ethanol is needed for a particular experiment, then what volume of ethanol should be measured out?

(a) 55.2 ml (b) 57.5 ml
(c) 69 ml (d) 47.9 ml

38. The volume of one mole of water at 277 K is 18 ml. One ml of water contains 20 drops. The number of molecules in one drop of water will be ($N_A = 6 \times 10^{23}$)

(a) $1.07 \times 10^{21}$ (b) $1.67 \times 10^{21}$
(c) $2.67 \times 10^{21}$ (d) $1.67 \times 10^{20}$

39. A given mixture consists only of pure substance X and pure substance Y. The total mass of the mixture is 3.72 g. The total number of moles is 0.06. If the mass of one mole of Y is 48 g and there is 0.02 mole of X in the mixture, then what is the mass of one mole of X?

(a) 90 g (b) 75 g
(c) 45 g (d) 180 g

40. Number of gas molecules present in 1 ml of gas at 0°C and 1 atm is called Loschmidt number. Its value is about

(a) $2.7 \times 10^{19}$ (b) $6 \times 10^{23}$
(c) $2.7 \times 10^{22}$ (d) $1.3 \times 10^{28}$

41. A quantity of 0.25 g of a substance when vaporized displaced 50 $cm^3$ of air at 0°C and 1 atm. The gram molecular mass of the substance will be

(a) 50 g (b) 100 g
(c) 112 g (d) 127.5 g

42. An amount of 6 moles of Cl atoms at STP occupies a volume of

(a) 134.4 L (b) 67.2 L
(c) 68.1 L (d) 136.2 L

43. While resting, an average 70 kg human male consumes 16.628 L of oxygen per hour at 27°C and 100 kPa. How many moles of oxygen are consumed by the 70 kg man while resting for 1 hour?

(a) 0.67 (b) 66.7
(c) 666.7 (d) 67.5

44. One molecule of haemoglobin will combine with four molecules of oxygen. If 1.0 g of haemoglobin combines with 1.642 ml of oxygen at body temperature (27°C) and a pressure of 760 torr, then what is the molar mass of haemoglobin?

(a) 6,00,000 (b) 1,50,000
(c) 15,000 (d) 60,000

45. A quantity of 2.0 g of a triatomic gaseous element was found to occupy a volume of 448 ml at 76 cm of Hg and 273 K. The mass of its each atom is

(a) 100 amu (b) $5.53 \times 10^{-23}$ g
(c) 33.3 g (d) 5.53 amu

46. The most abundant element dissolved in sea water is chlorine at a concentration of 19 g/kg of sea water. The volume of earth's ocean is $1.4 \times 10^{21}$ L. How many g-atoms of chlorine are potentially available from the oceans? The density of sea water is 1 g/ml ($N_A = 6 \times 10^{23}$).

(a) $7.5 \times 10^{20}$ (b) $27 \times 10^{21}$
(c) $27 \times 10^{24}$ (d) $7.5 \times 10^{19}$

47. From 2 mg calcium, $1.2 \times 10^{19}$ atoms are removed. The number of g-atoms of calcium left is (Ca = 40)

(a) $5 \times 10^{-5}$ (b) $2 \times 10^{-5}$
(c) $3 \times 10^{-5}$ (d) $5 \times 10^{-6}$

48. The number of g-molecules of oxygen in $6.023 \times 10^{24}$ CO molecules is

(a) 1 g-molecule (b) 0.5 g-molecule
(c) 5 g-molecules (d) 10 g-molecules

49. Equal masses of oxygen, hydrogen and methane are taken in identical conditions. What is the ratio of the volumes of the gases under identical conditions?

(a) 16 : 1 : 8 (b) 1 : 16 : 2
(c) 1 : 16 : 8 (d) 2 : 16 : 1

50. A pre-weighed vessel was filled with oxygen at NTP and weighed. It was then evacuated, filled with $SO_2$ at the same temperature and pressure, and again weighed. The weight of oxygen is

(a) the same as that of $SO_2$.
(b) $\frac{1}{2}$ that of $SO_2$.
(c) twice that of $SO_2$.
(d) $\frac{1}{4}$ that of $SO_2$.

## Average Molecular Mass

51. Molecular mass of dry air is

(a) less than moist air.
(b) greater than moist air.
(c) equal to moist air.
(d) may be greater or less than moist air.

52. At room temperature, the molar volume of hydrogen fluoride gas has a mass of about 50 g. The formula weight of hydrogen fluoride is 20. Therefore, gaseous hydrogen fluoride at room temperature is probably a mixture of

(a) $H_2$ and $F_2$ (b) HF and $H_2F_2$
(c) HF and $H_{2.5}F_{2.5}$ (d) $H_2F_2$ and $H_3F_3$

53. A gaseous mixture contains 70% $N_2$ and 30% unknown gas by volume. If the average molecular mass of gaseous mixture is 37.60, then the molecular mass of unknown gas is

(a) 42.2 (b) 60
(c) 40 (d) 50

54. The mass composition of universe may be given as 90% $H_2$ and 10% He. The average molecular mass of universe should be

(a) 2.20 (b) 2.10
(c) 3.80 (d) 3.64

55. A quantity of 10 g of a mixture of $C_2H_6$ and $C_5H_{10}$ occupy 4480 ml at 1 atm and 273 K. The percentage of $C_2H_6$ by mass in the mixture is

(a) 30% (b) 70%
(c) 50% (d) 60%

56. The density (in g/L) of an equimolar mixture of methane and ethane at 1 atm and 0°C is

(a) 1.03 (b) 2.05
(c) 0.94 (d) 1.25

57. '$n$' mol of $N_2$ and 0.05 mol of Ar is enclosed in a vessel of capacity 6 L at 1 atm and 27°C. The value of '$n$' is ($R = 0.08$ L atm mol$^{-1}$ K$^{-1}$)

(a) 0.25 (b) 0.20
(c) 0.05 (d) 0.1

58. A gaseous mixture contains 40% $H_2$ and 60% He by volume. What is the total number of moles of gases present in 10 g of such mixture?

(a) 5 (b) 2.5
(c) 3.33 (d) 3.125

59. A sample of ozone gas is found to be 40% dissociated into oxygen. The average molecular mass of sample should be

(a) 41.60 (b) 40
(c) 42.35 (d) 38.40

60. The vapour density of a sample of $SO_3$ gas is 28. Its degree of dissociation into $SO_2$ and $O_2$ is

(a) 1/7 (b) 1/6
(c) 6/7 (d) 2/5

## Percentage Composition

61. The commonly used pain reliever, aspirin, has the molecular formula $C_9H_8O_4$. If a sample of aspirin contains 0.968 g of carbon, then what is the mass of hydrogen in the sample?

(a) 0.717 g (b) 0.0717 g
(c) 8.000 g (d) 0.645 g

62. For $CuSO_4 \cdot 5H_2O$, which is the correct mole relationship?

(a) 9 × Mole of Cu = Mole of O
(b) 5 × Mole of Cu = Mole of O
(c) 9 × Mole of Cu = Mole of $O_2$
(d) Mole of Cu = 5 × Mole of O

63. The percentage of Fe(III) present in iron ore $Fe_{0.93}O_{1.00}$ is (Fe = 56)

(a) 94 (b) 6
(c) 21.5 (d) 15

64. A quantity of 5 g of a crystalline salt when rendered anhydrous lost 1.8 g of water. The formula mass of the anhydrous salt is 160. The number of molecules of water of crystallization in the salt is

(a) 3 (b) 5
(c) 2 (d) 1

65. Cortisone is a molecular substance containing 21 atoms of carbon per molecule. The mass percentage of carbon in cortisone is 69.98%. What is the molecular mass of cortisone?

(a) 180.05 (b) 360.1
(c) 312.8 (d) 205.8

66. A polystyrene of formula $Br_3C_6H_2(C_8H_8)_n$ was prepared by heating styrene with tribromobenzyl peroxide in the absence of air. It was found to contain 10.46% bromine by mass. The value of $n$ is (Br = 80)

(a) 20 (b) 21
(c) 19 (d) 22

67. A compound contains 36% carbon by mass. If each molecule contains two carbon atoms, the number of moles of compound in its 10 g is

(a) 66.67 (b) 0.15
(c) 0.30 (d) 1.5

68. The percentage of oxygen in a compound is 4%. Its minimum molecular mass will be

(a) 100 (b) 400
(c) 200 (d) 32

69. In Dumas method, 0.2 g of an organic nitrogenous compound gave 28 ml of $N_2$ (volume reduced to 0°C and 1 atm). What is the percentage of nitrogen by mass in the compound?

(a) 17.5 (b) 8.75
(c) 35.0 (d) 14.0

70. A quantity of 0.2 g of an organic compound containing C, H and O on combustion yielded 0.147 g $CO_2$ and 0.12 g water. The percentage of oxygen in it is

(a) 73.29% (b) 78.45%
(c) 83.23% (d) 89.50%

## Empirical and Molecular Formula

71. The empirical formula of an organic gaseous compound containing carbon and hydrogen is $CH_2$. The volume occupied by certain mass of this gas is exactly half of the volume occupied by the same mass of nitrogen gas under identical conditions. The molecular formula of the organic gas is

(a) $C_2H_4$ (b) $CH_2$
(c) $C_6H_{12}$ (d) $C_4H_8$

72. A compound has carbon, hydrogen and oxygen in 3 : 3 : 1 atomic ratio. If the number of moles in 1 g of the compound is $6.06 \times 10^{-3}$, then the molecular formula of the compound will be

(a) $C_3H_3O$
(b) $C_6H_6O_2$
(c) $C_9H_9O_3$
(d) $C_{12}H_{12}O_4$

73. A compound having the empirical formula, $C_3H_4O$ has a molecular weight of 170 ± 5. The molecular formula of the compound is

(a) $C_3H_4O$ (b) $C_6H_8O_2$
(c) $C_6H_{12}O_3$ (d) $C_9H_{12}O_3$

74. It was found from the chemical analysis of a gas that it has two hydrogen atoms for each carbon atom. At 0°C and 1 atm, its density is 1.25 g per litre. The formula of the gas would be

(a) $CH_2$ (b) $C_2H_4$
(c) $C_2H_6$ (d) $C_4H_8$

75. A quantity of 1.4 g of a hydrocarbon gives 1.8 g water on complete combustion. The empirical formula of hydrocarbon is

(a) CH (b) $CH_2$
(c) $CH_3$ (d) $CH_4$

EXERCISE I

76. An organic compound contains 40% carbon and 6.67% hydrogen by mass. Which of the following represents the empirical formula of the compound?

(a) $CH_2$ (b) $CH_2O$
(c) $C_2H_4O$ (d) $CH_3O$

77. A compound contains elements X and Y in 1 : 4 mass ratio. If the atomic masses of X and Y are in 1 : 2 ratio, the empirical formula of the compound should be

(a) $XY_2$ (b) $X_2Y$
(c) $XY_4$ (d) $X_4Y$

78. A compound contains equal masses of the elements A, B and C. If the atomic masses of A, B and C are 20, 40 and 60, respectively, then the empirical formula of the compound is

(a) $A_3B_2C$ (b) $AB_2C_3$
(c) ABC (d) $A_6B_3C_2$

79. A gaseous oxide contains 30.4% of nitrogen, one molecule of which contains one nitrogen atom. The density of the oxide relative to oxygen under identical conditions is about

(a) 0.69 (b) 1.44
(c) 0.35 (d) 2.88

80. Iron forms two oxides. If for the same mass of iron, the mass of oxygen combined in the first oxide is two-third of the mass of oxygen combined in the second oxide, then the ratio of valency of iron in first and second oxide is

(a) 1:1 (b) 2:3
(c) 3:2 (d) 2:5

## Stoichiometry

81. When a certain amount of octane, $C_8H_{18}$, is burnt completely, 7.04 g $CO_2$ is formed. What is the mass of $H_2O$ formed simultaneously?

(a) 1.62 g (c) 6.48 g
(c) 3.24 g (d) 2.28 g

82. If a rocket was fuelled with kerosene and liquid oxygen, then what mass of oxygen would be required for every litre of kerosene? Assume kerosene to have the average composition $C_{14}H_{30}$ and density 0.792 g/ml.

(a) 5.504 kg (b) 2.752 kg
(c) 1.376 kg (d) 3.475 kg

83. Air contains 20% $O_2$ by volume. What volume of air is needed at 0°C and 1 atm for complete combustion of 80 g methane?

(a) 10 L (b) 50 L
(c) 224 L (d) 1120 L

84. Acrylonitrile $C_3H_3N$ is the starting material for the production of a kind of synthetic fibre (acrylics). It can be made from propylene $C_3H_6$ by reaction with nitric oxide (NO).

$C_3H_6(g) + NO(g) \rightarrow C_3H_3N(g) + H_2O(g) + N_2(g)$ (Unbalanced)

How many grams of acrylonitrile may be obtained from 420 kg of propylene and excess NO?

(a) 265 kg (b) 530 kg
(c) 1060 kg (d) 795 kg

85. A quantity of 2.76 g of silver carbonate on being strongly heated yields a residue weighing (Ag = 108)

(a) 2.16 g (b) 2.48 g
(c) 2.32 g (d) 2.64 g

86. How many litres of detonating gas may be produced at 0°C and 1 atm from the decomposition of 0.1 mole of water by an electric current?

(a) 2.24 L (b) 1.12 L
(c) 3.36 L (d) 4.48 L

87. What mass of solid ammonium carbonate $H_2NCOONH_4$ when vaporized at 273°C will have a volume of 8.96 L at 760 mm of pressure. Assume that the solid completely decomposes as

$H_2NCOONH_4(s) \rightarrow CO_2(g) + 2NH_3(g)$

(a) 15.6 g (b) 5.2 g
(c) 46.8 g (d) 7.8 g

88. The minimum mass of sulphuric acid needed for dissolving 3 g of magnesium carbonate is

(a) 3.5 g (b) 7.0 g
(c) 1.7 g (d) 17.0 g

89. Samples of 1.0 g of Al are treated separately with an excess of sulphuric acid and an excess of sodium hydroxide. The ratio of the number of moles of the hydrogen gas evolved is

(a) 1 : 1 (b) 3 : 2
(c) 2 : 1 (d) 9 : 4

**90.** The minimum mass of water needed to slake 1 kg of quicklime, assuming no loss by evaporation, is

(a) 243.2 g (b) 642.8 g
(c) 160.7 g (d) 321.4 g

**91.** When 20 g $Fe_2O_3$ is reacted with 50 g of HCl, $FeCl_3$ and $H_2O$ are formed. The amount of unreacted HCl is (Fe = 56)

(a) 27.375 g (b) 22.625 g
(c) 30 g (d) 4.75 g

**92.** $SO_2$ gas is slowly passed through an aqueous suspension containing 12 g of $CaSO_3$ till the milkiness just disappears. What amount of $SO_2$ would be required?

(a) 6.4 mole (b) 0.3 mole
(c) 0.1 mole (d) 0.2 mole

**93.** A mixture of $N_2$ and $H_2$ is caused to react in a closed container to form $NH_3$. The reaction ceases before either reactant has been totally consumed. At this stage, 2.0 moles each of $N_2$, $H_2$ and $NH_3$ are present. The moles of $N_2$ and $H_2$ present originally were, respectively,

(a) 4 and 4 moles (b) 3 and 5 moles
(c) 3 and 4 moles (d) 4 and 5 moles

**94.** An ore contains 2.296% of the mineral argentite, $Ag_2S$, by mass. How many grams of this ore would have to be processed in order to obtain 1.00 g of pure solid silver? (Ag = 108)

(a) 1.148 g (b) 0.026 g
(c) 50 g (d) 2.296 g

**95.** A power company burns approximately 500 tons of coal per day to produce electricity. If the sulphur content of the coal is 1.5% by mass, then how many tons of $SO_2$ are dumped into the atmosphere every day?

(a) 15.0 (b) 7.5
(c) 30.0 (d) 18.75

## Limiting Reagent Based

**96.** An amount of $1.0 \times 10^{-3}$ moles of $Ag^+$ and $1.0 \times 10^{-3}$ moles of $CrO_4^{2-}$ reacts together to form solid $Ag_2CrO_4$. What is the amount of $Ag_2CrO_4$ formed? (Ag = 108, Cr = 52)

(a) 0.332 g (b) 0.166 g
(c) 332 g (d) 166 g

**97.** An amount of 0.3 mole of $SrCl_2$ is mixed with 0.2 mole of $K_3PO_4$. The maximum moles of KCl which may form is

(a) 0.6 (b) 0.5
(c) 0.3 (d) 0.1

**98.** Large quantities of ammonia are burned in the presence of a platinum catalyst to give nitric oxide, as the first step in the preparation of nitric acid.

$$NH_3(g) + O_2(g) \xrightarrow{Pt} NO(g) + H_2O(g)$$
(Unbalanced)

Suppose a vessel contains 0.12 moles of $NH_3$ and 0.14 of moles $O_2$. How many moles of NO may be obtained?

(a) 0.120 (b) 0.112
(c) 0.140 (d) 0.070

**99.** Equal masses of iron and sulphur are heated together to form FeS. What fraction of the original mass of excess reactant is left unreacted? (Fe = 56, S = 32)

(a) 0.22 (b) 0.43
(c) 0.86 (d) 0.57

**100.** Hydrogen cyanide, HCN, is prepared from ammonia, air and natural gas ($CH_4$) by the following process.

$$2NH_3(g) + 3O_2(g) + 2CH_4(g) \xrightarrow{Pt} 2HCN(g) + 6H_2O(g)$$

If a reaction vessel contains 11.5 g $NH_3$, 10.0 g $O_2$, and 10.5 g $CH_4$, then what is the maximum mass, in grams, of hydrogen cyanide that could be made, assuming the reaction goes to completion?

(a) 18.26 g (b) 5.625 g
(c) 17.72 g (d) 16.875 g

## Sequential and Parallel Reactions

**101.** What mass of carbon disulphide $CS_2$ can be completely oxidized to $SO_2$ and $CO_2$ by the oxygen liberated when 325 g of $Na_2O_2$ reacts with water?

(a) 316.67 g (b) 52.78 g
(c) 633.33 g (d) 211.11 g

**102.** An amount of 2 moles of $KClO_3$ is decomposed completely to produce $O_2$ gas. How many moles of butene $C_4H_8$ can be burnt completely by the $O_2$ gas produced?

(a) 0.5 (b) 1.0
(c) 2.0 (d) 3.0

**103.** On heating $KClO_3$ at a certain temperature, it is observed that one mole of $KClO_3$ yields one mole of $O_2$. What is the mole fraction of $KClO_4$ in the final solid mixture containing only KCl and $KClO_4$, the latter being formed by parallel reaction?

(a) 0.50 (b) 0.25
(c) 0.33 (d) 0.67

**104.** When 12 g graphite is burnt in sufficient oxygen, CO as well as $CO_2$ is formed. If the product contains 40% CO and 60% $CO_2$ by mass and none of the reactant is left, then what is the mass of oxygen gas used in combustion?

(a) 24.0 g (b) 21.33 g
(c) 23.8 g (d) 15.6 g

**105.** A mixture of 254 g of iodine and 142 g of chlorine is made to react completely to give a mixture of ICl and $ICl_3$. How many moles of each product are formed? (I = 127, Cl = 35.5)

(a) 0.1 mol of ICl and 0.1 mol of $ICl_3$.
(b) 1.0 mol of ICl and 1.0 mol of $ICl_3$.
(c) 0.5 mol of ICl and 0.1 mol of $ICl_3$.
(d) 0.5 mol of ICl and 1.0 mol of $ICl_3$.

## Percentage Based

**106.** A quantity of 4.35 g of a sample of pyrolusite ore, when heated with conc. HCl gave chlorine. The chlorine, when passed through potassium iodide solution liberated 6.35 g of iodine. The percentage of pure $MnO_2$ in the pyrolusite ore is (Mn = 55, I = 127)

(a) 40 (b) 50
(c) 60 (d) 70

**107.** How many grams of 90% pure $Na_2SO_4$ can be produced from 250 g of 95% pure NaCl?

(a) 640.6 g (b) 288.2 g
(c) 259.4 g (d) 320.3 g

**108.** A quantity of 10 g of a piece of marble was put into excess of dilute HCl acid. When the reaction was complete, 1120 cm$^3$ of $CO_2$ was obtained at 0°C and 1 atm. The percentage of $CaCO_3$ in the marble is

(a) 5% (b) 25%
(c) 50% (d) 2.5%

**109.** A 1.50 g sample of potassium bicarbonate having 80% purity is strongly heated. Assuming the impurity to be thermally stable, the loss in weight of the sample on heating is

(a) 3.72 g (b) 0.72 g
(c) 0.372 g (d) 0.186 g

**110.** Hydrazine $N_2H_4$ (used as a fuel in rocket system) can be produced according to the following reaction.

$$ClNH_2 + 2NH_3 \rightarrow N_2H_4 + NH_4Cl$$

When 1.0 kg $ClNH_2$ is reacted with excess of $NH_3$, 473 g of $N_2H_4$ is produced. What is the percentage yield?

(a) 76.12 (b) 67.21
(c) 26.17 (d) 16.72

**111.** Two successive reactions, A → B and B → C, have yields of 90% and 80%, respectively. What is the overall percentage yield for conversion of A to C?

(a) 90% (b) 80%
(c) 72% (d) 85%

**112.** Iodobenzene is prepared from aniline ($C_6H_5NH_2$) in a two-step process as shown here.

$$C_6H_5NH_2 + HNO_2 + HCl \longrightarrow C_6H_5N_2^+Cl^- + 2H_2O$$

$$C_6H_5N_2^+Cl^- + KI \rightarrow C_6H_5I + N_2 + KCl$$

In an actual preparation, 9.30 g of aniline was converted to 16.32 g of iodobenzene. The percentage yield of iodobenzene is (I = 127)

(a) 8% (b) 50%
(c) 75% (d) 80%

**113.** One mole of a mixture of CO and $CO_2$ requires exactly 20 g of NaOH in solution for complete conversion of all the $CO_2$ into $Na_2CO_3$. How many grams more of NaOH would it require for conversion into $Na_2CO_3$ if the mixture (one mole) is completely oxidized to $CO_2$?

(a) 60 g (b) 80 g
(c) 40 g (d) 20 g

**114.** When burnt in air, 14.0 g mixture of carbon and sulphur gives a mixture of $CO_2$ and $SO_2$ in the volume ratio of 2 : 1, the volume being measured at the same conditions of temperature and pressure. Moles of carbon in the mixture is

(a) 0.25 (b) 0.40
(c) 0.5 (d) 0.75

**115.** A mixture of NaI and NaCl on reaction with $H_2SO_4$ gave $Na_2SO_4$ equal to the weight of original mixture taken. The percentage of NaI in the mixture is (I = 127)

(a) 82.86 (b) 26.38
(c) 62.38 (d) 28.86

## Eudiometry

**116.** When 0.03 L of a mixture of hydrogen and oxygen was exploded, 0.003 L of oxygen remained. The initial mixture contains (by volume)

(a) 60% $O_2$ (b) 40% $O_2$
(c) 50% $O_2$ (d) 30% $O_2$

**117.** A volume of 100 ml of air containing only oxygen and nitrogen is a taken in a jar over water. NO is slowly passed till no more brown fumes appear in the gas jar. It is found that 42 ml of NO is required. The percentage of nitrogen in the air would be

(a) 42% (b) 79%
(c) 21% (d) 39.5%

**118.** A mixture of methane and ethylene in the ratio of a : b by volume occupies 30 ml. On complete combustion, the mixture yields 40 ml of $CO_2$. What volume of $CO_2$ would have been obtained if the ratio would have been b : a?

(a) 50 ml (b) 30 ml
(c) 40 ml (d) 60 ml

**119.** A volume of 200 ml of oxygen is added to 100 ml of a mixture containing $CS_2$ vapour and CO, and the total mixture is burnt. After combustion, the volume of the entire mixture is 245 ml. Calculate the volume of the oxygen that remains

(a) 67.5 ml (b) 125.0 ml
(c) 200.0 ml (d) 100.0 ml

**120.** A volume of 10 ml hydrogen requires 25 ml air for complete combustion. The volume per cent of $N_2$ in air is

(a) 20% (b) 80%
(c) 79% (d) 5%

**121.** A volume of 10 ml of gaseous $C_4H_x$ exactly requires 55 ml $O_2$ for complete combustion. The value of 'x' is

(a) 4 (b) 6
(c) 8 (d) 10

**122.** When 500 ml $CO_2$ gas is passed through red hot charcoal, the volume becomes 700 ml. The volume of $CO_2$ converted into CO is

(a) 200 ml (b) 300 ml
(c) 350 ml (d) 500 ml

**123.** The percentage by volume of $C_3H_8$ in a mixture of $C_3H_8$, $CH_4$ and CO is 36.5. The volume of $CO_2$ produced when 100 ml of the mixture is burnt in excess of $O_2$ is

(a) 153 ml (b) 173 ml
(c) 193 ml (d) 213 ml

**124.** A volume of 1 ml of a gaseous aliphatic compound $C_nH_{3n}O_m$ is completely burnt in an excess of oxygen. The contraction in volume (in ml) is

(a) $\left(1+\frac{1}{2}n-\frac{3}{4}m\right)$ (b) $\left(1+\frac{3}{4}n-\frac{1}{4}m\right)$

(c) $\left(1-\frac{1}{2}n-\frac{3}{4}m\right)$ (d) $\left(1+\frac{3}{4}n-\frac{1}{2}m\right)$

**125.** The explosion of a mixture consisting of one volume of a gas being studied and one volume of $H_2$ yielded one volume water vapour and one volume of $N_2$. The formula of gas being studied, is

(a) NO (b) $NO_2$
(c) $N_2O$ (d) $N_2O_3$

EXERCISE 1

**126.** A gaseous alkane is exploded with oxygen. The volume of $O_2$ for complete combustion to the volume of $CO_2$ formed is in 7:4 ratio. The molecular formula of alkane is

(a) $CH_4$ (b) $C_3H_8$
(c) $C_2H_6$ (d) $C_4H_{10}$

**127.** A volume $V$ of a gaseous hydrocarbon was exploded with an excess of oxygen. The observed contraction was $2.5V$, and on treatment with potash, there was a further contraction of $2V$. What is the molecular formula of the hydrocarbon?

(a) $C_2H_6$ (b) $C_3H_6$
(c) $C_4H_{12}$ (d) $C_2H_4$

**128.** A volume of 10 ml chlorine gas combines with 25 ml of oxygen gas to form 10 ml of a gaseous compound. If all the volumes are measured at the same pressure and temperature, then what is the molecular formula of compound formed?

(a) $Cl_2O$ (b) $Cl_2O_7$
(c) $ClO_2$ (d) $Cl_2O_5$

**129.** A volume of 10 ml of an oxide of nitrogen was taken in a eudiometer tube and mixed with hydrogen until the volume was 28 ml. On sparking, the resulting mixture occupied 18 ml. To this mixture, oxygen was added when the volume came to 27 ml and on explosion again, the volume fall to 15 ml. Find the molecular weight of the oxide of nitrogen originally taken in eudiometer tube. All measurements were made at STP.

(a) 22 (b) 44
(c) 88 (d) 176

**130.** $V_1$ ml of unknown gas (A) + $V_2$ ml of $O_2 \rightarrow (V_1 + V_2)$ ml of $CO_2$.

Gas 'A' may be

(a) CO
(b) ($CO + CO_2$) in equal proportion
(c) $C_{12}O_9$
(d) $C_3O_2$

---

## Concentration Terms

**131.** How many grams of solute should be added in 100 g water to get a solution of density 1.2 g/ml and strength 5% (w/v)?

(a) 5 g (b) 6 g
(c) 4.17 g (d) 4.35 g

**132.** An aqueous solution of glucose is 10% (w/v). The volume in which 1 mole of glucose is dissolved will be

(a) 18 L (b) 9 L
(c) 0.9 L (d) 1.8 L

**133.** A quantity of 50 g of water is saturated with HCl gas to get 75 ml of solution containing 40% HCl by mass. The density of solution formed is

(a) 1.11 g/ml (b) 0.4 g/ml
(c) 0.9 g/ml (d) 0.99 g/ml

**134.** The concentration of same aqueous solution of glucose is determined by two students—Sawan and Gautam. Sawan reported the concentration as 20% (w/w) and Gautam reported the concentration as 25% (w/v). If both the concentrations are correct, then the density of solution is

(a) 0.8 g/ml (b) 1.0 g/ml
(c) 1.25 g/ml (d) 1.33 g/ml

**135.** How much $Ca(NO_3)_2$, in mg, must be present in 50 ml of a solution with 2.35 ppm of Ca?

(a) 0.1175 (b) 770.8
(c) 4.7 (d) 0.48

**136.** The legal limit for human exposure to CO in the work place is 35 ppm. Assuming that the density of air is 1.3 g/L, how many grams of CO are in 1.0 L of air at the maximum allowable concentration?

(a) $4.55 \times 10^{-5}$ g (b) $3.5 \times 10^{-5}$ g
(c) $2.69 \times 10^{-5}$ g (d) $7.2 \times 10^{-5}$ g

**137.** What volume of 0.8 M-$AlCl_3$ solution should be mixed with 50 ml of 0.2 M-$CaCl_2$ solution to get a solution of chloride ion concentration equal to 0.6 M?

(a) 5.56 ml (b) 100 ml
(c) 50 ml (d) 4.89 ml

**138.** D5W refers to one of the solutions used as an intravenous fluid. It is 5% by mass solution of dextrose, $C_6H_{12}O_6$ in water. The density of D5W is 1.08 g/ml. The molarity of the solution is

(a) 0.3 M (b) 0.6 M
(c) 0.28 M (d) 0.26 M

**139.** How much $BaCl_2$ would be needed to make 250 ml of a solution having the same concentration of $Cl^-$ as one containing 3.78 g NaCl per 100 ml? (Ba = 137)

(a) 16.8 g (b) 67.2 g
(c) 33.6 g (d) 22.4 g

**140.** Upon heating a litre of semimolar HCl solution, 2.675 g of hydrogen chloride is lost and the volume of the solution shrinks to 750 ml. The molarity of the resultant solution is

(a) 0.569 M (b) 0.5 M
(c) 0.42 M (d) 1.707 M

**141.** A volume of 500 ml of a 0.1 M solution of $AgNO_3$ is added to 500 ml of 0.1 M solution of KCl. The concentration of nitrate ion in the resulting solution is

(a) 0.05 M (b) 0.1 M
(c) 0.2 M (d) Reduced to zero

**142.** In 1200 g solution, 12 g urea is present. If density of the solution is 1.2 g/ml, then the molarity of the solution is

(a) 0.2 M (b) 10 M
(c) 0.167 M (d) 12 M

**143.** Mole fraction of solute in an aqueous solution of NaOH is 0.1. If the specific gravity of the solution is 1.4, then the molarity of the solution is

(a) 6.93 (b) 0.1
(c) 71.4 (d) 0.14

**144.** What should be the density of an aqueous solution of urea (molar mass = 60 g/mol) such that the molality as well as molarity of the solution becomes equal to 1.0 unit ?

(a) 1.0 g/ml (b) 1.6 g/ml
(c) 1.06 g/ml (d) 1.16 g/ml

**145.** A quantity of 10 g of acetic acid is dissolved in 100 g of each of the following solvents. In which solvent, the molality of solution is maximum? Assume no any dissociation or association of acetic acid in the solvent.

(a) Water (b) Ethanol
(c) Benzene (d) Same in all solvents

**146.** A quantity of 10 g of acetic acid is dissolved in 100 g of each of the following solvents. In which solvent, the mole fraction of solute is maximum? Assume no any dissociation or association of acetic acid in the solvent.

(a) Water
(b) Ethanol
(c) Benzene
(d) Same in all solvents

**147.** An aqueous solution has urea and glucose in mass ratio 3 : 1. If the mass ratio of water and glucose in the solution is 10 : 1, then the mole fraction of glucose in the solution is

(a) $\frac{1}{110}$ (b) $\frac{9}{110}$
(c) $\frac{3}{110}$ (d) $\frac{100}{110}$

**148.** The volume strength of a sample of $H_2O_2$ is '9.08 vol'. The mass of $H_2O_2$ present in 250 ml of this solution is

(a) 0.4 g (b) 27.2 g
(c) 6.8 g (d) 108.8 g

**149.** What is the percentage of 'free $SO_3$' in a sample of oleum labelled as '104.5%'?

(a) 20% (b) 40%
(c) 60% (d) 80%

**150.** Which of the following percentage strength is not possible for a sample of oleum?

(a) 104% (b) 109%
(c) 118% (d) 127%

# EXERCISE II (JEE ADVANCED)

## Section A (Only one Correct)

1. A sample of clay contains 50% silica and 10% water. The sample is partially dried by which it loses 8 g water. If the percentage of silica in the partially dried clay is 52, then what is the percentage of water in the partially dried clay?

(a) 2.0% (b) 6.4%
(c) 10.4% (d) 2.4%

2. In the atomic weight determination, Dalton suggested the formula of water as HO and the composition of water as hydrogen = 12.5% and oxygen = 87.5% by weight. What should be the atomic weight of oxygen on H-scale, on the basis of this information?

(a) 16 (b) 8
(c) 14 (d) 7

3. The mercury content of a stream was believed to be above the minimum considered safe limit (1 part per billion, by mass). An analysis indicated that the concentration was 1.68 parts per billion. How many Hg atoms are present in 15 L of water, the density of which is 0.998 g/ml. (Hg = 200)

(a) $7.57 \times 10^{13}$ (b) $7.57 \times 10^{19}$
(c) $7.57 \times 10^{16}$ (d) $5.37 \times 10^{16}$

4. Assume that sodium atoms are spheres of radius 0.2 nm and that they are lined up side by side. How many miles, in length, is the line of atoms present in a 1.15 mg sample of sodium? ($N_A = 6 \times 10^{23}$)

(a) $1.2 \times 10^{10}$ (b) $1.2 \times 10^{8}$
(c) $7.5 \times 10^{8}$ (d) $7.5 \times 10^{6}$

5. The density of gold is 19.7 g/cm$^3$. The radius of gold atom is [Au = 197, $N_A = 6 \times 10^{23}$, $(10\pi)^{1/3} = 3.15$]

(a) $1.587 \times 10^{-8}$ m (b) $1.587 \times 10^{-9}$ m
(c) $1.587 \times 10^{-10}$ m (d) $1.587 \times 10^{-12}$ m

6. The average density of the universe as a whole is estimated as $3\times10^{-29}$ g/ml. If we assume that the entire mass is only H atoms, then what is the average volume of space that contains one H atom?

(a) 111.11 L (b) $1.8 \times 10^{-5}$ L
(c) 55.56 L (d) $3.6 \times 10^{-5}$ L

7. The waste of nuclear power plant contains $C^{12}$ and $C^{14}$ in the ratio of 4 : 1 by moles. What is the molecular mass of methane gas produced from this disposed waste? Given that the natural abundance of $C^{12}$ and $C^{14}$ are 98% and 2%, respectively.

(a) 15.998 (b) 16.0053
(c) 16 (d) 16.4

8. Two isotopes of an element Q are $Q^{97}$ (23.4% abundance) and $Q^{94}$ (76.6% abundance). $Q^{97}$ is 8.082 times heavier than $C^{12}$ and $Q^{94}$ is 7.833 times heavier than $C^{12}$. What is the average atomic weight of the element Q?

(a) 94.702 (b) 78.913
(c) 96.298 (d) 94.695

9. The $O^{18}/O^{16}$ ratio in some meteorites is greater than that used to calculate the average atomic mass of oxygen on earth. The average mass of an atom of oxygen in these meteorites is ______ that of a terrestrial oxygen atom?

(a) equal to (b) greater than
(c) less than (d) None of these

10. If the atomic mass were given by as 1/6th part and molecular mass as 1/12th part by mass of one atom of $C^{12}$ isotope, then what would be the molecular mass of water? Suppose atomic masses of hydrogen and oxygen on new scale are 1 and 16, respectively,

(a) 18 (b) 9
(c) 36 (d) Unpredictable

11. Assuming that 1,3,5-hexatriene has only pure double bonds and pure single bonds, how many grams of it contain one mole of double bonds?

(a) 13.3 g (b) 26.7 g
(c) 40 g (d) 80 g

12. In an experiment, it is found that 2.0769 g of pure X produces 3.6769 g of pure $X_2O_5$. The number of moles of X is

(a) 0.04 (b) 0.06
(c) 0.40 (d) 0.02

**13.** The volume occupied by 20 g water at 1.2 atm and 4°C is about

(a) 20 ml (b) $\frac{20 \times 0.082 \times 227}{18 \times 1.2}$ L

(c) $\frac{20 \times 0.082 \times 4}{18 \times 1.2}$ L (d) 20 L

**14.** A quantity of 2.3 g of a mixture of $NO_2$ and $N_2O_4$ has a pressure of 0.82 atm, at temperature TK in a container of volume $V$ litres such that the ratio, $T : V$ is 300 : 1 in magnitude. What is the degree of dissociation of $N_2O_4$?

(a) 0.17 (b) 0.33

(c) 0.67 (d) 0.70

**15.** When acetylene is passed through red hot metal tubes, some molecules trimerize to form benzene. The molecular mass of the gaseous mixture, when acetylene is passed through the tube is 60. The degree of trimerization of acetylene is

(a) 0.85 (b) 0.60

(c) 0.15 (d) 0.283

**16.** When a sample of hydrogen fluoride is cooled to 303 K, most of the molecules undergo dimerization. If the vapour density of such a sample is 18, then what per cent of total molecules in the sample are in dimer form? (F = 19)

(a) 88.89 (b) 80.0

(c) 20.0 (d) 11.11

**17.** Nitrogen (N), phosphorus (P) and potassium (K) are the main nutrients in plant fertilizers. According to an industry convention, the numbers on the label refer to the mass percent of N, $P_2O_5$ and $K_2O$ in that order. What is N : P : K ratio of a 30 : 10 : 10 fertilizer in terms of moles of each element expressed as x : y : 1.0? (N = 14, P = 31, K = 39)

(a) 10 : 0.66 : 1.0 (b) 20 : 0.66 : 1.0

(c) 8.4 : 1.3 : 1.0 (d) 16.8 : 1.3 : 1.0

**18.** A certain mixture of MnO and $MnO_2$ contains 66.67 mol percent of MnO. What is the approximate mass percent of Mn in it? (Mn = 55)

(a) 66.67 (b) 24.02

(c) 72.05 (d) 69.62

**19.** A sample of impure cuprous oxide contains 66.67% copper by mass. What is the percentage of pure $Cu_2O$ in the sample? (Cu = 63.5)

(a) 66.67 (b) 75

(c) 70 (d) 80

**20.** Sodium bicarbonate $NaHCO_3$ can be purified by dissolving it in hot water (60°C), filtering to remove insoluble impurities, cooling to 0°C to precipitate solid $NaHCO_3$, and then filtering to remove the solid, leaving soluble impurities in solution. Any $NaHCO_3$ that remains in the solution is not recovered. The solubility of $NaHCO_3$ in hot water at 60°C is 164 g/litre and is 69 g/litre in cold water at 0°C. What is the percentage yield of $NaHCO_3$, when it is purified by this method?

(a) 57.93% (b) 42.07%

(c) 69% (d) 100%

**21.** The mineral haematite is $Fe_2O_3$. Haematite ore contains unwanted material called gangue in addition to $Fe_2O_3$. If 5 kg of ore contains 2.78 kg of Fe, then what percentage of ore is gangue? (Fe = 56)

(a) 55.6% (b) 44.4%

(c) 20.6% (d) 79.4%

**22.** A sample of iron ore weighing 0.700 g is dissolved in nitric acid. The solution is then diluted with water, following with sufficient concentrated aqueous ammonia, to quantitative precipitation the iron as $Fe(OH)_3$. The precipitate is filtered, ignited and weighed as $Fe_2O_3$. If the mass of the ignited and dried precipitate is 0.541 g, then what is the mass percent of iron in the original iron ore sample? (Fe = 56)

(a) 27.0% (b) 48.1%

(c) 54.1% (d) 81.1%

**23.** The empirical formula of a compound is $CH_2O$. If 0.0833 moles of the compound contains 1.0 g of hydrogen, then its molecular formula should be

(a) $C_6H_{12}O_6$ (b) $C_5H_{10}O_5$

(c) $C_4H_8O_4$ (d) $C_3H_6O_3$

**24.** A hydrocarbon $C_nH_{2n}$ yields $C_nH_{2n+2}$ by reduction. In this process, the molar mass of the compound is raised by 2.38%. The value of $n$ is

(a) 8 (b) 4

(c) 6 (d) 5

**25.** A certain vitamin extracted from plant sources has carbon and hydrogen in 8 : 1 mass ratio. The percentage of oxygen is nearly 7.3. The compound gave no test for nitrogen or sulphur or any other element. What should be the empirical formula of the compound?

(a) $C_{30}H_{45}O_2$ (b) $C_{15}H_{23}O$

(c) $C_{29}H_{45}O_3$ (d) $C_{10}H_{15}O$

26. An unknown oxide of manganese is reacted with carbon to form manganese metal and $CO_2$. Exactly 31.6 g of the oxide, $Mn_xO_y$, yielded 13.2 g of $CO_2$. The simplest formula of the oxide is (Mn = 55)

(a) MnO (b) $MnO_2$
(c) $Mn_2O_3$ (d) $Mn_4O_6$

27. Assume that the atomic mass of oxygen is 7. A sample of 11 g of an oxide of uranium contains 10 g of uranium. Which of the following formula for the oxide is compatible with the data?

(a) Uranium oxide is UO and the atomic mass of U is 70.
(b) Uranium oxide is $U_3O_8$ and the atomic mass of U is 240.
(c) Uranium oxide is $UO_2$ and the atomic mass of U is 105.
(d) Uranium oxide is $U_3O_2$ and the atomic mass of U is 105.

28. A sample of protein was analysed for metal content and analysis revealed that it contained magnesium and titanium in equal amounts, by mass. If these are the only metallic species present in the protein and it contains 0.016% metal, by mass, the minimum possible molar mass of the protein is (Mg = 24, Ti = 48)

(a) 6,00,000 (b) 1,50,000
(c) 3,00,000 (d) 12,00,000

29. One mole of mixture of $N_2$, $NO_2$ and $N_2O_4$ has a mean molar mass of 55.4 g. On heating to a temperature, at which all the $N_2O_4$ is dissociated into $NO_2$, the mean molar mass tends to a lower value of 39.6 g. What is the mole ratio of $N_2$, $NO_2$ and $N_2O_4$ in the original mixture?

(a) 5:1:4 (b) 1:1:1
(c) 1:4:5 (d) 1:5:4

30. A protein isolated from a bovine preparation, was subjected to amino acid analysis. The amino acid present in the smallest amount was lysine, $C_6H_{14}N_2O_2$ and the amount of lysine was found to be 365 mg per 100 g protein. What is the minimum molecular mass of the protein?

(a) 40,000,000 (b) 40,000
(c) 40 (d) 4,00,000

31. Cupric ammonium sulphate was found to contain 27.03% water of crystallization by mass. Upon strongly heating, it gave cupric oxide corresponding to 19.89% of starting mass. Find the empirical formula of cupric ammonium sulphate. (Cu = 63.5)

(a) $CuSO_4 \cdot (NH_4)_2SO_4 \cdot 6H_2O$
(b) $CuSO_4 \cdot (NH_4)_2SO_4 \cdot 5H_2O$
(c) $CuSO_4 \cdot 2(NH_4)_2SO_4 \cdot 6H_2O$
(d) $CuSO_4 \cdot (NH_4)_2SO_4 \cdot 8H_2O$

32. A drug, marijuana, owes its activity to tetrahydrocannabinol, which contains 70 per cent as many carbon atoms as hydrogen atoms and 15 times as many hydrogen atoms as oxygen atoms. The number of moles in a gram of tetrahydrocannabinol is 0.00318. Determine its molecular formula.

(a) $CH_3O_2$ (b) $C_{21}H_{30}O_2$
(c) $C_{15}H_{30}O_2$ (d) $C_{70}H_{15}O$

33. How many millilitres (at 0°C and 1 atm) of hydrogen sulphide are needed to precipitate cupric sulphide completely from 100 ml of a solution containing 2.69 g of $CuCl_2$ in a 1 L solution? (Cu = 63.5)

(a) 448 (b) 4.48
(c) 22.4 (d) 44.8

34. When the hydrocarbon propane is burned in air, carbon dioxide and water are formed. If 0.15 mol of $CO_2$ is produced, then how many drops of water will be formed, assuming one drop is 0.05 $cm^3$ and contains $1.70 \times 10^{21}$ water molecules?

(a) $1.2 \times 10^{23}$ (b) 4
(c) 53 (d) 70

35. When a hydrocarbon is burnt completely, the ratio of masses of $CO_2$ and $H_2O$ formed is 44 : 27. The hydrocarbon is

(a) $CH_4$ (b) $C_2H_6$
(c) $C_2H_4$ (d) $C_2H_2$

36. An aqueous ammonium sulphate solution containing 50 moles of solute reacts with excess of calcium hydroxide. How many litres of a solution (specific gravity 0.85) containing 20% by mass of ammonia can be prepared using this reaction?

(a) 10.0 L (b) 8.5 L
(c) 20.0 L (d) 17.0 L

37. Specialized cells in the stomach release HCl to aid digestion. If they release too much, the excess can be neutralized by antacid tablets. Which of the following should be the more effective active ingredient of antacid tablets?

(a) $Mg(OH)_2$ (b) $Al(OH)_3$
(c) $Ca(OH)_2$ (d) $H_2SO_4$

**38.** A metal oxide has the formula $M_2O_3$. It can be reduced by hydrogen to give free metal and water. 0.1596 g of the metal oxide required 6 mg of hydrogen for complete reduction. The atomic mass of the metal is

(a) 111.60 (b) 159.60
(c) 79.80 (d) 55.80

**39.** If 0.250 g of an element M reacts with excess fluorine to produce 0.547 g of the hexafluoride, $MF_6$, the element should be (Cr = 52, Mo = 95.94, S = 32, Te = 127.6, F = 19)

(a) Cr (b) Mo
(c) S (d) Te

**40.** Fluorine reacts with uranium hexafluoride $UF_6$ as represented by the following equation.

$U(s) + 3F_2(g) \rightarrow UF_6(g)$

How many fluorine molecules are required to produce 2.0 mg of uranium hexafluoride $UF_6$ from an excess of uranium? The molar mass of $UF_6$ is 352.0 g $mol^{-1}$.

(a) $5.13 \times 10^{18}$ (b) $1.026 \times 10^{19}$
(c) $2.052 \times 10^{19}$ (d) $1.026 \times 10^{20}$

**41.** What is the total mass of the products formed, when 51 g of $H_2S$ is oxidized by oxygen to produce water and sulphur dioxide?

(a) 72 g (b) 27 g
(c) 123 g (d) 96 g

**42.** A quantity of 1.08 g of $Cr_2O_7^{2-}$ is reduced in an acidic solution by an excess of $SO_2$ to form $HSO_4^-$ and $Cr^{3+}$. What is the minimum number of moles of $H^+$ that must be present for this reaction to occur? (Cr = 52)

(a) 0.025 (b) 0.020
(c) 0.005 (d) 0.070

**43.** Diborane tetrachloride was treated with NaOH and the following reaction occurred.

$B_2Cl_4 + NaOH \rightarrow NaBO_2 + H_2O + H_2 + NaCl$

If 1362 ml of hydrogen gas is formed at STP, then how much $B_2Cl_4$ was consumed? (B = 11)

(a) 9.97 g (b) 9.84 g
(c) 0.0968 g (d) 23.57 g

**44.** What total volume, in litre at 727°C and 1 atm, could be formed by the decomposition of 16 g of $NH_4NO_3$?

Reaction: $2NH_4NO_3 \rightarrow 2N_2 + O_2 + 4H_2O(g)$.

(a) 57.47 L (b) 114.94 ml
(c) 41.78 L (d) 24.63 L

**45.** A compound of iron and chlorine is soluble in water. An excess of silver nitrate was added to precipitate all chloride ions as silver chloride. If a 127 mg sample of the compound gave 287 mg AgCl, then what is the formula of the compound? (Fe = 56, Ag = 108)

(a) $FeCl_2$ (b) $FeCl_3$
(c) FeCl (d) $FeCl_6$

**46.** From the following reactions,

$2CoF_2 + F_2 \rightarrow 2CoF_3$

$(CH_2)_n + 4n\ CoF_3 \rightarrow (CF_2)_n + 2n\ HF + 4n\ CoF_2$

calculate how much $F_2$ will be consumed to produce 1 kg of $(CF_2)_n$. (F = 19)?

(a) 1.52 kg (b) 2.04 kg
(c) 0.76 kg (d) 4.56 kg

**47.** An element 'A' reacts with the compound $BO_3$ to produce $A_3O_4$ and $B_2O_3$. The number of moles of $A_3O_4$ produced if 1 mole each of A and $BO_3$ are allowed to react is

(a) 3 (b) 1
(c) 1/3 (d) 2/3

**48.** A 1.50 g sample of type metal (an alloy of Sn, Pb, Cu and Sb) is dissolved in nitric acid, and metastannic acid, $H_2SnO_3$, precipitates. This is dehydrated by heating to tin (IV) oxide, which is found to weigh 0.50 g. What percentage of tin was in the original type metal sample? (Sn = 119)

(a) 33.33% (b) 26.27%
(c) 29.38% (d) 52.54%

**49.** An amount of 5 moles of A, 6 moles of B and excess amount of C are mixed to produce a final product D, according to the following reactions.

$A + 2B \rightarrow I$

$I + C \rightarrow B + D$

What is the maximum moles of D, which can be produced assuming that the products formed can also be reused in the reactions?

(a) 3 moles (b) 4.5 moles
(c) 5 moles (d) 6 moles

EXERCISE II

**50.** Hydrogen cyanide, HCN, can be made by a two-step process. First, ammonia is reacted with $O_2$ to give nitric oxide, NO.

$4NH_3(g) + 5O_2(g) \rightarrow 4NO(g) + 6H_2O(g)$

Then nitric oxide is reacted with methane, $CH_4$.

$2NO(g) + 2CH_4(g) \rightarrow 2HCN(g) + 2H_2O(g) + H_2(g)$

When 25.5 g of ammonia and 32.0 g of methane are used, how many grams of hydrogen cyanide can be produced?

(a) 1.5 (b) 2.0
(c) 40.5 (d) 54.0

**51.** To determine soluble (free) $SiO_2$ in a rock, an alkaline extraction was carried out, as a result of which there was found 1.52% of $SiO_2$ in the extract and also 1.02% of $Al_2O_3$. Considering that, apart from the free $SiO_2$, the extract also contained the $SiO_2$ that had passed into it from Kaolin ($2SiO_2 \cdot Al_2O_3$), the percentage of free $SiO_2$ in the rock being analysed is (Si = 28, Al = 27)

(a) 1.20 (b) 0.32
(c) 0.50 (d) 1.52

**52.** A sample of iron oxide has FeO and $Fe_2O_3$ in the mole ratio 2 : 1. It is partially oxidized to change this ratio to 1 : 2. The number of moles of FeO oxidized per mole of initial mixture is

(a) 0.2 (b) 0.333
(c) 0.4 (d) 0.5

**53.** When *x* g carbon is burnt with *y* g oxygen in a closed vessel, no residue is left behind. Which of the following statement is correct regarding the relative amounts of oxygen and carbon?

(a) $y/x$ must be less than 1.33.
(b) $y/x$ must be greater than 1.33.
(c) $y/x$ must be greater than 2.67.
(d) $y/x$ must lie between 1.33 and 2.67.

**54.** An amount of 1 mole of calcium cyanamide and 1 mole of water are allowed to react. The number of moles of ammonia produced is

(a) 3.0 (b) 2.0
(c) 1.0 (d) 0.67

**55.** An amount of 1 mole of $N_2$ and 4 moles of $H_2$ are allowed to react in a vessel and after reaction, water is added. Aqueous solution required 1 mole of HCl for complete reaction. Mole fraction of $H_2$ in the gas mixture after reaction is

(a) 1/6 (b) 5/6
(c) 1/3 (d) 2/3

**56.** A quantity of 5.08 g of iodine held in suspension in water is slowly acted upon by 460 ml of $H_2S$ measured at 0°C and 1 atm. What weight of sulphur will be liberated? (I = 127)

(a) 0.64 g (b) 0.657 g
(c) 1.297 g (d) 0.017 g

**57.** A quantity of 27.6 g of $K_2CO_3$ was treated by a series of reagent so as to convert all of its carbon to $K_2Zn_3[Fe(CN)_6]_2$. The mass of the product formed is (K = 39, Zn = 65.4, Fe = 56)

(a) 139.2 g (b) 11.6 g
(c) 69.6 g (d) 23.2 g

**58.** What is the volume required of a 20.0% HCl solution of density 1.20 g/ml to prepare 363.0 g of $AsCl_3$ according to the equations? (As = 75, Cl = 35.5)

$2KMnO_4 + 16\ HCl \rightarrow 2KCl + 2MnCl_2 + 5Cl_2 + 8H_2O$

$2As + 3Cl_2 \rightarrow 2AsCl_3$

(a) 2.56 l (b) 0.73 l
(c) 1.46 l (d) 2.92 l

**59.** Cyclohexanol is dehydrated to cyclohexene on heating with conc. $H_2SO_4$. If the yield of this reaction is 75%, then how much cyclohexene will be obtained from 100 g of cyclohexanol?

(a) 61.5 g (b) 82 g
(c) 109.3 g (d) 75 g

**60.** A sample of pure Cu (4.00 g) heated in a stream of oxygen for some time, gains in weight with the formation of black oxide of copper (CuO). The final mass is 4.90 g. What percent of copper remains unoxidized? (Cu = 64)

(a) 90% (b) 10%
(c) 20% (d) 80%

**61.** If the yield of chloroform obtainable from acetone and bleaching powder is 75%, then what mass of acetone is required for producing 30 g of chloroform?

(a) 40 g (b) 19.4 g
(c) 10.92 g (d) 14.56 g

62. Pure $FeS_2$ is burnt with 60% excess air. What is the percentage of $N_2$, by volume, in the gaseous mixture after the reaction? Air contains 20% $O_2$ and 80% $N_2$ by volume.

(a) 81.94 (b) 82.8
(c) 70.4 (d) 89.3

63. A 12 g sample of $CH_4$ and $C_2H_4$ yielded 35.2 g of $CO_2$ on complete oxidation. What was the mean molar mass of the original sample?

(a) 20.0 (b) 22.0
(c) 14.7 (d) 23.0

64. For a hydrocarbon, the ratio of volume $O_2$ used for complete combustion and the volume of $CO_2$ formed is independent to the number of carbon atoms present in the hydrocarbon. The hydrocarbon may be

(a) Alkane (b) Alkene
(c) Alkyne (d) Arene

65. A volume of 60 ml of a mixture of nitrous oxide and nitric oxide was exploded with excess hydrogen. If 38 ml of $N_2$ was formed, the volume of nitrous oxide in the original mixture is

(a) 16 ml (b) 44 ml
(c) 27 ml (d) 33 ml

66. A mixture is made equal volume of CO and air. A spark passed through so that all the oxygen is converted to carbon dioxide. What will be fractional decrease in the total volume of system assuming pressure and temperature remain constant? Air contains 20% oxygen by volume.

(a) 0.1 (b) 0.2
(c) 0.15 (d) 0.3

67. A mixture of formic acid and oxalic acid is heated with conc. $H_2SO_4$. The gaseous product is passed into KOH solution where the volume decreased by 1/6th. What was the molecular proportion of the organic acids, formic and oxalic acid in the mixture?

(a) 1 : 4 (b) 4 : 1
(c) 1 : 5 (d) 5 : 1

68. A volume of 50 ml of a gas mixed with 70 ml of oxygen gave after explosion 50 ml of $CO_2$ and after absorption by KOH, 45 ml of oxygen are left. What is the molecular formula of the gas?

(a) $CH_4$ (b) $C_2H_4$
(c) CO (d) $C_2H_2$

69. A human patient suffering from a duodenal ulcer may show a concentration of HCl of $80 \times 10^{-3}$ molar in gastric juice. If his stomach receives 3 L of gastric juice per day, how much medicine (antacid syrup) containing 2.6 g of $Al(OH)_3$ per 100 ml must he consumes per day to neutralize the acid?

(a) 27 ml (b) 80 ml
(c) 240 ml (d) 120 ml

70. When $V$ ml of 2.2 M – $H_2SO_4$ solution is mixed with $10V$ ml of water, the volume contraction of 2% takes place. The molarity of diluted solution is

(a) 0.2 M (b) 0.204 M
(c) 0.196 M (d) 0.224 M

71. A quantity of 23.6 g of succinic acid is dissolved in 500 ml of 0.1 M acetic acid solution. Assuming that neither acid is dissociated in solution, calculate the molarity of '–COOH' in the solution.

(a) 0.3 M (b) 0.5 M
(c) 0.9 M (d) 0.8 M

72. Chlorofluorocarbons such as $CCl_3F$ (M = 137.5) and $CCl_2F_2$ (M = 121) have been linked to ozone depletion in Antarctica. As of 2004, these gases were found in 275 and 605 parts per trillion ($10^{12}$), by volume. What are the concentrations of these gases under conditions typical of Antarctica stratosphere (200 K and 0.08 atm)? ($R$ = 0.08 l-atm/K-mol)

(a) $[CCl_3F] = 1.375 \times 10^{-12}$ mol $l^{-1}$, $[CCl_2F_2] = 3.025 \times 10^{-12}$ mol $l^{-1}$

(b) $[CCl_3F] = 2.75 \times 10^{-14}$ mol $l^{-1}$, $[CCl_2F_2] = 6.05 \times 10^{-14}$ mol $l^{-1}$

(c) $[CCl_3F] = 2.75 \times 10^{-10}$ mol $l^{-1}$, $[CCl_2F_2] = 6.05 \times 10^{-10}$ mol $l^{-1}$

(d) $[CCl_3F] = 1.375 \times 10^{-13}$ mol $l^{-1}$, $[CCl_2F_2] = 3.025 \times 10^{-12}$ mol $l^{-1}$

73. A quantity of 1 kg of 2 m urea solution is mixed with 2 kg of 4 m urea solution. The molality of the resulting solution is

(a) 3.33 m (b) 10 m
(c) 3.29 m (d) 5 m

74. A quantity of 1 kg of 1 m glucose solution is diluted to 5 kg. The molality of the diluted solution should be

(a) 0.2 m (b) 0.02 m
(c) 0.207 m (d) 0.175 m

75. A quantity of 500 g of an aq. urea solution having mole fraction of solute, 0.2 is diluted to 1500 g. The mole fraction of solute in the diluted solution is

(a) 0.05 (b) 0.067
(c) 0.6 (d) 0.1

76. A volume of 20 ml of 8.5% (w/v) $H_2O_2$ solution is diluted to 50 ml. A volume of 10 ml of the diluted solution is reacted with excess of an oxidant. It will cause liberation of ___ ml of ___ gas at 0°C and 1 atm.

(a) 56, $O_2$ (b) 112, $O_2$
(c) 224, $H_2$ (d) 224, $H_2$

77. A volume of 50 ml of '20 vol' $H_2O_2$ solution is mixed with 50 ml of '10 vol' $H_2O_2$ solution. The volume strength of the resulting solution is (assume neither expansion nor contraction in volume of solution, on mixing)

(a) '30 vol' (b) '10 vol'
(c) '15 vol' (d) '22.5 vol'

78. In 200 g of a sample of oleum labelled as 109.0%, 12 g water is added. The new labelling of the oleum sample is

(a) 106.0% (b) 103.0%
(c) 102.8% (d) 105.6%

79. When 200 g of an oleum sample labelled as 109% is mixed with 300 g of another oleum sample labelled as 118%, the new labelling of resulting oleum sample becomes

(a) 114.4% (b) 112.6%
(c) 113.5% (d) 127%

80. A sample of oleum is labelled as 112%. In 200 g of this sample, 18 g water is added. The resulting solution will contain

(a) 218 g pure $H_2SO_4$.
(b) 218 g $H_2SO_4$ and 6 g free $SO_3$.
(c) 212 g $H_2SO_4$ and 6 g free $SO_3$.
(d) 191.33 g $H_2SO_4$ and 26.67 g free $SO_3$.

## Section B (One or More than one Correct)

1. A quantity of 0.22 g of a gas occupies a volume of 112 ml at pressure of 1 atm and temperature of 273 K. The gas may be
(a) nitrogen dioxide
(b) nitrous oxide
(c) carbon dioxide
(d) propane

2. The number of hydrogen atoms in 0.9 g glucose, $C_6H_{12}O_6$, is same as
(a) 0.48 g hydrazine, $N_2H_4$
(b) 0.17 g ammonia, $NH_3$
(c) 0.30 g ethane, $C_2H_6$
(d) 0.03 g hydrogen, $H_2$

3. The composition of universe is approximately 90% hydrogen and 10% helium, by mass. It represents that

(a) there are 18 hydrogen atoms in the universe per atom of helium.
(b) there are 9 hydrogen atoms in the universe per atom of helium.
(c) there are 36 hydrogen atoms in the universe per atom of helium.
(d) the average molar mass of universe is 2.105 g per mole.

4. The vapour density of a sample of hydrogen fluoride gas is measured by an experiment as 20. It may represent that (F = 19)
(a) some molecules of hydrogen fluoride are dissociated.
(b) some molecules of hydrogen fluoride are in dimer form.
(c) all hydrogen fluoride molecules are in dimer form.
(d) some hydrogen fluoride molecules are in trimer form.

5. Which of the following statement(s) is/are correct for water?
(a) H and O are in 2 : 1 atomic ratio.
(b) H and O are in 2 : 1 mass ratio.
(c) H and O are in 1 : 8 mass ratio.
(d) Hydrogen and oxygen gases are combined in 2:1 volume ratio.

6. The atomic mass of a diatomic gaseous element is 19. Which of the following statement(s) is/are correct regarding the element?
   (a) The mass of one atom of the element is 19 amu.
   (b) The mass of $N_A$ molecules of the element is 38 g.
   (c) The volume of $N_A$ atoms of the element is 22.4 L at 0°C and 1 atm.
   (d) The volume of 2 g-molecules of the element is 44.8 L at 0°C and 1 atm.

7. Three isotopes of an element have mass numbers $M$, $(M + 1)$ and $(M + 2)$. If the mean mass number is $(M + 0.5)$, then which of the following ratio(s) may be accepted for $M$, $(M + 1)$ and $(M + 2)$ in the order ?
   (a) 1 : 1 : 1 (b) 4 : 1 : 1
   (c) 9 : 6 : 1 (d) 2 : 1 : 1

8. Which of the following statement(s) is/are correct about the Avogadro's number?
   (a) It is the number of atoms contained in one mole of atoms of any element.
   (b) It is the number of electrons required to deposit one mole of atoms of any metallic element from a solution of the metal salt.
   (c) It is the number of grams of any element which contains $6.022 \times 10^{23}$ atoms of that element.
   (d) It is the number of particles (atoms, molecules or ions) required to make one gram of the substance under consideration.

9. The non-stoichiometric compound, titanium monoxide, has a continuous range of composition from $Ti_{0.75}O$ to $TiO_{0.69}$. Which of the following is/are the correct regarding the possible composition of the compound? [Ti = 48]
   (a) The maximum percentage by mass of oxygen in the compound is 30.8.
   (b) The minimum percentage by mass of titanium in the compound is 69.2.
   (c) The minimum percentage by mass of oxygen in the compound is 18.7.
   (d) The minimum percentage by mass of titanium in the compound is 81.3.

10. Which of the following(s) is/are correct statement?
   (a) The empirical formula of all alkanes is same.
   (b) The empirical formula of all alkenes is same.
   (c) The empirical formula of all the members of any homologous series is same.
   (d) Two different compounds can have the same molecular formula.

11. Which of the following will have the composition (by mass) as similar as that of acetic acid?
   (a) Methyl formate, $HCOOCH_3$
   (b) Glucose, $C_6H_{12}O_6$
   (c) Formaldehyde, HCHO
   (d) Formic acid, HCOOH

12. Four groups of students are studying with different samples of alkali metal halides as given below.

   Group A: NaCl  Group B: NaBr
   Group C: KCl  Group D: KBr

   If all the four groups dissolved 0.1 moles of their salt in some water and then treated with the excess of acidified $AgNO_3$ solution, then which of the following statement(s) is/are correct regarding the mass of precipitate formed?
   (a) All the four groups will obtain the same mass of precipitate.
   (b) Group A and C will obtain the same mass of precipitate.
   (c) Group B and D will obtain the same mass of precipitate.
   (d) Group A and B will obtain the same mass of precipitate.

13. Which of the following is the incorrect conclusion regarding the following reaction.

   $2 H_2(g) + O_2(g) \rightarrow 2 H_2O(l)$

   (a) 2 mole of $H_2(g)$ will produce 2 mole of $H_2O(l)$.
   (b) 16 g of $O_2(g)$ will produce 18 g of $H_2O(l)$.
   (c) 2 litre of $O_2(g)$ at 25°C and 1 atm will produce 4 litre of $H_2O(l)$ at 25°C and 1 atm.
   (d) 2 molecules of $H_2O(l)$ is obtained from every 3 molecules of gaseous mixture of $H_2$ and $O_2$.

14. A quantity of 8 g $CH_4$ is mixed with 28 g $O_2$ and fired. Which of the following is correct about the combustion of $CH_4$ in this condition?
   (a) 1 g $CH_4$ will remain left unburned if carbon is quantitatively converted into $CO_2$.
   (b) 4 g $O_2$ will remain unused if carbon is quantitatively converted into CO.
   (c) Equal moles of CO and $CO_2$ are formed if none of the reactants is left and there is no other side reaction.
   (d) 18 g water will form in any possible condition.

15. The oxygen needed for complete combustion of 8 g $CH_4$ may be obtained from complete decomposition of
   (a) 2/3 mole of $KClO_3$
   (b) 1 mole of $H_2O_2$
   (c) 2 mole of $NaNO_3$ (up to 300°C)
   (d) 2 mole of $BaO_2$

16. A mixture of propane and benzene is burnt completely in excess of oxygen at 110°C. It results in the production of equal volumes of $CO_2(g)$ and steam (measured under identical pressure and temperature). Which of the following is correct regarding the original mixture?
   (a) The mole ratio of propane and benzene is 3 : 1.
   (b) The mass ratio of propane and benzene is 22 : 13.
   (c) The mole ratio of carbon and hydrogen atoms is 1 : 2.
   (d) The mass ratio of carbon and hydrogen atoms is 6 : 1.

17. A quantity of 6 g NaOH and 4.4 g $CO_2$ is allowed to react to form $Na_2CO_3$ or $NaHCO_3$ or both. Which of the following is correct statement regarding the reactions?
   (a) NaOH is the limiting reagent if there is no any formation of $NaHCO_3$.
   (b) NaOH is the limiting reagent if there is no any formation of $Na_2CO_3$.
   (c) Equal masses of $Na_2CO_3$ and $NaHCO_3$ are formed if none of the reactant is left.
   (d) The total mass of reaction mixture will be 10.4 g after the end of reaction, in any possible case.

18. When hydrocarbons are burnt completely in excess of oxygen gas, then
   (a) equal moles of $CO_2$ and $H_2O$ are formed from alkenes.
   (b) more moles of $H_2O$ than $CO_2$ are formed from alkanes.
   (c) more moles of $CO_2$ than $H_2O$ are formed from alkynes.
   (d) more moles of $CO_2$ than $H_2O$ are formed for any kind of hydrocarbon.

19. When hydrocarbons (alkanes, alkenes or alkynes) are burnt completely in excess of oxygen, then
   (a) for the same number of carbon atoms, more oxygen is consumed for alkanes.
   (b) for the same number of hydrogen atoms, more oxygen is consumed for alkynes.
   (c) for the same number of carbon atoms, more water is formed from alkynes.
   (d) for the same number of hydrogen atoms, more $CO_2$ is formed from alkynes.

20. A quantity of 12 g of magnesium is burnt completely in air ($O_2$ = 20% and $N_2$ = 80%, by volume). Which of the following is/are correct statement(s) regarding this combustion?
   (a) A minimum of 36 g air is needed if all Mg is converted into MgO only.
   (b) A minimum of 40 g air is needed if all Mg is converted into MgO only.
   (c) A minimum of 4.67 g air is needed if all Mg is converted into $Mg_3N_2$ only.
   (d) If air is consumed completely, then the total mass of products formed is 17.14 g.

21. A mixture contains NaCl and unknown chloride, MCl. When 1 g of this mixture is dissolved in water and excess of $AgNO_3$ solution is added to it, 2.567 g of white precipitate is obtained. In another experiment, 1 g of the same original mixture is heated to 300°C. Some vapours come out which are absorbed in acidified $AgNO_3$ solution by which 1.341 g of white precipitate is formed. The molecular mass of unknown chloride is
   (a) 53.4 (b) 58.5
   (c) 44.5 (d) 74.4

22. An amount of 0.15 moles of $K_2Cr_2O_7$ is required to oxidize a mixture of XO and $X_2O_3$ (total mass = 25.56 g) to form $XO_4^-$ and $Cr^{3+}$. If 0.21 moles of $XO_4^-$ is formed, then the correct information(s) is/are
   (a) Atomic mass of X = 100.
   (b) Moles of XO in original mixture = 0.06.
   (c) Moles of $Cr_2O_7^{2-}$ combined with $X_2O_3$ = 0.10.
   (d) Total moles of $H^+$ consumed = 0.99.

23. A volume of 10 ml of a mixture of $H_2$ and $O_2$ is exploded. If the final volume becomes 1 ml, the composition of original mixture may be
(a) 7 ml $H_2$, 3 ml $O_2$
(b) 6 ml $H_2$, 4 ml $O_2$
(c) 5 ml $H_2$, 5 ml $O_2$
(d) 3 ml $H_2$, 7 ml $O_2$

24. A definite volume of ammonia gas is passed through a series of electric sparks by which the volume becomes 90 ml. On washing with dilute orthophosphoric acid, the volume is reduced to 84 ml. Which of the following statement(s) is/are correct regarding the original ammonia sample?
(a) Its original volume was 45 ml.
(b) Its original volume was 48 ml.
(c) 12.5% of the original ammonia has decomposed.
(d) 87.5% of the original ammonia has decomposed.

25. To what extent must a given solution of concentration of 40 mg silver nitrate per ml be diluted to yield a solution of concentration of 16 mg silver nitrate per ml?
(a) Each ml should be diluted to 2.5 ml.
(b) To each ml of solution, 1.5 ml of water should be added.
(c) To 2.5 ml of solution, 2 ml of water should be added.
(d) To 1.5 ml of solution, 1.5 ml of water should be added.

26. An unknown volume of 40% (w/w) NaOH solution of specific gravity 1.6 is diluted until the specific gravity of the solution becomes 1.1. The strength of the resulting solution is
(a) 12.8% (w/v) (b) 10.67% (w/v)
(c) 11.6% (w/w) (d) 9.7% (w/w)

27. If a definite volume of '20 vol' $H_2O_2$ solution is diluted such that the volume of diluted solution becomes double than that of original volume, then
(a) the volume strength of diluted solution becomes '40 vol'.
(b) the molarity of solution becomes half of its initial molarity.
(c) the molality of solution becomes half of its initial molality.
(d) the maximum amount of $O_2$ gas obtainable from the solution remains the same.

28. A volume of 100 ml of M–NaCl solution, 100 ml of 2 M–$MgCl_2$ solution and 300 ml of 4 M–$Mg(NO_3)_2$ solution is mixed together and the mixture is diluted to 2 L. Which of the following is/are the correct final concentration of ions?
(a) $Na^+ = 0.05$ M (b) $Mg^{2+} = 0.7$ M
(c) $Cl^- = 0.15$ M (d) $NO_3^- = 1.2$ M

29. If the ratio of mole fractions of solute and solvent is unity, then the mass percent of solute is (Molar masses of solute and solvent are $X$ and $Y$, respectively)
(a) 50%
(b) $\frac{X}{X+Y} \times 100\%$
(c) $\frac{X}{Y} \times$ mass per cent of solvent
(d) $\frac{Y}{X} \times$ mass per cent of solvent

30. A quantity of 720 g water is added in 230 g ethanol at a certain temperature to get 1 L of solution. Which of the following is/are correct regarding the solution formed?
(a) The density of solution is 950 kg/m$^3$.
(b) The mole fraction of ethanol is 0.11.
(c) The molarity of solution is 5 M.
(d) The molality of solution is 6.94 M.

## Section C (Comprehensions)

### Comprehension I

**The first concept of atomic weight was given by Dalton. He defined that the absolute mass of an atom cannot be determined but we may compare the masses of atoms of different elements, perfectly, by knowing the chemical formula and percentage composition by mass of the compound formed by the elements concerned. The chemical or molecular formula can be determined with the help of Avogadro's hypothesis that is, under the similar conditions of pressure and temperature, equal volume of all the gases have equal number of molecules. Dalton defined the atomic weight of an element as the number of times by which one atom of the element is heavier than one atom of hydrogen. In order to determine the atomic weight of nitrogen, the following data are observed by experiments for a compound containing only nitrogen and hydrogen atoms.**

**Data I: The compound contains 88% nitrogen and 12% hydrogen by mass.**

**Date II: 10 ml of this gaseous compound exactly gives 10 ml nitrogen and 20 ml hydrogen on complete decomposition. (All volumes are at the same temperature and pressure)**

**1.** What is the molecular formula of the compound if both nitrogen and hydrogen are diatomic?

(a) $NH_3$ (b) $N_2H_4$
(c) $N_4H_2$ (d) $N_2H_2$

**2.** What is the atomic weight of nitrogen on this hydrogen scale?

(a) 14 (b) 14.67
(c) 14.33 (d) 13.67

**3.** What would be the molecular formula of the compound if nitrogen were triatomic and hydrogen were diatomic?

(a) $NH_3$ (b) $N_3H_4$
(c) $N_4H_3$ (d) $N_3H_2$

### Comprehension II

**A sample of hydrogen fluoride gas (only HF molecules) is collected in a vessel and left for some time. Then, a constant molar mass of the sample is experimentally determined as 34 g/mole. Assume that this abnormal molar mass is due to dimerization as well as trimerization of some HF molecules (no molecules in any other polymeric forms) and the mole ratio of monomeric and trimeric form of hydrogen fluoride molecules present is 4 : 1.**

**4.** What percentage of hydrogen fluoride molecules is dimerized?

(a) 50 (b) 58.8
(c) 76.47 (d) 17.65

**5.** What percentage of hydrogen fluoride molecules is trimerized?

(a) 40 (b) 58.8
(c) 76.47 (d) 17.65

**6.** What per cent of total molecules present in the final sample are $H_2F_2$ molecules?

(a) 10 (b) 40
(c) 50 (d) 58.8

## Comprehension III

**The vapour density of a gaseous mixture containing only Ar and $N_2O_4$ gases is 40. When the mixture is left for some time, the vapour density is decreased and finally becomes 37.5. It happened due to the dissociation of some $N_2O_4$ into $NO_2$. (Ar = 40)**

7. What is the degree of dissociation of $N_2O_4$?
   (a) 0.087 (b) 0.133
   (c) 0.067 (d) 0.0625

8. What is the initial mole ratio of Ar and $N_2O_4$?
   (a) 1 : 1 (b) 3 : 10
   (c) 1 : 3 (d) 1 : 5

9. What is the final mole ratio of Ar, $N_2O_4$ and $NO_2$?
   (a) 1 : 1 : 1 (b) 1 : 3 : 11
   (c) 45 : 137 : 26 (d) 4 : 13 : 3

## Comprehension IV

**When the Bayer's process is used for recovering aluminium from siliceous ores, some aluminium is always lost because of the formation of an unworkable mud having the following average formula: $3Na_2O{\cdot}3Al_2O_3{\cdot}5SiO_2{\cdot}5H_2O$. Since aluminium and sodium ions are always in excess in the solution from which this precipitate is formed, the precipitation of the silicon in the mud is complete. A certain ore contains 13% (by weight) Kaolin, $Al_2O_3{\cdot}2SiO_2{\cdot}2H_2O$ and 87% gibbsite, $Al_2O_3{\cdot}3H_2O$. (Al = 27, Si = 28)**

10. What per cent of the total aluminium in this ore is recoverable in the Bayer's process?
    (a) 80 (b) 90
    (c) 85 (d) 75

11. What is the percentage of silica present in the ore, by weight?
    (a) 2.82 (b) 3.02
    (c) 46.5 (d) 6.05

12. How many moles of $Al_2O_3$ are present per mole of ore?
    (a) 1.000 (b) 0.083
    (c) 0.222 (d) 0.242

## Comprehension V

**Vitamin C (M = 176) is a compound of C, H and O found in many natural sources, especially citrus fruits. When a 1.0 g sample of vitamin C is placed in a combustion chamber and burned, the following data is obtained.**

**Mass of $CO_2$ absorber after combustion = 85.35 g**
**Mass of $CO_2$ absorber before combustion = 83.85 g**
**Mass of $H_2O$ absorber after combustion = 37.96 g**
**Mass of $H_2O$ absorber before combustion = 37.55 g**

13. What is the percentage of carbon by weight in vitamin C ?
    (a) 66.67% (b) 40.9%
    (c) 20% (d) 60%

14. What is the percentage of hydrogen, by weight in vitamin C?
    (a) 4.55% (b) 41%
    (c) 20.5% (d) 9.11%

15. What is the empirical formula of vitamin C?
    (a) $CH_2O$ (b) $C_3H_4O_3$
    (c) $C_6H_8O_6$ (d) CHO

## Comprehension VI

**Figure shows a scheme for concentrating a dilute solution of NaOH.**

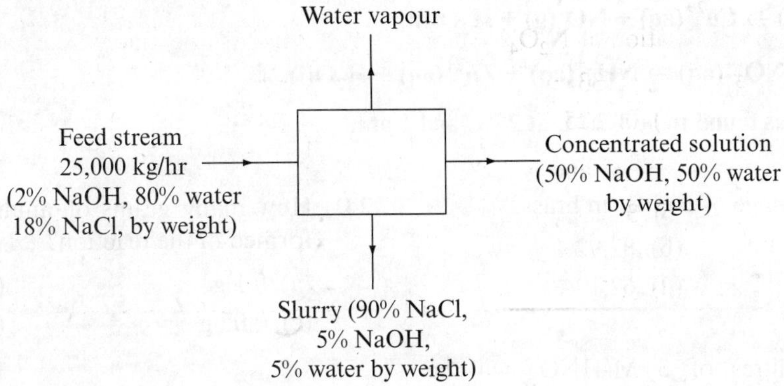


**16.** How much water is evaporated per hour?

(a) 5000 kg (b) 500 kg
(c) 19,500 kg (d) 20,000 kg

**17.** How much concentrated solution is obtained per hour?

(a) 5000 kg (b) 500 kg
(c) 19,500 kg (d) 20,000 kg

**18.** How much slurry is obtained per hour?

(a) 5000 kg (b) 500 kg
(c) 19,500 kg (d) 20,000 kg

---

## Comprehension VII

**A fuel mixture used in the early days of rocketry is composed of two liquids, hydrazine ($N_2H_4$) and dinitrogen tetraoxide ($N_2O_4$), which ignite on contact to form nitrogen gas and water vapour. The yield of $N_2$ gas is found to be less than its expected yield because some nitric oxide (NO) is also formed by a parallel reaction between the reactants. In an experiment, 96 g of $N_2H_4$ and 184 g of $N_2O_4$ are taken. It is found that 18 g of NO is formed.**

**19.** The limiting reagent is

(a) $N_2H_4$
(b) $N_2O_4$
(c) both the reactants will be used up completely.
(d) cannot predict, because the reactants are giving more than one reaction.

**20.** What is the highest percentage yield of $N_2$ that can be expected? The theoretical yield is the quantity of $N_2$ formed in the absence of parallel reaction.

(a) 96.67% (b) 90%
(c) 85.7% (d) 100%

**21.** What is the total mass of water vapour formed?

(a) 262 g (b) 140.2 g
(c) 108 g (d) 72 g

## Comprehension VIII

**A quantity of 1.9145 g of brass containing Cu and Zn reacts with 3 M-$HNO_3$ solution, the following reactions (unbalanced) take place.**

$$Cu(s) + HNO_3(aq) \rightarrow Cu^{2+}(aq) + NO_2(g) + H_2O(l)$$

$$Zn(s) + H^+(aq) + NO_3^-(aq) \rightarrow NH_4^+(aq) + Zn^{2+}(aq) + H_2O(l)$$

**The liberated $NO_2(g)$ was found to be 1.23 L at 27°C and 1 atm.**

22. What is the percentage of copper in brass?
(a) 17.08% (b) 82.92%
(c) 41.46% (d) 62.19%

23. How many millilitres of 3 M-$HNO_3$ will be required for complete reaction with brass?
(a) 33.33 ml (b) 37.5 ml
(c) 4.17 ml (d) 18.75 ml

24. How many grams of ammonium nitrate will be formed in the reaction?
(a) 0.1 g (b) 0.2 g
(c) 0.05 g (d) 0.025 g

## Comprehension IX

**Crude calcium carbide is made in an electric furnace by the following reaction.**

$$CaO + 3C \rightarrow CaC_2 + CO\uparrow$$

**The product contains 80% $CaC_2$ and 20% unreacted CaO.**

25. How much CaO is to be added to the furnace charge for each 1280 kg of pure $CaC_2$ produced?
(a) 1120 kg (b) 1440 kg
(c) 1152 kg (d) 1344 kg

26. How much CaO is to be added to the furnace charge for each 1280 kg of crude product?
(a) 1120 kg (b) 1440 kg
(c) 1152 kg (d) 1344 kg

27. What will be the volume of CO gas evolved, measured at 0°C and 1 atm, when 1280 kg of crude product is formed?
(a) 448 $m^3$ (b) 358.4 $m^3$
(c) 537.6 $m^3$ (d) 89.6 $m^3$

## Comprehension X

**A certain metal sulphide, $MS_n$ (where *n* is a small integer), is widely used as a high temperature lubricant. The substance is prepared by reaction of the metal pentachloride ($MCl_5$) with sodium sulphide ($Na_2S$). Heating the metal sulphide to 700°C in air gives the metal trioxide ($MO_3$) and sulphur dioxide ($SO_2$), which react with $Fe^{3+}$ ion under aqueous acidic conditions to give sulphate ion. Addition of aqueous $BaCl_2$ then forms a precipitate of $BaSO_4$. The chemical reactions (unbalanced) concerned are as follows.**

**(i) $MCl_5(s) + Na_2S(s) \rightarrow MS_n(s) + S(l) + NaCl(s)$**

**(ii) $MS_n(s) + O_2(g) \rightarrow MO_3(s) + SO_2(g)$**

**(iii) $SO_2(g) + Fe^{3+}(aq) \rightarrow Fe^{2+}(aq) + SO_4^{2-}(aq)$**

**(iv) $Ba^{2+}(aq) + SO_4^{2-}(aq) \rightarrow BaSO_4(s)$**

**Assume that you begin with 4.55 g of $MCl_5$ and that reaction (i) proceeds with 90% yield. After oxidation of $MS_n$ produced, oxidation of $SO_2$, and precipitation of $SO_4^{2-}$ ions, 6.99 g of $BaSO_4$ is obtained. (Ba = 137)**

**28.** How many moles of sulphur are present in the $MS_n$ sample?

(a) 0.01 (b) 0.02
(c) 0.03 (d) 0.04

**29.** Which of the following may be a permissible value of *n*?

(a) 3 (b) 2
(c) 4 (d) 6

**30.** If the value of *n* is 2, then the atomic weight of metal *M* is

(a) 95.5 (b) 232.5
(c) 125.8 (d) 187.6

## Comprehension XI

**Sixty millilitres of a mixture of equal volumes of chlorine and an oxide of chlorine were heated and then cooled back to the original temperature. The resulting gas mixture was found to have a volume of 75 ml. On treatment of caustic soda solution, the volume is contracted to 15 ml. Assuming that all measurements were made at the same temperature and pressure. The oxide of chlorine on heating decomposes quantitatively into oxygen and chlorine.**

**31.** What is the volume of chlorine in the original mixture?

(a) 15 ml (b) 30 ml
(c) 45 ml (d) 40 ml

**32.** What is the simplest formula of the oxide of chlorine?

(a) $ClO_2$ (b) $Cl_2O$
(c) $Cl_2O_3$ (d) $Cl_2O_5$

**33.** The gas finally present is

(a) $O_2$ (b) $Cl_2O$
(c) $Cl_2$ (d) $Cl_2O_5$

## Comprehension XII

**A volume of 18 ml of a gaseous mixture consisting of a gaseous organic compound A and just sufficient amount of oxygen is required for complete combustion yielding on burning 8 ml of $CO_2$, 12 ml of water vapour and 4 ml of $N_2$. All volumes are measured at the same temperature and pressure. The compound A contains only carbon, hydrogen and nitrogen.**

**34.** How many volumes of oxygen are required for complete combustion?

(a) 4 ml (b) 14 ml
(c) 7 ml (d) 11 ml

**35.** What is the molecular formula of the compound?

(a) $CH_5N$ (b) $C_2H_5N$
(c) $C_2H_6N_2$ (d) $C_4H_{10}N_2$

**36.** What volume of $H_2$ gas measured at the same temperature and pressure is needed for complete reduction of the same volume of compound A?

(a) 4 ml (b) 8 ml
(c) 28 ml (d) 14 ml

## Comprehension XIII

**A mixture of $CH_4$, $C_2H_4$ and $C_2H_2$ has a vapour density of 11.3 ml. When 10 ml of this mixture and 30 ml of oxygen are sparked together over aqueous KOH, the volume contracts to 5.5 ml and then disappears when pyrogallol is introduced. All volumes are measured under identical conditions of temperature, pressure and humidity.**

**37.** What was the volume ratio of $CH_4$, $C_2H_4$ and $C_2H_2$ in the original gaseous mixture?

(a) 2 : 2 : 3
(b) 3 : 3 : 4
(c) 4 : 3 : 3
(d) 1 : 1 : 3

**38.** If the mixture were not sparked over aqueous KOH, what was the total volume of resulting gases after sparking?

(a) 5.5 ml
(b) 25.5 ml
(c) 24.5 ml
(d) 21.5 ml

**39.** If the reactions were performed at 0°C and 1 atm, what should be the minimum mass of KOH present in the solution for complete conversion into $K_2CO_3$?

(a) 0.08 g
(b) 0.04 g
(c) 80 g
(d) 40 g

## Comprehension XIV

**Gaseous substances $(CHO)_p$ and $(COOH)_q$, when heated, decompose according to the following reactions.**

$$(CHO)_p \rightarrow CO + H_2$$

$$(COOH)_q \rightarrow CO_2 + H_2$$

**A volume of 10 ml of mixture containing $(CHO)_p$ and $(COOH)_q$ in 1:4 mole ratio was heated for complete decomposition. The resulting gases when passed through KOH, the volume decrease to $9/17^{th}$ of the volume of gaseous mixture passed. Remaining gases on complete combustion showed a contraction of 61 ml.**

**40.** The values of $p$ and $q$ are

(a) $p = 10, q = 6$
(b) $p = 6, q = 10$
(c) $p = 3, q = 5$
(d) $p = 5, q = 3$

**41.** The volume of $O_2$ used for combustion is

(a) 27 ml
(b) 20 ml
(c) 36 ml
(d) 60 ml

**42.** The increase in volume on decomposition of original mixture was

(a) 20 ml
(b) 46 ml
(c) 92 ml
(d) 61 ml

## Comprehension XV

**Once Tom and Jerry entered into a chemistry lab in which a chemist was preparing a solution of $H_2SO_4$ in a two litre container. He labelled the solution as 'density = 5.96 g/ml, 5m' which occupied half of the volume of the container. Tom saw the solution and a mischief came in his mind. As the chemist left the lab, Tom tried to throw the solution on Jerry. In doing so, some of the solution fell on the floor. Tom added water in the container to fill it up to the original level. The chemist returned back and got astonished to see the result of analysis that showed 'density = 0.4 g/ml and %w/w = 49'.**

**43.** What was the molarity of $H_2SO_4$ solution present initially in the container?

(a) 5 M
(b) 1.92 M
(c) 20 M
(d) 10 M

**44.** How many moles of $H_2SO_4$ had fallen down on the floor?

(a) 2
(b) 20
(c) 18
(d) 16

**45.** What volume of water was added to the solution by Tom?

(a) 1000 ml
(b) 900 ml
(c) 200 ml
(d) 100 ml

## Section D (Assertion–Reason)

The following questions consist of two statements. Mark the answer as follows.

(a) If both statements are CORRECT, and **Statement II** is the CORRECT explanation of **Statement I**.

(b) If both statements are CORRECT, and **Statement II** is CORRECT the CORRECT explanation of **Statement I**.

(c) If **Statement I** is CORRECT, but **Statement II** is INCORRECT.

(d) If **Statement I** is INCORRECT, but **Statement II** is CORRECT.

1. **Statement I:** The molecular mass of any substance is the sum of atomic masses of all the atoms present in each molecule of the substance.

   **Statement II:** The atomic as well as molecular masses are defined on the same carbon scale.

2. **Statement I:** The number of atoms in a given mass of dioxygen (oxygen) and trioxygen (ozone) gases is same.

   **Statement II:** The number of atoms depends on atomic mass, not on molecular mass.

3. **Statement I:** During a chemical reaction, the total moles remain constant.

   **Statement II:** During a chemical reaction, the total mass remains constant.

4. **Statement I:** For the reaction,

   $2A(g) + 3B(g) \rightarrow 4C(g) + D(g)$,

   the vapour density remains constant throughout the progress of reaction.

   **Statement II:** In all the gaseous chemical reactions, the vapour density remains constant.

5. **Statement I:** When any hydrocarbon is burnt, the moles of oxygen needed for complete combustion is always greater than the moles of hydrocarbon burnt.

   **Statement II:** Complete combustion of any substance requires more moles of oxygen than the moles of substance burnt.

6. **Statement I:** When 7.0 g nitrogen and 3.0 g hydrogen are allowed to react to form ammonia as a single product, 10.0 g ammonia is formed.

   **Statement II:** Chemical reactions follow the law of conservation of mass.

7. **Statement I:** The percentage yield of any product may be increased to more than 100% by adding more and more reactants to the reaction mixture.

   **Statement II:** Greater amount of reactants may result the production of greater amount of products.

8. **Statement I:** The mass ratio of reactants remains unchanged during the reaction, if they are taken in their stoichiometric amounts.

   **Statement II:** The mass ratio of products formed (in case of more than one products) is always independent from the relative masses of reactants taken.

9. **Statement I:** For the maximum yield of ammonia, the total amount of mixture of $N_2$ and $H_2$ should be taken in 1 : 3 mole ratio.

   **Statement II:** The yield of product becomes maximum when the reactants are taken in their stoichiometric amounts.

10. **Statement I:** Volumes of non-reacting gases are always additive.

    **Statement II:** Gases do not have their own volume.

11. **Statement I:** When a hydrocarbon is burnt and the products of combustion are cooled to the original temperature and pressure, a contraction in volume occurs.

    **Statement II:** The contraction in volume is solely due to the liquefaction of water vapours.

12. **Statement I:** Molarity and molality for very dilute aqueous solution is approximately equal.

    **Statement II:** For all aqueous solution, total mass of solvent is approximately equal to total volume of solution.

13. **Statement I:** Concentration of any solution is independent from the amount of solution, but it depends on the relative amount of solute and solvent.

    **Statement II:** Concentration of any solution has same magnitude in any unit to express concentration.

14. **Statement I:** For very dilute solutions, the strength of solution in w/w percent and in w/v percent have nearly equal value.
**Statement II:** For very dilute solution, the mass of solution becomes almost equal to the mass of solvent.

15. **Statement I:** One molar aqueous solution has always higher concentration than one molal.
**Statement II:** The molarity of a solution depends upon the density of the solution whereas molality does not.

## Section E (Column Match)

1. Match the columns.

| Column I Atomic masses | | | Column II Percentage composition of the heavier isotope |
|---|---|---|---|
| Isotope I | Isotope II | Average | |
| (A) $Z-1$ | $Z+2$ | $Z$ | (P) 33.33% by mole |
| (B) $Z+1$ | $Z+3$ | $Z+2$ | (Q) 50% by mole |
| (C) $Z$ | $3Z$ | $2Z$ | (R) % by mass depends on Z |
| (D) $Z-1$ | $Z+1$ | $Z$ | (S) 75% by mass |

2. Match the columns. ($N_A = 6 \times 10^{23}$)

| Column I | Column II |
|---|---|
| (A) 0.875 mole of $O_2$ gas | (P) 28 g |
| (B) 1.00 mole of $N_2$ gas | (Q) 22.4 L at 0°C and 1 atm |
| (C) 2.00 mole of $NaNO_3$ | (R) $1.20 \times 10^{24}$ atoms of nitrogen |
| (D) 0.4375 mole of $K_2SO_4$ | (S) $1.05 \times 10^{24}$ atoms of oxygen |
| | (T) 76.125 g |

3. Match the columns.

| Column I | Column II |
|---|---|
| (A) 3 mole of $Co(NH_3)_4SO_4$ | (P) 3 mole of S atom |
| (B) 1 mole of $FeKCo(NO_2)_6$ | (Q) 1 mole of Fe atom |
| (C) 1.5 mole of $[Fe(H_2O)_5SCN]SO_3$ | (R) 12 mole of O atoms |
| (D) 0.75 mole of $K_2Cu(SCN)_4$ | (S) 6 mole of N atoms |
| | (T) 1.5 mole of K atoms |

4. Match the columns.

| Column I (Average molecular mass) | Column II (Composition of gas mixture) |
|---|---|
| (A) $\frac{80}{3}$ | (P) $CH_4$ and $SO_3$ gases in 1:1 mole ratio |
| (B) $\frac{62}{3}$ | (Q) $CH_4$ and $SO_3$ gases in 1:1 mass ratio |
| (C) 48 | (R) $CH_4$ and $SO_3$ gases having H and O atoms in 8:3 ratio |
| (D) $\frac{112}{3}$ | (S) $CH_4$ and $C_2H_6$ gases having C and H atoms in 2:7 ratio |

5. When 1 mole of carbon reacts with 1 mole of oxygen producing 1 mole of $CO_2$, 100 kcal heat is released and when 1 mole of carbon reacts with 0.5 mole of oxygen producing 1 mole of CO, 25 kcal heat is released. Column I represents some amounts of carbon and oxygen which may react to form CO or $CO_2$ or both, in such a way that none of the reactant remain left, and Column II represents the heat released. Match the amounts with the corresponding heat released.

| Column I | Column II |
|---|---|
| (A) 36 g of C and 80 g of $O_2$ | (P) 125 kcal |
| (B) 12 g of C and 24 g of $O_2$ | (Q) 225 kcal |
| (C) 24 g of C and 48 g of $O_2$ | (R) 150 kcal |
| (D) 36 g of C and 64 g of $O_2$ | (S) 62.5 kcal |

**6.** Match the columns.

| Column I | Column II |
|---|---|
| (A) Amount of $O_2$ for complete combustion of 2 mole octane. | (P) 1100 g |
| (B) Amount of $CO_2$ produced when 300 g carbon combines with 800 g of oxygen. | (Q) 560 L of 273 K and 1 atm |
| (C) Amount of NaOH needed for complete neutralization of 1225 g $H_2SO_4$. | (R) 25 mole |
| (D) Amount of $N_2H_4$ formed from 50 mole $H_2$. | (S) $3.01 \times 10^{25}$ atoms |
| | (T) 800 g |

**7.** Match the columns

| Column I | Column II |
|---|---|
| (A) $N_2$ (3.5 g) + $H_2$ (1.0 g) $\rightarrow NH_3$ | (P) First reactant is the limiting reagent. |
| (B) $H_2$ (1.0 g) + $O_2$ (4.0 g) $\rightarrow H_2O$ | (Q) Second reactant is the limiting reagent. |
| (C) S (4.0 g) + $O_2$ (6.0 g) $\rightarrow SO_3$ | (R) Stoichiometric amounts of reactants. |
| (D) Fe (11.2 g) + $O_2$ (3.2 g) $\rightarrow Fe_2O_3$ | (S) Mass of reactants > Mass of product formed. |

**8.** Match the columns

| Column I<br>Compound | Column II<br>Relative amounts of products on complete combustion |
|---|---|
| (A) $CH_4$ | (P) Mole of $CO_2$ < Mole of $H_2O$ |
| (B) $C_2H_4$ | (Q) Mole of $CO_2$ = Mole of $H_2O$ |
| (C) $C_2H_2$ | (R) Mole of $CO_2$ > Mole of $H_2O$ |
| (D) $C_3H_8$ | (S) Mass of $CO_2$ > Mass of $H_2O$ |

**9.** Match the columns.

| Column I | Column II |
|---|---|
| (A) 200 ml of a mixture of 50% $H_2$, 40% $CH_4$ and 10% CO would evolve. The volume of $CO_2$ after combustion is | (P) 10 ml |
| (B) 100 ml of acetylene ($C_2H_2$) required oxygen for complete combustion is | (Q) 45 ml |
| (C) 10 ml of hydrogen sulphide ($H_2S$) required chlorine for complete decomposition | (R) 250 ml |
| (D) When a mixture of 30 ml of CO and 30 ml of $O_2$ was exploded, then the volume of gases produced due to explosion is | (S) 100 ml |

**10.** Match the columns.

| Column I<br>Masses of different components | Column II<br>Observation |
|---|---|
| (A) 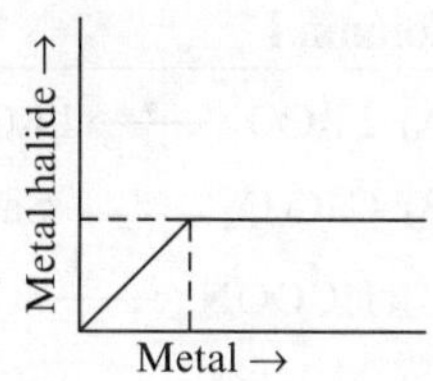 | (P) Metal is the limiting reagent. |
| (B) 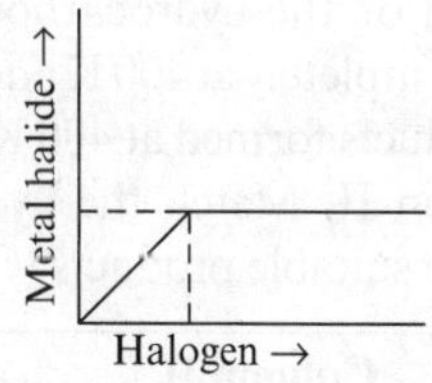 | (Q) Halogen is the limiting reagent. |
| (C) 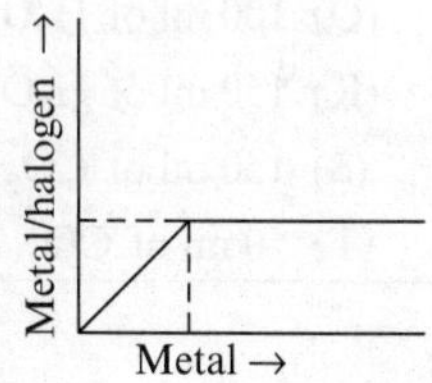 | (R) Metal and halogen are in stoichiometric amounts. |
| (D) 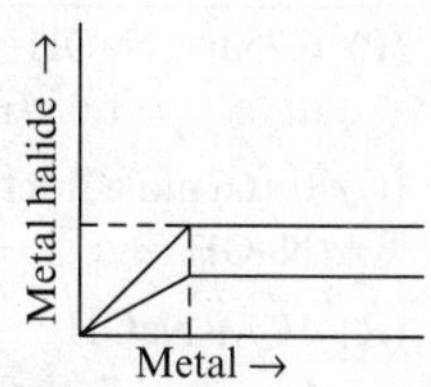 | (S) Metal is exhibiting a particular valency in the chloride formation. |
| | (T) Metal is exhibiting variable valency in the chloride formation. |

**11.** Match the columns.

| Section I (Gaseous organic compounds) | Section II (Volume of $O_2$ needed for complete combustion per volume of compound) |
|---|---|
| (A) $C_xH_{2x+2}$ | (P) $\frac{3x+1}{2}$ |
| (B) $C_xH_{2x+2}O$ | (Q) $\frac{3x}{2}$ |
| (C) $C_xH_{2x+3}N$ | (R) $\frac{3(2x+1)}{4}$ |
| (D) $C_xH_{2x+2}S$ | (S) $\frac{3(x+1)}{2}$ |

**12.** Column I consists of some decomposition reactions and Column II consists of some absorbent for the gases evolved in the reactions given in Column I. Match the gases evolved in Column I with the proper absorbent in Column II.

| Column I | Column II |
|---|---|
| (A) $Li_2CO_3 \xrightarrow{\Delta} Li_2O + CO_2\uparrow$ | (P) CaO |
| (B) $CaC_2O_4 \xrightarrow{\Delta} CaO + CO\uparrow + CO_2\uparrow$ | (Q) Ammonical CuCl |
| (C) $HCOONa \xrightarrow{\Delta} NaOH(s) + CO\uparrow$ | (R) $P_4O_{10}$ |
| (D) $2KHSO_3 \xrightarrow{\Delta} K_2SO_3 + H_2O\uparrow + SO_2\uparrow$ | (S) NaOH solution |

**13.** A volume of 50 ml of the hydrocarbons given in Column I is burnt completely at 400 K and 1 atm and the volumes of products formed at 400 K and 1 atm are given in Column II. Match the hydrocarbons (Column I) with the suitable products (Column II).

| Column I | Column II |
|---|---|
| (A) $CH_4$ | (P) 100 ml of $CO_2$ |
| (B) $C_2H_6$ | (Q) 100 ml of $H_2O$ |
| (C) $C_2H_4$ | (R) 150 ml of $H_2O$ |
| (D) $C_3H_4$ | (S) 150 ml of $CO_2$ |
| | (T) 50 ml of $CO_2$ |

**14.** Match the columns.

| Column I | Column II |
|---|---|
| (A) 400 g/L NaOH | (P) 6.25m – NaOH ($d_{solution}$ = 1.0 g/ml) |
| (B) 20% (w/w) NaOH | (Q) 0.166 mole fraction of NaOH ($d_{solution}$ = 1.3 g/ml) |
| | (R) 10 M-NaOH ($d_{solution}$ = 2 g/ml) |
| | (S) 7142.5 ppm ($d_{solution}$ = 1 g/ml) |

**15.** Match the columns.

| Column I | Column II |
|---|---|
| (A) 5m of NaOH solution ($d_{solution}$ = 0.6 g/ml) Molarity of solution is | (P) 16 M |
| (B) 250 ml of $H_2O_2$ solution provides 64 g of $O_2$. Molarity of $H_2O_2$ solution is | (Q) 1 M |
| (C) 100 ml of 1 M-$H_2SO_4$ solution ($d_{solution}$ = 1.5 g/ml) is mixed with 400 ml of water, density of final solution = 1.25 g/ml. Molarity of resulting solution is | (R) 2.5 M |
| (D) 100 ml of 6 M-NaCl solution is mixed with 100 ml of 17% (w/w) $AgNO_3$ solution ($d_{solution}$ = 8 g/ml). Molarity of $Ag^+$ ions in the resulting solution is | (S) 0.227 M |

## Section F (Subjective)

### Single-digit Integer Type

1. The density of mercury is 13.6 g/ml. The approximate diameter of an atom of mercury (in Å) assuming that each atom of mercury is occupying a cube of edge length equal to the diameter of the mercury atom is (Hg = 200)

2. Atoms of elements A, B and C combine to form a compound in the atomic ratio of 1 : 6 : 2. Atomic masses of A, B and C are 64, 9 and 16 amu, respectively. The maximum mass of the compound (in g) formed from 1.28 g of A, $3.0 \times 10^{23}$ atoms of B and 0.04 mole atom of C is

3. A compound which contains one atom of X and two atoms of Y for each three atoms of Z is made by mixing 5.0 g of X, $1.15 \times 10^{23}$ atoms of Y and 0.03 g-atoms of Z. If only 4.40 g of the compound results, then the value of atomic mass of Y divided by 10 is (The atomic masses of X and Z are 60 and 80, respectively.)

4. Recent controversial efforts to generate energy via 'cold fusion' of deuterium atoms have centred on the remarkable ability of palladium metal to absorb as much as 1120 times its own volume of deuterium gas at 1 atm and 0°C. The number of deuterium atoms per 10 atoms of Pd in a piece of fully saturated Pd metal is (Density of Pd = 11.8 g/ml and atomic mass of Pd = 106.2)

5. A solution contains 0.18 g/ml of a substance 'X', whose molecular mass is 64,000. It is found that 0.27 ml of oxygen at 760 mm and 300 K will combine with the amount of 'X' contained in 1 ml of the solution. The number of oxygen molecules combined with each molecule of 'X' is ($R$ = 0.08 L-atm/K-mol)

6. The number of ethoxy groups in an organic compound can be determined by the following reactions.

$R(OCH_2CH_3)_x + xHI \rightarrow R(OH)_x$
$+ xCH_3CH_2I$

$CH_3CH_2I + Ag^+ + H_2O \rightarrow CH_3CH_2OH$
$+ AgI(s) + H^+$

When 37 g of organic compound (molar mass = 176 g/mol) was treated as above, 148 g AgI was precipitated out. How many ethoxy groups are present in each molecule of the organic compound? (Ag = 108, I = 127)

7. A given sample of pure iron gains 10% of its weight on partially rusting to form $Fe_2O_3$. If the fraction of the iron converted to $Fe_2O_3$ is 'x', then the value of 30 times 'x' is (Fe = 56)

8. A sample of iron ore contains FeS and non-volatile inert impurity only. Roasting of this ore converts all FeS into $Fe_2O_3$ and a 4% loss in weight was observed. If the mass percent of FeS in the ore is '$x$', then the value of $\frac{x}{11}$ is (Fe = 56)

9. A volume of 50 ml of a gaseous mixture of hydrogen and hydrogen chloride was exposed to sodium amalgam. The volume is decreased to 40 ml. If 10 ml of the same mixture is mixed with 5 ml of gaseous ammonia and then exposed to water, then what will be the final volume (in ml) of gas left? All the volumes are measured at the same temperature and pressure.

10. A 1.174 g sample of special grade steel was treated appropriately with Chugaev's reagent by which nickel was precipitated as nickel dimethylglyoxime, $NiC_8H_{14}O_4N_4$. The dried precipitate weighed 0.2887 g. The percentage of nickel in the steel being analysed is (Ni = 58.7)

11. An amount of $2.5 \times 10^{-3}$ mole of an ion $A^{n+}$ exactly requires $1.5 \times 10^{-3}$ moles of $MnO_4^-$ for the oxidation of $A^{n+}$ to $AO_3^-$ in acid medium. What is the value of $n$?

12. A quantity of 1 g dry green algae absorbs $5.0 \times 10^{-3}$ moles of $CO_2$ per hour by photosynthesis. If the carbon atoms were all stored after photosynthesis as starch $(C_6H_{10}O_5)_n$, then how long (in hours) would it take for algae to increase its own weight by 81%, assuming that photosynthesis taking place at a constant rate?

13. Consider the production of tetraethyl lead according to the following reaction.

$4C_2H_5Cl + 4Na\text{–}Pb \rightarrow (C_2H_5)_4Pb + 4NaCl + 3Pb$

How many kilograms of ethyl chloride is required to produce enough tetraethyl lead (density = $\frac{12.96}{1.29}$ g/ml) needed for 500 litre of aviation fuel using 2 ml of tetraethyl lead per litre of fuel? (Pb = 208)

**14.** In one process of water proofing, a fibre is exposed to $(CH_3)_2SiCl_2$ vapour. The vapour reacts with hydroxyl groups on the surface of the fabric or with traces of water to form the waterproofing film $[(CH_3)_2SiO]_n$ by the following reaction:

$$n(CH_3)_2SiCl_2 + 2nOH^- \rightarrow 2nCl^- + nH_2O + [(CH_3)_2SiO]_n$$

Here, $n$ stands for a large integer. The waterproofing film is deposited on the fabric layer upon layer. Each layer is 3.7 Å thick (the thickness of the $(CH_3)_2SiO$ group). How much $(CH_3)_2SiCl_2$ (in g) is needed to waterproofing one side of a piece of fabric, 5.0 m by 4.0 m, with a film 200 layer thick? The density of film is $\frac{150}{129}$ g/ml. (Si = 28)

**15.** A magnesium ribbon, when burnt in air, left an ash containing MgO and $Mg_3N_2$. The ash was found to consume 0.6 mole of HCl, when it was taken in solution, according to the following reactions

$$MgO + 2HCl \rightarrow MgCl_2 + H_2O$$

$$Mg_3N_2 + 8HCl \rightarrow 3MgCl_2 + 2NH_4Cl$$

The solution so obtained was treated with excess of NaOH, when 0.1 mole of $NH_3$ was evolved. The mass (in g) of magnesium burnt is

**16.** A sample of $SF_5OF(g)$ was contained in a glass vessel at 117°C and a pressure of 380 mm. A quantity of $N_2F_4$ that was added brought the total pressure to 760 mm. The reaction that occurred produced a variety of products like $NF_3$, NO, $SiF_4$ (by the reaction with glass), $SF_6$, $SO_2F_2$, $SOF_4$, $SF_5ONF_2$ and $NO_2$. The yield of $SF_5ONF_2$ was 40 mole percent with respect to the reactant $SF_5OF$. All of the $SF_5OF$ and $N_2F_4$ were consumed in the reaction. What was the mass of $SF_5ONF_2$ produced (in g) if the volume of the vessel was 1.64 L? (F = 19)

**17.** An amount of 5 millimoles of $LiAlH_4$ was treated with 20 millimoles of $t$-butylalcohol. A total of 15 millimoles of hydrogen was evolved for the following reaction.

$$LiAlH_4 + 3(CH_3)_3COH \rightarrow Li[(CH_3)_3CO]_3AlH + 3H_2$$

The addition of an excess of another alcohol, methanol, to the above reaction mixture caused the fourth H atom of the $LiAlH_4$ to be replaced according to the following equation.

$$Li[(CH_3)_3CO]_3AlH + CH_3OH \rightarrow Li[(CH_3)_3CO]_3(CH_3O)Al + H_2$$

How many millimoles of $H_2$ was evolved due to the addition of $CH_3OH$?

**18.** To analyse cast iron for its sulphur content, a 6.4 g portion of the iron was weighed out for analysis and treated , where it was dissolved in hydrochloric acid, the hydrogen sulphide evolved from iron sulphide was distilled off and made to be absorbed by a solution of a cadmium salt, after which CdS was treated with an excess of a solution of $CuSO_4$, and the CuS precipitated formed was ignited. As a result, 0.795 g of an ignited CuO precipitate was obtained. Calculate the percentage content of sulphur in the cast iron. (Cu = 63.5)

**19.** A mixture containing 1.3 millimoles of $HNF_2$ gas and equal quantity of chlorine gas, was led into a flask containing 5.0 g of KF and allowed to stand for 18 hours at room temperature. The gas $ClNF_2$ (66.67% yield) and the solid KF-HCl were formed. If the volume percent of $ClNF_2$ in the gaseous mixture present after the reaction is $X$, then the value of $\frac{X}{10}$ is

**20.** In 1.885 g sample of a mixture of $MgSO_4 \cdot 7H_2O$ and $MgCl_2 \cdot 6H_2O$ containing some inert impurity was subjected to suitable treatment, as a result of which there were obtained 0.699 g of $BaSO_4$ and 0.888 g of $Mg_2P_2O_7$. The mass percentage of impurity is (Ba = 137, Mg = 24, P = 31)

## Four-digit Integer Type

**1.** A sample of ammonia contains only $H^1$ and $H^2$ isotopes of hydrogen in 4 : 1 ratio and $N^{14}$ and $N^{15}$ isotopes of nitrogen in 3 : 1 ratio. How many neutrons are present in 1.785 mg of ammonia? (Answer in the order $10^{18}$) ($N_A = 6 \times 10^{23}$)

**2.** The atomic ratio of $H^1$ to $H^3$ in a sample of water is $1 : 8 \times 10^{-8}$. How many $H^3$ atoms are present in 9.0 g of such water sample? (Answer in the order $10^{15}$) ($N_A = 6 \times 10^{23}$)

3. Assume that a polyethylene chain is truly linear. If a polymer chain had a molecular mass of $1\times10^6$, then what will be the length of one polyethylene molecule (in μm)? A carbon–carbon single bond length is 154 pm.

4. Chemical formula of a chelating agent versene is $C_2H_4N_2(C_2H_2O_2Na)_4$. If each mole of this compound could bind 1 mol of $Ca^{2+}$, then what would be the rating of pure versene, expressed as mg $CaCO_3$ bound per g of chelating agent? Here, $Ca^{+2}$ is expressed in terms of the amount of $CaCO_3$ it could form.

5. A polymeric substance, tetrafluoroethylene, can be represented by the formula $(C_2F_4)_x$, where $x$ is a large number. The material was prepared by polymerizing $C_2F_4$ in the presence of a sulphur-bearing catalyst that serves as a nucleus upon which the polymer grew. The final product was found to contain 0.012% S. What is the value of $x$, if each polymeric molecule contains one sulphur atom? Assume that the catalyst contributes a negligible amount to the total mass of the polymer. (F = 19, S = 32)

6. A compact car gets 20 miles per litre on the highway. Gasoline contains 84.0% carbon by mass and has a density of 0.80 g/ml. The mass of $CO_2$ produced (in g) during a 50 mile-trip is

7. A quantity of 2.0 g nitrate of univalent metal was heated with excess of previously ignited silica. A loss in weight of 1.08 g took place due to the total expulsion of the nitrate part of the salt as $N_2O_5$. The mass percentage of $NO_3^-$ group in the salt analysed is

8. A certain metal 'M' forms an insoluble oxalate complex $M_4O_3(C_2O_4)_3\cdot 12H_2O$. If 3.2 g of the complex is formed from 1 g of oxalic acid, then what is the atomic mass of M?

9. The maximum mass (in g) of $AlCl_3$, which may be formed from 321 g of a mixture of $Al_2O_3$ and HCl is (Al = 27)

10. Chlorine gas can be produced in the laboratory by the following reaction.

$K_2Cr_2O_7 + 14HCl \rightarrow 2KCl + 2CrCl_3 + 7H_2O + 3Cl_2$

If 75 g sample of $K_2Cr_2O_7$, i.e., s 98% pure is allowed to react with 365 ml of HCl solution having density of 1.2 g/ml and containing 28% HCl by mass, 'x' g of chlorine is produced. The value of '100$x$' is

11. A fluorine disposal plant was constructed to carry out the following reactions.

$2F_2 + 4NaOH \rightarrow 4NaF + 2H_2O + O_2$

$2NaF + CaO + H_2O \rightarrow CaF_2 + 2NaOH$

As the plant operated, excess lime was added to bring about complete precipitation of the fluorides as $CaF_2$. Over a period of operation, 1900 kg of fluorine was fed into the plant and 10,000 kg of lime was required. What was the percentage utilization of lime? (Ca = 40, F = 19)

12. A sample of chalk contained as impurity a form of clay which losses 20% if its weight as water on strong heating. A 5 g of chalk sample on heating shows a loss in weight by 1.636 g. The mass percentage of $CaCO_3$ in the chalk sample is (Ca = 40)

13. An impure sample of iron pyrite contains 28% iron, the impurity being silica. If 100 g of the sample is roasted to oxidize all the $FeS_2$ to $Fe_2O_3$, then what will be the mass of the roasted sample, in g? (Fe = 56)

14. Chlorine samples are prepared for analysis by using NaCl, KCl and $NH_4Cl$ separately or as mixture. What minimum volume (in ml) of 8.5%, by mass, $AgNO_3$ solution (specific gravity = 1.25) must be added to a sample of 10.7 g in order to ensure complete precipitation of chloride in every possible case?

15. A gas mixture contains $CH_4$ and $C_3H_6$. When this mixture undergo cracking into C(s) and $H_2$(g), the total number of moles of $H_2$(g) obtained is 42. If the total volume of the initial gas mixture at 1.5 atm and 27°C is 246.0 L, then what is the mole per cent of $CH_4$ gas in the initial mixture?

16. A solid mixture (5.02 g) containing lead nitrate and sodium nitrate was heated below 600°C until the weight of residue becomes constant. If the loss in weight is 1.4 g, then the amount of lead nitrate (in mg) in the mixture is (Pb = 208)

17. Octane is a component of gasoline. Complete combustion of octane leads to $CO_2$ and $H_2O$ while incomplete combustion produces CO and $H_2O$, which not only reduces the efficiency of the engine using the fuel but it is also toxic. In a certain test run, one gallon of octane is burned in an engine. The total mass of CO, $CO_2$ and $H_2O$ produced is 9.768 kg. Calculate the efficiency of the process, i.e., calculate the percentage of octane converted to $CO_2$. The density of octane is 2.28 kg/gallon.

**18.** A volume of 100 ml of water gas containing some $CO_2$ was mixed with 100 ml of oxygen and mixture exploded. The volume after explosion was 100 ml. On introducing NaOH, the volume was reduced to 52.5 ml. If the volume ratio of CO, $H_2$ and $CO_2$ in the original sample is ab : cd : 2, then the value of 'abcd' is

**19.** When 10 ml of acetic acid (density = 0.8 g/ml) is mixed with 40 ml of water (density = 1 g/ml) at a certain temperature, the final solution is found to have a density of $\frac{96}{98}$ g/ml. The percent change in total volume on mixing is (Answer as 'abcd' where the value of 'a' is 1 in case of increase in volume and 2 in case of decrease in volume, and 'bcd' is the magnitude of percentage change in volume)

**20.** The enzyme carbonic anhydrase catalyses the hydration of $CO_2$. The reaction $CO_2 + H_2O \rightarrow H_2CO_3$ is involved in the transfer of $CO_2$ from tissues to the lungs through the bloodstream. One enzyme molecule hydrates $10^6$ molecules of $CO_2$ per second. How many grams of $CO_2$ are hydrated in 1 hour by 10 ml of $10^{-6}$ M enzyme?

---

## Answer Keys — Exercise I

### Laws of Chemical Combinations

1. (a) 2. (a) 3. (a) 4. (c) 5. (b) 6. (c) 7. (c) 8. (a) 9. (d) 10. (b)

### Atomic Mass

11. (b) 12. (b) 13. (a) 14. (b) 15. (a) 16. (d) 17. (a) 18. (b) 19. (b) 20. (b)

### Molecular Mass

21. (b) 22. [illegible] 23. (c) 24. (d) 25. (c) 26. (a) 27. (c) 28. (a) 29. (c) 30. (d)
31. (b) 32. (d) 33. (b) 34. (d) 35. (b)

### Calculation of Mole

36. (a) 37. (c) 38. (b) 39. (a) 40. (a) 41. (c) 42. (c) 43. (a) 44. (d) 45. (b)
46. (a) 47. (c) 48. (c) 49. (b) 50. (b)

### Average Molecular Mass

51. (b) 52. (d) 53. (b) 54. (b) 55. (a) 56. (a) 57. (b) 58. (d) 59. (b) 60. (c)

### Percentage Composition

61. (b) 62. (a) 63. (d) 64. (b) 65. (b) 66. (c) 67. (b) 68. (b) 69. (a) 70. (a)

### Empirical and Molecular Formula

71. (d) 72. (c) 73. (d) 74. (b) 75. (b) 76. (b) 77. (a) 78. (d) 79. (b) 80. (b)

### Stoichiometry

81. (c) 82. (b) 83. (d) 84. (b) 85. (a) 86. (c) 87. (b) 88. (a) 89. (a) 90. (d)
91. (b) 92. (c) 93. (b) 94. (c) 95. (a)

### Limiting Reagent Based

96. (b) 97. (a) 98. (b) 99. (b) 100. (b)

### Sequential and Parallel Reactions

101. (b) 102. (a) 103. (b) 104. (c) 105. (b)

### Percentage Based

106. (b) 107. (d) 108. (c) 109. (c) 110. (a) 111. (c) 112. (d) 113. (a) 114. (c) 115. (d)

### Eudiometry

116. (b) 117. (b) 118. (a) 119. (b) 120. (b) 121. (b) 122. (a) 123. (b) 124. (d) 125. (c)
126. (c) 127. (a) 128. (d) 129. (b) 130. (d)

### Concentration Terms

131. (d) 132. (d) 133. (a) 134. (c) 135. (d) 136. (a) 137. (a) 138. (a) 139. (a) 140. (a)
141. (a) 142. (a) 143. (a) 144. (c) 145. (d) 146. (c) 147. (a) 148. (c) 149. (a) 150. (d)

## Answer Keys — Exercise II

### Section A (Only one Correct)

| | | | | | | | | | |
|---|---|---|---|---|---|---|---|---|---|
| 1. (b) | 2. (d) | 3. (c) | 4. (d) | 5. (c) | 6. (c) | 7. (d) | 8. (d) | 9. (b) | 10. (c) |
| 11. (b) | 12. (a) | 13. (a) | 14. (b) | 15. (a) | 16. (b) | 17. (a) | 18. (c) | 19. (b) | 20. (a) |
| 21. (c) | 22. (c) | 23. (a) | 24. (c) | 25. (a) | 26. (c) | 27. (a) | 28. (a) | 29. (a) | 30. (b) |
| 31. (a) | 32. (b) | 33. (d) | 34. (d) | 35. (b) | 36. (a) | 37. (b) | 38. (d) | 39. (b) | 40. (b) |
| 41. (c) | 42. (a) | 43. (b) | 44. (a) | 45. (a) | 46. (a) | 47. (c) | 48. (b) | 49. (c) | 50. (c) |
| 51. (b) | 52. (c) | 53. (d) | 54. (d) | 55. (b) | 56. (a) | 57. (b) | 58. (c) | 59. (a) | 60. (b) |
| 61. (b) | 62. (b) | 63. (a) | 64. (b) | 65. (a) | 66. (a) | 67. (b) | 68. (c) | 69. (c) | 70. (b) |
| 71. (c) | 72. (a) | 73. (c) | 74. (d) | 75. (a) | 76. (c) | 77. (c) | 78. (c) | 79. (a) | 80. (d) |

### Section B (One or More than one Correct)

| | | | |
|---|---|---|---|
| 1. (b), (c), (d) | 2. (a), (c) | 3. (c), (d) | 4. (c), (d) |
| 5. (a), (c), (d) | 6. (a), (b), (d) | 7. (b), (c) | 8. (a) |
| 9. (a), (b), (c) | 10. (b), (d) | 11. (a), (b), (c) | 12. (b), (c) |
| 13. (c), (d) | 14. (a), (b), (c) | 15. (a), (c), (d) | 16. (a), (b), (c), (d) |
| 17. (a), (d) | 18. (a), (b), (c) | 19. (a), (b), (d) | 20. (a), (d) |
| 21. (a) | 22. (b) | 23. (a), (b) | 24. (b), (d) |
| 25. (a), (b) | 26. (b), (d) | 27. (b), (d) | 28. (a), (b), (d) |
| 29. (b), (c) | 30. (a), (b), (c), (d) | | |

### Section C

**Comprehension I**

1. (b) 2. (b) 3. (b)

**Comprehension II**

4. (b) 5. (d) 6. (c)

**Comprehension III**

7. (a) 8. (b) 9. (c)

**Comprehension IV**

10. (b) 11. (d) 12. (a)

**Comprehension V**

13. (b) 14. (a) 15. (b)

**Comprehension VI**

16. (c) 17. (b) 18. (a)

**Comprehension VII**

19. (a) 20. (a) 21. (c)

**Comprehension VIII**

22. (b) 23. (b) 24. (a)

**Comprehension IX**

25. (b) 26. (c) 27. (b)

**Comprehension X**

28. (c) 29. (b) 30. (a)

**Comprehension XI**

31. (b) 32. (b) 33. (a)

**Comprehension XII**

34. (b) 35. (c) 36. (a)

**Comprehension XIII**

37. (c) 38. (d) 39. (a)

**Comprehension XIV**

40. (a) 41. (a) 42. (c)

**Comprehension XV**

43. (c) 44. (c) 45. (b)

## Section D (Assertion – Reason)

1. (a) 2. (a) 3. (d) 4. (c) 5. (c) 6. (d) 7. (d) 8. (b) 9. (a) 10. (d)
11. (c) 12. (c) 13. (c) 14. (d) 15. (d)

## Section E (Column Match)

1. A → P, R; B → Q, R; C → Q, S; D → Q, R
2. A → P, S; B → P, Q, R; C → R; D → S, T
3. A → P, R; B → Q, R, S; C → P, R; D → P, T
4. A → Q; B → S; C → P; D → R,
5. A → Q; B → S; C → P; D → R
6. A → Q, R, S, T; B → P, Q, R; C → R; D → Q, R, T
7. A → P, S; B → Q, S; C → R; D → Q, S
8. A → P, S; B → Q, S; C → R, S; D → P, S
9. A → Q, S; B → P, S; C → R, S; D → Q, T
10. A → S; B → R; C → P; D → Q
11. A → P; B → Q; C → R; D → S
12. A → P, S; B → P, Q, S; C → Q; D → P, R, S
13. A → Q, T; B → P, R; C → P, Q; D → Q, S
14. A → Q, R; B → P, R
15. A → R; B → P; C → S; D → Q

## Section F (Subjective)

### Single-digit Integer Type

1. (3) 2. (3) 3. (7) 4. (9) 5. (4) 6. (3) 7. (7) 8. (4) 9. (6) 10. (5)
11. (2) 12. (6) 13. (8) 14. (3) 15. (6) 16. (2) 17. (5) 18. (5) 19. (5) 20. (7)

### Four-digit Integer Type

1. (0471) 2. (0048) 3. (0011) 4. (0263) 5. (2667)
6. (6160) 7. (0062) 8. (0084) 9. (0267) 10. (5112)
11. (0028) 12. (0053) 13. (0080) 14. (0320) 15. (0020)
16. (3320) 17. (0080) 18. (1721) 19. (0002) 20. (1584)

# HINTS AND EXPLANATIONS

## EXERCISE I (JEE MAIN)

### Laws of Chemical Combinations

**1.** Mass of $CO_2$ formed = 27.5 g

From mass conservation, mass of water formed

$= 10 + 40 - 27.5 = 22.5$ g

**2.** Total mass of reactants taken

$= 10.8 + (50 \times 1.2) = 70.8$ g

Hence, mass of $H_2S$ gas released

$= 70.8 - 65.2 = 5.6$ g

Hence, volume of $H_2S$ gas released

$$= \frac{5.6}{1.4} = 4.0 \text{ L}$$

**3.** Mass of Iron(III) oxide left

$= (5.4 + 18.5) - (11.2 + 10.2) = 2.5$ g

**4.** Density of the liquid in bottle $= \frac{22.3}{15} = 1.487$ g/cm$^3$

**5.** Let the mass of ethanol in the solution be '$x$' g.

As, $V_{\text{solution}} = V_{\text{ethanol}} + V_{\text{water}}$.

$$55.0 = \frac{x}{0.8} + \frac{50-x}{1.0} \Rightarrow x = 20$$

Hence, mass per cent of water $= \frac{20}{50} \times 100 = 40$

**6.** In 100 g of weak liquor, the mass of caustic soda present = 4 g. As on evaporation, only loss in mass of water will occur, the final mass of solid must be 4 g. If '$x$' g water is evaporated, then

$$(100 - x) \times \frac{25}{100} = 4 \Rightarrow x = 84.$$

**7.** Mass of water molecules in 1 litre of steam = $0.0006 \times 1000 = 0.6$ g. As for $H_2O$ system, the most condensed state is liquid water, we may assume that in liquid water, the intermolecular space is negligible in comparison to the volume occupied by the liquid molecule. Hence, the volume occupied by the water molecules $= \frac{0.6}{1.0} = 0.6$ mL.

**8.** Mass of butter should be taken per day

$$= \frac{2.0 \times 10^{-3}}{5.5 \times 10^{-6}} = 363.6 \text{ g}$$

**9.** Aluminium has a variable valency.

**10.** In compund 'Y', 4 g of A will combine with $\frac{7}{3} \times 12$ = 28 g of B, and hence, 8 g of A will combine with $2 \times 28 = 56$ g of B.

### Atomic Mass

**11.** Atomic mass of the element

$$= \frac{8.0 \times 10^{-18}}{3.2 \times 10^{5}} \times 6 \times 10^{23} = 15$$

**12.**

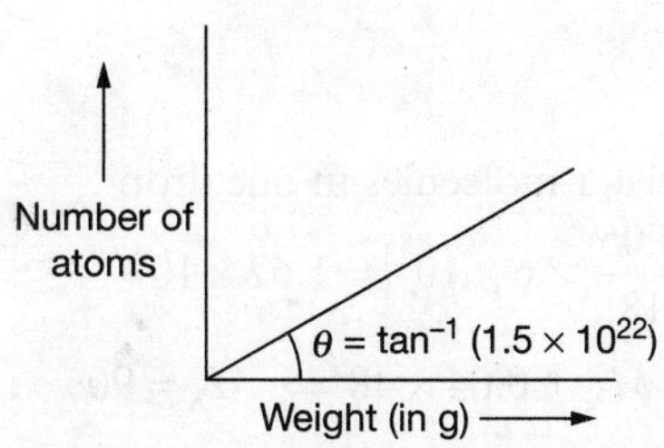


$$\text{Slope} = \frac{N_A}{\text{Atomic mass}} = 1.5 \times 10^{22}$$

$\Rightarrow$ Atomic mass = 40

**13.** Number of C-atoms $= \frac{3.0 \times 10^{-10}}{12} \times 6 \times 10^{23}$

$= 1.5 \times 10^{13}$

**14.** As the mole of atoms is same, the number of atoms is also same.

**15.** $\frac{4}{9} = \frac{40 / A_X}{60/16} \Rightarrow$ Atomic mass of X, $A_X = 24$

**16.** $\frac{1}{A_X + 12 + 3 \times 16} = \frac{1.11}{A_X + 2 \times 35.5}$

$\Rightarrow$ Atomic mass of X, $A_X = 40$.

**17.** $20.18 = \frac{20 \times 90 + x \times 21 + (10 - x) \times 22}{100} \Rightarrow x = 2$

18. Number of neutrons $= \frac{3}{1} \times 6.022 \times 10^{23} \times \frac{1}{6001}$

$= 3.01 \times 10^{20}$

19. Average atomic mass of C $= \frac{12 \times 98 + 14 \times 2}{100} = 12.04$

Number of $C^{14}$ atoms

$= \frac{2}{100} \times \frac{12}{12.04} \times 6.022 \times 10^{23} = 1.20 \times 10^{22}$

20. Average mass number $= 0.6 \times 35 + 0.4 \times 37 = 35.8$

## Molecular Mass

21. $20 \times 80 = N \times 32 \Rightarrow N = 50$

22. $\frac{w}{44} = \frac{20}{32} \Rightarrow w = 27.50$ g

23. $\frac{2 \times 10^{21}}{6 \times 10^{23}} \times 45 + \frac{3 \times 10^{21}}{6 \times 10^{23}} \times M_Q = 0.6 \Rightarrow M_Q = 90$

24. Volume of each virus particle

$= \pi r^2 l = \pi \left(\frac{170}{2} \times 10^{-8} \text{ cm}^2\right) \times (3000 \times 10^{-8} \text{ cm})$

$= 6.806 \times 10^{-17}$ cm$^3$

Mass of each virus particle

$= \frac{6.806 \times 10^{-17}}{12.5} = 5.4448 \times 10^{-18}$

$\therefore$ Molecular mass of virus

$= 5.4448 \times 10^{-18} \times 6 \times 10^{23} = 3.27 \times 10^{6}$

25. Volume of each molecule

$= \frac{\text{Mass}}{\text{Density}} = \frac{6 \times 10^{8}}{6 \times 10^{23} \times 1.1} = 9.09 \times 10^{-16}$ mL

26. If the atomicity of mercury is '$x$', then $200x$

$= 6.92 \times 29 \Rightarrow x \approx 1.$

27. Molecular mass $= 1.2 \times 30 = 36$

28. Molecular mass $= 7 \times 12 + 2 \times 16 + (9.96 \times 10^{-24} \times 6 \times 10^{23}) \approx 122$

29. New molecular mass of $H_2O = 2 \times 0.5 + 1 \times 20 = 21$

Percentage increase in molecular mass of $H_2O$

$= \frac{21 - 18}{18} \times 100 = 16.67\ \%$

30. Same mass of same element, hence same number of atoms.

31. Number of electrons

$= (2 \times 6 + 4 \times 8 + 2) \times \frac{4.4}{88} \times N_A = 2.3\ N_A$

32. Number of valence electrons

$= (2 \times 6 + 2) \times \frac{6.4}{32} \times N_A = 2.8\ N_A$

33. Number of $F^-$ ions $= \frac{4.2}{27 + 3 \times 19} \times 6.02 \times 10^{23} \times 3$

$= 9.03 \times 10^{22}$

34. Number of electrons lost $= \frac{13.5}{27} \times 6.02 \times 10^{23} \times 3$

$= 9.03 \times 10^{22}$

35. Number of iodide ions $= \frac{0.02}{166} \times 6 \times 10^{23}$

$= 7.23 \times 10^{19}$

## Calculation of Mole

36. $n = \frac{w}{M} = \frac{1}{153} = 0.00654$

37. $V = \frac{w}{d} = \frac{1.2 \times 46}{0.8} = 69$ mL

38. Volume of 18 g of water is 18 mL. Hence, the mass of one drop of water $= \frac{1}{20} = 0.05$ g.

Number of water molecules in one drop

$= \frac{0.05}{18} \times 6 \times 10^{23} = 1.67 \times 10^{21}$

39. $3.72 = 0.02 \times M_X + 0.04 \times 48 \Rightarrow M_X = 90$

40. Loschmidt number $= \frac{1}{22400} \times 6 \times 10^{23} = 2.688 \times 10^{19}$

41. $M = \frac{0.25}{50} \times 22400 = 112$ g/mol

**42.** 6 moles of chlorine atoms is 3 moles of chlorine molecules and hence, the volume $= 3 \times 22.7 = 68.1$ L

**43.** $n = \frac{PV}{RT} = \frac{100\times10^3\times16.628\times10^{-3}}{8.314\times300} = 0.67$

**44.** $\frac{1}{M}\times4 = \frac{1\times(1.642/1000)}{0.0821\times300} \Rightarrow M = 60000$

**45.** $\frac{2}{3A} = \frac{1\times(448/1000)}{0.0821\times273} \Rightarrow$ Atomic mass, A $= \frac{100}{3}$

Hence, mass of each atom $= \frac{100}{3}\times 1.66\times10^{-24}$

$= 5.53 \times 10^{-23}$ g

**46.** Number of g-atoms of chlorine

$= \frac{1.4\times10^{21}\times19}{35.5} = 7.49\times10^{20}$

**47.** Number of g-atoms left $= \frac{2\times10^{-3}}{40} - \frac{1.2\times10^{19}}{6.0\times10^{23}}$

$= 3\times10^{-5}$

**48.** Number of g-molecules of oxygen

$= \frac{6.022\times10^{24}}{6.022\times10^{23}}\times\frac{1}{2} = 5$

**49.** $V_{O_2} : V_{H_2} : V_{CH_4} = n_{O_2} : n_{H_2} : n_{CH_4}$

$= \frac{w}{32} : \frac{w}{2} : \frac{w}{16} = 1 : 16 : 2$

**50.** $\frac{w_{O_2}}{32} = \frac{w_{SO_2}}{64} \Rightarrow w_{O_2} = \frac{1}{2}\times w_{SO_2}$

## Average Molecular Mass

**51.** As the molar mass of $H_2O$ is less than that of $O_2$ and $N_2$, the molar mass of moist air is less than that of dry air.

**52.** The average molar mass of $H_2F_2$ and $H_3F_3$ may be 50.

**53.** $37.60 = \frac{70\times28+30\times M}{100} \Rightarrow M = 60$

**54.** $\frac{100}{M} = \frac{90}{2} + \frac{10}{4} \Rightarrow M = 2.105$

**55.** Average molar mass of the mixture

$= \frac{22400}{4480}\times10 = 50$

Now, $\frac{100}{50} = \frac{x}{30} + \frac{100-x}{70}$

$\Rightarrow$ Mass per cent of $C_2H_6$, $x = 30$

**56.** Average molar mass of the mixture $= \frac{16+30}{2} = 23$

Hence, the density of mixture $= \frac{23}{22.4} = 1.03$

**57.** $(n + 0.05) = \frac{1\times6}{0.08\times300} \Rightarrow n = 0.20$

**58.** Average molar mass of the mixture

$= \frac{40\times2+60\times4}{100} = 3.2$

Number of moles of gases present $= \frac{10}{3.2} = 3.125$

**59.** $0.4 = \frac{48-M}{\left(\frac{3}{2}-1\right)M} \Rightarrow M = 40$

**60.** $\alpha = \frac{80-56}{\left(\frac{3}{2}-1\right)56} = \frac{6}{7}$

## Percentage Composition

**61.** Mass of hydrogen $= \frac{8\times1}{9\times12}\times0.968 = 0.0717$ g

**62.** Moles of O-atoms = 9 × Moles of Cu

**63.** $x\times(+3) + (0.93 - x)\times(+2) + 1\times(-2) = 0$

$\Rightarrow x = 0.14$

Percentage of iron as Fe(III) $= \frac{0.14}{0.93}\times100$

$= 15.05\%$

**64.** $\frac{1.8}{18n} = \frac{5-1.8}{160} \Rightarrow n = 5$

**65.** $69.98 = \frac{12\times21}{M}\times100 \Rightarrow M = 360.1$

66. $10.46 = \frac{3\times 80}{314+108n} \times 100 \Rightarrow n \approx 19$

67. $36 = \frac{2\times 12}{M} \times 100 \Rightarrow M = \frac{200}{3}$

Hence, moles of compound $= \frac{10}{\left(\frac{200}{3}\right)} = 0.15$

68. $M_{\min} = \frac{100}{4} \times 16 = 400$

69. Mass of nitrogen in the compound

$= \frac{28}{22400} \times 28 = 0.035$ g

Percentage of nitrogen in the compound

$= \frac{0.035}{0.2} \times 100 = 17.5\%$

70. Mass per cent of carbon

$= \frac{12}{44} \times \frac{0.147}{0.2} \times 100 = 20.05$

Mass per cent of hydrogen

$= \frac{2}{18} \times \frac{0.12}{0.2} \times 100 = 6.67$

Hence, mass per cent of oxygen

$= 100 - (20.05 + 6.67) = 73.28\%$

## Empirical and Molecular Formula

71. $V_{\text{gas}} = \frac{1}{2} \times V_{\text{nitrogen}} \Rightarrow n_{\text{gas}} = \frac{1}{2} \times n_{\text{nitrogen}}$

$\Rightarrow \frac{w}{M_{\text{gas}}} = \frac{1}{2} \times \frac{w}{28} \Rightarrow M_{\text{gas}} = 56$

Now, $56 = 14 \times n \Rightarrow n = 4$

Hence, the molecular formula of compound

$= (CH_2)_4 = C_4H_8$

72. Molecular mass of the compound

$= \frac{1}{6.06\times 10^{-3}} \approx 165$

Empirical formula mass $= 3 \times 12 + 3 \times 1 + 1 \times 16 = 55$

Now, $n = \frac{165}{55} = 3$ and hence, the molecular formula $= (C_3H_3O)_3 = C_9H_9O_3$.

73. $n = \frac{170\pm 5}{12\times 3+4+16} \approx 3$

$\Rightarrow$ Molecular formula $= (C_3H_4O)_3 = C_9H_{12}O_3$

74. $1.25 \times 22.4 = (1 \times 12 + 2 \times 1) \times n \Rightarrow n = 2$

$\Rightarrow$ Molecular formula $= (CH_2)_2 = C_2H_4$

75. Mass per cent of hydrogen $= \frac{2}{18} \times \frac{1.8}{1.4} \times 100 = \frac{100}{7}$

and mass per cent of carbon $= 100 - \frac{100}{7} = \frac{600}{7}$

Now, atomic ratio of C and H

$= \frac{600/7}{12} : \frac{100/7}{1} = 1:2$

Hence, the empirical formula $= CH_2$

76. Mass per cent of oxygen

$= 100 - (40 + 6.67) = 53.33$

Now, $N_C : N_H : N_O = \frac{40}{12} : \frac{6.67}{1} : \frac{53.33}{16} = 1:2:1$

$\Rightarrow$ Empirical formula $= CH_2O$

77. $N_X : N_Y = \frac{1}{1} : \frac{4}{2} = 1:2$

$\Rightarrow$ Empirical formula $= XY_2$

78. $N_A : N_B : N_C = \frac{w}{20} : \frac{w}{40} : \frac{w}{60} = 6:3:2$

$\Rightarrow$ Empirical formula $= A_6B_3C_2$

79. Molecular mass of oxide $= \frac{100}{30.4} \times 14 = 46.05$

Hence, density of oxide relative to oxygen,

$\frac{d_{\text{oxide}}}{d_{\text{oxygen}}} = \frac{M_{\text{oxide}}}{M_{\text{oxygen}}} = 1.44$

80. Let the oxides be $Fe_2O_x$ and $Fe_2O_y$. From question,

$x \times 16 = \frac{2}{3} \times y \times 16 \Rightarrow x : y = 2:3$

## Stoichiometry

**81.** $C_8H_{18} + \frac{25}{2}O_2 \rightarrow 8CO_2 + 9H_2O$

$8 \times 44$ g  $9\times 18$ g

$\therefore 7.04$ g  $\frac{9\times18}{8\times44}\times7.04 = 3.24$ g

**82.** $C_{14}H_{30} + \frac{43}{2}O_2 \rightarrow 14\ CO_2 + 15\ H_2O$

198 g  $\frac{43}{2} \times 32$ g

$\therefore 792$ g  $\frac{43\times16}{198}\times792 = 2752$ g

**83.** $CH_4 + 2O_2 \rightarrow CO_2 + 2\ H_2O$

16 g  $2 \times 22.4$ L at 0°C and 1 atm

$\therefore 80$ g  $\frac{2\times22.4}{16}\times80 = 224$ L

Hence, the volume of air needed

$= \frac{100}{20}\times224 = 1120$ L

**84.**

$$\underset{\substack{42\text{ g}\\ \therefore 420\text{ kg}}}{C_3H_6(g)} + \frac{3}{2}NO(g) \rightarrow \underset{\substack{53\text{ g}\\ 530\text{ kg}}}{C_3H_3N(g)}$$

$$+\frac{3}{2}H_2O(g) + \frac{1}{4}N_2(g)$$

**85.** $2Ag_2CO_3 \rightarrow 4Ag + 2CO_2\uparrow + O_2\uparrow$

$2 \times 276$ g  $4 \times 108$ g

$\therefore 2.76$ g  $\frac{4\times108}{2\times276}\times2.76 = 2.16$ g

**86.** Detonating gas is the mixture of $H_2$ and $O_2$ gases.

$2H_2O \rightarrow 2H_2 + O_2$

2 mole  $(2 + 1) = 3$ mole $\Rightarrow 3\times 22.4$ L

$\therefore 0.1$ mole  $\frac{3\times22.4}{2}\times0.1 = 3.36$ L

**87.** Total moles of gases $= \frac{1\times8.96}{0.0821\times546} = 0.2$

Hence, the mass of ammonium carbonate

$= \frac{78}{3}\times0.2 = 5.2$ g

**88.** $MgCO_3 + H_2SO_4 \rightarrow MgSO_4 + CO_2 + H_2O$

84 g  98 g

$\therefore 3.0$ g  $\frac{98}{84}\times3 = 3.5$ g

**89.** $2Al + 3H_2SO_4 \rightarrow Al_2(SO_4)_3 + 3H_2$

$2Al + 2NaOH + 2H_2O \rightarrow 2NaAlO_2 + 3H_2$

**90.** $CaO + H_2O \rightarrow Ca(OH)_2$

56 g  18 g

$\therefore 1000$ g  $\frac{18}{56}\times1000 = 321.4$ g

**91.** $Fe_2O_3 + 6HCl \rightarrow 2FeCl_3 + 3H_2O$

160 g  $6\times36.5$ g

$\therefore 20$ g  $\frac{6\times36.5}{160}\times20 = 27.375$ g

Hence, the mass of unreacted HCl

$= 50 - 27.375 = 22.625$ g

**92.** $CaSO_3 + SO_2 + H_2O \rightarrow Ca(HSO_3)_2$

120 g  1 mole

**93.**

| $N_2$ | + $3H_2$ → | $2NH_3$ |
|---|---|---|
| $a$ mole | $b$ mole | 0 |
| $(a - x)$ mole | $(b - 3x)$ mole | $2x$ mole |
| = 2 mole | = 2 mole | = 2 mole |

Hence, $x = 1$, $a = 3$ and $b = 5$.

**94.** Molecular mass of $Ag_2S = 2 \times 108 + 32 = 248$

Now, $w \times \frac{2.296}{100} \times \frac{216}{248} = 1.0 \Rightarrow w = 50.0$ g

**95.** Mass of $SO_2 = 500 \times \frac{1.5}{100} \times \frac{64}{32} = 15.0$ g

HINTS AND EXPLANATIONS

## Limiting Reagent Based

**96.** $2Ag^+ + CrO_4^{2-} \rightarrow Ag_2CrO_4$

$10^{-3}$ mole (LR) $10^{-3}$ mole $\frac{332}{2} \times 10^{-3} = 0.166$ g

**97.** $3SrCl_2 + 2K_3PO_4 \rightarrow Sr_3(PO_4)_2 + 6KCl$

0.3 mole 0.2 mole 0.6 mole

**98.** $NH_3 + \frac{5}{4}O_2 \rightarrow NO + \frac{3}{2}H_2O$

For 0.12 mole of $NH_3$, 0.15 mole of $O_2$ is required. As $O_2$ is the only 0.14 mole, it is a limiting reagent.

**99.** $Fe + S \rightarrow FeS$

56 g 32 g

Hence, the excess reactant is sulphur and the mass fraction of sulphur remained $= \frac{56-32}{56} = 0.4285.$

**100.** $2NH_3(g) + 3O_2(g) + 2CH_4(g) \xrightarrow{Pt} 2HCN(g) + 6H_2O(g)$

$\frac{n_{NH_3}}{2} = \frac{11.5/17}{2} = 0.338; \frac{n_{O_2}}{3} = \frac{10/32}{3} = 0.104;$

$\frac{n_{CH_4}}{2} = \frac{10.5/16}{2} = 0.328$

Hence, $O_2$ is limiting reagent.

Mass of HCN formed $= \frac{2}{3} \times \frac{10}{32} \times 27 = 5.625$ g

## Sequential and Parallel Reactions

**101.** $2Na_2O_2 + 2H_2O \rightarrow 4NaOH + O_2$

$CS_2 + 3O_2 \rightarrow CO_2 + 2SO_2$

2 mole of $Na_2O_2 \equiv 1$ mole of $O_2 \equiv \frac{1}{3}$ mole $CS_2$

$2 \times 78$ g $\frac{1}{3} \times 76$ g

$\therefore 325$ g $\frac{76}{3\times2\times78} \times 325 = 52.78$ g

**102.** $2KClO_3 \rightarrow 2KCl + 3O_2$

$C_4H_8 + 6O_2 \rightarrow 4CO_2 + 4H_2O$

2 mole of $KClO_3 \equiv 3$ mole of $O_2 \equiv 0.5$ mole of $C_4H_8$

**103.** $2KClO_3(s) \rightarrow 2KCl(s) + 3O_2(g)$

$\frac{2}{3}$ mole $\frac{2}{3}$ mole 1 mole

$4KClO_3(s) \rightarrow 3KClO_4(s) + KCl(s)$

$\frac{1}{3}$ mole $\frac{1}{4}$ mole $\frac{1}{12}$ mole

Mole fraction of $KClO_4$

$= \frac{1/4}{1/4 + (2/3 + 1/12)} = 0.25$

**104.** $C + \frac{1}{2}O_2 \rightarrow CO$

$x$ mole $0.5x$ mole $x \times 28$ g

$C + O_2 \rightarrow CO_2$

$(1-x)$ mole $(1-x)$ mole $(1-x) \times 44$ g

From question, $[28x + 44(1-x)] \times \frac{40}{100} = 28x$

$\Rightarrow x = 0.5116$

$\therefore$ Mass of $O_2$ used $= [0.5x + (1-x)] \times 32 = 23.8$ g

**105.** $I_2 + Cl_2 \rightarrow 2ICl$ and $I_2 + 3Cl_2 \rightarrow 2ICl_3$

## Percentage Based

**106.** $MnO_2 + 4HCl \rightarrow MnCl_2 + Cl_2 + 2H_2O$

$Cl_2 + 2KI \rightarrow 2KCl + I_2$

1 mole of $I_2 \equiv 1$ mole of $Cl_2 \equiv 1$ mole of $MnO_2$

254 g 87 g

$\therefore 6.35$ g $\frac{87}{254} \times 6.35 = 2.175$ g

Hence, the percentage purity of pyrolusite ore

$= \frac{2.175}{4.3} \times 100 = 50\%$

**107.** $2NaCl \rightarrow Na_2SO_4$

$2 \times 58.5$ g 142 g

$\therefore 250 \times \frac{95}{100}$ g $\frac{142}{2\times58.5} \times (250\times0.95) = 288.248$ g

Hence, mass of 90 per cent pure $Na_2SO_4$ sample produced $= 288.248 \times \frac{100}{90} = 320.275$ g.

**108.** $CaCO_3 + 2HCl \rightarrow CaCl_2 + CO_2 + H_2O$

100 g 22400 $cm^3$ at 0°C and 1 atm

$\frac{100}{22400} \times 1120 = 5 \text{ g} \leftarrow$ 1120 $cm^3$

Hence, percentage of $CaCO_3$ in the marble

$= \frac{5}{10} \times 100 = 50\%$

**109.** $2KHCO_3 \rightarrow K_2CO_3 + CO_2\uparrow + H_2O\uparrow$

$2 \times 100$ g (44 + 18) = 62 g

$\therefore 1.50 \times \frac{80}{100} = 1.20 \text{ g}$ $\frac{62}{200} \times 1.20 = 0.372 \text{ g}$

**110.** Theoretical yield of $N_2H_4 = \frac{32}{51.5} \times 1000 = 621.36$ g.

Hence, the percentage yield

$= \frac{473}{621.36} \times 100 = 76.12\ \%$

**111.** Overall percentage yield $= 90 \times \frac{80}{100} = 72\ \%$

**112.** $C_6H_5NH_2 + HNO_2 + HCl$

$\rightarrow C_6H_5N_2^+Cl^- + 2H_2O$

$C_6H_5N_2^+Cl^- + KI \rightarrow C_6H_5I + N_2 + KCl$

Theoretical yield of $C_6H_5I = \frac{204}{93} \times 9.3 = 20.4$ g

Hence, the percentage yield $= \frac{16.32}{20.4} \times 100 = 80\%$

**113.** $2NaOH + CO_2 \rightarrow Na_2CO_3 + H_2O$

$2 \times 40$ g 1 mole

$\therefore 20$ g $\frac{1}{80} \times 20 = 0.25$ mole

On oxidation, the rest of 0.75 mole of CO will produce 0.75 mole of $CO_2$ on oxidation, which will require $0.75 \times 80 = 60$ g of NaOH.

**114.** $2a \times 12 + a \times 32 = 14$

$\Rightarrow$ Moles of carbon $= 2a = 0.5$

**115.** $2NaI \rightarrow Na_2SO_4$

$2 \times 150$ g 142 g

$\therefore x$ g $\frac{142}{300} \times x$ g

$2NaCl \rightarrow Na_2SO_4$

$2 \times 58.5$ g 142 g

$\therefore (100 - x)$ g $\frac{142}{117} \times (100 - x)$ g

From the question, $\frac{142}{300} \times x + \frac{142}{117} \times (100 - x) = 100$

Hence, the percentage of NaI in the mixture = 28.86%

## Eudiometry

**116.** $2H_2 + O_2 \rightarrow 2H_2O$

2V vol V vol

From question, $2V + V + 0.003 = 0.03$

$\Rightarrow V = 0.009$

$\therefore$ Percentage of oxygen

$= \frac{0.009 + 0.003}{0.03} \times 100 = 40\%$

**117.** $2NO + O_2 \rightarrow 2NO_2$

42 mL 21 mL

Hence, volume per cent of nitrogen = 79%

**118.** $CH_4 \rightarrow CO_2$ and $C_2H_6 \rightarrow 2CO_2$

$ax + bx = 30$ and $ax + 2bx = 40$

$\Rightarrow ax = 20, bx = 10 \Rightarrow a : b = 2 : 1$

For $b : a$ ratio, $CH_4 = 10$ mL and $C_2H_6 = 20$ mL

$\Rightarrow CO_2$ formed $= 10 + 2 \times 20 = 50$ mL

**119.** $CS_2 + 3O_2 \rightarrow CO_2 + 2SO_2$

V mL 3V mL V mL 2V mL

$2CO + O_2 \rightarrow 2CO_2$

(100 − V) mL $\frac{1}{2}$(100 − V) mL (100 − V) mL

Total contraction in volume,

$V + \frac{1}{2}(100 - V) = (100 + 200) - 245 \Rightarrow V = 10.$

Hence, the volume of $O_2$ remained

$= 200 - [3V + \frac{1}{2}(100 - V)] = 125$ mL

**120.** $2H_2 + O_2 \rightarrow 2H_2O$

10 mL 5 mL

Hence, $N_2 = 25 - 5 = 20$ mL $= \frac{20}{25} \times 100 = 80\%$

**121.** $C_4H_x + (4+\frac{x}{4})O_2 \rightarrow 4CO_2 + \frac{x}{2}H_2O$

10 mL $10(4+\frac{x}{4}) = 55 \Rightarrow x = 6$

**122.** $CO_2 + C(s) \rightarrow 2CO$

V mL 2V mL

Hence, increase in volume is

$2V - V = 700 - 500 \Rightarrow V = 200$

**123.** Volume of $CO_2$ formed

$= 3 \times 36.5 + (100 - 36.5) = 173$ mL

**124.** $C_nH_{3n}O_m + \left(\frac{7n}{4} - \frac{m}{2}\right)O_2 \rightarrow nCO_2 + \frac{3n}{2}H_2O(l)$

1 mL $\left(\frac{7n}{4} - \frac{m}{2}\right)$ mL $n$ mL 0

Hence, contraction in volume

$= \left(1 + \frac{7n}{4} - \frac{m}{2}\right) - n = \left(1 + \frac{3n}{4} - \frac{m}{2}\right)$ mL

**125.** $N_xO_y + H_2 \rightarrow N_2 + H_2O(g)$

1 vol 1 vol 1 vol 1 vol

Hence, $x = 2$ and $y = 1$.

**126.** $C_nH_{2n+2} + \left(\frac{3n+1}{2}\right)O_2 \rightarrow nCO_2 + (n+1)H_2O$

From question, $\left(\frac{3n+1}{2}\right) : n = 7 : 4$

$\Rightarrow n = 2 \Rightarrow$ Alkane $= C_2H_6$

**127.** $C_xH_y + \left(x + \frac{y}{4}\right)O_2 \rightarrow xCO_2 + \frac{y}{2}H_2O$

V ml $V\left(x + \frac{y}{4}\right)$ mL $Vx$ mL 0

Volume contraction, $\left[V + V\left(x + \frac{y}{4}\right)\right] - Vx = 2.5V$

and $Vx = 2V \Rightarrow x = 2$ and $y = 6 \Rightarrow C_2H_6$

**128.** $\frac{x}{2}Cl_2 + \frac{y}{2}O_2 \rightarrow Cl_xO_y$

10 mL $10\frac{y}{x}$ mL $= 25$ mL $\frac{20}{x}$ ml $= 10$ mL

Hence, $x = 2$ and $y = 5 \Rightarrow Cl_2O_5$

**129.** $N_xO_y + yH_2 \rightarrow \frac{x}{2}N_2 + yH_2O$

10 mL $10y$ mL $5x$ mL 0

Contraction in volume, $(10 + 10y) - 5x = 28 - 18$

$\Rightarrow x = 2y$

Volume of $H_2$ unreacted $= (28 - 10) - 10y$

$= (18 - 10y)$ mL

$2H_2 + O_2 \rightarrow 2H_2O$

$(18 - 10y)$ mL $(9 - 5y)$ mL 0

Contraction in volume, $27 - 15y = 27 - 15$

$\Rightarrow y = 1$ and $x = 2 \Rightarrow N_2O \Rightarrow$ Molecular mass $= 44$

**130.** $C_3O_2 + 2O_2 \rightarrow 3CO_2$

1 vol 2 vol 3 vol

## Concentration Terms

**131.** $\frac{w+100}{1.2} \times \frac{5}{100} = w \Rightarrow w = 4.35$ g

**132.** Volume of solution $= \frac{100}{10} \times 180 = 1800$ ml

**133.** $(75 \times d) \times \frac{60}{100} = 50 \Rightarrow d = 1.11$ g/mL

**134.** $d = \frac{25}{20} = 1.25$ g/mL

**135.** Mass of $Ca(NO_3)_2 = \frac{2.35}{10^6} \times 50 \times \frac{164}{40}$

$= 4.8175 \times 10^{-4}$ g

**136.** Mass of CO $= \frac{35}{10^6} \times 1.3 = 4.55 \times 10^{-5}$ g

**137.** $V \times 0.8 \times 3 + 50 \times 0.2 \times 2 = (V + 50) \times 0.6$

$\Rightarrow V = 5.56$ ml

**138.** Molarity $= \frac{5\times1000}{180\times100}\times1.08 = 0.3$ M

**139.** $\frac{w\times1000}{208\times250}\times2 = \frac{3.78\times1000}{58.5\times100}\times1 \Rightarrow w = 16.8$ g

**140.** Initial mass of HCl $= \frac{1}{2}\times36.5 = 18.25$ g

Final molarity of solution

$$= \frac{(18.25-2.675)\times1000}{36.5\times750} = 0.569 \text{ M}$$

**141.** $[NO_3^-] = \frac{500\times0.1}{1000} = 0.05$ M

**142.** Volume of solution $= \frac{1200}{1.2} = 1000$ mL

Hence, molarity of solution $= \frac{12}{60} = 0.2$ M

**143.** Mass of 1 mole solution = 0.1 × 40 + 0.9 × 18

= 20.2 g

Volume of solution $= \frac{20.2}{1.4}$ mL

Hence, molarity of solution

$$= \frac{0.1}{\left(\frac{20.2}{1.4}\right)}\times1000 = 6.93 \text{ M}$$

**144.** Mass of solvent (in g) = Volume of solution (in ml)

$(1000\times d - 1\times60) = 1000 \Rightarrow d = 1.06$ g/mL

**145.** Mass of solvent is same in all.

**146.** Mole of solvent is minimum for benzene.

**147.** $\chi_{glucose} = \frac{\frac{1}{180}}{\frac{1}{180}+\frac{3}{60}+\frac{10}{18}} = \frac{1}{110}$

**148.** Molarity of solution $= \frac{9.08}{11.35} = 0.8$ M

Hence, mass of $H_2O_2 = \frac{250\times0.8}{1000}\times34 = 6.8$ g

**149.** Percentage of free $SO_3 = \frac{80}{18}\times4.5 = 20\%$

**150.** Maximum limiting labelling of oleum is 122.5%

## EXERCISE II (JEE ADVANCED)

### Section A (Only one Correct)

1. $\frac{\text{Mass of silica}}{\text{Mass of all components other than water}}$ = Constant

or $\frac{50}{90} = \frac{52}{100-x} \Rightarrow$ Mass per cent of water in partially dried clay, $x = 6.4\%$

2. Atomic mass of oxygen $= \frac{87.5}{12.5} = 7$

3. Mass of mercury present

$$= \frac{1.68}{10^9}\times(15000\times0.998) = 2.515\times10^{-5} \text{ g}$$

Number of Hg-atoms $= \frac{2.515\times10^{-5}}{200}\times6\times10^{23}$

$= 7.545\times10^{16}$

4. Number of Na-atoms

$$= \frac{1.15\times10^{-3}}{23}\times6\times10^{23} = 3\times10^{19}$$

∴ Length of Na-atoms

$$= 3\times10^{19}\times(2\times0.2\times10^{-9})\times\frac{1}{1600}$$

$= 7.5\times10^6$ miles

5. $\frac{4}{3}\pi r^3 = \frac{197}{6\times10^{23}\times19.7}\text{cm}^3 \Rightarrow r = 1.587\times10^{-8}$ cm

6. Volume of one H-atom

$$= \frac{1}{6\times10^{23}\times3\times10^{-29}} = 5.56\times10^4 \text{ ml}$$

7. Average atomic mass of carbon

$$= \frac{4\times12+1\times14}{5} = 12.4$$

∴ Average molecular mass of $CH_4$

$= 12.4 + 4 \times 1 = 16.4$

8. Average atomic mass of $Q$

$$= \frac{23.4\times8.082\times12+76.6\times7.833\times12}{100}$$

$= 94.695$

9. Average atomic mass of oxygen in meteorites is greater.

10. Molecular mass of $H_2O$

$$= \frac{2\times M_{H\text{-atom}}}{\frac{1}{12}\times M_{C^{12}\text{-atom}}} + 1\times\frac{M_{O\text{-atom}}}{\frac{1}{12}\times M_{C^{12}\text{-atom}}}$$

$$= 2\times\left(\frac{2\times M_{H\text{-atom}}}{\frac{1}{6}\times M_{C^{12}\text{-atom}}} + \frac{1\times M_{O\text{-atom}}}{\frac{1}{6}\times M_{C^{12}\text{-atom}}}\right)$$

$= 2 \times (2 \times 1 + 1 \times 16) = 36$

11. $CH_2 = CH - CH = CH - CH = CH_2$

For 1 mole double bond, $\frac{1}{3}$ mole $= \frac{1}{3} \times 80 =$ 26.7 g compound is needed.

12. $\frac{2.0767 / A_x}{1.6/16} = \frac{2}{5} \Rightarrow A_x = 20.769\times2.5$

∴ Moles of $x = \frac{2.0769}{20.769\times2.5} = 0.04$

13. At 4°C, $H_2O$ is liquid.

14. $M = \frac{WRT}{PV} = \frac{2.3\times0.082}{0.82}\times\frac{300}{1} = 69$

Now, $\alpha = \frac{M_o - M}{(n-1).\,M} = \frac{92-69}{(2-1)\times69} = 0.33$

15. $\alpha = \frac{M_o - M}{(n-1).\,M} = \frac{26-60}{(\frac{1}{3}-1)\times60} = 0.85$

16. $\alpha = \frac{M_o - M}{(n-1).\,M} = \frac{20-36}{(\frac{1}{2}-1)\times36} = \frac{8}{9}$

∴ Percentage of total molecules in dimer form

$$= \frac{\alpha/2}{1-\alpha/2}\times100 = 80\%$$

17. $n_N : n_P : n_K = \frac{30}{14} : \frac{10}{142}\times2 : \frac{10}{94}\times2$

$= 10.07 : 0.66 : 1.00$

18. Mass per cent of Mn

$$= \frac{1\times55}{\frac{2}{3}\times71+\frac{1}{3}\times87}\times100 = 72.05\%$$

19. Mass per cent of $Cu_2O = \frac{143}{127}\times66.67 = 75.06\%$

20. Percentage yield of $NaHCO_3$

$$= \frac{164-69}{164}\times100 = 57.93\%$$

21. Mass of $Fe_2O_3$ in the ore $= \frac{160}{2\times56}\times2.78 = 3.97$ kg

∴ Percentage of gangue $= \frac{5-3.97}{5}\times100 = 20.6\%$

22. Mass of Fe $= \frac{2\times56}{160}\times0.541 = 0.3787$ g

∴ Percentage of Fe in the ore

$$= \frac{0.3787}{0.7}\times100 = 54.1\%$$

23. Let molecular formula $= C_x\,H_{2x}\,O_x$

As, $0.0833 \times 2x = 1 \Rightarrow x = 6$

⇒ Molecular formula $= C_6H_{12}O_6$

24. $\frac{(14n+2)-14n}{14n}\times100 = 2.38 \Rightarrow n = 6$

25. Mass per cent of C $= \frac{8}{9}\times92.7 = 82.4\%$

Mass per cent of H $= \frac{1}{9}\times92.7 = 10.3\%$

Now, $N_C : N_H : N_O = \frac{82.4}{12} : \frac{10.3}{1} : \frac{7.3}{16} = 30 : 45 : 2$

26. $Mn_x\,O_y + \frac{y}{2}\,C \rightarrow x\,Mn + \frac{y}{2}\,CO_2$

$(55x + 16y)$g $\qquad \frac{y}{2}\times44$ g

∴ 31.6 gm $\qquad \frac{22y}{55x+16y}\times31.6 = 13.2$

$\Rightarrow x : y = 2 : 3$

27. (a) $\frac{10}{70} : \frac{1}{7} = 1 : 1 \Rightarrow$ Formula = UO

(b) $\frac{10}{240} : \frac{1}{7} = 7 : 24 \Rightarrow$ Formula $= U_7O_{24}$

(c), (d) $\frac{10}{105} : \frac{1}{7} = 2 : 3 \Rightarrow$ Formula $= U_2O_3$

**28.** From Mg, $M_{min} = \frac{100}{0.08} \times 24 = 300000$

From Ti, $M_{min} = \frac{100}{0.08} \times 48 = 600000$

If M = 300000, the number of Ti-atoms in each molecule $= \frac{1}{2}$, which is not possible.

**29.** Let $N_2 = a$ mole, $NO_2 = b$ mole and $N_2O_4 = c$ mole.

$a + b + c = 1$ (1)

$28 \times a + 46 \times b + 92 \times c = 55.4$ (2)

$(a + b + 2c) \times 39.6 = 55.4$ (3)

Hence, $a = 0.5, b = 0.1, c = 0.4$

**30.** $M_{min} = \frac{100}{365 \times 10^{-3}} \times 146 = 40000$

**31.** $CuSO_4 \cdot x(NH_4)SO_4 \cdot y H_2O \xrightarrow{\Delta} CuO$

(159.5 + 132x + 18y) g 79.5 g

From question, $(159.5 + 132x + 18y) \times (159.5 + 132x + 18y) \times \frac{19.89}{100} = 79.5$ (1)

and $(159.5 + 132x + 18y) \times \frac{27.03}{100} = 18\, y$ (2)

Hence, $x = 1, y = 6$

**32.** Molecular formula of tetrahydrocannabinol = $C_{10.5x}H_{15x}O_x$

As $\frac{1.0}{157x} = 0.00318 \Rightarrow x = 2$

**33.** $CuCl_2 + H_2S \rightarrow CuS\downarrow + 2HCl$

$\frac{2.69}{134.5} = 0.02$ mole

Hence, volume of $H_2S$ needed

$= 0.02 \times 22400 = 448$ mL

**34.** $C_3H_8 + 5O_2 \rightarrow 3CO_2 + 4H_2O$

0.15 mole 0.20 mol

∴ Number of water droplets

$= \frac{0.2 \times 6 \times 10^{23}}{1.7 \times 10^{21}} = 70.6 = 70$

**35.** $C_xH_y + (x + \frac{y}{4}) O_2 \rightarrow xCO_2 + \frac{y}{2} H_2O$

$44x$ g $\frac{y}{2} \times 18 = 9y$ g

From question, $\frac{44x}{9y} = \frac{44}{27} \Rightarrow \frac{x}{y} = \frac{1}{3}$

**36.** $(NH_4)_2SO_4 + Ca(OH)_2 \rightarrow CaSO_4 + 2NH_3 + 2H_2O$

50 mole 100 mole

Now, $(V \times 0.85) \times \frac{20}{100} = 100 \times 17$

$\Rightarrow V = 10000$ mL

**37.** For 1 mole of HCl, 29 g of $Mg(OH)_2$ or 26 g of $Al(OH)_3$ or 37 g of $Ca(OH)_2$ is needed.

**38.** $M_2O_3 + 3H_2 \rightarrow 2M + 3H_2O$

(2A + 48)g 6 g

$\therefore 0.1596$ g $\frac{6}{2A+48} \times 0.1596 = 6 \times 10^{-3}$

$\Rightarrow A = 55.8.$

**39.** $M + 3F_2 \rightarrow MF_6$

$A$ g $(A + 6 \times 19)$g

$\therefore 0.25$ g $\frac{A+114}{A} \times 0.25 = 0.547$

$\Rightarrow A = 95.96$

**40.** $U + 3F_2 \rightarrow UF_6$

$3 \times 6.022 \times 10^{23}$ molecule 352 g

∴ Number of $F_2$ molecules needed

$= \frac{3 \times 6.022 \times 10^{23}}{352} \times 2 \times 10^{-3}$

$= 1.026 \times 10^{19}$

**41.** $H_2S + \frac{3}{2}O_2 \rightarrow H_2O + SO_2$

34 g $\frac{3}{2} \times 32$ g

$\therefore 51$ g $\frac{48}{34} \times 51 = 72$ g

∴ Total mass of product formed = 51 + 72 = 123 g

**42.** $Cr_2O_7^{2-} + 5H^+ + 3SO_2 \rightarrow 2Cr^{3+} + 3HSO_4^- + H_2O$

$\frac{1.08}{216} = 0.005$ mole $= 5 \times 0.005 = 0.025$ mole

**43.** $B_2Cl_4 + 6NaOH \rightarrow 2NaBO_2 + 2H_2O + H_2 + 4NaCl$

164 g 22700 mL at STP

∴ Mass of $B_2Cl_4$ needed $= \frac{164}{22700} \times 1362 = 9.84$ g

**44.** $2NH_4NO_3 \rightarrow 2N_2 + O_2 + 4H_2O(g)$

$2 \times 80$ g Total 7 mole gases

$\therefore 16$ g 0.7 mole

$\therefore$ Volume, $V = \frac{0.7 \times 0.082 \times 900}{1} = 57.4$ L

**45.** $FeCl_x + xAgNO_3 \rightarrow xAgCl \downarrow + Fe(NO_3)x$

$(56 + 35.5x)$g $x \times 143.5$ g

$\therefore 127$ mg $\frac{143.5x}{56+35.5x} \times 127 = 287 \Rightarrow x = 2$

**46.** 1 mole of $F_2 \equiv 2$ mole of Co $F_3 \equiv \frac{1}{2n}$ mole $(CF_2)_n$

$= 38$ g $= \frac{1}{2n} \times 50n = 25$ g

$\therefore$ Mass of $F_2$ needed $= \frac{38}{25} \times 1 = 1.52$ kg

**47.** $9A + 8BO_3 \rightarrow 3A_3O_4 + 4B_2O_3$

9 mole 3mole

(L.R.)

$\therefore$ 1 mole $\frac{3}{9} = \frac{1}{3}$ mole

**48.** Mass of tin $= \frac{119}{151} \times 0.5 = 0.394$ g

$\therefore$ Mass per cent of tin $= \frac{0.394}{1.5} \times 100 = 26.27\%$.

**49.** Overall reaction: A + B + C $\rightarrow$ D

5 moles 6 moles excess 5 moles

L.R.

**50.** $4NH_3 + 5O_2 \rightarrow 4NO + 6H_2O$

$\frac{25.5}{17} = 1.5$ mole 1.5 mole

$2NO + 2CH_4 \rightarrow 2HCN + 2H_2O$

1.5 mole 2 mole 1.5 mole

L.R. $= 1.5 \times 27 = 40.5$ g

**51.** $Al_2O_3 \cdot 2SiO_2 = 102 + 2 \times 60$

$\therefore SiO_2$ in combined state $= \frac{120}{102} \times 1.02 = 1.20$ g

Hence, percentage of free $SiO_2$

$= 1.52 - 1.20 = 0.32\%$

**52.** $2FeO \rightarrow Fe_2O_3$

$2a$ mole $a$ mole

$(2a - 2x)$mole $(a + x)$mole

From question, $\frac{a+x}{2a-2x} = \frac{2}{1} \Rightarrow x = \frac{3}{5}a$

Hence, moles of FeO oxidized per mole of mixture

$= \frac{2x}{3a} = \frac{2}{3} \times \frac{3}{5} = 0.4$

**53.** $C + \frac{1}{2}O_2 \rightarrow CO$

12 g 16 g

$\therefore x$ g $1.33x$ g

$C + O_2 \rightarrow CO_2$

12 g 32 g

$\therefore x$ g $2.67x$ g

Hence, $1.33x \le y \le 2.67x$

**54.** $Ca(NCN) + 3H_2O \rightarrow CaCO_3 + 2NH_3$

1 mole 1 mole $\frac{2}{3}$ mole

L.R.

**55.** $N_2 + 3H_2 \rightarrow 2NH_3$

$\frac{1}{2}$ mole $\frac{3}{2}$ mole 1 mole

$NH_3 + HCl \rightarrow NH_4Cl$

1 mole 1 mole (given)

Gases remained: $N_2 = 1 - \frac{1}{2} = \frac{1}{2}$ mole

$H_2 = 4 - \frac{3}{2} = \frac{5}{2}$ mole

**56.** $I_2 + H_2S \rightarrow 2HI + S$

$\frac{5.08}{254}$ $\frac{460}{22400}$ 0.02 mole

$= 0.02$ mole $= 0.0205$ mole $0.02 \times 32$

(L.R.) $= 0.64$ g

**57.** $12K_2CO_3 \rightarrow K_2Zn_3[Fe(CN)_6]_2$

$12 \times 138$ g 698.2 g

$\therefore 27.6$ g $\frac{698.2}{12 \times 138} \times 27.6 = 11.64$ g

**58.** 2 mole of $AsCl_3 \equiv 3$ mole of $Cl_2 \equiv \frac{16}{5} \times 3$ mole of HCl

$= 2 \times 181.5$ g $\qquad = \frac{48}{5} \times 36.5$ g

$\therefore 363$ g $\qquad 350.4$ g

$\therefore$ Volume of HCl solution needed

$$= \frac{350.4 \times \frac{100}{20}}{1.2} = 1460 \text{ mL}$$

**59.**

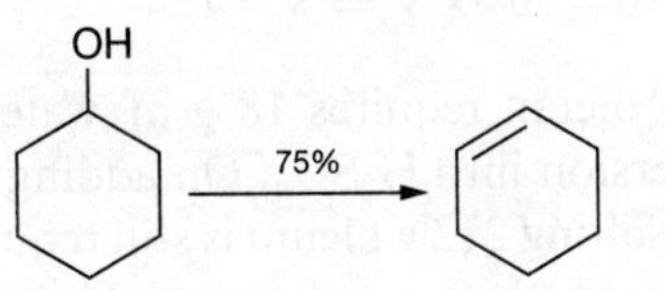


$(C_6H_{12}O)$ $\qquad$ $(C_6H_{10})$

100 g $\qquad$ $82 \text{ g} \times \frac{75}{100} = 61.5$ g

**60.** Cu $\rightarrow$ CuO

64 g $\qquad$ 80 g

$\therefore x$ g $\qquad \frac{80}{64} \times x = 1.25x$ g

Increase in mass, $1.25x - x - 4.90 - 4.00 \rightarrow x - 3.6$

$\therefore$ Percentage of copper unoxidized

$$= \frac{4 - 3.6}{4} \times 100 = 10\%$$

**61.** $2CH_3COCH_3 + 6Ca(OCl)Cl \xrightarrow{75\%} 2CHCl_3 + (CH_3\,COO)_2\,Ca + 3CaCl_2 + 2Ca(OH)_2$

$2 \times 58$ g $\qquad\qquad 2 \times 119.5$ g

$\therefore$ For 30 g $CHCl_3$, mass of $CH_3COCH_3$ needed

$$= \frac{2 \times 58}{2 \times 119.5} \times 30 \times \frac{100}{75} = 19.4 \text{ g}$$

**62.** $4FeS_2 \; + \; 11O_2 \rightarrow \; 2Fe_2O_3 + 8SO_2(g)$

4 mole (say) $\quad$ 11 mole $\qquad$ 8 mole

Final composition of gases:

$O_2 = 11 \times \frac{60}{100} = 6.6$ mole

$N_2 = (11 + 6.6) \times \frac{80}{20} = 70.4$ mole

$SO_2 = 8.0$ mole

$\therefore$ Total moles of gases $= 6.6 + 70.4 + 8.0 = 85$

$\therefore$ Volume per cent of $N_2 = \frac{70.4}{85} \times 100 = 82.82\%$

**63.** Let $CH_4 = x$ mole and $C_2H_4 = y$ mole

From question, $x \times 16 + y \times 28 = 12$ $\qquad$ (1)

and from $CO_2$, $(x + 2y) \times 44 = 35.2$ $\qquad$ (2)

Hence, $x = 0.4$ and $y = 0.2$

Now, $M_{av} = \frac{x + 16 + y \times 28}{x + y} = 20$

**64.** $C_xH_y + \left(x + \frac{y}{4}\right) O_2 \rightarrow xCO_2 + \frac{y}{2} H_2O$

From question, $\frac{(x + \frac{y}{4})}{x}$ = Independent of $x$.

For alkene, $y = 2x$.

**65.** $N_2O + H_2 \rightarrow N_2 + H_2O$

$x$ mL $\qquad x$ mL

$NO + H_2 \rightarrow \frac{1}{2} N_2 + H_2O$

$(60 - x)$ mL $\qquad \frac{1}{2}(60 - x)$ mL

From question, $x + \frac{1}{2}(60 - x) = 38 \Rightarrow x = 16$

**66.** $CO + \frac{1}{2}O_2 \rightarrow CO$

0.4V mL $\quad$ 0.2V mL $\quad$ 0.4V mL

Decrease in volume $= (0.4V + 0.2V) - 0.4V = 0.2V$

$\therefore$ Fractional decrease in volume $= \frac{0.2V}{2V} = 0.1$

**67.** $HCOOH \xrightarrow[\Delta]{\text{conc. } H_2SO_4} CO(g) + H_2O(l)$

$a$ mole $\qquad\qquad a$ mole

$H_2C_2O_4 \xrightarrow{\text{conc. } H_2SO_4} CO(g) + CO_2(g) + H_2O(l)$

$b$ mole $\qquad\qquad b$ mole $\quad b$ mole

From question, $(a + 2b) \times \frac{1}{6} = b \Rightarrow a : b = 4 : 1$

**68.** $O_2$ consumed $= 70 - 45 = 25$ mL

For any hydrocarbon, volume of $O_2$ used in combustion must be greater than the volume of hydrocarbon.

**69.** $3HCl \; + \; Al(OH)_3 \rightarrow \; AlCl_3 + \; 3H_2O$

3 mole $\qquad$ 78 g

$\therefore 3 \times 80 \times 10^{-3}$ mole $\frac{78}{3} \times 3 \times 80 \times 10^{-3} = 7.8 \times 0.8$ g

∴ Volume of antacid syrup needed

$$= \frac{100}{2.6} \times (7.8 \times 0.8) = 240 \text{ mL}$$

**70.** $V \times 2.2 = \left(11\,V \times \frac{98}{100}\right) \times M_{final} \Rightarrow M_{final} = 0.204 \text{ M}$

**71.** Moles of succinic acid $= \frac{23.6}{118} = 0.2$

∴ Final molarity of –COOH group

$$= \frac{0.2 \times 2}{500} \times 1000 + 0.1 \times 1 = 0.9 \text{ M}$$

**72.** $[CCl_3F] = \frac{275 \times 0.08}{0.08 \times 200} \times 10^{-12} = 1.375 \times 10^{-12}$ M

$[CCl_2F_2] = \frac{605 \times 0.08}{0.08 \times 200} \times 10^{-12} = 3.025 \times 10^{-12}$ M

**73.** $2 = \frac{x \times 1000}{60 \times (1000 - x)}$ and $4 = \frac{y \times 1000}{60 \times (2000 - y)}$

∴ Molality of resulting solution

$$= \frac{(x+y) \times 1000}{60 \times [3000 - (x+y)]} = 3.29 \text{ m}$$

**74.** $1 \times \left(\frac{1000}{1000 + 1 \times 180} \times 1000\right)$

$$= x \times \left(\frac{1000}{1000 + x \times 180} \times 5000\right) \Rightarrow x = 0.175 \text{ m}$$

**75.** $0.2 \times \left(\frac{50}{0.2 \times 60 + 0.8 \times 18}\right)$

$$= x \times \left(\frac{1500}{x \times 60 + (1-x) \times 18}\right) \Rightarrow x = 0.0508$$

**76.** Mass of $H_2O_2$ in 10 mL diluted solution

$$= \frac{1}{5} \times \left(20 \times \frac{8.5}{100}\right) = \frac{8.5}{25} \text{ g}$$

As 1 mole of $H_2O_2$ will produce 1 mole of $O_2$ on reaction with the oxidant, volume of $O_2$ gas produced at 0°C and 1 atm

$$= \frac{8.5/25}{34} \times 22400 = 224 \text{ mL}$$

**77.** $50 \times 20 + 50 \times 10 = 100 \times V \Rightarrow V = 15$

**78.** 200 g of 109% oleum requires 18 g of water for complete conversion into $H_2SO_4$. On adding 12 g of water, the resulting 212 g oleum is still requiring 6 g of water. Hence, the new labelling is

$$\left(100 + \frac{6}{212} \times 100\right) = 102.83\%$$

**79.** $200 \times 109 + 300 \times 118 = 500 \times x \Rightarrow x = 114.4$

**80.** Resulting oleum is still requiring (24 – 18) = 6 g of water.

$$H_2O + SO_3 \rightarrow H_2SO_4$$

18 g    80 g

∴ 6 g    26.67 g

Final composition: $SO_3 = 26.67$ g

and $H_2SO_4 = 218 - 26.6 = 191.33$ g

## Section B (One or More than one Correct)

**1.** Molar mass of gas $= \frac{0.22}{112} \times 22400 = 44$

**2.** Mole of H-atoms in glucose $= \frac{0.9}{180} \times 12 = 0.06$

Mole of H-atoms in hydrazine $= \frac{0.48}{32} \times 4 = 0.06$

Mole of H-atoms in ammonia $= \frac{0.17}{17} \times 3 = 0.03$

Mole of H-atoms in ethane $= \frac{0.30}{30} \times 6 = 0.06$

Mole of H-atoms in hydrogen $= \frac{0.03}{1} = 0.03$

**3.** Moles of H-atoms $= \frac{90}{1} = 90$

Moles of He-atoms $= \frac{10}{4} = 2.5$

∴ $N_H : N_{He} = 90 : 2.5 = 36 : 1$

And $M_{av} = \frac{100}{\frac{90}{2} + \frac{10}{4}} = 2.105$

**4.** $M = 20 \times 2 = 40$

Hence, either all HF molecules are in dimer form or some molecules may be in trimer form.

**5.** $2H_2 + O_2 \rightarrow 2H_2O$

HINTS AND EXPLANATIONS

6. Volume of $N_A$ molecules = 22.4 L at 0°C and 1 atm.

7. (b) $\frac{4\times M+1\times(M+1)-1(M+2)}{4+1+1}=M+0.5$

Similarly, try for other options.

8. Theoretical

9. For $Ti_{0.75}O = 0.75 \times 48 + 16 = 52$

$\therefore$ Percentage of Ti $= \frac{36}{52} \times 100 = 69.23\%$

and percentage of O = 30.8%

For $TiO_{0.69} = 48 + 16 \times 0.69 = 59.04$

$\therefore$ Percentage of Ti $= \frac{48}{59.04} \times 100 = 81.3\%$

and percentage of O = 18.7%

10. Isomers

11. Same empirical formula ⇒ Same composition

12. Group A and C ⇒ 0.1 mole AgCl

Group B and D ⇒ 0.1 mole AgBr

13. (c) $H_2O$ is in liquid state.

(d) $2H_2 + O_2 \rightarrow 2H_2O$

14. (a) $CH_4 + 2O_2 \rightarrow CO_2 + 2H_2O$

16 g 64 g

7 g ← 28 g

(b) $CH_4 + \frac{3}{2}O_2 \rightarrow CO + 2H_2O$

16 g 48 g

∴ 8 g 24 g

(c), (d) $\frac{1}{2}CH_4 + \frac{7}{8}O_2 \rightarrow \frac{1}{4}CO_2 + \frac{1}{4}CO + H_2O$

18 g

15. $CH_4 + 2O_2 \rightarrow CO_2 + 2H_2O$

16 g 2 mole

∴ 8 g 1 mole

$2KClO_3 \rightarrow 2KCl + 3O_2$

$2H_2O_2 \rightarrow 2H_2O + O_2$

$2NaNO_3 \rightarrow 2NaNO_2 + O_2$

$2BaO_2 \rightarrow 2BaO + O_2$

16. $C_3H_8 + 5O_2 \rightarrow 3CO_2 + 4H_2O(g)$

$x$ mole 3$x$ mole 4$x$ mole

$C_6H_6 + \frac{15}{2}O_2 \rightarrow 6CO_2 + 3H_2O(g)$

$y$ mole 6$y$ mole 3$y$ mole

From question, $3x + 6y = 4x + 3y \Rightarrow x : y = 3 : 1$

and $n_C : n_H = (3x + 6y) : (8x + 6y) = 1 : 2$

17. (a) $2NaOH + CO_2 \rightarrow Na_2CO_3 + H_2O$

2 × 40 g 44 g

8 g ← 4.4 g

But NaOH is only 6 g and hence, L.R.

(b) $NaOH + CO_2 \rightarrow NaHCO_3$

40 g 44 g

4 g ← 4.4 g

But NaOH is 6 g (excess). Hence, $CO_2$ is L.R.

18. $C_nH_{2n+2} + \left(\frac{3n+1}{2}\right)O_2 \rightarrow n\,CO_2 + (n+1)\,H_2O$

Alkane

$C_nH_{2n} + \frac{3n}{2}O_2 \rightarrow n\,CO_2 + n\,H_2O$

Alkene

$C_nH_{2n-2} + \left(\frac{3n-1}{2}\right)O_2 \rightarrow n\,CO_2 + (n-1)\,H_2O$

Alkyne

19. Reactions of question number 18.

20. (a) $Mg + \frac{1}{2}O_2 \rightarrow MgO$

$\frac{1}{2}$ mole $\frac{1}{4} \times 32 = 8$ g

∴ Mass of air needed = 8 + 1 × 28 = 36 g

(c) $3Mg + N_2 \rightarrow Mg_3N_2$

$\frac{1}{2}$ mole $\frac{1}{6}$ mole

∴ Mass of air needed $= \frac{1}{6} \times 28 + \frac{1}{24} \times 32 = 6$ g

(d) $Mg + \frac{1}{2}O_2 \rightarrow MgO$

$x$ mole $\frac{x}{2}$ mole

$3Mg + N_2 \rightarrow Mg_3N_2$

$\left(\frac{1}{2}-x\right)$ mole $\frac{1}{3}\left(\frac{1}{2}-x\right)$ mole

From question: $\frac{1}{3}\left(\frac{1}{2}-x\right) = 4 \times \frac{x}{2} \Rightarrow x = \frac{1}{14}$

∴ Mass of air needed $= \frac{x}{2} \times 32 + \frac{1}{3}\left(\frac{1}{2}-x\right) \times 28$

= 5.14 g

Final mass of products = 12 + 5.14 = 17.14 g

21. $NaCl \rightarrow AgCl$ and $MCl \rightarrow AgCl$

$x$ mole $\quad x$ mole $\quad y$ mole $\quad y$ mole

$\therefore (x + y) \times 143.5 = 2.567 \quad (1)$

From the second expression: $MCl \rightarrow AgCl$

$y$ mole $\quad y$ mole

$\therefore y \times 143.5 = 1.341 \quad (2)$

Hence, $x = \dfrac{1.226}{143.5}, y = \dfrac{1.341}{143.5}$

Now, $x \times 58.5 + y \times M = 1 \Rightarrow M = 53.5$

22. $5Cr_2O_7^{2-} + 6XO + 34H^+ \rightarrow 10Cr^{3+} + 6XO_4^- + 17H_2O$

$x$ mole $\quad \dfrac{6}{5}x$ mole $\quad \dfrac{6}{5}x$ mole

$4Cr_2O_7^{2-} + 3X_2O_3 + 26H^+ \rightarrow 8Cr^{3+} + 6XO_4^- + 13H_2O$

$(0.15 - x)$ mole $\quad \dfrac{3}{4}(0.15 - x)$ mole $\quad \dfrac{6}{4}(0.15 - x)$ mole

From question, $\dfrac{6}{5}x + \dfrac{6}{4}(0.15 - x) = 0.21$

$\Rightarrow x = 0.05.$

Now, $\dfrac{6}{5}x \times (A + 16) + \dfrac{3}{4}(0.15 - x) \times (2A + 48)$

$= 25.56 \Rightarrow A = 100.$

23. $2H_2 + O_2 \rightarrow 2H_2O$

L.R. may be $H_2$ or $O_2$.

24. Volume of undecomposed $NH_3 = 90 - 84 = 6$ mL

$2NH_3 \rightarrow N_2 + 3H_2$

V mL $\quad \dfrac{V}{2}$ mL $\quad \dfrac{3V}{2}$ mL

As $\dfrac{V}{2} + \dfrac{3V}{2} = 84 \Rightarrow V = 42$

Hence, initial volume of $NH_3 = 42 + 6 = 48$ mL

25. $V_1 \times 40 = V_2 \times 16 \Rightarrow \dfrac{V_2}{V_1} = \dfrac{2.5}{1}.$

26. $(V_1 \times 1.6) \times \dfrac{40}{100} = (V_1 + V_2) \times 1.1 \times \dfrac{x}{100}$

and $V_1 \times 1.6 + V_2 \times 1.0 = (V_1 + V_2) \times 1.1$

$\therefore$ Strength of final solution, $x = 9.7\%$ (w/w) and $10.67\%$ (w/v).

27. $V \times 20 = 2V \times S \Rightarrow S = 10$ vol.

28. For $Na^+$: $2000 \times M_{Na^+} = 100 \times 1 \Rightarrow M_{Na^+} = 0.05$ M

For $Mg^{2+}$: $2000 \times M_{Mg^{2+}} = 100 \times 2 + 300 \times 4$

$\Rightarrow M_{Mg^{2+}} = 0.7$ M.

For $Cl^-$: $2000 \times M_{Cl^-} = 100 \times 1 + 100 \times 2 \times 2$

$\Rightarrow M_{Cl^-} = 0.25$ M

For $NO_3^-$ : $2000 \times M_{NO_3^-} = 300 \times 4 \times 2$

$\Rightarrow M_{NO_3^-} = 1.2$ M

29. Mass of solute $= n \times x$ g

Mass of solvent $= n \times y$ g

$\therefore$ Mass per cent of solute $= \dfrac{x}{x + y} \times 100\%$

30. $d = \dfrac{950 \text{ gm}}{1000 \text{ cm}^3} = 950 \text{ kg/m}^3$

$X_{C_2H_{50}H} = \dfrac{230/46}{230/46 + 720/18} = \dfrac{1}{9}$

$M_{C_2H_{50}H} = 5$ M

and $M_{C_2H_{50}H} = \dfrac{5}{720} \times 1000 = 6.94$ M

## Section C (Comprehensions)

### Comprehension I

1. $N_xH_y \rightarrow \dfrac{x}{2}N_2 + \dfrac{y}{2}H_2$

10 mL $\quad 5x = 10 \quad 5y = 20$

$\therefore x = 2 \quad \therefore y = 4$

$\therefore$ Formula of compound $= N_2H_4$

2. $\dfrac{88/A_N}{12/1} = \dfrac{2}{4} \Rightarrow A_N = 14.67$

3. $N_xH_y \rightarrow \dfrac{x}{3}N_3 + \dfrac{y}{2}H_2$

10 mL $\quad \dfrac{10x}{3} = 10 \quad 5y = 20$

$\therefore x = 3 \quad \therefore y = 4$

$\therefore$ Formula of compound $= N_3H_4$

HINTS AND EXPLANATIONS

## Comprehension II

$$2HF \rightarrow H_2F_2 \qquad 3HF \rightarrow H_3F_3$$

$$a-(2x+3y) \quad x \qquad a-(3y+2x) \quad y$$

From question, $\dfrac{y}{a-(2x+3y)} = \dfrac{1}{4}$ (1)

and $[\{a-(2x+3y)\}+x+y]\times 34 = a\times 20$ (2)

From (1) and (2): $a = 17y = \dfrac{17}{5}x$

$\therefore$ Percentage of molecules in dimer form

$$= \frac{x}{a-x-2y}\times 100 = 50\%$$

Percentage of HF molecules trimerized

$$= \frac{3y}{a}\times 10 = 17.647\%$$

Percentage of HF molecules dimerized

$$= \frac{2x}{a}\times 100 = 58.82\%$$

## Comprehension III

Let the initial composition be Ar = $a$ mole and $N_2O_4 = b$ mole.

Now, $a\times 40 + b\times 92 = (a+b)\times 40\times 2$

$\Rightarrow a : b = 3 : 10$

$$N_2O_4 \rightarrow 2NO_2$$

$(b-x)$mole $\quad 2x$ mole

Now, $80\times(a+b) = (a+b+x)\times 37.5\times 2$

$\therefore$ Degree of dissociation of $N_2O_4 = \dfrac{x}{b} = 0.087$

And final mole ratio, $n_{Ar} : n_{N_2O_4} : n_{NO_2}$

$$= a : (b-x) : 2x = 45 : 137 : 26$$

## Comprehension IV

Ore = 100 g (say)

- Kaolia = 13 g ($Al_2O_3\cdot 2SiO_2\cdot 2H_2O$)
  - $Al = \dfrac{54}{258}\times 13 = 2.721$ g
  - $Si = \dfrac{56}{258}\times 13 = 2.822$ g
- Gibbsite = 87 g ($Al_2O_3\cdot 3H_2O$) $\Rightarrow Al = \dfrac{54}{156}\times 87 = 30.115$ g

Total mass of Al present

$$= 2.725 + 30.115 = 32.836 \text{ g.}$$

Mass of Al lost in mud $= \dfrac{6\times 27}{5\times 28}\times 2.822$

$$= 3.265 \text{ g}$$

$\therefore$ Percentage of Al recoverable

$$= \frac{(32.836-3.265)}{32.836}\times 100 = 90.06\%$$

Now, percentage of $SiO_2 = \dfrac{2\times 60}{258}\times 13 = 6.05$

## Comprehension V

Mass of $CO_2$ produced = 85.35 – 83.85 = 1.5 g

$\therefore$ Percentage of $C = \dfrac{12}{44}\times\dfrac{1.5}{1.0}\times 100 = 40.9\%$

Mass of $H_2O$ produced = 37.96 – 37.55 = 0.41 g

$\therefore$ Percentage of $H = \dfrac{2}{18}\times\dfrac{0.41}{1.0}\times 100 = 4.56\%$

Hence, percentage of oxygen

$$= 100 - (40.9 + 4.56) = 54.54\%$$

Now, $N_C : N_H : N_O = \dfrac{40.9}{12} : \dfrac{4.56}{1} : \dfrac{54.54}{16}$

$$= 3 : 4 : 3$$

## Comprehension VI

Let slurry = $x$ kg, concentrated solution = $y$ kg, water vapour = $z$ kg.

NaCl: $25000 \times \frac{18}{100} = x \times \frac{90}{100} \Rightarrow x = 5000$

NaOH: $25000 \times \frac{2}{100} = x \times \frac{5}{100} + y \times \frac{50}{100} \Rightarrow y = 500$

and $z = 25000 - (x + y) = 19500$

## Comprehension VII

$$2N_2H_4 \quad + \quad N_2O_4 \quad \rightarrow \quad 3N_2 \quad + \quad 4H_2O$$

$(3 - 0.1)$ $\quad (2 - 0.2)$ $\quad \frac{3}{2} \times 2.9$ $\quad 2 \times 2.9$ mole

$= 2.9$ mole $\quad = 1.8$ mole $\quad = 4.35$ mole

L.R.

$$N_2H_4 \quad + \quad 2N_2O_4 \quad \rightarrow \quad 6NO \quad + \quad 2H_2O$$

0.1 mole $\quad$ 0.2 mole $\quad \frac{18}{30} = 0.6$ mole $\quad$ 0.2 mole

Total $H_2O$ formed = 6 mole = 108 g

Theoretical yield of $N_2 = 4.5$ mole

$\therefore$ Percentage yield $= \frac{4.35}{4.5} \times 100 = 96.67\%$

## Comprehension VIII

$$Cu + 4HNO_3 \rightarrow Cu(NO_3)_2 + 2NO_2 + 2H_2O$$

0.025 mole $\quad \frac{1 \times 1.23}{0.082 \times 300} = 0.05$

$= 0.025 \times 63.5$

$= 1.5875$ g

$\therefore$ Percentage of Cu $= \frac{1.5875}{1.9145} \times 100 = 82.92\%$

and Zn present $= (1.9145 - 1.5875) = 0.327$ g

$= \frac{0.327}{65.4} = 0.005$ mole

$$Cu \quad + \quad 4HNO_3 \rightarrow Cu(NO_3)_2 + 2NO_2 + 2H_2O$$

0.025 mole $\quad$ 0.1 mole

$$4Zn + 10HNO_3 \rightarrow 4Zn(NO_3)_2 + NH_4NO_3 + 3H_2O$$

0.005 mole $\quad$ 0.0125 mole $\quad 0.00125 \times 80$

$= 0.1$ g

Now, moles of $HNO_3$, $\frac{V \times 3}{100} = 0.1125$

$\therefore V = 37.5$ mL

## Comprehension IX

**25.** $CaO + 3C \rightarrow CaC_2 + CO$

56 g $\quad$ 64 g

$\frac{56}{64} \times 1280 \leftarrow$ 1280 kg

$= 1120$ kg

And CaO unreacted $= \frac{20}{80} \times 1280 = 320$ kg

$\therefore$ Total CaO needed $= 1120 + 320 = 1440$ kg

**26.** Mass of $CaC_2$ formed $= 1280 \times \frac{80}{100} = 1024$ kg

Mass of CaO reacted $= \frac{56}{64} \times 1024 = 896$ kg

$\therefore$ Total CaO needed $= 986 + 1280 \times \frac{20}{100} = 1152$ kg

**27.** Moles of CO formed $= \frac{1}{64} \times (1024 \times 10^3) = 16 \times 10^3$

$\therefore$ Volume of CO produced $= 16 \times 10^3 \times 22.4$ L

$= 358.4\ m^3$

## Comprehension X

Moles of S = Moles of $BaSO_4 = \frac{6.99}{233} = 0.03$

$$MS_n + \left(\frac{2n+3}{2}\right) O_2 \rightarrow MO_3 + nSO_2$$

$$2MCl_5 + 5Na_2S \rightarrow 2MS_n + (5-2n)S + 10NaCl$$

2 mole of $MCl_5 \equiv 2 \times \frac{90}{100} = 1.8$ mole of $MS_n \equiv 1.8n$

Mole of $SO_2 \equiv 1.8n$ mole of $BaSO_4 \leftarrow 0.03$ mole

$$\frac{2}{1.8n} \times 0.03 = \frac{4.55}{A+35.5\times 5}$$

If $n = 2$, then $A = 95.5$.

## Comprehension XI

$Cl_2 = 30$ mL, $Cl_xO_y = 30$ mL

$$Cl_xO_y \rightarrow \frac{x}{2} Cl_2 + \frac{y}{2} O_2$$

30 mL 15$x$ mL 15$y$ mL

From the question, $30 + 15x + 15y = 75 \Rightarrow x + y = 3$

and $30 + 15x = 75 - 15 \Rightarrow x = 2, y = 1$.

## Comprehension XII

$$C_xH_yN_z + (x + \frac{y}{4})O_2 \rightarrow CO_2 + \frac{y}{2} H_2O + \frac{z}{2} N_2$$

V ml $\quad V(x + \frac{y}{4})$ mL $\quad Vx$ mL $\quad \frac{Vy}{2}$ mL $\quad \frac{Vz}{2}$ mL

$= (18 - V)$ mL $\quad = 8$ mL $\quad = 12$ mL $\quad = 4$ mL

$\therefore x = 2, y = 6, z = 2$ and $V = 4$

Now $C_2H_6N_2 + H_2 \rightarrow C_2H_8N_2$

4 mL 4 mL

## Comprehension XIII

Volume of $O_2$ used = 30 – 5.5 = 24.5 mL

Let the original mixture be $CH_4 = x$ mL, $C_2H_4 = y$ mL and $C_2H_2 = z$ mL.

From question, $x + y + z = 10$ (1)

$$\frac{16\times x+28\times y+26\times z}{x+y+z} = 11.3 \times 2 \quad (2)$$

$$CH_4 + 2O_2 \rightarrow CO_2 + 2H_2O$$

$x$ mL 2$x$ mL $x$ mL

$$C_2H_4 + 3O_2 \rightarrow 2CO_2 + 2H_2O$$

$y$ mL 3$y$ mL 2$y$ mL

$$C_2H_4 + \frac{5}{2} O_2 \rightarrow 2CO_2 + H_2O$$

$z$ mL $\frac{5z}{2}$ mL 2$z$ mL

$$2x + 3y + \frac{5z}{2} = 24.5 \quad (3)$$

$\therefore x = 4, y = 3, z = 3$

Volume of $CO_2$ formed $= x + 2y + 2z = 16$ mL

$\therefore$ Final volume in the absence of KOH

$= 5.5 + 16 = 21.5$ mL

$$CO_2 + 2KOH \rightarrow K_2CO_3 + H_2O$$

$\frac{16}{22400}$ mole $\quad 2 \times \frac{16}{22400} \times 56$

$= 0.08$ g

## Comprehension XIV

$$(CHO)P \rightarrow PCO + \frac{P}{2} H_2$$

2 mL 2$p$ mL $p$ mL

$$(COOH)_q \rightarrow qCO_2 + \frac{q}{2} H_2$$

8 mL 8$q$ mL 4$q$ mL

Volume of $CO_2$ formed $= 8q = \frac{8}{17}(3p + 12q)$

$\Rightarrow 5q = 3p$ (1)

HINTS AND EXPLANATIONS

$$H_2 + \frac{1}{2}O_2 \Rightarrow H_2O$$

$(p + 4q)$ mL $\frac{1}{2}(p + 4q)$ mL 0

$$CO + \frac{1}{2}O_2 \rightarrow CO_2$$

$2p$ mL $p$ mL $2p$ mL

$\therefore$ Contraction in volume $= \frac{3}{2}(p + 4q) + p = 61$

$\Rightarrow 5p + 12q = 122$ (2)

From equations (1) and (2), we get $p = 10$ and $q = 6$.

Volume of $O_2$ used in combustion $= \frac{1}{2}(p + 4q) + p = 27$ mL

The increase in volume in initial decomposition $= (3p - 2) + (12q - 8) = 92$ mL

## Comprehension XV

Original solution: V = 1000 mL, mass = 5960 g.

5 m solution $\Rightarrow$ 1000 g water + $5 \times 98$ g of $H_2SO_4$ = 1490 g

$\therefore$ Moles of $H_2SO_4 = \frac{5}{1490} \times 5960 = 20$ and molarity = 20 M.

Final solution: V = 1000 ml and mass = 400 g

Moles of $H_2SO_4 = \frac{\frac{49}{100} \times 400}{98} = 2$

$\Rightarrow$ Molarity = 2 M

$\therefore$ Moles of $H_2SO_4$ fallen down on the floor $= 20 - 2 = 18$

Now, $V \times 20 = 1000 \times 2$

$\Rightarrow$ Volume of solution left = 100 mL

$\therefore$ Volume of water added = 1000 – 100 = 900 mL

## Section D (Assertion–Reason)

1. If they were not on the same scale, the sum of atomic masses will not be equal to the molecular mass.
2. Same mass, same element and hence, same number of atoms.
3. Mass is conserved. Moles may or may not change.
4. In the given reaction, the total moles of gases is constant during the reaction.
5. $C_xH_y + (x + \frac{y}{4})O_2 \rightarrow x\,CO_2 + \frac{y}{2}H_2O$

   and $(x + \frac{y}{4}) > 1$

   $CO + \frac{1}{2}O_2 \rightarrow CO_2$

   and $n_{O_2} < n_{CO}$
6. $N_2$ is a limiting reagent. $H_2$ is in excess.
7. Percentage yield can never be more than 100%
8. $aP + bQ \rightarrow cR + dS$

   $t = 0$ $x$ mole $y$ mole

   $t = t$ $(x - z)$ mole $(y - \frac{b}{a}z)$ mole

   If $\frac{x}{y} = \frac{x - z}{y - \frac{b}{a}z} \Rightarrow \frac{x}{y} = \frac{a}{b}$

   and product will always form in $c : d$ mole ratio.
9. For maximum product formation, none of the reactant should be left.
10. Volume of non-reacting gases is additive only when all the volumes are given at the same P and T.
11. $C_xH_y + (x + \frac{y}{4})O_2 \rightarrow xCO_2 + \frac{y}{2}H_2O$

    Contraction in volume $= (1 + x + \frac{y}{4}) - x = (1 + \frac{y}{4})$ and the contraction is not due to $H_2O(l)$. It is due to the change in moles of gases.
12. For very dilute aqueous solution, the volume of solution is nearly equal to the mass of solvent because the density of water is almost equal to 1.0 g/mL.
13. On changing the unit, the numerical value of concentration may change.
14. w/w and w/v may be equal only when the density of the solution is 1.0 g/mL.
15. Mass is conservative but not the volume.

## Section E (Column Match)

**1.** (A) $Z = \frac{x(z-1)+(100-x)(z+2)}{100}$

$\therefore x = \frac{200}{3} \Rightarrow$ mole per cent of heavier isotope

$= \frac{100}{3}\%$

and mass per cent of heavier isotope

$= \frac{\frac{100}{3}\times(z+2)}{100\times z} \times 100 = \frac{100}{3}(1+\frac{2}{z})\%$

Similarly check other options.

**2.** (A) mass $= 0.875 \times 32 = 28$ g

Volume $= 0.875 \times 22.4$ L at 0°C and 1 atm.

Number of O-atoms $= 0.875 \times 2 \times 6.0 \times 10^{23}$

$= 1.05 \times 10^{24}$

Similarly check other options.

**3.** (A) 3 moles of $Co(NH_3)_4SO_4$ contains 3 moles of Co-atoms, 12 moles of N-atoms, 36 moles of H-atoms, 3 moles of S-atoms and 12 moles of O-atoms.

Similarly check other options.

**4.** $a$ mole of $CH_4$ + $b$ mole of $C_2H_6$

Mole ratio of C to H: $\frac{a+2b}{4a+6b} = \frac{2}{7} \Rightarrow \frac{a}{b} = \frac{2}{1}$

$\therefore M_{av} = \frac{2\times16+1\times30}{3} = \frac{62}{3}$

**5.** Given: $C + O_2 \rightarrow CO_2 + 100$ kcal

$C + \frac{1}{2}O_2 \rightarrow CO + 25$ kcal

(A) $3C + \frac{5}{2}O_2 \rightarrow CO + 2CO_2$

$\therefore$ Heat released $= 1 \times 25 + 2 \times 100 = 225$ kcal

(B) $C + \frac{3}{4}O_2 \rightarrow \frac{1}{2}CO + \frac{1}{2}CO_2$

$\therefore$ Heat released $= \frac{1}{2}\times 25 + \frac{1}{2}\times 100 = 62.5$ kcal

Similarly, check the other options.

**6** (A) $C_8H_{18} + \frac{25}{2}O_2 \rightarrow 8CO_2 + 9H_2O$

2 mole 25 mole

$= 25 \times 32 = 800$ g

$= 560$ L at 273 K and 1 atm

$= 25 \times 6.02 \times 10^{23} \times 2 = 3.01 \times 10^{25}$ atoms

Check similarly the other options.

**7.** $N_2 + 3H_2 \rightarrow 2NH_3$

$\frac{n_{N_2}}{1} = \frac{3.5}{28} = \frac{1}{8}$ and $\frac{n_{H_2}}{3} = \frac{1}{3}\times\frac{1}{2} = \frac{1}{6} > \frac{1}{8}$

Hence, $N_2$ is L.R.

As $H_2$ is in excess, the total mass of reactants taken is greater than the mass of $NH_3$ formed.

Similarly check other options.

**8.** (A) $CH_4 + 2O_2 \rightarrow CO_2 + 2H_2O$

1 mole 2 mole

$= 44$ g $= 2 \times 18 = 36$ g

Similarly check other options.

**9.** (A) On increasing the mass of metal, the mass of metal halide is increasing but it acquires a maximum value. It is possible only after the complete consumption of halogen. Hence, halogen is L.R.

**10.** (A) $CH_4 + 2O_2 \rightarrow CO_2 + 2H_2O$

80 mL 80 mL

$2CO + O_2 \rightarrow 2CO_2$

20 mL 20 mL

(B) $C_2H_2 + \frac{5}{2}O_2 \rightarrow 2CO_2 + H_2O$

100 mL 250 mL

(C) $H_2S + Cl_2 \rightarrow 2HCl + S$

10 mL 10 mL

(D) $2CO + O_2 \rightarrow 2CO_2$

30 mL 15 mL 30 mL

**11.** (A) $C_xH_{2x+2} + \frac{3x+1}{2} O_2 \rightarrow xCO_2 + (x+1)H_2O$

(B) $C_xH_{2x+2}O + \frac{3x}{2} O_2 \rightarrow xCO_2 + (x+1)H_2O$

(C) $C_xH_{2x+3}N + \frac{3(2x+1)}{4} O_2$

$\rightarrow xCO_2 + \left(\frac{2x+3}{2}\right) H_2O + \frac{1}{2} N_2$

(D) $C_xH_{2x+2}S + \frac{3(x+1)}{2} O_2$

$\rightarrow xCO_2 + (x+1) H_2O + SO_2$

**12.** $CO_2, SO_2 \rightarrow CaO, NaOH$

$CO \rightarrow$ Ammoniacal CuCl

$H_2O \rightarrow P_4O_{10}$

**13.** At 400 K and 1 atm, $H_2O$ will be in vapour form.

$CH_4 + 2O_2 \rightarrow CO_2 + 2H_2O(g)$

50 mL 50 mL 100 mL

**14.** (P) 1000 g of water + 6.25 × 40 g of NaOH = 1250 g solution

= 1250 mL solution.

$\therefore$ Concentration in g/L $= \frac{250}{1250} \times 1000 = 200$

and concentration in % (w/w) $= \frac{250}{1250} \times 100 = 20$

(Q) 1 mole solution $\equiv \frac{1}{6}$ mole of NaOH + $\frac{5}{6}$ mole of $H_2O$

$= \frac{1}{6} \times 40 + \frac{5}{6} \times 18 = \frac{130}{6}$ g solution

$\equiv \frac{130/6}{1.3} = \frac{100}{6}$ mL solution

$\therefore$ Concentration in g/L $= \frac{40/6}{100/6} \times 1000 = 400$

and concentration in % (w/w) $= \frac{40/6}{130/6} \times 100$

$= \frac{400}{13}$

**15.** (A) 1000 g of water + 5 × 40 g of NaOH = 1200 of solution

$\equiv \frac{1200}{0.6} = 2000$ mL solution

$\therefore$ Molarity $= \frac{5}{2000} \times 1000 = 2.5$ M

(D) $Ag^+ + Cl^- \rightarrow AgCl(s)$

$\frac{800 \times \frac{17}{100}}{170} = 0.8$ mole $\quad \frac{100 \times 6}{1000} = 0.6$ mole

Final 0.8 – 0.6 = 0.2 mole 0

$\therefore [Ag^+]_{final} = \frac{0.2}{200} \times 1000$

= 1 M

## Section F (Subjective)

### Single-digit Integer Type

**1.** $l^3 = \frac{200}{13.6 \times 6.022 \times 10^{23}}$ cm$^3 \Rightarrow l = 2.90$ Å

**2.** Moles of A-atoms $= \frac{1.28}{64} = 0.02$

Moles of B-atoms $= \frac{3.0 \times 10^{23}}{6.0 \times 10^{23}} 0.5$

Moles of C-atoms = 0.04

From the given atomic ratio, A and C are L.R.

$\therefore$ Maximum mass of $AB_6C_3$ formed

= 1.28 + (0.02 × 6 × 9) + (0.04 × 16) = 3.0 g

**3.** Moles of X-atoms $= \frac{5.0}{60} = 0.083$

Moles of Y-atoms $= \frac{1.15 \times 10^{23}}{6 \times 10^{23}} = 0.192$

Moles of Z-atoms = 0.03

Hence, L.R. is Z and 0.01 mole of $XY_2Z_3$ will be formed.

$0.01 = \frac{4.40}{60 + 2A_y + 3 \times 80} \Rightarrow A_y = 70$

**4.** $\frac{N_D}{N_{Pd}} = \frac{\left(\frac{1120V}{22400} \times N_A\right) \times 2}{\left(\frac{V \times 11.8}{106.2} \times N_A\right)} = \frac{9}{10}$

HINTS AND EXPLANATIONS

5. $$\frac{N_{O_2}}{N_X} = \frac{\left(\frac{1\times 0.27}{0.08\times 1000\times 300}\times N_A\right)\times 2}{\left(\frac{0.18}{64000}\times N_A\right)} = 4$$

6. $R(OEt)_x \equiv x$ Et I $\equiv x$ AgI

$\frac{37}{176}$ mole $\quad x\times\frac{37}{176} = \frac{148}{235} \Rightarrow x \approx 3$

7. $2Fe + \frac{3}{2}O_2 \rightarrow Fe_2O_3$

$2\times 56$ g $\quad \frac{3}{2}\times 32$ g

$\therefore x$ g $\quad \frac{3\times 16}{2\times 56}x$ g ($x$ = Fraction of iron rusted)

From the question, $\frac{3\times 16}{2\times 56}x = 0.1 \Rightarrow x = \frac{7}{30}$

8. $2FeS + \frac{7}{2}O_2 \rightarrow Fe_2O_3 + 2SO_2\uparrow$

$2\times 88$ g $\quad$ 160 g

From 176 g of FeS, only 160 g of $Fe_2O_3$ is produced. Hence, the loss in mass = 176 – 160 = 16 g

Now, for 16 g of loss in mass, the mass of FeS in ore = 176 g

$\therefore$ For 4 g of loss, mass of FeS = $\frac{176}{6}\times 4 = 44$ g

9. $Na - Hg + HCl(g) \rightarrow NaCl + Hg + \frac{1}{2}H_2(g)$

V mL $\quad \frac{V}{2}$ mL

From question, $V - \frac{V}{2} = 50 - 40 \Rightarrow V = 20$

Now, 10 mL of mixture will contain HCl(g)

$= \frac{20}{50}\times 10 = 4$ mL

and $H_2 = 10 - 4 = 6$ mL

$H_2 + NH_3(g) + HCl(g)$

$\rightarrow H_2 + NH_4Cl(g) \xrightarrow{\text{Water}} H_2(g)$

6 mL

10. Mass of nickel in $NiC_8H_{14}O_4N_4$

$= \frac{58.7}{288.7}\times 0.2887 = 0.0587.$

$\therefore$ Percentage of nickel in steel

$\frac{0.0587}{1.174}\times 100 = 5\%$

11. $5A^{n+} + (5-n)\ MnO_4^- + (10-8n)H^+ \rightarrow (5-n)\ Mn^{2+} + 5\ AO_3^- + (5-4n)\ H_2O$

5 mole $\quad (5-n)$ mole

$\therefore 2.5\times 10^{-3}$ mole $MnO_4^-$ $\frac{5-n}{5}\times 2.5\times 10^{-3}$

$= 1.5\times 10^{-3} \Rightarrow n = 2$

12. $6nCO_2 + 5nH_2O \rightarrow (C_6H_{10}O_5)_n + 6nO_2$

$6n$ mole $\quad 162n$ g

$\frac{6n}{162n}\times 0.81 \leftarrow 0.81$g

= 0.03 mole.

$\therefore$ Required time = $\frac{0.03}{5\times 10^{-3}} = 6$ hours.

13. Mass of tetraethyl lead needed

$= 500\times 2\times\frac{12.96}{1.29}$ g

$4C_2H_5Cl + 4NaPb \rightarrow (C_2H_5)_4Pb + 4NaCl + 3Pb$

$4\times 64.5$ g $\quad$ 324 g

$\therefore$ Mass of ethyl chloride needed

$= \frac{4\times 64.5}{324}\times\left(1000\times\frac{12.96}{1.29}\right) = 8000\text{ g} = 8\text{ kg}$

14. Mass of waterproofing film of $[(CH_3)_2SiO]_n$ needed

$= \frac{150}{129}\times(500\times 400\times 200\times 3.7\times 10^{-8})$

$= \frac{150\times 1.48}{129}$ g

$n(CH_3)_2SiCl_2 + 2nOH^- \rightarrow 2nCl^- + nH_2O + [(CH_3)_2SiO]_n$

$n\times 129$ g $\quad 74n$ g

$\frac{n\times 129}{74n}\times\frac{150\times 1.48}{129} = 3\text{ g} \leftarrow \frac{150\times 1.48}{129}$ g

15. Let $x$ moles of MgO and $y$ moles of $Mg_3N_2$ be formed. From the given reactions,

$2x + 8y = 0.6$ (1)

and $2y = 0.1$ (2)

$y = 0.05, x = 0.1$

$\therefore$ Mass of Mg burnt = $(x + 3y)\times 24 = 6$ g

16. Moles of $SF_5OF$ taken $= \frac{PV}{RT} = \frac{0.5\times1.642}{0.082\times390} = \frac{1}{39}$

$\therefore$ Mass of $SF_5ONF_2$ produced $= \frac{1}{39}\times\frac{40}{100}\times195 = 2$ g

17. $LiAlH_4 + 3(CH_3)_3COH \rightarrow Li[(CH_3)_3CO]_3\ AlH + 3H_2$

5 millimole 20 millimole 5 millimole 15 millimole
excess

$Li[(CH_3)_3CO]_3\ AlH + CH_3OH$

$\rightarrow Li[(CH_3)_3\ CO]_3\ [CH_3O]\ Al + H_2$

5 millimole excess 5 millimole

18. Cast iron $\xrightarrow{HCl} H_2S \xrightarrow{Cd^{2+}} CdS \xrightarrow{CuSO_4}$ CuS $\xrightarrow{\text{ignition}}$ CuO

Moles of S in iron sample = Moles of CuO

$= \frac{0.795}{79.5} = 0.01$

$\therefore$ Percentage of S in cast iron $= \frac{0.01\times32}{6.4}\times100 = 5\%$

19. $HNF_2(g) + Cl_2(g) + KF(s) \rightarrow ClNF_2(g) + KF\text{-}HCl(s)$

1.3 millimole = a 1.3 millimole = a excess

Final $a - \frac{2}{3}a$ $a - \frac{2}{3}a$ $\frac{2}{3}a$

$= \frac{a}{3}$ $= \frac{a}{3}$

$\therefore$ Volume per cent of $ClNF_2$ (g) in the gaseous mixture after the reaction $= \frac{2/3a}{4/3a}\times100 = 50\%$

20. Let the sample contains $x$ mole of $MgSO_4\cdot7H_2O$ and $y$ mole of $MgCl_2\cdot6H_2O$.

Now, $x = \frac{0.699}{233} = 3\times10^{-3}$

and $\frac{x}{2} + \frac{y}{2} = \frac{0.888}{222} = 4\times10^{-3} \Rightarrow y = 5\times10^{-3}$.

Now, the sum of masses of $MgSO_4\cdot7H_2O$ and $MgCl_2\cdot6H_2O$

$= 3\times10^{-3}\times246 + 5\times10^{-3}\times203 = 1.753$ g

## Four digit Integer type

1. Average molecular mass of $NH_3$

$= 1\times\left(\frac{3\times14+1\times15}{4}\right) + 3\times\left(\frac{4\times1+1\times2}{5}\right)$

$= 17.85$

and average number of neutrons per molecule

$= 1\times\left(\frac{3\times7+1\times8}{4}\right)+3\times\left(\frac{4\times0+1\times1}{5}\right) = 7.85$

$\therefore$ Total number of neutrons

$= \frac{1.785\times10^{-3}}{17.85}\times6\times10^{23}\times7.85$

$= 471\times10^{18}$

2. Number of $H^3$-atoms

$= \left(\frac{9.0}{18}\times6\times10^{23}\right)\times2\times\left(\frac{8\times10^{-8}}{1+8\times10^{-8}}\right) \approx 48\times10^{15}$

3. Number of $-CH_2-CH_2-$ units in a molecule $= \frac{10^6}{28}$

$\therefore$ Length of each polymer molecule

$= \left(2\times\frac{10^6}{28}-1\right)\times154\times10^{-12}$ m $= 11\times10^{-6}$ m

4. Rating $= \frac{100\times1000}{380} = 263.16$

5. Molecular mass of $(C_2F_4)x$

$\Rightarrow 100x = \frac{100}{0.012}\times32 \Rightarrow x = 2666.6$

6. Mass of carbon burnt $= \frac{50}{20}\times1000\times0.8\times\frac{84}{100}$

$= 1680$ g

$\therefore$ Mass of $CO_2$ produced $= \frac{44}{12}\times1680 = 6160$ g

7. Mole of $NO_3^-$ present $= \frac{1.08}{108}\times2 = 0.02$

$\therefore$ Mass percentage of $NO_3^-$ group

$= \frac{0.02\times62}{2}\times100 = 62\%$

8. $3H_2C_2O_4 \rightarrow M_4O_3(C_2O_4)_3\cdot12H_2O$

$3\times90$ g $(4A + 528)$ g

$\therefore$ 1 g $\frac{4A+528}{270}\times1 = 3.2 \Rightarrow A = 84$

**9.** $Al_2O_3 + 6HCl \rightarrow 2AlCl_3 + 3H_2O$

102 g $6 \times 36.5$ g $2 \times 133.5$ g

Total mass of reactant = $(102 + 6 \times 36.5) = 321$ g

$\therefore$ Mass of $AlCl_3$ formed = $2 \times 133.5 = 267$ g

**10.** $n_{K_2Cr_2O_7} = \dfrac{75 \times \frac{98}{100}}{294} = 0.25$

and $\dfrac{n_{HCl}}{14} = \dfrac{1}{14} \times \dfrac{365 \times 1.2 \times \frac{28}{100}}{36.5} = 0.24$

Hence, HCl is a limiting reagent.

$\therefore$ Mass of $Cl_2$ produced = $\dfrac{3 \times 71}{14} \times (0.24 \times 14)$

$= 51.12$ g

**11.** 2 mole of $F_2 \equiv 4$ mole of NaF $\equiv 2$ mole of CaO

$2 \times 38$ g $\qquad 2 \times 56$ g

$\therefore$ 1900 kg $\qquad \dfrac{2 \times 56}{2 \times 38} \times 1900 = 2800$ kg

$\therefore$ Percentage utilization of lime

$= \dfrac{2800}{10000} \times 100 = 28\%$

**12.** Let the sample contains $x$ g of $CaCO_3$.

$CaCO_3 \rightarrow CaO + CO_2\uparrow$

100 g $\qquad$ 44 g

$\therefore x$ g $\qquad \dfrac{44}{100} x$ g

Total loss in weight $\Rightarrow 0.44x + (5 - x) \times \dfrac{20}{100}$

$= 1.636 \Rightarrow x = 2.65$

$\therefore$ Percentage of $CaCO_3$ in chalk

$= \dfrac{2.65}{5} \times 100 = 53\%$

**13.** $2FeS_2 + \dfrac{11}{2} O_2 \rightarrow Fe_2O_3 + 4SO_2$

Mass of $Fe_2O_3$ formed = $XO_4^- \quad \times 28 = 40$ g.

Mass of $SiO_2$ present = $100 - \left(\dfrac{120}{56} \times 28\right) = 40$ g

$\therefore$ Final mass of roasted sample = $40 + 40 = 80$ g

**14.** Maximum moles of NaCl present = $\dfrac{10.7}{58.5}$

Maximum moles of KCl present = $\dfrac{10.7}{74.5}$

Maximum moles of $NH_4Cl$ present = $\dfrac{10.7}{53.5} = 0.2$

Hence, maximum moles of $Cl^-$ ion in the sample in every possible case = 0.2

$Cl^- \; + \; AgNo_3 \; \rightarrow \; AgCl\downarrow \; + \; NO_3^-$

0.2 mole 0.2 mole

$= 0.2 \times 170 = 34$ g

Hence, minimum mass of $AgNo_3$ needed = 34 g

or $(V_{min} \times 1.25) \times \dfrac{8.5}{100} = 34 \Rightarrow V_{min} = 320$ mL

**15.** Let $CH_4 = x$ mole, $C_2H_6 = y$ mole

From question: $2x + 3y = 42$ (1)

and $x + y = \dfrac{1.5 \times 246}{0.082 \times 300} = 15$ (2)

$\therefore x = 3$ and $y = 12$

Mole per cent of $CH_4 = \dfrac{3}{15} \times 100 = 20\%$

**16.** $2Pb(NO_3)_2 \xrightarrow{\Delta} 2PbO + 4NO_2\uparrow + O_2\uparrow$

$2 \times 332$ g $\qquad 2 \times 224$ g

$\therefore x$ g $\qquad \dfrac{224}{332} x$ g

$2NaNO_3 \xrightarrow{\Delta} 2NaNO_2 + O_2\uparrow$

$2 \times 85$ g $\qquad 2 \times 69$ g

$\therefore (5 - x)$ g $\qquad \dfrac{69}{85} \times (5.02 - x)$ g

From question, $\dfrac{224}{332} x + \dfrac{69}{85}(5.02 - x)$

$= (5.02 - 1.4) \Rightarrow x = 3.32$ g

**17.** Mass of octane burnt = 2.28 kg

$C_8H_{18} + \dfrac{25}{2} O_2 \rightarrow 8CO_2 + 9H_2O$

114 g $\qquad \dfrac{25}{2} \times 32$ g

$\therefore x$ g $\qquad \dfrac{25 \times 16}{114} x$ g

$$C_8H_{18} \quad + \quad \frac{17}{2} O_2 \rightarrow 8CO + 9H_2O$$

$$114 \text{ g} \qquad \frac{17}{2} \times 32 \text{ g}$$

$$\therefore (2280 - x) \text{ g} \qquad \frac{17 \times 16}{114} (2280 - x) \text{ g}$$

From the question,

$$2280 + \frac{25 \times 16}{114} x + \frac{17 \times 16}{114} (2280 - x) = 9768$$

$\Rightarrow x = 1824$ g

$\therefore$ Percentage of octane converted into $CO_2$

$$= \frac{1824}{2280} \times 100 = 80\%$$

**18.** Let CO = $x$ mL, $H_2$ = $y$ mL

$$CO \quad + \quad \frac{1}{2} O_2 \quad \rightarrow \quad CO_2$$

$$x \text{ mL} \qquad \frac{x}{2} \text{ mL} \qquad x \text{ mL}$$

$$H_2 \quad + \quad \frac{1}{2} O_2 \quad \rightarrow \quad H_2O$$

$$y \text{ mL} \qquad \frac{y}{2} \text{ mL} \qquad 0$$

Contraction in volume, $\frac{x}{2} + \frac{3y}{2} = (100 + 100) - 100$

or $x + 3y = 200$ (1)

and total volume of $CO_2$, $[100 - (x + y)] + x = 100 - 52.5$

$\therefore y = 52.5$ and $x = 42.5$

Hence, $V_{CO} : V_{H_2} : V_{CO_2} = 42.5 : 52.5 : 5 = 17 : 21 : 2.$

**19.** $M_{sol} = M_{acetic}$ acid + $M_{water}$

or $V_{sol} \times \frac{96}{98} = 10 \times 0.8 + 40 \times 1.0$

$\Rightarrow V_{sol} = 49$ mL

$\therefore$ Percentage decrease in volume

$$= \frac{50 - 49}{50} \times 100 = 2\%$$

**20.** Number of enzyme molecules

$$= \frac{10 \times 10^{-6}}{1000} N_A = 10^{-8} N_A$$

Hence, number of $CO_2$ molecules hydrated per hour = $10^{-8} N_A \times 10^6 \times 3600$

$\therefore$ Mass of $CO_2$ hydrated per hour

$$= \frac{36 N_A}{N_A} \times 44 = 1584 \text{ g}$$

CHAPTER 2

# Equivalent Concept

## EXERCISE I (JEE MAIN)

### Oxidation–Reduction

1. A compound contains X, Y and Z atoms. The oxidation states of X are $+a$, Y is $-b$ and Z is $-c$. The molecular formula of the compound is $X_mY_nZ_r$. Therefore, among the given relations, which one is correct?
   (a) $am + bn + cr = 0$ (b) $am + bn = cr$
   (c) $am + cr = bn$ (d) $bn + cr = am$

2. What is the oxidation state of Xe in $Ba_2XeO_6$?
   (a) 0 (b) +4
   (c) +6 (d) +8

3. When $K_2Cr_2O_7$ is converted into $K_2CrO_4$, the change in oxidation number of Cr is
   (a) 0 (b) 6
   (c) 4 (d) 3

4. The formula of brown ring complex is $[Fe(H_2O)_5(NO)]SO_4$. The oxidation state of iron is
   (a) +1 (b) +2
   (c) +3 (d) 0

5. In the reaction: $3Br_2 + 6CO_3^{2-} + 3H_2O \rightarrow 5Br^- + BrO_3^- + 6HCO_3^-$
   (a) Bromine is oxidized and carbonate is reduced.
   (b) Bromine is oxidized and water is reduced.
   (c) Bromine is both oxidized and reduced.
   (d) Bromine is neither oxidized nor reduced.

6. In which of the following compound, the oxidation state of sulphur is +7?
   (a) $Na_2S_2O_8$ (b) $H_2S_2O_7$
   (c) $H_2SO_4$ (d) None of these

7. In which of the following compound, iron has the lowest oxidation state?
   (a) $Fe(CO)_5$
   (b) $Fe_2O_3$
   (c) $K_4[Fe(CN)_6]$
   (d) $FeSO_4 \cdot (NH_4)_2SO_4 \cdot 6H_2O$

8. Which of the following have been arranged in the order of decreasing oxidation number of sulphur?
   (a) $H_2S_2O_7 > Na_2S_4O_6 > Na_2S_2O_3 > S_8$
   (b) $SO^{2+} > SO_4^{2-} > SO_3^{2-} > HSO_4^-$
   (c) $H_2SO_5 > H_2SO_3 > SCl_2 > H_2S$
   (d) $H_2SO_4 > SO_2 > H_2S > H_2S_2O_8$

9. The oxidation state of iron in oxygenated haemoglobin is
   (a) +1 (b) +2
   (c) +3 (d) Zero

10. The oxidation numbers of C in HCN and HNC, respectively, are
   (a) +2, +2 (b) +2, +4
   (c) +4, +4 (d) −2, −2

11. The oxidation number of carbon in carbon suboxide ($C_3O_2$) is
(a) +2/3 (b) +4/3
(c) +4 (d) −4/3

12. The oxidation states of the most electronegative element in the products of the reaction, $BaO_2$ with dil. $H_2SO_4$ are
(a) 0 and −1 (b) −1 and −2
(c) −2 and 0 (d) −2 and +1

13. The pair of compounds having metals in their highest oxidation state is
(a) $MnO_2$, $FeCl_3$
(b) $[MnO_4]^-$, $CrO_2Cl_2$
(c) $[Fe(CN)_6]^{3-}$, $[Co(CN)_3]$
(d) $[NiCl_4]^{2-}$, $[CoCl_4]^-$

14. The oxidation number of phosphorus in $Mg_2P_2O_7$ is
(a) +5 (b) −5
(c) +6 (d) −7

15. The sum of oxidation states of all carbon atoms in toluene molecule is
(a) −1 (b) −7/8
(c) −8/7 (d) −8

16. The oxidation number of K in $KO_2$ is
(a) +4 (b) +1
(c) +1/2 (d) −1/2

17. The oxidation state of chromium is +6 in
(a) $K_3CrO_8$ (b) $Cr_2O_3$
(c) $Cr_2(SO_4)_3$ (d) $CrO_5$

18. The oxidation states of N in aniline and nitrobenzene are, respectively,
(a) −3, +3 (b) −1, +5
(c) −3, +5 (d) −3, +1

19. Which of the following statements is true about oxidation state of S in $Na_2S_4O_6$?
(a) All S–atoms are in +2.5 state.
(b) All S–atoms are in +2 state.
(c) Two S–atoms are in 0 state and other two is in +5 state.
(d) Two S–atoms are in −1 state and other two is in +6 state.

20. The oxidation state of C in $C_6H_{12}O_6$ is equal to the oxidation state of C in
(a) HCOOH (b) HCHO
(c) $CH_4$ (d) CO

21. An oxide of iron contains 30% oxygen by mass. The oxidation state of iron in this oxide is (Fe = 56)
(a) +1 (b) +2
(c) +3 (d) +4

22. The strongest reducing agent is
(a) $H_2S$ (b) $H_2O$
(c) $H_2Se$ (d) $H_2Te$

23. During developing of an exposed camera film, one step involves in the following reaction.

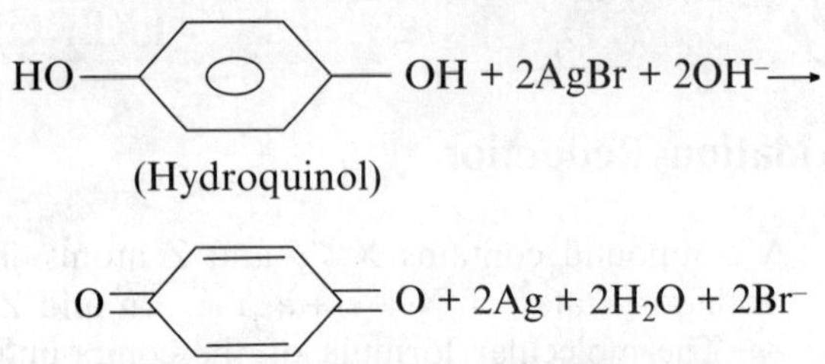


Which of the following best describes the role of hydroquinol?
(a) It acts as an acid.
(b) It acts as a reducing agent.
(c) It acts as an oxidant.
(d) It acts as a base.

24. A redox reaction is
(a) exothermic.
(b) endothermic.
(c) neither exothermic nor endothermic.
(d) either exothermic or endothermic.

25. The decomposition of $KClO_3$ to KCl and $O_2$ on heating is an example of
(a) intermolecular redox change
(b) intramolecular redox change
(c) disproportionation or auto redox change
(d) All the above

26. Which of the following reaction is non-redox?
(a) $2NaNO_3 \rightarrow 2NaNO_2 + O_2$
(b) $CaO + SiO_2 \rightarrow CaSiO_3$
(c) $Fe + H_2SO_4 \rightarrow FeSO_4 + H_2$
(d) $4Ag + 8CN^- + O_2 + 2H_2O \rightarrow 4[Ag(CN)_2]^- + 4OH^-$

**27.** Which of the following reaction is not a disproportionation reaction?

(a) $Br_2 + CO_3^{2-} + H_2O \rightarrow Br^- + BrO_3^- + HCO_3^-$
(b) $P_4 + OH^- + H_2O \rightarrow PH_3 + H_2PO_2^-$
(c) $H_2S + SO_2 \rightarrow S + H_2O$
(d) $H_2O_2 \rightarrow H_2O + O_2$

**28.** Which of the following reaction is an example of comproportionation reaction?

(a) $Cl_2 + OH^- \rightarrow Cl^- + ClO_3^- + H_2O$
(b) $CH_4 + O_2 \rightarrow CO_2 + H_2O$
(c) $H_2S + SO_2 \rightarrow S + H_2O$
(d) $NaOH + HCl \rightarrow NaCl + H_2O$

**29.** An oxide, $X_2O_3$ is oxidized to $XO_4^-$ by $Cr_2O_7^{2-}$ in acid medium. The number of moles of $X_2O_3$ oxidized per mole of $Cr_2O_7^{2-}$ is

(a) 3/4 (b) 3
(c) 3/2 (d) 2/3

**30.** The number of electrons involved in the reduction of nitrate ion to hydrazine is

(a) 8 (b) 7
(c) 5 (d) 3

**31.** In the disproportionation reaction (unbalanced),

$$Br_2 + OH^- \rightarrow Br^- + BrO_3^- + H_2O,$$

the ratio of $Br_2$ molecules undergoing oxidation and reduction is

(a) 5:1 (b) 1:5
(c) 2:3 (d) 3:2

**32.** For the process, $NO_3^- \rightarrow N_2O$, the number of $H_2O$ molecules needed for balancing in acid medium and the side in which it should be added are

(a) 2, right (b) 2, left
(c) 5, right (d) 5, left

**33.** In the process, $NO_2^- \rightarrow NH_3$, the number of $OH^-$ ions and the side in which they should be added in balancing are

(a) 7, right (b) 7, left
(c) 4, left (d) 5, right

**34.** For the redox reaction,

$Zn + NO_3^- \rightarrow Zn^{2+} + NH_4^+$

in basic medium, the coefficients of Zn, $NO_3^-$ and $OH^-$ in the balanced equation, respectively, are

(a) 4, 1, 7 (b) 7, 4, 1
(c) 4, 1, 10 (d) 1, 4, 10

**35.** The ratio of coefficients of $HNO_3$, $Fe(NO_3)_2$ and $NH_4NO_3$ in the following redox reaction,

$$Fe + HNO_3 \rightarrow Fe(NO_3)_2 + NH_4NO_3 + H_2O$$

respectively

(a) 10 : 1 : 4 (b) 10 : 4 : 1
(c) 4 : 10 : 1 (d) 4 : 1 : 10

## Equivalent Concept

**36.** An oxide of metal have 20% oxygen, the equivalent weight of oxide is

(a) 32 (b) 48
(c) 40 (d) 52

**37.** On heating in contact with tin, sulphurated hydrogen (V.D. = 17) is converted into hydrogen without change in volume. The equivalent weight of sulphur is

(a) 32 (b) 16
(c) 24 (d) 34

**38.** An element (X) having equivalent mass $E$ forms an oxide $X_mO_n$. The atomic mass of element should be

(a) $\dfrac{2En}{m}$ (b) $2mEn$
(c) $\dfrac{E}{n}$ (d) $En$

**39.** Equivalent weight of a metal is 18.67. When it reacts with chlorine, the mass of metal which will form 162.52 g of metal chloride is

(a) 143.83 g (b) 56 g
(c) 14.4 g (d) 5.6 g

**40.** How many grams of Mg would have to react in order to liberate 4 $N_A$ electrons?

(a) 12 g (b) 24 g
(c) 48 g (d) 96 g

41. Equivalent weight of $K_2CrO_4$ when it reacts with $AgNO_3$ to give $Ag_2CrO_4$ is
(a) infinite (b) $M$
(c) $\frac{M}{2}$ (d) $\frac{M}{3}$

42. The equivalent weight of $NaHC_2O_4$ in reaction with NaOH is
(a) 112 (b) 56
(c) 224 (d) 84

43. The equivalent weight of $NaHC_2O_4$ in reaction with HCl is
(a) 112 (b) 56
(c) 224 (d) 8

44. In a reaction, calcium phosphate is producing calcium hydrogen phosphate. The equivalent weight of calcium phosphate in this process is (Ca = 40, P = 31)
(a) 310 (b) 155
(c) 103.33 (d) 51.67

45. The equivalent weight of $MnSO_4$ is half of its molecular weight when it is converted to
(a) $Mn_2O_3$ (b) $MnO_2$
(c) $MnO_4^-$ (d) $MnO_4^{2-}$

46. Equivalent weight of $MnO_4^-$ in acidic, basic, neutral medium is in the ratio of
(a) 3 : 5 : 15 (b) 5 : 3 : 1
(c) 5 : 1 : 3 (d) 3 : 15 : 5

47. In the following reaction (unbalanced), equivalent weight of $As_2S_3$ is related to its molecular weight, *M*, by

$$As_2S_3 + H^+ + NO_3^- \rightarrow NO + H_2O + AsO_4^{3-} + SO_4^{2-}$$

(a) *M*/2 (b) *M*/4
(c) *M*/28 (d) *M*/24

48. In the following redox reactions, $NH_3$ appears either in reactant or product. In which case, the equivalent weight of $NH_3$ is maximum?
(a) $N_2 + 3H_2 \rightarrow 2NH_3$
(b) $4NH_3 + 5O_2 \rightarrow 4NO + 6H_2O$
(c) $2NH_3 + 2Na \rightarrow 2NaNH_2 + H_2$
(d) Equal in all cases

49. In the following unbalanced redox reaction,

$$Cu_3P + Cr_2O_7^{2-} \rightarrow Cu^{2+} + H_3PO_4 + Cr^{3+},$$

the equivalent weight of $H_3PO_4$ is
(a) M/3 (b) M/6
(c) M/7 (d) M/8

50. The equivalent weight of $Cl_2$ acting as oxidizing agent is
(a) 72 (b) 35.5
(c) 7.1 (d) 23.67

51. $H_2O_2$ disproportionates into $H_2O$ and $O_2$. The equivalent weight of $H_2O_2$ in this reaction is
(a) 34 (b) 17
(c) 68 (d) 8.5

52. The equivalent weight of $H_2SO_4$ in the reaction:

$$2KMnO_4 + 3H_2SO_4 + 10HCl \rightarrow 2MnSO_4 + K_2SO_4 + 5Cl_2 + 8H_2O, \text{ is}$$

(a) $\frac{M}{2}$ (b) $M$
(c) $\frac{3M}{10}$ (d) $\frac{3M}{5}$

53. What would be the equivalent weight of reductant in the reaction? (Fe = 56)

$$2[Fe(CN)_6]^{-3} + H_2O_2 + 2OH^- \rightarrow 2[Fe(CN)_6]^{4-} + 2H_2O + O_2$$

(a) 17 (b) 212
(c) 34 (d) 32

54. In the Haber's process, the equivalent weight of ammonia is
(a) M (b) M/3
(c) 2M/3 (d) 3M

55. Equivalent mass of a bivalent metal is 32.7. Molecular mass of its chloride is
(a) 68.2 (b) 103.7
(c) 136.4 (d) 166.3

56. In the reaction,

$$Zn + HNO_3 \rightarrow Zn(NO_3)_2 + NO + H_2O,$$

the equivalent weight of $HNO_3$ is
(a) *M* (b) 4*M*/3
(c) 8*M*/3 (d) 2*M*/3

EXERCISE I

**57.** Equivalent weight of $H_2SO_4$ in the reaction

$Mg + 2H_2SO_4 \rightarrow MgSO_4 + SO_2 + 2H_2O$, is

(a) 98 (b) 49
(c) 196 (d) 32.67

**58.** A metal carbonate on heating is converted to metal oxide and is reduced to 60% of its original weight. The equivalent weight of the metal is

(a) 5 (b) 25
(c) 60 (d) 70

**59.** A quantity of 0.298 g of the chloride of a metal yielded 0.348 g of the sulphate of the same metal. The equivalent weight of the metal is

(a) 12 (b) 20
(c) 39 (d) 41.5

**60.** $NH_3$ is oxidized to NO by $O_2$ in basic medium. The number of equivalents of $NH_3$ oxidized by 1 mole of $O_2$ is

(a) 4 (b) 5
(c) 6 (d) 7

**61.** The number of moles of $Cr_2O_7^{2-}$ needed to oxidize 0.136 equivalent of $N_2H_5^+$ through the reaction

$$N_2H_5^+ + Cr_2O_7^{2-} \rightarrow N_2 + Cr^{3+} + H_2O \text{ is}$$

(a) 0.023 (b) 0.091
(c) 0.136 (d) 0.816

**62.** A certain amount of a reducing agent reduces $x$ mole of $KMnO_4$ and y mole of $K_2Cr_2O_7$ in different experiments in acidic medium. If the change in oxidation state in reducing agent is same in both experiments, $x : y$ is

(a) 5 : 3 (b) 3 : 5
(c) 5 : 6 (d) 6 : 5

**63.** When a metal carbonate is treated with excess of dilute sulphuric acid, the weight of metal sulphate formed is 1.43 times the weight of carbonate reacted. What is the equivalent weight of metal?

(a) 23 (b) 20
(c) 39 (d) 12

**64.** How many grams of $H_2S$ will react with 6.32 g $KMnO_4$ to produce $K_2SO_4$ and $MnO_2$? (K = 39, Mn = 55)

(a) 4.08 g (b) 0.85 g
(c) 0.51 g (d) 2.04 g

**65.** The oxide of a metal contains 52.91% of the metal. If the formula of the metal oxide is $M_2O_3$, then what is the atomic mass of the metal?

(a) 8.99 (b) 26.96
(c) 17.97 (d) 53.93

**66.** The equivalent weight of an element is 25. If its specific heat is 0.085 cal/K-g, its exact atomic mass should be

(a) 75.29 (b) 75
(c) 50 (d) 50.8

**67.** The vapour density of metal chloride is 77. If its equivalent weight is 3, its atomic mass will be

(a) 3 (b) 6
(c) 9 (d) 12

**68.** One gram of the acid $C_6H_{10}O_4$ requires 0.768 g of KOH for complete neutralization. How many neutralizable hydrogen atoms are in this molecule?

(a) 4 (b) 3
(c) 2 (d) 1

**69.** A quantity of 1.878 g of a metal bromide when heated in a stream of hydrogen chloride gas is completely converted to the chloride weighing 1.00 g. The specific heat of the metal is 0.14 cal/°c-g. What is the molecular weight of the bromide? (Br = 80, Cl = 35.5)

(a) 45.54 (b) 125.54
(c) 285.54 (d) 205.54

**70.** Potassium sulphate is isomorphous with potassium chromate which contains 26.79% by mass of chromium. The atomic mass of chromium is

(a) 24 (b) 32
(c) 51.96 (d) 53.2

## Volumetric Analysis

71. Equal volumes of 10% (w/v) $H_2SO_4$ solution and 10% (w/v) NaOH solution are mixed. The nature of the resulting solution will be
(a) neutral (b) acidic
(c) basic (d) unpredictable

72. A quantity of 0.62 g of $Na_2CO_3 \cdot H_2O$ is added to 100 ml of 0.1 N-$H_2SO_4$ solution. The resulting solution would be
(a) acidic (b) alkaline
(c) neutral (d) buffer

73. The volume of 0.10 M-$AgNO_3$ should be added to 10.0 ml of 0.09 M-$K_2CrO_4$ to precipitate all the chromate as $Ag_2CrO_4$ is
(a) 18 ml (b) 9 ml
(c) 27 ml (d) 36 ml

74. What volume of 0.18 N-$KMnO_4$ solution would be needed for complete reaction with 25 ml of 0.21 N-$KNO_2$ in acidic medium?
(a) 57.29 ml (b) 11.67 ml
(c) 29.17 ml (d) 22.92 ml

75. A 0.1 M-$KMnO_4$ solution is used for the following titration. What volume of the solution will be required to react with 0.158 g of $Na_2S_2O_3$?

$$S_2O_3^{2-} + MnO_4^- + H_2O \rightarrow MnO_2(s) + SO_4^{2-} + OH^-$$

(a) 80 ml (b) 26.67 ml
(c) 13.33 ml (d) 16 ml

76. One litre of a solution contains 18.9 g of $HNO_3$ and one litre of another solution contains 3.2 g of NaOH. In what volume ratio must these solutions be mixed to obtain a neutral solution?
(a) 3:8 (b) 8:3
(c) 15:4 (d) 4:15

77. What volume of gaseous $NH_3$ at 0°C and 1 atm will be required to be passed into 30 ml of N–$H_2SO_4$ solution to bring down the acid strength of this solution to 0.2 N?
(a) 537.6 ml (b) 268.8 ml
(c) 1075.2 ml (d) 371.3 ml

78. A 26 ml of N-$Na_2CO_3$ solution is neutralized by the solutions of acids A and B in different experiments. The volumes of the acids A and B required were 10 ml and 40 ml, respectively. How many volumes of A and B are to be mixed in order to prepare 1 litre of normal acid solution?
(a) 179.4, 820.6 (b) 820.6, 179.4
(c) 500, 500 (d) 474.3, 525.7

79. A sample of an alloy of silver weighing 0.50 g and containing 90% silver was dissolved in conc. $HNO_3$ and silver was analysed by Volhard method. A volume of 25 ml of a KCNS solution was required for complete precipitation. The normality of KCNS solution is (Ag = 108)
(a) 4.167 (b) 0.167
(c) 3.136 (d) 0.125

80. A 0.5 g sample of $KH_2PO_4$ is titrated with 0.1 M NaOH. The volume of base required to do this is 25.0 ml. The reaction is represented as

$$H_2PO_4^- + OH^- \rightarrow HPO_4^{2-} + H_2O.$$

The percentage purity of $KH_2PO_4$ is (K = 39, P = 31)
(a) 68% (b) 34%
(c) 85% (d) 51%

81. A solution of $H_2O_2$ is titrated with a solution of $KMnO_4$. The reaction is

$$2MnO_4^- + 5H_2O_2 + 6H^+ \rightarrow 2Mn^{2+} + 5O_2 + 8H_2O$$

It requires 50 ml of 0.1 M-$KMnO_4$ to oxidize 10 ml of $H_2O_2$. The strength of $H_2O_2$ solution is
(a) 4.25% (w/v) (b) 8.5% (w/v)
(c) 0.85% (w/v) (d) 1.7% (w/v)

82. For the standardization of $Ba(OH)_2$ solution, 0.204 g of potassium acid phthalate was weighed which was then titrated with $Ba(OH)_2$ solution. The titration indicated equivalence at 25.0 ml of $Ba(OH)_2$ solution. The reaction involved is

$$KHC_8H_4O_4 + Ba(OH)_2 \rightarrow H_2O + K^+ + Ba^{2+} + C_8H_4O_4^{2-}$$

The molarity of the base solution is (K = 39)
(a) 0.04 M (b) 0.03 M
(c) 0.02 M (d) 0.01 M

**83.** A volume of 12.5 ml of 0.05 M selenium dioxide, $SeO_2$, reacted with exactly 25.0 ml of 0.1 M – $CrSO_4$. In this reaction, $Cr^{2+}$ is converted to $Cr^{3+}$. To what oxidation state the selenium is converted by the reaction?

(a) 0 (b) +1
(c) +2 (d) +4

**84.** The chromate ion may be present in waste water from a chrome planting plant. It is reduced to insoluble chromium hydroxide, $Cr(OH)_3$ by dithionation, in basic solution:

$$S_2O_4^{2-} + CrO_4^{2-} + H_2O + OH^- \rightarrow SO_3^{2-} + Cr(OH)_3$$

100 litre of water requires 522 g of $Na_2S_2O_4$. The molarity of $CrO_4^{2-}$ in waste water is

(a) 0.04 (b) 0.03
(c) 0.02 (d) 2.0

**85.** Calcium oxalate is insoluble in water. This property has been used to determine the amount of calcium ion in fluids such as blood. The calcium oxalate isolated from blood is dissolved in acid and titrated against a standard $KMnO_4$ solution. In one test, it is found that the calcium oxalate isolated from a 10 ml sample of blood requires 25 ml of 0.001 M-$KMnO_4$ for titration. The number of milligram of calcium per litre of blood is

(a) 0.25 (b) 0.50
(c) 0.80 (d) 0.40

**86.** How many grams of $KMnO_4$ will react with 50 ml of 0.2 M-$H_2C_2O_4$ solution in the presence of $H_2SO_4$?

(a) 1.58 g (b) 3.16 g
(c) 0.632 g (d) 0.79 g

**87.** What volume of 0.05 M-$Ca(OH)_2$ solution is needed for complete conversion of 10 ml of 0.1 M-$H_3PO_4$ into $Ca(H_2PO_4)_2$?

(a) 10 ml (b) 5 ml
(c) 20 ml (d) 40 ml

**88.** How many grams of oxalic acid crystals, $H_2C_2O_4 \cdot 2H_2O$ is needed to react completely with 100 ml of 0.4 M-$KMnO_4$ in the presence of $H_2SO_4$?

(a) 2.52 g (b) 12.6 g
(c) 25.2 g (d) 9.0 g

**89.** Borax has the formula $Na_2B_4O_7 \cdot 10H_2O$. It is a strong base in aqueous solution because $OH^-$ ions are produced by reaction with water.

$(B_4O_7^{2-} + 7H_2O \rightarrow 4H_3BO_3 + 2OH^-)$.

How many grams of borax is necessary to neutralize 25 ml of 0.2 M solution of hydrochloric acid? (B = 10.8)

(a) 0.4765 g (b) 0.953 g
(c) 9.53 g (d) 1.906 g

**90.** A volume of 50 ml of 0.1 M metal salt reacts completely with 25 ml of 0.1 M sodium sulphite. In the reaction, $SO_3^{2-}$ is oxidized to $SO_4^{2-}$. If the oxidation number of metal in the salt is +3, then what is its new oxidation number?

(a) 0 (b) +1
(c) 2 (d) 4

**91.** A 1 g sample of hydrogen peroxide solution containing $x$% of $H_2O_2$ by weight requires $x$ ml of $KMnO_4$ solution for complete oxidation under acidic conditions. What is the normality of $KMnO_4$ solution?

(a) 5.88 N (b) 58.8 N
(c) 0.0588 N (d) 0.588 N

**92.** One gram of ferrous oxalate dissolved in dil. $H_2SO_4$ is treated with $KMnO_4$ solution added in drops till a faint pink colour persists in the solution. If 60 ml of $KMnO_4$ solution is consumed, then calculate its molarity (Fe = 56).

(a) 0.694 M (b) 0.0694 M
(c) 0.294 M (d) 0.0294 M

**93.** Magnesium hydroxide is the white milky substance in milk of magnesia. What mass of $Mg(OH)_2$ is formed when 15 ml of 0.2 M-NaOH is combined with 12 ml of 0.15 M-$MgCl_2$?

(a) 0.087 g (b) 0.079 g
(c) 0.1044 g (d) 0.522 g

**94.** The formula weight of an acid is 82 amu. In a titration, 100 $cm^3$ of a solution of this acid containing 39.0 g of the acid per litre was completely neutralized by 95 $cm^3$ of aqueous solution of NaOH containing 40 g of NaOH in 1 L of solution. What is the basicity of the acid?

(a) 4 (b) 2
(c) 1 (d) 3

**95.** A quantity of 20 g of $H_3PO_4$ is dissolved in water and made up to 1 L. What is the normality of the solution, if titration against NaOH is carried only up to the second stage of neutralization?

(a) 0.408 (b) 0.204
(c) 0.612 (d) 0.102

96. A volume of 25 ml of 0.017 M-$HSO_3^-$ in strongly acidic solution required the addition of 16.9 ml of 0.01 M-$MnO_4^-$ for its complete oxidation. In neutral solution, 28.6 ml is required. Assign oxidation numbers of Mn in each of the products.

(a) 2, 4 (b) 3, 4
(c) 2, 3 (d) 3, 4

97. A quantity of 0.84 g of an acid (molecular mass = 150) was dissolved in water and the volume was made up to 100 ml. Twenty five millilitres of this solution required 28 ml of (N/10) NaOH solution for neutralization. The equivalent weight and basicity of the acid is

(a) 75, 2 (b) 150, 1
(c) 75, 4 (d) 150, 2

98. A quantity of 0.70 g of a sample of $Na_2CO_3 \cdot xH_2O$ was dissolved in water and the volume was made to 100 ml. Twenty millilitres of this solution required 19.8 ml of *N*/10 HCl for complete neutralization. The value of *x* is

(a) 2 (b) 1
(c) 4 (d) 10

99. The specific gravity of a given $H_2SO_4$ solution is 1.76. A quantity of 3.5 ml of the acid is diluted to 1.0 L and 25 ml of this diluted acid required 25.6 ml of *N*/10 ($f = 0.95$) NaOH solution for complete neutralization. The percentage strength (by mass) of the original acid solution is

(a) 61.6% (b) 77.38%
(c) 50% (d) 47.66%

100. A volume of 25 ml of (*N*/10)-$Na_2CO_3$ solution neutralizes 10 ml of a dilute $H_2SO_4$ solution. The volume of water that must be added to 400 ml of this $H_2SO_4$ solution in order to make it exactly *N*/10 is

(a) 1000 ml (b) 600 ml
(c) 500 ml (d) 400 ml

101. A volume of 10 ml of a $H_2SO_4$ solution is diluted to 100 ml. Twenty five millilitres of this diluted solution is mixed with 50 ml of 0.5 N-NaOH solution. The resulting solution requires 0.265 g $Na_2CO_3$ for complete neutralization. The normality of original $H_2SO_4$ solution is

(a) 12 N (b) 1.2 N
(c) 3 N (d) 0.275 N

102. The normality of a solution of a mixture containing HCl and $H_2SO_4$ is *N*/5. Twenty millilitres of this solution reacts with excess of $AgNO_3$ solution to give 0.287 g of silver chloride. The percentage of HCl in the mixture by mass is (Ag = 108)

(a) 42.69% (b) 57.31%
(c) 40% (d) 33.18%

103. A quantity of 0.10 g of anhydrous organic acid requires 25 ml of 0.10 N-NaOH for neutralization. A quantity of 0.245 g of the hydrated acid requires 50 ml of the same alkali. The number of moles of water of crystallization per equivalent of the anhydrous acid is

(a) 1.0 (b) 2.0
(c) 0.5 (d) 4.0

104. A volume of 100 ml of $H_2O_2$ is oxidized by 100 ml of 1 M-$KMnO_4$ in acidic medium ($MnO_4^-$ reduced to $Mn^{2+}$). A volume of 100 ml of same $H_2O_2$ is oxidized by 'V' ml of 1 M-$KMnO_4$ in basic medium ($MnO_4^-$ reduced to $MnO_2$). The value of 'V' is

(a) 500 (b) 100
(c) 33.33 (d) 166.67

105. A quantity of 1 g of metal carbonate was dissolved in 25 ml of normal HCl. The resulting liquid requires 50 ml of *N*/10 caustic soda solution to neutralize it completely. The equivalent weight of metal carbonate is

(a) 10 (b) 20
(c) 100 (d) 50

106. When 0.91 g of a mixture of $Na_2SO_4$ and $(NH_4)_2SO_4$ was boiled with 80 ml of 0.1 N-NaOH until no more $NH_3$ is evolved, the excess of NaOH required is 11.6 ml of 0.1 N-HCl. How many grams of $Na_2SO_4$ is present in the mixture?

(a) 0.594 g (b) 0.459 g
(c) 0.549 g (d) 0.945 g

107. A quantity of 10 g of a sample of silver, which is contaminated with silver sulphide, gave 11.2 ml of hydrogen sulphide at 0°C and 1 atm, on treatment with excess of hydrochloride acid. The amount of silver sulphide in the sample is (Ag = 108)

(a) 1.24 g (b) 124 mg
(c) $5 \times 10^{-3}$ mol (d) 62 mg

**108.** A 0.2 g sample of iron wire containing 98% iron is dissolved in acid to form ferrous ion. The solution requires 30 ml of $K_2Cr_2O_7$ solution for complete reaction. What is the normality of $K_2Cr_2O_7$ solution? (Fe = 56)

(a) 0.1167 N (b) 0.2333 N
(c) 0.0583 N (d) 0.167 N

**109.** One litre of a mixture of $O_2$ and $O_3$ at 0°C and 1 atm was allowed to react with an excess of acidified solution of KI. The iodine liberated requires 40 ml of $M/10$ sodium thiosulphate solution for titration. What is the mass percent of ozone in the mixture?

(a) 6.575% (b) 9.6%
(c) 93.425% (d) 90.4%

**110.** A 71 ml (specific gravity 1.1) of chlorine water is treated with an excess of KI. The liberated iodine requires 26 ml of 0.11 N-$Na_2S_2O_3$ (sodium thiosulphate or hypo) solution. What is the percentage of $Cl_2$ (by mass) in the chlorine water? Chlorine water is a solution of free chlorine in water.

(a) 0.26% (b) 0.13%
(c) 0.065% (d) 1.3%

**111.** Household bleach contains hypochlorite ion, which is formed when chlorine dissolves in water. To determine the concentration of hypochlorite in the bleach, the solution is first treated with a KI solution. The iodine liberated can be determined by titration with a standard thiosulphate solution. A 25 ml of certain household bleach requires 17.4 ml of a 0.02 M-$Na_2S_2O_3$ solution for titration. The mass of chlorine dissolved in one litre of the bleach solution is

(a) 0.1392 g (b) 0.494 g
(c) 9.88 g (d) 0.278 g

**112.** One gram of $Na_3AsO_4$ is boiled with excess of solid KI in the presence of strong HCl. The iodine evolved is absorbed in KI solution and titrated against 0.2 N hypo solution. Assuming the reaction to be

$$AsO_4^{3-} + 2H^+ + 2I^- \rightarrow AsO_3^{3-} + H_2O + I_2$$

The volume of thiosulphate hypo consumed is (As = 75)

(a) 48.1 ml (b) 38.4 ml
(c) 24.7 ml (d) 30.3 ml

**113.** $S_2O_3^{2-}$ ion is oxidized by $S_2O_8^{2-}$ ion, the products are $S_4O_6^{2-}$ and $SO_4^{2-}$ ions. What volume of 0.25 M thiosulphate solution would be needed to reduce 1 g of $K_2S_2O_8$? (K = 39)

(a) 36.92 ml (b) 32.69 ml
(c) 29.63 ml (d) 62.93 ml

**114.** $V_1$ ml of permanganate solution of molarity $M_1$ reacts exactly with $V_2$ ml of ferrous sulphate solution of molarity $M_2$, then

(a) $V_1M_1 = V_2M_2$ (b) $5V_1M_1 = V_2M_2$
(c) $V_1M_1 = 5V_1M_2$ (d) None of these

**115.** $x$ g of $KHC_2O_4$ requires 100 ml of 0.02 M-$KMnO_4$ in acidic medium. In another experiment, $y$ g of $KHC_2O_4$ requires 100 ml of 0.05 M-$Ca(OH)_2$. The ratio of $x$ and $y$ is

(a) 1 : 1 (b) 1 : 2
(c) 2 : 1 (d) 5 : 4

**116.** In the mixture of $NaHCO_3$ and $Na_2CO_3$, the volume of a given HCl required is $x$ ml with phenolphthalein indicator and further $y$ ml is required with methyl orange indicator. Hence, the volume of HCl for complete reaction of $NaHCO_3$ present in the original mixture is

(a) $2x$ (b) $y$
(c) $x/2$ (d) $(y - x)$

**117.** A volume of 25 ml of 0.107 M-$H_3PO_4$ was titrated with 0.115 M solution of NaOH to the end point identified by indicator bromocresol green. This requires 23.1 ml. The titration was repeated using phenolphthalein as indicator. This time 25 ml of 0.107 M-$H_3PO_4$ requires 46.2 ml of the 0.115 M-NaOH. What is the coefficient $n$ in the following reaction?

$$H_3PO_4 + nOH^- \rightarrow [H_{3-n}PO_4]^{n-} + nH_2O$$

(a) 1, 2 (b) 2, 1
(c) 3, 2 (d) 1, 3

**118.** Calculate the temporary and permanent hardness of water sample having the following constituents per litre.

$Ca(HCO_3)_2 = 162$ mg, $MgCl_2 = 95$ mg

$NaCl = 585$ mg, $Mg(HCO_3)_2 = 73$ mg

$CaSO_4 = 136$ mg

(a) 200 ppm, 150 ppm
(b) 100 ppm, 150 ppm
(c) 150 ppm, 200 ppm
(d) 150 ppm, 150 ppm

**119.** A volume of 100 L of hard water requires 5.6 g of lime for removing temporary hardness. The temporary hardness in ppm of $CaCO_3$ is

(a) 56 (b) 100
(c) 200 (d) 112

**120.** $RH_2$ (ion exchange resin) can replace $Ca^{2+}$ in hard water as follows.

$RH_2 + Ca^{2+} \rightarrow RCa + 2H^+$.

One litre of hard water after passing through $RH_2$ has pH = 2. Hence, hardness in ppm of $Ca^{2+}$ is

(a) 200 (b) 100
(c) 50 (d) 125

# EXERCISE II (JEE ADVANCED)

## Section A (Only one Correct)

**1.** Which of the following process is reduction?

(a) $CH_2=CH_2 \rightarrow \underset{OH}{\underset{|}{CH_2}}-\underset{OH}{\underset{|}{CH_2}}$

(b) $CH_3CH_2CH=CH-CH_2-CHO \rightarrow$
$CH_3CH_2CH=CH-CH_2-CH_2OH$

(c) $CH_3CHO \rightarrow CCl_3CHO$

(d) $Ag^+ + 2NH_3 \rightarrow [Ag(NH_3)_2]^+$

**2.** The oxidation number of sodium in sodium amalgam is

(a) +2 (b) +1
(c) −2 (d) zero

**3.** The oxidation state of molybdenum in its oxocomplex $[Mo_2O_4(C_2H_4)_2(H_2O)_2]^{2-}$ is

(a) +2 (b) +3
(c) +4 (d) +5

**4.** The oxidation state of boron in potassium tetrafluoroborate is

(a) +2 (b) +3
(c) +4 (d) −3

**5.** The oxidation state of bismuth in lithium bismuthate is

(a) +5 (b) +3
(c) +2 (d) +4

**6.** The compound of Xe and F is found to have 53.5% Xe. What is the oxidation number of Xe in this compound? (Xe = 131, F = 19)

(a) −4 (b) 0
(c) +4 (d) +6

**7.** The oxidation number of S in $(CH_3)_2SO$ is

(a) zero (b) +1
(c) +2 (d) +3

**8.** Sulphide ions react with $Na_4[Fe(NO)(CN)_5]$ to form a purple-coloured compound $Na_4[Fe(CN)_5(NOS)]$. In the reaction, the oxidation state of iron

(a) changes from +2 to +3.
(b) changes from +2 to +4.
(c) changes from +3 to +2.
(d) does not change.

**9.** The oxidation number of cobalt in $K[Co(CO)_4]$ is

(a) +1 (b) +3
(c) −1 (d) 0

**10.** Phosphorus has oxidation state of +3 in

(a) phosphorus acid
(b) orthophosphoric acid
(c) metaphosphoric acid
(d) pyrophosphoric acid

**11.** One gas bleaches the colour of the flowers by reduction while the other by oxidation. The gases are

(a) $CO, Cl_2$ (b) $H_2S, Br_2$
(c) $SO_2, Cl_2$ (d) $NH_3, SO_3$

**12.** In a reaction, $HNO_3$ is behaving as reducing agent. What should be its expected product?

(a) $H_2$ (b) $NO_2$
(c) $N_2O$ (d) $O_2$

**13.** Which of these substance is a good reducing agent?

(a) HI (b) KBr
(c) $FeCl_3$ (d) $KClO_3$

**14.** Which of the following ion cannot act as an oxidizing agent?

(a) $MnO_4^-$ (b) $CrO_4^{2-}$
(c) $I^-$ (d) $Fe^{3+}$

**15.** Which of the following reaction is redox?

(a) $Mg_3N_2 + 6H_2O \rightarrow 3Mg(OH)_2 + 2NH_3$

(b) $CaC_2 + 2H_2O \rightarrow Ca(OH)_2 + C_2H_2$

(c) $Ca(OCl)Cl + H_2O \rightarrow Ca(OH)_2 + Cl_2$

(d) $PCl_5 + 4H_2O \rightarrow H_3PO_4 + 5HCl$

**16.** During the oxidation of $Mn^{2+}$ to $MnO_4^-$ by $PbO_2$ in acid medium, the number of moles of acid consumed per mole of $Mn^{2+}$ ion is

(a) 4 (b) 1/2
(c) 2 (d) 1/4

17. During the oxidation of arsenite ion $AsO_3^{3-}$ to arsenate ion $AsO_4^{3-}$ in alkaline medium, the number of moles of hydroxide ions consumed per mole of arsenite ion is
(a) 2 (b) 3
(c) 2/3 (d) 3/2

18. $Cr(OH)_3 + ClO^- + OH^- \rightarrow ... + Cl^- + H_2O$. The missing ion is
(a) $Cr_2O_7^{2-}$ (b) $Cr^{3+}$
(c) $CrO_4^{2-}$ (d) $Cr_2O_3$

19. In a reaction, 4 moles of electrons is transferred to one mole of $HNO_3$. The possible product obtained due to reduction is
(a) 0.5 mole of $N_2$.
(b) 0.5 mole of $N_2O$.
(c) 1 mole of $NO_2$.
(d) 1 mole of $N_2O$.

20. The number of electrons lost per mole of ethanol in its oxidation into acetic acid is
(a) $4N_A$ (b) $2N_A$
(c) $6N_A$ (d) $8N_A$

21. For the process $CH_3CH_2OH \rightarrow CH_3COOH$, the number of $H^+$ ions needed for balancing and the side in which it should be added are, respectively
(a) 4, left (b) 4, right
(c) 2, Left (d) 2, right

22. In basic medium, $Cl_2$ disproportionates into $Cl^-$ and $ClO_x^-$. If there is loss and gain of one mole of electron per mole of $Cl_2$, then the value of $x$ is
(a) 3 (b) 1
(c) 2 (d) 4

23. An amount of 0.2 mole of $AO_3^-$ gains 1.2 mole of electron in a process. Assuming that there is no change in oxidation state of oxygen, determine the oxidation state of 'A' in product.
(a) +1 (b) −1
(c) 0 (d) +6

24. The equivalent weights of an element of variable valency are 21 and 14. The atomic mass of the element may be
(a) 35 (b) 42
(c) 70 (d) 126

25. When copper oxide is strongly heated with hydrogen, it reduces to copper. The loss in its weight is 14.9 g and the weight of water formed was 16.78 g. What is the equivalent weight of oxygen, taking the equivalent weight of hydrogen as 1.008?
(a) 8.000 (b) 7.989
(c) 8.064 (d) 16.00

26. A quantity of 1 g of metal ion, $M^{2+}$ was discharged by the passage of $1.81 \times 10^{22}$ electrons. The atomic mass of the metal is
(a) 33.27 (b) 99.81
(c) 66.54 (d) 133.08

27. Phosphoric acid has minimum equivalent weight when 1 mole of it reacts with
(a) 1 mole of NaOH (b) 2 moles of NaOH
(c) 3 moles of NaOH (d) 4 moles of NaOH

28. Equivalent weight of water in a neutralization reaction between dibasic acid and triacidic base is
(a) 9 (b) 18
(c) 6 (d) 3

29. Acetic acid on chlorination yields trichloroacetic acid. Its equivalent weight will be
(a) 60 (b) 40
(c) 20 (d) 10

30. Molecular masses of $NH_3$ and $N_2$ are $x_1$ and $x_2$, respectively. In the reaction,
$N_2 + 3H_2 \rightarrow 2NH_3$,
their equivalent weights are $y_1$ and $y_2$. Then $(y_1 - y_2)$ is
(a) $\left(\frac{2x_1 - x_2}{6}\right)$ (b) $(x_1 - x_2)$
(c) $(3x_1 - x_2)$ (d) $(x_1 - 3x_2)$

31. In the reaction,
$P_4 + NaOH + H_2O \rightarrow PH_3 + NaH_2PO_2$,
the equivalent weight of $P_4$ is
(a) $M$ (b) $M/3$
(c) $M/6$ (d) $2M/3$

32. In the reaction,
$Pb + PbO_2 + H_2SO_4 \rightarrow PbSO_4 + H_2O$,
the equivalent weight of $H_2SO_4$ is
(a) $M$ (b) $M/2$
(c) $2M$ (d) $M/4$

**33.** In the reaction,

$Cl_2 + NaOH \rightarrow NaCl + NaClO_3 + H_2O$,

the equivalent weight of $H_2O$ is

(a) $M$ (b) $3M/5$
(c) $6M/5$ (d) $M/2$

**34.** The equivalent weight of ozone behaving as an oxidizing agent is

(a) 48 (b) 24
(c) 16 (d) 32

**35.** In the reaction,

$MnO_2 + 4HCl \rightarrow MnCl_2 + Cl_2 + 2H_2O$,

the equivalent weight of HCl is

(a) $M$ (b) $M/2$
(c) $2M$ (d) $M/4$

**36.** In an acidic solution, $I^-$ changes to $I_2$. How many grams of $I_2$ is produced if, in the same process, 1.5 × $10^{22}$ electrons are used up to reduce $H_3AsO_4$ to $H_3AsO_3$? (I = 128, $N_A = 6 \times 10^{23}$)

(a) 1.6 g (b) 6.4 g
(c) 4.8 g (d) 3.2 g

**37.** An ion is reduced to the element when it absorbs 6 × $10^{20}$ electrons. The number of equivalents of the ion is

(a) 0.1 (b) 0.01
(c) 0.001 (d) 0.0001

**38.** In which of the following reactions, 1 g equivalent of $H_3PO_4$ reacts with 3 g equivalents of NaOH?

(a) $H_3PO_4 + NaOH \rightarrow NaH_2PO_4 + H_2O$
(b) $H_3PO_4 + 2NaOH \rightarrow Na_2HPO_4 + 2H_2O$
(c) $H_3PO_4 + 3NaOH \rightarrow Na_3PO_4 + 3H_2O$
(d) None of the above

**39.** A quantity of 8.6 g of an oxide of a metal reacts completely with hydrogen gas to yield 1.8 g of water. The equivalent weight of the metal is

(a) 23 (b) 37
(c) 78 (d) 35

**40.** A quantity of 20 g of an acid furnished 0.5 moles of $H_3O^+$ ions in its aqueous solution. The mass of 1 g equivalent of the acid will be

(a) 40 g (b) 20 g
(c) 10 g (d) 100 g

**41.** A quantity of 1.0 g of an acid when completely acted upon by magnesium gave 1.301 g of the anhydrous magnesium salt. The equivalent weight of the acid is

(a) 35.54 (b) 36.54
(c) 48 (d) 49

**42.** A quantity of 3.7 g of an oxide of a metal was heated with charcoal, and $CO_2$ so produced was absorbed in caustic soda solution whose weight increased by 1.0 g. The equivalent weight of the metal is

(a) 11 (b) 40.7
(c) 32.7 (d) 73.4

**43.** Which has the maximum number of equivalent per mole of the oxidant?

(a) $Zn(s) + VO^{2+}(aq) \rightarrow Zn^{2+}(aq) + V^{3+}(aq)$
(b) $Ag(s) + NO_2^-(aq) \rightarrow Ag^+(aq) + NO_2(g)$
(c) $Mg(s) + V^{4+}(aq) \rightarrow Mg^{2+}(aq) + V^{2+}(aq)$
(d) $I^-(aq) + IO_3^-(aq) \rightarrow I_3^-(aq)$

**44.** The number of moles of $KMnO_4$ that will be needed to react completely with one mole of ferrous oxalate in acidic solution is

(a) 3/5 (b) 2/5
(c) 4/5 (d) 1

**45.** The number of moles of $KMnO_4$ that will be needed to react completely with one mole of sulphite ion in acidic solution is

(a) 3/5 (b) 2/5
(c) 4/5 (d) 1

**46.** A certain amount of a reducing agent reduces $x$ mole of $MnO_2$ and $y$ mole of $K_2CrO_4$ in different reactions in acidic medium. If the changes in oxidation states of reducing agent in the reactions are in 1 : 2 ratio, respectively, then the ratio of $x$ and $y$ is

(a) 2:3 (b) 1:3
(c) 3:4 (d) 3:2

**47.** Dichloroacetic acid ($CHCl_2CO_2H$) is oxidized to $CO_2$, $H_2O$ and $Cl_2$ by 1.2 equivalents of an oxidizing agent. Same amount of the acid can neutralize 'X' moles of $NH_3$ to give ammonium dichloroacetate. The value of 'X' is

(a) 0.4 (b) 0.3
(c) 0.2 (d) 0.1

**48.** Two acids $H_2SO_4$ and $H_3PO_4$ are neutralized separately by the same amount of an alkali when sulphate and dihydrogen orthophosphate are formed, respectively. Find the ratio of the masses of $H_2SO_4$ and $H_3PO_4$.

(a) 1 : 1 (b) 1 : 2
(c) 2 : 1 (d) 2 : 3

**49.** A metal exhibits the valencies of 2 and 3. Its equivalent weight is 28 when it forms a metal oxide of formula MO. What mass of $H_2SO_4$ is needed for complete reaction with 4.8 g of $M_2O_3$?

(a) 8.82 g (b) 4.41 g
(c) 13.23 g (d) 11.03 g

**50.** The specific heat of a metal is 0.26. The chloride of the metal (always monomer) has its molecular mass 95. The volume of hydrogen gas that 1.2 g of the metal will evolve at 0°C and 1 atm, if it is allowed to react with excess of an acid, is

(a) 2.24 L (b) 1.12 L
(c) 0.56 L (d) 5.611 L

**51.** The vapour density of a volatile chloride of a metal is 74.6. If the specific heat of the metal is 0.55, then the atomic mass of the metal is

(a) 7.2 (b) 7.46
(c) 11.63 (d) 10

**52.** A quantity of 3 g of impure marble was treated with 200 ml of dil. HCl. After completion of the reaction a small quantity of the residue was left and 560 ml of a gas was evolved at 0°C and 1 atm. The normality of acid solution is

(a) 0.3 N (b) 0.125 N
(c) 0.25 N (d) 0.5 N

**53.** What will be present in the solution when 50 ml of 0.1M-HCl is mixed with 50 ml of 0.1 M-NaOH solution?

(a) 4.5 millimoles of $H^+$
(b) 0.05 millimoles of $OH^-$
(c) 0.1 M-NaCl
(d) $10^{-7}$ M of $H^+$ ion

**54.** A quantity of 5.88 g of $FeSO_4 \cdot (NH_4)_2SO_4 \cdot xH_2O$ was dissolved in 250 ml of its solution. Twenty millilitres of this solution requires 20 ml of $KMnO_4$ solution containing 3.16 g of 75% pure $KMnO_4$ dissolved per litre. The value of '$x$' is (K = 39, Mn = 55, Fe = 56)

(a) 3 (b) 4
(c) 6 (d) 7

**55.** The ratio of amounts of $H_2S$ needed to precipitate all the metal ions from 100 ml of 1 M-$AgNO_3$ and 100 ml of 1 M-$CuSO_4$ is

(a) 1:2 (b) 2:1
(c) zero (d) infinite

**56.** A volume of 100 ml of 0.1 M-$NaAl(OH)_2CO_3$ is neutralized by 0.25 N-HCl to form NaCl, $AlCl_3$ and $CO_2$. The volume of HCl required is

(a) 10 ml (b) 40 ml
(c) 100 ml (d) 160 ml

**57.** Purple of Cassius is prepared by reducing $AuCl_3$ to colloidal gold by $SnCl_2$. A 1 L solution containing 1.97 mg of gold per ml is prepared from 0.05 M solution of $AuCl_3$ by reduction with appropriate amount of 0.05 M-$SnCl_2$ solution, the resulting solution being diluted to 1 L with water. The volume of stannous chloride solution required, if its oxidation product is $SnCl_4$(aq), is (Au = 197)

(a) 300 ml (b) 500 ml
(c) 800 ml (d) 100 ml

**58.** The iodide content of a solution was determined by titration with cerium (IV) sulphate in the presence of HCl, in which $I^-$ is converted to ICl. A 250 ml sample of the solution required 20 ml of 0.05 N-$Ce^{4+}$ solution. What is the iodide concentration in the original solution in g/L? (I = 127)

(a) 0.508 (b) 0.254
(c) 0.762 (d) 0.127

**59.** A chemist is preparing to analyse samples that will contain no more than 0.5 g of uranium. His procedure calls for preparing the uranium as $U^{4+}$ ion and oxidizing it by $MnO_4^-$ in the following acid solution.

$$U^{4+} + MnO_4^- + H_2O \rightarrow UO_2^{2+} + Mn^{2+} + H_3O^+$$

If he wants to react the total $U^{4+}$ sample with a maximum of 50 ml of $KMnO_4$ solution, then what concentration does he choose? (U = 238)

(a) 0.0336 M (b) 0.0168 M
(c) 0.168 M (d) 0.0672 M

**60.** $KIO_3$ reacts with oxalic acid in solution to yield $K_2C_2O_4$, $CO_2$ and $I_2$. How many grams of oxalic acid will be required to react with one gram of $KIO_3$? (K = 39, I = 127)

(a) 1.262 g (b) 1.622 g
(c) 1.747 g (d) 1.022 g

61. What is the mass of oxalic acid, $H_2C_2O_4$, which can be oxidized to $CO_2$ by 100 ml of $MnO_4^-$ solution, 10 ml of which is capable of oxidizing 50 ml of 1.00 N – $I^-$ to $I_2$?
(a) 2.25 g (b) 52.2 g
(c) 25.2 g (d) 22.5 g

62. What volume of 0.2 M-$KMnO_4$ solution is needed for complete reaction with 26.56 gm $Fe_{0.9}O_{1.0}$, in acidic medium? (Fe = 56)
(a) 280 ml (b) $\frac{280}{9}$ ml
(c) $\frac{2800}{9}$ ml (d) 560 ml

63. A volume of 20 ml of M-$KMnO_4$ solution is diluted to 150 ml. In this solution, 50 ml of 10 M $H_2SO_4$ is added. 25 ml of this mixture is titrated with 20 ml of $FeC_2O_4$ solution. The molarity of $FeC_2O_4$ solution is
(a) 0.0416 (b) 0.208
(c) 0.625 (d) 0.125

64. A volume of 20 ml of M-$K_2Cr_2O_7$ solution is diluted to 200 ml. Twenty five millilitres of diluted solution is mixed with 50 ml of 4 M-$H_2SO_4$ solution. Thirty millilitres of this mixture is diluted to 150 ml. How many millilitres of 0.02 M-$H_2O_2$ solution is needed to titrate 15 ml of the diluted solution?
(a) 14 ml (b) 15 ml
(c) 30 ml (d) 45 ml

65. A polyvalent metal weighing 0.1 g and having atomic mass 51 reacted with dilute $H_2SO_4$ to give 43.9 ml of hydrogen at 0°C and 1 atm. The solution containing the metal in this lower oxidation state was found to require 58.8 ml of 0.1 N–permanganate for complete oxidation. What are the valencies of the metal?
(a) 2, 5 (b) 2, 4
(c) 3, 5 (d) 4, 5

66. A solution of $Na_2S_2O_3$ is standardized iodometrically against 0.1336 g of $KBrO_3$. This process required 40 ml of the $Na_2S_2O_3$ solution. What is the strength of the $Na_2S_2O_3$? (K = 39, Br = 80)
(a) 0.04 M (b) 0.02 M
(c) 0.05 M (d) 0.01 M

67. A sample of a metal carbonate $MCO_3$ was neutralized by 10 ml of 0.1 N-HCl and the resulting chloride gave 0.0517 g of phosphate, $M_3(PO_4)_2$. The equivalent weight of M is
(a) 20.03 (b) 40.06
(c) 51.7 (d) 8.62

68. A small amount of $CaCO_3$ completely neutralizes 525 ml of 0.1 N-HCl and no acid is left at the end. After converting all calcium chloride to $CaSO_4$, how much plaster of Paris can be obtained?
(a) 1.916 g (b) 5.827 g
(c) 3.57 g (d) 3.81 g

69. What volume of 0.40 M-$Na_2S_2O_3$ would be required to react with the $I_2$ liberated by adding excess of KI to 50 ml of 0.20 M $CuSO_4$?
(a) 12.5 ml (b) 25 ml
(c) 50 ml (d) 2.5 ml

70. To a 25 ml $H_2O_2$ solution, excess of acidified solution of potassium iodide was added. The iodine liberated required 20 ml of 0.3 N-$Na_2S_2O_3$ solution. The volume strength of $H_2O_2$ solution is
(a) 1.362 (b) 0.681
(c) 2.724 (d) 0.908

71. An unknown composition of a mixture of carbon disulphide and hydrogen sulphide was burnt in sufficient amount of oxygen. The resulting gases found to exert a pressure of 1.97 atm in a 20 l vessel at 400 K. The gaseous mixture required 2.8 M iodine solution and 250 ml of it was required to reach the end point forming $I^-$. Calculate the mole fraction of $CS_2$ in the original mixture.
(a) 0.2 (b) 0.4
(c) 0.6 (d) 0.8

72. One gram of a sample of $CaCO_3$ was strongly heated and the $CO_2$ liberated absorbed in 100 ml of 0.5 M-NaOH. Assuming 90% purity for the sample, how much ml of 0.5 M-HCl would be required to react with the solution of the alkali for the phenolphthalein end point?
(a) 73 ml (b) 41 ml
(c) 82 ml (d) 97 ml

73. A volume of 40 ml of 0.05 M solution of sodium sesquicarbonate ($Na_2CO_3 \cdot NaHCO_3 \cdot 2H_2O$) is titrated against 0.05 M-HCl solution. $x$ ml of HCl solution is used when phenolphthalein is the indicator and $y$ ml of HCl is used when methyl orange is the indicator in two separate titrations. Hence, $(y - x)$ is

(a) 80 ml (b) 30 ml
(c) 120 ml (d) 40 ml

74. A 100 ml mixture of $Na_2CO_3$ and $NaHCO_3$ is titrated against 1 M-HCl. If $V_1$ L and $V_2$ L are consumed when phenolphthalein and methyl orange are used as indicators, respectively, in two separate titrations, which of the following is true for molarities in the original solution?

(a) Molarity of $Na_2CO_3 = 20V_1$
(b) Molarity of $NaHCO_3 = 10(V_2 - 2V_1)$
(c) Molarity of $Na_2CO_3 = 10(V_2 + V_1)$
(d) Molarity of $NaHCO_3 = 10(V_2 - V_1)$

75. In the mysterious deserts of Egypt, large deposits of 'Trona' ($Na_2CO_3 \cdot NaHCO_3$) are found. If a sample of 'Trona' (containing same inert impurities) is dissolved in water and titration against 0.1 M-HCl, then which of the following readings are possible when $x$ and $y$ ml of HCl are required for titration against equal volumes of this solution, one using phenolphthalein and the other using methyl orange respectively as indicators?

(a) $x = 20, y = 20$ (b) $x = 10, y = 30$
(c) $x = 20, y = 40$ (d) $x = 20, y = 10$

## Section B (One or More than one Correct)

1. In the compound $NOClO_4$, the oxidation state of

(a) nitrogen is +1 (b) nitrogen is +3
(c) chlorine is +5 (d) chlorine is +7

2. Which of the following is/are peroxide(s)?

(a) $PbO_2$ (b) $H_2O_2$
(c) $SrO_2$ (d) $BaO_2$

3. Which of the following is a non-redox process?

(a) $SO_4^{2-} \rightarrow SO_3$ (b) $Cr_2O_7^{2-} \rightarrow CrO_4^{2-}$
(c) $PO_4^{3-} \rightarrow P_2O_7^{4-}$ (d) $C_2O_4^{2-} \rightarrow CO_2$

4. Which of the following compound does not decolourized an acidified solution of $KMnO_4$?

(a) $SO_2$ (b) $FeCl_3$
(c) $H_2O_2$ (d) $FeSO_4$

5. Which of the following statement(s) is/are true regarding the change $CN^- \rightarrow CNO^-$?

(a) Carbon is losing two electrons per atom.
(b) The oxidation state of carbon changes from +2 to +4.
(c) Oxidation state of nitrogen is not changing.
(d) Oxidation state of nitrogen changes from −3 to −1.

6. Substances which may be oxidized as well as reduced are

(a) HCl (b) HClO
(c) $HClO_3$ (d) $HClO_4$

7. A quantity of 15.8 g of $KMnO_4$ can be decolourized in acidic medium by (K = 39, Mn = 55, Fe = 56)

(a) 18.25 g HCl (b) 22.5 g $H_2C_2O_4$
(c) 32 g $SO_2$ (d) 38 g $FeSO_4$

8. When copper is treated with a certain concentration of nitric acid, nitric oxide and nitrogen dioxide are liberated in equal volumes according to the following equation.

$$X\text{Cu} + Y\text{HNO}_3 \rightarrow \text{Cu(NO}_3)_2 + \text{NO} + \text{NO}_2 + \text{H}_2\text{O}$$

The coefficients of $X$ and $Y$ are, respectively,

(a) 2 and 3 (b) 2 and 6
(c) 1 and 3 (d) 3 and 8

9. The equivalent volume of a gaseous substance is 5.6 L at 0°C and 1 atm. The substance may be

(a) $CH_4$ gas in combustion.
(b) $O_3$ gas as oxidizing agent.
(c) $H_2S$ gas as reducing agent.
(d) $CO_2$ formed from carbon.

10. A quantity of 0.5 g of a metal nitrate gave 0.43 g of metal sulphate.

(a) The equivalent weight of the metal is 38.
(b) The equivalent weight of the metal is 76.
(c) The atomic weight of metal may be 76.
(d) The atomic weight of metal may be 19.

**11.** A metal (M) forms a hydrated sulphate, isomorphous with $ZnSO_4 \cdot 7H_2O$. If the sulphate contains 20% metal, by weight, which of the following is/are correct for the metal?

(a) The atomic weight of metal is 24.
(b) The equivalent weight of the metal is 27.75.
(c) The metal is bivalent.
(d) The anhydrous metal sulphate contains 36.6% metal, by mass.

**12.** A metal forms two oxides. The higher oxide contains 20% oxygen, while 4.29 g of the lower oxide when converted to higher oxide, become 4.77 g. The equivalent weight of metal in

(a) lower oxide is 32.
(b) lower oxide is 64.4.
(c) higher oxide is 64.4.
(d) higher oxide is 32.

**13.** The specific heat of a metal is found to be 0.03. 10 g of the metal on treatment with nitric acid gave 18.9 g of pure dry nitrate. The correct statement(s) is/are

(a) The equivalent weight of the metal is 69.66.
(b) The atomic weight of the metal is 209.
(c) The metal is trivalent.
(d) The metal is an alkali metal.

**14.** A 100 ml mixture of $Na_2CO_3$ and $NaHCO_3$ is titrated against 1 M-HCl. If $V_1$ L and $V_2$ L are consumed when phenolphthalein and methyl orange are used as indicators, respectively, in two separate titrations, which of the following is true for molarities in the original solution?

(a) Molarity of $Na_2CO_3 = 20V_1$
(b) Molarity of $NaHCO_3 = 10\ (V_2 - 2V_1)$
(c) Molarity of $Na_2CO_3 = 10\ (V_2 + V_1)$
(d) Molarity of $NaHCO_3 = 10\ (V_2 - V_1)$

**15.** A volume of 20 ml of an aqueous solution of hydrated oxalic acid ($H_2C_2O_4 \cdot xH_2O$) containing 6.3 g per litre requires 40 ml of 0.05 M-NaOH solution for complete neutralization. Which of the following statement about the acid solution is/are correct?

(a) The value of $x$ is 2.
(b) The equivalent weight of anhydrous acid is 63.
(c) The molarity of acid solution is 0.1 M.
(d) 100 ml of the same acid solution requires 40 ml of 0.05 M-$KMnO_4$ solution for complete oxidation in the presence of $H_2SO_4$.

**16.** A bottle of oleum is labelled as 109%. Which of the following statement is/are correct for this oleum sample?

(a) It contains 40% of free $SO_3$ by weight.
(b) 1.0 g of this sample approximately requires 22.25 ml of 0.5 M-NaOH solution for complete neutralization.
(c) 0.5 g of this sample approximately requires 11.12 ml of 0.1 N-$Ba(OH)_2$ solution for complete neutralization.
(d) When 500 g water is added to 100 g of this sample, the resulting solution becomes $\left(\frac{109}{49}\right)$ m in $H_2SO_4$.

**17.** A quantity of 5.68 g of pure $P_4O_{10}$ is dissolved completely in sufficient water and the solution is diluted to 250 ml. Which of the following statement(s) is/are correct? (P = 31)

(a) The diluted solution has molarity 0.32 with respect to $H_3PO_4$.
(b) 25 ml of the diluted solution exactly requires 48 ml of 0.5 M-NaOH solution for complete neutralization.
(c) 15 ml of the diluted solution exactly requires 36 ml of 0.2 M-$BaCl_2$ solution for complete precipitation of phosphate.
(d) 40 ml of the diluted solution exactly requires 48 ml of 0.8 N-KOH solution for the first equivalent point.

**18.** An amount of 0.01 mole of $SO_2Cl_2$ is hydrolysed completely in sufficient water (no gas is allowed to escape out) and the solution is diluted to 200 ml. Which of the following statement is/are correct? (Ag = 108)

(a) The solution is 0.05 M in $H_2SO_4$.
(b) The solution is 0.1 M in HCl.
(c) A volume of 20 ml of the solution exactly requires 20 ml of 0.2 M NaOH solution for complete neutralization.
(d) When 100 ml of the solution is treated with excess of $AgNO_3$ solution, 1.435 g of AgCl will precipitate out.

19. A definite mass of $H_2O_2$ is oxidized by excess of acidified $KMnO_4$ and acidified $K_2Cr_2O_7$ in separate experiments. Which of the following is/are correct statements? (K = 39, Cr = 52, Mn = 55)
(a) Mass of $K_2Cr_2O_7$ used up will be greater than that of $KMnO_4$.
(b) Moles of $KMnO_4$ used up will be greater than that of $K_2Cr_2O_7$.
(c) Equal mass of oxygen gas is evolved in both the experiments.
(d) If equal volumes of both the solutions are used for complete reaction, then the molarities of $KMnO_4$ and $K_2Cr_2O_7$ solutions are in 6 : 5 ratio.

20. A quantity of 8.0 g of solid sulphur is first oxidized to $SO_2$ and then it is divided into two equal parts. One part is sufficient for complete decolourization of 200 ml of acidified $KMnO_4$ solution. Another part is oxidized to $SO_3$ and the $SO_3$ formed is sufficient for complete precipitation of all $BaCl_2$ present in 100 ml solution as $BaSO_4$. Which of the following statements is/are correct? (S = 32, Ba = 138)
(a) The molarity of $KMnO_4$ solution is 0.25.
(b) The molarity of $BaCl_2$ solution is 0.25.
(c) The weight of $BaSO_4$ precipitated out is 29.25 g.
(d) The same equivalents of $KMnO_4$ and $BaCl_2$ are reacted.

## Section C (Comprehensions)

### Comprehension I

**For the reaction: $MnBr_2 + PbO_2 + HNO_3 \rightarrow HMnO_4 + Pb(BrO_3)_2 + Pb(NO_3)_2 + H_2O$ (Atomic masses: Mn = 55, Br = 80, Pb = 208)**

1. The equivalent weight of $MnBr_2$ is
(a) 107.5 (b) 215
(c) 12.65 (d) 19.55

2. The equivalent weight of $PbO_2$ is
(a) 120 (b) 240
(c) 14.11 (d) 21.82

3. The equivalent weight of $HNO_3$ is
(a) 63 (b) 55.6
(c) 31.5 (d) 111.18

### Comprehension II

**It was found that 100 g of silver combined with all the chlorine in 56 g of arsenious chloride. The vapour density of arsenious chloride is 6.25 (air = 1). The specific heat of arsenic is 0.08. Given that one litre of air at 0°C and 1 atm weighs 1.3 g (Ag = 108).**

4. What is the exact atomic weight of arsenic?
(a) 74.94 (b) 24.98
(c) 80.00 (d) 182.47

5. What is the equivalent weight of arsenic in the arsenious chloride?
(a) 74.94 (b) 24.98
(c) 14.49 (d) 49.96

6. What is the molecular formula of arsenious chloride?
(a) $AsCl_3$ (b) $As_2Cl_6$
(c) $As_2Cl_5$ (d) $AsCl_5$

### Comprehension III

**A quantity of 0.4 g of oxygen and 4.0 g of a halogen combine separately with the same amount of metal.**

7. What is the equivalent weight of halogen if the element exhibits the same valency in both compounds?
(a) 40 (b) 80
(c) 20 (d) 160

8. What is the equivalent weight of halogen if the valency of element in the halide is twice that in oxide?
(a) 40 (b) 80
(c) 20 (d) 160

9. The atomic weight of the halogen can never have the value
(a) 40 (b) 80
(c) 20 (d) 160

## Comprehension IV

**$KMnO_4$ oxidizes $X^{n+}$ ion to $XO_3^-$ in acid solution. $2.5 \times 10^{-3}$ mole of $X^{n+}$ requires $1.5 \times 10^{-3}$ mole of $MnO_4^-$.**

10. What is the value of $n$?
    (a) 3 (b) 2
    (c) 1 (d) 4

11. What is the atomic mass of X, if the mass of 1 g-equivalent of $XCl_n$ in this reaction is 56? $XCl_n$ is the molecular form of $x^{+n}$. Assume no change in oxidation state of chlorine.
    (a) 71 (b) 112
    (c) 97 (d) 41

12. How many mole of $KMnO_4$ is needed per mole of $X^{n+}$ to oxidize it to $XO_3^-$ in strong basic medium?
    (a) 1 (b) 3
    (c) 0.6 (d) 2

## Comprehension V

**One gram of a moist sample of a mixture of potassium chloride and potassium chlorate was dissolved in water and made up to 250 ml. Twenty-five millilitres of this solution was treated with $SO_2$ to reduce the chlorate to chloride and excess $SO_2$ was removed by boiling. The total chloride was precipitated as silver chloride. The weight of the precipitate was 0.1435 g. In another experiment, 25 ml of the original solution was heated with 30 ml of 0.2 N solution of ferrous sulphate and unreacted ferrous sulphate required 37.5 ml of 0.08 N solution of an oxidizing agent for complete oxidation. (K = 39, Ag = 108)**

13. What is the molar ratio of the chlorate to chloride in the given mixture?
    (a) 1 : 1 (b) 1 : 2
    (c) 2 : 1 (d) 2 : 3

14. What is the mass percent of moisture present in the moist sample?
    (a) 1.0% (b) 1.5%
    (c) 1.75% (d) 3.5%

15. What is the mass percent of potassium chloride in the moist sample?
    (a) 37.25% (b) 61.25%
    (c) 3.725% (d) 74.5%

## Comprehension VI

**A forensic chemist needed to determine the concentration of HCN in the blood of a suspected homicide victim and decided to titrate a dilute sample of the blood with iodine using the following reaction.**

$$HCN(aq) + I_3^-(aq) \rightarrow ICN(aq) + 2I^-(aq) + H^+(aq)$$

**A diluted blood sample of volume 15.0 ml was titrated to the stoichiometric point with 5.0 ml of an $I_3^-$ solution. The molar concentration of the $I_3^-$ solution was determined by titrating it against arsenic (III) oxide, $As_4O_6$, which in solution forms arsenious acid, $H_3AsO_3$. A volume of 10.0 ml of the triiodide solution was needed to reach the stoichiometric point on a 0.1188 g sample of $As_4O_6$ in the following reaction.**

$$H_3AsO_3(aq) + I_3^-(aq) + H_2O(l) \rightarrow H_3AsO_4(aq) + 3I^-(aq) + 2H^+(aq)$$

**(Atomic mass of As = 75)**

16. What is the molar concentration of the triiodide solution?
    (a) 0.03 M (b) 0.12 M
    (c) 0.06 M (d) 0.00012 M

17. What is the molar concentration of HCN in the blood sample?
    (a) 0.04 M (b) 0.03 M
    (c) 0.12 M (d) 0.36 M

18. How many grams of HCN is present in the blood of victim if the total volume of blood present in the victim is 6.0 l?
    (a) 0.24 g (b) 6.48 g
    (c) 3.24 g (d) 2.16 g

## Comprehension VII

**In the presence of fluoride ion, $Mn^{2+}$ can be titrated with $MnO_4^-$, where both reactants being converted to a complex of Mn(III). A 0.458 g of sample containing $Mn_3O_4$ was dissolved and all manganese was converted to $Mn^{2+}$. Titration in the presence of fluoride ion consumed 30.0 ml of $KMnO_4$ that was 0.125 N against oxalate. (Mn = 55)**

19. The correct balanced reaction, assuming that the complex is $MnF_4^-$, is
    (a) $Mn^{2+} + MnO_4^- + H^+ + F^- \rightarrow MnF_4^- + H_2O$
    (b) $4Mn^{2+} + MnO_4^- + 8H^+ \rightarrow 5Mn^{3+} + 4H_2O$
    (c) $4Mn^{2+} + MnO_4^- + 8H^+ + 20F^- \rightarrow 5MnF_4^- + 4H_2O$
    (d) $Mn^{2+} + MnO_4^- + H^+ \rightarrow MnF_4^- + H_2O$

20. What is the percentage of $Mn_3O_4$ in the sample?
    (a) 50.00% (b) 40.00%
    (c) 62.50% (d) 75.00%

21. What is the normality of $KMnO_4$ solution against $Mn^{2+}$?
    (a) 0.125 N (b) 0.1 N
    (c) 0.01 N (d) 0.156 N

## Comprehension VIII

**Chromium exists as $FeCr_2O_4$ in the nature and it contains $Fe_{0.95}O_{1.00}$ as an impurity. To obtain pure chromium from $FeCr_2O_4$, the ore is fused with KOH and oxygen is passed through the mixture when $K_2CrO_4$ and $Fe_2O_3$ are produced. A quantity of 2 g of ore required 280 ml of $O_2$ at 0°C and 1 atm for complete oxidation of ore. $K_2CrO_4$ is then precipitated as $BaCrO_4$ after addition of Barium salt. To the remaining solution, 10 ml of 1 M-$K_4Fe(CN)_6$ is added when $Fe^{3+}$ ions reacts with it to form $KFe[Fe(CN)_6]$, after called 'Prussian Blue'. To determine excess of $K_4Fe(CN)_6$ in solution, 6 ml of 0.4 N-$Fe^{2+}$ is added when all the $K_4Fe(CN)_6$ is precipitated as $K_2Fe[Fe(CN)_6]$. (Fe = 56)**

22. What is the percentage of $Fe_{0.95}O_{1.00}$ in the ore?
    (a) 6.92% (b) 3.46%
    (c) 13.84% (d) 93.08%

23. What per cent of total iron present in the ore is in +2 state?
    (a) 77.53% (b) 97.73%
    (c) 78.41% (d) 87.9%

24. How many millimoles of Prussian blue is formed?
    (a) 8.9 (b) 8.8
    (c) 0.0088 (d) 7.85

## Comprehension IX

**Chile saltpeter, a source of $NaNO_3$ also contains $NaIO_3$. The $NaIO_3$ can be used as a source of iodine, produced in the following reactions.**

$$IO_3^- + 3HSO_3^- \rightarrow I^- + 3H^+ + 3SO_4^{2-}$$

**and**

$$5I^- + IO_3^- + 6H+ \rightarrow 3I_2 + 3H_2O$$

**One litre of Chile saltpeter solution containing 5.94 g $NaIO_3$ is treated with stoichiometric quantity of $NaHSO_3$. Now an additional amount of the same solution is added to the reaction mixture to bring about the second reaction. (I = 127)**

25. How many grams of $NaHSO_3$ is required in step I for complete reaction?
    (a) 9.36 g (b) 3.12 g
    (c) 6.24 g (d) 14.04 g

26. What additional volume of Chile saltpeter must be added in step II to bring in complete conversion of $I^-$ to $I_2$?
    (a) 1000 ml (b) 200 ml
    (c) 5000 ml (d) 400 ml

27. How many grams of $I_2$ can be produced per litre of Chile saltpeter?
    (a) 4.572 g (b) 2.286 g
    (c) 5.486 g (d) 3.810 g

## Comprehension X

**A volume of 50 ml of solution containing 1 g each of $Na_2CO_3$, $NaHCO_3$ and NaOH was treated with N-HCl.**

**28.** What will be the titre reading if only phenolphthalein is used as an indicator?

(a) 43.8 ml (b) 21.9 ml
(c) 34.4 ml (d) 57.9 ml

**29.** What will be the titre reading if only methyl orange is used as indicator from the very beginning?

(a) 67.7 ml (b) 55.8 ml
(c) 46.3 ml (d) 23.5 ml

**30.** What will be the titre reading if methyl orange is added after the first end point with phenolphthalein?

(a) 30.8 ml (b) 21.3 ml
(c) 33.2 ml (d) 51.9 ml

## Section D (Assertion–Reason)

The following questions consist of two statements. Mark the answer as follows.

**(a)** If both statements are CORRECT, and **Statement II** is the CORRECT explanation of **Statement I.**

**(b)** If both statements are CORRECT, and **Statement II** is NOT the CORRECT explanation of **Statement I.**

**(c)** If **Statement I** is CORRECT, but **Statement II** is INCORRECT.

**(d)** If **Statement I** is INCORRECT, but **Statement II** is CORRECT.

**1. Statement I:** $I^-$ can never act as an oxidizing agent.
**Statement II:** Oxidizing agent undergoes reduction.

**2. Statement I:** In propane, all carbon atoms are in the same oxidation state.
**Statement II:** The oxidation state is −8/3 per carbon atom.

**3. Statement I:** When $O_3$ reacts with KI, $O_3$ is reduced into $O_2$.
**Statement II:** There is no change in oxidation state of oxygen in this reaction.

**4. Statement I:** In $ClF_3$, chlorine has the oxidation number −1.
**Statement II:** Electron affinity of chlorine is greater than that of fluorine.

**5. Statement I:** The equivalent weight of any substance is its molecular weight divided by some factor, depending on the nature of the substance.
**Statement II:** The equivalent weight of any substance is always less than its molecular weight.

**6. Statement I:** The molecular weight of any substance is unique but the equivalent weight is not unique.
**Statement II:** Equivalent weight of any substance depends on its nature in the chemical reaction concerned.

**7. Statement I:** Equivalent weight of any element represents the parts by weight of the element which combines with or displaces 1 part by weight of hydrogen or 8 parts by weight of oxygen or 35.5 parts by weight of chlorine.
**Statement II:** The atomic weights of hydrogen, oxygen or chlorine are taken as reference for the determination of equivalent weights of all other elements.

**8. Statement I:** The number of g-equivalents of all the reactants reacted in any chemical reaction is always the same.
**Statement II:** In any chemical reaction, the total mass of reactants reacted is always equal to the total mass of products formed.

**9. Statement I:** Equal volumes of 0.3 M-$H_2SO_4$ solution and 0.2M–$H_3PO_4$ solution will require the same volume of the same NaOH solution for complete neutralization.
**Statement II:** $H_2SO_4$ is dibasic and $H_3PO_4$ is a tribasic acid.

**10. Statement I:** When a solution of $Na_2CO_3$ is titrated with HCl solution, the volume of acid solution required for the end point in case of methyl orange indicator is double than that required in case of phenolphthalein indicator.
**Statement II:** In case of phenolphthalein indicator, the sudden change in colour is observed when $Na_2CO_3$ is completely converted into $H_2CO_3$.

**11.** **Statement I:** When 10 ml of 0.5 M–$NaHCO_3$ solution is titrated with 0.25 M–HCl solution using phenolphthalein indicator, 20 ml of acid solution is consumed at the end point.

**Statement II:** End point cannot be detected when $NaHCO_3$ solution is titrated with HCl solution using phenolphthalein indicator.

**12.** **Statement I:** Equal volumes of 1 M–HCl solution and 1 M–NaOH solution is required for complete reaction with the same mass of $KHC_2O_4$.

**Statement II:** $KHC_2O_4$ is amphoteric and it can lose or gain one proton.

**13.** **Statement I:** The number of g-equivalents in the same mass of $KMnO_4$ is in 5 : 3 : 1 ratio in acid, neutral and strong basic medium, respectively.

**Statement II:** The oxidation state of Mn changes from +7 state to +2, +4 and +6 states, respectively, in acid, neutral and strong basic medium.

**14.** **Statement I:** When 20 ml of 0.5 M-$CuSO_4$ solution is treated with excess of KI solution, the liberated $I_2$ exactly requires 20 ml of 0.5 M-$Na_2S_2O_3$ solution for titration.

**Statement II:** For both the solutions ($CuSO_4$ and $Na_2S_2O_3$), their molarity and normality are equal.

**15.** **Statement I:** For the sequential reactions: A $\rightarrow$ B and B $\rightarrow$ C (both occurring completely), the number of g-equivalents of A and C must be equal.

**Statement II:** The relation between the number of g-equivalents of A and C depends on the equivalent weight of B in both the reactions.

## Section E (Column Match)

**1.** In **Column I**, some reactions are given and in **Column II**, the type of some reactions on the basis of oxidation and reduction processes are given. Match the reactions with their correct type.

| Column I | Column II |
|---|---|
| (A) $3CaO + 2P_2O_5 \rightarrow Ca_3(PO_4)_2$ | (P) Disproportionation |
| (B) $2Cu^+ \rightarrow Cu + Cu^{2+}$ | (Q) Comproportionation |
| (C) $NH_4NO_2 \rightarrow N_2 + 2H_2O$ | (R) Non-redox |
| (D) $CH_4 + 2O_2 \rightarrow CO_2 + 2H_2O$ | (S) Combustion |
| | (T) Redox |

**2.** Some compounds are given in **Column I**. Match them correctly with the terms given in **Column II**.

| Column I | Column II |
|---|---|
| (A) $SO_2$ | (P) Oxidizing agent |
| (B) $SO_3$ | (Q) Reducing agent |
| (C) $H_2O_2$ | (R) Undergoes disproportionation in air |
| (D) NaF | (S) Neither an oxidizing nor a reducing agent |

**3.** Match the columns.

| Column I (Oxidation number of underlined element) | Column II (Oxidation number) |
|---|---|
| (A) $\underline{N}H_4NO_3$ | (P) 0 |
| (B) $\underline{C}H_2O$ | (Q) −2 |
| (C) $\underline{Ni}(CO)_4$ | (R) +2 |
| (D) $\underline{Cl}_2O_5$ | (S) +5 |
| | (T) −3 |

**4.** In **Column I**, one mole of some oxidizing agents is given. Match them with the moles of the reducing agents needed for complete reaction, given in **Column II**.

| Column I | Column II |
|---|---|
| (A) $KMnO_4$ ($H^+$) | (P) 3.0 mole of $FeSO_4$ |
| (B) $KMnO_4$ ($OH^-$) | (Q) 0.5 mole of $I_2$ to $HIO_3$ |
| (C) $MnO_2$ ($H^+$) | (R) 1.0 mole of $K_2C_2O_4$ |
| (D) $K_2CrO_4$ ($H^+$) | (S) 1.5 mole of $K_2SO_3$ |

5. Some redox reactions are given in **Column I**. Match them with the correct mole ratio of oxidizing to reducing agents given in **Column II**.

| Column I | Column II |
|---|---|
| (A) $MnO_4^- + C_2O_4^{2-} \rightarrow MnO_2 + CO_2$ | (P) 2 : 1 |
| (B) $ClO^- + Fe(OH)_3 \rightarrow Cl^- + FeO_4^{2-}$ | (Q) 3 : 1 |
| (C) $HO_2^- + Cr(OH)_3^- \rightarrow CrO_4^{2-} + HO^-$ | (R) 2 : 3 |
| (D) $N_2H_4 + Cu(OH)_2 \rightarrow N_2O + Cu$ | (S) 3 : 2 |

6. Match the columns.

| Column I | Column II |
|---|---|
| (A) Equivalent volume of $Cl_2$ gas | (P) 5.6 L at 0°C and 1 atm |
| (B) Volume of $O_2$ needed for complete combustion of 5.6 L methane at 0°C and 1 atm | (Q) 11.2 L at 0°C and 1 atm |
| (C) Equivalent volume of $O_2$ gas | (R) 22.4 L at 0°C and 0.5 atm |
| (D) Equivalent volume of $SO_2$ gas as oxidizing agent | (S) 11.2 L at 0°C and 0.5 atm |

7. Match the columns.

| Column I (Process) | Column II (Equivalent weight of $Cl_2$) |
|---|---|
| (A) $Cl_2 \rightarrow Cl^-$ | (P) 71 |
| (B) $Cl_2 \rightarrow ClO_3^-$ | (Q) 35.5 |
| (C) $Cl_2 \rightarrow Cl^- + ClO_3^-$ | (R) 42.6 |
| (D) $Cl_2 \rightarrow Cl^- + ClO^-$ | (S) 7.1 |

8. Match the columns.

| Column I (Equivalent weight of HCl) | Column II (Chemical change) |
|---|---|
| (A) Greater than its molecular weight | (P) Neutralization reaction |
| (B) Equal to molecular weight | (Q) $MnO_2 + HCl \rightarrow MnCl_2 + Cl_2 + H_2O$ |
| (C) Less than molecular weight | (R) $HClO \rightarrow HCl$ |
| | (S) $HCl \rightarrow HClO_3$ |
| | (T) $Cu + HCl \rightarrow H_2[CuCl_4] + H_2$ |

9. Match the columns.

| Column I | Column II (Solution needed for complete reaction) |
|---|---|
| (A) 100 ml of 0.3 M – $H_2C_2O_4$ solution | (P) 100 ml of 0.3 M – KOH solution |
| (B) 50 ml of 0.6 M – $KHC_2O_4$ solution | (Q) 120 ml of 0.1 M – $KMnO_4$ solution in the presence of $H_2SO_4$ |
| (C) 50 ml of 0.6 M – HCl solution | (R) 60 ml of 0.1 M – $KMnO_4$ solution in the presence of $H_2SO_4$ |
| (D) 100 ml of 0.2 M – $H_3PO_4$ solution | (S) 100 ml of 0.6 M – KOH solution |

10. Match the columns.

| Column I | Column II (Solution needed for complete reaction) |
|---|---|
| (A) 50 ml of 0.5 M-$Na_2CO_3$ solution using methyl orange indicator. | (P) 50 ml of 0.5 M-$H_2SO_4$ solution |
| (B) 50 ml of 0.5 M-$Na_2CO_3$ solution using phenolphthalein indicator. | (Q) 50 ml of 0.5 M-HCl solution |
| (C) 50 ml of 0.5 M-$NaHCO_3$ solution using methyl orange indicator. | (R) 25 ml of 0.5 M-$H_2SO_4$ solution |
| (D) 50 ml of 0.5 M-NaOH solution using phenolphthalein indicator. | (S) 50 ml of 1.0 M-HCl solution |

## Section F (Subjective)

### Single-digit Integer Type

1. The value of $n$ in the following processes: $AO_4^{n-} + 2e \rightarrow HAO_n^{2-}$ is

2. $AO_2$ disproportionates into $AO_4^-$ and $A^{n+}$ ion. If the mole ratio of $AO_2$ undergone oxidation and reduction is 2 : 3, the value of $n$ is

3. A volume of 1.12 L dry chlorine gas at 0°C and 1 atm was passed over a heated metal when 5.55 g of chloride of the metal was formed. If the atomic mass of the metal is 40, then its valency is

4. The equivalent weight of $Br_2$ is 96 in the following disproportionation reaction.

   $Br_2 + OH^- \rightarrow Br^- + H_2O + ?$ (Oxidized product)

   The oxidation state of Br in the oxidized product is (Br = 80)

5. HCHO disproportionates to $HCOO^-$ and $CH_3OH$ in the presence of $OH^-$ (Cannizzaro's reaction).

   $$2HCHO + OH^- \rightarrow HCOO^- + CH_3OH$$

   If the equivalent weight of HCHO is $E$, then the value of $\frac{E}{10}$ is

6. When a solid element is reacted with chlorine, a gaseous chloride of vapour density 68.75 is formed. If this reaction is performed at constant temperature and pressure, the volume of the system reduces by one third. If the equivalent weight of the solid element is $E$, then the value of $\left(\frac{12}{31} \times E\right)$ is

7. $V$ litre of $SO_2$ at 0°C and 1 atm is required to reduce 16.9 g of $HClO_3$ to HCl. The number of moles in '5 V' litre of $SO_2$ at 273°C and 2 atm is

8. The approximate mass (in g) of $N_2H_4$ can be oxidized by 24 g of $K_2CrO_4$ is (Cr = 52)

   $$3N_2H_4 + 4CrO_4^{2-} + 4H_2O \rightarrow 3N_2 + Cr(OH)_4^- + 4OH^-$$

9. A sample of pure $KHC_2O_4.H_2C_2O_4.2H_2O$ requires 30 mol of NaOH for titration. How many moles of $KMnO_4$ will the same sample react with, in acid medium?

10. The basic solution of $Na_4XeO_6$ is powerful oxidants. How many millimoles of $Mn(NO_3)_2 \cdot 6H_2O$ reacts with 62.5 ml of a 0.04 M basic solution of $Na_4XeO_6$ that contains an excess of sodium hydroxide if the products include Xe and a solution of sodium permanganate? (Mn = 55)

11. A newly developed method for water treatment uses chlorine dioxide $ClO_2$ rather than $Cl_2$ itself. $ClO_2$ can be obtained by passing $Cl_2(g)$ into concentrated solution of sodium chlorite $NaClO_2$. NaCl is the other product. If this reaction has a 90% yield, then how many moles of $ClO_2$ are produced from 5 L of 2.0 M-$NaClO_2$?

12. A quantity of 1.245 g of $CuSO_4 \cdot xH_2O$ was dissolved in water and $H_2S$ was passed into it till CuS was completely precipitated. The $H_2SO_4$ produced in the filtrate required 10 ml of M-NaOH solution. Calculate the value of $x$. (Cu = 63.5)

13. A mixture of $CS_2$ and $H_2S$ when oxidized yields a mixture of $CO_2$, $SO_2$ and $H_2O(g)$, which exerts a pressure of 7.2 atm, when collected in 82.1 L vessel at 327°C. To oxidize $SO_2$ in the mixture, 7 L of 2 N – iodine was required. Moles of $CS_2$ in the mixture is

14. A mixture of Xe and $F_2$ was heated. A sample of white solid thus formed reacted with hydrogen to give 56 ml of Xe at 0°C and 1 atm and HF formed required 60 ml of 0.25 M-NaOH for complete neutralization. If the molecular formula of the solid formed is $XeF_x$, then the value of $x$ is

15. One litre of a sample of ozonized oxygen at 0°C and 1 atm on passing through a KI solution, liberated iodine which required 9 ml of a thiosulphate solution. A volume of 12 ml of a '5.675 volume' hydrogen peroxide solution liberated iodine from another iodide solution, which required 24 ml of the same thiosulphate solution. The volume percent of ozone in the ozonized oxygen sample is approximately

16. A certain mass of anhydrous oxalic acid is converted into $H_2O$, $CO_2$ and CO, on heating in the presence of $H_2SO_4$. The CO formed reacts completely with iodine pentoxide to liberate iodine. The iodine thus liberated required 200 ml of 0.2 N thiosulphate. The mass (in g) of oxalic acid taken was

17. When ammonium vanadate is heated with oxalic acid solution, a substance Z is formed. A sample of Z was treated with $KMnO_4$ solution in hot acidic solution. The resulting liquid was reduced with $SO_2$, the excess $SO_2$ boiled off and the liquid again titrated with same $KMnO_4$. The ratio of the volumes of $KMnO_4$ used in the two titrations was 5 : 1. What is the oxidation state of vanadium in substance Z? Given that $KMnO_4$ oxidizes all oxidation state of vanadium to vanadium (+5) and $SO_2$ reduces V (+5) to V (+4).

18. A solution of 0.2 g of a compound containing $Cu^{2+}$ and $C_2O_4^{2-}$ ions on titration with 0.02 M-$KMnO_4$ in the presence of $H_2SO_4$ consumes 22.6 ml of the oxidant. The resultant solution is neutralized with $Na_2CO_3$ acidified with dilute acetic acid and treated with excess KI. The liberated iodine requires 11.3 ml of 0.05 M-$Na_2S_2O_3$ for complete reduction. If the molar ratio of $Cu^{2+}$ to $C_2O_4^{2-}$ in the compound is 1 : $x$, then the value of $x$ is

19. A quantity of 1.0 g sample of $Fe_2O_3$ solid of 55.2 per cent purity is dissolved in acid and reduced by heating the solution with zinc dust. The resultant solution is cooled and made up to 100 ml. An aliquot of 25 ml of this solution requires 17 ml of 0.0167 M solution of an oxidant for titration. The number of electrons taken up by the oxidant in the reaction of the above titration is

20. A quantity of 1.16 g $CH_3(CH_2)_nCOOH$ was burnt in excess of air and the resultant gases ($CO_2$ and $H_2O$) were passed through excess NaOH solution. The resulting solution was divided into two equal parts. One part requires 50 ml of N-HCl for neutralization using phenolphthalein indicator. Another part required 80 ml of N-HCl for neutralization using methyl orange indicator. The value of $n$ is

## Four-digit Integer Type

1. A transition metal X forms an oxide of formula $X_2O_3$. It is found that only 50% of X atoms in this compound are in the +3 oxidation state. The only other stable oxidation states of X are +2 and +5. What percentage of X atoms is in the +2 oxidation state in this compound?

2. An amount of 0.1 moles of $OH^-$ ions is obtained from 8.50 g of hydroxide of a metal. What is the equivalent weight of the metal?

3. A quantity of 2.7 g of an alloy of copper and silver was dissolved in moderately conc. $HNO_3$ and excess of HCl was added to this solution when 2.87 g of a dry precipitate is formed. Calculate the percentage of copper in the alloy. (Cu = 63.5, Ag = 108)

4. Peroxides like oxides are basic. They form hydrogen peroxide upon treatment with an acid. What volume (in ml) of 0.25 M-$H_2SO_4$ solution is required to neutralize a solution that contains 7.2 g of $CaO_2$?

5. A volume of 30 ml of a solution containing 9.15 g per litre of an oxalate $K_xH_y(C_2O_4)_z.nH_2O$ is required for titrating 27 ml of 0.12 N-NaOH and 36 ml of 0.12 N – $KMnO_4$ separately. Assume all H-atoms are replaceable and $x$, $y$ and $z$ are in the simple ratio of g-atoms. The value of $xyzn$ is

6. A solution is made by mixing 200 ml of 0.1 M-$FeSO_4$, 200 ml of 0.1 M-$KMnO_4$ and 600 ml of 1 M-$HClO_4$. A reaction occurs in which $Fe^{2+}$ and $MnO_4^-$ convert to $Fe^{3+}$ and $Mn^{2+}$. If the molarity of $H^+$ ion in the final solution is '$x$' M, then the value of 1000$x$ is

7. The saponification number of fat or oil is defined as the number of mg of KOH required to saponify 1 g oil or fat. A sample of peanut oil weighing 1.5 g is added to 25.0 ml of 0.4 M-KOH. After saponification is complete, 8.0 ml of 0.25 M-$H_2SO_4$ is needed to neutralize excess of KOH. What is the saponification number of peanut oil?

8. A quantity of 1.6 g of pyrolusite ore was treated with 50 ml of 1.0 N-oxalic acid and some sulphuric acid. The oxalic acid left undecomposed was raised to 250 ml in a flask. A volume of 25 ml of this solution when titrated with 0.1 N-$KMnO_4$ required 32 ml of the solution. The percentage of available oxygen in the ore is

9. Calculate the amount (in mg) of $SeO_3^{2-}$ in solution, where 20 ml of M/40 solution of $KBrO_3$ was added to a definite volume of $SeO_3^{2-}$ solution. The bromine evolved was removed by boiling and excess of $KBrO_3$ was back titrated with 7.5 ml of M/25 solution of $NaAsO_2$. The reactions are (Se = 79)

$$SeO_3^{2-} + BrO_3^- + H^+ \rightarrow SeO_4^{2-} + Br_2 + H_2O$$

$$BrO_3^- + AsO_2^- + H_2O \rightarrow Br^- + AsO_4^{3-} + H^+$$

10. If 91 g of $V_2O_5$ is dissolved in acid and reduced to $V^{2+}$ by treatment with zinc metal, then how many grams of $I_2$ could be reduced by the resulting $V^{2+}$ solution, as it is oxidized to $V^{4+}$? (V = 51, I = 127)

$$V_2O_5 + 10H^+ + 6e^- \rightarrow 2V^{2+} + 5H_2O$$

$$V^{2+} + I_2 + H_2O \rightarrow 2I^- + VO^{2+} + 2H^+$$

11. A 200 ml sample of a citrus fruit drinks containing ascorbic acid (vitamin C) was acidified with $H_2SO_4$ and 10 ml of 0.025 M-$I_2$ was added. Some of the $I_2$ was reduced by the ascorbic acid to $I^-$. The excess of $I_2$ required 2.5 ml of 0.01 M-$Na_2S_2O_3$ for reduction. What was the vitamin C content of the drink in microgram vitamin per ml drink?

$$C_6H_8O_6 + I_2 \rightarrow C_6H_6O_6 + 2HI$$

$$5H_2O + S_2O_3^{2-} + 4I_2 \rightarrow 2SO_4^{2-} + 8I^- + 10H^+$$

12. A 0.2 g sample of chromite was fused with excess of $Na_2O_2$ and brought into solution according to the following reaction.

$$2Fe(CrO_2)_2 + 7Na_2O_2 \rightarrow 2NaFeO_2 + 4Na_2CrO_4 + 2Na_2O$$

The solution was acidified with dil. HCl and 1.96 g Mohr's salt (molar mass = 392 g/mol) was added. The excess of $Fe^{2+}$ required 40 ml of 0.05 N-$K_2Cr_2O_7$ for titration. What is the percent of Cr in sample? (Cr = 52, Fe = 56)

13. A 10 g mixture of $Cu_2S$ and CuS was treated with 400 ml of 0.4 M – $MnO_4^-$ in acid solution producing $SO_2$, $Cu^{2+}$ and $Mn^{2+}$. The $SO_2$ was boiled off and the excess of $MnO_4^-$ was titrated with 200 ml of 1 M – $Fe^{2+}$ solution. The percentage of CuS in original mixture is (Cu = 64)

14. A mixture containing $As_2S_3$ and $As_2S_5$ requires 20 ml of 0.05 N iodine for titration. The resulting solution is then acidified and excess of KI was added. The liberated iodine required 1.24 g hypo, $Na_2S_2O_3 \cdot 5H_2O$, for complete reaction. The reactions are as follows.

$$As_2S_3 + 2I_2 + 2H_2S \rightarrow As_2S_5 + 4H^+ + 4I^-$$

$$As_2S_5 + 4H^+ + 2I^- \rightarrow As_2S_3 + 2I_2 + 2H_2S$$

The mole percent of $As_2S_3$ in the original mixture is (As = 75)

15. The element Se dispersed in 2 ml sample of detergent for dandruff control was determined by suspending it in the warm ammonical solution that contained 45 ml of 0.02 M-$AgNO_3$.

$$6Ag^+ + 3Se(s) + 6NH_3 + 3H_2O \rightarrow 2Ag_2Se(s) + Ag_2SeO_3(s) + 6NH_4^+$$

The mixture was now treated with excess nitric acid which dissolved the $Ag_2SeO_3$ but not $Ag_2Se$. The $Ag^+$ from $Ag_2SeO_3$ and the excess of $AgNO_3$ consumed 10 ml of 0.01 N-KSCN in Volhard titration. How many milligrams of Se was contained per ml of sample? (Se = 80)

16. One gram of commercial $AgNO_3$ is dissolved in 50 ml of water. It is treated with 50 ml of a KI solution. The silver iodide thus precipitated is filtered off. Excess of KI is titrated with M/10-$KIO_3$ solution in the presence of 6 M-HCl till all iodide ions are converted into ICl. It requires 50 ml of M/10-$KIO_3$ solution. A 20 ml of the same stock solution of KI requires 30 ml of M/10-$KIO_3$ under similar conditions. The percentage of $AgNO_3$ in the sample is (Ag = 108)

Reaction: $KIO_3 + 2KI + 6HCl \rightarrow 3ICl + KCl + 3H_2O$

17. A 4.0 g sample containing $Fe_3O_4$, $Fe_2O_3$ and an inert impure substance is treated with excess of KI solution in the presence of dilute $H_2SO_4$. The entire iron is converted to $Fe^{2+}$ along with the liberation of iodine. The resulting solution is diluted to 100 ml. A 20 ml of dilute solution requires 11.2 ml of 0.5 M – $Na_2S_2O_3$ solution to reduce the iodine present. A 50 ml of the diluted solution, after complete extraction of iodine, requires 12.80 ml of 0.25 M – $KMnO_4$ solution in dilute $H_2SO_4$ medium for the oxidation of $Fe^{2+}$. The percentage of $Fe_2O_3$ in the original sample is

18. An aqueous solution containing 1.07 g $KIO_3$ was treated with an excess of KI solution. The solution was acidified with HCl. The liberated $I_2$ consumed 50 ml of thiosulphate solution to decolourize the blue starch-iodine complex. The molarity of the sodium thiosulphate solution is $x$ M, then the value of $1000x$ is (K = 39, I = 127)

EXERCISE II

**19.** Hydrogen peroxide solution (20 ml) reacts quantitatively with a solution of $KMnO_4$ (20 ml) acidified with dilute $H_2SO_4$. The same volume of the $KMnO_4$ solution is just decolourized by 10 ml of $MnSO_4$ in neutral medium simultaneously forming a dark brown precipitate of hydrated $MnO_2$. The brown precipitate is dissolved in 10 ml of 0.2 M sodium oxalate under boiling condition in the presence of dilute $H_2SO_4$. The strength of $H_2O_2$ solution in mg per 100 ml solution is

**20.** If 20 ml of 0.1 M solution of sodium sesquicarbonate ($Na_2CO_3 \cdot NaHCO_3$) is titrated against 0.05 M – HCl, using (i) phenolphthalein and (ii) methyl orange as indicators, then what difference in titre values (in ml) would be recorded?

## Answer Keys

**Exercise I**

### Oxidation–Reduction

| | | | | | | | | | |
|---|---|---|---|---|---|---|---|---|---|
| 1. (d) | 2. (d) | 3. (a) | 4. (a) | 5. (c) | 6. (d) | 7. (a) | 8. (a) | 9. (c) | 10. (a) |
| 11. (b) | 12. (b) | 13. (b) | 14. (a) | 15. (d) | 16. (b) | 17. (d) | 18. (a) | 19. (c) | 20. (b) |
| 21. (c) | 22. (d) | 23. (b) | 24. (d) | 25. (b) | 26. (b) | 27. (c) | 28. (c) | 29. (a) | 30. (b) |
| 31. (b) | 32. (c) | 33. (a) | 34. (c) | 35. (b) | | | | | |

### Equivalent Concept

| | | | | | | | | | |
|---|---|---|---|---|---|---|---|---|---|
| 36. (c) | 37. (b) | 38. (a) | 39. (b) | 40. (c) | 41. (c) | 42. (a) | 43. (a) | 44. (b) | 45. (b) |
| 46. (d) | 47. (c) | 48. (c) | 49. (d) | 50. (b) | 51. (a) | 52. (c) | 53. (a) | 54. (b) | 55. (c) |
| 56. (b) | 57. (a) | 58. (b) | 59. (c) | 60. (a) | 61. (a) | 62. (d) | 63. (d) | 64. (c) | 65. (b) |
| 66. (b) | 67. (d) | 68. (c) | 69. (c) | 70. (c) | | | | | |

### Volumetric Analysis

| | | | | | | | | | |
|---|---|---|---|---|---|---|---|---|---|
| 71. (c) | 72. (c) | 73. (a) | 74. (c) | 75. (b) | 76. (d) | 77. (a) | 78. (a) | 79. (b) | 80. (a) |
| 81. (a) | 82. (c) | 83. (a) | 84. (c) | 85. (a) | 86. (c) | 87. (a) | 88. (b) | 89. (b) | 90. (c) |
| 91. (d) | 92. (b) | 93. (a) | 94. (b) | 95. (a) | 96. (a) | 97. (a) | 98. (a) | 99. (b) | 100. (b) |
| 101. (a) | 102. (d) | 103. (c) | 104. (d) | 105. (d) | 106. (b) | 107. (b) | 108. (a) | 109. (a) | 110. (b) |
| 111. (b) | 112. (a) | 113. (c) | 114. (b) | 115. (b) | 116. (d) | 117. (a) | 118. (c) | 119. (b) | 120. (a) |

## Answer Keys

**Exercise II**

### Section A (Only one Correct)

| | | | | | | | | | |
|---|---|---|---|---|---|---|---|---|---|
| 1. (b) | 2. (d) | 3. (b) | 4. (b) | 5. (b) | 6. (d) | 7. (a) | 8. (d) | 9. (c) | 10. (a) |
| 11. (c) | 12. (d) | 13. (a) | 14. (c) | 15. (c) | 16. (c) | 17. (a) | 18. (c) | 19. (b) | 20. (a) |
| 21. (b) | 22. (b) | 23. (b) | 24. (b) | 25. (b) | 26. (c) | 27. (c) | 28. (b) | 29. (d) | 30. (a) |
| 31. (b) | 32. (a) | 33. (b) | 34. (b) | 35. (c) | 36. (d) | 37. (c) | 38. (d) | 39. (d) | 40. (a) |
| 41. (b) | 42. (c) | 43. (d) | 44. (a) | 45. (b) | 46. (c) | 47. (c) | 48. (b) | 49. (a) | 50. (b) |
| 51. (a) | 52. (c) | 53. (d) | 54. (c) | 55. (c) | 56. (d) | 57. (a) | 58. (b) | 59. (b) | 60. (a) |
| 61. (d) | 62. (a) | 63. (b) | 64. (b) | 65. (a) | 66. (b) | 67. (a) | 68. (d) | 69. (b) | 70. (a) |
| 71. (b) | 72. (c) | 73. (a) | 74. (b) | 75. (b) | | | | | |

## Section B (One or More than one Correct)

1. (b), (d) 2. (b), (c), (d)
3. (a), (b), (c) 4. (b)
5. (a), (b), (c) 6. (b), (c)
7. (a), (b) 8. (b)
9. (d) 10. (a), (c)
11. (b), (c), (d) 12. (b), (d)
13. (a), (b), (c) 14. (b)
15. (a), (d) 16. (a)
17. (a), (b), (c) 18. (a), (b), (c), (d)
19. (a), (b), (c), (d) 20. (a), (c), (d)

## Section C

### Comprehension I

1. (c) 2. (a) 3. (b)

### Comprehension II

4. (a) 5. (b) 6. (b)

### Comprehension III

7. (b) 8. (a) 9. (c)

### Comprehension IV

10. (b) 11. (c) 12. (b)

### Comprehension V

13. (a) 14. (b) 15. (a)

### Comprehension VI

16. (b) 17. (a) 18. (b)

### Comprehension VII

19. (c) 20. (a) 21. (b)

### Comprehension VIII

22. (a) 23. (b) 24. (b)

### Comprehension IX

25. (a) 26. (b) 27. (d)

### Comprehension X

28. (c) 29. (b) 30. (b)

## Section D (Assertion – Reason)

1. (a) 2. (d) 3. (b) 4. (d) 5. (c) 6. (a) 7. (c) 8. (b) 9. (a) 10. (c)
11. (d) 12. (a) 13. (a) 14. (a) 15. (d)

## Section E (Column Match)

1. A $\leftrightarrow$ R; B $\to$ P, T; C $\to$ Q, T; D $\to$ S, T
2. A $\to$ P, Q; B $\to$ P; C $\to$ P, Q, R; D $\to$ S
3. A $\to$ S, T; B $\to$ P, Q; C $\to$ P, Q, R; D $\to$ Q, S
4. A $\to$ Q; B $\to$ P, S; C $\to$ R; D $\to$ P, S
5. A $\to$ R; B $\to$ S; C $\to$ P; D $\to$ Q
6. A $\to$ Q, R; B $\to$ Q, R; C $\to$ P, S; D $\to$ P, S
7. A $\to$ Q; B $\to$ S; C $\to$ R; D $\to$ P
8. A $\to$ Q, T; B $\to$ P; C $\to$ R, S
9. A $\to$ Q, S; B $\to$ P, Q; C $\to$ P, R; D $\to$ S
10. A $\to$ P, S; B $\to$ Q, R; C $\to$ Q, R; D $\to$ Q,R

## Section F (Subjective)

### Single-digit Integer Type

1. (3) 2. (2) 3. (2) 4. (5) 5. (3) 6. (4) 7. (3) 8. (3) 9. (8) 10. (4)
11. (9) 12. (5) 13. (2) 14. (6) 15. (9) 16. (9) 17. (0) 18. (2) 19. (6) 20. (4)

### Four-digit Integer Type

1. (0033) 2. (0068) 3. (0020) 4. (0400) 5. (1322)
6. (0568) 7. (0224) 8. (0009) 9. (0127) 10. (0254)
11. (0132) 12. (0026) 13. (0080) 14. (0020) 15. (0024)
16. (0085) 17. (0040) 18. (0600) 19. (0340) 20. (0080)

# HINTS AND EXPLANATIONS

## EXERCISE I (JEE MAIN)

### Oxidation–Reduction

1. $m \times (+a) + n \times (-b) + r \times (-c) = 0 \Rightarrow ma = bn + cr$.

2. $2(+2) + x + 6(-2) = 0 \Rightarrow x = +8$.

3. Oxidation state of Cr in both compounds is +6.

4. In this compound, NO is present as $NO^+$.

5. Oxidation state of Br in $Br_2$, $Br^-$ and $BrO_3^-$ is 0, –1 and +5, respectively.

6. The maximum oxidation state of S in any of its compound is +6.

7. The oxidation state of Fe in $Fe(CO)_5$, $Fe_2O_3$, $K_4[Fe(CN)_6]$ and $FeSO_4.(NH_4)_2SO_4.6H_2O$ are 0, +3, +2 and +2, respectively.

8. (a) $H_2\overset{+6}{S_2}O_7$, $Na_2\overset{+2.5}{S_4}O_6$, $Na_2\overset{+2}{S_4}O_3$, $\overset{0}{S_8}$

   (b) $\overset{+4}{S}O^{2+}$, $\overset{+6}{S}O_4^{2-}$, $\overset{+4}{S}O_3^{2-}$, $H\overset{+4}{S}O_4^-$

   (c) $H_2\overset{+6}{S}O_5$, $H_2\overset{+4}{S}O_3$, $\overset{+2}{S}C_2$, $H_2\overset{-2}{S}$

   (d) $H_2\overset{+6}{S}O_4$, $\overset{+4}{S}O_2$, $H_2\overset{-2}{S}$, $H_2\overset{+6}{S_2}O_8$

9. On oxygenation, $Fe^{2+}$ present in haemoglobin oxidizes to $Fe^{3+}$.

10. $H^{+1}—{}^{-1}C^{+3}\equiv{}^{-3}N$ $\qquad$ $H^{+1}—{}^{-1}\overset{+}{N}{}^{-3}\equiv{}^{+3}\bar{C}$

    Oxidation state of C = +2 $\qquad$ Oxidation state of C = +2

11. $3x + 2(-2) = 0 \Rightarrow x = +\frac{4}{3}$.

12. $H_2SO_4 + BaO_2 \rightarrow BaSO_4^{-2} + H_2O_2^{-1}$

13. $\overset{+7}{Mn}O_4^-$, $\overset{+6}{Cr}O_2Cl_2$ These are highest oxidation states of Mn and Cr, respectively.

14. $2(+2) + 2x + 7(-2) = 0 \Rightarrow x = +5$.

15. Toluene $= C_7H_8 \Rightarrow 7x + 8(+1) = 0 \Rightarrow 7x = -8$.

16. Oxidation state of K in all of its compound is +1.

17. Oxidation state of Cr in $K_3CrO_8$ is +5. In $CrO_5$, it is +6. $K_3CrO_8$ has four peroxide ($O_2^{2-}$) linkage while $CrO_5$ has only two peroxide linkage.

18. 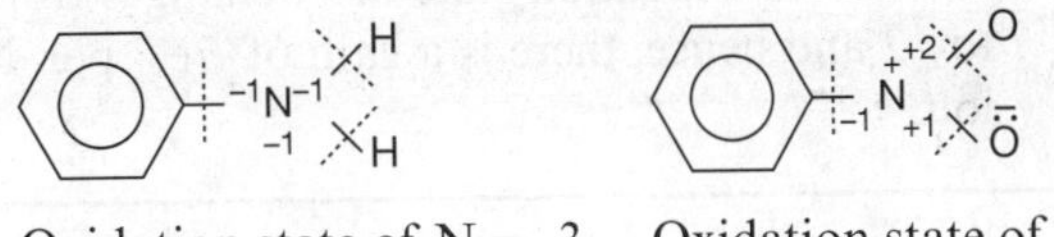

    Oxidation state of N = –3 $\qquad$ Oxidation state of N = +3

19. 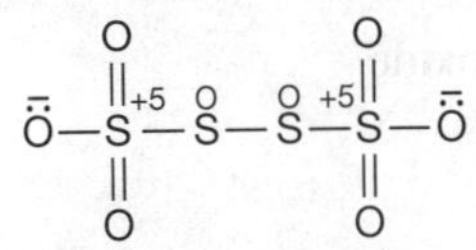

20. Empirical formula of $C_6H_{12}O_6$ and HCHO is same.

21. $\frac{N_{Fe}}{N_O} = \frac{\frac{70}{56}}{\frac{30}{16}} = \frac{2}{3} \Rightarrow$ Formula $= Fe_2O_3$

    ∴ Oxidation state of Fe = +3.

22. The reducing power of non-metal hydrides increase from top to bottom in a group

23. Hydroquinol undergoes removal of hydrogen, i.e., oxidation and hence, it acts as a reducing agent.

24. Redox reactions may involve absorption or release of heat.

25. Oxidation and reduction is occurring due to different elements and hence, it is not a disproportionation.

26. $\underset{Base}{CaO} + \underset{Acid}{\overset{+4}{Si}O_2} \rightarrow \underset{Salt}{Ca\overset{+4}{Si}O_3}$

27. $H_2S + SO_2 \rightarrow S + H_2O$

    Same element but belonging to different molecule undergoes oxidation and reduction and hence, it is not disproportionation. In fact, it is comproportionation.

28. Referer Question No. 27.

29. $3X_2O_3 + 4Cr_2O_7^{2-} + 26H^+ \rightarrow$
$6XO_4^- + 8Cr^{3+} + 13H_2O$
**Method II:** (After equivalent concept)
$n_{eq}$ of $X_2O_3 = n_{eq} Cr_2O_7^{-2}$
or $n \times 8 = 1 \times 6 \Rightarrow n_{X_2O_3} = \frac{3}{4}$

30. $2NO_3^- + 16H^+ + 14e^- \rightarrow N_2H_4 + 6H_2O$
$\therefore$ Number of $e^-$ per $NO_3^-$ ion $= \frac{14}{2} = 7.$
**Method II:** Oxidation state of N changes from +5 to –2 and hence, there is a gain of $7e^-$ per $NO_3^-$ ion.

31. Oxidation: $Br_2 + 12OH^- \rightarrow 2BrO_3^- + 6H_2O + 10e^-$
Reduction: $Br_2 + 2e^- \rightarrow 2Br^- 1 \times 5$
Net reaction:
$6Br_2 + 12OH^- \rightarrow 2BrO_3^- + 10Br^- + 6H_2O$

32. $2NO_3^- + 10H^+ + 8e^- \rightarrow N_2O + 5H_2O$

33. $NO_2^- + 5H_2O + 6e^- \rightarrow NH_3 + 7OH^-$

34. $4Zn + NO_3^- + 7H_2O \rightarrow 4Zn^{2+} + NH_4^+ + 10OH^-$

35. $4Fe + 10HNO_3 \rightarrow 4Fe(NO_3)_2 + NH_4NO_3 + 3H_2O$

## Equivalent Concept

36. $n_{eq}$ of oxygen $= n_{eq}$ of oxide
or $\frac{20}{8} = \frac{100}{E} \Rightarrow E = 40.$

37. Sulphurated hydrogen (g) + Sn(S) $\rightarrow$ Tin Sulphide(s) + $H_2(g)$
$x$ mole $\quad$ $x$ mole
$= x \times 34$ g $\quad$ $= x \times 2$ g
No change in volume means no change in mole of gases.
Mass of sulphur combined with $2x$ gm hydrogen $= 34x - 2x = 32x$ g.
$\therefore E_{sulphur} = \frac{32x}{2x} \times 1 = 16$

38. Atomic mass = Equivalent mass $\times$ Valency
$= E \times \frac{n}{m} \times 2$

39. $n_{eq}$ metal $= n_{eq}$ metal chloride
or $\frac{w}{18.67} = \frac{162.52}{18.67 + 35.5} \Rightarrow w = 56$ g.

40. $4N_A$ electrons means 4 equivalents. Hence, mass of Mg needed $= 4 \times 12 = 48$ gm

41. $K_2CrO_4$ is behaving as salt and hence, $E = \frac{M}{2}.$

42. $NaHC_2O_4$ is behaving as acid and hence, $E = \frac{M}{1}.$

43. $NaHC_2O_4$ is behaving as base and hence, $E = \frac{M}{1}.$

44. $Ca_3(PO_4)_2 + 2H^+ \rightarrow 2CaHPO_4 + Ca^{2+}$
Equivalent weight of $Ca_3(PO_4)_2 = \frac{M}{2}.$

45. Change in oxidation state of Mn should be 2 units.

46. $E_1 : E_2 : E_3 = \frac{M}{5} : \frac{M}{1} : \frac{M}{3} = 3:15:5$

47. $As_2S_3 + 20H_2O + 28e^- \rightarrow 2AsO_4^{3-} + 3SO_4^{2-} + 40H^+$

48. (a) $E = \frac{M}{3}$ (b) $E = \frac{M}{5}$ (c) $E = \frac{M}{1}$

49. $Cu_2P^{-3} \rightarrow H_3\overset{+5}{P}O_4$

50. $Cl_2 \rightarrow Cl^-$

51. $H_2O_2 \rightarrow H_2O + \frac{1}{2}O_2$
The reaction is balanced on the loss and gain of one electron per $H_2O_2$ molecule.

52. Reaction is balanced on the loss or gain of $10e^-$.

53. $H_2O_2$ is acting as reductant. Its equivalent weight is $\frac{34}{2} = 17.$

54. $N_2 + 3H_2 \rightarrow 2NH_3$
Reaction is balanced by the loss or gain of $6e^-$.

55. $MCl_2 = 32.7 \times 2 + 35.5 \times 2 = 136.4$

56. $3Zn + 8HNO_3 \rightarrow 3Zn(NO_3)_2 + 2NO + 4H_2O$
Reaction is balanced by the loss or gain of $6e^-$.

57. Reaction is balanced by the loss or gain of $2e^-$.

58. $n_{eq}$ metal carbonate $= n_{eq}$ metal oxide

or $\frac{100}{E+30} = \frac{60}{E+8} \Rightarrow E = 25$

59. $n_{eq}$ metal chloride $= n_{eq}$ metal sulphate

or $\frac{0.298}{E+35.5} = \frac{0.348}{E+48} \Rightarrow E = 39$

60. $n_{eq} NH_3 = n_{eq} O_2 = 1 \times 4 = 4$

61. $n_{eq} Cr_2O_7^{2-} = n_{eq} N_2H_5^+$

$\Rightarrow n \times 6 = 0.136$

$\Rightarrow n = 0.0226$

62. $x \times 5 = y \times 6 \Rightarrow x : y = 6 : 5$

63. $n_{eq}$ metal carbonate $= n_{eq}$ metal sulphate

or $\frac{w}{E+30} = \frac{1.43w}{E+48}$

$\therefore E = 11.86$

64. $n_{eq} H_2S = n_{eq} KMnO_4$

or $\frac{w}{34} \times 8 = \frac{6.32 \times 3}{158} \Rightarrow w = 0.51$ g

65. $n_{eq}$ metal $= n_{eq}$ oxygen

or $\frac{52.91}{E} = \frac{47.09}{8} \Rightarrow E = 8.99$

$\therefore$ Atomic mass of metal $= 8.99 \times 3 = 26.97$

66. Approximate atomic mass $= \frac{6.4}{0.085} = 75.29$

Now, Valency $= \frac{\text{Atomic mass}}{\text{Equivalent mass}} \approx \frac{75.29}{25}$

$\approx 3$ (integer)

$\therefore$ Exact atomic mass $= 25 \times 3 = 75$.

67. Valency $= \frac{2 \times \text{V.D.}}{E+35.5} = \frac{2 \times 77}{3+35.5} = 4$

$\therefore$ Atomic mass $= 3 \times 4 = 12$

68. $n_{eq} C_6H_{10}O_4 = n_{eq} KOH$

or $\frac{1}{146} \times \text{Basicity} = \frac{0.768}{56} \times 1$

$\Rightarrow$ Basicity of $C_6H_{10}O_4 = 2$

69. $n_{eq}$ metal bromide $= n_{eq}$ metal chloride

or $\frac{1.878}{E+80} = \frac{1.00}{E+35.5} \Rightarrow$ Equivalent mass of metal, $E = 15.18$.

Now, approximate atomic mass of metal $= \frac{6.4}{0.14} = 45.71$

$\therefore$ Valency $= \frac{45.71}{15.18} = 3$ (Integer)

$\therefore$ Exact atomic mass of metal bromide $= 15.18 \times 3 = 45.54$

$\therefore$ Molecular mass of metal bromide $= 45.54 + 3 \times 80 = 285.54$

70. The chemical formula of potassium sulphate is $K_2SO_4$ and hence, potassium chromate is $K_2CrO_4$.

Molecular mass $= 2 \times 39 + A + 4 \times 16 = 142 + A$

From question, $(142 + A) \times \frac{26.79}{100} = A \Rightarrow A = 51.96$.

## Volumetric Analysis

71. $n_{eq}$ of $H_2SO_4$ in $V$ ml solution $= \frac{V \times \frac{10}{100}}{49} = x$

$n_{eq}$ of NaOH in $V$ ml solution $= \frac{V \times \frac{10}{100}}{40} = y$

As $y > x$, the resulting solution should be basic.

72. $n_{eq}$ of $Na_2CO_3.H_2O = \frac{0.62}{124} \times 2 = 0.01$

and $n_{eq}$ of $H_2SO_4 = \frac{100 \times 0.1}{1000} = 0.01$.

Hence, 0.01 eq. of $Na_2SO_4$ will form and the resulting solution will be neutral.

73. $n_{eq} AgNO_3 = n_{eq} K_2CrO_4$

or $\frac{V \times 0.1}{1000} \times 1 = \frac{10 \times 0.09}{1000} \times 2 \Rightarrow V = 18$ ml

74. $n_{eq} KMnO_4 = n_{eq} KNO_2$

or $\frac{V \times 0.18}{1000} = \frac{25 \times 0.21}{1000} \Rightarrow V = 29.17$ ml

**75.** $n_{eq}\ KMnO_4 = n_{eq} Na_2S_2O_3$

or $\frac{V \times 0.1}{1000} \times 3 = \frac{0.158}{158} \times 8 \Rightarrow V = 26.67$ ml

**76.** $n_{eq}\ HNO_3 = n_{eq} NaOH$

or $\frac{V_1 \times 18.9}{63} \times 1 = \frac{V_2 \times 3.2}{40} \times 1 \Rightarrow V_1 : V_2 = 4:15$

**77.** $n_{eq}\ NH_3 = n_{eq} H_2SO_4$

or $\frac{V}{22400} \times 1 = \frac{30 \times (1-0.2)}{1000} \Rightarrow V = 537.6$ ml

**78.** $n_{eq}\ Na_2CO_3 = n_{eq}$ acid 'A'

or $\frac{26 \times 1}{1000} = \frac{10 \times N_A}{1000} \Rightarrow N_A = 2.6$ N

and $n_{eq}\ Na_2CO_3 = n_{eq}$ acid 'B'

or $\frac{26 \times 1}{1000} = \frac{40 \times N_B}{1000} \Rightarrow N_B = 0.65$ N

Now, $VN = V_A N_A + V_B N_B$

or $1000 \times 1 = V_A \times 2.6 + (1000 - V_A) \times 0.65$

$\Rightarrow V_A = 179.49$ ml.

**79.** $n_{eq}\ Ag = n_{eq} KCNS$

or $\frac{0.5 \times \frac{90}{100}}{108} = \frac{25 \times N}{1000} \Rightarrow N = 0.167$

**80.** $n_{eq} KH_2PO_4 = n_{eq} OH^-$

or $\frac{w}{136} \times 1 = \frac{25 \times 0.1}{1000} \Rightarrow w = 0.34$ g

$\therefore$ % Purity $= \frac{0.34}{0.5} \times 100 = 68\%$

**81.** $n_{eq} KMnO_4 = n_{eq} H_2O_2$

or $\frac{50 \times 0.1}{1000} \times 5 = \frac{w}{34} \times 2 \Rightarrow w = 0.425$ g

$\therefore$ Strength $= \frac{0.425}{10} \times 100 = 4.25\% (w/v)$

**82.** $n_{eq}\ KHC_8H_4O_4 = n_{eq} Ba(OH)_2$

or $\frac{0.204}{204} \times 1 = \frac{25 \times M}{1000} \times 2 \Rightarrow M = 0.02$

**83.** $n_{eq} SeO_2 = n_{eq} Cr^{2+}$

or $\frac{12.5 \times 0.05}{1000} \times (4 - x) = \frac{25 \times 0.1}{1000} \times 1 \Rightarrow x = 0$

**84.** $n_{eq} CrO_4^{-2} = n_{eq} S_2O_4^{2-}$

or $100 \times M \times 3 = \frac{522}{179} \times 2 \Rightarrow M = 0.02$

**85.** $n_{eq} CaC_2O_4 = n_{eq} KMnO_4$

or $n \times 2 = \frac{25 \times 0.001}{1000} \times 5$

$\therefore$ Moles of Ca = Moles of $CaC_2O_4 = 62.5 \times 10^{-6}$ in 10 ml of blood.

$\therefore$ Milligram of Ca per litre blood

$= \frac{62.5 \times 10^{-6} \times 40}{10} \times 1000 = 0.25$

**86.** $n_{eq} KMnO_4 = n_{eq} H_2C_2O_4$

or $\frac{w}{158} \times 5 = \frac{50 \times 0.2}{1000} \times 2 \Rightarrow w = 0.632$ g

**87.** $n_{eq} Ca(OH)_2 = n_{eq} H_3PO_4$

or $\frac{V \times 0.05}{1000} \times 2 = \frac{10 \times 0.1}{1000} \times 1 \Rightarrow V = 10$ ml

**88.** $n_{eq} H_2C_2O_4 \cdot 2H_2O = n_{eq} KMnO_4$

or $\frac{w}{126} \times 2 = \frac{100 \times 0.4}{1000} \times 5 \Rightarrow w = 12.6$ g

**89.** $n_{eq} Na_2B_4O_7.10H_2O = n_{eq} HCl$

or $\frac{w}{381.2} \times 2 = \frac{25 \times 0.2}{1000} \Rightarrow w = 0.953$ g

**90.** $n_{eq}$ metal salt $= n_{eq} Na_2SO_3$

or $\frac{50 \times 0.1}{1000} \times (3 - x) = \frac{25 \times 0.1}{1000} \times 2 \Rightarrow x = 2$

**91.** $n_{eq} H_2O_2 = n_{eq} KMnO_4$

or $\frac{1 \times \frac{x}{100}}{34} \times 2 = \frac{x \times N}{1000} \Rightarrow N = 0.588$

**92.** $n_{eq} FeC_2O_4 = n_{eq} KMnO_4$

or $\frac{1}{144} \times 3 = \frac{60 \times M}{1000} \times 5 \Rightarrow M = 0.0694$

**93.** $n_{eq}$ of NaOH $= \frac{15 \times 0.2}{1000} \times 1 = 3 \times 10^{-3}$

$n_{eq}$ of $MgCl_2 = \frac{12 \times 0.15}{1000} \times 2 = 3.6 \times 10^{-3}$

Hence, NaOH is a limiting reagent.

Now, the mass of $Mg(OH)_2$ formed $= 3 \times 10^{-3} \times 29 = 0.087$ g.

**94.** $n_{eq}$ acid $= n_{eq}$ NaOH

or $$\frac{\frac{39}{1000}\times 100}{82}\times n = \frac{\frac{40}{1000}\times 95}{40}\times 1 \Rightarrow n = 1.997 \simeq 2$$

**95.** Normality $= \frac{20}{98}\times 2 = 0.408$ N

**96.** **Acidic solution:**

$$\frac{25\times 0.017}{1000}\times 2 = \frac{16.9\times 0.01}{1000}\times(7-x) \Rightarrow x = 2$$

**Neutral Solution:**

$$\frac{25\times 0.017}{1000}\times 2 = \frac{28.6\times 0.01}{1000}\times(7-y) \Rightarrow y = 4$$

**97.** $n_{eq}$acid $= n_{eq}$NaOH

or $$\frac{0.84}{150}\times n = \frac{\left(\frac{28}{25}\times 100\right)\times\frac{1}{10}}{1000} \Rightarrow \text{Basicity, } N = 2$$

and equivalent weight of acid $= \frac{150}{2} = 75$

**98.** $n_{eq}Na_2CO_3\cdot xH_2O = n_{eq}HCl$

or $$\frac{0.70}{106+18x}\times 2 = \frac{\left(\frac{19.8}{20}\times 100\right)\times\frac{1}{10}}{1000} \Rightarrow x = 2$$

**99.** $n_{eq}H_2SO_4 = n_{eq}NaOH$

or $$\frac{w}{98}\times 2 = \frac{\left(\frac{25.6}{25}\times 1000\right)\times\frac{1}{10}\times 0.95}{1000}$$

$\Rightarrow w = 4.77$ g

$\therefore$ % Strength of acid $= \frac{w}{3.5\times 1.76}\times 100 = 77.38\%$

**100.** $n_{eq}Na_2CO_3 = n_{eq}H_2SO_4$

or $$\frac{25\times\frac{1}{10}}{1000} = \frac{10\times N}{1000} \Rightarrow N = 0.25$$

Now, for dilution, $400\times 0.25 = V\times 0.1$

$\Rightarrow V = 1000$ ml.

$\therefore$ Volume of water added $= 1000 - 400 = 600$ ml.

**101.** $n_{eq}H_2SO_4 = n_{eq}NaOH + n_{eq}Na_2CO_3$

or $$\frac{25\times N}{1000} = \frac{50\times 0.5}{1000} + \frac{0.265}{106}\times 2$$

$\therefore$ Normality of diluted solution, $N = 1.2$

Now, $V_1N_1 = V_2N_2 \Rightarrow 10\times N_1 = 100\times 1.2$

$\Rightarrow N_1 = 12N$

**102.** $n_{eq}HCl = n_{eq}AgCl$

or $$\frac{25\times N}{1000} = \frac{0.287}{143.5}\times 1 \Rightarrow N_{HCl} = 0.08$$

and $N_{H_2SO_4} = 0.2 - 0.08 = 0.12.$

$\therefore$ Mass per cent of

$$HCl = \frac{0.08\times 36.5}{0.08\times 36.5 + 0.12\times 49}\times 100 = 33.18\%$$

**103.** $n_{eq}$ anhydrous acid $= n_{eq}$ NaOH

or $$\frac{0.10}{E} = \frac{25\times 0.1}{1000} \Rightarrow E = 40$$

and $n_{eq}$ hydrated acid $= n_{eq}$NaOH

or $$\frac{0.245}{E+18x} = \frac{50\times 0.1}{1000} \Rightarrow x = 0.5$$

**104.** $$\frac{100\times 1}{1000}\times 5 = \frac{V\times 1}{1000}\times 3 \Rightarrow V = 166.67 \text{ ml}$$

**105.** $n_{eq}HCl = n_{eq}$metal carbonate + $n_{eq}$caustic soda

or $$\frac{25\times 1}{1000} = \frac{1}{E} + \frac{50\times\frac{1}{10}}{1000} \Rightarrow E = 50$$

**106.** $n_{eq}NaOH = n_{eq}(NH_4)_2SO_4 + n_{eq}HCl$

or $$\frac{80\times 0.1}{1000} = \frac{w}{132}\times 2 + \frac{11.6\times 0.1}{1000} \Rightarrow w = 0.451 \text{ g}$$

$\therefore$ Mass of $Na_2SO_4 = 0.91 - 0.451 = 0.459$ g

**107.** $n_{eq}Ag_2S = n_{eq}H_2S$

or $$\frac{w}{248}\times 2 = \frac{11.2}{22400}\times 2 \Rightarrow w = 0.124 \text{ g}$$

**108.** $n_{eq}Fe^{2+} = n_{eq}K_2Cr_2O_7$

or $$\frac{0.2\times\frac{98}{100}}{56}\times 1 = \frac{30\times N}{1000} \Rightarrow N = 0.1167$$

**109.** $n_{eq}\,O_3 = n_{eq}I_2 = n_{eq}Na_2S_2O_3$

or $$n\times 2 = \frac{40\times\frac{1}{10}}{1000}\times 1 \Rightarrow n_{O_3} = 2\times 10^{-3}$$

and total moles of gases $= \frac{1}{22.4} = 4.46\times 10^{-2}$

$\therefore$ Moles of $O_2 = 4.46\times 10^{-2} - 2\times 10^{-3} = 4.26\times 10^{-2}$

$\therefore$ Mass percent of

$$O_3 = \frac{2\times 10^{-3}\times 48}{2\times 10^{-3}\times 48 + 4.26\times 10^{-2}\times 32}\times 100 = 6.575\%$$

**110.** $n_{eq}\ Cl_2 = n_{eq} I_2 = n_{eq} Na_2S_2O_3$

or $\frac{w}{71} \times 2 = \frac{26 \times 0.15}{1000} \Rightarrow w = \frac{35.5 \times 26 \times 0.11}{1000}$ g

$\therefore$ Mass percent of $Cl_2 = \frac{w}{71 \times 1.1} \times 100 = 0.13\%$

**111.** $n_{eq} HClO = n_{eq} I_2 = n_{eq} Na_2S_2O_3$

or $n \times 2 = \frac{17.4 \times 0.02}{1000} \times 1 \Rightarrow n = \frac{0.174}{1000}$.

$Cl_2 + H_2O \rightarrow HCl + HClO$

$\therefore$ Mass of $Cl_2$ per litre $= \left(\frac{n}{25} \times 1000\right) \times 71 = 0.494$ g

**112.** $n_{eq} Na_3AsO_4 = n_{eq} I_2 = n_{eq}$ hypo

or $\frac{1}{208} \times 2 = \frac{V \times 0.2}{1000} \Rightarrow V = 48.1$ ml

**113.** $n_{eq} S_2O_3^{2-} = n_{eq} K_2S_2O_8$

or $\frac{V \times 0.25}{1000} \times 1 = \frac{1}{270} \times 2 \Rightarrow v = 29.63$ ml

**114.** $n_{eq} MnO_4^- = n_{eq} FeSO_4$

or $\frac{V_1 \times M_1}{1000} \times 5 = \frac{V_2 \times M_2}{1000} \times 1 \Rightarrow 5V_1M_1 = V_2M_2$

**115.** $n_{eq} KHC_2O_4 = n_{eq} KMnO_4$

or $\frac{x}{M} \times 2 = \frac{100 \times 0.02}{1000} \times 5$ (1)

$n_{eq} KHC_2O_4 = n_{eq} Ca(OH)_2$

or $\frac{y}{M} \times 1 = \frac{100 \times 0.5}{1000} \times 2$ (2)

$\therefore \frac{x}{y} = \frac{1}{2}$

**116.** Let $NaHCO_3 = a$ mole, $Na_2CO_3 =$ 6 mole.

In the presence of phenolphthalein, $n_{eq}$ HCl $= n_{eq} Na_2CO_3$

or $\frac{x \times N}{1000} = b \times 1$ (1)

In the presence of methyl orange,

$n_{eq}$ HCl $= n_{eq} NaHCO_3$ (original) +

$n_{eq} NaHCO_3$ (formed)

or $\frac{y \times N}{1000} = a \times 1 + b \times 1$ (2)

From (1) and (2), we get:

$V_{HCl}$ only forms $NaHCO_3$ (original) $= (y - x)$ ml.

**117.** Bromocresol green:

$\frac{25 \times 0.107}{1000} \times n = \frac{23.1 \times 0.115}{1000} \times 1 \Rightarrow n \simeq 1$

Phenolphthalein:

$\frac{25 \times 0.107}{1000} \times n = \frac{46.2 \times 0.115}{1000} \times 1 \Rightarrow n \simeq 2$

**118.** Temporary hardness is due to $Ca(HCO_3)_2$ and $Mg(HCO_3)_2$.

$n_{eq} CaCO_3 = n_{eq} Ca(HCO_3)_2 + n_{eq} Mg(HCO_3)_2$

or $\frac{w}{100} \times 2 = \frac{162 \times 10^{-3}}{162} \times 2 + \frac{73 \times 10^{-3}}{146} \times 2$

$\Rightarrow w = 150 \times 10^{-3}$ g

$\therefore$ Temporary hardness $= \frac{150 \times 10^{-3}}{1000} \times 10^6$

$= 150$ ppm.

Permanent hardness is due to $MgCl_2$ and $CaSO_4$.

$n_{eq} CaCO_3 = n_{eq} MgCl_2 + n_{eq} CaSO_4$

or $\frac{w}{100} \times 2 = \frac{95 \times 10^{-3}}{95} \times 2 + \frac{136 \times 10^{-3}}{136} \times 2$

$\Rightarrow w = 200 \times 10^{-3}$ g.

$\therefore$ Permanent hardness $= \frac{200 \times 10^{-3}}{1000} \times 10^6$

$= 200$ ppm.

**119.** Temporary hardness $= \left(\frac{\frac{5.6}{100} \times 1000}{56}\right) \times 200$

$= 100$ ppm.

**120.** $(Ca^{2+}) = \frac{10^2}{2} = 5 \times 10^{-3}$ M

$\therefore$ Hardness $= \left(\frac{5 \times 10^{-3}}{1000} \times 10^6\right) \times 40 = 200$ ppm

## EXERCISE II (JEE ADVANCED)

### Section A (Only one Correct)

1. Addition of hydrogen is reduction.

2. Sodium amalgam is an alloy (mixture of sodium and mercury).

3. $2x+4(-2)+2\times 0+2\times 0=-2 \Rightarrow x=+3$

4. $KBF_4:(+1)+x+4(-1)=0 \Rightarrow x=+3$

5. $LiBiO_2:(+1)+x+2(-2)=0 \Rightarrow x=+3$

6. $N_{Xe}:N_F=\frac{53.5}{131}:\frac{46.5}{19}=1:6 \Rightarrow XeF_6$

$\therefore$ Oxidation state of $Xe=+6$.

7. $H_3C-\overset{\overset{O}{||}}{S}-CH_3$ S is more $E_N$ than C, but less $E_N$ than O.

8. Oxidation state of Fe is +2 in both as $NO^-$ and $NOS^-$ are the ligands.

9. CO is a neutral oxide.

10. Phosphorus acid: $H_3PO_3$, oxidation state of P $=+3$.

Orthophosphoric acid: $H_3PO_4$, oxidation state of P $=+5$.

Metaphosphoric acid: $HPO_3$, oxidation state of P $=+5$.

Pyrophosphoric acid: $H_4P_2O_7$, oxidation state of P $=+5$.

11. Informative

12. Reducing agent undergoes oxidation. In $HNO_3$, H and N are in their maximum oxidation state and hence, oxidation is possible only for 'O'.

13. $I^-$ is a strong reducing agent.

14. Oxidizing agent must undergo reduction. $I^-$ can never be reduced.

15. In $Ca(OCl)Cl$, the oxidation state of 'Cl' is +1 and –1 but in the product $Cl_2$, it becomes zero.

16. $2Mn^{2+}+5PbO_2+4H^+ \rightarrow 2MnO_4^-+5Pb^{2+}+2H_2O$

17. $AsO_3^{3-}+2OH^- \rightarrow AsO_4^{3-}+H_2O+2e^-$

18. 'Cr' should be in +6 oxidation state in the product. For basic medium, the product should be $CrO_4^{2-}$.

19. $\overset{+5}{N}O_3^-+4e^- \rightarrow$ Oxidation state of N in the product should be +1.

20. $CH_3CH_2OH+H_2O \Rightarrow CH_3COOH+4H^++4e^-$

21. $CH_3CH_2OH+H_2O \Rightarrow CH_3COOH+4H^++4e^-$

22. The oxidation state of 'Cr' in $ClO_x^-$ must be +1 and hence, $x=1$.

23. 1 mole of $AO_3^-$ should gain 6 moles of electron.

$AO_3^-+6H^++6e^- \rightarrow A^{x+}+3H_2O$

From charge conservation, $(-1)+(+6)+(-6)=+x$ $\Rightarrow x=-1$.

24. $E=\frac{A}{V} \Rightarrow A=21V_1=14V_2$

As $V_1$ and $V_2$ must be integer, $A$ should be 42, 84, 126, etc.

25. Mass of oxygen present = 14.9 g and mass of hydrogen combined with oxygen to form water = 16.78 – 14.9 = 1.88 g.

$\therefore$ Equivalent weight of oxygen $=\frac{14.9}{1.88}\times 1.008=7.989$

26. $\underset{A\ gm}{M^{2+}}+\underset{2N_A e^-}{2e^-} \rightarrow M$

$\therefore 1\,g \rightarrow \frac{2N_A}{A}\times 1=1.81\times 10^{22} \Rightarrow A=66.54$.

27. For minimum equivalent weight, the basicity of acid should be maximum.

28. In the neutralization reactions, one mole of $H_2O$ is formed from 1 mole of $H^+$ and 1 mole of $OH^-$ ions and hence, equivalent weight of water = 18.

29. $CH_3COOH+3Cl^- \rightarrow CCl_3COOH+3H^++6e^-$

$\therefore E_{CH_3COOH}=\frac{60}{6}=10$

**30.** $\underset{\text{n-factor}=6}{N_2} + \underset{2}{3H_2} \rightarrow \underset{3}{2NH_3}$ (No. of $e^-$ involved = 6)

$\therefore E_{NH_3} = \frac{x_1}{3}$ and $E_{N_2} = \frac{x_2}{6}$.

**31.** $P_4 + 3NaOH + 3H_2O \rightarrow PH_3 + 3NaH_2PO_2$

The reaction is balanced by the loss and gain of 3 moles of electron per mole of $P_4$ and hence,

$E_{P_4} = \frac{M}{3}$.

**32.** $Pb + PbO_2 + 2H_2SO_4 \rightarrow 2PbSO_4 + 2H_2O$

The reaction is balanced by the loss and gain of 2 moles of electron per mole of Pb and hence,

$E_{H_2SO_4} = \frac{2M}{2} = M$.

**33.** $3Cl_2 + 6NaOH \rightarrow 5NaCl + NaClO_3 + 3H_2O$

The reaction is balanced by the loss and gain of 5 moles of electron per mole of $NaClO_3$ and hence,

$E_{H_2O} = \frac{3M}{5}$.

**34.** $O_3 + 2H^+ + 2e^- \rightarrow O_2 + H_2O$

$\therefore E_{O_3} = \frac{48}{2} = 24$.

**35.** The reaction is balanced by the loss and gain of 2 moles of electron per mole of $MnO_2$ and hence,

$E_{HCl} = \frac{4M}{2} = 2M$.

**36.** $n_{eq}\ I_2 = n_{eq} H_3AsO_4$

or $\frac{w}{254} \times 2 = \frac{1.5 \times 10^{22}}{6 \times 10^{23}} \Rightarrow w = 3.175\,\text{g}$.

**37.** $n_{eq} = \frac{6 \times 10^{20}}{6 \times 10^{23}} = 0.001$

**38.** 1 g equivalent always reacts with 1 g equivalent of other substance.

**39.** $n_{eq}$ metal oxide = $n_{eq}$ water

or $\frac{8.6}{E_M + 8} = \frac{1.8}{18} \times 2 \Rightarrow E_M = 35$

**40.** Mass of acid $= \frac{20}{0.5} \times 1 = 40\,\text{g}$

**41.** Let the acid be $H_nA$.

$n_{eq}$ acid = $n_{eq}$ magnesium salt

or $\frac{1.0}{1 + E_{A^{n-}}} = \frac{1.301}{12 + E_{A^{n-}}} \Rightarrow E_{A^{n-}} = 35.54$

$\therefore$ Equivalent weight of acid = 1 + 35.54 = 36.54.

**42.** $n_{eq}$ metal oxide = $n_{eq}\ CO_2$

or $\frac{3.7}{E_M + 8} = \frac{1.0}{44} \times 4$ $\qquad \left(\overset{\circ}{C} \rightarrow \overset{+4}{C}O_2\right)$

$\therefore E_M = 32.7$

**43.** $n_{eq} = n \times n$-factor and $n$-factor is maximum for

$IO_3^- \left(+5 \text{ to } -\frac{1}{3}\right)$.

**44.** $n_{eq}\ KMnO_4 = n_{eq} FeC_2O_4$

$(FeC_2O_4 \rightarrow Fe^{3+} + CO_2)$

or $n \times 5 = 1 \times 3 \Rightarrow n = \frac{3}{5}$.

**45.** $n_{eq}\ KMnO_4 = n_{eq}\ SO_3^{2-}$ $\qquad (SO_3^{2-} \rightarrow SO_4^{2-})$

or $n \times 5 = 1 \times 2 \Rightarrow n = \frac{2}{5}$.

**46.** $n_{eq}$R.A. $= n_{eq} MnO_2 \Rightarrow n \times 1 = x \times 2$

and $n_{eq}$R.A. $= n_{eq} K_2CrO_4 \Rightarrow n \times 2 = y \times 3$.

$\therefore x : y = 3 : 4$

**47.** $CHCl_2COOH + 3H_2O \rightarrow$
$2CO_2 + H_2O + Cl_2 + 6H^+ + 6e^-$

$n_{eq} CHCl_2COOH = n_{eq}$O.A.

or $n \times 6 = 1.2 \Rightarrow n = 0.2$.

Now, $n_{eq} CHCl_2COOH = n_{eq} NH_3$

or $0.2 \times 1 = x \times 1 \Rightarrow x = 0.2$.

**48.** $M_{H_2SO_4} : M_{H_3PO_4} = E_{H_2SO_4} : E_{H_3PO_4} = \frac{98}{2} : \frac{98}{1} = 1:2$

**49.** $28 = \frac{A_M}{2} \Rightarrow$ Atomic mass of metal, $A_M = 56$.

Now, $n_{eq} H_2SO_4 = n_{eq} M_2O_3$

or $\frac{w}{98} \times 2 = \frac{4.8}{160} \times 6 \Rightarrow w = 8.82$ g

**50.** Approximate atomic mass of metal $= \frac{6.4}{0.26} = 24.62$.

Let the metal chloride be $MCl_V$.

$24.62 + V \times 35.5 \approx 95 \Rightarrow V \approx 2$

$\therefore$ Exact atomic mass of metal $= 95 - 2 \times 35.5 = 24$.

Now, $n_{eq}$ metal = $n_{eq}\ H_2$

or $\frac{1.2}{24} \times 2 = \frac{V_{H_2}}{22.4\ \text{L}} \times 2 \Rightarrow V_{H_2} = 1.12$ L

**51.** Approximate atomic mass of metal $= \frac{6.4}{0.55} = 11.63.$

Let the metal chloride be $MCl_V$. Then,

$2 \times 74.6 \simeq 11.63 + V \times 35.5 \Rightarrow V \approx 4.$

Again, $2 \times 74.6 = A + 4 \times 35.5 \Rightarrow A = 7.2.$

**52.** $n_{eq} HCl = n_{eq} CO_2$

or $\frac{200 \times N}{1000} = \frac{560}{22400} \times 2 \Rightarrow N_{HCL} = 0.25$ N

**53.** The resulting solution becomes neutral.

**54.** $n_{eq} FeSO_4 \cdot (NH_4)_2 SO_4 \cdot xH_2O = n_{eq}$ $KMnO_4$

or $\frac{5.88}{284 + 18x} \times 1 = \frac{\left(\frac{20}{25} \times 250\right) \times \frac{3.16}{1000} \times \frac{75}{100}}{158} \times 5$

$\Rightarrow x = 6.$

**55.** $2Ag^+ + H_2S \rightarrow Ag_2S \downarrow + 2H^+$

$Cu^{2+} + H_2S \rightarrow CuS \downarrow + 2H$

$\therefore$ Ratio of amount of $H_2S = 1:2$

**56.** $NaAl(OH)_2 CO_3 + 4HCl \rightarrow$

$NaCl + AlCl_3 + CO_2 + 3H_2O$

$n_{eq} NaAl(OH)_2 CO_3 = n_{eq} HCl$

or $\frac{100 \times 0.1}{1000} \times 4 = \frac{V \times 0.25}{1000} \Rightarrow V = 160$ ml

**57.** $n_{eq} Au = n_{eq} SnCl_2$

or $\frac{1.97}{197} \times 3 = \frac{V \times 0.05}{1000} \times 2 \Rightarrow V = 300$ ml

**58.** $n_{eq}$ $I^- = n_{eq} Ce^{4+}$

or $\frac{250 \times M}{1000} \times 2 = \frac{20 \times 0.05}{1000} \Rightarrow M_{I^-} = \frac{1}{500}$M

$\therefore$ Concentration in $\frac{g}{L} = \frac{1}{500} \times 127 = 0.254$

**59.** $n_{eq} U^{4+} = n_{eq} KMnO_4$

or $\frac{0.5}{238} \times 2 = \frac{50 \times M}{1000} \times 5 \Rightarrow M = 0.0168$

**60.** $2IO_3^- + 6H_2C_2O_4 \rightarrow C_2O_4^{2-} + 10CO_2 + I_2 + 6H_2O$

$\frac{n_{KIO_3}}{2} = \frac{n_{H_2C_2O_4}}{6} \Rightarrow \frac{1}{214} = \frac{1}{3} \times \frac{w}{90} \Rightarrow w = 1.262$ g

**61.** $n_{eq} H_2C_2O_4 = n_{eq} MnO_4^- = n_{eq} I^-$

or $\frac{w}{90} \times 2 = \frac{500 \times 1.0}{1000} \Rightarrow w = 22.5$ g

**62.** $n_{eq} Fe_{0.9}O = n_{eq} KMnO_4$

or $\frac{26.56}{66.4} \times 0.7 = \frac{V \times 0.2}{1000} \times 5 \Rightarrow V = 280$ ml

**63.** $n_{eq} KMnO_4 = n_{eq} FeC_2O_4$

or $\frac{25 \times \left(\frac{20 \times 1}{200}\right)}{1000} \times 5 = \frac{20 \times M}{1000} \times 3$

$\Rightarrow M_{FeC_2O_4} = 0.208$ M

**64.** $20 \times 1 = 200 \times M_1 \Rightarrow M_1 = 0.1$ M

$20 \times 0.1 = 75 \times M_2 \Rightarrow M_2 = \frac{0.1}{3}$ M

$30 \times \frac{0.1}{3} = 150 \times M_3 \Rightarrow M_3 = \frac{1}{150}$ M

Now, $n_{eq} K_2Cr_2O_7 = n_{eq} H_2O_2$

or $\frac{15 \times \frac{1}{150}}{1000} \times 6 = \frac{V \times 0.02}{1000} \times 2 \Rightarrow V_{H_2O_2} = 15$ ml

**65.** $n_{eq}$ metal $= n_{eq}$ $H_2$

or $\frac{0.1}{51} \times x - \frac{43.9}{22400} \times 2 \Rightarrow x \simeq 2$

Now, $n_{eq} M^{x+} = n_{eq} MnO_4^-$

or $\frac{0.1}{51} \times (y - x) = \frac{58.8 \times 0.1}{1000} \Rightarrow y \simeq 5$

**66.** $n_{eq} Na_2S_2O_3 = n_{eq} KBrO_3$

or $\frac{40 \times M}{1000} \times 1 = \frac{0.1336}{167} \Rightarrow M_{Na_2S_2O_3} = 0.02$ M

**67.** $n_{eq} HCl = n_{eq} M_3 (PO_4)_2$

or $\frac{10 \times 0.1}{1000} = \frac{0.0517}{E_M + 31.67} \Rightarrow E_M = 20.03$

**68.** $n_{eq} HCl = n_{eq} CaSO_4 \cdot \frac{1}{2} H_2O$

or $\frac{525 \times 0.1}{1000} = \frac{w}{145} \times 2 \Rightarrow w = 3.81$ g

**69.** $Cu^{2+} + I^- \rightarrow Cu^+ + I_2$

$n_{eq} Na_2S_2O_3 = n_{eq} I_2 = n_{eq} Cu^{2+}$

or $\frac{V \times 0.4}{1000} \times 1 = \frac{50 \times 0.2}{1000} \times 1 \Rightarrow V_{Na_2S_2O_3} = 25$ ml

**70.** $n_{eq}H_2O_2 = n_{eq}I_2 = n_{eq}Na_2S_2O_3$

or $\frac{25 \times M}{1000} \times 2 = \frac{20 \times 0.3}{1000} \Rightarrow M_{H_2O_2} = \frac{6}{50}$ M

$\therefore$ Volume strength $= \frac{6}{50} \times 11.35 = 1.362$

**71.** $CS_2 + 3O_2 \rightarrow CO_2 + 2SO_2$
$x$ mole $\quad x$ mole $\quad 2x$ mole

$H_2S + \frac{3}{2}O_2 \rightarrow H_2O + SO_2$
$y$ mole $\quad y$ mole $\quad y$ mole

$3x + 2y = \frac{1.97 \times 20}{0.082 \times 400} = 1.2 \qquad (1)$

Now, $n_{eq}\ SO_2 = n_{eq}I_2$

or $(2x + y) \times 2 = \frac{250 \times 2.8}{1000} \times 2 \Rightarrow 2x + y = 0.7 \qquad (2)$

From (1) and (2), $x = 0.2$ ; $y = 0.3$

Now, $X_{CS_2} = \frac{x}{x + y} = 0.4$

**72.** $n_{CO_2}$ formed $= \frac{0.9}{100} = 9 \times 10^{-3}$

$n_{NaOH} = \frac{100 \times 0.5}{1000} = 50 \times 10^{-3}$

$2NaOH + CO_2 \rightarrow Na_2CO_3 + H_2O$
$18 \times 10^{-3} \quad 9 \times 10^{-3} \quad 9 \times 10^{-3}$

$\therefore$ Final solution contains NaOH = (50 – 18) × $10^{-3} = 32 \times 10^{-3}$ mole and $Na_2Co_3 = 9 \times 10^{-3}$ mole.

Now, $n_{eq}\ HCl = n_{eq}\ NaOH + n_{eq}\ Na_2CO_3$

or $\frac{V \times 0.5}{1000} \times 1 = 32 \times 10^{-3} \times 1 + 9 \times 10^{-3} \times 1$

$\Rightarrow V_{HCl} = 82$ ml

**73.** Phenolphthalein:

$\frac{x \times 0.05}{1000} \times 1 = \frac{40 \times 0.05}{1000} \times 1 \Rightarrow x = 40$

Methyl orange:

$\frac{y \times 0.05}{1000} \times 1 = \frac{40 \times 0.05}{1000} \times 3 \Rightarrow y = 120$

**74.** Let $M_{Na_2CO_3} = x$M ; $M_{NaHCO_3} = y$M

Now, $V_1 \times 1 \times 1 = \frac{100 \times x}{1000} \times 1$ (Phenolphthalein)

and $V_2 \times 1 \times 1 = \frac{100 \times x}{1000} \times 2 + \frac{100 \times y}{1000} \times 1$ (Methyl orange)

$\therefore x = 10V_1$ and $y = 10(V_2 - 2V_1)$

**75.** In presence of phenolphthalein, $n$-factor = 1.
In presence of methyl orange, $n$-factor = 3.

---

## Section B (One or More than one Correct)

**1.** $NO^+$ and $ClO_4^-$ ⇒, Oxidation state of N = +3, Cl = +7.

**2.** $PbO_2$ is an oxide.

**3.** (a) Oxidation state of S = +6 in both.

(b) Oxidation state of Cr = +6 in both.

(c) Oxidation state of P = +5 in both.

**4.** As $KMnO_4$ reduces, the compound must oxidize. $Fe^{+3}$ cannot oxidize.

**5.** $\bar{\text{C}}\equiv\text{N}$ Oxidation state of C = +2, N = –3

$:\ddot{O}=C=\ddot{N}:^- \longleftrightarrow :\ddot{\bar{O}}-C\equiv N$

O.S. of C= +4, O = –2, N = –3

**6.** 'Cl' should be in intermediate oxidation state.

**7.** $KMnO_4 = \frac{15.8}{158} \times 5 = 0.5$ eq

$HCl = \frac{18.25}{36.5} \times 1 = 0.5$ eq

$H_2C_2O_4 = \frac{22.5}{90} \times 2 = 0.5$ eq

$SO_2 = \frac{32}{69} \times 2 = 1.0$ eq

$FeSO_4 = \frac{38}{152} \times 1 = 0.25$ ec

**8.** $2Cu + 6HNO_3 \rightarrow$
$2Cu(NO_3)_2 + NO + NO_2 + 3H_2O$

**9.** Equivalent volume is the volume occupied by 1g-equivalent of the gas.

(a) 1g-equivalent of $CH_4 = \frac{1}{8}$ mole $= \frac{1}{8} \times 22.4$ L
$= 2.8$ L

(b) 1g-equivalent of $O_3 = \frac{1}{2}$ mole = 11.2 L

(c) 1g-equivalent of $H_2S = \frac{1}{2}$ mole = 11.2 L

(d) 1g-equivalent of $CO_2 = \frac{1}{4}$ mole = 5.6 L

**10.** $n_{eq}$ metal nitrate = $n_{eq}$ metal sulphate

or $\frac{0.5}{E_M+62} = \frac{0.43}{E_M+48} \Rightarrow E_M = 38$

**11.** The chemical formula of sulphate is $MSO_4 \cdot 7H_2O$.

$(A+222)\times\frac{20}{100} = A \Rightarrow A = 55.5$

As the valency of metal is 2, the equivalent weight $= \frac{55.5}{2} = 27.75$

**12.** Higher oxide: $\frac{80}{E_M} = \frac{20}{8} \Rightarrow E_M = 32.$

Lower oxide (4.29 g) → Higher oxide (4.77 g)

Mass of metal $= 4.77\times\frac{80}{100} = 3.816$ g

Mass of oxygen = 4.29 – 3.816

= 0.474 g

Lower oxide: $\frac{3.816}{E_M} = \frac{0.474}{8} \Rightarrow E_M = 64.4$

**13.** Atomic weight (approx.) $= \frac{6.4}{0.03} = 213.33$

Now, $\frac{10}{E_M} = \frac{18.9}{E_M+62} \Rightarrow E_M = 69.66$

Now, Valency $\approx \frac{213.33}{69.66} \approx 3$ (Integer value)

∴ Atomic weight (exactly) $= 69.66\times 3 = 208.98$

**14.** Let molarities of $Na_2CO_3$ and $NaHCO_3$ be $x$M and $y$M, respectively.

Phenolphthalein:

$\frac{100\times x}{1000}\times 1 = V_1\times 1\times 1 \Rightarrow x = 10V_1$

Methyl orange:

$\frac{100\times x}{1000}\times 2 + \frac{100\times y}{1000}\times 1 = (V_2\times 1)\times 1$

$\Rightarrow y = 10(V_2 - 2V_1)$

**15.** Molarity of oxalic acid solution $= \frac{6.3}{90+18x}$

Now, $\frac{20\times\left(\frac{6.3}{90+18x}\right)}{1000}\times 2 = \frac{40\times 0.05}{1000}\times 1 \Rightarrow x = 2$

$E_{H_2C_2O_4.2H_2O} = \frac{126}{2} = 63$ but $E_{H_2C_2O_4} = \frac{90}{2} = 45$

Molarity $= \frac{6.3}{126} = 0.05$ M

Now, $n_{eq}$ acid = $n_{eq}$ $KMnO_4$

or $\frac{100\times 0.05}{1000}\times 2 = \frac{V\times 0.05}{1000}\times 5 \Rightarrow V = 40$ ml.

**16.** (a) 109% oleum means 100 g oleum ($H_2SO_4 + SO_3$) exactly requires 9 g water to produce exactly 109 g of pure $H_2SO_4$.

$\underset{18\,g}{H_2O} + \underset{80\,g}{SO_3} \rightarrow H_2SO_4$

∴ 9 g → 40 g

∴ Percentage of free $SO_3$ = 40%

(b) 1 g of oleum contains 0.4 g of $SO_3$ and hence, 0.6 g of $H_2SO_4$.

Now, $n_{eq}SO_3 + n_{eq}H_2SO_4 = n_{eq}NaOH$

or $\frac{0.4}{80}\times 2 + \frac{0.6}{98}\times 2 = \frac{V\times 0.5}{1000}\times 1 \Rightarrow V = 44.49$ ml

(c) $n_{eq}\ SO_3 + n_{eq}H_2SO_4 = n_{eq}Ba(OH)_2$

or $\frac{2.0}{80}\times 2 + \frac{3.0}{98}\times 2 = \frac{V\times 0.1}{1000} \Rightarrow V = 111.22$ ml

(d) 100 g of oleum requires 9g of water to give 109 g of $H_2SO_4$. Hence, the final solution contains 109 g of $H_2SO_4$ and 491 g of water.

∴ Molality $= \frac{109/98}{491}\times 1000 = 2.265$ m.

**17.** (a) Moles of $P_4O_{10} = \frac{5.68}{284} = 0.02$

$P_4O_{10} \xrightarrow{\text{aqueous}} 4H_3PO_4$

∴ Molarity of $H_3PO_4$ solution

$= \frac{0.02\times 4}{250}\times 1000 = 0.32$ M

(b) $\frac{25\times 0.32}{1000}\times 3 = \frac{V\times 0.5}{1000}\times 1 \Rightarrow V = 48$ ml

(c) $\frac{15\times0.32}{1000}\times3=\frac{V\times0.2}{1000}\times2\Rightarrow V=36$ ml

(d) $\frac{40\times0.32}{1000}\times1=\frac{V\times0.8}{1000}\Rightarrow V=16$ ml

**18.** $\underset{0.01\text{mole}}{SO_2Cl_2}+2H_2O\rightarrow\underset{0.01\text{mole}}{H_2SO_4}+\underset{0.02\text{ mole}}{2HCl}$

(a) $[H_2SO_4]=\frac{0.01}{200}\times1000=0.05$ M

(b) $[HCl]=\frac{0.02}{200}\times1000=0.1$ M

(c) $\frac{20\times0.05}{1000}\times2+\frac{20\times0.1}{1000}\times1=\frac{V\times0.2}{1000}\times1$

$\Rightarrow V_{NaOH}=20$ ml.

(d) $\frac{100\times0.1}{1000}=\frac{w}{143.5}\Rightarrow W_{AgCl}=1.435$ g

**19.** $n_{eq}H_2O_2=n_{eq}KMnO_4=n_{eq}K_2Cr_2O_7=n_{eq}O_2$

or $n_{KMnO_4}\times5=n_{K_2Cr_2O_7}\times6$

$\Rightarrow n_{KMnO_4}>n_{K_2Cr_2O_7}$

or $\frac{W_{KMnO_4}}{158}\times5=\frac{W_{K_2Cr_2O_7}}{294}\times6$

$\Rightarrow W_{KMnO_4}<W_{K_2Cr_2O_7}$

or $\frac{V\times M_{KMnO_4}}{1000}\times5=\frac{V\times M_{K_2Cr_2O_7}}{1000}\times6$

$\Rightarrow\frac{M_{KMnO_4}}{M_{K_2Cr_2O_7}}=\frac{6}{5}$

**20.** Moles of $S=\frac{8.0}{32}=0.25$

(a) $\frac{0.25}{2}\times2=\frac{200\times M}{1000}\times5\Rightarrow M_{KMnO_4}=0.25$ M

(b) $M_{BaCl_2}=\frac{0.25/2}{100}\times1000=1.25$ M

(c) $M_{BaSO_4}=\frac{0.25}{2}\times234=29.25$ g

## Section C (Comprehensions)

### Comprehension I

**1.** $E_{MnBr_2}=\frac{215}{17}\simeq12.65$

**2.** $E_{PbO_2}=\frac{240}{2}=120$

**3.** $2MnBr_2+17PbO_2+30HNO_3\rightarrow$
$2HMnO_4+2Pb(BrO_3)_2+15Pb(NO_3)_2+14H_2O$

The reaction is balanced by the loss or gain of 34 electrons. Hence,

$E_{HNO_3}=\frac{30\times63}{34}=55.6$

### Comprehension II

**4.** Atomic weight (approx.) $=\frac{6.4}{0.08}=80$

$n_{eq}Ag=n_{eq}$ arsenious chloride

or $\frac{100}{108}\times1=\frac{56}{E_{As}+35.5}\Rightarrow E_{AS}=24.98$

Now, Valency $=\frac{80}{24.98}\simeq3$

$\therefore$ Exact atomic mass = 24.98 × 3 = 74.94

**5.** $E_{As}=24.98$

**6.** Molecular mass of arsenious chloride = 6.25 × (22.4 × 1.3) = 182.

Let the formula be $(AsCl_3)_x$.

$x(74.94+3\times35.5)=182\Rightarrow x=1.0$

$\therefore$ Molecular formula $=AsCl_3$

## Comprehension III

Let the oxide be $M_2O_x$ and the halide be $MX_y$.

For oxide: $\frac{0.4}{8}=\frac{w}{A}\times x$ (1)

For halide: $\frac{4.0}{E}=\frac{w}{A}\times y$ (2)

From (1) and (2), we get: $\frac{E}{80}=\frac{x}{y}$

7. $x=y\Rightarrow E=80$

8. $y=2x\Rightarrow E=40$

9. $A=E.n\Rightarrow$ As valency should be an integer, $A$ must be an integer multiple of 40.

## Comprehension IV

10. $n_{eq}KMnO_4=n_{eq}X^{n+}$

or $1.5\times10^{-3}\times5=2.5\times10^{-3}\times(5-n)\Rightarrow n=2.$

11. $E=\frac{M}{n\text{-factor}}$

$\therefore\ M=56\times(5-n)=168$

Hence, atomic mass of $X=168-2\times35.5=97.$

12. $n_{eq}KMnO_4=n_{eq}X^{n+}$

or $n\times1=1\times3\Rightarrow n=3$

## Comprehension V

Let the sample contains $x$ mole KCl and $y$ mole $KClO_3$.

$$\underset{\frac{y}{10}\text{mole}}{KClO_3}\xrightarrow{SO_2}\underset{\frac{y}{10}\text{mole}}{KCl}$$

Now, mole of AgCl formed,

$$\frac{x}{10}+\frac{y}{10}=\frac{0.1435}{143.5}\Rightarrow x+y=0.01 \quad (1)$$

For second experiment,

$n_{eq}KClO_3+n_{eq}O.A.=n_{eq}FeSO_4$

or $\frac{y}{10}\times6+\frac{37.5\times0.08}{1000}=\frac{30\times0.2}{1000}\Rightarrow y=5\times10^{-3}$

From Equation (1), we get: $x=5\times10^{-3}$

13. $n_{KClO_3}:n_{KCl}=y:x=1:1$

14. $m_{KCl}=x\times74.5=0.3725$ g

$m_{KClO_3}=y\times122.5=0.6125$ g

$\therefore\ m_{moisture}=1-(0.3725+0.6125)=0.015$ g, i.e., 1.5%

15. Mass percent of $KCl=\frac{0.3725}{1}\times100=37.25\%$

## Comprehension VI

16. $n_{eq}As_4O_6=n_{eq}I_3^-$

or $\frac{0.1188}{396}\times8=\frac{10\times M}{1000}\times2\Rightarrow M_{I_3^-}=0.12$ M

17. $n_{HCN}=nI_3^-$

or $\frac{15\times M}{1000}=\frac{5\times0.12}{1000}\Rightarrow M_{HCN}=0.04$ M

18. Mass of $HCN=(6\times0.04)\times27=6.48$ g

## Comprehension VII

19. $4Mn^{2+}+MnO_4^-+20F^-+8H^+\rightarrow5MnF_4^-+4H_2O$

20. Let the moles of $Mn_3O_4=x$. Hence, moles of $Mn^{2+}$ formed $=3x$.

Now, for $KMnO_4$ solution, molarity $= \frac{0.125}{5} = 0.025$ M.

Now, $n_{eq} KMnO_4 = n_{eq} Mn^{2+}$

or $\frac{30 \times 0.025}{1000} \times 4 = 3x \times 1 \Rightarrow x = 10^{-3}$

Now, mass of $Mn_3O_4 = x \times 229 = 0.229$ g

$\therefore$ Percentage of $Mn_3O_4 = \frac{0.229}{0.458} \times 100 = 50\%$

**21.** Normality $= 0.025 \times 4 = 0.1N$

## Comprehension VIII

Let the ore contains $x$ mole of $FeCr_2O_4$ and $y$ mole of $Fe_{0.95}O_{1.00}$.

Now, $n_{eq} FeCr_2O_4 + n_{eq} Fe_{0.95}O_{1.00} = n_{eq} O_2$

or $x \times 7 + y \times 0.85 = \frac{280}{22400} \times 4 \Rightarrow 7x + 0.85y = 0.05$ (1)

Now,

Moles of $K_4[Fe(CN)_6]$ taken $= \frac{10 \times 1}{1000} = 0.01$

$n_{eq} Fe^{2+} = n_{eq} K_4[Fe(CN)_6]$

or $\frac{6 \times 0.4}{1000} = n = 2 \Rightarrow n = 0.0012$.

$\therefore$ Moles of $K_4[Fe(CN)_6]$ reacted with

$Fe^{3+} = 0.01 - 0.0012 = 0.0088$

Now, $n_{eq} Fe^{3+} = n_{eq} K_4[Fe(CN)_6]$

or $(x + 0.95y) \times 3 = 0.0088 \times 3 \Rightarrow x + 0.95y = 0.0088$ (2)

From (1) and (2), we get:

$y = 2 \times 10^{-3}$, $x = 6.9 \times 10^{-3}$

**22.** Mass per cent of

$Fe_{0.95}O_{1.00} = \frac{2 \times 10^{-3} \times 69.2}{2} \times 100 = 6.92\%$

**23.** Let in $Fe_{0.95}O_{1.00}$, 'z' Fe-atom are in +2 state.

$z \times (+2) + (0.95 - z) \times (+3) = 2.00 \Rightarrow z = 0.85$

Hence, per cent of total iron in +2 state

$= \frac{x + y \times 0.85}{x + 0.95y} \times 100$

$= 97.73\%$

**24.** Moles of Prussian blue = moles of $Fe^{3+}$

$= x + 0.95y = 8.8 \times 10^{-3}$

## Comprehension IX

**25.** Moles of $NaHSO_3$ needed = 3× moles of $NaIO_3$

$= 3 \times \frac{5.94}{198} = 0.09$

$\therefore$ Mass of $NaHSO_3$ needed $= 0.09 \times 104 = 9.36$ g.

**26.** Moles of $SO_3^-$ needed in 2nd reaction.

$= \frac{1}{5} \times$ moles of $I^-$ formed in 1st reaction

$= \frac{1}{5} \times \frac{5.94}{198} = 6 \times 10^{-3}$

$\therefore$ Volume of solution required $= \frac{6 \times 10^{-3}}{5.94/198} = 0.2$ L

**27.** Mass of $I_2$ produced $= 6 \times 10^{-3} \times 3 \times 254 = 4.572$ g

$\therefore$ Mass of $I_2$ produced per litre of solution

$= \frac{4.572}{(1 + 0.2)} = 3.81$ g

## Comprehension X

**28.** $n_{eq} HCl = n_{eq} Na_2CO_3 + n_{eq} NaOH$

or $\frac{V \times 1}{1000} = \frac{1}{106} \times 1 + \frac{1}{40} \times 1 \Rightarrow V_{HCl} = 34.43$ ml

**29.** $n_{eq} HCl = n_{eq} Na_2CO_3 + n_{eq} NaHCO_3 + n_{eq} NaOH$

or $\frac{V \times 1}{1000} = \frac{1}{106} \times 2 + \frac{1}{84} \times 1 + \frac{1}{40} \times 1$

$\Rightarrow V_{HCl} = 55.77$ ml

30. $n_{eq}HCl = n_{eq}NaHCO_3$ formed $+ n_{eq}NaHCO_3$, present initially.

or $\frac{V \times 1}{1000} = \frac{1}{106} \times 1 + \frac{1}{84} \times 1 \Rightarrow V_{HCl} = 21.34$ ml

## Section D (Assertion–Reason)

1. $I^-$ can never be reduced.
2. The oxidation state of terminal C-atoms are –3 and middle C-atoms are –2.
3. $I^-$ can not be reduced and hence, reduction of $O_3$ occurs.

$O_3 + 2H^+ + 2e^- \rightarrow O_2 + H_2O$

4. Oxidation state of $Cl = +3$
5. *n*-factor may be fractional.
9. $V \times 0.3 \times 2 = V \times 0.2 \times 3$
10. In the presence of methyl orange, the colour change appears when $Na_2CO_3$ converts completely in $H_2CO_3$.
14. Copper converts from $Cu^{2+}$ to $Cu^+$.
15. $n_{eq}A = n_{eq}C$ only when n-factor of B is same in both reactions.

## Section E (Column Match)

4.

| Column I | Column II |
|---|---|
| (A) $n_{eq} = 1 \times 5$ | (P) $n_{eq} = 3 \times 1$ |
| (B) $n_{eq} = 1 \times 3$ | (Q) $n_{eq} = 0.5 \times 10$ |
| (C) $n_{eq} = 1 \times 2$ | (R) $n_{eq} = 1 \times 2$ |
| (D) $n_{eq} = 1 \times 3$ | (S) $n_{eq} = 1.5 \times 2$ |

5.

| Column I | Column II |
|---|---|

(A) $\underset{(O.A.)}{n_{eq}\,MnO_4^-} = \underset{(R.A.)}{n_{eq}\,C_2O_4^{2-}} \Rightarrow$

$n_1 \times 3 = n_2 \times 2 \Rightarrow n_1 : n_2 = 2:3$

(B) $\underset{(O.A.)}{n_{eq}\,ClO^-} = \underset{(R.A.)}{n_{eq}\,Fe(OH)_3} \Rightarrow$

$n_1 \times 2 = n_2 \times 3 \Rightarrow n_1 : n_2 = 3:2$

(C) $\underset{(O.A.)}{n_{eq}\,HO_2^-} = \underset{(R.A.)}{n_{eq}\,Cr(OH)_3} \Rightarrow$

$n_1 \times 2 = n_2 \times 4 \Rightarrow n_1 : n_2 = 2:1$

(D) $\underset{(R.A.)}{n_{eq}\,N_2H_4} = \underset{(O.A.)}{n_{eq}\,Cr(OH)_2} \Rightarrow$

$n_1 \times 6 = n_2 \times 2 \Rightarrow n_2 : n_1 = 3:1$

6. Equivalent volume is the volume of gas corresponding to 1 g-equivalent of the gas.

7.

(A) *n*-factor = 2

(B) *n*-factor = 10

(C) *n*-factor $= \frac{2 \times 10}{2 + 10} = \frac{5}{3}$

(D) *n*-factor $= \frac{2 \times 2}{2 + 2} = 1$

8.

(P) *n*-factor = 1

(Q) *n*-factor $= \frac{1}{2}$

(R) *n*-factor = 2

(S) *n*-factor = 6

(T) *n*-factor $= \frac{1}{2}$

9. Milliequivalents:

| Column I | Column II |
|---|---|
| (A) $100 \times 0.3 \times 2 = 60$ for P, Q, R, S | (P) $100 \times 0.3 \times 1 = 30$ |
| (B) $50 \times 0.6 \times 1 = 30$ for P, S $50 \times 0.6 \times 2 = 60$ for Q, R | (Q) $120 \times 0.1 \times 5 = 60$ |
| (C) $50 \times 0.6 \times 1 = 30$ for P, Q, R, S | (R) $60 \times 0.1 \times 5 = 30$ |
| (D) $100 \times 0.2 \times 3 = 60$ for P, Q, S | (S) $100 \times 0.6 \times 1 = 60$ |

10. Milliequivalents:

| Column I | Column II |
|---|---|
| (A) $50 \times 0.5 \times 2 = 50$ | (P) $50 \times 0.5 \times 2 = 50$ |
| (B) $50 \times 0.5 \times 1 = 25$ | (Q) $50 \times 0.5 \times 1 = 25$ |
| (C) $50 \times 0.5 \times 1 = 25$ | (R) $25 \times 0.5 \times 2 = 25$ |
| (D) $50 \times 0.5 \times 1 = 25$ | (S) $50 \times 1.0 \times 1 = 50$ |

## Section F (Subjective)

### Single-digit Integer Type

1. $AO_4^{n-} + (9-2n)H^+ + 2e^- \rightarrow HAO_n^{2-} + (4-n)H_2O$

From charge conservation,

$(-n)+(9-2n)+(-2) = -2 \Rightarrow n = 3.$

2. $5AO_2 + 4H^+ \rightarrow 2AO_4^- + 3A^{n+} + 2H_2O$

From charge conservation,

$(+4) = (-2) + 3n \Rightarrow n = 2.$

3. $n_{eq}Cl_2 = n_{eq}$ metal chloride

or $\frac{1.12}{22.4} \times 2 = \frac{5.55}{E+35.5} \Rightarrow$ Equivalent weight of metal, $E = 20$

$\therefore$ Valency $= \frac{A}{E} = \frac{40}{20} = 20$

4. $\frac{160}{96} = \frac{2 \times n}{2+n} \Rightarrow n = 10 \Rightarrow$ Oxidation state of Br in unknown product = 5

5. $E = \frac{30}{\left(\frac{2\times 2}{2+2}\right)} = 30$

6. $M(s) + Cl_2(g) \rightarrow Chloride(g)$

As the volume reduced by one-third, moles of chloride formed $= \frac{2}{3} \times$ Moles of $Cl_2$ reacted.

Mass of chorine used $= n \times 71$ g.

Mass of chloride formed

$= \frac{2}{3} n \times 68.75 \times 2 = \frac{275}{3} n$ g

$\therefore$ Mass of element used $= \frac{275}{3} n - 71n = \frac{62}{3} n$ g

Now, equivalent weight of element,

$E = \frac{62n/3}{71n} \times 35.5 = \frac{31}{3}$

7. $n_{eq}SO_2 = n_{eq}HClO_3$

or $\frac{V}{22.4} \times 2 = \frac{16.9}{84.5} \times 6 \Rightarrow V = 13.44$ L

Now $n = \frac{PV}{RT} = \frac{2 \times (5 \times 13.44)}{0.0821 \times 546} = 3$

8. $n_{eq}N_2H_4 = n_{eq}K_2CrO_4$

or $\frac{w}{32} \times 4 = \frac{24}{194} \times 3 \Rightarrow w = 2.97 \text{ g} = 3 \text{ g}$

9. $n_{eq}KHC_2O_4 \cdot H_2C_2O_4 \cdot 2H_2O = n_{eq}NaOH$

or $n \times 3 = 30 \times 1 \Rightarrow n = 10$

Now, $n_{eq}KHC_2O_4 \cdot H_2C_2O_4 \cdot 2H_2O = n_{eq}KMnO_4$

or $10 \times 4 = n \times 5 \Rightarrow n = 8$

10. $n_{eq}Mn(NO_3) \cdot 6H_2O = n_{eq}Na_4XeO_6$

or $n \times 5 = \frac{62.5 \times 0.04}{1000} \times 8 \Rightarrow n = 4 \times 10^{-3}$

**11.** $Cl_2 + 2NaClO_2 \Rightarrow 2ClO_2 + 2NaCl$

$n_{eq} NaClO_2 \times \frac{90}{100} = n_{eq} ClO_2$

or $5 \times 2 \times \frac{90}{100} = n \Rightarrow n = 9$

**12.** $n_{eq} CuSO_4 \cdot xH_2O = n_{eq} H_2SO_4 = n_{eq} NaOH$

or $\frac{1.245}{159.5 + 18x} \times 2 = \frac{10 \times 1}{1000} \times 1 \Rightarrow x = 4.97 \approx 5$

**13.** $\underset{x\,\text{mole}}{CS_2} + O_2 \rightarrow \underset{x\,\text{mole}}{CO_2} + \underset{2x\,\text{mole}}{2SO_2}$

$\underset{y\,\text{mole}}{H_2S} + \frac{3}{2}O_2 \rightarrow \underset{y\,\text{mole}}{H_2O(g)} + \underset{y\,\text{mole}}{SO_2}$

From the question, $3x + 2y = \frac{7.2 \times 82.1}{0.0821 \times 600} = 12$

and $(2x + y) \times 2 = 75 \times 2$

Hence, $x = 2$ and $y = 3$.

**14.** $XeF_x + \frac{x}{2}H_2 \rightarrow \underset{a\,\text{mole}}{Xe} + \underset{a\,x\,\text{mole}}{xHF}$

$a = \frac{56}{22400}$

and $ax \times 1 = \frac{60 \times 0.25}{1000} \times 1 \Rightarrow x = 6$

**15.** $n_{eq} H_2O_2 = n_{eq} I_2 = n_{eq} Na_2S_2O_3$

or $\frac{12 \times \frac{5.675}{11.35}}{1000} \times 2 = \frac{24 \times M}{1000} \times 1 \Rightarrow M = 0.5$

Now, $n_{eq} O_3 = n_{eq} I_2 = n_{eq} Na_2S_2O_3$

or $\frac{V}{22.4} \times 2 = \frac{9 \times 0.5}{1000} \times 1 \Rightarrow V = 0.504$ L

$\therefore$ Percentage of $O_3 = \frac{0.0504}{1} \times 100 = 5.04\% \approx 5\%$

**16.** $\underset{x\,\text{mole}}{H_2C_2O_4} \rightarrow H_2O + \underset{x\,\text{mole}}{CO} + CO_2$

$\underset{x\,\text{mole}}{5CO} + I_2O_5 \rightarrow 5CO_2 + \underset{\frac{x}{5}\text{mole}}{I_2}$

$n_{eq}\ I_2 = n_{eq} Na_2S_2O_3$

or $\frac{x}{5} \times 2 = \frac{200 \times 0.2}{1000} \Rightarrow x = 0.1$

$\therefore$ Mass of $H_2C_2O_4 = 0.1 \times 90 = 9$ g

**17.** Ammonium Vanadate $\xrightarrow[\text{acid}]{\text{oxalic}} \underset{\left(\overset{x+}{V}\right)}{Z} \xrightarrow{KMnO_4}$

$\overset{+5}{V} \xrightarrow{SO_2} \overset{+4}{V} \xrightarrow{KMnO_4} \overset{+5}{V}$

$n_{eq} Z = n_{eq} KMnO_4 \Rightarrow n \times (5 - x) = V_1 \times N$

$n_{eq} V^{4+} = n_{eq} KMnO_4 \Rightarrow n \times 1 = V_2 \times N$

From question, $\frac{V_1}{V_2} = \frac{5}{1} \Rightarrow x = 0$

**18.** Let $Cu^{2+}$ = a mole and $C_2O_4^{2-}$ = b mole.

$n_{eq} C_2O_4^{2-} = n_{eq} MnO_4^- \Rightarrow b \times 2 = \frac{22.6 \times 0.02}{1000} \times 5$

$n_{eq} Cu^{2+} = n_{eq} I_2 = n_{eq} Na_2S_2O_3 \Rightarrow$

$a \times 1 = \frac{11.3 \times 0.05}{1000} \times 1$

$\therefore a : b = 1 : 2$

**19.** Moles of $Fe_2O_3 = \frac{0.552}{160} \Rightarrow$

Moles of $Fe^{2+}$ formed $= 2 \times \frac{0.552}{160}$

Now, $n_{eq} Fe^{2+} = n_{eq}$ Oxidizing agent

or $\frac{2 \times 0.552}{160} \times 1 = \frac{\left(\frac{17}{25} \times 100\right) \times 0.0167}{1000} \times n_f$

$\Rightarrow$ n-factor, $n_f \approx 6$

**20.** $\underset{a\,\text{mole}}{CH_3(CH_2)_nCOOH} \rightarrow \underset{a(n+2)\text{mole}}{(n+2)CO_2} \xrightarrow[\text{excess}]{NaOH}$

$(n+2)Na_2CO_3 + x$ mole NaOH left

$a(n+2)$mole

In presence of phenolphthalein:

$$x \times 1 + \frac{a(n+2)}{2} \times 1 = \frac{50 \times 1}{1000} \quad (1)$$

In presence of methyl orange:

$$x \times 1 + \frac{a(n+2)}{2} \times 2 = \frac{80 \times 1}{1000} \quad (2)$$

From (1) and (2), $\frac{a(n+2)}{2} = 0.03$ and $a = \frac{1.16}{60 + 14n}$

$\therefore n = 4$

## Four-digit Integer Type

1. $1(+3)+x(+2)+1(1-x)(+5)+3\times(-2)=0$

$\Rightarrow x=\frac{2}{3}$

$\therefore$ Percentage of '$x$' in +2 state $=\frac{\frac{2}{3}}{2}\times 100\% = 33.33\%$

2. $n_{eq}OH^- = n_{eq}$ Metal hydroxide

or $0.1\times 1=\frac{8.50}{E+17}\Rightarrow E=68$

3. $n_{eq}Ag = n_{eq}AgCl$

or $\frac{w}{108}\times 1=\frac{2.87}{143.5}\times 1\Rightarrow w=2.16$ g

$\therefore$ Mass of $Cu = 2.7-2.16=0.54$ g

$\therefore$ Percentage of $Cu=\frac{0.54}{2.7}\times 100=20\%$

4. $n_{eq}H_2SO_4 = n_{eq}CaO_2$

or $\frac{V\times 0.25}{1000}\times 2=\frac{7.2}{72}\times 2\Rightarrow V=400$ ml

5. $n_{eq}NaOH = n_{eq}$ Oxalate

or $\frac{27\times 0.12}{1000}=\frac{30\times\frac{9.15}{M}}{1000}\times y$ (1)

$n_{eq}KMnO_4 = n_{eq}$ Oxalate

or $\frac{36\times 0.12}{1000}=\frac{30\times\frac{9.15}{M}}{1000}\times 2z$ (2)

From charge conservation,

$x+y=2z$ (3)

and molar mass, $M=39x+y+88z+18n$ (4)

Solving (1), (2), (3) and (4), we get: $x:y:z=1:3:2$ and $n=2$.

6.

$$\underset{\frac{200\times 0.1}{1000}M}{5Fe^{2+}} + \underset{\frac{200\times 0.1}{1000}M}{MnO_4^-} + \underset{\frac{600\times 1.0}{1000}M}{8H^+} \rightarrow 5Fe^{3+} + Mn^{2+} + 4H_2O$$

$$\left[H^+\right]_{final} = \frac{600}{1000}-\frac{8}{5}\times\frac{20}{1000}=\frac{568}{1000}M$$

7. Moles of KOH used in saponification

$=\frac{25\times 0.4}{1000}-\frac{8.0\times 0.25}{1000}\times 2=6\times 10^{-3}$

$\therefore$ Mass of KOH used $=6\times 10^{-3}\times 56=0.336$ g

$\therefore$ Saponification number $=\frac{0.336\times 10^3}{1.5}=224$

8. $n_{eq}$ Oxalic acid $= n_{eq}MnO_2 + n_{eq}KMnO_4$

or $\frac{50\times 1.0}{1000}=n_{MnO_2}\times 2+\frac{320\times 0.1}{1000}$

$\Rightarrow n_{MnO_2}=9\times 10^{-3}$

$MnO_2 \rightarrow MnO + \underset{\text{available oxygen}}{\text{'O'}}$

$\therefore$ Percentage of available oxygen

$=\frac{9\times 10^{-3}\times 16}{1.6}\times 100=9\%$

9. $n_{eq}SeO_3^{2-} = n_{eq}BrO_3^- \Rightarrow n\times 2=\frac{V_1\times 1/40}{1000}\times 5$ (1)

$n_{eq}AsO_2^- = n_{eq}BrO_3^-$

$\Rightarrow \frac{7.5\times\frac{1}{25}}{1000}\times 2=\frac{V_2\times\frac{1}{40}}{1000}\times 6$ (2)

and $V_1+V_2=20$ (3)

From (1), (2) and (3), we get:

$n=10^{-3}$

$\therefore$ Mass of $SeO_3^{2-}=10^{-3}\times 127$ g $=127$ mg

10. Moles of $V_2O_5=\frac{91}{182}=0.5$

$\therefore$ Moles of $V^{2+}$ formed $=0.5\times 2=1.0$

Now, $n_{eq}V^{2+} = n_{eq}I_2$

or $1.0\times 2=\frac{2}{254}\times 2\Rightarrow w=254$ g

11. $n_{eq}\ I_2 = n_{eq}C_6H_8O_6 + n_{eq}Na_2S_2O_3$

or $\frac{10\times 0.025}{1000}\times 2=\frac{w}{176}\times 2+\frac{2.5\times 0.01}{1000}\times 8$

$\Rightarrow w=0.0264$

$\therefore$ Vitamin C content $=\frac{0.0264}{200}\times 10^6=132$ mg/mL

12. $n_{eq}$ Mohr salt $= n_{eq}Na_2CrO_4 + n_{eq}K_2Cr_2O_7$

or $\frac{1.96}{392}\times 1 = 2a\times 3 + \frac{40\times 0.05}{1000}$

$\therefore$ Moles of $Fe(CrO_2)_2 = a = 5\times 10^{-4}$

$\therefore$ Mass of Cr – present $= \frac{2a\times 52}{0.2}\times 100\% = 26\%$

13. $n_{eq}CuS + n_{eq}Cu_2S + n_{eq}Fe^{2+} = n_{eq}MnO_4^-$

or $\frac{x}{96}\times 6 + \frac{10-x}{160}\times 8 + \frac{200\times 1}{1000}\times 1 = \frac{900\times 0.4}{1000}\times 5$

$\therefore x = 8.$

$\therefore$ Percentage of $CuS = \frac{x}{10}\times 100 = 80\%$

14. Let the mixture contains $x$ moles of $As_2S_3$ and $y$ moles of $As_2S_5$.

$n_{eq}As_2S_3 = n_{eq}I_2$

or $x\times 4 = \frac{20\times 0.05}{1000} \Rightarrow x = 2.5\times 10^{-4}$

Now, $n_{eq}As_2S_5 = n_{eq}I_2 = n_{eq}Na_2S_2O_3\cdot 5H_2O$

or $(x+y)\times 4 = \frac{1.24}{248}\times 1 \Rightarrow y = 1.0\times 10^{-3}$

$\therefore$ Mole percent of $As_2S_3 = \frac{x}{x+y}\times 100 = 20\%$

15. $3Se(s) + 6Ag^+ + 6NH_3 + 3H_2O \rightarrow 2Ag_2Se + Ag_2SeO_3 + 6NH_4^+$

| $3Se(s)$ | $6Ag^+$ | $2Ag_2Se$ | $Ag_2SeO_3$ |
|---|---|---|---|
| $x$ mole | $\frac{45\times 0.02}{1000} = 9\times 10^{-4}$ mole | 0 | 0 |
| 0 | $(9\times 10^{-4} - 2x)$ mole | $\frac{2x}{3}$ mole | $\frac{x}{3}$ mole |

From the question, $\left(9\times 10^{-4} - 2x\right) + 2\times\frac{x}{3} = \frac{10\times 0.01}{1000}$ $\left(n_{Ag^+} = n_{SCN^-}\right)$

$\therefore x = 6\times 10^{-4}$

$\therefore$ Mass of Se per ml $\frac{6\times 10^{-4}\times 80}{2}$ gm = 24 mg

16. $n_{eq}KI = n_{eq}KIO_3$

or $\frac{20\times M}{1000}\times 2 = \frac{30\times\frac{1}{10}}{1000}\times 4$

$\Rightarrow$ Molarity of KI solution = 0.3 M

Now, moles of KI taken $= \frac{50\times 0.3}{1000} = 15\times 10^{-3}$

and moles of KI reacted with

$KIO_3 = 2\times\frac{50\times\frac{1}{10}}{1000} = 10\times 10^{-3}$

$\therefore$ Moles of KI reacted with

$AgNO_3 = 15\times 10^{-3} - 10\times 10^{-3} = 5\times 10^{-3}$

= Moles of $AgNO_3$ present

$\therefore$ Percentage of $AgNO_3 = \frac{5\times 10^{-3}\times 170}{1}\times 100 = 85$

17. Let the original sample contains $x$ moles of $Fe_3O_4$ and $y$ mole of $Fe_2O_3$.

As, $n_{eq}Fe_3O_4 + n_{eq}Fe_2O_3 = n_{eq}I_2 = n_{eq}Na_2S_2O_3$

or $x\times 2 + y\times 2 = \frac{\left(\frac{11.2}{20}\times 100\right)\times 0.5}{1000}\times 1$

$\Rightarrow 2x + 2y = 28\times 10^{-3}$ (1)

Now, moles of $Fe^{2+}$ formed $= 3x + 2y$

As, $n_{eq}Fe^{2+} = n_{eq}KMnO_4$

$$\text{or } (3x+2y)\times 1 = \frac{\left(\frac{12.8}{50}\times 100\right)\times 0.25}{1000}\times 5 \quad (2)$$

$$\Rightarrow 3x+2y = 32\times 10^{-3}$$

From (1) and (2), we get:

$x = 4\times 10^{-3}$ and $y = 10\times 10^{-3}$.

Now, the percentage of $Fe_2O_3$ in sample

$$= \frac{y\times 160}{4}\times 100 = 40\%$$

**18.** $\underset{\frac{1.07}{214}=5\times 10^{-3}\text{ mole}}{IO_3^-} + 5I^- + 6H^+ \rightarrow \underset{15\times 10^{-3}\text{ mole}}{3I_2} + 3H_2O$

Now, $n_{eq}I_2 = n_{eq}Na_2S_2O_3$

$$\text{or } 15\times 10^{-3}\times 2 = \frac{50\times M}{1000}\times 1 \Rightarrow M = 0.6$$

**19.** $\underset{\frac{10\times 0.2}{1000}=2\times 10^{-3}\text{ mole}}{C_2O_4^{2-}} + \underset{2\times 10-3\text{ mole}}{MnO_2} + 4H^+ \rightarrow 2CO_2 + Mn^{2+} + 2H_2O$

$\underset{\frac{2}{5}\times 2\times 10^{-3}\text{ mole} = 0.8\times 10^{-3}\text{ mole}}{2MnO_4^-} + 3Mn^{2+} + 2H_2O \rightarrow \underset{2\times 10^{-3}\text{ mole}}{5MnO_2} + 4H^+$

$\underset{0.8\times 10^{-3}\text{ mole}}{2MnO_4^-} + 6H^+ + \underset{\frac{5}{2}\times 0.8\times 10^{-3}\text{ mole} = 2\times 10^{-3}\text{ mole} = 2\times 10^{-3}\times 34\text{ g}}{5H_2O_2} \rightarrow 2Mn^{2+} + 8H_2O + 5O_2$

$\therefore$ Mass of $H_2O_2$ per 100 ml of solution

$$= \frac{68\times 10^{-3}}{20}\times 100\text{ g} = 340\text{ mg}$$

**20.** In the presence of phenolphthalein,

$$\frac{20\times 0.1}{1000}\times 1 = \frac{V_1\times 0.05}{1000}\times 1 \Rightarrow V_1 = 40\text{ ml}$$

In the presence of methyl orange,

$$\frac{20\times 0.1}{1000}\times 3 = \frac{V_2\times 0.05}{1000}\times 1 \Rightarrow V_2 = 120\text{ ml}$$

$$\therefore V_2 - V_1 = 80\text{ ml}$$

CHAPTER 3

# Gaseous State

## EXERCISE I (JEE MAIN)

### Gaseous Parameters and Gas Laws

**1.** The pressure outside a jet plane flying at high altitude falls considerably below atmospheric pressure at sea level. Therefore, the air inside the cabin must be pressurized to protect the passengers. What is the pressure (in atmosphere) in the cabin if the barometer reading is 688 mm of Hg?

(a) 0.905 atm (b) 6.88 atm
(c) 9.05 atm (d) Data, insufficient

**2.** Dibutylphthalate is often used as a monometer fluid. It has a density of 1.047 g/ml. How many torrs are represented by 1 mm of this fluid?

(a) 0.077 torr (b) 1 torr
(c) 12.98 torr (d) 760 torr

**3.** A liquid of density 1.70 g/ml is filled in the barometer, in place of mercury. What should be the length of liquid column, if another barometer filled with mercury is measuring 75 cm pressure?

(a) 75 cm (b) 600 cm
(c) 9.375 cm (d) 6000 cm

**4.** A barometer is placed in open space of laboratory of weather bureau. If the level of mercury in the barometer falls considerably, then it will represent

(a) normal weather
(b) very slow speed of wind after some time.
(c) very high speed of wind after some time.
(d) very high atmospheric pressure.

**5.** At a constant temperature, a gas occupies a volume of 200 ml at a pressure of 0.720 bar. It is subjected to an external pressure of 0.900 bar. What is the resulting volume of the gas?

(a) 160 ml (b) 320 ml
(c) 80 ml (d) 400 ml

**6.** A volume of 2.5 L of a sample of a gas at 27°C and 1 bar pressure is compressed to a volume of 500 ml keeping the temperature constant, the percentage increase in pressure is

(a) 100% (b) 400%
(c) 500% (d) 80%

**7.** When we inflate cycle tubes, the volume of tube as well as the pressure of air inside the tube increases.

(a) It is an exceptional case of Boyle's law.
(b) It happens because air is not ideal gas.
(c) It happens because mass of air is not constant.
(d) It happens because external force is applied in inflating the tubes.

**8.** The value of Boyle's law constant (in S.I. unit) for 200 ml of gas at 1.2 atm is about

(a) 240 atm-ml (b) 0.24 atm-l
(c) 24.3 J (d) 0.24 J

**9.** In the outer space, the pressure recorded is $5 \times 10^{-4}$ torr. How much outer space could be compressed into 1 $dm^3$ box at a pressure of 1 atm?

(a) $1.52 \times 10^6$ $dm^3$ (b) $4.56 \times 10^6$ $dm^3$
(c) $2.28 \times 10^6$ $dm^3$ (d) $1.14 \times 10^6$ $dm^3$

10. What is the increase in volume, when the temperature of 600 ml of air increases from 27°C to 47°C under constant pressure?

(a) 50 ml (b) 60 ml
(c) 80 ml (d) 40 ml

11. The temperature of a certain mass of a gas is increased from 37°C to 38°C. The volume of the gas

(a) will decrease by $\frac{1}{273}$ of its volume at 37°C.
(b) will increase by $\frac{1}{273}$ of its volume at 273 K.
(c) will increase by $\frac{1}{273}$ of its volume at 310 K.
(d) will remain the same.

12. At 4°C, a gas occupies 200 ml and the same amount of gas at the same pressure occupies 300 ml at 144°C. The absolute zero from this data is

(a) −273°C (b) −276°C
(c) 276 K (d) −3°C

13. When a definite mole of different gases are heated from 20°C to 40°C at constant pressure, their volume

(a) becomes double.
(b) increases in the ratio of their molecular masses.
(c) increases to the same extent.
(d) decreases to the same extent.

14. A sample of gas at 35°C and 1 atmospheric pressure occupies a volume of 3.75 L. At what temperature should the gas be kept, if it is required to reduce the volume to 3.0 L at the same pressure?

(a) −26.6°C (b) 0°C
(c) 3.98°C (d) 28°C

15. A toy balloon can occupy 500 ml at 27°C. The maximum stretching capacity of the balloon is three times of the volume at 27°C. The temperature above which the balloon will burst, if pressure of the balloon does not change is

(a) 300 K (b) 900 K
(c) 625°C (d) 225°C

16. One day, when the temperature and pressure were 300 K and 760 mm, a mass of gas had a volume of 1200 ml. On the next day, the volume had changed to 1218 ml while the pressure was the same. What was the temperature on the next day?

(a) 546 K (b) 304.5 K
(c) 31.5 K (d) 300 K

17. If the pressure of a certain amount of a gas increases by 1% on heating by 1°C at constant volume, then its initial temperature must be

(a) 100 K (b) 100°C
(c) 250 K (d) 250°C

18. A gas has a volume of $V$ cm$^3$ at 10°C. If the pressure is doubled, at what temperature will the volume still be $V$ cm$^3$?

(a) 273°C (b) 300°C
(c) 283°C (d) 293°C

19. A quantity of 10 g of a gas at 1 atm pressure is cooled from 273°C to 273 K keeping its volume constant, the final pressure of the gas will be

(a) 273 atm (b) 0.5 atm
(c) 0.2 atm (d) 0.1 atm

20. A pre-weighed vessel was filled with oxygen at NTP and weighed. It was then evacuated, filled with $SO_2$ at the same temperature and pressure, and again weighed. The mass of oxygen will be

(a) the same as that of $SO_2$.
(b) half that of $SO_2$.
(c) twice that of $SO_2$.
(d) one-fourth of $SO_2$.

21. According to Avogadro's hypothesis, equal volumes of all gases under the same conditions of temperature and pressure will contain

(a) the same number of molecules.
(b) different number of molecules.
(c) the same number of molecules only if their molecular masses are equal.
(d) the same number of molecules if their densities are equal.

22. Four 1-L flasks are separately filled with the gas hydrogen, helium, oxygen and ozone at the same room temperature and pressure. The ratio of total number of atoms of these gases present in the different flasks would be

(a) 1 : 1 : 1 : 1 (b) 1 : 2 : 2 : 3
(c) 2 : 1 : 2 : 3 (d) 1 : 2 : 2 : 1

23. Reducing the pressure from 1.0 atm to 0.5 atm would change the number of molecules in one mole of ammonia to

(a) 75% of initial value.
(b) 50% of initial value.
(c) 25% of initial value.
(d) None of these

24. A closed container contains $N$ molecules at $P$ atm and $T$ K. If the absolute temperature is doubled,
(a) The number of molecules present in the container will become $2N$.
(b) The number of molecules present in the container will become $N/2$.
(c) gas pressure becomes $2P$.
(d) gas pressure becomes $P/2$.

25. An empty bulb weighs 120.0 g. When it is filled by $O_2$ gas, it weighs 120.5 gm. Now, the bulb is evacuated and refilled by an unknown gas to the same pressure and temperature. If the bulb now weighs 120.75 g, then the molecular mass of unknown gas will be
(a) 32 (b) 48
(c) $\frac{64}{3}$ (d) 96

## Ideal Gas Equation

26. Under what conditions will a pure sample of an ideal gas not only exhibit a pressure of 1 atm, also a concentration of 1 mole litre$^{-1}$?
(a) At STP
(b) When $V = 22.4$ L
(c) When $T \approx 12$ K
(d) Impossible under any condition.

27. The temperature of an ideal gas can be raised by
(a) decreasing the volume but keeping the quantity and pressure fixed.
(b) decreasing the quantity but keeping the pressure and volume fixed.
(c) decreasing the pressure but keeping the quantity and volume fixed.
(d) decreasing the pressure and volume but keeping the quantity fixed.

28. The approximate length of the side of a cube containing 850 million gas molecules at 0°C and 1 atm would be ($N_A = 6 \times 10^{23}$)
(a) $3.17 \times 10^{-4}$ cm (b) $3.17 \times 10^{-3}$ cm
(c) 1 mm (d) $5.25 \times 10^{-4}$ cm

29. The pressure of interstellar space is about $4.0 \times 10^{-22}$ atm. If the temperature is 3 K, then how many molecules are present per litre? ($R = 0.08$ L-atm/K-mol)
(a) 2000 (b) 1500
(c) 500 (d) 1000

30. The molar volume of $CO_2$ is maximum at
(a) 273 K and 1 atm (b) 546 K and 1 atm
(c) 273 K and 2 atm (d) 546 K and 2 atm

31. The volume occupied by 1.8 g $H_2O$ at 4°C and 760 mm pressure is
(a) 1.8 L (b) 1.8 ml
(c) 2.27 L (d) 2.98 ml

32. If air is pumped slowly but continuously into a metallic cylinder of strong wall, then what would happen to the air inside the cylinder?
(a) Temperature of air would increase.
(b) Pressure of air would increase.
(c) Pressure of air would decrease.
(d) Temperature and pressure of air would increase.

33. $V$ vs. $T$ curves at different pressures $P_1$ and $P_2$ for fixed amount of an ideal gas are shown below. Which one of the following is correct?
(a) $P_1 > P_2$
(b) $P_1 < P_2$
(c) $P_1 = P_2$
(d) Relation between $P_1$ and $P_2$ depends on the gas.

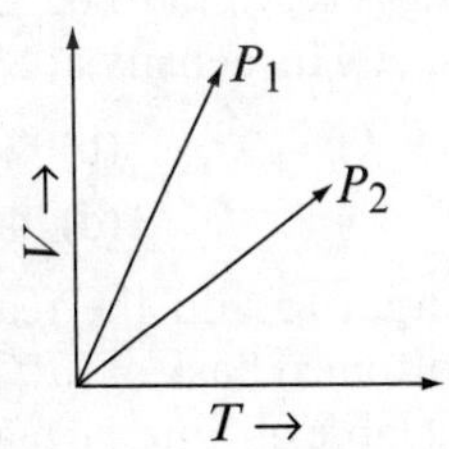


34. At which of the following four conditions, the density of an ideal gas will be maximum?
(a) 0°C and 0.1 atm.
(b) 0°C and 0.2 atm.
(c) 273°C and 0.1 atm.
(d) 273°C and 0.2 atm.

35. A vessel contains 1 mole of $O_2$ gas at a temperature $T$. The pressure of the gas is $P$. An identical vessel containing 1 mole of He gas at a temperature $2T$ has a pressure of
(a) $P/8$ (b) $P$
(c) $2P$ (d) $8P$

36. A 10 L bulb contains an ideal gas of molar mass 81 at 27°C. Some gas is removed from the bulb until the pressure of gas is decreased by 0.82 atm. The mass of gas removed is (R = 0.082 L-atm/K-mol)

(a) 27 g (b) 81 g
(c) 40.5 g (d) 54 g

37. A bulb with chlorine gas at ambient pressure contains 3.5 g at $t$°C. When the bulb is placed in a second thermostat at a temperature 30°C higher and the stopcock of the bulb temporarily opened (and closed again) to restore the initial pressure, the bulb is now found to contain 3.2 g of chlorine. What is the value of $t$°C?

(a) 47K (b) 14.1°C
(c) 320°C (d) 47°C

38. If the absolute temperature of an ideal gas having volume $V$ cm$^3$ is doubled and the pressure is reduced to half, the final volume of gas will be

(a) $0.25V$ (b) $0.50V$
(c) $2V^2$ (d) $4V$

39. A 0.5 dm$^3$ flask contains gas A and 1 dm$^3$ flask contains gas B at the same temperature. If density of A is twice that of B and the molar mass of A is half of B, then the ratio of pressure exerted by gases is

(a) $P_A/P_B = 2$ (b) $P_A/P_B = 1$
(c) $P_A/P_B = 4$ (d) $P_A/P_B = 3$

40. Density of a gas is found to be 5.46 g/dm$^3$ at 27°C at 2 bar pressure. Its density at STP will be

(a) 1 g/dm$^3$ (b) 2 g/dm$^3$
(c) 3 g/dm$^3$ (d) 4 g/dm$^3$

41. A student forgot to add the reaction mixture to the round-bottomed flask at 27°C but put it on the flame. After a lapse of time, he realized his mistake. Using a pyrometer, he found the temperature of the flask as 477°C. What fraction of air would have been expelled out?

(a) 5/3 (b) 3/5
(c) 2/5 (d) 3/2

42. Pressure of 1 g of an ideal gas A at 27°C is found to be 2 bar. When 2 g of another ideal gas B is introduced in the same flask at same temperature the pressure becomes 3 bar. What is the relationship between their molecular masses?

(a) $2M_B = M_A$ (b) $M_B = M_A$
(c) $M_B = 4M_A$ (d) $M_B = 2M_A$

43. A compound exists in the gaseous phase both as monomer (A) and dimer (B). The molecular mass of A is 48. In an experiment, 96 g of the compound was confined in a vessel of volume 33.6 L and heated to 546 K. What is the pressure developed if the compound exists as dimer to the extent of 50% by weight under these conditions?

(a) 2 atm (b) 4 atm
(c) 3 atm (d) 0.5 atm

44. A 2.24 L cylinder of oxygen at 0°C and 1 atm is found to develop a leakage. When the leakage was plugged, the pressure dropped to 570 mm of Hg. The number of moles of gas that escaped will be

(a) 0.025 (b) 0.050
(c) 0.075 (d) 0.09

45. A container contains certain gas of mass $m$ at high pressure. Some of the gas has been allowed to escape from the container. After some time, the pressure of the gas becomes half and its absolute temperature becomes two-third. The amount of the gas escaped is

(a) $2m/3$ (b) $m/2$
(c) $m/4$ (d) $m/6$

46. A perfectly expandable balloon filled with helium gas at 27°C and a pressure of 720 mm of Hg has a volume of 100 L. The balloon rises to an altitude where the pressure is 420 mm of Hg and the temperature –63°C. What is the change in the volume of the balloon?

(a) 16 L (b) 20 L
(c) 48 L (d) 120 L

47. There is 10 L of a gas at STP. Which of the following changes keep the volume constant?

(a) 273 K and 2.0 bar pressure.
(b) 546 K and 2.0 bar pressure.
(c) 546 K and 0.5 bar pressure.
(d) 273 K and 0.5 bar pressure.

48. A certain mass of an ideal gas at 9 atm and 30°C is first heated to 131°C at constant volume and then the amount of the gas is increased by 50% at constant volume and temperature. The final pressure of the gas becomes

(a) 9 atm (b) 4.5 atm
(c) 18 atm (d) 13.5 atm

49. A quantity of 10 g of an ideal gas is expanded at constant temperature until its pressure becomes half of the initial pressure. The quantity of 1.25 g of the rarefied gas occupies 50 ml at the same final pressure and temperature. The initial volume of the gas was

(a) 200 ml (b) 400 ml
(c) 800 ml (d) 25 ml

50. One mole of $N_2O_4(g)$ at 300 K is kept in a closed container under 1 atmospheric pressure. It is heated to 600 K when 20% by mass of $N_2O_4(g)$ decomposes to $NO_2(g)$. The resultant pressure is

(a) 1.2 atm (b) 2.4 atm
(c) 2.0 atm (d) 1.0 atm

## Dalton's Law

51. Which of the following mixture of gases at room temperature follows Dalton's law of partial pressure?

(a) $NH_3$, HCl (b) $H_2$, $O_2$
(c) NO, $O_2$ (d) All of these

52. The partial pressures of $N_2$, $O_2$ and $CO_2$ in a vessel are 38 cm of Hg, 190 torr and 0.5 atm, respectively. The total pressure of the mixture at the same temperature is

(a) 0.96 atm (b) 1.02 atm
(c) 1.64 atm (d) 1.25 atm

53. Assume that air is 21% of $O_2$ and 79% of $N_2$ by volume. If the barometric pressure is 740 mm, then the partial pressure of $O_2$ is closest to which one of the following?

(a) 155 mm (b) 310 mm
(c) 580 mm (d) 740 mm

54. Equal masses of ethane and hydrogen are mixed in an empty container at 25°C. The fraction of the total pressure exerted by hydrogen is

(a) 1 : 2 (b) 1 : 1
(c) 1 : 16 (d) 15 : 16

55. A mixture of hydrogen and oxygen at one bar pressure contains 20% by weight of hydrogen. The partial pressure of hydrogen will be

(a) 8 bar (b) 0.8 bar
(c) 4 bar (d) 6 bar

56. Quantities of 16 g of oxygen and 14 g of nitrogen are contained in a closed bottle. The pressure inside the bottle is 4 atm. Now, 8 g $O_2$ gas is removed from the bottle. What will be the new pressure inside the bottle?

(a) 2 atm (b) 4 atm
(c) 6 atm (d) 3 atm

57. The mass of pure $CH_4$ gas has to be mixed with 70 g of pure CO so that the partial pressure of CO is equal to the partial pressure of $CH_4$, is

(a) 40 g (b) 70 g
(c) 16 g (d) 28 g

58. A box of 1 L capacity is divided into two equal compartments by a thin partition, which is filled with 2 g hydrogen and 16 g methane, respectively. The pressure in each compartment is recorded as *P* atm. The total pressure when the partition is removed will be

(a) *P* atm (b) 2*P* atm
(c) *P*/2 atm (d) *P*/4 atm

59. A closed vessel contains equal number of nitrogen and oxygen molecules at a pressure of *P* mm. If nitrogen is removed from the system, then the pressure will be

(a) *P* mm (b) 2*P* mm
(c) *P*/2 mm (d) $P^2$ mm

60. A vessel is filled with a mixture of oxygen and nitrogen. At what ratio of partial pressures will the mass of gases be identical?

(a) $P(O_2) = 0.785\,P(N_2)$
(b) $P(O_2) = 8.75\,P(N_2)$
(c) $P(O_2) = 11.4\,P(N_2)$
(d) $P(O_2) = 0.875\,P(N_2)$

61. The volumes of the two vessels are in the ratio of 2 : 1. One contains nitrogen and the other oxygen at 800 mm and 680 mm pressure, respectively. Determine the resulting pressure when they are connected together.

(a) 760 mm (b) 670 mm
(c) 1140 mm (d) 1480 mm

62. What will be the pressure of the gas mixture when 0.5 L of $H_2$ at 0.8 bar and 2.0 L of oxygen at 0.7 bar are introduced in a 1 L vessel at 27°C?

(a) 3 bar (b) 1.8 bar
(c) 1 bar (d) 18 bar

63. A quantity of 4 g $H_2$ reacts with $9 \times 10^{23}$ $Cl_2$ molecules and forms HCl gas. If the total pressure of the system after reaction is 700 mm, then what is the partial pressure of HCl? Assume complete reaction. ($N_A = 6 \times 10^{23}$)

(a) 200 mm (b) 350 mm
(c) 300 mm (d) 600 mm

64. A vessel has $N_2$ gas saturated with water vapours at a total pressure of 1 atm. The vapour pressure of water is 0.3 atm. The contents of this vessel are completely transferred to another vessel having one-third of the capacity of original volume, at the same temperature. The total pressure of this system in the new vessel is

(a) 3.0 atm (b) 1 atm
(c) 3.33 atm (d) 2.4 atm

65. There is some liquid in a closed bottle. The amount of liquid is continuously decreasing. The vapour in the remaining part

(a) must be saturated.
(b) must be unsaturated.
(c) must be super saturated.
(d) there will be no vapour.

66. A sample of air contains only $N_2$, $O_2$ and $H_2O$. It is saturated with water vapour and pressure is 640 torr. The vapour pressure of water is 40 torr and the molar ratio of $N_2 : O_2$ is 3 : 1. The partial pressure of $N_2$ in the sample is

(a) 540 torr (b) 900 torr
(c) 480 torr (d) 450 torr

67. Oxygen gas is collected by downward displacement of water in a jar. The level of water inside the jar is adjusted to the height of water outside the jar. When the adjustment is made, the pressure exerted by the oxygen is

(a) Equal to the atmospheric pressure.
(b) Equal to the vapour pressure of oxygen at that temperature.
(c) Equal to atmospheric pressure plus aqueous tension at that temperature.
(d) Equal to atmospheric pressure minus aqueous tension at that temperature.

68. At 20°C, the vapour pressure of water is recorded as 22.57 mbar. What will be vapour pressure of water in the apparatus shown after the piston is lowered, thereby decreasing the volume of the gas above liquid to half of the original volume

(a) 45.14 mbar
(b) 22.57 mbar
(c) 11.28 mbar
(d) between 11.28 and 22.57 mbar

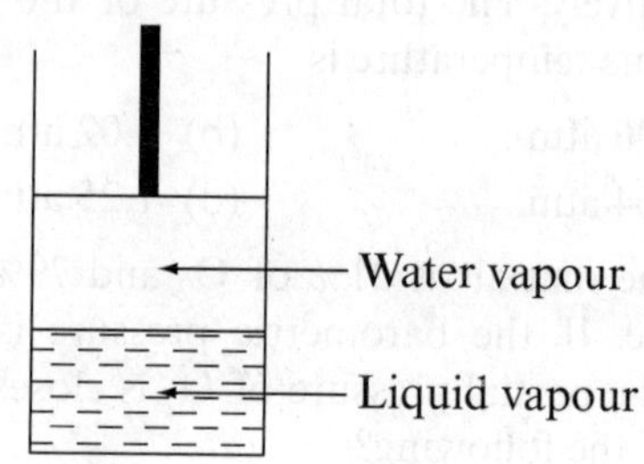


69. If the concentration of water vapour in the air is 1% and the total atmospheric pressure equals 1 atm, then the partial pressure of water vapour is

(a) 0.1 atm (b) 1 mm Hg
(c) 7.6 mm Hg (d) 100 atm

70. A volume of 190.0 ml of $N_2$ was collected in a jar over water at some temperature, water level inside and outside the jar standing at the same height. If barometer reads 740 mm Hg and aqueous tension at the temperature of the experiment is 20 mm Hg, the volume of the gas at 1 atm pressure and at the same temperature would be

(a) 185.0 ml (b) 180.0 ml
(c) 195.0 ml (d) 200 ml

## Graham's Law

**71.** If helium and methane are allowed to diffuse out of the container under similar conditions of temperature and pressure, then the ratio of rate of diffusion of helium to methane is

(a) 2 : 1 (b) 1 : 1
(c) 1 : 2 (d) 4 : 1

**72.** An unknown gas 'X' has rate of diffusion measured to be 0.88 times that of $PH_3$ at the same conditions of temperature and pressure. The gas may be

(a) $C_2H_6$ (b) CO
(c) $NO_2$ (d) $N_2O$

**73.** A quantity of 2 g of hydrogen diffuses from a container is 10 minutes. How many grams of oxygen would diffuse through the same container in the same time under similar conditions?

(a) 0.5 g (b) 4 g
(c) 6 g (d) 8 g

**74.** A gas with formula $C_nH_{2n+2}$ diffuses through the porous plug at a rate one-sixth of the rate of diffusion of hydrogen gas under similar conditions. The formula of gas is

(a) $C_2H_6$ (b) $C_{10}H_{22}$
(c) $C_5H_{12}$ (d) $C_6H_{14}$

**75.** A class consists of 7 rows. The teacher releases laughing gas $N_2O$ from the front row and a mischievous student releases a tear gas of molecular weight 176 from the rear end. The student of which row from the first will weep and smile simultaneously?

(a) Fifth (b) First
(c) Second (d) Third

**76.** A balloon is filled with $N_2O$ is pricked with a sharp point and quickly plunged into a tank of $CO_2$ under the same pressure and temperature. The balloon will

(a) be enlarged
(b) shrink
(c) remain unchanged in size
(d) collapse completely

**77.** A certain gas diffuses from two different vessels A and B. The vessel A has a circular orifice while vessel B has square orifice of length equal to the radius of the orifice of vessel A. The ratio of the rates of diffusion of the gas from vessel A to vessel B, assuming same temperature and pressure is

(a) $\pi : 1$ (b) $1 : \pi$
(c) 1 : 1 (d) $2 : \pi$

**78.** A volume of 180 ml of hydrocarbon takes 15 minute to diffuse. Under the same conditions, 120 ml of sulphur dioxide takes 20 minutes. The molecular weight of hydrocarbon is

(a) 16 (b) 32
(c) 48 (d) 64

**79.** In a glass tube of 1 m length and uniform cross section, $NH_3$(g) from one end and $BF_3$(g) from the other end are sent at a time. The distance from the ammonia end where the addition compound, $BF_3 \cdot NH_3$ will first form is (B = 11; F = 19)

(a) 66.67 cm (b) 33.33 cm
(c) 50.00 cm (d) 60.00 cm

**80.** For 10 minutes each, at 27°C, from two identical holes nitrogen and an unknown gas are leaked into a common vessel of 3 L capacity. The resulting pressure is 4.15 bar and the mixture contains 0.4 mole of nitrogen. The molar mass of the unknown gas is (R = 0.083 L-bar/K-mol)

(a) 112 g $mol^{-1}$ (b) 242 g $mol^{-1}$
(c) 224 g $mol^{-1}$ (d) 448 g $mol^{-1}$

## Kinetic Theory of Gases

**81.** The ratio of root mean square speed of $H_2$ at 50 K and that of $O_2$ at 800 K is

(a) 4 : 1 (b) 2 : 1
(c) 1 : 1 (d) 1 : 4

**82.** If only one gas molecule is taken in a container, the correct relation between average speed $c$, root mean square (RMS) speed $u$, and most probable speed $\alpha$, becomes

(a) $\alpha < c < u$ (b) $\alpha = c = u$
(c) $c < u < \alpha$ (d) $u < c < \alpha$

**83.** The average speed of an ideal gas molecule at 27°C is 0.3 m/s. The average speed at 927°C will be

(a) 0.6 m/s (b) 0.3 m/s
(c) 0.9 m/s (d) 3.0 m/s

**84.** The molecules of a given mass of a gas have RMS speed of 200 m/s at 300 K and 1,00,000 bar pressure. When the absolute temperature is doubled and the pressure is halved, the RMS speed of molecules will become

(a) 200 m/s (b) 400 m/s
(c) 100 m/s (d) $200\sqrt{2}$ m/s

**85.** At STP, the order of the RMS speed of molecules of $H_2$, $N_2$, $O_2$ and HBr gases is

(a) $H_2 > N_2 > O_2 > HBr$
(b) $HBr > O_2 > N_2 > H_2$
(c) $HBr > H_2 > O_2 > N_2$
(d) $N_2 > O_2 > H_2 > HBr$

**86.** Two gases X and Y have their molecular speed in ratio of 3 : 1 at certain temperature. The ratio of their molecular masses $M_x$:$M_y$ is

(a) 1 : 3 (b) 3 : 1
(c) 1 : 9 (d) 9 : 1

**87.** At what temperature will the total translational kinetic energy of 0.30 mole of He gas be the same as the total translational kinetic energy of 0.40 mol of Ar at 400 K?

(a) 533 K (b) 400 K
(c) 300 K (d) 266 K

**88.** The translational kinetic energy of $N$ molecules of $O_2$ is $x$ J at −123°C. Another sample of $O_2$ at 27°C has translational kinetic energy of $2x$ J. The latter sample contains

(a) $N$ molecules of $O_2$.
(b) $2N$ molecules of $O_2$.
(c) $N/2$ molecules of $O_2$.
(d) $N/4$ molecules of $O_2$.

**89.** Helium atom is two times heavier than a hydrogen molecule. At 298 K, the average kinetic energy of a helium atom is

(a) two times that of hydrogen molecule.
(b) same as that of a hydrogen molecules.
(c) four times that of a hydrogen molecules.
(d) half that of a hydrogen molecule.

**90.** The translational kinetic energy of $10^{20}$ molecules of nitrogen at a certain temperature is 0.63 J. The temperature is ($N_A = 6 \times 10^{23}$)

(a) 23°C (b) 303.1°C
(c) 30.1°C (d) 43.3°C

**91.** Consider three identical flasks with different gases.

Flask A: CO at 760 torr and 273 K.
Flask B: $N_2$ at 250 torr and 273 K.
Flask C: $H_2$ at 100 torr and 273 K.

In which flask will the molecules have the greatest average kinetic energy per mole?

(a) A (b) B
(c) C (d) same in all

**92.** Maxwell and Boltzmann have shown that the actual distribution of molecular velocities in all molecules in a given gas depends on

(a) temperature and pressure.
(b) temperature and molecular mass.
(c) volume and temperature.
(d) pressure and molecular mass.

**93.** A gas container observes Maxwellian distribution law of speed. If the number of molecules between the speed 5.0 and 5.1 km per sec at 298 K is $N$, then what would be number of molecules between this range of speed if the total number of molecules in the vessel are doubled?

(a) $2N$ (b) $N$
(c) $2N^2$ (d) $N^2/2$

EXERCISE I

**94.** If $X$ is the total number of collisions which a gas molecule registers with other molecules per unit time under particular conditions, then the collision frequency of the gas containing $N$ molecules per unit volume is

(a) $X/N$ (b) $NX$
(c) $2NX$ (d) $NX/2$

**95.** Which of the following gas molecule has the longest mean free path at the same pressure and temperature?

(a) $H_2$ (b) $N_2$
(c) $O_2$ (d) $Cl_2$

## Real Gases

**96.** Consider an ideal gas contained in a vessel. If the intermolecular interactions suddenly begin to act, then which of the following will happen?

(a) The pressure decreases.
(b) The pressure increases.
(c) The pressure remains unchanged.
(d) The gas collapses.

**97.** A real gas obeying van der Waals equation will resemble ideal gas if the constants

(a) $a$ and $b$ are small.
(b) $a$ is large and $b$ is small.
(c) $a$ is small and $b$ is large.
(d) $a$ and $b$ are large.

**98.** The behaviour of a real gas is usually depicted by plotting compressibility factor $Z$ versus $P$ at a constant temperature. At high temperature and high pressure, $Z$ is usually more than 1. This fact can be explained by van der Waals equation when

(a) the constant $a$ is negligible and not $b$.
(b) the constant $b$ is negligible and not $a$.
(c) both constants $a$ and $b$ are negligible.
(d) both constants $a$ and $b$ are not negligible.

**99.** The van der Waals equation for (½) mole of a gas is

(a) $\left(P+\frac{a}{V^2}\right)(V-b)=RT$

(b) $\left(P+\frac{a}{4V^2}\right)\left(V-\frac{b}{2}\right)=\frac{RT}{2}$

(c) $\left(P+\frac{a}{4V^2}\right)\left(\frac{V-b}{2}\right)=RT$

(d) $\left(P+\frac{a}{4V^2}\right)\left(\frac{V-b}{2}\right)=2RT$

**100.** The numerical value of $a$, the van der Waals constant, is maximum for

(a) He (b) $H_2$
(c) $O_2$ (d) $NH_3$

**101.** The molecular radius for a certain gas = 1.25 Å. What is the reasonable estimate of the magnitude of the van der Waals constant $b$, for the gas?

(a) $0.98 \times 10^{-2}$ litre/mole
(b) $1.43 \times 10^{-2}$ litre/mole
(c) $1.97 \times 10^{-2}$ litre/mole
(d) $3.33 \times 10^{-2}$ litre/mole

**102.** What is the pressure exerted by 32 g of methane in a 250 ml vessel at 300 K using van der Waals equation. Given that $a$ = 2.5 atm L mol$^{-2}$ and $b$ = 0.05 L mol$^{-1}$.

(a) 227.54 atm (b) 299.64 atm
(c) 168.4 atm (d) 328.4 atm

**103.** van der Waals constant $b$ for a gas is 0.0391 L mole$^{-1}$. What is the radius of the gas molecule?

(a) 3.8765 Å (b) 1.5708 Å
(c) 8.3675 Å (d) 5.8075 Å

**104.** When there is more deviation in the behaviour of a gas from the ideal gas equation $PV = nRT$?

(a) At high temperature and low pressure.
(b) At low temperature and high pressure.
(c) At high temperature and high pressure.
(d) At low temperature and low pressure.

**105.** The $PV$–$P$ isotherms of 1 mole of different gases at 273 K, if the limit of pressure tending to zero, converge to a value of $PV$ =

(a) 11.2 L atm (b) 22.4 L atm
(c) Zero (d) 22.4 L

## Liquefaction of Gases

**106.** Only the vapours of a liquid exist
(a) below boiling point.
(b) below critical temperature.
(c) below inversion temperature.
(d) above critical temperature.

**107.** Van der Waals constants $b$ and the corresponding values of critical temperature for three gases P, Q and R are given below

| Gas | Critical temperature | van der Waals constant $b$ |
|---|---|---|
| P | −200°C | 0.03 L/mol |
| Q | −100°C | 0.02 L/mol |
| R | +50°C | 0.01 L/mol |

Which of the gases are liquefiable at a temperature −110°C by application of increasing pressure?
(a) P, Q, R (b) P, Q
(c) Q, R (d) None of these

**108.** For a certain gas obeying van der Waals equation, the critical temperature is 305.5 K and the critical pressure is 48.2 atm. The critical volume is
(a) 0.098 L/mol (b) 0.195 L/mol
(c) 0.231 L/mol (d) 0.321 L/mol

**109.** The van der Waal's parameters for gases W, X, Y and Z are as follows.

| Gas | $a$ (atm $l^2$/mol$^2$) | $b$ (l/mol) |
|---|---|---|
| W | 4.0 | 0.027 |
| X | 8.0 | 0.030 |
| Y | 6.0 | 0.032 |
| Z | 12.0 | 0.027 |

Which one of these gases has the highest critical temperature?
(a) W (b) X
(c) Y (d) Z

**110.** The critical constants $P_C$ and $T_C$ for four gases are as follows.

| Gas | Critical temperature | Critical pressure |
|---|---|---|
| $O_2$ | 153 K | 50 bar |
| $Cl_2$ | 417 K | 76 bar |
| $H_2$ | 33.9 K | 13.5 bar |
| He | 5.12 K | 2.2 bar |

The gas with the smallest value of van der Waals constant $b$ is
(a) $O_2$ (b) $Cl_2$
(c) $H_2$ (d) He

**111.** Under critical conditions, the compressibility factor for a gas is
(a) 3/8 (b) 8/3
(c) 1 (d) 1/4

**112.** The behaviour of temporary gases like carbon dioxide approaches that of permanent gases, such as nitrogen, oxygen, etc., as we go
(a) below critical temperature.
(b) above critical temperature.
(c) above absolute zero.
(d) below absolute zero.

**113.** The van der Waals constants for three gases R, S and T are given below.

| Gas | $a$ (atm $L^2$/mol$^2$) | $b$ (l/mol) |
|---|---|---|
| R | 5.2 | 0.420 |
| S | 3.8 | 0.038 |
| T | 2.3 | 0.032 |

The correct sequence of placing the easily liquefiable gas last and the one difficult to liquefy first is
(a) R, S, T (b) S, T, R
(c) T, S, R (d) S, R, T

**114.** The correct order of the values of critical temperature, $T_C$; Boyle temperature, $T_B$ and inversion temperature, $T_i$ of a real gas is
(a) $T_C < T_B < T_i$ (b) $T_B < T_i < T_C$
(c) $T_B < T_C < T_i$ (d) $T_C < T_i < T_C$

**115.** The critical temperature and critical pressure of a gas obeying van der Waals equation are 27°C and 82.1 atm, respectively. Its van der Waals constant $b$ (in litres mol$^{-1}$) is
(a) 0.500 (b) 0.30
(c) 0.075 (d) 0.0375

EXERCISE I

# EXERCISE II (JEE ADVANCED)

## Section A (Only one Correct)

1. A manometer is connected to a gas containing bulb. The open arm reads 53.3 cm, whereas the arm connected to the bulb reads 15.6 cm. If the barometric pressure is 763 mm mercury, then what is the pressure of gas in atm?

(a) 1.05 atm (b) 1.5 atm
(c) 0.51 atm (d) 1.91 atm

2. A diver ascends quickly to the surface from the bottom of a lake of depth $H$ metre. During this period, he neither exhales nor inhales air. Assuming constant temperature, what would be the fractional increase in volume of his lungs? The atmospheric pressure is $7H$ metre of water.

(a) $\frac{1}{8}$ (b) $\frac{7}{8}$
(c) $\frac{8}{7}$ (d) $\frac{1}{7}$

3. A thin tube of uniform cross section has trapped air columns of lengths 46 and 44.5 cm above and below a middle pellet of mercury 5 cm in length (see figure) when held at an angle of 60° to the vertical. When placed horizontally, the columns are equal in length. The temperature is 27°C. The pressure of air columns, in cm Hg, when the tube is horizontal, is

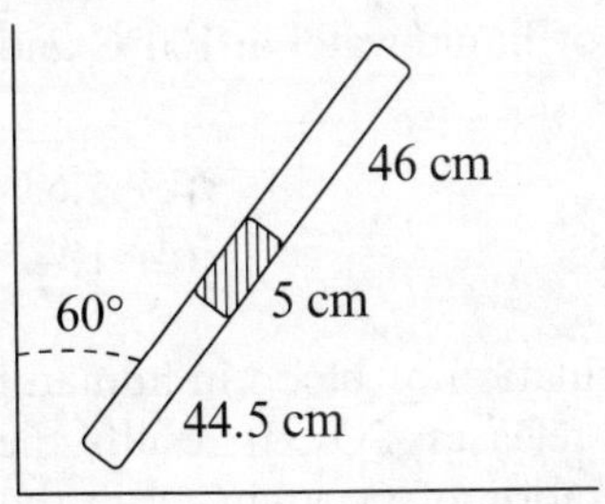


(a) 76.0 cm (b) 71.7 cm
(c) 73.2 cm (d) 75.4 cm

4. A 2 m long tube closed at one end is lowered vertically into water until the closed end is flushed with the water surface. See figure. Calculate the water level height in the tube, h. (Barometric pressure = 1 atm = 10 m of hydrostatic water head. Temperature = 25°C, density of water + 1.00 g/ml. Neglect water vapour pressure)

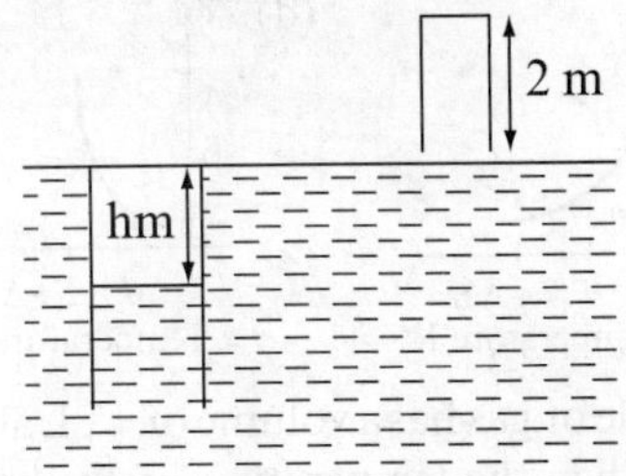


(a) 1.01 m (b) 0.29 m
(c) 1.71 m (d) 0.92 m

5. For which state of matter, the coefficient of cubic expansion is independent to chemical composition?

(a) Solid (b) Liquid
(c) Gas (d) All of these

6. When the volume of a fixed mass of a gas ($Y$–axis) is plotted against temperature in °C ($X$–axis) at constant pressure of 5 atm, the graph is straight line with slope 0.08 L/°C cutting the volume axis at $V_0$. The value of $V_0$ is

(a) 21.84 L (b) 0.08 L
(c) 19.04 L (d) 19.04 ml

7. At 0°C and a pressure of 1000 mm, a given weight of nitrogen occupies a volume of 1.0 L. At −100°C, the same weight of gas under the same pressure occupies a volume of 0.6 L. What is the value of absolute zero in degree celsius?

(a) −250°C (b) 250°C
(c) −273°C (d) 273°C

8. An ideal gas is initially at temperature $T$ and volume $V$. Its volume increases by $\Delta V$ due to an increase in temperature of $\Delta T$, pressure remaining constant. The quantity $\delta = \frac{\Delta V}{V \Delta T}$ varies with temperature as

(a) 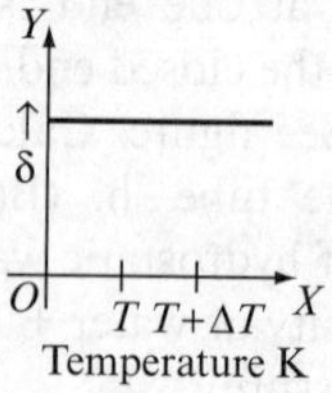


(b) 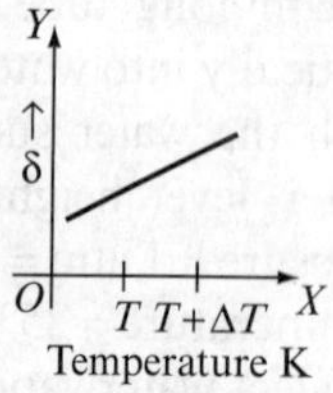


(c) 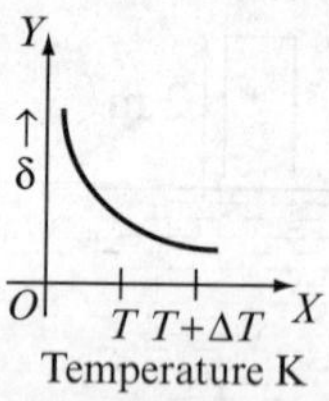


(d) 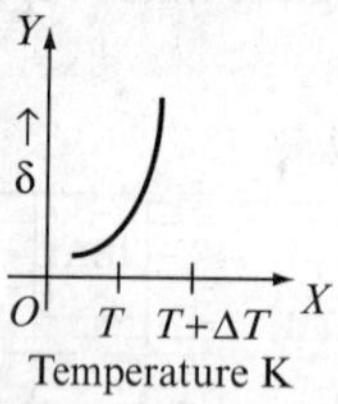


9. A sample of gas has a volume of $V_1$ L at temperature $t_1$°C. When the temperature of the gas is changed to $t_2$°C at constant pressure, the volume of the gas was found to increase by 10%. The percentage increase in temperature (in °C) is

(a) 10% (b) $(10 + 2730t_1^{-1})$
(c) 37.3% (d) $(0.1 + t_1^{-1})$

10. Two flasks A and B of 1 L capacity each contains $SO_2$ and $Br_2$ gases, respectively, maintained at 340 K and pressure of 1.5 atm. If the number of $Br_2$ molecules in flask B is $N$, then the total number of atoms in flask A will be

(a) $N$ (b) $2N$
(c) $N/2$ (d) $3N$

11. A volume of 1 litre of $N_2$ and 7/8 L of $O_2$ at the same temperature and pressure were mixed together. What is the relation between the masses of the two gases in the mixture?

(a) $M_{N_2} = 3M_{O_2}$ (b) $M_{N_2} = 8M_{O_2}$
(c) $M_{N_2} = M_{O_2}$ (d) $M_{N_2} = 16M_{O_2}$

12. The diameter of a balloon filled by 8 moles of He gas is 10 cm. If 7 moles of He gas effuses out in the night, then what should be the diameter of balloon in the next morning? Assume constant pressure and temperature for the gas.

(a) 5 cm (b) $\left(\frac{7}{8}\right)^{1/3} \times 5$ cm
(c) $\left(\frac{8}{7}\right)^{1/3} \times 5$ cm (d) 20 cm

13. A quantity of 22 g dry ice is taken in an open bottle of one litre capacity and the bottle is then closed properly. If the temperature of bottle is maintained at 27°C for long time, in which all the $CO_2$ is sublimed, then the pressure in bottle is (Ambient condition is 1 atm and 27°C)

(a) 13.315 atm (b) 12.315 atm
(c) 11.315 atm (d) 1.0 atm

14. The volume of a filled balloon of mass 20 g is 40 L. The density of gas filled in the balloon is 0.6 g/L and the density of air is 1.3 g/L. The balloon

(a) will not lift upward.
(b) will lift upward with pay load zero.
(c) will lift upward with pay load 6 g.
(d) will lift upwards with pay load 8 g.

15. If the intermolecular forces vanish away, the volume occupied by the molecules contained in 4.5 kg water at 0°C and 1 atm will be given by

(a) 4.5 $m^3$ (b) 5.6 $m^3$
(c) 11.2 $m^3$ (d) 11.2 litre

16. A mixture of CO and $CO_2$ has a density of 1.5 g/L at 27°C and 760 mm pressure. If 1 L of the mixture is exposed to alkali, then what would be the pressure of the remaining gas at the same volume and temperature?

(a) 533 mm (b) 473 mm
(c) 335 mm (d) 595 mm

17. By what factor does water expand when converted into vapour at 100°C and 1 atm pressure? The density of liquid water at 100°C and 1 atm is 0.96 g $cm^{-3}$?

(a) 815 (b) 1.63
(c) 1633 (d) 1042

18. The circulation of blood in human body supplies $O_2$ and releases $CO_2$. It results the variation in concentration of $O_2$ and $CO_2$ in the body. But in an average, blood contains 0.32 g of $O_2$ and 0.88 g of $CO_2$ per litre. What should be the volume occupied by $O_2$ and $CO_2$, as gas, at 1 atm and at body temperature 37°C, assuming 10 litre blood in human body?

(a) 2.545 L, 2.545 L (b) 5.09 L, 2.545 L
(c) 2.545 L, 5.09 L (d) 2.545 L, 7.635 L

EXERCISE II

**19.** A bottle of cold drink has 200 ml of liquid in which the concentration of $CO_2$ is 0.1 M. If $CO_2$ behaves as ideal gas, the volume of $CO_2$ at 0°C and 1 atm equivalent to the one in cold drink is

(a) 0.224 L (b) 0.448 L
(c) 0.112 L (d) 4.48 L

**20.** A plot of the density versus the pressure for air at 27°C has the limiting slope = $1.2 \times 10^{-5}$ kg m$^{-3}$Pa$^{-1}$. The composition of air (contains only $N_2$ and $O_2$) in volume ratio of $N_2$ and $O_2$ is (R = 8.0 J/K-mol)

(a) 80:20 (b) 90:10
(c) 50:50 (d) 60:40

**21.** A quantity of 4 g of oxygen occupies 10 L at a particular pressure and temperature. If the pressure of gas is doubled and absolute temperature is halved, in order to maintain constant volume,

(a) 3 g gas should be removed from the container.
(b) 3 g gas should be added in the container.
(c) 16 g gas should be added in the container.
(d) 12 g gas should be added in the container.

**22.** If at the top of a hill 2000 m above sea level, the atmospheric pressure is 50 cm of Hg and at the sea level, the atmospheric pressure is 74.5 cm of Hg, and you need as much oxygen to breath at sea level as on the top of the hill, then how much faster do you need to breathe at the top of the hill?

(a) 2.44 times (b) 1.49 times
(c) 5 times (d) 7 times

**23.** At 0°C the density of nitrogen at 1 atm is 1.25 kg/m$^3$. The nitrogen which occupied 1500 ml at 0°C and 1 atm was compressed at 0°C and 575 atm and the gas volume was observed to be 3.92 ml, in violation of Boyle's law. What was the final density of this non-ideal gas?

(a) 278 kg/m$^3$ (b) 378 kg/m$^3$
(c) 478 kg/m$^3$ (d) 578 kg/m$^3$

**24.** A spherical air bubble is rising from the depth of a lake where pressure is $P$ atm and temperature is $T$ K. The percentage increase in its radius when it comes to the surface of a lake will be (Assume temperature and pressure at the surface to be respectively $2T$ K and $P/4$)

(a) 100% (b) 50%
(c) 40% (d) 200%

**25.** An ideal gas expands according to the law $P^2V$ = constant. On expansion, the temperature

(a) decreases (b) increases
(c) remains constant (d) None of these

**26.** The ratio of universal gas constant and molar mass of gas is called molar gas constant. The value of molar gas constant is greater for

(a) He (b) $N_2$
(c) $H_2$ (d) Same for all

**27.** The pressure exerted on walls of a 3 L flask when 7 g of $N_2$ is introduced into it at 300 K, should be (assume ideal behaviour of gas)

(a) zero (b) 2.05 atm
(c) 4.10 atm (d) 207.85 atm

**28.** An amount of 1 mole of a gas is changed from its initial state (20 L, 2 atm) to final state (4 L, 10 atm), respectively. If the change can be represented by a straight line in $P$–$V$ curve, then the maximum temperature achieved by the gas in the process is ($R$ = 0.08 L-atm/K-mol)

(a) 900°C (b) 900 K
(c) 627 K (d) 1173°C

**29.** The drain cleaner called 'Drainex' contains small bits of aluminium which react with caustic soda to produce hydrogen. What volume of hydrogen at 27°C and 0.831 bar will be released when 0.15 g of aluminium reacts? (Al = 27)

(a) 250 ml (b) 150 ml
(c) 500 ml (d) 125 ml

**30.** A vessel of 5.0 L capacity contains 1.4 g nitrogen at 1800 K. Assuming that at this temperature, 40% of molecules are dissociated into atoms and the gas is ideal, what is the gas pressure?

(a) 2.07 atm (b) 1.476 atm
(c) 1.05 atm (d) 2.67 atm

**31.** A steel rigid vessel of volume 1 L is filled with a mixture of methane and oxygen at a total pressure of one atm at 27°C. The gas mixture is exploded. What would be the final pressure of the products at 127°C? Aqueous tension at 27°C = 0.04 atm.

(a) 3.13 atm (b) 1.33 atm
(c) 1.29 atm (d) 1.37 atm

**32.** A gaseous mixture of three gases A, B and C has a pressure of 10 atm. The total number of moles of all the gases is 10. If the partial pressures of A and B are 3.0 and 1.0 atm, respectively, and if C has molecular mass of 2.0, then what is the mass of C, in g, present in the mixture?

(a) 6 (b) 8
(c) 12 (d) 3

**33.** A quantity of 5.0 g of a mixture of He and another gas occupies a volume of 2.4 L at 300 K and 760 mm Hg. The gas freezes at 270 K. At 15 K, the pressure of the gas mixture is 19 mm Hg (at the same volume). What is the molecular mass of the gas?

(a) 96 (b) 4.0
(c) 48 (d) 192

**34.** A volume of 10 ml of ethane gas is mixed with 40 ml of oxygen gas in an eudiometer tube at 30°C and fired. When the resulting gases are cooled to 30°C, the volume of eudiometer becomes 26 ml. What is the vapour pressure of water at 30°C? Neglect the volume occupied by liquid water. Pressure is 1 atm and constant throughout.

(a) 1 atm (b) 29.23 mm Hg
(c) 26 mm Hg (d) 32.55 mm Hg

**35.** On a certain day, the vapour pressure is 24 mm of Hg for water vapour in air at 300 K. The saturated vapour pressure is 26.463 mm. How many moles of water vapour per litre of air would be required to saturate the air at this temperature?

(a) 0.1 (b) 0.0132
(c) $1.32 \times 10^{-4}$ (d) 7.60

**36.** A volume of 50 L of dry $N_2$ is passed through 36 g of water at 300 K. After the passage of gas, there is a loss of 1.20 g in water. The vapour pressure of water at 300 K is

(a) 0.033 mm Hg (b) 24.96 mm Hg
(c) 16.05 mm Hg (d) 38.73 mm Hg

**37.** Suppose the inhaled air has partial pressure of water vapour of 5 mm Hg and exhaled air is nearly saturated at body temperature (310 K) with water vapour. The mass of water lost per day by a person assuming that the normal man breaths 10,000 litre per day. Saturated vapour pressure of water at 310 K is 45 mm Hg.

(a) 20.68 g (b) 372.23 g
(c) 418.76 g (d) 46.53 g

**38.** If saturated vapours are compressed slowly and isothermally to half of the initial volume, then the pressure exerted by the vapours will

(a) become four times (b) become double
(c) remain unchanged (d) become half

**39.** Vapour is injected at a uniform rate in a closed vessel which was initially evacuated. The pressure in the vessel

(a) increases continuously.
(b) decreases continuously.
(c) first increases and then decreases.
(d) first increases and then becomes constant.

**40.** Some liquid is taken in an evacuated vessel and the vessel is sealed. Which of the following graph will correctly represent the achievement of equilibrium of liquid with its vapour at constant temperature? ($r_e$ = rate of evaporation, $r_c$ = rate of condensation)

(a)

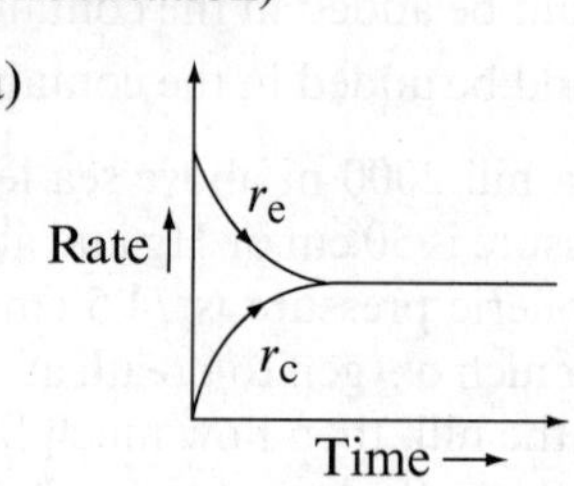


(b)

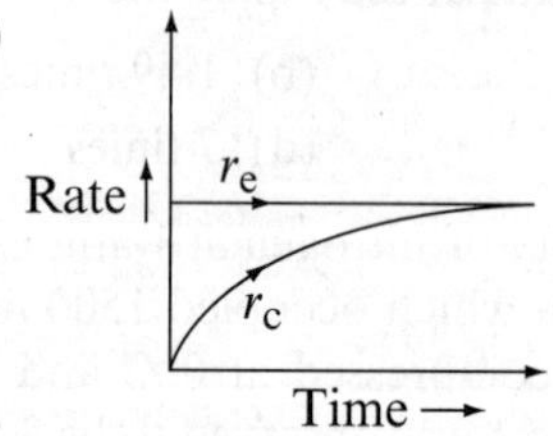


(c)

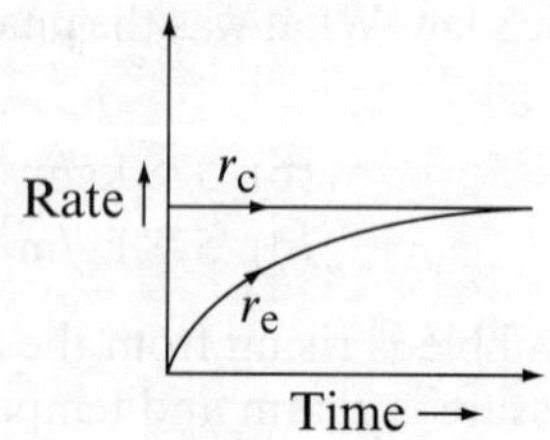


(d)

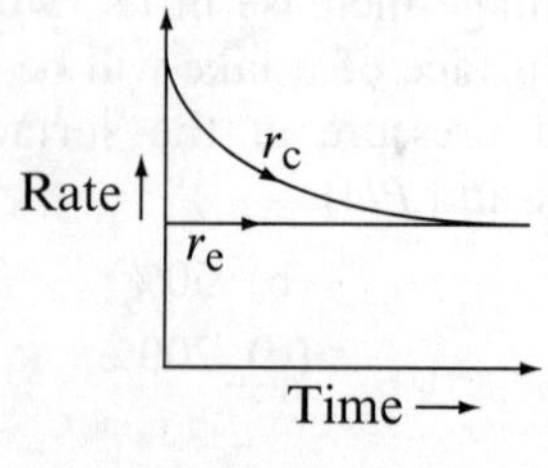

41. The rate of diffusion of two gases X and Y is in the ratio 1 : 5 and that of Y and Z is in the ratio of 1:6. The ratio of the rate of diffusion of Z with respect to X is
(a) 30/1 (b) 1/30
(c) 5/6 (d) 6/5

42. At room temperature, $NH_3(g)$ and HX(g) are passed into a 1 m long tube from two opposite ends at the same pressure. The formation of $NH_4X$ (s) will be closest from the injection of HX end when HX is
(a) HF (b) HCl
(c) HBr (d) HI

43. At 20°C, two balloons of equal volumes and porosity are filled to a pressure of 2 atm, one with 14 kg $N_2$ and other with 1 kg of $H_2$. The $N_2$ balloon leaks to a pressure of half atm in 1 hr. How long will it take for $H_2$ balloon to reach a pressure of half atm?
(a) 1 hour (b) 30 minutes
(c) 16 minutes (d) 3.74 hours

44. Which among the following has rate of effusion less than the moist air?
(a) He (b) Dry air
(c) $NH_3$ (d) Heavy hydrogen

45. A mixture of $CH_4$ and HBr in a vessel is allowed to effuse out through a small hole at the same temperature. What is the mole fraction of $CH_4$ if the initial rates of effusion are the same for both gases?
(a) 0.31 (b) 0.44
(c) 0.5 (d) 0.16

46. In a glass tube of uniform cross section, a mixture of HCl and He gases are sent from one end and a mixture of $NH_3$ and Ar gases are sent from the another end at the same time. The white fumes of $NH_4Cl$ will appear first
(a) at the middle of the tube.
(b) closer to $NH_3$ end.
(c) closer to HCl end.
(d) at the $NH_3$ end.

47. If Avogadro's number were to tend to infinity, the phenomenon of Brownian motion in gas molecules would
(a) remain completely unaffected.
(b) become more vigorous than that observed with the present finite value of Avogadro's number for all sizes of the Brownian particles.
(c) become more vigorous than that observed with the present finite value of Avogadro's number only for relatively large Brownian particles.
(d) become practically unobservable as the molecular impact would tend to balance one another for practically all sizes of Brownian particles.

48. A sample of gas in a closed container of fixed volume is at 250 K and 400 mm of Hg pressure. If the gas is heated to 375 K, then its pressure increases to 600 mm of Hg. By what factor will be the average speed of the molecules increases?
(a) 1.22 (b) 1.50
(c) 2.25 (d) 2.00

49. The RMS speed of oxygen molecules in a gas is $V$. If the absolute temperature is doubled and the oxygen molecules are dissociated into oxygen atoms, then the RMS speed will become
(a) $V$ (b) $\sqrt{2}\,V$
(c) $2V$ (d) $4V$

50. Three closed vessels A, B and C are at the same temperature and contain gases which obey the Maxwellian distribution of velocities. Vessel A contains only $O_2$, B only $N_2$, and C, a mixture of equal quantities of $O_2$ and $N_2$. If the average speed of the $O_2$ molecules in vessel A is $V_1$, then that of the $N_2$ molecules in vessel B is $V_2$, the average speed of the $O_2$ molecules in vessel C is
(a) $(V_1 + V_2)/2$ (b) $V_1$
(c) $\sqrt{V_1V_2}$ (d) $\sqrt{3kT/M}$

Here, $M$ is the mean molar mass of the mixture.

51. When temperature is increased, the difference between most probable velocity, RMS velocity and average velocity
(a) increases (b) decreases
(c) remains the same (d) None of these

52. Two flasks X and Y have capacity 1 and 2 L, respectively, and each of them contains 1 mole of a gas. The temperature of the flask is so adjusted that average speed of molecules in X is twice as those in Y. The pressure in flask X would be
(a) same as that in Y. (b) half of that in Y.
(c) twice of that in Y. (d) 8 times of that in Y.

EXERCISE II

53. When 1 g of an ideal gas A is introduced into an evacuated vessel at 300 K, the pressure was found to be 1 atm. Two grams of another ideal gas B is then added to A and the pressure is now found to be 1.5 atm. What is the ratio between the average speeds of A and B at the same temperature?
(a) 2 : 1
(b) 1 : 2
(c) 1 : 4
(d) 4 : 1

54. For two gases, A and B with molecular weights $M_A$ and $M_B$, it is observed that at a certain temperature $T$, the mean velocity of A is equal to the root mean square velocity of B. Thus, the mean velocity of A can be made equal to the mean velocity of B, if
(a) A is at temperature $T$, and B at $T'$, $T > T'$.
(b) A is lowered to a temperature $T' < T$ while B is at $T$.
(c) both A and B are raised to a higher temperature.
(d) both A and B are placed at lower temperature.

55. The number of gas molecules effusing per second unit area through an orifice on the wall of the container is $\frac{1}{4}.N^*.u_{av}$, where $N^*$ is the number of molecules per unit volume. What is this number under STP conditions for nitrogen gas?
(a) $1.53 \times 10^{24}$ m$^{-2}$s$^{-1}$
(b) $1.03 \times 10^{22}$ m$^{-2}$s$^{-1}$
(c) $3.05 \times 10^{27}$ m$^{-2}$s$^{-1}$
(d) $3.05 \times 10^{21}$ m$^{-2}$s$^{-1}$

56. The RMS speed of a sample of ozone is 'v' m/s at 27°C. When the temperature is increased to 327°C, all the ozone molecules are dissociated into oxygen molecules. What is the RMS speed of new gas sample?
(a) $\sqrt{\frac{2}{3}}v$ m/s
(b) $\frac{1}{\sqrt{3}}v$ m/s
(c) $\frac{1}{\sqrt{2}}v$ m/s
(d) $\sqrt{3}v$ m/s

57. Which of the following gas will have the highest value for translational kinetic energy per g at the same temperature?
(a) Methane
(b) Helium
(c) Nitrogen
(d) Same for all

58. If for two gases of molecular weights $M_A$ and $M_B$ at temperature $T_A$ and $T_B$, $T_AM_B = T_BM_A$, then which property has the same magnitude for both the gases?
(a) Density
(b) Pressure
(c) K.E. per mole
(d) RMS speed

59. At what temperature does the average translational K.E. of a molecule in a gas becomes equal to the K.E. of an electron accelerated through a potential density of 3 V?
(a) 232 K
(b) 2320 K
(c) 23,200 K
(d) 2,32,000 K

60. Which of the following statements is not true?
(a) The ratio of the mean speed to the RMS speed is independent of temperature.
(b) The square of mean speed of the molecules is equal to the mean square speed at the same temperature.
(c) Mean kinetic energy of the gas molecules at any given temperature is independent of the mean speed.
(d) The difference between RMS speed and mean speed at any temperature for different gases decreases as larger and yet larger molar masses are considered.

61. The number of collisions of Ar atoms with the walls of container per unit time, keeping the partial pressure of Ar constant,
(a) increases when the temperature increases.
(b) remains the same when $CO_2$ is added to the container at constant temperature.
(c) increases when $CO_2$ is added to the container at constant temperature.
(d) decreases when the average kinetic energy per molecule is decreased.

62. The ratio of the RMS speeds of $CH_4$ at $T$ and $SO_2$ at 300 K is 4 : 1. The average kinetic energy per mole of $CH_4$ is
(a) 3600 cal
(b) 1200 cal
(c) 900 cal
(d) 2400 cal

63. How many times the average speed of the molecules in a gas becomes when the temperature is raised from 27°C to 159°C?
(a) 1.2
(b) 1.44
(c) 5.89
(d) 2.43

64. In a closed flask of 5 L, 1.0 g $H_2$ is heated from 300 to 600 K. Which of the following statement is not correct?
(a) Pressure of the gas increases.
(b) The rate of collision increases.
(c) The number of moles of gas increases.
(d) The energy of gas molecules increases.

**65.** When the temperature of an ideal gas is increased at constant pressure,

(a) The collision number increases.
(b) The collision frequency increases.
(c) The mean free path increases.
(d) The number of molecules per unit volume increases.

**66.** Modern vacuum pumps permit the pressures down to $4.1 \times 10^{-14}$ atm to be reached at room temperature (27°C). Assuming that the gas exhausted is nitrogen, find the mean distance between the gas molecules remained at this pressure. Diameter of nitrogen molecule is 1.5 Å

(a) $1 \times 10^7$ m (b) $1 \times 10^{-7}$ m
(c) 0.01 cm (d) $1 \times 10^7$ cm

**67.** A vessel contains only two gas molecules. Which of the following relation is correct for RMS speed (u), average speed ($c$) and most probable speed ($\alpha$) for the molecules?

(a) $\alpha < c < u$
(b) $c < u$, $\alpha$ cannot be predicted
(c) $c = u$, $\alpha$ cannot be predicted
(d) $\alpha = c = u$

**68.** Average velocity of molecules of a gas in a container moving only in one dimension is

(a) $\sqrt{\dfrac{8RT}{\pi M}}$ (b) $\dfrac{1}{3} \cdot \sqrt{\dfrac{8RT}{\pi M}}$
(c) Zero (d) Infinite

**69.** The most probable kinetic energy of gas molecule is

(a) $kT/2$ (b) $3kT/2$
(c) $kT$ (d) $kT/4$

**70.** The assumptions of the kinetic theory of gases are most likely to be incorrect for gases under which of the following conditions?

(a) High temperature and high pressure.
(b) High temperature and low pressure.
(c) Low temperature and low pressure.
(d) Low temperature and high pressure.

**71.** Which of the following graph is correct for hydrogen gas at 0°C?

(a)
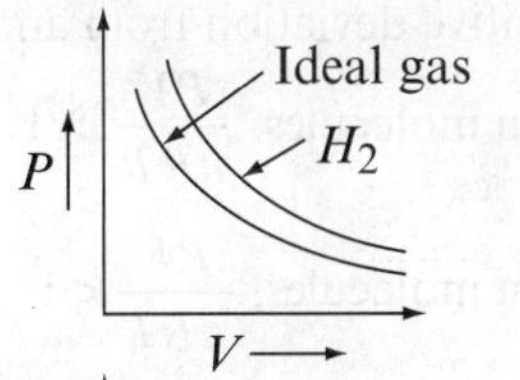


(b)
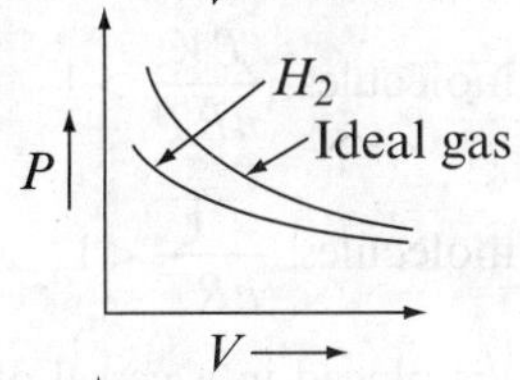


(c)
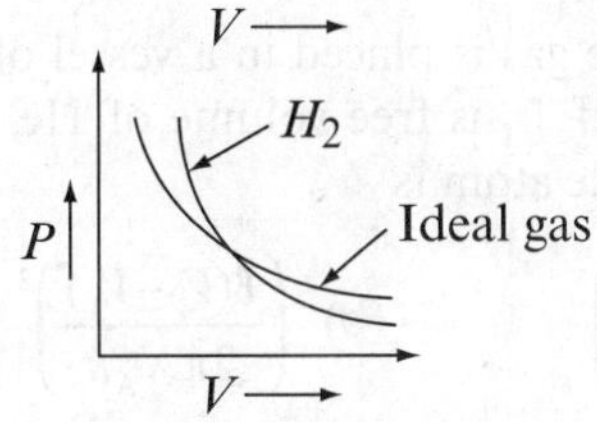


(d)
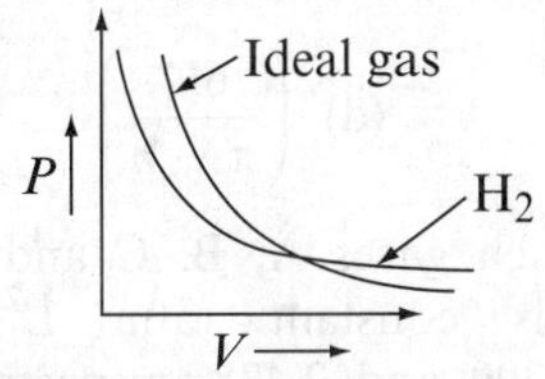


**72.** Which of the following graph is correct for real gases other than hydrogen and helium at 0°C?

(a)
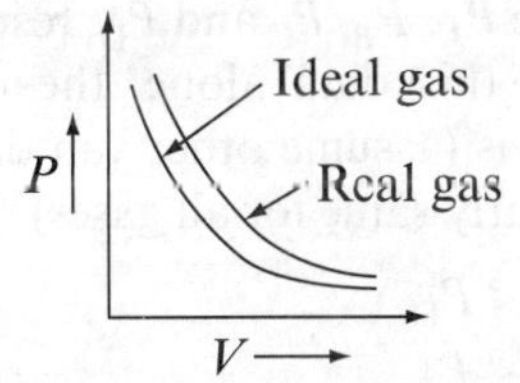


(b)
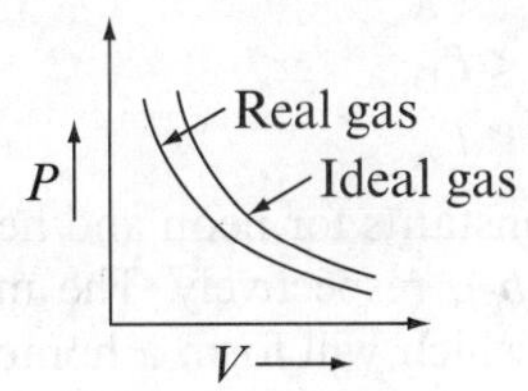


(c)
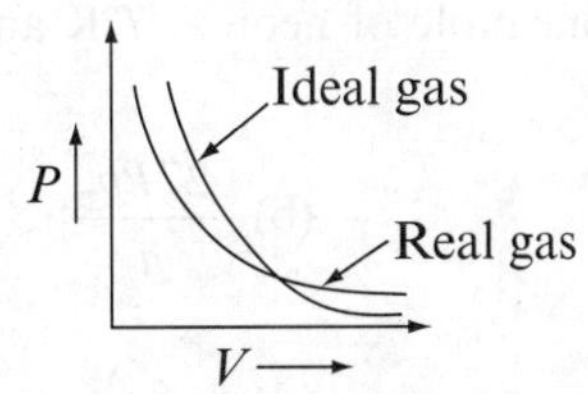


(d)
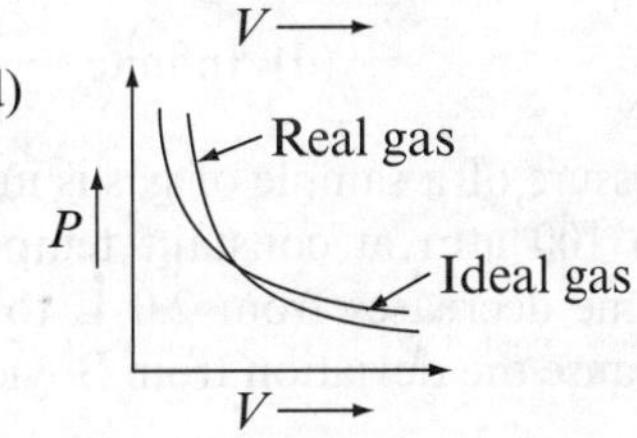

**73.** In the case of positive deviation from an ideal gas

(a) Interactions in molecules, $\frac{PV}{nRT} > 1$

(b) Interactions in molecules, $\frac{PV}{nRT} < 1$

(c) Finite size of molecules, $\frac{PV}{nRT} > 1$

(d) Finite size of molecules, $\frac{PV}{nRT} < 1$

**74.** $n$ moles of He gas is placed in a vessel of volume $V$ L at $T$ K. If $V_1$ is free volume of He, then the diameter of He atom is

(a) $\left(\frac{3V_1}{2\pi N_A n}\right)^{\frac{1}{3}}$ (b) $\left(\frac{3(V-V_1)}{2\pi N_A n}\right)^{\frac{1}{3}}$

(c) $\left(\frac{6(V-V_1)}{\pi N_A n}\right)^{\frac{1}{3}}$ (d) $\left(\frac{6V_1}{\pi N_A n}\right)^{\frac{1}{3}}$

**75.** One mole of each gases A, B, C and D with van der Waals constant (atm $L^2$ $mol^{-2}$) 1.348, 6.823, 4.390 and 2.438, respectively, are kept separately in four different vessels of equal volumes at identical temperature. Their pressures are observed to be $P_A$, $P_B$, $P_C$ and $P_D$, respectively. On the basis of this data alone, the order of pressure of gases is (assume other van der Waals constant to be nearly same for all gases)

(a) $P_A < P_B < P_C < P_D$

(b) $P_B < P_C < P_D < P_A$

(c) $P_B < P_A < P_C < P_D$

(d) $P_C < P_B < P_D < P_A$

**76.** Van der Waals constants for neon and helium are $(a_1, b_1)$ and $(a_2, b_2)$, respectively. The maximum moles of helium which will form a homogeneous mixture with one mole of neon at $T$ K and $P$ atm pressure is

(a) $\frac{4Pb_2^2}{a_2}$ (b) $\frac{27Pb_2^2}{a_2}$

(c) $\frac{4Pb_1^2}{a_1}$ (d) Infinite

**77.** When the pressure of a sample of gas is increased from 0.50 to 100 atm at constant temperature, then its volume decreases from 2.0 L to 13 ml. What could cause the deviation from Boyle's law?

(a) Volume of the gas molecules is a significant fraction of the volume of container at higher pressure.

(b) The force of attraction between the gas molecules is greater when the pressure is higher.

(c) The molecules are dimerized at higher pressure.

(d) The collision of the molecules on the walls of the container are no longer elastic at higher pressure.

**78.** At 320 K and 16 atm, the molar volume of ammonia gas is about 10% less than the molar volume of an ideal gas. The reason that the actual volume is less than the ideal volume is that

(a) the force of attraction between ammonia molecules is significant at this temperature and pressure.

(b) the volume occupied by the ammonia molecules themselves is a significant fraction of the volume of the container at this temperature and pressure.

(c) ammonia molecules move more slowly than predicted by the kinetic theory at this pressure and temperature.

(d) at 16.0 atm, ammonia molecules no longer move randomly.

**79.** The equation of state for a van der Waal gas can be expressed as $Z = 1 + \frac{B}{V_m} + \frac{C}{V_m^2} + \dots$. If the van der Waal constants $a$ and $b$ are 1.344 litre$^2$ atm/mol$^2$ and 0.03 litre mol$^{-1}$, respectively, then the value of second virial coefficient (in litre/mol) for oxygen gas at 0°C is

(a) –0.03 (b) +0.03

(c) –0.09 (d) +0.09

**80.** The equation of state for a gas is $P(V - nb) = nRT$, where $b$ and $R$ are constants. If the pressure and temperature are such that $V_m = 10$ b, then what is the compressibility factor of gas in this condition?

(a) $\frac{10}{9}$ (b) $\frac{9}{11}$

(c) $\frac{11}{10}$ (d) $\frac{10}{11}$

EXERCISE II

**81.** Under what pressure will carbon dioxide have the density of 2.2 g/litre at 300 K? For $CO_2$, $a = 3.6$ atm $L^2$ $mol^{-2}$ and $b = 0.05$ $mol^{-1}L$.

(a) 1.23 atm (b) 1.28 atm
(c) 2.46 atm (d) 0.64 atm

**82.** The virial form of van der Waals gas equation is $PV_m = RT\left(1+\frac{B}{V_m}+\frac{C}{V_m^2}+\ldots\right)$. The second virial coefficient of argon gas at 262.5 K is –1 L $mol^{-1}$. What is the density of argon gas at 262.5 K and 1 atm? Neglect all the terms after second term in the virial forms, under these conditions. ($R = 0.08$ L-atm/K-mol, Ar = 40)

(a) 2.0 g/L (b) 1.905 g/L
(c) 1.818 g/L (d) 1.964 g/L

**83.** Dieterici's equation of state is $P\cdot e^{a/VRT}(v-b) = RT$, where $V$ is the molar volume of gas. What is the pressure exerted by one mole of the gas in a vessel of 410.5 L capacity at 27°C, if $P \to 0$?

(a) 0.06 atm (b) Zero
(c) 60 atm (d) $6\times10^{-5}$ atm

**84.** One mole of a gas occupies 0.5 L at 27°C. The compressibility factor of gas at this temperature is 0.8. If $b = 0.04$ L $mol^{-1}$, then what is the value of Van der Waals constant $a$ for the gas? (R = 0.08 L-atm/K-mol)

(a) 3.44 atm $L^2$ $mol^{-2}$ (b) 2.87 atm $L^2$ $mol^{-2}$
(c) 4.14 atm $L^2$ $mol^{-2}$ (d) 3.82 atm $L^2$ $mol^{-2}$

**85.** At moderate pressure, the compressibility factor for a gas is given as $Z = 1 + 0.35P - \frac{168}{T}P$, where $P$ is in bar and $T$ is in Kelvin. What is the Boyle's temperature of the gas?

(a) 168 K (b) 480 K
(c) 58.8 K (d) 575 K

**86.** Above Boyle's temperature, a gas

(a) may be liquefied by suitable pressure.
(b) shows positive deviation at all pressure.
(c) shows negative deviation at low pressure and positive deviation at high pressure.
(d) shows positive deviation at low pressure and the deviation continuously increases with increase in temperature.

**87.** The expression for compressibility factor for one mole of a van der Waals gas at Boyle temperature is

(a) $1+\frac{b^2}{V(V-b)}$ (b) $1+\frac{b^2}{V^2}$
(c) $1+\frac{b}{V}$ (d) $1-\frac{b^2}{V^2}$

**88.** For an ideal gas, the critical temperature is

(a) 0°C (b) 0 K
(c) Infinite (d) Undefined

**89.** Critical temperatures of NO, $CO_2$ and $CCl_4$ are 178 K, 304 and 550 K, respectively. Which gas is more close to ideal behaviour at 600 K?

(a) NO (b) $CO_2$
(c) $CCl_4$ (d) Unpredictable

**90.** Gases do not liquefy above the critical temperature because above the critical temperature,

(a) the gases become ideal.
(b) the intermolecular attraction vanishes.
(c) the kinetic energy of molecules become so large that the attractive forces become unable to hold the molecules together.
(d) the repulsive forces dominates in the molecules.

**91.** At critical temperature, which of the following graphs is correct?

(a)
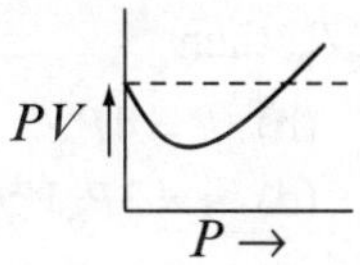


(b)
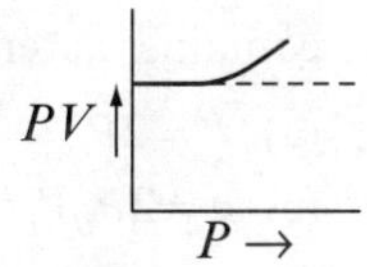


(c)
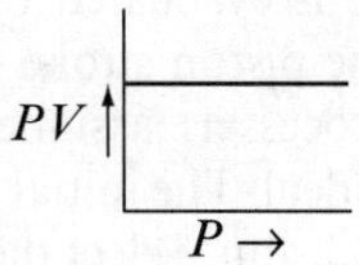


(d)
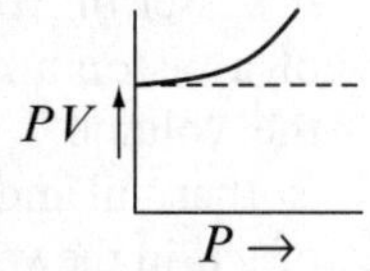


**92.** $CO_2$ and $CH_4$ have critical temperatures of 304 K and 190 K, respectively. The value of $P_C$ for $CO_2$ = 72 atm and $P_C$ for $CH_4$ = 45 atm. The ratio of $b_{CO_2} : b_{CH_4}$ is

(a) 2 : 3 (b) 1 : 1
(c) 2 : 1 (d) 3 : 2

**93.** A gas can never be liquefied at

(a) $T = T_C$ and $P = P_C$
(b) $T < T_C$ and $P = P_C$
(c) $T < T_C$ and $P < P_C$
(d) $T > T_C$ and $P >> P_C$

**94.** Consider the following statements.

1. Molecules of different ideal gases have the same translational kinetic energy at a given temperature.
2. The total translational kinetic energy for two moles of an ideal gas is equal to $3RT$.
3. At critical temperature, the intermolecular forces between gas molecules vanishes.
4. The gas with a larger value of the ratio of critical temperature to critical pressure will have larger value of excluded volume.

(a) 1, 3 and 4 are correct.
(b) 1, 2 and 3 are correct.
(c) 2, 3 and 4 are correct.
(d) 1, 2 and 4 are correct.

**95.** The critical pressure and temperature of a Van der Waal gas are 41.05 atm and 128 K, respectively. The critical volume of the gas is

(a) 0.096 L/mol (b) 0.192 L/mol
(c) 0.683 L/mol (d) 0.118 L/mol

## Section B (One or More than one Correct)

**1.** The value of Boyle's law constant for an ideal gas at a certain temperature depends on

(a) pressure of the gas.
(b) volume of the gas.
(c) amount of the gas.
(d) temperature of the gas.

**2.** Two identical containers, each of volume $V_o$ are joined by a small pipe of negligible volume. The containers contain identical gases at temperature $T_o$ and pressure $P_o$. One container is heated to temperature $2T_o$ while maintaining the other at the same temperature, $T_o$. The common pressure of the gas is $P$ and $n$ is the number of moles of gas in container at temperature $2T_o$, then

(a) $P = 2P_o$ (b) $P = 4P_o/3$
(c) $n = 2P_oV_o/3RT_o$ (d) $n = 3P_oV_o/2RT_o$

**3.** A vessel of volume 10 L is evacuated by means of a piston air pump. One piston stroke captures the volume 1 L. The process is assumed to be isothermal and the gas is ideal. The initial pressure of gas in the vessel was 24.2 atm. Select the correct statement(s) from the following.

(a) The pressure of gas remained in the vessel after first stroke is 22 atm.
(b) The pressure of gas remained in the vessel after second stroke is 20 atm.
(c) The number of strokes needed to reduce the pressure in the vessel $\eta$ times is $\dfrac{\ln \eta}{\ln 1.1}$.
(d) The pressure of gas remained in the vessel after '$n$' strokes is $24.2 \times \left(\dfrac{10}{11}\right)^n$ atm.

**4.** As a gas (insoluble in liquid) is bubbled through a liquid, part of the liquid vaporizes and these vapours are carried off with the gas. As a sample of dry nitrogen gas is bubbled through a liquid, it is found that under identical conditions of temperature and pressure, a definite volume of wet nitrogen gas weighs more than the equal volume of dry nitrogen gas. Hence, the liquid through which dry nitrogen gas was passed may be

(a) water (b) benzene
(c) $CCl_4$ (d) heavy water

**5.** A vessel contains a mixture of $H_2$ and $D_2$ gases. If a pin hole is made in the vessel, then

(a) the mole fraction of $H_2$ in the gas remaining in the vessel will increase with time.
(b) the mole fraction of $D_2$ in the gas remaining in the vessel will increase with time.
(c) the average molecular weight of gas remaining in the vessel will increase with time.
(d) the partial pressure of $H_2$ in the gas remaining in the vessel will increase with time.

**6.** One mole of helium and one mole of neon are taken in a vessel. Which of the following statements are correct?

(a) Molecules of helium strike the wall of vessel more frequently.
(b) Molecules of neon apply more average force per collision on the wall of vessel.
(c) Molecules of helium have greater average molecular speed.
(d) Helium exerts higher pressure than neon.

7. A container (cylindrical, base area $= 821\ cm^2$) fitted with frictionless, massless piston consist of five valves, such as I, II, III, IV and V. The distance of valves from the initial position of piston is 15, 30, 40, 45 and 50 cm, respectively. The initial height of piston from the base of container was 60 cm. These valves open automatically if pressure exceeds over 1.5, 2.2, 2.5, 4.4 and 4.8 atm, respectively. Under the given conditions (shown in the diagram), the system is in state of equilibrium. The piston is now compressed (moved downward) slowly and isothermally. Neglect the volumes of valve connectors.

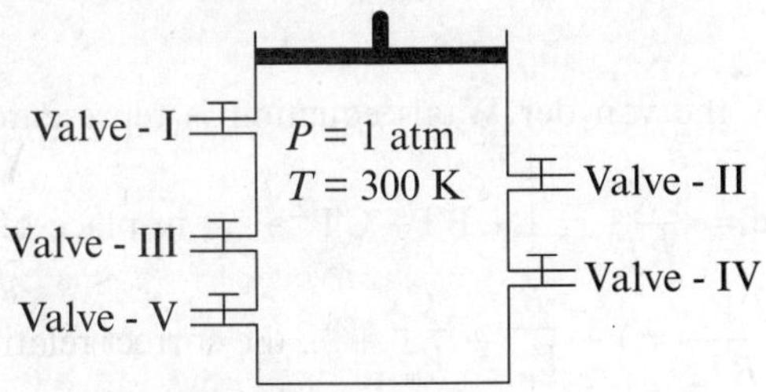


(a) Value - II will be opened first.
(b) As the piston crosses the value which will be opened first, the remaining number of moles in the container are $\frac{5}{3}$.
(c) Value – V will be the second value, which open.
(d) Number of moles will be zero as the piston crosses value – V.

8. Which of the following does not change during compression of a gas at constant temperature?
(a) Density of a gas.
(b) Molecular distance.
(c) Average speed of molecules.
(d) Number of collisions in a gas.

9. The RMS speeds of the molecules of two gases A and B are in the ratio $\sqrt{3} : 2$ at 27°C for A and 127°C for B. Gases A and B may be
(a) $N_2$ and CO (b) $O_2$ and $N_2H_4$
(c) $N_2O$ and $CO_2$ (d) $SO_2$ and $SO_3$

10. The volume of a gas is held constant while its temperature is raised. The pressure the gas exerts on the walls of container increases because
(a) the masses of the molecules increases.
(b) each molecule loses more kinetic energy when it strikes the wall.
(c) the molecules collide with the wall with relatively greater momentum.
(d) the molecules strike the walls more often.

11. Which of the following graphs are correct for ideal gases?

(a) $\frac{1}{dC}\cdot\left(\frac{dN}{N}\right)$

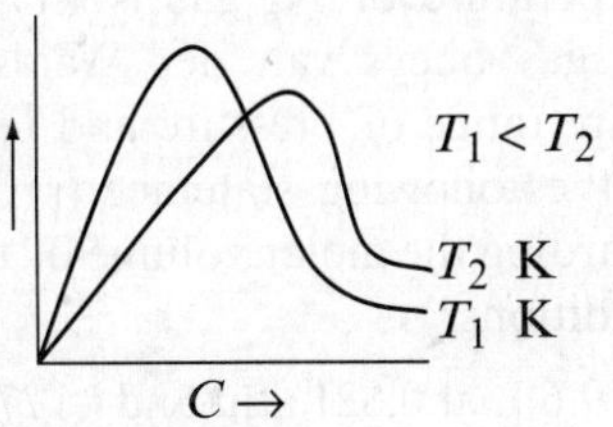


(b) $\frac{1}{dC}\cdot\left(\frac{dN}{N}\right)$

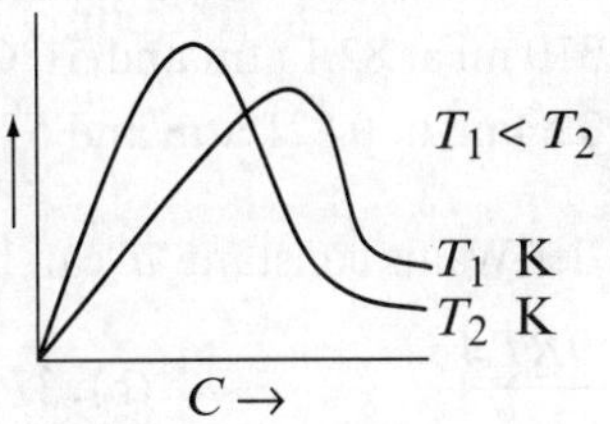


(c) $\frac{1}{dC}\cdot\left(\frac{dN}{N}\right)$

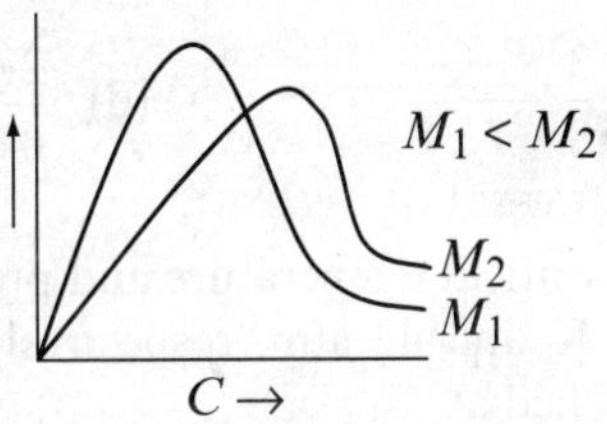


(d) $\frac{1}{dC}\cdot\left(\frac{dN}{N}\right)$

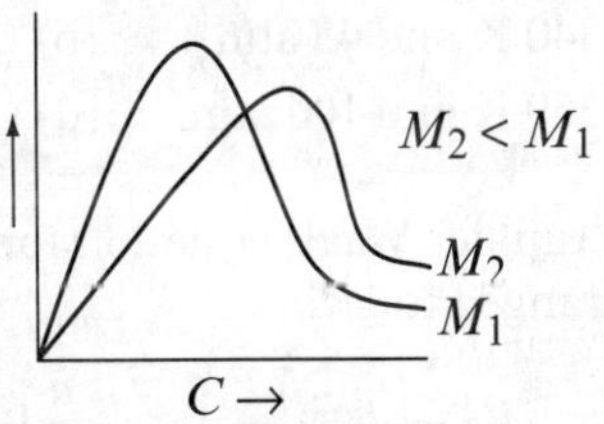


12. Which of the following statement(s) is/are true?
(a) Rise in compressibility factor with increasing pressure is due to *b*.
(b) Average velocity of molecules of an ideal gas in a container moving only in one dimension will be zero.
(c) The fraction of molecules having speeds in the range of $u$ to $u + du$ of a gas of molar mass $M$ at temperature $T$ is the same as that of the gas of molar mass $M/2$ at temperature $T/2$.
(d) Near critical condition, gases behave ideally.

13. At very high pressure, the compressibility factor of 1 mole of a van der Waals gas can be given as
(a) $Z = 1 + \frac{bP}{RT}$ (b) $Z = \frac{V}{V-b}$
(c) $Z = \frac{V+b}{V}$ (d) $Z = 1 - \frac{a}{VRT}$

**14.** $N_2$ gas behaves ideally at its Boyle temperature in the pressure range from 0 to 50 atm. The critical temperature of $N_2$ gas is −177°C. Assume that the gas obeys van der Waals equation in the entire range of pressure and temperature. Which of the following statement(s) must be incorrect regarding the molar volume of the gas at the given conditions?

(a) 9.6 L at 0.821 atm and −177°C.
(b) 40 L at 0.821 atm and 127°C.
(c) 310 ml at 82.1 atm and 51°C.
(d) 32.4 ml at 0.821 atm and 51°C.

**15.** van der Waals constant '*a*' can be expressed as

(a) $\frac{9RT_C V_C}{8}$ (b) $3P_C V_C^2$

(c) $\frac{3}{8}\cdot\frac{P_C V_C}{T_C}$ (d) $\frac{27}{64}\cdot\frac{R^2 T_C^2}{P_C}$

**16.** The critical temperature and pressure of a gas are 340 K and 42 atm, respectively. The gas can be liquefied at

(a) 340 K and 45 atm. (b) 340 K and 40 atm.
(c) 350 K and 100 atm. (d) 320 K and 42 atm.

**17.** The van der Waals equation for a real gas may be rearranged to give

$$V_m^3 - \left(b + \frac{RT}{P}\right)V_m^2 - \frac{a}{P}V_m - \frac{ab}{P} = 0$$

where $V_m$ is the molar volume of the gas. Indicate the correct statement(s) amongst the following.

(a) At temperature greater than $T_C$, there are three values of $V_m$, one real and two imaginary.
(b) At temperature $T_c$, the three real values of $V_m$ are identical.
(c) At temperature less than $T_c$ there are three real values of $V_m$.
(d) At any temperature, there are three values of $V_m$, Where one is real and two are imaginary.

**18.** For a real gas (van der Waals gas), behaving ideally, the pressure may be

(a) $a\cdot b\cdot V_m$ (b) $\frac{V_m}{a\cdot b}$

(c) $\frac{a}{V_m \cdot b}$ (d) $\frac{b}{a\cdot V_m}$

**19.** If the van der Waals equation is represented as $Z = \frac{PV_m}{RT} = 1 + B'P + C'P^2 + ...$, in place of $Z = \frac{PV_m}{RT} = 1 + \frac{B}{V_m} + \frac{C}{V_m^2} + ...$, the correct relation(s) is/are

(a) $B = B'RT$ (b) $B = B'/RT$
(c) $C = B^2 + C'(RT)^2$ (d) $C' = C/(RT)^2$

**20.** Which of the following statement(s) is/are correct?

(a) The physical state of the system at critical condition cannot be defined.
(b) All the physical properties of liquid and gaseous state of a substance become identical at critical condition.
(c) For a gas, there is no distinction between liquid and vapour state at critical condition.
(d) Easily liquefiable gases have higher Boyle's temperature while the gases difficult to liquefy have lower Boyle's temperature.

## Section C (Comprehensions)

### Comprehension I

**A pumped-up bicycle tire has a volume of about 1 litre and when properly inflated, it is at a total pressure of about 8 bar. A typical hand bicycle pump has a piston cross section area of 4 cm$^2$ and a stroke length of 25 cm. The atmospheric pressure is 1 bar ($g = 10$ ms$^{-2}$).**

1. How many strokes of the hand pump assuming no leakage or heating effects would be needed to inflate the tire?
 (a) 40 (b) 50
 (c) 80 (d) 100

2. What force would have to be applied to the hand pump to reach the final pressure?
 (a) 160 N (b) 320 N
 (c) 80 N (d) 640 N

3. If this force were exerted by the weight of an object placed on the pump handle, which is an unusual procedure, then what would be the mass of this object?
 (a) 16 kg (b) 3200 kg
 (c) 32 kg (d) 64 kg

### Comprehension II

**Recent communications with the inhabitants of Neptune have revealed that they have a Celsius type temperature scale, but it is based on the melting point (0°N) and boiling point (100°N) of their most common substance, hydrogen. Further communications have revealed that the Neptunians know about perfect gas behaviour and that they find in the limit of zero pressure, the value of $PV_m$ is 28 L – atm at 0°N and 40 L – atm at 100°N.**

4. What is the value of the absolute zero of temperature on their temperature scale?
 (a) 233.33°C (b) –233.33°C
 (c) –233.33°N (d) 233.33°N

5. What is the value of universal gas constant?
 (a) 0.082 L – atm /K – mol
 (b) 0.12 L – atm /K – mol
 (c) 0.4 L – atm /K – mol
 (d) Infinite

6. What is the volume occupied by 2 moles of a perfect gas at 2 atm and 66.67°N?
 (a) 36 L (b) 24.6 L
 (c) 20.0 L (d) 8.0 L

### Comprehension III

**A glass tube AD of uniform cross section of length 100 cm contains two columns of air (assume ideal behaviour) AB and CD, separately by a column of mercury (BC) of length 20 cm. When the tube is horizontal, AB = 20 cm and CD = 60 cm. When the tube is held vertically with the end 'A' up, the mercury column moves down 10 cm.**

7. Which of the following information regarding the tube is correct?
 (a) Its both ends are open.
 (b) Only end A is closed.
 (c) Only end D is closed.
 (d) Its both ends are closed.

8. What is the pressure of air in the air columns, when the tube was horizontal?
 (a) $P_{AB} = 3 \times P_{CD} = 37.5$ cm Hg
 (b) $P_{AB} = P_{CD} = 37.5$ cm Hg
 (c) $P_{CD} = 3 \times P_{AB} = 37.5$ cm Hg
 (b) $P_{AB} = P_{CD} = 20$ cm Hg

9. What will be the length of the gas column AB when the tube is held vertically with the end D up?
 (a) 30 cm (b) 10 cm
 (c) 15.33 cm (d) 13.88 cm

## Comprehension IV

**Chemical explosions are characterized by the instantaneous release of large quantities of hot gases, which set up a shock wave of enormous pressure (up to 70,000 atm) and velocity (up to 32,000 km/h). For example, the explosion of nitroglycerine ($C_3H_5N_3O_9$) releases four gases, such as A, B, C and D.**

$$n\,C_3H_5N_3O_9 \rightarrow a\,A(g) + b\,B(g) + c\,C(g) + d\,D(g)$$

**Assume that the explosion of 1 mol of nitroglycerine releases gases with a temperature of 1727°C and a volume of 2.9 L at 410.5 atm pressure.**

**10.** How many moles of hot gases are released by the explosion of 0.04 mole of nitroglycerine?

(a) 7.25 (b) 0.29
(c) 2.9 (d) 0.58

**11.** When the products released by explosion of 0.04 mole of nitroglycerine were placed in a 821 ml flask and the flask was cooled to –23°C, product 'A' gets solidified and the pressure inside the flask was 4.75 atm. How many moles of A were present and what is its likely identity? Neglect the volume of solid formed.

(a) 0.1 mol, $H_2$ (b) 2.5 mol, $H_2O$
(c) 0.1 mol, $H_2O$ (d) 0.1 mol, $CO_2$

**12.** When gases B, C and D were passed through a tube of powdered $Li_2O$, gas B reacted to form $Li_2CO_3$. The remaining gases, C and D, were collected in another 821 ml flask and found to have a pressure of 2.1 atm at 27°C. How many moles of B were present and what is its likely identity?

(a) 3.0 mol, $CO_2$ (b) 0.12 mol, $CO_2$
(c) 3.0 mol, CO (d) 0.12 mol, CO

**13.** When gases C and D were passed through a hot tube of powdered copper, gas C reacted to form CuO. The remaining gas D was collected in a third 821 ml flask and it was found to have a mass of 1.68 g and a pressure of 1.8 atm at 27°C. How many moles each of C and D were present and what are their likely identities?

(a) C : 0.01 mol $O_2$, D : 0.06 mol $N_2$
(b) C : 0.06 mol $O_2$, D : 0.01 mol $N_2$
(c) C : 0.02 mol $O_2$, D : 0.06 mol $N_2$
(d) C : 0.07 mol $O_2$, D : 0.06 mol $N_2$

## Comprehension V

**The apparatus shown consists of three temperature jacketed 1.642 L bulbs connected by stopcocks. Bulb A contains a mixture of $H_2O(g)$, $CO_2(g)$ and $N_2(g)$ at 27°C and a total pressure of 570 mm Hg. Bulb B is empty and held at a temperature of −73°C. Bulb C is also empty and is held at a temperature of −193°C. The stopcocks are closed and the volume of the lines connecting the bulbs is zero. $CO_2$ sublimes at −78°C and $N_2$ boils at −196°C. Neglect the volume of any solid, if formed.**

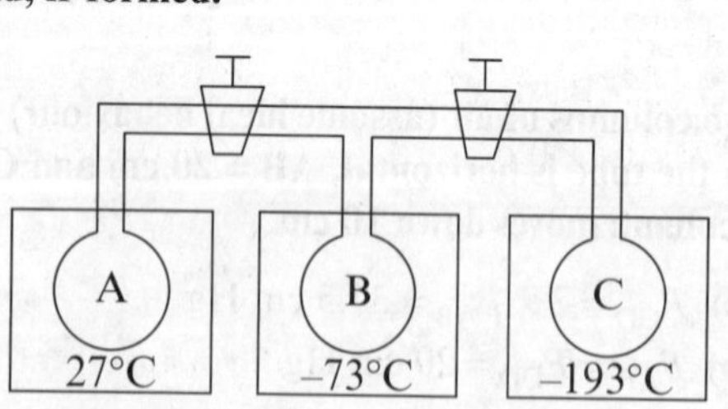


**14.** The stopcock between A and B is opened and the system is allowed to come to equilibrium. The pressure in A and B is now 0.21 atm. What do bulbs A and B contain?

(a) A: $CO_2(g)$,
B: $N_2(g)$, $H_2O(s)$
(b) A: $N_2(g)$,
B: $CO_2(g)$, $N_2(g)$, $H_2O(s)$
(c) A: $CO_2(g)$, $N_2(g)$,
B: $CO_2(g)$, $N_2(g)$, $H_2O(s)$
(d) A: $CO_2(g)$, $N_2(g)$, $H_2O(l)$,
B: $CO_2(g)$, $N_2(g)$

**15.** How many moles of $H_2O$ are in the system?

(a) 0.0075 (b) 0.015
(c) 0.05 (d) 0.035

16. Both stopcocks are opened and the system is again allowed to come at equilibrium. The pressure throughout the system is 22.8 mm Hg. What do bulbs A, B and C contain?
    (a) A: $N_2(g)$,
    B: $N_2(g)$, $H_2O(s)$,
    C: $N_2(g)$, $H_2O(s)$, $CO_2(s)$
    (b) A: $N_2(g)$,
    B: $N_2(g)$, $H_2O(s)$,
    C: $N_2(g)$, $CO_2(s)$
    (c) A: $N_2(g)$,
    B: $N_2(g)$, $H_2O(s)$, $CO_2(s)$,
    C: $N_2(g)$, $H_2O(s)$, $CO_2(s)$
    (d) A: $N_2(g)$,
    B: $N_2(g)$, $H_2O(s)$, $CO_2(g)$,
    C: $N_2(g)$, $H_2O(s)$, $CO_2(s)$

17. How many moles of $CO_2$ are in the system?
    (a) 0.0125 (b) 0.015
    (c) 0.0225 (d) 0.0375

## Comprehension VI

**A closed rigid bulb contains 0.01 mole of inert helium gas and a sample of solid white ammonium chloride, $NH_4Cl$. Assume that the volume of the solid $NH_4Cl$ is negligible compared to the volume of the bulb. The pressure of He is measured at 300 K and is found to be 114 mm Hg. The bulb is then heated to 600 K. All the $NH_4Cl$ decomposes to $NH_3(g)$ and HCl(g). The final total pressure in the bulb after complete decomposition of the solid is 908 mm Hg. Assume ideal behaviour for all gases.**

18. The approximate mass of $NH_4Cl$ present initially in the bulb is
    (a) 0.8 g (b) 1.6 g
    (c) 0.1 g (d) 2.4 g

19. The final partial pressure of $NH_3$ gas in the bulb is
    (a) 908 mm Hg (b) 454 mm Hg
    (c) 680 mm Hg (d) 340 mm Hg

20. The capacity (volume) of bulb is
    (a) 2.24 L (b) 1.642 L
    (c) 0.821 L (d) 3.284 L

## Comprehension VII

**A quantity of 5.0 kg liquid water is poured in an evacuated room of capacity 40 $m^3$ maintained at 300 K. Assume ideal gas behaviour for water vapour. The vapour pressure of water at 300 K is 0.04 atm ($R = 0.08$ L-atm/K-mol).**

21. What percentage of liquid water will vaporize?
    (a) 2.4% (b) 24%
    (c) 36% (d) 12%

22. What should be the minimum volume of the room, if no liquid water is to be present finally in the room?
    (a) 166.67 L (b) 16.67 L
    (c) 166.67 $m^3$ (d) 16.67 $m^3$

## Comprehension VIII

**The leak rate, $-\frac{dp}{dt}$, of a gas is directly proportional to the difference in pressure of the gas in the cylinder and the external pressure. Some ideal gas is taken in a cylinder of 10 L capacity at 20 atm and 300 K. The proportionality constant is 0.001 $sec^{-1}$. The constant external pressure is 1 atm (ln38 = 3.6, ln2 = 0.7, ln10 = 2.3).**

23. What is the pressure of gas in the cylinder after 1 hour?
    (a) 0 (b) 0.5 atm
    (c) 1.5 atm (d) 0.526 atm

24. How many balloons of capacity 2.0 $dm^3$ can be filled at 1 atm and 27°C by using the gas effused out in 1 hour?
    (a) 90 (b) 92
    (c) 93 (d) 97

25. In what time, the pressure of gas in the cylinder becomes 2 atm?
    (a) 2900 s (b) 2300 s
    (c) 3500 s (d) 3200 s

## Comprehension IX

**A U-shaped tube partly filled by mercury is taken as shown in the figure. The cross section area of closed arm is A $cm^2$ and that of open arm is 2A $cm^2$. A pin hole is now made at the top of the closed arm by which leakage of gas starts. The leak rate $-\frac{dP}{dt}$ is directly proportional to the pressure of remaining gas inside the tube over mercury. The proportionality constant is $Ks^{-1}$.**

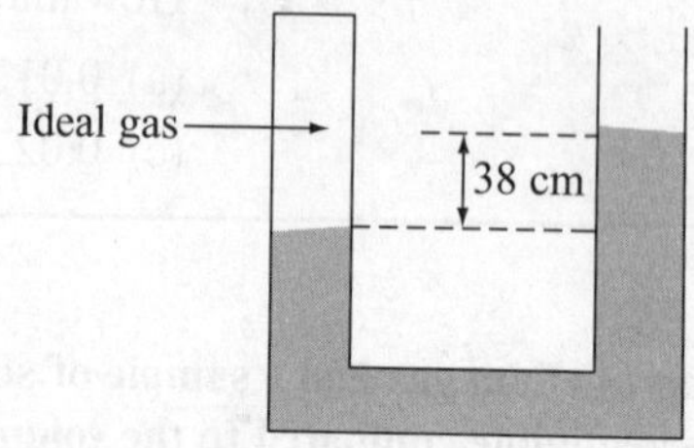


**26.** What will be the expression for pressures of gas above mercury in the closed arm at any time, $t$?

(a) 1.5 atm (b) $1.5\,e^{-kt}$ atm
(c) $1.5\,(e^{-kt} - 1)$ atm (d) $1.5\,(1 - e^{-kt})$ atm

**27.** What will be the increase in level of mercury in the closed arm at any time, $t$?

(a) $38\,(1 - e^{-kt})$ cm (b) $76\,(1 - e^{-kt})$ cm
(c) $(1 - e^{-kt})$ cm (d) $152(1 - e^{-kt})$ cm

**28.** What will be the decrease in level of mercury in the open arm at any time, $t$?

(a) $38\,(1 - e^{-kt})$ cm (b) $76\,(1 - e^{-kt})$ cm
(c) $(1 - e^{-kt})$ cm (d) $152\,(1 - e^{-kt})$ cm

## Comprehension X

**According to Maxwell's distribution law, the fraction of molecules per unit change in speed [velocity distribution function, $f(c)$], traveling with the speed from $C$ to $C + dC$ is given as: $f(c) = \frac{1}{dC}\cdot\left(\frac{dN}{N}\right) = 4\pi.\left(\frac{M}{2\pi RT}\right)^{3/2}.C^2.e^{-MC^2/2RT}$.**

**($R = 8.0$ J/K-mol, ln 2 = 0.7)**

**29.** At what temperature, the velocity distribution function for the oxygen molecules will have maximum value at the speed 400 m/s?

(a) 34.9°C (b) 47°C
(c) 320°C (d) –31.3°C

**30.** At what temperature, the RMS speed of hydrogen molecules exceeds their most probable speed by 400 m/s? ($\sqrt{6} = 2.45$)

(a) 127°C (b) 108.6°C
(c) 400°C (d) 381.6°C

**31.** At what temperature, the speeds of nitrogen gas molecules at $C_1 = 300$ m/s and $C_2 = 600$ m/s are associated with equal value of Maxwell distribution function $f(c)$?

(a) 337.5°C (b) 402°C
(c) 64.5°C (d) –95.75°C

**32.** At what speed of the molecules, the value of the Maxwell distribution function $f(c)$ for the temperature $T$ will be the same as that of the temperature $n$ times higher?

(a) $\sqrt{\frac{3RT\ln n}{M(n-1)}}$

(b) $\sqrt{\frac{3RT\,n\ln n}{M(n-1)}}$

(c) RMS speed

(d) Most probable speed

**33.** On increasing the temperature of a gas, the fraction of molecules having speed equal to most probable speed at that temperature

(a) increases
(b) decreases
(c) remains unchanged
(d) increases or decreases, depending on gas.

**34.** On increasing the molecular mass of gaseous sample at constant temperature, the fraction of molecules having speed equal to most probable speed at that molecular mass

(a) increases
(b) decreases
(c) remains unchanged
(d) increases or decreases, depending on gas.

**35.** What is the ratio of the number of molecules having speeds in the range of $2u_{mp}$ and $2u_{mp} + du$ to the number of molecules having speeds in the range of $u_{mp}$ and $u_{mp} + du$?

(a) $4.e^{-3}$ (b) $e^{-3}$
(c) $4.e^{3}$ (d) 2

## Comprehension XI

**Three flasks A, B and C have equal volumes. A is maintained at 300 K, B at 600 K and C at 900 K. Flask A contains $H_2$ gas, flask B contains $CH_4$ gas and flask C contains $O_2$ gas, and the masses of all gases are same. Assume ideal behaviour of gases.**

**36.** Flask containing greater number of molecules is

(a) A (b) B
(c) C (d) Same in all flask

**37.** Flask with greater number of collisions with the walls per unit area per unit time is

(a) A (b) B
(c) C (d) same for all

**38.** Flask with greater mean free path for the molecules ($r_{H_2} : r_{CH_4} : r_{O_2} = 1:2:2$) is

(a) A (b) B
(c) C (d) Same for all

**39.** Flask with greater total translational kinetic energy is

(a) A (b) B
(c) C (d) same for all

**40.** Flask with greater compressibility factor of gas is

(a) A (b) B
(c) C (d) same fo all

**41.** Flask with greater collision number of the molecules is

(a) A (b) B
(c) C (d) same for all

**42.** Flask with greater average speed for the molecules is

(a) A (b) B
(c) C (d) same for all

## Comprehension XII

**The equation of state for a gas is $P = \frac{RT}{V-b} - \frac{a}{T.V^2}$, where $V$ is the molar volume of the gas and $a$ and $b$ are constants different than van der Waals constant.**

**43.** The temperature above which the gas cannot be liquefied at any pressure is

(a) $\frac{8a}{Rb}$ (b) $\frac{a}{Rb}$
(c) $\frac{8a}{27Rb}$ (d) $\sqrt{\frac{8a}{27Rb}}$

**44.** The minimum pressure needed for liquefaction of gas at the temperature calculated above is

(a) $\frac{a}{27b^2}$ (b) $\sqrt{\frac{a}{27b^2}}$
(c) $\sqrt{\frac{aR}{216b^3}}$ (d) $\sqrt{\frac{aR}{216b^2}}$

**45.** The molar volume of the system at the temperature and pressure calculated above is

(a) $3b$ (b) $2b$
(c) $4b$ (d) zero

## Section D (Assertion – Reason)

The following questions consist of two statements. Mark

(a) If both statements are CORRECT, and **Statement II** is the CORRECT explanation of **Statement I**.

(b) If both statements are CORRECT, and **Statement II** is NOT the CORRECT explanation of **Statement I**.

(c) If **Statement I** is CORRECT, but **Statement II** is INCORRECT.

(d) If **Statement I** is INCORRECT, but **Statement II** is CORRECT.

1. **Statement I:** Under identical conditions of temperature and pressure, equal number of molecules of different gases occupies the same volume.

   **Statement II:** The distance between gas molecules is much larger than the actual dimensions of molecules.

2. **Statement I:** The ratio of volume and absolute temperature is constant for a fixed amount of gas at constant pressure.

   **Statement II:** At constant pressure, the volume of a given mass of a gas increases or decreases by $\frac{1}{273}$ of its volume at 0°C, for each 1°C rise or fall in temperature, respectively.

3. **Statement I:** At constant temperature, if pressure on the fixed mass of a gas is doubled, its density is doubled.

   **Statement II:** At constant temperature, the molecular mass of a gas is directly proportional to the density and inversely proportional to pressure.

4. **Statement I:** If $H_2$ and $Cl_2$ is enclosed separately in the same vessel exert pressures of 100 and 200 mm respectively, then their mixture in the same vessel at the same temperature will exert a pressure of 300 mm.

   **Statement II:** Dalton's law of partial pressure states that the total pressure is the sum of partial pressures.

5. **Statement I:** On increasing the surface area of any liquid at constant temperature, the rate of evaporation increases.

   **Statement II:** Increase in surface area at constant temperature increases the escaping tendency of molecule of liquid at surface.

6. **Statement I:** Graham's law is a limiting law, i.e., it is best applied at low pressures for gases.

   **Statement II:** Graham's law is valid only for ideal gases.

7. **Statement I:** At constant volume, the pressure of a fixed amount of an ideal gas is proportional to its absolute temperature.

   **Statement II:** Frequency of collisions and their impact both increases in proportion to the square root of absolute temperature.

8. **Statement I:** Absolute zero is a theoretically possible temperature at which the volume of the gas becomes zero.

   **Statement II:** The total kinetic energy of the molecules is zero at absolute zero.

9. **Statement I:** In kinetic theory of gases, the volume occupied by the gas molecules is considered negligible in comparison to the volume of gas.

   **Statement II:** The volume of an ideal gas should be only that volume which is available for the gas molecules.

10. **Statement I:** The average speed of gas molecules decreases with increase in molar mass of gas at constant temperature.

    **Statement II:** The average kinetic energy per mole of all the gases is same and it is independent of their molar masses.

11. **Statement I:** The total kinetic energy of vapours formed over liquid water in closed container A and B having free space 1 L and 2 L, respectively, over water at the same temperature is in the ratio 1 : 2, assuming ideal behaviour of vapours.

    **Statement II:** Vapour pressure of a pure substance depends only on temperature.

12. **Statement I:** On increasing the temperature, the fraction of molecules possessing the most probable speed increases.
**Statement II:** On increasing the temperature, the most probable speed of a gas increases.

13. **Statement I:** If the intermolecular attraction between gas molecules in a vessel of fixed volume vanishes at constant temperature, then the pressure of gas will increase.
**Statement II:** The speed at which the gas molecules are colliding at the walls will increase due to the disappearance of attractive forces.

14. **Statement I:** Excluded volume or co-volume equals to $(V - nb)$ for $n$ moles.
**Statement II:** Co-volume depends on the effective size of gas molecules.

15. **Statement I:** Critical temperature is the temperature at which the intermolecular attraction in gas molecules vanishes.
**Statement II:** At critical point, the density of substance in gaseous and liquid states becomes same.

16. **Statement I:** For a fixed amount of a gas, the product $PV$ is always constant.
**Statement II:** Gases deviate more from ideal behaviour at high pressure and low temperature.

17. **Statement I:** At 273 K, hydrogen and helium gases have positive slope at all pressure in PV vs. P graph.
**Statement II:** The Boyle temperature of hydrogen and helium gases is less than 273 K.

18. **Statement I:** For a Van der Waals gas, the constants 'a' and 'b' are independent of temperature.
**Statement II:** The gases obeying Van der Waals equation in all the conditions of pressure and temperature are called Van der Waals gases.

19. **Statement I:** At Boyle temperature, the compressibility factor of a real gas, $Z \geq 1$.
**Statement II:** All the gases tend to approach a value $Z = 1$, when the pressure of gas approaches to zero value at any temperature.

20. **Statement I:** Above critical temperature, a substance cannot exist in liquid state.
**Statement II:** Above critical temperature, the average kinetic energy of gas molecules is always greater than the maximum kinetic energy of a molecule possible in liquid state.

## Section E (Column Match)

1. Match the columns.

| Column I | Column II |
|---|---|
| (A) Boyle's law | (P) $\left(\frac{dP}{dV}\right)_T = -\frac{P}{V}$ |
| (B) Charle's law | (Q) $\left(\frac{dV}{dT}\right)_P = \frac{V}{T}$ |
| (C) Avogadro's law | (R) $\left[\frac{d(PV)}{dP}\right]_T = 0$ |
| (D) Graham's law | (S) $\left(\frac{dV}{dn}\right)_{P,T} = \frac{RT}{P}$ |
| | (T) $-\frac{dP}{dt} = \frac{k}{\sqrt{d}}$ |

2. Match the columns.

| Column I | Column II |
|---|---|
| (A) $H_2$ gas at 300 K | (P) Maximum translational K.E. per mol. |
| (B) $CH_4$ gas at 400 K | (Q) Maximum translational K.E. per gram. |
| (C) $O_2$ gas at 900 K | (R) Minimum translational K.E. per mol. |
| (D) He gas at 600 K | (S) Minimum translational K.E. per gram. |

**3.** Van der waals constant for three different gases are given as follows.

| Gas | $a$ (atm L$^2$/mol$^2$) | $b$ (L/mol) |
|---|---|---|
| X | 3.0 | 0.025 |
| Y | 10.0 | 0.030 |
| Z | 6.0 | 0.035 |

Which of the following is correctly matched?

| Column I | Column II |
|---|---|
| (A) X | (P) Maximum critical temperature |
| (B) Y | (Q) Maximum critical volume |
| (C) Z | (R) Least deviation from ideal behaviour |
| | (S) Maximum critical pressure |

**4.** Match the description in Column I with graph in Column II for $n$ moles of an ideal gas at constant temperature.

| Column I | Column II |
|---|---|
| (A) $\frac{P}{V}$ vs. $P$ | (P) 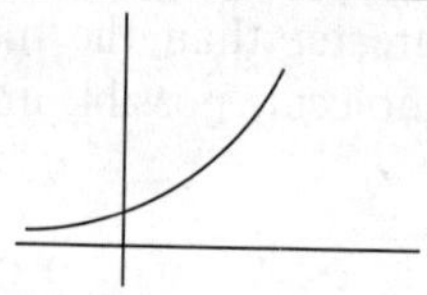 |
| (B) $\frac{P}{V}$ vs. $V$ | (Q) 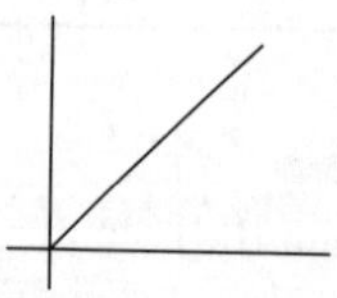 |
| (C) $\frac{V}{P}$ vs. $\frac{1}{P^2}$ | (R) 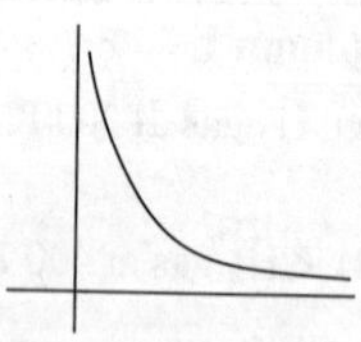 |
| (D) $\frac{P}{V}$ vs. $\log P$ | (S) 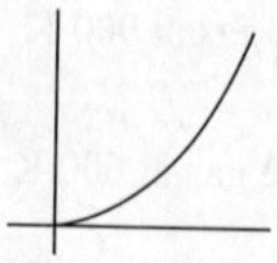 |

**5.** A volume of 6 L $H_2O$ is placed in a closed evacuated room of volume 827 L at the temperature of 300 K. The density of liquid water at 300 K is 1.0 g/ml. The vapour pressure of water at 300 K is 22.8 mm Hg. Neglect the change in volume of liquid water by vaporization.

| Column I | Column II |
|---|---|
| (A) Mass of water vapour formed (in g) | (P) 6 |
| (B) Moles of water vapour formed | (Q) 18 |
| (C) Approximately mass of liquid water left (in kg) | (R) 3 |
| (D) Total moles of atoms in vapour form | (S) 1 |

**6.** Van der Waals constant $a$ and $b$ for a gas are 4.0 atm L$^2$/mol$^2$ and 0.05 L/mol, respectively. If 80 g of the gas (molecular mass = 16) is placed in a 10 L vessel at 300 K, then

| Column I | Column II |
|---|---|
| (A) Pressure correction (in atm) | (P) 0.0625 |
| (B) Ideal volume of gas (in L) | (Q) 1 |
| (C) Actual volume occupied by molecules (in L) | (R) 9.75 |
| (D) Volume correction (in L) | (S) 0.25 |

**7.** Match Column I with Column II for a van der Waals gas.

| Column I | Column II |
|---|---|
| (A) Free volume | (P) Depends upon nature of gas. |
| (B) Critical temperature | (Q) Constant for a particular gas. |
| (C) Boyle's temperature | (R) Depends upon pressure for a particular gas. |
| (D) Compressibility factor | (S) Depends upon the radius of gas molecule. |

**8.** A container contains 2 moles of an ideal gas at 27°C. It is heated to 327°C. Match Column I (parameters) with Column II (increases by a factor).

| Column I | Column II |
|---|---|
| (A) Mean free path, $\lambda$ | (P) 1 (at constant volume) |
| (B) Collision number, $Z_1$ | (Q) $\sqrt{2}$ (at constant volume) |
| (C) Collision frequency, $Z_{11}$ | (R) $\frac{1}{\sqrt{2}}$ (at constant pressure) |
| | (S) 2 (at constant pressure) |

**9.** Match the columns.

| Column I (Pressure exerted by gas in figures) | Column II (Consider atmospheric pressure equal to 1 atm and density of mercury = 13.6 g/ml, if needed) |
|---|---|
| (A) 1.50 atm | (P) 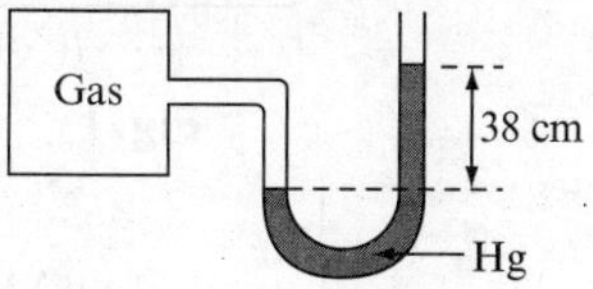  |
| (B) 1.75 atm | (Q) 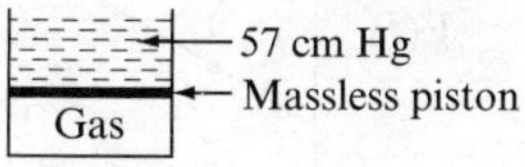  |
| (C) 0.50 atm | (R) 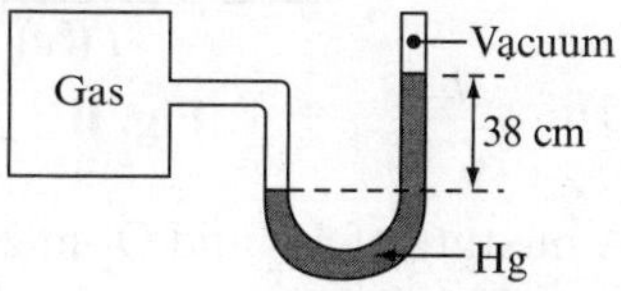  |
| | (S) 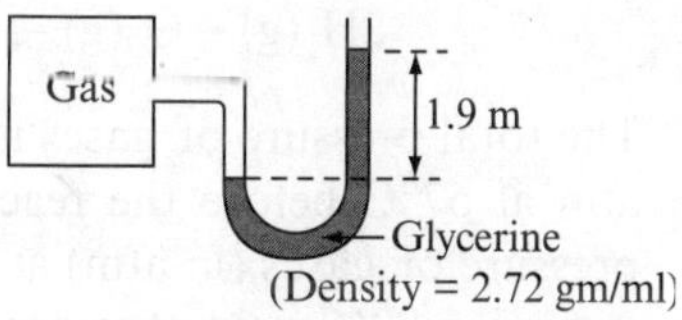  |

**10.** The critical temperature and pressure of a Van der Waals gas is –177°C and 20 atm, respectively. At Boyle's temperature, the gas behaves ideally up to 50 atm. If 0.2 moles of this gas is taken at the temperature and pressure given in Column I, then match with the expected volume of system in Column II ($R$ = 0.08 L-atm/K-mol).

| Column I | Column II |
|---|---|
| (A) −177°C, 20 atm | (P) 821 ml |
| (B) 51°C, 6.48 atm | (Q) 28.8 ml |
| (C) 77°C, 7.0 atm | (R) 760 ml |
| (D) 27°C, 6.0 atm | (S) 800 ml |
| (E) 51°C, 64.8 atm | (T) 85 ml |

## Section F (Subjective)

### Single-digit Integer Type

**1.** A gas is confined in a graduated cylinder enclosed by a movable piston. When the temperature of the room is 27°C, the volume of the gas at atmospheric pressure is found to be 45 ml. The cylinder is then immersed in a bath containing a mixture of liquid and solid benzene, maintained at the normal freezing point of benzene. The piston moves to maintain constant atmospheric pressure. The volume of the gas at the freezing point of benzene is 42 ml. The freezing point of benzene (in °C) is

**2.** A swimming pool (Figure) is conical in shape of diameter 20 m. To measure the depth of the pool, a pipe line having cross sectional area of $\pi$ mm$^2$ has been introduced through the base centre to the depth of the pool. The other end of the pipeline is connected with the gas cylinder. When gases are passed through the pipeline, it has been measured that the volume of each bubble across the pipeline at the surface of the pool is $2\pi$ mm$^3$. Assume the temperature of water to be same as that of atmosphere and 1 atm pressure is equal to 10 m of water column. If the water holding capacity of the swimming pool is $V$ m$^3$, then the value of $\frac{3V}{100\,\pi}$ is

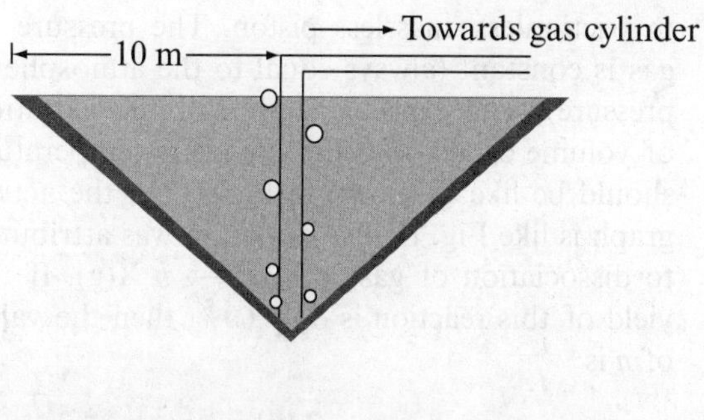

**3.** An iron meteorite was analysed for its isotopic argon content. The amount of $Ar^{36}$ was 1.911 $mm^3$ (STP) per kg of meteorite. If each $Ar^{36}$ atom had been formed by a single cosmic event, then how many such events must there have been per kg of meteorite? If the answer is $X$, then the value of $\frac{X}{10^{16}}$ is ($R = 8.4$ J/K-mol, $N_A = 6.0 \times 10^{23}$)

**4.** What will be the temperature difference (in °C) needed in a hot air balloon to lift $\frac{10.0}{8.314}$ kg of mass? Assume that the volume of the balloon is 91.0 $m^3$, the temperature of the ambient air is 17°C, the pressure is 1.0 bar, and air is an ideal gas with an average molar mass of 29 g/mole.

**5.** A vessel of volume 3.0 L contains ideal gas at the temperature 0°C. After a portion of the gas has been let out, the pressure in the vessel decreases by 0.8 atm (the temperature remaining constant). The gas density under the normal condition (0°C and 1 atm) is 1.25 g/L. The mass of released gas in (g) is

**6.** Two identical vessels are connected by a tube with a valve letting the gas to pass from one vessel into the other if the pressure difference $\Delta P \geq 2.0$ atm. Initially, there was a vacuum in one vessel while the other contained ideal gas at a temperature 300 K and pressure 4.0 atm. Then both vessels were heated to a temperature 600 K. Up to what value will be the pressure (in atm) in the first vessel (which had a vacuum initially) increase?

**7.** A 5.0 L box contains 28.4 g of mixture of gases $C_xH_8$ and $C_xH_{10}$. The pressure of the gas mixture at 300 K is 2.46 atm. The analysis of the gas mixture shows that carbon is 84.5% by mass. The value of $x$ is

**8.** An ideal gas, $X_n$, is taken in a container connected to frictionless, massless piston. The pressure of gas is constant (always equal to the atmospheric pressure). The expected graph for the variation of volume of gas with the change in temperature should be like as shown in Fig. I, but the actual graph is like Fig. II. The deviation was attributed to dissociation of gas as $X_n(g) \rightarrow n\,X(g)$. If the yield of this reaction is only 60%, then the value of $n$ is

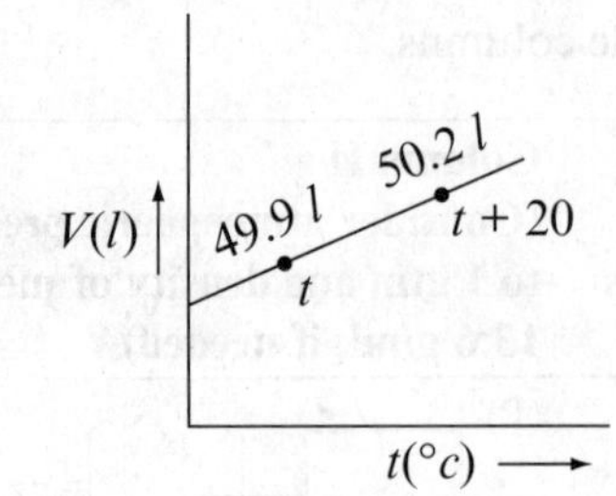


**Fig. I**

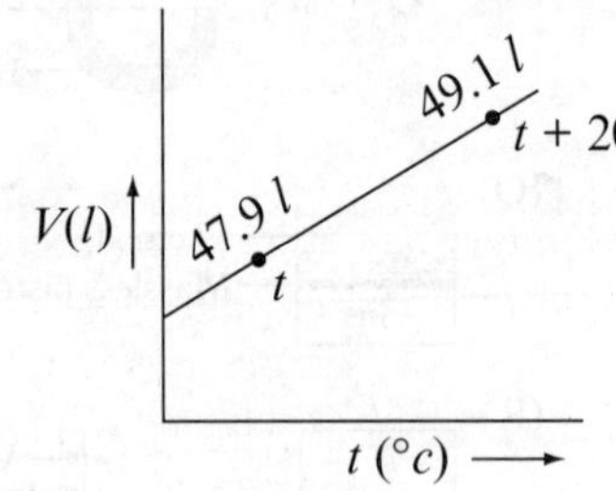


**Fig. II**

**9.** A mixture of $H_2$ and $O_2$ in 2 : 1 mole ratio is used to prepare water vapour by the following reaction.

$$2H_2(g) + O_2(g) \rightarrow 2H_2O(g)$$

The total pressure of gases in the container is 4.5 atm at 57°C before the reaction. The final total pressure of gases (in atm) at 127°C after reaction assuming 80% yield of water vapour is

**10.** What would be the final pressure of $O_2$ in the following experiment? A collapsed polyethylene bag of 30 L capacity is partially blown up by the addition of 10 L of $N_2$ at 0.9 atm at 298 K. Subsequently, enough $O_2$ is pumped into the bag so that at 298 K and external pressure of 1.1 atm, the bag contains full 30 L. If the answer is $x$ atm, then the value of $10x$ is

**11.** Pure water vapour is trapped in a vessel of volume 10 ml. The relative humidity is 40%. The vapour is compressed slowly and isothermally. The volume of vapour (in ml) at which first drop of liquid will form is

**12.** A container with a volume of 20.0 L holds $N_2$ (g) and $H_2O$ (l) at 300 K. The pressure is found to be 1.0 atm. The water is then split into hydrogen and oxygen by electrolysis. After the reaction is complete, the pressure is 1.86 atm. What mass (in g) of water was present initially in the container? The aqueous tension of water at 300 K is 0.04 atm. ($R = 0.08$ L-atm/K-mol)

13. Find the number of diffusion steps required to separate the isotopic mixture initially containing some mass of hydrogen gas and 1 mole of deuterium gas in a 3 litre container maintained at 24.63 atm and 300 K to the final mass ratio of hydrogen to deuterium gas equal to 1 : 4.

14. The equation of state for a certain gas is $P = \frac{RT}{V_m - b} - \frac{a}{V_m}$, where $a$ and $b$ are constants distinct from zero. Ascertain whether the gas has a critical point or not. Answer as **1** if the gas has critical point and as **2**, if no critical point.

15. The Van der Waals constants $a$ and $b$ for a gas of molar mass 164.2 g/mol are 4.105 atm-$L^2$/$mol^2$ and 0.1 L/mol, respectively. The density (in kg/$m^3$) of the gas at 2 atm and 500 K is

## Four-digit Integer Type

1. A faulty barometer has some trapped air above the mercury column due to which it does not give the correct reading. When the atmospheric pressure is 760 mm Hg, the barometer reads 750 mm Hg and when the pressure is 800 mm Hg, the barometer reads 770 mm Hg. The actual pressure (in mm Hg) when the barometer reads 760 mm Hg is

2. Liquefied natural gas (LNG) is mainly methane. A 10.0 $m^3$ tank is constructed to store LNG at –164°C and 1 atm pressure, under which conditions its density is 416 kg/$m^3$. The volume (in $m^3$) of a storage tank capable of holding the same mass of LNG as a gas at 27°C and 1.642 atm pressure is

3. An unspecified quantity of an ideal gas was at initial pressure of 5 atm and temperature of 303 K. The gas is expanded at 303 K until the volume has increased by 60% of the initial value. Next, the quantity of the gas in the vessel is increased by 20% of the initial value while the volume is maintained constant. Finally, the temperature is adjusted at constant volume until the gas pressure is again 5 atm. What is the final temperature in Kelvin?

4. A cylinder contains 64 g of an ideal gas (M = 64) at 27°C and 3 atm. In transportation, the cylinder fell down and a dent was created, i.e., the effective volume of the cylinder decreases. But the valve attached to the cylinder cannot keep the pressure greater than 3 atm, so 8 g of gas was leaked out. The volume of the cylinder before and after the dent was $a$ L and $b$ L, respectively. If the valve was pretty strong, the pressure of gas in the cylinder after incurring a dent was $\frac{cd}{7}$ atm. Temperature remained constant during this process. The value of $abcd$ is (R = 0.08 L-atm/K-mol)

5. A vertical cylinder closed from both ends is equipped with an easily moving piston dividing the volume into two parts, each containing one mole of air. In equilibrium at 320 K, the volume of the upper part is 4.0 times greater than that of the lower part. At what temperature (in kelvin), the volume of upper part becomes 3.0 times than that of lower part?

6. An LPG (liquefied petroleum gas) cylinder weighs 14.8 kg when empty. When full, it weighs 28.8 kg and shows a pressure of 35 atm. In the course of use at 300 K, the weight of the full cylinder is reduced to 23.2 kg. Assume LPG to be $n$-butane with normal boiling point of 273 K. If

   $ab = \frac{1}{100} \times$ the volume (in litre) of the gas used up at the normal usage conditions, and

   $cd$ = final pressure (in atm) inside the cylinder

   then the value of $abcd$ is (R = 0.08 L-atm/K-mol)

7. At room temperature, the following reaction goes to completion.

   $$2\,NO + O_2 \rightarrow 2\,NO_2 \rightarrow N_2O_4$$

   Dimer $N_2O_4$ at 262 K is solid. A 750 ml flask and a 250 ml flask are separated by a stopcock. At 300 K, the nitric oxide in the longer flask exerts a pressure of 1.6 atm and the smaller contains oxygen at 1.2 atm. The gases are mixed by opening the stopcock. After the completion of reaction, the flasks are cooled to 200 K. Neglect the vapour pressure of dimer ($R$ = 0.08 L-atm/K-mol)

   $a$ = 1 (if the gas remaining at 200 K is NO) or 2 (if the gas remaining at 200 K is $O_2$) or 3 (if the gas remaining at 200 K is $NO_2$) or 4 (if the gas remaining at 200 K is $N_2O_4$)

   $bc$ = millimole of gas remaining at 200 K

   $d$ = 10 × the pressure (in atm) of gas remaining at 200 K.

   The value of $abcd$ is

**8.** A barometer tube contains a mixture of air and saturated water vapour in the space above the mercury column. If it reads 70 cm when the actual atmospheric pressure is 76 cm of Hg, then the saturation vapour pressure at room temperature is 1.0 cm of Hg. The tube is now lowered in the reservoir till the space above the mercury column is reduced to half of its original volume. The reading of barometer (in cm) is (Assume that the temperature remains constant)

**9.** Temperature and relative humidity of air are 27°C and 75% on a certain day. Vapour pressure of water at 27°C and 7°C are 24.0 mm and 8.4 mm Hg, respectively. If the fraction of the mass of water vapour that will condense, if the temperature falls to 7°C is $x$, then the value of $1000x$ is

**10.** Suppose you are a manger of a fish drying unit. You have to dry 200 kg of fish containing 36% water on wet basis. The drier specification is mentioned in the figure.

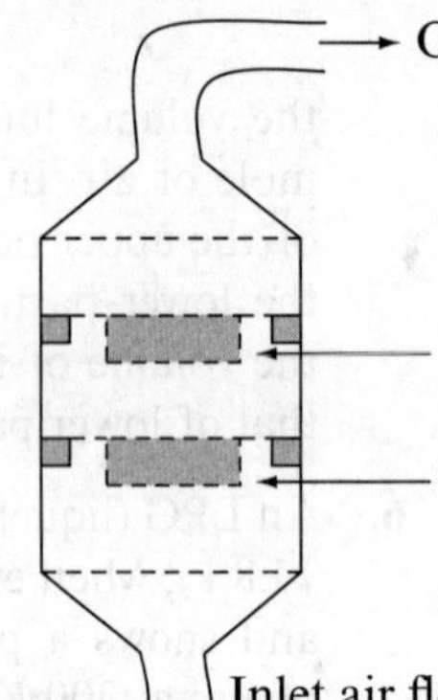


Saturation vapour pressure at 227°C = 38 mm of Hg
Saturation vapour pressure at 127°C = 19 mm of Hg
What is the process time of the drier (in hr)? ($R$ = 0.08 L-atm/K-mol)

**11.** A diver at a depth of 10 m exhales some air by which a bubble of air of volume 24.0 ml is formed. The bubble catches an organism which survives on the exhaled air trapped in the bubble. The organism just inhales the air at the rate of 0.05 millimoles/min and exhales nothing. The atmospheric pressure is 1 atm and the temperature of water is throughout 300 K. The density of water is 1.013 g/ml ($R$ = 0.08 L-atm/K-mol, g = 10 ms$^{-2}$),

$ab$ = volume (in ml) of bubble when it reaches the surface after 10 minutes,

$cd$ = average rate (in $10^{-5}$ mole/min) at which the organism should inhale air so that the volume of bubble remains the same at the depth and at the surface.

The value of $abcd$ is

**12.** A container is divided into two parts in the ratio 1 : 3, by a fixed semipermeable membrane (SPM), which permits the movement of $NO_2$ gas only. At $T$ K, Chamber – I contains only $N_2O_4$ gas at 20 mm Hg pressure and Chamber – II contains small amount of liquid water, but it is sufficient to maintain equilibrium with vapours at any temperature. Now, both the chambers are heated to '$1.2T$' K at which 50% of $N_2O_4$ is dissociated into $NO_2$. The vapour pressure of water at $T$ K and '$1.2T$' K are 20 and 30 mm Hg, respectively. If $ab$ and $cd$ are the final pressures (in mm Hg) of gases in chambers I and II, respectively, then the value of $abcd$ is

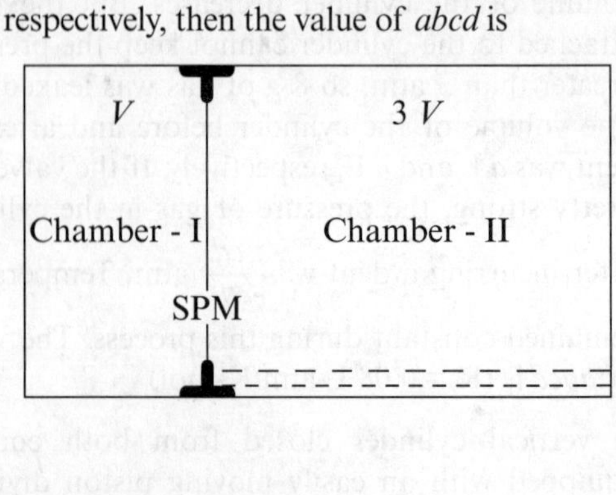


EXERCISE II

13. During one of his adventure, Chacha Chaudhary got trapped in an underground cave which was sealed two hundred years back. The air inside the cave was poisonous, having some amount of CO (5.0%, by volume) in addition to $O_2$ and $N_2$. The safe limit of CO in the atmosphere is less than 0.001% by volume. Sabu, being huge, could not enter into the cave, so in order to save Chacha Choudhary, he started sucking the poisonous air out of the cave by mouth. Each time, he fills his lungs with the cave air and exhaled it out in the surroundings. In the meantime, fresh air from surrounding effused into the cave till the pressure becomes again 1 atm. If each time, Sabu sucked out half of the air present in the cave, then how many times does Sabu need to suck out air in order to safe Chacha Chaudhary?

14. A 50 litre vessel is equally divided into three parts with the help of two stationary (fixed) semi-permeable membranes (SPM). The vessel contains 60 g $H_2$ in the left chamber, 160 g $O_2$ in the middle and 140 g $N_2$ in the right chamber. The left SPM allows transfer of only $H_2$ gas while the right SPM allows the transfer of both $H_2$ and $N_2$. If the ratio of final total pressures in the three chambers is $x : y : 5$, then the value of $(5 \times x \times y)$ is

15. A very long rectangular box is divided into $N$ equal compartments with $(N - 1)$ fixed semipermeable membranes (SPM) numbered from 1 to $(N - 1)$ as shown in the figure. The gases are initially present in only the first compartment and they can pass through only those SPM whose number is less than or equal to their subscript. For example, gas $A_1$ can pass through only first SPM, gas $A_2$ can pass only through first and second SPM, and so on. If initially all gases have same moles and after a long time, the ratio of partial pressures of gas $A_4$ in third compartment to that of gas $A_{N-1}$ in first compartment is 3, then the value of $N$ is

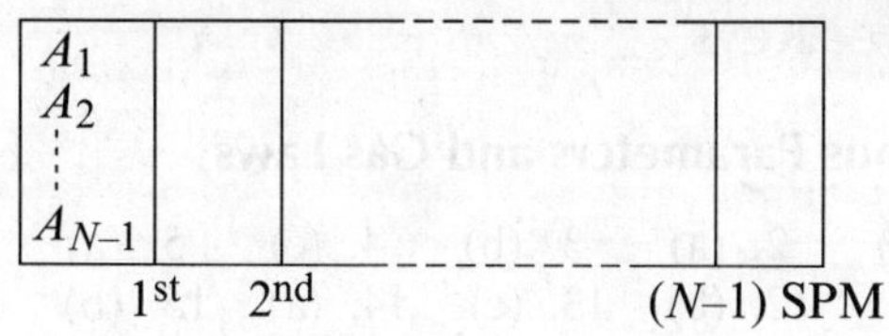


16. A volume of 100 ml of hydrogen were confined in a diffusion tube and exposed to air and at equilibrium a volume of 26 ml of air was measured in the tube. Again, when 100 ml of an unknown gas were placed in the same tube and exposed to air, 130 ml of air were measured in the tube at the equilibrium. The molecular mass of the unknown gas is

17. An argon atom is released from the surface of the earth to travel upwards at 300 K. Assuming that it undergoes no collisions with other molecules, how long (in metre) will it travel before coming to rest? Atomic mass of Ar = 40, g = 10 $ms^{-2}$, $R = 8.4$ J/K-mol).

18. The second virial coefficient of methane can be approximate by the empirical equation,

$$B = a + b \cdot e^{-C/T^2}$$

where $a = -0.2\ \text{bar}^{-1}$, $b = 0.22\ \text{bar}^{-1}$, and $c = 950K^2$. What is the value of the Boyle temperature of methane (in K)? ($ln1.1 = 0.095$)

19. For a Van der Waals gas, the critical pressure and temperature are 73.89 atm and 27°C, respectively. What is the volume (in ml) occupied only by the molecules in a sample of gas containing its 24 moles?

20. The density of a gas (in g/litre) depends on pressure (in atm) at 300 K as

$$d = 4P + 0.02P^2 + 0.001P^3 + \ldots$$

The molar mass of gas (in g/mol) is ($R$ = 0.08 L-atm/K-mol)

## Answer Keys — Exercise I

### Gaseous Parameters and Gas Laws

1. (a) 2. (a) 3. (b) 4. (c) 5. (a) 6. (b) 7. (c) 8. (c) 9. (a) 10. (d)
11. (b) 12. (b) 13. (c) 14. (a) 15. (b) 16. (b) 17. (a) 18. (d) 19. (b) 20. (b)
21. (a) 22. (c) 23. (d) 24. (c) 25. (b)

### Ideal Gas Equation

26. (c) 27. (b) 28. (a) 29. (d) 30. (b) 31. (b) 32. (b) 33. (b) 34. (b) 35. (c)
36. (a) 37. (d) 38. (d) 39. (c) 40. (c) 41. (b) 42. (c) 43. (a) 44. (a) 45. (c)
46. (b) 47. (b) 48. (c) 49. (a) 50. (b)

### Dalton's Law

51. (b) 52. (d) 53. (a) 54. (d) 55. (b) 56. (d) 57. (a) 58. (a) 59. (c) 60. (d)
61. (a) 62. (b) 63. (d) 64. (d) 65. (b) 66. (d) 67. (d) 68. (b) 69. (c) 70. (b)

### Graham's Law

71. (a) 72. (d) 73. (d) 74. (c) 75. (a) 76. (c) 77. (a) 78. (a) 79. (a) 80. (d)

### Kinetic Theory of Gases

81. (c) 82. (b) 83. (a) 84. (d) 85. (a) 86. (c) 87. (a) 88. (a) 89. (b) 90. (c)
91. (d) 92. (b) 93. (a) 94. (d) 95. (a)

### Real Gases

96. (a) 97. (a) 98. (a) 99. (b) 100. (d) 101. (c) 102. (c) 103. (b) 104. (b) 105. (b)

### Liquefaction of Gases

106. (d) 107. (c) 108. (b) 109. (d) 110. (d) 111. (a) 112. (b) 113. (c) 114. (a) 115. (d)

## Answer Keys — Exercise II

### Section A (Only one Correct)

1. (b) 2. (d) 3. (d) 4. (c) 5. (c) 6. (c) 7. (a) 8. (c) 9. (b) 10. (d)
11. (c) 12. (a) 13. (a) 14. (d) 15. (b) 16. (c) 17. (c) 18. (c) 19. (b) 20. (a)
21. (d) 22. (b) 23. (c) 24. (a) 25. (b) 26. (c) 27. (b) 28. (b) 29. (a) 30. (a)
31. (b) 32. (c) 33. (a) 34. (b) 35. (c) 36. (b) 37. (b) 38. (c) 39. (d) 40. (b)
41. (a) 42. (d) 43. (c) 44. (b) 45. (a) 46. (c) 47. (b) 48. (a) 49. (c) 50. (b)
51. (a) 52. (d) 53. (a) 54. (b) 55. (c) 56. (d) 57. (b) 58. (d) 59. (c) 60. (b)
61. (b) 62. (a) 63. (a) 64. (c) 65. (c) 66. (a) 67. (b) 68. (c) 69. (a) 70. (d)
71. (a) 72. (d) 73. (c) 74. (b) 75. (b) 76. (d) 77. (a) 78. (a) 79. (a) 80. (a)
81. (a) 82. (a) 83. (a) 84. (a) 85. (b) 86. (b) 87. (a) 88. (d) 89. (a) 90. (c)
91. (a) 92. (b) 93. (d) 94. (d) 95. (a)

## Section B (One or More than one Correct)

1. (c), (d) 2. (b), (c) 3. (a), (b), (c), (d) 4. (b), (c)
5. (b), (c) 6. (a), (b), (c) 7. (b), (c) 8. (c)
9. (a), (b), (c) 10. (c), (d) 11. (a), (d) 12. (a), (b), (c)
13. (a), (b) 14. (a), (b), (c), (d) 15. (a), (b), (d) 16. (a), (d)
17. (a), (b), (c) 18. (c) 19. (a), (c) 20. (a), (b), (c), (d)

## Section C

### Comprehension I

1. (c) 2. (b) 3. (c)

### Comprehension II

4. (c) 5. (b) 6. (a)

### Comprehension III

7. (d) 8. (b) 9. (d)

### Comprehension IV

10. (b) 11. (c) 12. (b) 13 (a)

### Comprehension V

14. (c) 15. (b) 16. (a) 17. (c)

### Comprehension VI

18. (a) 19. (d) 20. (b)

### Comprehension VII

21. (b) 22. (c)

### Comprehension VIII

23. (c) 24. (b) 25. (a)

### Comprehension IX

26. (b) 27. (b) 28. (a)

### Comprehension X

29. (b) 30. (a) 31. (c) 32. (b)
33. (b) 34. (a) 35. (a)

### Comprehension XI

36. (a) 37. (a) 38. (c) 39. (a)
40. (d) 41. (a) 42. (a)

### Comprehension XII

43. (d) 44. (c) 45. (a)

## Section D (Assertion – Reason)

1. (a) 2. (a) 3. (c) 4. (d) 5. (c) 6. (c) 7. (b) 8. (b) 9. (a) 10. (a)
11. (a) 12. (d) 13. (a) 14. (d) 15. (d) 16. (d) 17. (a) 18. (b) 19. (b) 20. (a)

## Section E (Column Match)

1. $A \rightarrow P, R; B \rightarrow Q; C \rightarrow S; D \rightarrow T$
2. $A \rightarrow Q, R; B \rightarrow S; C \rightarrow P; D \rightarrow Q$
3. $A \rightarrow R; B \rightarrow P, S; C \rightarrow Q$
4. $A \rightarrow S; B \rightarrow R; C \rightarrow Q; D \rightarrow P$
5. $A \rightarrow Q; B \rightarrow S; C \rightarrow P; D \rightarrow R$
6. $A \rightarrow Q; B \rightarrow R; C \rightarrow P; D \rightarrow S$
7. $A \rightarrow P, R, S; B \rightarrow P, Q, S; C \rightarrow P, Q, S; D \rightarrow P, R, S$
8. $A \rightarrow P, S; B \rightarrow Q, R; C \rightarrow Q$
9. $A \rightarrow P, S; B \rightarrow Q; C \rightarrow R$
10. $A \rightarrow Q; B \rightarrow S; C \rightarrow P; D \rightarrow R; E \rightarrow T$

## Section F (Subjective)

### Single-digit Integer Type

1. (7) 2. (5) 3. (5) 4. (4) 5. (3) 6. (3) 7. (4) 8. (6) 9. (4) 10. (8)
11. (4) 12. (9) 13. (4) 14. (2) 15. (8)

### Four-digit Integer Type

1. (0775) 2. (3900) 3. (0404) 4. (8724) 5. (0450)
6. (2421) 7. (1254) 8. (0065) 9. (0500) 10. (0320)
11. (3610) 12. (1836) 13. (0013) 14. (0140) 15. (0015)
16. (0050) 17. (9450) 18. (0100) 19. (0250) 20. (0096)

# HINTS AND EXPLANATIONS

## EXERCISE I (JEE MAIN)

### Gaseous Parameters and Gas Laws

1. $P = \frac{688}{760} = 0.905$ atm

2. $\rho_1 h_1 = \rho_2 h_2$
   or $1.047 \times 1 = 13.6 \times h_2 \Rightarrow h_2 = 0.077$ mmHg.

3. $\rho_1 h_1 = \rho_2 h_2$
   or $13.6 \times 75 = 1.7 \times h_2 \Rightarrow h_2 = 600$ cm

4. As the pressure in that area is low, there must be storm.

5. $P_1V_1 = P_2V_2 \Rightarrow 0.72 \times 200 = 0.90 \times V_2$
   $\Rightarrow V_2 = 160$ ml.

6. $P_1V_1 = P_2V_2 \Rightarrow 1 \times 2500 = P_2 \times 500 \Rightarrow P_2 = 5$ bar.
   $\therefore$ % Increase in pressure $= \frac{5-1}{1} \times 100 = 400\%$

7. Theory based

8. Boyle's law constant $= PV = \left(1.2 \times 1.013 \times 10^5 \frac{\text{N}}{\text{m}^2}\right)$
   $\times \left(200 \times 10^{-6} \text{m}^3\right)$
   $= 24.3$ J

9. $P_1V_1 = P_2V_2 \Rightarrow 5 \times 10^{-4} \times V_1 = 760 \times 1$
   $\Rightarrow V_1 = 1.52 \times 10^6$ dm$^3$.

10. $\frac{V_1}{T_1} = \frac{V_2}{T_2} \Rightarrow \frac{600}{300} = \frac{V_2}{320} \Rightarrow V_2 = 640$ ml.
    $\therefore$ Increase in volume $= 640 - 600 = 40$ ml.

11. Theory based

12. $\frac{V_1}{T_1} = \frac{V_2}{T_2} \Rightarrow \frac{200}{x+4} = \frac{300}{x+144} \Rightarrow x = 276.$
    $\therefore$ Absolute zero $= -276$°C.

13. $V \propto T$

14. $\frac{V_1}{T_1} = \frac{V_2}{T_2} \Rightarrow \frac{3.75}{308} = \frac{3.0}{T_2} \Rightarrow T_2 = 246.4$ K $= -26.6$°C.

15. $\frac{V_1}{T_1} = \frac{V_2}{T_2} \Rightarrow \frac{V}{300} = \frac{3V}{T_2} \Rightarrow T_2 = 900$ K.

16. $\frac{V_1}{T_1} = \frac{V_2}{T_2} \Rightarrow \frac{1200}{300} = \frac{1218}{T_2} \Rightarrow T_2 = 304.5$ K.

17. $\frac{P_1}{T_1} = \frac{P_2}{T_2} \Rightarrow \frac{100}{T} = \frac{101}{T+1} \Rightarrow T = 100$ K.

18. $\frac{P_1}{T_1} = \frac{P_2}{T_2} \Rightarrow \frac{P}{283} = \frac{2P}{T_2} \Rightarrow T_2 = 566$ K $= 293$°C.

19. $\frac{P_1}{T_1} = \frac{P_2}{T_2} \Rightarrow \frac{1}{546} = \frac{P_2}{273} \Rightarrow P_2 = 0.5$ atm.

20. $n_{O_2} = n_{SO_2} \Rightarrow \frac{m_{O_2}}{32} = \frac{m_{SO_2}}{64} \Rightarrow m_{O_2} = \frac{1}{2} \times m_{SO_2}$

21. Theory based

22. The number of molecules will be the same. Hence, the ratio of number of atoms = 2 : 1 : 2 : 3.

23. Number of molecules in a mole will remain equal to Avogadro's number.

24. The number of molecules in a closed container will not change, but the pressure of gas becomes double ($P \propto$ T when 'n' and 'V' are constant).

25. $n_{O_2} = n_{gas} \Rightarrow \frac{0.5}{32} = \frac{0.75}{\text{M}} \Rightarrow \text{M} = 48.$

## Ideal Gas Equation

**26.** $PV = nRT \Rightarrow P = \frac{n}{V}RT = CRT$

$\therefore$ $1\text{ atm} = 1\frac{\text{mol}}{\text{litre}} \times 0.0821\frac{\text{L-atm}}{\text{K.mol}} \times \text{T} \Rightarrow \text{T} \simeq 12.2\text{ K}.$

**27.** Theory based

**28.** Volume of cube, $l^3 = n \times 22.4\text{ L}$

$$= \frac{850 \times 10^6}{6 \times 10^{23}} \times 22.4 \times 10^3\text{ cm}.$$

$\therefore\ l = 3.17 \times 10^{-4}$ cm.

**29.** $PV = nRT$

or $4 \times 10^{-22} \times 1 = \frac{N}{6 \times 10^{23}} \times 0.08 \times 3 \Rightarrow N = 1000.$

**30.** $V = \frac{nRT}{P} \Rightarrow$ For $V_{\max}$, T should be high and P should be low.

**31.** At 4°C, $H_2O$ is a liquid with density = 1.0 g/ml.

Hence, $V = \frac{M}{d} = \frac{1.8}{1.0} = 1.8$ ml.

**32.** As metallic cylinder is a good conductor of heat, the temperature will remain constant. But the pressure will increase due to increase in mole.

**33.** $V = \frac{nR}{P} \cdot \text{T}$

Slope will be smaller at higher 'P'.

**34.** $d = \frac{PM}{RT} \Rightarrow$ For maximum density, 'P' should be high and 'T' should be low.

**35.** $PV = nRT \Rightarrow P \propto \text{T} \Rightarrow \left(\frac{P}{T}\right)_{v_2} = \left(\frac{P}{T}\right)_{\text{He}}$

$\Rightarrow \frac{P}{T} = \frac{P_{\text{He}}}{2T} \Rightarrow P_{\text{He}} = 2$

**36.** $n_{\text{removed}} = n_{\text{taken}} - n_{\text{remained}}$

or $\frac{w}{81} = \frac{P \times 10}{0.082 \times 300} - \frac{(P - 0.82) \times 10}{0.082 \times 300} \Rightarrow w = 27\text{ g}.$

**37.** $PV = \frac{w}{M}RT \Rightarrow w_1T_1 = w_2T_2$

or $3.5 \times (273 + t) = 3.2 \times (303 + t) \Rightarrow t = 47°\text{C}.$

**38.** $\frac{P_1V_1}{T_1} = \frac{P_2V_2}{T_2} \Rightarrow \frac{P \times V}{T} = \frac{\frac{P}{2} \times V_2}{2T} \Rightarrow V_2 = 4V.$

**39.** $d = \frac{PM}{RT} \Rightarrow \frac{PM}{d} = \text{Constant} \Rightarrow \frac{P_AM_A}{d_A} = \frac{P_BM_B}{d_B}$

or $\frac{P_A \times \frac{M_B}{2}}{2 \times d_B} = \frac{P_B \times M_B}{d_B} \Rightarrow \frac{P_A}{P_B} = \frac{4}{1}.$

**40.** $\frac{d_1T_1}{P_1} = \frac{d_2T_2}{P_2} \Rightarrow \frac{5.46 \times 300}{2} = \frac{d_2 \times 273}{1}$

$\Rightarrow d_2 = 3.0\text{ gm/dm}^3.$

**41.** $n_1T_1 = n_2T_2 \Rightarrow n_1 \times 300 = n_2 \times 750 \Rightarrow \frac{n_2}{n_1} = \frac{2}{5}.$

$\therefore$ Fraction of air escaped $= 1 - \frac{n_2}{n_1} = \frac{3}{5}.$

**42.** $2 \times V = \frac{1}{M_A} \times RT$ and $3 \times V = \left(\frac{1}{M_A} + \frac{2}{M_B}\right) \times RT$

$\therefore\ \frac{M_A}{M_B} = \frac{1}{4}.$

**43.** $P = 33.6 = \left(\frac{48}{48} + \frac{48}{96}\right) \times 0.0821 \times 546 \Rightarrow P = 2\text{ atm}.$

**44.** $n_{\text{escaped}} = n_{\text{taken}} - n_{\text{remained}}$

$$= \frac{1 \times 2.24}{0.0821 \times 273} - \frac{\frac{570}{760} \times 2.24}{0.0821 \times 273} = 0.025.$$

**45.** $\frac{n_1T_1}{P_1} = \frac{n_2T_2}{P_2} \Rightarrow \frac{n_1 \times T}{P} = \frac{n_2 \times \frac{2}{3}T}{\frac{P}{2}} \Rightarrow n_2 = \frac{3}{4}n_1$

$\therefore$ Mass of gas escaped $= m - \frac{3}{4}m = \frac{m}{4}.$

46. $\frac{P_1V_1}{T_1}=\frac{P_2V_2}{T_2} \Rightarrow \frac{720\times100}{300}=\frac{420\times V_2}{210} \Rightarrow V_2=120$ L.

$\therefore$ Increase in volume $=120-100=20$ L.

47. $V=\frac{nRT}{P} \Rightarrow$ For constant volume, $\frac{T}{P}$ should be the same.

48. $\frac{P_1}{n_1T_1}=\frac{P_2}{n_2T_2} \Rightarrow \frac{9}{n\times303}=\frac{P_2}{1.5n\times404} \Rightarrow P_2=18$ atm.

49. $\frac{P_1V_1}{m_1}=\frac{P_2V_2}{m_2} \Rightarrow \frac{P\times V_1}{10}=\frac{\frac{P}{2}\times50}{1.25}$

$\therefore V_1=200$ ml.

50. $N_2O_4 \rightarrow 2NO_2$

| Initial | 1 mole | 0 |
|---|---|---|
| Final | 1 – 0.2 = 0.8 | 2 × 0.2 = 0.4 |

Now, $\frac{P_1}{n_1T_1}=\frac{P_2}{n_2T_2} \Rightarrow \frac{1}{1\times300}=\frac{P_2}{1.2\times600}$
$\Rightarrow P_2=2.4$ atm.

## Dalton's Law

51. $H_2$ and $O_2$ gases are unreactive in normal condition.

52. $P_{total}=P_{N_2}+P_{O_2}+P_{CO_2}=\frac{38}{76}+\frac{190}{760}+0.5$
$-1.25$ atm.

53. $P_{O_2}=X_{O_2}\cdot P_{total}=\frac{21}{100}\times740=155.4$ mmHg.

54. $\frac{P_{H_2}}{P_{total}}=X_{H_2}=\frac{n_{H_2}}{n_{H_2}+n_{C_2H_6}}=\frac{\frac{w}{2}}{\frac{w}{2}+\frac{w}{30}}=\frac{15}{16}$.

55. $P_{H_2}=X_{H_2}\cdot P_{total}=\frac{\frac{20}{2}}{\frac{20}{2}+\frac{80}{32}}\times1=0.8$ bar.

56. $\frac{n_1}{P_1}=\frac{n_2}{P_2} \Rightarrow \frac{\frac{16}{32}+\frac{14}{28}}{4}=\frac{\frac{8}{32}+\frac{14}{28}}{P_2} \Rightarrow P_2=3$ atm.

57. $P_{CO}=P_{CH_4} \Rightarrow n_{CO}=n_{CH_4} \Rightarrow \frac{70}{28}=\frac{w}{16} \Rightarrow w=40$ g.

58. $n_{total}=n_{H_2}+n_{CH_4} \Rightarrow \frac{P_f\times2V}{RT}=\frac{P\times V}{RT}+\frac{P\times V}{RT}$
$\Rightarrow P_f=P.$

59. $\frac{P_1}{n_1}=\frac{P_2}{n_2} \Rightarrow \frac{P}{n+n}=\frac{P_2}{n} \Rightarrow P_2=\frac{P}{2}$ mm.

60. $\frac{P_{O_2}}{P_{N_2}}=\frac{n_{O_2}}{n_{N_2}}=\frac{w/32}{w/28}=\frac{7}{8}$

$\therefore P_{O_2}=0.875\times P_{N_2}$.

61. $n_{total}=n_{N_2}+n_{O_2} \Rightarrow \frac{P.3V}{RT}=\frac{800\times2V}{RT}+\frac{680\times V}{RT}$

$\therefore P=760$ mm.

62. $n_{total}=n_{H_2}+n_{O_2}$

or $\frac{P\times1}{RT}=\frac{0.8\times0.5}{RT}+\frac{0.7\times2.0}{RT} \Rightarrow P=1.8$ bar.

63. $H_2+Cl_2 \Rightarrow 2HCl$

| Initial mole | $\frac{4}{2}=2$ | $\frac{9\times10^{23}}{6\times10^{23}}=1.5$ | 0 |
|---|---|---|---|
| Final moles | 0.5 | 0 | 3.0 |

Now, $P_{HCl}=X_{HCl}\times P_{total}=\frac{3.0}{3.5}\times700=600$ mm

64. For $N_2$ gas, $P_1V_1=P_2V_2 \Rightarrow (1-0.3)\times V=P_2\times\frac{V}{3}$

$\Rightarrow P_2=2.1$ atm.

$\therefore$ Final pressure $=P_{N_2}+V.P.=2.1+0.3=2.4$ atm.

65. Rate of evaporation is greater than the rate of condensation.

66. $P_{N_2} = X_{N_2} \times P_{total} = \frac{3}{4} \times (640-40) = 450$ Torr.

67. $P_{O_2}$ + Vapour pressure of water = External pressure.

68. Vapour pressure depends only on temperature.

69. $P_{H_2O} = \frac{1}{100} \times 760$ mmHg = 7.6 mmHg.

70. $P_1V_1 = P_2V_2 \Rightarrow (740-20) \times 190 = 760 \times V_2$
$\Rightarrow V_2 = 180$ mm.

## Graham's Law

71. $\frac{r_{H_e}}{r_{CH_4}} = \sqrt{\frac{M_{CH_4}}{M_{H_e}}} = \sqrt{\frac{16}{4}} = \frac{2}{1}$.

72. $\frac{r_X}{r_{PH_3}} = \sqrt{\frac{M_{PH_3}}{M_X}} \Rightarrow 0.88 = \sqrt{\frac{34}{M_X}} \Rightarrow M_X = 43.9.$

Gas may be $N_2O(M = 44)$

73. $\frac{r_{O_2}}{r_{H_2}} = \sqrt{\frac{M_{H_2}}{M_{O_2}}} \Rightarrow \frac{n_{O_2}/t}{n_{H_2}/t} = \sqrt{\frac{2}{32}} \Rightarrow n_{O_2} = \frac{1}{4} \times n_{H_2}$.

or $\frac{w_{O_2}}{32} = \frac{1}{4} \times \frac{2}{2} \Rightarrow w_{O_2} = 8$ g.

74. $\frac{r_{C_nH_{2n+2}}}{r_{H_2}} = \sqrt{\frac{M_{H_2}}{M_{C_nH_{2n+2}}}} \Rightarrow \frac{1}{6} \sqrt{\frac{2}{14n+2}} \Rightarrow n = 5.$

75. $\frac{r_{N_2O}}{r_{tear\ gas}} = \sqrt{\frac{M_{tear\ gas}}{M_{N_2O}}} \Rightarrow \frac{x/t}{8-x/t} = \sqrt{\frac{176}{44}}$
$\Rightarrow x = 5.33$

76. $M_{N_2O} = M_{CO_2} \Rightarrow r_{N_2O} = r_{CO_2}$

77. $\frac{r_A}{r_B} = \frac{\text{Area}_A}{\text{Area}_B} = \frac{\pi r^2}{r^2} = \frac{\pi}{1}$.

78. $\frac{r_{HC}}{r_{SO_2}} = \sqrt{\frac{M_{SO_2}}{M_{HC}}} \Rightarrow \frac{180/15}{120/20} = \sqrt{\frac{64}{M_{HC}}}$
$\Rightarrow M_{HC} = 16.$

79. $\frac{r_{NH_3}}{r_{BF_3}} = \sqrt{\frac{M_{BF_3}}{M_{NH_3}}} \Rightarrow \frac{x/t}{(100-x)/t} = \sqrt{\frac{68}{17}}$
$\Rightarrow x = 66.67$ cm.

80. $n_{total} = \frac{PV}{RT} = \frac{4.15 \times 3}{0.083 \times 300} = 0.5$

$\therefore n_{gas} = 0.5 - n_{N_2} = 0.5 - 0.4 = 0.1.$

Now, $\frac{r_{N_2}}{r_{gas}} = \sqrt{\frac{M_{gas}}{M_{N_2}}} \Rightarrow \frac{0.1/t}{0.4/t} = \sqrt{\frac{M_{gas}}{28}}$
$\Rightarrow M_{gas} = 448$

## Kinetic Theory of Gases

81. $\frac{u_{rms,H_2}}{u_{rms,O_2}} = \frac{\sqrt{\frac{3R \times 50}{2}}}{\sqrt{\frac{3R \times 800}{32}}} = \frac{1}{1}$

82. $\alpha = c = u$

83. $u_{av} \propto \sqrt{T} \Rightarrow \frac{u_2}{u_1} = \sqrt{\frac{T_2}{T_1}} \Rightarrow \frac{u_2}{0.3} = \sqrt{\frac{1200}{300}}$
$\Rightarrow u_2 = 0.6$ m/s

84. $u_{rms} \propto \sqrt{T} \Rightarrow u_2 = \sqrt{2} \times u_1$

85. $u_{rms} \propto \frac{1}{\sqrt{M}} \Rightarrow H_2 > N_2 > O_2 > HBr.$

86. $u \propto \frac{1}{\sqrt{M}} \Rightarrow \frac{u_1}{r_2} = \sqrt{\frac{M_2}{M_1}} \Rightarrow \frac{3}{1} = \sqrt{\frac{M_2}{M_1}} \Rightarrow \frac{M_1}{M_2} = \frac{1}{9}.$

87. $E_T = \frac{3}{2}nRT \Rightarrow \frac{3}{2}\times 0.3\times RT = \frac{3}{2}\times 0.4\times R\times 400$

$\Rightarrow T = \frac{1600}{3}$

88. $x = \frac{3}{2}\times\frac{N}{N_A}\times R\times 150$ and $2x = \frac{3}{2}\times\frac{N^1}{N_A}\times R\times 300$

$\therefore N^1 = N$

89. $Av.\text{K.E.} \propto T$

90. $E_T = \frac{3}{2}nRT \Rightarrow 0.63 = \frac{3}{2}\times\frac{10^{20}}{6\times 10^{23}}\times 8.314\times T$

$\Rightarrow T = 303.1 = \text{K} = 30.1°\text{C}$

91. $Av.\text{K.E.}\ \alpha\ T$

92. Theory based

93. Fraction of molecule remains the same.

94. $Z_{11} = Z_1.\frac{N^*}{2} = X.\frac{N^*}{2}$

95. $\lambda = \frac{RT}{\sqrt{2}\pi\sigma^2.P.N_A} \Rightarrow \lambda\ \alpha\ \frac{1}{\sigma^2}$

$\therefore$ Smaller the size of molecule, larger is $\lambda$.

## Real Gases

96. Theory based

97. Theory based

98. $Z = 1 + \frac{P.b}{RT}$

99. $\left(P + \frac{a.n^2}{V^2}\right)(V - nb) = nRT \Rightarrow \left(P + \frac{a}{4V^2}\right)\left(V - \frac{b}{2}\right)$

$= \frac{RT}{2}.$

100. Greater the intermolecular forces, greater is 'a'.

101. $b = 4\times$ Volume of a molecule $\times N_A$

$= 4\times\frac{4}{3}\pi\times\left(1.25\times 10^{-8}\ \text{cm}\right)^3\times 6.022\times 10^{23}$

$= 19.7\,\text{cm}^3/\text{mol} = 1.97\times 10^{-2}\,\text{L/mol}.$

102. $P = \frac{nRT}{V - nb} - \frac{an^2}{v^2} = \frac{2\times 0.0821\times 300}{0.25 - 2\times 0.05} - \frac{2.5\times 2^2}{(0.25)^2}$

$= 168.4$ atm.

103. $b = 4\times\frac{4}{3}\pi r^3 N_A$

or, $0.0391\times 10^3 = 4\times\frac{4}{3}\pi r^3\times 6.022\times 10^{23}$

$\Rightarrow r = 1.57\times 10^{-8}$ cm

104. Theory based

105. When $P \to 0$, gases behave ideally. $PV = nRT$.

## Liquefaction of Gases

106. Theory based

107. $\underset{P}{-200°C} < \underset{Q}{-110°C} < \underset{R}{-100°C} < +50°C$

Gases cannot be liquefied above critical temperature ($T_c$).

108. $P_C V_C = \frac{3}{8}R.T_C \Rightarrow V_C = \frac{3}{8}\times\frac{0.0821\times 305.5}{48.2}$

$= 0.195$ L/mol.

109. $T_C = \frac{8a}{27Rb}$ and $\frac{a}{b}$ is maximum for Z.

110. $T_C = \frac{8a}{27Rb}$ and $P_C = \frac{a}{27b^2} \Rightarrow b = \frac{R.T_c}{8.P_C}$ and $\frac{T_C}{P_C}$ is minimum for He.

111. $Z_C = \frac{P_C.V_C}{R.T_C} = \frac{\frac{a}{27b^2}\times 3b}{R\times\frac{80}{27Rb}} = \frac{3}{8}.$

112. Gases cannot be liquefied above Tc.

113. Gases with higher 'a' value are easily liquefiable.

114. $T_c = \frac{8}{27}.\frac{a}{Rb}$; $T_B = \frac{a}{Rb}$; $T_i = 2.\frac{a}{Rb}$.

115. $b = \frac{R.T_c}{8.P_c} = \frac{0.0821\times 300}{8\times 82.1} = 0.0375$ L/mol.

## EXERCISE II (JEE ADVANCED)

### Section A (Only one Correct)

1. $P_{gas} + 15.6 = 53.3 + 76.3 \Rightarrow P_{gas} = 114$ cm Hg = 1.5 atm

2. Fractional increase $= \frac{V_2 - V_1}{V_1} = \frac{V_2}{V_1} - 1 = \frac{P_1}{P_2} - 1$

$$= \frac{H+7H}{7H} - 1 = \frac{1}{7}$$

3. 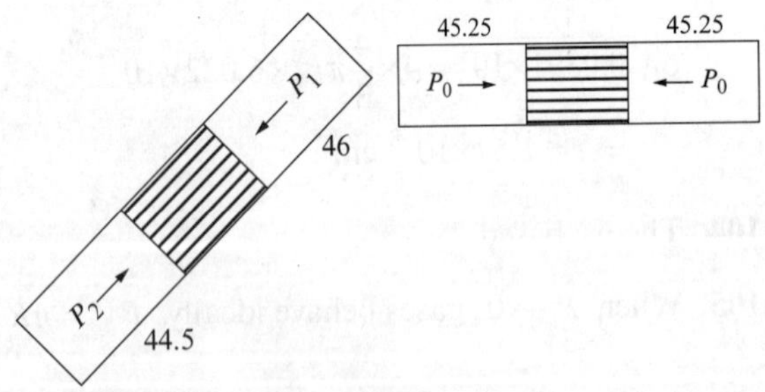


$P_0 \times 45.25 = P_1 \times 46 = P_2 \times 44.5$

$P_1 + 5\sin 30° = P_2$

$$\frac{P_0 \times 45.25}{46} + 5 \times \frac{1}{2} = \frac{P_0 \times 45.25}{44.5}$$

$\Rightarrow P_0 = 75.4$ cm Hg

4. $P_1V_1 = P_2V_2 \Rightarrow 10 \times 2A = (10 + h) \times hA$
$\Rightarrow h = 1.71$ m

5. Theory based

6. $\frac{dv}{dt} = \frac{V_0}{273} = 0.08 \Rightarrow V_0 = 21.84$ L

7. $\frac{V_1}{T_1} = \frac{V_2}{T_2} \Rightarrow \frac{1.0}{x+0} = \frac{0.6}{x+(-100)}$
$\Rightarrow x = 250 \Rightarrow 0$ K $= -250$°C

8. $\frac{V_1}{T_1} = \frac{V_2}{T_2} \Rightarrow \frac{V}{T} = \frac{V+\Delta V}{T+\Delta T}$

$$\Rightarrow \frac{\Delta V}{V.\Delta T} = \frac{1}{T} \Rightarrow y = \frac{1}{x}$$

9. $\frac{V_1}{T_1} = \frac{V_2}{T_2} \Rightarrow \frac{V_1}{273+t_1} = \frac{1.1V_1}{273+t_2}$

$\therefore$ Percentage increase in temperature

$$= \frac{t_2 - t_1}{t_1} \times 100 = (10 + \frac{2730}{t_1})\ \%$$

10. Number of $SO_2$ molecules $= N \Rightarrow$ Number of atoms = 3N

11. $\frac{n_{N_2}}{n_{O_2}} = \frac{V_{N_2}}{V_{O_2}} \Rightarrow \frac{m_{N_2}/28}{m_{O_2}/32} = \frac{1}{7/8} \Rightarrow \frac{m_{N_2}}{m_{O_2}} = \frac{1}{1}$

12. $\frac{V_1}{n_1} = \frac{V_2}{n_2} \Rightarrow \frac{4/3\pi(10/2)^3}{8}$

$$= \frac{4/3\pi(d/2)^3}{1} \Rightarrow d = 5 \text{ cm}$$

13. $P = P_{CO_2} + P_{air} = \frac{0.5\times 0.0821\times 300}{1} + 1 = 13.315$ atm

14. Weight of filled balloon,
W = 20 g + 40 × 0.6 = 44 g
Weight of displaced air,
B = 40 × 1.3 = 52 g
$\therefore$ Balloon will lift upward with pay load = 52 – 44 = 8 g

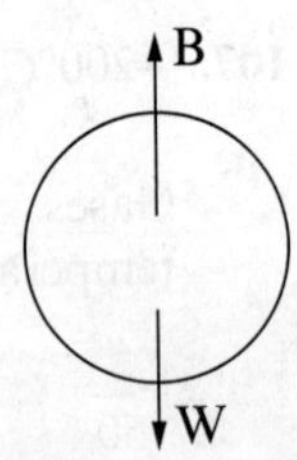

**15.** Water will behave like ideal gas on disappearance of intermolecular forces.

$$V = \frac{nRT}{P} = \frac{4.5\times10^3}{18}\times(22.4\times10^{-3})\ \text{m}^3 = 5.6\ \text{m}^3$$

**16.** $$d = \frac{m}{v} \Rightarrow 1.5 = \frac{n_{co}\times 28 + n_{co_2}\times 44}{(n_{co}+n_{co_2})\times\frac{0.0821\times300}{1}}$$

$$\Rightarrow n_{co} = \frac{7.055}{8.945}\ n_{co_2}$$

Alkali will absorb all $CO_2$. Hence, final pressure is due to CO.

$$P_{co} = \frac{n_{co}}{n_{co}+n_{co_2}}\times P_{total} = \frac{7.055}{7.055+8.945}\times 760\ \text{mm}$$

$$= 335.1\ \text{mm}$$

**17.** $$\frac{V_{\text{water vapour}}}{V_{\text{water}}} = \frac{\left(\frac{1\times0.0821\times373}{1}l\right)}{\left(\frac{18}{0.96}ml\right)} = 1633.24$$

**18.** $$V_{O_2} = \frac{\frac{3.2}{32}\times0.0821\times310}{1} = 2.5451\ \text{L}$$

$$V_{CO_2} = \frac{\frac{8.8}{44}\times0.0821\times310}{1} = 5.0902\ \text{L}$$

**19.** $$n_{CO_2} = \frac{200\times0.1}{1000} = 0.02$$

$$\therefore V_{CO_2} = 0.02\times22.4 = 0.448\ \text{L}$$

**20.** $$\frac{d(d)}{d\rho} = \frac{M}{RT} = 1.2\times10^{-5}\ \text{Kg m}^{-3}\ \text{Pa}^{-1}$$

$$\Rightarrow \frac{M}{8.314\times300} = 1.2\times10^{-5}$$

$$\therefore M_{air} = 0.03\ \text{Kg/mol} = 30\ \text{gm/mol}$$

$$\text{Now, } 30 = \frac{n_{N_2}\times28 + n_{O_2}\times32}{n_{N_2}+n_{O_2}} \Rightarrow n_{N_2}:n_{O_2} = 1:1$$

**21.** $$\frac{m_1T_1}{P_1} = \frac{m_2T_2}{P_2} \Rightarrow \frac{4\times T}{P} = \frac{m_2\times\frac{T}{2}}{2P}$$

$$\Rightarrow m_2 = 16\ \text{gm}$$

Hence, (16 – 4) = 12 gm gas should be added.

**22.** $$\frac{74.5}{50} = 1.49\ \text{times}$$

**23.** $$V_1d_1 = V_2d_2 \Rightarrow 1500\times1.25 = 3.92\times d_2$$

$$\Rightarrow d_2 = 478.3\ \text{kg/mol}$$

**24.** $$\frac{P_1V_1}{T_1} = \frac{P_2V_2}{T_2} \Rightarrow \frac{P\times\frac{4}{3}\pi r_1^3}{T} = \frac{\frac{P}{4}\times\frac{4}{3}\pi r_2^3}{2T}$$

$$\Rightarrow r_2 = 2r_1$$

$$\therefore \%\ \text{Increase in radius} = \frac{r_2-r_1}{r_1}\times100 = 100\%$$

**25.** $$\text{Constant} = P^2V = \left(\frac{nRT}{V}\right)^2 V \Rightarrow \frac{T^2}{V} = \text{Constant}$$

∴ On expansion, temperature will increase.

**26.** $$r = \frac{R}{M} \Rightarrow r_{n_2} > r_{H_e} > r_{N_2}$$

**27.** $$P\times3 = \frac{7}{28}\times0.0821\times300 \Rightarrow 2.0525\ \text{atm}$$

**28.**

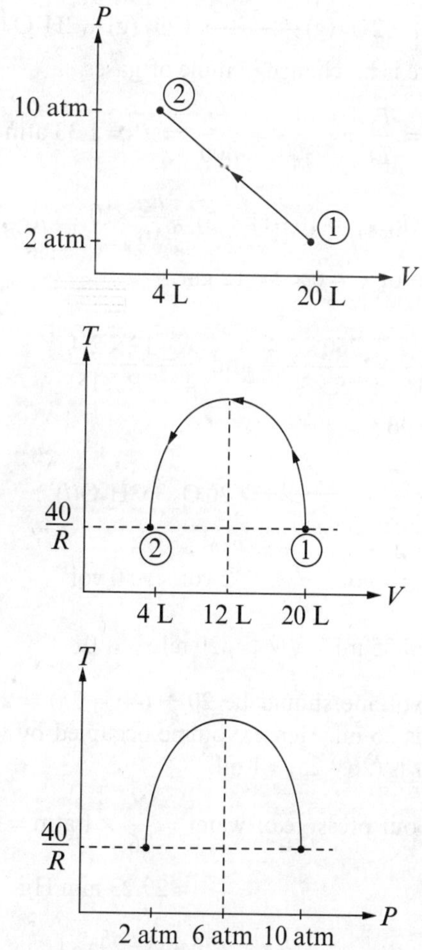


For $T_{max}$, $V = 12$ L and $P = 6$ atm and hence,

$$T_{max} = \frac{6\times12}{1\times0.08} = 900\ \text{K}.$$

HINTS AND EXPLANATIONS

29. $2Al + 2NaOH + 2H_2O \rightarrow 2NaAlO_2 + 3H_2$
2 mole 3 mole

$\therefore \frac{0.15}{27}$ mole $\frac{3}{2}\times\frac{0.15}{27}$ mole

$\therefore V_{H_2} = \frac{1.5\times0.15}{27}\times\frac{0.0831\times300}{0.831} = 0.25L$

30. $N_2 \rightarrow 2N$

Initial mole $a = \frac{1.4}{28}$ 0

Final mole $a - 0.4a$ $2\times0.4a$
$= 0.6a$ $= 0.8a$

Final total moles $= 0.6a + 0.8a = 1.4\times\frac{1.4}{28} = 0.07$

$\therefore P = \frac{0.07\times0.0821\times1800}{5} \simeq 2.07$ atm

31. $CH_4 (g) + 2O_2 (g) \xrightarrow{127°C} CO_2 (g) + 2H_2O (g)$
As there is no change in mole of gases,

$\frac{P_1}{T_1} = \frac{P_2}{T_2} \Rightarrow \frac{1}{300} = \frac{P_2}{400} \Rightarrow P_2 = 1.33$ atm

32. $P_c = X_c P_{total} \Rightarrow 10 - (1 + 3) = \frac{n_C}{10}\times10 \Rightarrow n_c = 6$

$\therefore$ Mass of $C = 6\times2 = 12$ gm

33. $\frac{x}{4}+\frac{5-x}{M} = \frac{760\times2.4}{R\times300}$ and $\frac{x}{4} = \frac{19\times2.4}{R\times15}$

$\therefore M = 96$

34. $C_2H_6 + \frac{7}{2}O_2 \longrightarrow 2CO_2 + 3H_2O(l)$

1 vol $\frac{7}{2}$ vol 2 vol 0 vol

$\therefore$ 10 ml 35 ml 20 ml 0

Final volume should be 20 + (40 – 35) = 25 ml but it is 26 ml. Hence, volume occupied by water vapour is (26 – 25) = 1 ml.

$\therefore$ Vapour pressure of water $= \frac{1}{26}\times1$ atm $= \frac{760}{26}$

$= 29.23$ mm Hg

35. $n_{H_2O}$ vapour needed $= \frac{(26.463-24)\times1}{760\times0.0821\times300}$

$= 1.32\times10^{-4}$

36. $P = \frac{1.2\times0.0821\times300}{18\times50}\times760 = 24.96$ mm Hg

37. Mass of water lost per day $= \frac{\Delta P.V}{RT}\times M$

$= \frac{\frac{(45-5)}{760}\times10000}{0.0821\times310}\times18 = 372.23$ gm

38. Vapour pressure is a function of temperature only

39. After achievement of equilibrium with its liquid form which will form on continuous injection of vapour, the pressure due to vapours become constant.

40. Rate of evaporation will remain constant throughout because neither surface area nor temperature are changing

41. $\frac{r_x}{r_y} = \frac{1}{5}$ and $\frac{r_y}{r_z} = \frac{1}{6} \Rightarrow \frac{r_z}{r_x} = \frac{30}{1}$

42. Smaller the rate of diffusion of HX, more closer to the HX end, $NH_4X$ will form.

43. $\frac{r_{N_2}}{r_{H_2}} = \sqrt{\frac{M_{H_2}}{M_{N_2}}} \Rightarrow \frac{\Delta P/60}{\Delta P/t} = \sqrt{\frac{2}{28}}$

$\Rightarrow t = 16.04$ min

44. $M_{dry\ air} > M_{moist\ air}$

45. $\frac{r_{CH_4}}{r_{HBr}} = \frac{P_{CH_4}}{P_{HBr}}\sqrt{\frac{M_{HBr}}{M_{CH_4}}} \Rightarrow \frac{1}{1} = \frac{n_{CH_4}}{n_{HBr}}\sqrt{\frac{81}{16}}$

$\Rightarrow \frac{n_{CH_4}}{n_{HBr}} = 0.4$

$\therefore X_{CH_4} = \frac{n_{CH_4}}{n_{CH_4}+n_{HBr}} = 0.31$

46. As HCl will diffuse slowly, white fumes will form closer to HCl end.

47. In gases, the intermolecular distance is much higher than the size of molecules.

48. $\frac{u_{av,2}}{u_{av,1}} = \sqrt{\frac{T_2}{T_1}} = \sqrt{\frac{375}{250}} = 1.22$

49. $\frac{u_{rms,O}}{u_{rms,O_2}} = \frac{\sqrt{\frac{3R\times2T}{16}}}{\sqrt{\frac{3RT}{32}}}\ \frac{2}{1} \Rightarrow u_{rms,o} = 2V$

**50.** Average speed for a gas depends on temperature and it is independent from the presence of other gas.

**51.** Difference in any two kind of speed, $\Delta u = K \times \sqrt{T}$

Now, $\dfrac{d(\Delta u)}{dT} = \dfrac{K}{2\sqrt{T}}$

$\Rightarrow$ On increasing temperature, $\Delta u$ decreases.

**52.** $\dfrac{u_{av,\,x}}{u_{av,\,y}} = \dfrac{2}{1} = \sqrt{\dfrac{T_X}{T_Y}} \Rightarrow \dfrac{T_X}{T_Y} = \dfrac{4}{1}$

Now, $\dfrac{P_X}{P_Y} = \dfrac{nRT_X/V_X}{nRT_Y/V_Y} = \dfrac{T_X}{T_Y} \times \dfrac{V_Y}{V_X} = \dfrac{4}{1} \times \dfrac{2}{1} = \dfrac{8}{1}$

**53.** $1 \times V = \dfrac{1}{M_A} \times R \times T \Rightarrow \dfrac{M_B}{M_A} = \dfrac{4}{1}$

$0.5 \times V = \dfrac{2}{M_B} \times RT$

$\therefore \dfrac{u_{av,\,A}}{u_{av,\,B}} = \sqrt{\dfrac{M_B}{M_A}} = \dfrac{2}{1}$

**54.** $u_{av,\,A} = u_{av,\,B} \Rightarrow \sqrt{\dfrac{8RT_A}{\pi M_A}} = \sqrt{\dfrac{3RT}{M_B}} \Rightarrow \dfrac{M_B}{M_A} = \dfrac{3\pi}{8}$

Now, $u_{av,\,A} = u_{av,\,B} \Rightarrow \sqrt{\dfrac{8RT_A}{\pi M_A}} = \sqrt{\dfrac{8RT_B}{\pi M_B}}$

$\Rightarrow \dfrac{T_A}{T_B} = \dfrac{M_A}{M_B} = \dfrac{8}{3\pi} < 1$

**55.** $\dfrac{1}{4} \cdot N^*.u_{av} = \dfrac{1}{4} \times \dfrac{6\times10^{23}}{22.4\times10^{-3}} \times \sqrt{\dfrac{8\times8.314\times273}{\pi\times28\times10^{-3}}}$

$= 3.05 \times 10^{27} m^{-2} s^{-1}$

**56.** $\dfrac{u_{rms,O_2}}{u_{rms,O_3}} = \sqrt{\dfrac{3R\times600}{32} \times \dfrac{48}{3R\times300}} = \sqrt{3}$

$\Rightarrow u_{rms,O_2} = \sqrt{3}$ v m/s

**57.** Average translational K.E. per gm $= \dfrac{3}{2}\dfrac{RT}{M}$

**58.** $\dfrac{T_A}{M_A} = \dfrac{T_B}{M_B} \Rightarrow u_{rms} = \sqrt{\dfrac{3RT}{M}}$ = Same for both

**59.** $\dfrac{3}{2}$ KT = qV $\Rightarrow \dfrac{3}{2} \times \dfrac{8.314}{6.022\times10^{23}} \times T$

$= 1.602 \times 10^{-19} \times 3$

$\Rightarrow T = 23207.2$ K

**60.** $u^2_{rms} \neq u^2_{av}$

**61.** $Z_w = \dfrac{1}{4} \cdot N^*.\,u_{av} = \dfrac{1}{4} \times \dfrac{P.N_A}{RT} \times \sqrt{\dfrac{8RT}{\pi M}}$

$\Rightarrow Z_w \propto \dfrac{1}{\sqrt{T}}$

**62.** $\dfrac{u_{rms,\,CH_4}}{u_{rms,\,SO_2}} = \sqrt{\dfrac{3R\times T}{16} \times \dfrac{64}{3R\times300}} = \dfrac{4}{1}$

$\Rightarrow T = 1200$ K

$\therefore$ Average K.E. per mole $= \dfrac{3}{2}RT = \dfrac{3}{2} \times 2 \times 1200$

$= 3600$ cal

**63.** $u_{av} \propto \sqrt{T} \Rightarrow \dfrac{u_2}{u_1} = \sqrt{\dfrac{432}{300}} = 1.2$

**64.** Mole of gas cannot change.

**65.** Collision number, $Z_1 = \sqrt{2}\,\pi\sigma^2.u_{av}.N^*$

$= \sqrt{2}\,\pi\sigma^2 . \sqrt{\dfrac{8RT}{\pi M}} \times \left(\dfrac{PN_A}{RT}\right)$

$\therefore Z_1 \propto \dfrac{1}{\sqrt{T}}$

Collision frequency, $Z_{11} = \dfrac{1}{\sqrt{2}}\pi\sigma^2 u_{av}.N^{*2}$

$= \dfrac{1}{\sqrt{2}}\pi\sigma^2 . \sqrt{\dfrac{8RT}{\pi M}} \times \left(\dfrac{PN_A}{RT}\right)^2$

$\therefore Z_{11} \propto \dfrac{1}{T^{3/2}}$

Mean free path, $\lambda = \dfrac{1}{\sqrt{2}\pi\sigma^2 N^*} = \dfrac{RT}{\sqrt{2}\pi\sigma^2.PN_A}$

$\Rightarrow \lambda \propto T$

**66.** $\lambda = \dfrac{RT}{\sqrt{2}\pi\sigma^2 \times PN_A}$

$= \dfrac{8.314\times300}{\sqrt{2}\pi\times(1.5\times10^{-10})^2\times(4.1\times10^{-14}\times1.013\times10^5)\times(6.022\times10^{23})}$

$= 1.0 \times 10^7$ m

**67.** Theory based

**68.** Velocity is a vector quality.

**69.** $\dfrac{dN}{N} = 4\pi\left(\dfrac{m}{2\pi KT}\right)^{3/2} \times u^2 \times e^{-\frac{mu^2}{2KT}} \times du$

$= \dfrac{2}{\sqrt{\pi}}\left(\dfrac{1}{KT}\right)^{3/2} \times \sqrt{E} \times e^{-E/KT} \times dE$

For most probable K.E., $\frac{d(dN/N)}{dE} = 0$

$\Rightarrow E = \frac{1}{2}KT$

**70.** Deviation from ideal behavior is maximum at low temperature and high pressure.

**71.** $Z > 1$ for $H_2$ at 0°C at all pressure.

**72.** $Z < 1$ at low pressure and $Z > 1$ at high pressure.

**73.** Theory based

**74.** $V_1 = V - nb \Rightarrow b = \frac{V - V_1}{n}$

$$\Rightarrow 4\times\frac{\pi}{6}d^3 \times N_A = \frac{V - V_1}{n}$$

$$\therefore d = \left(\frac{3(V - V_1)}{2\pi n N_A}\right)^{1/3}$$

**75.** $P_i = P + \frac{an^2}{V^2} \Rightarrow P = \left(P_i + \frac{an^2}{V^2}\right)$

Greater the value of '$a$', smaller will be '$P$'.

**76.** Gaseous mixture is always homogeneous.

**77.** $P_1V_1 = P_2V_2 \Rightarrow 0.5 \times 2000 = 100 \times V_2$

$\Rightarrow V_2 = 10$ ml < 13 ml

As the real volume is greater than ideal, the volume occupied by the molecule is significant.

**78.** When attractive forces are dominant, $V_{real} < V_{ideal}$.

**79.** $B = b - \frac{a}{RT} = 0.03 - \frac{1.344}{0.0821\times 273} = -0.03$ l/mol

**80.** $Z = \frac{PV_m}{RT} = \frac{V_m}{V_m - b} = \frac{10b}{10b - b} = \frac{10}{9}$

**81.** $\left(P + \frac{an^2}{V^2}\right)(V - nb) = nRT$ may be expressed as

$$\left(P + \frac{a.d^2}{M^2}\right)\left(\frac{M}{d} - b\right) = RT \text{ as } d = \frac{m}{v} = \frac{n\times m}{v}$$

Now, $\left[P + \frac{3.6\times(2.2)^2}{(44)^2}\right]\left[\frac{44}{2.2} - 0.05\right] = 0.0821 \times 300$

$\Rightarrow P = 1.226$ atm

**82.** $B = b - \frac{a}{RT} = -1.0$ L/mol

Now, $PV_m = RT\left(1 + \frac{B}{V_m}\right)$ and $d = \frac{M}{V_m}$

Hence, $\frac{PM}{d} = RT\left(1 + \frac{B\times d}{M}\right)$

or, $\frac{1\times 40}{d} = 0.08 \times 262.5\left[1 + \frac{(-1.0)\times d}{40}\right]$

$\therefore d = 2.005$ g/L

**83.** When $P \to 0$, $V \to \infty$ and hence $e^{a/VRT} \to 1$ and $(V - b) \to V$. Hence, $P = \frac{RT}{V} = \frac{0.0821\times 300}{410.5} =$ 0.06 atm

**84.** For a van der Waals gas, $Z = \frac{V_m}{V_m - b} - \frac{a}{V_m RT}$

or, $0.8 = \frac{0.5}{0.5 - 0.04} - \frac{a}{0.5\times 0.08\times 300}$

$\Rightarrow a = 3.44$ atm L$^2$/mol$^2$

**85.** At Boyle's temperature, $\frac{dz}{dp} = 0 \Rightarrow T = \frac{168}{0.35} = 480$ K

**86.** Theory based.

The initial slope of $Z$ vs. $P$ curve increases with increase in temperature, above Boyle's temperature, only upto $2 \times T_B$. Then, the slope starts decreasing.

**87.** For van der Waals gas, $Z = \frac{V_m}{V_m - b} - \frac{a}{V_m \times RT}$

At Boyle's temperature, $Z = \frac{V_m}{V_m - b} - \frac{a}{V_m \times R \times \left(\frac{a}{Rb}\right)}$

$$= 1 + \frac{b^2}{V_m(V_m - b)}$$

**88.** Ideal gas can never be liquified.

**89.** For ideal behavior, Boyle's temperature should be closer to 600 K.

**90.** Theory based

**91.** $T_c < T_B$

**92.** $\frac{T_c}{P_c} = \frac{8a/27Rb}{a/27b^2} = \frac{8b}{R}$

$$\therefore \frac{\left(\frac{T_c}{P_c}\right)_{CO_2}}{\left(\frac{T_c}{P_c}\right)_{CH_4}} = \frac{b_{CO_2}}{b_{CH_4}} = \frac{304/72}{190/45} = \frac{1}{1}$$

93. At $T > T_c$, the gas can never be liquified.

94. Theory based

95. $P_cV_c = \frac{3}{8}RT_c \Rightarrow V_c = \frac{3}{8} \times \frac{0.0821 \times 128}{41.05}$

$= 0.096$ L/mol

## Section B (One or More than one Correct)

1. Boyle's law constant $= PV = nRT$

2. $\frac{P_0V_0}{RT_0} + \frac{P_0V_0}{RT_0} = \frac{PV_0}{R \times 2T_0} + \frac{PV_0}{RT_0} \Rightarrow P = \frac{4}{3}P_0$

and $n = \frac{\frac{4}{3}P_0 \times V_0}{R \times 2T_0} = \frac{2P_0V_0}{3RT_0}$

3. $P_f = P_i\left(\frac{V}{V + \Delta V}\right)^n$

(a) $P_f = 24.2 \times \left(\frac{10}{10+1}\right)^1 = 22$ atm

(b) $P_f = 24.2 \times \left(\frac{10}{10+1}\right)^2 = 20$ atm

(c) $\frac{P}{\eta} = P\left(\frac{10}{10+1}\right)^n \Rightarrow n = \frac{\ln \eta}{\ln 1.1}$

(d) $P_f = 24.2 \times \left(\frac{10}{10+1}\right)^n$ atm

4. As the average molar mass increases, the molar mass of vapours must be greater than that of $N_2$.

5. $r_{H_2} > r_{D_2}$

6. (a) Number of molecules colliding at the wall per unit time per unit area, $Z_w = \frac{1}{4}.u_{av}.N^*$

$N^*$ is same for both but $u_{av,\,He} > u_{av,\,Ne}$

(b) Average force per collision ∝ Change in momentum $\propto \sqrt{M}$

7. **At valve – I:** $P_1V_1 = P_2V_2 \Rightarrow 1 \times 60\text{ A} = P_2 \times 45\text{ A}$
$\Rightarrow P_2 = 1.33$ atm $< 1.5$ atm

Hence, valve – I will not open.

**At valve – II:** $P_1V_1 = P_2V_2 \Rightarrow 1 \times 60\text{ A} = P_2 \times 30\text{ A}$
$\Rightarrow P_2 = 2$ atm $< 2.2$ atm

Hence, valve – II will not open.

**At valve – III:** $P_1V_1 = P_2V_2 \Rightarrow 1 \times 60\text{ A} = P_2 \times 20\text{ A}$
$\Rightarrow P_2 = 3$ atm $> 2.5$ atm

Hence, valve – III will open first. As the piston will reach at valve – III, the gas will come out till the pressure of gas becomes 2.5 atm. Now,

$$2.5 \times \frac{20 \times 821}{1000} = n \times 0.0821 \times 300$$

$\Rightarrow$ Moles of gas remained, $n = \frac{5}{3}$

**At valve – IV:** $P_1V_1 = P_2V_2 \Rightarrow 2.5 \times 20\text{ A} = P_2 \times 15\text{ A}$
$\Rightarrow P_2 = 3.33$ atm $< 4.4$ atm

Hence, valve - IV will not open.

**At valve – V:** $P_1V_1 = P_2V_2 \Rightarrow 2.5 \times 20\text{ A} = P_2 \times 10\text{ A}$
$\Rightarrow P_2 = 5$ atm $> 4.8$ atm

Hence, valve – V will open until the gas pressure becomes 4.8 atm.

8. $u_{av} \propto \sqrt{T}$

9. $\frac{u_{\text{rms, A}}}{u_{\text{rms, B}}} = \sqrt{\frac{3R \times 300}{M_A} : \frac{M_B}{3R \times 400}} = \frac{\sqrt{3}}{2}$

$\Rightarrow M_A = M_B$

10. On increasing the temperature at constant volume, the average speed of molecules as well as number of molecular collisions at wall increases.

11. Theory based

12. Theory based

13. At very high pressure, $\left(P + \frac{a}{V^2}\right) \simeq P$

$\Rightarrow Z = 1 + \frac{b.P}{RT}$

and $Z = \frac{PV}{RT} \Rightarrow \frac{P}{RT} = \frac{Z}{V} \Rightarrow Z = \frac{V}{V-b}$

**14.** $T_c = 273 + (-177) = 96$ K

$\Rightarrow T_B = \frac{27}{8} \times 96 = 324\text{K} = 51°\text{C}$

(a) $Z = \frac{P.V_m}{RT} = \frac{0.821 \times 9.6}{0.0821 \times 96} = 1$

But at $T = T_c$, $Z < 1$ at low pressure

(b) $Z = \frac{P.V_m}{RT} = \frac{0.821 \times 40}{0.0821 \times 400} = 1$

But at $T > T_B$, $Z > 1$ at all pressure

(c) $Z = \frac{P.V_m}{RT} = \frac{82.1 \times 0.310}{0.0821 \times 324} = 0.96 < 1$

But at $T = T_B$ and $P > 50$ atm, $Z > 1$

(d) $Z = \frac{P.V_m}{RT} = \frac{0.821 \times 32.4 \times 10^{-3}}{0.0821 \times 324} = 10^{-3} < 1$

But at $T = T_c$ and $P < 50$ atm, $Z = 1$

**15.** a. $\dfrac{9R \times \frac{8a}{27Rb} \times 36}{8} = a$

b. $3 \times \frac{a}{27b^2} \times (3b)^2 = a$

c. $\frac{3}{8} \times \dfrac{\frac{a}{27b^2} \times 36}{\frac{8a}{27Rb}} \neq a$

d. $\frac{27}{64} \times \dfrac{R^2 \times \left(\frac{8a}{27Rb}\right)^2}{\frac{a}{27b^2}} = a$

**16.** Theory based

**17.** Theory based

**18.** Real gas may behave ideally at Boyle's temperature.

$$P = \frac{RT}{V_m} = \frac{R \times \frac{a}{Rb}}{V_m} = \frac{a}{b.Vm}$$

**19.** $Z = \frac{PV_m}{RT} = 1 + B'.P + C'.P^2 + \ldots.$ (1)

$Z = \frac{PV_m}{RT} = 1 + \frac{B}{Vm} + \frac{C}{V_m^2} + \ldots.$ (2)

Or, $P = \frac{RT}{V_m}\left(1 + \frac{B}{V_m} + \frac{C}{V_m^2} + \ldots.\right)$

Substituting this value in Equation (1), we get:

$$Z = 1 + B'.\left\{\frac{RT}{V_m}\left(1 + \frac{B}{V_m} + \frac{C}{V_m^2} + \ldots\right)\right\} + C'.\left\{\frac{RT}{V_m}\left(1 + \frac{B}{V_m} + \frac{C}{V_m^2} + \ldots\right)\right\}^2 + \ldots$$

$$= 1 + \frac{B'RT}{V_m} + \frac{B'RT.B + C'(RT)^2}{V_m^2} + \ldots.$$

Comparing this with Equation (2), we get:

$B'RT = B$ and $B'RT.B + C.(RT)^2 = C$

**20.** Theory based

## Section C (Comprehensions)

### Comprehension – I

**1.** Number of strokes $= \frac{(8 \text{ bar}) \times (1000 \text{ cm}^3)}{(1 \text{ bar}) \times (25 \times 4 \text{ cm}^3)} = 80$

**2.** $F = P.A = \left(8 \times 10^5 \frac{N}{m^2}\right) \times (4 \times 10^{-4} \text{m}^2) = 320$ N

**3.** $m = \frac{F}{g} = \frac{320}{10} = 32$ kg

**Comprehension – II**

**4.** $PV_m = RT$ (Let $0K = -x°N$)

$28 = R(0 + x)$ $x = 233.33$

$40 = R(100 + x)$ $\therefore 0K = -233.33°N$

**5.** $R = \dfrac{28}{x} = 0.12$ L – atm/K-mol

**6.** $V = \dfrac{nRT}{P} = \dfrac{2 \times 0.12 \times (66.67 + 233.33)}{2} = 36$ L

---

**Comprehension – III**

**7.** **A**

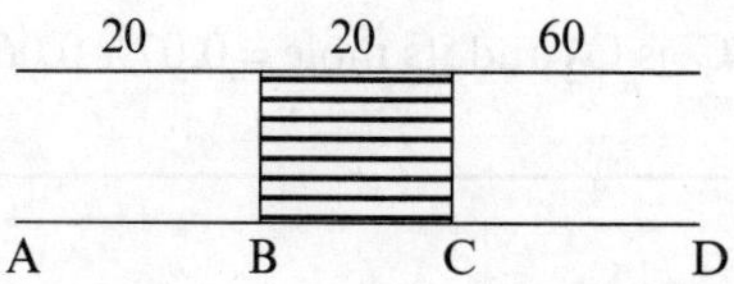


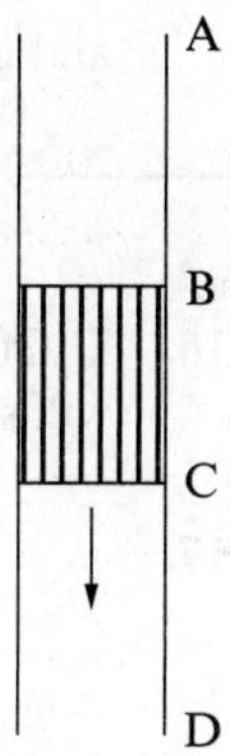


If both ends are open, then mercury will fall down.

**B**

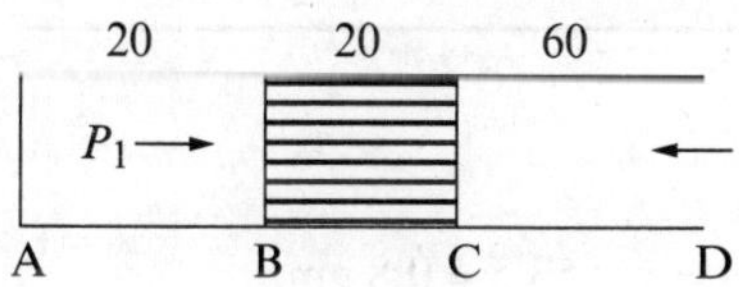


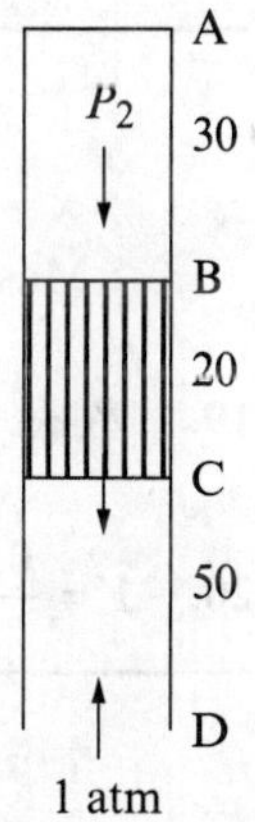


For AB column:

$P_1 = 1$ atm $= 76$ cm Hg

$V_1 = 20$ A cm$^3$

$P_2 = 1$ atm – 20 cm Hg

$= 76 - 20$

$= 56$ cm Hg

$V_2 = 30$ A cm$^3$

Now, $P_1V_1 \neq P_2V_2$

Hence, only end A is not closed.

**C**

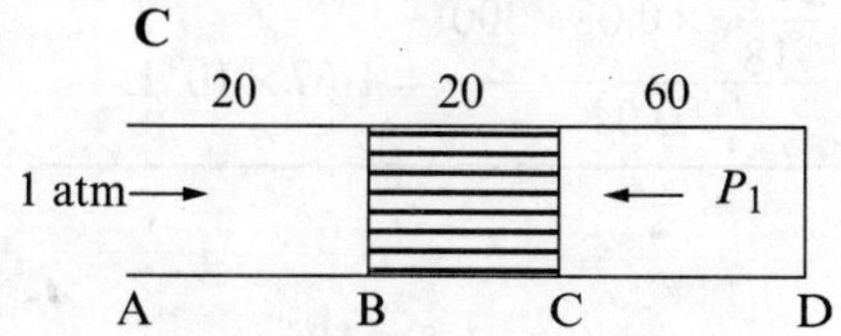


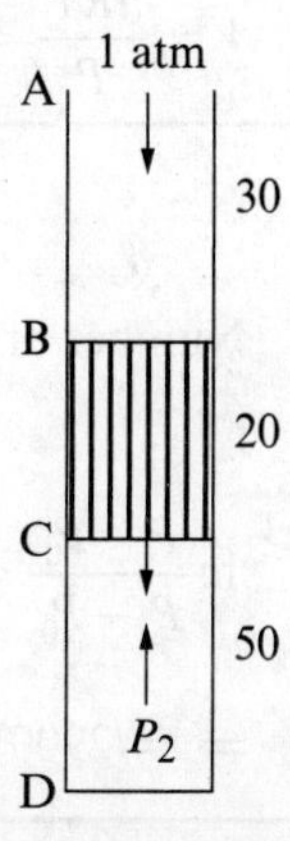


For CD column:

$P_1 = 1$ atm $= 76$ cm Hg

$V_1 = 60$ A cm$^3$

$P_2 = 1$ atm + 20 cm Hg

$= 76 + 20$

$= 96$ cm Hg

$V_2 = 50$ A cm$^3$

As $P_1V_1 \neq P_2V_2$, only end D is not closed.

Hence, both the ends are closed.

**8.**

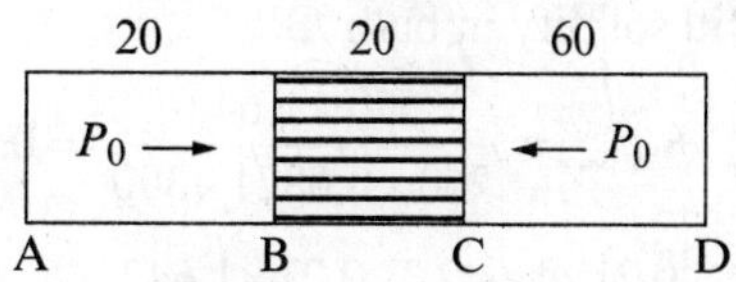


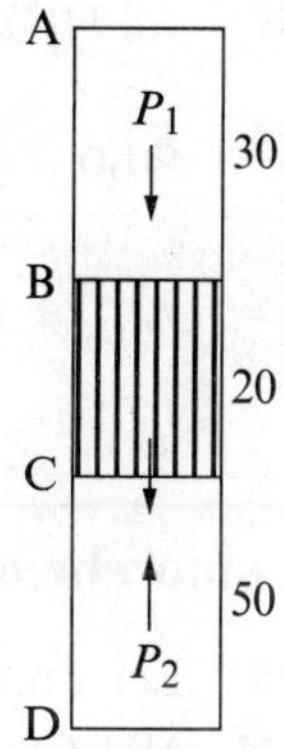


For AB column:

$P_0 \times 20 = P_1 \times 30$

For CD column:

$P_0 \times 60 = P_2 \times 50$

As $P_1 + 20$ cm Hg $= P_2$

or, $\dfrac{20P_0}{30} + 20 \text{ cm Hg} = \dfrac{60P_0}{50} \Rightarrow P_0 = 37.5$ cm Hg

**9.** **D**

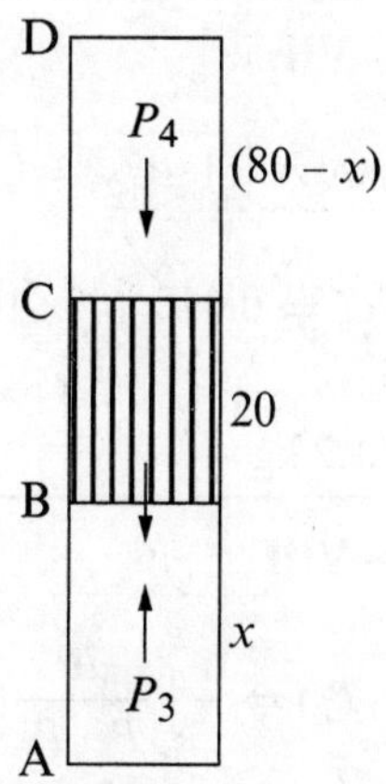


$P_4 + 20 \text{ cm Hg} = P_3$

or, $\dfrac{60P_0}{(80 - x)} + 20 = \dfrac{20P_0}{x} \Rightarrow x = 13.88$

## Comprehension – IV

10. Total moles of product gases $= \frac{PV}{RT}$

$= \frac{410.5 \times 2.9}{0.0821 \times 2000} = 7.25$

$\therefore$ Moles of gases per 0.04 mole of nitroglycerine $= 0.04 \times 7.25 = 0.29$

11. Moles of gases except A $= \frac{4.75 \times 0.821}{0.0821 \times 250} = 0.19$

'A' must be $H_2O$ because it solidifies at –23°C and its mole = 0.29 – 0.19 = 0.10.

12. Moles of gases C and D $= \frac{2.1 \times 0.821}{0.0821 \times 300} = 0.07$

$\therefore$ Mole of gas 'B', which is $CO_2$ = 0.19 – 0.07 = 0.12

13. The gas remained, D must be $N_2$ and its mole $= \frac{1.8 \times 0.821}{0.0821 \times 300} = 0.06$

and gas 'C' is $O_2$ and its mole = 0.07 – 0.06 = 0.01.

## Comprehension – V

14. All $H_2O(g)$ will solidify in bulb 'B'.

15. $n_{H_2O} + n_{CO_2} + n_{N_2} = \frac{570 \times 1.642}{760 \times 0.0821 \times 300} = 0.05$

$n_{CO_2} + n_{N_2} = \frac{0.21 \times 1.642}{0.0821 \times 300} + \frac{0.21 \times 1.642}{0.0821 \times 200} = 0.035$

$\therefore n_{H_2O} = 0.05 - 0.035 = 0.015$

16. $H_2O(g)$ will solidify in 'B' as well as 'C' but $CO_2(g)$ will solidify only in 'C'.

17. $n_{N_2} = \frac{22.8 \times 1.642}{760 \times 0.0821}\left[\frac{1}{300} + \frac{1}{200} + \frac{1}{80}\right] = 0.0125$

$\therefore n_{CO_2} = 0.035 - 0.0125 = 0.0225$

## Comprehension VI

18. Let $x$ mole, $NH_4Cl$ was present initially.

$NH_4Cl(s) \rightarrow NH_3(g) + HCl(g)$

$x$ mole $\quad x$ mole

Now, $PV = nRT$

$114 \times V = 0.01 \times R \times 300$

and $908 \times V = (0.01 + 2x) \times R \times 600$

$\therefore x \approx 0.015$

$\therefore$ Mass of $NH_4Cl = x \times 53.5 \approx 0.8$ gm

19. $P_{NH_3} = \frac{908 - 114 \times 2}{2}$ 340 mm Hg

20. $V = \frac{0.01 \times 0.0821 \times 300 \times 760}{114} = 1.642$ L

## Comprehension – VII

21. Water will vaporize till $P_{H_2O} = 0.04$ atm

Now, $PV = nRT \Rightarrow 0.04 \times (40 \times 10^3) = \frac{w}{18} \times 0.08 \times 300$

or $w = 1200 = 1.2$ kg

$\therefore$ Percentage of water vaporized $= \frac{1.2}{5} \times 100 = 24\%$

22. $V = \frac{nRT}{P} = \frac{\frac{5000}{18} \times 0.08 \times 300}{0.04} = 1.67 \times 10^5$ L

## Comprehension – VIII

23. $\frac{-dP}{dt} = K(P - P_0) \Rightarrow -\int_{P_1}^{P_2} \frac{dP}{P - P_0} = K\int_0^t dt$

$\Rightarrow \ln\frac{P_1 - P_0}{P_2 - P_0} = Kt$

or, $\ln\frac{20 - 1}{P_2 - 1} = 0.001 \times 3600 = \ln 38 \Rightarrow P_2 = 1.5$ atm

24. Number of balloons $= \frac{(20 - 1.5) \times 10}{1 \times 2} = 92.5 \approx 92$

25. $\ln\frac{P_1 - P_0}{P_2 - P_0} = Kt \Rightarrow \ln\frac{20 - 1}{2 - 1} = 0.001 \times t$

$\Rightarrow t = 2900$ sec

## Comprehension – IX

**26.** $-\dfrac{dP}{dt} = \text{K.P} \Rightarrow -\displaystyle\int_{1.5\,\text{atm}}^{P} \dfrac{dP}{P} = K\int_0^t dt$

$\Rightarrow P = (1.5\text{ atm}).e^{-Kt}$

**27.**

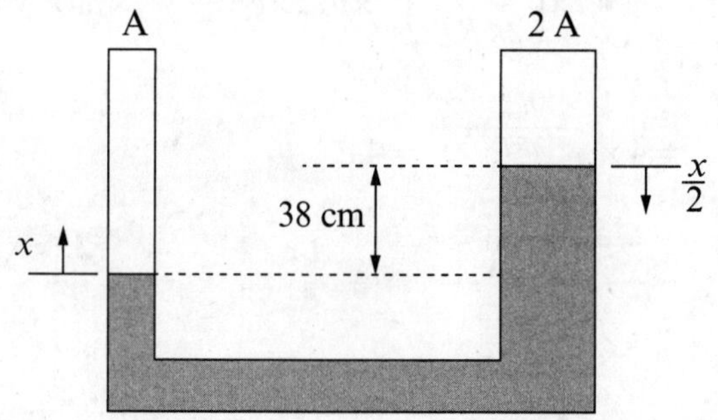


The pressure of gas in closed arm,

$P = 1\text{ atm} + \left(38 - \dfrac{3x}{2}\right)\text{cm}$

$= \left(114 - \dfrac{3x}{2}\right)\text{cm Hg}$

Now, $\left(114 - \dfrac{3x}{2}\right) = 114 \times e^{-kt}$

$\Rightarrow x = 76\,(1 - e^{-kt})\text{ cm Hg}$

**28.** $\dfrac{x}{2} = 38(1 - e^{-kt})\text{ cm Hg}$

## Comprehension – X

**29.** $u_{\text{mp}} = \sqrt{\dfrac{2RT}{M}} \Rightarrow T = \dfrac{M \times u_{\text{mp}}^2}{2R}$

$= \dfrac{(32\times10^{-3})\times(400)^2}{2\times8} = 320\text{K} = 47°\text{C}$

**30.** $u_{\text{rms}} - u_{\text{mp}} = 400\text{ m/s} \Rightarrow \sqrt{\dfrac{3RT}{M}} - \sqrt{\dfrac{2RT}{M}} = 400\text{ m/s}$

or, $T = \left(\dfrac{400}{\sqrt{3}-\sqrt{2}}\right)^2 \times \dfrac{M}{R}$

$= \dfrac{(400)^2}{3+2-2\sqrt{6}} \times \dfrac{2\times10^{-3}}{8} = 400\text{ K} = 127°\text{C}$

**31.**

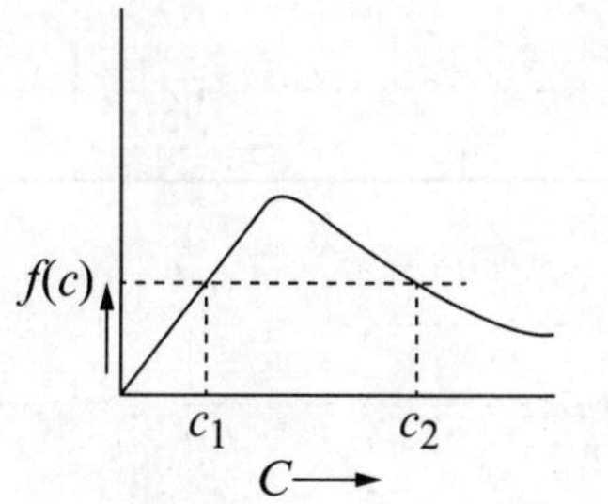


$4\pi\left(\dfrac{M}{2\pi RT}\right)^{3/2} \times C_1^{\,2} \times e^{\frac{-MC_1^2}{2RT}}$

$= 4\pi\left(\dfrac{M}{2\pi RT}\right)^{3/2} \times C_2^{\,2} \times e^{\frac{-MC_2^2}{2RT}}$

or, $\dfrac{C_1^2}{C_2^2} = e^{\frac{M(C_1^2 - C_2^2)}{2RT}}$

or, $\dfrac{M(C_1^2 - C_2^2)}{2RT} = 2\ln\left(\dfrac{C_1}{C_2}\right)$

$\therefore T = \dfrac{M(C_1^2 - C_2^2)}{4R.\ln\left(\dfrac{C_1}{C_2}\right)} = \dfrac{(28\times10^{-3})(300^2 - 600^2)}{4\times8\times\ln\left(\dfrac{300}{600}\right)}$

$= 337.5\text{ K} = 64.5°\text{C}$

**32.**

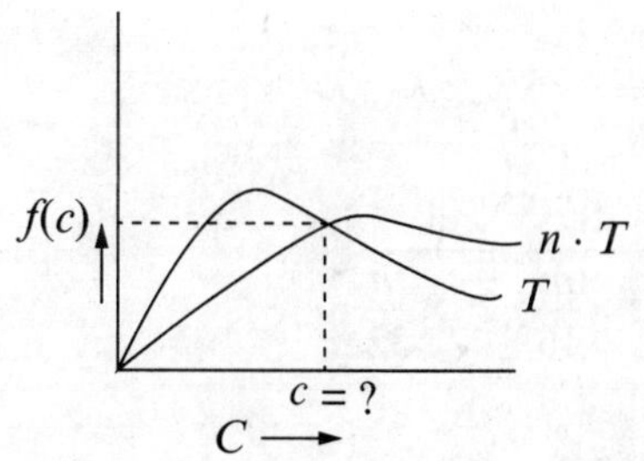


$4\pi\left(\dfrac{M}{2\pi RT}\right)^{3/2} \times c^2 \times e^{-\frac{MC^2}{2RT}}$

$= 4\pi\left(\dfrac{M}{2\pi RT.n}\right)^{3/2} \times c^2 \times e^{-\frac{MC^2}{2RT.n}}$

or, $n^{3/2} = e^{\frac{MC^2}{2RT}\left(1-\frac{1}{n}\right)}$

or, $\dfrac{3}{2}\ln n = \dfrac{MC^2}{2RT} \times \dfrac{n-1}{n}$

$\therefore C = \sqrt{\dfrac{3nRT\ln n}{M(n-1)}}$

**33.**

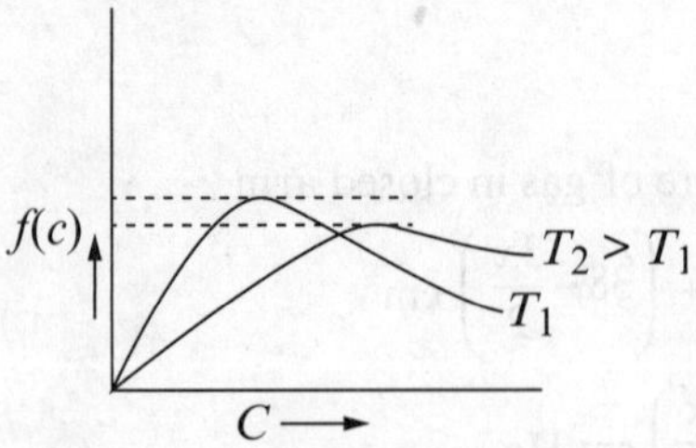


**34.**

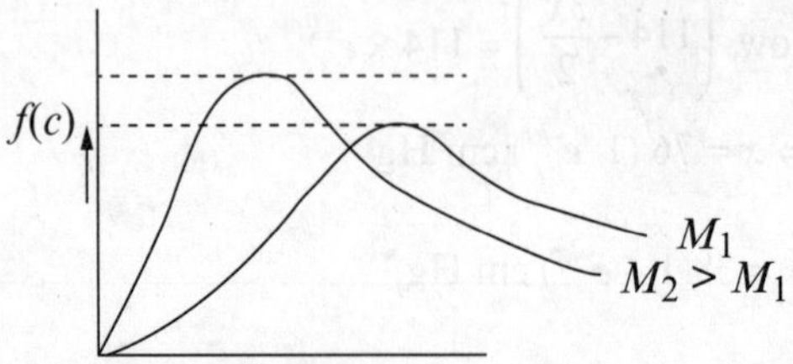


**35.** $\dfrac{(dN)_1}{(dN)_2} = \dfrac{4\pi \times \left(\dfrac{M}{2\pi RT}\right)^{3/2} \times (2u_{mp})^2 \times e^{-\frac{M(2u_{mp})^2}{2RT}} \times du \times N}{4\pi \times \left(\dfrac{M}{2\pi RT}\right)^{3/2} \times u_{mp}^2 \times e^{-\frac{M \times u_{mp}^2}{2RT}} \times du \times N}$

$= 4 \times e^{-\frac{Mu_{mp}^2}{2RT}(4-1)} = 4.e^{-3}$

## Comprehension – XI

**36.** $n_A = \dfrac{m}{2}, n_B = \dfrac{m}{16}, n_C = \dfrac{m}{32}$

**37.** $Z_W = \dfrac{1}{4} \times u_{av} \times N^* = \dfrac{1}{4}\sqrt{\dfrac{8RT}{\pi M}} \times N^*$

$\Rightarrow Z_W \propto \dfrac{N^*}{\sqrt{M}}$

**38.** $\lambda = \dfrac{1}{\sqrt{2}\pi\sigma^2 N^*} \Rightarrow \lambda_A : \lambda_B : \lambda_C$

$= \dfrac{1}{1^2 \times \dfrac{m}{2}} : \dfrac{1}{2^2 \times \dfrac{m}{16}} : \dfrac{1}{2^2 \times \dfrac{m}{32}}$

$= 1 : 2 : 4$

**39.** $E_{total} = \dfrac{3}{2}$nRT and $n_{max}$ for A

**40.** $Z = 1$ for all (Ideal behaviour)

**41.** $Z_1 = \sqrt{2}\,\pi\,\sigma^2\, u_{av} \times N^*$

$Z_A : Z_B : Z_C = 1^2 \times \dfrac{1}{\sqrt{2}} \times \dfrac{m}{2} : 2^2 \times \dfrac{1}{\sqrt{16}} \times \dfrac{m}{16} : 2^2 \times \dfrac{1}{\sqrt{32}} \times \dfrac{m}{32}$

$= \dfrac{1}{2\sqrt{2}} : \dfrac{1}{64} : \dfrac{1}{128\sqrt{2}}$

**42.** $u_{av} \propto \dfrac{1}{\sqrt{M}}$

## Comprehension – XII

For critical point, $\dfrac{dP}{dV} = 0$ and $\dfrac{d^2P}{dV^2}$

Now, $\dfrac{dP}{dV} = 0 \Rightarrow -\dfrac{RT}{(V-b)^2} + \dfrac{2a}{T.V^3} = 0$

$\Rightarrow \dfrac{RT}{(V-b)^2} = \dfrac{2a}{T.V^3}$ (1)

and $\dfrac{d^2P}{dV^2} = 0 \Rightarrow \dfrac{2RT}{(V-b)^3} - \dfrac{6a}{T.V^4} = 0$

$\Rightarrow \dfrac{2RT}{(V-b)^3} = \dfrac{3a}{T.V^4}$ (2)

From (1) ÷ (2) : $V - b = \dfrac{2V}{3} \Rightarrow V_C = 3b$

Eq 1 : $\dfrac{RT}{(3b-b)^2} = \dfrac{2a}{T.(3b)^3} \Rightarrow T_C = \sqrt{\dfrac{8a}{27Rb}}$

HINTS AND EXPLANATIONS

and $P_C = \frac{RT}{V-b} - \frac{a}{T.V^2} = \sqrt{\frac{aR}{216b^3}}$

**43.** $T_C = \sqrt{\frac{8a}{27Rb}}$

**44.** $P_C = \sqrt{\frac{aR}{216b^3}}$

**45.** $V_C = 3b$

## Section D (Assertion – Reason)

1. Avogadro's hypothesis is valid only for gases due to large intermolecular distance.
2. Charle's law
3. $d = \frac{PM}{RT}$ but $M$ is independent from $d$, $P$ or $T$.
4. $H_2$ and $Cl_2$ are reactive gases.
5. Escaping tendency increases only on increasing the energy of molecules.
6. Graham's law is valid for ideal as well as non-ideal gases.
7. Theory based
8. Theory based
9. Volume of ideal gas should be the total volume minus the volume occupied by gas molecules.
10. As the average K.E. is same, increase in mass decreases their speed.
11. Total K.E. = $\frac{3}{2}$nRT

    As the pressure exerted by the vapour is same in both but volume is in 1 : 2 ratio, the moles is also in 1 : 2 ratio.
12. 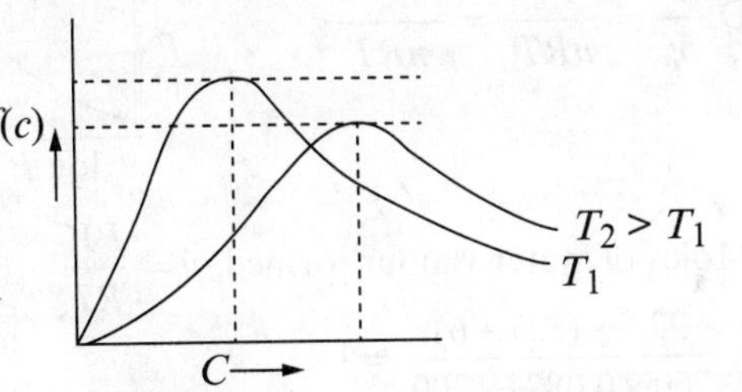

13. Concept based
14. Excluded volume is 'nb'.
15. $T_C < T_B$ and hence, attractive forces are dominant.
16. $PV$ is constant at constant temperature.
17. Above Boyle's temperature, gases show positive deviation.
18. Theory based
19. Theory based
20. K.E. of molecules is the function of T.

## Section E (Column Match)

**1.** Boyle's law : $PV = K \Rightarrow \left(\frac{dP}{dV}\right)_T = -\frac{K}{V^2} = -\frac{P}{V}$

And $\left[\frac{d(PV)}{dP}\right]_T = 0$

Charle's law $\frac{V}{T} = K \Rightarrow \left(\frac{dV}{dT}\right)_P = K = \frac{V}{T}$

Avogadro's law : $\frac{V}{n} = K = \frac{RT}{P}$

$\Rightarrow \left(\frac{dV}{dn}\right)_{P,T} = \frac{RT}{P}$

Graham's law $r = \left(-\frac{dP}{dt}\right) \propto \frac{1}{\sqrt{d}}$

**2.** Average translational K.E. per mole = $\frac{3}{2}$RT

Average translational K.E. per gram = $\frac{3}{2}\frac{RT}{M}$

**3.** $T_C = \frac{8a}{27Rb}$ $\left(\frac{a}{b}\right)_X = 120$, $\left(\frac{a}{b}\right)_Y = 333.33$, $\left(\frac{a}{b}\right)_Z = 171.4$

$V_C = 3b$

$P_C = \frac{a}{27b^2}$, $\left(\frac{a}{b^2}\right)_X = 4800$, $\left(\frac{a}{b^2}\right)_Y = 11111.11$, $\left(\frac{a}{b^2}\right)_Z = 4898$

**4.** A. $\frac{P}{V} = \frac{P}{\left(\frac{nRT}{P}\right)} = \frac{P^2}{nRT}$

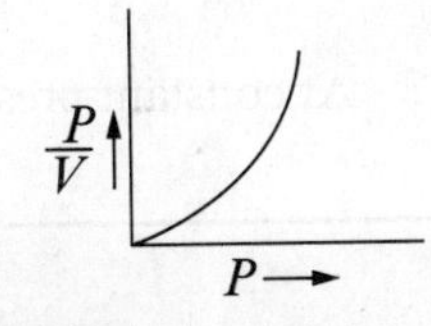

B. $\frac{P}{V} = \frac{nRT/V}{V} = \frac{nRT}{V^2}$

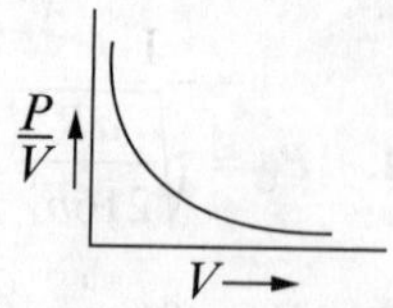


C. $\frac{V}{P} = \frac{nRT}{P^2}$

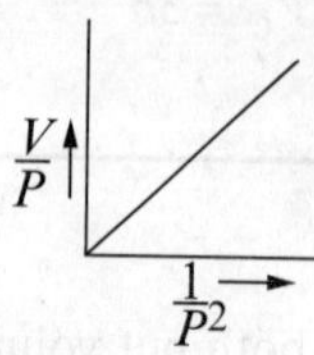


D. $\frac{P}{V} = \frac{P^2}{nRT} = \frac{10^{2\log P}}{nRT}$

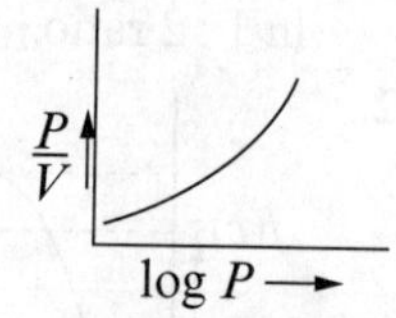


**5.** Moles of water vapour formed, $n = \frac{PV}{RT}$

$= \frac{22.8\times(827-6)}{760\times0.0821\times300} = 1$

**6.** Pressure correction $= \frac{a.n^2}{V^2} = \frac{4\times5^2}{10^2} = 1$ atm

Ideal volume $= V - nb = 10 - 5\times0.05 = 9.75$ L

Volume occupied by molecules $= \frac{nb}{4} = \frac{0.25}{4}$

$= 0.0625$ L

Volume correction $= nb = 0.25$L

**7.** Theory based

**8.** (A) $\lambda = \frac{1}{\sqrt{2}\pi\sigma^2 N^*} = \frac{RT}{\sqrt{2}\pi\sigma^2 PN_A}$

At constant volume, $N^*$ = constant $\Rightarrow \frac{\lambda_2}{\lambda_1} = 1$

At constant pressure, $\lambda \propto T \Rightarrow \frac{\lambda_2}{\lambda_1} = 2$

(B) $Z_1 = \sqrt{2}\,\pi\sigma^2 \times u_{av} \times N^*$

$= \sqrt{2}\,\pi\sigma^2 \times \sqrt{\frac{8RT}{\pi M}}\left(\frac{PN_A}{RT}\right)$

At constant volume : $Z_1 \propto \sqrt{T} \Rightarrow \frac{Z_{1,2}}{Z_{1,1}} = \sqrt{2}$

At constant pressure : $Z_1 \propto \frac{1}{\sqrt{T}} \Rightarrow \frac{Z_{1,2}}{Z_{1,1}} = \frac{1}{\sqrt{2}}$

(C) $Z_{11} = \frac{1}{\sqrt{2}}\pi\sigma^2 . u_{av} . N^{*2} = \frac{1}{\sqrt{2}}\pi\sigma^2 \sqrt{\frac{8RT}{\pi M}}\left(\frac{PN_A}{RT}\right)^2$

At constant volume : $Z_{11} \propto \sqrt{T} \Rightarrow \frac{Z_{11,2}}{Z_{11,1}} = \sqrt{2}$

At constant pressure : $Z_{11} \propto \frac{1}{(T)^{3/2}} \Rightarrow \frac{Z_{11,2}}{Z_{11,1}}$

$= \frac{1}{2\sqrt{2}}$

**9.** A. $P = 1$ atm + 38 cm Hg = 1.5 atm

Q. $P = 1$ atm + 57 cm Hg = 1.75 atm

R. $P = 38$ cm Hg = 0.5 atm

S. $P = 1$ atm + 1.9 m glycerine $= 1 + \frac{190\times2.72}{13.6\times76}$

$= 1.5$ atm

**10.** $T_C = 273 + (-177) = 96$ K $= -177$°C

$T_B = \frac{27}{8}\times96 = 324$ K $= 51$°C

**A.** $V_i = \frac{0.2\times0.08\times96}{20}\times10^3 = 76.8$ ml

But $V_{real} < V_{ideal}$ in given condition $\Rightarrow V_r < 76.8$ ml

**B.** $Z = 1 \Rightarrow V_r = V_i = \frac{0.2\times0.08\times324}{6.48}\times10^3 = 800$ ml

**C.** Above Boyle's temperature, $Z > 1$

$\therefore V_r > V_i = \frac{0.2\times0.08\times350}{7}\times10^3 = 800$ ml

**D.** Below Boyle's temperature, $Z < 1$

$\therefore V_r < V_i = \frac{0.2\times0.08\times300}{6}\times10^3 = 800$ ml

**E.** $T = T_B$ but $P > 50$ atm $\Rightarrow Z > 1$

$\therefore V_r > V_i = \frac{0.2\times0.08\times324}{64.8}\times10^3 = 80$ ml

## Section F (Subjective)

### Single-digit Integer Type

**1.** $\frac{V_1}{T_1} = \frac{V_2}{T_2} \Rightarrow \frac{45}{300} = \frac{42}{T_2} \Rightarrow T_2 = 280$ K $= 7$°C

**2.**

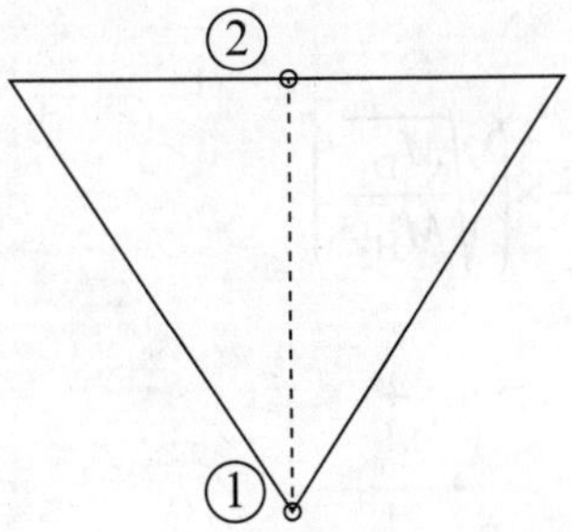


$P_1 = 1 \text{ atm} + hm \text{ water} = (10 + h) \text{ m water}$

$V_1 = \frac{4}{3}\pi\ (1 \text{ mm})^3 \qquad A = \pi r^2 = \pi \text{ mm}^2$

$\therefore r = 1 \text{ mm}$

$P_2 = 1 \text{ atm} = 10 \text{ m water}$

$V_2 = 2\pi \text{ mm}^3$

Now, $P_1V_1 = P_2V_2 \Rightarrow (10 + h) \times \frac{4}{3}\pi = 10 \times 2\pi$

$\Rightarrow h = 5 \text{ m}$

Hence, water holding capacity of pool, $V = \frac{1}{3}\pi r^2 h$

$= \frac{1}{3}\pi (10 \text{ m})^2 \times 5 \text{ m} = \frac{500\pi}{3} \text{ m}^3$

**3.** Number of cosmic events = Number of Ar-atoms

$= \frac{1.911 \times 10^{-6} l}{22.7 l} \times 6 \times 10^{23} = 5.05 \times 10^{16}$

**4.** For lifting of balloon, $B > W$

or, $(V \times \rho_{\text{outside air}} \times g) > (V \times \rho_{\text{inside air}} + m_{\text{additional}})g$

or, $V(\rho_{\text{outside air}} - \rho_{\text{inside air}}) > m_{\text{additional}}$

or, $91\left(\frac{1 \times 29}{0.08314 \times 290} - \frac{1 \times 29}{0.08314 \times T}\right) > \frac{10}{8.314}$

$\therefore T > 293.22$

Hence, difference in temperature, $\Delta T > 3.22 \approx 4$ K.

**5.** $n_{\text{released gas}} = n_{\text{taken}} - n_{\text{remained}}$

or, $\frac{m}{22.4 \times 1.25} = \frac{P \times 3}{0.0821 \times 273} - \frac{(p - 0.8) \times 3}{0.0821 \times 273}$

$\Rightarrow m = 3$ gm

**6.**

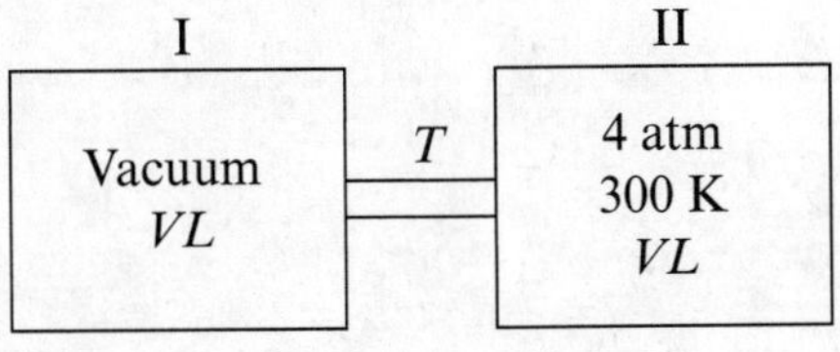


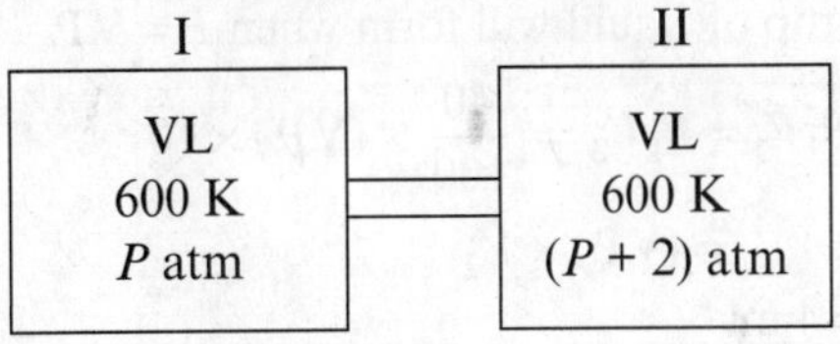


Final moles of gases in vessel I and II = Initial mole in vessel II

or, $\frac{P \times V}{R \times 600} + \frac{(P+2) \times V}{R \times 600} = \frac{4 \times V}{R \times 300} \Rightarrow P = 3 \text{ atm}$

**7.** Let the mixture contains $a$ moles $C_xH_8$ and $b$ moles $C_xH_{10}$.

Now, $a \times (12x + 8) + 6 \times (12x + 10) = 28.4$ (1)

$a + b = \frac{PV}{RT} = \frac{2.46 \times 5}{0.082 \times 300} = 0.5$ (2)

and $28.4 \times \frac{84.5}{100} = a \times 12x + b \times 12x$ (3)

On solving, $x \approx 4$

**8.** From $PV = nRT$, $V = \frac{nR}{P}.T$

As the pressure is constant, the change in slope is only due to change in moles.

| | $X_n$ | $\rightarrow$ | $nX$ |
|---|---|---|---|
| Initial | $a$ mole | | o |
| Final | $a - 0.6a$ $= 0.4a$ | | $0.6axn$ |

Now, $\frac{a}{0.4a + 0.6an} = \frac{(50.2 - 49.9)/20}{(49.1 - 47.9)/20} \Rightarrow n = 6$

**9.**

| | $2H_2$ | + | $O_2$ | $\rightarrow$ | $2H_2O$ |
|---|---|---|---|---|---|
| | 2a | | a | | 0 |
| Final | 2a – 1.6a = 0.4a | | a – 0.8a = 0.2a | | 1.6a |

Now, $\frac{P}{nT} = \frac{R}{V} = \text{Constant} \Rightarrow \frac{P_1}{n_1T_1} = \frac{P_2}{n_2T_2}$

$\Rightarrow \frac{4.5}{3a \times 330} = \frac{P_2}{2.2a \times 400} \quad \therefore P_2 = 4 \text{ atm}$

**10.** $n_{\text{total}} = n_{O_2} + n_{N_2}$

or, $\frac{1.1 \times 30}{RT} = \frac{Po_2 \times 30}{RT} + \frac{0.9 \times 10}{RT}$

$\Rightarrow P_{O_2} = 0.8 \text{ atm}$

**11.** $P_{H_2O} = \frac{40}{100} \times (\text{V.P.})$

HINTS AND EXPLANATIONS

First drop of liquid will form when $P = \text{V.P.}$

Now, $P_1V_1 = P_2V_2 \Rightarrow \frac{40}{100} \times (\text{V.P.}) \times 10$

$= (\text{V.P.}) \times V_2$

$\therefore V_2 = 4 \text{ ml}$

**12.** $H_2O(l) \rightarrow H_2(g) + \frac{1}{2} O_2(g)$

| | $H_2O(l)$ | $H_2(g)$ | $\frac{1}{2}O_2(g)$ |
|---|---|---|---|
| | $a$ mole | 0 | 0 |
| Final | 0 | $a$ mole | 0.5 $a$ mole |

$\Delta \text{P.V} = \Delta n.\text{RT}$

or, $(1.86 - 0.96) \times 20 = (1.5a) \times 0.08 \times 300 \Rightarrow a = 0.5$

$\therefore$ Mass of water present initially $= 0.5 \times 18 = 9$ gm

**13.** Initial total moles $= \frac{PV}{RT} = \frac{24.63 \times 3}{0.0821 \times 300} = 3$

$\therefore$ Initial mole of $H_2 = 3 - 1 = 2$

Final mole ratio, $\frac{n_{H_2}}{n_{D_2}} = \frac{1/2}{4/4} = \frac{1}{2}$

Now, $\frac{n^f{}_{H_2}}{n^f{}_{D_2}} = \frac{n^i{}_{H_2}}{n^i{}_{D_2}} \times \left(\sqrt{\frac{M_{D_2}}{M_{H_2}}}\right)^n$

or, $\frac{1}{2} = \frac{2}{1} \times \left(\sqrt{\frac{4}{2}}\right)^n \Rightarrow n = 4$

**14.** For critical point, $\frac{dP}{dV_m} = 0$ and $\frac{d^2P}{dV_m^2} = 0$

On solving, $b = 0$. But $b \neq 0$ from question. Hence, the gas does not have critical condition.

**15.** Boyle's temperature:

$T_B = \frac{a}{Rb} = \frac{4.105}{0.0821 \times 0.1} = 500 \text{ K}$

Hence, at 500 K, the gas will behave ideally.

Not, $d = \frac{PM}{RT} = \frac{2 \times 164.2}{0.0821 \times 500} = 8 \text{ g/L} = 8 \text{ kg/m}^3$

## Four-digit Integer Type

**1.**

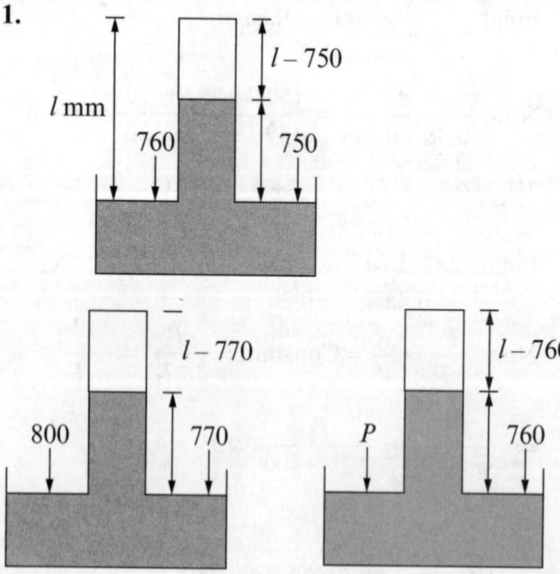


$(760 - 750) \times (l - 750) = (800 - 770) \times (l - 770) =$
$(P - 760) \times (l - 760)$

$\therefore P = 775$ mm Hg

**2.** Mass of LNG $= 10 \text{ m}^3 \times 416 \text{ kg/m}^3 = 416 \times 10^4$ gm

$\therefore$ Moles of $CH_4 = \frac{916 \times 10^4}{16} = 26 \times 10^4$

Now, $V = \frac{nRT}{P} = \frac{26 \times 10^4 \times 0.021 \times 300}{1.692} = 3.9 \times 10^6 \text{ L}$

$= 3900 \text{ m}^3$

**3.** $\frac{V_1}{n_1T_1} = \frac{V_2}{n_2T_2} \Rightarrow \frac{V}{n \times 303} = \frac{1.6V}{1.2n \times T_2} \Rightarrow T_2 = 404 \text{ K}$

**4.** $V_{\text{initial}} = a = \frac{nRT}{P} = \frac{64 \times 0.08 \times 300}{64 \times 3} = 8 \text{ L}$

$V_{\text{initial}} = b = \frac{nRT}{P} = \frac{(64 - 8) \times 0.08 \times 300}{64 \times 3} = 7 \text{ L}$

$P = \frac{nRT}{V} = \frac{64 \times 0.08 \times 300}{64 \times 7} = \frac{24}{7} \text{ atm}$

**5.** 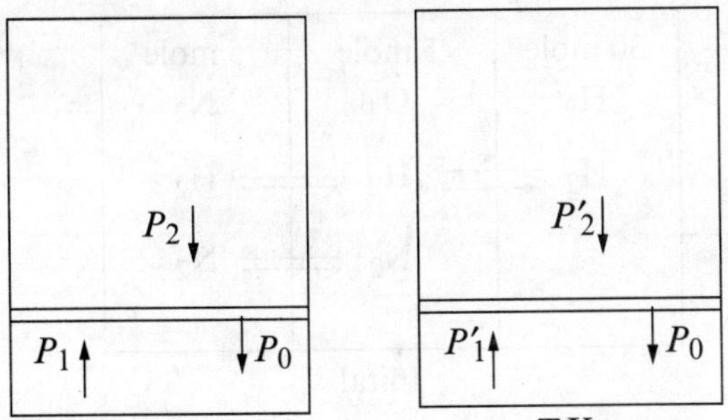


$P_2 + P_0 = P_1$ $\qquad$ $P'_2 + P_0 = P'_1$

$P_0 = P_1 - P_2 = P'_1 - P'_2$

or,

$$\frac{n\times R\times 320}{\left(\frac{V}{5}\right)} - \frac{n\times R\times 320}{\left(\frac{4V}{5}\right)} = \frac{n\times R\times T}{\left(\frac{V}{4}\right)} - \frac{n\times R\times T}{\left(\frac{3V}{4}\right)}$$

$\therefore T = 450$ K

**6.** Mass of gas used = 28.8 – 23.2 = 5.6 kg

Volume of gas used up,

$$V = \frac{nRT}{P} = \frac{(5.6\times10^3)\times0.08\times300}{56\times1}$$

= 2400 L

Now, $\frac{P_1}{M_1} = \frac{P_2}{M_2} \Rightarrow \frac{35}{(28.8-14.8)} = \frac{P_2}{(23.2-14.8)}$

$\Rightarrow P_2 = 21$ atm

**7.** $n_{NO} = \frac{1.6\times0.75}{0.08\times3.00} = 0.05$

$n_{O_2} = \frac{1.2\times0.25}{0.08\times3.00} = 0.0125$

| | $2NO$ | + $O_2$ | → $2NO_2$ | → $N_2O_4$ |
|---|---|---|---|---|
| | 0.05 | 0.0125 | 0 | 0 |
| Final | $\frac{-0.025}{0.025}$ | $\frac{-0.0125}{0}$ | $\frac{0}{0}$ | $\frac{0.0125}{0.0125}$ |

But at 200 K, $N_2O_4$ is solid. Hence, the only gas is NO.

Millimoles of NO remained = 0.025 × 1000 = 25

Now, $P = \frac{0.025\times0.08\times200}{(0.75+0.25)} = 0.4$ atm

**8.** 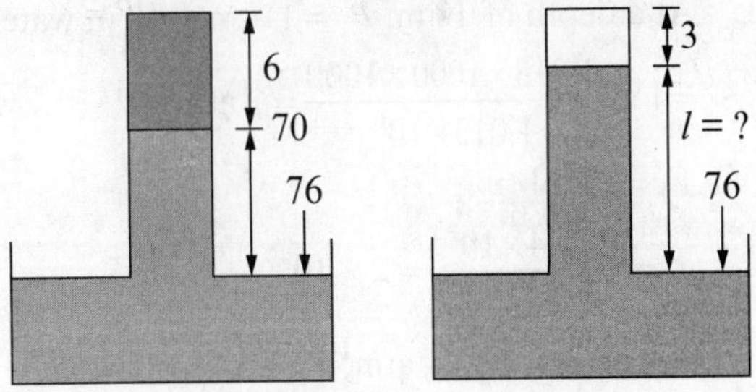


$P_1 = (6 - 1) = 5$ cm Hg $\qquad$ $P_2 = ?$

$V_1 = 6$ A cm$^3$ $\qquad$ $V_2 = 3$ A cm$^3$

$$P_2 = \frac{5\times6A}{3A} = 10 \text{ cm Hg}$$

Hence, final total pressure in table above mercury = 10 + 1 = 11 cm Hg

∴ Barometer reading = 76 – 11 = 65 cm

**9.** Mass of water vapour present initially,

$$m_1 = \frac{\left(\frac{756}{100}\times\frac{24}{760}\right)\times V}{R\times300}\times18$$

Mass of water vapour finally remained,

$$m_2 = \frac{\frac{8.4}{760}\times V}{R\times280}\times18$$

$\therefore$ Fraction of water condensed $= \frac{m_1 - m_2}{m_1} = 0.5$

**10.** Let the process time = $t$ min

Mass of water vapour in inlet air,

$$m_1 = \frac{\left(\frac{20}{100}\times\frac{38}{760}\right)\times\left(10t\times10^3\right)}{0.08\times500}\times18 \text{ gm} = 2.5\,t \text{ gm}$$

Mass of water vapour in outlet air,

$$m_2 = \frac{\left(\frac{80}{100}\times\frac{19}{760}\right)\times\left(10t\times10^3\right)}{0.08\times400}\times18 \text{ gm} = 6.25\,t \text{ gm}$$

From question, $m_1 + 200 \text{ kg}\times\frac{36}{100} = m_2$

$\Rightarrow t = 19200$

**11.** At a depth of 10 m, $P_1 = 1$ atm + 10 m water

$= 1 + \dfrac{1.013\times1000\times1000}{1.013\times10^6} = 2$ atm

$V_1 = 24$ ml

$n_1 = \dfrac{2\times24\times10^{-3}}{0.08\times300} = 2\times10^{-3}$

At surface, $P_2 = 1$ atm, $V_2 = ?$,

$n_2 = 2\times10^{-3} - 0.05\times10^{-3}\times10$

$= 1.5\times10^{-3}$

Now, $\dfrac{P_1V_1}{n_1} = \dfrac{P_2V_2}{n_2} \Rightarrow \dfrac{2\times24}{2\times10^{-3}} = \dfrac{1\times V_2}{1.5\times10^{-3}} \Rightarrow V_2$

$= 36$ ml

For volume remaining uncharged, $\dfrac{P_1}{n_1} = \dfrac{P_2}{n_2}$

or, $\dfrac{2}{2\times10^{-3}} = \dfrac{1}{2\times10^{-3} - r\times10} \Rightarrow r = 10^{-4}$ mol/min

**12.**

| | $N_2O_4$ | $\rightarrow$ | $2NO_2$ |
|---|---|---|---|
| Initial mole | $\dfrac{20\times V}{RT}$ | | 0 |
| Final mole | $\dfrac{20\times V}{RT} - \dfrac{10V}{RT}$ | | $\dfrac{20V}{RT}$ |

$= \dfrac{10V}{RT}$

$NO_2$ will effuse through SPM till its pressure becomes same in both chamber and hence, mole ratio of $NO_2$ in chamber-I and II should be 1 : 3.

Final moles in chamber-I = $\dfrac{10V}{RT}$ of $N_2O_4$ and $\dfrac{5V}{RT}$ of $NO_2$

Final moles in chamber-II except $H_2O$ vapour = $\dfrac{15V}{RT}$ of $NO_2$

$\therefore$ Pressure of gas in chamber-I = $\dfrac{\dfrac{15V}{RT}\times R\times1.2T}{V}$ = 18 mm

and pressure of gases in chamber-II = $\dfrac{\dfrac{15V}{RT}\times R\times1.2T}{3V} + 30 = 36$ mm

**13.** $\dfrac{5}{2^n} \le 0.01 \Rightarrow n \ge 12.28$

**14.**

| 30 mole $H_2$ | 5 mole $O_2$ | 5 mole $N_2$ |
|---|---|---|
| $H_2 \rightleftharpoons$ | $H_2 \rightleftharpoons$ | $H_2$ |
| | $N_2 \rightleftharpoons$ | $N_2$ |

Inital

| $H_2$ = 10 mole | $H_2$ = 10 mole | $H_2$ = 10 mole |
|---|---|---|
| | $O_2$ = 5 mole | |
| | $N_2$ = 2.5 mole | $N_2$ = 2.5 mole |

Final

$P_1 : P_2 : P_3 = 10 : 17.5 : 12.5 = 4 : 7 : 5$

**15.**

| | | | | |
|---|---|---|---|---|
| $A_1 = \frac{n}{2}$ | $A_1 = \frac{n}{2}$ | | | |
| $A_2 = \frac{n}{3}$ | $A_2 = \frac{n}{3}$ | $A_2 = \frac{n}{3}$ | | |
| $A_3 = \frac{n}{4}$ | $A_3 = \frac{n}{4}$ | $A_3 = \frac{n}{4}$ | $A_3 = \frac{n}{4}$ | |

$\therefore \dfrac{P_{A_4}}{P_{A_{N-1}}} = \dfrac{n/5}{n/N} = 3 \Rightarrow N = 15$

**16.** $\dfrac{r_{H_2}}{r_{air}} = \sqrt{\dfrac{M_{air}}{M_{H_2}}} \Rightarrow \dfrac{100/t}{26/t} = \sqrt{\dfrac{M_{air}}{2}}$

and $\dfrac{r_{gas}}{r_{air}} = \sqrt{\dfrac{M_{air}}{M_{H_2}}} \Rightarrow \dfrac{100/t}{130/t} = \sqrt{\dfrac{M_{air}}{M_{gas}}}$

$\therefore M_{gas} = 50$

**17.** $\dfrac{3}{2}$ kT = mgh $\Rightarrow \dfrac{3}{2}\times\dfrac{8.4}{N_A}\times 300 = \dfrac{(40\times10^{-3})}{N_A}\times 10\times h$

$\therefore h = 9450$ m

**18.** At Boyle's temperature, the second virial coefficient is zero.

$B = a + b\,.\,e^{c/T^2} = 0$

or, $e^{-c/T^2} = -\dfrac{a}{b} \Rightarrow e^{-950/T^2} = -\dfrac{-0.02}{0.22} \Rightarrow T = 100$ k

**19.** $P_c = 73.89 \text{ atm} = \dfrac{a}{27b^2}$ and $T_c = 300 \text{ k} = \dfrac{8a}{27Rb}$

$\therefore \dfrac{300}{73.89} = \dfrac{8b}{R} \Rightarrow b = 0.0416 \text{ l mol}^{-1}$

$b = 4 \times$ Volume occupied by molecules per mole

$\therefore$ Volume occupied by molecules in 24 moles

$= \dfrac{b}{4} \times 24$

$= 0.25 \text{ L} = 250 \text{ ml}$

**20.** $\dfrac{d(d)}{dP} = 4 + 0.02 \times 2P + \ldots$

$\therefore \left(\dfrac{d(d)}{dP}\right)_{P \to 0} = 4 = \dfrac{M}{RT}$

$\Rightarrow M = 4 \times 0.08 \times 300 = 96$

# CHAPTER 4 Thermodynamics

## EXERCISE I (JEE MAIN)

### Basics

1. A thermodynamic property is that physical quantity
   (a) which is used in thermochemistry.
   (b) which obeys all the laws of thermodynamics.
   (c) which is used to define a state of the system.
   (d) which is used in measuring thermal change.

2. Which of the following is not a thermodynamic property of a system?
   (a) $H$ (b) $P$
   (c) $E$ (d) $w$

3. Which of the following is the best container for gas during isothermal process?
   (a) Glass vessel (b) Wood vessel
   (c) Copper vessel (d) Thermos flask

4. The internal energy of an ideal gas increases during an isothermal process when the gas is
   (a) expanded by adding more molecules to it.
   (b) expanded by adding more heat to it.
   (c) expanded against zero pressure.
   (d) compressed by doing work on it.

5. Maximum work can a gas do, if it is allowed to expand isothermally against
   (a) vacuum.
   (b) high pressure of surrounding.
   (c) low pressure of surrounding.
   (d) atmospheric pressure.

6. The internal energy change when a system goes from state A to B is 40 kJ/ mol. If the system goes from A to B by a reversible path and returns to state A by an irreversible path, then what would be the net change in internal energy?
   (a) 40 kJ (b) >40 kJ
   (c) <40 kJ (d) Zero

7. A system is said to be in thermodynamic equilibrium with the surroundings, if
   (a) it is only in thermal equilibrium with the surrounding.
   (b) it is in both thermal and chemical equilibrium with the surrounding.
   (c) it is in thermal, chemical as well as mechanical equilibrium with the surrounding.
   (d) it is in thermal and mechanical equilibrium, but not in chemical equilibrium with the surrounding.

8. If a closed system has adiabatic boundaries, then at least one boundary must be
   (a) permeable (b) imaginary
   (c) movable (d) fixed

9. Which of the following pair does show the extensive properties?
   (a) Temperature and pressure.
   (b) Viscosity and surface tension.
   (c) Refractive index and specific heat.
   (d) Volume and heat capacity.

EXERCISE I

10. Which of the following statement is correct?
    (a) Heat is thermodynamic property of system.
    (b) Work is thermodynamic property of system.
    (c) Work done by a conservative force is path function.
    (d) Heat involved in chemical reaction is a path independent physical quantity.

11. Which of the following statement is incorrect?
    (a) Only a state function may be expressed as difference in its value at two states in any process.
    (b) A process cannot be defined on the basis of initial and final states of the system.
    (c) In a cyclic process, the internal energy of the system remains throughout constant.
    (d) During irreversible process, the equation $PV = nRT$ is not applicable to ideal gas.

12. For an isothermal process, the essential condition is
    (a) $\Delta T = 0$ (b) $\Delta H = 0$
    (c) $\Delta U = 0$ (d) $dT = 0$

13. Which of the following statement is correct?
    (a) An ideal gas always obeys the equation $PV^{\gamma} =$ constant in adiabatic process.
    (b) An ideal gas always obeys the equation $PV^{x} =$ constant in polytropic process.
    (c) In a polytropic process, the heat capacity of the system remains same throughout.
    (d) In all the cyclic process, $w_{net}$ by the system is non-zero.

14. Which of following is incorrect about reversible process?
    (a) System always remains in thermodynamic equilibrium.
    (b) The process is extremely slow.
    (c) The process may be reversed at any stage only by making infinitesimally small change in opposite direction.
    (d) Reversible processes may be performed in finite time.

15. The law of equipartition of energy is applicable to the system whose constituents are
    (a) in random motion.
    (b) in orderly motion.
    (c) moving with constant speed.
    (d) in rest.

## First Law of Thermodynamics

16. A system absorbs 20 kJ heat and does 10 kJ of work. The internal energy of the system
    (a) increases by 10 kJ (b) decreases by 10 kJ
    (c) increases by 30 kJ (d) decreases by 30 kJ

17. The volume of a system becomes twice its original volume on the absorption of 300 cal of heat. The work done on the surrounding was found to be 200 cal. What is $\Delta U$ for the system?
    (a) 500 cal (b) 300 cal
    (c) 100 cal (d) −500 cal

18. A system absorbs 100 kJ heat in the process shown in the figure. What is $\Delta U$ for the system?

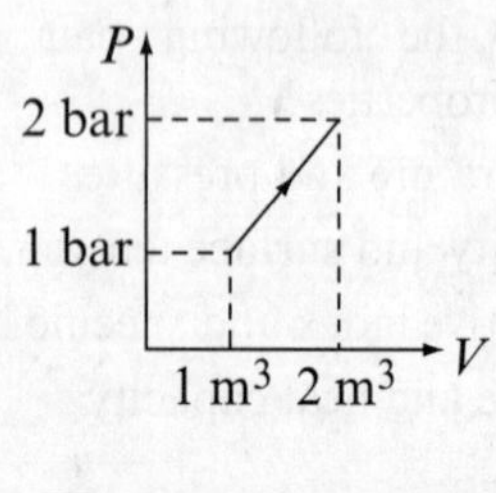


    (a) −50 kJ (b) +50 kJ
    (c) +150 kJ (d) −150 kJ

19. In a given process on an ideal gas, $dw = 0$ and $dq < 0$. Then for the gas,
    (a) the temperature will decrease.
    (b) the volume will increase.
    (c) the pressure will remain constant.
    (d) the temperature will increase.

20. Five moles of an ideal gas is expanded isothermally from 5 $dm^3$ to 5 $m^3$ at 300 K. Which of the following is incorrect about the gas?
    (a) No heat is absorbed or rejected by the gas.
    (b) There is no change in internal energy of the gas.
    (c) There is no change in enthalpy of the gas.
    (d) Pressure of the gas will decrease by 1000 times.

**21.** One mole of an ideal gas at 300 K is expanded isothermally from an initial volume of 1 L to 10 L. The change in internal energy, $\Delta U$, for the gas in this process is

(a) 163.7 cal (b) Zero
(c) 1381.1 cal (d) 9 L-atm

**22.** Five moles of an ideal gas expand isothermally and reversibly from an initial pressure of 100 atm to a final pressure of 1 atm at 27°C. The work done by the gas is ($\ln 100 = 4.6$)

(a) 2760 cal (b) 6000 cal
(c) 0 (d) 13,800 cal

**23.** The work done in the isothermal reversible expansion of argon gas at 27°C from 4 L to 16 L was equal to 4200 cal. What is the amount of argon subjected to such an expansion? (Ar = 40, $\ln 4 = 1.4$)

(a) 5.0 g (b) 20.0 g
(c) 200.0 g (d) 48.1 g

**24.** The minimum work which must be done to compress 16 g of oxygen isothermally at 300 K from a pressure of $1.01325 \times 10^3$ N/m$^2$ to $1.01325 \times 10^5$ N/m$^2$ is ($\ln 100 = 4.6$, $R = 8.3$ J/K-mol)

(a) 5727 J (b) 11.454 kJ
(c) 123.255 kJ (d) 1232.55 J

**25.** For a reversible process at $T = 300$ K, the volume of the ideal gas is increased from 1 L to 10 L. If the process is isothermal, then $\Delta H$ of the process is

(a) 11.47 kJ (b) 4.98 kJ
(c) 0 (d) −11.47 kJ

**26.** The magnitude of work done by one mole of a Van der Waals gas during its isothermal reversible expansion from volume $V_1$ to $V_2$ at temperature $T$ K, is

(a) $RT\ln\left(\frac{V_2}{V_1}\right)$

(b) $RT\ln\left(\frac{V_2-b}{V_1-b}\right)$

(c) $RT\ln\left(\frac{V_2-b}{V_1-b}\right)+a\left(\frac{1}{V_2}-\frac{1}{V_1}\right)$

(d) $RT\ln\left(\frac{V_2-b}{V_1-b}\right)-a\left(\frac{1}{V_2}-\frac{1}{V_1}\right)$

**27.** An ideal gas undergoes isothermal expansion from (10 atm, 1 L) to (1 atm, 10 L) either by path–I (infinite stage expansion) or by path–II (first against 5 atm and then against 1 atm). The value of $\left(\frac{q_{\text{path-I}}}{q_{\text{path-II}}}\right)$ is

(a) $\frac{2.303}{1.3}$ (b) $\frac{1.3}{2.303}$
(c) $\frac{1.0}{13\times 2.303}$ (d) $13\times 2.303$

**28.** An ideal gas is expanded irreversibly from 5 L to 10 L against a constant external pressure of 1 bar. The value of heat involved ($q$) in this isenthalpic process is

(a) 0 (b) +500 J
(c) +5 J (d) −500 J

**29.** The work done in an adiabatic change of fixed amount of an ideal gas depends on change in

(a) volume (b) pressure
(c) temperature (d) density

**30.** In the reversible adiabatic expansion of an ideal monoatomic gas, the final volume is 8 times the initial volume. The ratio of final temperature to initial temperature is

(a) 8 : 1 (b) 1 : 4
(c) 1 : 2 (d) 4 : 1

**31.** One mole of monoatomic ideal gas at $T$ K is expanded from 1 L to 2 L adiabatically under a constant external pressure of 1 atm. The final temperature of the gas in Kelvin is

(a) $T$ (b) $\frac{T}{2^{\left(\frac{5}{3}-1\right)}}$
(c) $T-\frac{2}{3\times 0.0821}$ (d) $T+\frac{3}{2\times 0.0821}$

**32.** Two moles of an ideal gas $\left(C_{v,m}=\frac{5}{2}R\right)$ was compressed adiabatically against constant pressure of 2 atm, which was initially at 350 K and 1 atm. The work done on the gas in this process is

(a) $250R$ (b) $500R$
(c) $125R$ (d) $300R$

EXERCISE I

33. Two moles of an ideal gas ($\gamma = 1.4$) was allowed to expand reversibly and adiabatically from 1 L, 527°C to 32 L. The molar enthalpy change of the gas is

(a) $-4200R$ (b) $-2100R$
(c) $-1500R$ (d) $-3000R$

34. Equal moles of He, $H_2$, $CO_2$ and $SO_3$ gases are expanded adiabatically and reversibly from the same initial state to the same final volume. The magnitude of work is maximum for (Assume ideal behaviour of gases and all the degree of freedoms are active.)

(a) He (b) $H_2$
(c) $CO_2$ (d) $SO_3$

35. An ideal monoatomic gas initially at 300 K expands adiabatically into vacuum to double its volume. The final temperature of gas is

(a) 300 K (b) $300 \times (0.5)^{2/3}$ K
(c) $300 \times (2)^{2/3}$ K (d) 600 K

36. Temperature of one mole of an ideal gas is increased by one degree at constant pressure. Work done by the gas is

(a) $R$ (b) $2R$
(c) $R/2$ (d) $3R$

37. What is the change in internal energy when a gas contracts from 325 ml to 125 ml at a constant pressure of 2 bar, while at the same time being cooled by removing 124 J heat?

(a) −524 J (b) −84 J
(c) −164 J (d) +84 J

38. Two moles of an ideal gas [$C_{v,m}$ (/JK$^{-1}$mol$^{-1}$) = 20 + 0.01$T$(/K)] is heated at constant pressure from 27°C to 127°C. The amount of heat absorbed by the gas is

(a) 1662.8 J (b) 4700 J
(c) 6362.8 J (d) 3037.2 J

39. A monoatomic gas expands isobarically. The percentage of heat supplied that increases the thermal energy and that involved in doing work for expansion is

(a) 50 : 50 (b) 60 : 40
(c) 40 : 60 (d) 75 : 25

40. The maximum high temperature molar heat capacity at constant volume to be expected for acetylene which is a linear molecule is

(a) 9 cal/deg-mole (b) 12 cal/deg-mole
(c) 19 cal/deg-mole (d) 14 cal/deg-mole

41. The molar heat capacity of water in equilibrium with ice at constant pressure is

(a) zero (b) infinity
(c) 40.45 kJ/K-mol (d) 75.48 J/K-mol

42. For which of the following ideal gas, $C_{v,m}$ is independent of temperature?

(a) He (b) $H_2$
(c) CO (d) $SO_2$

43. If one mole of a monoatomic gas ($\gamma = 5/3$) is mixed with one mole of a diatomic gas ($\gamma = 7/5$), then the value of $\gamma$ for the mixture is

(a) 1 (b) 1.5
(c) 2 (d) 3.0

44. When an ideal diatomic gas is a heated at a constant pressure, the fraction of heat energy supplied which increases the internal energy of the gas is

(a) $\frac{2}{5}$ (b) $\frac{3}{5}$
(c) $\frac{5}{7}$ (d) $\frac{3}{7}$

45. A gas is heated at constant pressure. The fraction of heat absorbed used in doing work is

(a) $\frac{1}{\gamma}$ (b) $1-\frac{1}{\gamma}$
(c) $\gamma - 1$ (d) $\gamma$

46. An ideal gas undergoes a process in which its pressure and volume are related as $PV^n$ = constant, where $n$ is a constant. The molar heat capacity for the gas in this process will be zero if

(a) $n = \gamma$ (b) $n = \gamma - 1$
(c) $n = \gamma + 1$ (d) $n = 1 - \gamma$

47. An ideal gas ($\gamma = 1.5$) undergoes a change in state such that the magnitude of heat absorbed by the gas is equal to the magnitude of work done on the gas. The molar heat capacity of the gas in this process is

(a) $2R$ (b) $R$
(c) $3R$ (d) $1.5R$

**48.** A quantity of 70 calories of heat is required to raise the temperature of 2 mole of an ideal gas at constant pressure from 40°C to 50°C. The amount of heat required to raise the temperature of the same gas through the same range at constant volume is

(a) 90 calorie (b) 70 calorie
(c) 50 calorie (d) 30 calorie

**49.** During an adiabatic process, the pressure of a gas is found to be proportional to the cube of its absolute temperature. The Poisson's ratio of gas is

(a) 3/2 (b) 7/2
(c) 5/3 (d) 9/7

**50.** A diatomic ideal gas initially at 273 K is given 100 cal heat due to which system did 210 J work. The molar heat capacity of the gas for the process is (1 cal = 4.2 J)

(a) $\frac{3}{2}R$ (b) $\frac{5}{2}R$
(c) $\frac{5}{4}R$ (d) $5R$

**51.** An ideal monoatomic gas undergoes a reversible process, where $\frac{P}{V}$ = constant, from (2 bar, 273 K) to 4 bar. The value of $\frac{\Delta U}{w}$ for this process is

(a) +3.0 (b) −3.0
(c) −1.5 (d) +1.5

**52.** The work done on one mole of an ideal gas in the reversible process: $PV^3$ = constant, from (1 atm, 300 K) to $2\sqrt{2}$ atm is

(a) $150R$ (b) $300R$
(c) $75R$ (d) $600R$

**53.** 2 moles of an ideal monoatomic gas undergoes reversible expansion from (4 L, 400 K) to 8 L such that $TV^2$ = constant. The change in enthalpy of the gas is

(a) $-1500R$ (b) $-3000R$
(c) $+1500R$ (d) $+3000R$

**54.** Two moles of helium gas undergoes a cyclic process as shown in the figure. Assuming ideal behaviour of gas, the magnitude of net work done by the gas in this cyclic process is

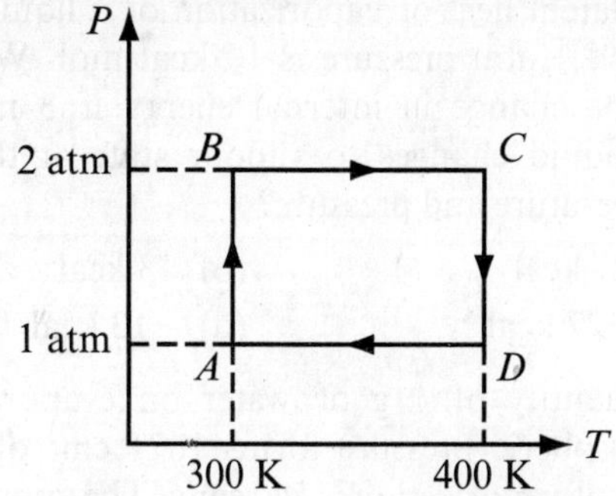


(a) 0 (b) $100R$ ln2
(c) $100R$ ln4 (d) $200R$ ln4

**55.** One mole of an ideal gas undergoes the following cyclic process.

(i) Isochoric heating from $(P_1, V_1, T_1)$ to double temperature.
(ii) Isobaric expansion to double volume.
(iii) Linear expansion (on $P$–$V$ curve) to $(P_1, 8V_1)$.
(iv) Isobaric compression to initial state.

If $T_1$ = 300 K, then the magnitude of net work done by the gas in the cyclic process is

(a) 2400 cal (b) 1200 cal
(c) 4800 cal (d) 3600 cal

**56.** One mole of a non-ideal gas undergoes a change of state (2 atm, 3 L, 95 K) → (4 atm, 5 L, 245 K) with a change in internal energy, $\Delta U$= 30.0 L-atm. The change in enthalpy ($\Delta H$) of the process in L-atm is

(a) 40.0
(b) 42.3
(c) 44.0
(d) undefined, because pressure is not constant.

**57.** The normal boiling point of water is 100°C. At 100°C

(a) the average kinetic energy of molecules in vapour is greater than that of liquid water.
(b) the average potential energy of molecules in vapour is greater than that of liquid water.
(c) the molar internal energy is same for vapour and liquid water.
(d) water boils at any pressure.

58. The latent heat of vaporization of a liquid at 500 K and 1 atm pressure is 10 kcal/mol. What will be the change in internal energy if 3 moles of the liquid changes to vapour state at the same temperature and pressure?
(a) 27 kcal (b) 13 kcal
(c) −27 kcal (d) −13 kcal

59. A quantity of 1 g of water on evaporation at atmospheric pressure forms 1671 $cm^3$ of steam. Heat of vaporization is 540 cal/g. The approximate increase in internal energy is (1L – atom = 24 cal)
(a) 250 cal (b) 500 cal
(c) 1000 cal (d) 1500 cal

60. At 500 kbar and $T$ K, the densities of graphite and diamond are 2.0 and 3.0 g/$cm^3$, respectively. The value of $(\Delta H - \Delta U)$ for the conversion of 1 mole of graphite into diamond at 500 kbar and $T$ K is
(a) 100 kJ (b) −100 kJ
(c) 1000 kJ (d) −1000 kJ

## Second and Third Law of Thermodynamics, Entropy

61. The efficiency of the reversible heat engine is $\eta_r$ and that of irreversible heat engine is $\eta_i$. Which of the following relation is correct?
(a) $\eta_r = \eta_1$
(b) $\eta_r > \eta_1$
(c) $\eta_r < \eta_1$
(d) $\eta_1$ may be less than, greater than or equal to $\eta_r$, depending on the gas.

62. For which of the following process, $\Delta S$ is negative?
(a) $H_2(g) \rightarrow 2H(g)$
(b) $N_2$ (g, 1 atm) $\rightarrow$ $N_2$ (g, 8 atm)
(c) $2SO_3(g) \rightarrow 2SO_2(g) + O_2(g)$
(d) C (Diamond) $\rightarrow$ C (Graphite)

63. $\Delta S$ will be highest for which of the following reaction ?
(a) $Ca(s) + 1/2\ O_2(g) \rightarrow CaO(s)$
(b) $CaCO_3(s) \rightarrow CaO(s) + CO_2(g)$
(c) $C(s) + O_2(g) \rightarrow CO_2(g)$
(d) $N_2(g) + O_2(g) \rightarrow 2NO(g)$

64. Entropy of universe decreases during
(a) crystallization of sucrose from unsaturated solution.
(b) rusting of iron.
(c) melting of ice at 0°C and 1 atm.
(d) vaporization of camphor above normal sublimation temperature and 1 atm.

65. Change in entropy is negative for
(a) Bromine (l) $\rightarrow$ Bromine (g)
(b) $C(s) + H_2O\ (g) \rightarrow CO(g) + H_2(g)$
(c) $N_2$ (g, 10 atm) $\rightarrow$ $N_2$ (g, 1 atm)
(d) Fe (1 mol, 400 K) $\rightarrow$ Fe (1 mol, 300 K)

66. When a substance is heated, its entropy increases. The increase will be maximum at
(a) 0°C
(b) the melting point
(c) the boiling point
(d) 100°C

67. An isolated system comprises of the liquid in equilibrium with vapours. At this stage, the molar entropy of the vapour is
(a) less than that of liquid.
(b) more than that of liquid.
(c) equal to zero.
(d) equal to that of liquid.

68. According to second law of thermodynamics, heat is partly converted into useful work and part of it
(a) becomes electrical energy.
(b) is always wasted.
(c) increases the weight of the body.
(d) becomes K.E.

69. Choose the substance which has higher possible entropy (per mole) at a given temperature.
(a) Solid carbon dioxide
(b) Nitrogen gas at 1 atm.
(c) Nitrogen gas at 0.01 atm.
(d) Nitrogen gas at 0.00001 atm.

70. The change that does not increase entropy is
(a) evaporation of liquid
(b) condensation
(c) sublimation
(d) melting of solid

**71.** Ammonium chloride when dissolved in water leads to cooling sensation. The dissolution of $NH_4Cl$ at constant temperature is accompanied by

(a) increase in entropy.
(b) decrease in entropy.
(c) no change in entropy.
(d) no change in enthalpy.

**72.** When the value of entropy is greater, then the ability for work is

(a) maximum (b) minimum
(c) medium (d) None of these

**73.** When one mole of an ideal gas is compressed to half of its initial volume and simultaneously heated to twice its temperature, the change in entropy is

(a) $C_{V,m} \ln 2$ (b) $C_{P,m} \ln 2$
(c) $R \ln 2$ (d) $(C_{V,m} - R) \ln 2$

**74.** What is the entropy change when 3.6 g of liquid water is completely converted into vapours at 373 K? The molar heat of vaporization is 40.85 kJ/mol.

(a) 219.0 J/K (b) 2.190 J/K
(c) 21.90 J/K (d) 109.5 J/K

**75.** The entropy change in the fusion of one mole of a solid melting at 300 K (latent heat of fusion, 2930 J/mol) is

(a) 9.77 J/K-mol (b) 10.73 J/K-mol
(c) 2930 J/K-mol (d) 108.5 J/K-mol

**76.** Oxygen gas weighing 64 g is expanded from 1 atm to 0.25 atm at 30°C. What is the entropy change, assuming the gas to be ideal? ($\ln 4 = 1.4$, $R = 8.3$ J/K-mol)

(a) 23.24 J/K (b) 34.86 J/K
(c) 46.48 J/K (d) 11.62 J/K

**77.** Two moles of an ideal monoatomic gas is heated from 27°C to 627°C, reversibly and isochorically. The entropy of gas

(a) increases by $2R \ln 3$
(b) increases by $3R \ln 3$
(c) decreases by $2R \ln 3$
(d) decreases by $3R \ln 3$

**78.** One mole of an ideal monoatomic gas undergoes adiabatic free expansion from 2 to 20 dm$^3$, 300 K. The value of $\Delta S$ for the gas is

(a) 0 (b) $+R \ln 10$
(c) $-R \ln 10$ (d) $+1.5\, R \ln 10$

**79.** The standard molar entropy of an ideal gas $\left(\gamma = \frac{4}{3}\right)$ is 2.5 cal/K-mol at 25°C and 1 bar. The standard molar entropy of the gas at 323°C and 1 bar is ($\ln 2 = 0.7$)

(a) undefined (b) $4R \ln 2$
(c) 5.6 cal/K-mol (d) 8.1 cal/K-mol

**80.** Given the following entropy values (in J/K-mol) at 298 K and 1 atm $H_2(g) = 130.6$, $Cl_2(g) = 223.0$ and $HCl(g) = 186.7$. The entropy change (in J/K-mol) for the reaction $H_2(g) + Cl_2(g) \rightarrow 2HCl(g)$ is

(a) +540.3 (b) +727.0
(c) −166.9 (d) +19.8

## Gibbs Free Energy Function

**81.** The $\Delta G$ in the process of melting of ice at −15°C and 1 atm, is

(a) less than zero.
(b) greater than zero.
(c) equal to zero.
(d) None of these

**82.** For a system in equilibrium, $\Delta G = 0$ under conditions of constant

(a) temperature and pressure.
(b) temperature and volume.
(c) pressure and volume.
(d) energy and volume.

**83.** One mole of ice is converted into water at 273 K and 1 atm. The entropies of $H_2O$ (s) and $H_2O$ (l) are 38.0 and 58.0 J/K-mol, respectively. The enthalpy change for the conversion is

(a) 5460 kJ/mol (b) 5460 J/mol
(c) −5460 J/mol (d) 20 J/mol

**84.** The solubility of NaCl(s) in water at 298 K is about 6 moles per litre. Suppose you add 1 mole of NaCl(s) to a litre of water. For the reaction:

$NaCl(s) + H_2O \rightarrow NaCl(aq)$

(a) $\Delta G > 0, \Delta S > 0$ (b) $\Delta G < 0, \Delta S > 0$
(c) $\Delta G > 0, \Delta S < 0$ (d) $\Delta G < 0, \Delta S < 0$

85. The values of $\Delta G$ are very important in metallurgy. The $\Delta G$ values for the following reactions at 1000 K are given as follows.

$S_2(s) + 2O_2(g) \rightarrow 2SO_2(g); \Delta G = -544$ kJ

$2Zn(s) + S_2(s) \rightarrow 2ZnS(s); \Delta G = -293$ kJ

$2Zn(s) + O_2(g) \rightarrow 2ZnO(s); \Delta G = -480$ kJ

The $\Delta G$ for the reaction

$2ZnS(s) + 3O_2(g) \rightarrow 2ZnO(s) + 2SO_2(g)$

will be

(a) −357 kJ (b) −731 kJ
(c) −773 kJ (d) −229 kJ

86. The following reaction is at equilibrium at 298 K

$2NO$ (g, 0.00001 bar) + $Cl_2$ (g, 0.01 bar) $\rightleftharpoons$ 2NOCl (g, 0.01 bar).

$\Delta G^o$ for the reaction is

(a) −45.65 kJ (b) −28.53 kJ
(c) −22.82 kJ (d) −57.06 kJ

87. What is the free energy change, $\Delta G$, when 1.0 mole of water at 100°C and 1 atm pressure is converted into steam at 100°C and 1 atm pressure?

(a) 540 cal (b) −9800 cal
(c) 9800 cal (d) 0 cal

88. The enthalpy and entropy change for a chemical reaction are −2500 cal and +7.4 cal/K, respectively. The nature of reaction at 298 K is

(a) spontaneous
(b) reversible
(c) never occurring
(d) non-spontaneous

89. A spontaneous reaction is impossible if

(a) both $\Delta H$ and $\Delta S$ are negative.
(b) $\Delta H$ and $\Delta S$ are positive.
(c) $\Delta H$ is negative and $\Delta S$ is positive.
(d) $\Delta H$ is positive and $\Delta S$ is negative.

90. For a reversible reaction, if $\Delta G^o = 0$, the equilibrium constant of the reaction should be equal to

(a) zero (b) 1
(c) 2 (d) 10

# EXERCISE II (JEE ADVANCED)

## Section A (Only one Correct)

1. Which of the following gas possess the largest internal energy?
   (a) 2 moles of He gas occupying 1000 L at 300 K.
   (b) 56 kg of nitrogen at 107 Pa and 300 K.
   (c) 8 g of oxygen at 8 atm and 300 K.
   (d) 2000 mole of argon occupying 40000 L at 900 K.

2. A gas expands reversibly in a piston cylinder device from $V_1$ to $V_2$, the process being described by $PV = a + bV$, where $P$ is in $Nm^{-2}$, $V$ is in $m^3$ and $a$ and $b$ are constants. The work done in the process (w) is
   (a) $a\log_e\left(\frac{V_1}{V_2}\right) + b(V_2 - V_1)$
   (b) $-a\log_e\left(\frac{V_2}{V_1}\right) - b(V_2 - V_1)$
   (c) $-a\log_e\left(\frac{V_1}{V_2}\right) - b(V_2 - V_1)$
   (d) $a\log_e\left(\frac{V_2}{V_1}\right) + b(V_2 - V_1)$

3. An average human produces about 10 MJ of heat each day through metabolic activity. If a human body were an isolated system of mass 80 kg with the heat capacity of water, what temperature rise would the body experience? Heat capacity of water = 4.2 J/K-g.
   (a) 29.76°C (b) 2.976 K
   (c) $2.976 \times 10^4$°C (d) 0.029°C

4. The heat capacity of liquid water is 75.6 J/K-mol, while the enthalpy of fusion of ice is 6.0 kJ/mol. What is the smallest number of ice cubes at 0°C each containing 9.0 g of water needed to cool 500 g of liquid water from 20°C to 0°C?
   (a) 1 (b) 7
   (c) 14 (d) 21

5. An insulated container of gas has two chambers separated by an insulating partition. One of the chambers has volume $V_1$ and contains an ideal gas at pressure $P_1$ and temperature $T_1$. The other chamber has volume $V_2$ and it contains the same ideal gas at pressure $P_2$ and temperature $T_2$. If the partition is removed without doing any work on the gas, the final equilibrium temperature of the gas in the container will be
   (a) $\frac{T_1T_2(P_1V_1 + P_2V_2)}{P_1V_1T_2 + P_2V_2T_1}$ (b) $\frac{P_1V_1T_1 + P_2V_2T_2}{P_1V_1 + P_2V_2}$
   (c) $\frac{P_1V_1T_2 + P_2V_2T_1}{P_1V_1 + P_2V_2}$ (d) $\frac{T_1T_2(P_1V_1 + P_2V_2)}{P_1V_1T_1 + P_2V_2T_2}$

6. The work involved ($w$) in an isothermal expansion of $n$ moles of an ideal gas from an initial pressure of '$P$' atm to final pressure of 1 atm in number of steps such that in every step, the constant external pressure exactly 1 atm less than the initial pressure of gas is maintained, is given as
   (a) $-nRT\sum_{i=1}^{i=P-1}\left(\frac{1}{P+1-i}\right)$
   (b) $-nRT\sum_{i=1}^{i=P}\left(\frac{1}{P+1-i}\right)$
   (c) $-nRT\sum_{i=1}^{i=P}\left(\frac{i}{P+1-i}\right)$
   (d) $-nRT\sum_{i=1}^{i=P-1}\left(\frac{i}{P+1-i}\right)$

7. An ideal gaseous sample at initial state ($P_o$, $V_o$, $T_o$) is allowed to expand to volume $2V_o$ using two different processes. For the first process, the equation of process is $2PV^2 = K_1$ and for the second process, the equation of the process is $PV = K_2$.
   (a) Magnitude of work done in the first process will be greater than that in the second process.
   (b) Magnitude of work done in the second process will be greater than that in the first process.
   (c) Work done in both the processes cannot be compared without knowing the relation between $K_1$ and $K_2$.
   (d) First process is impossible.

8. If a triatomic non-linear gas is heated isothermally, then what percentage of the heat energy is used to increase the internal energy?
(a) zero (b) 60%
(c) 50% (d) 100%

9. A student is calculating the work done by 2 mole of an ideal gas in a reversible isothermal expansion shown in the figure. By mistake he calculated the area of the shaded area in the $PV$ graph shown, as work and answered the magnitude of work equal to 49.26 L-atm. What is the correct magnitude of work done by the gas in L-atm? ($R$ = 0.0821 L-atm/K-mol)
(a) 49.26 (b) 98.52
(c) 78.63 (d) 34.14

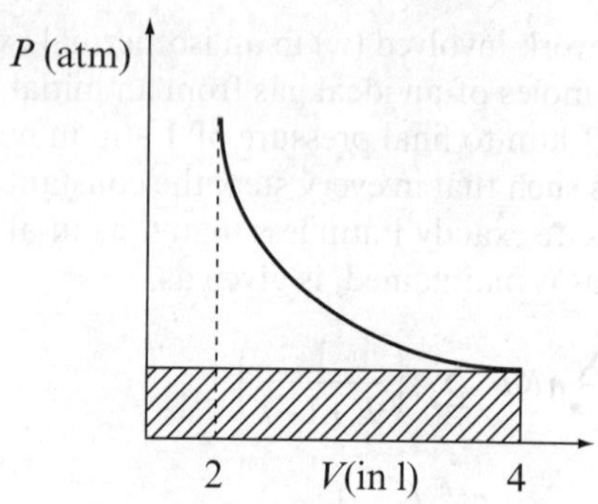


10. In a certain polytropic process, the volume of argon was increased 4 times. Simultaneously, the pressure decreased 8 times. The molar heat capacity of argon in this process, assuming the gas to be ideal, is
(a) −4.2 J/K-mol (b) +4.2 J/K-mol
(c) 12.47 J/K-mol (d) 20.78 J/K-mol

11. The molar heat capacity at 25°C should be close to 27 J/K-mol for all of the given elements except
(a) Pt (b) W
(c) Ne (d) Sr

12. The pressure and density of a diatomic gas ($\gamma$ = 7/5) change from ($P_1$, $d_1$) to ($P_2$, $d_2$) adiabatically. If $d_2/d_1$ = 32, then what is the value of $P_2/P_1$?
(a) 32 (b) 64
(c) 128 (d) 256

13. How many times a diatomic gas should be expanded adiabatically so as to reduce the RMS speed to half?
(a) 8 (b) 16
(c) 32 (d) 64

14. Work done by a sample of an ideal gas in a process A is double the work done in another process B. The temperature rises through the same amount in the two processes. If $C_A$ and $C_B$ are the molar heat capacities for the two processes, then
(a) $C_A = C_B$ (b) $C_A > C_B$
(c) $C_A < C_B$ (d) Both, undefined

15. One mole of oxygen is heated from 0°C at constant pressure till its volume is increased by 10%. The specific heat of oxygen, under these conditions, is 0.22 cal/g-K. The amount of heat required is
(a) $32 \times 0.22 \times 27.3 \times 4.2$ J
(b) $16 \times 0.22 \times 27.3 \times 4.2$ J
(c) $\frac{32 \times 0.22 \times 27.3}{4.2}$ J
(d) $\frac{16 \times 0.22 \times 27.3}{4.2}$ J

16. Heat energy of 743 J is needed to raise the temperature of 5 moles of an ideal gas by 2 K at constant pressure. How much heat energy is needed to raise the temperature of the same mass of the gas by 2 K at constant volume? ($R$ = 8.3 J/K-mol)
(a) 826 J (b) 743 J
(c) 660 J (d) 600 J

17. When an ideal gas at pressure $P$, temperature $T$ and volume, $V$, is isothermally compressed to $V/n$, its pressure becomes $P_i$. If the gas is compressed adiabatically to $V/n$, its pressure becomes $P_a$. The ratio of $P_i/P_a$ is
(a) 1 (b) $n$
(c) $n^{\gamma}$ (d) $n^{(1-\gamma)}$

18. If the ratio $C_p/C_v = \gamma$, the change in internal energy of the mass of a gas, when volume changes from $V$ to $2V$ at constant pressure $P$, is
(a) $\frac{R}{\gamma - 1}$ (b) $PV$
(c) $\frac{PV}{\gamma - 1}$ (d) $\frac{\gamma PV}{\gamma - 1}$

19. A thermally insulated vessel containing a gas whose molar mass is equal to 40 and ratio of specific heats is 1.5, moves with a velocity 100 m/s. Find the gas temperature increment (in °C) resulting from the sudden stoppage of the vessel.
(a) 12 (b) 24
(c) 36 (d) 48

**20.** With what minimum pressure must a given volume of an ideal gas ($\gamma = 1.4$), originally at 400 K and 100 kPa, be adiabatically compressed in order to raise its temperature up to 700 K?

(a) 708.9 kPa (b) 362.5 kPa
(c) 1450 kPa (d) 437.4 kPa

**21.** Two rigid adiabatic vessels A (volume = 4 L) and B (volume = 6 L), which initially contains two gases at different temperatures are connected by a pipe of negligible volume. The vessel A contains 2 moles of Ne gas ($C_{P,m}$ = 5 cal/K-mol) at 300 K and vessel B contains 3 moles of $SO_2$ gas ($C_{P,m}$ = 8 cal/K-mol) at 400 K. What is the final pressure (in atm) when the valve is opened and 12 kcal heat is supplied through it to the vessels? ($R$ = 0.08 L-atm/K-mol)

(a) 3.5 (b) 7
(c) 35 (d) 70

**22.** A container of volume 1 $m^3$ is divided into two equal parts by a partition. One part has an ideal diatomic gas at 300 K and the other part has vacuum. The whole system is isolated from the surrounding. When the partition is removed, the gas expands to occupy the whole volume. Its final temperature will be

(a) 300 K (b) 227.5°C
(c) 455 K (d) 455°C

**23.** If all degree of freedom of a three-dimensional N-atomic gaseous molecules is excited, then $C_P/C_V$ ratio of gas should be

(a) 1.33 (b) $1+\frac{1}{3N-3}$
(c) $1+\frac{1}{N}$ (d) $1+\frac{1}{3N-2}$

**24.** One mole of an ideal gas undergoes a reversible process $T = T_0 + \alpha V$, where $T_0$ and $\alpha$ are constants. If its volume increases from $V_1$ to $V_2$, then the amount of heat transferred to the gas is

(a) $C_{P,m}RT_0\ln\left(\frac{V_2}{V_1}\right)$

(b) $\alpha C_{P,m}\frac{V_2-V_1}{RT_0}\ln\left(\frac{V_2}{V_1}\right)$

(c) $\alpha C_{P,m}(V_2-V_1)+RT_0\ln\left(\frac{V_2}{V_1}\right)$

(d) $RT_0\ln\left(\frac{V_2}{V_1}\right)-\alpha C_{P,m}(V_2-V_1)$

**25.** The average degree of freedom per molecule for a gas is 6. The gas performs 25 J of work when it expands at constant pressure. What is the amount of heat absorbed by the gas?

(a) 25 J (b) 50 J
(c) 75 J (d) 100 J

**26.** One mole of an ideal monoatomic gas is heated in a process $PV^{-5/2}$ = constant. By what amount heat is absorbed in the process in 26°C rise in temperature?

(a) 100 J (b) 180 J
(c) 200 J (d) 208 J

**27.** What is the heat capacity of the ideal monoatomic gas undergoing the process shown in the figure?

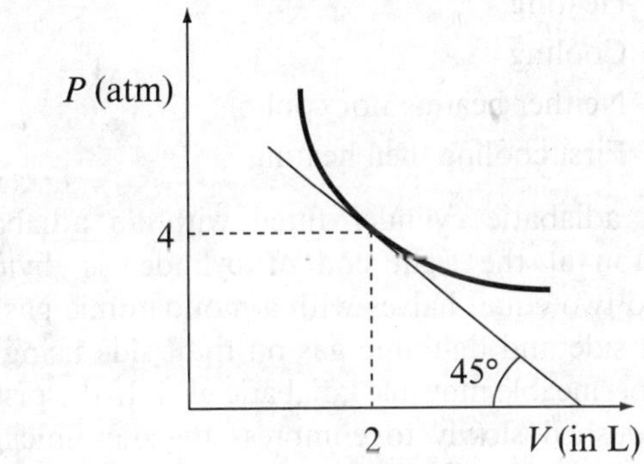


(a) 1.5$R$ (b) 2.5$R$
(c) 3.5$R$ (d) 1.33$R$

**28.** A sample of ideal gas is compressed from initial volume of $2V_o$ to $V_o$ using three different processes as follows.

(1) Reversible isothermal
(2) Reversible adiabatic
(3) Irreversible adiabatic under a constant external pressure.

Then

(a) Final temperature of gas will be highest at the end of second process.
(b) Magnitude of enthalpy change of sample will be highest in isothermal process.
(c) Final temperature of gas will be highest at the end of third process.
(d) Final pressure of gas will be highest at the end of second process.

29. An ideal gas expands from volume $V_1$ to $V_2$. This may be achieved by any of the three processes, such as isobaric, isothermal and adiabatic. Which of the following statement is not true for $\Delta E$?

(a) $\Delta E$ is the least in the adiabatic expansion.

(b) $\Delta E$ is the greatest in the adiabatic expansion.

(c) $\Delta E$ is the greatest in the isobaric process.

(d) $\Delta E$ in isothermal process lies in between the value obtained under isobaric and adiabatic process.

30. Inversion temperature is defined as the temperature above which a gas gets warm up and below which, the gas become cooler, when expanded adiabatically. Boyle temperature for a gas is 20°C. What will happen to the gas if it is adiabatically expanded at 50°C?

(a) Heating

(b) Cooling

(c) Neither heating nor cooling.

(d) First cooling then heating.

31. An adiabatic cylinder fitted with an adiabatic piston at the right end of cylinder is divided into two equal halves with a monoatomic gas on left side and diatomic gas on right side using an impermeable movable adiabatic wall. If the piston is pushed slowly to compress the diatomic gas to $\frac{3}{4}$ th of its original volume. The ratio of new volume of monoatomic gas to its initial volume would be

(a) $\left(\frac{4}{3}\right)^{\frac{25}{21}}$ (b) $\left(\frac{7}{5}\right)^{\frac{3}{4}}$

(c) $\left(\frac{3}{4}\right)^{\frac{21}{25}}$ (d) $\frac{3}{4}$

32. If four identical samples of an ideal gas initially at the same state ($P_o$, $V_o$, $T_o$) are allowed to expand to double their volumes by the following processes

Process I: Isothermal reversible process

Process II: Reversible process $P^2V$ = Constant

Process III: Reversible adiabatic process

Process IV: Irreversible adiabatic expansion against constant external pressure.

If the final states of gases are shown by different points in the graph represented, then match each point with the correct process.

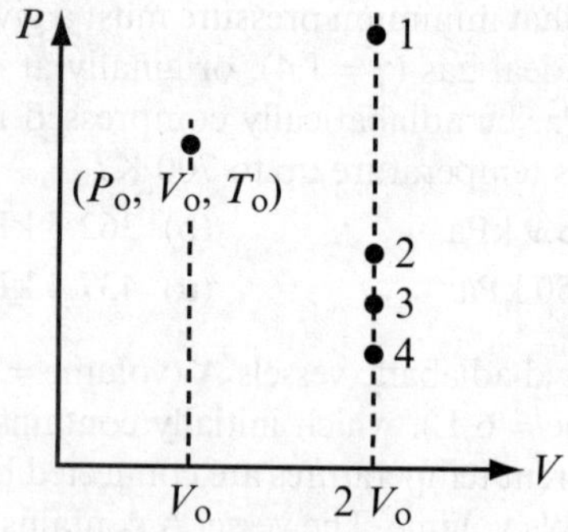


(a) I – 1, II – 2, III – 3, IV – 4

(b) I – 2, II – 1, III – 4, IV – 3

(c) I – 4, II – 3, III – 2, IV – 1

(d) I – 2, II – 1, III – 3, IV – 4

33. For an ideal gas subjected to different processes as shown in the graphs, select the graph which will involve the greatest amount of heat exchange if the initial and final temperatures are same in all?

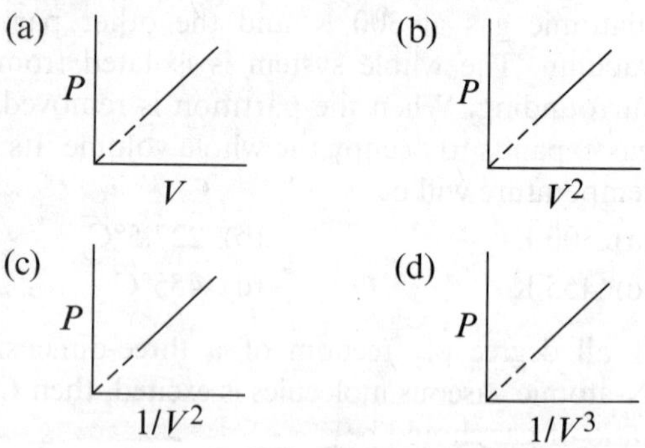


34. A thermodynamic process is shown in the following figure. The pressure and volumes corresponding to some points in the figure are $P_A = 3 \times 10^4$ Pa, $P_B = 8 \times 10^4$ Pa, $V_A = 2 \times 10^{-3}$ m$^3$, $V_D = 5 \times 10^{-3}$ m$^3$. In the process AB, 600 J of heat is added to the system and in BC, 200 J of heat is added to the system. The change in internal energy of the system in the process AC would be

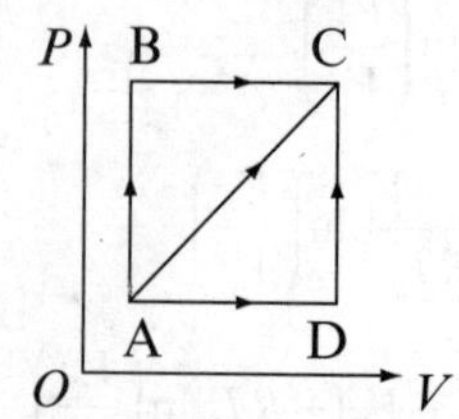


(a) 560 J (b) 800 J

(c) 600 J (d) 640 J

**35.** Two moles of an ideal monoatomic gas undergoes a cyclic process ABCA as shown in the figure. What is the ratio of temperature at B and A?

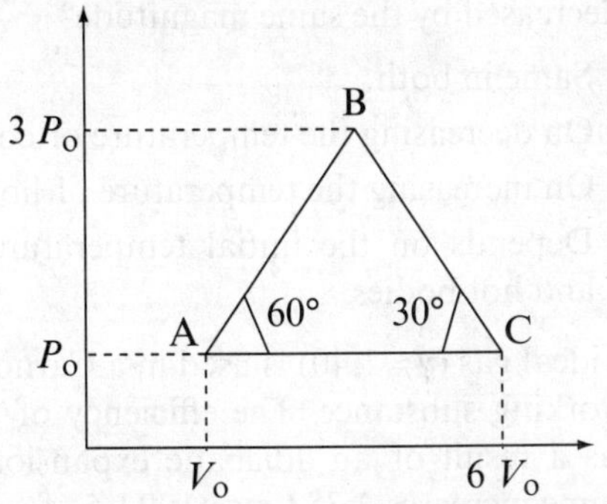


(a) 1 : 1 (b) 3 : 2
(c) 27 : 4 (d) 9 : 2

**36.** One mole of a certain ideal gas is contained under a weightless piston of a vertical cylinder at a temperature $T$. The space over the piston opens into the atmosphere. What work has to be performed in order to increase the gas volume isothermally under the piston $\eta$ times by slowly raising the piston? The friction of the piston against the cylinder walls is negligibly small.

(a) $RT(\eta - 1 - \ln \eta)$
(b) $RT(1 - \eta + \ln \eta)$
(c) $RT \ln \eta$
(d) $-RT \ln \eta$

**37.** A piston can freely move inside a horizontal cylinder closed from both ends. Initially, the piston separates the inside space of the cylinder into two equal parts each of volume $V_o$, in which an ideal gas is contained under the same pressure $P_o$ and at the same temperature. What work has to be performed in order to increase the volume of one part of gas isothermally η times when compared to that of the other by slowly moving the piston?

(a) $P_oV_o \ln \eta$

(b) $P_oV_o \ln \dfrac{(\eta+1)^2}{4\eta}$

(c) $P_oV_o \ln \dfrac{(\eta-1)^2}{4\eta}$

(d) $2P_oV_o \ln \eta$

**38.** Three samples A, B and C of the same ideal gas ($\gamma$ = 1.5) have equal volumes and temperatures. The volume of each sample is doubled, the process being isothermal for A, adiabatic for B and isobaric for C. If the final pressures are equal for the three samples, then the ratio of their initial pressures is

(a) 1 : 1 : 1 (b) 2 : 2 : 1
(c) $2 : 2\sqrt{2} : 1$ (d) $1 : 2\sqrt{2} : 2$

**39.** A cylinder with thermally insulated walls contains an insulated portion which can slide freely. The partition divides the cylinder into two chambers containing equal moles of the same gas, the initial pressure, temperature and volume being $P_o$, $T_o$ and $V_o$, respectively. By means of heating the coil, heat is supplied slowly to the gas in one chamber (A) until its pressure becomes $27P_o/8$. If the value of $\gamma$ is 1.5, then find the heat supplied to the gas in chamber A.

(a) $5P_oV_o/4$ (b) $19P_oV_o/2$
(c) $15P_oV_o/2$ (d) $15P_oV_o/4$

**40.** A portion of helium gas in a vertical cylindrical container is in thermodynamic equilibrium with the surroundings. The gas is confined by a movable heavy piston. The piston is slowly elevated by a distance $H$ from its equilibrium position and then kept in the elevated position long enough for the thermal equilibrium to be re-established. After that, the container is insulated and then the piston is released. After the piston comes to rest, what is the new equilibrium position of the piston with respect to initial position?

(a) The piston ends up 0.4$H$ above its initial position.
(b) The piston ends up 0.6$H$ above its initial position.
(c) The piston ends at its initial position.
(d) The piston ends up 0.4$H$ below its initial position.

**41.** Three lawn chairs, one made up of aluminium (heat capacity = 0.90 J/K-g), one of iron (heat capacity = 0.45 J/K-g) and one of tin (heat capacity = 0.60 J/K-g) are painted of the same colour. On a sunny day, which chair will be hotter to sit?

(a) Iron chair (b) Tin chair
(c) Aluminium chair (d) All are same

**42.** One mole of a real gas is subjected to a process from (2 bar, 30 L, 300 K) to (2 bar, 40 L, 500 K). The molar heat capacity of gas at constant volume and constant pressure are 25 and 40 J/K-mol, respectively. What is the change in internal energy of the gas in this process?

(a) 5000 J (b) 6000 J
(c) 8000 J (d) 10,000 J

**43.** A real gas is subjected to an adiabatic process from (2 bar, 40 L, 300 K) to (4 bar, 30 L, 300 K) against a constant pressure of 4 bar. The enthalpy change for the process is

(a) zero (b) 6000 J
(c) 8000 J (d) 10,000 J

**44.** The equation of state for one mole of a gas is $PV = RT + BP$, where $B$ is a constant, independent of temperature. The internal energy of fixed amount of gas is the function of temperature only. If one mole of the above gas is isothermally expanded from 12 L to 22 L at a constant external pressure of 1 bar at 400 K, then the change in enthalpy of the gas is approximately (B = 2 L/mol)

(a) 0 (b) −3.32 J
(c) −332 J (d) −166 J

**45.** One mole of a real gas is subjected to heating at constant volume from ($P_1$, $V_1$, $T_1$) state to ($P_2$, $V_1$, $T_2$) state. Then it is subjected to irreversible adiabatic compression against constant external pressure of $P_3$ atm, till the system reaches final state ($P_3$, $V_2$, $T_3$). If the constant volume molar heat capacity of real gas is $C_V$, then the correct expression for $\Delta H$ from State 1 to State 3 is

(a) $C_V(T_3 - T_1) + (P_3V_1 - P_1V_1)$
(b) $C_V(T_2 - T_1) + (P_3V_2 - P_1V_1)$
(c) $C_V(T_2 - T_1) + (P_3V_1 - P_1V_1)$
(d) $C_P(T_2 - T_1) + (P_3V_1 - P_1V_1)$

**46.** The efficiency of a Carnot cycle is 1/6. On decreasing the temperature of the sink by 65°C, the efficiency increases to 1/3. The temperature of source is

(a) 117°C (b) 52°C
(c) 390°C (d) 1755°C

**47.** A Carnot engine operates between 327°C and 117°C. If it absorbs 120 cal heat per cycle from the source, then the heat rejected per cycle to the sink is

(a) 120 cal (b) 42 cal
(c) 78 cal (d) 90 cal

**48.** In which case will the efficiency of a Carnot cycle be higher, when the hot body temperature is increased by $\Delta T$ or when the cold body temperature is decreased by the same magnitude?

(a) Same in both.
(b) On decreasing the temperature of cold body.
(c) On increasing the temperature of hot body.
(d) Depends on the initial temperature of cold and hot bodies.

**49.** An ideal gas ($\gamma = 1.40$) is used in a Carnot cycle as a working substance. The efficiency of the cycle, if as a result of an adiabatic expansion the gas volume increases 2.75 times, is [$(1.5)^{2.5} = 2.75$]

(a) $\frac{100}{3}\%$ (b) $\frac{200}{3}\%$
(c) 50% (d) 25%

**50.** What is the efficiency of a cycle consisting of two isochoric and two adiabatic lines, if the volume of the ideal gas changes 10 times within the cycle? The working substance is nitrogen [$(10)^{0.4} = 2.5$].

(a) 40% (b) 25%
(c) 60% (d) 75%

**51.** A reversible heat engine absorbs 40 kJ of heat at 500 K and performs 10 kJ of work rejecting the remaining amount to the sink at 300 K. The entropy change for the universe is

(a) −80 J/K (b) 100 J/K
(c) 20 J/K (d) 180 J/K

**52.** Molar heat capacity of $CD_2O$ (deuterated form of formaldehyde) vapour at constant pressure is vapour 14 cal/K-mol. The entropy change associated with the cooling of 3.2 g of $CD_2O$ vapour from 1000 K to 900 K at constant pressure is (assume ideal gas behaviour for $CD_2O$) [ln 0.9 = −0.1]

(a) +0.14 cal/K (b) – 0.14 cal/K
(c) −1.4 cal/K (d) +1.4 cal/K

**53.** An amount of 2 mole of CO was in the small cubical container of length ($l = a$) at 57°C. The gas was taken out completely from the small container and put into large container ($l = 2a$) at the same temperature. What is the change in entropy, if it is assumed that the process is reversible and gas is behaving as an ideal gas?

(a) 34.58 J/K (b) – 34.58 J/K
(c) −11.53 J/K (d) +11.53 J/K

**54.** The change in entropy accompanying the heating of one mole of helium gas ($C_{v,m} = 3R/2$), assumed ideal, from a temperature of 250 K to a temperature of 1000 K at constant pressure is (ln 2 = 0.7)

(a) 4.2 cal/K (b) 7.0 cal/K
(c) 2.1 cal/K (d) 3.5 cal/K

**55.** The molar entropy of a constant volume sample of neon at 500 K if it is 46.2 cal/K-mol at 250 K is (ln 2 = 0.7)

(a) 2.1 cal/K-mol (b) 44.1 cal/K-mol
(c) 48.3 cal/K-mol (d) 46.2 cal/K-mol

**56.** A sample of perfect gas that initially occupies 15.0 L at 300 K and 1.0 bar is compressed isothermally. To what volume must the gas be compressed to reduce its entropy by 5.0 J/K? [ln 0.36 = −1.0, ln 2.7 = 1.0]

(a) 5.4 L (b) 8.22 L
(c) 40.5 L (d) 5.56 L

**57.** A system undergoes a process in which the entropy change is +5.51 $JK^{-1}$. During the process, 1.50 kJ of heat is added to the system at 300 K. The correct information regarding the process is

(a) the process thermodynamically reversible.
(b) the process is thermodynamically irreversible.
(c) the process may or may not be thermodynamically reversible.
(d) the process must be isobaric.

**58.** Consider a reversible isentropic expansion of 1 mole of an ideal monoatomic gas from 27°C to 927°C. If the initial pressure of gas was 1 bar, then the final pressure of gas becomes

(a) 4 bar (b) 8 bar
(c) 32 bar (d) 0.25 bar

**59.** A quantity of 1.6 g helium gas is expanded adiabatically 3.0 times and then compressed isobarically to the initial volume. Assume ideal behaviour of gas and both the processes to bereversible. The entropy change of the gas in this process is (ln 3 = 1.1)

(a) −1.1 cal/K (b) +1.1 cal/K
(c) −2.2 cal/K (d) +2.2 cal/K

**60.** The entropy change of 2.0 moles of an ideal gas whose adiabatic exponent $\gamma = 1.50$, if as a result of a certain process, the gas volume increased 2.0 times while the pressure dropped 4.0 times, is (ln 2 = 0.7)

(a) −11.64 J/K (b) +11.64 J/K
(c) −34.92 J/K (d) +34.92 J/K

**61.** Each of the vessels 1 and 2 contain 1.2 moles of gaseous helium. The ratio of the volume of vessels is $V_2/V_1 = 2.0$, and the ratio of the absolute temperature of helium in them is $T_1/T_2 = 2.0$. Assuming the gas to be ideal, find the difference of gas entropies in these vessels, $S_2 - S_1$ (ln 2 = 0.7).

(a) 0.84 cal/K (b) 4.2 cal/K
(c) −0.84 cal/K (d) −4.2 cal/K

**62.** One mole of an ideal gas with the adiabatic exponent 'γ' goes through a polytropic process as a result of which the absolute temperature of the gas increases τ-fold. The polytropic constant equals *n*. The entropy increment of the gas in this process is

(a) $\frac{(n-\gamma)R}{(n-1)(\gamma-1)}\ln\tau$ (b) $\frac{(n-1)(\gamma-1)}{(n-\gamma)R}\ln\tau$

(c) $\frac{(n-\gamma)R}{(\gamma-1)}\ln\tau$ (d) $\frac{(n-\gamma)R}{(n-1)}\ln\tau$

**63.** The expansion process of 2.0 moles of argon proceeds so that the gas pressure increases in direct proportion to its volume. The entropy change of the gas in this process if its volume increases 2.0 times, is (ln 2 = 0.7)

(a) +11.2 cal/K (b) −11.2 cal/K
(c) +5.6 cal/K (d) −5.6 cal/K

**64.** An ideal gas with the adiabatic exponent γ goes through a process $P = P_o - \alpha V$, where $P_o$ and α are positive constants and *V* is the volume. At what volume will the gas entropy have the maximum value?

(a) $\frac{\gamma.P_o}{\alpha(\gamma-1)}$ (b) $\frac{\gamma.P_o}{\alpha(\gamma+1)}$

(c) $\frac{\alpha.P_o}{\gamma+1}$ (d) $\frac{\alpha.P_o}{\gamma-1}$

**65.** One mole of an ideal gas undergoes a reversible process in which the entropy of the gas changes with absolute temperature *T* as $S = aT + C_{v,m}\ln T$, where *a* is a positive constant. If $T = T_o$ at $V = V_{o}$, then the volume dependence of the gas on temperature in this process is

(a) $T = T_o + \ln V$ (b) $T = T_o + \frac{R}{a}.\ln\frac{V_o}{V}$

(c) $T = T_o + \frac{R}{a}.\ln\frac{V}{V_o}$ (d) $V = V_o + \frac{a}{R}.\ln\frac{T}{T_o}$

66. At very low temperature, the heat capacity of crystals is equal to $C = aT^3$, where $a$ is a constant. Find the entropy of a crystal as a function of temperature in this temperature interval.

(a) $S = \frac{a.T^3}{3}$ (b) $S = aT^3$

(c) $\frac{a.T^2}{2}$ (d) $\frac{a.T}{3}$

67. The entropy change accompanying the transfer of 12,000 J of heat from a body A at 327°C to a body B at 127°C is

(a) −10.0 J/K (b) +10.0 J/K
(c) −57.8 J/K (d) +57.8 J/K

68. A piece of alloy weighing 4 kg and at a temperature of 800 K is placed in 4 kg of water at 300 K. If the specific heat capacity of water is 1.0 cal/K-g and that of alloy is 4 cal/K-g, then the $\Delta S_{mix}$ is ($\ln 2 = 0.7$, $\ln 3 = 1.1$, $\ln 7 = 1.95$)

(a) +3.33 kcal/K (b) −1.0 kcal/K
(c) +1.0 kcal/K (d) +1.33 kcal/K

69. Two blocks of copper metal are of the same size (heat capacity = C) but at different temperatures $T_1$ and $T_2$. These blocks are brought together and allowed to attain thermal equilibrium. The entropy change of system is

(a) $C\cdot\ln\left[\frac{(T_2-T_1)^2}{4T_1T_2}+1\right]$

(b) $C\cdot\ln\left[\frac{(T_2-T_1)^2}{4T_1T_2}\right]$

(c) $C\cdot\ln\left[\frac{(T_2+T_1)^2}{4T_1T_2}+1\right]$

(d) $C\cdot\ln\left[\frac{(T_2+T_1)^2}{4T_1T_2}-1\right]$

70. Assuming ideal gas behaviour, the $\Delta S$ for the isothermal mixing of 0.8 mole $N_2$ and 0.2 mole of $O_2$ is ($\ln 2 = 0.7$, $\ln 10 = 2.3$)

(a) +0.96 cal/K (b) +0.32 cal/K
(c) −0.96 cal/K (d) −0.32 cal/K

71. If all the following gases are in monoatomic form, then which of the following has greater entropy?

(a) *H* (b) N
(c) O (d) Cl

72. Which of the following would be expected to have the largest entropy per mole?

(a) $SO_2Cl_2(s)$ (b) $SO_2Cl_2(l)$
(c) $SO_2Cl_2(g)$ (d) $SO_2(g)$

73. The enthalpy of formation steadily changes from −17.89 Kcal/mol to −49.82 Kcal/mol as we go from $CH_4$, $C_2H_6$ to $C_8H_{18}$. The value of $\Delta G$ however shows opposite trend from −12.12 Kcal/mol for $CH_4$ to 4.14 Kcal/mol for $C_8H_{18}$. Why?

(a) As the number of carbon atoms increases, the number of possible isomers increases. This reverses the expected trend of $\Delta G$ values.
(b) The increase in the number of C–C bonds in relation to the number of C–H bonds modifies the trend of $\Delta G$ values in relation to $\Delta H$ values.
(c) In the formation of $C_nH_{2n+2}$ from $n$ carbon atoms and $(n + 1)$ hydrogen molecules there is a large decrease in entropy. This is reflected in the $\Delta G$ values.
(d) No simple reason is possible.

74. An amount of 5 mole $H_2O(l)$ at 100°C and 1 atm is converted into $H_2O(g)$ at 100°C and 5 atm. $\Delta G$ for the process is

(a) zero (b) 1865 ln 5 cal
(c) 3730 ln 5 cal (d) −3730 ln 5 cal

75. Heat liberated by a given amount of an ideal gas undergoing reversible isothermal process is 1200 cal at 300 K. What is the Gibbs free energy change of the gas in this process?

(a) zero (b) +1200 cal
(c) −1200 cal (d) 4 cal

76. A reaction at 300 K with $\Delta G^\circ = -1743$ J consists of 3 moles of A(g), 6 moles of B(g) and 3 moles of C(g). If A, B and C are in equilibrium in one liter vessel, then the reaction should be ($\ln 2 = 0.7$, $R = 8.3$ J/K-mol)

(a) $A + B \rightleftharpoons C$ (b) $A \rightleftharpoons B + 2C$
(c) $2A \rightleftharpoons B + C$ (d) $A + B \rightleftharpoons 2C$

77. The vapour pressures of water and ice at −10°C are 0.28 and 0.26 Pa, respectively. What is the molar free energy change for the process?

$H_2O$ (l, −10°C, 0.28 Pa) $\rightarrow$ $H_2O$ (s, −10°C, 0.26 Pa)

(a) $R\times 263\times\ln\frac{14}{13}$ (b) $R\times 263\times\ln\frac{13}{14}$

(c) $R\times 10\times\ln\frac{13}{14}$ (d) $R\times 10\times\ln\frac{14}{13}$

EXERCISE II

78. A definite mass of a monoatomic ideal gas at 1 bar and 27°C expands against vacuum from 1.2 $dm^3$ to 2.4 $dm^3$. The change in free energy of the gas, $\Delta G$, is ($R = 0.08$ bar- L/K-mol, ln 2 = 0.7)

(a) 0 (b) −64 bar-l
(c) +84 J (d) −84 J

79. For a reaction A $\rightleftharpoons$ B carried out at 27°C, the ratio of equilibrium concentrations of the product to reactant changes by a factor of $e^4$ for every

(a) 1.2 kcal rise in $\Delta G°$.
(b) 1.2 kcal fall in $\Delta G°$.
(c) 2.4 kcal rise in $\Delta G°$.
(d) 2.4 kcal fall in $\Delta G°$.

80. The vapour pressure of water is 0.04 atm at 27°C. The free energy change for the following process is

$H_2O$ (g, 0.04 atm, 27°C) $\rightarrow$ $H_2O$ (*l*, 0.04 atm, 27°C)

(a) 0 (b) $R \times 300 \times \ln\frac{1}{25}$
(c) $R \times 300 \times \ln 25$ (d) $300R$

## Section B (One or More than One Correct)

1. Which of the following physical quantities are both independent from amount and path
(a) Internal energy
(b) Temperature
(c) Molar heat exchanged at constant volume.
(d) Molar heat exchanged at constant pressure.

2. Which of the following sets does show the intensive properties?
(a) Temperature and pressure.
(b) Viscosity and surface tension.
(c) Refractive index and specific heat.
(d) Volume and heat capacity.

3. For an isolated system, the wall/boundary separating the system from surrounding must be
(a) rigid (b) impermeable
(c) adiabatic (d) diathermal

4. An ideal gas is taken reversibly from state A ($P$, $V$) to the state B (0.5 $P$, 2 $V$) along a straight line in $PV$ diagram. Which of the following statement(s) is/are correct regarding the process?
(a) The work done by gas in the process A to B exceeds the work that would be done by it if the same change in state were performed isothermally.
(b) In the $T$–$V$ diagram, the path AB becomes a part of parabola.
(c) In the $P$–$T$ diagram, the path AB becomes a part of hyperbola.
(d) On going from A to B, the temperature of the gas first increases to a maximum value and then decreases.

5. $\Delta E$ must be zero for
(a) cyclic process
(b) isothermal ideal gas expansion
(c) isothermal real gas expansion
(d) isothermal change in physical state

6. Select the correct option(s).
(a) Molar internal energy is an intensive property.
(b) Ideal gases produce more work in an isothermal reversible expansion as compared to Van der Waals gases, assuming ($V >> nb$).
(c) Reversible process can be reversed at any point in the process by making infinitesimal change.
(d) Less heat is absorbed by the gas in the reversible isothermal expansion as compared to irreversible isothermal expansion to the same final volume.

7. Which of the following statement(s) is/are true?
(a) $C_{v,m}$ is independent of temperature for a perfect gas.
(b) If neither heat nor matter can enter or leave a system, the system must be isolated.
(c) $\oint dF = 0$, where $F = U, H, G, S$.
(d) A process in which the final temperature equals to the initial temperature must be an isothermal process.

**8.** Which of the following statement(s) is/are incorrect?

(a) $\Delta E = q + w$ for every thermodynamic system at rest in the absence of external field.

(b) A thermodynamic process is specified by specifying the initial and final state of the system.

(c) Adiabatic free expansion of any gas is also isothermal.

(d) For every cyclic process, the final state of the surroundings is the same as the initial state of surroundings.

**9.** Which of the following statement(s) is/are true for ideal gas?

(a) $(-w_{rev}) - (-w_{irr}) > 0$ for isothermal expansion.

(b) $w_{rev} > w_{irr}$ for isothermal compression.

(c) Heat rejected in isothermal process is greater in $(n-1)$th stage as compared to $n$th stage work.

(d) Arrow shown in the graph below indicates the decreasing value of $\gamma$.

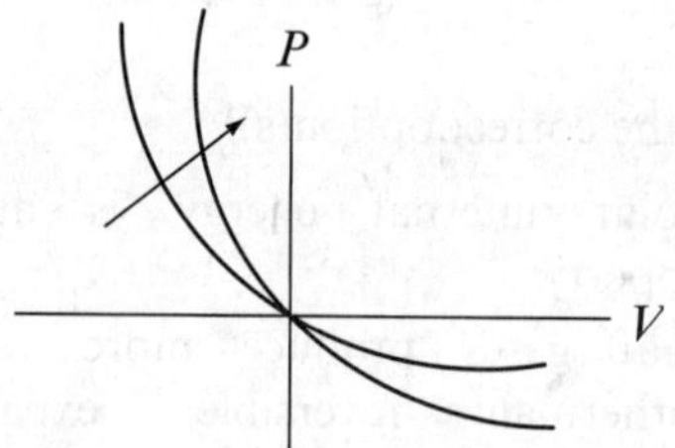


**10.** An amount of 4 moles of an ideal monoatomic gas expands adiabatically and reversibly by which its temperature decreases from 47°C to 17°C. Which of the following is/are true?

(a) $q = 0$ (b) $\Delta H = 0$

(c) $\Delta H = -600$ cal (d) $\Delta U = -600$ cal

**11.** Which of the following statement(s) is/are correct regarding ideal gas?

(a) If all the gas molecules are assumed to be rigid spheres of negligible volume, then the only possible molecular motion is translational.

(b) In polyatomic molecules, the motion of each atom is independent to the motion of other atoms of the same molecule and the net motion of the molecule is the resultant of motion of all the constituent atoms.

(c) Kinetic and potential energies are independent modes to describe the energy of a particle.

(d) All polyatomic molecules have two rotational degrees of freedom.

**12.** Two moles of an ideal gas ($C_{v,m} = 1.5R$) is subjected to the following changes in states.

$$A\ (500\ K,\ 5\ bar) \xrightarrow[\text{Expansion}]{\text{Reversible isothermal}} B \xrightarrow{\text{Isochoric cooling}} C(250\,K, 1\,bar) \xrightarrow[\text{compression}]{\text{Single stage adiabatic}} D\ (3\ bar)$$

The correct statement(s) is/are

(a) The pressure at B is 2.0 bar.

(b) The temperature at D is 450 K.

(c) $\Delta H_{CD} = 1000R$

(d) $\Delta U_{BC} = 375R$

**13.** Which of the following is true for reversible adiabatic process involving an ideal gas?

(a) Gas with higher $\gamma$ has high magnitude of slope in a $P$ ($y$-axis) vs. $T$ ($x$-axis) curve.

(b) Gas with higher $\gamma$ has high magnitude of slope in a $V$ ($y$-axis) vs. $T$ ($x$-axis) curve.

(c) Gas with higher $\gamma$ has high magnitude of slope in a $P$ ($y$-axis) vs. $V$ ($x$-axis) curve.

(d) Gas with higher $\gamma$ has low magnitude of slope in a $P$ ($y$-axis) vs. $T$ ($x$-axis) curve.

**14.** For a process to occur under adiabatic conditions, the essential condition(s) is/ are

(a) $\Delta T = 0$ (b) $\Delta P = 0$

(c) $q = 0$ (d) $\Delta U = 0$

**15.** For a process to occur under isothermal conditions, the essential condition(s) is/ are

(a) $dT = 0$ (b) $dP = 0$

(c) $q = 0$ (d) $dU = 0$

**16.** Assume ideal gas behaviour for all the gases considered and vibrational degrees of freedom to be active. Separate equimolar samples of Ne, $O_2$, $SO_2$ and $CH_4$ gases were subjected to a two-step process as mentioned. Initially, all are at the same temperature and pressure.

**Step I:** All undergo reversible adiabatic expansion to attain the same final volume, which is double the original volume thereby causing decrease in their temperature.

**Step II:** After Step I, all are given appropriate amount of heat isochorically to restore the original temperature.

Which of the following is/are correct statement(s)?

(a) Due to Step I only, the decrease in temperature will be maximum for Ne.

(b) During Step II, heat given will be maximum for $CH_4$.

(c) There will be no change in internal energy for any gas after both the steps of process are completed.

(d) The $P$–$V$ graph of $CH_4$ and $SO_2$ will be the same for overall process.

**17.** Which of the following statement(s) is/are incorrect?

(a) The specific heat capacity of substance is greater in the solid state than in liquid state.

(b) The specific heat capacity of a substance is greater in gaseous state than in the liquid state.

(c) The latent heat of vaporization of a substance is greater than that of fusion.

(d) The internal energy of an ideal gas is a function of its temperature.

**18.** For which of the following gases, the difference in specific heats at constant pressure and at constant volume is equal to 0.04545 cal/gm?

(a) $N_2$ (b) $N_2O$

(c) CO (d) $CO_2$

**19.** A liquid is adiabatically expanded from state – I to state – II, suddenly by a single step, as shown in the figure then

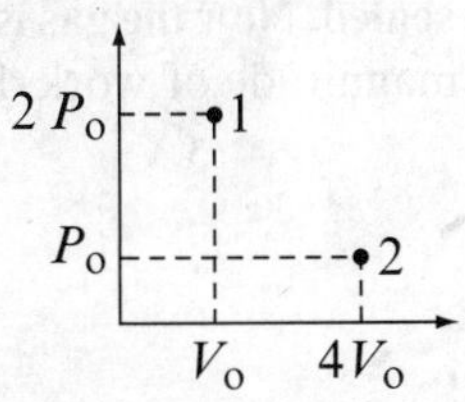


(a) $\Delta H = \frac{2\gamma P_o V_o}{\gamma - 1}$ (b) $\Delta U = \frac{3P_o V_o}{\gamma - 1}$

(c) $\Delta H = -P_o V_o$ (d) $\Delta U = -3P_o V_o$

**20.** Which of the following statement(s) is/are incorrect?

(a) For a closed system at rest with no fields, the sum of $q + w$ has the same value for every process that goes from a given state 1 to a given state 2.

(b) If a closed system at rest in the absence of external fields undergoes an adiabatic process that has $w = 0$, then the temperature of system must remain constant.

(c) A change in state from state 1 to state 2 produces a greater increase in entropy of the system when carried out irreversible than when carried out reversibly.

(d) The change in entropy of the system for an adiabatic process in a closed system must be zero.

**21.** Which of the following statement(s) is/are incorrect?

(a) Combustion of methane gas in an adiabatic rigid container will cause no change in temperature of the system.

(b) It is possible to have both adiabatic reversible and adiabatic irreversible processes between two states.

(c) For a reaction involving only ideal gases and occurring at constant temperature, there will not be any change in the internal energy of the system.

(d) $P$–$V$ work is always non-zero when there is some change in the volume of system.

**22.** Which of the following statement(s) is/ are correct regarding reversible adiabatic and isothermal processes for an ideal gas, starting from the same initial state to the same final volume?

(a) In expansion, more work is done by the gas in isothermal process.

(b) In compression, more work is done on the gas in adiabatic process.

(c) In expansion, the final temperature of gas will be higher in isothermal process.

(d) In compression, the final temperature of gas will be higher in adiabatic process.

**23.** A heat engine is operating in between 500 K and 400 K. If the engine absorbs 100 J heat per cycle, then which of the following is impossible amount of heat rejected by the engine per cycle?

(a) 75 J (b) 85 J

(c) 90 J (d) 70 J

**24.** For the cyclic process given below, which of the following relations are correct?

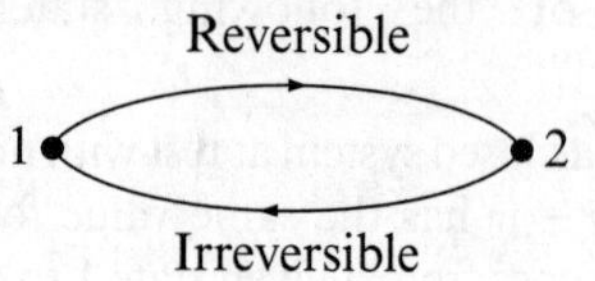


(a) $\Delta S = S_2 - S_1 = \int_1^2 \frac{\delta q_{rev}}{T}$

(b) $\Delta S = S_1 - S_2 = \int_2^1 \frac{\delta q_{irr}}{T}$

(c) $\Delta S_{cycle} = 0 = \int_1^2 \frac{\delta q_{rev}}{T} + \int_2^1 \frac{\delta q_{irr}}{T}$

(d) $\Delta S_{cycle} = 0 > \left( \int_1^2 \frac{\delta q_{rev}}{T} + \int_2^1 \frac{\delta q_{irr}}{T} \right)$

**25.** For which of the following process $\Delta S^o_{sys}$ is positive?

(a) The dissolution of ammonium nitrate in water.
(b) The expansion of a sample of chlorine into a larger volume.
(c) The decomposition of mercuric oxide.
(d) The rusting of iron.

**26.** Any process will be spontaneous at constant pressure and temperature when

(a) $\Delta S_{system}$ = +ve (b) $\Delta S_{univ.}$ = +ve
(c) $\Delta G_{sys}$ = –ve (d) $\Delta G_{univ.}$ = +ve

**27.** Which of the following statement(s) is/are incorrect?

(a) Entropy is a measure of unavailable energy.
(b) Decrease in free energy of the system represents the unavailable energy of the system.
(c) Complete conversion of heat into work is not possible in a reversible isothermal expansion of ideal gas.
(d) Complete conversion of heat into work is not possible in an irreversible isothermal expansion of ideal gas.

**28.** Which of the following processes are expected to be spontaneous at low temperature and non-spontaneous at high temperature?

(a) Dissociation of $PCl_5(g)$ into $PCl_3(g)$ and $Cl_2(g)$.
(b) Formation of $H_2O(l)$ from $H_2(g)$ and $O_2(g)$.
(c) Formation of $H_2O(g)$ from $H_2(g)$ and $O_2(g)$.
(d) Freezing of water.

**29.** The normal boiling point of a liquid is 350 K and $\Delta H_{vap}$ is 35 kJ/mol. Assume that $\Delta H_{vap}$ is independent from temperature and pressure. The correct statement(s) is/are

(a) $\Delta S_{vap}$ > 100 J/K-mol at 350 K and 0.5 atm.
(b) $\Delta G_{vap}$ < 0 at 350 K and 0.5 atm.
(c) $\Delta S_{vap}$ > 100 J/K-mol at 350 K and 2.0 atm.
(d) $\Delta G_{vap}$ > 0 at 350 K and 2.0 atm.

**30.** Select the correct statement(s).

(a) In a reversible process, $\Delta G$ is always zero in a closed system.
(b) In a reversible process, $\Delta S_{univ}$ is always zero in a closed system.
(c) In a reversible process, $\Delta S_{sys}$ is always zero in a closed system.
(d) In a reversible process, $\Delta S_{sys}$ is always zero in an isolated system.

## Section C (Comprehensions)

### Comprehension I

**An amount of 4 mole of hydrogen gas is taken in a vessel at STP and the vessel is sealed. Now the gas is cooled to −50°C.**

**1.** What is the change in internal energy of gas?

(a) −1400 cal (b) −600 cal
(c) −1000 cal (d) Zero

**2.** What is the change in enthalpy of gas?

(a) −1400 cal (b) −600 cal
(c) −1000 cal (d) Zero

**3.** What is the magnitude of work done by the gas?

(a) 600 J
(b) 600 cal
(c) 1000 cal
(d) Zero

## Comprehension II

**An ideal gas ($C_P/C_v = \gamma$) is expanded so that the amount of heat transferred to the gas is equal to the decrease in its internal energy.**

**4.** What is the molar heat capacity of gas in this process?

(a) $C_{v,m}$ (b) $-C_{v,m}$
(c) $C_{P,m}$ (d) $-C_{P,m}$

**5.** What is the relation between $T$ and $V$ of gas in this process?

(a) $T \cdot V^{2(\gamma-1)} = \text{Constant}$
(b) $T \cdot V^{\gamma-1} = \text{Constant}$
(c) $T \cdot V^{(\gamma-1)/2} = \text{Constant}$
(d) $T \cdot V^{\gamma} = \text{Constant}$

**6.** What is the magnitude of work performed by one mole of the gas when its volume increases 8 times if the initial temperature of the gas is 300 K? $C_{v,m}$ for the gas is $1.5R$.

(a) 900 cal (b) 450 cal
(c) 1247.7 cal (d) 623.8 cal

## Comprehension III

**As a result of the isobaric heating by $\Delta T = 72$ K, one mole of a certain ideal gas obtains an amount of heat $Q = 1.60$ kJ.**

**7.** The work performed by the gas is

(a) 8.60 kJ (b) 0.60 kJ
(c) 16.60 kJ (d) 4.60 kJ

**8.** The increment of its internal energy (in kJ) is

(a) 1.0 (b) 1.6
(c) 2.2 (d) 2.0

**9.** The value of $\gamma$ for the gas is

(a) 0.6 (b) 0.16
(c) 1.6 (d) 2.2

## Comprehension IV

**An ideal gas whose adiabatic exponent equals $\gamma$ is expanded according to the law $P = \alpha V$, where $\alpha$ is a constant. The initial volume of the gas is equal to $V_0$. As a result of expansion, the volume increases 4 times.**

**10.** What is the in the increment of the internal energy of the gas?

(a) $\dfrac{15\alpha V_0^2}{\gamma-1}$ (b) $\dfrac{\alpha V_0^2}{\gamma-1}$

(c) $\dfrac{15\alpha V_0^2}{2}$ (d) $\dfrac{15\alpha(\gamma+1)V_0^2}{2(\gamma-1)}$

**11.** What is the magnitude of work performed by the gas?

(a) $\dfrac{15\alpha V_0^2}{\gamma-1}$ (b) $\dfrac{\alpha V_0^2}{\gamma-1}$

(c) $\dfrac{15\alpha V_0^2}{2}$ (d) $\dfrac{15\alpha(\gamma+1)V_0^2}{2(\gamma-1)}$

**12.** What is the molar heat capacity of the gas in the process?

(a) $\dfrac{R(\gamma+1)}{2(\gamma-1)}$

(b) $\dfrac{\alpha V_0^2}{\gamma-1}$

(c) $\dfrac{R(\gamma-1)}{2(\gamma+1)}$

(d) $\dfrac{15\alpha(\gamma+1)V_0^2}{2(\gamma-1)}$

## Comprehension V

**One mole of an ideal gas whose adiabatic exponent equals $\gamma$ undergoes a process in which the gas pressure relates to the temperature as $p = aT^{\alpha}$, where $a$ and $\alpha$ are constants.**

**13.** What is the magnitude of work performed by the gas if its temperature is increased by $\Delta T$?

(a) $R(1-\alpha)\cdot\Delta T$ (b) $R\cdot\alpha\cdot\Delta T$
(c) $R\cdot\Delta T/(1-\alpha)$ (d) $R\cdot\Delta T/\alpha$

**14.** What is the molar heat capacity of the gas in the process?

(a) $\dfrac{R(1+\alpha)}{\gamma-1}$ (b) $\dfrac{R\alpha}{\gamma-1}$

(c) $\dfrac{R}{\gamma-1}+R(1-\alpha)$ (d) $\dfrac{R}{\gamma-1}+\dfrac{R}{1-\alpha}$

**15.** At what value of $\alpha$ will the heat capacity be negative?

(a) $\alpha < -1$ (b) $\alpha < 0$

(c) $\alpha > \dfrac{\gamma}{\gamma-1}$ (d) $\alpha < \gamma$

## Comprehension VI

**An ideal gas with the adiabatic exponent $\gamma$ undergoes a process in which its internal energy relates to the volume as $U = a.V^{\alpha}$, where $a$ and $\alpha$ are constants.**

**16.** What is the magnitude of work performed by the gas to increase its internal energy by $\Delta U$?

(a) $\dfrac{R.\Delta U}{\alpha(\gamma-1)}$ (b) $\dfrac{\Delta U}{\alpha(\gamma-1)}$

(c) $\dfrac{\Delta U.(\gamma-1)}{\alpha}$ (d) $\dfrac{\Delta U\cdot\alpha}{(\gamma-1)}$

**17.** What amount of heat is to be transferred to this gas to increase its internal energy by $\Delta U$?

(a) $\Delta U\left(1+\dfrac{\gamma-1}{\alpha}\right)$ (b) $\Delta U\left(1-\dfrac{\gamma-1}{\alpha}\right)$

(c) $\Delta U\left(1+\dfrac{\alpha}{\gamma-1}\right)$ (d) $R\cdot\Delta U\left(1+\dfrac{\gamma-1}{\alpha}\right)$

**18.** What is the molar heat capacity of the gas in this process?

(a) $\dfrac{R}{\alpha(\gamma-1)}$ (b) $\dfrac{R}{\alpha+\gamma-1}$

(c) $\dfrac{R}{\gamma-1}+\dfrac{R}{\alpha}$ (d) $\dfrac{R}{\gamma-1}-\dfrac{R}{\alpha}$

## Comprehension VII

**The adiabatic compression of a mixture of 2 volumes of hydrogen and 1 volume of oxygen from 320 ml to 10 ml caused explosion. The initial pressure and temperature of the gaseous mixture were 1 atm and 27°C.**

**19.** What is the pressure of the gaseous mixture at the moment of explosion?

(a) 32 atm (b) 128 atm
(c) 1/32 atm (d) 1/128 atm

**20.** What is the temperature of the gaseous mixture at the moment of explosion?

(a) 1200 K (b) 300 K
(c) 75 K (d) 3023.8 K

**21.** What is the magnitude of maximum work performed by the gaseous mixture before explosion?

(a) 145.87 J (b) 243.12 J
(c) 2400 J (d) 0

## Comprehension VIII

**A cylindrical container of volume 44.8 L is containing equal number of moles of an ideal monoatomic gas in two sections A and B, separated by an adiabatic frictionless piston as shown in the figure. The initial temperature and pressure of gas in both section is 27.3 K and 1 atm. Now, gas in section A is slowly heated till the volume of section B becomes one-eighth of the initial volume.**

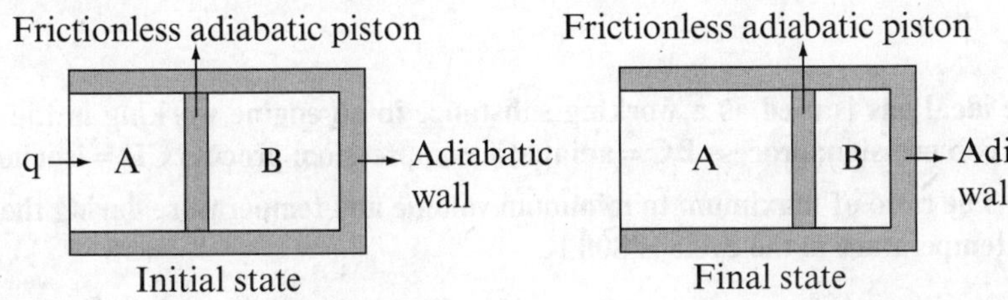


**22.** What will be the final pressure of gas in Section B?

(a) 2 atm (b) 8 atm
(c) 16 atm (d) 32 atm

**23.** What is the final temperature of gas in Section A?

(a) 1638 K (b) 6988 K
(c) 3274 K (d) 51 K

**24.** The change in enthalpy of gas in Section A is

(a) 48.3 kcal (b) 80.53 kcal
(c) 4.83 kcal (d) 8.05 kcal

---

## Comprehension IX

**The vapour pressure of water at 353 K is 532 mm Hg. The external pressure on $H_2O$ (l) taken in a cylinder fitted with frictionless movable piston initially containing 0.9 L (= 0.9 kg) of $H_2O$ (l) at 353 K is increased to 1 atm at the constant temperature. Then, heat is supplied keeping the pressure constant till 0.45 L of $H_2O$ (l) is evaporated to form $H_2O$ (g) at 373 K. Assume the internal energy of liquid to be dependent only on temperature. Answer the following questions by carefully observing the diagrams and the data provided.**

**Specific heat of $H_2O$ (l) = 4.2 J/°C-g**

**$\Delta H_{vap}$ at 373 K and 1 atm = +40 kJ/mol**

**1 L-atm = 100 J**

**$R$ = 8 J/K-mol**

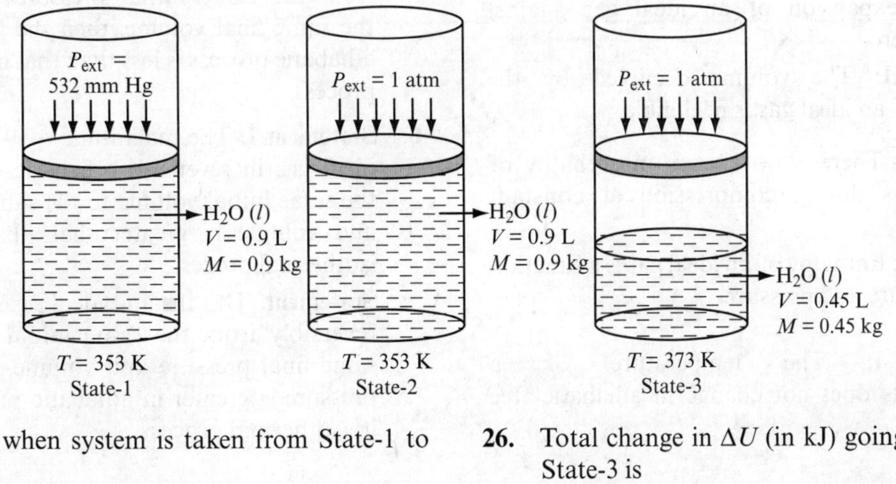


**25.** $\Delta H$ (in J), when system is taken from State-1 to State-2, is

(a) zero (b) 0.27
(c) 27 (d) 90

**26.** Total change in $\Delta U$ (in kJ) going from State-1 to State-3 is

(a) 75.6 (b) 1075.6
(c) 1001 (d) 74.6

27. Total change in enthalpy (in kJ) going from State-1 to State-3 is
(a) 75.6 (b) 1075.6
(c) 1001 (d) 74.6

28. What is the magnitude of work done (in J) in going State-1 to State-3?
(a) Zero (b) 74.6
(c) 90 (d) 31.5

## Comprehension X

**One mole of a monoatomic ideal gas is used as a working substance in an engine working in the following reversible cycle, process AB = isobaric compression; process BC = adiabatic compression; process CD = isochoric heating; process DA = isothermal expansion. The ratio of maximum to minimum volume and temperature during the cycle is $8\sqrt{2}$ and 4, respectively. The maximum temperature in the cycle is 800 K.**

29. The temperature of gas at state 'C' is
(a) 800 K (b) 200 K
(c) 400 K (d) 100 K

30. The change in internal energy of gas in the process BC is
(a) 2.49 kJ (b) −2.49 kJ
(c) 4.16 kJ (d) −4.16 kJ

## Section D (Assertion – Reason)

The following questions consist of two statements. Mark the answer as follows.

**(a)** If both statements are CORRECT, and **Statement II** is the CORRECT explanation of **Statement I.**

**(b)** If both statements are CORRECT, and **Statement II** is NOT the CORRECT explanation of **Statement I.**

**(c)** If **Statement I** is CORRECT, but **Statement II** is INCORRECT.

**(d)** If **Statement I** is INCORRECT, but **Statement II** is CORRECT.

1. **Statement I:** The heat absorbed during the isothermal expansion of an ideal gas against vacuum is zero.
**Statement II:** The volume occupied by the molecules of an ideal gas is negligible.

2. **Statement I:** There is no change in enthalpy of an ideal gas during compression at constant temperature.
**Statement II:** Enthalpy of an ideal gas is a function of temperature and pressure.

3. **Statement I:** The temperature of a non-ideal gas does not change in adiabatic free expansion.
**Statement II:** In adiabatic free expansion of ideal as well non-ideal gas, $q = w = \Delta U = 0$.

4. **Statement I:** Due to adiabatic expansion, the temperature of an ideal gas always decreases.
**Statement II:** For an adiabatic process, $\Delta U = w$.

5. **Statement I:** The magnitude of work done by an ideal gas in adiabatic reversible expansion from the same initial pressure and volume to the same final volume is less than that in isothermal reversible process.
**Statement II:** If an ideal gas is expanded reversibly from the same initial pressure and volume to the same final volume, then the final pressure in adiabatic process is less than that in the isothermal process.

6. **Statement I:** The magnitude of work done by an ideal gas in reversible adiabatic expansion from the same initial volume to the same final pressure and volume is greater than that in reversible isothermal process.
**Statement II:** If an ideal gas is expanded reversibly from the same initial volume to the same final pressure and volume, then the initial pressure is greater in adiabatic process than that in isothermal process.

7. **Statement I:** The final temperature of an ideal gas in adiabatic expansion is less in reversible expansion than in irreversible expansion against a constant external pressure.

   **Statement II:** The magnitude of work done by an ideal gas in adiabatic expansion is more in reversible process than that in irreversible process.

8. **Statement I:** The final temperature of an ideal gas in adiabatic compression is less in reversible compression than in irreversible compression against a constant external pressure.

   **Statement II:** The magnitude of work done on an ideal gas in adiabatic compression is more in irreversible process than that in reversible process.

9. **Statement I:** For the same mole of $SO_2$ and $O_2$ gases, more heat is needed for $SO_2$ gas than $O_2$ gas for the same increase in the temperature of the gas at constant pressure.

   **Statement II:** $SO_2$ gas, being triatomic, has higher heat capacity than the diatomic $O_2$ gas.

10. **Statement I:** Decrease in free energy causes spontaneous reaction.

    **Statement II:** Spontaneous reactions are invariably exothermic.

11. **Statement I:** Many endothermic reactions that are not spontaneous at room temperature become spontaneous at high temperature.

    **Statement II:** Entropy of the system increases with increase in temperature.

12. **Statement I:** An exothermic process, non-spontaneous at high temperature, may become spontaneous at low temperature.

    **Statement II:** With decrease in temperature, randomness (entropy) decreases.

13. **Statement I:** A reaction which is spontaneous and accompanied by decrease of randomness must be exothermic.

    **Statement II:** All exothermic reactions are accompanied by decrease of randomness.

14. **Statement I:** The endothermic reactions are favoured at lower temperatures and the exothermic reactions are favoured at higher temperatures.

    **Statement II:** When a system in equilibrium is disturbed by changing the temperature, it will tend to adjust itself so as to overcome the effect of change.

15. **Statement I:** For an irreversible cyclic process in a closed system, $\Delta S_{surr} = 0$.

    **Statement II:** Entropy is a state function.

## Section E (Column Match)

1. Match the columns

| Column I (Process) | Column II (Type of boundary possible) |
|---|---|
| (A) Isochoric | (P) Rigid |
| (B) Isothermal | (Q) Non-rigid |
| (C) Isobaric | (R) Impermeable |
| (D) Adiabatic | (S) Diathermal |
| | (T) Insulated |

2. Match the columns

| Column I | Column II |
|---|---|
| (A) Change in internal energy | (P) Independent from the quantity of system |
| (B) Heat | (Q) Depends on the quantity of system |
| (C) Work done by a gas | (R) State function |
| (D) Molar heat capacity at constant volume | (S) Path function |

3. Match the columns

| Column I (Ideal gas) | Column II (Related equations) |
|---|---|
| (A) Reversible isothermal process | (P) $w = nRT \ln \frac{P_2}{P_1}$ |
| (B) Reversible adiabatic process | (Q) $w = nC_{V,m}(T_2 - T_1)$ |
| (C) Irreversible adiabatic process | (R) $PV = nRT$ |
| (D) Irreversible isothermal process | (S) $w = -\int_{V_1}^{V_2} P_{ext} \cdot dV$ |

**4.** Match the columns (Given process does not include chemical reaction and phase change).

| Column I (Relation) | Column II (Applicable to) |
|---|---|
| (A) $\Delta H = \Delta U + \Delta(PV)$ | (P) Any matter undergoing any process. |
| (B) $\Delta H = n \cdot C_{P,m} \cdot \Delta T$ | (Q) Isochoric process involving any substance. |
| (C) $q = \Delta U$ | (R) Ideal gas, under any process. |
| | (S) Ideal gas under isothermal process. |

**5.** Match the columns (assume $C_{v,m}$ is independent of temperature).

| Column I | Column II |
|---|---|
| (A) $dU = n \cdot C_{v,m} \cdot dT$ | (P) Ideal gas |
| (B) $dU = n \cdot C_{v,m} \cdot dT + n^2 a \dfrac{dV}{V^2}$ | (Q) Van der Waals gas |
| (C) $dU = n^2 a \dfrac{dV}{V^2}$ | (R) Van der Waals gas in isothermal process |
| | (S) Van der Waals gas in isochoric process |

**6.** Match the process of Column I with entropy or enthalpy change in Column II.

| Column I | Column II |
|---|---|
| (A) $N_2(g) + O_2(g) \rightarrow 2NO(g)$ | (P) $\Delta S \approx 0$ |
| (B) $2KI(aq) + HgI_2(aq) \rightarrow K_2[HgI_4](aq)$ | (Q) $\Delta S < 0$ |
| (C) $PCl_5(g) \rightarrow PCl_3(g) + Cl_2(g)$ | (R) $\Delta H > 0$ |
| (D) $NH_3(g) + HCl(g) \rightarrow NH_4Cl(s)$ | (S) $\Delta H < 0$ |

**7.** Match the columns.

| Column I | Column II |
|---|---|
| (A) Reversible isothermal expansion of an ideal gas | (P) $\Delta S_{sys} = \Delta S_{surr} = \Delta S_{total} = 0$ |
| (B) Reversible adiabatic expansion of an ideal gas | (Q) $\Delta H = \Delta U = \Delta S_{total} = 0$ |
| (C) Adiabatic free expansion | (R) $\Delta S_{total} > 0$ |
| | (S) $q = 0$ |

**8.** Match the columns.

| Column I | Column II |
|---|---|
| (A) For the process: $A(l) \rightarrow A(s)$, $\Delta H$ and $\Delta V$ may be | (P) −ve, +ve |
| (B) $A_2(s) + B_2(g) \rightarrow C_2(s) + D_2(s)$, $\Delta H$ and $\Delta G$ may be | (Q) +ve, −ve |
| (C) $A_2(g) \rightarrow B(g) + C(g)$, $E_{a(forward)} = 50$ kJ/mol, $E_{a(backward)} = 40$ kJ/mol $\Delta H$ and $\Delta G$ at very high temperature are | (R) +ve, +ve |
| (D) $A(g) \rightarrow B(g)$ $\Delta H$ and $\Delta G$ at very low temperature are | (S) −ve, −ve |

**9.** Match the columns.

| Column I | Column II |
|---|---|
| (A) Fusion at melting point | (P) $\Delta G = 0$ |
| (B) Vaporization at boiling point | (Q) $\Delta G < 0$ |
| (C) Condensation at triple point | (R) $\Delta S > 0$ |
| (D) Melting at normal boiling point | (S) $\Delta H \approx \Delta U$ |

**10.** Match Column I (graph) with the physical quantity in Column II (area shown in the graph represents the magnitude of).

| Column I | Column II |
|---|---|
| (A) 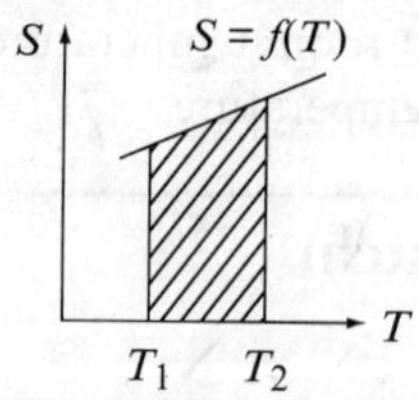  | (P) $q$ |
| (B) 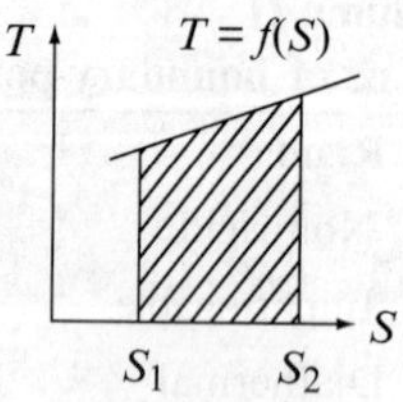  | (Q) $w$ |
| (C) 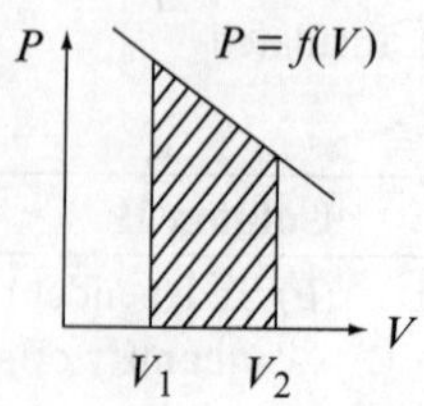  | (R) $(\Delta G)_T$ |
| (D) 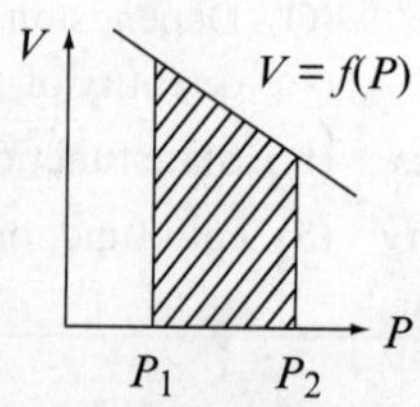  | (S) $(\Delta G)_P$ |

## Section F (Subjective)

### Single-digit Integer Type

1. As 1 mole of liquid water is heated from 288 to 298 K, it expands slightly against the atmosphere. Given coefficient of cubic expansion of water is 0.0002/°C; density of water = 1.0 g $cm^{-3}$ and external pressure = 1 bar. If the magnitude of work for this process is $X$ J, then the value of $\frac{X}{6\times10^{-4}}$ is

2. The final volume (in L) of one mole of an ideal gas initially at 27°C and 8.21 atm pressure, if it absorbs 420 cal of heat during a reversible isothermal expansion is (ln 2 = 0.7)

3. A balloon containing air at 27°C and 1 bar initially is filled with air further, isothermally and reversibly till the pressure is 4 bar. If the initial diameter of the balloon is 1 m and the pressure at each stage is proportional to diameter of balloon, the magnitude of work done (in $10^7$ J) is

4. A quantity of 8 g oxygen gas is expanded isothermally at 27°C from 2 $dm^3$ to 8 $dm^3$ at a constant external pressure of 4 bar. If the magnitude of work done in this process is used in lifting a body of mass 40 kg, the height (in metre) up to which the body can be lifted is ($g$ = 10 $ms^{-2}$)

5. An amount $Q$ of heat is added to a monoatomic ideal gas in a process in which the gas performs a work $Q$/2 on its surrounding. The molar heat capacity of gas (in cal/K-mol) for the process is

6. What amount of heat (in J) is to be transferred to nitrogen in the isobaric heating process for that gas to perform the work 2.0 J?

7. Ten moles of a certain ideal gas at a temperature 300 K was cooled isochorically so that the gas pressure reduced 2.0 times. Then, as a result of the isobaric process, the gas expanded till its temperature got back to the initial value. The total amount of heat absorbed (in Kcal) by the gas in this process is

8. Three moles of an ideal gas being initially at a temperature 273 K was isothermally expanded 5.0 times its initial volume and then isochorically heated so that the pressure in the final state became equal to that in the initial state. The total amount of heat transferred to the gas during the process equals 80 kJ. The value of $C_{v,m}$ for this gas (in cal/K-mol) is (ln 5 = 1.6)

9. The number of degrees of freedom of molecules in a gas whose molar heat capacity is 29 J/mol-K in the process $PT$ = constant is

10. A quantity of 56 g of nitrogen gas is enclosed in a rigid vessel at a temperature 300 K. The amount of heat (in kcal) has to be transferred to the gas to increase the root mean square velocity of its molecules 2.0 times is

11. A gas consisting of rigid diatomic molecules was expanded in a polytropic process so that the rate of collisions of the molecules against the vessel's wall did not change. The molar heat capacity of the gas (in cal/K-mol) in this process is

12. The increase in Gibbs free energy (in kJ) of 13 g of ethanol (density = 0.78 g $cm^{-3}$), when the pressure is increased isothermally from 1 bar to 3001 bar, is

13. At 298 K, the standard enthalpy of combustion of sucrose is –5737 kJ $mol^{-1}$ and the standard Gibb's energy of the reaction is –6333 kJ $mol^{-1}$. The additional (non-expansion) work (in kJ) that may be obtained by raising the temperature to 29°C is

14. The diamonds are formed from graphite under very high pressure. Calculate the equilibrium pressure (in $10^4$ bar) at which graphite is converted into diamond at 25°C. The densities of graphite and diamond are 2.4 and 3.6 g/$cm^3$, respectively, and are independent of pressure. $\Delta G^\circ$ for the conversion of graphite into diamond is 5.0 kJ/mol.

15. The standard entropy change (in cal/K-mol) for the reaction: X $\rightleftharpoons$ Y, if the value of $\Delta H^\circ$ = 7.5 kcal/mol and $K_C = e^{-10}$ at 300 K.

EXERCISE II

## Four-digit Integer Type

1. One mole of a Van der Waals gas expands reversibly and isothermally at 27°C from 2 L to 20 L. The magnitude of work done (in J) if $a = 1.42 \times 10^{12}$ dynes $cm^4$/mole and $b = 30$ ml/mole.

2. One mole of an non-ideal gas undergoes a change of state from (2.0 atm, 3.0 L, 95 K) to (4.0 atm, 5.0 L, 245 K) with a change in internal energy, $\Delta U$ = 30.0 L-atm. The enthalpy change ($\Delta H$) of the process in L-atm is

3. The internal energy change in the conversion of 1 mole of the calcite form of $CaCO_3$ to the aragonite form is +0.21 kJ. The enthalpy change (in J) in the conversion at 2.7 bar is (The densities of the solids calcite and aragonite are 2.7 g $cm^{-3}$ and 3.0 g $cm^{-3}$, respectively.)

4. A system undergoes a certain change in state by path I and the corresponding heat absorbed and work done are 10 kcal and 0 erg, respectively. For the same change in state by path II, the respective quantities are 11 kcal and 0.5 $w_{max}$, where $w_{max}$ represents the work done if the changes were reversibly carried out. The magnitude of $w_{max}$, in J, is (1 cal = 4.2 J)

5. The internal energy of a gas is given by $U = 1.5\,PV$. It expands from 100 to 200 $cm^3$ against a constant pressure of $1.0 \times 10^5$ Pa. The heat absorbed (in KJ) by the gas in the process is

6. The internal energy of a monoatomic ideal gas is 1.5 $nRT$. One mole of helium is kept in a cylinder of cross section 8.5 $cm^2$. The cylinder is closed by a light frictionless piston. The gas is heated slowly in a process during which a total of 42 J heat is given to the gas. If the temperature rises through 2°C, then find the distance moved by the piston (in cm). Atmospheric pressure = 100 kPa.

7. Only at extremely high pressure does $\Delta H - \Delta U$ for condensed state reactions becomes significantly different than zero. Determine the pressure (in bar) at which $\Delta H - \Delta U$ is equal to – 1.0 kJ for the reaction, C(graphite) $\rightarrow$ C(diamond). For graphite and diamond, the densities are 2.4 and 3.6 g/ml, respectively.

8. The melting point of a certain substance is 70°C, its normal boiling point is 450°C, its enthalpy of fusion is 30 cal/g, its enthalpy of vaporization is 45 cal/g, and its specific heat is 0.215 cal/g-K. The heat required (in cal) to convert 10 g of the substance from the solid state at 70°C to vapour at 450°C is

9. When an electric current of 0.50 A from a 12 V supply is passed for $\frac{1805}{6}$ sec through a resistance in thermal contact with water maintained at 1 atm and 373 K, it is found that 0.9 g of water is vaporized. The molar internal energy change (in kJ/mol) of water is (Take $8.314 \times 373 = 3100$)

10. A volume of 100 ml of a liquid contained in an adiabatic container at a pressure of 1 bar. The pressure is steeply increased to 100 bar by which the volume of liquid is decreased by 1 ml. The change in enthalpy, $\Delta H$, of the liquid is (Answer as 'abcd', where $a = 1$, if $\Delta H$ is +ve and $a = 2$, if $\Delta H$ is – ve, and 'bcd' is the magnitude of $\Delta H$, in J)

11. Pressure over 1000 ml of a liquid is gradually increased from 1 bar to 1001 bar under adiabatic conditions. If the final volume of the liquid is 990 ml and there is linear variation of volume with pressure, the value of $\Delta U$ of the process is (Answer as 'abcd', where $a = 1$, if $\Delta U$ is +ve and $a = 2$, if $\Delta U$ is – ve, and 'bcd' is the magnitude of $\Delta U$, in J)

12. One mole of an ideal gas ($\gamma = 1.4$) is expanded isothermally at 27°C till its volume is doubled. It is then adiabatically compressed to its original volume. The magnitude of total work done (in J) by the gas is ($\ln 2 = 0.7$, $2^{0.4} = \frac{4}{3}$, R = 8.3 J/K-mol)

13. Consider a classroom that is roughly 8.21 m × 10 m × 3 m. Initially, $T = 290$ K and $P = 1$ atm. There are 50 people in the class, each losing energy to the room at the average rate of 166 W. Assume that the walls, ceiling, floor, and furniture are perfectly insulated and do not absorb any heat. Also assume that all the doors and windows are tightly closed to prevent any exchange of air from surrounding. How long (in sec) will the physical chemistry examination last if the professor (Mr. Neeraj Kumar) has foolishly agreed to dismiss the class when the air temperature in the room reaches body temperature, 310 K? For air, $C_{P,m} = 7R/2$ ($R = 0.0821$ L-atm/K-mol = 8.3 J/K-mol).

EXERCISE II

**14.** Three moles of an ideal gas ($C_{P,m}$ = 2.5R) and 2 moles of another ideal gas ($C_{P,m}$ = 3.5R) are taken in a vessel and compressed reversibly and adiabatically. In this process, the temperature of gaseous mixture increased from 300 K to 400 K. The increase in internal energy of gaseous mixture (in cal) is

**15.** One mole of an ideal monoatomic gas initially at 1200 K and 64 atm is expanded to a final state at 300 K and 1 atm. To achieve the above change, a reversible path is constructed that involve an adiabatic expansion in the beginning followed by an isothermal expansion to the final state. The magnitude of net work done by the gas (in cal) is (ln 2 = 0.7)

**16.** Calculate $\Delta S_{univ}$ (in J/K) for the chemical reaction: C(Graphite) + $2H_2$(g) → $CH_4$(g); $\Delta H^\circ_{300} = -75.0$ kJ. The standard entropies of C (Graphite), $H_2$(g) and $CH_4$(g) are 6.0, 130.6 and 186.2 J/K-mol, respectively.

**17.** An athlete in the weight room lifts a 50 kg mass through a vertical distance of 2.0 m. The mass is allowed to fall through the 2.0 m distance while coupled to an electrical generator. The electrical generator produces an equal amount of electrical work, which is used to produce aluminium by Hall electrolytic process.

$Al_2O_3$(Solution) + 3C(Graphite) → 2 Al(l) + 3CO(g); $\Delta G^\circ$ = 600 kJ

How many times must the athlete lift the 50 kg mass to provide sufficient Gibbs energy to produce 27 g Al? ($g = 10$ m/s$^2$)

**18.** How much energy (in kJ) is available for sustaining muscular and nervous activity from the combustion of 1.0 mole of glucose molecules under standard conditions at 37°C (body temperature)? The standard entropy and enthalpy of reaction are +200 J/K-mol and –2808 kJ/ mol, respectively.

**19.** Select the conditions which represent the criteria for spontaneity of a process in a closed system, from the following given conditions.

| Serial Number | Conditions |
|---|---|
| 1. | $(dG)_{P,T} > 0$ |
| 2. | $(\Delta S)_{universe} > 0$ |
| 3. | $(dU)_{S,V} > 0$ |
| 4. | $(dH)_{S,P} < 0$ |
| 5. | $(dS)_{U,V} < 0$ |
| 6. | $(dS)_{H,P} > 0$ |

(Answer by adding the serial numbers of the correct conditions.)

**20.** Consider the following phase transitions.

(i) $H_2O$ (s, 1 atm, 273 K) → $H_2O$ (*l*, 1 atm, 273 K)
(ii) $H_2O$ (s, 1 atm, 300 K) → $H_2O$ (*l*, 1 atm, 300 K)
(iii) $H_2O$ (s, 1 atm, 200 K) → $H_2O$ (*l*, 1 atm, 200 K)
(iv) $H_2O$ (s, 0.5 atm, 273 K) → $H_2O$ (*l*, 0.5 atm, 273 K)
(v) $H_2O$ (s, 2 atm, 273 K) → $H_2O$ (*l*, 2 atm, 273 K)
(vi) $C_6H_6$ (*l*, 1 atm, 353 K) → $C_6H_6$ (g, 1 atm, 353 K)
(vii) $C_6H_6$ (*l*, 1 atm, 400 K) → $C_6H_6$ (g, 1 atm, 400 K)
(viii) $C_6H_6$ (*l*, 1 atm, 300 K) → $C_6H_6$ (g, 1 atm, 300 K)
(ix) $C_6H_6$ (*l*, 2 atm, 323 K) → $C_6H_6$ (g, 2 atm, 323 K)

The normal freezing point of water is 273 K and the normal boiling point of benzene is 353 K.

Now, a four-digit number 'abcd' is defined as follows.

Digit 'a': Number of phase transitions for which $\Delta S_{total} = 0$.

Digit 'b': Number of phase transitions for which $\Delta S_{total} > 0$.

Digit 'c': Number of phase transitions for which $\Delta S_{total} < 0$.

Digit 'd': Number of phase transitions for which $\Delta H > 0$.

Determine the number 'abcd'.

## Answer Keys

**Exercise I**

### Basics

1. (c) 2. (d) 3. (c) 4. (a) 5. (b) 6. (d) 7. (c) 8. (c) 9. (d) 10. (d)
11. (c) 12. (d) 13. (c) 14. (d) 15. (a)

### First Law of Thermodynamics

16. (a) 17. (c) 18. (a) 19. (a) 20. (a) 21. (b) 22. (d) 23. (c) 24. (a) 25. (c)
26. (c) 27. (a) 28. (b) 29. (c) 30. (b) 31. (c) 32. (b) 33. (b) 34. (d) 35. (a)
36. (a) 37. (b) 38. (c) 39. (b) 40. (c) 41. (b) 42. (a) 43. (b) 44. (c) 45. (b)
46. (a) 47. (b) 48. (d) 49. (a) 50. (d) 51. (b) 52. (a) 53. (a) 54. (c) 55. (a)
56. (c) 57. (b) 58. (a) 59. (b) 60. (b)

### Second and Third Law of Thermodynamics, Entropy

61. (b) 62. (b) 63. (b) 64. (a) 65. (d) 66. (c) 67. (b) 68. (b) 69. (d) 70. (b)
71. (a) 72. (b) 73. (d) 74. (c) 75. (a) 76. (a) 77. (b) 78. (b) 79. (d) 80. (d)

### Gibbs Free Energy Function

81. (b) 82. (a) 83. (b) 84. (b) 85. (b) 86. (a) 87. (d) 88. (a) 89. (d) 90. (b)

## Answer Keys

**Exercise II**

### Section A (Only one Correct)

1. (d) 2. (b) 3. (a) 4. (c) 5. (a) 6. (a) 7. (b) 8. (a) 9. (d) 10. (a)
11. (c) 12. (c) 13. (c) 14. (b) 15. (a) 16. (c) 17. (d) 18. (c) 19. (a) 20. (b)
21. (c) 22. (a) 23. (b) 24. (c) 25. (d) 26. (b) 27. (c) 28. (c) 29. (b) 30. (b)
31. (c) 32. (b) 33. (a) 34. (a) 35. (c) 36. (a) 37. (b) 38. (c) 39. (b) 40. (a)
41. (a) 42. (b) 43. (c) 44. (c) 45. (c) 46. (a) 47. (c) 48. (b) 49. (a) 50. (c)
51. (c) 52. (b) 53. (a) 54. (b) 55. (c) 56. (a) 57. (b) 58. (c) 59. (c) 60. (a)
61. (c) 62. (a) 63. (a) 64. (b) 65. (c) 66. (a) 67. (b) 68. (c) 69. (a) 70. (a)
71. (d) 72. (c) 73. (c) 74. (c) 75. (b) 76. (c) 77. (b) 78. (d) 79. (d) 80. (a)

### Section B (One or More than one Correct)

1. (b), (c), (d) 2. (a), (b), (c) 3. (a), (b), (c) 4. (a), (b), (d)
5. (a), (b) 6. (a), (b), (c) 7. (c) 8. (a), (b), (c), (d)
9. (a), (c) 10. (a), (c) 11. (a), (b), (c) 12. (a), (b), (c)
13. (c), (d) 14. (c) 15. (a) 16. (a), (b), (c)
17. (a), (b) 18. (b), (d) 19. (c), (d) 20. (b), (c), (d)
21. (a), (b), (c), (d) 22. (a), (b), (c), (d) 23. (a), (d) 24. (a), (d)
25. (a), (b), (c) 26. (b), (c) 27. (b), (c), (d) 28. (b), (c), (d)
29. (a), (b), (d) 30. (b), (d)

## Section C

### Comprehension I

1. (c) 2. (a) 3. (d)

### Comprehension II

4. (b) 5. (c) 6. (a)

### Comprehension III

7. (b) 8. (a) 9. (c)

### Comprehension IV

10. (a) 11. (c) 12. (a)

### Comprehension V

13. (a) 14. (c) 15. (c)

### Comprehension VI

16. (c) 17. (a) 18. (c)

### Comprehension VII

19. (b) 20. (a) 21. (b)

### Comprehension VIII

22. (d) 23. (a) 24. (b)

### Comprehension IX

25. (c) 26. (c) 27. (b) 28. (b)

### Comprehension X

29. (c) 30. (a)

## Section D (Assertion – Reason)

1. (b) 2. (c) 3. (d) 4. (d) 5. (a) 6. (a) 7. (a) 8. (a) 9. (a) 10. (c)
11. (a) 12. (a) 13. (c) 14. (d) 15. (d)

## Section E (Column Match)

1. A → P, R, S, T; B → P, Q, R, S; C → Q, R, S, T; D → P, Q, R, T
2. A → Q; B → Q, S; C → Q, S; D → P
3. A → P, R, S; B → Q, R, S; C → Q, R, S; D → R, S
4. A → P, Q, R, S; B → R, S; C → Q
5. A → P, S; B → Q, R, S; C → R
6. A → P, R; B → Q, S; C → R; D → Q, S
7. A → Q; B → P, S; C → R, S
8. A → P, S; B → P, R, S; C → Q; D → R, S
9. A → P, S, R; B → P, R; C → P; D → Q, R, S
10. A → S; B → P; C → Q; D → R

## Section F (Subjective)

### Single-digit Integer Type

1. (6) 2. (6) 3. (1) 4. (6) 5. (6) 6. (7) 7. (3) 8. (5) 9. (3) 10. (9)
11. (6) 12. (5) 13. (8) 14. (3) 15. (5)

### Four-digit Integer Type

1. (5713) 2. (0044) 3. (0209) 4. (8400) 5. (0025)
6. (0020) 7. (6000) 8. (1567) 9. (0033) 10. (1990)
11. (1500) 12. (0332) 13. (0500) 14. (1900) 15. (3120)
16. (0169) 17. (0300) 18. (2870) 19. (0012) 20. (2349)

# HINTS AND EXPLANATIONS

## EXERCISE I (JEE MAIN)

### Basics

1. Theory based
2. Work cannot be defined at a particular state of system.
3. For isothermal process, the container must be a good conductor of heat.
4. Internal energy of ideal gas depends on the temperature as well as quantity of gas.
5. Greater the external pressure, greater will be the magnitude of work done by the gas.
6. Internal energy is a state function.

$\Delta U_{net} = \Delta U_{AB} + \Delta U_{BA} = (U_B - U_A) + (U_A - U_B) = 0.$

7. Theory based
8. A closed system should exchange energy with the surrounding. If heat exchange is not possible, then energy must be exchanged as work.
9. Theory based
10. Hess's law of constant heat summation.
11. During cyclic process, the internal energy of system may change but $\Delta U_{net} = 0$.
12. 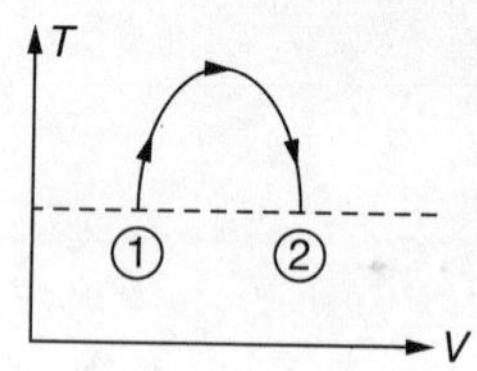


For the process, $\Delta T = 0$ but during the process, temperature is changing and hence, the process is not isothermal.

13. $PV^\gamma$ = Constant or $PV^x$ = Constant, are valid only when the gas is ideal, the process is reversible and $C_{v,m}$ is temperature independent.

For a cyclic process, $w_{net}$ may or may not be zero.

14. A true reversible process takes infinite time.
15. Theory based

### First Law of Thermodynamics

16. $\Delta U = q + w = (+20) + (-10) = +10$ J.
17. $\Delta U = q + w = (+300) + (-200) = 100$ cal.

(Work is done by the system)

18. $W = -1 \times \frac{1+2}{2} = -1.5 \text{ bar} \cdot \text{m}^3 = \frac{-1.5 \times 1000 \times 100}{1000}$ kJ

$= -150$ kJ

and $q = +100$ kJ

$\therefore \Delta U = q + w = -50$ kJ

19. As the system rejects heat, the temperature will decrease.
20. As the system is doing work but its internal energy is not changing, the system must absorb heat.
21. $\Delta U = n.C_{v,m}\ \Delta T = 0$ (as $\Delta T = 0$)
22. $w = -nRT \ln \frac{P_1}{P_2} = -5 \times 2 \times 300 \times \ln \frac{100}{1}$

$= -13800$ cal.

23. $w = -nRT \ln \frac{V_2}{V_1} \Rightarrow -4200 = -\frac{w}{40} \times 2 \times 300 \times \ln \frac{16}{4}$

$\therefore w = 200$ g.

24. For minimum work in compression, the process must be reversible.

$w = -nRT \cdot \ln \frac{P_1}{P_2} = -\frac{16}{32} \times 8.3 \times 300 \times \ln \frac{1.01325 \times 10^3}{1.01325 \times 10^{25}}$

$= +5727$ J

**25.** $\Delta H = n \cdot C_{P,m} \cdot \Delta T = 0$ (as $\Delta T = 0$)

**26.** $W = -\int_{V_1}^{V_2} P_{ext} \cdot dv = -\int_{v_2}^{v_2} P \cdot dv$ (For reversible process)

$$= -\int_{v_1}^{v_2}\left(\frac{nRT}{V-nb} - \frac{an^2}{V^2}\right) \cdot dV$$

$$= -\left[nRT \cdot \ln\frac{V_2 - nb}{V_1 - nb} + an^2\left(\frac{1}{V_2} - \frac{1}{V_1}\right)\right]$$

**27.** For both paths, $\Delta U = 0$, as $\Delta T = 0$.

$$\frac{q_{path-I}}{q_{path-II}} = \frac{-w_{path-I}}{-w_{path-II}} = \frac{PV \cdot \ln\frac{V_2}{V_1}}{(P \cdot \Delta V)_1 + (P \cdot \Delta V)_2}$$

$$= \frac{10 \times 1 \times \ln\left(\frac{10}{1}\right)}{5(2-1) + 1(10-2)} = \frac{10 \times 2.303}{13}$$

**28.** $\Delta H = 0 \Rightarrow \Delta U = 0 \Rightarrow q = -w$

and $w = -P_{ert}(V_2 - V_1) = -1$ bar (10 L – 5 L)

$= -5$ Lbar $= -500$ J

**29.** For adiabatic process:

$q = 0 \Rightarrow w = \Delta U = n \cdot C_{v,m} \cdot \Delta T$

**30.** $T_1 \cdot V_1^{\gamma-1} = T_2 \cdot V_2^{\gamma-1} \Rightarrow \frac{T_2}{T_1} = \left(\frac{V_1}{V_2}\right)^{\gamma-1} = \left(\frac{1}{8}\right)^{5/3-1}$

$= \frac{1}{4}$

**31.** $q = 0 \Rightarrow \Delta U = w \Rightarrow n \cdot C_{v,m} \cdot (T_2 - T_1)$

$= -P_{ert} \cdot (V_2 - V_1)$

or $1 \times \frac{3}{2} R \times (T_2 - T) = -1 \text{ atm} \times (2 \text{ L} - 1 \text{ L})$

$\therefore T_2 = T - \frac{2}{3 \times 0.0821}$

**32.** $w = \Delta U \Rightarrow -P_{ext}(V_2 - V_1) = n \cdot C_{v,m}(T_2 - T_1)$

or $-P_2\left(\frac{nRT_2}{P_2} - \frac{nRT_1}{P_1}\right) = n \times \frac{5}{2} R(T_2 - T_1)$

or $-\left(T_2 - T_1 \cdot \frac{P_2}{P_1}\right) - \frac{5}{2}(T_2 - T_1)$

or $-\left(T_2 - 350 \times \frac{2}{1}\right) = \frac{5}{2}(T_2 - 350) \Rightarrow T_2 = 450$ K.

Now, $w = \Delta U = n \cdot C_{v,m}(T_2 - T_1)$

$= 2 \times \frac{5}{2} R \times (450 - 350) = 500R$

**33.** $T_1 V_1^{\gamma-1} = T_2 \cdot V_2^{\gamma-1} \Rightarrow T_2 = T_1\left(\frac{V_1}{V_2}\right)^{\gamma-1}$

$= 800 \times \left(\frac{1}{32}\right)^{7/5-1} = 200$ K

$\therefore \Delta H = n \cdot C_{P,m} \cdot (T_2 - T_1) = 1 \times \frac{7}{2} R \times (200 - 800)$

$= -2100$ R

**34.**

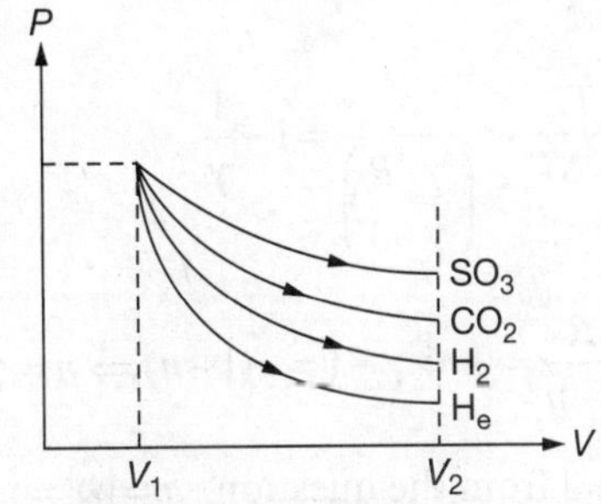


**35.** Adiabatic free expansion of an ideal gas is also isothermal.

**36.** $w = -P \cdot \Delta V = -nR \cdot \Delta T = -1 \times R \times 1 = -R$

**37.** $q = -124$ J and

$w = -P \cdot \Delta V = -2$ bar (125 – 325) ml

$= +400$ bar-ml

$= \frac{400}{1000} \times 100 = 40$ J.

$\therefore \Delta U = q + w = (-124) + 40 = -84$ J.

**38.** $q_P = n \cdot \int_{T_1}^{T_2} C_{P,m} \cdot dT = 2 \times \int_{T_1=300K}^{T_2=400K} (28.314 + 0.01T) \cdot dT$

$$= 2\left[28.314(T_2 - T_1) + \frac{0.01}{2}\left(T_2^2 - T_1^2\right)\right] = 6362.8 \text{ J}$$

**39.** $\frac{\Delta U}{-w} = \frac{n.C_{v,m}.\ \Delta T}{nR.\Delta T} = \frac{3}{2}$

**40.** $C_{v,m} = 3 \times \frac{1}{2} R + 2 \times \frac{1}{2} R + (3 \times 4 - 5) \times R = \frac{19}{2} R$

**41.** For isothermal changes, $C_m = \frac{q}{N \cdot \Delta t}$ = Infinite.

**42.** Monoatomic gases have only translational degree of freedom.

**43.** $\gamma_{mix} = 1 + \frac{(n_1 + n_2) \cdot R}{n_1 \cdot C_{v,m,1} + n_2 \cdot C_{v,m,2}}$

$= 1 + \frac{(1+1) \times R}{1 \times \frac{R}{\frac{5}{3}-1} + 1 \times \frac{R}{\frac{7}{5}-1}}$

$= 1.5$

**44.** $\frac{\Delta U}{q_P} = \frac{C_{v,m}}{C_{p,m}} = \frac{\frac{5}{2}R}{\frac{7}{2}R} = \frac{5}{7}$

**45.** $\frac{-w}{q_p} = \frac{nR \cdot \Delta T}{n \cdot C_{p,m} \cdot \Delta T} = \frac{R}{\left(\frac{\gamma \cdot R}{\gamma - 1}\right)} = 1 - \frac{1}{\gamma}$

**46.** $C_m = \frac{R}{\gamma - 1} + \frac{R}{1-n} = 0 \Rightarrow \gamma - 1 = -(1-n) \Rightarrow n = \gamma.$

**47.** $q = \Delta U - w$ and from the question, $q = |w| = w$

or $2q = \Delta U$

or $2 \cdot n \cdot C_m \cdot \Delta T = n \cdot C_{v,m} \cdot \Delta T \Rightarrow C_m = \frac{C_{v,m}}{2}$

$= \frac{1}{2} \times \frac{R}{1.5-1} = R$

**48.** $C_{p,m} = \frac{q_p}{n \cdot \Delta T} = \frac{70}{2 \times 10} = \frac{7}{2}$ cal

$\therefore C_{v,m} = \frac{7}{2} - 2 = \frac{3}{2}$ cal

Now, $q_v = n \cdot C_{v,m} \cdot \Delta T$

$= 2 \times \frac{3}{2} \times 10 = 30$ cal

**49.** $P \propto T^3 \Rightarrow P \propto \left(\frac{PV}{nR}\right)^3 \Rightarrow PV^{3/2}$ = Constant

$\Rightarrow \gamma = 3/2$

**50.** $\Delta U = q + w = 100 + \left(-\frac{210}{4.2}\right) = 50 \text{ cal} = n \cdot c_{v,m} \cdot \Delta T$

and $q = 100 \text{ cal} = n \cdot c_m \cdot \Delta T$

$\therefore \frac{50}{100} = \frac{n \cdot C_{v,m} \cdot \Delta T}{n \cdot C_m \cdot \Delta T} \Rightarrow C_m = 2C_{v,m} = 2 \times \frac{5}{2}R = 5R$

**51.** $P \cdot V^{-1} = K \Rightarrow x = -1$

Now, $\frac{\Delta U}{w} = \frac{n \cdot C_{v,m} \cdot \Delta T}{\left(\frac{n \cdot R \cdot \Delta T}{x-1}\right)} = -2 \cdot \frac{C_{v,m}}{R} = -2 \times \frac{\frac{3}{2}R}{R}$

$= -3.$

**52.** $P_1 = 1$ atm, $V_1 = V$ L, $T_1 = 300$ K

$P_2 = 2\sqrt{2}$ atm, $V_2 = V_1 \cdot \left(\frac{P_1}{P_2}\right)^{1/3} = V_1 \left(\frac{1}{2^{3/2}}\right)^{1/3} = \frac{V_1}{\sqrt{2}}$

; $T_2 = \frac{P_2 V_2 \cdot T_1}{P_1 V_1}$

or $T_2 = \frac{2\sqrt{2}}{1} \times \frac{1}{\sqrt{2}} \times 300 = 600$ K

Now, $w = \frac{-nR(T_2 - T_1)}{1-x} = -\frac{1 \times R \times (600-300}{1-3}$

$= +150R.$

**53.** $T_1 V_1^2 = T_2 V_2^2 \Rightarrow T_2 = T_1 \cdot \left(\frac{V_1}{V_2}\right)^2 = 400 \times \left(\frac{4}{8}\right)^2$

$= 100$ K

Now, $\Delta H = n \cdot C_{p,m} \cdot (T_2 - T_1) = 2 \times \frac{5}{2} R \times (100 - 400)$

$= -1500R$

**54.** $w_{AB} = -nRT \cdot \ln \frac{P_1}{P_2} = -2 \times R \times 300 \times \ln \frac{1}{2}$

$= +600R \cdot \ln 2$

$w_{BC} = -nR(T_2 - T_1) = -2 \times R \times (400 - 300)$

$= -200R$

$w_{CD} = -nRT \ln \frac{P_1}{P_2} = -2 \times R \times 400 \times \ln \frac{2}{1}$

$= -800R \cdot \ln 2$

$w_{DA} = -nR(T_2 - T_1) = -2 \times R \times (300 - 400)$

$= +200R$

$\therefore w_{net} = w_{AB} + w_{BC} + w_{CD} + w_{DA} = -200R \cdot \ln 2$

$= -100R \cdot \ln 4$

**55.**

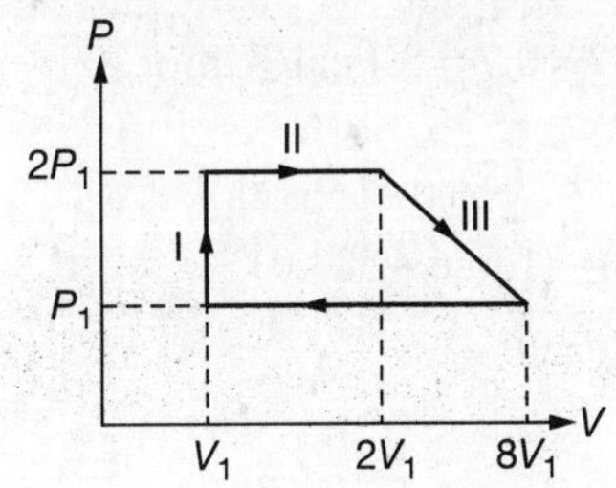


$$w = -\left(P_1V_1 + \frac{1}{2} \times P_1 \times 6V_1\right) = -4P_1V_1$$
$$= -4nRT_1 = -4 \times 1 \times 2 \times 300 \text{ cal}$$
$$= -2400 \text{ cal}$$

**56.** $\Delta H = \Delta U + \Delta(PV) = \Delta U + (P_2V_2 - P_1V_1)$
$= 30.0 \text{ L-atm} + (4 \times 5 - 2 \times 3) \text{ L-atm}$
$= 44.0 \text{ L-atm}$

**57.** As the intermolecular distance increases against intermolecular attraction, the potential energy increases.

**58.** $q = n \cdot L_m = 3 \times 10 = 30 \text{ kcal}$

$$w = -P(V_{\text{vap}} - V_{\text{liq}}) \approx -P \cdot V_{vap} = -nRT$$
$$= -3 \times \frac{2}{1000} \times 500$$
$$= -3 \text{ kcal}$$
$$\therefore \Delta U = q + w = 27 \text{ kcal}$$

**59.** $q = 540 \text{ cal}$
$$w = -P\left(V_{vap} - V_{liq}\right) = -1 \text{ atm}(1671 - 1) \text{ cm}^3$$
$$= -\frac{1670}{1000} \text{L-atm} = -1.670 \times 101.3 \text{ J}$$
$$= -\frac{1.670 \times 101.3}{4.184} \text{ cal} \approx -40 \text{ cal}$$
$$\therefore \Delta U = q + w \approx 500 \text{ cal}$$

**60.** $\Delta H - \Delta U = P \cdot \Delta V = P(V_{\text{diamond}} - V_{\text{graphite}})$
$$= 500 \times 10^3 \times 10^5 \frac{\text{N}}{\text{m}^2}\left(\frac{12}{3.0} - \frac{12}{2.0}\right) \times 10^{-6} \text{m}^3$$
$$= -100 \text{ kJ}$$

## Second and Third Law of Thermodynamics, Entropy

**61.** Reversible heat engine is always more efficient than irreversible heat engine.

**62.** Entropy increases with increase in mole of gases as well as volume of gas (decrease in pressure).

**63.** Entropy increases with increase in mole of gases.

**64.** In all spontaneous process, entropy of universe increases but for system at equilibrium, entropy of universe becomes constant.

**65.** Entropy decreases on decreasing temperature.

**66.** Entropy: Solid < Liquid << Gas

**67.** Entropy: Solid < Liquid << Gas

**68.** Efficiency is never 100%.

**69.** Entropy of gas is greater and it increases with the increase in volume (decrease in pressure).

**70.** Entropy: Solid < Liquid << Gas

**71.** As $\Delta H = +\text{ve}$ but the process is spontaneous, it is possible only when $\Delta S$ is also +ve.

**72.** Entropy measures unavailable energy, i.e., energy which can't be converted into useful work.

**73.** $\Delta S = n \cdot C_{v,m} \cdot \ln\frac{T_2}{T_1} + nR \cdot \ln\frac{V_2}{V_1}$
$$= 1 \times C_{v,m} \times \ln 2 + 1 \times R \times \ln\frac{1}{2} = (C_{v,m} - R) \cdot \ln 2$$

**74.** $\Delta S = \frac{\Delta H_{vap}}{B.P.} = \frac{\frac{40.85}{18} \times 3.6 \times 1000}{373} = 21.9 \text{ J/K}$

**75.** $\Delta S = \frac{\Delta H_{fus}}{M.P.} = \frac{2930}{300} = 9.77 \text{ J/K.mol}$

**76.** $\Delta S = nR \cdot \ln\frac{P_1}{P_2} = \frac{64}{32} \times 8.3 \times \ln\frac{1}{0.25} = 23.24 \text{ J/K}$

**77.** $\Delta S = n \cdot C_{v,m} \cdot \ln\frac{T_2}{T_1} = 2 \times R \times \ln\frac{900}{300} = 2R \cdot \ln 3$

**78.** $\Delta S = nR\ln\frac{V_2}{V_1} = 1 \times R \times \ln\frac{20}{2} = R \cdot \ln 10$

(Adiabatic free expansion is also isothermal)

79. $S_2 - S_1 = n \cdot C_{p,m} \cdot \ln \frac{T_2}{T_1}$

or $S_2 - 2.5 = 1 \times 4R \cdot \ln \frac{596}{298} = 4R \cdot \ln 2$

$\therefore S_2 = 2.5 + 4 \times 2 \times 0.7 = 8.1$ cal/K.mol

80. $\Delta_r S = (2 \times S_{m,HCl}) - (S_{m,H_2} + S_{m,Cl_2})$
$= (2 \times 186.7) - (130.6 + 223.0)$
$= +19.8$ J/K.mol

## Gibbs Free Energy Function

81. Melting of ice at $-15°$ C and 1 atm is a non-spontaneous process and hence, $\Delta G = +$ve.

82. $\Delta G_{T,P} = 0$

83. $\Delta S = S_{water} - S_{ice} = 58.0 - 38.0 = 20$ J/K.mol

Now, $\Delta H = T \cdot \Delta S = 273 \times 20 = 5460$ J/mol.

84. Amount is less than saturation and hence, NaCl will dissolve spontaneously.

85. $\Delta_r G = [(-480) + (-544)] - [(-293)] = -731$ kJ

86. $\Delta G^\circ = -RT \cdot \ln K_P^\circ = -RT \cdot \ln \frac{P_{NOCl}^2}{P_{NO}^2 \cdot P_{Cl_2}}$

$= -8.314 \times 298 \times \ln \frac{(0.01)^2}{(0.00001)^2 \times 0.01}$

$= +45.65$ kJ

87. For equilibrium condition, $\Delta G = 0$

88. $\Delta G = \Delta H - T \cdot \Delta S = (-2500) - 298 \times (+7.4) = -$ve
$\Rightarrow$ Spontaneous

89. $\Delta H = +ve$, $\Delta S = -ve$, then $\Delta G = +ve$ always

90. $\Delta G° = RT \cdot \ln K_{eq} = 0 \Rightarrow K_{eq} = 1$

---

# EXERCISE II (JEE ADVANCE)

## Section A (Only one Correct)

1. $U = n \times \frac{f}{2} R \times T$

For larger $U$, $n, f, T$, should be higher.

2. $w = -\int_{V_1}^{V_2} P \cdot dv = -\int_{V_1}^{V_2} \left( \frac{a}{V} + b \right) dv$

$= -a \ln \frac{V_2}{V_1} - b(V_2 - V_1)$

3. $q = m.s.\Delta T$
$\Rightarrow 10 \times 10^6 = 80 \times (4.2 \times 10^3) \times \Delta T$
$\Rightarrow \Delta T = 29.76\ K$
$= 29.76°\ C$

4. Heat lost by water = Heat gained by ice

or, $500 \times \frac{75.6}{18} \times 20 = (N \times 9) \times \frac{6000}{18}$

$\Rightarrow N = 14$

5. Let $T_1 > T_2$. Now, heat lost by gas (1) = Heat gained by gas (2)

or, $n_1 \cdot C_m \cdot (T_1 - T_f) = n_2 \cdot C_m \cdot (T_f - T_2)$

or, $\frac{P_1 V_1}{RT_1} \cdot (T_1 - T_f) = \frac{P_2 V_2}{RT_2} (T_f - T_2)$

$\Rightarrow T_f = \frac{T_1 T_2 (P_1 V_1 + P_2 V_2)}{P_1 V_1 T_2 + P_2 V_2 T_1}$

6. $w = -P_{ext} (V_2 - V_1)$

$= -P_2 \left( \frac{nRT}{P_2} - \frac{nRT}{P_1} \right)$

$= -nRT \left( 1 - \frac{P_2}{P_1} \right) = -nRT \left( 1 - \frac{P_1 - 1}{P_1} \right)$

$= -nRT \times \frac{1}{P_1}$

Now, $w_{total} = w_1 + w_2 + \ldots + w_f$

$$= \left(-nRT \times \frac{1}{P}\right) + \left(-nRT \times \frac{1}{P-1}\right) + \cdots + \left(-nRT \times \frac{1}{2}\right)$$

$$= -nRT \sum_{i=1}^{i=P-1} \left(\frac{1}{P+1-i}\right)$$

**7.** $w_1 = -\int_{V_1}^{V_2} P \cdot dV = -\int_{V_0}^{2V_0} \frac{K_1}{2V^2} \cdot dV = +\frac{K_1}{2}\left[\frac{1}{2V_0} - \frac{1}{V_0}\right] = -\frac{K_1}{4V_0}$

$$= -\frac{2P_0V_0^2}{4V_0} = -\frac{P_0V_0}{2} = -0.5P_0V_0$$

$$w_2 = -\int_{V_1}^{V_2} P \cdot dV = -\int_{V_0}^{2V_0} \frac{K_1}{V} \cdot dV = -K_2 \ln \frac{2V_0}{V_0} = -P_0V_0 \times 0.7$$

**8.** As $\Delta T = 0$, $\Delta U = 0$

**9.** Area $= P_2 \times \Delta V \Rightarrow P_2 \times 4 = 49.26$ *L*-atom

Now, correct work, $w = -nRT \cdot \ln \frac{V_2}{V_1} = -P_2V_2 \cdot \ln \frac{4}{2}$

$= -49.26 \times 0.693 = -34.137$ *L*-atom

**10.** $PV^x = \text{Constant}$

$$\Rightarrow P_1V_1^x = P_2V_2^x$$

$$\Rightarrow \frac{P_1}{P_2} = \left(\frac{V_2}{V_1}\right)^x \Rightarrow 8 = 4^x \Rightarrow x = \frac{3}{2}$$

Now, $C_m = C_{v,m} + \frac{R}{1-x} = \frac{3R}{2} + \frac{R}{1-\frac{3}{2}} = -\frac{R}{2}$

**11.** Dulong and Petit's law is applicable only for solid element. (Molar heat capacity $\approx$ 6.4 cal/K-mol = 26.8 J/K-mol).

**12.** $P_1V_1^\gamma = P_2V_2^\gamma$

$$\Rightarrow \frac{P_1}{d_1^\gamma} = \frac{P_2}{d_2^\gamma}$$

$$\Rightarrow \frac{P_2}{P_1} = \left(\frac{d_2}{d_1}\right)^\gamma = (32)^{7/5} = 128$$

**13.** $\frac{U_{rms},2}{U_{rms},1} = \frac{1}{2} = \sqrt{\frac{T_2}{T_1}} \Rightarrow \frac{1}{4}$

Now, $T.V^{r-1} = \text{Constant}$

$$\Rightarrow \frac{T_2}{T_1} = \left(\frac{V_1}{V_2}\right)^{r-1} \Rightarrow \frac{1}{4} = \left(\frac{V_1}{V_2}\right)^{7/5-1}$$

$\therefore V_2 = 32\ V_1$

**14.** $|w_A| = 2 \times |w_B|$ But $\Delta U_A = \Delta U_B$, Hence $q_A > q_B$ or, $(C_A \cdot \Delta T) > (C_B \cdot \Delta T) \Rightarrow C_A > C_B$

**15.** $q = n \cdot C_m \cdot \Delta T = 1 \times (0.22 \times 32) \times (273 \times 1.1 - 273) \times 4.2$ J

**16.** $\frac{q_V}{q_P} = \frac{n \cdot C_{V,m} \cdot \Delta T}{n \cdot C_{P,m} \cdot \Delta T} = \frac{1}{\gamma}$

$$\Rightarrow q_V = \frac{1}{\left(\frac{74.3}{66}\right)} \times 793 = 660 \text{ J}$$

$\left(C_{P,m} = \frac{743}{5 \times 2} = 74.3 \Rightarrow C_{V,m} = 74.3 - 8.3 = 66.0\right)$

**17.** Isothermal: $P.V = P_i \times \frac{V}{n} \Rightarrow P_i = n \cdot P$

Adiabatic: $P.V^r = P_a \times \left(\frac{V}{n}\right)^\gamma \Rightarrow P_a = n^\gamma \cdot P$

$$\therefore \frac{P_i}{P_a} = \frac{n \cdot P}{n^\gamma \cdot P} = n^{1-\gamma}$$

**18.** $\Delta U = n \cdot C_{V,m} \cdot (T_2 - T_1) = n \cdot \frac{R}{\gamma - 1} \cdot \left(\frac{P \cdot 2V}{nR} - \frac{P \cdot V}{nR}\right) = \frac{PV}{\gamma - 1}$

**19.** K.E. $= \Delta U \Rightarrow \frac{1}{2} \times \frac{(n \times 40)}{1000} \times (100)^2 = n \times \frac{8.314}{(1.5-1)} \times \Delta T$

$\therefore \Delta T = 12.03\ K$

**20.** For minimum pressure, compression should be irreversible.

$$\Delta U = w \Rightarrow n \cdot \frac{R}{\gamma - 1} \cdot (T_2 - T_1) = -P_{ext}(V_2 - V_1)$$

$$= -P_2\left(\frac{nRT_2}{P_2} - \frac{nRT_1}{P_1}\right)$$

$$\text{or, } \frac{T_2 - T_1}{\gamma - 1} = -\left(T_2 - T_2 \cdot \frac{P_2}{P_1}\right)$$

$$\Rightarrow \frac{700 - 400}{1.4 - 1} = -\left(700 - 400 \times \frac{P_2}{100}\right)$$

$\therefore P_2 = 362.5$ kPa

**21.** $q = (n \cdot C_{V,m} \cdot \Delta T)_{Ne} + (n \cdot C_{V,m} \cdot \Delta T)_{SO_3}$

or, $12 \times 10^3 = 2 \times 3 \times (T_f - 300) + 3 \times 6 \times (T_f - 400)$

$\Rightarrow T_f = 875$ K

Now, $P_{\text{final}} = \frac{nRT}{V} = \frac{5 \times 0.08 \times 875}{10} = 35$ atm

**22.** Free expansion is isothermal.

**23.** $C_{V,m} = 3 \times \frac{1}{2} R + 3 \times \frac{1}{2} R + (3N - 6) \times R = (3N - 3)R$

$$\therefore \gamma = \frac{C_P}{C_V} = 1 + \frac{R}{C_{V,m}} = 1 + \frac{1}{3N - 3}$$

**24.** $C_m = C_{V,m} + \frac{P \cdot dV}{dT} = C_{V,m} + \frac{P}{\alpha} = C_{V,m} + \frac{RT}{V \cdot \alpha}$

$$= C_{V,m} + \frac{R(T_0 + \alpha V)}{V\alpha}$$

$$= C_{P,m} + \frac{RT_0}{V\alpha}$$

Now, $q = \int_{T_1}^{T_2} C_m \cdot dT = \int_{V_1}^{V_2} C_m \cdot (\alpha \cdot dV)$

$$= \int_{V_1}^{V_2} \left(C_{P,m} \cdot \alpha + \frac{RT_0}{V}\right) dV$$

$$= \alpha \cdot C_{P,m}(V_2 - V_1) + RT_0 \cdot \ln \frac{V_2}{V_1}$$

**25.** $w = -25\text{ J} = -nR \cdot \Delta T$

$$\Delta U = n \cdot C_{V,m} \cdot \Delta T = n \cdot \left(\frac{6}{2} R\right) \cdot \Delta T = 75J$$

$\therefore q = \Delta U - w = 100$ J

**26.** $C_m = C_{V,m} + \frac{R}{1 - x} = \frac{3R}{2} + \frac{R}{1 - \frac{5}{2}} = \frac{5R}{6}$

$$\therefore q = n \cdot C_m \cdot \Delta T = 1 \times \frac{5R}{6} \times 26 = 180 \cdot 14J$$

**27.** $PV^x = K \Rightarrow \frac{dP}{dV} = -x \cdot \frac{P}{V} \Rightarrow -1 = -x \times \frac{4}{2} \Rightarrow x = \frac{1}{2}$

$$\therefore C_m = C_{V,m} + \frac{R}{1 - x} = \frac{3R}{2} + \frac{R}{1 - \frac{1}{2}} = 3.5R$$

**28.**

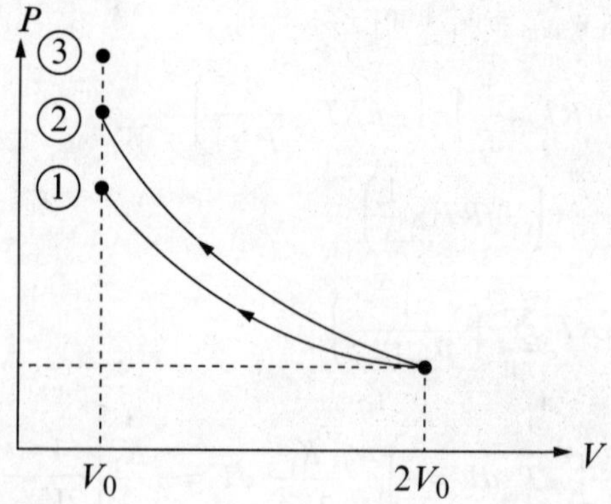


**29.**

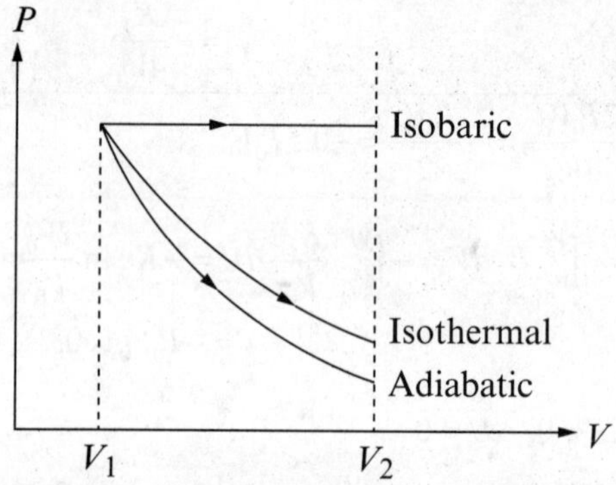


$\Delta E_{\text{adiabatic}}$ = Negative

$\Delta E_{\text{isothermal}} = 0$

$\Delta E_{\text{isobaric}}$ = Positive

**30.** Boyle temperature, $T_B = 20 + 273 = 293$ K

Inversion temperature, $T_i = 2 \times T_B = 586\text{ K} = 313°\text{ C} > 50°\text{ C}$

**31.**

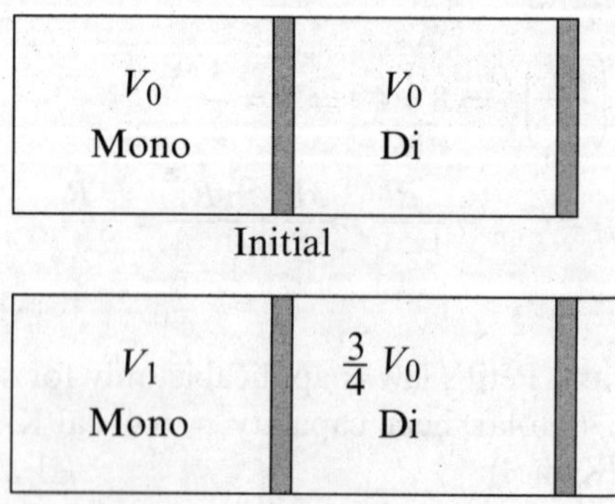


Monoatomic : $P_1 \cdot V_0^{5/3} = P_2 \cdot V_1^{5/2}$

Diatomic : $P_1 \cdot V_0^{7/5} = P_2 \cdot \left(\frac{3}{4} V_0\right)^{7/5}$

$$\therefore \frac{V_1}{V_0} = \left(\frac{3}{4}\right)^{\frac{21}{25}}$$

**32.** 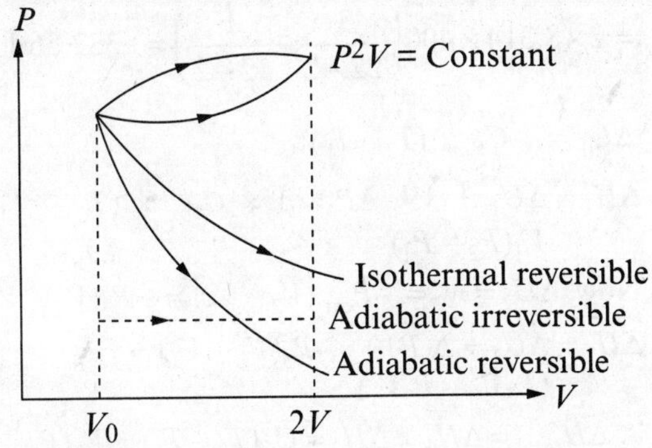


**33.** For greater heat exchange, heat capacity should be high.

$$C_m = C_{V,m} + \frac{R}{1-x} \text{ for } PV^x = \text{Constant}$$

**34.** $q_{ABC} = 600 + 200 = 800$ J

$$w_{AB} = 0 \text{ and } w_{BC} = -\left(8\times10^4 \times \frac{N}{m^2}\right)(5\times10^{-3} - 2\times10^{-3})m^2$$

$$= -240 \text{ J}$$

$$\therefore \Delta U_{AC} = \Delta U_{ABC} = q_{ABC} + w_{ABC}$$

$$= 800 + (-240) = 560 \text{ J}$$

**35.** 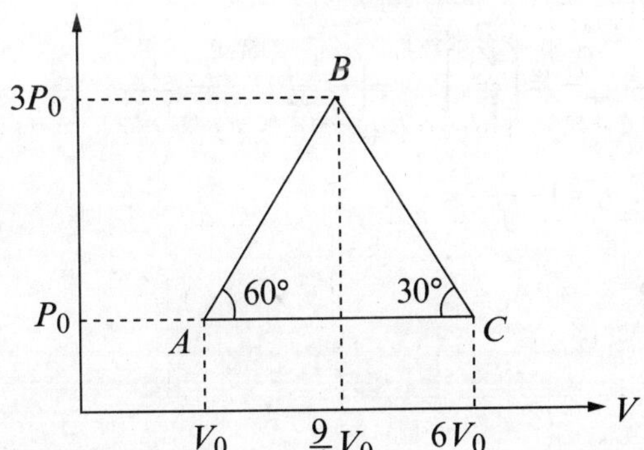


$AB: P = \sqrt{3}V + C_1$

$$P_0 = \sqrt{3}V_0 + C_1 \quad (1)$$

$$\text{and } 3P_0 = \sqrt{3}V_B + C_1 \quad (2)$$

$$BC: P = -\frac{1}{\sqrt{3}}V + C_2$$

$$P_0 = -\frac{1}{\sqrt{3}}\cdot 6V_0 + C_2 \quad (3)$$

$$3P_0 = -\frac{1}{\sqrt{3}}\cdot V_B + C_2 \quad (4)$$

From equation (1), (2), (3) and (4), $V_B = \frac{9}{4}V_0$

$$\text{Now, } \frac{T_B}{T_A} = \frac{3P_0\cdot\frac{9}{4}V_0}{P_0\cdot V_0} = \frac{27}{4}$$

**36.** 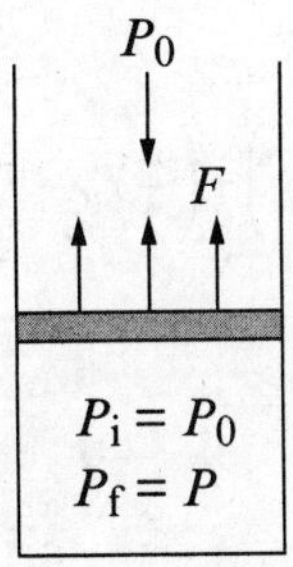


$$dw = F\cdot dx = (P_0 - P)A\cdot dx = (P_0 - P)\cdot dV$$

$$\therefore w = \int_V^{\eta\cdot V}\left(P_0 - \frac{nRT}{V}\right)dV$$

$$= P_0(\eta V - V) - RT\cdot\ln\frac{V\cdot\eta}{V}$$

$$P_0V(\eta-1) - \text{RT}.\ln\eta = \text{RT}[\eta-1-\ln\eta]$$

**37.** 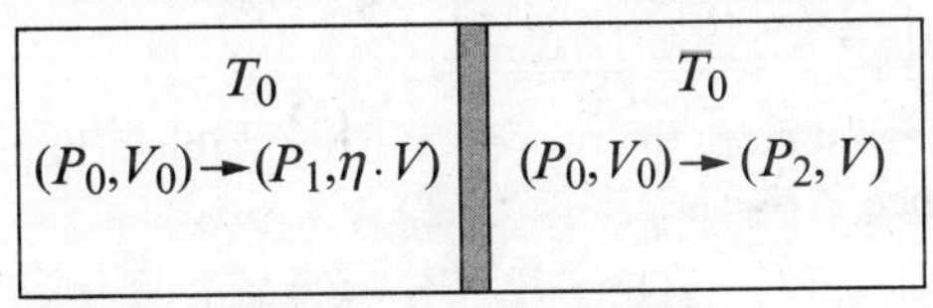


Work performed on the piston

$$= -\left[\int_{V_0}^{\eta\cdot V} P_1\cdot dV + \int_{V_0}^{V} P_2\cdot dV\right] \text{ and } (V + \eta\cdot V) = 2V_0$$

$$= P_0V_0\cdot\ln\frac{(\eta+1)^2}{4\eta}$$

**38.** Isothermal : $P_A\cdot V = P\cdot(2V) \Rightarrow P_A = 2P$

Adiabatic : $P_B\cdot V^{1.5} = P\cdot(2V)^{1.5} \Rightarrow P_A = 2\sqrt{2}\,P$

Isobaric : $P_C = P$

$\therefore P_A : P_B : P_C = 2 : 2\sqrt{2} : 1$

**39.** 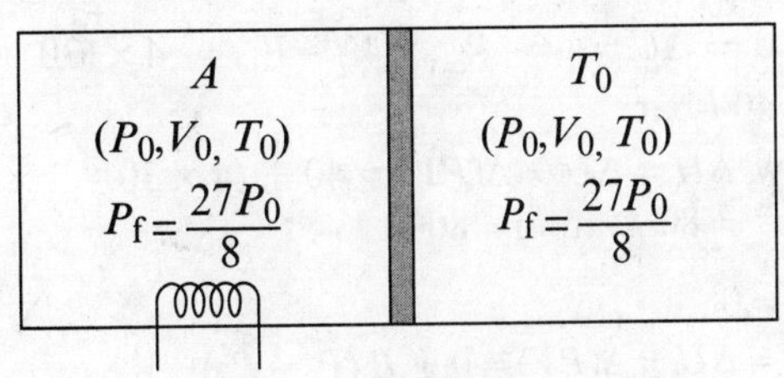


$$\text{Chamber } B: P_0\cdot V_0^\gamma = \frac{27P_0}{8}\cdot V_B^\gamma$$

HINTS AND EXPLANATIONS

$$\therefore\ V_B = \frac{4V_0}{9} \Rightarrow T_B = \frac{3}{2}T_0$$

and $V_A = 2V_0 - \frac{4V_0}{9} = \frac{14V_0}{9} \Rightarrow T_A = \frac{21}{4}T_0$

Now, $q_A = \Delta U_A + \Delta U_B = n \cdot C_{V,m} \cdot (T_A - T_0)$

$$+ n \cdot C_{V,m} \cdot (T_B - T_0)$$

$$= \frac{P_0V_0}{R \cdot T_0} \cdot 2R \cdot \left[\left(\frac{21}{4}T_0 - T_0\right) + \left(\frac{3}{2}T_0 - T_0\right)\right]$$

$$= \frac{19}{2} \cdot P_0V_0$$

**40.**

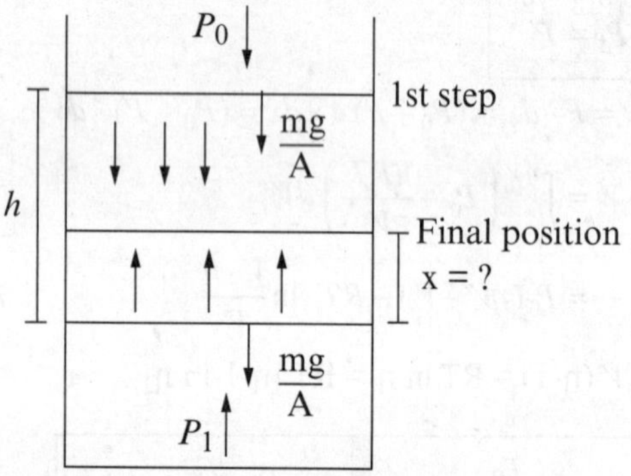


After 1st step, the process is irreversible adiabatic. Hence, $\Delta U = w$

$$n \cdot C_{V,m} \cdot (T_2 - T_1) = -P_{ext}(V_2 - V_1)$$

or, $n \cdot \frac{3}{2} R\left(\frac{P_1 \cdot V_2}{nR} - \frac{P_1V_1}{nR}\right) = -P_1[V_2 - (V_1 + A \cdot H)]$

$\therefore\ V_2 = V_1 + 0.4\ H.A \Rightarrow x = 0.4\ H$

(The final pressure of gas after 2nd step will remain same as initial, beginning of processes.)

**41.** Smaller the heat capacity larger is $\Delta T$.

**42.** $\Delta H = n \cdot C_{P,m} \cdot (T_2 - T_1) = 1 \times 40 \times (500 - 300)$

$= 8000$ J

$\therefore\ \Delta U = \Delta H - P \cdot \Delta V = 8000 - 2(40 - 30) \times 100$
$= 6000$ J

**43.** $q = 0 \Rightarrow \Delta U = w = -P_{ext} \cdot (V_2 - V_1) = -4 \times (30 - 40)$
$= 40$ $l$-bar

Now, $\Delta H = \Delta U + \Delta(PV) = 40 + (4 \times 30 - 2 \times 40)$
$= 80$ $L$-atom $= 8000$ J

**44.** $\Delta U = 0$

$\Delta H = \Delta U + \Delta(PV) = 0 + B(P_2 - P_1)$

$$= B \cdot \frac{RT}{V_2 - B} - \frac{RT}{V_1 - B}$$

$$= 2 \times 8.314 \times 400\left(\frac{1}{22-2} - \frac{1}{12-2}\right) = -332.56 \text{ J}$$

**45.** $\Delta U_1 = n \cdot C_{V,m} \cdot (T_2 - T_1)$

$\Delta H_1 = \Delta U_1 + V \cdot \Delta P = 1 \times C_{V,m} \times (T_2 - T_1) + V_1(P_2 - P_1)$

Now, $\Delta U_2 = w_2 = -P_{ext}(V_2 - V_1) = -P_3(V_2 - V_1)$

$\Delta H_2 = \Delta U_2 + \Delta(PV) = -P_3(V_2 - V_1) + (P_3V_2 - P_2V_1)$

$\therefore\ \Delta H_{total} = \Delta H_1 + \Delta H_2 = C_V(T_2 - T_1) + V_1(P_3 - P_1)$

**46.** $\eta = 1 - \frac{T_C}{T_H}$

$\frac{1}{6} = 1 - \frac{T_C}{T_H}$ and $\frac{1}{3} = 1 - \frac{T_C - 65}{T_H} \Rightarrow T_H = 390$ K

$= 117°$ C

**47.** $\frac{|q_{rej}|}{q_{abs}} = \frac{T_C}{T_H} \Rightarrow |q_{rej}| = \frac{390}{600} \times 120 = 78$ cal

**48.** $\left(1 - \frac{T_C - \Delta T}{T_H}\right) > \left(1 - \frac{T_C}{T_H + \Delta T}\right)$

**49.** $T_H \cdot V_2^{\gamma-1} = T_C \cdot V_3^{\gamma-1}$

$$\Rightarrow \frac{T_C}{T_M} = \left(\frac{V_2}{V_3}\right)^{\gamma-1} = \left(\frac{1}{2.75}\right)^{1.4-1} = \frac{1}{1.5}$$

$$\therefore\ \eta = 1 - \frac{T_C}{T_H} = \frac{1}{3}$$

**50.**

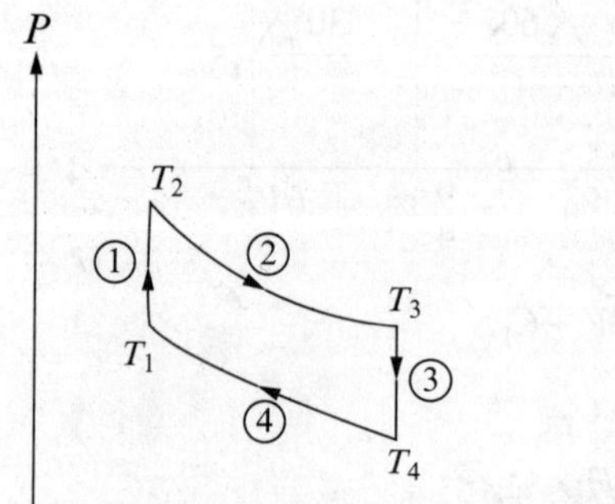


$T_2 \cdot V_1^{\gamma-1} = T_3 \cdot V_2^{\gamma-1}$

and $T_1 \cdot V_1^{\gamma-1} = T_4 \cdot V_2^{\gamma-1}$

$$\therefore\ \frac{T_2}{T_1} = \frac{T_3}{T_4} \Rightarrow \frac{T_2 - T_1}{T_1} = \frac{T_3 - T_4}{T_4}$$

$$\eta = 1 - \frac{|q_{rej}|}{q_{abs}} = 1 - \frac{n \cdot C_{V,m}(T_3 - T_4)}{n \cdot C_{V,m}(T_2 - T_1)} = 1 - \frac{T_4}{T_1}$$

$$= 1 - \left(\frac{V_1}{V_2}\right)^{\gamma - 1}$$

$$= 1 - \left(\frac{1}{10}\right)^{\frac{7}{5} - 1} = -0.6$$

**51.** $\Delta S_{unit} = \Delta S_{Source} + \Delta S_{Heat\ engine} + \Delta S_{Sink}$

$$= -\frac{40 \times 10^3}{500} + 0 + \frac{30 \times 10^3}{300} = +20 J/K$$

**52.** $\Delta S = n \cdot C_{P,m} \cdot \ln \frac{T_2}{T_1}$

$$= \frac{3.2}{32} \times 14 \times \ln \frac{900}{1000} = -0.14 \text{ cal/K}$$

**53.** $\Delta S = nR \cdot \ln \frac{V_2}{V_1}$

$$= 2 \times 8.314 \times \ln \frac{(2a)^3}{a^3} = 34.58 \text{ J/K}$$

**54.** $\Delta S = n \cdot C_{P,m} \cdot \ln \frac{T_2}{T_1}$

$$= 1 \times \frac{5R}{2} \times \ln \frac{1000}{250} = 7.0 \text{ cal/K}$$

**55.** $\Delta S = n \cdot C_{V,m} \cdot \ln \frac{T_2}{T_1}$

$$S_{500\,K} - 46.2 = 1 \times \frac{3R}{2} \times \ln \frac{500}{250}$$

$$\therefore S_{500\,K} = 48.3 \text{ Ccal/K-mol}$$

**56.** $\Delta S = nR \cdot \ln \frac{V_2}{V_1}$

$$\Rightarrow -5.0 = \frac{10^5 \times (15 \times 10^{-3})}{300} \times \ln \frac{V_2}{15} \Rightarrow V_2 = 5.4 \text{ L}$$

**57.** $\Delta S_{Surr} = -\frac{1.5 \times 10^3}{300} = -5 \text{ J/K}$

Now, $\Delta S_{unit} = \Delta S_{Sys} + \Delta S_{Surr} = 5.51 + (-5)$

$= + 0.51$ J/K

Hence, the process is irreversible.

**58.** $\Delta S = n \cdot C_{P,m} \cdot \ln \frac{T_2}{T_1} + nR \cdot \ln \frac{P_1}{P_2}$

or, $0 = n \times \frac{5}{2} R \times \ln \frac{1200}{300} + nR \times \ln \frac{1}{P_2}$

$\Rightarrow P_2 = 32$ bar

**59.** $\Delta S = \Delta S_{adiabatic} + \Delta S_{isobaric} = 0 + n \cdot C_{P,m} \cdot \ln \frac{T_2}{T_1}$

$$= \frac{1.6}{4} \times \frac{5R}{2} \times \ln \frac{1}{3} = -2.2 \text{ cal/K}$$

**60.** $\Delta S = n \cdot C_{V,m} \cdot \ln \frac{T_2}{T_1} + nR \cdot \ln \frac{V_2}{V_1}$

$$= n \cdot C_{V,m} \cdot \ln \frac{P_2}{P_1} + n \cdot C_{P,m} \cdot \ln \frac{V_2}{V_1}$$

$$= 2 \times \frac{R}{1.5 - 1} \times \ln \frac{1}{4} + 2 \times \frac{1.5R}{1.5 - 1} \times \ln 2$$

$$= -11.64 \text{ J/K}$$

**61.** $S_2 - S_1 = n \cdot C_{V,m} \cdot \ln \frac{T_2}{T_1} + nR \cdot \ln \frac{V_2}{V_1}$

$$= 1.2 \times \frac{3R}{2} \times \ln \frac{1}{2} + 1.2 \times R \times \ln 2$$

$$= -0.84 \text{ cal/K}$$

**62.** $\Delta S = n \cdot C_{V,m} \cdot \ln \frac{T_2}{T_1} + nR \cdot \ln \frac{V_2}{V_1}$

$$= 1 \times \frac{R}{\gamma - 1} \times \ln \frac{T_2}{T_1} + 1 \times R \times \ln \left(\frac{T_1}{T_2}\right)^{1/n-1}$$

(as $T.V^{n-1}$ = Constant)

$$R \cdot \ln \frac{T_2}{T_1} \left[\frac{1}{\gamma - 1} - \frac{1}{n - 1}\right]$$

$$= \frac{(n - \gamma)R}{(n-1)(\gamma - 1)} \cdot \ln \tau$$

**63.** $\Delta S = n \cdot C_{V,m} \cdot \ln \frac{P_2}{P_1} + n \cdot C_{P,m} \cdot \ln \frac{V_2}{V_1}$

$$= 2 \times \frac{3R}{2} \times \ln 2 + 2 \times \frac{5R}{2} \times \ln 2$$

$$= + 11.2 \text{ cal/K}$$

**64.** $dS = \frac{n \cdot C_{V,m} \cdot dT + P \cdot dV}{T}$

For maximum entropy, $\frac{dS}{dV} = 0$

or, $n \cdot C_{V,m}\dfrac{dT}{dV}+P=0$ (1)

Now, $P=\dfrac{RT}{V}=P_0-\alpha V \Rightarrow \dfrac{dT}{dV}=\dfrac{1}{R}(P_0-2\alpha V)$ (2)

From (1) and (2),

$$1\times\frac{R}{\gamma-1}\times\frac{1}{R}(P_0-2\alpha V)+(P_0-\alpha V)=0$$

$$\therefore V=\frac{\gamma\cdot P_0}{\alpha(\gamma+1)}$$

**65.** $dS=C_{V,m}\cdot\dfrac{dT}{T}+P\cdot\dfrac{dV}{T}=a\cdot dT+C_{V,m}\cdot\dfrac{1}{T}\cdot dT$

or, $\int_{V_0}^{V}\dfrac{R}{V}\cdot dV=a\cdot\int_{T_0}^{T}dT \Rightarrow R\cdot\ln\dfrac{V}{V_0}=a(T-T_0)$

$$\therefore T=T_0+\frac{R}{a}\cdot\ln\frac{V}{V_0}$$

**66.** $\int_0^S dS=\int_0^T C\cdot\dfrac{dT}{T}=\int_0^T T^3\cdot\dfrac{dT}{T}\Rightarrow S=\dfrac{aT^3}{3}$

**67.** $\Delta S=\Delta S_A+\Delta S_B=-\dfrac{12000}{600}+\dfrac{12000}{400}=+10$ J/K

**68.** Heat lost by alloy = Heat gained by water

or $4\times4\times(800-T)=4\times1.0\times(T-300)$

$\Rightarrow T=700$ K

(As date is not given for vaporization of water)

Now, $\Delta S_{mix}=\Delta S_{alloy}+\Delta S_{water}$

$$=4\times4\times\ln\frac{700}{800}+4\times1\times\ln\frac{700}{300}$$

= 1.0 K cal/K

**69.** Final temperature of both blocks $=\dfrac{T_1+T_2}{2}$

$\therefore \Delta S=\Delta S_1+\Delta S_2=$

$$C\cdot\ln\frac{(T_1+T_2)/2}{T_1}+C\cdot\ln\frac{(T_1+T_2)/2}{T_2}$$

$$=C\cdot\ln\frac{(T_1+T_2)^2}{4T_1T_2}$$

**70.** $\Delta S=-R[n_1\cdot\ln x_1+n_2\cdot\ln x_2]$

$=-R[0.8\times\ln0.8+0.2\times\ln0.2]$

= + 0.96 Cal/K.

**71.** Larger molar mass, greater is the molar entropy.

**72.** Greater the number of atoms, greater is the molar entropy.

**73.** $nC(s)+(n+1)H_2(g)\rightarrow C_nH_{2n+2}(g)$

with increase in *n*, the decrease in entropy increases.

**74.** $H_2O$ (*l*, 1 atm, 100°C) $\xrightarrow{(1)}$ $H_2O$ (*g*, 1 atm, 100°C) $\xrightarrow{(2)}$ $H_2O$ (*g*, 5 atm, 100°C)

$\Delta G_1=0$ and $\Delta G_2=nRT\ln\dfrac{P_2}{P_1}=5\times2\times373\times\ln\dfrac{5}{1}$

= 3730 ln 5 Cal

**75.** $q=\Delta U-w=0-\left(-nRT\cdot\ln\dfrac{P_1}{P_2}\right)=nRT\cdot\ln\dfrac{P_1}{P_2}$

$=-\Delta G$

$\therefore \Delta G=-q=-(-1200)=+1200$ cal

**76.** $\Delta G^\circ=-RT\cdot\ln K_{eq}$

$\Rightarrow -1743=-8.3\times300\times\ln K_{eq}$

$\therefore K_{eq}=2$

Now, (a) $K_{eq}=\dfrac{3}{3\times6}$

(b) $K_{eq}=\dfrac{6\times3^2}{3}$

(c) $K_{eq}=\dfrac{6\times3}{3^2}$

(d) $K_{eq}=\dfrac{3^2}{3\times6}$

**77.**

$$\begin{array}{ccc} H_2O(l,-10°C,0.28\,Pa) & \longrightarrow & H_2O(s,-10°C,0.26\,Pa) \\ \downarrow \Delta G_1=0 & & \uparrow \Delta G_3=0 \\ H_2O(g,-10°C,0.28\,Pa) & \xrightarrow{\Delta G_2} & H_2O(g,-10°C,0.26\,Pa) \end{array}$$

$$\Delta G_2=nRT\ln\frac{P_2}{P_1}=1\times R\times263\times\ln\frac{0.26}{0.28}$$

$$\therefore \Delta G=\Delta G_1+\Delta G_2+\Delta G_3=263\,R\ln\frac{13}{14}$$

**78.** Free expansion is isothermal

$$\Delta G=nRT\ln\frac{P_2}{P_1}=nRT\ln\frac{V_2}{V_1}$$

$$=10^5\times(1.2\times10^{-3})\times\ln\frac{1.2}{2.4}=-84\text{ J}$$

**79.** $\Delta G_1° = -RT \cdot \ln K_1$ and $\Delta G_2° = -RT \cdot \ln K_2$

Now, $\Delta G_2° - \Delta G_1° = -RT[\ln K_2 - \ln K_1]$

$= -RT[\ln e^4]$

$= -2 \times 300 \times 4 = -2400$ cal

**80.** At 0.04 atom, the system is in equilibrium.

## Section B (One or More than One Correct)

1. Theory based
2. Theory based
3. For isolated system, there should not be any mass and energy transfer with surroundings.
4.

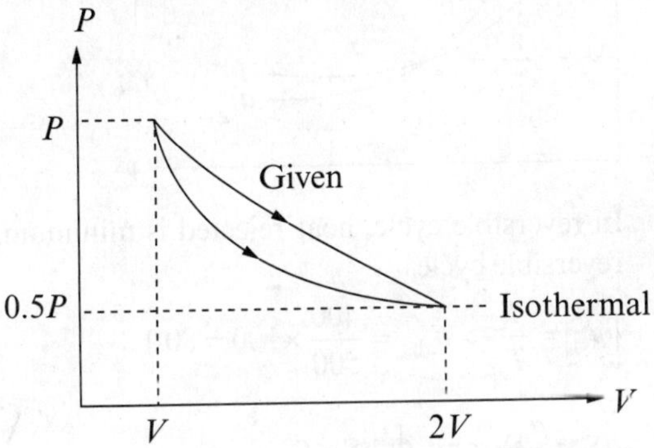


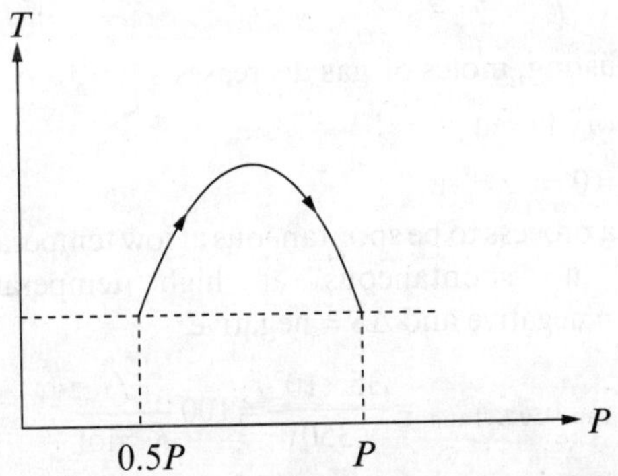


5. The internal energy of real gas may change on changing the volume of gas. Change in physical state also changes the physical state.
6. Theory based
7. Theory based
8. Option (c) should be changed with (c) adiabatic free expansion of any gas is also isothermal.
9. $w_{rev} - w_{irr} = (-P \cdot dV) - (-P_{ext} \cdot dV)$

$= (P_{ext} - P) \cdot dV =$ negative, always and

$q_{rev} - q_{irr} =$ positive, always

10. $q = 0$

$$w = \Delta U = n \cdot C_{V,m} \cdot (T_2 - T_1) = 4 \times \frac{3}{2} R \times (290 - 320)$$
$$= -360 \text{ cal}$$

$$\Delta H = \gamma \cdot \Delta U = \frac{5}{3} \times (-360) = -600 \text{ cal}$$

11. Theory based

12. Process $BC$: $\frac{P_B}{T_B} = \frac{P_C}{T_C} \Rightarrow \frac{P_B}{500} = \frac{1}{250} \Rightarrow P_B = 2$ bar

and $\Delta U_{BC} = n \cdot C_{V,m}(T_C - T_B)$

$= 2 \times 1.5 \text{R} \times (250 - 500)$

$= -750$ R

Process $CD$: $\Delta U = w$

$\Rightarrow n \cdot C_{V,m}(T_D - T_C)$

$= -P_{ext}(V_D - V_C)$

or, $n \times 1.5R \times (T_D - T_C)$

$= -P_D\left(\frac{nRT_D}{P_D} - \frac{nRT_C}{P_C}\right)$

$\Rightarrow T_D = 450$ K

and $\Delta H_{CD} = n \cdot C_{P,m} \cdot (T_D - T_C)$

$= 2 \times 2.5\, R \times (450 - 250)$

$= 1000$ R

13. $PV^{\gamma} = K_1$

$\Rightarrow P = K_1 \cdot V^{-\gamma}$

$\Rightarrow \frac{dP}{dV} = K_1 \cdot (-\gamma).V^{-\gamma-1}$

$= -\gamma \cdot \frac{P}{V}$

The gas having higher $\gamma$ will have higher magnitude of slope of $P$ vs. $V$ curve.

Now, $n \cdot \frac{R}{\gamma - 1} \cdot dT = -P \cdot dV = -\frac{nRT}{V} \cdot dV$

or, $\frac{dV}{dT} = -\frac{1}{\gamma - 1} \cdot \frac{V}{T}$

Gas having higher $\gamma$ will have lower magnitude of slope of $V$ vs. $T$ curve.

Now, $n \cdot C_{V,m} \cdot dT = -P \cdot dV = -[nRdT - V \cdot dP]$

or, $n \cdot C_{P,m} \cdot dT = -V \cdot dP \Rightarrow \frac{n \cdot \gamma R}{\gamma - 1} \cdot dT = \frac{nRT}{P} \cdot dP$

$\therefore \frac{dP}{dT} = \frac{\gamma}{\gamma - 1} \cdot \frac{P}{T}$

Gas having higher $\gamma$ will have lower magnitude of slope of $P$ vs. $T$ curve.

**14.** $q = 0$

**15.** $dT = 0$

**16.**

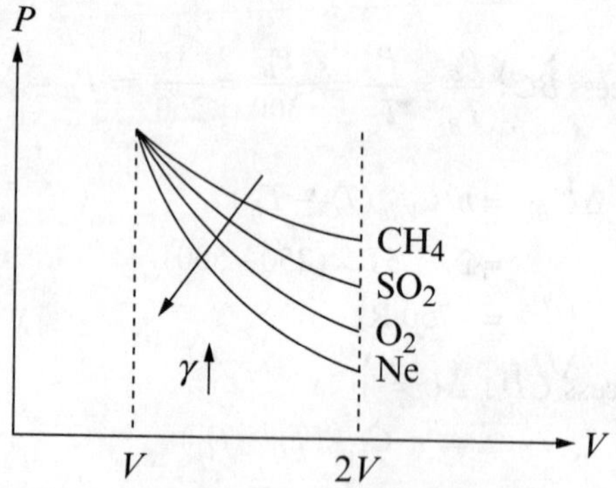


Here, $\gamma$ decreases on increasing degree of freedoms. As final pressure is minimum for Ne, its final temperature is minimum (decrease in temperature is maximum).

Now, for overall process, $\Delta T = 0 \Rightarrow \Delta U_{total} = 0$

or, $\Delta U_I + \Delta U_{II} = 0 \Rightarrow (0 + w_I) + (q_{II} + 0) = 0$

$\therefore q_{II} = -w_I =$ maximum for $CH_4$

**17.** Theory based

**18.** $C_{P,m} - C_{V,m} = R$

$\Rightarrow S_P - S_V = \frac{R}{M} = 0.04545$

$\therefore M = 44$ gm/mol

**19.** $q = 0 \Rightarrow \Delta U = w = -P_0(4V_0 - V_0) = -3P_0V_0$

Now, $\Delta H = \Delta U + \Delta(PV) = (-3P_0V_0) + (P_0 \cdot 4V_0 - 2P_0 \cdot V_0)$

$= -P_0 \cdot V_0$

**20.** Theory based

**21.** Theory based

**22.**

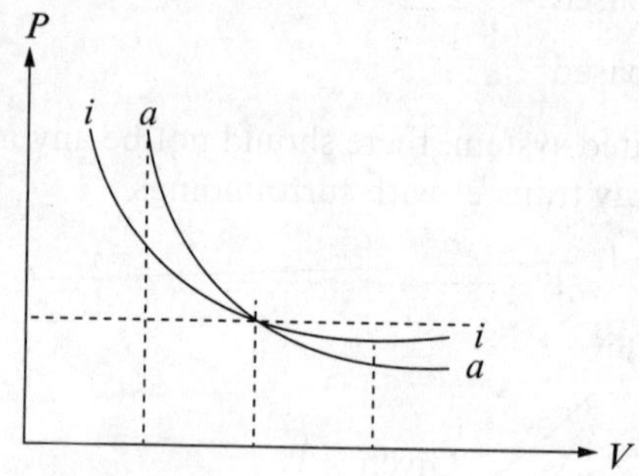


**23.** In reversible cycle, heat rejected is minimum. For reversible cycle,

$|q_{rej}| = \frac{T_C}{T_H} \times q_{abs} = \frac{400}{500} \times 100 = 80$ J

**24.** $dS = \frac{q_{rev}}{T}$ and $\oint dS = 0$

**25.** In rusting, moles of gas decreases.

**26.** Theory based.

**27.** $\Delta U = 0 \Rightarrow q = -w$

**28.** For a process to be spontaneous at low temperature and non-spontaneous at high temperature, $\Delta H =$ negative and $\Delta S =$ negative.

**29.** $(\Delta S_{Vap})_{350\,K, 1\,atm} = \frac{35 \times 10^3}{350} = 100 \frac{J}{K\ mol}$

On increasing pressure at constant temperature entropy decreases.

$(\Delta G_{Vap})_{350\,K, 1\,atm} = 0$

On increasing pressure at constant temperature energy increases.

**30.** Theory based.

## Section C (Comprehensions)

### Comprehension I

**1.** $\Delta U = n \cdot C_{V,m} \cdot (T_2 - T_1) = 4 \times \frac{5}{2} R \times (-50 - 0)$

$= -1000$ cal

**2.** $\Delta H = \gamma \cdot \Delta U = \frac{7}{5} \times (-1000) = -1400$ cal

**3.** $w = 0$ ($V =$ Constant)

## Comprehension II

4. $q = -\Delta U \Rightarrow n \cdot C_{V,m} \cdot \Delta T = -n \cdot C_{V,m} \cdot \Delta T$

$\Rightarrow C_m = -C_{V,m}$

5. $C_m = C_{V,m} + \dfrac{R}{1-x}$

$\Rightarrow -C_{V,m} = C_{V,m} + \dfrac{R}{1-x} \Rightarrow -2 \cdot \dfrac{R}{\gamma-1} = \dfrac{R}{1-x}$

$\therefore\ x = \dfrac{\gamma+1}{2}$

Now, $T \cdot V^{x-1} = \text{Constant} \Rightarrow T \cdot V^{(\gamma-1)/2} = \text{Constant}$

6. $\dfrac{T_2}{T_1} = \left(\dfrac{V_1}{V_2}\right)^{\gamma-1/2} \Rightarrow \dfrac{T_2}{300} = \left(\dfrac{1}{8}\right)^{\left(\frac{5}{3}-1\right)/2} \Rightarrow T_2 = 150\text{ K}$

$\therefore\ w = -\dfrac{nR(T_2 - T_1)}{1-x} = -\dfrac{1 \times 2(150-300)}{1-\left(\frac{5}{3}+1\right)/2}$

$= -900\text{ cal}$

## Comprehension III

7. $w = -nR \cdot \Delta T = -1 \times 8.314 \times 72 = -598.6\text{ J}$

$= -0.6\text{ kJ}$

8. $\Delta U = q + w = 1.6 + (-0.6) = 1.0\text{ kJ}$

9. $\gamma = \dfrac{\Delta H}{\Delta U} = \dfrac{1.6}{1.0} = 1.6$

## Comprehension IV

10. $V_2 = 4\ V_0 \Rightarrow P_2 = 4\ P_0$

As $\dfrac{P_1V_1}{T_1} = \dfrac{P_2V_2}{T_2}$

$\Rightarrow \dfrac{P_0V_0}{T_0} = \dfrac{4P_0 \cdot 4P_0}{T_2}$

$\Rightarrow T_2 = 16T_0$

Now, $\Delta U = n \cdot C_{V,m} \cdot \Delta T = n \times \dfrac{R}{\gamma-1} \times (16.T_0 - T_0)$

$= \dfrac{15P_0V_0}{\gamma-1} = \dfrac{15\alpha V_0^2}{\gamma-1}$

11. $w = \dfrac{nR(T_2 - T_1)}{1-x} = \dfrac{nR \times 15T_0}{1-(-1)} = \dfrac{15\alpha V_0^2}{2}$

12. $C_m = C_{V,m} + \dfrac{R}{1-x} = \dfrac{R}{\gamma-1} + \dfrac{R}{1-(-1)} = \dfrac{(\gamma+1)R}{2(\gamma-1)}$

## Comprehension V

13. $P = a \cdot T^{\alpha} = a \cdot \left(\dfrac{PV}{nR}\right)^{\alpha}$

$\Rightarrow P \cdot V^{\frac{\alpha}{\alpha-1}} = \text{Constant}$

$\therefore\ w = \dfrac{nR \cdot \Delta T}{x-1} = \dfrac{1 \times R \cdot \Delta T}{\dfrac{\alpha}{\alpha-1} - 1} = R(\alpha-1) \cdot \Delta T$

14. $C_m = C_{V,m} + \dfrac{R}{1-x} = \dfrac{R}{\gamma-1} + \dfrac{R}{1-\dfrac{\alpha}{\alpha-1}}$

$= \left[\dfrac{1}{\gamma-1} + (1-\alpha)\right] . R$

15. $\dfrac{1}{\gamma-1} + (1-\alpha) < 0$

$\Rightarrow \alpha > \dfrac{\gamma}{\gamma-1}$

## Comprehension VI

**16.** $U = \alpha \cdot V^{\alpha} = n \cdot C_{V,m} \cdot T$

$\Rightarrow T \cdot V^{-\alpha} = \text{Constant}$

As $T \cdot V^{x-1} = \text{Constant}$

$\Rightarrow x - 1 = -\alpha$

Now, $w = \dfrac{n \cdot R \cdot \Delta T}{x-1} = \dfrac{(\gamma-1)\cdot \Delta U}{-\alpha}$

**17.** $q = \Delta U - w = \Delta U + \dfrac{(\gamma-1)\cdot \Delta U}{\alpha} = \Delta U\left[1 + \dfrac{\gamma-1}{\alpha}\right]$

**18.** $C_m = \dfrac{R}{\gamma-1} + \dfrac{R}{1-x} = \dfrac{R}{\gamma-1} + \dfrac{R}{\alpha}$

## Comprehension VII

**19.** $P_1V_1^{\gamma} = P_2V_2^{\gamma}$

$\Rightarrow P_2 = 1 \times \left(\dfrac{320}{10}\right)^{7/5} = 128 \text{ atm}$

**20.** $\dfrac{P_1V_1}{T_1} = \dfrac{P_2V_2}{T_2}$

$\Rightarrow \dfrac{1 \times 320}{300} = \dfrac{128 \times 10}{T_2}$

$\Rightarrow T_2 = 1200 \text{ K}$

**21.** $w = n \cdot C_{V,m} \cdot \Delta T = \dfrac{1 \times 0.32}{0.082 \times 300} \times \left(\dfrac{5}{2} \times 8.314\right) \times (1200 - 300)$

$= 243.3 \text{ J}$

## Comprehension VIII

**22.** $P_1V_1^{\gamma} = P_2V_2^{\gamma} \Rightarrow P_2 = 1 \times (8)^{5/3} = 32 \text{ atm}$

**23.** For $A$: $P_1 = 1$ atm, $P_2 = 32$ atm

$V_1 = VL$, $V_2 = V + \dfrac{7}{8}V = \dfrac{15}{8}V\ L$

$T_1 = 27.3$ K; $T_2 = ?$

Now, $\dfrac{P_1V_1}{T_1} = \dfrac{P_2V_2}{T_2}$

$\Rightarrow T_2 = 1638 \text{ K}$

**24.** $\Delta H_A = n \cdot C_{P,m} \cdot \Delta T = \dfrac{1 \times 22.4}{0.082 \times 27.3} \times \left(\dfrac{5}{2} \times 2\right) \times (1638 - 27.3)$

$= 80535 \text{ cal}$

## Comprehension IX

**25.** $\Delta U = 0$

$\Delta H = \Delta U + V \cdot \Delta P = 0.9L \times \left(1 - \dfrac{532}{760}\right) \text{atm} = 0.27$ $L$-atm $= 27$ J

**26.** $\Delta U_{1\to 2} = 0$

$\Delta U_{2\to 3} = \Delta U$ for temperature increase + $\Delta U$ for vaporization of water.

$= m.s. \cdot \Delta T + (q + w)$

$= \left(900 \times \dfrac{4.2}{1000} \times 20\right) + \left(\dfrac{450}{18} \times 40\right) + \left(-\dfrac{450}{18} \times \dfrac{8}{1000} \times 373\right)$

$= 1001 \text{ kJ}$

**27.** $\Delta H_{1\to 3} = \Delta H_{1\to 2} + q_{2\to 3}$

$= 27\text{J} + \left(\dfrac{450}{18} \times 40 \text{ kJ} + 900 \times \dfrac{4.2}{1000} \times 20 \text{ kJ}\right)$

$= 1075.573 \text{ kJ}$

**28.** $w_{1\to 2} = 0$

$w_{2\to 3} = -\dfrac{450}{18} \times \dfrac{8}{1000} \times 373 = -74.6 \text{ kJ}$

## Comprehension X

**29.** $T_A > T_B < T_C < T_D = T_A$

From question : $\dfrac{T_A}{T_B} = 4$ and $T_A = 800$ K

$\Rightarrow T_B = 200$K

Also, $V_A > V_B > V_C = V_D$

From equation : $\dfrac{V_A}{V_C} = 8\sqrt{2}$ and $\dfrac{V_A}{V_B} = \dfrac{T_A}{T_B} = 4$

For process BC : $T_B \cdot V_B^{\gamma-1} = T_C \cdot V_C^{\gamma-1}$

$$\Rightarrow T_C = T_B \cdot \left(\frac{V_B}{V_C}\right)^{r-1}$$

$$= 200 \times \left(\frac{8\sqrt{2}}{4}\right)^{\frac{5}{3}-1} = 400 \text{ K}$$

**30.** $\Delta U_{BC} = C_{V,m} \cdot (T_C - T_B) = 1 \times \frac{3}{2} R \times (400 - 200)$

$= 2.4942$ kJ

## Section D (Assertion – Reason)

1. $q = 0$ because $\Delta U = 0$ and $w = 0$
2. Enthalpy of ideal gas is independent from pressure.
3. For non-ideal gas, $U = f(T, V)$
4. In adiabatic free expansion, $\Delta T = 0$
5.

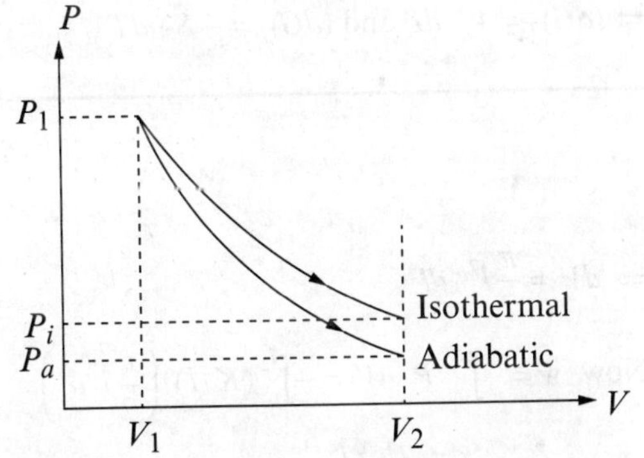


6.

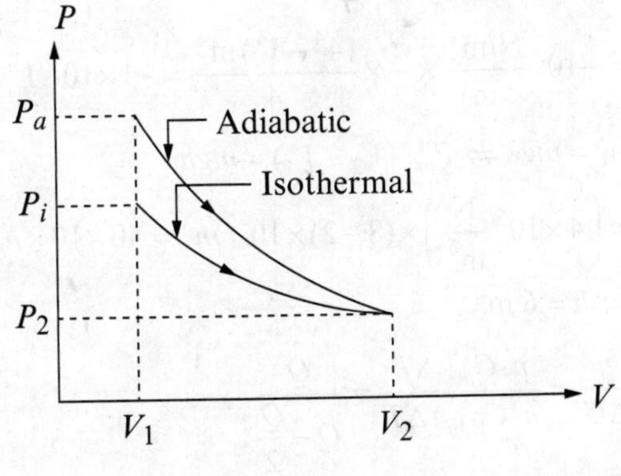


7. Magnitude of work in adiabatic process depends on change in temperature.
8. Magnitude of work in adiabatic process depends on change in temperature.
9. $q_P = n \cdot C_{P,m} \cdot \Delta T$
10. Endothermic reactions may also be spontaneous.
11. At high temperature, process may become entropy driven.
12. At low temperature, process may become enthalpy driven.
13. If $\Delta G$ = negative and $\Delta S$ = negative, $\Delta H$ must be negative and $\Delta G = \Delta H - T \cdot \Delta S$.
14. Theory based
15. $\Delta S_{\text{sys}} = 0$ but $\Delta S_{\text{univ}} = +\text{ve}$

## Section E (Column Match)

1. Theory based
2. Theory based
3. Theory based
4. For ideal gas, $H = f(T)$ but in general, $H = f(T, P)$
5. $U = f(T,V) \Rightarrow dU = \left(\frac{\partial U}{\partial T}\right)_V \cdot dT + \left(\frac{\partial U}{\partial V}\right)_T \cdot dV$

$\left(\frac{\partial U}{\partial T}\right)_V = n \cdot C_{V,m}$ and $\left(\frac{\partial U}{\partial V}\right)_T = T \cdot \left(\frac{\partial P}{\partial T}\right)_V - P$

6. $N_2(g) + O_2(g) \rightarrow 2NO(g)$ ; $\Delta H$ = Positive, $\Delta S \approx 0$

$2KI(aq) + HgI_2(aq) \rightarrow K_2[HgI_4](aq)$ ;
$\Delta H$ = Negative, $\Delta S \approx$ Negative

$PCl_5(g) \rightarrow PCl_3(g) + Cl_2$ ; $\Delta H$ = Positive, $\Delta S \approx$ Positive

$NH_3(g) + HCl(g) \rightarrow NH_9Cl(s)$; $\Delta H$ = Negative, $\Delta S$ = Negative

7. (A) $\Delta H = 0, \Delta U = 0, \Delta S_{total} = 0, \Delta S_{Sys}$ = Positive

(B) $q = 0, \Delta S_{Sys} = 0, \Delta S_{total} = 0, \Delta S_{surr} = 0$
(C) $q = 0, \Delta S_{surr} = 0, \Delta S_{total}$ = Positive.

8. (A) $\Delta H$ = Negative, $\Delta V = \pm Ve$
(B) $\Delta H = \pm Ve, \Delta S = -Ve, \Delta G = \Delta H - T, \Delta S$
$= +Ve$, if $\Delta H = +Ve$
$= \pm Ve$, if $\Delta H = -Ve$
(C) $\Delta H = Ea_f - Ea_b = 10$ kJ / mol $= +Ve$
$\Delta S = +Ve$
$\therefore \Delta G = \Delta H - T\Delta S = -Ve$, at high temperature
(D) $\Delta H = +Ve, \Delta S \approx 0 \Rightarrow \Delta G \approx \Delta H$

9. (A) Solid $\rightleftharpoons$ Liquid ; $\Delta G = 0, \Delta S$ = Positive, $\Delta V \approx 0$
$\Rightarrow \Delta H \approx \Delta U$
(B) Liquid $\rightleftharpoons$ Vapour : $\Delta G = 0, \Delta S$ = Positive,
(C) Triple point is equilibrium condition.
(D) Melting at boiling point is spontaneous.

10. $dG = V \cdot dP - S \cdot dT$

$\Rightarrow (dG)_T = V \cdot dP$ and $(dG)_P = -S \cdot dT$

## Section F (Subjective)

### Single-digit Integer Type

1. $w = -P_{ext}(V_2 - V_1)$

$= -P_{ext}[V_0(1 + \gamma \cdot t_2) - V_0(1 + \gamma \cdot t_1)]$

$= -P_{ext} \cdot V_0 \cdot \gamma \cdot (t_2 - t_1)$

$-\left(10^5 \frac{N}{m^2}\right) \times (18 \times 10^{-6} m^3) \times \frac{0.0002}{°C} \times 10°C$

$= 0.0036$ J

2. $q = \Delta U - w = 0 - \left(-nRT \ln \frac{V_2}{V_1}\right)$

or, $420 = 1 \times 2 \times 300 \times \ln \frac{V_2}{\left(\frac{1 \times 0.082 \times 300}{8.21}\right)}$

$\Rightarrow V_2 = 6\ L$

3. $P = K \cdot l$

For initial condition, 1 bar $= K \times 1$ m
$\Rightarrow K = 1$ bar/m

And, $V = \frac{4}{3}\pi r^3 = \frac{\pi}{6} l^3$

$\Rightarrow dV = \frac{\pi}{2} l^2 \cdot dl$

Now, $w = -\int_{V_1}^{V_2} P \cdot dV = -\int_{l_1}^{l_2} (K \cdot l) \cdot \left(\frac{\pi}{2} l^2 dl\right)$

$= -\frac{K\pi}{2} \cdot \frac{l_2^4 - l_1^4}{4}$

$= -10^5 \frac{N/m^2}{m} \times \frac{\pi}{2} \times \frac{(4^4 - 1^4)m^4}{4} \simeq -1 \times 10^7$ J

4. $-w = mgh \Rightarrow P_{ext}(V_2 - V_1) = mgh$

$= \left(4 \times 10^5 \frac{N}{m^2}\right) \times (8 - 2) \times 10^{-3} m^3 = 40 \times 10 \times h$

$\Rightarrow h = 6$ m

5. $\frac{q}{\Delta U} = \frac{n \cdot C_m \cdot \Delta T}{n \cdot C_{V,m} \cdot \Delta T} = \frac{Q}{Q - \frac{Q}{2}}$

$\Rightarrow \frac{C_m}{\frac{3}{2}R} = 2$

$\Rightarrow C_m = 6$ cal/K mole

**6.** $\frac{q}{w} = \frac{n \cdot C_{P,m} \cdot \Delta T}{-nR \cdot \Delta T}$

$\Rightarrow \frac{q}{-2} = \frac{\frac{7}{2}R}{-R}$

$\Rightarrow q = 7\text{ J}$

**7.** $q = q_1 + q_2 = \Delta U_1 + \Delta H_2 = n \cdot C_{V,m} \cdot \left(\frac{300}{2} - 300\right) + n \cdot C_{P,m} \cdot \left(300 - \frac{300}{2}\right)$

$= n \cdot R \cdot \frac{300}{2} = 10 \times 2 \times \frac{300}{2} = 3000\text{ cal}$

**8.** State I $\xrightarrow{\text{Isothermal}}$ State II $\xrightarrow{\text{Isochoric}}$

| | | |
|---|---|---|
| $T_1 = 273$ K | $T_2 = 273$ K | $T_3 = 5 \times 273$ K |
| $V_1 = V$ | $V_2 = 5$ V | $V_3 = 5$ V |
| $P_1 = P_0$ | $P_2 = \frac{P}{5}$ | $P_3 = P$ |

$q_{\text{total}} = nRT \cdot \ln\frac{V_2}{V_1} + n \cdot C_{V_1m} \cdot (T_3 - T_2)$

or, $80 \times 10^3 = 3 \times 8.314 \times 273 \times \ln 5 + 3 \times C_{V,m} \times 4 \times 273$

$\therefore\ C_{V,m} \approx 21$ J/K-mol = 5 cal/K-mol

**9.** $PT$ = Constant $\Rightarrow P \cdot V^{1/2}$ = Constant

Now, $C_m = C_{V_1m} + \frac{R}{1-x} \Rightarrow 29 = f \times \frac{8.314}{2} + \frac{8.314}{1-\frac{1}{2}}$

$\Rightarrow f \approx 3$

**10.** $\frac{u_2}{u_1} = 2 = \sqrt{\frac{T_2}{300}} \Rightarrow T_2 = 1200\text{ K}$

$\Rightarrow T_2 = 1200$ K

Now, $q_V = n \cdot C_{V,m} \cdot (T_2 - T_1)$

$= \frac{56}{28} \times \left(\frac{5}{2} \times 2\right) \times (1200 - 300)$

$= 9000$ cal

**11.** $Z_w = \frac{1}{4} \cdot u_{aV} \cdot N^* = \frac{1}{4} \cdot \sqrt{\frac{8RT}{\pi M}} \cdot \frac{PN_A}{RT}$ = Constant

or, $\frac{P}{\sqrt{T}}$ = Constant $\Rightarrow \text{P} \cdot \text{V}^{-1}$ = Constant

$\therefore\ C_m = C_{V_1m} + \frac{R}{1-x} = \frac{5}{2}R + \frac{R}{1-(-1)} = 3R$

$= 6$ cal/K mol

**12.** $\Delta G = V \cdot \Delta P$

$= \left(\frac{13}{0.78} \times 10^{-6}\,\text{m}^3\right)(3001-1) \times 10^5\ \frac{\text{N}}{\text{m}^2}$

$= 5000$ J

**13.** $\Delta G_1 = \Delta H - T_1 \cdot \Delta S$ and $\Delta G_2 = \Delta H - T_2 \cdot \Delta S$

$\therefore (-\Delta G_2) - (-\Delta G_1) = (T_2 - T_1) \cdot \Delta S = (T_2 - T_1) \times \frac{\Delta H - \Delta G_1}{T_1}$

$= (302 - 298) \times \frac{(-5737) - (-6333)}{298} = 8\text{ kJ}$

**14.** Graphite $\rightleftharpoons$ Diamond; $\Delta G° = 5.0$ kJ $\quad P = 1$ bar

$\Delta G = 0 \quad P = ?$

Now, $\Delta G_2 - \Delta G_1 = (V_P - V_G)(P_2 - P_1)$

or, $0 - 5000 = \left[\left(\frac{12}{3.6} - \frac{12}{2.4}\right) \times 10^{-6}\right] \times (P_2 - 10^5)$

$\Rightarrow 3 \times 10^9\ \frac{\text{N}}{\text{m}^2}$

**15.** $\Delta G° = -RT \cdot \ln K_{eq} = -2 \times 300 \times \ln(e^{-10}) = +6000$ cal

Now, $\Delta S° = \frac{\Delta H° - \Delta G°}{T} = \frac{7500 - 6000}{300}$

$= 5$ cal/k-mol

## Four-digit Integer Type

1. $w = -\left[nRT\ln\dfrac{V_2 - nb}{V_1 - nb} + an^2\left(\dfrac{1}{V_2} - \dfrac{1}{V_1}\right)\right]$

$$= -\left[\begin{array}{l} 1\times 8.314\times 300\times \ln\left(\dfrac{20-1\times 0.03}{2-1\times 0.03}\right) \\ +1.42\times 10^{12}\times 1^2\times\left(\dfrac{1}{20}-\dfrac{1}{2}\right)\times 10^{-10}\end{array}\right]$$

$= -5713.16$ J

2. $\Delta H = \Delta U + \Delta(PV) = 30 + (4\times 5 - 2\times 3) = 44$ $L$-atm

3. Calcite → Aragonite

$\Delta H = \Delta U + P\cdot\Delta V$

$\Delta V = 210\text{ J} + \left(2.7\times 10^5\,\dfrac{\text{N}}{\text{m}^2}\right)\left(\dfrac{100}{3} - \dfrac{100}{2.7}\right)\times 10^{-6}\,\text{m}^3$

$= 209$ J

4. $\Delta U = (q + w)_{\text{path I}} = (q + w)_{\text{path II}}$

or, $10\times 10^3\times 4.2\text{ J} + 0 = (11\times 10^3\times 4.2\text{ J}) + (-0.5\,w_{\max})$

$\therefore w_{\max} = 8400$ J

5. $H = U + PV = 2.5\,PV$

$\therefore \Delta H = 2.5\times 10^5\,\dfrac{\text{N}}{\text{m}^2}\times(200-100)\times 10^{-3}\,\text{m}^3$

$= 25$ kJ

6. $q = \Delta U - w = 1.5\,nR\cdot\Delta T + P_{\text{ext}}\cdot A\cdot\Delta l$

$42 = 1.5\times 1\times 8.314\times 2 + 100\times 10^3\times 8.5\times\times 10^{-4}\times\Delta l$

$\therefore \Delta l \approx 0.2\ m$

7. $\Delta H - \Delta U = P(V_D - V_G)$

$\Rightarrow -1000 = P\left[\dfrac{12}{3.6} - \dfrac{12}{2.4}\right]\times 10^{-6}$

$\therefore P = 6000\times 10^5\ Pa$

8. Solid (70°C) $\xrightarrow{1}$ Liquid (70°C) $\xrightarrow{2}$ Liquid (450°C) $\xrightarrow{3}$ Vapour (450°C)

$q = q_1 + q_2 + q_3 = 30\times 10 + 10\times 0.215\times 380 + 10\times 45 = 1567$ cal

9. $q = 12\times 0.5\times\dfrac{1805}{6} = 1805$ J

$w = -P\cdot(V_g - V_l) = -P\cdot V_g = -nRT$

$= -\dfrac{0.9}{18}\times 8.314\times 373 = -155$ J

$\therefore \Delta U = q + w = 1805 + (-155) = 1650$ J (for 0.9 $g$)

$= \dfrac{1650}{0.9}\times 18\times 10^{-3} = 33$ kJ

10. $q = \Delta U - w = 0 \Rightarrow \Delta U = w = -P_{\text{ext}}(V_2 - V_1)$

$= -100(-1) = 100$ bar-ml

Now, $\Delta H = \Delta U + \Delta PV = 100 + (100\times 99 - 1\times 100)$

$= 9900$ bar-ml $= 990$ J

11. $q = \Delta U - w = 0 \Rightarrow \Delta U = w$

$= \dfrac{1}{2}\times 10\times 1000$ bar-ml

$= 500$ J

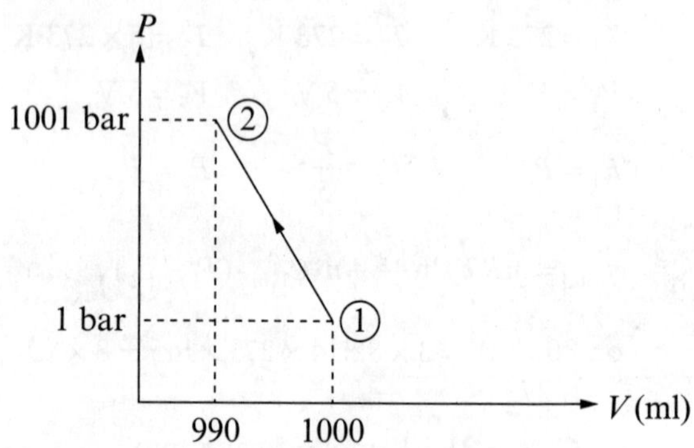


12. For adiabatic process:

$T_2 = T_1\cdot\left(\dfrac{V_1}{V_2}\right)^{\gamma-1} = 300\times(2)^{1.4-1} = 400$ K

Now, $w = w_1 + w_2$

$= \left(-nRT\cdot\ln\dfrac{V_2}{V_1}\right) + [nC_{V,m}\cdot(T_2 - T_1)$

$= [-1\times 8.3\times 300\times\ln 2] + \left[1\times\dfrac{8.3}{1.4-1}\times(400-300)\right]$

$= 332$ J

13. $q = n\cdot C_{V,m}\cdot\Delta T$

or, $50\times 166\times$

$t = \dfrac{1\times 8.21\times 10\times 2.9\times 10^3}{0.0821\times 290}\times\dfrac{5\times 8.3}{2}\times 20$

$\therefore t = 500$ sec

14. $\Delta U = 3\times 1.5\,R\times 100 + 2\times 2.5\,R\times 100 = 1900$ cal

**15.**

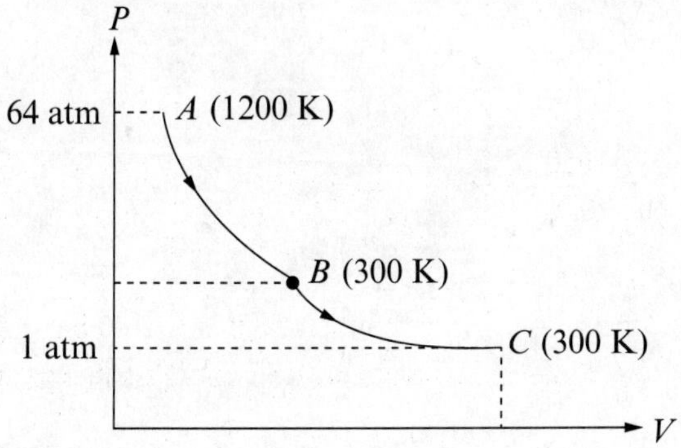


Path AB (Adiabatic):

$$P_2 = P_1 \cdot \left(\frac{T_1}{T_2}\right)^{\frac{1}{\frac{1}{\gamma}-1}} = 64 \times \left(\frac{1200}{300}\right)^{\frac{1}{\frac{3}{5}-1}} = 2 \text{ atm}$$

Now, $w_{\text{total}} = w_{AB} + w_{BC}$

$$= n \cdot C_{V,m} \cdot \Delta T + \left(-nRT \ln \frac{P_B}{P_C}\right)$$

$$= 1 \times \frac{3}{2} R \times (300 - 1200) - 1 \times R \times 300 \times \ln \frac{2}{1}$$

$$= -3120 \text{ cal}$$

**16.** $\Delta S° = S_{CH_4} - (S_{grap} + 2 \times S_{H_2})$

$= 186.2 - (6.0 + 2 \times 130.6)$

$= -81$ J/K

Now, $\Delta G° = \Delta H° - T \cdot \Delta S° = -T \cdot \Delta S_{univ}$

or, $(-75 \times 10^3) - 300 \times (-81) = -300 \times \Delta S_{univ}$

$\Rightarrow \Delta S_{univ} = 169$ J/K

**17.** $(-mgh) \times N = (-\Delta G°)$

or $50 \times 10 \times 2 \times N = \dfrac{600 \times 10^3}{2 \times 27} \times 27 \Rightarrow N = 300$

**18.** $(-\Delta G) = -(\Delta H - T \cdot \Delta S)$

$= \left[(-2808) - 310 \times \dfrac{200}{1000}\right] = 2870$ kJ

**19.** $2 + 4 + 6 = 12$

**20.** $a = 2$ (i, vi)

$b = 3$ (ii, v, vii)

$c = 4$ (iii, iv, viii, ix)

$d = 9$ (all)

CHAPTER 5

# Thermochemistry

## EXERCISE I (JEE MAIN)

### Basics

1. The word standard in molar enthalpy change implies
   (a) temperature 298 K and pressure 1 atm.
   (b) any temperature and pressure 1 atm.
   (c) any temperature and pressure 1 bar.
   (d) any temperature and pressure.

2. The enthalpy change in a reaction does not depend upon
   (a) the state of reactants and products.
   (b) the nature of the reactants and products.
   (c) different intermediate steps in the reaction.
   (d) initial and final enthalpy of the system.

3. A quantity of 1.6 g sample of $NH_4NO_3$ is decomposed in a bomb calorimeter. The temperature of the calorimeter decreases by 6.0 K. The heat capacity of the calorimeter system is 1.25 kJ/K. The molar heat of decomposition for $NH_4NO_3$ is
   (a) 7.5 kJ/mol (b) −600 kJ/mol
   (c) −375 kJ/mol (d) 375 kJ/mol

4. The difference between enthalpies of reaction at constant pressure and constant volume for the reaction

$$2C_6H_6(l) + 15O_2(g) \rightarrow 12CO_2(g) + 6H_2O(l)$$

   at 298 K in kJ is
   (a) −7.43 (b) +3.72
   (c) −3.72 (d) +7.43

5. Among the following the reaction for which $\Delta H = \Delta E$ is
   (a) $PCl_5(g) \rightarrow PCl_3(g) + Cl_2(g)$
   (b) $H_2(g) + Cl_2(g) \rightarrow 2HCl(g)$
   (c) $C_2H_5OH(l) + 3O_2(g) \rightarrow 2CO_2(g) + 3H_2O(g)$
   (d) $C_2H_4(g) + H_2(g) \rightarrow C_2H_6(g)$

6. The heat capacity of bomb calorimeter is 500 J/°C. A 2°C rise in temperature has been observed on the combustion of 0.1 g of methane. What is the value of $\Delta E$ per mole of methane?
   (a) 1 kJ (b) 160 kJ
   (c) −160 kJ (d) −1 kJ

7. For the reaction $C(s) + \frac{1}{2}O_2(g) \rightarrow CO(g)$,
   (a) $\Delta H > \Delta E$ (b) $\Delta H < \Delta E$
   (c) $\Delta H = \Delta E$ (d) $|\Delta H| > |\Delta E|$

8. The enthalpy of formation of methane(g) at constant pressure is −18,500 cal/mol at 27°C. The enthalpy of formation at constant volume would be
   (a) −19,700 cal (b) −17,300 cal
   (c) −18,498.8 cal (d) −18,500 cal

9. When 0.18 g of glucose was burnt in a bomb calorimeter, the temperature rose by 4 K. The heat capacity of the calorimeter system is ($\Delta_C H^\circ = -2.8 \times 10^6$ J/mol glucose)
   (a) 700 J/K (b) 700 kJ/K
   (c) 126 J/K (d) 7.0 kJ/K

10. Geological conditions are sometimes so extreme that quantities neglected in normal laboratory experiments take on an overriding importance. For example, consider the formation of diamond under geophysically typical conditions. The density of graphite is 2.4 g/cm$^3$ and that of diamond is 3.6 g/cm$^3$ at a certain temperature and 500 kbar. By how much does $\Delta U_{trans}$ differs from $\Delta H_{trans}$ for the graphite to diamond transition?
(a) 83.33 kJ/mol (b) 0.83 kJ/mol
(c) $8.33 \times 10^7$ kJ/mol (d) 83.33 J/mol

11. Study the following thermochemical equations.
$A \rightarrow B; \Delta H = +100$ kcal
$B \rightarrow C; \Delta H = -80$ kcal
The correct order of enthalpies of formation of A, B and C is
(a) $A < B < C$ (b) $A < C < B$
(c) $C < A < B$ (d) $B < C < A$

## Enthalpy of Formation

12. The standard enthalpy of formation of a substance
(a) is always positive.
(b) is always negative.
(c) is zero.
(d) may be positive, negative or zero.

13. The enthalpy of formation of ammonia gas is −46.0 kJ/mol. The enthalpy change for the reaction
$2NH_3(g) \rightarrow N_2(g) + 3H_2(g)$ is
(a) 46.0 kJ (b) 92.0 kJ
(c) 23.0 kJ (d) −92.0 kJ

14. The enthalpy of formation of HCl(g) from the following reaction
$H_2(g) + Cl_2(g) \rightarrow 2HCl(g) + 44$ kcal is
(a) −44 kcal mol–1 (b) −22 kcal mol–1
(c) 22 kcal mol–1 (d) −88 kcal mol–1

15. Formation of ozone from oxygen is an endothermic process. In the upper atmosphere, ultraviolet is the source of energy that drives the reaction. Assuming that both the reactions and the products of the reaction are in standard states, the standard enthalpy of formation of ozone from the following information
$3O_2(g) \rightarrow 2O_3(g), \Delta H^\circ = 286$ kJ, is
(a) +143 kJ/mol (b) −143 kJ/mol
(c) +286 kJ/mol (d) −286 kJ/mol

16. The $\Delta_f H^\circ$ for $CO_2(g)$, CO(g) and $H_2O(g)$ are −393.5, −110.5 and −241.8 kJ mol$^{-1}$, respectively. The standard enthalpy change (in kJ) for the reaction
$CO_2(g) + H_2(g) \rightarrow CO(g) + H_2O(g)$ is
(a) 524.1 (b) 41.2
(c) –262.5 (d) –41.2

17. The value of $\Delta_f H^\circ$ of $U_3O_8(s)$ is –853.5 kJ mol$^{-1}$. $\Delta H^\circ$ for the reaction
$3UO_2(s) + O_2(g) \rightarrow U_3O_8(s)$
is −76.00 kJ. The value of $\Delta_f H^\circ$ of $UO_2(s)$ is
(a) −259.17 kJ/mol (b) −310.17 kJ/mol
(c) +259.17 kJ/mol (d) 930.51 kJ/mol

18. Given enthalpy of formation of $CO_2(g)$ and CaO(s) are −94.0 kJ and −152 kJ, respectively, and the enthalpy of the reaction
$CaCO_3(s) \rightarrow CaO(s) + CO_2(g)$
is 42 kJ. The enthalpy of formation of $CaCO_3(s)$ is
(a) −42 kJ/mol (b) −202 kJ/mol
(c) +202 kJ/mol (d) −288 kJ/mol

19. The standard enthalpies of formation of $NH_3(g)$, CuO(s) and $H_2O(l)$ are −46, −155 and −285 kJ/mol, respectively. The enthalpy change when 6.80 g of $NH_3$ is passed over cupric oxide is
(a) −59.6 kJ (b) +59.6 kJ
(c) −298 kJ (d) −119.2 kJ

20. The standard enthalpies of formation of $H_2O(l)$, $Li^+(aq)$ and $OH^-(aq)$ are −285.8, −278.5 and −228.9 kJ/mol, respectively. The standard enthalpy change for the below reaction is
$2Li(s) + 2H_2O(l) \rightarrow 2Li^+(aq) + 2OH^-(aq) + H_2(g)$
(a) +443.2 kJ (b) −443.2 kJ
(c) −221.6 kJ (d) +221.6 kJ

## Enthalpy of Combustion

**21.** The standard enthalpy of combustion of solid boron is numerically equal to

(a) $\frac{1}{2}\Delta_f H^\circ_{B_2O_3(s)}$ (b) $\Delta_f H^\circ_{B_2O_3(s)}$

(c) $-\Delta_f H^\circ_{B_2O_3(s)}$ (d) $-\frac{1}{2}\Delta_f H^\circ_{B_2O_3(s)}$

**22.** The heat evolved in the combustion of glucose $C_6H_{12}O_6$ is −680 kcal/mol. The mass of $CO_2$ produced, when 170 kcal of heat is evolved in the combustion of glucose is

(a) 45 g (b) 66 g
(c) 11 g (d) 44 g

**23.** The standard molar enthalpy of formation of $CO_2$ is equal to

(a) zero
(b) the standard molar enthalpy of combustion of gaseous carbon.
(c) the sum of standard molar enthalpies of formation of CO and $O_2$.
(d) the standard molar enthalpy of combustion of carbon (graphite).

**24.** Heat evolved in the complete combustion of 1.026 kg sucrose at constant pressure will be

$C_{12}H_{22}O_{11}(s) + 12O_2(g) \rightarrow 12CO_2(g) + 11H_2O(l); \Delta H = -5.65 \times 10^3$ kJ

(a) $1.695 \times 10^4$ kJ (b) $5.65 \times 10^3$ kJ
(c) $1.883 \times 10^3$ kJ (d) $3.22 \times 10^4$ kJ

**25.** Enthalpies of combustion of $CH_4$, $C_2H_4$ and $C_2H_6$ are −890, −1411 and −1560 kJ/mole, respectively. Which has the highest fuel value (heat produced per gram of the fuel)?

(a) $CH_4$ (b) $C_2H_6$
(c) $C_2H_4$ (d) All are same

**26.** The enthalpy of combustion of methane is −890 kJ. The volume of methane at 0°C and 1 atm to be burnt to produce 2670 kJ heat is

(a) 33.6 L (b) 67.2 L
(c) 7.47 L (d) 11.2 L

**27.** Enthalpies of combustion of $CH_4$, $C_2H_6$ and $C_3H_8$ are −210, −370 and −526 kcal/mol, respectively. Enthalpy of combustion of hexane can be predicted as

(a) −840 kcal mol$^{-1}$ (b) −684 kcal mol$^{-1}$
(c) −1000 kcal mol$^{-1}$ (d) −500 kcal mol$^{-1}$

**28.** For a specific work, on an average a person requires 5616 kJ of energy. How many kilograms of glucose must be consumed if all the required energy has to be derived from glucose only? $\Delta H$ for combustion of glucose is −2808 kJ mol$^{-1}$.

(a) 0.720 kg (b) 0.36 kg
(c) 0.18 kg (d) 1.0 kg

**29.** Benzene burns in oxygen according to the following reactions.

$$C_6H_6(l) + \frac{15}{2}O_2(g) \rightarrow 3H_2O(l) + 6CO_2(g)$$

If the standard enthalpies of formation of $C_6H_6(l)$, $H_2O(l)$ and $CO_2(g)$ are 11.7, −68.1 and −94 kcal/mole, respectively, the amount of heat that will liberate by burning 780 g of benzene is

(a) 7800 kcal (b) 780 kcal
(c) 78 kcal (d) 608.4 kcal

**30.** The enthalpy of combustion at 25°C of $H_2(g)$, cyclohexane(l) and cyclohexene(l) are −241, −3920 and −3800 kJ/mol, respectively. The enthalpy of hydrogenation of cyclohexene(l) is

(a) −121 kJ/mol (b) +121 kJ/mol
(c) −242 kJ/mol (d) +242 kJ/mol

**31.** The enthalpy change involved in the oxidation of glucose is −2880 kJ/mol. Twenty five per cent of this energy is available for muscular work. If 100 kJ of muscular work is needed to walk 1 km, what is the maximum distance that a person will be able to walk after eating 120 g of glucose?

(a) 19.2 km (b) 9.6 km
(c) 2.4 km (d) 4.8 km

**32.** A geyser, operating on LPG (liquefied petroleum gas) heats water flowing at the rate of 3.0 litres per minute, from 27°C to 77°C. If the heat of combustion of LPG is 40,000 J/g, then how much fuel (in g) is consumed per minute? (Specific heat capacity of water is 4200 J/kg-K)

(a) 15.25 (b) 15.50
(c) 15.75 (d) 16.00

33. For the allotropic change represented by the equation C(graphite) → C(diamond); $\Delta H = +1.9$ kJ. If 6.0 g of diamond and 6.0 g of graphite is burnt completely in separate experiments, then the heat liberated will be higher by

(a) 1.9 kJ in case of graphite.
(b) 1.9 kJ in case of diamond.
(c) 0.95 kJ in case of diamond.
(d) 0.95 kJ in case of graphite.

## Enthalpy of Neutralization

34. Equal volumes of one molar hydrochloric acid and one molar sulphuric acid are neutralized completely by dilute NaOH solution by which $X$ and $Y$ kcal of heat are liberated, respectively. Which of the following is true?

(a) $X = Y$ (b) $2X = Y$
(c) $X = 2Y$ (d) $4X = Y$

35. Enthalpy of neutralization of oxalic acid is −25.4 kcal/mol using strong base, NaOH. Enthalpy change for the process $H_2C_2O_4(aq) \rightarrow 2H^+(aq) + C_2O_4^{2-}(aq)$ is about

(a) 2.0 kcal (b) −11.7 kcal
(c) 1.0 kcal (d) 4.0 kcal

36. A solution of 500 ml of 2 M-KOH is added to 500 ml of 2 M-HCl and the mixture is well shaken. The rise in temperature $T_1$ is noted. The experiment is again performed using 250 ml of KOH solution and 500 ml of HCl solution and rise in temperature $T_2$ is again noted. Which of the following is correct?

(a) $T_1 = T_2$
(b) $T_1$ is 2 times as larger as $T_2$.
(c) $T_2$ is twice larger as $T_1$.
(d) $T_1$ is 1.5 times as larger as $T_2$.

37. Under identical conditions, how many millilitres of 1 M-KOH and 2 M-$H_2SO_4$ solutions are required to produce a resulting volume of 100 ml with the highest rise in temperature?

(a) 80, 20 (b) 20, 80
(c) 60, 40 (d) 50, 50

38. Enthalpy of neutralization of the reaction between $CH_3COOH(aq)$ and NaOH(aq) is −13.2 kcal/eq and that of the reaction between $H_2SO_4(aq)$ and KOH(aq) is −13.7 kcal/eq. The enthalpy of dissociation of $CH_3COOH(aq)$ is

(a) −0.5 kcal $eq^{-1}$ (b) +0.5 kcal $eq^{-1}$
(c) −26.9 kcal $eq^{-1}$ (d) +13.45 kcal $eq^{-1}$

39. Enthalpy of neutralization of $H_3PO_3$ by NaOH is −106.68 kJ/mol. If the enthalpy of neutralization of HCl by NaOH is −55.84 kJ/mol, then the $\Delta H_{\text{ionization}}$ of $H_3PO_3$ into its ions is

(a) 50.84 kJ/mol (b) 5 kJ/mol
(c) 10 kJ/mol (d) 2.5 kJ/mol

## Hess's Law

40. Given that $C(s) + O_2(g) \rightarrow CO_2(g); \Delta H^o = -X$ kJ

$2CO(g) + O_2(g) \rightarrow 2CO_2(g): \Delta H^o = -Y$ kJ

The enthalpy of formation of carbon monoxide will be

(a) $(2X - Y)/2$ (b) $(Y - 2X)/2$
(c) $2X - Y$ (d) $Y - 2X$

41. The standard heat of combustion of propane is −2220.1 kJ/mol. The standard heat of vaporization of liquid water is 44 kJ/mol. What is the $\Delta H^o$ of the following reaction?

$C_3H_8(g) + 5O_2(g) \rightarrow 3CO_2(g) + 4H_2O(g)$

(a) −2220.1 kJ (b) − 2044.1 kJ
(c) −2396.1 kJ (d) −2176.1 kJ

**42.** Calculate $\Delta_f H$ for $ZnSO_4(s)$ from the following data

$ZnS(s) \rightarrow Zn(s) + S$ (rhombic), $\Delta H_1 = 44$ kcal/mol

$2ZnS(s) + 3O_2(g) \rightarrow 2ZnO(s) + 2SO_2(g)$, $\Delta H_2 = -221.88$ kcal/mol

$2SO_2(g) + O_2(g) \rightarrow 2SO_3(g)$, $\Delta H_3 = -46.88$ kcal/mol

$ZnSO_4(s) \rightarrow ZnO(s) + SO_3(g)$, $\Delta H_4 = 55.1$ kcal/mol

(a) −233.48 kcal/mol (b) −343.48 kcal/mol
(c) −434.84 kcal/mol (d) −311.53 kcal/mol

**43.** The value of $\Delta H_{sol}$ of anhydrous copper (II) sulphate is −66.11 kJ. Dissolution of 1 mole of blue vitriol, [Copper (II) sulphate pentahydrate] is followed by the absorption of 11.5 kJ of heat. The enthalpy of dehydration of blue vitriol is

(a) −77.61 kJ (b) +77.61 kJ
(c) −54.61 kJ (d) +54.61 kJ

**44.** The data below refers to gas phase reaction at constant pressure at 25°C.

$CH_3–CH_3 \rightarrow CH_3–CH_2 + H$: $\Delta H_1 = +420$ kJ mol$^{-1}$

$CH_3–CH_2 \rightarrow CH_2 = CH_2 + H$: $\Delta H_2 = +168$ kJ mol$^{-1}$

From these data, the enthalpy change $\Delta H$ for the reaction $2CH_3–CH_2 \rightarrow CH_3–CH_3 + CH_2 = CH_2$ is

(a) +250 kJ (b) +588 kJ
(c) −252 kJ (d) −588 kJ

**45.** Study the following thermochemical data.

$S + O_2 \rightarrow SO_2$: $\Delta H = -298.2$ kJ

$SO_2 + ½\, O_2 \rightarrow SO_3$: $\Delta H = -98.2$ kJ

$SO_3 + H_2O \rightarrow H_2SO_4$: $\Delta H = -130.2$ kJ

$H_2 + ½\, O_2 \rightarrow H_2O$: $\Delta H = -287.3$ kJ

The enthalpy of formation of $H_2SO_4$ at 298 K will be

(a) −433.7 kJ (b) −650.3 kJ
(c) +320.5 kJ (d) −813.9 kJ

**46.** Based on the following thermochemical equations,

$H_2O(g) + C(s) \rightarrow CO(g) + H_2(g)$: $\Delta H = 131$ kJ

$CO(g) + ½\, O_2(g) \rightarrow CO_2(g)$: $\Delta H = -282$ kJ

$H_2(g) + ½\, O_2(g) \rightarrow H_2O(g)$: $\Delta H = -242$ kJ

$C(s) + O_2(g) \rightarrow CO_2(g)$: $\Delta H = X$ kJ

The value of 'X' will be

(a) −393 kJ (b) −655 kJ
(c) +393 kJ (d) +655 kJ

**47.** Enthalpies of solution of $BaCl_2(s)$ and $BaCl_2 \cdot 2H_2O(s)$ are −20.6 kJ/mol and 8.8 kJ/mol, respectively. $\Delta H$ hydration of $BaCl_2(s)$ to $BaCl_2 \cdot 2H_2O(s)$ is

(a) −29.4 kJ (b) −11.8 kJ
(c) 29.6 kJ (d) 11.8 kJ

**48.** The dissolution of $CaCl_2 \cdot 6H_2O$ in a large volume of water is endothermic to the extent of 3.5 kcal/mol. For the reaction, $CaCl_2(s) + 6H_2O(l) \rightarrow CaCl_2 \cdot 6H_2O(s)$; $\Delta H$ is −23.2 kcal. The heat of solution of anhydrous $CaCl_2$ in large quantity of water will be

(a) −26.7 kcal mol$^{-1}$ (b) −19.7 kcal mol$^{-1}$
(c) 19.7 kcal mol$^{-1}$ (d) 26.7 kcal mol$^{-1}$

**49.** For the given two processes

(i) $½P_4(s) + 3Cl_2(g) \rightarrow 2PCl_3(l)$: $\Delta H = -635$ kJ

(ii) $PCl_3(l) + Cl_2(g) \rightarrow PCl_5(s)$: $\Delta H = -137$ kJ

the value of $\Delta_f H$ of $PCl_5(s)$ is

(a) 454.5 kJ mol$^{-1}$ (b) −454.5 kJ mol$^{-1}$
(c) −772 kJ mol$^{-1}$ (d) −498 kJ mol$^{-1}$

**50.** Calculate $\Delta_f H^\circ$ for aqueous chloride ion from the following data.

$\frac{1}{2}H_2(g) + \frac{1}{2}Cl_2(g) \rightarrow HCl(g)$: $\Delta_f H^\circ = -92.4$ kJ

$HCl(g) + nH_2O(l) \rightarrow H^+(aq) + Cl^-(aq)$: $\Delta H^\circ = -74.8$ kJ

$\Delta_f H^\circ (H^+, aq.) = 0.0$ kJ

(a) 0.0 (b) +83.6 kJ
(c) +167.2 kJ (d) −167.2 kJ

51. Tungsten carbide is very hard and is used to make cutting tools and rock drills. What is the enthalpy of formation (in kJ/mol) of tungsten carbide? The enthalpy change for this reaction is difficult of measure directly, because the reaction occurs at 1400°C. However, the enthalpies of combustion of the elements and of tungsten carbide can be measured easily.

$2W(s) + 3O_2(g) \rightarrow 2WO_3(s)$: $\Delta H = -1680.6$ kJ

$C(Graphite) + O_2(g) \rightarrow CO_2(g)$: $\Delta H = -393.5$ kJ

$2WC(s) + 5O_2(g) \rightarrow 2WO_3(s) + 2CO_2(g)$: $\Delta H = -2391.6$ kJ

(a) −38.0 (b) −76.0
(c) −19.0 (d) −1233.8

52. Diborane is a potential rocket fuel which undergoes combustion according to the following reaction.

$B_2H_6(g) + 3O_2(g) \rightarrow B_2O_3(s) + 3H_2O(g)$

From the following data, calculate the enthalpy change for the combustion of diborane.

$2B(s) + 3/2\ O_2(g) \rightarrow B_2O_3(s)$: $\Delta H = -1273$ kJ/mol

$H_2(g) + 1/2\ O_2(g) \rightarrow H_2O(l)$: $\Delta H = -286$ kJ/mol

$H_2O(l) \rightarrow H_2O(g)$: $\Delta H = 44$ kJ/mol

$2B(s) + 3H_2(g) \rightarrow B_2H_6(g)$: $\Delta H = 36$ kJ/mol

(a) −2167 kJ/mol (b) −1478 kJ/mol
(c) −2035 kJ/mol (d) −1999 kJ/mol

53. Calculate the enthalpy of formation (in kcal/mol) of anhydrous $Al_2Cl_6$ from the following data.

$2Al(s) + 6HCl(aq) \rightarrow Al_2Cl_6(aq) + 3H_2(g)$: $\Delta H = -239.760$ kcal

$H_2(g) + Cl_2(g) \rightarrow 2HCl(g)$: $\Delta H = -44$ kcal

$HCl(g) + aq \rightarrow HCl(aq)$: $\Delta H = -17.315$ kcal

$Al_2Cl_6(s) + aq \rightarrow Al_2Cl_6(aq)$: $\Delta H = -153.690$ kcal

(a) −371.76 (b) −321.960
(c) −218.07 (d) −525.45

54. Which of the following salts shall cause more cooling when one mole of the salt is dissolved in the same amount of water? (Integral heat of solution at 298 K is given for each solute.)

(a) $KNO_3$: $\Delta H = 35.4$ kJ/mol
(b) NaCl: $\Delta H = 5.35$ kJ/mol
(c) KOH: $\Delta H = -55.6$ kJ/mol
(d) HBr: $\Delta H = -83.3$ kJ/mol

55. When 1 mole of Na(s) is dissolved in large volume of water at 298 K and 1 bar, 184 kJ/mol heat is released. When 1 mole of $Na_2O(s)$ is dissolved in large volume of water at 298 K and 1 bar, 238 kJ/mol is released. If the enthalpy of formation of water is −286 kJ/mol, then the enthalpy of formation of sodium oxide is

(a) +54 kJ/mol (b) +156 kJ/mol
(c) −416 kJ/mol (d) −130 kJ/mol

56. The factor of $\Delta G$ values is important in metallurgy. The $\Delta G$ values for the following reactions at 800°C are given as follows.

$S_2(s) + 2O_2(g) \rightarrow 2SO_2(g)$: $\Delta G = -544$ kJ

$2Zn(s) + S_2(s) \rightarrow 2ZnS(s)$: $\Delta G = -293$ kJ

$2Zn(s) + O_2(g) \rightarrow 2ZnO(s)$: $\Delta G = -480$ kJ

The $\Delta G$ for the reaction $2ZnS(s) + 3O_2(g) \rightarrow 2ZnO(s) + 2SO_2(g)$ will be

(a) −357 kJ (b) −731 kJ
(c) −773 kJ (d) −229 kJ

57. For the following reaction,

$C_6H_{12}O_6(s) + 6O_2(g) \rightarrow 6CO_2(g) + 6H_2O(l)$

which of the following expression is incorrect?

(a) $\Delta H^\circ_{reaction} = 6\Delta_f H^\circ(CO_2, g) + 6\Delta_f H^\circ(H_2O, l) - 6\Delta_f H^\circ(C_6H_{12}O_6, s)$

(b) $\Delta G^\circ_{reaction} = 6\Delta_f G^\circ(CO_2, g) + 6\Delta_f G^\circ(H_2O, l) - 6\Delta_f G^\circ(C_6H_{12}O_6, s)$

(c) $\Delta S^\circ_{reaction} = 6S^\circ(CO_2, g) + 6S^\circ(H_2O, l) - 6S^\circ(C_6H_{12}O_6, s)$

(d) $\Delta S^\circ_{reaction} = 6S^\circ(CO_2, g) + 6S^\circ(H_2O, l) - S^\circ(C_6H_{12}O_6, s) - 6S^\circ(O_2, g)$

58. Calculate the free energy change for the reaction: $H_2(g) + Cl_2(g) \rightarrow 2HCl(g)$ by using the following data at the reaction temperature of 27°C.

Bond enthalpies (kJ/mol):
H – H = 435; Cl – Cl = 240; H – Cl = 430

Entropies (J/K-mol):
$H_2 = 130$; $Cl_2 = 222$; HCl = 186

(a) −185 kJ (b) −20 kJ
(c) −179 kJ (d) −191 kJ

59. The $\Delta G^o$ values for the hydrolysis of creatine phosphate (creatine-P) and glucose-6-phosphate (G-6-P) are

(i) Creatine-P + $H_2O$ → Creatine + P; $\Delta G^o$ = −29.2 kJ

(ii) G-6-P + $H_2O$ → G + P; $\Delta G^o$ = −12.4 kJ

The value of $\Delta G^0$ for the reaction G-6-P + Creatine → G + Creatine-P is

(a) +16.8 kJ (b) −16.8 kJ
(c) −41.6 kJ (d) +41.6 kJ

60. Calculate the standard free energy change for the ionization HF(aq) → $H^+$(aq) + $F^-$(aq) from the following data.

HF(aq) → HF(g): $\Delta G^o$ = 23.9 kJ

HF(g) → H(g) + F(g): $\Delta G^o$ = 555.1 kJ

H(g) → $H^+$(g) + e: $\Delta G^o$ = 1320.2 kJ

F(g) + e → $F^-$(g): $\Delta G^o$ = −347.5 kJ

$H^+$(g) + $F^-$(g) $\xrightarrow{aq.}$ $H^+$(aq) + $F^-$(aq): $\Delta G^o$ = −1513.6 kJ

(a) −38.1 kJ (b) +38.1 kJ
(c) −1489.7 kJ (d) −1513.6 kJ

61. Calculate the standard free energy of the reaction at 27°C for the combustion of methane using the given data $CH_4(g) + 2O_2(g) \rightarrow CO_2(g) + 2H_2O(l)$.

| **Species** | **$CH_4$(g)** | **$O_2$(g)** | **$CO_2$(g)** | **$H_2O$(l)** |
|---|---|---|---|---|
| **$\Delta_f H^o$ / (kJ mol$^{-1}$)** | −74.5 | 0 | −393.5 | −286.0 |
| **$S^o$ / (JK$^{-1}$ mol$^{-1}$)** | 186 | 205 | 216 | 70 |

(a) −891.0 kJ/mol (b) −240 kJ/mol
(c) −819 kJ/mol (d) −963 kJ/mol

## Bond Enthalpy

62. $\Delta_f H^o$ for $NF_3$(g) is −113 kJ/mol. Bond energy for the N–F bond is 273.5 kJ/mol. The bond energies of $N_2$ and $F_2$, if their magnitudes are in the ratio 6:1, are respectively,

(a) 822.6, 137.1 kJ/mol
(b) 979.8, 163.3 kJ/mol
(c) 943.33, 157.22 kJ/mol
(d) 762.6, 127.1 kJ/mol

63. Among the following, for which reaction the heat of reaction represents bond energy of HCl?

(a) HCl(g) → H(g) + Cl(g)
(b) 2HCl(g) → $H_2$(g) + $Cl_2$(g)
(c) HCl(g) → ½ $H_2$(g) + ½ $Cl_2$(g)
(d) HCl(g) → $H^+$(g) + $Cl^-$(g)

64. 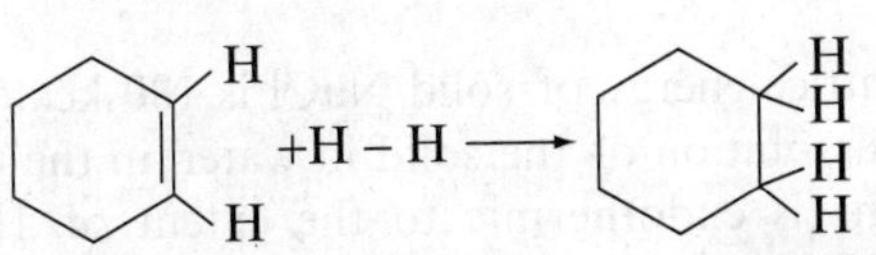

The bond energies (in kJ/mol) at 25°C are C–C = 346, C–H = 413, H–H = 437, C = C; 611. From these data, the value of $\Delta H$ at 25°C for the above reaction is

(a) −289 kJ mol$^{-1}$ (b) −124 kJ mol$^{-1}$
(c) 124 kJ mol$^{-1}$ (d) 289 kJ mol$^{-1}$

65. Heat absorbed in the reaction $H_2$(g) + $Cl_2$(g) → 2HCl(g) is 182 kJ. Bond energies of H–H = 430 kJ/mole and Cl–Cl = 242 kJ/mole. The H–Cl bond energy is

(a) 763 kJ/mole (b) 245 kJ/mole
(c) 336 kJ/mole (d) 154 kJ/mole

66. $\Delta H$ for the reaction 2C(s) + $3H_2$(g) → $C_2H_6$(g) is −20.24 kcal/mol. The bond energies (in kcal/mol) of C–C, C–H and H–H are 63, 85.6 and 102.6, respectively. The enthalpy of sublimation of C(s) is

(a) 124.3 kcal/mol (b) 185.3 kcal/mol
(c) 158.3 kcal/mol (d) 211.5 kcal/mol

67. Sublimation energy of Ca is 121 kJ/mol. Dissociation energy of $Cl_2$ is 242.8 kJ/mol, the total ionization energy of Ca(g) → $Ca^{2+}$(g) is 2422 kJ/mol and electron affinity of Cl is −355 kJ/mol. Lattice energy of $CaCl_2$ is −2430.8 kJ/mol. What is $\Delta H$ for the process Ca(s) + $Cl_2$(g) → $CaCl_2$(s)?

(a) −355 kJ mol$^{-1}$ (b) +3550 kJ mol$^{-1}$
(c) −35.5 kJ mol$^{-1}$ (d) −1720 kJ mol$^{-1}$

68. The following are various $\Delta H$ values (kJ per mol), $\Delta_f H$ (NaCl) = – 411.2; $\Delta_f H$ (Na, g) = 107.3; $\Delta_f H$ (Cl, g) = 121.7; $\Delta_i H$ (Na, g) = 495.4; $\Delta_{eg} H$ (Cl, g) = – 348.5. The lattice enthalpy of NaCl(s) is

(a) 495.4 kJ (b) 107.3 kJ
(c) 411.2 kJ (d) 787.1 kJ

69. The enthalpy of hydrogenation of benzene is −49.8 kcal/mol while its resonance energy is 36.0 kcal/mol. The enthalpy of hydrogenation of cyclohexene is

(a) −4.6 kcal (b) −28.6 kcal/mol
(c) −85.8 kcal/mol (d) −13.8 kcal/mol

70. The enthalpy of hydrogenation of benzene is –51.0 kcal/mol. If enthalpy of hydrogenation of 1, 4-cyclohexadiene and cyclohexene is –58 kcal/mol and –29 kcal/mol, respectively, then what is the resonance energy of benzene?

(a) 29 kcal/mole (b) 36 kcal/mole
(c) 58 kcal/mole (d) 7 kcal/mole

71. Use the following data to calculate the enthalpy of hydration for caesium iodide and caesium hydroxide, respectively.

| Compound | Lattice energy (kJ/mol) | $\Delta H_{Solution}$ (kJ/mol) |
|---|---|---|
| CsI | +604 | +33 |
| CsOH | +724 | –72 |

(a) −571 kJ/mol and −796 kJ/mol
(b) 637 kJ/mol and 652 kJ/mol
(c) −637 kJ/mol and −652 kJ/mol
(d) 571 kJ/mol and 796 kJ/mol

72. From the following thermochemical equations, find out the bond dissociation enthalpy of $CH_3$–H bond.

$CH_3I(g) \rightarrow CH_3(g) + I(g)$: $\Delta H = 54.0$ kcal

$CH_4(g) + I_2(s) \rightarrow CH_3I(g) + HI(g)$: $\Delta H = 29.0$ kcal

$HI(g) \rightarrow H(g) + I(g)$: $\Delta H = 79.8$ kcal

$I_2(s) \rightarrow 2I(g)$: $\Delta H = 51.0$ kcal

(a) 125.2 kcal/mol (b) 91.7 kcal/mol
(c) 101.9 kcal/mol (d) 111.8 kcal/mol

73. Estimate the average S–F bond energy in $SF_6$. The values of standard enthalpy of formation of $SF_6$(g), S(g) and F(g) are −1100, 275 and 80 kJ/mol, respectively.

(a) 183.33 kJ/mol (b) 309.17 kJ/mol
(c) 366.37 kJ/mol (d) 345 kJ/mol

74. The enthalpy of atomization of $PH_3$(g) is +954 kJ/mol and that of $P_2H_4$ is +1.488 MJ/mol. The bond energy of the P–P bond is

(a) 318 kJ/mol (b) 372 kJ/mol
(c) 216 kJ/mol (d) 534 kJ/mol

75. The standard molar enthalpies of formation of cyclohexane (l) and benzene(l) at 298 K are −156 and +49 kJ/mol, respectively. The standard enthalpy of hydrogenation of cyclohexene (l) at 298 K is −119 kJ/mol. Use these data to estimate the magnitude of the resonance energy of benzene.

(a) 152 kJ/mol (b) 250 kJ/mol
(c) 12 kJ/mol (d) 86 kJ/mol

76. Calculate the magnitude of resonance energy of $CO_2$ from the following data (in kJ/mol).

Bond energies: C=O = 539.0, O=O = 498.0

Heat of sublimation of C(s) = 718.0

Heat of combustion of C(s) = −393.0

(a) 255 (b) 531
(c) 138 (d) 247

77. AB, $A_2$ and $B_2$ are diatomic molecules. If the bond enthalpies of $A_2$, AB and $B_2$ are in the ratio 2:2:1 and enthalpy of formation AB from $A_2$ and $B_2$ is −100 kJ $mole^{-1}$, then what is the bond energy of $A_2$?

(a) 200 kJ $mol^{-1}$ (b) 100 kJ $mol^{-1}$
(c) 300 kJ $mol^{-1}$ (d) 400 kJ $mol^{-1}$

78. The lattice energy of solid NaCl is 180 kcal/mol. The dissolution of the solid in water, in the form of ions is endothermic to the extent of 1kcal/mol. If the solvation energies of $Na^+$ and $Cl^-$ ions are in the ratio 6:5, then what is the enthalpy of hydration of sodium ion?

(a) −85.6 kcal/mol (b) −97.6 kcal/mol
(c) 82.6 kcal/mol (d) 100 kcal/mol

**79.** The enthalpy of atomization of graphite is 698.6 kJ/mol and the mean bond enthalpy of C–C bond in diamond is 348.4 kJ/mol. The enthalpy of conversion of graphite into diamond is

(a) +1.2 kJ/mol (b) +1.8 kJ/mol
(c) −1.2 kJ/mol (d) −1.8 kJ/mol

**80.** The lattice energy of $Na_2CO_3(s)$ is 205 kJ/mol and the hydration energies of $Na^+$ and $CO_3^{2-}$ ions are −80 and −40 kJ/mol, respectively. The correct prediction about the solubility of $Na_2CO_3$ in water is (Assume that $\Delta H$ is independent of temperature)

(a) increases with increase in temperature.
(b) decreases with increase in temperature.
(c) unaffected by the change in temperature.
(d) first increases and then decreases with increase in temperature.

# EXERCISE II (JEE ADVANCED)

## Section A (Only one Correct)

1. In a flask, colourless $N_2O_4$ is in equilibrium with brown coloured $NO_2$. At equilibrium, when the flask is heated at 373 K, the brown colour deepens and on cooling it becomes less coloured. The change in enthalpy for this reaction is
   (a) negative (b) positive
   (c) zero (d) unpredictable

2. $2MnO_4^- + 16H^+ + 10Cl^- \rightarrow 2Mn^{2+} + 5Cl_2(g) + 8H_2O$

   The above reaction is endothermic and hence, the actual temperature of the reaction vessel (isolated from the surrounding) may be different from that expected. Given that the initial temperature of the reaction vessel was used in the calculations, how would this affect the predicted value of moles of $Cl_2$ ($n$) according to the equation $n = PV/RT$?
   (a) It would be greater than the actual value.
   (b) It would be less than the actual value.
   (c) It would be the same as the actual value.
   (d) This cannot be determined from the information given.

3. The molar heat capacities of A, B and C are in the ratio 1:2:3. The enthalpy change for the reaction A + B → C at temperature $T_1$ is $\Delta H_1$. Assuming that the heat capacities do not change with temperature, the enthalpy change, $\Delta H_2$, at temperature $T_2$ ($T_2 > T_1$) will be
   (a) greater than $\Delta H_1$.
   (b) equal to $\Delta H_1$.
   (c) less than $\Delta H_1$.
   (d) greater or less than $\Delta H_1$, depending on the values of $T_2$ and $T_1$.

4. The specific heats of iodine vapours and solid are 0.031 and 0.055 cal/g, respectively. If the enthalpy of sublimation of iodine is 24 cal/g at 200°C, then the enthalpy of sublimation of iodine at 250°C should be
   (a) 24 cal/g (b) 22.8 cal/g
   (c) 26.4 cal/g (d) 25.2 cal/g

5. A quantity that cannot be directly measured is
   (a) heat of formation of $H_2O(l)$.
   (b) heat of formation of $CH_4(g)$.
   (c) latent heat of fusion of ice.
   (d) heat of combustion of ethyl alcohol.

6. Which of the following gas will liberate maximum heat on combustion per gram?
   (a) Methane (b) Ethane
   (c) Acetylene (d) Ethylene

7. Ethyl chloride is prepared by the reaction of ethylene with hydrogen chloride as

   $C_2H_4(g) + HCl(g) \rightarrow C_2H_5Cl(g): \Delta H = -72.3$ kJ

   What is the value of $\Delta U$ (in kJ) if 70 g of ethylene and 73 g of HCl are allowed to react at 25°C?
   (a) −69.8 (b) −180.75
   (c) −174.5 (d) −139.6

8. Reactions involving gold have been of particular interest to a chemist. Consider the following reactions.

   $Au(OH)_3 + 4HCl \rightarrow HAuCl_4 + 3H_2O: \Delta H = -28$ kcal

   $Au(OH)_3 + 4HBr \rightarrow HAuBr_4 + 3H_2O: \Delta H = -36.8$ kcal

   In an experiment, there was absorption of 0.44 kcal when one mole of $HAuBr_4$ was mixed with 4 moles of HCl. What is the percentage conversion of $HAuBr_4$ into $HAuCl_4$?
   (a) 0.5% (b) 0.6%
   (c) 5% (d) 50%

EXERCISE II

9. When carbon is burnt in a definite amount of oxygen, the product will be CO, if excess amount of carbon is present and the product will be $CO_2$ if excess amount of $O_2$ is present. The enthalpies of formation of CO(g) and $CO_2$(g) are −75 and −95 kcal/mol, respectively. In which of the following case, the amount of heat evolved will be maximum?
(a) 10 moles of carbon and 4.5 moles of $O_2$.
(b) 24 g of carbon and 64 g of $O_2$.
(c) 4 moles of carbon and 3.5 moles of $O_2$.
(d) 30 g of carbon and 80 g of $O_2$.

10. The standard molar enthalpies of formation of trinitrotoluene(l), $CO_2$(g) and $H_2O$(l) are 65, −395 and −285 kJ/mol, respectively. The density of trinitrotoluene is 1.816 g/ml. Trinitrotoluene can be used as rocket fuel, with the gases resulting from its combustion streaming out of the rocket to give the required thrust. What is the enthalpy density for the combustion reaction of trinitrotoluene?
(a) −28.34 MJ/L (b) −28.34 kJ/L
(c) 27.30 MJ/L (d) −8.59 MJ/L

11. When 1 g equivalent of strong acid reacts with strong base, the heat released is 13.5 kcal. When 1 g equivalent $H_2A$ is completely neutralized against strong base, 13 kcal is released. When 1 g equivalent $B(OH)_2$ is completely neutralized against strong acid, 10 kcal heat is released. What is the enthalpy change when 1 mole of $H_2A$ is completely neutralized by $B(OH)_2$?
(a) −27 kcal (b) −10 kcal
(c) −20 kcal (d) −19 kcal

12. The enthalpy change for the reaction, NaOH(aq) + HCl(aq) → NaCl(aq) + $H_2O$(l) is −57 kJ. Predict the value of the enthalpy change in the following reaction.
$Ba(OH)_2(aq) + H_2SO_4(aq) \rightarrow BaSO_4(s) + 2H_2O(l)$
(a) −57 kJ (b) −76 kJ
(c) −114 kJ (d) −200 kJ

13. The enthalpy of neutralization of a strong monobasic acid by a strong monoacidic base is −13,700 cal. A certain monobasic weak acid is 10% ionized in a molar solution. If the enthalpy of ionization of the weak acid is +400 cal/mole, then what is the enthalpy of neutralization of one molar solution of the weak acid?
(a) −13,700 cal (b) −13,340 cal
(c) −13,660 cal (d) −13,300 cal

14. In biological cells that have a plentiful supply of $O_2$, glucose is oxidized completely to $CO_2$ and $H_2O$ by a process called aerobic oxidation. Muscle cells may be deprived of $O_2$ during vigorous exercise and in that case, one molecule of glucose is converted to two molecules of lactic acid, $CH_3CH(OH)COOH$, by a process called anaerobic glycolysis.
$C_6H_{12}O_6(s) + 6O_2(g) \rightarrow 6CO_2(g) + 6H_2O(l)$; $\Delta H° =$ −2880 kJ/mol
$C_6H_{12}O_6(s) \rightarrow 2CH_3CH(OH)COOH(s)$; $\Delta H° =$ +2530 kJ/mol
Which of the following statements is true regarding aerobic oxidation and anaerobic glycolysis with respect to energy change as heat?
(a) Aerobic oxidation has biological advantage over anaerobic glycolysis by 5410 kJ/mol.
(b) Aerobic oxidation has biological advantage over anaerobic glycolysis by 350 kJ/mol
(c) Anaerobic glycolysis has biological advantage over aerobic oxidation by 5410 kJ/mol.
(d) Anaerobic glycolysis has biological advantage over aerobic oxidation by 350 kJ/mol.

15. The intermediate $SiH_2$ is formed in the thermal decomposition of silicon hydrides. Calculate $\Delta H_f°$ of $SiH_2$ from the following reactions.
$Si_2H_6(g) + H_2(g) \rightarrow 2SiH_4(g)$: $\Delta H° = -11.7$ kJ/mol
$SiH_4(g) \rightarrow SiH_2(g) + H_2(g)$: $\Delta H° = +239.7$ kJ/mol
$\Delta_f H°$, $Si_2H_6(g) = +80.3$ kJ mol$^{-1}$
(a) 353 kJ/mol (b) 321 kJ/mol
(c) 198 kJ/mol (d) 274 kJ/mol

16. Study the following thermodynamic data given by E. H. P. Cordfunke, A. S. Booji and M. Y. Furkalionk.
(i) $DyCl_3(s) \rightarrow DyCl_3$ (aq., in 4.0 M-HCl): $\Delta H^0 =$ −180.06 kJmol$^{-1}$
(ii) Dy(s) + 3HCl(aq, 4.0 M) → $DyCl_3$ (aq, in 4.0 M-HCl) + 3/2 $H_2$(g): $\Delta H° =$ −699.43 kJ mol$^{-1}$
(iii) ½ $H_2$(g) + ½ $Cl_2$(g) → HCl (aq, 4.0 M): $\Delta H° =$ −158.31 kJ mol$^{-1}$
What is $\Delta H_f^0$ of $DyCl_3$(s) from these data?
(a) −248.58 kJ mol$^{-1}$ (b) −994.30 kJ mol$^{-1}$
(c) −3977.2 kJ mol$^{-1}$ (d) −1469.2 kJ mol$^{-1}$

**17.** The enthalpies of formation of $SO_2$(g), $H_2O$(l), HCl(g) and $H_2SO_4$(l) are −70.97, −68.32, −22.1 and −188.84 (kcal mol$^{-1}$). The enthalpies of solution of $SO_2$, $H_2SO_4$ and HCl in dilute solution are −8.56, −22.05 and −17.63 (kcal mol$^{-1}$), respectively. What is the enthalpy change for the reaction?

$SO_2(aq) + Cl_2(g) + 2H_2O(l) \rightarrow H_2SO_4(aq) + 2HCl(aq)$

(a) −74.18 kcal (b) −78.97 kcal
(c) −43.71 kcal (d) −87.14 kcal

**18.** The enthalpy of neutralization of a strong acid by a strong base is −57.32 kJ mol$^{-1}$. The enthalpy of formation of water is −285.84 kJ mol$^{-1}$. The enthalpy of formation of aqueous hydroxyl ion is

(a) +228.52 kJ/mol (b) −114.26 kJ/mol
(c) −228.52 kJ/mol (d) +114.2 kJ/mol

**19.** $\Delta_f H°$ for $CO_2$(g) and $H_2O$(l) are −94.0 and −68.0 kcal/mol. $\Delta_f H°$ for the propanoic acid is one-third of the enthalpy of combustion of the acid and has the same sign as the latter. The value of $\Delta_f H°$ of propanoic acid is

(a) −364.5 kcal/mol (b) −729.0 kcal/mol
(c) −121.5 kcal/mol (d) −243.0 kcal/mol

**20.** The enthalpy change when *x* g of phenol dissolves in *y* g of $CHCl_3$ is given below at a certain temperature.

| *x* | *Y* | $\Delta H$ (kcal) |
|---|---|---|
| 0.470 | 120 | −0.02 |
| 1.410 | 140 | −0.03 |

Calculate the enthalpy of dilution per mole of phenol, when the second solution is diluted to the concentration of the first by addition of $CHCl_3$.

(a) −4.0 kcal/mol (b) −2.0 kcal/mol
(c) +4.0 kcal/mol (d) +2.0 kcal/mol

**21.** Consider the following equations.

$CH_3CH(OH)CHClCOOH + 2KOH$
$\rightarrow CH_3CHCHCOOK + KCl + 2H_2O;$
\ /
O

$\Delta H = -14.7$ kcal

$CH_3CH(OH)CHClCOOK + KOH$
$\rightarrow CH_3CHCHCOOK + KCl + H_2O;$
\ /
O

$\Delta H = -2.7$ kcal

$\Delta H$ for the neutralization of HCl and NaOH is −13.75 kcal/eq. The enthalpy of ionization of $CH_3CH(OH)CHClCOOH$ is

(a) +1.75 kcal/mol (b) +3.5 kcal/mol
(c) +0.875 kcal/mol (d) −1.75 kcal/mol

**22.** The enthalpies of formation of FeO(s) and $Fe_2O_3$(s) are −65.0 and −197.0 kcal/mol, respectively. A mixture of the two oxides contains FeO and $Fe_2O_3$ in the mole ratio 2:1. If by oxidation it is changed in to a 1:2 mole ratio mixture, then how much of thermal energy will be released per mole of the initial mixture?

(a) 13.4 kcal (b) 67 kcal
(c) 47.2 kcal (d) 81 kcal

**23.** Calculate the heat effect produced when a solution of 1 mole of ethanol in 5 moles of water is mixed with a solution of 2 moles of ethanol in 20 moles of water.

$C_2H_5OH + 5H_2O \rightarrow C_2H_5OH\ (5H_2O)$: $\Delta H$ = −1120 cal

$C_2H_5OH + 10H_2O \rightarrow C_2H_5OH\ (10H_2O)$: $\Delta H$ = −1760 cal

$C_2H_5OH + 25/3H_2O \rightarrow C_2H_5OH\ (25/3H_2O)$: $\Delta H$ = −1650 cal

(a) 310 cal is absorbed.
(b) 310 cal is released.
(b) 4640 cal is absorbed.
(d) 4640 cal is released.

**24.** A solution was prepared by dissolving 7.45 g of KCl in 200 g of $H_2O$ in a calorimeter with a water equivalent of 25 g. The temperature of the water had reduced from 28°C to 25°C. The value of $\Delta H$ (in kJ/mol) for dissolving KCl in water is (Specific heat capacity of water is 4.2 J/K-g)

(a) +2.52 (b) +2.835
(c) +25.2 (d) +28.35

**25.** In an ice calorimeter, a chemical reaction is allowed to occur in thermal contact with an ice-water mixture at 0°C. Any heat liberated by the reaction is used to melt some ice and the volume change of the ice–water mixture indicates the amount of melting. When solutions containing 1.0 millimole each of $AgNO_3$ and NaCl were mixed in such a calorimeter, both solutions having been pre-cooled to 0°C, 0.20 g of ice melted. Assuming complete reaction in this experiment, what is $\Delta H$ for the reaction $Ag^+ (aq) + Cl^- (aq) \rightarrow AgCl(s)$? Latent heat of fusion of ice at 0°C is 80 cal/g.

(a) −16 kcal (b) +16 kcal
(c) −16 cal (d) +16 cal

26. A volume of 4.0 L of a mixture of ethane and methane gases on complete combustion at 300 K produced 6.0 L of carbon dioxide. Find out the amount of heat evolved on burning 1 L of gaseous mixture. The heats of combustion of ethylene and methane are −1573 and −890 kJ per mole, respectively, at 300 K.

(a) 4926 kJ (b) 50 kJ
(c) 1231.5 kJ (d) 200 kJ

27. A quantity of 0.3 g of carbon was converted into $CO_2$ producing 2400 cal of heat. When 0.6 g of carbon was converted to CO, 1400 cal of heat were produced. What will be the heat produced when 0.7 g of CO is converted to $CO_2$?

(a) 1700 cal (b) 1633.33 cal
(c) 700 cal (d) 1000 cal

28. Assume that for a domestic hot water supply, 160 kg of water per day must be heated from 10°C to 60°C and gaseous fuel propane, $C_3H_8$, is used for this purpose. What volume of propane gas at STP would have to be used for heating domestic water, with efficiency of 40%? Heat of combustion of propane is −500 kcal/mol and specific heat capacity of water is 1.0 cal/K-g.

(a) 896 L (b) 908 L
(c) 896 $m^3$ (d) 908 $m^3$

29. As a 0.1 mole sample of solid $NH_4Cl$ was dissolved in 50 ml of water, the temperature of the solution decreased. A small electrical immersion heater restored the temperature of the system by passing 0.125 A from a 15 V power supply for a period of 14 min. The value of $\Delta H$ for the process $NH_4Cl(s) \rightarrow NH_4Cl(aq)$ is

(a) −15.75 kJ (b) +15.75 kJ
(c) −787.5 J (d) +787.5 J

30. The thermochemical equation for the dissociation of hydrogen gas into atoms may be written as $H_2 \rightarrow 2H$; $\Delta H = 432.0$ kJ. What is the ratio of the energy yield on combustion of hydrogen atoms to steam to the yield on combustion of an equal mass of hydrogen molecules to steam? Heat of formation of steam is −240.0 kJ/mol.

(a) 2.80 (b) 1.80
(c) 0.8 (d) 2.40

31. The most exothermic 'ordinary' chemical reaction for a given mass of reactants is $2H(g) \rightarrow H_2(g)$, where $\Delta E = 103$ kcal. The theoretical decrease in mass on combination of 2.0 moles of hydrogen atoms to form 1.0 mole of hydrogen molecules, assuming that the energy is released only due to decrease in mass of the system is

(a) $4.8 \times 10^{-12}$ kg (b) $4.8 \times 10^{-12}$ g
(c) $2.4 \times 10^{-12}$ kg (d) $9.6 \times 10^{-12}$ kg

32. Calculate the enthalpy of formation (in kcal/mol) of gaseous HCl using the following data.

| Substance | $NH_3(g)$ | HCl(g) | $NH_4Cl(s)$ |
|---|---|---|---|
| Heat of formation | −11 | X | −75 |
| Heat of solution | −8.5 | −17.5 | +3.9 kcal |

$NH_3$ (aq) + HCl (aq) → $NH_4Cl$ (aq): $\Delta H = -12$ kcal

(a) −44.2 (b) −22.1
(c) −11.05 (d) −28.7

33. Two bars of different metals are heated to 60°C and then immersed in identical insulated containers each containing 200 g of water at 20°C. Will the metal with higher or lower atomic mass cause a greater temperature rise in water?

(a) Lower atomic mass
(b) Higher atomic mass
(c) Same for both
(d) Cannot be predicted

34. From the following data, calculate the enthalpy change (in kJ/mol) for the combustion of cyclopropane(g) at 298 K. The enthalpy of formation of $CO_2(g)$, $H_2O(l)$ and propene(g) are −394, −286 and 20 kJ/mol, respectively. The enthalpy of isomerization of cyclopropane(g) to propene(g) is −33 kJ/mol.

(a) −2073 (b) −2093
(c) −2060 (d) −2027

35. The reaction of zinc metal with hydrochloric acid was used to produce 1.5 moles of hydrogen gas at 298 K and 1 atm pressure. The magnitude of work done in pushing back the atmosphere is

(a) 596 cal (b) 894 cal
(c) 447 cal (d) 298 cal

36. The molar enthalpy of vaporization of benzene at its boiling point (353 K) is 7.4 kcal/mol. The molar internal energy change of vaporization is

(a) 7.4 kcal/mol (b) 8.106 kcal/mol
(c) 6.694 kcal/mol (d) 62.47 kcal/mol

**37.** Determine the standard enthalpy of the reaction $C_3H_8(g) + H_2(g) \rightarrow C_2H_6(g) + CH_4(g)$, using the given enthalpies under standard conditions.

| Compound | $H_2(g)$ | $CH_4(g)$ | $C_2H_6(g)$ | C (Graphite) |
|---|---|---|---|---|
| $\Delta_c H^0$ (kJ/mol) | −285.8 | −890.0 | −1560.0 | −393.5 |

The standard enthalpy of formation of $C_3H_8(g)$ is −103.8 kJ/mol.

(a) −55.7 kJ (b) +55.7 kJ (c) −2060.4 kJ (d) +2060.4 kJ

**38.** The enthalpy of formation of KCl(s) from the following data is

(i) $KOH(aq) + HCl(aq) \rightarrow KCl(aq) + H_2O(l)$: $\Delta H = -13.7$ kcal

(ii) $H_2(g) + ½ O_2(g) \rightarrow H_2O(l)$: $\Delta H = -68.4$ kcal

(iii) $½ H_2(g) + ½ Cl_2(g) + aq \rightarrow HCl(aq)$: $\Delta H = -39.3$ kcal

(iv) $K(s) + ½ O_2(g) + ½ H_2(g) + aq \rightarrow KOH(aq)$: $\Delta H = -116.5$ kcal

(v) $KCl(s) + aq \rightarrow KCl(aq)$: $\Delta H = +4.4$ kcal

(a) +105.5 kcal/mol (b) −105.5 kcal/mol
(c) −13.7 kcal/mol (d) −18.1 kcal/mol

**39.** Calculate $\Delta H$ for the following reaction at 298 K,

$Fe_2O_3(s) + 3CO(g) \rightarrow 2Fe(s) + 3CO_2(g)$

from the following thermochemical equations.

(i) $3Fe_2O_3(s) + CO(g) \rightarrow 2Fe_3O_4(s) + CO_2(g)$: $\Delta H = -46.4$ kJ

(ii) $FeO(s) + CO(g) \rightarrow Fe(s) + CO_2(g)$: $\Delta H = 9.0$ kJ

(iii) $Fe_3O_4(s) + CO(g) \rightarrow 3FeO(s) + CO_2(g)$: $\Delta H = -41.0$ kJ

(a) −24.8 kJ (b) +24.8 kJ
(c) −17.97 kJ (d) +17.97 kJ

**40.** Calculate the enthalpy of formation (in kcal/mol) of HI(g) from the following data.

$H_2(g) + Cl_2(g) \rightarrow 2HCl(g)$: $\Delta H = -44.20$ kcal

$HCl(g) + aq \rightarrow HCl(aq)$: $\Delta H = -17.31$ kcal

$HI(g) + aq \rightarrow HI(aq)$: $\Delta H = -19.21$ kcal

$KOH(aq) + HCl(aq) \rightarrow KCl(aq)$: $\Delta H = -13.74$ kcal

$KOH(aq) + HI(aq) \rightarrow KI(aq)$: $\Delta H = -13.67$ kcal

$Cl_2(g) + 2KI(aq) \rightarrow 2KCl(aq) + I_2(s)$: $\Delta H = -52.42$ kcal

(a) +19.21 (b) −32.87
(c) +5.94 (d) +7.82

**41.** Calculate the enthalpy of formation of $I_2O_5(s)$ from the following data.

(i) $I_2O_5(s) + H_2O(l) \rightarrow 2HIO_3(aq)$: $\Delta H = +4.0$ kJ

(ii) $KI(aq) + 3HClO(aq) \rightarrow HIO_3(aq) + 2HCl(aq) + KCl(aq)$: $\Delta H = -322.0$ kJ

(iii) $NaOH(aq) + HClO(aq) \rightarrow NaOCl(aq) + {}_2O(l)$: $\Delta H = -44.0$ kJ

(iv) $NaOH(aq) + HCl(aq) \rightarrow NaCl(aq) + H_2O(l)$: $\Delta H = -57.0$ kJ

(v) $2NaOH(aq) + Cl_2(g) \rightarrow NaOCl(aq) + NaCl(aq) + H_2O(l)$: $\Delta H = -100.0$ kJ

(vi) $2KI(aq) + Cl_2(g) \rightarrow 2KCl(aq) + I_2(s)$: $\Delta H = -224.0$ kJ

(vii) $H_2(g) + 1/2\ O_2(g) \rightarrow H_2O(l)$: $\Delta H = -285.0$ kJ

(viii) $1/2H_2(g) + 1/2\ Cl_2(g) \rightarrow HCl(g)$: $\Delta H = -92.0$ kJ

(ix) $HCl(g) + aq \rightarrow HCl(aq)$: $\Delta H = -75.0$ kJ

(a) −169.0 kJ (b) −173.0 kJ
(c) −165.0 kJ (d) −177.0 kJ

**42.** Calculate proton affinity of $NH_3(g)$ from the following data.

$\Delta H_{dissociation}$ $H_2 = 218$ kJ mole$^{-1}$

$\Delta H_{dissociation}$ $Cl_2 = 124$ kJ mole$^{-1}$

$\Delta H_f^0$ of $NH_3(g) = -46$ kJ mole$^{-1}$

$\Delta H_f^0$ of $NH_4Cl(s) = -314$ kJ mole$^{-1}$

Ionization energy of $H = 1310$ kJ mole$^{-1}$

E.A. of Cl(g) = −348 kJ mole$^{-1}$

Lattice energy of $NH_4Cl(s) = -683$ kJ mole$^{-1}$

(a) −818 kJ mole$^{-1}$ (b) −718 kJ mole$^{-1}$
(c) −318 kJ mole$^{-1}$ (d) −418 kJ mole$^{-1}$

**43.** The polymerization of ethylene to linear polyethylene is represented by the reaction

$nCH_2 = CH_2 \rightarrow (-CH_2-CH_2-)n$

where $n$ has a large integral value. Given that the average enthalpies of bond dissociation for C=C and C–C at 298 K are +590 and +331 kJ/mol, respectively, the enthalpy of polymerization per mole of ethylene at 298 K is

(a) −72 kJ (b) +259 kJ
(c) −259 kJ (d) −849 kJ

**44.** The enthalpy of formation of liquid methyl alcohol in kJ/mol, using the following data (in kJ/mol).

Heat of vaporization of liquid methyl alcohol = 38.

Heat of formation of gaseous atoms from the elements in their standard states: H, 218; C, 715; O, 249.

Average bond energies: C–H, 415; C–O, 356; O–H, 463.

(a) −190 (b) −702
(c) −626 (d) −266

**45.** The strain energy (in kJ/mol) of cyclopropane from the following data is

$\Delta_f H$ [$C_3H_6$(g)] = 53.0 kJ/mol; $\Delta_f H$ [C(g)] = 715.0 kJ/mol; $\Delta_f H$ [H(g)] = 218.0 kJ/mol

B.E. (C–C) = 356.0 kJ/mol; B.E. (C–H) = 408.0 kJ/mol.

(a) 770 kJ (b) 116 kJ
(c) 240 kJ (d) 346 kJ

**46.** Standard enthalpy of formation of gaseous ethane, ethene and benzene from gaseous atoms are −2839, −2275 and −5506 kJ/mol, respectively. The bond enthalpy of C–H bond is 412 kJ/mol. The magnitude of resonance energy of benzene compared with one Kekule structure is

(a) 24 kJ (b) 52 kJ
(c) 2524 kJ (d) 152 kJ

**47.** Find the bond energy of S–S bond from the following data.

$C_2H_5 - S - C_2H_5(g)$: $\Delta H^\circ_f = -148$ kJ,

$C_2H_5 - S - S - C_2H_5(g)$: $\Delta H^\circ_f = -202$ kJ,

S(g): $\Delta H^\circ_f = 222$ kJ

(a) 276 kJ/mol (b) 128 kJ/mol
(c) 168 kJ/mol (d) 222 kJ/mol

**48.** Given the bond dissociation enthalpy of $CH_3$–H bond as 103 kcal/mol and the enthalpy of formation of $CH_4$(g) as −18 kcal/mol, find the enthalpy of formation of methyl radical. The dissociation energy of $H_2$(g) into H (atoms) is 103 kcal/mol.

(a) −33.5 kcal/mol (b) 33.5 kcal/mol
(c) 18 kcal/mol (d) −9 kcal/mol

**49.** Calculate the enthalpy of the following homogeneous gaseous reaction

$$CH_3COCH_3 + 2O_2 \rightarrow CH_3COOH + CO_2 + H_2O$$

from the following data.

Bond energies (kJ/mol): C–H = 414; C–C = 348; C=O = 580; C–O = 354; O=O = 610; O–H = 462;

Magnitude of resonance energies (kJ/mol): COOH = 118; $CO_2$ = 140.

(a) 348 kJ (b) 168 kJ
(c) −168 kJ (d) −348 kJ

**50.** What is the enthalpy change for the isomerization reaction?

$CH_2$=CII–$CH_2$ CH–CH–CH=$CH_2$ (A) $\xrightarrow[\Delta]{NaNH_2}$ $CH_2$=CH–CH=CH–CH =CH–$CH_3$ (B)

Magnitude of resonance energies of A and B are 50 and 70 kJ/mol, respectively.

Enthalpies of formation of A and B are –2275.2 and – 2839.2 kJ/mol, respectively.

(a) −584 kJ (b) −564 kJ
(c) −544 kJ (d) −20 kJ

**51.** The bond enthalpies of C–C, C=C and C≡C bonds are 348, 610 and 835 kJ/mol, respectively at 298 K and 1 bar. The enthalpy of polymerization per mole of 2-Butyne at 298 K and 1 bar as shown below is

$nCH_3$–C≡C–$CH_3$(g) → –($CH_2$–CH=CH–$CH_2$)$_n$ – (g)

(a) −123 kJ (b) −132 kJ
(c) −139 kJ (d) −37 kJ

**52.** The enthalpies of combustion of formaldehyde and paraformaldehyde (a polymer of formaldehyde) are −134 and −732 kcal/mol, respectively. The enthalpy of polymerization per mole of paraformaldehyde is −72 kcal. The molecular formula of paraformaldehyde is

(a) $CH_2O$ (b) $C_6H_{12}O_6$
(c) $C_3H_6O_3$ (d) $C_4H_8O_4$

53. Boron exists in different allotropic forms. All allotropic form contains icosahedral units (icosahedral is a regular shape with 12 corners and 20 faces) with boron atoms at all 12 corners and all bonds are equivalent. Calculate the heat evolved at constant pressure (in kJ/mole) of boron atom undergoing the above change if the bond dissociation enthalpy of B–B bond is 300 kJ/mol.

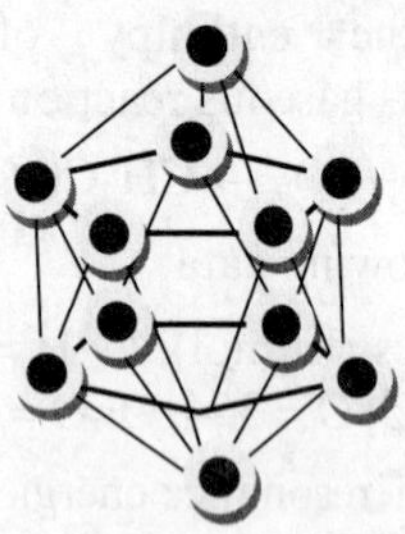

(a) 500 (b) 750
(c) 1500 (d) 900

54. Butane exists in various conformations in nature. At any given instant, the probability that a given butane molecule is in anti, gauche, eclipsed and fully eclipsed conformation is 0.7, 0.2, 0.06 and 0.04, respectively. If the molar enthalpy of combustion of natural butane is −690 kcal/mol at 25°C, then calculate the enthalpy of combustion of butane if all the butane molecules are in gauche conformation. The potential energy of gauche form is 2 kcal/mol higher than anti form but it is 3 kcal/mol less than eclipsed form and 5.5 kcal/mol less than fully eclipsed form.

(a) −690 kcal/mol (b) −689 kcal/mol
(c) −691 kcal/mol (d) −692 kcal/mol

55. For an ionic solid $MX_2$, where X is monovalent, the enthalpy of formation of the solid from M(s) and $X_2$(g) is 1.5 times the electron gain enthalpy of X(g). The first and second ionization enthalpies of the metal (M) are 1.2 and 2.8 times of the enthalpy of sublimation of M(s). The bond dissociation enthalpy of $X_2$(g) is 0.8 times the first ionization enthalpy of metal and it is also equal to one-fifth of the magnitude of lattice enthalpy of $MX_2$. If the electron gain enthalpy of X(g) is −96 kcal/mol, then what is the enthalpy of sublimation (in kcal/mol) of the metal (M)?

(a) 41.38 (b) 52.5
(c) 48.0 (d) 38.27

## Section B (One or More than one Correct)

1. Which of the following is/are endothermic reaction(s)?
(a) Combustion of methane.
(b) Decomposition of water.
(c) Dehydrogenation of ethane to ethylene.
(d) Conversion of graphite to diamond.

2. The enthalpy change for the following process would be expected to be a negative for
(a) $Na^+(g) + e^- \rightarrow Na(g)$
(b) $F(g) + e^- \rightarrow F^-(g)$
(c) $Na^+(g) + F^-(g) \rightarrow NaF(s)$
(d) $H_2O(l) \rightarrow H_2O(g)$

3. For which of the following substance(s), the standard enthalpy of formation is/are zero?
(a) C (graphite) (b) $O_3$(g)
(c) $I_2$(g) (d) $Br_2$(l)

4. Which of the following is/are endothermic compound(s)?
(a) NO(g) (b) $CO_2$(g)
(c) CO(g) (d) $NH_3$(g)

5. Among the following the reaction for which $\Delta H = \Delta E$, is
(a) $PCl_5(g) \rightarrow PCl_3(g) + Cl_2(g)$
(b) $H_2(g) + Cl_2(g) \rightarrow 2HCl(g)$
(c) $C_2H_5OH(l) + 3O_2(g) \rightarrow 2CO_2(g) + 3H_2O(g)$
(d) $C_2H_4(g) + H_2(g) \rightarrow C_2H_6(g)$

6. $\Delta H$ of which of the following reaction(s) may be directly taken as enthalpy of combustion of the concerned substance?
(a) $C(s) + O_2(g) \rightarrow CO_2(g)$
(b) $2C_2H_6(g) + 7O_2(g) \rightarrow 4CO_2(g) + 6H_2O(l)$
(c) $C_3H_8(g) + 7/2\, O_2(g) \rightarrow 3CO(g) + 4H_2O(l)$
(d) $1/2\, N_2(g) + 1/2\, O_2(g) \rightarrow NO(g)$

7. Which of the following reaction(s) is/are endothermic?
(a) $N_2(g) + O_2(g) \rightarrow 2NO(g)$
(b) $CaCO_3(s) \rightarrow CaO(s) + CO_2(g)$
(c) $2CO(g) + O_2(g) \rightarrow 2CO_2(g)$
(d) $C_2H_4(g) + H_2(g) \rightarrow C_2H_6(g)$

**8.** A volume of 100 ml of 0.5 N-$H_2SO_4$ solution is neutralized with 200 ml of 0.2 M-$NH_4OH$ in a constant pressure calorimeter which resulted 1.4°C rise in temperature. The heat capacity of the calorimeter system is 1.5 kJ/°C. Some useful thermochemical equations are as follows:

$HCl + NaOH \rightarrow NaCl + H_2O + 57$ kJ

$CH_3COOH + NH_4OH \rightarrow CH_3COONH_4 + H_2O +$ 48.1 kJ

Which of the following statements are correct?

(a) Enthalpy of neutralization of HCl vs. $NH_4OH$ is −52.5 kJ/mol.
(b) Enthalpy of dissociation (ionization) of $NH_4OH$ is 4.5 kJ/mol.
(c) Enthalpy of dissociation (ionization) of $CH_3COOH$ is 4.6 kJ/mol.
(d) $\Delta H$ for $2H_2O(l) \rightarrow 2H^+(aq) + 2OH^-(aq)$ is 114 kJ.

**9.** From the following data at 25°C, which of the following statement(s) is/are correct?

$\frac{1}{2}H_2(g) + \frac{1}{2}O_2(g) \rightarrow OH(g): \Delta H^\circ = 42$ kJ

$H_2(g) + \frac{1}{2}O_2(g) \rightarrow H_2O(g): \Delta H^\circ = -242$ kJ

$H_2(g) \rightarrow 2H(g): \Delta H^\circ - 436$ kJ

$O_2(g) \rightarrow 2O(g): \Delta H^\circ = 495$ kJ

(a) $\Delta_r H^\circ$ for the reaction $H_2O(g) \rightarrow 2H(g) + O(g)$ is 925.5 kJ.
(b) $\Delta_r H^\circ$ for the reaction $OH(g) \rightarrow H(g) + O(g)$ is 502 kJ.
(c) Enthalpy of formation of H(g) is −218 kJ/mol.
(d) Enthalpy of formation of OH(g) is 42 kJ/mol.

**10.** Which of the following molecules will have different values of standard molar enthalpy of formation, one calculated using bond energy concept and other calculated calorimetrically?

(a) $C_2H_6$
(b) 1,3-Butadiene
(c) 1,4-Cyclohexadiene
(d) $N_2O$

**11.** Consider the following isomerization process.

$CH_2{=}CH{-}CH_2{-}CH{=}CH_2(g)$
$\rightarrow CH_2{=}CH{-}CH{=}CH{-}CH_3(g)$

Which of the following statement(s) is/are true regarding this process?

(a) The process is exothermic.
(b) Enthalpy change of reaction = Resonance enthalpy of product.
(c) The magnitude of enthalpy of combustion of product is less than that of reactant.
(d) The magnitude of enthalpy of hydrogenation of product is greater than that of reactant.

**12.** α-maltose can be hydrolysed to glucose according to the following reaction.

$\alpha\text{-}C_{12}H_{22}O_{11}(aq) + H_2O(l) \rightarrow 2C_6H_{12}O_6(aq)$

The standard enthalpy of formation of $H_2O(l)$, $C_6H_{12}O_6(aq)$ and $\alpha\text{-}C_{12}H_{22}O_{11}(aq)$ are −285, −1263 and −2238 kJ/mol, respectively. Which of the following statement(s) is/are true?

(a) The hydrolysis reaction is exothermic.
(b) Heat liberated in combustion of 1.0 mole of α-maltose is smaller than the heat liberated in combustion of 2.0 mole of glucose neglecting the effect of hydration.
(c) Increase in temperature will increase the degree of hydrolysis of α-maltose.
(d) Enthalpy of reaction will remain the same even if solid α-maltose is taken in the reaction.

**13.** In certain areas where coal is cheap, artificial gas is produced for house hold use by the 'water gas' reaction.

$C(s) + H_2O(g) \rightarrow H_2(g) + CO(g)$

Assume that coke is 100% carbon. Given $\Delta H_C$ (kcal/mol) at 25°C: $H_2(g) = -68.0$; $CO(g) = -68.0$; $C(s) = -94.0$. Select the correct option(s)from the following.

(a) The maximum heat obtainable at 25°C from the combustion of 1.2 kg of coke is 9400 kcal.
(b) The maximum heat obtainable at 25°C from burning water gas produced from 1.2 kg of coke is 13,600 kcal.
(c) The maximum heat obtainable at 25°C from burning 1.2 kg of water gas is 5440 kcal.
(d) The maximum heat obtainable at 25°C from burning water gas or burning the same mass of coke will be same.

**14.** At 300 K, the standard enthalpies of formation of $C_6H_5COOH(s)$, $CO_2(g)$ and $H_2O(l)$ are −408, −393 and −286 kJ/mol, respectively. The enthalpy of combustion of benzoic acid(s) at 300 K is

(a) −3201 kJ/mol, at constant pressure.
(b) −3199.75 kJ/mol, at constant pressure.
(c) −3201 kJ/mol, at constant volume.
(d) −3199.75 kJ/mol, at constant volume.

15. For the reaction, $2NO(g) + O_2(g) \rightarrow 2NO_2(g)$; $\Delta H = -35.0$ kcal at 300 K. If 6.0 moles of NO reacts with 3.0 moles of $O_2$ at constant pressure of 1.0 atm and temperature 300 K to form $NO_2$, then which of the following statement(s) is/are correct?
(a) The magnitude of work done by the system is 1.8 kcal.
(b) The amount of heat released by the reaction is 35 kcal.
(c) The internal energy of system decreased by 103.2 kcal.
(d) The internal energy of system decreased by 106.8 kcal.

## Section C (Comprehensions)

### Comprehension I

**Study the following thermochemical equations.**

**(a)** $N_2O(g) + 3H_2(g) \rightarrow H_2O(l) + N_2H_4(l)$: $\Delta H = -76$ kcal
**(b)** $4N_2(g) + 3H_2O(l) \rightarrow 2NH_3(g) + 3N_2O(g)$: $\Delta H = +240$ kcal
**(c)** $N_2H_4(l) + H_2O(l) \rightarrow 2NH_3(g) + 1/2\ O_2(g)$: $\Delta H = +36$ kcal
**(d)** $H_2O(l) \rightarrow H_2(g) + 1/2\ O_2(g)$: $\Delta H = +68$ kcal

1. The enthalpy of combustion (in kcal/mol) of liquid $N_2H_4$ is
(a) −129 (b) −138
(c) −147 (d) +147

2. The enthalpy of formation (in kcal/mol) of liquid $N_2H_4$ is
(a) −11 (b) +44
(c) +11 (d) +13

3. Ammonia can be obtained according to reaction (b) or (c). Which of the following is correct regarding these reactions, for the same mass of ammonia formed?
(a) More mass of reactants is needed for reaction (c).
(b) Less mass of reactants is needed for reaction (b).
(c) More heat is needed for reaction (b).
(d) More heat is needed for reaction (c).

### Comprehension II

**The integral enthalpy of solution of one mole of $H_2SO_4$ in $n$ mole of water is given by the equation $\Delta H = -\frac{75n}{(n+1.8)}$ kJ/mol. Determine $\Delta H$ for the following solutions.**

4. ***Solution I:*** 1 mole of $H_2SO_4$ dissolved in 5 moles of water.
(a) −64.66 kJ
(b) −66.18 kJ
(c) −208.33 kJ
(d) −55.15 kJ

5. ***Solution II:*** 1 mole of $H_2SO_4$ dissolved in 10 moles of water.
(a) −69.44 kJ
(b) −63.56 kJ
(c) −416.67 kJ
(d) −58.59 kJ

6. ***Solution III:*** 1 mole of $H_2SO_4$ dissolved in large excess of water.
(a) Indeterminate (b) zero
(c) −750 kJ (d) −75 kJ

7. ***Solution IV:*** Solution I + 5 moles of water.
(a) −63.56 kJ (b) −55.15 kJ
(c) +8.41 kJ (d) −8.41 kJ

8. ***Solution V:*** Solution II + large excess of water.
(a) Indeterminate (b) −63.56 kJ
(c) −11.44 kJ (d) −75 kJ

## Comprehension III

**Study the following thermochemical equations**

$$2H_2(g) + O_2(g) \rightarrow 2H_2O(g): \Delta H = -483.636 \text{ kJ}$$

$$3H_2(g) + O_3(g) \rightarrow 3H_2O(g): \Delta H = -868.2 \text{ kJ}$$

$$H_2(g) + H_2O_2(g) \rightarrow 2H_2O(g): \Delta H = -347.33 \text{ kJ}$$

**9.** Which oxidizing agent will generate the greatest amount of energy for 1 mole of $H_2(g)$?

(a) $O_2(g)$ (b) $O_3(g)$
(c) $H_2O_2(g)$ (d) All are same

**10.** Which oxidizing agent will generate the greatest amount of energy for 1 g of oxidizing agent?

(a) $O_2(g)$ (b) $O_3(g)$
(c) $H_2O_2(g)$ (d) All are same

**11.** Which of the reaction will generate the greatest amount of energy on a total mass basis of reactants?

(a) $H_2$–$O_2$ (b) $H_2$–$O_3$
(c) $H_2$–$H_2O_2$ (d) All are same

## Comprehension IV

**The thermal effects of reactions in liquid $NH_3$ at −33°C were measured by observing the quantity of liquid $NH_3$ vaporized by the process of interest. The heat of vaporization of $NH_3$ at −33°C is 320.0 cal/g. When 0.98 g of $NH_4Br$ was dissolved in 20 g of liquid $NH_3$, 0.25 g of $NH_3$ was vaporized (Br = 80).**

**12.** The molar heat of solution of $NH_4Br$ in liquid $NH_3$ at this concentration is

(a) +80.0 cal (b) – 80.0 cal
(c) – 8.0 kcal (d) +8.0 kcal

**13.** When 0.49 g of $NH_4Br$ was dissolved in 20 g of liquid $NH_3$ containing an equimolar amount of $KNH_2$, 0.80 g of ammonia was vaporized. The $\Delta H$ for the reaction

$NH_4^+ (NH_3, l) + NH_2^- (NH_3, l) \rightarrow 2NH_3(l)$

at 240 K is

(a) −256.0 cal (b) +256.0 cal
(c) +51.2 kcal (d) −51.2 kcal

## Comprehension V

**Use the data (all values are in kJ per mole at 25°C) given below to answer the following.**

**Enthalpy of formation of $CH_3CN$ = +88.0**
**Enthalpy of formation of $C_3H_8$ = −85.0**
**Enthalpy of sublimation of graphite = 719.0**
**Enthalpy of dissociation of nitrogen = 948.0**
**Enthalpy of dissociation of hydrogen = 435.0**
**Bond enthalpies: C–H = 414.0; C–N = 378.0; N–H = 426.0**

**14.** The bond enthalpy of C–C bond (kJ/mol) is

(a) 250.0 (b) 335.0
(c) 223.33 (d) 248.5

**15.** The bond enthalpy of C≡N bond (kJ/mol) is

(a) 987.5 (b) 811.5
(c) 899.5 (d) 890.0

**16.** The enthalpy of hydrogenation of $CH_3CN$ (kJ/mol) is

(a) −288.5 (b) +288.5
(c) −89.5 (d) +89.5

## Comprehension VI

**An intimate mixture of hydrogen gas and a theoretical amount of air at 27°C and a total pressure of 1 atm is exposed in a closed rigid vessel. If the process occurs under adiabatic condition, then using the following data, answer the following questions.**

**Given: (i) $C_{P,m} = 6.9$ cal/K-mol (ii) $C_{P,m} = 8.2$ cal/K-mol (iii) $\Delta_f H$ [$H_2O(g)$] = −55.85 kcal (iv) Air contains 20% $O_2$ and 80% $N_2$, by volume**

17. The values of $C_{P,m}$ of $N_2(g)$ and $H_2O(g)$ (in cal/K-mol) should be
(a) 6.9, 6.9 (b) 6.9, 8.2
(c) 8.2, 8.2 (d) 8.2, 6.9

18. What will be the maximum temperature (approximately) attained if the process occurs in adiabatic container?
(a) 3800 K (b) 3500 K
(c) 2540 K (d) 300 K

19. What will be the final pressure (approximately)?
(a) 10.86 atm (b) 12.67 atm
(c) 14.78 atm (d) 1.0 atm

20. If at initial temperature, $T_1$, $E_1$ is the internal energy and at higher temperature $T_2$, $E_2$ is the internal energy, then
(a) $E_1 > E_2$ (b) E1 < E2
(c) E1 = E2 (d) Unpredictable

## Comprehension VII

**When 0.1 mole of $C_8H_{18}(l)$ at 300 K is completely burned at constant pressure in stoichiometric amount of oxygen gas at 300 K, yielding as products gaseous $H_2O$, CO and $CO_2$ at 800 K, the process yielding 87.3 kcal of heat to the surrounding.**

**$\Delta_f H$ (kcal/mol): $C_8H_{18}(l) = -74.0$, $CO_2(g) = -94.0$, $CO(g) = -26.5$, $H_2O(g) = -58.0$**

**Molar heat capacity at constant pressure (cal/K-mol): CO(g) = 7.0, $CO_2(g) = 8.0$, $H_2O(g) = 6.0$**

**Assume that all $\Delta_r H$ are independent of temperature.**

21. The value of $\Delta_c H$ of $C_8H_{18}(l)$ is (in kcal/mol) assuming the formation of $H_2O(g)$
(a) +1200.0 (b) +660.0
(c) −1200.0 (d) −660.0

22. The value of $\Delta_r H$ for the reaction $C_8H_{18}(l) + \frac{17}{2} O_2(g) \rightarrow 8CO(g) + 9H_2O(g)$ is (in kcal/mol)
(a) +1200.0 (b) +660.0
(c) −1200.0 (d) −660.0

23. How many moles of $CO_2$ are produced?
(a) 0.1 (b) 0.8
(c) 0.4 (d) 0.05

24. How many moles of $H_2O$ are produced?
(a) 0.1 (b) 0.9
(c) 0.45 (d) 1.8

25. What is the magnitude of work done by the system?
(a) 1.045 kcal (b) 2.09 kcal
(c) 0.96 kcal (d) 5.7 kcal

## Comprehension VIII

**A factory is producing methanol based on the following reaction.**

$$CO + 2H_2 \rightarrow CH_3OH; \Delta H = -100 \text{ R}$$

**Hydrogen and carbon monoxide are obtained by the following reaction.**

$$CH_4 + H_2O \rightarrow CO + 3H_2$$

**Three units of factory, namely reformer (for the production of $H_2$ and CO), reactor (for the production of methanol from CO and $H_2$) and separator (to separate $CH_3OH$ from CO and $H_2$) are schematically shown in the following figure.**

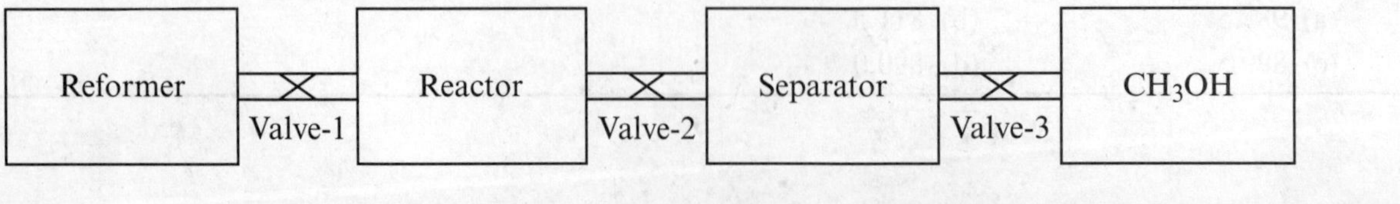

**The flow of methanol from Valve-3 is 1000 mol/sec. The factory is so designed that 2/3rd of the CO is converted to $CH_3OH$. Assume that the reformer reaction goes to completion.**

26. What is the flow of CO and $H_2$ at Valve-1?
(a) CO = 3000 mol/sec; $H_2$ = 9000 mol/sec
(b) CO = 1500 mol/sec; $H_2$ = 3000 mol/sec
(c) CO = 1000 mol/sec; $H_2$ = 2000 mol/sec
(d) CO = 1500 mol/sec; $H_2$ = 4500 mol/sec

27. What is the flow of CO and $H_2$ at Valve-2?
(a) CO = 500 mol/sec; $H_2$ = 1000 mol/sec
(b) CO = 500 mol/sec; $H_2$ = 2500 mol/sec
(c) CO = 500 mol/sec; $H_2$ = 2000 mol/sec
(d) CO = 500 mol/sec; $H_2$ = 1500 mol/sec

28. Amount of energy released in methanol reactor in one minute is
(a) 12,000 kcal (b) 1200 kcal
(c) 6000 kcal (d) 600 kcal

## Comprehension IX

**When 12 g of carbon reacted with oxygen to form CO and $CO_2$ at 298 K and constant pressure, 53.2 kcal of heat was liberated and no carbon remained. Heat of formation of CO and $CO_2$ are −26.0 and −94.0 kcal/mole, respectively.**

29. Moles of CO formed is
(a) 0.4 (b) 0.5
(c) 0.6 (d) 1.0

30. Mass of oxygen reacted with carbon is
(a) 16 g (b) 32 g
(c) 24 g (d) 22.4 g

## Comprehension X

**The bond dissociation enthalpy of the first H–S bond in hydrogen sulphide is 376.0 kJ/mol. The enthalpies of formation of $H_2S$(g) and S(g) are −20.0 and 277.0 kJ/mol, respectively. The bond dissociation enthalpy of H–H bond is 436.0 kJ/mol.**

31. The enthalpy of formation of the free radical HS is
(a) 138 kJ/mol (b) −138 kJ/mol
(c) −10 kJ/mol (d) 357 kJ/mol

32. The bond dissociation enthalpy of the free radical HS is
(a) 138 kJ/mol (b) 276 kJ/mol
(c) 357 kJ/mol (d) 376 kJ/mol

## Comprehension XI

**For the reaction at 25°C, $X_2O_4(l) \rightarrow 2XO_2(g)$, $\Delta E° = 2.1$ kcal and $\Delta S° = 20$ cal/K.**

33. The value of $\Delta G°$ for the reaction is
(a) −2.7 kcal
(b) −9.25 kcal
(c) +2.7 kcal
(d) +9.25 kcal

34. The reaction is
(a) Spontaneous and exergonic.
(b) Non-spontaneous and exergonic.
(c) Spontaneous and endergonic.
(d) Non-spontaneous and endergonic.

## Section D (Assertion – Reason)

The following questions consist of two statements. Mark the answer as follows.

(a) If both statements are CORRECT, and **Statement II** is the CORRECT explanation of **Statement I.**
(b) If both statements are CORRECT, and **Statement II** is NOT the CORRECT explanation of **Statement I.**
(c) If **Statement I** is CORRECT, but **Statement II** is INCORRECT.
(d) If **Statement I** is INCORRECT, but **Statement II** is CORRECT.

1. **Statement I:** On increasing the temperature, the enthalpy of reaction may increase, decrease or remain constant.

   **Statement II:** On changing the temperature, the enthalpies of reactants and products may change to same or different extent depending on their heat capacities.

2. **Statement I:** Standard enthalpy of isomerization of an enantiomer into the other is zero.

   **Statement II:** The two enantiomers of any chiral compound have the same enthalpy of formation.

3. **Statement I:** Heat evolved in the neutralization of either 1 mole of HCl or 1 mole of $H_2SO_4$ with NaOH is same.

   **Statement II:** Both HCl and $H_2SO_4$ are strong acids.

4. **Statement I:** When a salt is dissolved in water, the temperature of solution decreases.

   **Statement II:** Solubility of any salt in water may be endothermic or exothermic.

5. **Statement I:** If 5 ml of an acid solution is completely neutralized by adding 5 ml of a base solution, then the temperature of solution increases by $x$°C. If 20 ml of the same acid solution is completely neutralized by adding 20 ml of the same base solution, then the temperature of the solution increases by $4x$°C.

   **Statement II:** Heat liberated in the second case will be four times the heat liberated in the first case.

6. **Statement I:** The solubility of any gas in any liquid is an exothermic process.

   **Statement II:** All the gases are highly soluble in any liquid.

7. **Statement I:** For all the salts completely soluble in water, the magnitude of sum of the enthalpies of hydration of ions is greater than the magnitude of lattice enthalpy of the salt.

   **Statement II:** If the magnitude of sum of enthalpies of hydration of ions is less than the lattice enthalpy of the salt, then the salt is completely insoluble in water.

8. **Statement I:** The magnitude of enthalpy of combustion of diamond is greater than that of graphite.

   **Statement II:** Graphite is thermodynamically more stable form of carbon than diamond.

9. **Statement I:** The magnitude of enthalpy of combustion of 2-Butene is less than that of 1-Butene.

   **Statement II:** 2-Butene is thermodynamically more stable than 1-Butene.

10. **Statement I:** Enthalpy changes are positive when $Na_2SO_4 \cdot 10H_2O$, $CuSO_4 \cdot 5H_2O$ and salt like NaCl, KCl, etc., which do not form hydrates is dissolved in water. But enthalpy changes are negative when anhydrous salts capable of forming hydrates are dissolved in water.

    **Statement II:** The difference in the behaviour is due to large difference in the molecular masses of hydrated and anhydrous salts. The substance with large molecular mass usually shows positive enthalpy changes on dissolutions.

## Section E (Column Match)

**1.** Match the columns.

| Column I (Reaction) | Column II (Process) |
|---|---|
| (A) $C(s) + \frac{1}{2} O_2(g) \rightarrow CO(g)$ | (P) Combustion |
| (B) $CO(g) + \frac{1}{2} O_2(g) \rightarrow CO_2(g)$ | (Q) Neutralization |
| (C) $NaOH(aq) + HCl(aq) \rightarrow NaCl(aq) + H_2O(l)$ | (R) Formation |
| (D) $H_2(g) + \frac{1}{2} O_2(g) \rightarrow H_2O(l)$ | (S) Used in fuel cell |

**2.** Match the columns.

| Column I (Reaction) | Column II (Relation) |
|---|---|
| (A) $H_2(g) + Cl_2(g) \rightarrow 2HCl(g)$ | (P) $\Delta H = \Delta U + RT$ |
| (B) $2N_2O(g) + O_2(g) \rightarrow 2NO_2(g)$ | (Q) $\Delta H = \Delta U$ |
| (C) $H_2(g) + I_2(s) \rightarrow 2HI(g)$ | (R) $\Delta H = \Delta U - 2RT$ |
| (D) $N_2(g) + 3H_2(g) \rightarrow 2NH_3(g)$ | (S) $\Delta H = \Delta U - RT$ |

**3.** Column I contains some chemical reactions and Column II contains some $\Delta H$ values (in kJ). Match the correct $\Delta H$ in Column II for the chemical reaction in Column I with the help of the following thermochemical equations given.

$H^+(aq) + OH^-(aq) \rightarrow H_2O(l)$; $\Delta H = -57.3$ kJ

$\Delta H_{Solution}$ of $HA(g) = -70.7$ kJ/mol

$\Delta H_{Solution}$ of $BOH(g) = 20$ kJ/mol

$\Delta H_{Ionization}$ of $HA = 15$ kJ/mol and BOH is a strong base.

| Column I (Chemical reactions) | Column II ($\Delta H$ values in kJ) |
|---|---|
| (A) $HA(aq) + BOH(aq) \rightarrow BA(aq) + H_2O$ | (P) −42.3 |
| (B) $HA(g) + BOH(g) \rightarrow BA(aq) + H_2O$ | (Q) −93 |
| (C) $HA(g) \rightarrow H^+(aq) + A^-(aq)$ | (R) −55.7 |
| (D) $B^+(aq) + OH^-(aq) \rightarrow OH(aq)$ | (S) 0 |

**4.** Carefully observe the given diagrams which indicate the standard enthalpy of formation of different states of one mole of Mg and two moles of Cl atoms and match the entries in Column I and II provided.

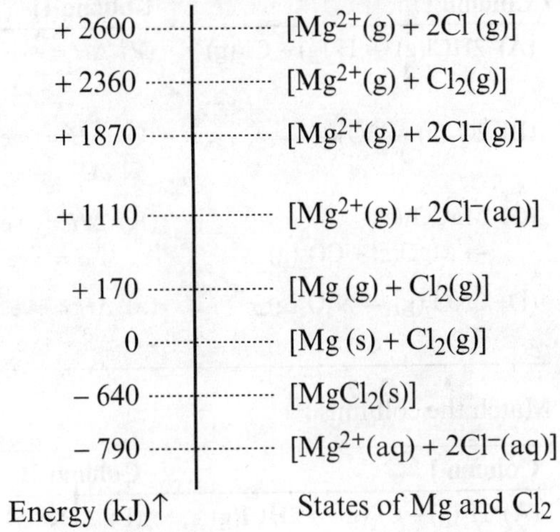


| Column I | Column II |
|---|---|
| (A) $\Delta_f H\,[Mg^{2+}(aq)]$ | (P) −1900 kJ/mol |
| (B) $\Delta_f H\,[Cl^-(aq)]$ | (Q) +460 kJ/mol |
| (C) $\Delta H_{Hydration}\,[Mg^{2+}(g)]$ | (R) −625 kJ/mol |
| (D) Lattice enthalpy of $MgCl_2(s)$ | (S) +2510 kJ/mol |

**5.** Match the columns.

| Column I | Column II ($\Delta_r H$ is also known as) |
|---|---|
| (A) $C\,(Graphite) + O_2(g) \rightarrow CO_2(g)$ | (P) $\Delta H_{formation}$ |
| (B) $C\,(Graphite) \rightarrow C(g)$ | (Q) $\Delta H_{combustion}$ |
| (C) $HCl(aq) + OH^-(aq) \rightarrow Cl^-(aq) + H_2O(l)$ | (R) $\Delta H_{atomization}$ |
| | (S) $\Delta H_{neutralization}$ |

**6.** Match the columns.

| Column I | Column II |
|---|---|
| (A) Sb(s) | (P) $\Delta_f H^\circ = +ve$, $\Delta_f S^\circ = +ve$ |
| (B) $O_3(g)$ | (Q) $\Delta_f H^\circ = 0$, $\Delta_f S^\circ = 0$ |
| (C) $I_2(g)$ | (R) $\Delta_f H^\circ = +ve$, $\Delta_f S^\circ = -ve$ |
| (D) CO(g) | (S) $\Delta_f H^\circ = -ve$, $\Delta_f S^\circ = +ve$ |

7. Match the columns.

| Column I | Column II |
|---|---|
| (A) $2HCl(g) \rightarrow H_2(g) + Cl_2(g)$ | (P) $\Delta H = +ve$, $\Delta S = +ve$ |
| (B) $2O_3(g) \rightarrow 3O_2(g)$ | (Q) $\Delta H = -ve$, $\Delta S = +ve$ |
| (C) $MgCO_3(s) \rightarrow MgO(s) + CO_2(g)$ | (R) $\Delta H = -ve$, $\Delta S = -ve$ |
| (D) $2NO_2(g) \rightarrow N_2O_4(g)$ | (S) $\Delta H = +ve$, $\Delta S \approx +ve$ |

8. Match the columns

| Column I | Column II |
|---|---|
| (A) $H_2(g) + Cl_2(g) \rightarrow 2HCl(g)$ | (P) $\Delta H = \Delta U$ |
| (B) $2C(s) + O_2(g) \rightarrow 2CO(g)$ | (Q) $\Delta H > \Delta U$ |
| (C) $PCl_5(g) \rightarrow PCl_3(g) + Cl_2(g)$ | (R) $\Delta H < \Delta U$ |
| (D) $C_2H_4(g) + 3O_2(g) \rightarrow 2CO_2(g) + 2H_2O(l)$ | (S) $\lvert\Delta H\rvert > \lvert\Delta U\rvert$ |
| | (T) $\lvert\Delta H\rvert < \lvert\Delta U\rvert$ |

9. Match the columns.

| Column I | Column II |
|---|---|
| (A) $C(Diamond) + O_2(g) \rightarrow CO_2(g)$ | (P) $\Delta_r H^\circ = \Delta_c H^\circ$ |
| (B) $C(Graphite) + O_2(g) \rightarrow CO_2(g)$ | (Q) $\Delta_r H^\circ = \Delta_f H^\circ$ |
| (C) $H_2(g) \rightarrow 2H(g)$ | (R) $\Delta_r H^\circ = \Delta_{atomization} H^\circ$ |
| (D) $CH_4(g) \rightarrow C(g) + 4H(g)$ | (S) $\Delta_r H^\circ = \Delta_{bond} H^\circ$ |

10. Some thermochemical details are given as follows.

$3A(g) \rightarrow A_3(g)$, $\Delta H_{300} = -100$ kJ

$A(l) \rightarrow A(g)$, $\Delta_{vap} H_{300} = +25$ kJ/mol

$A_3(l) \rightarrow A_3(g)$, $\Delta_{vap} H_{400} = +50$ kJ/mol

The standard boiling points of A(l) and $A_3(l)$ are 300 K and 400 K, respectively.

Molar heat capacities at constant pressure (in J/K-mol): A(l) = 40; A(g) = 20; $A_3(l)$ = 50; $A_3(g)$ = 30

Match the columns on the basis of these details.

| Column I | Column II |
|---|---|
| (A) $A(l) \rightarrow A(g)$ $\Delta_{vap} H_{400}$ | (P) −84 kJ/mol |
| (B) $A_3(l) \rightarrow A_3(g)$ $\Delta_{vap} H_{300}$ | (Q) +23 kJ/mol |
| (C) $3A(l) \rightarrow A_3(l)$ $\Delta H_{300}$ | (R) +52 kJ/mol |
| (D) $3A(l) \rightarrow A_3(l)$ $\Delta H_{400}$ | (S) −77 kJ/mol |

## Section F (Subjective)

### Single-digit Integer Type

1. Volumes of 50 ml of 1 M-NaOH and 50 ml of 1 M-HCl both at the same temperature were mixed in a calorimeter of very small heat capacity. A temperature rise of 411 K was recorded. In a second experiment, the HCl was replaced by 50 ml of 1 M-HCOOH. The temperature rise was 321 K. The enthalpy of ionization (in kcal/mol) of HCOOH is

2. The heat evolved on combustion of 1 g of starch, $(C_6H_{10}O_5)_x$ into $CO_2(g)$ and $H_2O(l)$ is 4.6 kcal. Heat of formation of $CO_2(g)$ and $H_2O(l)$ are −94.2 and −68.4 kcal/mol, respectively. The magnitude of standard enthalpy of formation of 1 g of starch (in kcal) is

3. In order to get maximum calorific output, a burner should have an optimum fuel to oxygen ratio which corresponds to 3 times as much oxygen as required theoretically for complete combustion of the fuel. A burner which has been adjusted for methane as fuel (with 'X' L/hr of methane and '6X' L/hr of oxygen) is to be adjusted for butane. In order to get same calorific output, what should be the supply of oxygen (in L/hr) with 1 L/hr supply of butane? Assume that losses due to incomplete combustion, etc., are the same for both fuels and that the gases behave ideally. Enthalpies of combustion; methane = −800 kJ/mol; butane = −3120 kJ/mol.

4. The enthalpy of neutralization of monobasic acid in decinormal solution by a dilute solution of KOH is −12,200 cal. The enthalpy of neutralization of strong acid by strong base is −13,700 cal. Assuming that the acid is 25% dissociated in decinormal solution, the enthalpy of dissociation of the acid (in kcal/mole) is

5. A volume of 1.642 L sample of a mixture of methane gas and oxygen measured at 298 K and 1.192 atm, was allowed to react at constant pressure in a calorimeter which together with its content had a heat capacity of 1260 cal/K. The complete combustion of methane to carbon dioxide and water caused a temperature rise in calorimeter 0.667 K. The volume percent of methane in original mixture is (Given the heat of combustion of methane is −210 kcal/mole)

6. A solution of 6.3 g of haemoglobin (molar mass = 64,000 g/mol) in 25 ml of solution shows a temperature rise of 0.03°C for complete oxygenation. Each mole of haemoglobin binds 4 moles of oxygen. If the heat capacity of the solution is 4.2 J/K-ml, then the amount of heat released per mole of oxygen bound (in kJ) is

7. Two solutions, initially at 25°C, were mixed in an insulated bottle. One contained 200 ml of 0.4 M weak monoprotic acid solution. The other contained 100 ml of a solution having 0.5 mole NaOH per litre. After mixing, the temperature rose to 26°C. Assume that the densities of both the solutions are 1.0 g/ml and that their specific heat capacities are 1.0 cal/g-K. The amount of heat evolved (in kcal) in the neutralization of 1 mole of the acid is

8. In solid $NH_3$, each $NH_3$ molecule has six other $NH_3$ molecules as nearest neighbours. The enthalpy of sublimation of $NH_3(s)$ at its melting point is 30.4 kJ/mol and estimated value of enthalpy of sublimation of $NH_3(s)$, where there were no hydrogen bonds is 15.4 kJ/mol. What is the average strength of hydrogen bonds in solid $NH_3$ in kJ/mol?

9. When 3.0 g graphite is burnt in limited supply of oxygen at 298 K and 1 bar, 7.5 kcal heat is released. No solid is left and the gaseous product formed is not absorbed in aqueous KOH solution. When 4.0 g graphite is burnt in excess supply of oxygen at 298 K and 1 bar, 32 kcal heat is released. No solid residue is left and the gaseous product formed is absorbed completely in aqueous KOH solution. When 4.0 g $CO_2(g)$ is decomposed completely into CO(g) and $O_2(g)$ at 29 K and 1 bar, the enthalpy of system increases (in kcal) by

10. The enthalpies of neutralization of a weak acid HA and a weak acid HB by NaOH are −6900 cal/equivalent and −2900 cal/equivalent, respectively. When one equivalent of NaOH is added to a solution containing one equivalent of HA and one equivalent of HB, the enthalpy change was −3900 cal. If the base is distributed between HA and HB in the ratio 1 : $x$, the value of 'x' is

## Four-digit Integer Type

1. The heat of total cracking of hydrocarbons $\Delta H_{TC}$ is defined as $\Delta H$ at 298.15 K and 101.325 kPa for the following process.

$$C_nH_m + \left(2n - \frac{m}{2}\right) H_2(g) \rightarrow nCH_4(g)$$

The values of $\Delta H_{TC}$ is −65.2 kJ for $C_2H_6$ and −87.4 kJ for $C_3H_8$. Calculate $\Delta H$ (in kJ) for

$$CH_4(g) + C_3H_8(g) \rightarrow 2C_2H_6(g)$$

2. An athlete takes 20 breaths per minute at room temperature. The air inhaled in each breath is 164.2 ml which contains 20% oxygen by volume, while exhaled air contains 10% oxygen by volume. Assuming that all the oxygen consumed is used for converting glucose into carbon dioxide and water, how much heat is produced (in kJ) in the body in one hour? Body temperature is 310 K and enthalpy of combustion of glucose is −3100 kJ/mol at 310 K.

3. The heat of combustion of glycogen is about 432 kJ/mol of carbon. Assume that the average rate of heat loss by an adult male is 150 W. If we were to assume that all the heat comes from the oxidation of glycogen, how many units of glycogen (1 mole carbon per unit) must be oxidized per day to provide for this heat loss?

4. Only gases remains after 15.0 g of carbon is treated with 20 L of air at 380 K and 8.21 atm pressure. (Assume 19% by volume oxygen, 80% nitrogen, 1% carbon dioxide). Determine the amount of heat evolved (in kcal) under constant pressure. Enthalpies of formation of $CO_2(g)$ and $CO(g)$ are −96.0 and −26.0 kcal/mol, respectively.

5. Ethanol was oxidized to acetic acid in a catalyst chamber at 18°C. What should be rate of removal of heat (in kcal/hr) to maintain the reaction chamber at 18°C with the feed rate of 2.3 kg ethanol per hour, along with excess oxygen to the system at 18°C, with a 40 mole percent yield based on ethanol? The enthalpies of formation of $H_2O(l)$, $C_2H_5OH(l)$ and $CH_3COOH(l)$ are, respectively −68, −66 and −118 kcal/mol.

6. The carbon dioxide exhaled in the breath of astronaut is often removed from the spacecrafts by reaction with lithium hydroxide.

$$2LiOH(s) + CO_2(g) \rightarrow Li_2CO_3(s) + H_2O(l)$$

Assume that each astronaut requires 2100 kcal of energy per day. Further assume that this energy is obtained only from the combustion of glucose into $CO_2(g)$ and $H_2O(l)$. The standard enthalpies of formation of glucose, $CO_2$ and water are −1280, −395 and −285 kJ/mol, respectively. The minimum mass (in g) of LiOH required per astronaut per day to react completely with all the $CO_2$ produced is

7. The reversible reaction

$$Na_2SO_4 \cdot 10H_2O \rightarrow Na_2SO_4 + 10H_2O;\ \Delta H = +16.1\ \text{kcal}$$

goes completely to the right at temperature above 32.4°C and remains completely on the left below this temperature. This system has been used in some solar houses for heating at night with the energy absorbed from the sun's radiation during the day. How many litres of fuel gas could be saved per night by the reversal of the dehydration of a fixed charge of 100 kg $Na_2SO_4\cdot10H_2O$? Assume that the fuel value of the gas is 10,000 kcal/m$^3$.

8. Ethanol can undergo decomposition to form two sets of products as follows.

$$C_2H_5OH(g) \rightarrow \begin{cases} C_2H_4(g) + H_2O(g); \Delta H = +45.0\ \text{kJ} \\ CH_3CHO(g) + H_2O(g); \Delta H = +72.0\ \text{kJ} \end{cases}$$

If the molar ratio of $C_2H_4(g)$ to $CH_3CHO(g)$ in the product is 8:1, the heat absorbed in decomposition of 2.5 mole of ethanol(g) (in kJ) is

9. A slice of banana weighing 2.5 g was burnt in a bomb calorimeter and produced a temperature rise of 3.0 K. In the same calorimeter, combustion of a 0.305 g sample of benzoic acid produced a temperature rise of 4.0 K. The heat of combustion of benzoic acid at constant volume is −800 kcal/mol. If an average banana weighs 125 g, then how many calories can be obtained from one average banana?

10. The enthalpy of formation of liquid water at 25°C is −286 kJ. Given $C_P$ = 75.4 J/K–mol for $H_2O(l)$ and 33.4 J/K–mol for $H_2O(g)$ and that the molar enthalpy of vaporization of liquid water at 125°C is 40.8 kJ/mol. The enthalpy of dissociation (in kJ/mol) of $H_2O(g)$ into $H_2$ and $O_2$ gases at 25°C is

11. For the hypothetical reaction, $2B(g) \rightarrow B_2(g)$; $\Delta C_P$ [/JK$^{-1}$] = $2.0 \times 10^{-2}$ T[/K] and $\Delta H_{300}$ = −4.0 kJ/mol. Estimate the absolute temperature at which $\Delta H = 0$ for this reaction.

12. The value of heat of combustion per $CH_2$ unit of cyclopropane is −697 kJ/mol. Calculate its strain energy. Given that the heats of formation of C(g), H(g), carbon dioxide and water are 715, 218, −393 and −285 kJ/mol, respectively, and C–C and C–H bond energies are 356 and 408 kJ/mol, respectively.

13. The hydrogen bond between $F^-$ and CHCOOH is very strong and its strength may be analysed by setting up a Born–Haber cycle with the following data (in kJ/mol). Lattice energy of $KF.CH_3COOH$, 734; enthalpy of vaporization of $CH_3COOH$, 20; enthalpy of solution of KF, 35; solvation energy of $K^+(g)$, −325; solvation energy of $F^-$, −389; enthalpy of formation of $KF.CH_3COOH(s)$ from KF(s) and $CH_3COOH(l)$, −25. Find the energy of the hydrogen bond between $F^-$ and $CH_3COOH$ in the gas phase (in kJ/mol).

EXERCISE II

**14.** Find the bond enthalpy (in kJ/mol) of 'three centre two electron bond' in $B_2H_6$ from the following data. $\Delta_f H^\circ$ [$BH_3$(g)] = 100 kJ/mol; $\Delta_f H^\circ$ [$B_2H_6$(g)] = 36 kJ/mol; $\Delta H_{\text{Atomization}}$ [B(s)] = 565 kJ/mol; $\Delta H_{\text{Atomization}}$ [$H_2$(g)] = 436 kJ/mol.

**15.** Calculate the enthalpy change (in kcal) for the reaction $XeF_4 \rightarrow Xe^+ + F^- + F_2 + F$. The average Xe – F bond enthalpy is 34 kcal/mol, first ionization enthalpy of Xe is 279 kcal/mol, electron gain enthalpy of fluorine is −85 kcal/mol and bond dissociation enthalpy of $F_2$ is 38 kcal/ mol.

**16.** Calculate $\Delta H_{\text{vap}}$ [$CH_3COOH$(l)] in kJ/mol from the following data.

$\Delta H_{\text{Solution}}$ [KF.$CH_3COOH$(s)] in glacial acetic acid = −3 kJ/mol

$\Delta H_{\text{Solution}}$ [KF(s)] in glacial acetic acid = +35 kJ/mol

The strength of H-bond between $F^-$(g) and $CH_3COOH$(g) = +46 kJ/mol

Lattice enthalpy of KF.$CH_3COOH$(s) = +734 kJ/mol

Lattice enthalpy of KF(s) = +797 kJ/mol

**17.** Estimate $\Delta_f H^\circ$ [Pyridine(l)] (in kJ/mol) from the given data.

| Compound | $\Delta_f H^\circ$ (kJ/mol) |
|---|---|
| $CH_2 = N - CH_3$(g) | +44 |
| $CH_3 - NH - CH_3$(g) | −18 |
| 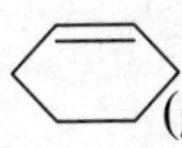 (g) | −37 |
| 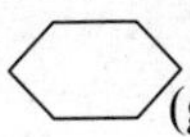 (g) | −156 |
| 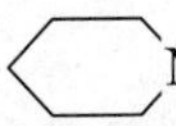 N – H (g) | −50 |

Magnitude of resonance energy of pyridine = 125 kJ/mol

$\Delta H_{\text{vap, 298 K}}$ [Pyridine(l)] = 40 kJ/mol

**18.** The standard molar enthalpies of formation of $IF_3$(g) and $IF_5$(g) are −470 kJ and −847 kJ, respectively. Valence shell electron-pair repulsion theory predicts that $IF_5$(g) is square pyramidal in shape in which all I – F bonds are equivalent while $IF_3$(g) is T-shaped (based on trigonal–bipyramidal geometry) in which I – F bonds are of different lengths. It is observed that the axial I – F bonds in $IF_3$ are equivalent to the I – F bonds in $IF_5$. Calculate the equatorial I – F bond strength (in kJ/mol) in $IF_3$. Some other details given are as follows.

$I_2(s) \rightarrow I_2(g)$: $\Delta H$ = 62 kJ

$F_2(g) \rightarrow 2F(g)$: $\Delta H$ = 155 kJ

$I_2(g) \rightarrow 2I(g)$: $\Delta H$ = 149 kJ

**19.** The standard molar enthalpies of formations of $H_2O$(l) and $H_2O_2$(l) are −286 and −188 kJ/mol, respectively. Molar enthalpies of vaporization of $H_2O$ and $H_2O_2$ are 44 and 53 kJ, respectively. The bond dissociation enthalpy of $O_2$(g) is 498 kJ/mol. Calculate the bond dissociation enthalpy (in kJ/ mol) of O – O bond in $H_2O_2$, assuming that the bond dissociation enthalpy of O – H bond is same in both $H_2O$ and $H_2O_2$.

**20.** Enthalpy of the reaction $Ag^+$(aq) + $Br^-$ (aq) → AgBr(s) is −84.54 kJ. Magnitude of enthalpies of formation of $Ag^+$(aq) and $Br^-$(aq) is in 8:9 ratio but their signs are opposite. Enthalpy of formation of AgBr is −99.54 kJ/mol. The magnitude of enthalpy of formation of $Ag^+$(aq) (in kJ/mol) is

## Answer Keys

### Exercise I

### Basics

1. (c) 2. (c) 3. (d) 4. (a) 5. (b) 6. (c) 7. (b) 8. (b) 9. (a) 10. (a)
11. (b)

### Enthalpy of Formation

12. (d) 13. (b) 14. (b) 15. (a) 16. (b) 17. (a) 18. (d) 19. (a) 20. (b)

### Enthalpy of Combustion

21. (a) 22. (b) 23. (d) 24. (a) 25. (a) 26. (b) 27. (c) 28. (b) 29. (a) 30. (a)
31. (d) 32. (c) 33. (c)

### Enthalpy of Neutralization

34. (b) 35. (a) 36. (d) 37. (a) 38. (b) 39. (b)

### Hess's Law

40. (b) 41. (b) 42. (a) 43. (b) 44. (c) 45. (d) 46. (a) 47. (a) 48. (b) 49. (b)
50. (d) 51. (a) 52. (c) 53. (b) 54. (a) 55. (c) 56. (b) 57. (c) 58. (d) 59. (a)
60. (b) 61. (c)

### Bond Enthalpy

62. (c) 63. (a) 64. (b) 65. (b) 66. (a) 67. (a) 68. (d) 69. (b) 70. (b) 71. (a)
72. (d) 73. (b) 74. (c) 75. (a) 76. (b) 77. (d) 78. (b) 79. (b) 80. (a)

---

## Answer Keys

### Exercise II

### Section A (Only one Correct)

1. (b) 2. (b) 3. (b) 4. (d) 5. (b) 6. (a) 7. (d) 8. (c) 9. (a) 10. (a)
11. (d) 12. (d) 13. (b) 14. (a) 15. (d) 16. (b) 17. (a) 18. (c) 19. (c) 20. (b)
21. (a) 22. (a) 23. (b) 24. (d) 25. (a) 26. (b) 27. (a) 28. (b) 29. (b) 30. (a)
31. (a) 32. (b) 33. (a) 34. (b) 35. (b) 36. (c) 37. (a) 38. (b) 39. (a) 40. (c)
41. (a) 42. (b) 43. (a) 44. (d) 45. (b) 46. (b) 47. (a) 48. (b) 49. (d) 50. (d)
51. (a) 52. (b) 53. (b) 54. (c) 55. (a)

### Section B (One or More than one Correct)

1. (b), (c), (d) 2. (a), (b), (c) 3. (a), (d) 4. (a)
5. (b) 6. (a) 7. (a), (b) 8. (a), (b), (d)
9. (a), (d) 10. (b), (d) 11. (a), (b), (c) 12. (a), (b)
13. (a), (b), (c) 14. (a), (d) 15. (c)

## Section C

**Comprehension I**

1. (c) 2. (c) 3. (c)

**Comprehension II**

4. (d) 5. (b) 6. (d) 7. (d) 8. (c)

**Comprehension III**

9. (c) 10. (b) 11. (b)

**Comprehension IV**

12. (c) 13. (d)

**Comprehension V**

14. (b) 15. (c) 16. (a)

**Comprehension VI**

17. (b) 18. (a) 19. (a) 20. (c)

**Comprehension VII**

21. (c) 22. (d) 23. (c) 24. (b) 25. (b)

**Comprehension VIII**

26. (d) 27. (b) 28. (a)

**Comprehension IX**

29. (c) 30. (d)

**Comprehension X**

31. (a) 32. (c)

**Comprehension XI**

33. (a) 34. (a)

## Section D (Assertion – Reason)

1. (a) 2. (a) 3. (d) 4. (d) 5. (d) 6. (c) 7. (c) 8. (a) 9. (a) 10. (c)

## Section E (Column Match)

1. A → R; B → P; C → Q; D → P, R, S
2. A → Q; B → S; C → P; D → R
3. A → P, B → Q; C → R; D → S
4. A → Q; B → R; C → P; D → S
5. A → P, Q; B → P, R; C → S
6. A → Q; B → R; C → P; D → S
7. A → S; B → Q; C → P; D → R
8. A → P; B → Q, T; C → Q, S; D → R, S
9. A → P; B → P, Q; C → R, S; D → R
10. A → Q; B → R; C → S; D → P

## Section F (Subjective)

**Single-digit Integer Type**

1. (3) 2. (1) 3. (5) 4. (2) 5. (5) 6. (8) 7. (6) 8. (5) 9. (6) 10. (3)

**Four-digit Integer Type**

1. (0043) 2. (0400) 3. (0030) 4. (0085) 5. (2400)
6. (0216) 7. (0500) 8. (0120) 9. (0075) 10. (0241)
11. (0700) 12. (0120) 13. (0060) 14. (0455) 15. (0292)
16. (0021) 17. (0085) 18. (0272) 19. (0142) 20. (0120)

# HINTS AND EXPLANATIONS

## EXERCISE I (JEE MAIN)

### Basics

1. For standard state, the pressure should be 1 bar, but temperature, any specified temperature.

2. Enthalpy is a state function.

3. As the temperature is decreased, the reaction must be endothermic.

   Heat absorbed in reaction = Heat released by calorimeter

   $= c\cdot\Delta T = 1.25\times 6 = 7.5\text{ kJ}$

   $\therefore \Delta H = +\frac{7.5}{1.6}\times 80 = +375\text{ kJ/mol}$

4. $\Delta H - \Delta E = \Delta n_g \cdot RT = (12-15)\times\frac{8.314}{1000}\times 298$

   $= -7432.7\text{ J}$

5. For $\Delta H = \Delta E,\ \Delta n_g = 0$

6. Heat released by reaction $= c\cdot\Delta T = 500\times 2 = 1000\text{ J}$

   $\therefore \Delta E = -\frac{1000}{0.1}\times 16 = -160\text{ kJ/mol}$

7. $\Delta n_g = 1 - \frac{1}{2} = \frac{1}{2}$

   $\Delta H = \Delta E + \Delta n_g RT$ and $\Delta H, \Delta E$ are – ve, hence $\Delta H < \Delta E$.

8. $CH_4(g) + 2O_2(g) \rightarrow CO_2(g) + 2H_2O(l);$

   $\Delta n_g = 1 - 3 = -2$

   Now,

   $\Delta E = \Delta H - \Delta n_g RT = (-18500) - (-2)\times 2\times 300$

   $= -17300\text{ cal/mol}$

9. Heat released by reaction $= c\cdot\Delta T$

   or $\frac{2.8\times 10^6}{180}\times 0.18 = c\times 4 \Rightarrow c = 700\text{ J/K}$

10. $\Delta H - \Delta U = P\cdot\Delta V = P[V_D - V_G]$

    $= 500\times 10^3\times 10^5\frac{\text{N}}{\text{m}^2}\left[\frac{12}{3.6} - \frac{12}{2.4}\right]\times 10^{-6}\text{ m}^3$

    $= -8.33\times 10^4\text{ J/mol}$

11. 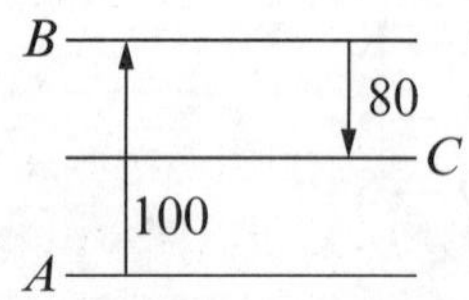


### Enthalpy of Formation

12. $\Delta_f H° = 0$, +ve or –ve

13. $\Delta H = 2\times 46.0 = 92\text{ kJ}$

14. $\Delta_f H_{HCl(g)} = -\frac{44}{2} = -22\text{ kcal/mol}$

15. $\Delta_f H°_{O_3(g)} = -\frac{286}{2} = 143\text{ kJ/mol}$

16. $$\Delta_r H° = \sum \Delta_f H°_{\text{Products}} - \sum \Delta_f H°_{\text{Reactants}}$$

    $$= \left[\Delta_f H°_{co(g)} + \Delta_f H°_{H_2O(g)}\right] - \left[\Delta_f H°_{co_2(g)} + \Delta_f H°_{H_2(g)}\right]$$

    $$= [(-110.5) + (-241.8)] - [(-393.5) + 0] = 41.2\text{ kJ}$$

**17.** $\Delta_r H^\circ = \sum \Delta_f H^\circ{}_{Products} - \sum \Delta_f H^\circ{}_{Reactants}$

$= \left[\Delta_f H^\circ{}_{U_3O_8(s)}\right] - \left[3\times\Delta_f H^\circ{}_{UO_2(s)} + \Delta_f H^\circ{}_{O_2(g)}\right]$

or, $-76.0 = (-853.5) - \left[3\times\Delta_f H^\circ{}_{UO_2(s)} + 0\right]$

$\therefore\ \Delta_f H^\circ{}_{UO_2(s)} = -259.17$ kJ/mol

**18.** $\Delta H = \sum \Delta_f H_{Products} - \sum \Delta_f H_{Reactants}$

or, $42 = [(-152)+(-74)] - \left[\Delta_f H_{CaCO_3(s)}\right]$

$\therefore\ \Delta_f H_{CaCO_3(s)} = -288$ kJ/mol

**19.**

$2NH_3(g) + 3CuO(s) \rightarrow 3Cu(s) + N_2(g) + 3H_2O(l)$

$\Delta_r H = \sum \Delta_f H_{Products} - \sum \Delta_f H_{Reactants}$

$= [0+0+3\times(-285)] - [2\times(-46)+3\times(-155)]$

$= -298$ kJ

$\therefore\ \Delta H_{required} = -\frac{298}{2\times17}\times6.8 = 59.6$ kJ

**20.** $\Delta rH = \Sigma\Delta_f H_{products} - \Sigma\Delta_f H_{Reactants}$

$= \left[2\times\Delta_f H_{Li^+(aq)} + 2\times\Delta_f H_{OH^-(aq)} + \Delta_f H_{H_2(g)}\right]$

$-\left[2\times\Delta_f H_{Li(s)} + 2\times\Delta_f H_{H_2O(l)}\right]$

$= [2\times(-278.5) + 2\times(-228.9) + 0]$

$- [2\times0 + 2\times(-285.8)]$

$= -443.2$ KJ

## Enthalpy of Combustion

**21.** $2B(s) + \frac{3}{2}O_2(g) \longrightarrow B_2O_3(s)$

$\Delta_r H^\circ = \Delta_f H^\circ{}_{B_2O_3(s)} = 2\times\Delta_C H^\circ{}_{B(s)}$

**22.** $C_6H_{12}O_6(s) + 6O_2(g) \longrightarrow 6CO_2(g) + 6H_2O(l)$;

$\Delta H = -680$ Kcal

For 680 Kcal, 6 × 44 gm $CO_2$ is produced

Hence, for 170 Kcal, mass of $CO_2$ produced

$= \frac{6\times44}{680}\times170 = 66$ gm

**23.** $C(Graphite) + O_2(g) \rightarrow CO_2(g)$;

$\Delta_f H^\circ{}_{co_2(g)} = \Delta_c H^\circ{}_{C(Graphite)}$

**24.** $\frac{5.65\times10^3}{342}\times1026 = 1.695\times10^4$ kJ

**25.** For $CH_4$ : $\frac{890}{16} = 55.625$ kJ/gm

For $C_2H_4$ : $\frac{1411}{28} = 50.393$ kJ/gm

For $C_2H_6$ : $\frac{1560}{30} = 52.0$ kJ/gm

**26.** Moles of $CH_4$ burned $= \frac{2670}{890} = 3$

$\therefore$ Volume of $CH_4 = 3\times22.4 = 67.2$ l

**27.** Increase in $|\Delta_c H|$ per unit increase in $CH_2$ is

$\frac{(370-210)+(526-370)}{2} = 158$

$\therefore$ Expected $\Delta_c H$ of

$C_6H_{14} = (-210) + 5\times(-158) = -1000$ kcal/mol

**28.** Mass of glucose needed $= \frac{5616}{2808}\times130 = 360$ g

**29.** $\Delta_c H^\circ{}_{Benzene} = \Delta_r H$

$= \sum \Delta_f H_{Products} - \sum \Delta_f H_{Reactants}$

$= 7800$ kcal

**30.**  $(l) + H_2(g) \longrightarrow$ 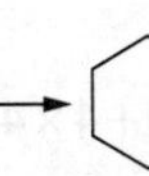 $(l)$

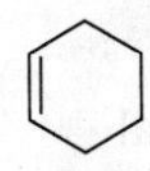 $(l) + H_2(g) \longrightarrow$ 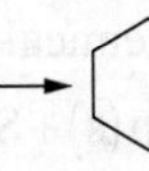 $(l)$

$\Delta H = \Sigma\Delta_C H_{Reactants} - \Sigma\Delta_C H_{Products}$

$= [(-3800) + (-241)] - [(-3920)]$

$= -121$ KJ/mol

31. Heat released from 120 gm glucose

$$= \frac{2880}{180} \times 120 = 1920 \text{ kJ}$$

$$\therefore \text{Distance } \frac{1920}{100} \times \frac{25}{100} = 4.8 \text{km}$$

32. Heat released by constitution of LPG = Heat absorbed by water

$w \times 40000 = 3 \times 4200 \times 50 \Rightarrow w = 15.75$ gm

33.

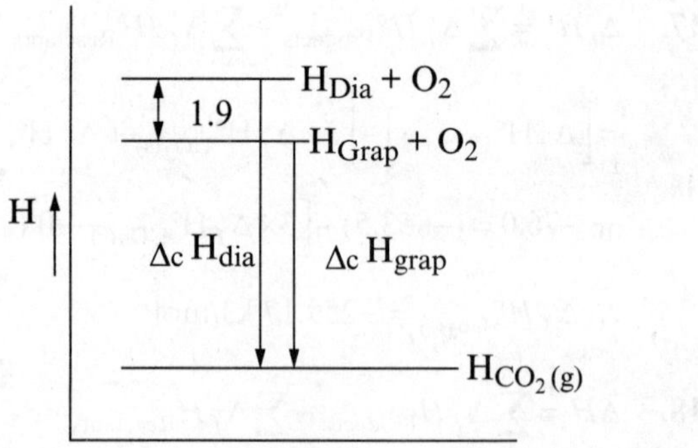


Heat liberated is greater for diamond by $\frac{1.9}{12} \times 6 = 0.95$ kJ

## Enthalpy of Neutralization

34. Moles of $H^+$ from $H_2SO_4$ = 2 × Moles of $H^+$ from HCl

35. $\Delta H = (-25.4) - 2 \times (-13.7) = 2.0$ kcal/mol

36. $q = (1000 \times d) \times s \times T_1$

$\therefore \frac{q}{2} = (1000 \times d) \times s \times T_2$ (KOH is limiting reagent)

$\therefore T_1 = 1.5 T_2$

37. For maximum rise in temperature, both should react completely.

$V \times 1 = (100 - V) \times 2 \times 2 \Rightarrow V = 80 = V_{KOH}$

38. $\Delta H = (-13.2) - (-13.7) = +0.5$ kcal / eq.

39. $\Delta H = (-106.68) - 2 \times (-55.84) = 5$ kJ/mol

## Hess's Law

40. $C(s) + O_2(g) \rightarrow CO_2(g); \Delta H° = -x$ KJ

$-\frac{1}{2} \times [2CO(g) + O_2(g) \rightarrow 2CO_2(g); \Delta H° = -y \text{ kJ}]$

---

$C(s) + \frac{1}{2} O_2(g) \rightarrow CO(g); \Delta H° = (-x) - \frac{1}{2}(-y) = \frac{y - 2x}{2}$

41. $\Delta H° = (-2220.1) + 4 \times 44 = -2044.1$ kJ

42. Given thermo chemical equations are:

(a) $ZnS(s) \rightarrow Zn(s) + S(s); \Delta H = 44$ kcal

(b) $2ZnS(s) + 3O_3(g) \rightarrow 2ZnO(s) + 2SO_2(g);$

$\Delta H = 221.88$ kcal

(c) $2SO_2(g) + O_2(g) \rightarrow 2SO_3(g);$

$\Delta H = 46.88$ kcal

(d) $ZnSO_4(s) \rightarrow ZnO(s) + SO_3(g);$

$\Delta H = 55.1$ kcal

Required thermochemical equation is

$Zn(s) + S(s) + 2O_2(g) \rightarrow ZnSO_4(s); \Delta H = ?$

From $\frac{1}{2}(b) + \frac{1}{2}(c) - (a) - (d);$

$\Delta H = \frac{1}{2}(-221.88) + \frac{1}{2}(-46.88) - (44) - (55.1)$

$= -233.48$ kcal

43. Given T.C.E. are

(a) $CuSO_4(s) + aq \rightarrow CuSO_4(aq); \Delta H = -66.11$ kJ

(b) $CuSO_4 \cdot 5H_2O(s) + aq \rightarrow CuSO_4(aq);$

$\Delta H = +11.5$ kJ

Required T.C.E. is

$$CuSO_4 \cdot 5H_2O(s) \rightarrow CuSO_4(s); \Delta H = ?$$

From $(b)-(a); \Delta H = 11.5-(-66.11) = 77.61$ kJ

**44.** Given T.C.E. are

(a) $CH_3-CH_3 \rightarrow CH_3-CH_3+H;$

$\Delta H = 420$ kJ

(b) $CH_3-CH_2 \rightarrow CH_2=CH_2+H;$

$\Delta H = 168$ kJ

Required T.C.E. is

$2CH_3-CH_2 \rightarrow CH_3-CH_3+CH_2=CH_2;$

$\Delta H = ?$

From $(b)-(a): \Delta H = 168-420 = -252$ kJ

**45.** Adding all the given T.C.E.,

$\Delta_f H_{H_2SO_4} = -813.9$ kJ/mol

**46.** Adding all the given T.C.E.: X = –393

**47.** $BaCl_2(s)+aq \rightarrow BaCl_2(aq);$

$\Delta H = -20.6$ kJ

$BaCl_2 \cdot 2H_2O(s)+aq \rightarrow BaCl_2(aq);$

$\Delta H = 8.8$ kJ

Hence, for the reaction,

$BaCl_2(s)+2H_2O(l) \rightarrow BaCl_2 \cdot 2H_2O(s),$

$\Delta II - (-20.6)-(8.8)$

$= -29.4$ kJ

**48.** $-23.2 = \Delta_{sol} H_{CaCl_2(s)} - \Delta_{sol} H_{CaCl_2(s).6H_2O(s)}$

$\therefore \Delta_{sol} H_{cacl_2(s)} = (-23.2)+(3.5) = -19.7$ kcal/mol

**49.** $\Delta_f H_{PCl_5(s)} = \frac{1}{2} \times (i) + (ii) = \frac{1}{2}(-635)+(-137)$

$= -454.5$ kJ

**50.** (a) $\frac{1}{2}H_2(g)+\frac{1}{2}Cl_2(g) \longrightarrow HCl(g);$

$\Delta H° = -92.4$ KJ

(b) $HCl(g)+nH_2O(l) \longrightarrow H^+(aq)+Cl^-(aq);$

$\Delta H° = -74.8$ KJ

(c) $\frac{1}{2}H_2(g)+nH_2O(l) \longrightarrow H^+(aq); \Delta H° = 0$

Required T.C.E. is

$\frac{1}{2}Cl_2(g)+nH_2O(l) \longrightarrow Cl^-(aq); \Delta H° = ?$

From (a) + (b) – (c); $\Delta H° = -167.2$ KJ

**51.** (a) $2W(s) + 3O_2(g) \rightarrow 2WO_3(s); \Delta H = -1680.6$ KJ

(b) $C(s) + O_2(g) \rightarrow CO_2(g); \Delta H = -393.5$ KJ

(c) $2WC(s) + 5O_2(g) \rightarrow 2WO_3(s) + 2CO_2(g);$ $\Delta H = -2391.6$ KJ

Required T.C.E. is $W(s) + C(s) \rightarrow WC(s); \Delta H = ?$

From $\frac{1}{2}(a)+(b)-\frac{1}{2}(c); \Delta H = -38.0$ KJ

**52.** From (i) + 3(ii) + 3(iii) – (iv); $\Delta H = -2035$ KJ

**53.** (a) $2Al(s)+6HCl(aq) \rightarrow Al_2Cl_6(aq)+3H_2(g);$

$\Delta H = -259.76$ kcal

(b) $H_2(g)+Cl_2(g) \rightarrow 2HCl(g); \Delta H = -44$ kcal

(c) $HCl(g)+aq \rightarrow HCl(aq); \Delta H = -17.315$ Kcal

(d) $Al_2Cl_6(s)+aq \rightarrow Al_2Cl_6(aq);$

$\Delta H = -153.69$ kCal

Required T.C.E. is

$2Al(s)+3Cl_2(g) \rightarrow Al_2Cl_6(s); \Delta H = ?$

From $(a)+3(b)+6(c)-(d);$

$\Delta H = (-239.76)+3(-44)+6(-17.315)-(-153.69)$

$= -321.96$ kcal

**54.** For more cooling, heating absorbed on dissolution should be high.

**55.** (a) $Na(s)+H_2O(l) \rightarrow NaOH(aq)+\frac{1}{2}H_2(g);$

$\Delta H = -184$ kJ

(b) $Na_2O(s)+H_2O(l) \rightarrow 2NaOH(aq);$

$\Delta H = -238$ kJ

(c) $H_2(g)+\frac{1}{2}O_2(g) \rightarrow H_2O(l);$

$\Delta H = -286$ kJ

Required T.C.E. is $2Na(s)+\frac{1}{2}O_2(g) \rightarrow Na_2O(s);$

$\Delta H = ?$

From $2(a)-(b)+(c);$

$\Delta H = 2(-184)-(-238)+(-286) = -416$ kJ

56. From $(a)-(b)+(c)$;
$\Delta H = (-544)-(-293)+(-480) = -731$ kJ

57. $\Delta(\text{S.F.}) = \sum(\text{S.F.})_{\text{Products}} - \sum(\text{S.F.})_{\text{Reactants}}$

58. From $(ii)-(i)$;
$\Delta G° = (-12.4)-(-29.2) = +16.8$ kJ

59. From (a) + (b) + (c) + (d) + (e)
Now

$$\Delta G° = (23.9)+(555.1)+(1320.2)+(-347.5)+(-1513.6)$$
$$= +38.1 \text{ kJ}$$

60. $$\Delta_r H° = \left[\Delta_f H_{co_2(g)} + 2\times\Delta_f H_{H_2O(l)}\right] - \left[\Delta_f H_{CH_4(g)} + 2\times\Delta_f H_{O_2(g)}\right]$$
$$= \left[(-393.5)+2(-286)\right]-\left[(-74.5)+2x0\right] = -891 \text{ kJ}$$

$$\Delta_r S° = \left[S_{co_2(g)} + 2\times S_{H_2o(l)}\right] - \left[\Delta_f S_{CH_4(g)} + 2\times S_{O_2(g)}\right]$$
$$= [216+2\times 70]-[186+2\times 205] = -240 \text{ J/K}$$

Now,
$$\Delta G° = \Delta H° - T\cdot\Delta S° = (-891)-300\times\left(\frac{-240}{1000}\right)$$
$$= -819 \text{ kJ}$$

61. $\Delta H = \sum(\text{B.E.})_{\text{Reactants}} - \sum(\text{B.E.})_{\text{Products}}$
$= (435+240)-(2\times 430) = -185$ KJ
$\Delta S = \sum S_{\text{Products}} - \sum S_{\text{Reactants}}$
$= (2\times 186)-(130+222) = 20$ J/K
Now, $\Delta G = \Delta H - T\cdot\Delta S = (-185)-300\times\frac{20}{1000}$
$= -191$ kJ

## Bond Enthalpy

62. $\frac{1}{2}N_2(g)+\frac{3}{2}F_2(g)\rightarrow NF_3(g); \Delta H = -113$ kJ
Now, $-113 = \left[\frac{1}{2}\times x + \frac{3}{2}\times\frac{x}{6}\right]-(3\times 273.5)$
$\therefore\ x = 943.33$ kJ

63. Only one bond cleavage should involve in this reaction.

64. $\Delta H = (\text{B.E.}_{\text{c-c } \pi\text{bond}} + \text{B.E.}_{\text{H-H}}) - (2\times\text{B.E.}_{\text{C-H}})$
$= \left[(611-346)+437\right]-(2\times 413) = -124$ kJ

65. $+182 = (430+242)-(2\times\text{B.E.}_{\text{H-Cl}})$
$\therefore\ \text{B.E.}_{\text{H-Cl}} = 245$ kJ/mol

66. $-20.24 = \left[2\times\Delta_{\text{sol}}H_{c(s)} + 3\times 102.6\right]-[63+6\times 85.6]$
$\therefore\ \Delta_{\text{sub}}H_{c(s)} = 124.28$ kcal/mol

67. $\Delta H = 121+242.8+2422-2\times(355)-2430.8$
$= -355$ kJ

68. $-411.2 = 107.3+121.7+495.4-348.5-(\text{L.E.})$
$\therefore$ L.E. = 787.1 kJ

69. Required $\Delta H = -\left[\frac{49.8+36.0}{3}\right] = -28.6$ kcal/mol

70. $|\text{R.E.}| = 29\times 3-51 = 36$ kcal/mol

71. $\Delta_{\text{Lattice}}H + \Delta_{\text{Hydration}}H = \Delta_{\text{solution}}H$

For CSI : $\Delta_{\text{Hyd}}H = 33-604 = -571$ kJ
For CsOH : $\Delta_{\text{Hyd}}H = (-72)-724 = -796$ kJ

72. From (a) + (b) + (c) – (d),
$CH_3-H(g)\rightarrow CH_3(g)+H(g); \Delta H = 54+29+79.8-51$
$= 111.8$ kcal/mol

73. $SF_6(g)\rightarrow S(g)+6F(g)$
$\Delta H = (275+6\times 80)-(-1100) = 6\times\text{B.E.}_{\text{S-F}}$
$\therefore\ \text{B.E.}_{\text{S-F}} = 309.17$ kJ/mol

**74.** $PH_3(g) \rightarrow P(g) + 3H(g); \Delta H = 954$ kJ

$\therefore$ $3 \times B.E._{P-H} = 954 \Rightarrow B.E._{P-H} = 318$ kJ/mol

$P_2H_4(g) \rightarrow 2P(g) + 4H(g); \Delta H = 1488$ kJ

$\therefore$ $B.E._{P-P} + 4 \times B.E._{P-H} = 1488$

$\Rightarrow B.E._{P-P} = 216$ kJ/mol

**75.**  (l) + $3H_2(g) \rightarrow$ 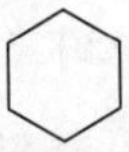 (l);

$\Delta H = (-156) - (49 + 0) = -205$ kJ

$\therefore$ Magnitude of resonance energy $= 3 \times 119 - 205 = 152$ kJ

**76.** $C(s) + O_2(g) \rightarrow CO_2(g); \Delta H = -393$ kJ

Now, $(-393) = [718 + 498] - 2 \times 539 - |R.E.|_{CO_2}$

$\therefore$ $|R.E.|_{CO_2} = 531$ kJ/mol

**77.** $\frac{1}{2}A_2(g) + \frac{1}{2}B_2(g) \rightarrow AB(g); \Delta H = -100$ kJ

Now, $(-100) = \left[\frac{1}{2} \times x + \frac{1}{2} \times \frac{x}{2}\right] - x \Rightarrow x = 400$

**78.** $\Delta_{Lattice}H + \Delta_{Hydration}H = \Delta_{Solution}H$

or, $180 + (6x + 5x) = 1 \Rightarrow x = -16.27$

$\therefore$ $\Delta_{Hyd}H_{Na^+} = 6x = -97.64$ kcal/mol

**79.** $C(\text{Graphite}) \rightarrow C(\text{Diamond})$

$\Delta H = 698.6 - 2x348.4 = 1.8$ kJ/mol

(In diamond, there is 2 C – C bonds per C-atom)

**80.** $\Delta_{Solution}H = 205 + [(-80) \times 2 + (-40)] = +5$ kJ/mol

As the solubility is endothermic, it should increase with increase in temperature.

---

## EXERCISE II (JEE ADVANCE)

### Section A (Only one Correct)

1. $N_2O_4 \rightleftharpoons \underset{\text{Brown}}{2NO_2}$

   On heating, colour deepens means reaction is endothermic ($\Delta H = +ve$).

2. As the process is endothermic but no heat is absorbed from surrounding, the temperature of the system will decrease, As initial temperature is used, the calculated mole will be lower.

3. $\frac{\Delta H_2 - \Delta H_1}{T_2 - T_1} = \Delta(C_P) = (3x) - (x + 2x) = 0$

   $\Rightarrow \Delta H_2 = \Delta H_1$

4. $I_2(s) \rightarrow I_2(g)$;

   $\Delta H_1 = 24$ cal/gm at $T_1 = 473$ K

   $\Delta H_2 = ?$ at $T_2 = 523$ K

   $\frac{\Delta H_2 - \Delta H_1}{T_2 - T_1} = C_P, I_2(g) - C_P, I_2(s)$

   $\Rightarrow \frac{\Delta H_2 - 24}{523 - 473} = 0.055 - 0.031$

   $\therefore$ $\Delta H_2 = 25.2$ cal/gm

5. For direct measurement, reaction must occur directly in the conditions to measure heat.

6. Greater the mass per cent of hydrogen, greater is the calorific value.

7. For the reaction:

   $\Delta U = \Delta H - \Delta n_g \cdot RT = (-72.3) - (-1) \times \frac{8.314}{1000} \times 298$

   $= -69.8$ kJ

   As HCl is limiting reagent, for the given amount,

   $\Delta U = 2 \times (-69.8) = -139.6$ kJ

**8.** $HAuBr_4 + 4HCl \rightarrow HAuCl_4 + 4HBr;$

$\Delta H = (-28) - (-36.8)$

$= 8.8$ kcal

$\therefore$ Percentage reaction $= \frac{0.44}{8.8} \times 100 = 5\%$

**9.** (a) $\underset{9.0}{C} + \underset{4.5}{\frac{1}{2}O_2} \rightarrow CO;$

$\Delta H = -75$ kcal

Heat evolved $= 75 \times 9 = 675$ kcal

(b) $\underset{2}{C} + \underset{2}{O_2} \rightarrow CO_2;$

$\Delta H = -95$ kcal

Heat evolved $= 95 \times 2 = 190$ kcal

(c) $4C + 3.5O_2 \rightarrow CO + 3CO_2$

Heat evolved $= 75 \times 1 + 95 \times 3 = 360$ kcal

(d) $\underset{2.5}{C} + \underset{2.5}{O_2} \rightarrow CO_2$

$\Delta H = -95$ kcal

Heat evolved $= 95 \times 2.5 = 237.5$ kcal

**10.**

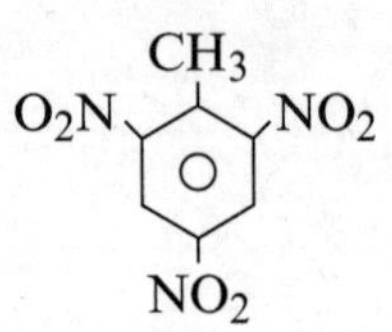

$(1) + \frac{21}{4} O_2(g) \rightarrow 7CO_2(g) + \frac{5}{2} H_2O(l) + \frac{3}{2} N_2(g)$

$\Delta H = 7 \times (-395) + \frac{5}{2} \times (-285) - (65)$

$= -3542.5$ kJ/mol

$= -\frac{3542.5}{227} \times 1.816$ kJ/mol

$= -28.34$ kJ/mol $= -28.34$ MJ/L

**11.** For $H_2A$, enthalpy of ionization

$= 2 \times (13.5 - 13) = 1$ kcal/mol

For $B(OH_2)$, enthalpy of ionization

$= 2 \times (13.5 - 10) = 7$ kcal/mol

$\therefore$ Required $\Delta H = 2 \times 13.5 - 1 - 7 = -19$ kcal

**12.** If $BaSO_4$ were water soluble, then

$\Delta H_{expected} = 2 \times (-57) = -114$ kJ

**13.** Required $\Delta H = -(13700 - 400 \times 0.9) = -13340$ cal

**14.** Aerobic oxidation results release of energy and hence, it is biologically benefical by (2880 + 2530 = 5410 kJ/mol)

**15.** (a) $Si_2H_6(g) + H_2(g) \rightarrow 2SiH_4(g);$

$\Delta H = -11.7$ kcal

(b) $SiH_4(g) \rightarrow SiH_2(g) + H_2(g);$

$\Delta H = +239.7$ kcal

(c) $2Si(s) + 3H_2(g) \rightarrow Si_2H_6(g);$

$\Delta H = +80.3$ kcal

Required thermochemical equation is

$Si(s) + H_2(g) \rightarrow SiH_2(g)$

From (b) $+ \frac{1}{2}$(a) $+ \frac{1}{2}$(c): $\Delta H = +274$ kcal/mol

**16.** Required thermochemical equation is

$Dy(s) + \frac{3}{2}Cl_2(g) \rightarrow DyCl_3(s)$

From (ii) + 3 × (iii) – (i), we get:

$\Delta H = (-699.43) + 3(-158.31) - (-180.06)$

$= -994.3$ kJ/mol

**17.**

$\Delta H = [(-188.84) + (-22.05) + 2(-22.1) + 2(-17.63)]$

$-[(-70.97) + (-8.56) + 0 + 2(-68.32)] = -74.18$ kcal

**18.** $H^+(aq) + OH^-(aq) \rightarrow H_2O(l)$

$\Delta H = \Delta_f H_{H_2O(l)} - \left[\Delta_f H_{H^+(aq)} + \Delta_f H_{OH^-(aq)}\right]$

or, $-57.32 = (-285.84) - \left[0 + \Delta_f H_{OH^-(aq)}\right]$

$\therefore \Delta_f H_{OH^-(aq)} = -228.52$ kJ/mol

**19.**

$CH_3CH_2COOH(l) + \frac{7}{2}O_2(g) \rightarrow 3CO_2(g) + 3H_2O(l)$

$\Delta_C H_{CH_3CH_2COOH(l)} = \left[3 \times \Delta_f H_{CO_2(g)} + 3 \times \Delta_f H_{H_2O(l)}\right]$

$-\left[\Delta_f H_{CH_3CH_2COOH(l)} + \frac{7}{2} \times \Delta_f H_{O_2(g)}\right]$

or, $3 \times \Delta_f H_{CH_3CH_2COOH(l)}$

$= [3(-94) + 3(-68)] - \left[\Delta_f H_{CH_3CH_2COOH(l)} + O\right]$

$\therefore \Delta_f H_{CH_3CH_2COOH(l)} = -121.5$ kcal/mol

**20.** phOH(solution II) → phOH(solution I)

$$\Delta H=\left(\frac{-0.02}{0.47}\times 94\right)-\left(\frac{-0.03}{1.410}\times 94\right)=-2\text{ kcal/mol}$$

**21.** $\Delta H_{\text{required}}=[(-14.7)-(-2.7)]-[-13.75]$
$=+1.75$ kcal/mol

**22.** $2FeO+\frac{1}{2}O_2\rightarrow Fe_2O_3;\Delta H=(-197)-2(-65)$

$=-67$ kcal

| | | |
|---|---|---|
| Initial | $2a$ | $a$ |
| Final | $2a-2x$ | $a+x$ |

$$\frac{2a-2x}{a+x}=\frac{1}{2}\ \Rightarrow\ x=\frac{3}{5}a$$

and heat released = $67x$ kcal

∴ Heat released per mole of initial mixture

$$=\frac{67x}{3a}=13.4\text{ kcal}$$

**23.** $3C_2H_5OH+25H_2O\rightarrow 3C_2H_5OH(25H_2O);$

$\Delta H_{\text{theo}}=(-1120)+2(-1760)$
$=-4640$ kcal

$\Delta H_{\text{exp}}=3(-1650)=-4950$ cal

As experimentally, more heat is released means the mixing is exothermic by (4950 – 4640) = 310 cal.

**24.** Heat absorbed in solubility = Heat released from solution

$=\text{m.s.}\Delta T$
$=(200+25)\times 4.2\times 3$
$=2835$ J

$$\therefore\ \Delta H=+\frac{2835}{7.45}\times 74.5=+28350\text{ J}$$

**25.** Heat released by reaction = Heat gained by ice
$=\text{m.L}=0.2\times 80=16$ cal

$$\Delta H=-\frac{16}{10^{-3}}=-16\times 10^3\text{ cal}$$

**26.** let $C_2H_6=x$L, then $CH_4=(4-x)$L

Volume of $CO_2$ produced, $2x+(4-x)=6\Rightarrow x=2$

∴ Total heat evolved

$$=\left(\frac{-1}{2}\times 1573+\frac{1}{2}\times 890\right)\times\frac{1}{0.0821\times 300}=50\text{ kJ}$$

**27.** $C+O_2\rightarrow CO_2;\Delta H=-\frac{2400}{0.3}\times 12=-96000$ cal

$C+\frac{1}{2}O_2\rightarrow CO;\Delta H=-\frac{1400}{0.6}\times 12=-28000$ cal

Now, $CO+\frac{1}{2}O_2\rightarrow CO_2;$
$\Delta H=-(-96000)-(-28000)=-68000$ cal

$$\therefore\ \text{Heat produced}=\frac{68000}{28}\times 0.7=1700\text{ cal}$$

**28.** Heat liberated from propane = Heat absorbed by water

or, $n\times 500\times 10^3\times\frac{40}{100}=160\times 10^3\times 1\times 50$

$\Rightarrow\ n=40$

**29.** $q=v.i.t=15\times 0.125\times(14\times 60)\text{ J}=1575\text{ J}$

$$\therefore\ \Delta H=\frac{1575}{0.1}\times 1=15750\text{ J/mol}$$

**30.** $H_2(g)+\frac{1}{2}O_2(g)\rightarrow H_2O(g);\Delta H_1=-240$ kJ

$2H(g)+\frac{1}{2}O_2(g)\rightarrow H_2O(g);$

$\Delta H_2=-(240+432)=-672$ kJ

$$\therefore\ \frac{\Delta H_2}{\Delta H_1}=\frac{-672}{-240}=2.8$$

**31.** $$\Delta m=\frac{E}{C^2}=\frac{103\times 10^3\times 4.2}{(3\times 10^8)^2}=4.8\times 10^{-12}\text{ kg}$$

**32.** $-12=[(-75)+(+3.9)]$
$-[(-11)+(-8.5)+x+(-17.5)]$

∴ $X=-22.1$ kcal/mol

**33.** Dulong and petit's law : Atomic mass × Specific heat ≅ 6.4 for greater temperature rise, heat lost should be high.

**34.** $\triangle(g)+\frac{9}{2}O_2(g)\rightarrow 3CO_2(g)+3H_2O(l)$

$$\Delta_C H_{cyclopropane}=\left[3\times\Delta_f H_{CO_2(g)}+3\times\Delta_f H_{H_2O(l)}\right]-\left[\Delta_f H_{cyclopropane}+\frac{9}{2}\times\Delta_f H_{O_2}(g)\right]$$
$$=\left[3\times(-394)+3\times(-286)\right]-\left[\{(33)+(20)\}+\frac{9}{2}\times 0\right]$$
$$=-\ 2093\text{ kJ/mol}$$

**35.** $w=-P.V_{H_2}=-nRT=-1.5\times 2\times 298=-894\text{ cal}$

**36.** $w=-nRT=-1\times 2\times 353=-706\text{ cal}$

$$\therefore\ \Delta U=q+w=7.4+\left(-\frac{706}{1000}\right)=6.694\text{ kcal/mol}$$

**37.** $C_3H_8(g)+5O_2(g)\rightarrow 3CO_2(g)+4H_2O(l)$

$$\Delta_C H_{C_3H_3(g)}=\left[3\times\Delta_f H_{CO_2(g)}+4\times\Delta_f H_{H_2O(l)}\right]-\left[\Delta_f H_{C_3H_3(g)}+5\times\Delta_f H_{O_2(g)}\right]$$
$$=\left[3(-393.5)+4(-285.8)\right]-(-103.8)=-2219.9\text{ kJ}$$

$$\Delta_r H_{required}=-\left[\Delta_C H_{C_2H_6(g)}+\Delta_C H_{CH_4(g)}\right]+\left[\Delta_C H_{C_3H_8(g)}+\Delta_C H_{H_2(g)}\right]$$
$$=-\left[(-1560.0)+(-890.0)\right]+\left[(-2219.9)+(-285.8)\right]$$
$$=-55.7\text{ kJ}$$

**38.** The required thermochemical equation is

$K(s)+\frac{1}{2}Cl_2(g)\rightarrow KCl(s);\ \Delta H=?$

From (iv) + (iii) – (v) + (i) – (ii): we get,

$$\Delta H=(-116.5)+(-39.3)-(4.4)+(-13.7)-(-68.4)$$
$$=-105.5\text{ Kcal}$$

**39.** From $\frac{1}{3}\times(i)+2\times(ii)+\frac{2}{3}\times(iii)$: we get,

$$\Delta H_{Required}=\frac{1}{3}(-46.4)+2(9.0)+\frac{2}{3}(-41)$$
$$=-24.8\text{ kJ}$$

**40.** For the reaction, $\frac{1}{2}H_2(g)+\frac{1}{2}I_2(s)\rightarrow HI(g)$;

$$\Delta H_{Required}=\frac{1}{2}(-44.20)-\frac{1}{2}(-52.42)+(-17.31)-(-19.21)+(-13.74)-(-13.67)$$
$$=5.94\text{ kcal}$$

**41.** The required thermochemical equation is

$I_2(s)+\frac{5}{2}O_2(g)\rightarrow I_2O_5(s)$

From 2 × (ii) + 6 × (v) + 5 × (vii) – (i) – 6 × (iii) – 6 × (iv) – (vi) – 10 × (viii) – 10 × (ix), we get,

$$\Delta H_{Required}=2(-322)+6(-100)+5(-255)-(4.0)-6(-44)-6(-57)-(-224)-10(-92)-10(-75)$$
$$=-169\text{ kJ}$$

**42.** Given thermochemical equations are

(i) $H_2(g)\rightarrow 2H(g);\Delta H=218\text{ kJ}$

(ii) $Cl_2(g)\rightarrow 2Cl(g);\Delta H=124\text{ kJ}$

(iii) $\frac{1}{2}N_2(g)+\frac{3}{2}H_2(g)\rightarrow NH_3(g);\Delta H=-46\text{ kJ}$

(iv) $\frac{1}{2}N_2(g)+2H_2(g)+\frac{1}{2}Cl_2(g)\rightarrow NH_4Cl(s);$ $\Delta H=-314\text{ kJ}$

(v) $H(g)\rightarrow H^+(g)+e^-;\Delta H=1310\text{ kJ}$

(vi) $Cl(g)+e^-\rightarrow Cl^-(g);\Delta H=-348\text{ kJ}$

(vii) $NH_4Cl(s)\rightarrow NH_4^+(g)+Cl^-(g);\Delta H=683\text{ kJ}$

Required thermochemical equations are

$NH_3(g)+H^+(g)\rightarrow NH_4^+(g);\Delta H=?$

From $(vii)+(iv)-(iii)-\frac{1}{2}(ii)-\frac{1}{2}(i)-(v)-(vi)$

$$\Delta H_{required}=(683)+(-314)-(-46)-\frac{1}{2}\times(124)-\frac{1}{2}\times(218)-(1310)-(-348)$$
$$=-718\text{ kJ/mol}$$

**43.** In such polymerization, one sigma bond is formed on cleavage of one pi bond.

$$\Delta H_{required} = (B.E._{C-C\pi bond}) - (B.E._{C-C\sigma bond})$$
$$= (590 - 331) - (331) = -72 \text{ kJ/mole}$$

**44.** Required thermochemical equation is

$$C(S) + 2H_2(g) + \frac{1}{2}O_2(g) \rightarrow CH_3OH(l)$$
$$\Delta H = [715 + 4\times 218 + 249] - [3\times 415 + 356 + 463] - 38$$
$$= -266 \text{ kJ}$$

**45.** $3C(s) + 3H_2(g) \rightarrow C_3H_6(g); \Delta H_{exp} = 53 \text{ kJ}$

$$\Delta H_{theo} = [3\times 715 + 6\times 218] - [3\times 356 + 6\times 408$$
$$= -63 \text{ kJ}$$
$$\therefore \text{ Strain energy} = \Delta H_{exp} - \Delta H_{theo} = 116 \text{ kJ}$$

**46.** $2C(g) + 6H(g) \rightarrow C_2H_6(g)$

$$\Delta H = -2839 = [0+0] - [B.E_{C-C} + 6\times 412]$$
$$\Rightarrow B.E_{C-C} = 367 \text{ kJ}$$

and, $2C(g) + 4H(g) \rightarrow C_2H_4(g)$

$$\Delta H = -2275 = [0+0] - [B.E_{C=C} + 4\times 412]$$
$$\Rightarrow B.E._{C=C} = 627 \text{ kJ}$$

Now, $6C(g) + 6H(g) \rightarrow C_6H_6(g)$

$$\Delta H = -5506$$
$$= [0+0] - [3\times 367 + 3\times 627 + 6\times 412] - \text{R.E.}$$
$$\therefore \text{ R.E.} = 52 \text{ kJ/mol}$$

**47.**

$$C_2H_5-S-C_2H_5(g) + S(g) \rightarrow C_2H_5-S-S-C_2H_5(g)$$
$$\Delta H = (-202) - [(-143) + 222] = -276 \text{ kJ}$$
$$B.E._{S-S} = 276 \text{ kJ/mol}$$

**48.** $CH_4(g) \rightarrow CH_3(g) + H(g)$

$$103 = \left[\Delta_f H_{CH_3(g)} + \frac{103}{2}\right] - [-18]$$
$$\Rightarrow \Delta_f H_{CH_3(g)} = 33.5 \text{ kcal/mol}$$

**49.** $\Delta H_{required} = [6\times 414 + 2\times 348 + 580 + 2\times 610]$

$$-\begin{bmatrix} 3\times 414 + 348 + 580 + 354 + 462 \\ +118 + 2\times 580 + 140 + 2\times 462 \end{bmatrix}$$
$$= -348 \text{ KJ}$$

**50.** $\Delta H = 50 - 70 = -20 \text{ kJ}$

**51.** 
$$\Delta H = (B.E._{\pi \text{ bond in } C=C}) - (B.E._{\sigma \text{ bond in } C-C})$$
$$= (835 - 610) - (348) = -123 \text{ kJ}$$

**52.** $n\,HCHO(g) \rightarrow (HCHO)_2$

$$\Delta H = -72 = n\times(-134) - (-732) \Rightarrow n = 6$$
$$\therefore \text{ Molecular formula} = (HCHO)_6 = C_6H_{12}O_6$$

**53.** There is 2.5 B-B bond per B atom. Hence, $\Delta H = -2.5\times 300 = -750$ kJ/mole of Boron.

**54.** Let the enthalpy of combustion of gauche form be $-x$ kcal/mol.

Now,

$$690 = 0.7\times(x-2) + 0.2\times x + 0.06$$
$$\times(x+3) + 0.04\times(x+5.5)$$
$$\therefore x = 691$$

**55.** $M(s) + X_2(g) \rightarrow MX_2(s); \Delta H = 1.5\times \Delta_{eg}H_{x(g)}$

From Born–Hafer Cycle, we get:

$$\Delta H = \Delta_{sub}H_{M(s)} + \Delta_{i_1}H_{M(g)} + \Delta_{i_2}H_{M(g)} + \Delta_{Bond}H_{X_2(g)}$$
$$+2\times \Delta_{eg}H_{X(g)} + \Delta_{latlice}H_{MX_2(s)}$$

or, $1.5\times(-96) = \Delta_{sub}H_{M(s)} + 1.2\times \Delta_{sub}H_{M(s)}$

$$+2.8\times \Delta_{sub}H_{M(s)} + 0.8\times(1.2\times \Delta_{sub}H_{M(s)})$$
$$+2(-96) + 5\times[-0.8\times 1.2\times \Delta_{sub}H_{M(s)}]$$
$$\Delta_{sub}H_{M(s)} = 41.38 \text{ kcal/mol}$$

## Section B (One or More than one Correct)

1. Combustion is exothermic. Decomposition or elimination are endothermic. Graphite is more stable form.
2. Conversion of liquid into gas is endothermic.
3. $\Delta_f H^\circ = 0$ for elements in their reference state.
4. Endothermic compounds have +ve $\Delta_f H^\circ$.
5. For $\Delta H = \Delta E, \Delta n_g = 0$
6. One mole of the substance should burn completely.
7. $\Delta_f H_{NO(g)} = +ve$
8. Heat released in reaction = Heat gained by calorimeter system
$= 1.5 \times 1.4 = 2.1$ kJ

$$n_{eq}(H_2SO_4) = \frac{100 \times 0.5}{1000} = 0.05$$

$$n_{eq}(NH_4OH) = \frac{200 \times 0.2}{1000} = 0.04 \quad \text{(Limiting reagent)}$$

$$\Delta_{neut}H_{NH_4OH}\text{(By strong acid)} = -\frac{2.1}{0.04} = -52.5 \text{ kJ/eq}$$
$$= -52.5 \text{ kJ/mole}$$

$$\Delta_{diss}H_{NH_4OH} = (-52.5) - (-57) = 4.5 \text{ kJ/mol}$$

$$\Delta_{diss}H_{CH_3COOH} = (57 - 48.1) - 4.5 = 4.4 \text{ kJ/mol}$$

9. (a) $\Delta_r H = \left(436 + \frac{1}{2} \times 495\right) - (-242) = 925.5 \text{ kJ}$

(b) $\Delta_r H = \left(\frac{1}{2} \times 436 + \frac{1}{2} \times 495\right) - (-42) = 423.5 \text{ kJ}$

(c) $\Delta_f H_{H(g)} = \frac{1}{2} \times 436 = 218 \text{ kJ / mol}$

(d) $\Delta_f H_{OH(g)} = 42 \text{ kJ/mol}$

10. Resonance occurs in 1, 3-Butadiene and $N_2O$.
11. Resonance occurs in product but not in reactant.
12. (a)
$\Delta_r H = [2 \times (-1263)] - [(-2238) + (-285)] = -3 \text{ kJ}$

(b) $-3 \text{ KJ} = \Delta_C H_{\alpha\text{-maltose}} - 2 \times \Delta_C H_{glucose}$

13. (a) $n_{C(s)} = \frac{1.2 \times 1000}{12} = 100$

$\therefore$ Maximum obtainable heat $= 100 \times 94 = 9400$ cal

(b) Heat released $= 100 \times 68 + 100 \times 68 = 13600$ cal

(c) Heat released $= \frac{13600}{100} \times \left(\frac{1200}{30}\right) = 5440$ cal

14. $C_6H_5COOH(s) + \frac{15}{2}O_2(g) \rightarrow 7CO_2(g) + 3H_2O(l)$

$$\Delta H = [7(-393) + 3(-286)] - [-408]$$
$$= -3201 \text{ kJ/mol}$$

and $\Delta U = \Delta H - \Delta n_g.RT$

$$= (-3201) - \left(7 - \frac{15}{2}\right) \times \frac{8.314}{1000} \times 300$$
$$= -3199.75 \text{ kJ/mol}$$

15. $\Delta H = q = 3 \times (-35) = -105 \text{ kcal}$

$$\Delta U = \Delta H - \Delta n_g.RT$$
$$= \left[(-35) - (2-3) \times \frac{2}{1000} \times 300\right] \times 3$$
$$= -103.2 \text{ kcal}$$

and $w = u - q = (-103.2) - (105) = 1.8$ kcal

## Section C (Comprehensions)

### Comprehension I

1. From $-\frac{3}{4} \times a - \frac{1}{4} \times b + \frac{1}{4} \times c - \frac{9}{4} \times d,$

$$\Delta H_{required} = -\frac{3}{4} \times (-76) - \frac{1}{4} \times (240) + \frac{1}{4} \times (36) - \frac{9}{4}(68)$$
$$= -147 \text{ kcal}$$

2. From $\frac{3}{4} \times a + \frac{1}{4} \times b - \frac{1}{4} \times c + \frac{1}{4} \times d,$

$$\Delta H_{required} = \frac{3}{4} \times (-76) + \frac{1}{4} \times (240) - \frac{1}{4} \times (36) + \frac{1}{4}(68)$$
$$= 11 \text{ kcal/mol}$$

3. Given data based

## Comprehension II

4. $\Delta H = -\frac{75\times5}{5+1\cdot8} = -55\cdot15$ kJ

5. $\Delta H = -\frac{75\times10}{10+1\cdot8} = -63\cdot56$ kJ

6. $\Delta H = -\frac{75}{1+\frac{1\cdot8}{n}} = -\frac{75}{1+\frac{1\cdot8}{\infty}} = -75$ kJ

7. $\Delta H = (-63.56)-(-55.15) = -8.41$ kJ

8. $\Delta H = (-75)-(-63.56) = -11.44$ kJ

## Comprehension III

9. $\Delta H_1 = -\frac{483\cdot636}{2} = -241\cdot818$ kJ/mol $H_2$

$\Delta H_2 = -\frac{868\cdot2}{3} = -289\cdot4$ kJ/mol $H_2$

$\Delta H_3 = -347.33$ kJ/mol $H_2$

10. $\Delta H_1 = -\frac{483\cdot636}{32} = -15\cdot11$ kJ/gm $O_2$

$\Delta H_2 = -\frac{868\cdot2}{48} = -18\cdot09$ kJ/gm $O_3$

$\Delta H_3 = -\frac{347\cdot33}{34} = -10\cdot22$ kJ/gm $H_2O_2$

11. $\Delta H_1 = -\frac{483\cdot636}{36} = -13\cdot43$ kJ/gm reactant

$\Delta H_2 = -\frac{868\cdot2}{54} = -16\cdot08$ kJ/gm reactant

$\Delta H_3 = -\frac{347\cdot33}{36} = -9\cdot65$ kJ/gm reactant

## Comprehension IV

12. Heat released = 0.25 × 320 = 80 cal

∴ Molar enthalpy of solution

$= -\frac{80}{0\cdot98}\times98 = -8000\,\text{cal}$

13. Heat released = 0.80 × 320 = 256 cal

$\therefore\ \Delta_r H = -\frac{256}{0\cdot49}\times98 = -51200\,\text{cal}$

## Comprehension V

16. $2C(S)+\frac{3}{2}H_2(g)+\frac{1}{2}N_2(g)\rightarrow CH_3CN(g)$;

$\Delta H = 88 = [2\times719+\frac{3}{2}\times435+\frac{1}{2}\times948]$

$-[3\times414+B.E._{C-C}+B.E._{C\equiv N}]\rightarrow 1$

$3C(s)+4H_2(g)\rightarrow C_3H_8(g)$

$\Delta H = -85 = [3\times719+4\times435]$

$-[2\times B.E._{C-C}+8\times414]\rightarrow 2$

From (1) and (2), we get:

$B.E._{C-C} = 335$ kJ/mol and

$B.E._{C\equiv N} = 899.5$ kJ/mol

Now, $CH_3CN(g)+2H_2(g)\rightarrow CH_3CH_2NH_2(g)$,

$\Delta H = [3\times414+335+899.5+2\times435]$

$-[5\times414+335+378+2\times426]$

$= -288.5$ kJ/mol

## Comprehension VI

**17.** $C_{p,m,N_2(g)} < C_{p,m,H_2O(g)}$

**18.** $H_2(g)+\frac{1}{2}O_2(g) \rightarrow H_2O(g);$

$\Delta H = -55{\cdot}85$ kcal

$\Delta U = -56{\cdot}0$ kcal

Let $x$ mole $H_2$ be burnt. $\frac{x}{2}$ mol $O_2(g)$ is needed and hence, $\frac{x}{2}\times 4 = 2\times$ mol $N_2$ is also present.

Now,
Heat released from reaction = Heat gained by $H_2O(g)$ and $N_2(g)$

$56.0 \times 10^3 = x \times 6.2 \times (T_2 - 300) + 2x \times 4.9 \times (T_2 - 300)$

$\therefore T_2 = 3800$ K

**19.** $\frac{p_1}{n_1T_1} = \frac{p_2}{n_2T_2} \Rightarrow \frac{1}{x+\left(\frac{x}{2}+2x\right)\times 300}$

$= \frac{p_2}{(x+2x)\times 3800}$

$\therefore p_2 = 10.86$ atm

**20.** $q = 0,\ w = 0 \Rightarrow \Delta E = 0$

## Comprehension VII

**21.** $C_8H_{18}(l)+\frac{25}{2}O_2(g) \rightarrow 8CO_2(g)+9H_2O(g)$

$\Delta_c H = [8 \times (-94) + 9\times(-58)] - [-74]$

$= -1200$ kcal/mol

**22.** $\Delta H_{required} = [8 \times (-26.5) + 9\times(-58)] - [-74]$

$= -660$ kcal/mol

**23.** Let $x$ mole of $C_8H_{18}$ be converted into $CO_2$. As temperature is increased, some heat is absorbed by product gases.

Now, $[x\times 1200+(0{\cdot}1-x)\times 660]-\frac{1}{1000}$

$\times[8x\times 8\times 500+8(0{\cdot}1-x)\times 7{\cdot}0$

$\times 500+0{\cdot}9\times 6{\cdot}0\times 500] = 87{\cdot}3$

$\therefore x = 0.05$

Moles of $CO_2$ formed $= 0.05 \times 8 = 0.4$

**24.** Moles of $H_2O$ formed $= 9x + (0.1 - x) \times 9 = 0.9$

**25.** $w = -p(v_2 - v_1) = -p\left(\frac{n_2RT_2}{p} - \frac{n_1RT_1}{p}\right)$

$= -R(n_2T_2 - n_1T_1)$

$= -2\times[\{0{\cdot}05\times 8+(0{\cdot}1-0{\cdot}05)\times 8+0{\cdot}9\}\times 800$

$-\left\{0{\cdot}05\times\frac{25}{2}+(0{\cdot}1-0{\cdot}05)\times\frac{17}{2}\right\}\times 300]$

$= -2090$ cal

## Comprehension VIII

**26.** $CO+2H_2 \xrightarrow{\frac{2}{3}\text{rd}} \underset{1000\,\text{mol}}{CH_3OH}$

$= \frac{3}{2}\times 1000$

$= 1500$ mol

In reformer, CO and $H_2$ is forming in 1 : 3 ratio.

**27.** CO = 1500 − 1000 = 500 mole

$H_2$ = 4500 − 2000 = 2500 mole

**28.** Heat produced in 1 min = 1000 × 100 R × 60 $= 1.2 \times 10^7$ cal

## Comprehension IX

**29.** Let $x$ mole C be converted into CO. Hence,

$x \times 26 + (1 - x) \times 94 = 53.2 \Rightarrow x = 0.6$

Hence, moles of C formed = 0.6

**30.** $O_2$ consumed $= \left[\frac{x}{2} + (1-x)\right]32 = 22 \cdot 4\,\text{gm}$

## Comprehension X

Given thermochemical equations are

(a) $H_2S\,(g) \rightarrow H\,(g) + H\,S\,(g);\ \Delta H = 376.0$ kcal

(b) $H_2\,(g) + S(s) \rightarrow H_2\,S\,(g);\ \Delta H = -20.0$ kcal

(c) $S\,(s) \rightarrow S\,(g);\ \Delta H = 277.0$ kcal

(d) $H_2\,(g) \rightarrow 2H\,(g);\ \Delta H = 436.0$ kcal

**31.** $\frac{1}{2}H_2(g) + S(s) \rightarrow HS(g)$

From $a + b - \frac{1}{2} \times d$, we get:

$\Delta H_{required} = 376 + (-20) - \frac{1}{2} \times 436 = 138$ kJ / mol

**32.** $HS(g) \rightarrow H(g) + S(g)$

From (d) − (a) − (b) + (c)

$\Delta H_{required} = 436 - 376 - (-20) + 277 = 357$ kJ/mol

## Comprehension XI

**33.** $\Delta H° = \Delta E° + \Delta n_g \cdot RT = 2.1 + 2 \times \frac{2}{1000} \times 298$

$= 3.292$ kcal

Now, $\Delta G° = \Delta H° - T \cdot \Delta S° = 3.292 - \frac{298}{1000} \times 20$

$= -2.668$ kcal

**34.** Spontaneous as $\Delta G° = -$ve

## Section D (Assertion – Reason)

1. Theory based
2. Theory based
3. $H_2SO_4$ is dibasic but HCl is monobasic.
4. Information based
5. Heat liberated will be four times but as quantity is also four times, the change in the temperature will be same.
6. Solubility is exothermic but all gases are not highly soluble in all liquid.
7. If $|\Delta_{Hydration}\,H| < |\Delta_{lattice}\,H|$, the salt dissolves partially and the extent depends on the difference in two values.
8.

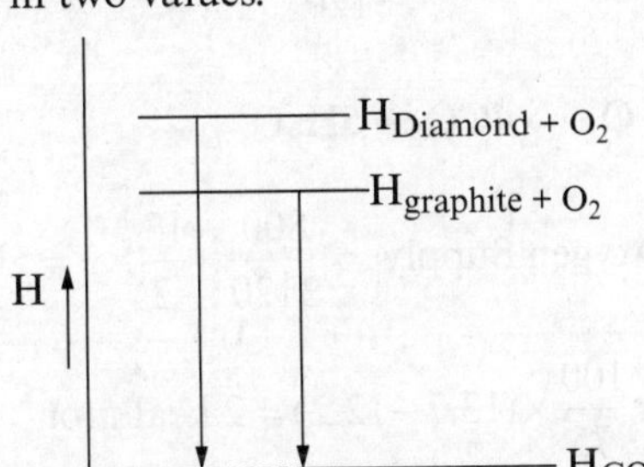


9.

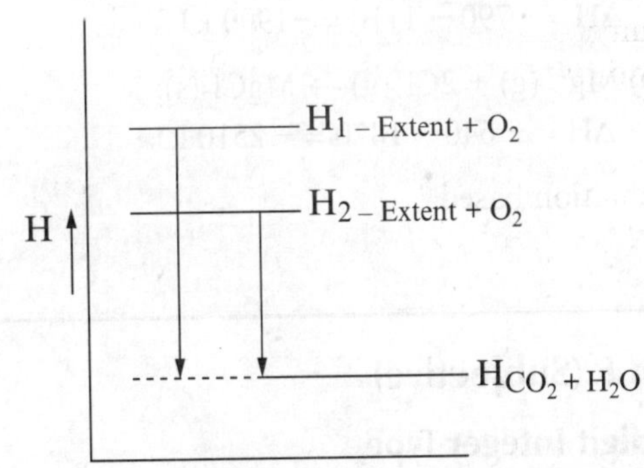


10. Theory based

## Section E (Column Match)

1. $C \rightarrow CO, \Delta H \neq \Delta H_{combustion}$
2. (A) $\Delta n_g = 0$ (B) $\Delta n_g = -1$ (C) $\Delta n_g = 1$ (D) $\Delta n_g =-2$
3. (A) $\Delta H = (-57.3) + 15 = -42.3$ kJ
   (B) $\Delta H = -42.3 - 70.7 + 20 = -93.0$ kJ
   (C) $\Delta H = -70.7 + 15 = -55.7$ kJ
   (D) $\Delta H = 0$
4. (A)

$$\begin{array}{ccc} Mg(s) & \longrightarrow & Mg^{2f}(aq); \\ \downarrow +Cl_2(g),\ +1110 & -(1110+790) & \uparrow +Cl_2(g),\ (2360-1110) \\ Mg^{2+}(g) + 2Cl^-(aq) & \longrightarrow & Mg^{2+}(aq) + 2Cl^-(aq) \end{array}$$

or = 1110 – (1110 + 790) + (2360 – 1110)
= 460 kJ/mol

(B)
$\frac{1}{2}Cl_2(g) \rightarrow \quad Cl^-(ag); \Delta H \quad = \frac{1}{2}(1110-2360)$
$+\frac{1}{2}Mg^{2+}(g) \quad +\frac{1}{2}Mg^{2+}(g) \quad = -652$ KJ/mol

(C) $Mg^{2+}(g) + 2Cl^-(aq) \rightarrow Mg^{2+}(aq) + 2Cl^-(aq)$;
$\Delta H = -790 - 1110 = -1900$ kJ

(D) $Mg^{2+}(g) + 2Cl^-(g) \rightarrow MgCl_2(s)$;
$\Delta H = -640 - 1870 = -2510$ kJ

5. Definition based
6. $3O_2(g) \rightarrow 2O_2(g)$
7. Theory based
8. $\Delta n_g = 0 \Rightarrow \Delta H = \Delta U$
   $\Delta n_g = +ve \Rightarrow \Delta H > \Delta U$
   $\Rightarrow$ If $\Delta H = -ve$, then $|\Delta H| < |\Delta U|$
   $\Rightarrow$ If $\Delta H = +ve$, then $|\Delta H| > |\Delta U|$
   $\Delta n_g = -ve \Rightarrow \Delta H < \Delta U$
   $\Rightarrow$ If $\Delta H = -ve$, then $|\Delta H| > |\Delta U|$
   $\Rightarrow$ If $\Delta H = +ve$, then $|\Delta H| < |\Delta U|$
9. Definition based

**(10)** (A) $\Delta H_{400} = \Delta H_{300} + [c_{p,A(g)} - c_{p,A(l)}] \times [T_2 - T_1]$
$= 25 + (20-40) \times \frac{1}{1000} \times (400-300)$
$= +23$ kJ/mol

(B) $\Delta H_{300} = \Delta H_{400} + [c_{p,A_3(g)} - c_{p,A_3(l)}] \times [T_2 - T_1]$
$= 50 + (30-50) \times \frac{1}{1000} \times (300-400)$
$= +52$ kJ/mol

(C) $\Delta H_{300} = 3 \times 25 - 100 - 52 = -77$ kJ/mol

(D)
$\Delta H_{400} = \Delta H_{300} + [c_{p,A_3(l)} - 3 \times c_{p,A_3(l)}] \times [T_2 - T_1]$
$= (-77) + (50 - 3 \times 40) \times \frac{1}{1000} \times (400-300)$
$= -84$ kcal/mol

---

## Section F (Subjective)

### Single-digit Integer Type

1. For HCl : $13.7 \times 0.05 = c \times 411$ (1)
   For HCOOH : $q \times 0.05 = c \times 321$ (2)
   From (2) ÷ (1) $\Rightarrow q = 10.7$ kcal
   $\therefore$ Enthalpy of ionisation of HCOOH = 13.7 − 10.7 = 3.0 kcal/mol
2. $(C_6H_{10}O_5)_x + 6x\,O_2(g) \rightarrow 6x\,CO_2(g) + 5x\,H_2O(l)$
   $\Delta H = -4.6 \times 162x = [6x(-94.2) + 5x(-68.4)] - [\Delta_f H_{(C_6H_{10}O_5)_x} + 0]$
   $\therefore \Delta_f H_{(C_6H_{10}O_5)_x} = -162$ kcal/mol $= -1$ kcal/gm
3. $1 \times 800 = a \times 3120 \Rightarrow a = \frac{800}{3120}$ L/hr Butane
   $C_4H_{10} + \frac{13}{2}O_2 \rightarrow 4CO_2 + 5H_2O$
   $\therefore$ Rate of Oxygen Supply $= \frac{800}{3120} \times \frac{13}{2} \times 3 = 5$ L/hr
4. $\Delta H_{required} = \frac{100}{75} \times (13.7 - 12.2) = 2$ kcal/mol

5. Total moles of gases $= \dfrac{1.192 \times 1.642}{0.0821 \times 298} = 0.08$

Now,

$n_{CH_4} \times 210 \times 10^3 = 1260 \times 0.667 \Rightarrow n_{CH_4} = 0.004$

$\therefore$ Volume per cent of $CH_4 = \dfrac{0\cdot004}{0\cdot08} \times 100 = 5\%$

6. Heat released by $\dfrac{6\cdot3}{64000}$ mole haemoglobin

$= 25 \times 4.2 \times 0.03 = 3.15\,\text{J}$

$\therefore$ Heat released per mole haemoglobin

$= \dfrac{3\cdot15 \times 64000}{6\cdot3} = 32000\,\text{J}$

$\therefore$ Heat released per mole $O_2 = \dfrac{32000}{4} = 8000\,\text{J}$

7. Heat released $= 300 \times 1.0 \times 1.0 \times (26 - 25) = 300\,\text{cal}$

Now,

$n_{HA} = \dfrac{200 \times 0\cdot4}{1000} = 0\cdot08$

$n_{NaOH} = \dfrac{100 \times 0\cdot5}{1000} = 0\cdot05$

Hence, NaOH is a limiting reagent.

$\therefore \Delta_{neut}H = -\dfrac{300}{0\cdot05} \times 1 = -6000$ cal/mol

8. There is 3 H-bond per $NH_3$ molecule because for each bond two $NH_3$ molecules are required.

$\therefore$ Strength of H-bond

$= \dfrac{30.4 - 15.4}{3} = 5.0$ kcal/mol

9. $C(s) + \frac{1}{2}O_2(g) \rightarrow CO(g);$

$\Delta H_1 = -\dfrac{7\cdot5}{3} \times 12 = -30$ kcal/mol

$C(s) + O_2(g) \rightarrow CO_2(g);$

$\Delta H_2 = -\dfrac{32}{4} \times 12 = -96$ kcal/mol

Now, $CO_2(g) \rightarrow CO(g) + \frac{1}{2}O_2(g);$

$\Delta H = \Delta H_1 - \Delta H_2 = +66$ kcal/mol

For $4\cdot0$ gm $CO_2$, $\Delta H = \dfrac{66}{44} \times 4 = 6$ kcal

10. $(1 - a) \times 6900 + a \times 2900 = 3900 \Rightarrow a = \dfrac{3}{4}$

$\therefore n_{eq(HA)} : n_{eq(HB)} = (1 - a) : a = \dfrac{1}{4} : \dfrac{3}{4} = 1:3$

## Four-digit Integer Type

1. $C_2H_6 + H_2 \rightarrow 2\,CH_4$ ; $\Delta H = -65.2$ kJ

$C_3H_8 + 2H_2 \rightarrow 3\,CH_4$ ; $\Delta H = -87.4$ kJ

Hence, for $CH_4\,(g) + C_3H_8\,(g) \rightarrow 2\,C_2H_6\,(g)$;

$\Delta H = (-87.4) - 2 \times (-65.2) = +\,43$ kJ

2. Moles of $O_2$ consumed

$= \dfrac{\left\{\dfrac{164.2}{1000} \times \left(\dfrac{20-10}{100}\right) \times 20 \times 60\right\} \times 1}{0\cdot0821 \times 310} = \dfrac{24}{31}$

$C_6H_{12}O_6 + 6O_2 \rightarrow 6CO_2 + 6HO_2;$

$\Delta H = -3100$ kJ

$\therefore$ Heat produced in body per hr

$= \dfrac{3100}{6} \times \dfrac{24}{31} = 400\,\text{kJ}$

3. Number of glycogen units oxidized per day

$= \dfrac{150 \times 60 \times 60 \times 24}{432 \times 10^3} = 30$

4. Moles of $C = \dfrac{15}{12} = 1.25$

Moles of $O_2 = \dfrac{\left(20 \times \dfrac{19}{100}\right) \times 8.21}{0\cdot0821 \times 380} = 1$

$1.25\ C + O_2 \rightarrow 0.5CO + 0.75\ CO_2$

$\therefore$ Heat produced $= 0.5 \times 26 + 0.75 \times 96 = 85$ kcal

**5.** $C_2H_5OH(l) + O_2(g) \rightarrow CH_3COOH(g) + H_2O(l)$

$\Delta H = [(-118)+(-68)] - [(-66)+0] = -120$ kcal

Hence, rate of heat removal

$= \frac{120}{46} \times 2.3\times10^3 \times \frac{40}{100}$

$= 2400$ kcal/mol

**6.** $C_6H_{12}O_6(s) + 6\,O_2(g) \rightarrow 6\,CO_2(g) + 6\,H_2O(l)$

$\Delta H = [6\times(-395) + 6\times(-285)] - [(-1280) + 0] = -2800$ kJ

Moles of $CO_2$ released per astronaut

$= \frac{6}{2800}\times 2100 = 4.5$

$\therefore$ Mass of LiOH required $= 4.5\times2\times24 = 216$ gm

**7.** $\frac{16.1}{322}\times100\times10^3 = V\times10000 \Rightarrow V = 0.5\text{m}^3 = 500\,\text{L}$

**8.** Total heat absorbed

$= \left(\frac{8}{9}\times45+\frac{1}{9}\times72\right)\times2.5 = 120\,\text{KJ}$

**9.** For banana: $q = c\times3.0$ (1)

For benzoic acid: $\frac{800}{122}\times0.305 = c\times4.0$ (2)

From (1) and (2), $q = 1.5$ kacl for 2.5 gm banana

$\therefore$ Heat obtained per banana $= \frac{1\cdot5}{2\cdot5}\times125 = 75$ kcal

**10.** $H_2(g)+\frac{1}{2}O_2(g)\rightarrow H_2O(l); \Delta H_{298} = -286$ kJ

$H_2O(l)\rightarrow H_2O(g);\ \Delta H_{398} = 40\cdot8$ kJ

$\Delta H_{298} = \Delta H_{398} + \Delta C_p\cdot\Delta T$

$= 40.8+\frac{33.4-75.4}{1000}\times(298-398)$

$= 45$ kJ

$\therefore\ H_2O(g)\rightarrow H_2(g)+\frac{1}{2}O_2(g); \Delta H_{298} = -[45-286]$

$= 241$ kJ

**11.** $\Delta H_2 - \Delta H_1 = \int_{T_1}^{T_2} \Delta C_p\cdot dT$

or, $0-(-4000) = \int_{300}^{T}(2\times10^{-2}T)dT$

$= \frac{2\times10^{-2}}{2}(T^2-300^2)$

$\therefore$ T = 700 K

**12.** $3C(s) + 3H_2(g) \rightarrow C_3H_6(g)$

$\Delta H_{theo} = (3\times715 + 6\times218) - (3\times356 + 6\times408) = -63$ kJ

$\Delta H_{exp} = [3\times(-393) + 3\times(-285)] - [3\times(-697)] = 57$ kJ

$\therefore$ Strain energy $= 57 - (-63) = 120$ kJ/mol

**13.** $KF.CH_3COOH(s)\rightarrow K^+(g)+F.^-CH_3COOH(g);$ $\Delta H = 734$ kJ

$CH_3COOH(l)\rightarrow CH_3COOH(g); \Delta H = 20$ kJ

$KF(s)\rightarrow K^+(aq)+F^-(aq); \Delta H = 35$ kJ

$K^+(g)\rightarrow K^+(aq); \Delta H = -325$ kJ

$F^-(g)\rightarrow F^-(aq); \Delta H = -389$ kJ

$KF(s)+CH_3COOH(l)\rightarrow KF_,CH_3COOH(s);$ $\Delta H = -25$ kJ

Required:

$F^-(g)+CH_3COOH(g)\rightarrow F^-CH_3COOH(g);$

$\Delta H = (-389)+734-20-35+(-325)+(-25)$

$= -60$ kJ/mol

**14.** $B(s)+\frac{3}{2}H_2(g)\rightarrow BH_3(g)$

$\Delta H = 100 = \left(565+\frac{3}{2}\times436\right)-(3\times B.E._{B-H})$

$\therefore$ $B.E._{B-H} = 373$ kJ/mol

$2B(s) + 3H_2(g)\rightarrow B_2H_6(g)$

$\Delta H = 36 = [2\times565+3\times436]-[4\times373+2\times B.E._{3c-2e}]$

$\therefore$ $B.E._{3c-2e} = 455$ kJ/mol

**15.** $XeF_4 \rightarrow Xe^+ + F^- + F_2 + F$

$\Delta H = (4\times34)+279+(-85)+(-38) = 292$ kcal

**16.** (a) $KF.CH_3COOH(s) \rightarrow K^+(ACOH) + F.^-(ACOH) + CH_3COOH(l) = 3$ kJ

(b) $KF(s) \rightarrow K^+(ACOH) + F^-(ACOH)$; $\Delta H = 35$ kJ

(c) $F.^-CH_3COOH(g) \rightarrow F^-(g) + CH_3COOH(g)$; $\Delta H = 46$ kJ

(d) $KF.CH_3COOH(s) \rightarrow K^+(g) + F.^-CH_3COOH(g)$; $\Delta H = 734$ kJ

(e) $KF(s) \rightarrow K^+(g) + F^-(g)$; $\Delta H = 797$ kJ

Required: $CH_3COOH(l) \rightarrow CH_3COOH(g)$

From (c) − (a) + (d) − (e) + (b), we get:

$\Delta H = 46 - (-3) + 734 - 797 + 35 = 21$ kJ/mol

**17.** 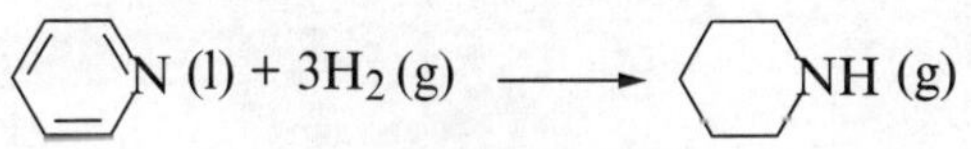


$\Delta H = (-50) - [\Delta_f H_{py(l)} + 0] = (40 + 125) + [2\times\{(-156) - (-37)\} + \{(-18) - 44\}]$

$\therefore \Delta_f H_{py(l)} = 85$ kJ/mol

**18.** $\frac{1}{2}I_2(s) + \frac{5}{2}F_2(g) \rightarrow IF_5(g)$; $\Delta H = -847$ kJ

$-847 = \frac{1}{2}\times(62+149) + \frac{5}{2}\times155 - 5\times B.E._{I-F}$

$B.E._{I-F} = 268$ kJ/mol

$\frac{1}{2}I_2(s) + \frac{3}{2}F_2(g) \rightarrow IF_3(g)$; $\Delta H = -470$ kJ

$-470 = \frac{1}{2}(62+149) + \frac{3}{2}\times155 - \left[2\times268 + B.E._{I-F(eq)}\right]$

$B.E._{I-F(eq)} = 272$ kJ/mol

**19.** $H_2(g) + \frac{1}{2}O_2(g) \rightarrow H_2O(l)$; $\Delta H = -286$ kJ

$-286 = \left(B.E._{H-H} + \frac{1}{2}\times498\right) - 2\times B.E._{O-H} - 44$ (1)

$H_2(g) + O_2(g) \rightarrow H_2O_2(l)$; $\Delta H = -188$ kJ

$-188 = (B.E._{H-H} + 498) - (2\times B.E._{O-H} + B.E._{O-O}) - 53$ (2)

From (1) − (2), we get: $B.E._{O-O} = 142$ kJ/mol

**20.** (a) $Ag^+(aq) + Br^-(aq) \rightarrow AgBr(s)$; $\Delta H = -84.54$ kJ

(b) $Ag(s) \rightarrow Ag^+(aq)$; $\Delta H = -8x$ kJ

(c) $\frac{1}{2}Br_2(l) \rightarrow Br^-(aq)$; $\Delta H = 9x$ kJ

(d) $Ag(s) + \frac{1}{2}Br_2(l) \rightarrow AgBr(s)$; $\Delta H = -99.54$ kJ

As (a) + (b) + (c) = (d), we get:

$(-84.54) + (-8x) + 9x = -99.54$

$\Rightarrow x = -15$

$\therefore \Delta_f H_{Ag^+(aq)} = -8x = 120$ kJ/mol

CHAPTER 6

# Chemical Equilibrium

## EXERCISE I (JEE MAIN)

### Basic

1. The reaction which proceeds towards completion in the forward direction is
   (a) $Fe_2O_3 + 6HCl \rightleftharpoons 2FeCl_3 + 3H_2O$
   (b) $SnCl_4 + Hg_2Cl_2 \rightleftharpoons SnCl_2 + 2HgCl_2$
   (c) $NH_3 + H_2O + NaCl \rightleftharpoons NH_4Cl + NaOH$
   (d) $2CuI + I_2 + 4K^+ \rightleftharpoons 2Cu^{2+} + 4KI$

2. The concentration of a pure solid or liquid phase is not included in the expression of equilibrium constant because
   (a) solid and liquid concentrations are independent of their quantities.
   (b) solid and liquids react slowly.
   (c) solid and liquids at equilibrium do not interact with gaseous phase.
   (d) the molecules of solids and liquids cannot migrate to the gaseous phase.

3. For the reversible reaction, $N_2(g) + 3H_2(g) \rightleftharpoons 2NH_3(g)$ at 500°C, the value of $K_p$ is $1.44 \times 10^{-5}$ when partial pressure is measured in atmospheres. The corresponding value of $K_c$, with concentration in mole litre$^{-1}$ is
   (a) $\dfrac{1.44\times10^{-5}}{(0.082\times500)^{-2}}$ (b) $\dfrac{1.44\times10^{-5}}{(8.314\times773)^{-2}}$
   (c) $\dfrac{1.44\times10^{-5}}{(0.082\times773)^{2}}$ (d) $\dfrac{1.44\times10^{-5}}{(0.082\times773)^{-2}}$

4. When two reactants, A and B are mixed to give products C and D, the reaction quotient $Q$ at the initial stages of the reaction
   (a) is zero.
   (b) decreases with time.
   (c) is independent of time.
   (d) increases with time.

5. At constant temperature, the equilibrium constant ($K_p$) for the decomposition reaction $N_2O_4 \rightleftharpoons 2NO_2$ is expressed by $K_p = \dfrac{4x^2P}{1-x^2}$, where $P$ = total pressure at equilibrium, $x$ = extent of decomposition. Which one of the following statements is true?
   (a) $K_p$ increases with increase of $P$.
   (b) $K_p$ increases with increase of $x$.
   (c) $K_p$ increases with decrease of $x$.
   (d) $K_p$ remains constant with change in $P$ and $x$.

6. For the reaction $CaCO_3(s) \rightleftharpoons CaO(s) + CO_2(g)$, the value of $K_P$ is
   (a) $P_{CO_2}$ (b) $\dfrac{P_{CO_2}}{P_{CaCO_3}}$
   (c) $\dfrac{[CaO][CO_2]}{[CaCO_3]}$ (d) $\dfrac{P_{CaCO_3}}{P_{CaO}P_{CO_2}}$

7. $XeF_6 + H_2O \rightleftharpoons XeOF_4 + 2HF$; equilibrium constant = $K_1$.

$XeO_4 + XeF_6 \rightleftharpoons XeOF_4 + XeO_3F_2$; equilibrium constant = $K_2$.

The equilibrium constant for the following reaction will be:

$XeO_4 + 2HF \rightleftharpoons XeO_3F_2 + H_2O$

(a) $\frac{K_1}{K_2}$ (b) $K_1 + K_2$

(c) $\frac{K_2}{K_1}$ (d) $K_2 - K_1$

8. $K_P$ for formation of ethane from hydrogen and ethylene is $5.5 \times 10^{18}$ atm$^{-1}$ and $K_P$ for formation of ethylene from hydrogen and acetylene is $5 \times 10^{26}$ atm$^{-1}$ at 323 K. What is $K_P$ for the reaction between hydrogen and acetylene to form ethane at 323 K?

(a) $2.75 \times 10^{45}$ atm$^{-2}$ (b) $1.1 \times 10^{-8}$

(c) $9.09 \times 10^{7}$ (d) $3.63 \times 10^{-46}$ atm$^{2}$

9. For a reversible reaction $A + B \rightleftharpoons C$, if the concentrations of the reactants are doubled at a definite temperature, then equilibrium constant will

(a) be doubled (b) be halved

(c) be one fourth (d) remain same

10. The value of $\frac{K_C}{K_P}$ the below for reaction is

$CO(g) + \frac{1}{2}O_2(g) \rightleftharpoons CO_2(g)$

(a) $\frac{1}{\sqrt{RT}}$ (b) $\sqrt{RT}$

(c) $1/RT$ (d) 1

11. The equilibrium constant for the reaction $N_2(g) + O_2(g) \rightleftharpoons 2NO(g)$ is $K_1$ and the equilibrium constant for the reaction $NO(g) \rightleftharpoons \frac{1}{2}N_2(g) + \frac{1}{2}O_2(g)$ is $K_2$ both at the same temperature. The value of $K_1$ and $K_2$ are related as

(a) $K_1 = \left(\frac{1}{K_2}\right)^2$ (b) $K_1 = K_2^2$

(c) $K_2 = \left(\frac{1}{K_1}\right)^2$ (d) $K_2 = K_1^2$

12. Equilibrium constants for some reactions are given. In which of the following case does the reaction go farthest to completion?

(a) $K = 10^2$ (b) $K = 10^{-2}$

(c) $K = 10$ (d) $K = 1$

13. For a reversible reaction, the rate constants for the forward and backward reactions are 0.16 and $4 \times 10^4$, respectively. What is the value of equilibrium constant of the reaction?

(a) $0.25 \times 10^6$ (b) $2.5 \times 10^5$

(c) $4 \times 10^{-6}$ (d) $4 \times 10^{-4}$

14. The following reaction has an equilibrium constant $K_C$ equal to $3.07 \times 10^{-4}$ at 24°C.

$2NOB(g) \rightleftharpoons 2NO(g) + Br_2(g)$

The correct set of concentrations at which the rate of forward reaction is greater than that of backward reaction is

(a) [NOBr] = 0.06 M, [NO] = 0.015 M, [$Br_2$] = 0.01 M

(b) [NOBr] = 0.15 M, [NO] = 0.025 M, [$Br_2$] = 0.014 M

(c) [NOBr] = 0.18 M, [NO] = 0.012 M, [$Br_2$] = 0.02 M

(d) [NOBr] = 0.045 M, [NO] = 0.0105 M, [$Br_2$] = 0.01 M

15. $NO_2$ is involved in the formation of smog and acid rain. It is formed importantly as:

$NO(g) + O_3(g) \rightleftharpoons NO_2(g) + O_2(g)$; $K_C = 6.0 \times 10^{34}$

The air over a metropolitan city contained $1.0 \times 10^{-5}$ M-NO, $1.0 \times 10^{-6}$ M-$O_3$, $2.5 \times 10^{-4}$ M-$NO_2$ and $8.2 \times 10^{-3}$ M-$O_2$. These data suggest that

(a) more of NO and $O_3$ tend to be formed.

(b) more of $NO_2$ and $O_2$ tend to be formed.

(c) more of $NO_2$ and $O_3$ tend to be formed.

(d) no tendency to change because the reaction is at equilibrium.

16. For the reaction $I_2(g) \rightleftharpoons 2I(g)$, $K_C = 1.0 \times 10^{-2}$ mol lit$^{-1}$. What volume of the vessel should be taken so that at equilibrium 1 mole of $I_2$ and 0.5 mole of 'I' are present at equilibrium?

(a) 25 L (b) 0.04 L

(c) 0.25 L (d) 5 L

**17.** The equilibrium constant for the reaction $3C_2H_2 \rightleftharpoons C_6H_6$ is 4.0 at $T$ K. If the equilibrium concentration of $C_2H_2$ is 0.5 M, then the concentration of $C_6H_6$ at equilibrium is

(a) 0.5 M (b) 1.5 M
(c) $5 \times 10^{-2}$ M (d) 0.25 M

**18.** The equilibrium mixture for the reaction $2H_2S(g) \rightleftharpoons 2H_2(g) + S_2(g)$ has 1 mole of $H_2S$, 0.20 mole of $H_2$ and 0.80 mole of $S_2$ in a 2 L vessel. The value of $K_C$ for the reaction is

(a) 0.16 M (b) 0.008 M
(c) 0.016 M (d) 0.032 M

**19.** A gaseous mixture contains 0.30 moles of CO, 0.10 moles of $H_2$, and 0.03 moles of $H_2O$ vapour and an unknown amount of $CH_4$ per litre. This mixture is in equilibrium at 1200 K.

$CO(g) + 3H_2(g) \rightleftharpoons CH_4(g) + H_2O(g); K_C = 3.9$

What is the concentration of $CH_4$ in this mixture?

(a) 0.39 M (b) 0.039 M
(c) 0.78 M (d) 0.078 M

**20.** For the reaction $2NOCl(g) \rightleftharpoons 2NO(g) + Cl_2(g)$, $\Delta H° = 18$ kcal and $\Delta S° = 30$ cal/K at 300 K. The equilibrium constant $K_P°$ for the reaction at 300 K is

(a) $e^{15}$ (b) $e^{-15}$
(c) $e^{-18}$ (d) $e^{-12}$

**21.** For the gas phase reaction $2NO(g) \rightleftharpoons N_2(g) + O_2(g)$; $\Delta H = -43.5$ kcal. Which one of the following is true for the reaction $N_2(g) + O_2(g) \rightleftharpoons 2NO(g)$?

(a) $K$ is independent of $T$.
(b) $K$ decreases as $T$ decreases.
(c) $K$ increases as $T$ decreases.
(d) $K$ varies with addition of NO.

**22.** The equilibrium constant for the reaction $N_2(g) + O_2(g) \rightleftharpoons 2NO(g)$ is $4.0 \times 10^{-4}$ at 2000 K. In the presence of a catalyst, the equilibrium is attained 10 times faster. Therefore, the equilibrium constant in the presence of the catalyst at 2000 K is

(a) $4 \times 10^{-3}$ (b) $4 \times 10^{-5}$
(c) $4 \times 10^{-4}$ (d) Unpredictable

**23.** What is the approximate value of $\log K_P$ for the following reaction?

$N_2(g) + 3H_2(g) \rightleftharpoons 2NH_3(g)$ at 25°C.

The standard enthalpy of formation of $NH_3(g)$ is $-40.0$ kJ/mol and the standard entropies of $N_2(g)$, $H_2(g)$ and $NH_3(g)$ are 191, 130 and 192 $JK^{-1} mol^{-1}$, respectively.

(a) 0.04 (b) 7.05
(c) 8.6 (d) 3.73

**24.** If $K_1$ and $K_2$ are the equilibrium constants for a reversible reaction at $T_1$ K and $T_2$ K temperature, respectively ($T_1 < T_2$) and the reaction takes place with neither heat evolution nor absorption, then

(a) $K_1 > K_2$ at high temperature.
(b) $K_1 < K_2$ at high temperature.
(c) $K_1 = K_2$ only at high temperature.
(d) $K_1 = K_2$ at any temperature.

**25.** From the following data,

(i) $H_2(g) + CO_2(g) \rightleftharpoons H_2O(g) + CO(g)$; $K_{2000\,K} = 4.4$

(ii) $2H_2O(g) \rightleftharpoons 2H_2(g) + O_2(g)$; $K_{2000\,K} = 5.31 \times 10^{-10}$

(iii) $2CO(g) + O_2(g) \rightleftharpoons 2CO_2(g)$; $K_{1000\,K} = 2.24 \times 10^{22}$

show whether reaction (iii) is exothermic or endothermic or thermal.

(a) Exothermic (b) Endothermic
(c) Thermal (d) Cannot say

**26.** In the decomposition equilibrium of a certain alkaline earth metal carbonate, the partial pressure of $CO_2$ becomes a hundred fold when the temperature is increased from 400 K to 500 K. What is the mean value of $\Delta H°$ for the given range of temperature?

(a) 18.4 kcal/mol
(b) 13.84 kcal/mol
(c) 14.83 kcal/mol
(d) 10.83 kcal/mol

**27.** The equilibrium constants for the reaction $A_2 \rightleftharpoons 2A$ at 500 K and 1000 K are $1 \times 10^{-10}$ and $1 \times 10^{-5}$, respectively. The reaction is

(a) exothermic (b) very slow
(c) very fast (d) endothermic

**28.** The activation energies for the forward and reverse elementary reactions in the system $A \rightleftharpoons B$ are 10.303 and 8.000 kcal, respectively at 500 K. Assuming the pre-exponential factor to be the same for both the forward and reverse steps, the equilibrium constant of the reaction at 500 K is

(a) 1.00 (b) 10.0
(c) 100 (d) 0.1

29. $\Delta G°$ for the reaction $X + Y \rightleftharpoons C$ is – 4.606 kcal at 1000 K. The equilibrium constant for the reverse mode of the reaction is

(a) 100 (b) 10
(c) 0.01 (d) 0.1

30. For a gaseous equilibrium $2A(g) \rightleftharpoons 2B(g) + C(g)$, $K_P$ has a value 1.8 at 700 K. The value of $K_C$ for the equilibrium $2B(g) + C(g) \rightleftharpoons 2A(g)$ at that temperature is about

(a) 0.031 (b) 32
(c) 57.4 (d) 103.3

## Application of Equilibrium Constant

31. The amounts of 0.8 mol of $PCl_5$ and 0.2 mole of $PCl_3$ are mixed in a 1 L flask. At equilibrium, 0.4 mole of $PCl_3$ is present. The equilibrium constant for the reaction, $PCl_5(g) \rightleftharpoons PCl_3(g) + Cl_2(g)$ will be

(a) 0.05 mol $L^{-1}$ (b) 0.13 mol $L^{-1}$
(c) 0.013 mol $L^{-1}$ (d) 0.60 mol $L^{-1}$

32. One mole of pure $PCl_5$ is placed in an evacuated container and maintained at 250°C. The equilibrium is established at total pressure of 2 atm. What is the partial pressure of chlorine at equilibrium?

$PCl_5(g) \rightleftharpoons PCl_3(g) + Cl_2(g)$: $K_P = 1.6$ atm

(a) 0.4 atm (b) 0.67 atm
(c) 0.80 atm (d) 0.64 atm

33. For the reaction $PCl_5(g) \rightleftharpoons PCl_3(g) + Cl_2(g)$, if initial moles of $PCl_5$ is 'x', α is the degree of dissociation and $P$ is the total pressure at equilibrium, then $P_{PCl_3} \cdot P^{-1}$ is equal to

(a) $\frac{x}{1+x}$ (b) $\frac{\alpha x}{\alpha + x}$

(c) $\frac{\alpha}{1+\alpha}$ (d) $\frac{\alpha}{x + \alpha x}$

34. In a 5.76 L vessel, 0.5 moles of $H_2$ gas and 0.5 moles of $I_2$ vapours are allowed to react to form HI (g) at 447°C, then the total pressure of gases at equilibrium would be ($R = 0.08$ L-atm/K-mol)

(a) 20 atm (b) 10 atm
(c) 5 atm (d) 1 atm

35. In a closed tube, HI(g) is heated at 440°C up to establishment of equilibrium. If it dissociates into $H_2(g)$ and $I_2(g)$ up to 22%, then the dissociation constant is

(a) 0.282 (b) 0.0796
(c) 0.0199 (d) 1.99

36. For the reaction $H_2(g) + I_2(g) \rightleftharpoons 2HI(g)$, the value of equilibrium constant is 9.0. The degree of dissociation of HI will be

(a) 0.5 (b) 0.33
(c) 0.4 (d) 0.67

37. At 444°C, HI is 30% dissociated. If initially 3 moles of HI are taken, then the number of moles of HI at equilibrium is

(a) 0.9 (b) 2.1
(c) 0.45 (d) 1.8

38. One mole of $N_2O_4$ (g) at 300 K is kept in a closed container under one atm. It is heated to 600 K when 20% by mass of $N_2O_4(g)$ decomposes to $NO_2(g)$. The resultant pressure is

(a) 1.2 atm (b) 2.4 atm
(c) 2.0 atm (d) 1.0 atm

39. For the reaction $N_2O_4(g) \rightleftharpoons 2NO_2(g)$, the relation between the degree of dissociation of $N_2O_4(g)$ at pressure, $P$ with its equilibrium constant $K_P$ is

(a) $\alpha = \frac{K_P/P}{4 + K_P/P}$ (b) $\alpha = \frac{K_P}{4 + K_P}$

(c) $\alpha = \left[\frac{K_P/P}{4 + K_P/P}\right]^{1/2}$ (d) $\alpha = \left[\frac{K_P}{4 + K_P}\right]^{1/2}$

40. At total pressure $P_1$ atm and $P_2$ atm, $N_2O_4$ is dissociated to an extent of 33.33% and 50.00%, respectively. The ratio of pressures $P_1$ and $P_2$ is

(a) 3:8 (b) 2:1
(c) 8:3 (d) 1:2

41. At 0°C and 1 atm pressure, 1 L of $N_2O_4$ decomposes to $NO_2$ according to the equation $N_2O_4(g) \rightleftharpoons 2NO_2$ (g). To what extent has the decomposition proceeded when the original volume is 25% less than that of existing volume?

(a) 0.67 (b) 0.33
(c) 0.25 (d) 0.75

**42.** Forty percent of a mixture of 0.2 mol of $N_2$ and 0.6 mol of $H_2$ reacts to give $NH_3$ according to the equation

$N_2(g) + 3H_2(g) \rightleftharpoons 2NH_3(g)$

at constant temperature and pressure. Then, the ratio of the final volume to the initial volumes of gases is

(a) 4 : 5 (b) 5 : 4
(c) 7 : 10 (d) 8 : 5

**43.** An amount of 4 moles of $NH_3$ gas is introduced with a previously evacuated 1 L container in which it is partially dissociated at high temperature as

$2NH_3(g) \rightleftharpoons N_2(g) + 3H_2(g)$

At equilibrium, 2 mole of $NH_3(g)$ remained. The value of $K_C$ for the reaction is

(a) $1.5M^2$ (b) $6.75M^2$
(c) $0.44M^{-2}$ (d) $2.25M^2$

**44.** In a closed container maintained at 1 atm pressure and 25°C, 2 moles of $SO_2(g)$ and 1 mole of $O_2(g)$ were allowed to react to form $SO_3(g)$ under the influence of a catalyst.

$2SO_2(g) + O_2(g) \rightleftharpoons 2SO_3(g)$

At equilibrium, it was found that 50% of $SO_2(g)$ was converted to $SO_3(g)$. The partial pressure of $O_2(g)$ at equilibrium will be

(a) 0.17 atm (b) 0.5 atm
(c) 0.33 atm (d) 0.20 atm

**45.** One mole of ethanol is treated with one mole of ethanoic acid at 25°C. One-fourth of the acid changes into ester at equilibrium. The equilibrium constant for the reaction will be

(a) 1/9 (b) 4/9
(c) 9 (d) 9/4

**46.** One mole each of A and B and 3 moles each of C and D are placed in 1 L flask. If equilibrium constant is 2.25 for the reaction A + B $\rightleftharpoons$ C + D, then equilibrium concentrations of A and C will be in the ratio

(a) 2:3 (b) 3:2
(c) 1:2 (d) 2:1

**47.** For the reaction $XY_2(g) \rightleftharpoons XY(g) + Y(g)$, the reaction is started with initial pressure of $XY_2$, 600 mm Hg. The total pressure for gases at equilibrium is 800 mm Hg. Assuming that volume and temperature of the system remains constant, the value of $K_P$ is

(a) 100 atm (b) 100 mm Hg
(c) 0.01 atm (d) 400 mm Hg

**48.** For the reaction A + B $\rightleftharpoons$ C + D, the initial concentration of A and B is equal, but the equilibrium concentration of C is twice that of equilibrium concentration of A. The equilibrium constant is

(a) 4 (b) 9
(c) 1/4 (d) 1/9

**49.** $I_2 + I^- \rightleftharpoons I_3^-$. This reaction is set up in aqueous medium. We start with 1 mol of $I_2$ and 0.5 mol of $I^-$ in 1 L flask. After equilibrium is reached, excess of $AgNO_3$ gave 0.25 mol of yellow precipitate. The equilibrium constant is

(a) 1.33 (b) 2.66
(c) 0.375 (d) 0.75

**50.** The equilibrium constant for the mutarotation, α-D–glucose $\rightleftharpoons$ β-D–glucose is 1.8. What percent of the α-form remains under equilibrium?

(a) 35.7 (b) 64.3
(c) 55.6 (d) 44.4

**51.** For the reaction,

$2NOBr(g) \rightleftharpoons 2NO(g) + Br_2(g)$,

the ratio $\frac{K_P}{P}$, where $P$ is the total pressure of gases at equilibrium and $P_{Br_2} = \frac{P}{9}$ at a certain temperature is

(a) $\frac{1}{9}$ (b) $\frac{1}{81}$
(c) $\frac{1}{27}$ (d) $\frac{1}{3}$

**52.** The progress of the reaction A $\rightleftharpoons$ $n$B with time is represented by the graph given below.

The value of $n$ is

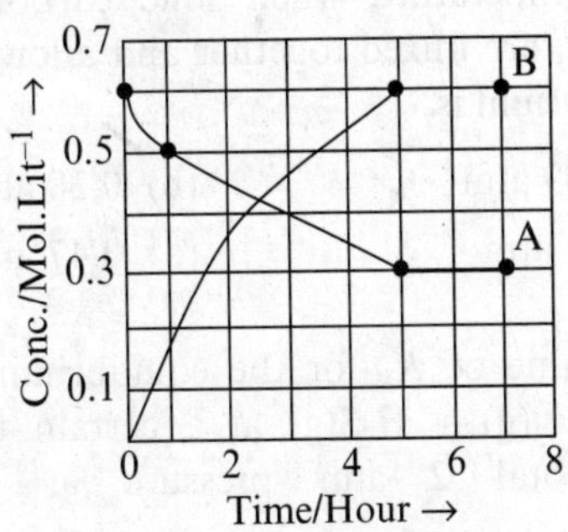


(a) 1 (b) 2
(c) 3 (d) 4

53. The value of $K_P$ for the reaction $N_2O_4(g) \rightleftharpoons 2NO_2$ is 640 mm at 775 K. The percentage dissociation of $N_2O_4$ at equilibrium pressure of 160 mm is

(a) $\frac{100}{\sqrt{2}}$ (b) 50
(c) $\frac{50}{\sqrt{2}}$ (d) $10\sqrt{2}$

54. At 127°C and 1 atm pressure, $PCl_5(g)$ is partially dissociated into $PCl_3(g)$ and $Cl_2(g)$ as $PCl_5(g) \rightleftharpoons PCl_3(g) + Cl_2(g)$. The density of the equilibrium mixture is 3.5 g/L. Percentage dissociation of $PCl_5$ is ($R = 0.08$ L-atm/K-mol, P = 31, Cl = 35.5)

(a) 46.28 (b) 86.16
(c) 15.86 (d) 53.72

55. At $T$ K, a compound $AB_2(g)$ dissociates according to the reaction $2AB_2(g) \rightleftharpoons 2AB(g) + B_2(g)$, with a degree of dissociation 'x' which is small compared with unity. The expression for 'x' in terms of the equilibrium constant, $K_P$ and the total pressure $P$ is

(a) $\frac{K_P}{P}$ (b) $(K_P)^{1/3}$
(c) $\left(\frac{2K_P}{P}\right)^{1/3}$ (d) $\left(\frac{K_P}{P}\right)^{1/3}$

## Heterogeneous Equilibrium

56. A quantity of 34 g sample of $BaO_2$ is heated to 1000 K in a closed and rigid evacuated vessel of 8.21 L capacity. What percentage of peroxide is converted into oxide? (Ba = 138)

$2BaO_2\ (s) \rightleftharpoons 2BaO(s) + O_2\ (g);\ K_P = 0.5$ atm

(a) 20% (b) 50%
(c) 75% (d) 80%

57. For the reaction $NH_2COONH_4(s) \rightleftharpoons 2NH_3(g) + CO_2(g)$, $K_P = 3.2 \times 10^{-5}$ atm$^3$. The total pressure of the gaseous products when sufficient amount of reactant is allowed to achieve equilibrium is

(a) 0.02 atm (b) 0.04 atm
(c) 0.06 atm (d) 0.095 atm

58. At a certain temperature, $K_P$ for the dissociation of solid $CaCO_3$ is $4.5 \times 10^{-2}$ atm and for the reaction $C(s) + CO_2(g) \rightleftharpoons 2CO(g)$, the $K_P$ value is 2.0 atm. The pressure of carbon monoxide at this temperature, when solid carbon CaO and $CaCO_3$ are mixed together and allowed to attain equilibrium is

(a) 0.09 atm (b) 0.30 atm
(c) 2.1 atm (d) 0.47 atm

59. The value of $K_P$ for the equilibrium $NH_4HS(s) \rightleftharpoons NH_3(g) + H_2S(g)$ at a certain temperature and total 2 atm pressure at equilibrium will be

(a) 0.25 atm$^2$ (b) 1.0 atm$^2$
(c) 4.0 atm$^2$ (d) 1.0 atm

60. Iron fillings and water were placed in a 5 L vessel and sealed. The tank was heated to 1000°C. Upon analysis, the tank was found to contain 1.2 g of $H_2(g)$ and 54.0 g of $H_2O(g)$. If the reaction is represented as $3Fe(s) + 4H_2O(g) \rightleftharpoons Fe_3O_4(s) + 4H_2(g)$, then the value of equilibrium constant is

(a) 0.2 (b) 0.04
(c) 0.008 (d) 0.0016

61. At 1000°C and a pressure of 16 atm, the equilibrium constant of the reaction

$CO_2(g) + C(s) \rightleftharpoons 2CO(g)$

is such that for every nine moles of CO, there is one mole of $CO_2$. For what pressure of the mixture, is the ratio CO : $CO_2$ = 4 : 1? The temperature remains 1000°C.

(a) 40.5 atm (b) 81 atm
(c) 33.75 atm (b) 6.7 atm

62. At a certain temperature, the value of $K_P$ for the reaction: $2CO(g) \rightleftharpoons CO_2(g) + C(\text{graphite})$ is 0.1 atm$^{-1}$. What is the ratio of partial pressures of CO and $CO_2$ at equilibrium, taking the total pressure to be 1.1 atm?

(a) 9:1 (b) 10:1
(c) 1:10 (d) 1:9

63. For the equilibrium $LiCl{\cdot}3NH_3(s) \rightleftharpoons LiCl{\cdot}NH_3(s) + 2NH_3(g)$; $K_P = 9$ atm$^2$ at 27°C. A 8.21 L vessel contains 0.1 mole of $LiCl.NH_3(s)$. How many moles of $NH_3(g)$ should be added to the flask at this temperature to derive the backward reaction for completion?

(a) 0.8 (b) 1.0
(c) 1.2 (d) 1.1

**64.** The reaction $ZnO(s) + CO(g) \rightleftharpoons Zn(g) + CO_2(g)$ has an equilibrium constant of 1 atm at 1500 K. The equilibrium partial pressure of zinc vapour in a reaction vessel if an equimolar mixture of CO and $CO_2$ is brought into contact with solid ZnO at 1500 K and the equilibrium is achieved at 1 atm is

(a) 0.68 atm (b) 0.76 atm
(c) 0.24 atm (d) 0.5 atm

**65.** If 0.3 moles of hydrogen gas and 2.0 moles of sulphur solid are heated to 87°C in a 2.0 L vessel, then what will be the partial pressure of $H_2S$ gas at equilibrium? (Given: $R = 0.08$ L-atm/K-mol)

$H_2(g) + S(s) \rightleftharpoons H_2S(g)$: $K_C = 0.08$

(a) 0.32 atm (b) 0.43 atm
(c) 0.62 atm (d) 0.48 atm

## Le-Chatelier's Principle

**66.** Consider the following equilibrium in a closed container.

$N_2O_4(g) \rightleftharpoons 2NO_2(g)$

At a fixed temperature, the volume of the reaction container is halved. For this change, which of the following statement holds true regarding the equilibrium constant ($K_P$) and degree of dissociation ($\alpha$)?

(a) Neither $K_P$ nor $\alpha$ changes.
(b) Both $K_P$ and $\alpha$ changes
(c) $K_P$ changes, but $\alpha$ does not change.
(d) $K_P$ does not change, but $\alpha$ changes.

**67.** Which of the following hypothetical reactions is favoured by the increase of temperature as well as pressure?

(a) $A(s) + 2B(g) \rightleftharpoons C(g) + D(g)$: $\Delta H = +ve$
(b) $A(s) + 2B(g) \rightleftharpoons 2C(g) + D(g)$: $\Delta H = +ve$
(c) $2A(g) + B(g) \rightleftharpoons 2C(g) + D(s)$: $\Delta H = +ve$
(d) $2A(s) + 2B(g) \rightleftharpoons 2C(g) + 2D(g)$: $\Delta H = -ve$

**68.** In the system $AB(s) \rightleftharpoons A(g) + B(g)$, doubling the quantity of AB(s) would

(a) increase the amount of A to double its value.
(b) increase the amount of B to double its value.
(c) increase the amounts of both A and B to some extent.
(d) cause no changes in the amounts of A and B.

**69.** In a system $A(s) \rightleftharpoons 2B(g) + 3C(g)$, if the concentration of C at equilibrium is increased by a factor of 2, then it will cause the equilibrium concentration of B to decrease by

(a) two times the original value.
(b) one half of its original value.
(c) $2\sqrt{2}$ times its original value.
(d) 8 times its original value.

**70.** The reaction $MgCO_3(s) \rightleftharpoons MgO(s) + CO_2(g)$ is in progress. If the number of moles of MgO in the vessel is doubled at an instance,

(a) the reaction quotient $Q$ is halved.
(b) the reaction quotient $Q$ is doubled.
(c) the moles of $CO_2$ present at equilibrium is halved.
(d) the partial pressure of $CO_2$ in the vessel remains unchanged.

**71.** When pressure is applied to the equilibrium system Ice $\rightleftharpoons$ water, which of the following phenomenon will happen?

(a) More ice will be formed.
(b) Ice will sublime.
(c) More water will be formed.
(d) Equilibrium will not disturb.

**72.** The equilibrium $SOCl_2(g) \rightleftharpoons SO_2(g) + Cl_2(g)$ is attained at 25°C in a closed rigid container and helium gas is introduced. Which of the following statements is correct?

(a) Concentration of $SO_2$ is increased.
(b) More $Cl_2$ is formed.
(c) Concentrations of all change.
(d) Concentrations will not change.

**73.** When calcium acetate is dissolved in water, heat is evolved. If the temperature is raised, then the solubility of calcium acetate

(a) increases.
(b) decreases.
(c) is not affected.
(d) first increases and then decreases.

**74.** The Haber's process for the manufacture of ammonia is usually carried out at about 500°C. If a temperature of about 250°C was used instead of 500°C then,

(a) no ammonia would be formed at all.
(b) the percentage of ammonia in the equilibrium mixture would be too low.
(c) a catalyst would be of no use at all at this temperature.
(d) the percentage of ammonia in the equilibrium mixture would be too high.

**75.** Densities of diamond and graphite are 3.5 and 2.4 g/ml, respectively. The increase in pressure (at constant temperature) at the equilibrium in C (Diamond) $\rightleftharpoons$ C (Graphite) will

(a) favour the forward reaction.
(b) favour the backward reaction.
(c) have no effect.
(d) increases the equilibrium constant.

**76.** In a flask, colourless $N_2O_4$ is in equilibrium with brown coloured $NO_2$. At equilibrium, when the flask is heated at 100°C the brown colour deepens and on cooling, it becomes less coloured. The change in enthalpy, $\Delta H$, for the system is

(a) negative (b) positive
(c) zero (d) undefined

**77.** For the chemical reaction $3X(g) + Y(g) \rightleftharpoons X_3Y(g)$, the amount of $X_3Y$ at equilibrium is affected by

(a) temperature and pressure.
(b) temperature only.
(c) pressure only.
(d) temperature, pressure and catalyst.

**78.** What will the change on increasing the pressure at equilibrium water $\rightleftharpoons$ water vapour?

(a) The boiling point of water will increase.
(b) The boiling point of water will decrease.
(c) The boiling point of water will not change.
(d) Boiling point is not related with pressure.

**79.** In equilibrium:

$SCN^-(aq) + Fe^{+3}(aq) \rightleftharpoons [Fe(SCN)^{2+}](aq)$

Colourless Yellow Deep red

If thiocyanate ions are added in equilibrium mixture

(a) The solution becomes colourless.
(b) The yellow colour of solution deepens.
(c) The red colour of the solution deepens.
(d) Concentration of $[Fe(SCN)]^{2+}$ ion will decrease.

**80.** An aqueous solution of volume 500 ml, when the reaction $2Ag^+(aq) + Cu(s) \rightleftharpoons Cu^{2+}(aq) + 2Ag(s)$ reached equilibrium, the concentration of $Cu^{2+}$ ions was $x$ M. To this solution, 500 ml of water is added. At the new equilibrium, the concentration of $Cu^{2+}$ ions would be

(a) $2x$ M.
(b) $x$ M.
(c) between $x$ and $0.5x$ M.
(d) less than $0.5x$ M.

**81.** $PCl_5(g) \rightleftharpoons PCl_3(g) + Cl_2(g)$. In the above reaction, the partial pressure of $PCl_3$, $Cl_2$ and $PCl_5$ are 0.3, 0.2 and 0.6 atm, respectively. If partial pressure of $PCl_3$ and $Cl_2$ was increased twice at the new equilibrium, then what will be the new partial pressure of $PCl_5$ (in atm)?

(a) 0.3 (b) 1.2
(c) 2.4 (d) 0.15

**82.** In the equilibrium mixture $H_2(g) + I_2(g) \rightleftharpoons 2HI(g)$, the mole ratio of gases are $\sqrt{2}:\sqrt{2}:10$, respectively. What would be the effect on the mole ratio on adding 5 mole of He gas at constant pressure?

(a) No change
(b) The new molar ratio becomes 1 : 1 : 5.
(c) The new molar ratio becomes 2 : 2 : 5.
(d) The new molar ratio becomes $2:2:5\sqrt{2}$.

**83.** One mole of He (g, colourless) is added to the equilibrium mixture containing $N_2O_4$ (g, colourless) and $NO_2$ (g, red-brown) present in a cylinder piston arrangement maintained at constant pressure and temperature. It will result in

(a) no change in colour of the equilibrium mixture.
(b) lightening the colour of the equilibrium mixture.
(c) darkening the colour of the equilibrium mixture.
(d) no effect on the equilibrium composition.

**84.** Consider the formation of $SO_3$ according to the reaction $2SO_2(g) + O_2(g) \rightleftharpoons 2SO_3(g)$; $\Delta H = -198$ kJ. Which of the following may change the value of equilibrium constant of the above reaction?

(a) Adding He gas to reaction vessel.
(b) Adding more of $O_2$ to the reaction vessel.
(c) Increasing the temperature.
(d) Doubling the volume of reaction vessel.

**85.** For the given chemical equation $2X(g) + Y(g) \rightleftharpoons 2Z(g) + 8$ kcal, which combination of pressure and temperature gives the highest yield of 'Z' at equilibrium?

(a) 1000 atm and 500°C
(b) 500 atm and 500°C
(c) 500 atm and 100°C
(d) 1000 atm and 100°C

# EXERCISE II (JEE ADVANCED)

## Section A (Only one Correct)

1. The approach to the following equilibrium was observed kinetically from both directions.

$$PtCl_4^{2-} + H_2O \rightleftharpoons Pt(H_2O)Cl_3^- + Cl^-$$

At 25°C, it was found that

$$-\frac{d[PtCl_4^{2-}]}{dt} = (3.9 \times 10^{-5}\ s^{-1})[PtCl_4^{2-}] - (2.1 \times 10^{-3}\ L\ mol^{-1}s^{-1})[Pt(H_2O)Cl_3^-][Cl^-]$$

The value of $K_{eq}$ (equilibrium constant) for the complexation of the fourth $Cl^-$ by Pt (II) is

(a) 53.8 mol $L^{-1}$ (b) 0.018 mol $L^{-1}$
(c) 53.8 L $mol^{-1}$ (d) 0.018 L $mol^{-1}$

2. The complexion of $Fe^{2+}$ with the chelating agent dipyridyl has been studied kinetically in both the forward and reverse directions.

$Fe^{2+} + 3\text{ dipy} \rightleftharpoons [Fe(dipy)_3]^{2+}$

For this reaction, the rates of forward and reverse reactions are $(1.45 \times 10^{13}\ M^{-3}s^{-1})[Fe^{2+}][dipy]^3$ and $(1.22 \times 10^{-4}\ s^{-1})[Fe(dipy)_3^{2+}]$, at 25°C. What is the stability constant of the complex?

(a) $1.77 \times 10^9$ (b) $8.4 \times 10^{-18}$
(c) $1.18 \times 10^{17}$ (d) $5.65 \times 10^{-10}$

3. The rate constant for the forward reaction: $A(g) \rightleftharpoons 2B(g)$ is $1.5 \times 10^{-3}\ s^{-1}$ at 300 K. If $10^{-5}$ moles of 'A' and 100 moles of 'B' are present in a 10 litre vessel at equilibrium, then the rate constant of the backward reaction at this temperature is

(a) $1.5 \times 10^{-3}\ M^{-1}s^{-1}$ (b) $1.5 \times 10^{-1}\ M^{-1}s^{-1}$
(c) $1.5 \times 10^{-11}\ M^{-1}s^{-1}$ (d) $1.5 \times 10^{-12}\ M^{-1}s^{-1}$

4. $CuSO_4 \cdot 5H_2O(s) \rightleftharpoons CuSO_4.3H_2O(s) + 2H_2O(g)$, $K_P$ for this equilibrium is $1.0 \times 10^{-4}\ atm^2$ at 25°C. What is the maximum pressure of water vapour (moisture) in the atmosphere, below which the pentahydrate is efflorescent?

(a) 7.60 mm (b) 0.01 mm
(c) 0.076 mm (d) 760 mm

5. Ammonia at a pressure of 10 atm and $CO_2$ at a pressure of 20 atm are introduced into an evacuated chamber. If $K_P$ for the reaction $NH_2COONH_4(s) \rightleftharpoons 2NH_3(g) + CO_2(g)$ is 2020 $atm^3$, then the total pressure after a long time is

(a) less than 30 atm (b) more than 30 atm
(c) equal to 30 atm (d) unpredictable

6. The value of $K_P$ for the process: $CuSO_4 \cdot 5H_2O(s) \rightleftharpoons CuSO_4 \cdot 3H_2O(s) + 2H_2O(g)$ is $1.21 \times 10^{-4}\ atm^2$ at certain temperature. If aqueous tension at that temperature is 40 torr, then at what relative humidity of air will $CuSO_4 \cdot 5H_2O$ effloresces?

(a) Above 40.8% (b) Below 40.8%
(c) Above 20.9% (d) Below 20.9%

7. The value of $K_P$ for the reaction: $H_2(g) + I_2(s) \rightleftharpoons 2HI(g)$ is $6.4 \times 10^{-4}$ atm. On close observation, it is found that the partial pressure of iodine present in vapour state is $1.6 \times 10^{-4}$ atm, at the same temperature. The value of $K_P$ for the reaction: $H_2(g) + I_2(g) \rightleftharpoons 2HI(g)$ is

(a) 0.25 (b) 4
(c) $1.024 \times 10^{-7}$ (d) $9.76 \times 10^6$

8. For the reaction $H_2(g) + I_2(g) \rightleftharpoons 2HI(g)$, the rate law expression is

(a) $-\frac{1}{2}\cdot\frac{d[HI]}{dt} = K_{-1}[HI]^2 - K_1[H_2][I_2]$

(b) $-\frac{1}{2}\cdot\frac{d[HI]}{dt} = \frac{K_1[HI]^2}{K_{-1}[H_2][I_2]}$

(c) $-\frac{1}{2}\cdot\frac{d[HI]}{dt} = K_1[H_2][I_2] - K_{-1}[HI]^2$

(d) $-\frac{1}{2}\cdot\frac{d[HI]}{dt} = K_1K_{-1}[H_2][I_2]$

9. At 200°C, $PCl_5$ dissociates as $PCl_5(g) \rightleftharpoons PCl_3(g) + Cl_2(g)$. It was found that the equilibrium vapours are 62 times as heavy as hydrogen. The percentage dissociation of $PCl_5$ at 200°C is

(a) 34.1% (b) 70.3%
(c) 40.5% (d) 68.1%

10. Under what pressure must an equimolar mixture of $Cl_2$ and $PCl_3$ be placed at 250°C in order to obtain 75% conversion of $PCl_3$ into $PCl_5$? Given that $PCl_3(g) + Cl_2(g) \rightleftharpoons PCl_5(g)$; $K_P = 2\ atm^{-1}$.

(a) 12 atm (b) 6 atm
(c) 15 atm (d) 30 atm

**11.** In the reaction $N_2(g) + 3H_2(g) \rightleftharpoons 2NH_3(g)$ + Heat. One mole of $N_2$ reacts with three moles of $H_2$. If at equilibrium '$x$' moles of $N_2$ combined, then the value of '$x$' in terms of $K_P$ and the total pressure of gases at equilibrium $P$ is ($x << 1$)

(a) $\frac{3P.\sqrt{3.K_P}}{8}$ (b) $\frac{8P}{3\sqrt{3.K_P}}$

(c) $\frac{3\sqrt{3.K_P}}{8P}$ (d) $\frac{8}{P}\sqrt{\frac{K_P}{3}}$

**12.** For the reaction $X_2(g) + Y_2(g) \rightleftharpoons 2XY(g)$, 2 moles of '$X_2$' was taken in a 2 L vessel and 3 moles of '$Y_2$' was taken in a 3 L vessel. Both vessels were then connected. At equilibrium, the concentration of '$XY$' is 0.7 M. The equilibrium concentrations of '$X_2$' and '$Y_2$' would be

(a) 0.65 M, 0.65 M (b) 0.30 M, 0.30 M
(c) 0.25 M, 1.25 M (d) 0.05 M, 0.25 M

**13.** The equilibrium constant for the reaction $N_2(g) + O_2(g) \rightleftharpoons 2NO(g)$ is 0.09 at 3500 K. The fraction of equimolar mixture of $N_2$ and $O_2$ converted into NO is

(a) 0.13 (b) 0.23
(c) 0.30 (d) 0.357

**14.** At a certain temperature, the equilibrium constant of the reaction $N_2(g) + O_2(g) \rightleftharpoons 2NO(g)$ is 0.0004. Assuming air to be a mixture of four volumes of nitrogen with one volume of oxygen, the percentage of nitric oxide by volume, in the gas produced by allowing air to reach equilibrium at this temperature is

(a) 3.0% (b) 0.8%
(c) 0.04% (d) 8.0%

**15.** When a mixture of $N_2$ and $H_2$ in the volume ratio of 1 : 5 is allowed to react at 700 K and $10^3$ atm pressure, 0.4 mole fraction of $NH_3$ is formed at equilibrium. The value of $K_P$ for the below reaction

$$N_2(g) + 3H_2(g) \rightleftharpoons 2NH_3(g)$$

(a) $2.6 \times 10^{-5}$ atm$^{-2}$ (b) $2.6 \times 10^{-4}$ atm$^{-2}$
(c) $2.6 \times 10^{3}$ atm$^{-2}$ (d) $5.1 \times 10^{-3}$ atm$^{-2}$

**16.** An amount of 16 moles $H_2$ and 4 moles of $N_2$ is confined in a vessel of volume one litre. The vessel is heated to a constant temperature until the equilibrium is established. At equilibrium, the pressure was found to be 9/10th of the initial pressure. The value of $K_C$ for the reaction $N_2(g) + 3H_2(g) \rightleftharpoons 2NH_3(g)$ is

(a) 8100 (b) $6.07 \times 10^{-4}$
(c) 1647.75 (d) $8.99 \times 10^{-5}$

**17.** An amount of 3 moles of $N_2$ and some $H_2$ is introduced into an evacuated vessel. The reaction starts at $t = 0$ and equilibrium is attained at $t = t_1$. The amount of ammonia at $t = 2t_1$ is found to be 34 g. It is observed that $\frac{w(N_2)}{w(H_2)} = \frac{14}{3}$ at $t = \frac{t_1}{3}$ and $t = \frac{t_1}{2}$. The only correct statement is

(a) $w(N_2) + w(H_2) + w(NH_3) = 118$ g at $t = t_1$
(b) $w(N_2) + w(H_2) + w(NH_3) = 102$ g at $t = 2t_1$
(c) $w(N_2) + w(H_2) + w(NH_3) = 50$ g at $t = t_1/3$
(d) $w(N_2) + w(H_2) + w(NH_3)$ cannot be predicted

**18.** Starting with 2 moles of $SO_2$ and 1 mol of $O_2$ in 1 L flask, the equilibrium mixture required 0.4 moles of $MnO_4^-$ for complete reaction in acidic medium. The value of $K_C$ for the reaction: $2SO_2(g) + O_2(g) \rightleftharpoons 2SO_3(g)$ is

(a) 2 (b) 0.5
(c) 0.25 (d) 4

**19.** A quantity of 60 g of $CH_3COOH$ and 46 g of $CH_3CH_2OH$ reacts in 5 L flask to form 44 g of $CH_3COOC_2H_5$ at equilibrium. On taking 120 g of $CH_3COOH$ and 46 g of $CH_3CH_2OH$, $CH_3COOC_2H_5$ formed at equilibrium is

(a) 44 g (b) 29.33 g
(c) 66 g (d) 58.67 g

**20.** Two moles of an equimolar mixture of two alcohols $R_1$–OH and $R_2$–OH are esterified with one mole of acetic acid. If only 80% of the acid is consumed till equilibrium and the quantities of ester formed under equilibrium are in the ratio 3 : 2. What is the value of equilibrium constant for the esterification of $R_1$–OH?

(a) 0.48 (b) 2.2
(c) 0.32 (d) 3.69

**21.** For the reaction $2NO(g) + Cl_2(g) \rightleftharpoons 2NOCl(g)$, NO and $Cl_2$ are initially taken in mole ratio of 2 : 1. The total pressure at equilibrium is found to be 1 atm. If the moles of NOCl are one-fourth of that of $Cl_2$ at equilibrium, the value of $K_P$ for the reaction is

(a) $\frac{13}{36}$ (b) $\frac{13}{256}$

(c) $\frac{13}{512}$ (d) $\frac{13}{128}$

22. When S in the form of $S_8$ is heated at 900 K, the initial pressure of 1 atm falls by 30% at equilibrium. This is because of conversion of some $S_8(g)$ to $S_2(g)$. The value of $K_P$ for the reaction is
(a) 0.011 $atm^{-3}$ (b) 2.96 $atm^3$
(c) 1.71 $atm^3$ (d) 204.8 $atm^3$

23. The Deacon reaction is the oxidation of HCl by $O_2$ in the following reaction

$$HCl(g) + \frac{1}{4}O_2(g) \rightleftharpoons \frac{1}{2}Cl_2(g) + \frac{1}{2}H_2O(g)$$

at a pressure of 730 mm and with an initial mixture containing 8% of HCl and 92% of $O_2$, the degree of decomposition of the HCl is 0.08. What is the equilibrium partial pressure of oxygen?
(a) 671.6 mm (b) 659.92 mm
(c) 537.28 mm (d) 670.43 mm

24. The equilibrium constant for the reaction: $H_3BO_3$ + Glycerin $\rightleftharpoons$ ($H_3BO_3$ + Glycerin complex) is 0.90. How much glycerin should be added to 1 L of 0.10 M-$H_3BO_3$ solution, so that 60% of the $H_3BO_3$ is converted to boric acid–glycerin complex?
(a) Infinite (b) 1.73 M
(c) 0.10 M (d) 2.27 M

25. The process $2A(g) \rightleftharpoons A_2(g)$ has $K_P = 8 \times 10^8$ $atm^{-1}$. If 'A' atoms are taken at 1 atm pressure, then what should be the equilibrium pressure of 'A'?
(a) 0 (b) $2.5 \times 10^{-5}$ atm
(c) $4 \times 10^4$ atm (d) $1.25 \times 10^{-5}$ atm

26. When $CO_2(g)$ is dissolved in water, then the following equilibrium is established

$$CO_2(aq) + 2H_2O(l) \rightleftharpoons H_3O^+(aq) + HCO_3^-(aq)$$

for which the equilibrium constant is $3.8 \times 10^{-7}$. If the pH of solution is 6.0, then what would be the ratio of concentration of $HCO_3^-(aq)$ to $CO_2(aq)$?
(a) $3.8 \times 10^{-13}$ (b) 6.0
(c) 0.38 (d) 13.4

27. For the equilibrium $A(g) \rightleftharpoons nB(g)$, the equilibrium constant $K_P$ is related with the degree of dissociation α, and the total pressure of gases at equilibrium $P$ is

(a) $\dfrac{(n\alpha)^n.P^{n-1}}{(1-\alpha).[1+(n-1)\alpha]^{n-1}}$

(b) $\dfrac{(n\alpha)^{n-1}.P^{n-1}}{(1-\alpha).[1+(n-1)\alpha]^{n-1}}$

(c) $\dfrac{\alpha^{n-1}.P^{n-1}}{(1-\alpha).[1+(n-1)\alpha]}$

(d) $\dfrac{n\alpha.P^{n-1}}{(1-\alpha)^{n-1}.[1+(n-1)\alpha]}$

28. The reaction $A(g) + B(g) \rightleftharpoons C(g) + D(g)$ occurs in a single step. The rate constant of forward reaction is $2.0 \times 10^{-3}$ $mol^{-1}$ L $s^{-1}$. When the reaction is started with equimolar amounts of A and B, it is found that the concentration of A is twice that of C at equilibrium. The rate constant of the backward reaction is
(a) $5.0 \times 10^{-4}$ $mol^{-1}$ L $s^{-1}$
(b) $8.0 \times 10^{-3}$ $mol^{-1}$ L $s^{-1}$
(c) $1.25 \times 10^{2}$ $mol^{-1}$ L $s^{-1}$
(d) $2.0 \times 10^{3}$ $mol^{-1}$ L $s^{-1}$

29. Find the concentration of manomeric dichloroacetic acid in a $CCl_4$ solution which contains 0.0129 g of the acid in 100 ml of solution. The dissociation constant of the dimeric acid is $5.0 \times 10^{-4}$. Assume that the acids are unionized in $CCl_4$ solution.
(a) $5.0 \times 10^{-4}$ M (b) $2.5 \times 10^{-4}$ M
(c) $1.0 \times 10^{-3}$ M (d) $3.9 \times 10^{-4}$ M

30. In an evacuated rigid vessel of volume $V$ litre, one mole of solid ammonium carbonate ($NH_2CONH_4$) is taken and the vessel is heated to $T$ K. The equilibrium total pressure of gases is found to be $P$ atm. The percentage dissociation of solid into $NH_3(g)$ and $CO_2(g)$ is
(a) $\dfrac{100RT}{3PV}\%$ (b) $\dfrac{RT}{3PV}\%$
(c) $\dfrac{100PV}{3RT}\%$ (d) $\dfrac{300PV}{RT}\%$

31. For the equilibrium $SrCl_2 \cdot 6H_2O(s) \rightleftharpoons SrCl_2 \cdot 2H_2O(s) + 4H_2O(g)$, $K_P = 8.1 \times 10^{-7}$ $atm^4$ at 27°C. If 1.642 L of air saturated with water vapour at 27°C is exposed to a large quantity of $SrCl_2 \cdot 2H_2O(s)$, then what mass of water vapour will be absorbed? Saturated vapour pressure of water at 27°C = 30.4 torr.
(a) 12 mg (b) 6.67 mg
(c) 9 mg (d) 48 mg

EXERCISE II

**32.** Two systems $PCl_5(g) \rightleftharpoons PCl_3(g) + Cl_2(g)$ and $COCl_2(g) \rightleftharpoons CO(g) + Cl_2(g)$ are simultaneously in equilibrium in a vessel at constant volume. If some CO is introduced into the vessel, then at the new equilibrium, the moles of

(a) $PCl_5$ increases.
(b) $PCl_5$ remains unchanged.
(c) $PCl_5$ decreases.
(d) $Cl_2$ increases.

**33.** In the system, $LaCl_3(s) + H_2O(g) + \text{heat} \rightleftharpoons LaClO(s) + 2HCl(g)$, equilibrium is established. More water vapour is added to disturb the equilibrium. If the pressure of water vapour at new equilibrium is double that of at initial equilibrium, the factor by which pressure of HCl is changed is

(a) 2 times (b) $\sqrt{2}$ times
(c) $\frac{1}{\sqrt{2}}$ times (d) 4 times

**34.** Indicate the correct answer out of the following for the below reaction.

$NH_4Cl + H_2O \rightleftharpoons NH_4OH + HCl$

(a) The reaction is retarded by the addition of KOH.
(b) The reaction is favoured by the addition of $NH_4OH$.
(c) The reaction is retarded by the addition of hydrogen ion.
(d) The reaction is favoured by the addition of hydrogen ion.

**35.** The two equilibria $AB \rightleftharpoons A^+ + B^-$ and $AB + B^- \rightleftharpoons AB_2^-$ are simultaneously maintained in a solution with equilibrium constants $K_1$ and $K_2$, respectively. The ratio of $A^+$ to $AB_2^-$ in the solution is

(a) directly proportional to the concentration of B.
(b) inversely proportional to the concentration of B.
(c) directly proportional to the square of the concentration of B.
(d) inversely proportional to the square of the concentration of B.

**36.** Solid $NH_4HS$ dissociates into $NH_3$ and $H_2S$ at a certain temperature, the equilibrium pressure is $P$ atm. If now, $NH_3$ is pumped into the system so that its partial pressure becomes $P$ atm, then what will be the partial pressure (in atm) of $H_2S$?

(a) 0.5 $P$ (b) 0.25 $P$
(c) 0.33 $P$ (d) 0.67 $P$

**37.** In a closed rigid vessel, the following equilibrium partial pressures are measured, where $N_2$ = 100 mm, $H_2$ = 400 mm and $NH_3$ = 1000 mm. Now, nitrogen is removed from the vessel until the pressure of hydrogen at equilibrium is equal to 700 mm. The new equilibrium partial pressure of $N_2$ is

(a) 11.94 mm (b) 200 mm
(c) 18.66 mm (d) 43.78 mm

**38.** Suppose the gas phase isomerization reactions: $A \rightleftharpoons B$, $A \rightleftharpoons C$ and $B \rightleftharpoons C$ achieve equilibrium simultaneously at a fixed temperature. The equilibrium mole fraction of 'A' in terms of equilibrium constants, $K_1$, $K_2$ and $K_3$ (respectively) is

(a) $\frac{K_1}{K_1+K_2+K_3}$ (b) $\frac{1}{1+K_1+K_2}$
(c) $\frac{1}{K_1+K_2+K_3}$ (d) $\frac{1}{K_1+K_2}$

**39.** Methanol ($CH_3OH$) can be prepared from CO and $H_2$ as

$CO(g) + 2H_2(g) \rightleftharpoons CH_3OH(g)$; $K_P = 6.23 \times 10^{-3}$ at 500 K.

What total pressure is required to convert 25% of CO to $CH_3OH$ at 500 K, if CO and $H_2$ comes from the following reaction?

$CH_4(g) + H_2O(g) \rightarrow CO(g) + 3H_2(g)$

(a) 20.48 bar (b) 21 bar
(c) 10.24 bar (d) 5.12 bar

**40.** An amount of 1 mole each of A and D is introduced in 1 L container. Simultaneously the following two equilibria are established.

$A \rightleftharpoons B + C$: $K_C = 10^6$ M and
$B + D \rightleftharpoons A$: $K_C = 10^{-6}\ M^{-1}$

The equilibrium concentration of A will be

(a) $10^{-6}$ M (b) $10^{-3}$ M
(c) $10^{-12}$ M (d) $10^{-4}$ M

**41.** For a reversible reaction $A \underset{K_2}{\overset{K_1}{\rightleftharpoons}} B$, the initial molar concentration of A and B are $a$ M and $b$ M, respectively. If $x$ M of A is reacted till the achievement of equilibrium, then $x$ is

(a) $\frac{K_1a - K_2b}{K_1 + K_2}$ (b) $\frac{K_1a - K_2b}{K_1 - K_2}$
(c) $\frac{K_1a - K_2b}{K_1K_2}$ (d) $\frac{K_1a + K_2b}{K_1 + K_2}$

**42.** For the given reaction $2A(s) + B(g) \rightleftharpoons C(g) + 2D(s) + E(s)$, the extent of reaction of B was found to be 20% at 300 K and 24% at 500 K. The rate of backward reaction

(a) increases with increase in pressure and temperature.
(b) increases with increase in pressure and decrease in temperature.
(c) depends on temperature only and decreases with increases in temperature.
(d) depends on temperature only and increases with increase in temperature.

**43.** Steam decomposes at high temperature according to the following equation.
$2H_2O(g) \rightleftharpoons 2H_2(g) + O_2(g)$; $\Delta H^\circ = 240$ kJ/mole and $\Delta S^\circ = 50$ JK$^{-1}$/mole
The temperature at which the equilibrium constant ($K_P^\circ$) becomes 1.0 is

(a) 4.8 K  (b) 4800 K
(c) 480 K  (d) impossible

**44.** $\Delta_f G^\circ$ for the formation of HI(g) from its gaseous elements is – 2.303 kcal/mol at 500 K. When the partial pressure of HI is 10 atm and of $I_2(g)$ is 0.001 atm, then what must be the partial pressure of hydrogen be at this temperature to reduce the magnitude of $\Delta G$ for the reaction to zero?

(a) 1000 atm  (b) 10000 atm
(c) 100 atm  (d) 31.63 atm

**45.** When α-D-glucose is dissolved in water, it undergoes a partial conversion to β-D-glucose. This conversion is called mutarotation and it stops when 64.0% of the glucose is in the β-form. Assuming that equilibrium has been attained, what is $\Delta G^\circ$ for the reaction α-D-glucose $\rightleftharpoons$ β-D-glucose, at this experimental temperature?

(a) $-RT\log_{10}(1.6)$  (b) $-RT\log_{10}(1.78)$
(c) $-RT\log_e(1.78)$  (d) $-RT\log_e(1.6)$

**46.** At chemical equilibrium, a reaction is

(a) spontaneous in forward direction.
(b) spontaneous in backward direction.
(c) spontaneous in both direction.
(d) spontaneous in neither direction.

**47.** Rate of disappearance of the reactant 'A' in the reversible reaction $A \rightleftharpoons B$ at two temperatures is given as

$$-\frac{d[A]}{dt} = (2.0 \times 10^{-3}\ s^{-1})[A] - (5.0 \times 10^{-4}\ s^{-1})[B]$$

(at 27°C)

$$-\frac{d[A]}{dt} = (8.0 \times 10^{-2}\ s^{-1})[A] - (4.0 \times 10^{-3}\ s^{-1})[B]$$

(at 127°C)

The enthalpy of reaction in the given temperature range is

(a) $-\frac{2.303\times8.314\times300\times400}{100}\cdot\log(50)$ J/mol
(b) $-\frac{2.303\times8.314\times300\times400}{100}\cdot\log(5)$ J/mol
(c) $\frac{2.303\times8.314\times300\times400}{100}\cdot\log(50)$ J/mol
(d) $\frac{2.303\times8.314\times300\times400}{100}\cdot\log(5)$ J/mol

**48.** A certain gas 'A' polymerizes to a very small extent at a given temperature as $nA(g) \rightleftharpoons A_n(g)$. The reaction is started with one mole of 'A' in a container of capacity $V$. Which of the following is the correct value of $\frac{PV}{RT}$, at equilibrium?

(a) $1-\frac{(n-1).K_C}{V^{n-1}}$  (b) $\frac{(n-1).K_C}{V^{n-1}}$
(c) $1-\frac{n.K_C}{V^{n-1}}$  (d) $1-\frac{n.K_C}{V^n}$

**49.** $PCl_5$ (molecular mass = $M$) dissociates into $PCl_3$ and $Cl_2$ as $PCl_5(g) \rightleftharpoons PCl_3(g) + Cl_2(g)$. If the total pressure of the system at equilibrium is $P$ and the density is 'd' at temperature $T$ K. The degree of dissociation of $PCl_5$ may be represented as

(a) $\frac{PM}{dRT}$  (b) $\frac{PM}{dRT}-1$
(c) $\frac{dRT}{PM}-1$  (d) $\frac{dRT}{PM}$

EXERCISE II

**50.** A 250 ml flask and 100 ml flask are separated by a stopcock. At 350 K, the nitric oxide in the larger flask exerts a pressure of 0.4 atm, and the smaller one contains oxygen at 0.8 atm. The gases are mixed by opening the stopcock. The reactions occurring are as follows.

$2NO + O_2 \rightarrow 2NO_2 \rightleftharpoons N_2O_4$

The first reaction is complete while the second one is at equilibrium. Assuming all the gases to behave ideally, calculate the $K_P$ for the second reaction if the total pressure is 0.3 atm.

(a) 3.5 $atm^{-1}$ (b) 0.87 $atm^{-1}$
(c) 0.07 $atm^{-1}$ (d) 7.0 $atm^{-1}$

**51.** If 'a' is the fraction of ammonia present by volume in an equilibrium mixture made from one volume of $N_2$ and three volumes of $H_2$ and $P$ is the total pressure, then

(a) $\frac{a}{1-a} \alpha P$ (b) $\frac{a}{(1-a)^2} \alpha P$
(c) $\frac{a}{(1+a)^2} \alpha P$ (d) $\frac{a}{1+a} \alpha P$

**52.** At 525 K, $PCl_5(g)$ is 80% dissociated at a pressure of 1 atm. Now, sufficient quantity of an inert gas at constant pressure is introduced into the above reaction mixture to produce inert gas partial pressure of 0.9 atm. What is the percentage dissociation of $PCl_5(g)$ when equilibrium is re-established?

(a) 97.3% (b) 80%
(c) 65.6% (d) 4.7%

**53.** For the reaction: $CuSO_4.3H_2O(s) \rightleftharpoons CuSO_4.H_2O(s) + 2H_2O(g)$; $\Delta H = 3360$ cal. The dissociation pressure is $7 \times 10^{-3}$ atm at 27°C. What will be the dissociation pressure at 127°C (ln2 = 0.7)?

(a) $9.8 \times 10^{-3}$ atm (b) $1.4 \times 10^{-2}$ atm
(c) $1.4 \times 10^{-3}$ atm (d) $9.8 \times 10^{-2}$ atm

**54.** Carbon monoxide in water gas reacts with steam according to the following reaction.

$CO(g) + H_2O(g) \rightleftharpoons CO_2(g) + H_2(g)$; $K_{eq} = 0.3333$

If two volumes of water gas (containing 1:1 of CO and $H_2$) are mixed with five volumes of steam, then the volume ratio of the four gases present at equilibrium is

(a) 1 : 1 : 1 : 1 (b) 1 : 3 : 1 : 1
(c) 1 : 15 : 1 : 5 (d) 1 : 9 : 1 : 3

**55.** If for the equilibria $NH_2COONH_4(s) \rightleftharpoons N_2 + H_2 + CO + O_2$, the value of $K_P$ at 800 K is $27 \times 2^{\lambda/2}$ and the equilibrium pressure is 22 atm. The value of λ is

(a) 21 (b) 22
(c) 11 (d) 12

**56.** Solid ammonium carbamate dissociates as: $NH_2COONH_4(s) \rightleftharpoons 2NH_3(g) + CO_3(g)$. In a closed vessel, solid ammonium carbamate is in equilibrium with its dissociation products. At equilibrium ammonia is added such that the partial pressure of $NH_3$ at new equilibrium now equals the original total pressure. The ratio of total pressure at new equilibrium to that of original total pressure is

(a) 1 : 1 (b) 27 : 31
(c) 31 : 27 (d) 3 : 4

**57.** The value of equilibrium constant for the following reaction at 300 K and constant pressure is

A(g) + B(g) $\rightleftharpoons$ C(g) + D(g) + E(g); $\Delta E^o = 30$ kcal and $\Delta S^o - 100$ cal/K.

(a) $e$ (b) $\frac{1}{e}$
(c) $e^2$ (d) $\frac{1}{e^2}$

**58.** $\Delta_f G^o$ are 30.426 kJ/mol for *trans*-1,2-dichloroethene and 22.112 kJ/mol for *cis*-1,2-dichloroethene, at 27°C. The molar ratio of *trans* to *cis* isomers at equilibrium at 27°C, is [ln28 = 3.33]

(a) 10 : 3 (b) 3 : 10
(c) 28 : 1 (d) 1 : 28

**59.** The equilibrium constant for the reaction CO(g) + $H_2O(g) \rightleftharpoons CO_2(g) + H_2(g)$ is 3.0 at 500 K. In a 2.0 L vessel, 60 g of water gas [equimolar mixture of CO(g) and $H_2$(g)] and 90 g steam is initially taken. What is the equilibrium concentration of $H_2$(g)?

(a) 1.75 M (b) 3.5 M
(c) 1.5 M (d) 0.75 M

**60.** The equilibrium $NH_4HS(s) \rightleftharpoons NH_3(g) + H_2S(g)$ is achieved at the equilibrium pressure of 'X' bar at $T$ K. The value of $\Delta_r G^o$ for the reaction is

(a) $-RT$ lnX (b) $-2RT$ lnX
(c) $-2RT$ (lnX – ln2) (d) $-2RT$ ln(2X)

## Section B (One or More than one Correct)

1. When $NH_4HS(s)$ is vaporized in an empty vessel and maintained at 20°C, the equilibrium is established $NH_4HS(s) \rightleftharpoons NH_3(g) + H_2S(g)$ and the total pressure of gases at equilibrium is 0.4 atm. When $NH_4HS(s)$ is vaporized in the presence of $NH_3(g)$ in the same vessel at 20°C, the partial pressure of $NH_3(g)$ at equilibrium is 0.5 atm. Which of the following is/are correct statement(s)?
   (a) At second equilibrium, the final partial pressure of $H_2S(g)$ is 0.08 atm.
   (b) In second experiment, the initial pressure of $NH_3(g)$ was 0.42 atm.
   (c) In the presence of $NH_3(g)$, the extent of dissociation of $NH_4HS(s)$ is decreased.
   (d) At second equilibrium, the mole fraction of $H_2S(g)$ is 0.5.

2. Which of the following reaction(s) have $K_P = K_C$?
   (a) $H_2(g) + I_2(g) \rightleftharpoons 2HI(g)$
   (b) $N_2(g) + O_2(g) \rightleftharpoons 2NO(g)$
   (c) $2NO(g) + Cl_2(g) \rightleftharpoons 2NOCl(g)$
   (d) $2SO_2(g) + O_2(g) \rightleftharpoons 2SO_3(g)$

3. For dissociation of a gas $N_2O_5$ as $N_2O_5(g) \rightleftharpoons 2NO_2(g) + \frac{1}{2}O_2(g)$. The reaction is performed at constant temperature and volume. If $D$ is the vapour density of equilibrium mixture, $P_o$ is initial pressure of $N_2O_5(g)$ and $M$ is molecular mass of $N_2O_5$, then the correct information(s) at the equilibrium is/are
   (a) the total pressure of gases at equilibrium is $\frac{P_o \cdot M}{2D}$.
   (b) the degree of dissociation of $N_2O_5(g)$ is $\frac{M-2D}{3D}$.
   (c) the partial pressure of $N_2O_5(g)$ at equilibrium is $\frac{(5D-M)\cdot P_o}{3D}$.
   (d) the partial pressure of $O_2(g)$ at equilibrium is $\frac{(M-2D)\cdot P_o}{3D}$.

4. Which of the following is/are correct statement(s)?
   (a) The dissociation of $CaCO_3$ is suppressed at high pressure.
   (b) The apparent molecular mass of $PCl_5$ shows lower value on dissociation.
   (c) Low pressure is favourable for melting of ice.
   (d) Combination of hydrogen atoms to form hydrogen molecule is favourable at high temperature.

5. What is/are true about equilibrium state?
   (a) Catalyst has no effect on equilibrium state.
   (b) Equilibrium constant is independent of initial concentration of reaction.
   (c) The reaction ceases at equilibrium.
   (d) Equilibrium constant is independent of pressure.

6. A chemical system is in equilibrium. Addition of a catalyst would result in
   (a) increase in the rate of forward reaction.
   (b) increase in the rate of reverse reaction.
   (c) a new reaction pathway to reaction.
   (d) increase the amount of heat evolved.

7. A cylinder fitted with a movable piston contains liquid water in equilibrium with water vapour at 25°C. Which operation results in a decrease in the equilibrium vapour pressure?
   (a) Moving the piston downwards a short distance.
   (b) Removing a small amount of vapour.
   (c) Removing a small amount of the liquid water.
   (d) Dissolving salt in the water.

8. The condition suitable for forming atomic chlorine from molecular chlorine is
   (a) Low temperature (b) Low pressure
   (c) High temperature (d) High pressure

9. The position of equilibrium will shift in the given direction by the addition of inert gas at constant pressure in which of the following case(s)?
   (a) $N_2(g) + 3F_2(g) \rightleftharpoons 2NF_3(g)$; forward direction
   (b) $COCl_2(g) \rightleftharpoons CO(g) + Cl_2(g)$; forward direction
   (c) $CO(g) + 2H_2(g) \rightleftharpoons CH_3OH(g)$; backward direction
   (d) $2C(s) + O_2(g) \rightleftharpoons 2CO(g)$; backward direction

10. Two gases A and B, one being the dimer of the other, are at equilibrium. Decreasing of pressure at constant temperature or increasing of temperature at constant pressure favours the formation of more of B. The reaction could not be represented by
   (a) $2A \rightleftharpoons B + q$ calories
   (b) $2A \rightleftharpoons B - q$ calories
   (c) $2B \rightleftharpoons A - q$ calories
   (d) $2B \rightleftharpoons A + q$ calories

EXERCISE II

**11.** The equilibrium constant for some reactions are given below against each of the reaction.

(i) $2N_2 + 5O_2 \rightleftharpoons 2N_2O_5$; $K = 5 \times 10^{-27}$
(ii) $N_2 + O_2 \rightleftharpoons 2NO$; $K = 2 \times 10^{-15}$
(iii) $N_2 + 2O_2 \rightleftharpoons 2NO_2$; $K = 1.5 \times 10^{-29}$

Which of the following statement is correct?
(a) The least stable oxide is $NO_2$.
(b) The most stable oxide is NO.
(c) The stability order is $N_2O_5 > NO_2 > NO$.
(d) The stability order is $NO_2 > NO > N_2O_5$.

**12.** For the gaseous reaction: $CO + H_2O \rightleftharpoons CO_2 + H_2$, the following thermodynamical data are given.

$\Delta H^\circ_{300\,K} = -41.0\ \text{kJ mol}^{-1}$;
$\Delta S^\circ_{300\,K} = -0.04\ \text{kJ K}^{-1}\ \text{mol}^{-1}$

$\Delta H^\circ_{1200\,K} = -33.0\ \text{kJ mol}^{-1}$;
$\Delta S^\circ_{1200\,K} = -0.03\ \text{kJ K}^{-1}\ \text{mol}^{-1}$

Assuming partial pressure of each component is 1 bar, the direction of spontaneous reaction is/are
(a) forward at 300 K.
(b) forward at 1200 K.
(c) backward at 300 K.
(d) backward at 1200 K.

**13.** van 't Hoff equations show the effect of temperature on equilibrium constants $K_C$ and $K_P$. $K_P$ and $K_C$ varies with temperature according to which of the following relations?

(a) $\log \dfrac{K_{P_2}}{K_{P_1}} = \dfrac{\Delta H}{2.303R}\left[\dfrac{T_1 - T_2}{T_1T_2}\right]$

(b) $\log \dfrac{K_{P_2}}{K_{P_1}} = \dfrac{\Delta H}{2.303R}\left[\dfrac{T_2 - T_1}{T_1T_2}\right]$

(c) $\log \dfrac{K_{C_2}}{K_{C_1}} = \dfrac{\Delta U}{2.303R}\left[\dfrac{T_2 - T_1}{T_1T_2}\right]$

(d) $\log \dfrac{K_{C_2}}{K_{C_1}} = \dfrac{\Delta U}{2.303R}\left[\dfrac{T_1 - T_2}{T_1T_2}\right]$

**14.** For a chemical reaction at the state of equilibrium, which of the following statement(s) is/are correct?
(a) None of the variables like temperature, pressure or volume appear to change.
(b) Reaction system has maximum stability with minimum energy content.
(c) Addition of catalyst disturbs the point of equilibrium.
(d) Addition of inert gas at equilibrium for constant volume system will move the equilibrium in the direction, where more number of moles of gases is present.

**15.** Solubility of a solute in a solvent (say, water) is dependent on the temperature as given by $S = A.e^{-\Delta H/RT}$, where $\Delta H$ is the heat of reaction: solute $+ H_2O \rightleftharpoons$ solution. For a given solution, variation of log $S$ with temperature is shown graphically. The solute should be

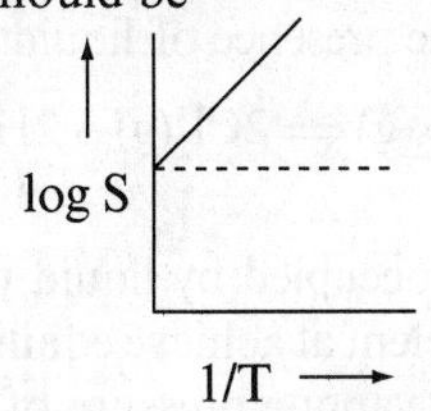


(a) $CuSO_4 \cdot 5H_2O$ (b) NaCl
(c) Sucrose (d) CaO

**16.** At chemical equilibrium, the
(a) enthalpy of system is minimum.
(b) entropy of system is maximum.
(c) free energy of system is minimum.
(d) free energy of system is zero.

**17.** At –10°C, the solid compound $Cl_2(H_2O)_8$ is in equilibrium with gaseous chlorine, water vapour and ice. The partial pressure of the two gases in equilibrium with a mixture of $Cl_2(H_2O)_8$ and ice are 0.20 atm for $Cl_2$ and 0.001 atm for water vapour. The processes may be represented as follows.

(i) $Cl_2(H_2O)_8(s) \rightleftharpoons Cl_2(g) + 8H_2O(g)$; $K_{P_1}$

(ii) $Cl_2(H_2O)_8(s) \rightleftharpoons Cl_2(g) + 8H_2O(s)$; $K_{P_2}$

Identify the correct statement among the following regarding the processes.
(a) The value of $K_{P_1}$ is $2.0 \times 10^{-25}\ \text{atm}^9$.
(b) The value of $K_{P_2}$ is 0.2 atm.
(c) The vapour pressure of ice is 0.001 atm at 263 K.
(d) Process (i) must be exothermic.

**18.** In an evacuated vessel of capacity 112 L, 4 moles of Ar(g) and 5 moles of $PCl_5(g)$ were introduced and the temperature is maintained at 273°C. At equilibrium, the total pressure of the mixture was found to be 4.8 atm.

(a) The degree of dissociation of $PCl_5(g)$ into $PCl_3(g)$ and $Cl_2(g)$ is 0.6.

(b) $K_P$ for the reaction: $PCl_5(g) \rightleftharpoons PCl_3(g) + Cl_2(g)$ is 2.7 atm.

(c) At equilibrium, the total moles of gases are 12.

(d) On removing Ar(g) from the equilibrium mixture at constant pressure and temperature, the extent of dissociation of $PCl_5$ will increase.

**19.** In a vessel of 1.0 L capacity, $O_2(g)$ at 0.25 atm pressure and HCl(g) at 1.0 atm pressure are allowed to react in the presence of liquid water at 57°C.

$4HCl(g) + O_2(g) \rightleftharpoons 2Cl_2(g) + 2H_2O(g)$; $K_P = 5.0 \times 10^{12}\ atm^{-1}$.

The volume occupied by liquid water is negligible but it is sufficient at achieve equilibrium with water vapour. The vapour pressure of water at 57°C is 0.4 atm. Select the correct statement(s) regarding the equilibrium mixture.

(a) The partial pressure of water vapour at any stage of reaction is 0.4 atm.

(b) The partial pressure of $Cl_2(g)$ at equilibrium is 0.5 atm.

(c) The partial pressure of $O_2(g)$ at equilibrium is $5.0 \times 10^{-4}$ atm.

(d) The partial pressure of HCl(g) at equilibrium is $2.0 \times 10^{-3}$ atm.

**20.** In the following equilibrium $N_2O_4(g) \rightleftharpoons 2NO_2(g)$, when 5 moles of each is taken and the temperature is kept at 298 K, the total pressure was found to be 20 bar. Given $\Delta_f G^o[N_2O_4(g)] = 100$ kJ/mol; $\Delta_f G^o[NO_2(g)] = 50$ kJ/mol.

(a) $K_P$ for the reaction is 1.0 bar.

(b) $\Delta G^o$ for the reaction is zero at 298 K.

(c) Under the given condition, the reaction is mainly occurring in backward direction.

(d) The mixture taken is at equilibrium.

**21.** A gaseous substance $AB_2(g)$ converts to AB(g) in the presence of solid A as $AB_2(g) + A(s) \rightleftharpoons 2AB(g)$. The initial pressure and equilibrium pressure are 0.7 and 0.95 bar, respectively. Now, the equilibrium mixture is expanded reversibly and isothermally till the gas pressure falls to 0.4 bar. Then, which of the following statements is correct?

(a) The volume percent of AB(g) at final equilibrium is nearly 27%.

(b) The volume percent of $AB_2(g)$ at final equilibrium is nearly 32.5%.

(c) $K_P$ for the equilibrium is 5/9.

(c) The ratio of equilibrium pressure of $AB_2(g)$ at initial and final equilibrium is more than 2 but less than 3.

**22.** An amount of 1 mole of $PCl_3(g)$ and 1 mole of $PCl_5(g)$ is taken in a vessel of 10 L capacity maintained at 400 K. At equilibrium, the moles of $Cl_2$ is found to be 0.004.

(a) $K_C$ for the reaction: $PCl_5(g) \rightleftharpoons PCl_3(g) + Cl_2(g)$ is 0.0004 M.

(b) $K_P$ for the reaction: $PCl_3(g) + Cl_2(g) \rightleftharpoons PCl_5(g)$ is $0.0004 \times (0.082 \times 400)$ atm.

(c) If $PCl_3(g)$ is added to the equilibrium mixture, $\Delta G$ at the new equilibrium becomes greater than the $\Delta G$ at old equilibrium.

(d) After equilibrium is achieved, the moles of $PCl_3$ are doubled and moles of $Cl_2$ are halved simultaneously, then the partial pressure of $PCl_5$ remains unchanged.

**23.** A reaction at 300 K with $\Delta G^o = -1743$ J consists of 3 moles of A(g), 6 moles of B(g) and 3 moles of C(g). If A, B and C are in equilibrium in 1 L container, then the reaction may be (ln 2 = 0.7, $R = 8.3$ J/K-mol)

(a) $A + B \rightleftharpoons C$  (b) $A \rightleftharpoons B + 2C$

(c) $2A \rightleftharpoons B + C$  (d) $A + B \rightleftharpoons 2C$

**24.** The equilibrium between gaseous isomers A, B and C can be represented as follows.

$A(g) \rightleftharpoons B(g)$; $K_1 = ?$
$B(g) \rightleftharpoons C(g)$; $K_2 = 0.4$
$C(g) \rightleftharpoons A(g)$; $K_3 = 0.6$

If one mole of A is taken in a closed vessel of volume 1 L, then

(a) [A] + [B] + [C] = 1 M at any moment of reaction.

(b) concentration of C is 1.2 M at equilibrium.

(c) the value of $K_1$ is 1/0.24.

(d) isomer A is thermodynamically least stable.

25. Consider the equilibrium $HgO(s) + 4I^-(aq) + H_2O(l) \rightleftharpoons HgI_4^{2-}(aq) + 2OH^-(aq)$, which changes will decrease the equilibrium concentration of $HgI_4^{2-}(aq)$?
(a) Addition of 0.1 M-HI(aq).
(b) Addition of HgO(s).
(c) Addition of $H_2O(l)$.
(d) Addition of KOH(aq).

## Section C (Comprehensions)

### Comprehension I

**Equilibrium constants are given for the following reactions at 27°C.**

$$SrCl_2 \cdot 6H_2O(s) \rightleftharpoons SrCl_2 \cdot 2H_2O(s) + 4H_2O\ (g);\ K_P = 2.56 \times 10^{-10}\ atm^4$$

$$Na_2HPO_4 \cdot 12H_2O(s) \rightleftharpoons Na_2HPO_4 \cdot 7H_2O(s) + 5H_2O\ (g);\ K_P = 2.43 \times 10^{-13}\ atm^5$$

$$Na_2SO_4 \cdot 10H_2O(s) \rightleftharpoons Na_2SO_4\ (s) + 10\ H_2O(g);\ K_P = 1.024 \times 10^{-27}\ atm^{10}$$

**The vapour pressure of water at 27°C is 0.04 atm.**

1. Which is the most effective drying agent at 27°C?
(a) $SrCl_2{\cdot}2H_2O$, (b) $Na_2HPO_4{\cdot}7H_2O$
(c) $Na_2SO_4$ (d) All, equally

2. At what relative humidity will $Na_2SO_4.10H_2O(s)$ be efflorescent when exposed to the air at 27°C?
(a) Below 50% (b) Above 50%
(c) Above 5% (d) Below 5%

3. At what relative humidity will $Na_2HPO_4.7H_2O(s)$ deliquescent when exposed to the air at 27°C?
(a) Below 75% (b) Above 75%
(c) Above 7.5% (d) Below 7.5%

### Comprehension II

**An amount of 0.20 moles of CO taken in a 2.463 L flask is maintained at 750 K along with a catalyst so that the following reaction can take place $CO(g) + 2H_2(g) \rightleftharpoons CH_3OH(g)$. Hydrogen is introduced until the total pressure of the system is 7.5 atm at equilibrium and 0.1 mole of methanol is formed.**

4. For the reaction, $K_P$ is
(a) 0.16 $atm^{-2}$ (b) 6.25 $atm^{-2}$
(c) 0.04 $atm^{-2}$ (d) 1.56 $atm^{-2}$

5. For the reaction, $K_C$ is [Given: $(2.463)^2 = 6.07$]
(a) 6.07 $M^{-2}$ (b) 607 $M^{-2}$
(c) 151.75 $M^{-2}$ (d) 2428 $M^{-2}$

6. What is the final pressure if the same amount of CO and $H_2$ as before used, but with no catalyst so that the reaction does not take place?
(a) 18.75 atm (b) 10.0 atm
(c) 15.0 atm (d) 12.5 atm

### Comprehension III

**In the esterification reaction,**

$$C_2H_5OH(l) + CH_3COOH(l) \rightleftharpoons CH_3COOC_2H_5(l) + H_2O(l)$$

**an equimolar mixture of alcohol and acid taken initially, yields under equilibrium, the ester with mole fraction = 0.333.**

7. The equilibrium constant is
(a) 10 (b) 4
(c) 100 (d) 16

8. What fraction of the alcohol would be esterified if the initial mole fraction of the alcohol is 0.333? [Given: $\sqrt{12} = 3.45$]
(a) 0.15 (b) 0.85
(c) 0.28 (d) 0.72

**9.** If in this esterification, initially equal moles of alcohol and acid are taken. At equilibrium, unreacted alcohol and acid, on treatment of sodium, produced $H_2$ which occupied 44.8 L at 0°C and 1 atm. The percentage esterification of acid and alcohol is

(a) 25% (b) 33.3%
(c) 50% (d) 66.7%

## Comprehension IV

**At 444°C, HI undergoes dissociation to the extent of 22.22% at equilibrium**

**10.** What is the equilibrium constant of the reaction $2HI(g) \rightleftharpoons H_2(g) + I_2(g)$ at 444°C?

(a) 50 (b) 0.02
(c) 0.143 (d) 3.14

**11.** What is the percentage dissociation of HI at 444°C if the experiment is started with one mole of HI(g) and one mole of $I_2$(g)?

(a) Less than 22.22%
(b) More than 22.22%
(c) Equal to 22.22%
(d) Nothing can be said

**12.** What is the percentage dissociation of HI at 444°C, if the experiment is started with one mole of HI(g) and 1 mole of He(g)?

(a) Less than 22.22%
(b) More than 22.22%
(c) Equal to 22.22%
(d) Nothing can be said

## Comprehension V

**Le Chatelier performed an experiment in science academy. He introduced excess of solid $NH_4HS$ into a closed 5.0 L vessel containing $NH_3$ gas at a partial pressure of *P* mm and at temperature 300 K. Due to it, dissociation of the solid takes place and at equilibrium, he measured the partial pressure of $NH_3$ gas equal to 625 mm and the total pressure of gases equal to 725 mm.**

**13.** The value of *P* is

(a) 725 (b) 625
(c) 525 (d) 100

**14.** The value of $K_P$ for the reaction $NH_3(g) + H_2S(g) \rightleftharpoons NH_4HS(s)$, is

(a) $6.25 \times 10^4$ mm$^2$ (b) $1.6 \times 10^{-5}$ mm$^{-2}$
(c) $1.6 \times 10^{-3}$ mm$^{-2}$ (d) $4.53 \times 10^5$ mm$^2$

**15.** What would have been the partial pressure of $NH_3$, if excess solid $NH_4HS$ had been taken in an evacuated vessel of the same volume at the same temperature?

(a) 1000 mm (b) 750 mm
(c) 500 mm (d) 250 mm

**16.** The minimum mass of $NH_4HS$(s) needed to establish equilibrium in the same evacuated vessel at the same temperature is

(a) $\dfrac{250 \times 5}{0.0821 \times 300}$ g

(b) $\dfrac{250 \times 5 \times 51}{0.0821 \times 300}$ g

(c) $\dfrac{250 \times 5 \times 51}{0.0821 \times 300 \times 760}$ g

(d) $\dfrac{250 \times 5}{0.0821 \times 300 \times 760}$ g

## Comprehension VI

**For the system $A(g) + B(g) \rightleftharpoons C(g) + D(g)$, $\Delta H$ for the forward reaction is −24.942 kJ/mol. The activation energies of forward and backward reactions are in the ratio of 2 : 3. Assume that the pre-exponential factor is the same for the forward and backward reactions ($A_f = A_b = 1$).**

17. What is the equilibrium constant for the reaction at 27°C?
   (a) $e^4$ (b) $e^{-4}$
   (c) $e^{10}$ (d) $e^{-10}$

18. What is the ratio of rate constant of backward to forward direction at 27°C?
   (a) $e^4$ (b) $e^{-4}$
   (c) $e^{10}$ (d) $e^{-10}$

19. Which of the following statement is correct at 27°C?
   (a) The rate constant of forward reaction is $e^{30}$.
   (b) The rate constant of forward reaction is $e^{-30}$.
   (c) The rate constant of backward reaction is $e^{30}$.
   (d) The rate constant of backward reaction is $e^{-30}$.

## Comprehension VII

**The degree of dissociation of $NH_3$ in a vessel $V_1$ (initially evacuated) at $T$ K at an equilibrium pressure of 50 atm is same as that of $SO_3$ (g) at a total equilibrium pressure equal to $K_P$ for**

$$2SO_3(g) \rightleftharpoons 2SO_2(g) + O_2(g)$$

**A vessel $V_2$ (capacity 100 L and maintained at $T$ K) initially contains equimolar mixture of $NH_3$, $N_2$, $H_2$ and Ne. The total mass of the mixture containing four gases is 0.134 kg. At equilibrium, $N_2$ was found to be 52.24% by mass. (Ne = 20).**

20. The degree of dissociation of $NH_3$ at $T$ K in the vessel of volume $V_1$ is
   (a) 0.33 (b) 0.50
   (c) 0.67 (d) 0.75

21. The value of $K_P$ for the reaction: $2NH_3(g) \rightleftharpoons N_2(g) + 3H_2(g)$ at $T$ K is
   (a) 2700 $atm^2$ (b) 2500 $atm^2$
   (c) $3.7 \times 10^{-4}$ $atm^2$ (d) $4.0 \times 10^{-4}$ $atm^2$

22. The total pressure of gases at equilibrium in the vessel $V_2$ is
   (a) $\sqrt{\frac{2700 \times 8}{2.5}}$ atm (b) $\sqrt{\frac{2700 \times 8}{2.5 \times 7}}$ atm
   (c) $\sqrt{\frac{2700 \times 81}{2.5 \times (3.5)^3}}$ atm (d) $\sqrt{\frac{2700 \times 81}{3.5 \times (2.5)^3}}$ atm

## Comprehension VIII

**In a vessel, the equilibria $N_2(g) + 3H_2(g) \rightleftharpoons 2NH_3(g)$ and $N_2(g) + 2H_2(g) \rightleftharpoons N_2H_4(g)$ are achieved simultaneously. Initially, the vessel contains $N_2$ and $H_2$ in molar ratio of 9 : 13. The equilibrium pressure is $7P_o$ in which due to ammonia, the pressure is $P_o$ and due to hydrogen, the pressure is $2P_o$.**

23. The value of $K_P$ for the reaction $N_2(g) + 3H_2(g) \rightleftharpoons 2NH_3(g)$ is
   (a) $20\ P_o^2$ (b) $\frac{20\ P_o^2}{3}$
   (c) $\frac{1}{20\ P_o^2}$ (d) $\frac{3}{20\ P_o^2}$

24. The value of $K_P$ for the reaction $N_2(g) + 2H_2(g) \rightleftharpoons N_2H_4(g)$ is
   (a) $20\ P_o^2$ (b) $\frac{20\ P_o^2}{3}$
   (c) $\frac{1}{20\ P_o^2}$ (d) $\frac{3}{20\ P_o^2}$

## Section D (Assertion–Reason)

The following questions consist of two statements. Mark the answer as follows.

(a) If both statements are CORRECT, and **Statement II** is the CORRECT explanation of **Statement I.**

(b) If both statements are CORRECT, and **Statement II** is NOT the CORRECT explanation of **Statement I.**

(c) If **Statement I** is CORRECT, but **Statement II** is INCORRECT.

(d) If **Statement I** is INCORRECT, but **Statement II** is CORRECT.

1. **Statement I:** For the reaction $A(g) \rightleftharpoons B(g)$, equilibrium moles of A and B are, respectively, 'a' and 'b' in 1 L container. If 5 moles of A and 3 moles of B are added, then reaction must move in forward direction.

   **Statement II:** Even if the amount of reactant added to a system at equilibrium is more than the amount of product added at the same time, the equilibrium can shift in any direction.

2. **Statement I:** Total number of moles in a closed system at new equilibrium is less than the old equilibrium if some amount of a substance is removed from the system (reaction: $A(g) \rightleftharpoons B(g)$).

   **Statement II:** The number of moles of the substance which is removed, is partially compensated as the system reaches to the new equilibrium.

3. **Statement I:** A net reaction can occur only if a system is not at equilibrium.

   **Statement II:** All reactions occur to reach a state of equilibrium.

4. **Statement I:** A catalyst does not influence the value of equilibrium constant.

   **Statement II:** Catalyst influence the rates of both forward and backward reactions by the same factor.

5. **Statement I:** Le Chatelier's principle predicts that an increase in temperature favours an endothermic process.

   **Statement II:** An endothermic process is one that absorbs heat and hence, tends to minimize the temperature increase.

6. **Statement I:** For a gaseous reversible reaction, $K_P$ can be equal to or less than or even greater than the value of $K_C$.

   **Statement II:** The relation between $K_P$ and $K_C$ depends on the change in the number of moles of gaseous reactants and products as well as temperature.

7. **Statement I:** The system $N_2O_4(g) \rightleftharpoons 2NO_2(g)$ is said to be at equilibrium when the intensity of brown colour does not change with time.

   **Statement II:** $N_2O_4(g)$ is colourless while $NO_2(g)$ is coloured.

8. **Statement I:** On cooling in a freezing mixture, the colour of the following mixture turns to pink from deep blue for the reaction:

   $$\underset{\text{Blue}}{Co(H_2O)_6^{2+}(aq.)} + 4Cl^-(aq.) \rightleftharpoons \underset{\text{Pink}}{CoCl_4^{2-}(aq.)} + 6H_2O(l)$$

   **Statement II:** The reaction is endothermic.

9. **Statement I:** NaCl solution can be purified by passing excess of hydrogen chloride gas through NaCl solution.

   **Statement II:** Increase in chloride ion concentration in the solution results in the precipitation of NaCl.

10. **Statement I:** Decrease in volume of the system at constant temperature at equilibrium results increase in molar concentration of all the gaseous components involved in the reaction $A(g) + B(s) \rightleftharpoons 3C(g)$.

    **Statement II:** Decrease in volume of the system at constant temperature at equilibrium results increase in moles of A(g) and decrease in moles of C(g) but no change in the moles of B(s) in the reaction $A(g) + B(s) \rightleftharpoons 3C(g)$.

## Section E (Column Match)

**1.** Match the columns.

| Column I | Column II |
|---|---|
| (A) $Q = K_{eq}$ | (P) Reaction is near to completion. |
| (B) $Q < K_{eq}$ | (Q) Reaction is not at equilibrium. |
| (C) $Q > K_{eq}$ | (R) Reaction is fast in forward direction. |
| (D) $K_{eq} >>> 1$ | (S) Reaction at equilibrium. |

**2.** Match the columns.

| Column I | Column II |
|---|---|
| (A) $H_2(g) + I_2(g) \rightleftharpoons 2HI(g)$ | (P) $K_P = K_C\,(RT)$ |
| (B) $N_2(g) + 3H_2(g) \rightleftharpoons 2NH_3(g)$ | (Q) $K_P = K_C\,(RT)^2$ |
| (C) $PCl_5(g) \rightleftharpoons PCl_3(g) + Cl_2(g)$ | (R) $K_P = K_C\,(RT)^{-2}$ |
| (D) $NH_4HS(s) \rightleftharpoons NH_3(g) + H_2S(g)$ | (S) $K_P = K_C$ |

**3.** Match the columns.

| Column I | Column II |
|---|---|
| (A) $H_2(g) + I_2(g) \rightleftharpoons 2HI(g)$ | (P) Unaffected by inert gas addition. |
| (B) $CaCO_3(s) \rightleftharpoons CaO(s) + CO_2(g)$ | (Q) Forward shift by rise in pressure and backward shift by inert gas addition. |
| (C) $N_2(g) + 3H_2(g) \rightleftharpoons 2NH_3(g)$ | (R) Unaffected by increase in pressure. |
| (D) $PCl_5(g) \rightleftharpoons PCl_3(g) + Cl_2(g)$ | (S) Backward shift by rise in pressure and forward shift by inert gas addition. |

**4.** Match the columns.

| Column I | Column II |
|---|---|
| (A) $2NH_3(g) \rightleftharpoons N_2(g) + 3H_2(g)$ | (P) Extent of reaction will not increase with increase in pressure. |
| (B) $2HI(g) \rightleftharpoons H_2(g) + I_2(g)$ | (Q) Concentration of product will increase with increase in temperature. |
| (C) $2(CH_3COOH)(g) \rightleftharpoons (CH_3COOH)_2(g)$ | (R) Increase in volume will increase moles of reactant. |
| | (S) Introduction of inert gas at constant pressure will shift the equilibrium in the product side. |

**5.** Match the columns.

| Column I | Column II |
|---|---|
| (A) $2NH_3(g) \rightleftharpoons N_2(g) + 3H_2(g)$ | (P) Introduction of hydrogen gas at constant volume shift equilibrium back. |
| (B) $PCl_5(g) \rightleftharpoons PCl_3(g) + Cl_2(g)$ | (Q) $\alpha$ increases on increasing temperature. |
| (C) $2HI(g) \rightleftharpoons H_2(g) + I_2(g)$ | (R) On adding inert gas at constant volume, equilibrium state does not change. |
| (D) $NH_4HS(s) \rightleftharpoons NH_3(g) + H_2S(g)$ | (S) On increasing the pressure, concentration of all reactants increases. |
| | (T) $\alpha$ is independent of equilibrium pressure. |

6. For an endothermic reaction: $4A(g) + B_2(g) \rightleftharpoons 2A_2B(g)$

**Column I**

(A)

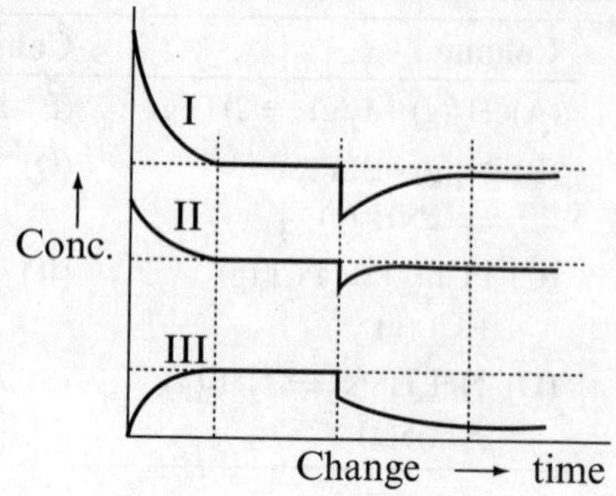


(B)

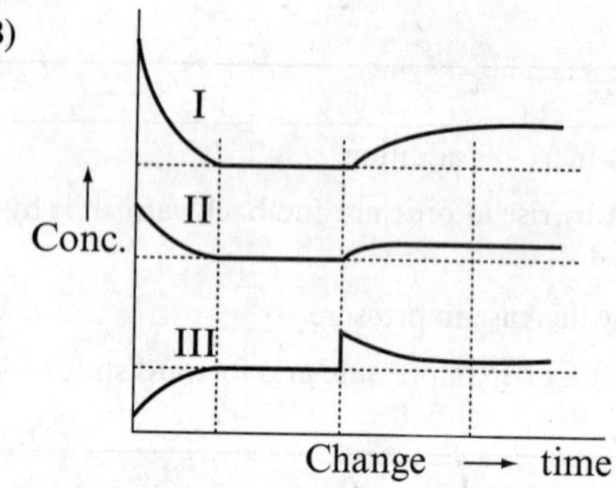


(C)

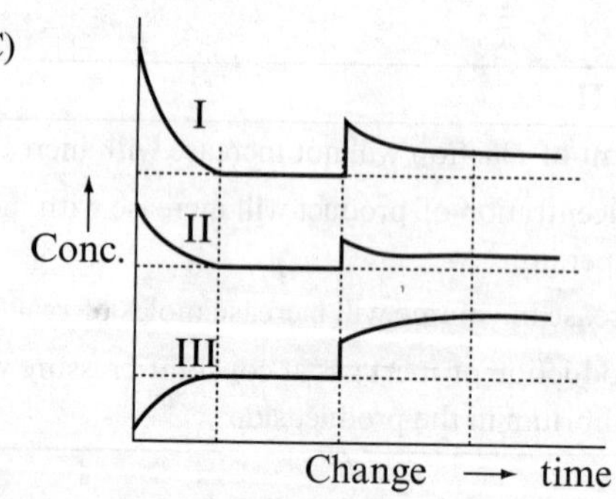


(D)

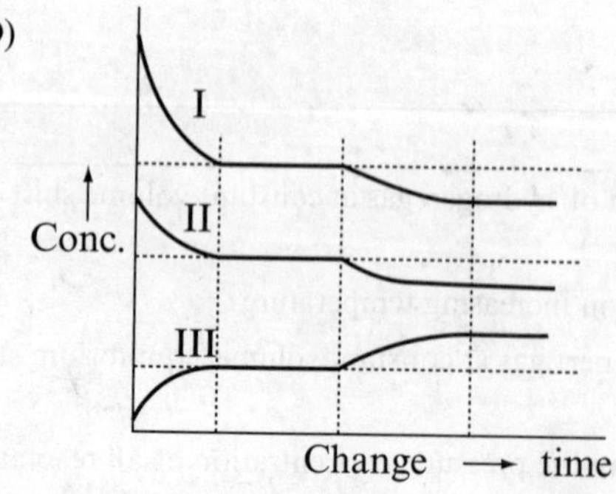


**Column II**

(P) Increase in temperature.

(Q) Increase in pressure.

(R) Addition of $A_2B$ at equilibrium.

(S) Addition of inert gas at constant pressure.

(T) Increase in volume.

**7.** Match the columns.

| Column I | Column II |
|---|---|
| (A) $PCl_5(g) \rightleftharpoons PCl_3(g) + Cl_2(g)$ | (P) $K_P > K_C$ at 300 K. |
| (B) $2NO_2(g) \rightleftharpoons N_2O_4(g)$ | (Q) $K_P < K_C$ at 300 K. |
| (C) $H_2(g) + I_2(g) \rightleftharpoons 2HI(g)$ | (R) $P$ is increased, reaction moves to right side. |
| (D) $N_2(g) + 3H_2(g) \rightleftharpoons 2NH_3(g)$ | (S) Inert gas added at constant volume, equilibrium will not be affected. |
| | (T) Inert gas added at constant pressure, equilibrium will shift right. |

**8.** Match the following for the reaction: $2A(g) \rightleftharpoons 3B(g) + C(s)$; $\Delta H = -\text{ve}$.

| Column I | Column II |
|---|---|
| (A) Increase in total pressure at equilibrium. | (P) Increase in moles of A. |
| (B) Increase in volume at equilibrium. | (Q) Increase in moles of B. |
| (C) Addition of He(g) at constant pressure. | (R) Increase in moles of C. |
| (D) Increase in temperature at constant volume. | (S) Increase in molar concentration of A. |
| | (T) Increase in molar concentration of B. |

**9.** For the reaction $CaCO_3(s) \rightleftharpoons CaO(s) + CO_2(g)$; $K_P = 2.463$ atm at 900 K. The reaction is performed in a rigid vessel of 15.0 L capacity maintained at 900 K, starting with the following amounts of $CaCO_3(s)$ (given in Column I). Match the correct information given in Column II for the amount given in Column I.

| Column I | Column II |
|---|---|
| (A) 100 g | (P) 50% dissociation of $CaCO_3$ |
| (B) 50 g | (Q) 100% dissociation of $CaCO_3$ |
| (C) 25 g | (R) Addition of small amount of $CO_2(g)$ may result shift of equilibrium in backward direction. |
| | (S) Addition of 10 g of $CaCO_3(s)$ may result in increasing moles of $CO_2(g)$. |

**10.** Match the columns.

| Column I (Reactions) | Column II (Effect on addition of water) |
|---|---|
| (A) $A(aq) \rightleftharpoons 2B(aq) + H_2O(l)$ | (P) Moles of 'A' increases. |
| (B) $3A(aq) \rightleftharpoons 2B(aq)$ | (Q) Moles of 'B' increases. |
| (C) $2A(aq) + H_2O(l) \rightleftharpoons B(aq)$ | (R) Molar concentration of 'A' decreases. |
| (D) $2A(aq) \rightleftharpoons 3B(aq)$ | (S) Molar concentration of 'B' decreases. |

## Section F (Subjective)

### Single-digit Integer Type

**1.** The percent dissociation of $H_2S(g)$ if 0.1 mole of $H_2S$ is kept in 0.4 L vessel at 1000 K for the reaction $2H_2S(g) \rightleftharpoons 2H_2(g) + S_2(g)$; $K_C = 1.0 \times 10^{-6}$.

**2.** $COF_2(g)$ passed over catalyst at 1000°C comes to equilibrium $2COF_2(g) \rightleftharpoons CO_2(g) + CF_4(g)$. Analysis of the equilibrium mixture (after quick cooling to freeze the equilibrium) shows that 500 ml of the equilibrium mixture (STP) contains 300 ml (STP) of $(COF_2 + CO_2)$ taking the total pressure to be 10 atm. The value of $K_P$ for the reaction is

3. The dissociation constant of $PCl_5$ is 8 atm at 273°C. What pressure (in atm) will be developed when 62.55 g of this substance is vaporized at 273°C in 4480 ml vessel originally full of chlorine gas at 0°C and 1 atm pressure?

4. A mixture of equimolar quantities of ethyl alcohol and acetic acid is prepared. Immediately after mixing, 10 ml of mixture was neutralized by 10 ml of N-NaOH. When the mixture reaches the equilibrium, 10 ml requires 5 ml of N-NaOH for neutralization. What is the equilibrium constant for the formation of ester?

5. The degree of dissociation of HI at a particular temperature of 0.8. Calculate the volume (in litre) of 1.6 M-$Na_2S_2O_3$ solution required to neutralize the iodine present in an equilibrium mixture of a reaction when 2 moles each of $H_2$ and $I_2$ are heated in a closed vessel of 2 L capacity.

6. It was found that when 1.0 mole of dichloroacetic acid and 4.0 moles of amylene were heated together at 363 K to equilibrium state, 0.5 moles of ester were formed and the total volume of the mixture was 700 ml. In another experiment, 1.0 mole of the acid was heated at the same temperature with some moles of amylene and the equilibrium mixture now occupied a volume of 720 ml. If the amount of ester at equilibrium is 0.6 mole in the second experiment, then how many moles of amylene was taken initially? Amylene is one of the pentene.

7. The minimum mass (in g) of $CaCO_3$ required to establish the equilibrium:

$$CaCO_3(s) \rightleftharpoons CaO(s) + CO_2(g),\ K_C = 0.05\ M$$

at a certain temperature in a 1.0 L container is

8. To 500 ml of 0.9 M-$AgNO_3$ solution was added 500 ml of 1.0M-$Fe^{2+}$ solution and the reaction is allowed to achieve equilibrium at 25°C.

$$Ag^+(aq) + Fe^{2+}(aq) \rightleftharpoons Fe^{3+}(aq) + Ag(s)$$

For 30 ml of the solution, 25 ml of 0.06 M–$KMnO_4$ was required for oxidation of $Fe^{2+}$ present. The equilibrium constant (in $M^{-1}$) for the reaction at 25°C is

9. A vessel of 2.5 L was filled with 0.01 mole of $Sb_2S_3$ and 0.01 mole of $H_2$ to attain equilibrium at 440°C as

$$Sb_2S_3(s) + 3H_2(g) \rightleftharpoons 2Sb(s) + 3H_2S(g)$$

After equilibrium, the $H_2S$ formed was analysed by dissolving it in water and treating with excess of $Pb^{2+}$ to give 1.19 g of PbS as precipitate. The value of $K_C$ of the reaction at 440°C is (Pb = 206)

10. At 25°C, 560 g of deuterium oxide, $D_2O$ ($d$ = 1.10 g/ml) and 504 g $H_2O$ ($d$ = 0.997 g/ml) are mixed. The volumes are additive. Fifty percent of the $H_2O$ reacts to form HDO. The value of $K_C$ at 25°C for the reaction $H_2O + D_2O \rightleftharpoons 2HDO$ is

11. The following equilibria are established on mixing two gases $A_2$ and C.

$$3A_2(g) \rightleftharpoons A_6(g);\ K_P = 1.6\ atm^{-2}$$

$$A_2(g) + C(g) \rightleftharpoons A_2C(g);\ K_P = \text{'}x\text{'}\ atm^{-1}$$

When $A_2(g)$ and C(g) are mixed in 2 : 1 molar ratio, the total pressure of gases at equilibrium is found to be 1.4 atm and partial pressure of $A_6(g)$, 0.2 atm. The value of '4$x$' is

12. A 8.28 g sample of IBr(g) is placed in a container of capacity 164.2 ml and heated to 500 K. The equilibrium pressure of $Br_2(g)$ in the system is 4.0 atm. The value of $K_P$ for the reaction

$2IBr(g) \rightleftharpoons I_2(g) + Br_2(g)$ is (Given: Atomic masses are I = 127, Br = 80)

13. When 1.0 mole of $H_2(g)$ and 3.0 moles of $I_2$ vapours are allowed to react, '$x$' moles of HI(g) is formed at equilibrium. Addition of a further 2.0 moles of $H_2(g)$ gave an additional '$x$' moles of HI(g) at new equilibrium. The value of equilibrium constant for the reaction $H_2(g) + I_2(g) \rightleftharpoons 2HI(g)$ is

14. When $N_2O_5(g)$ is heated to 600 K, it dissociates as $N_2O_5(g) \rightleftharpoons N_2O_3(g) + O_2(g)$; $K_C$ = 2.5 M. Simultaneously, $N_2O_3(g)$ decomposes as $N_2O_3(g) \rightleftharpoons N_2O(g) + O_2(g)$. When initially 4.0 moles of $N_2O_5(g)$ is taken in a 2.0 L flask and allowed to attain equilibrium, the equilibrium concentration of $O_2(g)$ is found to be 2.5 M. The equilibrium concentration of $N_2O(g)$ (in M) is

15.

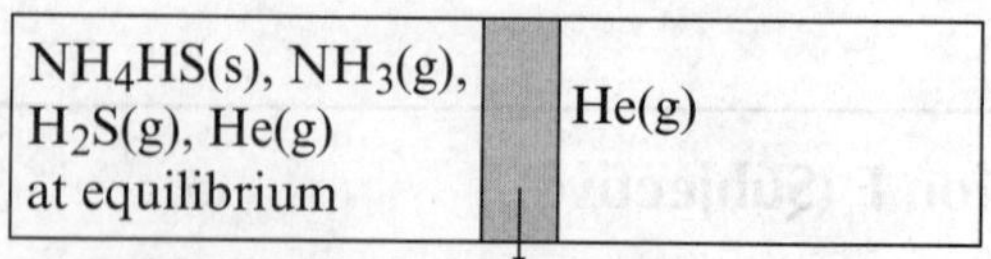


Fixed SPM, which allows only He(g) to cross it.

The entire system is at equilibrium at 300 K. The volume of each chamber is 82.1 L. The total pressure in left chamber is 4 atm and in right chamber, 2 atm. $NH_3(g)$ and $H_2S(g)$ are obtained only from the dissociation of $NH_4HS(s)$. The value of $K_P$ (in $atm^2$) for the reaction $NH_4HS(s) \rightleftharpoons NH_3(g) + H_2S(g)$ is

## Four-digit Integer Type

1. The equilibrium: *p*–xyloquinone + methylene white $\rightleftharpoons$ *p*–xylohydroquinone + methylene blue, may be studied conveniently by observing the difference in colour between methylene blue and methylene white. One millimole of methylene blue was added to 1.00 L of solution that was 0.24 M in *p*–xylohydroquinone and 0.012 M in *p*–xyloquinone. It was then found that 4.0% of the added methylene blue was reduced to methylene white. What is the equilibrium constant for the above reaction? The equation is balanced with 1 molecule of each of the four substances.

2. The theoretically computed equilibrium constant for the polymerization of formaldehyde to glucose in aqueous solution is $6.4 \times 10^{19}$ $M^{-5}$. If 1 M solution of glucose were taken, what would be the equilibrium concentration of formaldehyde in milligram per litre?

3. A container of capacity *V* L contains an equilibrium mixture that consists of 2 moles each of $PCl_5$, $PCl_3$ and $Cl_2$ (all as gases). The pressure is 30.3975 kPa and temperature is *T* K. A certain amount ('*x*' mole) of $Cl_2$(g) is now introduced keeping the pressure and temperature constant, until the equilibrium volume becomes 2*V* L. The value of '90*x*' is

4. The standard reaction enthalpy of the reaction $Zn(s) + H_2O(g) \rightleftharpoons ZnO(s) + H_2(g)$ is + 223 kJ/mol and the standard reaction Gibb's functions is +33 kJ $mol^{-1}$ at 1520 K. Assuming that both $\Delta H^o$ and $\Delta S^o$ remain constant, estimate the minimum temperature (in Kelvin) above which the equilibrium constant becomes greater than one.

5. The diamonds are formed from graphite under very high pressure. Given that the densities of graphite and diamond are, respectively, 2.4 and 3.6 g/cm$^3$ and are independent of pressure. $\Delta_f G^o$ values for graphite and diamond are zero and 3.0 kJ/mol, respectively. If the equilibrium pressure at which graphite is converted into diamond at 25°C is *P* bar, then the value of 0.01*P* is

6. For the equilibrium $NiO(s) + CO(g) \rightleftharpoons Ni(s) + CO_2(g)$, $\Delta G^o$(cal/mol$^{-1}$) = – 5320 – 5.6*T* (K). The temperature (in Kelvin) at which the gaseous mixture at equilibrium contains 400 ppm of CO by mole is [ln10 = 2.3, ln2 = 0.7)

7. Assume that the decomposition of $HNO_3$ can be represented as

$$4HNO_3(g) \rightleftharpoons 4NO_2(g) + 2H_2O(g) + O_2(g)$$

and that at a given temperature of 400 K and pressure of 30 atm, the reaction approaches equilibrium. At equilibrium, the partial pressure of $HNO_3$(g) is 2 atm. The value of $K_C$ (in M$^3$) for the reaction at 400 K is (R = 0.082 L-atm/K-mol)

8. Gaseous nitrosyl chloride (NOCl) and $N_2$ are taken in a flask, sealed and heated to some temperature where the total pressure would have been 1.0 bar had not the following equilibrium been established

$$2NOCl(g) \rightleftharpoons 2NO(g) + Cl_2(g)$$

But the actual pressure was found to be 1.2 bar. Now, into the equilibrium mixture, some $Cl_2$ gas was introduced at constant volume and temperature so that the total pressure would have been 8.3 bar had no further reaction occurred but the actual pressure was found to be 8.2 bar. The equilibrium constant $K_P$ (in bar) for the decomposition reaction under the given experimental condition is (Answer by multiplying the $K_P$ value with 10)

9. Three ideal gases A, B and D were taken in a vessel of constant volume in molar ratio of 1 : 2 : 3, respectively, causing the following reaction.

$$A(g) + 2B(g) \rightleftharpoons C(g)$$

After a long time, when the equilibrium is established, the total pressure was 5/6th of the initial total pressure. At this point, the volume was reduced to half and a catalyst was added starting the following reaction.

$$2C(g) + D(g) \rightleftharpoons 2F(g)$$

If at the second equilibrium, moles of A and C are equal and the ratio of equilibrium total pressure at second equilibrium to the equilibrium total pressure at the first equilibrium is '*x* : 100', then the value of '*x*' is

10. An amount of 0.2 mole of each $A_2(g)$ and $B_2(g)$ is introduced in a sealed flask and heated to 2000 K where the following equilibrium is established.

$A_2(g) + B_2(g) \rightleftharpoons 2\,AB(g)$

At equilibrium, the moles of AB is 0.3. At this stage, 0.1 mole of $C_2(g)$ is added and a new equilibrium is also established as follows.

$A_2(g) + C_2(g) \rightleftharpoons 2\,AC\,(g)$

At the new equilibrium, the moles of AB becomes 0.24. What is the equilibrium constant for the second reaction?

11. At 27°C, the rate of forward reaction at time $t$ is $e^4$ times greater than that of reverse reaction for the reaction $A + B \rightleftharpoons P$. The magnitude of free energy change involved at that time (in cal) is

12. An amount of 5.0 moles each of 'A', 'B' and 'D' is added to a 1.0 L container.

$A(g) + B(g) \rightleftharpoons C(g);\ K_C = 4 \times 10^{10}\ M^{-1}$

$A(g) + D(g) \rightleftharpoons E(g);\ K_C = 10^{10}\ M^{-1}$

If at equilibrium, the moles of 'B' is '$x$', then the value of '$150x$' is

13. For the reaction $Br_2(l) + Cl_2(g) \rightleftharpoons 2BrCl(g);\ K_P = 1$ atm. In a closed container of volume 164 L, initially 10 moles of $Cl_2(g)$ are present at 27°C. What minimum mass (in g) of $Br_2(l)$ must be introduced into this container so that the above equilibrium is maintained at a total pressure of 2.25 atm. Vapour pressure of $Br_2(l)$ at 27°C is 0.25 atm. Assume that volume occupied by the liquid is negligible ($R = 0.082$ L-atm/K-mol, Atomic mass of Br = 80).

14. In a 10.0 L container, an equilibrium was established between $SO_3$, $SO_2$ and $O_2$ gases, by starting with $SO_3$ only. The density of equilibrium mixture was found to be 16 g/litre at a temperature of $\frac{900}{0.0821}$ K. If the degree of dissociation of $SO_3$ is 40%, then the value of $K_P$ (in atm) for the reaction $2SO_3(g) \rightleftharpoons 2SO_2(g) + O_2(g)$ is

15. Equimolar mixture of two gases $A_2$ and $B_2$ is taken in a rigid vessel at constant temperature 300 K. The gases achieve equilibrium as follows.

$A_2(g) \rightleftharpoons 2A(g),\ K_P = x$ atm

$B_2(g) \rightleftharpoons 2B(g),\ K_P = y$ atm

$A_2(g) + B_2(g) \rightleftharpoons 2AB(g),\ K_P = 2$

If the initial pressure in the container was 2 atm and the final pressure at equilibrium is 2.75 atm in which the partial pressure of AB(g) is 0.5 atm, ththe value of $y : x$ is ($y > x$)

## Answer Keys

**Exercise I**

### Basic

1. (a) 2. (a) 3. (d) 4. (d) 5. (d) 6. (a) 7. (c) 8. (a) 9. (d) 10. (b)
11. (a) 12. (a) 13. (c) 14. (c) 15. (b) 16. (a) 17. (a) 18. (c) 19. (b) 20. (b)
21. (b) 22. (c) 23. (d) 24. (d) 25. (a) 26. (a) 27. (d) 28. (d) 29. (d) 30. (b)

### Application of Equilibrium Constant

31. (b) 32. (c) 33. (c) 34. (b) 35. (c) 36. (c) 37. (b) 38. (b) 39. (c) 40. (c)
41. (b) 42. (a) 43. (b) 44. (d) 45. (a) 46. (a) 47. (b) 48. (a) 49. (a) 50. (a)
51. (b) 52. (b) 53. (a) 54. (b) 55. (c)

### Heterogeneous Equilibrium

56. (b) 57. (c) 58. (b) 59. (b) 60. (d) 61. (a) 62. (b) 63. (c) 64. (c) 65. (a)

### Le-Chatelier's Principle

66. (d) 67. (c) 68. (d) 69. (c) 70. (d) 71. (c) 72. (d) 73. (b) 74. (a) 75. (b)
76. (b) 77. (a) 78. (a) 79. (c) 80. (d) 81. (c) 82. (a) 83. (c) 84. (c) 85. (d)

## Answer Keys

**Exercise II**

### Section A (Only one Correct)

1. (c) 2. (c) 3. (c) 4. (a) 5. (c) 6. (d) 7. (b) 8. (a) 9. (d) 10. (a)
11. (a) 12. (d) 13. (a) 14. (b) 15. (a) 16. (b) 17. (b) 18. (a) 19. (d) 20. (d)
21. (b) 22. (b) 23. (d) 24. (b) 25. (b) 26. (c) 27. (a) 28. (b) 29. (d) 30. (c)
31. (a) 32. (c) 33. (b) 34. (c) 35. (d) 36. (b) 37. (a) 38. (b) 39. (c) 40. (a)
41. (a) 42. (a) 43. (b) 44. (a) 45. (c) 46. (d) 47. (d) 48. (a) 49. (b) 50. (a)
51. (b) 52. (a) 53. (b) 54. (d) 55. (a) 56. (c) 57. (b) 58. (d) 59. (a) 60. (c)

### Section B (One or More than one Correct)

1. (a), (b), (c) 2. (a), (b) 3. (a), (b), (c) 4. (a), (b) 5. (a), (b), (d)
6. (a), (b), (c) 7. (d) 8. (b), (c) 9. (b), (c) 10. (a), (b), (d)
11. (a), (b) 12. (a), (d) 13. (b), (c) 14. (a), (b) 15. (d)
16. (a), (b), (c) 17. (a), (b), (c) 18. (a), (c) 19. (a), (b), (c), (d) 20. (a), (b)
21. (b), (c) 22. (a), (d) 23. (c) 24. (a), (c), (d) 25. (c), (d)

### Section C

**Comprehension I**

1. (c) 2. (d) 3. (c)

**Comprehension II**

4. (a) 5. (b) 6. (d)

**Comprehension III**

7. (b) 8. (b) 9. (d)

**Comprehension IV**

10. (b) 11. (a) 12. (c)

**Comprehension V**

13. (c) 14. (b) 15. (d) 16. (c)

**Comprehension VI**

17. (c) 18. (d) 19. (d)

**Comprehension VII**

20. (c) 21. (a) 22. (c)

**Comprehension VIII**

23. (a) 24. (d)

### Section D (Assertion – Reason)

1. (d) 2. (a) 3. (a) 4. (a) 5. (a) 6. (a) 7. (b) 8. (c) 9. (a) 10. (c)

### Section E (Column Match)

1. $A \rightarrow S; B \rightarrow Q, R; C \rightarrow Q; D \rightarrow P$
2. $A \rightarrow S; B \rightarrow R; C \rightarrow P; D \rightarrow Q$
3. $A \rightarrow P, R; B \rightarrow S; C \rightarrow Q; D \rightarrow S$
4. $A \rightarrow P, Q, S; B \rightarrow P, Q; C \rightarrow R$
5. $A \rightarrow P, Q, R, S; B \rightarrow Q, R, S; C \rightarrow P, Q, R, S, T; D \rightarrow Q, R$
6. $A \rightarrow S, T; B \rightarrow R; C \rightarrow Q; D \rightarrow P$
7. $A \rightarrow P, S, T; B \rightarrow Q, R, S; C \rightarrow S; D \rightarrow Q, R, S$
8. $A \rightarrow P, S, T; B \rightarrow Q, R; C \rightarrow Q, R; D \rightarrow P, S$
9. $A \rightarrow P, R; B \rightarrow Q, R; C \rightarrow Q, S$
10. $A \rightarrow Q, R, S; B \rightarrow P, R, S; C \rightarrow P, R, S; D \rightarrow Q, R, S$

### Section F (Subjective)

**Single-digit Integer Type**

1. (2) 2. (4) 3. (7) 4. (1) 5. (2) 6. (5) 7. (5) 8. (5) 9. (1) 10. (4)
11. (6) 12. (4) 13. (4) 14. (1) 15. (1)

**Four-digit Integer Type**

1. (0480) 2. (0015) 3. (0600) 4. (1784) 5. (0180)
6. (0532) 7. (0032) 8. (0032) 9. (0170) 10. (0018)
11. (2400) 12. (0250) 13. (0800) 14. (0016) 15. (0008)

# HINTS AND EXPLANATIONS

## EXERCISE I (JEE MAIN)

### Basic

1. $FeCl_3$ will produce $Fe(OH)_3$ and HCl on hydrolysis.

2. [Solid] or [Pure liquid] $= \frac{n}{V} = \frac{w/M}{w/d} = \frac{d}{M} = \frac{\text{Density}}{\text{Molar mass}}$
Thus, it is independent from their quantity.

3. $\Delta n_g = 2 - (1 + 3) = -2$
$K_p = K_c.(RT)^{\Delta n_g} = K_c.(RT)^{-2}$
$\therefore K_c = \frac{K_p}{(RT)^{-2}} = \frac{1.44\times10^{-5}}{(0.0821\times773)^{-2}}$

4. At the initial stage means time after start of reaction, not at $t = 0$.

5. Equilibrium constant is a function of temperature only.

6. $K_p = \frac{1.P_{co_2}}{1} = P_{co_2}$

7. Required reaction may be obtained by subtracting the first reaction from second and hence, $K = \frac{K_2}{K_1}$.

8. $C_2H_4 + H_2 \rightleftharpoons C_2H_6$; $K_1 = 5.5\times10^{18}$ atm$^{-1}$
$C_2H_2 + H_2 \rightleftharpoons C_2H_4$; $K_2 = 5\times10^{26}$ atm$^{-1}$
$\therefore C_2H_2 + 2H_2 \rightleftharpoons C_2H_6$;
$K_{required} = K_1.K_2 = 2.75\times10^{45}$ atm$^{-2}$

9. $K_{eq}$ is independent from concentration.

10. $K_p = K_c.(RT)^{\Delta n_g} \Rightarrow \frac{K_c}{K_p} = (RT)^{-\Delta n_g} = (RT)^{-\left(-\frac{1}{2}\right)} = \sqrt{RT}$

11. $K_2 = (K_1)^{-1/2} = \frac{1}{\sqrt{K_1}} \Rightarrow K_1.K_2^2 = 1$

12. Greater the value of $K_{eq}$, greater is the extent of reaction.

13. $K_{eq} = \frac{K_f}{K_b} = \frac{0.16}{4\times10^4} = 4\times10^{-6}$

14. For $r_f > r_b$ means the net reaction in forward direction, the reaction quotient $Q$ should be less than $K_{eq}$.

15. $Q = \frac{[NO_2][O_2]}{[NO][O_3]} = \frac{(2.5\times10^{-4})\times(8.2\times10^{-3})}{(1.0\times10^{-5})\times(1.0\times10^{-6})} = 2.05\times10^5 < K_C$
Hence, the net reaction is in forward direction.

16. $K_C = \frac{[I]^2}{[I_2]} \Rightarrow 10^{-2} = \frac{\left(\frac{0.5}{V}\right)^2}{\left(\frac{1}{V}\right)} \Rightarrow V = 25$ L

17. $K = \frac{[C_6H_6]}{[C_2H_2]^3} \Rightarrow 4 = \frac{[C_6H_6]}{(0.5)^3} \Rightarrow [C_6H_6] = 0.5$ M

18. $K_C = \frac{[H_2]^2[S_2]}{[H_2S]^2} = \frac{\left(\frac{0.2}{2}\right)^2\times\left(\frac{0.8}{2}\right)}{\left(\frac{1}{2}\right)^2} = 0.016$ M

19. $K_C = \frac{[CH_4][H_2O]}{[CO][H_2]^3} \Rightarrow 3.9 = \frac{[CH_4]\times0.03}{0.3\times(0.1)^3}$
$\Rightarrow [CH_4] = 0.039$ M

20. $\Delta G° = \Delta H° - T.\Delta S° = -RT.\ln K_p°$
or, $18 - 300\times\frac{30}{1000} = -\frac{2}{1000}\times300\times\ln K_p°$
or, $\ln K_p° = -15 \Rightarrow K_p° = e^{-15}$

21. $N_2(g) + O_2(g) \rightleftharpoons 2NO(g)$; OH = +43.5 Kcal
For endothermic reactions, $K_{eq}$ increases with the increase in temperature.

22. $K_{eq}$ is independent from catalyst.

23. For the given reaction, $\Delta H° = 2\times(-40) = -80$ KJ
and $\Delta S° = (2\times192) - (191 + 3\times130) = -197$ J/K
Now, $\Delta G° = \Delta H° - \Delta S° = -2.303\,RT\log K°_p$
$\therefore \log K_p = \frac{(-80\times1000) - 298\times(-197)}{(-2.303\times8.314\times298)} = 3.73$

**24.** $\Delta H° = O \Rightarrow K_{eq}$ is independent from $T$

**25.** From (i) and (ii),

$2CO(g) + O_2(g) \rightleftharpoons 2CO_2(g)$

$$K_{2000} = \frac{1}{(4.4)^2 \times 5.31 \times 10^{-10}} = 9.73 \times 10^7$$

$K_{1000} = 2.24 \times 10^{22}$

On increasing temperature, $K_{eq}$ is decreased and hence, $\Delta H° = -ve$.

**26.** $MCO_3(S) \rightleftharpoons MO(S) + CO_2(g)$: $K_p = P_{CO_2}$

$$\text{Now, } \ln\frac{K_2}{K_1} = \frac{\Delta H°}{R}\left(\frac{1}{T_1} - \frac{1}{T_2}\right)$$

$$\text{or, } 2.303 \times \log 100 = \frac{\Delta H°}{2}\left(\frac{1}{400} - \frac{1}{500}\right)$$

$\Rightarrow \Delta H° = 18424$ cal

**27.** On increasing temperature, $K_{eq}$ is increasing and hence, the reaction is endothermic.

**28.** $\Delta H = \varepsilon a_f - \varepsilon a_b = 10.303 - 8.000 = 2.303$ kcal

$$\text{Now, } K_{eq} = \frac{A_f}{A_b} \cdot e^{-\Delta H/RT} = 1 \times e^{-\frac{2.303 \times 10^3}{2 \times 500}} = 0.1$$

**29.** $\Delta G° = -2.303RT \cdot \log K_{eq}$

$$\text{or, } -4.606 = -2.303 \times \frac{2}{1000} \times 1000 \times \log K_{eq}$$

$\Rightarrow K_{eq} = 10$

$$\text{For reverse reaction, } K'_{eq} = \frac{1}{10} = 0.1$$

**30.** $K_P = K_C \cdot (RT)^{\Delta Hg}$

or $1.8 = K_C \times (0.0821 \times 700)' \Rightarrow K_C = 0.031$

$$\text{Hence, for reverse reaction, } K'_C = \frac{1}{K_C} = 31.93$$

## Application of Equilibrium Constant

**31.**

| | $PCl_5$ | $\rightleftharpoons$ $PCl_3$ | + $Cl_2$ |
|---|---|---|---|
| | 0.8 mole | 0.2 mole | 0 |
| Equilibrium | $0.8 - x$ | $0.2 + x$ | $x$ |
| | $= 0.6$ | $= 0.4$ | $= 0.2$ |

$\therefore x = 0.2$

$$\text{Now, } K_C = \frac{[PCl_3][Cl_2]}{[PCl_5]} = \frac{\left(\frac{0.4}{1}\right) \times \left(\frac{0.2}{1}\right)}{\left(\frac{0.6}{1}\right)} = 0.133 \text{ M}$$

**32.**

| | $PCl_5$ | $\rightleftharpoons$ $PCl_3$ | + $Cl_2$ |
|---|---|---|---|
| | 1 mole | 0 | 0 |
| Equilibrium | $(1 - x)$ mole | $x$ mole | $x$ mole |
| Equilibrium partial pressure ($P = 2$ atm) | $\frac{1-x}{1+x} \times P$ | $\frac{x}{1+x} \times P$ | $\frac{x}{1+x} \times P$ |

$$K_p = \frac{\left(\frac{x}{1+x} \cdot P\right)\left(\frac{x}{1+x} \cdot P\right)}{\left(\frac{1-x}{1+x} \cdot P\right)} = \frac{x^2 \cdot P}{1 - x^2}$$

$$\text{or } 1.6 = \frac{x^2 \times 2}{1 - x^2} \Rightarrow x = \frac{2}{3}$$

$$\Rightarrow P_{Cl_2} = \frac{x}{1+x} \times P = 0.8 \text{ atm}$$

**33.**

| | $PCl_5$ | $\rightleftharpoons$ $PCl_3$ | + $Cl_2$ |
|---|---|---|---|
| | $X$ mole | 0 | 0 |
| Equilibrium | $x(1 - \alpha)$ | $x \cdot \alpha$ | $x \cdot \alpha$ |

$$\text{E} \quad P_{PCl_3} = \frac{x \cdot \alpha}{x(1+\alpha)} \times P \Rightarrow P_{PCl_3} \cdot P^{-1} = \frac{\alpha}{1+\alpha}$$

**34.** $H_2(g) + I_2(g) \rightleftharpoons 2HI(g)$

As $\Delta n_g = 0$, moles of system is not changing.

$$P = \frac{nRT}{V} = \frac{(0.5 + 0.5) \times 0.08 \times 720}{5.76} = 10 \text{ atm}$$

**35.**

| | $2HI(g)$ | $\rightleftharpoons$ $H_2(g)$ | + $I_2(g)$ |
|---|---|---|---|
| Equilibrium | $1 - \alpha\frac{\alpha}{2}$ | $\frac{\alpha}{2}$ | |
| | $= 0.78$ | $= 0.11$ | $= 0.11$ |

$$K = \frac{0.11 \times 0.11}{(0.78)^2} = 0.0199$$

**36.** $K = \dfrac{\dfrac{\alpha}{2}\cdot\dfrac{\alpha}{2}}{(1-\alpha)^2} = \dfrac{1}{9} \Rightarrow \alpha = 0.4$

**37.** $n_{HI} = 3 - 3 \times \dfrac{30}{100} = 2.1$

**38.**

| | $N_2O_4$ | $\rightleftharpoons$ | $2NO_2$ |
|---|---|---|---|
| | 1 mole | | 0 |
| Final | $1 - 0.2$ | | $2 \times 0.2$ |
| | $= 0.8$ mole | | $= 0.4$ mole |

Now, $\dfrac{P_1}{n_1T_1} = \dfrac{P_2}{n_2T_2} \Rightarrow \dfrac{1}{1 \times 300} = \dfrac{P_2}{1.2 \times 600}$

$\Rightarrow P_2 = 2.4$ atm

**39.**

| | $N_2O_4$ | $\Rightarrow$ | $2NO_2$ |
|---|---|---|---|
| Equilibrium | $1 - \alpha$ | | $2\alpha$ |

$$K_p = \frac{P_{NO_2}^2}{p_{N_2O_4}} = \frac{\left(\dfrac{2\alpha}{1+\alpha}\cdot P\right)^2}{\left(\dfrac{1-\alpha}{1+\alpha}\cdot P\right)} = \frac{4\alpha^2 \cdot P}{1-\alpha^2}$$

$$\therefore \alpha = \sqrt{\frac{Kp/4p}{1 + Kp/4p}}$$

**40.** $K_p = \dfrac{4\alpha^2 . P}{1-\alpha^2} = \text{Constant}$

$$\therefore \frac{4 \times \left(\dfrac{1}{3}\right)^2 \times P_1}{1 - \left(\dfrac{1}{3}\right)^2} = \frac{4 \times \left(\dfrac{1}{2}\right)^2 \times P_2}{1 - \left(\dfrac{1}{2}\right)^2} \Rightarrow \frac{P_1}{P_2} = \frac{8}{3}$$

**41.**

| | $N_2O_4$ | $\rightleftharpoons$ | $2NO_2$ |
|---|---|---|---|
| Initial mole | $a$ | | 0 |
| Mole at Equilibrium | $a - x$ | | $2x$ |

From question, $(a + x) \times \dfrac{75}{100} = a \Rightarrow \alpha = \dfrac{x}{a} = \dfrac{1}{3}$

**42.**

| | $N_2$ | + | $3H_2$ | $\rightleftharpoons$ | $2NH_3$ |
|---|---|---|---|---|---|
| | 0.2 mole | | 0.6 mole | | 0 |
| Equilibrium | $0.2 - 0.2 \times \dfrac{40}{100}$ | | $0.6 - 0.08 \times 3$ | | $2 \times 0.08$ |
| | $= 0.12$ | | $= 0.36$ | | $= 0.16$ |

$$\therefore \frac{V_{\text{final}}}{V_{\text{initial}}} = \frac{n_{\text{final}}}{n_{\text{initial}}} = \frac{0.64}{0.80} = \frac{4}{5}$$

**43.**

| | $2NH_3$ | $\rightleftharpoons$ | $N_2$ | + | $3H_2$ |
|---|---|---|---|---|---|
| | 4 mole | | 0 | | 0 |
| Equilibrium | $4 - 2x$ | | $x$ | | $3x$ |
| | $= 2$ | | $= 1$ | | $= 3$ |

$\therefore x = 1$

$$\text{Now, } K_C = \frac{[N_2][H_2]^3}{[NH_3]^2} = \frac{\dfrac{1}{1} \times \left(\dfrac{3}{1}\right)^3}{\left(\dfrac{2}{1}\right)^2} = 6.75 \text{ M}^2$$

**44.**

| | $2SO_2$ | + | $O_2$ | $\rightleftharpoons$ | $2SO_3$ |
|---|---|---|---|---|---|
| | 2 mole | | 1 mole | | 0 |
| Equilibrium | $2 - 2 \times \dfrac{50}{100}$ | | $1 - \dfrac{1}{2}$ | | 1 |
| | $= 1$ | | $= 0.5$ | | |

$$\therefore P_{O_2} = \frac{\frac{1}{2}0.5}{2.5} \times 1 = 0.2 \text{ atm}$$

**45.**

| | $CH_3COOH$ + | $C_2H_5OH$ $\rightleftharpoons$ | $CH_3COOC_2H_5$ + | $H_2O$ |
|---|---|---|---|---|
| | 1 | 1 | 0 | 0 |
| Equilibrium | $1 - \dfrac{1}{4}$ | $1 - \dfrac{1}{4}$ | $\dfrac{1}{4}$ | $\dfrac{1}{4}$ |
| | $= \dfrac{3}{4}$ | $\dfrac{3}{4}$ | | |

$$K = \frac{\dfrac{1}{4} \times \dfrac{1}{4}}{\dfrac{3}{4} \times \dfrac{3}{4}} = \frac{1}{9}$$

**46.** At equilibrium, $[A]=[B]$ and $[C]=[D]$ and hence

$$K=\frac{[C][D]}{[A][B]}=\frac{[C]^2}{[A]^2}\Rightarrow\frac{[C]}{[A]}=\sqrt{2.25}$$

$$=1.5\Rightarrow\frac{[A]}{[C]}=\frac{2}{3}$$

**47.**

| | $XY_2$ | $\rightleftharpoons$ $XY$ | + $Y$ |
|---|---|---|---|
| | 600 mm | 0 | 0 |
| Equilibrium | $(600-x)$ mm | $x$ mm | $x$ mm |

From Question: $(600-x)+x+x=800\Rightarrow x=200$

$$\therefore Kp=\frac{x.x}{(600-x)}=100\text{ mm Hg}$$

**48.**

| | $A$ | + $B$ | $\rightleftharpoons$ $C$ | + $D$ |
|---|---|---|---|---|
| Initial | cM | cM | 0 | 0 |
| Equilibrium | $(c-x)$M | $(c-x)$M | $x$ M | $x$ M |

From Question: $x=2\times(c-x)\Rightarrow x=\frac{2}{3}c$

$$\therefore K=\frac{x.x}{(c-x)(c-x)}=4$$

**49.** At equilibrium, mole of $I^-$ – mole of AgI (yellow) precipitate

$=0.25$

| | $I_2$ | + $I^-$ | $\rightleftharpoons$ $I_3^-$ |
|---|---|---|---|
| | 1 mole | 0.5 mole | 0 |
| Equilibrium | $1-x$ | $0.5-x$ | $x$ |
| | $=0.75$ | $=0.25$ | $=0.25$ |

$\therefore x=0.25$

$$\therefore K=\frac{\left(\frac{0.25}{1}\right)}{\left(\frac{0.75}{1}\right)\times\left(\frac{0.25}{1}\right)}=1.33$$

**50.**

| | α – D – Glucose $\rightleftharpoons$ | $b$ – D – Glucose |
|---|---|---|
| | $a$ | 0 |
| Equilibrium | $a-x$ | $x$ |

$$K=\frac{x}{a-x}=1.8\Rightarrow x=\frac{9}{14}a$$

$$\therefore \%\text{ of }\alpha-\text{D}-\text{Glucose remained}=\frac{a-x}{a}\times100$$

$$=35.7\%$$

**51.**

| | $2NOBr$ $\rightleftharpoons$ | $2NO$ | + $Br_2$ |
|---|---|---|---|
| | $P_0$ | 0 | 0 |
| Equilibrium | $P_0-2x$ | $2x$ | $x=\frac{P}{9}$ |

$$P_{total}=(P_0-2x)+2x+x=P_0+x$$

$$=P\Rightarrow P_0=\frac{8P}{9}$$

$$\therefore Kp=\frac{(2x)^2\times x}{(P_0-2x)^2}=\frac{\left(\frac{2P}{9}\right)^2\times\frac{P}{9}}{\left(\frac{8P}{9}-\frac{2P}{9}\right)^2}=\frac{P}{81}$$

**52.**

| | A $\rightleftharpoons$ | nB |
|---|---|---|
| | 0.6M | 0 |
| Equilibrium | $0.6-x$ | $nx$ |
| | $=0.3$ M | $=0.6$ |
| | $\therefore x=0.3$ | $\therefore n=2$ |

**53.** $$K_p=\frac{4\alpha^2\cdot P}{1-\alpha^2}\Rightarrow 640=\frac{4\alpha^2\times160}{1-\alpha^2}\Rightarrow\alpha=\frac{1}{\sqrt{2}}$$

**54.** $$M_{mix}=\frac{dRT}{P}=\frac{3.5\times0.08\times400}{1}=112$$

$$\text{Now, }\alpha=\frac{M_{PCl_5}-M_{mix}}{(n-1)\cdot M_{mix}}=\frac{208.5-112}{(2-1)\times112}=0.8616$$

**55.**

| | $2AB_2$ $\rightleftharpoons$ | $2AB$ | + $B_2$ |
|---|---|---|---|
| | 1 | 0 | 0 |
| Equilibrium | $1-x$ | $x$ | $\frac{x}{2}$ |

$$K_p = \frac{\left(\frac{x}{1+\frac{x}{2}} \times P\right)^2 \cdot \left(\frac{x/2}{1+x/2} \cdot P\right)}{\left(\frac{1-x}{1+x/2} \cdot P\right)^2}$$

$$= \frac{x^3 \cdot P}{2(1-x)^2\left(1+\frac{x}{2}\right)}$$

As $x << 1$, $K_p = \frac{x^3 \cdot P}{2} \Rightarrow x = \left(\frac{2Kp}{P}\right)^{1/3}$

## Heterogeneous Equilibrium

**56.** $2BaO_2(s) \rightleftharpoons 2BaO(s) + O_2(g)$

$\frac{34}{170}$ mole = 0.2

Equilibrium $0.2x$ $x$ $\frac{x}{2}$

Now, $K_p = P_{O_2} \Rightarrow 0.5 = \frac{\frac{x}{2} \times 0.0821 \times 1000}{8.21}$

$\Rightarrow x = 0.1$

$\therefore$ % of $BaO_2$ decomposed $= \frac{0.1}{0.2} \times 100 = 50\%$

**57.** $K_p = (2p)^2 . p \Rightarrow P = \left(\frac{3.2 \times 10^{-5}}{4}\right)^{1/3} = 0.02$

$\therefore$ Total pressure $= 2P + P = 0.06$ atm

**58.** $CaCO_3(s) \rightleftharpoons CaO(s) + CO_2(g)$;

$K_p = 4.5 \times 10^{-2}$ atm

$C(s) + CO_2(g) \rightleftharpoons 2CO(g)$; $K_p = 2.0$ atm

On adding, $CaCO_3(s) + C(s) \rightleftharpoons CaO(s) + 2CO(g)$;

$Kp = 9 \times 10^{-2}$ atm$^2$

or $P_{CO}{}^2 = 9 \times 10^{-2} \Rightarrow P_{CO} = 0.3$ atm

**59.** $K = \frac{[H_2]^4}{[H_2O]^4} = \left(\frac{1.2/2}{54/18}\right)^4 = 0.0016$

**60.** $CO_2(g) + C(s) \rightleftharpoons 2CO(g)$

Given: $a$ mole $9a$ mole

Required: $b$ mole $4b$ mole

Now, $K_p = \frac{\left(\frac{9a}{10a} \times 16\right)^2}{\left(\frac{a}{10a} \times 16\right)} = \frac{\left(\frac{4b}{5b} \times P\right)^2}{\left(\frac{b}{5b} \times P\right)}$

$\Rightarrow P = 40.5$ atm

**62.** $2CO \rightleftharpoons CO_2 + C(s)$

Equilibrium $x$ atm $y$ atm

From question, $x + y = 1.1$ and $\frac{y}{x^2} = 0.1$

$\therefore \frac{x}{y} = \frac{10}{1}$

**63.** Any reaction of heterogeneous equilibrium may complete with respect to solid or pure liquid, maintaining equilibrium with gaseous components.

$LiCl.NH_3(s) + 2NH_3(g) \rightleftharpoons LiCl.3NH_3(s)$; $K_p' = \frac{1}{9 \text{ atm}^2}$

0.1 mole $a$ mole 0

Equilibrium

0 $(a - 0.2)$ mole 0.1 mole

Now, $K_p' = \frac{1}{P^2_{NH_3}} \Rightarrow P_{NH_3} = 3$ atm

$$= \frac{(a-0.2) \times 0.0821 \times 30}{8.21}$$

$\therefore a = 1.2$

**64.** $ZnO(s) + CO(g) \rightleftharpoons Zn(g) + Co_2(g)$; $K_p = 1$ atm

$a$ atm 0 $a$ atm

Equilibrium $a - x$ $x$ $a + x$

$$K_p = \frac{P_{Zn} \cdot P_{CO_2}}{P_{CO}} \Rightarrow 1 = \frac{x(a+x)}{a-x}$$

and $(a-x) + x + (a+x) = P_{total} = 1$ atm

$\therefore x = 0.24$

**65.** $H_2(g) \quad + \quad S(s) \quad \rightleftharpoons H_2S(g)$

| | 0.3 mole | 2 mole | 0 |
|---|---|---|---|
| Equilibrium | $(0.3-x)$ mole | $(2-x)$ mole | $x$ mole |

$$H_C = \frac{x/2}{(0.3-x)/2} = 0.08 \Rightarrow 0.022$$

Now, $P_{H_2S} = \frac{x \times 0.08 \times 360}{2} = 0.32$ atm

## Le-Chatelier's Principle

**66.** $K_p$ is a function of temperature only. On reducing the volume, equilibrium will shift backward and hence, $\alpha$ will decrease.

**67.** Increase in temperature favours endothermic direction and increase in pressure favours the direction of decreases in volume (moles of gases).

**68.** The molar concentration of $AB$(s) will not change on changing its amount and hence, equilibrium will not be affected.

**69.** $K = [B]^2[C]^3 = \text{Constant}$

Now, $[B]^2[C]^3 = [B]^2_{new} \times (2[C])^3 \Rightarrow [B]_{new} = \frac{[B]}{2\sqrt{2}}$

**70.** MgO is solid.

**71.** Increase in pressure will shift the equilibrium in the direction of decrease in volume and hence, more water will form.

**72.** There is no effect of adding inert gas constant volume.

**73.** Exothermic reactions are unfavoured by increasing temperature.

**74.** At 250°C, $N_2$ gas will not react.

**75.** $V_{Diamond} < V_{Graphite}$

On increasing pressure, equilibrium will shift toward diamond.

**76.** $N_2O_4 \rightleftharpoons 2NO_2$

On increasing temperature, the reaction will shift forward and hence, $\Delta H = +ve$.

**77.** Catalyst has no effect on equilibrium.

**78.** Equilibrium will shift towards water. In order to maintain equilibrium, the temperature should be increased.

**79.** Addition of reactant will shift the equilibrium in forward direction.

**80.** If equilibrium were not disturbed, $[Cu^{2+}]_{final} = \frac{x}{2}$.

But as equilibrium will shift backward, its concentration will decrease further.

**81.** $Kp = \frac{P_{PCl_3} \cdot P_{Cl_2}}{P_{PCl_5}} = \text{constant}$

If $P'_{PCl_3} = 2 \times P_{PCl_3}$

and $P'_{Cl_2} = 2 \times P_{Cl_2}$,

then $P'_{PCl_5} = 4 \times P_{PCl_5}$

**82.** $\Delta n_g = 0$, hence no effect.

**83.** Equilibrium will shift towards $NO_2$.

**84.** $K_{eq}$ is a function of temperature only.

**85.** For forward reaction (exothermic), temperature should be low. As mole of gas is decreasing, the pressure should be high.

## EXERCISE II (JEE ADVANCED)

### Section A (Only one Correct)

1. $K_{eq}$ for the reaction in backward direction

$$= \frac{K_b}{K_f} = \frac{2.1\times10^{-3}\,\text{L mol}^{-1}\text{s}^{-1}}{3.9\times10^{-5}\,s^{-1}} = 53.846\ \text{L mol}^{-1}$$

2. Stability constant,

$$K = \frac{K_f}{K_b} = \frac{1.45\times10^{13}}{1.22\times10^{-9}} = 1.1885\times10^{17}$$

3. $$K_{eq} = \frac{K_f}{K_b} = \frac{[B]^2}{[A]}$$

$$\Rightarrow \frac{1.5\times10^{-3}}{K_b} = \frac{(100/10)^2}{(10^{-5}/10)}$$

$$\Rightarrow K_b = 1.5\times10^{-11}\ M^{-1}\ S^{-1}$$

4. For pent hydrate to be efflorescent, $Q < K_p$

or, $P^2_{H_2O} < 10^{-4}\,\text{atm}^2$

$$\Rightarrow P_{H_2O} < 10^{-2}\,\text{atm} = 7.6\ \text{mm}$$

5. $Q_P = P^2_{NH_3}\times P_{CO_2} = 10^2\times20 = 2000\ \text{atm}^3 < K_P$

Hence, the reaction should shift forward. But as solid $NH_2COONH_4$ is not present initially, the pressure will remain at 30 atm.

6. $Q_P < K_P \Rightarrow P^2_{H_2O} < K_p$

$$\Rightarrow \left(\frac{40}{760}\times\frac{R.H}{100}\right)^2 < 1.21\times10^{-4}$$

$\therefore$ R.H. < 20.9 %

7. $H_2(g) + I_2(s) \rightleftharpoons 2HI(g)$ ; $K_p = 6.4\times10^{-4}$ atm

$I_2(s) \rightleftharpoons I_2(g)$ ; $K_p = 1.6\times10^{-4}$ atm

$$\therefore H_2(g) + I_2(g) \rightleftharpoons 2HI(g)\ ;\ K_P = \frac{6.4\times10^{-4}}{1.6\times10^{-4}} = 4$$

8. Net rate of reaction of HI,

$$-\frac{1}{2}\cdot\frac{d[\text{HI}]}{dt} = r_b - r_f = \text{K}_{-1}[\text{HI}]^2 - \text{K}_1[\text{H}_2][\text{I}_2]$$

9. $$\alpha = \frac{M_0 - M}{(n-1)\cdot M} = \frac{208.5-124}{(2-1)\times124} = 0.681$$

10. $$PCl_3(g) + Cl_2(g) \rightleftharpoons PCl_5(g)$$

| | $PCl_3(g)$ | $Cl_2(g)$ | $PCl_5(g)$ |
|---|---|---|---|
| Initial partial pressure | $P_0$ | $P_0$ | 0 |
| Equilibrium partial pressure | $P_0 - 0.75\,P_0$ | $P_0 - 0.75\,P_0$ | $0.75\,P_0$ |
| | $-0.25\,P_0$ | $-0.25\,P_0$ | |

Now, $$K_P = \frac{P_{PCl_5}}{P_{PCl_3}\times P_{Cl_2}}$$

$$\Rightarrow 2 = \frac{0.75\,P_0}{0.25\,P_0\times0.25\,P_0}$$

$$\Rightarrow P_0 = 6\ \text{atm}$$

$\therefore$ Initial total pressure of mixture $= 2P_0 = 12$ atm

11. $$N_2(g) + 3H_2(g) \rightleftharpoons 2NH_3(g)$$

| | $N_2(g)$ | $3H_2(g)$ | $2NH_3(g)$ |
|---|---|---|---|
| Initial moles | 1 | 3 | 0 |
| Moles at equilibrium | $1-x$ | $3-3x$ | $2x$ |

Total moles of gases $= (1-x) + (3-3x) + 2x$
$= 4 - 2x$

Equilibrium partial pressure

$$\frac{1-x}{4-2x}\times P \qquad \frac{3-3x}{4-2x}\times P \qquad \frac{2x}{4-2x}\times P$$

Now, $$K_P = \frac{\left(\frac{2x}{4-2x}\times P\right)^2}{\left(\frac{1-x}{4-2x}\times P\right)\left(\frac{3-3x}{4-2x}\times P\right)^3}$$

$$= \frac{4x^2}{27(1-x)^4}\times\frac{(4-2x)^2}{P^2} \approx \frac{4x^2\times16}{27\times P^2}$$

$$\therefore x = \sqrt{\frac{27\,K_P\cdot P^2}{64}} = \frac{3P\cdot\sqrt{3\,K_P}}{8}$$

12. $$X_2 + Y_2 \rightleftharpoons 2XY$$

| | $X_2$ | $Y_2$ | $2XY$ |
|---|---|---|---|
| Initial moles | 2 | 3 | 0 |
| Final moles | $2-x$ | $3-x$ | $2x$ |

$[XY] = \frac{2x}{5} = 0.7$

$\Rightarrow x = 1.75$

$\therefore [X_2] = \frac{2-x}{5} = 0.05$ M

and $[Y_2] = \frac{3-x}{5} = 0.25$ M

**13.** $N_2 + O_2 \rightleftharpoons 2NO$

Equilibrium moles $1-x \quad 1-x \quad 2x$

$0.09 = \frac{(2x)^2}{(1-x)(1-x)}$

$\Rightarrow x = 0.13$

**14.** $N_2 + O_2 \rightleftharpoons 2NO$

Initial moles $4a \quad a \quad 0$

Equilibrium moles $4a-x \quad a-x \quad 2x$

Now, $0.0004 = \frac{(2x^2)}{(4a-x)(a-x)} \approx \frac{4x^2}{4a \cdot a}$

$\Rightarrow \frac{x}{a} = 0.02$

$\therefore$ Per cent of NO $= \frac{2x}{5a} \times 100 = 0.8\%$

**15.** $N_2 + 3H_2 \rightleftharpoons 2NH_3$

Initial moles $1 \quad 5 \quad 0$

Moles at equilibrium $1-x \quad 5-3x \quad 2x$

Total moles $= (1-x) + (5-3x) + 2x = 6 - 2x$

From question, $\frac{2x}{6-2x} = 0.4$

$\Rightarrow x = \frac{6}{7}$

$K_P = \frac{(2x)^2}{(1-x)\times(5-3x)^3} \times \left(\frac{6-2x}{P}\right)^2$

$= 2.6 \times 10^{-4}$ atm$^{-2}$

**16.** $N_2 + 3H_2 \rightleftharpoons 2NH_3$

Initial moles $4 \quad 16 \quad 0$

Moles at equilibrium $4-x \quad 16-3x \quad 2x$

Total moles $= (4-x) + (16-3x) + 2x = 20 - 2x$

From question, $20 \times \frac{9}{10} = 20 - 2x$

$\Rightarrow x = 1$

Now, $K_C = \frac{(2x)^2}{(4-x)(16-3x)^3} \cdot V^2$

$= 6.07 \times 10^{-4}$ M$^{-2}$

**17.** If reactants are taken in stoichiometric amount, then their mass ratio does not change at any stage of reaction. For 3 mole $N_2$, there should be 9 mole $H_2$. Hence, at any stage, $m_{N_2} + m_{H_2} + m_{NH_3} = 3 \times 28 + 9 \times 2 = 102$ gm.

**18.** $2SO_2 + O_2 \rightleftharpoons 2SO_3$

Initial moles $2 \quad 1 \quad 0$

Moles at equilibrium $2-2x \quad 1-x \quad 2x$

From question $n_{eq}\ SO_2 = n_{eq}\ MnO_4^-$.

or, $(2-2x) \times 2 = 0.4 \times 5$

$\Rightarrow x = 0.5$

$\therefore K_C = \frac{(2x)^2}{(2-2x)^2 \times (1-x)} = 2$ M$^{-1}$.

**19.** $CH_3COOH + C_2H_5OH \rightleftharpoons CH_3COOC_2H_5 + H_2O$

Case I $\frac{60}{60} = 1$ mole $\quad \frac{46}{46} = 1$ mole $\quad 0 \quad 0$

Moles at Equ. $1-x \quad 1-x \quad x \quad x$; $x = \frac{44}{88} = 0.5$

Case II $\frac{120}{60} = 2$ mole $\quad \frac{46}{46} = 1$ mole $\quad 0 \quad 0$

Moles at Equ. $2-y \quad 1-y \quad y \quad y$

$K_{eq} = \frac{x \cdot x}{(1-x)\cdot(1-x)} = \frac{y \cdot y}{(2-y)\cdot(1-y)}$

$\Rightarrow y = \frac{2}{3}$

$\therefore$ Mass of $CH_3COOC_2H_5$ at equilibrium $= \frac{2}{3} \times 88 =$

**20.** $R_1OH + CH_3COOH \rightleftharpoons CH_3COOR_1 + H_2O$

Initial moles $1 \quad 1 \quad 0 \quad 0$

Equ. moles $1-x \quad 1-(x+y) \quad x \quad x+y$

$$R_2OH + CH_3COOH \rightleftharpoons CH_3COOR_2 + H_2O$$

| | $R_2OH$ | $CH_3COOH$ | $CH_3COOR_2$ | $H_2O$ |
|---|---|---|---|---|
| Initial moles | 1 | 1 | 0 | 0 |
| Equ. moles | $1-y$ | $1-(x+y)$ | $y$ | $x+y$ |

From question, $x + y = 0.8$ and $\frac{x}{y} = \frac{3}{2}$

$\therefore\ x = 0.48$ and $y = 0.32$

Now, $K_1 = \frac{x\cdot(x+y)}{(1-x)[1-(x+y)]} = \frac{0.48\times 0.8}{0.52\times 0.2} = 3.69$

**21.** $$2NO(g) + Cl_2(g) \rightleftharpoons 2NOCl(g)$$

| | $2NO(g)$ | $Cl_2(g)$ | $2NOCl(g)$ |
|---|---|---|---|
| Initial partial pressure | $2P_0$ | $P_0$ | 0 |
| Equ. partial pressure | $2P_0 - 2x$ | $P_0 - x$ | $2x$ |

From question, $(2P_0 - 2x) + (P_0 - x) + 2x = 1$

$\Rightarrow 3P_0 - x = 1$ (1)

and $2x = \frac{1}{4}(P_0 - x)$ (2)

From (1) and (2), $P_0 = 9x$ and $x = \frac{1}{26}$

$\therefore\ K_P = \frac{(2x)^2}{(2P_0-2x)^2(P_0-x)} = \frac{13}{256}\ \text{atm}^{-1}$

**22.** $S_8(g) \rightleftharpoons 4\,S_2(g)$

| | $S_8(g)$ | $4S_2(g)$ |
|---|---|---|
| Initial partial pressure | 1 atm | 0 |
| Equ. partial pressure | $1 - 0.3$ | $4 \times 0.3$ |
| | = 0.7 atm | = 1 .2 atm |

$\therefore\ K_P = \frac{(1.2)^4}{0.7} = 2.96\ \text{atm}^3$

**23.** $HCl(g) + \frac{1}{4}O_2(g) \rightleftharpoons \frac{1}{2}Cl_2(g) + \frac{1}{2}H_2O(g)$

Initial partial pressure $730 \times \frac{8}{100}$ $\quad 730 \times \frac{92}{100}$

= 58.4 mm $\quad$ = 671.6 mm

Equilibrium partial pressure $58.4 - 58.4 \times 0.08$ $\quad 671.6 - \frac{58.4\times 0.08}{4}$

= 670.432 mm

**24.** $$H_3BO_3 + \text{Glycerin} \rightleftharpoons \text{complex}$$

| | $H_3BO_3$ | Glycerin | complex |
|---|---|---|---|
| Initial concent. | 0.1 | $a$ M | 0 |
| Equ. Concert | $0.1 - 0.06$ | $(a - 0.06)$ M | 0.06 M |
| | = 0.04 M | | |

Now, $K_{eq} = 0.9 = \frac{0.06}{(0.04)\times(a-0.06)}$

$\Rightarrow a = 1.73$ M

**25.** $$2A(g) \rightleftharpoons A_2(g);\ K_P = 8 \times 10^8\ \text{atm}^{-1}$$

| | $2A(g)$ | $A_2(g)$ |
|---|---|---|
| Initial partial pressure | 1 atm | 0 |
| Partial pressure on complete reaction | 0 | 0.5 atm |
| Equilibrium partial pressure | $2x$ atm | $0.5 - x \approx 0.5$ atm |

Now, $8\times 10^8 = \frac{0.5}{P_A^2}$

$\Rightarrow P_A = 2.5 \times 10^{-5}$ atm

**26.** $K_{eq} = 3.8 \times 10^{-7}\ \frac{10^{-6}\times[HCO_3^-]}{[CO_2]}$

$\Rightarrow \frac{[HCO_3^-]}{[CO_2]} = 0.38$

**27.** $$A(g) \rightleftharpoons nB(g)$$

| | $A(g)$ | $nB(g)$ |
|---|---|---|
| Initial mole | 1(say) | 0 |
| Equilibrium mole | $1-\alpha$ | $n\alpha$ |

Total moles = $1 - \alpha + n\alpha = 1 + \alpha(n-1)$

Now $K_P = \frac{P_B^n}{P_A} = \frac{\left[\frac{n\alpha}{1+\alpha(n-1)}\cdot P\right]^n}{\left[\frac{1-\alpha}{1+\alpha(n-1)}\cdot P\right]}$

$= \frac{(n\alpha)^n \cdot P^{n-1}}{(1-\alpha)\cdot[1+\alpha(n-1)]^{n-1}}$

**28.** $$A + B \rightleftharpoons C + D$$

| | A | B | C | D |
|---|---|---|---|---|
| Initial moles | $a$ | $a$ | 0 | 0 |
| Equilibrium moles | $a-x$ | $a-x$ | $x$ | $x$ |

From question, $[A] = 2[C] \Rightarrow a - x = 2x \Rightarrow a = 3x$

Now, $K_{eq} = \frac{K_f}{K_b} = \frac{x\cdot x}{(a-x)\cdot(a-x)}$

$\Rightarrow \frac{2\times 10^{-3}}{K_b} = \frac{x\cdot x}{2x\cdot 2x}$

$\therefore\ K_b = 8 \times 10\ -3\ \text{mol}^{-1}\ \text{L S}^{-1}$

**29.** $(Cl_2CHCOOH)_2 \rightleftharpoons 2Cl_2CHOOH$

Initial Conc. $\dfrac{0.0129/258}{100/1000}$ 0

$= 5 \times 10^{-4}$ M

Equ. Conc. $5 \times 10^{-4} - x$ $2x$

Now, $K_{eq} = 5 \times 10^{-4} = \dfrac{(2x)^2}{(5\times10^{-4} - x)}$

$\Rightarrow x = 1.95 \times 10^{-4}$

$\therefore [Cl_2CHOOH] = 3.90 \times 10^{-4}$ M

**30.** $NH_2CONH_4(s) \rightleftharpoons 2NH_3(g) + CO_2(g)$

| | | | |
|---|---|---|---|
| Initial moles | 1 | 0 | 0 |
| Equ. moles | $1-\alpha$ | $2\alpha$ | $\alpha$ |

From question, $3\alpha = \dfrac{P\cdot V}{RT}$

$\therefore$ Percentage dissociation of solid $= 100\alpha$ %

$= \dfrac{100\cdot PV}{3RT}\%$

**31.** $P_{H_2O,\,eq} = (K_P)^{1/4} = (8.1 \times 10^{-7}) = 0.03$ atm

$P_{H_2O,\,eq}$, actual $= \dfrac{30.4}{760} = 0.04$ atm

$\therefore$ Mass of water vapour absorbed

$= \dfrac{(0.09 - 0.03)\times 1.642}{0.0821\times 300}\times 18$

$= 0.012$ gm

**32.** Addition of CO will shift second reaction backward. Decrease in $Cl_2$ will shift the first reaction forward.

**33.** $\dfrac{P_{H_2O}(g)}{P^2_{HCl(g)}} = \dfrac{P_{H_2O}(g)\times 2}{P^2_{HCl(g),\,new}}$

$\Rightarrow P_{HCl\,(g),\,new} = \sqrt{2}\times P_{HCl\,(g)}$

**34.** Ionic form of the reaction is

$NH_4^+ + H_2O \rightleftharpoons NH_4OH + H^+$

**35.** $K_1 = \dfrac{[A^+][B^-]}{[AB]}$ and $K_2 = \dfrac{[AB_2^-]}{[AB][B^-]}$

Now, $\dfrac{[A^+]}{[AB_2^-]} = \dfrac{K_1}{K_2}\cdot[B^-]^2$

**36.** $NH_4HS(s) \rightleftharpoons NH_3\,(g) + H_2S(g)$

| | | |
|---|---|---|
| Equ. partial pressure | $\dfrac{P}{2}$ atm | $\dfrac{P}{2}$ atm |
| New Equ.partial pressure | $P$ atm | $P'$ atm |

Now, $\dfrac{P}{2}\times\dfrac{P}{2} = P\times P'$

$\Rightarrow P' = 0.25$ P

**37.** $N_2 + 3H_2 \rightleftharpoons 2NH_3$

| | $N_2$ | $3H_2$ | $2NH_3$ |
|---|---|---|---|
| Equ. partial pressure | 100 mm | 400 mm | 1000 mm |
| New equ. partial pressure | $100 - a + x$ | $400 + 3x = 700$ mm | $1000 - 2x = 800$ mm |

$\therefore x = 100$ mm

Now, $K_P = \dfrac{1000^2}{100\times 400^3} = \dfrac{800^2}{P_{N_2}\times 700^3}$

$\Rightarrow P_{N_2} = 11.94$ mm

$K_1 = \dfrac{[B]}{[A]},\ K_2 = \dfrac{[C]}{[A]}$

Now, $X_A = \dfrac{[A]}{[A]+[B]+[C]} = \dfrac{[A]}{[A]+K_1[A]+K_2[A]}$

$= \dfrac{1}{1+K_1+K_2}$

**39.** CO and $H_2$ are initially in 1 : 3 mole ratio, as they are formed by 2nd reaction.

$CO + 2H_2 \rightleftharpoons CH_3OH$

| | CO | $2H_2$ | $CH_3OH$ |
|---|---|---|---|
| Initial moles | 1 | 3 | 0 |
| Equilibrium moles | $1 - 0.25 = 0.75$ | $3 - 0.25\times 2 = 2.5$ | 0.25 |

Total moles = 0.75 + 2.5 + 0.25 = 3.5

Now, $K_P = \dfrac{0.25}{0.75\times(2.5)^2}\times\left(\dfrac{P}{3.5}\right)^{-2} = 6.23\times 10^{-3}$

$\therefore P = 10.24$ bar

**40.** $A \rightleftharpoons B + C;\ K_1 = 10^6$

| | A | B | C |
|---|---|---|---|
| Initial moles | 1 | 0 | 0 |
| Equilibrium moles | $1 - x + y$ | $x - y$ | $x$ |

$B + D \rightleftharpoons A;\ K_2 = 10^{-6}$

| | B | D | A |
|---|---|---|---|
| | $x$ | 1 | 1 |
| Equilibrium moles | $x - y$ | $1 - y$ | $1 + y - x$ |

As $K_1 >> 1$, we may assume $x \approx 1$

Now, $K_2 = \frac{(1+y-x)}{(x-y)(1-y)} \approx \frac{y}{(1-y)\cdot(1-y)}$

As $K_2 << 1$, we may assume $y << 1$

$$K_2 = \frac{y}{(1-y)(1-y)} \approx y$$

$\therefore$ [A] $= 1 - x + y \approx y = 10^{-6}$ M

**41.**

| | A | $\rightleftharpoons$ | B |
|---|---|---|---|
| Initial | $a$ M | | $b$ M |
| Equilibrium | $(a-x)$ M | | $(b+x)$ M |

Now, $K_{eq} = \frac{K_1}{K_2} = \frac{b+x}{a-x}$

$\therefore\ x = \frac{K_1a - K_2b}{K_1 + K_2}$

**42.** $r_b = K_b \cdot P_{C(g)}$

**43.** $\Delta G° = -2.303RT \cdot \ln K_p°$

$\Delta H° - T\cdot\Delta S°$

or $-2.303 \times 8.314 \times T \times \ln 1.0 = 240 \times 10^3 - T \times 50$

$\therefore\ T = 4800\ K$

**44.** $\Delta G° = -2.303RT \cdot \ln K_{eq}$

$\Rightarrow -2.303 \times 10^3 = -2.303 \times 2 \times 500 \times \ln K_{eq}$

$\therefore\ K_{eq} = 10$

Now, $K_{eq} = \frac{P_{HI}}{P_{H_2}^{1/2} \times P_{I_2}^{1/2}}$

$\Rightarrow \frac{10}{P_{H_2}^{1/2} \times (0.001)^{1/2}}$

$\therefore\ P_{H_2} = 1000$ atm

**45.** $\Delta G° = -RT \cdot \ln\frac{[B]}{[\alpha]} = -RT \cdot \ln\frac{64}{36} = -RT \cdot \ln 1.78$

**46.** There is no net change at equilibrium.

**47.** $K_{eq}$ at 27°C, $K_1 = \frac{2\times10^{-3}}{2\times10^{-4}} = 4$

and $K_{eq}$ at 127°C, $K_2 = \frac{8\times10^{-2}}{4\times10^{-3}} = 20$

Now, $\ln\frac{K_2}{K_1} = \frac{\Delta H}{R}\left(\frac{1}{T_1} - \frac{1}{T_2}\right)$

or, $2.303 \log\frac{20}{4} = \frac{\Delta H}{R}\left(\frac{1}{300} - \frac{1}{400}\right)$

$\therefore\ \Delta H = 2.303 \times 8.314 \times 1200 \log(5)$ J/mol

**48.**

| | $n$A | $\rightleftharpoons$ | $A_n$ |
|---|---|---|---|
| Initial moles | 1 | | 0 |
| Equilibrium moles | $1-x$ | | $\frac{x}{n}$ |

Now, $K_C = \frac{\left(\frac{x/n}{V}\right)}{\left(\frac{1-x}{V}\right)^n} = \frac{x\cdot V^{n-1}}{(1-x)^n \cdot n}$

$\approx \frac{x\cdot V^{n-1}}{n}$ (as $x << 1$)

Now, total moles $= (1-x) + \frac{x}{n} = 1 + x\cdot\left(\frac{1}{n} - 1\right)$

$= 1 + \frac{n\cdot KC}{V^{n-1}}\cdot\left(\frac{1-n}{n}\right) = 1 + \frac{(1-n)\cdot K_C}{V^{n-1}}$

**49.** $\alpha = \frac{M - M_{mix}}{(n-1)\cdot M_{mix}} = \frac{M - M_{mix}}{(2-1)\cdot M_{mix}} = \frac{M}{M_{mix}} - 1$

$= \frac{M}{\left(\frac{dRT}{P}\right)} - 1 = \frac{PM}{dRT} - 1$

**50.** $n_{NO} = \frac{0.4\times250}{RT} = \frac{100}{RT}$ $\quad n_{O_2} = \frac{0.8\times100}{RT} = \frac{80}{RT}$

| | 2 NO + | $O_2$ | $\rightarrow$ 2 $NO_2$ $\rightleftharpoons$ | $N_2O_4$ |
|---|---|---|---|---|
| Initial moles | $\frac{100}{RT}$ | $\frac{80}{RT}$ | 0 | 0 |
| Final moles | 0 | $\frac{30}{RT}$ | $\frac{100}{RT} - x$ | $\frac{x}{2}$ |

From question, $\frac{30}{RT} + \frac{100}{RT} - x + \frac{x}{2} = \frac{0.3\times350}{RT}$

$\therefore\ x = \frac{50}{RT}$

Now, $K_P$ of second reaction

$= \frac{P_{N_2O_4}}{P_{NO_2}^2} = \frac{\frac{x}{2}}{\left(\frac{100}{RT} - x\right)^2} \times \left(\frac{0.3}{\frac{0.3\times350}{RT}}\right)^{-1}$

$= 3.5$ atm

**51.**

| | $N_2$ + | $3H_2$ $\rightleftharpoons$ | $2NH_3$ |
|---|---|---|---|
| Initial moles | 1 | 3 | 0 |
| Equilibrium moles | $1-x$ | $3-3x$ | $2x$ |

From question, $\frac{2x}{4-2x} = a$

$$\Rightarrow x = \frac{2a}{1+a}$$

$$\text{Now, } K_P = \frac{(2x)^2}{(1-x)(3-3x)^3} \cdot \left(\frac{P}{4-2x}\right)^{-2}$$

$$= \frac{4x^2 \cdot (4-2x)^2}{27(1-x)^4 \cdot P^2}$$

$$\text{or, } \sqrt{K_P} = \frac{2x \cdot (4-2x)}{\sqrt{27}(1-x)^2 \cdot P} = \frac{2 \cdot \left(\frac{2a}{1+a}\right) \cdot \left(4 - \frac{4a}{1+a}\right)}{\sqrt{27} \cdot \left(1 - \frac{2a}{1+a}\right)^2 \cdot P}$$

$$= \frac{32a}{\sqrt{27} \cdot (1-a)^2 \cdot P}$$

$$\therefore \frac{a}{(1-a)^2} \alpha P$$

**52.** $K_P = \dfrac{\alpha^2 \cdot P}{1-\alpha^2}$

$$\Rightarrow \frac{(0.3)^2 \times 1}{1-(0-3)^2} = \frac{\alpha^2 \times 0.1}{1-\alpha^2}$$

$$\Rightarrow \alpha = 0.973$$

**53.** $K_P = P^2$

$$\text{Now, } \ln \frac{K_2}{K_1} = \frac{\Delta H}{R}\left(\frac{1}{T_1} - \frac{1}{T_2}\right)$$

$$\text{or, } \ln \frac{P_2^2}{(7 \times 10^{-3})^2} = \frac{3360}{2}\left(\frac{1}{300} - \frac{1}{400}\right)$$

$$\Rightarrow P_2 = 1.4 \times 10^{-2} \text{ atm}$$

**54.** $CO(g) + H_2(g) \rightleftharpoons CO_2(g) + H_2(g)$

| | CO(g) | $H_2(g)$ | $CO_2(g)$ | $H_2(g)$ |
|---|---|---|---|---|
| Initial | 1 | 5 | 0 | 1 |
| Equilibrium | $1-x$ | $5-x$ | $x$ | $1-x$ |

$$\text{Now, } K_{eq} = \frac{1}{3} = \frac{x \cdot (1+x)}{(1-x) \cdot (5-x)}$$

$$\Rightarrow x = \frac{1}{2}$$

**55.** $NH_2COONH_4(s) \rightleftharpoons N_2 + 3H_2 + CO + \frac{1}{2}O_2$

| | $N_2$ | $3H_2$ | CO | $\frac{1}{2}O_2$ |
|---|---|---|---|---|
| Equilibrium partial pressure | $\frac{22}{5.5} = 4$ | $\frac{3 \times 22}{5.5} = 12$ | $\frac{22}{5.5} = 4$ | $\frac{22}{2 \times 5.5} = 2$ |

$$\text{Now, } K_p = 4 \times (12)^3 \times 4 \times (2)^{1/2} = 27 \times (2)^{10.5}$$

**56.** $NH_2COONH_4(s) \rightleftharpoons 2NH_3(g) + CO_2(g)$

| | $2NH_3(g)$ | $CO_2(g)$ |
|---|---|---|
| Equ. partial pressure | $2P_0$ | $P_0$ |
| New Equ. partial pressure | $3P_0$ | $P_0$ |

$$\text{Now, } K_P = (2P_0)^2 \cdot P_0 = (3P_0)^2 \cdot P$$

$$\Rightarrow P = \frac{4}{9} P_0$$

$$\text{Now, } \frac{3P_0 + P}{3P_0} = \frac{31}{27}$$

**57.** $\Delta H° = \Delta E° + \Delta n_g \cdot RT$

$$= (+30) + (3-2) \times \frac{2}{1000} \times 300 = +30.6 \text{ K cal}$$

$$\text{Now, } \Delta G° - RT \cdot \ln K_{eq} = \Delta H° - T \cdot \Delta S°$$

$$\text{or, } -2 \times 300 \times \ln K_{eq} = 30.6 \times 10^3 - 300 \times 100$$

$$\Rightarrow \ln K_{eq} =$$

$$\therefore K_{eq} = \frac{1}{e}$$

**58.** $trans \rightleftharpoons Cis$; $\Delta G° = 22.112 - 30.426 = -8.314$ KJ

$$\text{Now, } \Delta G° = -RT \cdot \ln \frac{[Cis]}{[trans]}$$

$$\Rightarrow -8.314 \times 10^3 = -8.314 \times 300 \times \ln \frac{[Cis]}{[trans]}$$

$$\therefore \frac{[Cis]}{[trans]} = \frac{28}{1}$$

**59.** $CO(g) + H_2O(g) \rightleftharpoons CO_2(g) + H_2(g)$

| | CO(g) | $H_2O(g)$ | $CO_2(g)$ | $H_2(g)$ |
|---|---|---|---|---|
| Initial moles | 2 | 5 | 0 | 2 |
| Equilibrium moles | $2-x$ | $5-x$ | $x$ | $2+x$ |

$$\text{Now, } K_{eq} = 3.0 = \frac{x \cdot (2+x)}{(2-x)(5-x)}$$

$$\Rightarrow x = 1.5$$

$$\therefore [H_2] = \frac{2+x}{2} = 1.75 \text{ M}$$

**60.** $NH_4HS(s) \rightleftharpoons NH_3(g) + H_2S(g)$

$$\frac{X}{2} \text{ bar} \qquad \frac{X}{2} \text{ bar}$$

$$\therefore \Delta G° = -RT \cdot \ln K_P°$$

$$= -\text{RT} \ln\left(\frac{X}{2}, \frac{X}{2}\right) = -2RT(\ln \times - \ln 2)$$

HINTS AND EXPLANATIONS

## Section B (One or More than one Correct)

1. $NH_4HS(s) \rightleftharpoons NH_3(g) + H_2S(g)$

   Equ. partial pressure 0.2 atm 0.2 atm

   Second Equ. partial pressure 0.5 atm $P$ atm

   Now, $K_P = 0.2 \times 0.2 = 0.5 \times P \Rightarrow P = 0.08$

2. $\Delta ng = 0$

3. $$N_2O_5(g) \rightleftharpoons 2NO_2(g) + \frac{1}{2}O_2(g)$$

   Initial partial pressure $P_0$ 0 0

   Equ. partial pressure $P_0(1-\alpha)$ $2\alpha P_0$ $\frac{\alpha P_0}{2}$

   and $\alpha = \frac{M-2D}{\left(\frac{5}{2}-1\right)\cdot 2D} = \frac{M-2D}{3D}$

   Total equilibrium pressure $= P_0(1-\alpha) + 2\alpha P_0 + \frac{\alpha P_0}{2} = P_0\left(1+\frac{3\alpha}{2}\right)$

4. Le Chatelier's principle
5. Theory based
6. Theory based
7. Vapour pressure of a particular liquid system depends only on temperature.
8. $Cl_2(g) \rightleftharpoons 2Cl(g)$ $T\uparrow$ $P\downarrow$
9. Addition of insert gas at constant pressure shifts the equilibrium in the direction of increase in moles of gases.
10. Decrease in pressure favors the reaction is the direction of increase in moles of gas and hence, $B$ should be monomer.
11. $\Delta_f G° : NO_2 > N_2O_5 > NO$
12. At $300K : \Delta G° = (-41) - 300 \times (-0.04)$

    $= -29$ KJ/mol

    Hence, the reaction is spontaneous in forward direction.

    At 1200 K : $\Delta G° = (-33) - 1200 \times (-0.03)$

    $= +3$ KJ/mol

    Hence, the reaction is spontaneous in backward direction.

13. Theory based.
14. Theory based.
15. $S = A \cdot e^{-\Delta H/RT}$

    $\Rightarrow \ln s = \ln A - \frac{\Delta H}{RT}$

    Positive slope represents that $\Delta H$ = negative.

16. Theory based.
17. $K_{P_2} = P_{Cl_2(g)} = 0.2$ atm

    $K_{P_1} = P_{Cl_2(g)} \cdot P^8_{H_2O(g)} = 0.2 \times (0.001)^8 = 2 \times 10^{-25}$ atm$^9$

    $P_{H_2O(g)}$ = Vapour pressure of ice.

18. $$PCl_5(g) \rightleftharpoons PCl_3(g) + Cl_2(g)$$

    Initial moles 5 0 0

    Moles at equilibrium $5-x$ $x$ $x$

    From question, $(5-x)+x+x+4 = \frac{4.8 \times 112}{0.0821 \times 546}$

    $\Rightarrow x = 3$

    $\therefore \alpha = \frac{x}{5} = 0.6$

    and $K_P = \frac{x \cdot x}{5-x} \cdot \left(\frac{4.8}{12}\right) = 1.8$ atm

19. $$4HCl(g) + O_2(g) \rightleftharpoons 2Cl_2(g) + 2H_2O(g)$$

    | | $4HCl(g)$ | $O_2(g)$ | $2Cl_2(g)$ | $2H_2O(g)$ |
    |---|---|---|---|---|
    | Initial partial pressure | 1.0 atm | 0.25 atm | 0 | 0.4 atm |
    | On completion | 0 | 0 | 0.5 atm | 0.4 atm |
    | Equ. partial pressure | $4x$ atm | $x$ atm | 0.5 atm | 0.4 atm |

    $K_P = 5 \times 10^{12} = \frac{(0.5)^2(0.4)^2}{(4x)^4 \times x}$

    $\Rightarrow x = 5 \times 10^{-4}$

20. $\Delta G° = 2 \times \Delta_f G°_{NO_2(g)} - \Delta_f G°_{N_2O_4(g)} = 0$

    $\Rightarrow K_P° = 1$

    Now, $\Delta G = \Delta G° + RT \cdot \ln Q = 0 + RT \cdot \ln \frac{P^2_{NO_2}}{P_{N_2O_4}}$

    $= RT \cdot \ln \frac{10^2}{10}$ = positive.

HINTS AND EXPLANATIONS

**21.** $AB_2\,(g) + A(s) \rightleftharpoons 2\,AB(g)$

| | | |
|---|---|---|
| Initial partial pressure | 0.7 bar | 0 |
| Equ. partial pressure | $(0.7 - x)$ bar | $2x$ bar |
| Second equ. partial pressure | $y$ bar | $(0.4 - y)$ bar |

From question, $(0.7 - x) + 2x = 0.95$

$\Rightarrow x = 0.25$

$$\therefore\ K_P = \frac{(2x)^2}{(0.7-x)} = \frac{(0.5)^2}{0.45} = \frac{5}{9}$$

Now, $\frac{5}{9} = \frac{(0.4-y)^2}{y}$

$\Rightarrow y = 0.13$

$\therefore$ At second equilibrium, the volume per cent of

$$AB_2 = \frac{0.13}{0.4}\times 100 = 32.5\%$$

**22.** $PCl_5(g) \rightleftharpoons PCl_3(g) + Cl_2(g)$

| | | | |
|---|---|---|---|
| Initial moles | 1 | 1 | 0 |
| Equ. moles | $1 - x$ | $1 + x$ | $x$ |
| | $\approx 1$ | $\approx 1$ | $= 0.004$ |

$$\therefore\ K_C = \frac{1\times 0.004}{1}\times\left(\frac{1}{10}\right) = 0.0004\ \text{M}$$

**23.** $\Delta G^\circ = -RT\cdot \ln K_P{}^\circ$

$\Rightarrow -1743 = -\,8.3 \times 300 \times \ln K_P{}^\circ$

$\therefore\ K_P{}^\circ = 2$

**24.** $K_1 = \frac{1}{K_2 \times K_3} = \frac{1}{0.24}$

As $\Delta n_g = 0$, $[A] + [B] + [C] = 1$ M

**25.** Addition of water will shift the reaction in the direction of increase in mole of *aq* species.

## Section C (Comprehensions)

### Comprehension I

**1.** For $SrCl_2 \cdot 2H_2O(s)$, $P_{H_2O} = (2.56 \times 10^{-10})^{1/4} = 0.004$ atm

For $Na_2HPO_4 \cdot 7H_2O$ $P_{H_2O} = (2.43 \times 10^{-13})^{1/5} = 0.003$ atm

For $Na_2SO_4(s)$, $P_{H_2O} = (1.024 \times 10^{-27})^{1/10} = 0.002$ atm

As $P_{H_2O}$ is minimum for $Na_2SO_4(s)$, it is the best dehydrating agent.

**2.** For $Na_2SO_4(s)$, $10H_2O(s)$ to be efflorescent, $P_{H_2O} < 0.002$ atm

or $\left(0.04\times\frac{R.H.}{100}\right) < 0.002$

$\Rightarrow R.H. < 5\%$

**3.** $Na_2HPO_4 \cdot 7H_2O(s)$ to be deliquescent, $P_{H_2O} >$ 0.003 atm

or, $\left(0.04\times\frac{R.H.}{100}\right) > 0.003$

$\Rightarrow R.H. > 7.5\%$

### Comprehension II

**4.** $CO(g) + 2H_2(g) \rightleftharpoons CH_3OH(g)$

| | | | |
|---|---|---|---|
| Initial moles | 0.2 | $a$(say) | 0 |
| Moles at equ. | $0.2 - x$ | $a - 2x$ | $x$ |
| | $= 0.1$ | $= a - 0.2$ | $= 0.1$ |

Total moles $= 0.1 + (a - 0.2) + 0.1 = \frac{7.5\times 2.463}{0.0821\times 750}$

$\Rightarrow a = 0.3$

Now, $K_P = \frac{0.1}{0.1\times(0.1)^2}\times\left(\frac{7.5}{0.3}\right)^{-2} = 0.16\ \text{atm}^{-2}$

**5.** $K_C = \frac{K_P}{(RT)^{\Delta n_g}} = \frac{0.16}{(0.0821\times 750)^{-2}} = 607\ \text{M}^{-2}$

**6.** $P = \frac{(0.2+0.3)\times 0.0821\times 750}{2.463} = 12.5$ atm

## Comprehension III

7. $C_2H_5OH + CH_3COOH \rightleftharpoons CH_3COOC_2H_5 + H_2O$

Initial moles

$a \quad a \quad 0 \quad 0$

Equilibrium moles

$a-x \quad a-x \quad x \quad x$

From question, $\frac{x}{2a} = 0.333 = \frac{1}{3}$

$\Rightarrow x = \frac{2}{3}a$

Now, $K_{eq} = \frac{x \cdot x}{(a-x)\cdot(a-x)} = 4$

8. $C_2H_5OH + CH_3COOH \rightleftharpoons CH_3COOC_2H_5 + H_2O$

Initial moles

$\frac{a}{3} \quad \frac{2a}{3} \quad 0 \quad 0$

Equilibrium moles

$\frac{a}{3}-x \quad \frac{2a}{3}-x \quad x \quad x$

Now, $K_{eq} = 4 = \frac{x \cdot x}{\left(\frac{a}{3}-x\right)\left(\frac{2a}{3}-x\right)}$

$\Rightarrow x = 0.2833\, a$

$\therefore$ Fraction of alcohol reacted $= \frac{x}{a/3} = 0.85$

9. Solution of $0.7 = \frac{x}{a} \times 100 = 66.67\%$

## Comprehension IV

10.

| | $2HI(g)$ | $\rightleftharpoons$ | $H_2(g) + I_2(g)$ | |
|---|---|---|---|---|
| Initial moles | 1(say) | | 0 | 0 |
| Equilibrium moles | $1 - 0.2222 = 0.7778$ | | 0.1111 | 0.1111 |

$\therefore K_{eq} = \frac{0.1111 \times 0.1111}{(0.7778)^2} = \frac{1}{49} \approx 0.02$

11. In the presence of $I_2(g)$, the extent of dissociation of HI will decrease.

12. Addition of He($g$) will not affect thequilibrium.

## Comprehension V

13.

| | $NH_4HS\,(s) \rightleftharpoons$ | $NH_3\,(g)$ | $+ H_2S\,(g)$ |
|---|---|---|---|
| Initial partial pressure | | $P$ mm | 0 |
| Equilibrium partial pressure | | $(P + x)$mm | $x$ mm |

From question, $P + x = 625$ and $(P + x) + x = 725$

$\therefore x = 100$ and $P = 525$

14. $K_P = (P + x) \cdot x = 625 \times 100\ mm^2$

$\therefore K'_P$ (required) $= \frac{1}{K_P} = 1.6 \times 10^{-5}\ mm^{-2}$.

15. $P_{NH_3} = P_{H_2S} = \sqrt{K_P} = 250$ mm

16. Minimum mass of $NH_4HS\,(s)$ needed.

$= \frac{\frac{250}{760} \times 5.0}{0.0821 \times 300} \times 51$ gm

## Comprehension VI

17. $K_{eq} = \frac{A_f}{A_b} \cdot e^{-\Delta H/RT} = e^{\frac{-(-24.942\times10^3)}{8.314\times300}} = e^{10}$

18. $\frac{K_b}{K_f} = \frac{1}{K_{eq}} = e^{-10}$

$\Delta H = Ea_f - Ea_b = Ea_f - \frac{3}{2}Ea_f \Rightarrow Ea_f = 2\cdot(-\Delta H)$

Now, $K_f = A_f \cdot e^{-Ea_f/RT} = 1\times e^{\frac{-2\times24.942\times10^3}{8.314\times300}} = e^{-20}$

and $K_b = e^{-30}$

## Comprehension VII

20. $2SO_3 \rightleftharpoons 2SO_2 + O_2$

| | | | |
|---|---|---|---|
| Equilibrium moles | $1-\alpha$ | $\alpha$ | $\frac{\alpha}{2}$ |

Now. $K_P = \frac{\alpha^2\cdot\frac{\alpha}{2}}{(1-\alpha)^2}\cdot\left(\frac{K_P}{1+\frac{\alpha}{2}}\right)'$

$\Rightarrow \alpha = \frac{2}{3}$

21. $2NH_3 \rightleftharpoons N_2 + 3H_2$

| | | | |
|---|---|---|---|
| Equilibrium moles | $1-\alpha$ | $\frac{\alpha}{2}$ | $\frac{3\alpha}{2}$ |
| | $=\frac{1}{3}$ | $=\frac{1}{3}$ | $=1$ |

Now, $K_P = \frac{\frac{1}{3}\times1^3}{\left(\frac{1}{3}\right)^2}\times\left(\frac{50}{\left(\frac{5}{3}\right)}\right)^2 = 2700\text{ atm}^2$

22. $n_{\text{initial}} \times (17 + 28 + 2 + 20) = 134 \Rightarrow n_{\text{initial}} = 2$

| | $2NH_3$ | $\rightleftharpoons N_2$ | $+\ 3H_2$ |
|---|---|---|---|
| Initial moles | 2 | 2 | 2 |
| Moles at equilibrium | $2-2x$ | $2+x$ | $2+3x$ |
| | $=1.0$ | $=\frac{134\times0.5224}{28}$ | $=3.5$ |

$\therefore x = 0.5$

Now, $K_P = 2700 = \frac{2.5\times(3.5)^3}{(1.0)^2}\times\left(\frac{P}{9}\right)^2$

$\Rightarrow P = \sqrt{\frac{2700\times81}{2.5\times(3.5)^3}}$ atm

## Comprehension VIII

$N_2 + 3H_2 \rightleftharpoons 2NH_3;\ K_{P_1}$

| | | | |
|---|---|---|---|
| Initial partial pressure | $9P$ | $13P$ | 0 |
| Equilibrium partial pressure | $9P-x-y$ | $13P-3x-2y$ | $2x$ |

$N_2 + 2H_2 \rightleftharpoons N_2H_4;\ K_{P_2}$

| | | | |
|---|---|---|---|
| Initial partial pressure | $9P$ | $13P$ | 0 |
| Equilibrium partial pressure | $9P-y-x$ | $13P-2y-3x$ | $y$ |

From question, $P_{NH_3} = 2x = P_0 \Rightarrow x = \frac{P_0}{2}$

$P_{H_2} = 13P - 3x - 2y = 2P_0$

and $P_{\text{total}} = (9P - x - y) + (13P - 3x - 2y) + 2x + y = 7P_0$

$\therefore\ y = \frac{3P_0}{2}$ and $P = \frac{P_0}{2}$

$K_{P_1} = \frac{P^2_{NH_3}}{P_{N_2}\cdot P^3_{H_2}} = \frac{P_0^2}{\frac{5P_0}{2}\times(2P_0)^3} = \frac{1}{20P_0^2}$

$\therefore K_P\text{ (required)} = 20\,P_0^2$

24. $K_{P_2} = \frac{P_{N_2H_4}}{P_{N_2}\cdot P^2_{H_2}} = \frac{\left(\frac{3P_0}{2}\right)}{\frac{5P_0}{2}\times(2P_0)^2} = \frac{3}{20P_0^2}$

## Section D (Assertion–Reason)

1. Direction of shifting of equilibrium will depend on relative values of *a* and *b*.
2. Equilibrium opposes the changes.
3. Theory based
4. Theory based
5. Theory based
6. $K_P = K_C \cdot (RT)^{\Delta n_g}$
7. Theory based
8. Exothermic direction is favoured on lowering temperature.
9. $NaCl(s) \rightleftharpoons Na^+(aq) + Cl^-(aq)$
10. On decreasing the volume, moles of A(*g*) as well as B(*s*) will increase.

## Section E (Column Match)

1. Theory based
2. $K_P = K_C \cdot (RT)^{\Delta n_g}$
3. Le Chatelier's principle
4. Le Chatelier's principle
5. Le Chatelier's principle
6. Le Chatelier's principle
7. Le Chatelier's principle
8. Le Chatelier's principle
9. $P_{CO_2} = K_P = 2.463$ atm

   $\therefore n_{CO_2}$ at equilibrium $= \dfrac{2.463\times 15}{0.0821\times 900} = 0.5$

   (A) % of $CaCO_3$ decomposed
   $= \dfrac{0.5}{1.0}\times 100 = 50\%$ (Eqn)

   (B) % of $CaCO_3$ decomposed
   $= \dfrac{0.5}{0.5}\times 100 = 100\%$ (Eqn)

   (C) % of $CaCO_3$ decomposed $= 100\%$ (non-Eqn)
10. Le Chatelier's principle

## Section F (Subjective)

### Single-digit Integer Type

1. $2H_2S(g) \rightleftharpoons 2H_2(g) + S(g);\ K_e = 10^{-6}$

| | $2H_2S(g)$ | $2H_2(g)$ | $S(g)$ |
|---|---|---|---|
| Initial moles | 0.1 | 0 | 0 |
| Equilibrium moles | $0.1 - x \simeq 0.1$ | $x$ | $\frac{x}{2}$ |

Now,

$$K_C = 10^{-6} = \frac{x^2 \times \frac{x}{2}}{(0.1)^2} \times \left(\frac{1}{0.4}\right)$$

$\Rightarrow x = 2\times 10^{-3}$

$\therefore$ Percentage dissociation $= \dfrac{2\times 10^{-3}}{0.1}\times 100 = 2\%$

2. $V_{CF_4} = 500 - 300 = 200\text{ ml} = V_{CO_2}$

$\therefore V_{COF_2} = 500 - 2\times 200 = 100$ ml

Hence, $P_{CF_4} = P_{CO_2} = \dfrac{200}{500}\times 10 = 4$ atm

$P_{COF_2} = \dfrac{100}{500}\times 10 = 2$ atm

$K_p = \dfrac{4\times 4}{2^2} = 4$

3. Initial: $n_{PCl_5} = \dfrac{62.55}{208.5} = 0.3$ and $n_{Cl_2} = \dfrac{4.48}{22.4} = 0.2$

| | $PCl_5 \rightleftharpoons$ | $PCl_3$ | $+\ Cl_2$ |
|---|---|---|---|
| Initial moles | 0.3 | 0 | 0.2 |
| Equilibrium moles | $0.3 - x$ | $x$ | $0.2 + x$ |

Now,

$$K_p = \frac{x\cdot(0.2+x)}{(0.3-x)}\cdot\left(\frac{p}{0.5+x}\right) = \frac{x(0.2+x)}{(0.3-x)}\times\frac{RT}{V}$$

or, $8 = \dfrac{x(0.2+x)}{0.3-x}\times\dfrac{0.0821\times 546}{4.48}$

$\Rightarrow x = 0.2$

$\therefore$ Final pressure $= \dfrac{(0.5+x)\times 0.0821\times 546}{4.48}$

$= 7$ atm

**4.** NaOH is used to neutralize acetic acid. From the given data, half of the acid taken is neutralize.

| | $CH_3COOH + C_2H_5OH \rightleftharpoons$ | | $CH_3COOC_2H_5 + H_2O$ | |
|---|---|---|---|---|
| Initial moles | $a$ | $a$ | 0 | 0 |
| Equilibrium moles | $a-\frac{a}{2}$ | $a-\frac{a}{2}$ | $\frac{a}{2}$ | $\frac{a}{2}$ |

$$\therefore K_{eq} = \frac{\frac{a}{2}\times\frac{a}{2}}{\frac{a}{2}\times\frac{a}{2}} = 1$$

**5.**

| | $2HI \rightleftharpoons$ | $H_2$ + | $I_2$ |
|---|---|---|---|
| Equilibrium moles | $1-0.8 = 0.2$ | 0.4 | 0.4 |

$$\therefore K_{eq} = \frac{0.4\times 0.4}{(0.2)^2} = 4$$

| Now, | $H_2$ + | $I_2 \rightleftharpoons$ | $2HI$ |
|---|---|---|---|
| Initial moles | 2 | 2 | 0 |
| Equilibrium moles | $2-x$ | $2-x$ | $2x$ |

$$K_{eq} = \frac{1}{4} = \frac{(2x)^2}{(2-x)(2-x)}$$

$\Rightarrow x = 0.4$

Now, $n_{eq}$ of $I_2 = n_{eq}$ of $Na_2S_2O_3$

or, $(2-x)\times 2 = V\times(1.6\times 1)$

$\Rightarrow V = 2$ L

**6.**

| | $Cl_2CHCOOH$ + | $C_5H_{10} \rightleftharpoons$ | $Cl_2CHCOOC_5H_{11}$ |
|---|---|---|---|
| I: Initial moles | 1 | 4 | 0 |
| Equilibrium moles | $1-x$ | $4-x$ | $x = 0.5$ |
| II: Initial moles | 1 | $a$ | 0 |
| Equilibrium moles | $1-y$ | $a-y$ | $y = 0.6$ |

Now, $K_{eq} = \dfrac{0.5}{0.5\times 3.5}\times 0.7 = \dfrac{0.6}{0.4\times(a-0.6)}\times 0.72$

$\Rightarrow a = 5$

7. $n_{CO_2}$ at equilibrium = 0.05

∴ Minimum mass of $CaCO_3$ needed = 0.05 × 100 = 5 gm

8. $$Ag^+(aq) + Fe^{2+}(aq) \rightleftharpoons Fe^{3+}(aq) + Ag(s)$$

Initial moles $\frac{500\times0.9}{1000} = 0.45$ $\frac{500\times1.0}{1000} = 0.50$ 0 0

Equilibrium moles $0.45 - x$ $0.50 - x$ $x$ $x$

Now, $n_{eq}\ Fe^{2+} = n_{eq}\ MnO_4^-$

or, $\frac{(0.50-x)}{1000}\times30\times1 = \frac{25\times0.06}{1000}\times5$

$\Rightarrow x = 0.25$

$\therefore K_{eq} = \frac{x}{(0.45-x)(0.5-x)} = 5\ M^{-1}$

9. $$Sb_2S_3(s) + 3H_2(g) \rightleftharpoons 2Sb(s) + 3H_2S(g)$$

Initial moles 0.01 0.01 0 0

Equ. moles $0.01 - x$ $0.01 - 3x$ $2x$ $3x = \frac{1.19}{238}$

$= 5\times10^{-3}$ $= 5\times10^{-3}$

Now, $K_c = \frac{(5\times10^{-3})^3}{(5\times10^{-3})^3} = 1$

10. $$H_2O + D_2O \rightleftharpoons 2HDO$$

Initial moles 28 28 0

Equ. moles $28 - 14 = 14$ $28 - 14 = 14$ $2\times14 = 28$

$K_C = \frac{(28)^2}{14\times14} = 4$

11. $$3A_2(g) \rightleftharpoons A_6(g),\ K_{p_1} = 1.6\ atm^{-2}$$

Initial partial pressure $2P_0$ 0

Equilibrium partial pressure $2P_0 - 3a - b$ $a$

$$A_2(g) + C(g) \rightleftharpoons A_2C(g),\ K_{p_2} = x\ atm^{-1}$$

Initial partial pressure $2P_0$ $P_0$ 0

Equilibrium partial pressure $2P_0 - b - 3a$ $P_0 - b$ $b$

From question, $a = 0.2$,

$\frac{P_{A_6}}{P_{A_2}^{\ 3}} = 1.6 \Rightarrow \frac{0.2}{P_{A_2}^{\ 3}} = 1.6$

$\Rightarrow P_{A_2} = 0.5 = 2P_0 - 3a - b$

and $(2P_0 - 3a - b) + a + (P_0 - b) + b = 1.4$

$\Rightarrow P_0 = 0.7$ and $b = 0.3$

Now, $K_{p_2} = \frac{b}{(2P_0 - 3a - b)(P_0 - b)} = \frac{0.3}{0.5\times.4}$

$= 1.5\ atm^{-1}$

12. Initial partial pressure of IBr (g)

$= \frac{8.28}{207}\times\frac{0.0821\times500}{0.1642} = 10$

$$2IBr(g) \rightleftharpoons I_2(g) + Br_2(g)$$

Initial partial pressure 10 0 0

Equilibrium partial pressure $10 - 2x$ $x$ $x = 4$

$\therefore K_p = \frac{4\times4}{(2)^2} = 4$

13. $$H_2(g) + I_2(g) \rightleftharpoons 2HI(g)$$

I: Initial moles 1 3 0

Equilibrium moles $1-\frac{x}{2}$ $3-\frac{x}{2}$ $x$

II: Initial moles 3 3 0

Equilibrium moles $3 - x$ $3 - x$ $2x$

Now, $K_{eq} = \frac{(x)^2}{\left(1-\frac{x}{2}\right)\left(3-\frac{x}{2}\right)} = \frac{(2x)^2}{(3-x)(3-x)}$

$\Rightarrow x = \frac{3}{2}$

$\therefore K_{eq} = 4$

**14.** $N_2O_5(g) \rightleftharpoons N_2O_3(g) + O_2(g);\ K_{C_1} = 2.5$ M

| | | | |
|---|---|---|---|
| Initial moles | 4 | 0 | 0 |
| Equilibrium moles | $4-x$ | $x-y$ | $x+y$ |

$N_2O_3(g) \rightleftharpoons N_2O(g) + O_2(g);\ K_{C_2}$

| | | | |
|---|---|---|---|
| | $x$ | 0 | 0 |
| Equilibrium moles | $x-y$ | $y$ | $x+y$ |

From question, $[O_2] = \frac{x+y}{2} = 2.5$

$\Rightarrow x + y = 5$

And $K_{C_1} = 2.5 = \frac{(x-y)\times 5}{4-x}\times\frac{1}{2} = \frac{(5-2y)\times 5}{(y-1)}\times\frac{1}{2}$

$\Rightarrow y = 2$

$\therefore [N_2O] = \frac{y}{2} = 1$ M

**15.** In left chamber, $P_{H_e} = 2$ atm

$\therefore P_{NH_3} = P_{H_2 3} = \frac{4-2}{2} = 1$ atm

$\therefore K_p = 1\times 1 = 1\ \text{atm}^2$

## Four-digit Integer Type

**1.** P-xyloquinone + M.W $\rightleftharpoons$ P-xylohydroquinone + M.B.

| | | | | |
|---|---|---|---|---|
| Initial conc. | 0.012 M | 0 | 0.24 M | $10^{-3}$ M |
| Equ. con. | $0.012 + 4\times10^{-5}$ | $4\times10^{-5}$ M | $0.24 - 4\times10^{-5}$ | $10^{-3} - \frac{4}{100}\times10^{-3}$ |
| | $\approx 0.012$ | | $\approx 0.24$ M | $0.96\times10^{-3}$m |

$\therefore K_{eq} = \frac{0.24\times0.96\times10^{-3}}{0.012\times4\times10^{-5}} = 480$

**2.** $6HCHO \rightleftharpoons C_6H_{12}0_6;\ K_{eq} = 6.4\times10^{19}$

| | | |
|---|---|---|
| Initial conc. | 0 | 1 M |
| Equ. con. | $6x$ | $1-x = 1$ M |

Now, $K_{eq} = 6.4\times10^{19} = \frac{1}{[HCHO]^6}$

$\Rightarrow [HCHO] = 5\times10^{-4}$ M

$= 5\times10^{-4}\times 30$ g/L

$= 15$ mg/L

**3.** $PCl_5 \rightleftharpoons PCl_3 + Cl_2$

| | | | |
|---|---|---|---|
| Initial equ. moles | 2 | 2 | 2 |
| Moles on adding $Cl_2$ | 2 | 2 | $2+x$ |
| Final Equ. moles | $2+y$ | $2-y$ | $2+x-y$ |

From question, $(2 + y) + (2 - y) + 2 + (x - y) = 2\times 6$

or, $x - y = 6$

and $K_c = \frac{2\times2}{2}\times\frac{1}{V} = \frac{(2-y)\times(2+x-y)}{(2+y)}\times\frac{1}{2V}$

Or, $4 - \frac{(8-x)\times8}{(x-4)} \Rightarrow x = \frac{20}{3}$

**4.** $K^\circ_{eq} = 1 \Rightarrow \Delta G^\circ = 0$

$\Rightarrow T = \frac{\Delta H^\circ}{\Delta S^\circ} = \frac{223\times10^3}{\left(\frac{223-33}{1520}\times10^3\right)} = 1784$ K

**5.** Graphite $\rightleftharpoons$ Diamond; $\Delta G^\circ = (3.0 - 0)$ kJ/mol; $P_1 = 1$ bar

$\Delta G = 0 \quad P_2 = P$

Now, $\Delta(\Delta G) = \Delta V\cdot\Delta P$

or, $(\Delta G - \Delta G^\circ) = (V_{Dia} - V_{Gra})(P_2 - P_1)$

or, $(0 - 3.0\times10^3) = \left[\left(\frac{12}{3.6} - \frac{12}{2.4}\right)\times10^{-6}\right]\left[P_2 - 10^5\right]$

$\therefore P_2 = 1.8\times10^9\ Pa = 1.8\times10^4$ bar

**6.** $K^\circ_P = \frac{P_{CO_2}}{P_{CO}} = \frac{10^6}{400}$

Now, $\Delta G^\circ = -5320 - 5.6\,T = -RT\ln K^\circ_P$

$$= -2 \times T \times \ln \frac{10^6}{400}$$

$\therefore T = 532$ k

7. $4HNO_3 (g) \rightleftharpoons 4NO_2 (g) + 2H_2O (g) + O_2 (g)$

| | | | | |
|---|---|---|---|---|
| Initial partial pressure | $P_0$ | 0 | 0 | 0 |
| Equ. partial pressure | $P_0 - 4x$ | $4x$ | $2x$ | $x$ |

From question, $P_0 - 4x = 2$ atm

and $(P_0 - 4x) + 4x + 2x + x = 30$ atm

$\therefore P_0 = 18$ atm and $x = 4$ atm

$$\text{Now, } K_p = \frac{(4x)^4 \times (2x)^2 \times x}{(P_o - 4x)^4} = 2^{20}\,\text{atm}^3$$

$$\text{and } K_c = \frac{K_p}{(RT)^{\Delta n_g}} = \frac{2^{20}}{(0.08 \times 400)^3} = 32\,\text{M}^3$$

8. Initial: $P_{NOcl} = P_0$ bar and $P_{N_2} = (1 - P_0)$ bar

$$2NOCl \rightleftharpoons 2NO + Cl_2$$

| | | | |
|---|---|---|---|
| Initial partial pressure | $P_0$ | 0 | 0 |
| Eqn. partial pressure | $P_0 - 2x$ | $2x$ | $x = 1.2 - 1.0 = 0.2$ |
| Par. pre. on adding $Cl_2$ | $P_0 - 2x$ | $2x$ | $x + (8.3 - 1.2)$ |
| New Equ.partial pre. | $P_0 - 2x + 2y$ | $2x - 2y$ | $7.1 + x - y$ |

From question, $y = 8.3 - 8.2 = 0.1$

$$\text{Now, } K_p = \frac{(2x)^2 \times x}{(P_o - 2x)^2} = \frac{(2x - 2y)^2 \times (7.1 + x - y)}{(P_0 - 2x + 2y)^2}$$

$$\text{or, } \frac{0.4^2 \times 0.2}{(P_0 - 0.4)^2} = \frac{0.2^2 \times 7.2}{(P_0 - 0.2)^2} \Rightarrow P_0 = 0.5$$

and $K_p = 3.2$

9. Initial equilibrium:

$$A + 2B \rightleftharpoons C$$

| | | | |
|---|---|---|---|
| Initial partial pressure | $P_0$ | $2P_0$ | O |
| Equ. partial pressure | $P_0 - x$ | $2P_0 - 2x$ | $x$ |

From question, $(P_0 - x) + (2P_0 - 2x) + x + 3P_0$

$$= \frac{5}{6} \times 6P_0$$

$$\text{or, } x = \frac{P_0}{2}$$

Second equilibrium:

$$A + 2B \rightleftharpoons C$$

| | | | |
|---|---|---|---|
| Initial partial pressure | $2P_0$ | $4P_0$ | 0 |
| Equ. partial pressure | $2P_0 - y$ | $4P_0 - 2y$ | $y - 2z$ |

$$2C + D \rightleftharpoons 2F$$

| | | | |
|---|---|---|---|
| | $y$ | $6P_0$ | 0 |
| Equ. partial pressure | $y - 2z$ | $6P_0 - z$ | $2z$ |

From question, $2P_0 - y = y - 2z$

Now for the first reaction,

$$K_p = \frac{x}{(P_o - x)(2P_0 - 2x)^2} = \frac{y - 2z}{(2P_0 - y)(4P_0 - 2y)^2}$$

$$\text{or, } \frac{1}{P_0^2} = \frac{1}{(4P_0 - 2y)^2} \Rightarrow y = \frac{3}{2} P_0 \text{ and } z = \frac{P_0}{2}$$

Total equilibrium pressure:

First equilibrium $= 5P_0$

Second equilibrium $= (2P_0 - y) + (4P_0 - 2y) + (y - 2z) + (6P_0 - z) + 2z$
$= 3.5P_0$

10. $A_2 + B_2 \rightleftharpoons 2AB$

| | | | |
|---|---|---|---|
| Initial moles | 0.2 | 0.2 | 0 |
| Equ. moles | $0.2 - x = 0.05$ | $0.2 - x = 0.05$ | $2x = 0.3$ |

$$K_{eq} = K_1 = \frac{(0.3)^2}{0.05 \times 0.05} = 36$$

After adding $C_2$:

$$A_2 + B_2 \rightleftharpoons 2AB$$

| | | | |
|---|---|---|---|
| Initial moles | 0.2 | 0.2 | 0 |
| Equ. moles | $0.2 - y - z = 0.08 - z$ | $0.2 - y = 0.08$ | $2y = 0.24$ |

$$A_2 + C_2 \rightleftharpoons 2AC$$

| | | | |
|---|---|---|---|
| Initial moles | 0.2 | 0.1 | 0 |
| Equ. moles | $0.2 - z - y$ | $0.1 - z$ | $2z$ |

$$\text{Now, } K_1 = 36 = \frac{(0.24)^2}{(0.08 - z) \times 0.08} \Rightarrow z = 0.06$$

$$\therefore K_{eq} = K_2 = \frac{(0.12)^2}{0.02 \times 0.04} = 18$$

11. $\Delta G = \Delta G° + \text{RT.} \ln Q = -\text{RT.} \ln \dfrac{K_f}{K_b} + \text{RT.} \ln \dfrac{[\text{Product}]}{[\text{Reactants}]}$

$$= RT.\ln\frac{K_b[\text{Product}]}{K_f[\text{Reactants}]} = RT.\ln\frac{r_b}{r_f}$$

$$= 2 \times 300 \times \ln\frac{1}{e^4} = -2400 \text{ cal}$$

**12.** $A + B \rightleftharpoons C;\ K_1 = 4 \times 10^{10}$

| | A | B | C |
|---|---|---|---|
| Initial moles | 5 | 5 | 0 |
| Equ. moles | $5-(x+y)$ | $5-x$ | $x$ |

$A + D \rightleftharpoons C;\ K_2 = 10^{10}$

| | A | D | C |
|---|---|---|---|
| Initial moles | 5 | 5 | 0 |
| Equ. moles | $5-(x+y)$ | $5-y$ | $y$ |

As $K_1$ and $K_2$ are very large, $(x + y) = 5$ (1)

and $\frac{K_1}{K_2} = 4 = \frac{x}{5-x} \times \frac{5-y}{y} \Rightarrow \frac{x}{y} = 2$ (2)

From (1) and (2), $x = \frac{10}{3}$

$\therefore$ Moles of B at equilibrium $= 5 - x = \frac{5}{3}$

**13.** $Br_2\,(l) + Cl_2\,(g) \rightleftharpoons 2BrCl\,(g);\ K_p = 1$ atm

| | $Br_2$ | $Cl_2$ | BrCl |
|---|---|---|---|
| Initial moles | $x$ | 10 | 0 |
| Equ. moles | $\approx 0$ | $10-x$ | $2x$ |

$Br_2\,(l) \rightleftharpoons Br_2\,(g);\ K_p = 0.25$ atm

| | $Br_2$ (l) | $Br_2$ (g) |
|---|---|---|
| Initial moles | $y$ | 0 |
| Equ. moles | $\approx 0$ | $y$ |

From question: $\frac{y \times 0.082 \times 300}{164} = 0.25 \Rightarrow y = \frac{5}{3}$

and $\frac{(10 - x + 2x) \times 0.082 \times 300}{164} = 2.00 \Rightarrow x = \frac{10}{3}$

$\therefore$ Minimum mass of $Br_2(l) = (x + y) \times 160$ gm $=$ 800 gm

**14.** $2SO_3 \rightleftharpoons 2SO_2 + O_2$

| | $SO_3$ | $SO_2$ | $O_2$ |
|---|---|---|---|
| Initial moles | 1 (say) | 0 | 0 |
| Equ. moles | $1-0.4=0.6$ | 0.4 | 0.2 |

$\therefore M_{av} = \frac{1 \times 80}{1.2}$

Now, $d = \frac{PM}{RT} \Rightarrow 16 = \frac{p \times \frac{80}{1.2}}{0.0821 \times \frac{920}{0.0821}}$

$\Rightarrow p = 216$ atm

$\therefore K_p = \frac{(0.4)^2 \times 0.2}{(0.6)^2} \times \frac{216}{1.2} = 16$ atm

**15.** $A_2 \rightleftharpoons 2A;\ K_1 = x$ atm

| | $A_2$ | A |
|---|---|---|
| Initial partial pressure | 1 atm | 0 |
| Equ. partial pressure | $1-(x+z)$ | $2x$ |

$B_2 \rightleftharpoons 2B;\ K_2 = y$ atm

| | $B_2$ | B |
|---|---|---|
| Initial partial pressure | 1 atm | 0 |
| Equ.partial pressure | $1-(y+z)$ | $2y$ |

$A_2 + B_2 \rightleftharpoons 2AB;\ K_3 = 2$

| | $A_2$ | $B_2$ | AB |
|---|---|---|---|
| Initial partial pressure | 1 | 1 | 0 |
| Equ. partial pressure | $1-(x+z)$ | $1-(y+z)$ | $2z-0.5$ |

(1)

From question, $[1-(x+z)] + 2x + [1-(y+z)] + 2y + 2z = 2.75$

$\therefore x + y = 0.75$ (2)

Now, $K_3 = \frac{(0.5)^2}{(0.75-x)(0.75-y)} = 2$

$\Rightarrow x = 0.25$ or $0.50$

$y = 0.50$ or $0.25$

$$\therefore \frac{K_2}{K_1} = \frac{\frac{(2y)^2}{1-(y+z)}}{\frac{(2x)^2}{1-(x+z)}} = \frac{(2y)^2 \times (0.75-x)}{(2x)^2 \times (0.75-y)} = \frac{1}{8} \text{ or } \frac{8}{1}$$

CHAPTER 7

# Ionic Equilibrium

## EXERCISE I (JEE MAIN)

### Basics

1. When rain is accompanied by a thunderstorm, the collected rain water will have a pH value
   (a) depending on the amount of dust in air.
   (b) slightly lower than that of rain water without thunderstorm.
   (c) slightly higher than that when the thunderstorm is not there.
   (d) uninfluenced by occurrence of thunderstorm.

2. The pH of water is 7.0 at 25°C. If water is heated to 70°C, then
   (a) pH will decrease and the sample becomes acidic.
   (b) pH will increase but the sample will remain neutral.
   (c) pH will remain constant as 7.
   (d) pH will decrease but the sample will remain neutral.

3. The degree of dissociation of water at 25°C is $1.8 \times 10^{-7}$% and the density is 1.0 g $cm^{-3}$. The ionic product for water is
   (a) $1.0 \times 10^{-14}$ (b) $1.8 \times 10^{-16}$
   (c) $1.0 \times 10^{-16}$ (d) $1.0 \times 10^{-8}$

4. The degree of dissociation of pure water at 25°C is found to be $1.8 \times 10^{-9}$. The dissociation constant, $K_d$ of water, at 25°C is
   (a) $10^{-14}$ (b) $1.8 \times 10^{-16}$
   (c) $5.56 \times 10^{-13}$ (d) $1.8 \times 10^{-14}$

5. What is the pH of a neutral solution at 37°C, where $K_w$ equals $2.5 \times 10^{-14}$? (log 2 = 0.3)
   (a) 7.0 (b) 13.6
   (c) 6.8 (d) 6.6

6. At 40°C, the density of heavy water is 1.02 g/ml and its ionic product is $5.1 \times 10^{-15}$. Which of the following if the only incorrect information regarding heavy water at 40°C?
   (a) The molar concentration of heavy water is 51 M.
   (b) The dissociation constant of heavy water is $10^{-16}$.
   (c) Its degree of dissociation is $10^{-8}$.
   (d) The molal concentration of heavy water is 50 m.

7. The ionic product of water is $1.0 \times 10^{-14}$ at 25°C. Assuming the density of water independent from change in temperature, the ionic product of water at 50°C will be
   (a) $2.0 \times 10^{-14}$ (b) $5.0 \times 10^{-15}$
   (c) $5.9 \times 10^{-14}$ (d) $1.0 \times 10^{-14}$

8. The hydronium ion concentration in an aqueous solution of $H_2SO_4$ is $2.0 \times 10^{-4}$ M at 25°C. The hydroxide ion concentration in the solution is
   (a) 0 (b) $2.0 \times 10^{-4}$ M
   (c) $5 \times 10^3$ M (d) $5 \times 10^{-11}$ M

9. The pH of an aqueous solution of sodium chloride at 60°C is
(a) 7.0 (b) > 7.0
(c) < 7.0 (d) 0

10. The number of hydronium ions in 1 ml of an aqueous solution of pH 12.0 at 25°C is
(a) 0.01 (b) $10^{-12}$
(c) $6.02 \times 10^8$ (d) $6.02 \times 10^{11}$

## Strong Acids and Bases

11. The pH of $4.0 \times 10^{-4}$ M-$HNO_3$ solution is (log 2 = 0.3)
(a) 4.6 (b) 3.4
(c) 3.6 (d) 4.0

12. The pH of 0.005 M-NaOH solution is (log 2 = 0.3)
(a) 2.3 (b) 2.7
(c) 11.3 (d) 11.7

13. How many grams of HCl should be dissolved in sufficient water to get 500 ml of an aqueous solution of pH, 2.0?
(a) 0.01 (b) 0.005
(c) 0.1825 (d) 0.365

14. What is the pH of $10^{-7}$ M-HCl solution at 25°C?
(a) 7.0 (b) 6.70
(c) 6.62 (d) 6.79

15. What mass of NaOH should be dissolved in sufficient water to get 20 $m^3$ of an aqueous solution of pH 7.3 at 25°C?
(a) 0.16 g (b) $1.6 \times 10^{-4}$ g
(c) 0.04 g (d) 0.12 g

16. What is the pH of solution made by mixing equal volumes of 0.1 N-$H_2SO_4$, 0.1 N-$HNO_3$, 0.1 N-HCl?
(a) 1 (b) 2
(c) 3 (d) 4

17. The following five solutions of KOH were prepared as first, 0.1 mole in 1 L; second, 0.2 mole in 2 L; third, 0.3 mole in 3 L; fourth, 0.4 mole in 4 L; fifth, 0.5 mole in 5 L. The pH of resultant solution, when all these solutions are mixed is
(a) 2 (b) 1
(c) 13 (d) 7

18. At 90°C, the hydronium ion concentration in pure water is $10^{-6}$ M. If 100 ml of 0.5 M-NaOH solution is mixed with 250 ml of 0.2 M-$HNO_3$ solution at 90°C, then pH of the resulting solution will be
(a) 7.0 (b) 6.0
(c) 8.0 (d) 0.85

19. Three solutions of strong electrolytes, 25 ml of 0.1 M-HX, 25 ml of 0.1 M-$H_2Y$ and 50 ml of 0.1 N-$Z(OH)_2$ are mixed, the pOH of the resulting solution is
(a) 1.6 (b) 7.0
(c) 12.4 (d) 11.6

20. What will be the percentage error in measuring hydrogen ion concentration in a $10^{-6}$ M-HCl solution on neglecting the contribution of water at 25°C?
(a) 5% (b) 9.8%
(c) 1.98% (d) 0.98%

## Weak Acids and Bases

21. When 0.05 moles of the following acids are dissolved in 1000 ml of $H_2O$, the $[H^+]$ will be greatest in
(a) $HNO_2$: $pK_a = 3.0$
(b) HCOOH: $pK_a = 3.75$
(c) HCN: $pK_a = 9.4$
(d) $CH_3COOH$: $pK_a = 4.75$

22. The concentration of acetate ions in 1 M acetic acid ($K_a = 2 \times 10^{-5}$) solution containing 0.1 M-HCl is
(a) $2 \times 10^{-1}$ M
(b) $2 \times 10^{-3}$ M
(c) $2 \times 10^{-4}$ M
(d) $4.4 \times 10^{-3}$ M

**23.** The dissociation constants of formic and acetic acids are $1.77 \times 10^{-4}$ and $1.75 \times 10^{-5}$, respectively. Which of the following statement is correct?

(a) Formic acid is 3.18 times stronger than acetic acid at equal concentration.
(b) Acetic acid is 3.18 times stronger than formic acid at equal concentrations.
(c) Formic acid is 10.11 times stronger than acetic acid at equal concentrations.
(d) Formic acid is 10.11 times stronger than acetic acid at different concentrations.

**24.** The dissociation constant of acetic acid is 0.000018 and that for cyanoacetic acid is 0.0036 at 298 K. What would be the ratio of volumes of the two acid solutions, each containing equal moles of the acids, so that the solutions becomes isohydric?

(a) 1 : 1 (b) $1 : \sqrt{200}$
(c) 1 : 200 (d) 200 : 1

**25.** The p*K*a of acetylsalicylic acid (aspirin) is 3.5. The pH of gastric juice in human stomach is about 2 to 3 and the pH in the small intestine is about 8. Aspirin will be

(a) unionized in the small intestine as well as in the stomach.
(b) completely ionized in the small intestine as well as in the stomach.
(c) ionized in the stomach and almost unionized in the small intestine.
(d) ionized in the small intestine and almost unionized in the stomach.

**26.** The active ingredient in aspirin is acetyl salicylic acid

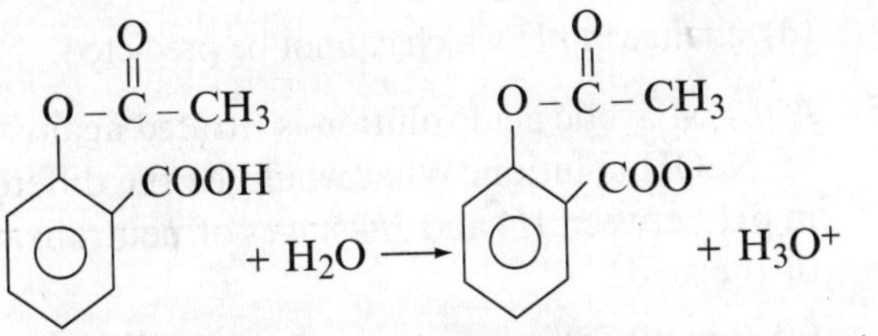

with $K_a = 4.0 \times 10^{-9}$. The pH of the solution obtained by dissolving two aspirin tablets (containing 0.36 g of acetyl salicylic acid in each tablet) in 250 ml of water is (log 2 = 0.3)

(a) 5.1 (b) 8.9
(c) 10.2 (d) 5.25

**27.** For weak electrolyte, AB, the degree of ionization would be ($V$ = volume of solution having 1 mole of electrolyte and $K$ is the ionization constant of the electrolyte)

(a) $\frac{K}{V^2}$ (b) $K.V$
(c) $\frac{K}{V}$ (d) $\sqrt{K \cdot V}$

**28.** What would be the pH of an ammonia solution if the pH of acetic acid solution of same strength is 3.2? The dissociation constants of ammonia and acetic acid are same.

(a) 3.2 (b) 3.8
(c) 10.2 (d) 10.8

**29.** Isohydric solutions are the solutions having the same concentration of hydronium ion. If 0.2 M-HA solution is isohydric with $4 \times 10^{-4}$ M-HCl solution, then $K_b$ of $A^-$ is

(a) $8 \times 10^{-7}$ (b) $1.25 \times 10^{-8}$
(c) $1.25 \times 10^{-6}$ (d) $8 \times 10^{7}$

**30.** If p$K_b$ for fluoride ion at 25°C is 10.3, the ionization constant of hydrofluoric acid in water at this temperature is (log 2 = 0.3)

(a) $2 \times 10^{-4}$ (b) $2 \times 10^{-3}$
(c) $2 \times 10^{-5}$ (d) $5 \times 10^{-11}$

**31.** *n*-coproic acid $C_5H_{11}COOH$ found in coconut and palm oil is used in making artificial flavours and it has solubility in water equal to 11.6 g/L. The saturated solution has pH = 3.0. The $K_a$ of acid is

(a) $10^{-6}$ (b) $10^{-5}$
(c) $2 \times 10^{-5}$ (d) $2 \times 10^{-6}$

**32.** The dissociation constant of formic acid is 0.00024. The hydrogen ion concentration in 0.002 M-HCOOH solution is nearly

(a) $6.93 \times 10^{-4}$ M (b) $4.8 \times 10^{-7}$ M
(c) $5.8 \times 10^{-4}$ M (d) $1.4 \times 10^{-4}$ M

**33.** Calculate the pH of 0.02 M – HA solution. The value of $K_a$ for HA = $2 \times 10^{-12}$ (log 2 = 0.3, log 3 = 0.48).

(a) 6.65 (b) 6.70
(c) 6.85 (d) 6.52

**34.** How much water must be added to 300 ml of 0.2 M solution of $CH_3COOH$ for the degree of dissociation of the acid to double? The value of $K_a$ for acetic acid = $1.8 \times 10^{-5}$.

(a) 1200 ml (b) 300 ml
(c) 600 ml (d) 900 ml

35. A solution has initially 0.1 M-HCOOH and 0.2 M-HCN. The value of $K_a$ for HCOOH = $2.56 \times 10^{-5}$, $K_a$ of HCN = $9.6 \times 10^{-10}$. The only incorrect statement for the solution is (log 2 = 0.3)

(a) $[H^+] = 1.6 \times 10^{-3}$ M
(b) $[HCOO^-] = 1.6 \times 10^{-3}$ M
(c) $[CN^-] = 1.2 \times 10^{-7}$ M
(d) pOH = 2.8

## Polyprotic Acids and Bases

36. What is the pH of $4 \times 10^{-3}$ M–$Y(OH)_2$ solution assuming the first dissociation to be 100% and second dissociation to be 50%, where Y represents a metal cation? (log 2 = 0.3, log 3 = 0.48)

(a) 11.78 (b) 11.22
(c) 2.22 (d) 2.78

37. The species present in solution when $CO_2$ is dissolved in water is

(a) $CO_2$, $H_2CO_3$, $HCO_3^-$, $CO_3^{2-}$
(b) $H_2CO_3$, $CO_3^{2-}$
(c) $CO_3^{2-}$, $HCO_3^-$
(d) $CO_2$, $H_2CO_3$

38. An aqueous solution is prepared by dissolving 0.1 mole $H_2CO_3$ in sufficient water to get 100 ml solution at 25°C. For $H_2CO_3$, $K_{a1} = 4.0 \times 10^{-6}$ and $K_{a2} = 5.0 \times 10^{-11}$. The only incorrect equilibrium concentration is

(a) $[H^+] = 6.32 \times 10^{-4}$ M
(b) $[HCO_3^-] = 2 \times 10^{-3}$ M
(c) $[CO_3^{2-}] = 5 \times 10^{-11}$ M
(d) $[OH^-] = 5 \times 10^{-12}$ M

39. Ascorbic acid (vitamin C) is a diprotic acid, $H_2C_6H_6O_6$. What is the pH of a 0.10 M solution? The acid ionization constants are $K_{a1} = 9.0 \times 10^{-5}$ and $K_{a2} = 1.6 \times 10^{-12}$. (log 2 = 0.3, log 3 = 0.48).

(a) 3.52 (b) 2.52
(c) 1.52 (d) 2.48

40. The pH of 0.1 M – $N_2H_4$ solution is (For $N_2H_4$, $K_{b1} = 3.6 \times 10^{-6}$, $K_{b2} = 6.4 \times 10^{-12}$, log 2 = 0.3, log 3 = 0.48)

(a) 3.22 (b) 2.72
(c) 10.78 (d) 11.22

## Buffer Solutions

41. The dissociation constant of a weak acid HX is $10^{-5}$. The buffer HX + NaX can be best used to maintain the pH in the range

(a) 9–11 (b) 2–4
(c) 11–13 (d) 4–6

42. A physician wishes to prepare a buffer solution at pH = 3.58 that efficiently resists a change in pH yet contains only small concentrations of the buffering agents. Which one of the following weak acid together with its sodium salt would be best to use?

(a) *m*-chloro benzoic acid ($pK_a = 3.98$)
(b) *p*-chlorocinnamic acid ($pK_a = 4.41$)
(c) 2,5-dihydroxy benzoic acid ($pK_a = 2.97$)
(d) acetoacetic acid ($pK_a = 3.58$)

43. pH of 0.01 M-$(NH_4)_2SO_4$ and 0.02 M-$NH_4OH$ buffer ($pK_a$ of $NH_4^+ = 9.26$) is

(a) 9.26 + log 2 (b) 9.26 – log 2
(c) 4.74 + log 2 (d) 9.26

44. The addition of sodium acetate to acetic acid solution will cause

(a) increase in its pH value.
(b) decrease in its pH value.
(c) no change in pH value.
(d) change in pH which cannot be predicted.

45. A 0.1 M acetic acid solution is titrated against 0.1 M-NaOH solution. What would be the difference in pH between 1/4 and 3/4 stages of neutralization of the acid?

(a) 2 log(0.75) (b) 2 log(0.25)
(c) log 3 (d) 2 log 3

46. An amount of 0.1 mole of $CH_3NH_2$ ($K_b = 5 \times 10^{-4}$) is mixed with 0.08 mole of HCl and diluted to one litre. What will be the $H^+$ concentration in the solution?

(a) $1.25 \times 10^{-4}$ M (b) $8 \times 10^{-11}$ M
(c) $1.6 \times 10^{-11}$ M (d) $2 \times 10^{-3}$ M

**47.** A volume of 10 ml of a strong acid solution of pH = 2.0 are mixed with 990 ml of a buffer solution of pH = 4.0. The pH of the resulting solution will be

(a) 4.2 (b) 6.0
(c) 4.002 (d) 4.0

**48.** An amount of 0.15 mole of pyridinium chloride has been added into 500 ml of 0.2 M pyridine solution. Calculate pH of the resulting solution assuming no change in volume. The value of $K_b$ for pyridine = $1.5 \times 10^{-9}$ (log 2 = 0.3, log 0.3 = 0.48).

(a) 9.0 (b) 5.0
(c) 8.64 (d) 5.36

**49.** A volume of 20 ml of 0.8 M-HCN solution is mixed with 80 ml of 0.4 M-NaCN solution. Calculate the pH of the resulting solution. The value of $K_a$ of HCN = $2.5 \times 10^{-10}$ (log 2 = 0.3).

(a) 9.9 (b) 9.3
(c) 4.1 (d) 4.7

**50.** The base imidazole has a $K_b$ of $1.0 \times 10^{-7}$ at 25°C. In what volumes should 0.02 M-HCl and 0.02 M imidazole be mixed to make 120 ml of a buffer at pH = 7?

(a) 60 ml, 60 ml (b) 40 ml, 80 ml
(c) 30 ml, 90 ml (d) 20 ml, 100 ml

## Hydrolysis of Salts

**51.** Separate solutions of NaW, NaX, NaY and NaZ, each of concentrations 0.1 M has pH 7.0, 9.0, 10.0 and 11.0, respectively, at 25°C. The strongest acid among these is

(a) HW (b) HX
(c) HY (d) HZ

**52.** If the pH of 0.001 M potassium propionate solution be 8.0, then the dissociation constant of propionic acid will be

(a) 10–3 (b) 10–2
(c) 10–2.5 (d) 10–5

**53.** The correct order of increasing [$OH^-$] in the following aqueous solution is

(a) 0.01 M-$NaHCO_3$ < 0.01 M-NaCN < 0.01 M-KCl
(b) 0.01 M-KCl < 0.01 M-NaCN < 0.01 M-$NaHCO_3$
(c) 0.01 M-KCl < 0.01 M-$NaHCO_3$ < 0.01 M-NaCN
(d) 0.01 M-NaCN < 0.01 M-KCl < 0.01 M-$NaHCO_3$

**54.** The pH of solutions of both ammonium acetate and sodium chloride is 7 due to

(a) hydrolysis in both case.
(b) the former hydrolyses and not the latter.
(c) no hydrolysis in both.
(d) hydrolysis of the latter but not the former.

**55.** For the titration of a dibasic weak acid $H_2A$ $\left(p^{K_{a(2)}} - p^{K_{a(1)}} \geq 2\right)$ with a strong base, pH versus volume of the base graph is as shown in the figure. The value of $p^{K_{a(1)}}$ and $p^{K_{a(2)}}$ may be equal to the pH values corresponding to the points.

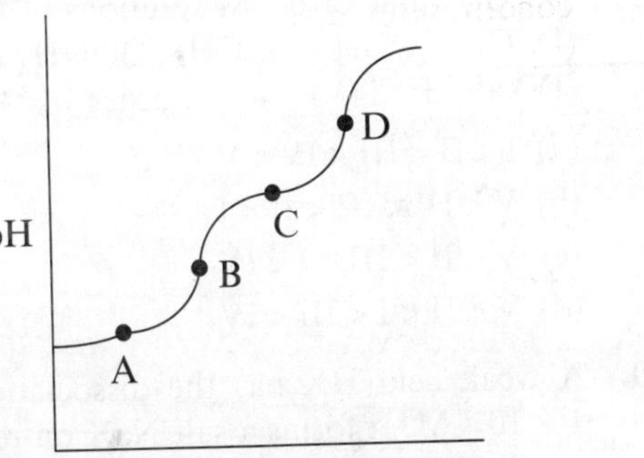


(a) B and D, respectively
(b) A and B, respectively
(c) C and D, respectively
(d) A and C, respectively

**56.** A salt of strong acid and a weak base is dissolved in water. Its hydrolysis in solution is

(a) not affected by heating.
(b) increased by adding the strong acid.
(c) suppressed by adding strong acid.
(d) suppressed by dilution.

57. The curve in the figure shows the variation of pH during the course of titration of a weak acid, HA with a strong base (NaOH). At which point in the titration curve is the concentration of the acid may be equal to that of its conjugate base?

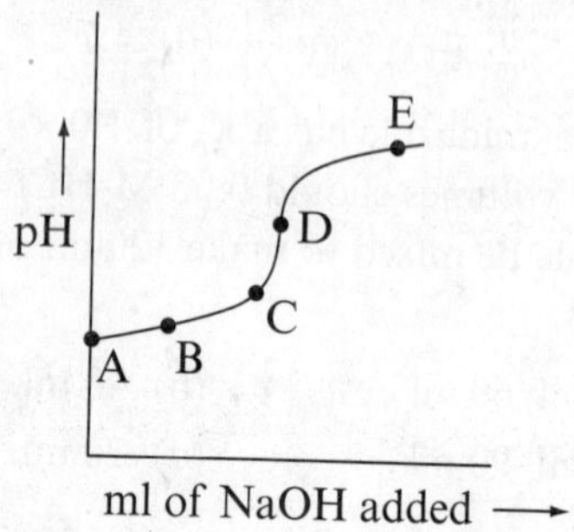


(a) Point D (b) Point E
(c) Point C (d) Point B

58. The pH of 0.1 M solution of the following compounds increases in the order
(a) $NaCl < NH_4Cl < NaCN < HCl$
(b) $HCl < NH_4Cl < NaCl < NaCN$
(c) $NaCN < NH_4Cl < NaCl < HCl$
(d) $HCl < NaCl < NaCN < NH_4Cl$

59. The pH value of 0.1 M solutions of $CH_3COONa$ (I), $CH_3COOH$ (II), $CH_3COONH_4$ (III), NaOH (IV) and HCl (V) is in the order
(a) I < II < III < IV < V
(b) V < IV < III < II < I
(c) V < II < III < I < IV
(d) V < II < I < III < IV

60. A weak acid HX has the dissociation constant $1 \times 10^{-5}$ M. It forms a salt NaX on reaction with alkali. The percentage hydrolysis of 0.1 M solution of NaX is
(a) 0.0001% (b) 0.01%
(c) 0.1% (d) 0.15%

61. The pH of an aqueous solution of 1.0 M ammonium formate assuming complete dissociation is ($pK_a$ of formic acid = 3.8 and $pK_b$ of ammonia = 4.8)
(a) 7.0 (b) 7.5
(c) 6.5 (d) 4.3

62. What is the pH of a 0.50 M aqueous NaCN solution? The value of $pK_b$ of $CN^-$ is 4.70 (log 2 = 0.3).
(a) 3.5 (b) 11.5
(c) 4.7 (d) 9.3

63. The pH at the equivalence point when a solution of 0.01 M-$CH_3COOH$ is titrated with a solution of 0.01 M-NaOH, is ($pK_a$ of $CH_3COOH$ = 4.7, log 5 = 0.7)
(a) 8.2 (b) 9.4
(c) 8.35 (d) 10.5

64. The acid ionization constant of $Zn^{2+}$ is $2.0 \times 10^{-10}$. What is the pH of 0.001 M solution of $ZnCl_2$? (log 2 = 0.3)
(a) 9.7 (b) 4.85
(c) 6.35 (d) 3.35

65. The addition of ammonium chloride to acetic acid solution will cause
(a) increase in its pH value.
(b) decrease in its pH value.
(c) no change in pH value.
(d) change in pH which cannot be predicted.

## Indicators

66. The indicator constant for an acidic indicator, HIn is $5 \times 10^{-6}$ M. This indicator appears only in the colour of acidic form when $\frac{[In^-]}{[HIn]} \le \frac{1}{20}$ and it appears only in the colour of basic form when $\frac{[HIn]}{[In^-]} \le \frac{1}{40}$. The pH range of indicator is (log 2 = 0.3)
(a) 4.3 – 6.3 (b) 4.0 – 6.6
(c) 4.0 – 6.9 (d) 3.7 – 6.6

67. For the indicator thymol blue, the value of pH is 2.0 when half of the indicator is present in the unionized form. The percentage of the indicator in the unionized form in a solution of $4.0 \times 10^{-3}$ M hydrogen ion concentration is
(a) 40% (b) 28.6%
(c) 71.4% (d) 60%

**68.** A certain sample of rainwater gives a yellow colour with methyl red [pH range 4.2 (red) – 6.2 (yellow)] and a yellow colour with phenol red [pH range 6.4 (yellow) – 8.0 (red)]. What is the approximate pH of the water? Is the rainwater acidic, neutral or basic?

(a) 6.3, acidic (b) 6.1, acidic
(c) 6.5, acidic (d) 6.3, basic

**69.** An acid type indicator, HIn differs in colour from its conjugate base ($In^-$). The human eye is sensitive to colour differences only when the ratio [$In^-$]/[HIn] is greater than 10 or smaller than 0.1. What should be the minimum change in the pH of the solution to observe a complete colour change ($K_a = 1.0 \times 10^{-5}$)?

(a) 0.0 (b) 1.0
(c) 2.0 (d) 5.0

**70.** The range of most suitable indicator which should be used for titration of NaX (0.1 M, 10 ml) with 0.1 M HCl should be ($K_b$ of $X^- = 10^{-6}$)

(a) 2–3 (b) 3–5
(c) 6–8 (d) 8–10

## Solubility

**71.** If ionization of $X_aY_b$ takes place then, the number of $Y^{-a}$ ions will be equal to

(a) *b*/*a* times of $X^{+b}$
(b) *a*/*b* times of $X^{+b}$
(c) *b*/*a* times of $X^{a+}$
(d) equal to $X^{+b}$

**72.** The solubility of sparingly soluble salt $A_3B_2$ (molar mass = 'M' g/mol) in water is 'x' g/L. The ratio of molar concentration of $B^{3-}$ to the solubility product of the salt is

(a) $\frac{108\,x^5}{M^5}$ (b) $\frac{x^4}{108\,M^4}$
(c) $\frac{x^4}{54\,M^4}$ (d) $\frac{M^4}{54\,x^4}$

**73.** The solubility product of $Zn(OH)_2$ is $10^{-14}$ at 25°C. What would be the concentration $Zn^{+2}$ ion in 0.1 M-$NH_4OH$ solution which is 50% ionized?

(a) $2 \times 10^{-13}$ (b) $4 \times 10^{-12}$
(c) $4 \times 10^{-8}$ (d) $2 \times 10^{-11}$

**74.** In which of the following, the solubility of AgCl will be maximum?

(a) 0.1 M-$AgNO_3$ (b) Water
(c) 0.1 M-$NH_3$(aq) (d) 0.1 M-NaCl

**75.** What is the equilibrium constant of the following reaction? $Fe(OH)_3(s) + 3H_3O^+ \rightleftharpoons Fe^{3+} + 6H_2O$? $K_{sp}$ of $Fe(OH)_3 = 4 \times 10^{-38}$

(a) $2.5 \times 10^{-5}$ (b) $4.0 \times 10^4$
(c) $4.0 \times 10^{-4}$ (d) $4 \times 10^{-80}$

**76.** The solubility product of AgCl is $1.0 \times 10^{-10}$. The equilibrium constant of the reaction

$AgCl(s) + Br^- \rightleftharpoons AgBr(s) + Cl^-$

is 200 and that of the reaction

$2AgBr(s) + S^{2-} \rightleftharpoons Ag_2S(s) + 2Br^-$

is $1.6 \times 10^{24}$. What is the $K_{sp}$ of $Ag_2S$?

(a) $3.2 \times 10^{16}$ (b) $1.56 \times 10^{-49}$
(c) $3.95 \times 10^{-25}$ (d) $3.13 \times 10^{-17}$

**77.** What is the solubility product of $Al(OH)_3$ in water? Given:

$Al(OH)_4^-(aq) \rightleftharpoons Al^{3+}(aq) + 4OH^-(aq)$
$K = 1.3 \times 10^{-34}$

$Al(OH)_3(s) + OH^-(aq) \rightleftharpoons Al\,(OH)_4^-(aq)$; $K = 38.5$

(a) $3.1 \times 10^{-35}$ (b) $5 \times 10^{-33}$
(c) $6.1 \times 10^{-33}$ (d) $5 \times 10^{-34}$

**78.** A recent investigation of the complexation of $SCN^-$ with $Fe^{3+}$ led to values of 125, 20 and 1.0 for $K_1$, $K_2$ and $K_3$, respectively. What is the dissociation constant of $Fe(SCN)_3$ into its simplest ions on the basis of these data?

(a) $2.5 \times 10^3$ (b) $4.0 \times 10^{-4}$
(c) 1.0 (d) $8.0 \times 10^{-3}$

**79.** Solubility of $BaF_2$ in a solution of $Ba(NO_3)_2$ will be represented by the concentration term

(a) [$Ba^{2+}$] (b) [$F^-$]
(c) 0.5 [$F^-$] (d) 2 [$NO_3^-$]

80. How many times the solubility of $CaF_2$ is decreased in $4 \times 10^{-3}$ M-KF(aq) solution as compared to pure water at 25°C. Given: $K_{sp}(CaF_2) = 3.2 \times 10^{-11}$

(a) 50 (b) 100
(c) 500 (d) 1000

81. The solubility of $A_2X_3$ is $y$ mol dm$^{-3}$. It solubility product is

(a) $6y^4$ (b) $64y^4$
(c) $36y^4$ (d) $108y^5$

82. For a sparingly soluble salt $A_pB_q$, the relationship of its solubility product ($L_S$) with its solubility ($S$) is

(a) $L_S = S^{p+q}.p^p.q^q$ (b) $L_S = S^{p+q}.p^q.q^p$
(c) $L_S = S^{pq}.p^p.q^q$ (d) $L_S = S^{pq}.(pq)^{p+q}$

83. $Ag^+ + NH_3 \rightleftharpoons [Ag(NH_3)^+]$; $K_1 = 1.6 \times 10^3$

$[Ag(NH_3)^+] + NH_3 \rightleftharpoons [Ag(NH_3)_2^+]$; $K_2 = 6.8 \times 10^3$

The formation constant of $[Ag(NH_3)_2^+]$ is

(a) $1.08 \times 10^7$ (b) $6.08 \times 10^6$
(c) $1.08 \times 10^3$ (d) $1.08 \times 10^5$

84. Solubility product constant ($K_{sp}$) of salts of types MX, $MX_2$ and $M_3X$ at temperature, $T$ are $4.0 \times 10^{-8}$, $3.2 \times 10^{-14}$ and $2.7 \times 10^{-15}$, respectively. Solubilities (in M) of the salts at temperature, $T$, are in the order

(a) $MX > MX_2 > M_3X$
(b) $M_3X > MX_2 > MX$
(c) $MX_2 > M_3X > MX$
(d) $MX > M_3X > MX_2$

85. The solubility of AgCl in water is 0.001435 g per litre at 15°C. The solubility product of AgCl is (Ag = 108, Cl = 35.3)

(a) $10^{-5}$ (b) $10^{-10}$
(c) $2 \times 10^{-10}$ (d) $10^{-9}$

86. The solubility of $Li_3Na_3(AlF_6)_2$ is 0.0744 g per 100 ml at 298 K. Calculate the solubility product of the salt (Atomic masses: Li = 7, Na = 23, Al = 27, F = 19).

(a) $2.56 \times 10^{-22}$ (b) $2 \times 10^{-3}$
(c) $7.46 \times 10^{-19}$ (d) $3.46 \times 10^{-12}$

87. The solubility product of $CaF_2$ is $1.08 \times 10^{-10}$. What mass of $CaF_2$ will dissolve in 500 ml water in order to make a saturated solution? (Ca = 40, F = 19)

(a) $3 \times 10^{-4}$ g (b) $1.17 \times 10^{-2}$ g
(c) 1.17 mg (d) $3 \times 10^{-3}$ g

88. The solubility product of $Mg(OH)_2$ is $9.0 \times 10^{-12}$. The pH of an aqueous saturated solution of $Mg(OH)_2$ is (log 1.8 = 0.26, log 3 = 0.48)

(a) 3.58 (b) 10.42
(c) 3.88 (d) 6.76

89. The molar solubility of $Zn(OH)_2$ in 1 M ammonia solution at room temperature is ($K_{sp}$ of $Zn(OH)_2 = 1.6 \times 10^{-17}$; $K_{stab}$ of $Zn(NH_3)_4^{2+} = 1.6 \times 10^{10}$)

(a) $4 \times 10^{-3}$ M (b) $1.58 \times 10^{-6}$ M
(c) $4 \times 10^{-9}$ M (d) $2.56 \times 10^{-7}$ M

90. Assuming no change in volume, calculate the minimum mass of NaCl necessary to dissolve 0.01 mole of AgCl in 100 L solution. The value of $K_{sp}$ of AgCl = $2.0 \times 10^{-10}$ and $K_f$ of $AgCl_2^- = 2.5 \times 10^5$.

(a) 117 g (b) 11.7 kg
(c) 58.5 kg (d) 585 g

91. The solubility product of $AgC_2O_4$ at 25°C is $2.3 \times 10^{-11}$ M$^3$. A solution of $K_2C_2O_4$ containing 0.15 moles in 500 ml water is shaken at 25°C with excess of $Ag_2CO_3$ till the following equilibrium is reached.

$$Ag_2CO_3 + K_2C_2O_4 \rightleftharpoons Ag_2C_2O_4 + K_2CO_3$$

At equilibrium, the solution contains 0.035 mole of $K_2CO_3$. Assuming the degree of dissociation of $K_2C_2O_4$ and $K_2CO_3$ to be equal, calculate the solubility product of $Ag_2CO_3$.

(a) $2.3 \times 10^{-11}$ M$^3$ (b) $7.0 \times 10^{-10}$ M$^3$
(c) $3.0 \times 10^{-13}$ M$^3$ (d) $7.0 \times 10^{-12}$ M$^3$

92. For the reaction $[Ag(CN)_2]^- \rightleftharpoons Ag^+ + 2CN^-$, the equilibrium constant, at 25°C, is $4.0 \times 10^{-19}$. Calculate the silver ion concentration in a solution which was originally 0.10 molar in KCN and 0.03 molar in $AgNO_3$.

(a) 0 (b) 0.03 M
(c) $3 \times 10^{-19}$ M (d) $7.5 \times 10^{-18}$ M

**93.** A sample of AgCl was treated with 5.00 ml of 2.0 M $Na_2CO_3$ solution to give $Ag_2CO_3$. The remaining solution contained 0.00355 g of $Cl^-$ ions per litre. The solubility product of AgCl is ($K_{sp}$ of $Ag_2CO_3$ is $8.0 \times 10^{-12}$).

(a) $2 \times 10^{-10}$ (b) $1 \times 10^{-10}$
(c) $4 \times 10^{-10}$ (d) $8 \times 10^{-10}$

**94.** Given $Ag(NH_3)_2^+ \rightleftharpoons Ag^+ + 2NH_3$, $K_c = 7.2 \times 10^{-8}$ and $K_{sp}$ of AgCl = $1.8 \times 10^{-10}$ at 298 K. If ammonia is added to a water solution containing excess of AgCl(s) only, then calculate the concentration of the complex in 1.0 M aqueous ammonia.

(a) 1.0 M (b) 0.091 M
(c) 0.0455 M (d) 0.023 M

**95.** The solubility of $Pb(OH)_2$ in water is $6.0 \times 10^{-6}$ M. The solubility of $Pb(OH)_2$ in a buffer solution of pH = 8 is

(a) 8.64 M (b) 2.16 × 10−16 M
(c) 8.64 × 10−16 M (d) $8.64 \times 10^{-4}$ M

**96.** The silver ion concentration in a 0.2 M solution of $Ag(NH_3)_2NO_3$ is ($K_{diss} = 6.8 \times 10^{-8}$, $1.5^3 = 3.4$)

(a) 0.2 M (b) 1.5 × 10−3 M
(c) 1.16 × 10−4 M (d) $6.8 \times 10^{-8}$ M

**97.** The formation constant of $Cu(NH_3)_4^{2+}$ is $1.25 \times 10^{12}$. What will be the equilibrium concentration of $Cu^{2+}$ if 0.0125 moles of Cu is oxidized and put into 1.0 L of 0.25 M–$NH_3$ solution?

(a) $2.5 \times 10^{-11}$ M (b) $2.5 \times 10^{-13}$ M
(c) $4 \times 10^{-12}$ M (d) $6.25 \times 10^{-12}$ M

**98.** The simultaneous solubilities of AgSCN and AgBr are, respectively ($K_{sp}$ of AgSCN = $1 \times 10^{-12}$, $K_{sp}$ of AgBr = $2.1 \times 10^{-13}$),

(a) $9.09 \times 10^{-7}$ M, $1.909 \times 10^{-7}$ M
(b) $1.909 \times 10^{-7}$ M, $9.09 \times 10^{-7}$ M
(c) $9.09 \times 10^{-6}$ M, $1.909 \times 10^{-7}$ M
(d) $1.1 \times 10^{-6}$ M, $2.1 \times 10^{-7}$ M

**99.** The solubility of AgCN in a buffer solution of pH = 3.0 is ($K_{sp}$ of AgCN = $1.2 \times 10^{-16}$; $K_a$ of HCN = $4.8 \times 10^{-10}$)

(a) 1.58 × 10−5 M (b) 2.0 × 10−5 M
(c) $1.58 \times 10^{-4}$ M (d) $2.5 \times 10^{-9}$ M

**100.** The solubility of $PbCl_2$ when it is 80% ionized is

(a) 25% less than the solubility of $PbCl_2$ when it is 100% ionized.
(b) 50% less than the solubility of $PbCl_2$ when it is 100% ionized.
(c) More than the solubility of $PbCl_2$ when it is 100% ionized.
(d) is equal to the solubility of $PbCl_2$ when it is 100% ionized.

## Precipitation

**101.** The solubility products of $Mg(OH)_2$, $Cd(OH)_2$, $Al(OH)_3$ and $Zn(OH)_2$ are $4 \times 10^{-11}$, $8 \times 10^{-6}$, $8.5 \times 10^{-23}$ and $1.8 \times 10^{-14}$, respectively. The cation that will precipitate first as hydroxide, on adding limited quantity of $NH_4OH$ in a solution containing equimolar amount of metal cations, is

(a) $Al^{3+}$ (b) $Zn^{2+}$
(c) $Mg^{2+}$ (d) $Cd^{2+}$

**102.** Silver ions are slowly added in a solution with $[Br^-] = [Cl^-] = [CO_3^{2-}] = [AsO_4^{3-}] = 0.1$ M. Which of the following compound will precipitate first?

(a) AgBr ($K_{sp} = 5 \times 10^{-13}$)
(b) AgCl ($K_{sp} = 1.8 \times 10^{-10}$)
(c) $Ag_2CO_3$ ($K_{sp} = 8.1 \times 10^{-12}$)
(d) $Ag_3PO_4$ ($K_{sp} = 1 \times 10^{-22}$)

**103.** The $K_{sp}$ of $Ag_2CrO_4$ = $1.2 \times 10^{-11}$. What concentration of $Ag^+$ ion in aqueous solution will just fail to give a precipitate of $Ag_2CrO_4$ with a solution in which $[CrO_4^{-2}] = 3 \times 10^{-4}$ M?

(a) $10^{-3}$ M (b) $10^{-1}$ M
(c) $10^{-4}$ M (d) $2 \times 10^{-4}$ M

**104.** A 0.1 mole of $AgNO_3$ is dissolved in 1 L of 1 M-$NH_3$. If 0.01 mole of NaCl is added to this solution, will AgCl(s) precipitate? The value of $K_{sp}$ for AgCl = $1.8 \times 10^{-10}$ and $K_{stab}$ for $Ag(NH_3)_2^+$ = $1.6 \times 10^7$.

(a) Yes
(b) No
(c) Addition of NaCl in any amount can never result in precipitation.
(d) Addition of even smaller amount of NaCl may result in precipitation.

**105.** In 500 ml of $2.5 \times 10^{-5}$ M – $AgNO_3$ solution, 2000 ml of $5.0 \times 10^{-2}$ M – NaCl solution is added. The mass of precipitate of AgCl formed is ($K_{sp}$ of AgCl = 2 $\times 10^{-10}$, Ag = 108)

(a) 1.794 g (b) 1.794 mg
(c) $5 \times 10^{-6}$ g (d) $1.25 \times 10^{-2}$ g

**106.** The solubility product of $PbI_2$ is $7.2 \times 10^{-9}$. The maximum mass of NaI which may be added in 500 ml of 0.005 M-$Pb(NO_3)_2$ solution without any precipitation of $PbI_2$ is (I = 127)

(a) 0.09 g (b) $1.2 \times 10^{-3}$ g
(c) $6 \times 10^{-4}$ g (d) $1.08 \times 10^{-5}$ g

**107.** What is the minimum mass of NaBr which should be added in 200 ml of 0.0004 M-$AgNO_3$ solution just to start the precipitation of AgBr? The value of $K_{sp}$ of AgBr = $4 \times 10^{-13}$ (Br = 80).

(a) $1.0 \times 10^{-9}$ g (b) $2 \times 10^{-10}$ g
(c) $2.06 \times 10^{-8}$ g (d) $1.03 \times 10^{-7}$ g

**108.** A sample of hard water contains 0.005 mole of $CaCl_2$ per litre. What is the minimum concentration of $Al_2(SO_4)_3$ which must be exceeded for removing $Ca^{2+}$ ions from this water sample? The solubility product of $CaSO_4$ is $2.4 \times 10^{-5}$.

(a) $4.8 \times 10^{-3}$ M (b) $1.6 \times 10^{-3}$ M
(c) 0.0144 M (d) $2.4 \times 10^{-3}$ M

**109.** To 100 ml of a solution, which contains $8.32 \times 10^{-3}$ g lead ions, $10^{-4}$ moles of $H_2SO_4$ is added. How much lead remains in the solution unprecipitated? $K_{sp}$ of $PbSO_4 = 1.6 \times 10^{-7}$. (Pb = 208)

(a) $4 \times 10^{-4}$ g (b) $2.67 \times 10^{-4}$ g
(c) $2 \times 10^{-4}$ g (d) $4.16 \times 10^{-3}$ g

**110.** An aqueous solution of a metal bromide $MBr_2$ (0.04 M) is saturated with $H_2S$. What is the minimum pH at which MS will precipitate? The value of $K_{sp}$ for MS = $6.0 \times 10^{-21}$; concentration of saturated $H_2S$ = 0.1M, $K_1 = 10^{-7}$ and $K_2 = 1.5 \times 10^{-13}$ for $H_2S$.

(a) 1.0 (b) 1.3
(c) 13.0 (d) 0.7

**111.** An amount of 0.1 millimole of $CdSO_4$ is present in 10 ml acid solution of 0.08 M-HCl. Now $H_2S$ is passed to precipitate all the $Cd^{2+}$ ions. What would be the pH of solution after filtering off precipitate, boiling off $H_2S$ and making the solution 100 ml by adding water?

(a) 3.0 (b) 2.0
(c) 4.0 (d) 2.22

**112.** A solution contains a mixture of $Ag^+$ (0.10 M) and $Hg_2^{2+}$ (0.10 M), which are to be separated by selective precipitation. Calculate the maximum concentration of iodide ion at which one of them gets precipitated almost completely. What per cent of that metal ion is precipitated before the start of precipitation of second metal ion? $K_{sp}(AgI) = 8.5 \times 10^{-17}$ and $K_{sp}(Hg_2I_2) = 2.5 \times 10^{-26}$.

(a) $5 \times 10^{-13}$ M, 99.83%
(b) $8.5 \times 10^{-16}$ M, 99.83%
(c) $2.5 \times 10^{-25}$ M, 100%
(d) $5 \times 10^{-13}$ M, 98.3%

**113.** The solubility of $CaCO_3$ is 7 mg/litre. Calculate the solubility product of $BaCO_3$ from this information and from the fact that when $Na_2CO_3$ is added slowly to a solution containing equimolar concentration of $Ca^{2+}$ and $Ba^{2+}$, no precipitate of $CaCO_3$ is formed until 90% of $Ba^{2+}$ has been precipitated as $BaCO_3$.

(a) $4.9 \times 10^{-8}$ (b) $4.9 \times 10^{-9}$
(c) $4.9 \times 10^{-10}$ (d) $7 \times 10^{-4}$

**114.** Small amount of freshly precipitated magnesium hydroxides are stirred vigorously in a buffer solution containing 0.25 M of $NH_4Cl$ and 0.05 M of $NH_4OH$. $[Mg^{2+}]$ in the resulting solution is ($K_b$ for $NH_4OH = 2.0 \times 10^{-5}$ and $K_{sp}$ of $Mg(OH)_2 = 8.0 \times 10^{-12}$)

(a) $4 \times 10^{-6}$ M (b) $2 \times 10^{-6}$ M
(c) 0.5 M (d) 2.0 M

**115.** The solubility of metal sulphide in saturated solution of $H_2S$ (concentration = 0.1 M) can be represented as follows.

$$MS(s) + 2H^+(aq) \rightleftharpoons M^{2+}(aq) + H_2S(aq);$$

$$K_{eq} = \frac{[M^{2+}][H_2S]}{[H^+]^2}$$

The values of $K_{eq}$ for the metal sulphides, MnS, ZnS, CoS and PbS are $3 \times 10^{10}$, $3 \times 10^{-2}$, 3 and $3 \times 10^{-7}$, respectively. If the concentration of each metal ion in a saturated solution of $H_2S$ is 0.01 M, then which metal sulphide(s) will precipitate at $[H^+] = 1.0$ M?

(a) MnS, ZnS, CoS
(b) PbS
(c) PbS, ZnS, CoS
(d) PbS, ZnS

EXERCISE I

# EXERCISE II (JEE ADVANCED)

## Section A (Only one Correct)

1. Liquid ammonia ionizes to slight extent. At −50°C, its self-ionization constant, $K = [NH_4^+][NH_2^-] = 10^{-30}$ M$^2$. How many amide ions are present per ml of pure liquid ammonia? ($N_A = 6 \times 10^{23}$)
   (a) $10^{-15}$ (b) $10^{-18}$
   (c) $6 \times 10^5$ (d) $6 \times 10^8$

2. For a sample of pure water,
   (a) pH increases and pOH decreases with increase in temperature.
   (b) pH decreases and pOH increases with increase in temperature.
   (c) both pH and pOH increases with increase in temperature.
   (d) both pH and pOH decrease with increase in temperature.

3. The pH at which water is maximum dissociated at 25°C, is
   (a) 2 (b) 7
   (c) 10 (d) 14

4. What will be the resultant pH when 200 ml of an aqueous solution of HCl (pH = 2.0) is mixed with 300 ml of an aqueous solution of NaOH (pH = 12.0)?
   (a) 2.7 (b) 11.3
   (c) 3.7 (d) 10.3

5. When 20 ml of 0.2 M-DCl solution is mixed with 80 ml of 0.1 M-NaOD solution, pH of the resulting solution becomes 13.6. The ionic product of heavy water $D_2O$ is
   (a) $10^{-15}$ (b) $10^{-16}$
   (c) $4 \times 10^{-15}$ (d) $4 \times 10^{-16}$

6. The equilibrium constant of $T_2O$ (Tritium is an isotope of H) differ from those of $H_2O$ at 298 K. Let at 298 K, pure $T_2O$ has pT (like pH) 7.60. What is the pT of a solution prepared by adding 100 ml of 0.4 M-TCl to 400 ml of 0.2 M-NaOT? (log 2 = 0.3)
   (a) 1.1 (b) 0.08
   (c) 6.5 (d) 14.1

7. The acid ionization constant of $Zn^{2+}$ is $2.0 \times 10^{-10}$. What is the basic dissociation constant of $Zn(OH)^{+}$?
   (a) $5 \times 10^{-5}$ (b) $2.0 \times 10^4$
   (c) $2 \times 10^{-10}$ (d) $5 \times 10^9$

8. The dissociation constant of $NH_3$ at 27°C from the following data:
   $NH_3 + H^+ \rightleftharpoons NH_4^+$; $\Delta H° = -52.21$ kJ/mol; $\Delta S° = +1.6\ JK^{-1}\ mol^{-1}$
   $H_2O \rightleftharpoons H^+ + OH^-$; $\Delta H° = 54.70$ kJ/mol; $\Delta S° = -76.3\ JK^{-1}\ mol^{-1}$
   Given: R = 8.3 J/K-mol
   (a) $e^{10}$ (b) $e^{-10}$
   (c) $e^{-8}$ (d) $e^{-9}$

9. How many moles of acetic acid should be added to 100 ml of 0.6 M formic acid solution such that the percentage dissociation of formic acid remains unchanged? The value of $K_a$ for acetic acid = $1.8 \times 10^{-5}$ and $K_a$ for formic acid = $2.4 \times 10^{-4}$.
   (a) 0.8 (b) 0.08
   (c) 8.0 (d) 0.6

10. The dissociation constant of a weak monoprotic acid is numerically equal to the dissociation constant of its conjugate base. What is the pH of 0.1M solution of this acid?
    (a) 7.0 (b) 6.0
    (c) 8.0 (d) 4.0

11. The ionization constant of $NH_4^+$ in water is $5.6 \times 10^{-10}$ mol $L^{-1}$ at 25ºC. The rate constant for the reaction of $NH_4^+$ and $OH^-$ to form $NH_3$ and $H_2O$ at 25ºC is $3.4 \times 10^{10}$ L mol$^{-1}$ s$^{-1}$. The rate constant for proton transfer from water to $NH_3$ at 25ºC is
    (a) $6.07 \times 10^5\ s^{-1}$ (b) $6.07 \times 10^{-18}\ s^{-1}$
    (c) $1.65 \times 10^{-6}\ s^{-1}$ (d) $1.65 \times 10^{-19}\ s^{-1}$

12. A solution is prepared in which 0.1 mole each of HCl, $CH_3COOH$ and $CHCl_2COOH$ is present in a litre. If the ionization constant of $CH_3COOH$ is $10^{-5}$ and that of $Cl_2CHCOOH$ is 0.15, then the pH of solution is (log 2 = 0.3, log 3 = 0.48)
    (a) 1.18 (b) 0.82
    (c) 1.0 (d) 0.95

13. A solution contains 4.25 g ammonia per 250.0 ml of solution. Electrical conductivity measurement at 25°C shows that 0.40% of the ammonia has reacted with water. The pH of the solution is (log 2 = 0.3)

(a) 11.6 (b) 2.4
(c) 12.6 (d) 10.6

14. Morphine ($C_{17}H_{19}NO_3$) is administered medically to relieve pain. It is a naturally occurring base or alkaloid. What is the pH of a 0.0025 M solution of morphine at 25°C? The base-ionization constant $K_b$ is $1.6 \times 10^{-6}$ at 25°C. (log 2 = 0.3)

(a) 4.2 (b) 9.8
(c) 3.7 (d) 10.3

15. Saccharin ($K_a = 2 \times 10^{-12}$) is a weak acid represented by the formula HSac. A $4 \times 10^{-4}$ mole amount of saccharin is dissolved in 200 ml water of pH 3.0. Assuming no change in volume, the concentration of $Sac^-$ ions in the resulting solution at equilibrium is

(a) $4 \times 10^{-12}$ M (b) $2 \times 10^{-12}$ M
(c) $8 \times 10^{-13}$ M (d) $6.32 \times 10^{-8}$ M

16. Dissociation constants of acids HA and HB are $2.0 \times 10^{-4}$ and $5 \times 10^{-5}$, respectively. The $[H^+]$ in the resulting solution obtained by mixing 20 ml of 0.5 M-HA solution and 30 ml of 0.2 M-HB solution is

(a) $1.05 \times 10^{-2}$ M (b) $6.78 \times 10^{-3}$ M
(c) $1.05 \times 10^{-3}$ M (d) $6.78 \times 10^{-2}$ M

17. At 25°C, the dissociation constants of acid HA and base BOH in aqueous solution is same. The pH of 0.01 M solution of HA is 5.0. The pH of 0.1 M solution of BOH is

(a) 5.0 (b) 9.0
(c) 9.5 (d) 8.5

18. An aqueous solution initially contains 0.01 M-$RNH_2$ ($K_b = 2.0 \times 10^{-6}$) and $10^{-4}$ M-NaOH. The final concentration of $OH^-$ in the solution is about

(a) $10^{-4}$ M (b) $2.0 \times 10^{-4}$ M
(c) $3.0 \times 10^{-4}$ M (d) $1.414 \times 10^{-4}$ M

19. What will be the effect of adding 100 ml of 0.001 M-HCl solution to 100 ml of a solution having 0.1 M-HA? The acid dissociation constant of HA is $10^{-5}$.

(a) The degree of dissociation of HA will decrease but the pH of solution remains unchanged.
(b) The degree of dissociation of HA remains unchanged but the pH of solution decreases.
(c) Neither degree of dissociation nor pH of solution will change.
(d) The degree of dissociation as well as pH of solution will decrease.

20. Fear or excitement, generally cause one to breathe rapidly and it results in the decrease of concentration of $CO_2$ in blood. In what way, it will change the pH of blood?

(a) pH will increase (b) pH will decrease
(c) No change (d) pH becomes 7.0

21. An amount of 0.16 g of $N_2H_4$ is dissolved in water and the total volume is made up to 500 ml. What is the percentage of $N_2H_4$ that has reacted with water in this solution? The value of $K_b$ for $N_2H_4 = 4.0 \times 10^{-6}$.

(a) 0.02% (b) 0.014%
(c) 2% (d) 2.82%

22. Water in equilibrium with air contains $4.4 \times 10^{-5}$% $CO_2$. The resulting carbonic acid ($H_2CO_3$) gives the solution a hydronium ion concentration of $2.0 \times 10^{-6}$ M about 20 times greater than that of pure water. What is the pH of the solution at 298 K? (log 4.4 = 0.64, log 2 = 0.3)

(a) 5.36 (b) 5.70
(c) 8.30 (d) 5.64

23. What is the pH of $6.67 \times 10^{-3}$ M aqueous solution of $Al(OH)_3$ if its first dissociation is 100%, second dissociation is 50% and the third dissociation is negligible.

(a) 2 (b) 12
(c) 11 (d) 3

24. The only incorrect information related with 0.09 M solution of ($NH_2CH_2CH_2NH_2$) ethylenediamine (en) is ($K_{b1} = 8.1 \times 10^{-5}$, $K_{b2} = 7.0 \times 10^{-8}$, log 3 = 0.48, log 7 = 0.85)

(a) pH = 11.44
(b) $[enH^+] = 2.7 \times 10^{-3}$ M
(c) $[enH_2^{2+}] = 7.0 \times 10^{-8}$ M
(d) $[H^+] = 2.7 \times 10^{-3}$ M

**25.** Calculate $[S^{2-}]$ in a solution originally having 0.1 M – HCl and 0.2 M – $H_2S$. For $H_2S$, $K_{a1} = 1.4 \times 10^{-7}$ and $K_{a2} = 1.0 \times 10^{-14}$.

(a) 0.1 M (b) $2.8 \times 10^{-20}$ M
(c) $2.8 \times 10^{-22}$ M (d) $1.4 \times 10^{-20}$ M

**26.** For a tribasic acid, $H_3A$, $K_{a1} = 2 \times 10^{-5}$, $K_{a2} = 5 \times 10^{-9}$ and $K_{a3} = 4 \times 10^{-12}$. The value of $\frac{[A^{3-}]}{[H_3A]}$ at equilibrium in an aqueous solution originally having 0.2 M – $H_3A$ is

(a) $5 \times 10^{-17}$ (b) $5 \times 10^{-9}$
(c) $1 \times 10^{-17}$ (d) $2 \times 10^{-22}$

**27.** $H_3A$ is a weak tribasic acid with $K_{a1} = 10^{-5}$, $K_{a2} = 10^{-9}$ and $K_{a3} = 10^{-13}$. The value of pX of 0.1 M – $H_3A$ solution, where $pX = -\log_{10}X$ and $X = \frac{[A^{3-}]}{[HA^{2-}]}$, is

(a) 5.0 (b) 4.0
(c) 9.0 (d) 10.0

**28.** To 20 ml of 0.1 M – NaOH solution, 3 ml of 1 M acetic acid solution is added. Is the solution now neutral, acidic or alkaline? How much more of the acetic acid solution we add to produce a change of pH = 0.3 unit? ($pK_a$ for $CH_3COOH$ = 4.74, log 2 = 0.3)

(a) acidic, 2 ml (b) alkaline, 1 ml
(c) acidic, 1 ml (d) neutral, 2 ml

**29.** A volume of 18 ml of acetic acid mixture and sodium acetate required 6 ml of 0.1 M-NaOH for neutralization of the acid and 12 ml of 0.1 M-HCl reaction with salt separately. If $pK_a$ of acetic acid is 4.75, then what is the pH of the mixture? (log 2 = 0.3)

(a) 5.05 (b) 4.45
(c) 4.15 (d) 5.35

**30.** What is the ratio of pH of a solution containing 1 mole of $CH_3COONa$ and 1 mole of HCl per litre and the other solution containing 1 mole of $CH_3COONa$ and 1 mole of $CH_3COOH$ per litre?

(a) 1 : 2 (b) 2 : 1
(c) $pK_a$ : 2 (d) Zero

**31.** To a solution of acetic acid, solid sodium acetate is gradually added. When '*x* g' of the salt has been added, the pH has a certain value. When total '*y* g' of the salt has been added, the pH has been further raised by 0.6 units. What is the ratio of *x* : *y*? (log 3.98 = 0.6)

(a) 3.98 : 1 (b) 1 : 3.98
(c) 2 : 3.98 (d) 3.98 : 2

**32.** Two buffers, X and Y of pH 4.0 and 6.0, respectively, are prepared from acid HA and the salt NaA. Both the buffers are 0.50 M in HA. What would be the pH of the solution obtained by mixing equal volumes of the two buffers? The value of $K_a$ for HA = $1.0 \times 10^{-5}$ (log 5.05 = 0.7).

(a) 5.0 (b) 4.3
(c) 4.7 (d) 5.7

**33.** The buffer capacity (β) for a weak acid (A) – conjugate base (B) buffer is defined as the number of moles of strong acid or base needed to change the pH of 1 L of solution by 1 pH unit, where $\beta = \frac{2.303(C_A + C_B)K_a[H^+]}{([H^+] + K_a)^2}$. Under what condition will a buffer best resist a change in pH?

(a) $pH = 3\,pK_a$ (b) $2\,pH = pK_a$
(c) $pH = pK_a$ (d) $pH = 2\,pK_a$

**34.** A 40.0 ml solution of weak base, BOH is titrated with 0.1 N – HCl solution. The pH of the solution is found to be 10.0 and 9.0 after adding 5.0 ml and 20.0 ml of the acid, respectively. The dissociation constant of the base is (log 2 = 0.3)

(a) $2 \times 10^{-5}$ (b) $1 \times 10^{-5}$
(c) $4 \times 10^{-5}$ (d) $5 \times 10^{-5}$

**35.** How many grams of NaOH should be added in 500 ml of 2 M acetic acid solution to get a buffer solution of maximum buffer capacity?

(a) 20.0 (b) 10.0
(c) 40.0 (d) 30.0

**36.** A 0.28 g sample of an unknown monoprotic organic acid is dissolved in water and titrated with a 0.1 M sodium hydroxide solution. After the addition of 17.5 ml of base, a pH of 5.0 is recorded. The equivalence point is reached when a total of 35.0 ml of NaOH is added. The molar mass of the organic acid is

(a) 160 (b) 80
(c) 40 (d) 120

37. Equilibrium constant for the acid ionization of $Fe^{3+}$ to $Fe(OH)^{2+}$ and $H^+$ is $9.0 \times 10^{-3}$. What is the maximum pH, which could be used so that at least 90% of the total $Fe^{3+}$ in a dilute solution exists as $Fe^{3+}$? (log 3 = 0.48)

(a) 3.0 (b) 1.08
(c) 1.92 (d) 2.04

38. A student was given 0.01 mole of a weak organic acid and told to determine the $K_a$ of the acid. He prepared 100 ml of an aqueous solution containing the sample. The volume of 50 ml of this solution was then titrated with NaOH to the equivalence point. The titrated solution was then mixed with the other 50 ml of the solution and the pH was determined. A value of 4.80 was obtained for the mixed solution. What is the approximate value of $pK_a$ for the acid?

(a) 5.10 (b) 4.50
(c) 4.80 (d) 4.20

39. What is the aqueous ammonia concentration of a solution prepared by dissolving 0.15 mole of $NH_4^+CH_3COO^-$ in 1 L of water? Given: $K_a$ ($CH_3COOH$) = $1.8 \times 10^{-5}$; $K_b$ ($NH_4OH$) = $1.8 \times 10^{-5}$.

(a) $8.3 \times 10^{-4}$ M (b) 0.15 M
(c) $5.52 \times 10^{-3}$ M (d) $3.8 \times 10^{-4}$ M

40. A volume of 2.5 ml of $\frac{2}{5}$ M weak monoacidic base ($K_b = 1 \times 10^{-12}$ at 25°C) is titrated with $\frac{2}{15}$ M-HCl in water at 25°C. The concentration of $H^+$ at equivalence point is ($K_w = 1 \times 10^{-14}$ at 25°C)

(a) $3.7 \times 10^{-13}$ M (b) $3.2 \times 10^{-7}$ M
(c) $3.2 \times 10^{-2}$ M (d) $2.7 \times 10^{-2}$ M

41. A volume of 10 ml of 0.1 M tribasic acid, $H_3A$ is titrated with 0.1 M-NaOH solution. What is the ratio (approximate value) of $\frac{[H_3A]}{[A^{3-}]}$ at the second equivalent point? Given: $K_1 = 7.5 \times 10^{-4}$; $K_2 = 10^{-8}$; $K_3 = 10^{-12}$

(a) $10^{-4}$ (b) $10^{-3}$
(c) $10^{-7}$ (d) $10^{-6}$

42. The equilibrium carbonate ion concentration after equal volumes of 0.7 M-$Na_2CO_3$ and 0.7 M-HCl solutions are mixed, is ($K_{a1}$ and $K_{a2}$ for $H_2CO_3$ are $4.9 \times 10^{-6}$ and $4.0 \times 10^{-11}$, respectively)

(a) 0.7 M (b) 0.35 M
(c) 0.002 M (d) 0.001 M

43. Calcium lactate is a salt of weak acid and represented as $Ca(Lac)_2$. A saturated solution of $Ca(Lac)_2$ contains 0.125 mole of salt in 0.50 L solution. The pOH of this is 5.60. Assuming complete dissociation of the salt, calculate $K_a$ of lactate acid (log 2.5 = 0.4).

(a) $1.25 \times 10^{-11}$ (b) $8.0 \times 10^{-4}$
(c) $3.2 \times 10^{-17}$ (d) $4 \times 10^{-5}$

44. A volume of 50 ml of a solution which is 0.05 M in the acid HA ($pK_a$ = 3.80) and 0.08 M in HB ($pK_a$ = 8.20) is titrated with 0.2 M-NaOH solution. The pH of solution at the first equivalent point is (log 2 = 0.3, log 1.6 = 0.2)

(a) 6.0 (b) 9.2
(c) 4.8 (d) 5.9

45. Calculate the degree of hydrolysis of 0.005 M-$K_2CrO_4$. For $H_2CrO_4$, $K_{a_1}$ = infinite, $K_{a_2} = 5 \times 10^{-7}$.

(a) 0.002 (b) 0.02
(c) 0.2 (d) 0.005

46. When glycinium hydrochloride ($NH_2CH_2COOH.HCl$) is titrated against NaOH, pH at the first half equivalence point is 2.40 and the pH at second half equivalence point is 9.60. The pH at first equivalence point is

(a) 2.40 (b) 9.60
(c) 6.00 (d) 7.20

47. An acid–base indicator has $K_a = 3.0 \times 10^{-5}$. The acid form of the indicator is red and the basic form is blue. The $[H^+]$ required to change the indicator from 75% blue to 75% red is

(a) $8 \times 10^{-5}$ M (b) $9 \times 10^{-5}$ M
(c) $1 \times 10^{-5}$ M (d) $3.33 \times 10^{-5}$ M

48. An acid base indicator which is a weak acid has a $pK_a$ value = 5.5. At what concentration ratio of sodium acetate to acetic acid would the indicator show a colour half way between those of its acid and conjugate base forms? The $pK_a$ value of acetic acid = 4.75 [Antilog (0.75) = 5.62, Antilog (0.79) = 6.3, Antilog (0.69) = 4.93].

(a) 4.93 : 1 (b) 6.3 : 1
(c) 5.62 : 1 (d) 2.37 : 1

**49.** A volume of 224 ml of $CO_2(g)$ at 1 atm and 0°C was passed in 1 L of NaOH solution of unknown molarity. The resulting solution when titrated with 1.0 M-HCl solution requires 30 ml for the phenolphthalein end point. The molarity of NaOH solution used is

(a) 0.04 M (b) 0.02 M
(c) 0.03 M (d) 0.08 M

**50.** The correct increasing order of solubility of the following substances in g/100 ml is $PbSO_4$ ($K_{sp} = 2 \times 10^{-9}$), ZnS ($K_{sp} = 1 \times 10^{-22}$), AgBr ($K_{sp} = 4 \times 10^{-13}$), $CuCO_3$ ($K_{sp} = 1 \times 10^{-8}$). (Atomic masses: Pb = 208, Zn = 65, Ag = 108, Br = 80, Cu = 63)

(a) $PbSO_4 < ZnS < AgBr < CuCO_3$
(b) $PbSO_4 < CuCO_3 < AgBr < ZnS$
(c) $ZnS < AgBr < CuCO_3 < PbSO_4$
(d) $ZnS < AgBr < PbSO_4 < CuCO_3$

**51.** Calculate $K_{form}$ for $HgCl_4^{2-}$ if the concentration of $Hg^{2+}$ is $1.6 \times 10^{-17}$ M in a solution prepared by dissolving 0.10 mole of $Hg(NO_3)_2$ in 1 L of solution containing 0.9 mole of NaCl.

(a) $10^{17}$ (b) $10^{16}$
(c) $10^{18}$ (d) $6.25 \times 10^{16}$

**52.** Unexposed silver halides are removed from photographic film when they react with sodium thiosulphate to form the complex ion $Ag(S_2O_3)_2^{3-}$. What amount of $Na_2S_2O_3$ is needed to prepare 1 L of a solution that dissolves 0.1 moles of AgBr by the formation of $Ag(S_2O_3)_2^{3-}$? $K_{sp}$ of AgBr = $4.0 \times 10^{-13}$ and $K_f$ of $Ag(S_2O_3)_2^{3-}$ = $1.6 \times 10^{12}$.

(a) 3.25 g (b) 3.25 moles
(c) 0.325 moles (d) 6.5 moles

**53.** A saturated solution of silver benzoate ($AgOCOC_6H_5$) has pH of 8.6. $K_a$ for benzoic acid is $5.0 \times 10^{-5}$. The value of $K_{sp}$ for silver benzoate is (log 2 = 0.3)

(a) $8.0 \times 10^{-2}$ (b) $6.4 \times 10^{-3}$
(c) $6.4 \times 10^{-4}$ (d) 0.282

**54.** The solubility product of $Co(OH)_3$ is $2.7 \times 10^{-43}$. The pH of saturated solution of $Co(OH)_3$ is about

(a) 7.0 (b) 11.0
(c) 3.0 (d) 3.48

**55.** In an attempted determination of the solubility product constant of $Tl_2S$, the solubility of this compound in pure $CO_2$ free water was determined as $2.0 \times 10^{-6}$ M. Assume that the dissolved sulphide hydrolyses almost completely to $HS^-$ and that the further hydrolysis to $H_2S$ can be neglected, what is the computed $K_{sp}$? For $H_2S$, $K_{a1} = 1.4 \times 10^{-7}$, $K_{a2} = 1.0 \times 10^{-14}$

(a) $6.4 \times 10^{-23}$ (b) $1.6 \times 10^{-23}$
(c) $3.2 \times 10^{-17}$ (d) $3.2 \times 10^{-24}$

**56.** Calculate the formation constant for the reaction of a tripositive metal ion with thiocyanate ions to form the monocomplex if the total metal concentration in the solution is $2 \times 10^{-3}$ M, the total $SCN^-$ concentration is $1.51 \times 10^{-3}$ M and the free $SCN^-$ concentration is $1.0 \times 10^{-5}$ M.

(a) $7.55 \times 10^4$ (b) $3 \times 10^5$
(c) $3.33 \times 10^{-6}$ (d) $1.5 \times 10^5$

**57.** After solid $SrCO_3$ was equilibrated with a buffer at pH 8.6, the solution was found to have $[Sr^{2+}]$ = $2.0 \times 10^{-4}$ M, what is the $K_{sp}$ of $SrCO_3$? ($K_{a2}$ for $H_2CO_3 = 5.0 \times 10^{-11}$, log 2 = 0.3)

(a) $4.0 \times 10^{-8}$ (b) $8.0 \times 10^{-8}$
(c) $\frac{4}{51} \times 10^{-8}$ (d) $\frac{2}{51} \times 10^{-8}$

**58.** What is the solubility of MnS in pure water, assuming hydrolysis of $S^{2-}$ ions? $K_{sp}$ of MnS = $2.5 \times 10^{-10}$, $K_{a1} = 1 \times 10^{-7}$ and $K_{a2} = 1 \times 10^{-14}$ for $H_2S$. ($0.63^3 = 0.25$)

(a) $6.3 \times 10^{-4}$ M (b) $2.5 \times 10^{-4}$ M
(c) $6.3 \times 10^{-3}$ M (d) $1.58 \times 10^{-5}$ M

**59.** An amount of 0.10 moles of AgCl(s) is added to one litre of water. Next, the crystals of NaBr are added until 75% of the AgCl is converted to AgBr(s), the less soluble silver halide. What is $Br^-$ at this point? $K_{sp}$ of AgCl = $2 \times 10^{-10}$ and $K_{sp}$ of AgBr = $4 \times 10^{-13}$.

(a) 0.075 M (b) 0.025 M
(c) $1.5 \times 10^{-4}$ M (d) 0.027 M

**60.** An amount of 0.01 moles of solid AgCN is rendered soluble in 1 L by adding just sufficient excess cyanide ion to form $Ag(CN)_2^-$ and the concentration of free cyanide ion is $2.5 \times 10^{-7}$ M. Determine $[Ag^+]$ in the solution neglecting hydrolysis of cyanide ion. The value of $K_{diss}$ for $Ag(CN)_2^-$ = $1.0 \times 10^{-20}$.

(a) $6.25 \times 10^{-9}$ M (b) $1.6 \times 10^{-9}$ M
(c) $1.6 \times 10^{-7}$ M (d) $6.25 \times 10^{-7}$ M

**61.** An amount of 2.0 M solution of $Na_2CO_3$ is boiled in a closed container with excess of $CaF_2$. Very little amount of $CaCO_3$ and NaF are formed. If the solubility product of $CaCO_3$ is '$x$' and the molar solubility of $CaF_2$ is '$y$', the molar concentration of $F^-$ in the resulting solution after equilibrium is attained is

(a) $\sqrt{\frac{2y}{x}}$ (b) $\frac{8y^3}{x}$

(c) $\sqrt{\frac{8y^3}{x}}$ (d) $\sqrt{\frac{4y^3}{x}}$

**62.** Solid $BaF_2$ is added to a solution containing 0.1 mole of $Na_2C_2O_4$ solution (1 L) until equilibrium is reached. If the $K_{sp}$ of $BaF_2$ and $BaC_2O_4$ is $10^{-6}$ mol$^3$ L$^{-3}$ and $10^{-10}$ mol$^2$ L$^{-2}$, respectively, then find the equilibrium concentration of $Ba^{2+}$ in the solution. Assume that addition of $BaF_2$ does not cause any change in volume.

(a) 0.2 M (b) $4 \times 10^{-6}$ M
(c) $2.5 \times 10^{-5}$ M (d) $2.5 \times 10^{-6}$ M

**63.** What is the solubility of solid zinc hydroxide at a pH of 13? Given that

$Zn(OH)_2(s) \rightleftharpoons Zn(OH)_2(aq)$: $K_1 = 10^{-6}$ M

$Zn(OH)_2(aq) \rightleftharpoons Zn(OH)^+(aq) + OH^-(aq)$: $K_2 = 10^{-7}$ M

$Zn(OH)^+(aq) \rightleftharpoons Zn^{2+}(aq) + OH^-(aq)$: $K_3 = 10^{-4}$ M

$Zn(OH)_2(aq) + OH^-(aq) \rightleftharpoons Zn(OH)_3^-(aq)$: $K_4 = 10^3$ M$^{-1}$

$Zn(OH)_3^-(aq) + OH^-(aq) \rightleftharpoons Zn(OH)_4^{2-}(aq)$: $K_5 = 10$ M$^{-1}$

(a) $10^{-17}$ M (b) $10^{-6}$ M
(c) $10^{-4}$ M (d) $2 \times 10^{-4}$ M

**64.** On dissolving a substance in water, it can produce either an ionic or molecular solution. From the following data, predict what type of solution is likely to be formed by $CaCl_2$ and $HgCl_2$ at 25°C?

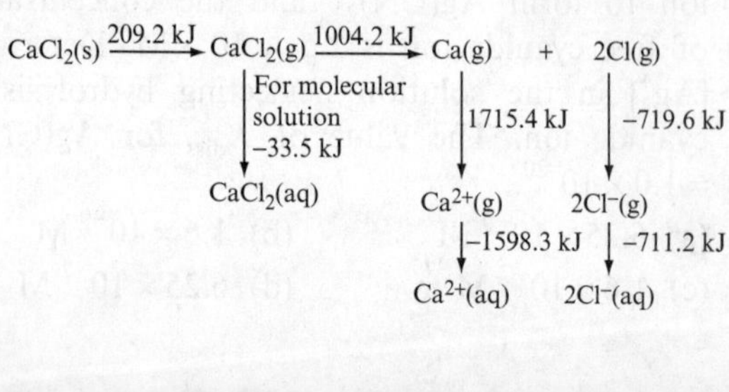


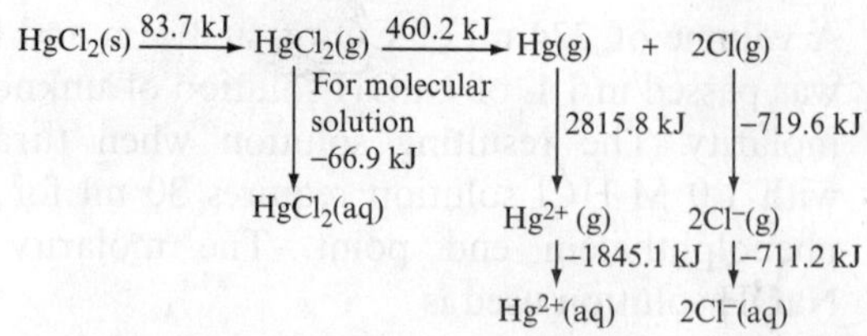


The maximum contribution of $T\Delta S$ in dissolution process is 30 kJ at 25°C

(a) both ionic.
(b) both molecular.
(c) $CaCl_2$ is ionic but $HgCl_2$ is molecular.
(d) $CaCl_2$ is molecular but $HgCl_2$ is ionic.

**65.** A volume of 250 ml of saturated clear solution of $CaC_2O_4$(aq) requires 6.0 ml of 0.001 M-$KMnO_4$ in acidic medium for complete oxidation of $C_2O_4^{2-}$ ions. What is the value of $K_{sp}$ for $CaC_2O_4$?

(a) $3.6 \times 10^{-9}$ (b) $6 \times 10^{-5}$
(c) $5.76 \times 10^{-10}$ (d) $1.44 \times 10^{-8}$

**66.** $Sr^{2+}$ forms a very unstable complex with $NO_3^-$. A solution that was 0.001 M-$Sr(ClO_4)_2$ and 0.05 M-$KNO_3$ was found to have only 75% of its strontium in the uncomplexed $Sr^{2+}$ form, the balance being $Sr(NO_3)^+$. What is the value of $K_f$ for complexation?

(a) 6.67 (b) 0.15
(c) 60 (d) 26.67

**67.** The concentration of $CH_3COO^-$ ion in a solution prepared by adding 0.1 mole of $CH_3COOAg$(s) in 1 L of 0.1 M-HCl solution is [Given: $K_a(CH_3COOH) = 10^{-5}$; $K_{sp}(AgCl) = 10^{-10}$; $K_{sp}(CH_3COOAg) = 10^{-8}$]

(a) $10^{-3}$ M (b) $10^{-2}$ M
(c) $10^{-1}$ M (d) 1 M

**68.** Among the different types of salts that have nearly the same solubility product constant $K_{sp}$ but much smaller than one, the most soluble salt is that which

(a) produces maximum number of ions.
(b) produces minimum number of ions.
(c) produces high charge on ions.
(d) produces low charges on ions.

**69.** When excess oxalic acid is added to $CaCl_2$ solution, $CaC_2O_4$ is precipitated and the solution still contains some unprecipitated $Ca^{2+}$. The reason is

(a) $CaC_2O_4$ is a soluble salt.
(b) oxalic acid does not ionize at all.
(c) the solution becomes acidic and hence, ionization of $H_2C_2O_4$ is suppressed.
(d) the solution becomes basic and hence, ionization of $H_2C_2O_4$ increases.

**70.** At what pH, is the solubility of $Zn(OH)_2$ minimum? What is the minimum solubility?

$Zn(OH)_2(s) \rightleftharpoons Zn^{2+}(aq) + 2OH^-(aq)$;
$K_{sp} = 1.2 \times 10^{-17}$

$Zn(OH)_2(s) + 2OH^-(aq) \rightleftharpoons Zn(OH)_4^{2-}(aq)$;
$K_f = 0.12$

(a) 10.0, $2.4 \times 10^{-9}$ M (b) 4.0, $2.4 \times 10^{-9}$ M
(c) 10.0, $1.2 \times 10^{-9}$ M (d) 10.0, $1.32 \times 10^{-9}$ M

**71.** At what minimum pH will $10^{-3}$ M – $Al(OH)_3$ go into solution ($V = 1$ L) as $Al(OH)_4^-$ and at what maximum pH, it will be dissolved as $Al^{3+}$? Given: log 2 = 0.3

$Al(OH)_4^- \rightleftharpoons Al^{3+} + 4OH^-$ ; $K_{eq} = 1.6 \times 10^{-34}$
$Al(OH)_3 \rightleftharpoons Al^{3+} + 3OH^-$ ; $K_{eq} = 8.0 \times 10^{-33}$

(a) 9.3, 4.7 (b) 4.7, 9.3
(c) 9.3, 9.3 (d) 4.3, 9.3

**72.** A 0.1 M solution of $[Cu(NH_3)_4]^+$ is stirred with an excess of potassium cyanide sufficient to convert all the ammonium complex to the corresponding cuprocyanide complex $[Cu(CN)_4]^{-3}$ and in addition to provide the solution with an excess of $CN^-$ equal to 0.2 M. Calculate the maximum pH of the solution when the final solution is treated with hydrogen sulphide to maintain $[H_2S] = 0.1$ M and the precipitation of cuprous sulphide is prevented. The instability constant for $[Cu(CN)_4]^{-3}$ is $6.4 \times 10^{-15}$, $K_{a,overall}$ of $H_2S = 1.6 \times 10^{-21}$, $K_{sp}$ of $Cu_2S = 2.56 \times 10^{-27}$.

(a) 4.0 (b) 10.0
(c) 10.8 (d) 3.2

**73.** A particular water sample has 136 ppm $CaSO_4$. What percentage of water by mass must be evaporated in a container before solid $CaSO_4$ begins to deposit. Assume that the solubility of $CaSO_4$ does not change with temperature in the range 0°C to 100°C. The value of $K_{sp}$ for $CaSO_4 = 1.6 \times 10^{-5}$.

(a) 80% (b) 50%
(c) 75% (d) 60%

**74.** There exist an equilibrium between solid $BaSO_4$, $Ba^{2+}$ and $SO_4^{2-}$ ions in aqueous medium. Now, if equilibrium is disturbed by the addition of (a) $BaNO_3$ and (b) $K_2SO_4$ in different experiments and the equilibrium is re-achieved, then match the columns given below. (Thick line represents the initial equilibrium and the dotted line represents the approach of system towards equilibrium).

(a) Addition of $BaNO_3$.
(b) Addition of $K_2SO_4$.
(c)

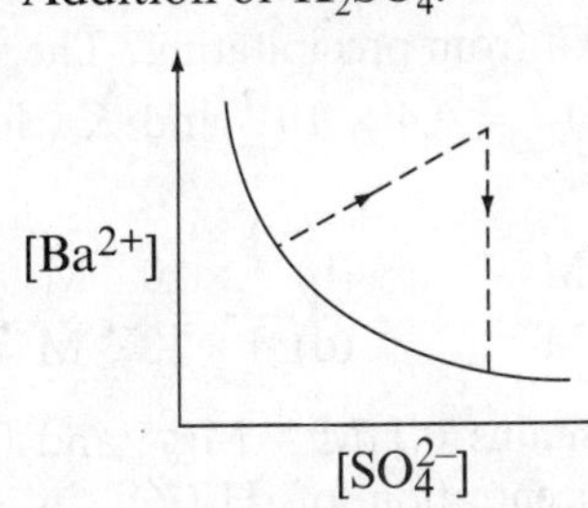


(d)

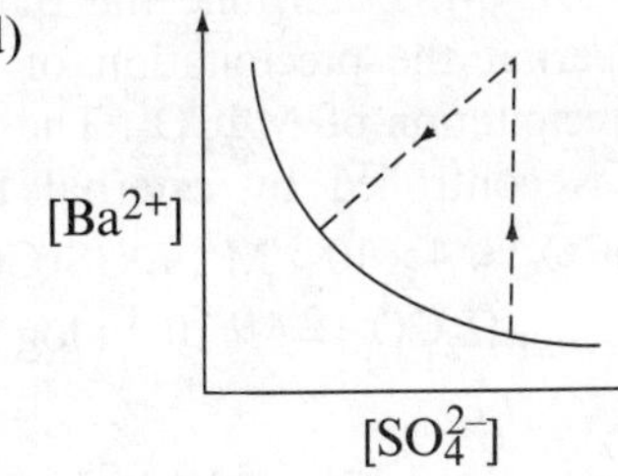


(e)

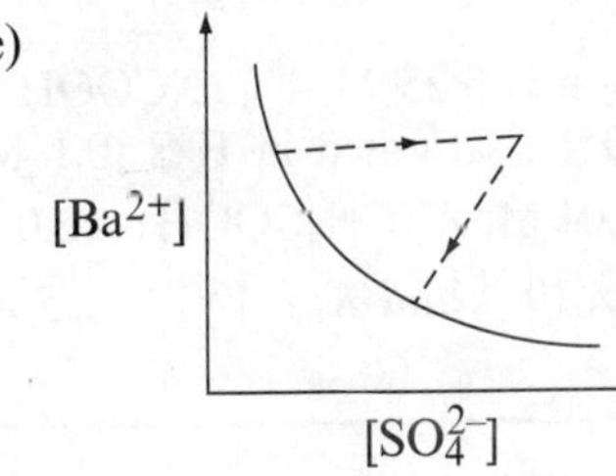


(f)

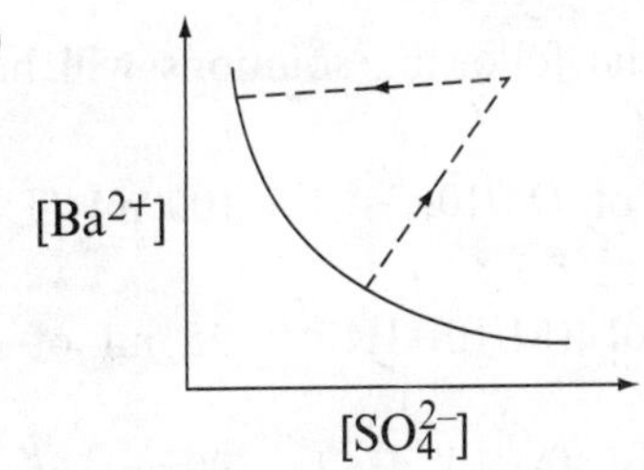


(a) ac, bd (b) ad, be
(c) ac, be (d) ae, bd

**75.** A volume of 1.0 L of solution which was in equilibrium with solid mixture of AgCl and $Ag_2CrO_4$ was found to contain $1 \times 10^{-4}$ moles of $Ag^+$ ions, $1.0 \times 10^{-6}$ moles of $Cl^-$ ions and $8.0 \times 10^{-4}$ moles of $CrO_4^{2-}$ ions. The $Ag^+$ ions are added slowly to the above mixture (keeping the volume constant) till $8.0 \times 10^{-7}$ moles of AgCl got precipitated. How many moles of $Ag_2CrO_4$ were precipitated simultaneously?

(a) $7.68 \times 10^{-4}$ (b) $4.8 \times 10^{-4}$
(c) $8.0 \times 10^{-4}$ (d) $7.68 \times 10^{-5}$

76. What concentration of free $CN^-$ must be maintained in a solution that is 1.8 M-$AgNO_3$ and 0.16 M-NaCl to prevent AgCl from precipitating? The value of $K_f$ for $Ag(CN)_2^-$ = $6.4 \times 10^{17}$ and $K_{sp}$ for AgCl = $1.8 \times 10^{-10}$.

(a) $2.5 \times 10^{-9}$ M (b) $5 \times 10^{-5}$ M
(c) $2.5 \times 10^{-5}$ M (d) $1 \times 10^{-4}$ M

77. A solution contains 0.1 M – $Mg^{2+}$ and 0.1 M – $Sr^{2+}$. The concentration of $H_2CO_3$ in solution is adjusted to 0.05 M. Determine the pH range which would permit the precipitation of $SrCO_3$ without any precipitation of $MgCO_3$. The $H^+$ ion concentration is controlled by external factors. Given: $K_{sp}(MgCO_3) = 4 \times 10^{-8}$ $M^2$; $K_{sp}(SrCO_3) = 9 \times 10^{-10}$ $M^2$; $K_{a,overall}(H_2CO_3) = 5 \times 10^{-17}$; log 2 = 0.3; log 3 = 0.48.

(a) 4.78 to 5.6 (b) 4.6 to 5.78
(c) 5.78 to 6.4 (d) 5.22 to 5.4

78. A buffer solution is 0.25 M – $CH_3COOH$ + 0.15 M – $CH_3COONa$, saturated in $H_2S$ (0.1 M) and has $[Mn^{2+}]$ = 0.04 M, $K_a(CH_3COOH) = 2.0 \times 10^{-5}$, $K_a(H_2S) = 1.0 \times 10^{-21}$ and $K_{sp}(MnS) = 2.5 \times 10^{-13}$. Which buffer component should be increased in concentration and to which minimum value to just start precipitation of MnS?

(a) $CH_3COOH$, 1.25 M
(b) $CH_3COONa$, 0.25 M
(b) $CH_3COONa$, 1.25 M
(d) $CH_3COONa$, 0.50 M

79. To 0.35 L of 0.1 M – $NH_3$ 0.15 L of 0.1 M-$MgCl_2$ is added. What minimum mass of $(NH_4)_2SO_4$ should be added to cause the $Mg(OH)_2$ to re-dissolve? The value of $K_{sp}$ for $Mg(OH)_2 = 1.2 \times 10^{-11}$, $K_b$ for $NH_3 = 2.0 \times 10^{-5}$.

(a) 4.62 g (b) 2.31 g
(c) 9.24 g (d) 1.155 g

80. A volume of 500 ml of 0.01 M – $AgNO_3$ solution, 250 ml of 0.02 M – NaCl solution and 250 ml of 0.02 M – NaBr solution are mixed. The final concentration of bromide ion in the solution is ($K_{sp}$ of AgCl and AgBr are $10^{-10}$ and $5 \times 10^{-13}$, respectively.)

(a) 0.01 M (b) 0.02 M
(c) 0.005 M (d) $2.5 \times 10^{-5}$ M

## Section B (One or More than one Correct)

1. Which of the following solutions will have pH close to 1.0?

(a) 100 ml of (M/10) HCl + 100 ml of (M/10) NaOH.
(b) 55 ml of (M/10) HCl + 45 ml of (M/10) NaOH.
(c) 10 ml of (M/10) HCl + 90 ml of (M/10) NaOH.
(d) 75 ml of (M/5) HCl + 25 ml of (M/5) NaOH.

2. Which of the following relation(s) is/are true for alkaline solution?

(a) $pH > \frac{pK_w}{2}$ (b) pH > pOH
(c) $pOH < \frac{pK_w}{2}$ (d) pH < pOH

3. Which of the following expression(s) for the degree of dissociation of weak monobasic acid in aqueous solution is/are incorrect?

(a) $\sqrt{\frac{K_a}{C}}$ (b) $\frac{K_a}{K_a + [H^+]}$
(c) $\frac{[H^+]}{K_a + [H^+]}$ (d) $\frac{1}{1+10^{(pK_a - pH)}}$

4. Which of the following will occur if a 0.1 M solution of a weak acid is diluted to 0.01 M at constant temperature?

(a) $[H^+]$ will decrease.
(b) pH will increase.
(c) Percentage ionization will increase.
(d) $K_a$ will increase.

5. The relation $pK_a = 14 - pK_b$ is true for which of the following pair(s)?

(a) $MeNH_3^+$, $CH_3NH_2$
(b) $CH_3COOH$, $CH_3COO^-$
(c) $H_3O^+$, $OH^-$
(d) $Na^+$, NaOH

EXERCISE II

6. Which would decrease the pH of 25 ml of a 0.01 M solution of hydrochloric acid? The addition of
   (a) magnesium metal
   (b) 25 ml 0.02 M of HCl
   (c) 25 ml 0.005 M of HCl
   (d) 50 ml water

7. Which of the following result(s) is/are correct for the equilibrium state in a solution originally having 0.1 M-$CH_3COOH$ and 0.1 M-HCl? The value of $K_a$ for $CH_3COOH = 1.8 \times 10^{-5}$.
   (a) $[H^+] = 0.1$ M
   (b) $[CH_3COO^-] = 1.8 \times 10^{-5}$ M
   (c) Degree of dissociation of acetic acid $= 1.8 \times 10^{-4}$
   (d) $[H^+]$ from water $= 10^{-13}$ M

8. The equilibrium constant for the ionization of $RNH_2(g)$ in water as
   $$RNH_2(g) + H_2O(l) \rightleftharpoons RNH_3^+(aq) + OH^-(aq)$$
   is $10^{-6}$ at 25°C. Which of the following is/are correct?
   (a) pH of the solution is 11.0 at partial pressure of $RNH_2(g) = 1$ bar.
   (b) Forward reaction is favoured by the addition of HCl(aq).
   (c) Forward reaction is favoured by the addition of $H_2O(l)$.
   (d) Forward reaction is favoured by the addition of $RNH_2(g)$.

9. Which of the following processes will increase $[OH^-]$ in $NH_4OH$ solution?
   (a) Addition of HCl solution.
   (b) Addition of water.
   (c) Addition of $NH_4Cl$ solution.
   (d) Addition of solid $NH_4OH$.

10. Which of the following statement(s) is/are correct?
    (a) The pH of $10^{-8}$ M-HCl solution is 8.
    (b) The conjugate base of $H_2PO_4^-$ is $HPO_4^{2-}$.
    (c) Autoprotolysis constant of water increases with the increase in temperature.
    (d) When a solution of a weak monoprotic acid is titrated against a strong base, at half neutralization point, pH = $pK_a/2$.

11. If $K_1$ and $K_2$ are the first and second ionization constants of $H_2CO_3$ and $K_1 >> K_2$, then the incorrect relation(s) is/are
    (a) $[H^+] = [HCO_3^-]$
    (b) $[H^+] = \sqrt{K_1.[H_2CO_3]}$
    (c) $K_2 = [CO_3^{2-}]$
    (d) $[H^+] = 2.[CO_3^{2-}]$

12. Which of the following may act as buffer in aqueous solution?
    (a) $NH_4Cl + NH_4OH$
    (b) $CH_3COOH + CH_3COONa$
    (c) $CH_3COONa + NaCl$
    (d) Borax + Boric acid

13. A buffer solution can be prepared from a mixture of
    (a) $NH_4Cl$ and NaOH in 2 : 1 mole ratio.
    (b) $CH_3COONa$ and HCl in 1 : 1 mole ratio.
    (c) $CH_3COONa$ and HCl in 2 : 1 mole ratio.
    (d) $CH_3COONa$ and HCl in 1 : 2 mole ratio.

14. Which of the following statements is/are correct?
    (a) A buffer solution may contain a weak acid and its conjugate base.
    (b) A buffer solution shows little change in pH on the addition of a small amount of acid or base.
    (c) A buffer solution can be prepared by mixing a solution of ammonium acetate and acetic acid.
    (d) The addition of solid potassium cyanide to water increases the pH of water.

15. A 2.5 g impure sample containing weak monoacidic base (Molecular weight = 45) is dissolved in 100 ml water and titrated with 0.5-M HCl at 25°C. When 1/5th of the base was neutralized, the pH was found to be 9 and at equivalent point, the pH of solution is 4.5 (log 2 = 0.3).
    (a) $K_b$ of base is less than $10^{-6}$.
    (b) Concentration of salt at equivalent point is 0.25 M.
    (c) Volume of HCl used at equivalent point is 100 ml.
    (d) Mass percentage of base in given sample is 80%.

**16.** $H_2CO_3$ ionizes as follows.

$H_2CO_3 \rightleftharpoons H^+ + HCO_3^-$; $K_1 = 4.0 \times 10^{-6}$

$HCO_3^- \rightleftharpoons H^+ + CO_3^{2-}$; $K_2 = 5.0 \times 10^{-11}$

Which of the following information is correctly related to 0.5 M – $Na_2CO_3$ solution?

(a) The degree of hydrolysis of $Na_2CO_3$ is 0.02.

(b) The pH of solution is 2.0.

(c) The pOH of solution is 2.0.

(d) $[H_2CO_3] = 2.5 \times 10^{-9}$ M

**17.** The amino acid glycine ($NH_2CH_2COOH$) is basic because of its $-NH_2$ group and acidic because of its –COOH group. By a process equivalent to base dissociation, glycine can acquire an additional proton to form $\overset{\oplus}{N}H_3CH_2COOH$. The resulting cation may be considered to be a diprotic acid, since one proton from the –COOH group and one from the $-\overset{\oplus}{N}H_3$ group may be lost. The p$K_a$ values for these processes are 2.22 and 9.78, respectively. For a 0.01 M solution of neutral glycine (log 1.7 = 0.22, log 6 = 0.78),

(a) the pH is 6.0.

(b) percentage of the glycine in the cationic form is 0.017%.

(c) the pOH is 6.0.

(d) percentage of the glycine in the anionic form is 0.017%.

**18.** A certain indicator (an organic dye) has p$K_a$ = 5. For which of the following titrations may it be suitable?

(a) Acetic acid against NaOH.

(b) Aniline hydrochloride against NaOH.

(c) Sodium acetate against HCl.

(d) Barium hydroxide against oxalic acid.

**19.** Which of the following statement(s) is/are incorrect?

(a) When water is added to KCN, a reaction occurs because the $CN^-$ ion is a strong proton acceptor.

(b) No reaction occurs when KCl is dissolved in water because neither $K^+$ nor $Cl^-$ can remove or accept proton from water.

(c) When water is added to sodium acetate, an acidic solution is produced because of the formation of free acetic acid from the action of water on $CH_3COO^-$.

(d) $(NH_4)_2CO_3$ smells strongly of $NH_3$ because the strong base $CO_3^{2-}$ is able to remove protons from $NH_4^+$ ion.

**20.** An aqueous solution contains 0.02 M-$FeCl_2$ and 0.05 M-$FeCl_3$. The solubility products are 8 × $10^{-16}$ for $Fe(OH)_2$ and 4 × $10^{-28}$ for $Fe(OH)_3$. Identify the correct option(s) among the following regarding the precipitation of metal hydroxides.

(a) At pH = 9.0, neither $Fe(OH)_2$ nor $Fe(OH)_3$ will precipitate.

(b) At pH = 6.0, neither $Fe(OH)_2$ nor $Fe(OH)_3$ will precipitate.

(c) If pH of the solution is in between 5.3 and 7.3, only $Fe(OH)_3$ will precipitate but not $Fe(OH)_2$.

(d) If pOH of the solution is in between 6.7 and 8.7, only $Fe(OH)_2$ will precipitate but not $Fe(OH)_3$.

## Section C (Comprehensions)

### Comprehension I

**Acetic acid tends to form dimer due to the formation of intermolecular hydrogen bonding.**

$$2\ CH_3COOH \rightleftharpoons (CH_3COOH)_2$$

**The equilibrium constant for this reaction is $1.5 \times 10^2\ M^{-1}$ in benzene solution and $3.6 \times 10^{-2}\ M^{-1}$ in water. In benzene, monomer does not dissociate and however, in water, monomer dissociates simultaneously with acid dissociation constant $2.0 \times 10^{-5}$ M. Dimer does not dissociate in benzene as well as water.**

1. The molar ratio of dimer to monomer for 0.1 M acetic acid in benzene is equal to
   (a) 150 : 1 (b) 1 : 150
   (c) 5 : 2 (d) 2 : 5

2. The molar ratio of dimer to monomer for 0.1 M acetic acid in water (neglecting the dissociation of acetic acid in water) is equal to
   (a) 250 : 1 (b) 1 : 250
   (c) 9 : 2500 (d) 2500 : 9

3. The pH of 0.1 M acetic acid solution in water, considering the simultaneous dimerization of acid is
   (a) 1 (b) 2.85
   (c) 5.7 (d) 3.42

### Comprehension II

**The dissociation constant of acetic acid is $2.0 \times 10^{-5}$. An aqueous solution of acetic acid is prepared by dissolving 0.7 millimole acetic acid in sufficient watcr to get 10 $m^3$ of solution at 25°C.**

4. What is the pH of the solution? (log 2 = 0.3, log 7 – 0.85, log 17 = 1.23, $\sqrt{4.49}$ = 2.1)
   (a) 7.15 (b) 6.77
   (c) 6.85 (d) 7.0

5. What is the concentration of unionized acetic acid in the solution?
   (a) 0 (b) $4.9 \times 10^{-10}$ M
   (c) $7 \times 10^{-8}$ M (d) $2.45 \times 10^{-10}$ M

### Comprehension III

**The concentration of hydrogen ion in a 0.2 M solution of formic acid is $8.0 \times 10^{-3}$ M. To this solution, formate is added so as to adjust the concentration of sodium formate to 1.0 M. The degree of dissociation of sodium formate is 0.8 (log 2 = 0.3).**

6. What will be the pH of the final solution?
   (a) 4.1 (b) 2.1
   (c) 2.9 (d) 4.9

7. The dissociation constant of formic acid is
   (a) $2.4 \times 10^{-4}$ (b) $3.2 \times 10^{-4}$
   (c) $3.2 \times 10^{-5}$ (d) $1.6 \times 10^{-4}$

### Comprehension IV

**A buffer solution contains 0.8 M-$NH_4OH$ and 0.2 M-$NH_4Cl$. The value of $K_a$ fo $NH_4^+ = 5.0 \times 10^{-10}$ (log 2 = 0.3, log 3 = 0.48).**

8. The pH of the solution is
   (a) 9.9 (b) 8.7
   (c) 9.6 (d) 4.1

9. The pH of the resulting solution on adding 0.15 mole HCl in 500 ml of the buffer solution is
   (a) 9.3 (b) 0.48
   (c) 0.52 (d) 9.9

10. The pH of the resulting solution on adding 0.5 mole HCl in 500 ml of the buffer solution is
    (a) 9.3 (b) 0.30
    (c) 0.70 (d) 9.9

## Comprehension V

**Potash alum is $KAl(SO_4)_2 \cdot 12H_2O$. As a strong electrolyte, it is considered to be 100% dissociated into $K^+$, $Al^{3+}$ and $SO_4^{2-}$. The solution is acidic due to the hydrolysis of $Al^{3+}$, but not so acidic as might be expected because the $SO_4^{2-}$ can sponge up some of the $H_3O^+$ by forming $HSO_4^-$. Given a solution made by dissolving 11.85 g of $KAl(SO_4)_2 \cdot 12H_2O$ in enough water to make 100 $cm^3$ of solution. What is $[H_3O^+]$ of the solution if (K = 39, Al = 27)**

11. None of the ion is hydrolysing.
   (a) $10^{-7}$ M
   (b) less than $10^{-7}$ M
   (c) more than $10^{-7}$ M
   (d) 0.0

12. Only $Al^{3+}$ is hydrolysing and its first hydrolysis constant is $1.4 \times 10^{-5}$ M.
   (a) $1.87 \times 10^{-3}$ M
   (b) $6.24 \times 10^{-4}$ M
   (c) 0.09 M
   (d) $6.32 \times 10^{-7}$ M

13. Only $SO_4^{2-}$ is hydrolysing and acid dissociation constant of $HSO_4^-$ in water is $1.25 \times 10^{-2}$.
   (a) $1.26 \times 10^{-13}$ M
   (b) $6.32 \times 10^{-7}$ M
   (c) $4.47 \times 10^{-7}$ M
   (d) $1.58 \times 10^{-8}$ M

14. Both, $Al^{3+}$ and $SO_4^{2-}$ are hydrolysing.
   (a) $2.97 \times 10^{-4}$ M
   (b) 0.0118 M
   (c) $5.43 \times 10^{-6}$ M
   (d) $9.35 \times 10^{-4}$ M

## Comprehension VI

**A 0.010 M solution of $PuO_2(NO_3)_2$ was found to have a pH of 3.80 [Antilog (–3.8) = $1.6 \times 10^{-4}$].**

15. What is the hydrolysis constant, $K_b$, for $PuO_2^{2+}$?
   (a) $2.56 \times 10^{-6}$
   (b) $3.2 \times 10^{-8}$
   (c) $5.12 \times 10^{-4}$
   (d) $2.56 \times 10^{-8}$

16. What is the value of $K_b$ for $PuO_2 \cdot OH^+$?
   (a) $1.0 \times 10^{-6}$
   (b) $3.9 \times 10^{-8}$
   (c) $3.9 \times 10^{-9}$
   (d) $3.9 \times 10^{-10}$

## Comprehension VII

**A volume of 100 ml of 0.1 M-$H_3PO_4$ solution is titrated with 0.5 M-NaOH solution till the second equivalence point. Then 10 ml of 0.5 M-HCl solution is mixed in the resulting solution. The dissociation constants of $H_3PO_4$ are $10^{-3}$, $10^{-8}$ and $10^{-13}$.**

17. pH at the second equivalence point is
   (a) 13.0 (b) 8.0
   (c) 5.5 (d) 10.5

18. pH of the solution after adding HCl is
   (a) 8.0 (b) 1.0
   (c) 6.0 (d) 10.5

19. The solubility product of base $A(OH)_2$ is $4.0 \times 10^{-30}$. Its solubility in the final solution is
   (a) $10^{-10}$ M
   (b) $4.0 \times 10^{-16}$ M
   (c) $4.0 \times 10^{-18}$ M
   (d) $4.0 \times 10^{-22}$ M

## Comprehension VIII

**Human blood has a narrow pH range of 7.3–7.4, which must be maintained for metabolic processes to function properly. To keep the pH in this range, it requires a delicate balance between the concentrations of the conjugate acid–base pairs, making up to the buffer system. The main buffer is a carbonic acid-hydrogen carbonate system, which involves the following equilibria.**

$$CO_2(g) \rightleftharpoons CO_2(aq)$$
$$CO_2(aq) + H_2O(l) \rightleftharpoons H_2CO_3(aq)$$
$$H_2CO_3(aq) + H_2O(l) \rightleftharpoons HCO_3^-(aq) + H_3O^+(aq)$$

**At the body temperature, the p$K_a$ for carbonic acid is 6.40. However, the normal concentration of $CO_2$(g) in the lungs maintains a ratio of $HCO_3^-$(aq) and $H_2CO_3$(aq) in blood plasma at about 8 : 1. The $H_2CO_3$ concentration in blood is largely controlled by breathing and respiration and $HCO_3^-$ concentration is largely controlled by excretion in urine.**

**If blood pH rises above 7.4, thena potentially life-threatening condition called alkalosis can result. It normally occurs in climbers suffering from oxygen deficiency at high altitude or in patients who are hyperventilating from severe anxiety (log 2 = 0.3).**

**20.** The pH of blood at body temperature is

(a) 6.4 (b) 7.4
(c) 7.3 (d) 7.2

**21.** The maximum permissible value of $\frac{[H_2CO_3]}{[HCO_3^-]}$ in human blood to just prevent alkalosis is

(a) 0.1 (b) 10
(c) 8.0 (d) 0.125

**22.** Which of the following statement is correct?

(a) One way to treat alkalosis can be to get the patient to breathe more quickly so that the amount of $CO_2$ exhaled increases and the blood becomes deficient in $CO_2$.
(b) One way to treat alkalosis is to get the patient breathe into a bag so that the exhaled $CO_2$ is re-inhaled.
(c) The pH of blood is independent of the concentration of $CO_2$.
(d) Alkalosis cannot be controlled by breathing and respiration.

---

## Comprehension IX

**The first and second dissociation constants of $H_2CO_3$ are $4.0 \times 10^{-6}$ and $2.5 \times 10^{-11}$, respectively. In a particular titration, 50 ml of 1.0 M-$Na_2CO_3$ is taken in a flask and 1.0 M-HCl solution is added drop by drop. Determine the pH of the resulting solution on adding the following volume of HCl solution. Assume thatvolume is additive. (log 2 = 0.3, log 3 = 0.48).**

**23.** 0 ml?

(a) 1.7 (b) 12.3
(c) 3.7 (d) 10.3

**24.** 25 ml?

(a) 10.6 (b) 10.3
(c) 10.9 (d) 3.4

**25.** 50 ml?

(a) 10.6 (b) 5.4
(c) 8.0 (d) 6.0

**26.** 75 ml?

(a) 10.6 (b) 8.6
(c) 8.0 (d) 5.4

**27.** 100 ml?

(a) 2.94 (b) 11.06
(c) 5.4 (d) 5.88

## Comprehension X

**Amino acid glycine ($NH_2$–$CH_2$–COOH) exists as a zwitterion in aqueous solution. The $K_a$ and $K_b$ values of glycine are $1.6 \times 10^{-10}$ ($pK_a = 9.8$) and $2.5 \times 10^{-12}$ ($pK_b = 11.6$), respectively. The $K_a$ and $K_b$ values are for zwitterion of amino acid with the following structure [$\overset{+}{N}H_3 - CH_2 - COO^-$].**

**28.** What is the value of $K_b$ for $-NH_2$ group in glycine?

(a) $4 \times 10^{-3}$
(b) $1.6 \times 10^{-10}$
(c) $6.25 \times 10^{-5}$
(d) $2.5 \times 10^{-12}$

**29.** An aqueous solution of glycine has pH

(a) nearly 7
(b) nearly 7.9
(c) nearly 6.1
(d) nearly 11.5

## Comprehension XI

**To find the $K_{sp}$ of $AgBrO_3$, a student prepared 1 L of a just saturated solution by adding $AgBrO_3$ in water at 27°C. He found that a copper wire left in the solution overnight became covered with silver and $Cu^{2+}$ ions were also formed in the solution. The wire was cleaned, dried and found to weigh 6.35 mg less that its original weight.**

$$Cu(s) + 2Ag^+(aq) \rightleftharpoons Cu^{2+}(aq) + 2Ag(s); \Delta G° = -120 \text{ kJ}$$

$$(R = 8.0 \text{ J/K-mol}; Cu = 63.5)$$

**30.** The number of moles of Cu reacted was

(a) $1.0 \times 10^{-4}$ (b) 1.0
(c) 0.10 (d) $1.0 \times 10^{-3}$

**31.** The concentration of $Ag^+$ in the original saturated solution was

(a) $1.0 \times 10^{-2}$ (b) $1.0 \times 10^{-4}$
(c) $2.0 \times 10^{-4}$ (d) $5.0 \times 10^{-5}$

**32.** The $K_{sp}$ for $AgBrO_3$ is

(a) $1.0 \times 10^{-8}$ (b) $3.0 \times 10^{-9}$
(c) $1.6 \times 10^{-4}$ (d) $4.0 \times 10^{-8}$

## Comprehension XII

**The solubility product of AgCN is $1.0 \times 10^{-16}$ and the formation constant of $Ag(CN)_2^-$ is $1.5 \times 10^{17}$.**

**33.** The solubility of AgCN in 0.02 M-KCN solution, assuming no complex formation, is

(a) $1.0 \times 10^{-8}$ M (b) $5.0 \times 10^{-15}$ M
(c) 0.02 M (d) $5.0 \times 10^{-14}$ M

**34.** The solubility of AgCN in 0.02 M-KCN solution, assuming complex formation, is

(a) 0.3 M (b) 0.02 M
(c) $1.33 \times 10^{-19}$ M (d) $1.87 \times 10^{-2}$ M

**35.** At what [$CN^-$] in the solution, the solubility of AgCN will be minimum?

(a) $3.65 \times 10^{-9}$ M
(b) $1.33 \times 10^{-17}$ M
(c) $2.58 \times 10^{-9}$ M
(d) $5.48 \times 10^{-8}$ M

## Section D (Assertion–Reason)

The following questions consist of two statements. Mark the answer as follows.

(a) If both statements are CORRECT, and **Statement II** is the CORRECT explanation of **Statement I**.

(b) If both statements are CORRECT, and **Statement II** is NOT the CORRECT explanation of **Statement I**.

(c) If **Statement I** is CORRECT, but **Statement II** is INCORRECT.

(d) If **Statement I** is INCORRECT, but **Statement II** is CORRECT.

1. **Statement I:** An aqueous solution having pH 6.8 must be acidic.

   **Statement II:** An aqueous solution having $[H^+] > \sqrt{K_w}$ must be acidic.

2. **Statement I:** The pH of $10^{-7}$ M-NaOH solution is in between 7.0 to 7.3 at 25°C.

   **Statement II:** Due to common ion effect, ionization of water is reduced.

3. **Statement I:** $10^{-4}$ M-HCl solution is more acidic compared to 1 M-HCOOH solution ($K_a = 10^{-3}$ M).

   **Statement II:** HCl is stronger acid as compared to HCOOH.

4. **Statement I:** When an aqueous solution of weak base like BOH is diluted, $[OH^-]$ increases.

   **Statement II:** On dilution, the degree of dissociation of weak electrolytes increases.

5. **Statement I:** The pH of acidic buffer solution always increases on dilution.

   **Statement II:** The pH of any aqueous solution of acid always increases on dilution.

6. **Statement I:** An aqueous solution containing a weak acid (HA) and its conjugate base ($A^-$) may behave as buffer solution.

   **Statement II:** At 25°C, the pH of such buffer solution must be less than 7.0.

7. **Statement I:** Aqueous solutions of NaCl and $CH_3COONH_4$ are neutral (pH = 7.0 at 25°C).

   **Statement II:** Both the salts do not undergo hydrolysis.

8. **Statement I:** In general, phenolphthalein is used as an indicator for the titration of weak acid against strong base.

   **Statement II:** At equivalent point, solution is basic and the pH range of phenolphthalein is about 8.0 to 10.0.

9. **Statement I:** Phenolphthalein does not show any change in colour during the titration of $NaHCO_3$ with HCl.

   **Statement II:** In the presence of phenolphthalein, HCl does not react with $NaHCO_3$.

10. **Statement I:** Moles of $Sr^{2+}$ furnished by sparingly soluble substance $Sr(OH)_2$ decreases due to dilution.

    **Statement II:** Solubility product of $Sr(OH)_2$ is not affected by dilution.

## Section E (Column Match)

1. Match the columns.

| Column I | Column II |
|---|---|
| (A) NaCl | (P) True electrolyte. |
| (B) HCl | (Q) Strong electrolyte in water. |
| (C) $CH_3COOH$ | (R) Potential electrolyte. |
| (D) $CrCl_3 \cdot 3NH_3$ | (S) Weak electrolyte in water. |
| | (T) Non-electrolyte in water. |

2. Match the columns.

| Column I (pH value) | Column II (Solution) |
|---|---|
| (A) 5.0 | (P) pH at which $Mg(OH)_2$ just start precipitating from $2.0 \times 10^{-3}$ M-$Mg^{2+}$ solution ($K_{sp}$ of $Mg(OH)_2 = 2 \times 10^{-6}$ $M^3$). |
| (B) 9.0 | (Q) Maximum pH at which 0.1 mole of $Al(OH)_3$ dissolves completely as $Al^{3+}$ in 1 L solution ($K_{sp}$ of $Al(OH)_3 = 1 \times 10^{-28}$ $M^4$). |
| (C) 12.5 | (R) pH at which 0.1 M-$CH_3COOH$ dissociates 1000/11% ($K_a$ of $CH_3COOH = 1.0 \times 10^{-5}$ M). |
| (D) 6.0 | (S) pH of 0.001 M-HA (weak acid) solution (p$K_a$ of HA = p$K_b$ of $A^-$). |
| | (T) pH of $3 \times 10^{-5}$ M-$CaA_2$ solution ($CaA_2$ dissociates completely in water; $K_a$ of HA = $5 \times 10^{-9}$). |

3. Match the columns.

| Column I | Column II |
|---|---|
| (A) NaCl | (P) Cationic hydrolysis |
| (B) $Na_2CO_3$ | (Q) Anionic hydrolysis |
| (C) $NH_4Cl$ | (R) pH = 7.0 at 25°C |
| (D) $CH_3COONH_4$ | (S) pH < 7.0 at 25°C |
| | (T) pH > 7.0 at 25°C |

4. For a tribasic acid, $H_3A$, $K_{a1} = 10^{-4}$, $K_{a2} = 10^{-8}$ and $K_{a3} = 10^{-12}$. Match the pH (Column II) of the resulting solution (Column I) at 25°C.

| Column I | Column II |
|---|---|
| (A) Equimolar mixture of $H_3A$ and $NaH_2A$. | (P) 12.0 |
| (B) Equimolar mixture of $NaH_2A$ and $Na_2HA$. | (Q) 8.0 |
| (C) Equimolar mixture of $Na_2HA$ and $Na_3A$. | (R) 4.0 |
| (D) Equimolar mixture of $H_3A$ and NaOH. | (S) 6.0 |
| (E) Equimolar mixture of $NaH_2A$ and NaOH. | (T) 10.0 |

5. Match the columns.

| Column I | Column II |
|---|---|
| (A) $\frac{pK_{a,H_2O} + pK_{b,H_2O}}{2}$ at 25°C ($d_{water}$ = 1.0 g/ml) | (P) $\frac{pK_w}{2}$ |
| (B) pH of $CH_3COONH_4$(aq) ($K_{a,CH_3COOH} = K_{b,NH_4OH}$) | (Q) 16 – log 1.8 |
| (C) pH of pure water at 320 K. | (R) $-\log K_{a,H_2O}$ |
| | (S) < 7.0 |

## Section F (Subjective)

### Single-digit Integer Type

1. The ionic product of heavy water ($D_2O$) is $1.0 \times 10^{-16}$ at 7°C. The $P^D$ value of pure heavy water at 7°C is

2. The self-ionization constant for pure formic acid, $K = [HCOOH_2^+][HCOO^-]$ has been estimated as $10^{-6}$ $M^2$ and the density of formic acid is 1.15 g/$cm^3$ at room temperature. If '$x$ %' of formic acid molecules in pure formic acid is converted to formate ions, then the value of '1000$x$' is

3. An aqueous solution contains 10% ammonia by mass and has a density of 0.85 g/ml. If $[H_3O^+]$ in this solution is '$x$M', then the value of '$x \times 10^{12}$' is ($K_a$ for $NH_4^+ = 5.0 \times 10^{-10}$ M)

EXERCISE II

**4.** Boric acid $B(OH)_3$ is used as a mild antiseptic. What is the pH of a 0.0025 M aqueous solution of boric acid? The hydrogen ion arises principally from the following reaction.
$B(OH)_3(aq) + H_2O(l) \rightleftharpoons B(OH)_4^-(aq) + H^+(aq)$; $K = 4.0 \times 10^{-10}$

**5.** The average concentration of $SO_2$ in the atmosphere over a city on a certain day is 10 ppm, when the average temperature is 298 K. Given that the solubility of $SO_2$ in water at 298 K is 1.28 gm litre$^{-1}$ and the p$K_a$ of $H_2SO_3$ is 2.0, estimate the pH of rain on that day.

**6.** An artificial fruit beverage contains 30.0 g of tartaric acid ($H_2C_4H_4O_6$) and 18.8 g of its salt, potassium hydrogen tartrate per litre. What is the pH of the beverage? For tartaric acid, $K_{a1} = 5.0 \times 10^{-4}$, $K_{a2} = 4 \times 10^{-9}$ (log 2 = 0.3).

**7.** When 0.05 mole of HCl is added in 200 ml of a buffer solution, the pH of the solution decreases by 0.05 unit. What is the buffer capacity of the solution?

**8.** A solution of weak acid was titrated with base NaOH. The equivalence point was reached when 36.12 ml of 0.1 M-NaOH have been added. Now, 18.06 ml of 0.1 M-HCl was added to the titrated solution, the pH was found to be 5.0. The p$K_a$ of acid is

**9.** A volume of 30 ml of 0.06 M solution of the protonated form of an amino acid, methionine ($H_2A^+$) is titrated with 0.09 M-NaOH. The pH of the resulting solution after addition of 20 ml of base, is (p$K_{a1}$ = 2.28, p$K_{a2}$ = 9.72)

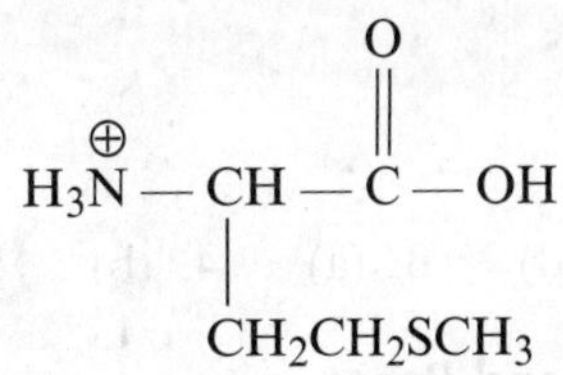

(Methionine cation)

**10.** The pH range of a basic indicator (InOH) is 3.4 – 4.6. Determine the ratio $[In^+]/[InOH]$ above which the solution appears only in the colour of $In^+$? (log 2 = 0.3)

## Four-digit Integer Type

**1.** An aqueous solution of aniline of concentration 0.2 M is prepared. How many milligrams of NaOH should be added in 500 ml of this solution so that anilinium ion concentration in the solution becomes $10^{-8}$ M? $K_b$ of $C_6H_5NH_2 = 4.0 \times 10^{-10}$.

**2.** How many milligram of sodium acetate should be added in 500 ml of 0.2 M acetic acid solution in order to make the $[H^+]$ in the solution $4 \times 10^{-4}$ M. $K_a$ of $CH_3COOH = 1.8 \times 10^{-5}$.

**3.** For $H_2SO_4$, $K_{a1}$ = infinite and $K_{a2} = 1.2 \times 10^{-2}$. The molarity of $H_2SO_4$ solution of pH 2.0 is '$x$M'. The value of $1.7 \times 10^5 x$ is

**4.** The pH of blood stream is maintained by a proper balance of $H_2CO_3$ and $NaHCO_3$ concentrations. What volume of 5 M-$NaHCO_3$ solution should be mixed with a 10 ml sample of blood which is 2 M in $H_2CO_3$ in order to maintain a pH of 7.4? The value of $K_a$ for $H_2CO_3$ in blood is $8.0 \times 10^{-7}$ (log 2 = 0.3).

**5.** The hydronium ion concentration (in millimole per litre) in a solution containing 1.8 g $NaHSO_4$ per 100 ml is ($K_a$ for $HSO_4^-$ is $4.0 \times 10^{-2}$)

**6.** The acid ionization of hydrated aluminium ion is

$Al(H_2O)_6^{3+}(aq) + H_2O(l) \rightleftharpoons Al(H_2O)_5OH^{2+}(aq) + H_3O^+(aq)$; $K_a = 1.0 \times 10^{-5}$

How many milligrams of $AlCl_3$ should be dissolved in sufficient water to get 400 ml of solution of pH 3.0?

**7.** A weak acid-type indicator was found to be 60% dissociated at pH = 9.18. What will be the percentage dissociation at pH = 9.0? (log 2 = 0.3, log 3 = 0.48)

**8.** A saturated solution of iodine in water contains 0.254 g of $I_2$ in 1 L. More than this can dissolve in a KI solution because of the following equilibrium.

$I_2(aq) + I^-(aq) \rightleftharpoons I_3^-(aq)$

A 0.1 M-KI solution actually dissolved 12.7 g of iodine per litre, most of which is converted to $I_3^-$. Assuming that the concentration of $I_2$ in all saturated solutions is the same, calculate the equilibrium constant for the above reaction (Take: $0.96 \times 5.1 = 4.9$, atomic mass of iodine = 127).

**9.** The solubility product ($K_{sp}$) of $Ca(OH)_2$ at 25°C is $3.2 \times 10^{-5}$. A 500 ml of saturated solution of $Ca(OH)_2$ is mixed with equal volume of 1.6 M-NaOH. How much $Ca(OH)_2$ (in milligrams) is precipitated?

**10.** Most ordinary soaps are sodium salt of long chain fatty acids and are soluble in water. Soaps of divalent cations such as $Ca^{2+}$ are only slightly soluble and are often seen in the common soap is calcium palmitate, $Ca[CH_3(CH_2)_{14}COO]_2$. A handbook of chemistry lists the solubility of this soap as 0.0055 g per 100 ml at 25°C. If sufficient sodium soap is used to produce a final concentration of palmitate ion equal to 0.10 M in a water sample having 40 ppm $Ca^{2+}$ initially, then how many milligrams of calcium palmitate would precipitate in a bowl containing 10 L of this water sample?

---

## Answer Keys

**Exercise I**

### Basics

1. (b) 2. (d) 3. (a) 4. (b) 5. (c) 6. (c) 7. (c) 8. (d) 9. (c) 10. (c)

### Strong Acids and Bases

11. (b) 12. (d) 13. (c) 14. (d) 15. (d) 16. (a) 17. (c) 18. (b) 19. (c) 20. (d)

### Weak Acids and Bases

21. (a) 22. (c) 23. (a) 24. (c) 25. (d) 26. (a) 27. (d) 28. (d) 29. (b) 30. (a)
31. (b) 32. (c) 33. (a) 34. (d) 35. (d)

### Polyprotic Acids and Bases

36. (a) 37. (a) 38. (a) 39. (b) 40. (c)

### Buffer Solutions

41. (d) 42. (d) 43. (d) 44. (a) 45. (d) 46. (b) 47. (d) 48. (b) 49. (a) 50. (b)

### Hydrolysis of Salts

51. (a) 52. (d) 53. (c) 54. (b) 55. (d) 56. (c) 57. (d) 58. (b) 59. (c) 60. (b)
61. (c) 62. (b) 63. (a) 64. (c) 65. (b)

### Indicators

66. (c) 67. (c) 68. (a) 69. (c) 70. (b)

### Solubility

71. (a) 72. (d) 73. (b) 74. (c) 75. (b) 76. (b) 77. (b) 78. (b) 79. (c) 80. (b)
81. (d) 82. (a) 83. (a) 84. (d) 85. (b) 86. (c) 87. (b) 88. (b) 89. (a) 90. (b)
91. (d) 92. (d) 93. (b) 94. (c) 95. (d) 96. (d) 97. (d) 98. (a) 99. (a) 100. (c)

### Precipitation

101. (a) 102. (a) 103. (d) 104. (b) 105. (b) 106. (a) 107. (c) 108. (b) 109. (d) 110. (a)
111. (b) 112. (a) 113. (c) 114. (c) 115. (b)

## Answer Keys — Exercise II

### Section A (Only one Correct)

1. (c) 2. (d) 3. (b) 4. (b) 5. (a) 6. (d) 7. (a) 8. (b) 9. (a) 10. (d)
11. (a) 12. (b) 13. (a) 14. (b) 15. (a) 16. (b) 17. (c) 18. (b) 19. (c) 20. (a)
21. (c) 22. (b) 23. (b) 24. (d) 25. (b) 26. (a) 27. (d) 28. (c) 29. (a) 30. (a)
31. (b) 32. (d) 33. (c) 34. (a) 35. (a) 36. (b) 37. (b) 38. (c) 39. (a) 40. (d)
41. (c) 42. (d) 43. (b) 44. (d) 45. (a) 46. (c) 47. (a) 48. (c) 49. (a) 50. (c)
51. (a) 52. (c) 53. (b) 54. (a) 55. (a) 56. (b) 57. (c) 58. (a) 59. (c) 60. (b)
61. (c) 62. (c) 63. (d) 64. (c) 65. (a) 66. (a) 67. (b) 68. (a) 69. (c) 70. (a)
71. (c) 72. (b) 73. (c) 74. (b) 75. (a) 76. (b) 77. (a) 78. (c) 79. (b) 80. (d)

### Section B (One or More than one Correct)

1. (d) 2. (a), (b), (c) 3. (c) 4. (a), (b), (c)
5. (a), (b) 6. (b) 7. (a), (b), (c), (d) 8. (a), (b), (c), (d)
9. (d) 10. (b), (c) 11. (d) 12. (a), (b), (d)
13. (a), (c) 14. (a), (b), (c), (d) 15. (b), (c) 16. (a), (c), (d)
17. (a), (b), (d) 18. (c) 19. (c) 20. (c)

### Section C

**Comprehension I**

1. (c) 2. (c) 3. (b)

**Comprehension II**

4. (c) 5. (b)

**Comprehension III**

6. (a) 7. (b)

**Comprehension IV**

8. (a) 9. (a) 10. (c)

**Comprehension V**

11. (a) 12. (a) 3. (d) 14. (a)

**Comprehension VI**

15. (a) 16. (c)

**Comprehension VII**

17. (d) 18. (a) 19. (c)

**Comprehension VIII**

20. (c) 21. (a) 22. (b)

**Comprehension IX**

23. (b) 24. (a) 25. (c)
26. (d) 27. (a)

**Comprehension X**

28. (c) 29. (c)

**Comprehension XI**

30. (a) 31. (c) 32. (d)

**Comprehension XII**

33. (b) 34. (d) 35. (c)

### Section D (Assertion – Reason)

1. (d) 2. (a) 3. (d) 4. (d) 5. (d) 6. (c) 7. (c) 8. (a) 9. (c) 10. (d)

## Section E (Column Match)

1. A → P, Q; B → Q, R; C → R, S; D → T
2. A → Q, S; B → T; C → P; D → R
3. A → R; B → Q, T; C → P, S; D → P, Q, R
4. A → R; B → Q; C → P; D → S; E → T
5. A → Q, R; B → P; C → P, S

## Section F (Subjective)

### Single-digit Integer Type

1. (8) 2. (4) 3. (1) 4. (6) 5. (2) 6. (3) 7. (5) 8. (5) 9. (6) 10. (4)

### Four-digit Integer Type

1. (0160) 2. (0369) 3. (1100) 4. (0080) 5. (0060)
6. (5340) 7. (0050) 8. (0960) 9. (0740) 10. (5500)

# HINTS AND EXPLANATIONS

## EXERCISE I (JEE MAIN)

### Basics

**1.** Dissociation of water into ions will increase

**2.** Dissociation of water into ions will increase and hence $P^H$ will decrease. Water will remain neutral.

**3.**
$$H_2O \rightleftharpoons H^+ + OH^-$$
$$CM = \frac{1000}{18}M \quad 0 \quad 0$$
Equilibrium $C(1-\alpha)M \quad C\alpha M \quad C\alpha M$
$$K_w = [H^+][OH^-] = C\alpha \cdot C\alpha = \alpha^2 \cdot C^2$$
$$= (1.8\times10^{-9})^2 \times \left(\frac{1000}{18}\right)^2$$
$$= 1.0\times10^{-14}$$

**4.**
$$K_d = \frac{[H^+][OH^-]}{[H_2O]}$$
$$= \frac{\alpha^2 \cdot C}{1-\alpha} \approx \alpha^2 \cdot C = (1.8\times10^{-9})^2 \times \frac{1000}{18}$$
$$= 1.8\times10^{-16}$$

**5.** $[H^+] = \sqrt{K_w} \Rightarrow P^H = -\log(2.5\times10^{-14})^{1/2} = 6.8$

**6.** (a) $[D_2O] = \frac{d}{M} = \frac{1.02\times10^3 \text{ g/L}}{20 \text{ g/mol}} = 51 \text{ M}$

(b) $K_d = K_w/[D_2O] = \frac{5.1\times10^{-15}}{51} = 10^{-15}$

(c) $[D^+] = [OD^-] = \sqrt{K_w}$

$$\therefore \alpha = \frac{\sqrt{K_w}}{C} = \frac{\sqrt{5.1\times10^{-15}}}{51} = \frac{10^{-8}}{\sqrt{51}}$$

(d) Molality $= \frac{1000}{20} = 50$ m

**7.**
$$\ln\frac{K_2}{K_1} = \frac{\Delta H}{R}\left(\frac{1}{T_1} - \frac{1}{T_2}\right)$$
or, $\ln\frac{K_2}{10^{-14}} = \frac{13.7\times10^3}{2}\left(\frac{1}{298} - \frac{1}{323}\right)$

$\Rightarrow K_2 = 5.9\times10^{-14}$

**8.** $[OH^-] = \frac{K_w}{[H_3O^+]} = \frac{10^{-14}}{2\times10^{-4}} = 5\times10^{-11}$ M

**9.** NaCl solution is neutral and at 60°C, $P^H < 7$

**10.** No. of $H_3O^+$ ions $\frac{1\times10^{-12}}{1000}\times6.02\times10^{23} = 6.02\times10^8$

### Strong Acids and Bases

**11.** $P^H = -\log(4\times10^{-4}) = 3.4$

**12.** $P^{OH} = -\log(0.005) = 2.3$

$\therefore P^H = 14 - 2.3 = 11.7$

**13.** $M = \frac{500\times10^{-2}}{1000}\times36.5 = 0.1825$ gm

**14.** As the acid is very dilute, the contribution of water should be taken. Let $H_2O$ dissociates only upto $x$M in presence of HCl.
$$H_2O \rightleftharpoons H^+ + OH^-$$
Equilibrium $(x+10^{-7})$ M $\quad x$M

Now, $[H^+][OH^-] = K_w \Rightarrow (x+10^{-7})\cdot x = 10^{-14}$

$\Rightarrow x = 6.18\times10^{-8}$

$\therefore [H^+] = (x+10^{-7})$ M $= 1.618\times10^{-7}$ M

$\Rightarrow P^H = 6.79$

**15.** $P^H$ is close to 7.0 and hence, contribution of water is considered. let water dissociates only upto '$x$' M is presence of CM-NaOH.
$$H_2O \rightleftharpoons H^+ + OH^-$$
Equilibrium $x$M $\quad (x+C)$ M

Now, $x = 10^{-7.3}$ and $x + C = \frac{10^{-14}}{x} = 10^{-6.7}$

$\therefore C = 10^{-6.7} - 10^{-7.3} = 2 \times 10^{-7} - 5 \times 10^{-8}$
$= 1.5 \times 10^{-7}$ M
$\therefore$ Mass of NaOH $= (20 \times 10^3) \times 1.5 \times 10^{-7} \times 40$
$= 0.12$ gm

**16.** $[H^+]_{final} = \dfrac{V \times 0.1 + V \times 0.1 \times V \times 0.1}{3V}$
$= 0.1\,N = 0.1\,M$
$\therefore P^H = -\log(0.1) = 1.0$

**17.** As all have 0.1 M concentration, $[KOH]_{final} = 0.1$ M
$\therefore P^{OH} = -\log(0.1) = 1.0$ and $P^H = 13$

**18.** $n_{OH^-}$ taken $= \dfrac{100 \times 0.5}{1000} = 0.05$

$n_{H^+}$ taken $= \dfrac{250 \times 0.2}{1000} = 0.05$

Hence, resulting solution is neutral $P^H = -\log(10^{-6})$
$= 6.0$

**19.** $n_{eq}$ of $H^+ = \dfrac{25 \times 0.1}{1000} \times 1 + \dfrac{25 \times 0.1}{1000} \times 2 = 7.5 \times 10^{-3}$

$n_{eq}$ of $OH^- = \dfrac{50 \times 0.1}{1000} = 5 \times 10^{-3}$

$\therefore n_{eq}$ of $H^+$ left $= 2.5 \times 10^{-3}$

$\Rightarrow [H^+] = \dfrac{2.5 \times 10^{-3}}{100} \times 1000 = 0.025$ M

$\therefore P^H = -\log(0.025) = 1.6 \Rightarrow P^{OH} = 14 - 1.6 = 12.4$

**20.** On neglecting the contribution of water, $[H^+] = 10^{-6}$ M
When contribution of water is considered,

$$H_2O \rightleftharpoons H^+ + OH^-$$

Equilibrium $\quad (x + 10^{-6})$ M $\quad x$M

Now, $(x + 10^{-6})\cdot x = 10^{-14} \Rightarrow x = 9.9 \times 10^{-9}$

$\therefore [H^+]_2 = (x + 10^{-6})$ M $= 1.0099 \times 10^{-6}$ M

Now, % error in$[H^+] = \dfrac{[H^+]_2 - [H^+]}{[H^+]_2}[H^+] \times 100\%$
$= 0.98\%$

## Weak Acids and Bases

**21.** Smaller $P^{Ka}$, stronger acid, greater $[H^+]$

**22.**

$$CH_3COOH \rightleftharpoons CH_3COO^- + H^+$$

| | $CH_3COOH$ | $CH_3COO^-$ | $H^+$ |
|---|---|---|---|
| | 1M | 0 | 0.1 M |
| Equilibrium | $1 - x$ | $x$ | $0.1 + x$ |
| | $\simeq$ 1M | | $\simeq$ 0.1M |

Now, $K_a = \dfrac{[CH_3COO^-][H^+]}{[CH_3COOH]} \Rightarrow 2 \times 10^{-5} = \dfrac{x \times 0.1}{1}$

$\Rightarrow x = 2 \times 10^{-4}$ M

**23.** $\dfrac{[H^+]_{HCOOH}}{[H^+]_{CH_3COOH}} = \dfrac{(\sqrt{K_a \cdot C})_{HCOOH}}{(\sqrt{K_a \cdot C})_{CH_3COOH}}$

$= \sqrt{\dfrac{1.77 \times 10^{-4}}{1.75 \times 10^{-5}}} = 3.18$

**24.** $[H^+]_{CH_3COOH} = [H^+]_{CH_2(CN)COOH}$

or, $\sqrt{0.000018 \times \dfrac{n}{V_1}} = \sqrt{0.0036 \times \dfrac{n}{V_2}} \Rightarrow \dfrac{V_1}{V_2} = \dfrac{1}{200}$

**25.** $HA \rightleftharpoons H^+ + A^-$
low $P^H \Rightarrow$ High $[H^+]$
$\Rightarrow$ Equilibrium in backward direction
High $P^H \Rightarrow$ low $[H^+]$
$\Rightarrow$ Equilibrium in forward direction

**26.** Conc. of aspirin $= \dfrac{2 \times 0.36/180}{0.250} = 0.016\,M$

Now, $[H^+] = \sqrt{K_a \cdot C} = \sqrt{4 \times 10^{-9} \times 0.016}$
$= 8 \times 10^{-6}$ M

$\therefore P^H = 5.1$

**27.** $\alpha = \sqrt{\dfrac{K}{C}} = \sqrt{\dfrac{K}{1/V}} = \sqrt{K \cdot V}$

**28.** $P^H_{CH_3COOH} = P^{OH}_{NH_3} = 3.2 \Rightarrow P^H_{NH_3} = 14 - 3.2 = 10.8$

**29.** $(\sqrt{K_a \cdot C})_{HA} = [H^+]_{HCl}$
or, $\sqrt{K_a \times 0.2} = 4 \times 10^{-4} \Rightarrow K_a = 8 \times 10^{-7}$

$\therefore K_b$ of $A^- = \dfrac{10^{-14}}{8 \times 10^{-7}} = 1.25 \times 10^{-8}$

**30.** $P^{K_b} = 10.3 = -\log K_b \Rightarrow K_b = 5\times10^{-11}$

$\therefore K_{a(\text{HF})} = \dfrac{K_w}{K_b(F^-)} = \dfrac{10^{-14}}{5\times10^{-11}} = 2\times10^{-4}$

**31.** $[C_5H_{11}COOH] = \dfrac{11.6}{116} = 0.1\text{ M}$

Now, $[H^+] = \sqrt{K_a\cdot C} \Rightarrow 10^{-3} = \sqrt{K_a\times0.1}$

$\Rightarrow K_a = 10^{-5}$

**32.** $\alpha = \sqrt{\dfrac{K_a}{C}} = \sqrt{\dfrac{0.00024}{0.002}} = 0.346\text{ or }34.6\%$

Hence, α can not be neglected.

$K_a = \dfrac{\alpha^2\cdot C}{1-ga} \Rightarrow 0.00024 = \dfrac{\alpha^2\times0.002}{1-\alpha} \Rightarrow \alpha = 0.29$

Now, $[H^+] = C\alpha = 0.002\times0.29 = 5.8\times10^{-4}$

**33.** $[H^+] = \sqrt{K_a\cdot C} = \sqrt{2\times10^{-12}\times0.02} = 2\times10^{-7}\text{ M}$

As $[H^+]$ is very small, contribution of $H^+$ from water must be considered.

$$HA \rightleftharpoons H^+ + A^-$$

Equilibrium $(0.02-x)\text{M}$ $(x+y)\text{ M}$ $x\text{M}$

$$H_2O \rightleftharpoons H^+ + OH^-$$

Equilibrium $(x+y)\text{ M}$ $y\text{M}$

Now, $K_a = \dfrac{[H^+][A^-]}{[HA]}$

$\Rightarrow 2\times10^{-12} = \dfrac{(x+y)\cdot x}{(0.02-x)} \approx \dfrac{(x+y)\cdot x}{0.02}$

or, $4\times10^{-14} = (x+y)\cdot x$ (1)

and $K_w = [H^+][OH^-] \Rightarrow 10^{-14} = (x+y)\cdot y$ (2)

From (1) + (2), $(x+y) = \sqrt{5\times10^{-14}}\text{ M} = [H^+]$

$\therefore P^H = -\log(5\times10^{-14})^{1/2} = 6.65$

**34.** $\alpha_2 = 2\times\alpha_1 \Rightarrow \sqrt{\dfrac{K_a}{n}\times V_2} = 2\times\sqrt{\dfrac{K_a}{n}\times V_1}$

$\Rightarrow V_2 = 4\times V_1 = 4\times300 = 1200\text{ ml}$

$\therefore V_{\text{water added}} = 1200 - 300 = 900\text{ ml}$

**35.**

$$HCN \rightleftharpoons H^+ + CN^-$$

Equilibrium $(0.2-x)\text{M}$ $(x+y)\text{M}$ $x\text{M}$

$9.6\times10^{-10} = \dfrac{(x+y)\cdot x}{(0.2-x)} \approx \dfrac{y\cdot x}{0.2}$

$$HCOOH \rightleftharpoons H^+ + HCOO^-$$

Equilibrium $(0.1-y)\text{M}$ $(x+y)\text{M}$ $y\text{M}$

$2.56\times10^{-5} = \dfrac{(x+y)\cdot y}{(0.1-y)} \approx \dfrac{y\cdot y}{0.1}$

$\therefore y = 1.6\times10^{-3}$ and $x = 1.2\times10^{-7}$

Option (d) may be answered without solving because solution is acidic.

## Polyprotic Acids and Bases

**36.**

$$Y(OH)_2 \rightleftharpoons Y(OH)^+ + OH^-$$

$4\times10^{-3}\text{ M}$ 0 0

Final 0 $4\times10^{-3}\text{ M}$ $4\times10^{-3}\text{ M}$

$$Y(OH)^+ \rightleftharpoons Y^{2+} + OH^-$$

$4\times10^{-3} - 2\times10^{-3} = 2\times10^{-3}\text{ M}$ $2\times10^{-3}\text{ M}$ $4\times10^{-3} + 2\times10^{-3} = 6\times10^{-3}\text{ M}$

$\therefore P^{OH} = -\log(6\times10^{-3}) = 2.22 \Rightarrow P^H = 11.78$

**37.** All are correct answers, but if only one is to select, then (a), because it have more species.

**38.**

$$H_2CO_3 \rightleftharpoons H^+ + HCO_3^-$$

$\dfrac{0.1}{100}\times1000\text{ M}$ 0 0

Equilibrium $(1-x)\text{ M}$ $(x+y)\text{ M}$ $(x-y)\text{ M}$

$$HCO_3^- \rightleftharpoons H^+ + CO_3^{2-}$$

Equilibrium $(x-y)\text{ M}$ $(x+y)\text{ M}$ $y\text{M}$

Now, $K_{a_1} = \dfrac{[H^+][HCO_3^-]}{[H_2CO_3]}$

$\Rightarrow 4\times10^{-6} = \dfrac{(x+y)\cdot(x-y)}{(1-x)} \approx \dfrac{x\cdot x}{1}$

$\therefore x = 2\times10^{-3}$

and $K_{a_2} = \dfrac{[H^+][CO_3^{2-}]}{[HCO_3^-]}$

$\Rightarrow 5\times10^{-11} = \dfrac{(x+y)\cdot y}{(x-y)} \approx \dfrac{x\cdot y}{x}$

$\therefore y = 5\times10^{-11}$

$\therefore [H^+] = (x+y) \approx x = 2\times10^{-3}$ M

$[HCO_3^-] = (x-y) \approx x = 2\times10^{-3}$ M

$[CO_3^-] = y = 5\times10^{-11}$ M

$$[OH^-] = \frac{K_w}{[H^+]} = \frac{10^{-14}}{2\times10^{-3}} = 5\times10^{-12}\text{ M}$$

**39.** For. $P^H$, 2nd dissociation may be neglected.

$[H^+] = \sqrt{K_{a_1}\cdot C} = \sqrt{9\times10^{-5}\times0.1} = 3\times10^{-3}$ M

$\therefore P^H = -\log(3\times10^{-3}) = 2.52$

**40.** For $P^{OH}$, 2nd protonation may be neglected

$[OH^-] = \sqrt{K_{b_1}\cdot C} = \sqrt{3.6\times10^{-6}\times0.1} = 6\times10^{-4}$ M

$\therefore P^{OH} = -\log(6\times10^{-4}) = 3.22 \Rightarrow P^H = 10.78$

## Buffer Solutions

**41.** $P^{Ka} = 5 \Rightarrow P^H$ range = 4 to 6

**42.** For maximum buffer capacity, $P^H = P^{Ka}$

**43.** $P^H = P^{K_a(NH_4^+)} + \log\dfrac{[NH_3]_0}{[NH_4^+]_0}$

$= 9.26 + \log\dfrac{0.02}{0.01\times2} = 9.26$

**44.** Sodium acetate is basic in nature.

**45.** $CH_3COOH + OH^- \rightleftharpoons CH_3COO^- + H_2O$

$a$ mole $\quad x$ mole $\quad 0$

Final $(a-x)$mole $\quad 0 \quad x$ mole

$P^H = P^{K_a} + \log\dfrac{x}{a-x}$

For $\dfrac{1}{4}th$ neutralization,

$P_1^H = P^{K_a} + \log\dfrac{a/4}{a-a/4} = P^{K_a} + \log\dfrac{1}{3}$

For $\dfrac{3}{4}th$ neutralization,

$P_2^H = P^{K_a} + \log\dfrac{3a/4}{a-3a/4} = P^{K_a} + \log 3$

$\therefore \Delta P^H = P_2^H - P_1^H = 2\log 3$

**46.** $CH_3NH_2 + H^+ \rightleftharpoons CH_3NH_3^+$

| | $CH_3NH_2$ | $H^+$ | $CH_3NH_3^+$ |
|---|---|---|---|
| | 0.1 mole | 0.08 mole | 0 |
| Final | 0.02 mole | 0 | 0.08 mole |

$P^{OH} = P^{K_b} + \log\dfrac{[CH_3NH_3^+]_0}{[CH_3NH_2]_0}$

$= -\log(5\times10^{-4}) + \log\dfrac{0.08}{0.02}$

$\therefore [OH^-] = \dfrac{5\times10^{-4}}{4} \Rightarrow [H^+] = \dfrac{10^{-14}}{[OH^-]} = 8\times10^{-11}$ M

**47.** On adding acid, $P^H$ should decrease slightly. As none of the option is slightly less than 4.0, answer is 4.0

**48.** $[OH^-] = \dfrac{K_6\cdot[\text{Pyridine}]}{[\text{Pyridinium ion}]} = \dfrac{1.5\times10^{-9}\times0.2}{\left(\dfrac{0.15}{500}\times1000\right)}$

$= 1\times10^{-9}$ M

$\therefore P^{OH} = -\log(10^{-9}) = 9.0 \Rightarrow P^H = 5.0$

**49.** $P^H = P^{K_a} + \log\dfrac{[CN^-]_0}{[HCN]_0}$

$= -\log(2.5\times10^{-10}) + \log\dfrac{80\times0.4/100}{20\times0.8/100}$

$= 9.9$

**50.** $\underset{\text{(Imidazole)}}{B} + H^+ \rightleftharpoons BH^+$

$\dfrac{V\times0.02}{120}$ M $\quad \dfrac{(120-V)\times0.02}{120}$ M $\quad 0$

Final $\dfrac{(2V-120)\times0.02}{120}$ M $\quad 0 \quad \dfrac{(120-V)\times0.02}{120}$ M

$P^{OH} = P^{K_b} + \log\dfrac{[BH^+]_0}{[B]_0} \Rightarrow 7 = 7 + \log\dfrac{120-V}{2V-120}$

$\Rightarrow V = 80$

### Hydrolysis of Salts

**51.** Stronger the acid, smaller is $P^H$.

**52.** Potassium propionate is the salt of strong base and weak acid and hence.

$$P^H = 7 + \frac{1}{2}(P^{K_a} + \log C)$$

or, $8 = 7 + \frac{1}{2}(P^{k_a} + \log 0.001)$

$\Rightarrow P^{K_a} = 5 \Rightarrow K_a = 10^{-5}$

**53.** The order of acidic strength is HCl > $H_2CO_3$ > HCN

**54.** NaCl does not hydrolysis but $CH_3COONH_4$ undergoes cationic as well as anionic hydrolyisis.

**55.** Point A: Buffer $H_2A + HA^-$; $P^H = P^{K_a} + \log\frac{[HA^-]_0}{[H_2A]_0}$

Point B: 1st equivalent point $HA^-$;

$$P^H = \frac{1}{2}(P^{K_{a1}} + P^{K_{a2}})$$

Point C: Buffer $HA^- + A^{2-}$; $P^H = P^{K_{a2}} + \log\frac{[A^{2-}]_0}{[HA^-]_0}$

Point D: 2nd equivalent point, $A^{2-}$;

$$P^H = 7 + \frac{1}{2}(P^{K_{a1}} + \log C)$$

**56.** $B^+ + H_2O \rightleftharpoons BOH + H^+$; $\Delta H$ = positive

**57.** Point B : Buffer HA + $A^-$

**58.** NaCl = Neutral, $NH_4Cl$ = Acidic, NaCN = Basic, HCl = Acid

**59.** $CH_3COONa$ = Basic, $CH_3COOH$ = Weak acid, $CH_3COONH_4$ = Neutral, NaOH = strong base, HCl = strong acid

**60.** $h = \sqrt{\frac{K_h}{C}} = \sqrt{\frac{K_w}{K_a \cdot C}} = \sqrt{\frac{10^{-14}}{10^{-5} \times 0.1}} = 10^{-4}$ or 0.01%

**61.** $P^H = 7 + \frac{1}{2}(P^{K_a} - P^{K_b}) = 7 + \frac{1}{2}(3.8 - 4.8) = 6.5$

**62.** $P^{K_b}$ of $CN^- = 4.70 \Rightarrow P^{K_a}$ of HCN = 9.30

Now, $P^H = 7 + \frac{1}{2}(P^{K_a} + \log C)$

$= 7 + \frac{1}{2}(9.30 + \log 0.5) = 11.5$

**63.** Equal volumes of both will consume and hence,

$[CH_3COONa] = \frac{0.01}{2} = 0.005$

Now, $P^H = 7 + \frac{1}{2}(P^{K_a} + \log C)$

$= 7 + \frac{1}{2}(4.7 + \log 0.005) = 8.2$

**64.** $Zn^{2+} + H_2O \rightleftharpoons Zn(OH)^+ + H^+$;

Equilibrium $(0.001 - x)$M $\quad x$M $\quad x$M

$K_a = 2 \times 10^{-10}$

Now, $2 \times 10^{-10} = \frac{x \cdot x}{(0.001 - x)} \approx \frac{x^2}{0.001}$

$\Rightarrow x = \sqrt{2 \times 10^{-13}}$

$\therefore P^H = -\log x = -\log(2 \times 10^{-13})^{1/2} = 6.35$

**65.** Ammonium chloride is acidic in nature.

### Indicators

**66.** $P^{K_{In}} = -\log K_{In} = -\log(5 \times 10^{-6}) = 5.3$

Now, $P^H = P^{K_{In}} + \log\frac{[In^-]}{[HIn]}$

Only for acidic form, $P^H \le \left(5.3 + \log\frac{1}{20}\right) = 4.0$

Only for basic form, $P^H \ge (5.3 + \log 40) = 6.9$

**67.** $P^H = P^{K_{In}} + \log\frac{[In^-]}{[HIn]}$

$\Rightarrow 2 = P^{k_{In}} + \log\frac{C}{C} \Rightarrow P^{K_{In}} = 2$

Now, $P^H = P^{K_{In}} + \log\frac{[In^-]}{[HIn]}$

$\Rightarrow -\log(4 \times 10^{-3}) = 2 + \log\frac{100 - x}{x}$

$\therefore x = 71.4$

68. Yellow colour with methyl red means $P^H > 6.2$

Yellow colour with phenol red means $P^H < 6.4$

69. For complete colour change, $P^H$ should change from, $P^H \le \left(P^{k_{In}} + \log\frac{1}{10}\right)$ to $P^H \ge \left(P^{k_{In}} + \log 10\right)$ and hence, minimum $P^H$ change should be of 2.0 units.

70.

| | $X^-$ | + | $H^+$ | $\rightleftharpoons$ | HX |
|---|---|---|---|---|---|
| | $\frac{10\times 0.1}{20}$ M | | $\frac{10\times 0.1}{20}$ M | | 0 |
| Final | 0 | | 0 | | 0.05 M |

(10 ml HCl solution is needed for equivalent point)

At equivalent point $[H^+] = \sqrt{K_a \times C}$

$= \sqrt{10^{-6} \times 0.05}$

$\therefore P^H = -\log(5\times 10^{-8})^{1/2} = 3.65$

**Solubility**

71. $X_aY_b \rightleftharpoons aX^{b+} + bY^{a-}$

$\quad a$ mole $\quad b$ mole

$\therefore$ Moles of $Y^{a-} = \frac{b}{a} \times$ moles of $X^{b+}$

72. $A_3B_2(s) \rightleftharpoons 3A^{2+} + 2B^{3-}$ (Solubility $= \frac{x}{M}$ mol/l)

$\quad \frac{3x}{M} \quad \frac{3x}{M}$

$$K_{sp} = [A^{2+}]^3[B^{3-}]^2 = \left(\frac{3x}{M}\right)^3 \cdot \left(\frac{2x}{M}\right)^2 = \frac{108x^5}{M^5}$$

$$\therefore \frac{[B^{3-}]}{K_{sp}} = \frac{2x/M}{108x^5/M^5} = \frac{M^4}{54x^4}$$

73. $[OH^-] = 0.1\times\frac{50}{100} = 0.05$ M

$$\therefore [Zn^{2+}] = \frac{K_{sp}}{[OH^-]^2} = \frac{10^{-14}}{(0.05)^2} = 4\times 10^{-12} \text{ M}$$

74. $Ag^+$ form complex with $NH_3$

75. $Fe(OH)_3(s) \rightleftharpoons Fe^{3+}(aq) + 3OH^-(aq)$;

$K_{sp} = [Fe^{3+}][OH^-]^3$

Now, for given reaction,

$Fe(OH)_3(s) + 3H_3O^+ \rightleftharpoons Fe^{3+} + 6H_2O$

$$K_{eq} = \frac{[Fe^{3+}]}{[H_3O^+]^3} \times \frac{[OH^-]^3}{[OH^-]^3} = \frac{K_{sp}}{K_w^3}$$

$$= \frac{4\times 10^{-38}}{(10^{-14})^3} = 4\times 10^4$$

76. (a) $AgCl(s) \rightleftharpoons Ag^+ + Cl^-$; $K_1 = 10^{-10}$

(b) $AgCl(s) + Br^- \rightleftharpoons AgBr(s) + Cl^-$; $K_2 = 200$

(c) $2AgBr(s) + S^{2-} \rightleftharpoons Ag_2S(s) + 2Br^-$; $K_3 = 1.6\times 10^{24}$

From 2(a) – 2(b) – 2(c); $Ag_2S(s) \rightleftharpoons 2Ag^+ + S^{2-}$;

$$K = \frac{(10^{-10})^2}{(200)^2 \times 1.6\times 10^{24}} = 1.5625\times 10^{-49}$$

77. Adding both reactions, $K_{sp} = 1.3\times 10^{-34} \times 38.5 \approx 5\times 10^{-33}$

78. $K_{formation} = K_1 \cdot K_2 \cdot K_3 = 125\times 20\times 1 = 2500$

$$\therefore K_{dissociation} = \frac{1}{K_{formation}} = 4\times 10^{-4}$$

79. $BaF_2(s) \rightleftharpoons Ba^{2+} + 2F^-$

$\quad (S + C)$ M $\quad 2S$ M

where S = solubility of $BaF_2$ in CM-$Be(NO_3)_2$ solution

Hence, Solubility, $S = \frac{[F^-]}{2}$

80. $CaF_2(s) \rightleftharpoons Ca^{2+} + 2F^-$

$\quad S_1M \quad 2S_1M$

Now, $K_{sp} = S_1 \cdot (2S_1)^2$

$$\Rightarrow S_1 = \left(\frac{K_{sp}}{4}\right)^{1/3} = \left(\frac{3.2\times 10^{-11}}{4}\right)^{1/3} = 2\times 10^{-4} \text{ M}$$

Now, in KF solution,

$$S_2 = \frac{K_{sp}}{[F^-]^2} = \frac{3.2\times 10^{-11}}{(4\times 10^{-3})^2} = 2\times 10^{-6} \text{ M}$$

$$\text{Now, } \frac{S_2}{S_1} = \frac{2\times 10^{-6}}{2\times 10^{-4}} = \frac{1}{100}$$

**81.** $A_2X_3(s) \rightleftharpoons 2A^{3+} + 3X^{2-}$

$2y$M $3y$M

$K_{sp} = [A^{3+}]^2 [X^{2-}]^3 = (2y)^2 \cdot (3y)^3 = 108y^5$

**82.** $A_pB_q(s) \rightleftharpoons pA^{2+} + qB^{P-}$

$p \cdot s$ M $q \cdot s$ M

Solubility product, $L_s = (p \cdot s)^p \cdot (q \cdot s)^q$

$= p^p \cdot q^q \cdot s^{(p+q)}$

**83.** $K_f = K_{f_1} \cdot K_{f_2} = 1.6\times10^3 \times 6.8\times10^3 = 1.08\times10^7$

**84.** For MX : Solubility $= \sqrt{K_{sp}} = \sqrt{4\times10^{-8}}$

$= 2\times10^{-4}$M

For $MX_2$ : Solubility $= \left(\dfrac{K_{sp}}{4}\right)^{1/3} = \left(\dfrac{3.2\times10^{-14}}{4}\right)^{1/3}$

$= 2\times10^{-5}$M

For $M_3X$ : Solubility $= \left(\dfrac{K_{sp}}{27}\right)^{1/4} = \left(\dfrac{2.7\times10^{-15}}{27}\right)^{1/3}$

$= 1\times10^{-4}$M

∴ Order of solubility = $MX_2 < M_3X < MX$

**85.** $S = \dfrac{0.001435}{143.5} = 10^{-5}$M

∴ Solubility product, $K_{sp} = S^2 = 10^{-10}$

**86.** Solubility, $S = \dfrac{0.0744/372}{100/1000} = 2\times10^{-3}$ M

$Li_3Na_3(AlF_6)_2\,(s) \rightleftharpoons 3Li^+ + 3Na^+ + 2AlF_6^{3-}$

3S 3S 2S

$K_{sp} = [Li+]^3\,[Na^+]^3\,[AlF_6^{3-}]^2 = (3S)^3 \cdot (3S)^3 \cdot (2S)^2$

$= 7.46496 \times 10^{-19}$

**87.** For $CaF_2$ : $K_{sp} = 4S^3$

$\Rightarrow S = \left(\dfrac{K_{sp}}{4}\right)^{1/3} = \left(\dfrac{1.08\times10^{-10}}{4}\right)^{1/3}$

$= 3 \times 10^{-4}$ M

∴ Mass of $CaF_2$ needed $= \left(\dfrac{500\times3\times10^{-4}}{1000}\right)\times 78$

$= 1.17 \times 10^{-2}$ gm

**88.** $Mg(OH)_2\,(s) \rightleftharpoons Mg^{2+} + 2OH^-$

SM 2SM

Now, $K_{sp} = 4S^3 \Rightarrow S = \left(\dfrac{9\times10^{-12}}{4}\right)^{1/3}$ M

Now, $[OH^-] = 2S = 2\times\left(\dfrac{9\times10^{-12}}{4}\right)^{1/3}$

$= (18\times10^{-12})^{1/3}$M

∴ $P^{OH} = -\log(18 \times 10^{-12})^{1/3} = 3.58 \Rightarrow P^H = 10.42$

**89.** $Zn(OH)_2(s) + 4NH_3 \rightleftharpoons Zn(NH_3)_4^{2+} + 2OH^-$

1 M 0 0

Equilibrium (1 – 4S)M SM 2SM

$K_{eq} = \dfrac{[Zn[NH_3]_4^{2+}[OH^-]^2}{[NH_3]^4} \times \dfrac{[Zn^{2+}]}{[Zn^{2+}]}$

$= K_{sp} \cdot K_{stab} = 1.6 \times 10^{-17} \times 1.6 \times 10^{10} = 2.56 \times 10^{-7}$

Now, $2.56\times10^{-7} = \dfrac{S\times(2S)^2}{(1-4S)^4} \approx \dfrac{4S^3}{1}$

$\Rightarrow S = 4 \times 10^{-3}$M

**90.** $AgCl(s) + Cl^- \rightleftharpoons AgCl_2^-$

aM 0

Final $\left(a - \dfrac{0.01}{100}\right)$M $10^{-4}$ M

Now, $K_{eq} = \dfrac{[AgCl_2^-]}{[Cl^-]} \times \dfrac{[Ag^+][Cl^-]}{[Ag^+][Cl^-]} = K_{sp} \cdot K_f$

$= 2\times10^{-10} \times 2.5\times10^5 = 5\times10^{-5}$

∴ $5\times10^{-5} = \dfrac{10^{-4}}{(a-10^{-4})} \Rightarrow a \approx 2$M

Hence, the minimum mass of NaCl added = $(100 \times 2) \times 58.5 = 11700$ gm

**91.** $Ag_2CO_3(s) + C_2O_4^{2-} \rightleftharpoons Ag_2C_2O_4(s) + CO_3^{2-}$

Initial $\dfrac{0.15}{500}\times1000$ 0

$= 0.3$ M

Final $0.3 - x$ $x = \dfrac{0.035}{500}\times1000$

$= 0.23$ M $= 0.07$ M

Now, $K_{eq} = \frac{[CO_3^{2-}]}{[C_2O_4^{2-}]} \times \frac{[Ag^+]^2}{[Ag^+]^2} = \frac{K_{sp}(Ag_2CO_3)}{K_{sp}(Ag_2C_2O_4)}$

or, $\frac{0.07}{0.23} = \frac{K_{sp}(Ag_2CO_3)}{2.3\times10^{-11}}$

$\Rightarrow K_{sp}(Ag_2CO_3) = 7\times10^{-12}$

**92.** $Ag^+ + 2CN^- \rightleftharpoons Ag(CN)_2^-$;

$K = \frac{1}{4\times10^{-19}} = 2.5\times10^{18}$

| | $Ag^+$ | $2CN^-$ | $Ag(CN)_2^-$ |
|---|---|---|---|
| | 0.03 M | 0.1 M | 0 |
| 100% | 0 | (0.1 – 0.06)M = 0.04 M | 0.03 M |
| Eqn. | $x$M | $(0.04 + 2x)$M ≈ 0.04 M | $(0.03 - x)$M ≈ 0.03 M |

Now, $2.5\times10^{18} = \frac{0.03}{x\times(0.04)^2}$

$\Rightarrow [Ag^+] = x = 7.5\times10^{-18}$ M

**93.** $2AgCl(s) + CO_3^{2-} \rightleftharpoons Ag_2CO_3(s) + 2Cl^-$

| | $CO_3^{2-}$ | $Cl^-$ |
|---|---|---|
| | 2.0 M | 0 |
| Equ. | $(2 - x)$M ≈ 2M | $2x = \frac{0.00355}{35.5}$ ∴ $x = 5\times10^{-5}$ |

Now, $K_{eq} = \frac{[Cl^-]^2}{[CO_3^{2-}]} \times \frac{[Ag^+]^2}{[Ag^+]^2} = \frac{K_{sp}^2(AgCl)}{K_{sp}(Ag_2CO_3)}$

or, $\frac{(5\times10^{-5})^2}{2} = \frac{K_{sp}^2(AgCl)}{8\times10^{-12}}$

$\Rightarrow K_{sp}(AgCl) = 1\times10^{-10}$

**94.** $AgCl(s) + 2NH_3(aq) \rightleftharpoons Ag(NH_3)_2^+(aq) + Cl^-(aq)$

| | $NH_3$ | $Ag(NH_3)_2^+$ | $Cl^-$ |
|---|---|---|---|
| | 1.0 M | 0 | 0 |
| Eqn. | $(1 - 2x)$M | $x$M | $x$M |

Now, $K_{eq} = \frac{[Ag(NH_3)_2^+][Cl^-]}{[NH_3]^2} \times \frac{[Ag^+]}{[Ag^+]} = \frac{K_{sp}}{K_c}$

$= \frac{1.8\times10^{-10}}{7.2\times10^{-8}} = \frac{1}{400}$

Now, $\frac{x\cdot x}{(1-2x)^2} = \frac{1}{400} \Rightarrow x = \frac{1}{22} = 0.045$M

**95.** For saturated solution of $Pb(OH)_2$ in water, $K_{sp} = 4.\ S^3$

$= 4\times(6\times10^{-6})^3 = 8.64\times10^{-16}$

Now, for the given buffer, $[OH^-] = 10^{-6}$ M

∴ Solubility, $S = [Pb^{2+}] = \frac{K_{sp}}{[OH^-]^2} = \frac{8.64\times10^{-16}}{(10^{-6})^2}$

$= 8.64\times10^{-4}$ M

**96.** $Ag(NH_3)_2^+ \rightleftharpoons Ag^+ + 2NH_3$; $K_{eq} = 6.8\times10^{-8}$

| | $Ag(NH_3)_2^+$ | $Ag^+$ | $2NH_3$ |
|---|---|---|---|
| | 0.2 M | 0 | 0 |
| Eqn | $0.2 - x$ = 0.2 M | $x$M | $2x$M |

Now, $6.8\times10^{-8} = \frac{x\times(2x)^2}{0.2}$

$\Rightarrow x = 1.5\times10^{-3}$ M $= [Ag^+]$

**97.** $Cu^{2+} + 4NH_3 \rightleftharpoons Cu(NH_3)_4^{2+}$;

$K_{eq} = 1.25\times10^{12}$

| | $Cu^{2+}$ | $4NH_3$ | $Cu(NH_3)_4^{2+}$ |
|---|---|---|---|
| | 0.0125 M | 0.25 M | 0 |
| 100% | 0 | 0.25 – 0.05 = 0.20 M | 0.0125 M |
| Equ. | $x$M | $0.2 + 4x$ ≈ 0.2 M | $0.0125 - x$ ≈ 0.0125 M |

Now, $1.25\times10^{12} = \frac{0.0125}{x\times(0.2)^4} \Rightarrow x = 6.25\times10^{-12}$

**98.** $AgSCN(s) \rightleftharpoons Ag^+ + SCN^-$

$(x + y)$M $\quad x$M

$AgBr(s) \rightleftharpoons Ag^+ + Br^-$

$(x + y)$M $\quad y$M

Now, $(x + y)\cdot x = 1\times10^{-12}$ (1) $\quad x = 9.09\times10^{-7}$

$(x + y)\cdot y = 2.1\times10^{-13}$ (2) $\quad y = 1.909\times10^{-7}$

**99.** $AgCN(s) \rightleftharpoons Ag^+ + CN^-$; $K_{sp} = 1.2\times10^{-16}$

$CN^- + H^+ \rightleftharpoons HCN$; $k_{eq} = \frac{1}{K_a} = \frac{1}{4.8\times10^{-10}}$

Adding both reactions,

$AgCN(s) + H^+ \rightleftharpoons Ag^+ + HCN$;

$K_{eq} = \frac{K_{sp}}{K_a} = \frac{1}{4\times10^6}$

$10^{-3} \quad$ SM $\quad$ SM

Now, $\frac{1}{4\times10^6} = \frac{S\cdot S}{10^{-3}} \Rightarrow 1.58\times10^{-5}$ M

**100.** $PbCl_2(s) \rightleftharpoons Pb^{2+} + 2Cl^-$

$S_1M \quad 2S_1M$

When it is 80% ionized, then

$PbCl_2(s) \rightleftharpoons Pb^{2+} + 2Cl^-$

$0.8\ S_2\ M \quad 1.6\ S_2$

Now, $K_{sp} = [Pb^{2+}][Cl^-] = \text{constant}$

or, $S_1 \times (2S_1)^2 = (0.8S_2) \times (1.6S_2)^2 \Rightarrow S_1 = 0.512S_2$

$\Rightarrow S_2 > S_1$

**Precipitation**

**101.** $[OH^-]_{min}$ to start ppt. of $Mg(OH)_2 = \sqrt{\dfrac{K_{sp}}{[Mg^{2+}]}}$

$= \sqrt{\dfrac{4\times10^{-11}}{x}}$

For $Cd(OH)_2$, $[OH^-]_{min} = \sqrt{\dfrac{8\times10^{-6}}{x}}$

For $Al(OH)_3$, $[OH^-]_{min} = \left(\dfrac{8.5\times10^{-23}}{x}\right)^{1/3}$

For $Zn(OH)_2$, $[OH^-]_{min} = \sqrt{\left(\dfrac{1.8\times10^{-14}}{x}\right)}$

Considering significant concentration of metal ion ($x$ is not extremely small), $[OH^-]$ needed is minimum for $Al^{3+}$ and hence, it will precipitate first.

**102.** $[Ag^+]_{min}$ for the start of precipitation of

$AgBr = \dfrac{K_{sp}(AgBr)}{[Br^-]} = \dfrac{5\times10^{-13}}{0.1} = 5\times10^{-12}\ M$

For $Cl^-$ : $[Ag^+]_{min} = \dfrac{1.8\times10^{-10}}{0.1} = 1.8\times10^{-9}\ M$

For $CO_3^{2-}$ : $[Ag^+]_{min} = \sqrt{\dfrac{8.1\times10^{-12}}{0.1}} = 9\times10^{-6}\ M$

For $AsO_4^{3-}$ : $[Ag^+]_{min} = \left(\dfrac{1\times10^{-22}}{0.1}\right)^{1/3} = 1\times10^{-7}\ M$

As [Ag+] needed is minimum for AgBr, it will precipitate first.

**103.** $Ag_2CrO_4(s) \rightleftharpoons 2Ag^+ + CrO_4^{2-}$

To prevent ppt (backward reaction),

$Q \le K_{sp}$

or, $[Ag^+]^2 \times 3\times10^{-4} \le 1.2\times10^{-11}$

$\therefore [Ag^+]_{max} = 2\times10^{-4}\ M$

**104.** $Ag^+ + 2NH_3 \rightleftharpoons Ag(NH_3)_2^+$; $K_{stab} = 1.6\times10^7$

| | $Ag^+$ | $2NH_3$ | $Ag(NH_3)_2^+$ |
|---|---|---|---|
| | 0.1 M | 1M | 0 |
| 100% | 0 | $(1-0.2)$ M $= 0.8$ M | 0.1 M |
| Equ. | $x$M | $0.8 + 2x \approx 0.8$ M | $0.1 - x \approx 0.1$ M |

Now, $1.6\times10^7 = \dfrac{0.1}{x\times(0.8)^2} \Rightarrow x = 9.76\times10^{-9}\ M$

Now, on adding NaCl, $Q = [Ag^+][Cl^-]$

$= 9.76\times10^{-9}\times0.01$

$= 9.76\times10^{-11} < K_{sp}$

Hence, no precipitation of AgCl.

**105.** $Ag^+ + Cl^- \rightleftharpoons AgCl(s)$; $K_{eq} = \dfrac{1}{K_{sp}} = \dfrac{1}{2\times10^{-10}}$

| | $Ag^+$ | $Cl^-$ |
|---|---|---|
| | $\dfrac{500\times2.5\times10^{-5}}{2500} = 5\times10^{-6}$ M | $\dfrac{2000\times5\times10^{-2}}{2500} = 4\times10^{-2}$ M |
| 100% | 0 | $(4\times10^{-2} - 5\times10^{-6}) \approx 4\times10^{-2}$ M |
| Equ. | $x$M | $(4\times10^{-2} + x) \approx 4\times10^{-2}$ M |

Now, $\dfrac{1}{2\times10^{-10}} = \dfrac{1}{x\times4\times10^{-10}} \Rightarrow x = 5\times10^{-9}$

Hence, $Ag^+$ ppt $= (5\times10^{-6} - 5\times10^{-9}) \approx 5\times10^{-6}$ M

$\therefore$ Mass of AgCl ppt

$= \left(\dfrac{2500\times5\times10^{-6}}{1000}\right)\times143.5 = 1.79\times10^{-2}$ gm

**106.** $PbI_2(s) \rightleftharpoons Pb^{2+} + 2I^-$

To prevent ppt, $Q \le K_{sp}$

or, $0.005\times[I^-]^2 \le 7.2\times10^{-9}$

$\therefore [I^-]_{max} = 1.2\times10^{-3}$ M

Hence, maximum mass of NaI

$$= \left(\frac{500 \times 1.2 \times 10^{-3}}{1000}\right) \times 150 = 0.09 \text{ gm}$$

**107.** $AgBr(s) \rightleftharpoons Ag^+ + Br^-$

For precipitation, $Q > K_{sp}$

or, $0.0004 \times [Br^-] > 4 \times 10^{-13}$

$\therefore [Br^-] > 10^{-9}$ M

Hence, minimum mass of NaBr needed

$$= \left(\frac{200 \times 10^{-9}}{1000}\right) \times 103 = 2.06 \times 10^{-8} \text{ gm}$$

**108.** $CaSO_4(s) \rightleftharpoons Ca^{2+} + SO_4^{2-}$

To start ppt, $Q > K_{sp}$

or, $0.005 \times [SO_4^{2-}] > 2.4 \times 10^{-5}$

$\therefore [SO_4^{2-}] > 4.8 \times 10^{-3}$ M

$$\therefore [Al_2(SO_4)_3]_{min} = \frac{4.8 \times 10^{-3}}{3} = 1.6 \times 10^{-3} \text{M}$$

**109.** $Pb^{2+} + SO_4^{2-} \rightleftharpoons PbSO_4(s)$ ;

$$K_{eq} = \frac{1}{K_{sp}} = \frac{1}{1.6 \times 10^{-7}}$$

$$\frac{8.32 \times 10^{-3}/200}{100/1000} \qquad \frac{10^{-4}}{100/1000}$$

$= 4 \times 10^{-4}$ $\qquad = 10^{-3}$

100% $\quad 0 \quad (10^{-3} - 4 \times 10^{-4})$ $= 6 \times 10^{-4}$ M

Equ. $\quad x$M $\quad (6 \times 10^{-4} + x)$M

As $[Pb^{2+}][SO_4^{2-}] = K_{sp} \Rightarrow x \cdot (6 \times 10^{-4} + x) = 1.6 \times 10^{-7}$

$\Rightarrow x = 2 \times 10^{-4}$ M

$\therefore$ Mass of lead unprecipitated

$$= \left(\frac{100 \times 2 \times 10^{-4}}{1000}\right) \times 208$$

$= 4.16 \times 10^{-3}$ gm

**110.** $M^{2+} + H_2S \rightleftharpoons MS(s) + 2H^+$

For ppt. of MS(s), $Q < k_{eq}$

$$\text{or, } \frac{[H^+]}{[M^{2+}][H_2S]} < \frac{K_{a_1} \cdot K_{a_2}}{K_{sp}}$$

$$\text{or, } \frac{[H^+]^2}{0.04 \times 0.1} < \frac{10^{-7} \times 1.5 \times 10^{-13}}{6 \times 10^{-21}}$$

$\therefore [H^+]_{max} = 0.1 \text{ M} \Rightarrow P^H_{min} = 1.0$

**111.** $Cd^{2+} + H_2S \rightleftharpoons CdS(s) + 2H^+$

$0.1 \times 10^{-3}$ mole $\qquad \frac{10 \times 0.08}{1000}$ $= 0.8 \times 10^{-3}$ mole

Final $\quad 0 \quad 0.8 \times 10^{-3} + 0.2 \times 10^{-3}$ $= 10^{-3}$ mole

$$\therefore [H^+]_{final} = \frac{10^{-3}}{100} \times 1000 = 0.01 \text{ M} \Rightarrow P^H = 2.0$$

**112.** $[I^-]_{min}$ for ppt of AgI

$$= \frac{K_{sp}}{[Ag^+]} = \frac{8.5 \times 10^{-17}}{0.1} = 8.5 \times 10^{-16} \text{ M}$$

$[I^-]_{min}$ for ppt of $Hg_2I_2$

$$= \sqrt{\frac{K_{sp}}{[Hg_2^{2+}]}} = \sqrt{\frac{2.5 \times 10^{-26}}{0.1}} = 5 \times 10^{-13} \text{M}$$

Hence, AgI will precipitate first. The $[I^-]$ to start precipitation of other metal ion $= 5 \times 10^{-13}$ M

Now, $[Ag^+]_{left}$ when $Hg_2I_2$ start precipitating

$$= \frac{K_{sp}}{[I^-]} = \frac{8.5 \times 10^{-17}}{5 \times 10^{-13}} = 1.7 \times 10^{-4} \text{ M}$$

Hence, % of $Ag^+$ precipitate

$$= \frac{0.1 - 1.7 \times 10^{-4}}{0.1} \times 100 = 99.83\%$$

**113.** $K_{sp}$ of $CaCO_3 = \left(\frac{7 \times 10^{-3}}{100}\right)^2 = 4.9 \times 10^{-9}$

Let initial $[Ca^{2+}] = [Ba^{2+}] = x$M. From question final $[Ba^{2+}] = x \times \frac{10}{100} = 0.1xM$. But $[CO_3^{2-}]$ is same for both. Hence,

$$\frac{K_{sp}(BaCO_3)}{K_{sp}(CaCO_3)} = \frac{[Ba^{2+}][CO_3^{2-}]}{[Ca^{2+}][CO_3^{2-}]}$$

$$\Rightarrow \frac{K_{sp}(BaCO_3)}{4.9 \times 10^{-9}} = \frac{0.1x}{x} \therefore K_{sp}(BaCO_3) = 4.9 \times 10^{-10}$$

**114.**

$$[OH^-] = K_b \cdot \frac{[NH_4OH]}{[NH_4^+]} = 2 \times 10^{-5} \times \frac{0.05}{0.25} = 4 \times 10^{-6} \text{ M}$$

$$\therefore [Mg^{2+}] = \frac{K_{sp}}{[OH^-]^2} = \frac{8 \times 10^{-12}}{(4 \times 10^{-6})^2} = 0.5 \text{ M}$$

**115.** For ppt, $Q > K_{eq}$

For the metal sulphides,

$$Q = \frac{[M^{2+}][H_2S]}{[H^+]^2} = \frac{0.01 \times 0.1}{(1.0)^2} = 10^{-3} > K_{eq} \text{ for PbS only}$$

Hence, only Pbs will precipitate.

# EXERCISE II (JEE ADVANCED)

## Section A (Only one Correct)

**1.** $[NH_2^-] = \sqrt{10^{-30}} = 10^{-15}\,M$

$\therefore$ Number of $NH_2^-$ ions per ml

$= \dfrac{10^{-15}\times 1}{1000} \times (6\times10^{23}) = 6\times10^{5}$

**2.** On increasing temperature, the dissociation of water will increase. It will result increase in $[H^+]$ and as well as in $[OH^-]$ and hence, decreases in $P^H$ and as well as $P^{OH}$.

**3.** For maximum dissociation, $[H^+] = [OH^-]$.

**4.** $[H^+] = 10^{-2}\,M \Rightarrow n_{H^+} = \dfrac{200\times10^{-2}}{1000} = 2\times10^{-3}$

$[OH^-] = 10^{-2}\,M \Rightarrow n_{OH^-} = \dfrac{300\times10^{-2}}{1000} = 3\times10^{-3}$

$\therefore$ Moles of excess $OH^-$ remained $= 1\times10^{-3}$

$[OH^-] = \dfrac{1\times10^{-3}}{500} \times 1000 = 2\times10^{-3}\,M$

$\therefore P^{OH} = -\log(2\times10^{-3}) = 2.7 \Rightarrow P^H = 11.3$

**5.** $[OD^-]_{excess} = \dfrac{80\times0.1 - 20\times0.2}{100} = 0.04\,M$

$\therefore P^{OD} = -\log(0.04) = 1.4$

Now, $P^{Kw}$ of $D_2O = P^D + P^{OD} = 13.6 + 1.4 = 15$

$\therefore Kw = 1\times10^{-15}$

**6.** $[OT^-]_{excess} = \dfrac{400\times0.2 - 100\times0.4}{500} = 0.08\,M$

$\therefore P^{OT} = -\log(0.08) = 1.1$

Now, $PT = P^{Kw} - P^{OT} = 2\times7.60 - 1.1 = 14.1$

**7.** $K_b = \dfrac{Kw}{Ka} = \dfrac{10^{-14}}{2\times10^{-10}} = 5\times10^{-5}$

**8.** $NH_3 + H_2O \rightleftharpoons NH_4^+ + OH^-$

$\Delta H° = (-52.21) + (54.70) = 2.49$ kJ

$\Delta S° = 1.6 + (-76.3) = -74.7$ J/K

Now, $\Delta H° = -RT.\ln K_{eq}$

or, $2490 - 300\times(-74.7) = -8.3\times300\times\ln K_{eq}$

$\therefore K_{eq} = e^{-10}$

**9.** $[H^+]_{HCOOH} = [H^+]_{CH_3COOH}$

or, $\sqrt{2.4\times10^{-4}\times0.6} = \sqrt{1.8\times10^{-5}\times C} \Rightarrow C = 8\,M$

$\therefore$ Moles of $CH_3COOH$ added $= \dfrac{100\times8}{1000} = 0.8$

**10.** $K_a(HA) = K_b(A^-) = \sqrt{Kw} = 10^{-7}$

Now, $[H^+] = \sqrt{10^{-7}\times0.1} = 10^{-4}\,M \Rightarrow P^H = 4.0$

**11.** $NH_4^+ + OH^- \xrightleftharpoons[K_b = ?]{K_f = 3.4\times10^{10}\,M^{-1}S^{-1}} NH_3 + H_2O$

Given : $NH_4^+ \rightleftharpoons NH_3 + H^+;\ K_1 = 5.6\times10^{-10}\,M$

and $H_2O \rightleftharpoons H^+ + OH^-;\ K_2 = 1.0\times10^{-14}\,M^2$

$\therefore NH_4^+ + OH^- \rightleftharpoons NH_3 + H_2O;\ K_{eq} = \dfrac{K_1}{K_2}$

Now, $\dfrac{3.4\times10^{10}}{K_b} = \dfrac{5.6\times10^{-10}}{10^{-14}}$

$\Rightarrow K_b = 6.07\times10^{5}\,S^{-1}$

**12.** $\underset{0.1-x}{CH_3COOH} \rightleftharpoons \underset{x}{CH_3COO^-} + \underset{0.1+x+y\approx0.1+y}{H^+}$

$\underset{0.1-y}{Cl_2CHCOOH} \rightleftharpoons \underset{y}{Cl_2CHCOO^-} + \underset{0.1+x+y\approx0.1+y}{H^+}$

$0.15 = \dfrac{y\times(0.1+y)}{(0.1-y)} \Rightarrow y = 0.05$

$\therefore [H^+] = 0.1 + x + y \approx 0.1 + y = 0.15\,M$

$\therefore P^H = -\log(0.15) = 0.82$

**13.** $[OH^-] = \dfrac{0.4}{100} \times \dfrac{4.25/17}{250/1000} = 0.004\,M$

$P^{OH} = -\log(0.004) = 2.4$

$\therefore P^H = 14 - 2.4 = 11.6$

**14.** $[OH^-] = \sqrt{K_b\times C} = \sqrt{1.6\times10^{-6}\times0.0025}$

$= \sqrt{4\times10^{-9}}\,M$

$P^{OH} = -\log\sqrt{4\times10^{-9}} = 4.2 \Rightarrow P^H = 9.8$

15. $[HSaC]_O = \frac{4\times10^{-4}}{200/1000} = 2\times10^{-3}M$

and $P^H = 3.0 \Rightarrow [H^+] = 10^{-3}$ M

Now, $2\times10^{-12} = \frac{[SaC^-]\times10^{-3}}{2\times10^{-3}}$

$\Rightarrow [SaC^-] = 4\times10^{-12}M$

16. $[HA]_O = \frac{20\times0.5}{50} = 0.2$ M

$[HB]_O = \frac{30\times0.2}{50} = 0.12$ M

$\underset{0.2-x}{HA} \rightleftharpoons \underset{x+y}{H^+} + \underset{x}{A^-}$

$\underset{0.12-y}{HB} \rightleftharpoons \underset{x+y}{H^+} + \underset{y}{B^-}$

Now, $2\times10^{-4} = \frac{(x+y)\cdot x}{0.2-x} \approx \frac{(x+y)\cdot x}{0.2}$

$\therefore (x+y)\cdot x = 4\times10^{-5}$ (1)

and, $5\times10^{-5} = \frac{(x+y)\cdot y}{(0.12-y)} \approx \frac{(x+y)\cdot y}{0.12}$

$\therefore (x+y)\cdot y = 6\times10^{-6}$ (2)

From (1) and (2), $[H^+] = x + y = 6.78\times10^{-3}$ M

17. $10^{-5} = \sqrt{K_a\times0.01} \Rightarrow K_a = 10^{-8}$

Now, $[OH^-] = \sqrt{10^{-8}\times0.1} = 10^{-4.5} \Rightarrow P^{OH} = 4.5$

$\therefore P^H = 9.5$

18. $\underset{0.01-x}{RNH_2} + H_2O \rightleftharpoons \underset{x}{RNH_3^+} + \underset{x+10^{-4}}{OH^-}$

$2\times10^{-6} = \frac{x(x+10^{-4})}{0.01-x} \Rightarrow x = 10^{-4}$

$\therefore [OH^-] = 2\times10^{-4}$ M

19. As on adding HCl, $[H^+]$ is not changing and will remain unchanged.

20. $\underset{\downarrow \text{Decrease}}{Co_2} + H_2O \underset{\leftarrow \text{Shiftleft}}{\rightleftharpoons} HCo_3^- + H^+$

As $[H^+]$ decreases, $P^H$ increases.

21. $[W_2H_4]_O = \frac{0.16/32}{500/1000} = 0.01$ M

$\therefore \propto = \sqrt{\frac{4\times10^{-6}}{0.01}} = 0.02$ or 2%

22. $P^H = -\log(2\times10^{-6}) = 5.70$

23. $[OH-] = 6.67\times10^{-3} + \frac{6.67\times10^{-3}}{2} + 0 \approx 10^{-2}$ M

$\therefore P^{OH} = -\log(10^{-2}) = 2.0 \Rightarrow P^H = 12.0$

24. $\underset{(0.09-x)M}{en} + H_2O \rightleftharpoons \underset{(x-y)M}{enH^+} + \underset{(x+y)M}{OH^-}$; $K_{b_1} = 8.1\times10^{-5}$

$\underset{(x-y)M}{enH^+} + H_2O \rightleftharpoons \underset{yM}{enH_2^{2+}} + \underset{(x+y)M}{OH^-}$; $K_{b_2} = 7.0\times10^{-8}$

Now, $8.1\times10^{-5} = \frac{(x-y)(x+y)}{(0.09-x)} K \approx \frac{x\cdot x}{0.09}$

$\Rightarrow x = 2.7\times10^{-3}$

and $7.0\times10^{-8} = \frac{y\cdot(x+y)}{(x-y)} \approx \frac{y\cdot x}{x} \Rightarrow y = 7.0\times10^{-8}$

$\therefore [en\ H^+] = (x-y) \approx xM = 2.7\times10^{-3}$ M

$[enH_2^{2+}] = y = 7.0\times10^{-8}M$

$\therefore [OH^-] = (x+y) = x = 2.7\times10^{-3}$ M

and $P^{OH} = -\log(2.7\times10^{-3}) = 2.56 \Rightarrow P^H = 11.44$

25. $K_{a_1}\cdot K_{a_2} = \frac{[H^+]^2[5^{2-}]}{[H_2S]}$

or, $(1.4\times10^{-7})\times(1.0\times10^{-14}) = \frac{(0.1)^2\times[5^{2-}]}{0.2}$

$\Rightarrow [5^{2-}] = 2.8\times10^{-20}M$

26. $[H^+] = \sqrt{0.2\times2\times10^{-5}} = 2\times10^{-3}M$

Now, $(2\times10^{-5})\times(5\times10^{-9})\times(4\times10^{-12})$

$= \frac{(2\times10^{-3})^3\times[A^{3+}]}{[H_3A]}$

$\therefore \frac{[A^{3-}]}{[H_3A]} = 5\times10^{-17}$

27. $[H^+] = \sqrt{0.1\times10^{-5}} = 10^{-3}M$

Now, $K_{a_3} = \frac{[H^+][A^{3-}]}{[HA^{2-}]} \Rightarrow \frac{[A^{3-}]}{[HA^{2-}]} = \frac{10^{-13}}{10^{-3}} = 10^{-10}$

$\therefore P^X = 10$

28. $\underset{\substack{2\text{ m mol}\\ \text{Final } 0}}{OH^-} + \underset{\substack{3\text{ m mol}\\ 1\text{ m mol}}}{ACOH} \rightleftharpoons \underset{\substack{0\\ 2\text{ m mol}}}{ACO^-} + H_2O$

$p^H = 4.74 + \log\frac{2}{1} = 5.04$ (Acidic)

Addition of 1 ml ACOH will decrease $P^H$ by 0.3 unit.

HINTS AND EXPLANATIONS

**29.** Millimoles of ACOH = 6 × 0.1 = 0.6

Millimoles of $ACO^-$ = 12.× 0.1 = 1.2

$\therefore\ p^H = 4.75 + \log\frac{1.2}{0.6} = 5.05$

**30.** 1st solution will finally have 1 mole of $CH_3COOH$.

$\therefore\ P_1^H = \frac{1}{2}P^{Ka}$

and for 2nd solution, $P_2^H = P^{Ka}$

**31.** $P_2^H = P_1^H + 0.6$

$p^{K_a} + \log\frac{y/M}{C} = P^{K_a} + \log\frac{x/M}{C} + \log 3.98$

$\therefore \frac{y}{x} = 3.98$

**32.** $4.0 = 5.0 + \log\frac{C_1}{0.5} \Rightarrow C_1 = 0.05\text{ M}$

$6.0 = 5.0 + \log\frac{C_2}{0.5} \Rightarrow C_2 = 5.0\text{ M}$

Now, final $P^H = 5.0 + \log\frac{V\times 0.05 + V\times 5.0}{V\times 0.5 + V\times 0.5} = 5.7$

**33.** For maximum $\beta, \frac{\alpha\beta}{\alpha[H^+]} = 0 \Rightarrow P^H = P^{K_a}$

**34.**

$$\begin{array}{lccc} & BOH & + \ H^+ & \rightleftharpoons B^+ + H_2O \\ & \frac{40\times C}{40+V}M & \frac{V\times 0.1}{40+V}M & 0 \\ \text{Final} & \frac{40C-0.1V}{40+V} & 0 & \frac{0.1V}{40+V} \end{array}$$

$H^+$ must be a limiting reagent because both $P^H$ are above >.0.

Now, $P^{OH} = P^{K_b} + \log\frac{0.1V}{40C - 0.1V}$

$14 - 10 = P^{K_b} + \log\frac{0.1\times 5}{40C - 0.1\times 5}$ (1)

$14 - 9 = P^{K_b} + \log\frac{0.1\times 20}{40C - 0.1\times 20}$ (2)

$\therefore K_b = 2\times 10^{-5}$

**35.** For maximum buffer capacity:

$\frac{[ACOH]}{[NaOH]} = \frac{2}{1} \Rightarrow [NaOH] = \frac{2\text{ M}}{2} = 1\text{ M}$

$\therefore$ Mass of NaOH added $= \frac{500\times 1}{1000}\times 40 = 20$ gm

**36.** $n_{HA} = n_{OH^-} \Rightarrow \frac{0.28}{M} = \frac{35\times 0.1}{1000} \Rightarrow M = 80$

**37.** $\underset{0.9x\text{M}}{Fe^{3+}} + H_2O \rightleftharpoons \underset{0.1x\text{M}}{Fe(OH)^{2+}} + \underset{?}{H^+}$

$9\times 10^{-3} = \frac{0.1x\times[H^+]}{0.9x} \Rightarrow [H^+] = 0.081 \Rightarrow P^H = 1.08$

**38.** In final solution: $[HA] = [A^-]$

**39.** $[NH_3] = \sqrt{\frac{K_w}{K_a\cdot K_b}}\times C = 8.33\times 10^{-4}$ M

**40.** For equivalence point, $2.5\times\frac{2}{5} = V_{Hcl}\times\frac{2}{15}$

$\therefore V_{HCl} = 7.5$ ml

$$\begin{array}{lccc} & BOH & + \ H^+ & \rightleftharpoons B^+ + H_2O;\ K_{aq} = \frac{10^{-12}}{10^{-14}} = 10^2 \\ & \frac{2.5\times\frac{2}{5}}{10}M & \frac{7.5\times\frac{2}{15}}{10}M & 0 \\ & 0 & 0 & 0.1\text{M} \\ \text{Eqn.} & x\text{M} & \text{XM} & (0.1-x)\text{M} \end{array}$$

$100 = \frac{0.1-x}{x\cdot x} \Rightarrow x = 2.7\times 10^{-2}\text{M} = (H^+)$

**41.** At 2nd equation point:

$P^H = \frac{1}{2}(P^{K_{a2}} + P^{K_{a3}}) = \frac{8+12}{12} = 10$

Now, $K_{a_1}\cdot K_{a_2}\cdot K_{a_3} = \frac{[H^+][A^{3-}]}{[H_3A]}$

or $7.5\times 10^{-4}\times 10^{-8}\times 10^{-12} = \frac{(10^{-10})^3\times[A^{3-}]}{[H_3A]}$

$\therefore \frac{[H_3A]}{[A^{3-}]} = \frac{10^{-6}}{7.5} = 1.33\times 10^{-7}$

**42.**

$$\begin{array}{lccc} & CO_3^{2-} & + \ H^+ & \rightleftharpoons HCO_3^-;\ K = \frac{1}{4\times 10^{-11}} \\ & 0.35\text{ M} & 0.35\text{ M} & 0 \\ \text{100\% run} & 0 & 0 & 0.35\text{ M} \end{array}$$

For $HCO_3^-$ solution,

$[H^+] = \sqrt{K_{a_1}, K_{a_2}} = 1.4\times 10^{-8}$M

Now, $K_{a_2} = \frac{[H^+][CO_3^{2-}]}{[HCO_3^-]}$

$\therefore [CO_3^{2-}] = \frac{4\times 10^{-11}\times 0.35}{1.4\times 10^{-8}} = 10^{-3}$M

**43.** $[LaC^-]_0 = \frac{0.125}{0.5} \times 2 = 0.5$ M

Now, $P^{OH} = 7 - \frac{1}{2}(P^{K_a} + \log C)$

$5.6 = 7 - \frac{1}{2}(P^{K_a} + \log 0.5)$

$\therefore P^{K_a} = 3.1 \Rightarrow K_a = 8 \times 10^{-4}$

**44.** As HA is stronger acid, it will react first. For first equivalent point, $V_{NaOH} \times 0.2 = 50 \times 0.05 \Rightarrow V_{NaOH} = 12.5$ ml

At first equivalent point,

$$[A^-] = \frac{50 \times 0.05}{62.5} = 0.04 \text{ M}$$

$$[HB] = \frac{50 \times 0.08}{62.5} = 0.064 \text{ M}$$

Now, $\underset{\substack{0.04-x \\ \approx 0.04}}{A^-} + \underset{\substack{0.064-x \\ \approx 0.064}}{HB} \rightleftharpoons \underset{x}{B^-} + \underset{x}{HA}$;

$$K_{eq} = \frac{K_a(HB)}{K_a(HA)} = \frac{10^{-8.2}}{10^{-3.8}} = 10^{-4.4} = 4 \times 10^{-5}$$

$$4 \times 10^{-5} = \frac{x \cdot x}{0.04 \times 0.064} \Rightarrow x = 3.2 \times 10^{-4}$$

Now, $K_a(HA) = \frac{[H^+][A^-]}{[HA]}$

$$\Rightarrow 1.6 \times 10^{-4} = \frac{[H^+] \times 0.04}{3.2 \times 10^{-4}}$$

$\therefore [H^+] = 1.28 \times 10^{-6} \Rightarrow P^H = 5.9$

**45.** $\underset{0.005-x}{CrO_4^{2-}} + H_2O \rightleftharpoons \underset{x}{HCrO_4^-} + \underset{x}{OH^-}$;

$$K_h = \frac{Kw}{K_{a_2}} = 2 \times 10^{-8}$$

$$2 \times 10^{-8} = \frac{x \cdot x}{0.005 - x} \approx \frac{x^2}{0.005} \Rightarrow x = 10^{-5}$$

$$\therefore h = \frac{10^{-5}}{0.005} = 0.002$$

**46.** $P^{K_{a_1}} = 2.40$ and $P^{K_{a_2}} = 9.60$

$\therefore$ Required $pH = \frac{1}{2}(2.40 + 9.60) = 6.00$

**47.** $\underset{\text{Red}}{HA} \rightleftharpoons H^+ + \underset{\text{Blue}}{A^-}$

$$[H^+] = \frac{K_a \cdot [HA]}{[A^-]}$$

$$\therefore [H^+]_{\text{required}} = [H^+]_2 - [H^+]_1 = K_a\left(\frac{75}{25} - \frac{25}{75}\right)$$

$$= 8 \times 10^{-5} \text{ M}$$

**48.** $5.5 = 4.75 + \log\frac{[ACo^-]_O}{[ACOH]_O}$

$$\therefore \frac{[ACo^-]_O}{[ACOH]_O} = \frac{5.62}{1}$$

**49.** $n_{CO_2} = \frac{224}{22400} = 0.01$ and $n_{H^+}$ used $= \frac{30 \times 1}{1000} = 0.03$

$$\underset{0.01 \text{ mole (L.R.)}}{CO_2 + OH^-} \longrightarrow \underset{0.01 \text{ mole or less}}{HCO_3^-}$$

Hence, moles of $H^+$ used should be 0.01 or less and titration of $HCO_3^-$ and H+ should not be detected by phenolphthalein. Hence, $OH^-$ must be in excess.

$$\underset{0.01 \text{ mole}}{CO_2} + \underset{0.02 \text{ mole}}{2OH^-} \longrightarrow \underset{0.01 \text{ mole}}{CO_3^{2-}} + H_2O$$

Thus, 0.01 mole of $CO_3^{2-}$ will require only 0.01 mole of $H^+$ in the presence of phenolphthalein. As the mole of $H^+$ used is 0.03, 0.02 mole $OH^-$ must be present in excess.

Hence, total moles of $OH^-$ used = 0.02 + 0.02 = 0.04,

$$\therefore [NaOH]_{\text{used}} = \frac{0.04}{1} = 0.04 \text{ M}$$

**50.** $PbSO_4 = \sqrt{2 \times 10^{-9}} \times \frac{304}{10} \simeq 1.36 \times 10^{-3}$ g/100ml

$ZaS = \sqrt{10^{-22}} \times \frac{97}{10} = 9.7 \times 10^{-11}$ g/100ml

$AgBr = \sqrt{4 \times 10^{-13}} \times \frac{188}{10} = 1.19 \times 10^{-5}$ g/100ml

$CuCo_3 = \sqrt{10^{-8}} \times \frac{123}{10} = 1.23 \times 10^{-3}$ g/100ml

**51.**

| | $Hg^{2+}$ | $+ 4Cl^- \rightleftharpoons$ | $HgCl_4^{2-}$ |
|---|---|---|---|
| | 0.1 M | 0.9 M | 0 |
| 100% ram | 0 | 0.5M | 0.1M |
| Eqn. | $1.6 \times 10^{-17}$ M | 0.5M | 0.1M |

$$\therefore K_{form} = \frac{0.1}{1.6 \times 10^{-17} \times (0.5)^4} = 10^{17}$$

**52.** $AgBr(s) + 2S_2O_3^{2-} \rightleftharpoons Ag(S_2O_3)_2^{3-} + Br^-$

0.1 aM 0 0

Eqn/final 0 a−0.2 0.1 0.1

$$K_{eq} = 4\times10^{-13}\times1.6\times10^{12} = \frac{0.1\times0.1}{(a-0.2)^2} \Rightarrow a = 0.325$$

**53.** $Tl_2S(S) \rightleftharpoons 2Tl^+ + S^{2-}$

2SM x

$S^{2-} + H_2O \rightleftharpoons HS^- + OH^-$; $K_h = \frac{Kw}{K_{a_2}}$

xM SM SM

$$\frac{10^{-14}}{10^{-14}} = \frac{2\times10^{-6}\times2\times10^{-6}}{x} \Rightarrow x = 4\times10^{-12}$$

$$\therefore K_{sp} = (2\times2\times10^{-6})^2\times4\times10^{-12} = 6.4\times10^{-23}$$

**56.** $M^{3+} + SCN^- \rightleftharpoons M(SCN)^{2+}$

$2\times10^{-3}$ M $1.51\times10^{-3}$ M 0

Eqn. $2\times10^{-3}-x$ $1.51\times10^{-3}-x$ xM

$=5\times10^{-4}$ M $=1.0\times10^{-5}$ M $=1.5\times10^{-3}$ M

$$\therefore K_f = \frac{1.5\times10^{-3}}{5\times10^{-4}\times1\times10^{-5}} = 3\times10^5$$

**57.** $SrCO_3(s) \rightleftharpoons Sr^{2+} + CO_3^{2-}$

$2\times10^{-4}$ M $2\times10^{-4}-x$

$CO_3^{2-} + H_2O \rightleftharpoons HCO_3^- + OH^-$

$(2\times10^{-4}-x)$M xM $4\times10^{-6}$ M

$$\text{Now, } \frac{10^{-14}}{5\times10^{-11}} = \frac{x\times4\times10^{-6}}{(2\times10^{-4}-x)} \Rightarrow x = \frac{0.01}{51}$$

$$\therefore K_{sp} = (2\times10^{-4})\times(2\times10^{-4}-x) = \frac{4}{51}\times10^{-8}$$

**58.** $MnS(S) \rightleftharpoons Mn^{2+} + S^{2-}$

SM (S−x)M

$S^{2-} + H_2O \rightleftharpoons HS^- + OH^-$

(S−x)M xM xM

$$\frac{10^{-14}}{10^{-14}} = \frac{x\cdot x}{(S-x)} \text{ and } 2.5\times10^{-10} = S\cdot(S-x)$$

$\therefore S = 6.3\times10^{-4}$ M

**59.** $AgCl(s) + Br^-(aq) \rightleftharpoons AgBr(s) + Cl^-(aq)$

0.1−0.075 xM 0.075 0.075M

$$K_{eq} = \frac{[Cl^-]}{[Br^-]} = \frac{K_{sp}(AgCl)}{K_{sp}(AgBr)} \Rightarrow \frac{0.075}{[Br^-]} = \frac{2\times10^{-10}}{4\times10^{-13}}$$

$\therefore [Br^-] = 1.5\times10^{-4}$ M

**60.** $[Ag(CN)_2^-] = 0.01$M

$$K_{diss} = \frac{[Ag^+][CN^-]^2}{[Ag^+(CN)_2^-]} \Rightarrow 1\times10^{-20} = \frac{[Ag^+]\times(2.5\times10^{-7})^2}{0.01}$$

$\therefore [Ag^+] = 1.6\times10^{-9}$ M

**61.** $[CO_3^{2-}] = 2.0$ M

$$\text{Now, } \frac{K_{sp}(CaCO_3)}{K_{sp}(CaF_2)} = \frac{[CO_3^{2-}]}{[F^-]^2}$$

$$\Rightarrow \frac{x}{4y^3} = \frac{2}{[F^-]} \Rightarrow (F^-) = \sqrt{\frac{8y^3}{x}}$$

**62.** $BaF_2(s) + C_2O_4^{2-}(aq) \Rightarrow BaC_2O_4(s) + 2F^-(aq)$

(0.1−x)M 2xM

$$K_{eq} = \frac{[F^-]}{[C_2O_4^{2-}]} = \frac{K_{sp}(BrF_2)}{K_{sp}(BrC_2O_4)} = \frac{10^{-6}}{10^{-10}} = 10^4$$

$\therefore x \approx 0.1 \Rightarrow [F^-] = 0.2$ M

$$\therefore [Ba^{2+}] = \frac{10^{-6}}{(0.2)^2} = 2.5\times10^{-5}\text{ M}$$

**63.** $S = [Zn(OH)_2(aq)] + [Zn(OH)^+] + [Zn^{2+}] + [Zn(OH)_3^-] + [Zn(OH)_4^{2-}]$

$$= K_1 + \frac{K_2\cdot K_1}{[OH^-]} + \frac{K_3\cdot K_2K_1}{[OH^-]^2} + K_4K_1[OH^-] + K_5K_4K_1[OH^-]^2$$

$$= 10^{-6} + \frac{10^{-7}\times10^{-6}}{0.1} + \frac{10^{-4}\times10^{-7}\times10^{-6}}{(0.1)^2} + 10^3\times10^{-6}\times0.1 + 10\times10^3\times10^{-6}\times(0.1)^2$$

$$= 10^{-6} + 10^{-12} + 10^{-15} + 10^{-4} + 10^{-4} \approx 2\times10^{-4}\text{ M}$$

**64.** For molecular solubility of $CaCl_2$,

$\Delta H_{solution} = 209.2 + (-33.5) = 175.7$ KJ > 30 KJ

For ionic solubility of $CaCl_2$,

$\Delta H_{solution} = 209.2 + 1004.2 + 1715.4 - 1598.3 - 719.6 - 711.2 = -100.3$ KJ

For molecular solubility of $HgCl_2$,

$\Delta H_{solution} = 83.7 - 66.9 = 16.8$ KJ < 30 KJ

For ionic solubility of $HgCl_2$,

$\Delta H_{solution} = 83.7 + 460.2 + 2815.8 - 1845.1 - 719.6 - 711.2$

$= 83.3$ KJ > 30 KJ

Hence, $CaCl_2$ is ionic and $HgCl_2$ is molecular solubility.

**65.** $$[C_2O_4^{2-}] = \frac{6\times0.001\times5}{250\times2} = 6\times10^{-5}\text{M}$$

$$\therefore K_{sp} = (6\times10^{-5})^2 = 3.6\times10^{-9}$$

**66.** $$\underset{\substack{0.001\text{M}\\ \text{Eqn. } 0.001-x\\ =0.001\times\frac{75}{100}}}{Sr^{2+}} + \underset{\substack{0.05\text{M}\\ 0.05-x\\ \approx 0.05\text{M}}}{NO_3^-} \rightleftharpoons \underset{\substack{0\\ x}}{Sr(NO_3)^+}$$

$\therefore x = 0.00025$

$$\text{Now, } K_f = \frac{2.5\times10^{-4}}{7.5\times10^{-4}\times0.05} = \frac{20}{3}$$

**67.** $$\underset{0.1\,\text{Mole}}{ACOAg(s)} + \underset{0.1\text{M}}{H^+} + \underset{0.1\text{M}}{Cl^-} \rightleftharpoons ACOH + AgCl(s)$$

$$K_{eq} = \frac{[ACOH]}{[H^+][Cl^-]}\times\frac{[ACO^-]}{[ACO^-]}\times\frac{[Ag^+]}{[Ag^+]}$$

$$= \frac{K_{sp}[ACOAg]}{K_{sp}(AgCl)\times K_a}$$

$$= \frac{10^{-8}}{10^{-10}\times10^{-5}} = 10^7 \Rightarrow \text{Almost complete reaction}$$

$$\therefore [ACOH] \simeq 0.1\text{M}, [H^+] = \sqrt{\frac{0.1}{10^7}} = 10^{-4}\text{ M}$$

$$\text{and } [ACO^-] = \frac{Ka\times[ACOH]}{[H^+]} = 0.01\text{ M}$$

**68.** $A_xB_y(s) \rightleftharpoons xA^{y+} + yB^{x-}$

$$K_{sp} = x^x\cdot y^y\cdot s^{x+y} \Rightarrow s = \left(\frac{K_{sp}}{x^x\cdot y^y}\right)^{\frac{1}{x+y}}$$

As $K_{sp} << 1$, greater the value of $(x + y)$, greater is $s$.

**69.** Theory based

**70.** $$s = [Zn^{2+}] + [Zn(OH)_4^-] = \frac{K_{sp}}{[OH^-]^2} + K_f[OH^-]^2$$

$$\text{For maximum or minimum S, } \frac{dS}{d[OH^-]} = 0$$

$$\text{or, } \frac{-2K_{sp}}{[OH^-]^3} + 2\,K_f[OH^-] = 0$$

$$\Rightarrow [OH^-] = \left(\frac{K_{sp}}{K_f}\right)^{\frac{1}{4}} = 10^{-4}\text{ M}$$

$\therefore P^H = 10$ and $S_{min} = 2.4\times10^{-9}$ M

**71.** $Al(OH)_3(s) + OH^- \rightleftharpoons Al(OH)_4^-;$

From question $\quad ? \quad\quad 10^{-3}$ M

$$K = \frac{8\times10^{-33}}{1.6\times10^{-34}} = 50$$

$$50 = \frac{10^{-3}}{[OH^-]} \Rightarrow [OH^-] = 2\times10^{-5}\text{M} \Rightarrow P^H = 9.30$$

As the calculated $[OH^-]$ is minimum $OH^-$, $P^H$ is minimum.

$$Al(OH)_3(s) \rightleftharpoons \underset{10^{-3}\text{ M}}{Al^{3+}} + 3OH^-;\ K = 8\times10^{-33}$$

$$8\times10^{-33} = 10^{-3}\times[OH^-]^3 \Rightarrow [OH^-] = 2\times10^{-10}\text{M}$$

$\Rightarrow P^H = 4.30$

As the calculated $[OH^-]$ is maximum $OH^-$, $P^H$ is maximum.

**72.** From the question,

$[Cu(CN)_4^{3-}] = 0.1$ M and $[CN^-] = 0.2$ M

$$\therefore [Cu^+] = \frac{K_{Instab}\cdot[Cu(CN)_4^{3-}]}{[CN^-]^4} = \frac{6.4\times10^{-15}\times0.1}{(0.2)^4}$$

$$= 4\times10^{-13}\text{ M}$$

$$\text{Now, } [S^{2-}] = \frac{K_{sp}(Cu_2S)}{[Cu^+]^2} = \frac{2.56\times10^{-27}}{(4\times10^{-13})^2}$$

$$= 1.6\times10^{-2}\text{ M}$$

$$\therefore [H^+] = \sqrt{\frac{K_a\times[H_2S]}{[S^{2-}]}} = \sqrt{\frac{1.6\times10^{-21}\times0.1}{1.6\times10^{-2}}} = 10^{-10}\text{ M}$$

and $P^H = 10.0$

**73.** $$S = \sqrt{1.6\times10^{-5}} = 4\times10^{-3}\text{M} = \frac{4\times10^{-3}\times136}{10^3}\times10^6$$

$= 4\times136$ ppm

For increase in concentration 4 times, volume should be $\frac{1}{4}^{\text{th}}$. Hence, 75% water should be evaporated.

**74.** During precipitation, the concentration of both $Ba^{2+}$ and $SO_4^{2-}$ ions will decrease.

**75.** $K_{sp}(AgCl) = 10^{-4}\times10^{-6} = 10^{-10}$

$K_{sp}(Ag_2CrO_4) = (10^{-4})^2\times8\times10^{-4} = 8\times10^{-12}$

After precipitation of AgCl, find the concentration of $Cl^-$. $[Cl^-]_{final} = 1.0\times10^{-6} - 8\times10^{-7}\times2\times10^{-7}$ M

$$\text{Now, } \frac{[CrO_4^{2-}]_{final}}{[Cl^-]^2_{final}} = \frac{K_{sp}(AgCrO_4)}{K^2_{sp}(AgCl)}$$

$$\frac{[CrO_4^{2-}]_{final}}{(2\times10^{-7})^2} = \frac{8\times10^{-12}}{(10^{-10})^2} \Rightarrow [CrO_4^{2-}]_{final} = 3.2\times10^{-5}\text{ M}$$

Hence, moles of $Ag_2CrO_4$ precipitated

$$= 8\times10^{-4} - 3.2\times10^{-5} = 7.68\times10^{-4}$$

**76.** To prevent precipitation of AgCl, the concentration of $Ag^+$ needed in solution $= \dfrac{K_{sp}(AgCl)}{(Cl^-)}$

$$= \frac{1.8\times10^{-10}}{0.16} << 1.8\text{ M}$$

Hence, almost all $Ag^+$ ion must form complete with $CN^-$ ions.

$$\underset{\frac{1.8\times10^{-10}}{0.16}\text{M}}{Ag^+} + \underset{\text{CM}}{2CN^-} \rightleftharpoons \underset{1.8\text{ M}}{Ag(CN)_2^-};\ K_f = 6.4\times10^{17}$$

**77.** $[CO_3^{2-}] = K_a(\text{overall})\cdot\dfrac{[H_2CO_3]}{[H^+]^2}$

To prevent precipitation of $MCO_3$,

or, $[M^{2+}][CO_3^{2-}] \le K_{sp}$

or, $[M^{2+}]\cdot\dfrac{K_a\cdot[H_2CO_3]}{[H^+]^2} \le K_{sp}$

$$\therefore [H^+] \ge \sqrt{\frac{[M^{2+}]\cdot K_a[H_2CO_3]}{K_{sp}}}$$

For $MgCO_3$: $[H^+] \ge \sqrt{\dfrac{0.1\times5\times10^{-17}\times0.05}{9\times10^{-8}}}$

$$= 2.5\times10^{-6}\text{M}$$

$\therefore P^H \le 5.6$

For $SrCO_3$: $[H^+] \ge \sqrt{\dfrac{0.1\times5\times10^{-17}\times0.05}{9\times10^{-10}}}$

$$= \frac{5\times10^{-5}}{3}\text{ M}$$

$\therefore P^H \le 4.78$

For precipitation of $SrCO_3$ without any precipitation of $MgCO_3$, the $P^H$ range should be 4.78 to 5.6

**78.** $Mn^{2+}(aq) + H_2S(aq) \rightleftharpoons MnS(s) + 2H^+(aq)$

To just start precipitation of MnS, $Q < K_{eq}$

or, $\dfrac{[H^+]^2}{[Mn^{2+}][H_2S]} < \dfrac{K_a(H_2S)}{K_{sp}(MnS)}$

or, $\dfrac{[H^+]^2}{0.04\times0.1} < \dfrac{1.0\times10^{-21}}{2.5\times10^{-13}} \Rightarrow [H^+] < 4\times10^{-6}$ M

Now, in the given buffer, $[H^+] = \dfrac{K_a\cdot[CH_3COOH]_O}{[CH_3COO^-]_O}$

$$= \frac{2\times10^{-5}\times0.25}{0.15} = 3.33\times10^{-5} > 4\times10^{-6}\text{ M}$$

Hence, no precipitation. To start precipitation $[H^+]$ should decrease and hence, $CH_3COONa$ should be added.

Now, $4\times10^{-6} = \dfrac{2\times10^{-5}\times0.25}{[CH_3COONa]_O}$

$\therefore [CH_3COONa]_O = 1.25$ M

**79.** $Mg(OH)_2(S) + 2NH_4^+ \rightleftharpoons Mg^{2+} + 2NH_4OH$

To re-dissolve $Mg(OH)_2$, $Q \le K_{eq}$

or, $\dfrac{[Mg^{2+}][NH_4OH]^2}{[NH_4^+]^2} \le \dfrac{K_{sp}}{K_b^2}$

or, $\dfrac{\left(\frac{0.15\times0.1}{0.5}\right)\left(\frac{0.35\times0.1}{0.5}\right)^2}{\left(\frac{n}{0.5}\right)^2} \le \dfrac{1.2\times10^{-11}}{(2.0\times10^{-5})^2}$

$\therefore n \ge 0.035$

Hence, minimum mass of $(NH_4)_2SO_4$ needed

$$= \frac{0.035}{2}\times132 = 2.31\text{ gm}$$

**80.** $Ag^+ + Cl^- \rightleftharpoons AgCl(S)$

| | $Ag^+$ | $Cl^-$ |
|---|---|---|
| | $\frac{500\times0.01}{1000} = 5\times10^{-3}$ M | $\frac{250\times0.02}{1000} = 5\times10^{-3}$ M |
| Eq | $5\times10^{-3}-(x+y)$ M | $(5\times10^{-3}-x)$ M |

$Ag^+ + Br^- \rightleftharpoons AgBr(S)$

| $Ag^+$ | $Br^-$ |
|---|---|
| $= 5\times10^{-3}$ M | $= 5\times10^{-3}$ M |
| $5\times10^{-3}-(x+y)$ M | $(5\times10^{-3}-y)$ M |

As both reactions will tend towards completion, $(x + y) = 5\times10^{-3}$

Now, $[Ag^+](5\times10^{-3} - x) = 10^{-10}$ (1)

$\Rightarrow [Ag^+]\cdot y = 10^{-10}$

and $[Ag^+](5\times10^{-3} - y) = 5\times10^{-13}$ (2)

From (1) ÷ (2), $\dfrac{y}{5\times10^{-3} - y} = 200 \Rightarrow y = \dfrac{1}{201}$

$\therefore [Br^-] = 5\times10^{-3} - y \approx 2.5\times10^{-5}$ M

## Section B (One or More than one Correct)

1. (a) Complete neutralization $\Rightarrow P^H = 7.0$

(b) $[H^+]_{final} = \frac{55\times0.1-45\times0.1}{100} = 0.01\,M$

$\Rightarrow P^H = 2.0$

(c) $OH^-$ is in excess.

(d) $[H^+]_{final} = \frac{75\times\frac{1}{5}-25\times\frac{1}{5}}{100} = 0.1\,M \Rightarrow P^H = 1.0$

2. For basic solution: $[H^+] < [OH^-]$ and $[H^+] < \sqrt{Kw}$

$\therefore P^H > P^{OH}$ or $P^H > \frac{P^{kw}}{2}$ or $P^{OH} < \frac{P^{kw}}{2}$

3. $\underset{C(1-\alpha)M}{HA} \rightleftharpoons \underset{C\alpha M}{H^+} + \underset{C\alpha M}{A^-}$

$$K_a = \frac{C\alpha\cdot C\alpha}{C(1-\alpha)} = \frac{\alpha^2\cdot C}{1-\alpha} \approx \alpha^2\cdot C \Rightarrow \alpha = \sqrt{\frac{K_a}{C}}$$

$$\text{Now, } K_a = \frac{[H^+]\cdot C\alpha}{C(1-\alpha)} = \frac{[H^+]\cdot\alpha}{1-\alpha} \Rightarrow \alpha = \frac{K_a}{K_a+[H^+]}$$

$$= \frac{1}{1+\frac{[H^+]}{K_a}} = \frac{1}{1+10^{(P^{Ka}-P^H)}}$$

4. Dilution results in increased degree of dissociation but decrease in concentrations of all active components.

5. Relation is valid only for conjugate pairs.

6. $P^H$ may decrease only on increasing $[H^+]$.

7. $\underset{\substack{(0.1-x)\,M\\ \approx 0.1\,M}}{CH_3COOH} \rightleftharpoons \underset{xM}{CH_3COO^-} + \underset{\substack{(0.1+x)\,M\\ \approx 0.1\,M}}{H^+}$

$$\text{Now, } 1.8\times10^{-5} = \frac{x\times0.1}{0.1} \Rightarrow x = 1.8\times10^{-5}$$

$$\text{and } \alpha = \frac{x}{0.1} = 1.8\times10^{-4}$$

$$\text{Now, } [H^+]_{\text{from water}} = [OH^-] = \frac{Kw}{[H^+]_{acid}} = 10^{-13}\text{ M}$$

8. $\underset{1\,bar}{RNH_2(g)} + H_2O(l) \rightleftharpoons \underset{xM}{RNH_3^+(aq)} + \underset{xM}{OH^-(aq)}$

$$10^{-6} = \frac{x\cdot x}{1} \Rightarrow x = 10^{-3} \Rightarrow P^{OH} = 3.0 \Rightarrow P^H = 11.0$$

9. $NH_4OH(aq) \rightleftharpoons NH_4^+ + OH^-$

Addition of solid $NH_4OH$ will increase $NH_4OH(aq)$ concentration and hence, $[OH^-]$ will increase.

10. (c) $H_2O + H_2O \rightleftharpoons H_3O^+ + OH^-$; $\Delta H° = +ve$

(d) $HA + OH^- \rightleftharpoons A^- + H_2O$

| | HA | $OH^-$ | $A^-$ |
|---|---|---|---|
| | $a$ | $\frac{a}{2}$ | 0 |
| Final | $\frac{a}{2}$ | | $\frac{a}{2}$ |

$$P^H = P^{K_a} + \log\frac{a/2}{a/2} = P^{K_a}$$

11. $[H^+] >> [CO_3^{2-}]$

12. Theory based

13. $NH_4Cl + NaOH \rightleftharpoons NH_4OH + NaCl$

| | $NH_4Cl$ | NaOH | $NH_4OH$ | |
|---|---|---|---|---|
| | $a$ | $b$ | 0 | |
| For buffer: | $(a-b)$ | $\approx0$ | $b$ | $\Rightarrow a>b$ |

$CH_3COONa + HCl \rightleftharpoons CH_3COOH + NaCl$

| | $CH_3COONa$ | HCl | $CH_3COOH$ | |
|---|---|---|---|---|
| | $a$ | $b$ | 0 | |
| For buffer: | $(a-b)$ | $\approx0$ | $b$ | $\Rightarrow a>b$ |

14. KCN is a salt of weak acid (HCN) and strong base (KOH).

15. $BOH + H^+ \rightleftharpoons B^+ + H_2O$

| | BOH | $H^+$ | $B^+$ |
|---|---|---|---|
| | $a$ mole | $b$ mole | 0 |
| $\frac{1}{5}$th run | $a-\frac{a}{5}$ | 0 | $\frac{a}{5}$ |
| Equivalent point | $\approx0$ | $\approx0$ | $a$ mole |

$$\text{Now, for 1/5th reaction, } P^{OH} = P^{K_b} + \log\frac{[B^-]}{[BOH]}$$

$$\text{or, } (14-9) = P^{K_b} + \log\frac{a/5}{4a/5}$$

$$\therefore P^{K_b} = 5.6 \Rightarrow K_b = 2.5\times10^{-6}$$

$$\text{At equivalent point: } P^H => -\frac{1}{2}(P^{K_b} + \log C)$$

$$\text{or, } 4.5 => -\frac{1}{2}(5.6 + \log C) \Rightarrow C = 0.25\text{ M}$$

Now, $n_{\text{HCl used}} = n_{B^+\text{ formed}}$

$$\text{or, } \frac{V\times0.5}{1000} = \frac{(100+V)\times0.25}{1000} \Rightarrow V_{HCl} = 100\text{ ml}$$

Finally, $n_{\text{BOH takes}} = n_{\text{Hcl used for equivalent point}}$

or, $\frac{w}{45} = \frac{100 \times 0.5}{1000} \Rightarrow w = 2.25$ gm

$\therefore$ Percentage purity of base $= \frac{2.25}{2.5} \times 100 = 90\%$

**16.** $CO_3^{2-} + H_2O \rightleftharpoons HCO_3^- + OH^-$ ;

$(0.5-x)$ M $\quad (x-y)$M $\quad (x+y)$M

$$K_{h_1} = \frac{Kw}{K_{a_2}} = 2 \times 10^{-4}$$

$HCO_3^- + H_2O \rightleftharpoons H_2CO_3 + OH^-$ ;

$(x-y)$M $\quad y$M $\quad (x+y)$M

$$K_{h_2} = \frac{Kw}{K_{a_1}} = 2.5 \times 10^{-9}$$

Now, $2 \times 10^{-4} = \frac{(x-y)\cdot(x+y)}{(0.5-x)} \approx \frac{x \cdot x}{0.5} \Rightarrow x = 10^{-2}$

and $2.5 \times 10^{-9} = \frac{y\cdot(x+y)}{(x-y)} \approx \frac{y \cdot x}{x} \Rightarrow y = 2.5 \times 10^{-9}$

Now, $h = \frac{x}{0.5} = 0.02$

$P^{OH} = -\log(x+y) \approx -\log(10^{-2}) = 2.0 \Rightarrow P^H = 12.0$

and $[H_2CO_3] = y = 2.5 \times 10^{-9}$ M

**17.** $\overset{+}{N}H_3CH_2COOH \xrightarrow{P^{K_{a1}} = 2.22} \overset{+}{N}H_3CH_2COO^-$

$\xrightarrow{P^{Ka2} = 9.78} NH_2CH_2COO^-$

$$P^H = \frac{1}{2}(P^{K_{a1}} + P^{K_{a2}}) = \frac{1}{2}(2.22 + 9.78) = 6.0$$

Now, $K_{a_1} = \frac{[\overset{\oplus}{N}H_3\,CH_2COO^-][H^+]}{[\overset{\oplus}{N}H_3\,CH_2COOH]}$

$\Rightarrow 10^{-2.22} = \frac{0.01 \times 10^{-6}}{[\overset{\oplus}{N}H_3\,CH_2COOH]}$

$\therefore [\overset{\oplus}{N}H_3\,CH_2COOH] = 10^{-5.78} = 1.7 \times 10^{-6}$ M

% of glycine in cationic form

$= \frac{1.7 \times 10^{-6}}{0.01} \times 100 = 0.017\%$

**18.** At equivalent point, the solution should be acidic.

**19.** Sodium acetate solution is basic.

**20.** For precipitation of $Fe(OH)_2$,

$[OH^-]_{min} = \sqrt{\frac{8 \times 10^{-16}}{0.02}} = 2 \times 10^{-7}$

$\Rightarrow P^{OH}_{max} = 6.7 \Rightarrow P^H_{min} = 7.3$

For precipitation of $Fe(OH)_3$,

$[OH^-]_{min} = \left(\frac{4 \times 10^{-28}}{0.05}\right)^{1/3} = 2 \times 10^{-9}$

$\Rightarrow P^{OH}_{max} = 8.7 \Rightarrow P^H_{min} = 5.3$

## Section C (Comprehensions)

### Comprehension I

**1.** $K = 1.5 \times 10^2 = \frac{[\text{Dimer}]}{[\text{Monomer}]^2} = \frac{x}{(0.1-2x)^2}$

$\Rightarrow x = \frac{5}{120}$

$\therefore \frac{[\text{Dimer}]}{[\text{Monomer}]} = \frac{x}{0.1-2x} = \frac{5}{2}$

**2.** $K = 3.6 \times 10^{-2} = \frac{(\text{Dimer})}{[\text{Monomer}]^2} = \frac{x}{(0.1-2x)}$

$\Rightarrow x = 3.6 \times 10^{-4}$

$\therefore \frac{(\text{Dimer})}{[\text{Monomer}]} = \frac{x}{0.1-2x} \Rightarrow \frac{x}{0.1} = \frac{9}{2500}$

**3.** $[H^+] \approx \sqrt{0.1 \times 2 \times 10^{-5}} = \sqrt{2} \times 10^{-3}$ M

(Dimerization is negative as Q. 2)

$\therefore P^H = 2.85$

## Comprehension II

4. $[CH_3COOH]_O = \frac{0.7\times10^{-3}}{10\times10^{3}} = 7\times10^{-8} M$

The solution is so dilute that we may assume almost complete dissociation of acid.

$\therefore [H^+]_{acid} \approx 7\times10^{-8}$ M

Now, $H_2O \rightleftharpoons \underset{(7\times10^{-8}+x)M}{H^+} + \underset{xM}{OH^-}$

$10^{-14} = (7\times10^{-8}+x)\cdot x \Rightarrow x = 7\times10^{-8}$

$\therefore p^H = -\log(7\times10^{-8}+x) = 6.85$

5. $CH_3COOH \rightleftharpoons CH_3COO^- + H^+$

Eqn. $yM$ $\quad 7\times10^{-8}M \quad 7\times10^{-8}+x = 14\times10^{-8}$ M

Now, $2.0\times10^{-5} = \frac{7\times10^{-8}\times14\times10^{-8}}{y}$

$\therefore y = 4.9\times10^{-10}$ M

## Comprehension III

7. $K_a = \frac{(8\times10^{-3})^2}{0.2} = 3.2\times10^{-4}$

6. $K_a = 3.2\times10^{-4} = \frac{[H^+]\times(1.0\times0.8)}{0.2}$

$\therefore [H^+] = 8\times10^{-5} \Rightarrow p^H = 4.1$

## Comprehension IV

8. $p^H = p^{Ka}(NH_4^+) + \log\frac{[NH_3]_O}{[NH_4^+]}$

$= 9.3 + \log\frac{0.8}{0.2} = 9.9$

9. $NH_4OH + H^+ \rightleftharpoons NH_4^+ + H_2O$

| | $NH_4OH$ | $H^+$ | $NH_4^+$ |
|---|---|---|---|
| | 0.8M | 0.3M | 0.2M |
| Final | 0.5M | ≈0 | 0.5M |

$p^H = p^{K_a(NH_4^+)} + \log\frac{[NH_3]_O}{[NH_4^+]_O} = 9.3 + \log\frac{0.5}{0.5} = 9.3$

10. $H^+$ added in excess. Final $[H^+] = 1.0 - 0.8 = 0.2$ M

$\therefore p^H = -\log(0.2) = 0.7$

## Comprehension V

11. No hydrolysis $\Rightarrow p^H = 7.0$

12. Concentration of

$KAl(SO_4)_2.12H_2O = \frac{11.85/474}{100/1000} = 0.25$ M

$Al^{3+} + H_2O \rightleftharpoons Al(OH)^{2+} + H^+$

$(0.25-x)M \quad xM \quad xM$

$1.4\times10^{-5} = \frac{x\cdot x}{(0.25-x)} \simeq \frac{x^2}{0.25}$

$\Rightarrow x = 1.87\times10^{-3} M = [H^+]$

13. $SO_4^{2-} + H_2O \rightleftharpoons HSO_4^- + OH^-$

$(0.5-x)M \quad x \quad x$

$\frac{10^{-19}}{1.25\times10^{-2}} = \frac{x\cdot x}{(0.5-x)} \simeq \frac{x^2}{0.5} \Rightarrow x = 6.32\times10^{-7}$ M

$\therefore [H^+] = \frac{10^{-14}}{6.32\times10^{-7}} = 1.58\times10^{-8}$ M

14. $Al^{3+} + SO_4^{2-} + H_2O \rightleftharpoons Al(OH)^{2+} + HSO_4^-$

| | $Al^{3+}$ | $SO_4^{2-}$ | $Al(OH)^{2+}$ | $HSO_4^-$ |
|---|---|---|---|---|
| | 0.25M | 0.5M | 0 | 0 |
| Eqn. | (0.25–x)M | (0.5–x)M | x | x |

$\frac{1.4\times10^{-5}}{1.25\times10^{-2}} = \frac{x\cdot x}{(0.25-x)(0.5-x)} \simeq \frac{x^2}{0.25\times0.5}$

$\therefore x = 1.18\times10^{-2}$

Now, $1.4\times10^{-5} = \frac{1.18\times10^{-2}\times[H^+]}{0.25}$

$\therefore [H^+] = 2.97\times10^{-4}$ M

## Comprehension VI

**15.** $PuO_2^{2+} + H_2O \rightleftharpoons PuO_2(OH)^+ + H^+$

$0.01-x \approx 0.01$ $\quad x \quad$ $x=1.6\times10^{-4}$

$\therefore K_h = \dfrac{x\cdot x}{0.01-x} \simeq \dfrac{x^2}{0.1} = 2.56\times10^{-6}$

**16.** $K_a = \dfrac{Kw}{K_b} \Rightarrow K_b = 3.9\times10^{-9}$

## Comprehension VII

**17.** $p^H = \dfrac{1}{2}(P^{K_{a2}} + P^{K_{a3}}) = \dfrac{1}{2}(8+13) = 10.5$

**18.** For 2nd equivalent point,

$n_{NaoH} = 2\times n_{H_3PO_4}$

$\dfrac{V\times0.5}{1000} = 2\times\dfrac{100\times0.1}{1000} \Rightarrow V = 40 \text{ ml}$

After adding HCl,

$HPO_4^{2-} + H^+ \rightleftharpoons H_2O_4^-$

10 millimole $\quad$ 5 millimole $\quad$ 0

Final $\quad$ 5 $\quad \approx 0$ $\quad$ 5

$\therefore P^H = 8 + \log\dfrac{5}{5} = 8.0$

**19.** $S = \dfrac{K_{sp}}{[OH^-]^2} = \dfrac{4.0\times10^{-30}}{(10^{-6})^2} = 4.0\times10^{-18}$ M

## Comprehension VIII

**20.** $P^H = P^{K_a} + \log\dfrac{[HCO_3^-]_O}{[H_2CO_3]_O} = 6.4 + \log\dfrac{8}{1} = 7.3$

**21.** $7.4 - 6.4 + \log\dfrac{[HCO_3^-]_O}{[H_2CO_3]_O} \Rightarrow \dfrac{[H_2CO_3]_O}{[HCO_3^-]_O} = \dfrac{1}{10}$

**22.** More $CO_2$ should dissolve in solution.

## Comprehension IX

**23.** $pH = 7 + \dfrac{1}{2}(P^{K_{a2}} + \log C) = 7 + \dfrac{1}{2}(10.6 + \log 1) = 12.3$

**24.** $CO_3^{2-} + H^+ \rightleftharpoons HCO_3^-$

$\dfrac{50\times1}{75}$ M $\quad \dfrac{25\times1}{75}$ M $\quad$ 0

Final $\quad \dfrac{25}{75}$ M $\quad \approx 0 \quad \dfrac{25}{75}$ M

$pH = P^{K_{a2}} + \log\dfrac{[CO_3^{2-}]_O}{[HCO_3^-]_O} = 10.6 + \log\dfrac{1/3}{1/3} = 10.6$

**25.** $CO_3^{2-} + H^+ \rightleftharpoons HCO_3^-$

$\dfrac{50\times1}{100}$ M $\quad \dfrac{50\times1}{100}$ M $\quad$ 0

Final $\quad \approx 0 \quad \approx 0 \quad \dfrac{50}{100}$ M

$\therefore pH = \dfrac{1}{2}(P^{K_{a1}} + P^{K_{a2}}) = \dfrac{1}{2}(5.4 + 10.6) = 8.0$

**26.** $CO_3^{2-} + H^+ \rightleftharpoons HCO_3^-$

$\dfrac{50\times1}{125}$ M $\quad \dfrac{75\times1}{125}$ M $\quad$ 0

Final $\quad 0 \quad \dfrac{25}{125}$ M $\quad \dfrac{50}{125}$ M

$HCO_3^- + H^+ \rightleftharpoons H_2CO_3$

$\dfrac{50}{125}$ M $\quad \dfrac{25}{125}$ M $\quad$ 0

Final $\quad \dfrac{25}{125}$ M $\quad =0 \quad \dfrac{25}{125}$ M

$\therefore pH = P^{K_a} + \log\dfrac{[HCO_3^-]_O}{[H_2CO_3]_O}$

$= 5.4 + \log\dfrac{25/125}{25/125} = 5.4$

**27.** $[H_2CO_3] = \dfrac{50\times10}{150} = \dfrac{1}{3}$ M

$\therefore pH = \dfrac{1}{2}(P^{K_{a1}} + \log C) = \dfrac{1}{2}(5.4 + \log\dfrac{1}{3}) = 2.94$

### Comprehension X

28. $\overset{+}{N}H_3-CH_3-COOH \underset{K_{a_1}=4\times10^{-3}}{\overset{K_{b_2}=2.5\times10^{-12}}{\rightleftharpoons}}$

$\overset{+}{N}H_3-CH_3-COO^- \underset{K_{b_1}=6.25\times10^{-5}}{\overset{K_{a_2}=1.6\times10^{-10}}{\rightleftharpoons}}$

$NH_2-CH_2-COO^-$

Required $K_b = K_{b_1} = 6.25\times10\times10^{-5}$

29. $\therefore \text{pH} = \frac{1}{2}(P^{K_{a1}} + P^{K_{a2}}) = \frac{1}{2}(2.4+9.8) = 6.1$

### Comprehension XI

30. Moles of Cu reacted $= \frac{6.35\times10^{-3}}{63.5} = 10^{-4}$

$\therefore [Ag^+] = 2 \times$ Mole of Cu reacted $= 2\times10^{-4}$ M

31. $\ln K_{eq} = \frac{\Delta G^\circ}{-RT} = \frac{-120\times10^3}{-8.0\times300} = 50 \Rightarrow K_{eq} >>> 1$

32. $K_{sp} = [Ag^+][BRO_3^-] = (2\times10^{-4})^2 = 4\times10^{-8}\ M^2$

### Comprehension XII

33. $S = \frac{K_{sp}[AgCN]}{[CN^-]} = \frac{1.0\times10^{-16}}{0.02} = 5.0\times10^{-15}$ M

34. $AgCN(s) + \underset{(0.02-S)M}{CN^-} \rightleftharpoons \underset{SM}{Ag(CN)_2^-}$

$K_{eq} = K_{sp}\times K_f = 15$

$15 = \frac{s}{0.02-s} \Rightarrow s = \frac{0.3}{16} = 1.875\times10^{-2}$ M

35. $s = [Ag^+]+[Ag(Cu)_2^-]; \frac{K_{sp}}{[Cu^-]} + K_f\cdot K_{sp}\cdot[Cu^-]$

5M

For minimum solubility: $\frac{ds}{d[Cu^-]} = 0$

or, $-\frac{K_{sp}}{[Cu^-]^2} + K_f\cdot K_{sp} = 0$

$\Rightarrow [Cu^-] = \sqrt{\frac{1}{K_f}} = 2.58\times10^{-9}$ M

## Section D (Assertion–Reason)

1. For acidic solution, pH < 7.0 at 25° C.
2. If there were no common ion effect, $P^H$ should lie in between 7.0 and 7.3
3. $[H^+]_{HCl} = 10^{-4}\ M < [H^+]_{HCOOH} = 3.16\times10^{-2}$ M
4. Dilution results decrease in concentration of BOH (aq), $B^+$ (aq) as well as $OH^-$ (aq).
5. The pH of buffer remains constant on slight dilution but for acidic solution, the dilution results in the decrease in $[H^+]$ and hence, increase in pH.
6. pH of buffer containing $H^A$ and $A^-$ may be less than, greater than or equal to 7.0.
7. $K_a(CH_3COOH) = K_b(NH_4OH)$ and hence, $CH_3COONH_4$ solution is also neutral. But, it undergoes hydrolysis.
8. Theory based
9. Reaction occurs but at equivalent point, pH will be less than 7.
10. As dilution does not change the concentration of ions in saturated solution, the mole of ions will increase.

## Section E (Column Match)

1. True electrolytes produce ions in pure liquid form as well as in solution. Potential electrolytes are molecular in pure liquid state but it produces ions in solution.

2. (P) [OH–] to just start precipitation $= \sqrt{\dfrac{K_{sp}}{[Mg^{2+}]}}$

$$= \sqrt{\frac{2\times10^{-6}}{2\times10^{-3}}} = 10^{-1.5}$$

$$\therefore \; p^{OH} = 1.5 \Rightarrow P^{H} = 12.5$$

(Q) $[OH^-]_{max} = \left(\dfrac{K_{sp}}{[Al^{3+}]}\right)^{1/3} = \left(\dfrac{10^{-28}}{0.1}\right)^{1/3}$

$$= 10^{-9} \text{ M}$$

$$\therefore \; p^{OH}_{min} = 9 \Rightarrow p^{H}_{max} = 5.0$$

(R) $CH_3COOH \rightleftharpoons CH_3COO^- + H^+$

$$\left(0.1 - 0.1\times\frac{10}{11}\right)\text{M} \quad 0.1\times\frac{10}{11}\text{M} \quad ?$$

$$= \frac{0.1}{11}\text{M} \quad = \frac{1}{11}\text{M}$$

Now, $K_a = 10^{-5} = \dfrac{\frac{1}{11}\times[H^+]}{\frac{0.1}{11}} \Rightarrow [H^+] = 10^{-6}$ M

$$\therefore \; p^{H} = 6.0$$

(S) $[H^+] = \sqrt{K_a \cdot C} = \sqrt{10^{-7}\times 0.001} = 10^{-5}$ M

$$\Rightarrow pH = 5.0$$

(T) $A^- + H_2O \rightleftharpoons HA + OH^-$;

$$(6\times10^{-5} - x)\text{M} \quad x\text{M} \quad x\text{M}$$

$$\frac{Kw}{K_a} = 2\times10^{-6}$$

$$2\times10^{-6} = \frac{x\cdot x}{(6\times10^{-5} - x)} \Rightarrow x = 1\times10^{-5}$$

$$\therefore \; p^{OH} = 5 \Rightarrow pH = 9$$

3. Theory based

4. (A) $pH = P^{K_{a1}} + \log\dfrac{[H_2A^-]_O}{[H_3A]_O} = 4.0$

(B) $pH = P^{K_{a2}} + \log\dfrac{[HA^{2-}]_O}{[H_2A^-]_O} = 8.0$

(C) $pH = P^{K_{a3}} + \log\dfrac{[A^{3-}]_O}{[HA^{2-}]_O} = 12.0$

(D) $pH = \dfrac{1}{2}(P^{K_{a1}} + P^{K_{a2}}) = \dfrac{1}{2}(4+8) = 6.0$

(E) $pH = \dfrac{1}{2}(P^{K_{a2}} + P^{K_{a3}}) = \dfrac{1}{2}(8+12) = 10.0$

5. (A) $K_a(H_2O) = K_b(H_2O) = \dfrac{Kw}{[H_2O]} = \dfrac{10^{-14}}{\left(\frac{1000}{18}\right)}$

$$= 1.8\times10^{-16}$$

(C) On increasing temperature, Kw increases and hence, $P^{Kw}$ decreases.

## Section F (Subjective)

### Single-digit Integer Type

1. $[D^+] = \sqrt{K_W} = \sqrt{10^{-16}} = 10^{-8}$ M

$$\Rightarrow P^{D} = -\log[D^+] = 8$$

2. $[HCOOH]_O = \dfrac{1.15\times10^3}{46} = 25 \text{ M} = C$

$$HCOOH + HCOOH \rightleftharpoons HCOOH_2^+ + HCOO^-$$

$$C - x \quad C - x \quad x \quad x$$

From given data: $x = \sqrt{K} = 10^{-3}$ M

∴ Percentage of HCOOH molecules converted into $HCOO^-$

$$= \frac{x}{C}\times100 = \frac{10^{-3}}{25}\times100 = 4\times10^{-3}$$

3. $[NH_3]_O = \dfrac{10/17}{100/0.85}\times10^3 = 5$ M

$$[OH^-] = \sqrt{K_b \cdot C} = \sqrt{\frac{Kw}{K_a(NH_4^+)} \cdot C}$$

$$= \sqrt{\frac{10^{-14}}{5\times10^{-10}} \times 5} = 10^{-2}\,M$$

$$\therefore [H_3O^+] = \frac{Kw}{[OH^-]} = \frac{10^{-14}}{10^{-2}} = 10^{-12}\,M$$

4. $[H^+] = \sqrt{K_a \cdot C} = \sqrt{4\times10^{-10} \times 0.0025} = 10^{-6}\,M$

$\therefore P^H = -\log 10^{-6} = 6$

5. $[H_2SO_3]_O = \frac{1.28}{64} = 0.02\,M$

$$H_2SO_3 \rightleftharpoons H^+ + HSO_3^-$$

$(0.02-x)$ M $\quad xM \quad xM$

Now, $K_a = 10^{-2} = \frac{x \cdot x}{(0.02-x)}$

$\Rightarrow x = 0.01\,M \Rightarrow pH = -\log x = 2$

6. $pH = P^{K_{a1}} + \log\frac{[HC_4H_4O_6^-]_O}{[H_2C_4H_4O_6]_O}$

$= 3.3 + \log\frac{18.8/18.8}{30/150} = 3$

7. Buffer capacity,

$$\beta = -\frac{[H^+]\,\text{added}}{\Delta P^H} = -\frac{0.05/0.2}{(-0.05)} = 5$$

8.

$$HA \quad + \quad OH^- \rightleftharpoons A^- \quad + H_2O$$

$a$ millimole $\quad 36.12\times0.1$ millimole $\quad 0$

Equ.point $\simeq 0 \quad \simeq 0 \quad 3.612$ millimole

$$A^- \quad + \quad H^+ \rightleftharpoons HA$$

3.612 mmole $\quad 18.06\times0.1$ mmole $\quad 0$

Final 1.806 mmole $\quad \simeq 0 \quad 1.806$ mmole

$$\therefore pH = P^{K_a} + \log\frac{[A^-]_O}{[HA]_O} = 5 + \log\frac{1.806}{1.806} = 5$$

9.

$$H_2A^+ \quad + \quad OH^- \rightleftharpoons HA \quad + H_2O$$

$30\times0.06$ mmole $\quad 20\times0.09$ mmole

Final $\simeq 0 \quad \simeq 0 \quad 1.8$ mmole

$$\therefore pH = \frac{1}{2}(P^{K_{a1}} + P^{K_{a2}}) = \frac{1}{2}(2.28+9.72) = 6$$

10. For appearance of only $In^+$ colour,

$$\log\frac{[In^+]}{[In\,OH]} = \frac{4.6-3.4}{2} = 0.6 = \log 4$$

$$\therefore \frac{[In^+]}{[In\,OH]} = 4$$

## Four-digit Integer Type

1. $$C_6H_5NH_2 + H_2O \rightleftharpoons C_6H_5NH_3^+ + OH^-$$

$0.2\,M \quad 0 \quad CM$

Eqn. $(0.2-x)$ M $\quad xM \quad (C+x)$ M

$\simeq 0.2\,M \quad =10^{-8}\,M \quad \simeq CM$

Now, $K_b = \frac{[C_6H_5NH_3^+][OH^-]}{C_6H_5NH_2}$

$\Rightarrow 4\times10^{-10} = \frac{10^{-8}\times C}{0.2}$

$\therefore C = 8\times10^{-3}\,M$

Now, mass of NaOH added

$= \frac{8\times10^{-3}}{1000} \times 500 \times 40 = 0.16\,gm = 160\,mg$

2. $K_a = \frac{[H^+][CH_3COO^-]}{[CH_3COOH]}$

$\Rightarrow 1.8\times10^{-5} = \frac{4\times10^{-4}\times[CH_3COO^-]}{0.2}$

$\therefore [CH_3COO^-] = 9\times10^{-3}\,M$

$\therefore$ Mass of $CH_3COONa$ added

$\frac{9\times10^{-3}\times500}{1000} \times 82 = 0.369\,gm = 369\,mg$

3. $$H_2SO_3 \rightleftharpoons H^+ + HSO_4^-$$

$CM \quad 0 \quad 0$

Final $\simeq 0 \quad CM \quad CM$

HINTS AND EXPLANATIONS

$$HSO_4^- \rightleftharpoons H^+ + SO_4^{2-}$$

Fianl $(C-x)$M  $(C+x)$M $=0.01$ M  $x$M

Now, $1.2\times10^{-2} = \dfrac{0.01\times x}{C-x} \Rightarrow x = \dfrac{6}{11}C$

and $C + x = 0.01 \Rightarrow C = \dfrac{0.01\times11}{17}$ M

**4.** $P^H = P^{K_a} + \log\dfrac{[NaHCo_3]_O}{[H_2Co_3]_O}$

or, $7.4 = 6.1 + \log\dfrac{V\times5}{10\times2} \Rightarrow V = 80$ ml

**5.** $[HSO_4^-]_O = \dfrac{1.8/120}{100/1000} = 0.15$ M

$$HSO_4^- \rightleftharpoons H^+ + SO_4^{2-}$$

$(0.15-x)$M  $x$M  $x$M

$4\times10^{-2} = \dfrac{x\cdot x}{0.15-x}$

$\Rightarrow x = 6\times10^{-2}$ M = 60 millimole/L

**6.** $Al(H_2O)_6^{3+} + H_2O \rightleftharpoons Al(H_2O)_5^{OH^{2+}} + H_3O^+$

$(C-x)$M  $x$M  $x$M $=10^{-3}$M $C\approx0.1$M

Now, $10^{-5}=$

$\therefore$ Mass of $Al(OH)_3$ added = 400 × 0.1 × 133.5 = 5.34 gm = 5340 mg

**7.** $9.18 = P^{K_a} + \log\dfrac{60}{40}$ (1)

$9.00 = P^{K_a} + \log\dfrac{x}{100-x}$ (2)

$\therefore\ x = 50$

**8.** $I_2 + I^- \rightleftharpoons I_3^-$

$\dfrac{12.7}{254}$ $=0.05$M  0.1M  0

Eqn. $(0.05-x)$M  $0.1-x$  $x$

$= \dfrac{0.254}{254} = 0.001$

$\therefore\ x = 0.049$

Now, $K_c = \dfrac{x}{(0.05-x)(0.1-x)}$

$= \dfrac{0.049}{0.001\times0.051} = 960$

**9.** Concentration of $Ca(OH)_2$ in its saturated solution

$= \left(\dfrac{K_{sp}}{4}\right)^{1/3} = \left(\dfrac{3.2\times10^{-5}}{4}\right)^{1/3} = 0.02$ M

Now, $Ca^{2+} + 2OH^- \rightleftharpoons Ca(OH)_2(s)$

0.02M  0.04M

On adding NaOH

0.01 M  (0.02 + 0.8) M = 0.82 M

Equ  $(0.01 - x)$M = 0  $(0.82 - 2x)$ M = 0.8M

$\therefore\ [Ca^{2+}]_{left} = \dfrac{3.2\times10^{-5}}{(0.8)^2} = 5\times10^{-5} << 0.01$

$\therefore$ $Ca(OH)_2$ precipitated = 0.01 mole = 0.01 × 74 = 0.74 gm = 740 mg

**10.** $S = \dfrac{0.0055/550}{100/1000} = 10^{-4}$ M

$\therefore$ $K_{sp}$ of $Ca(pam)_2 = 4S^3 = 4\times10^{-12}$ M

Now, $Ca^{2+} + 2\,pam^- \rightleftharpoons Ca(pam)_2(s)$

$\dfrac{40/40}{10^6}\times10^2$ $=10^{-3}$ M  0.1M

Final =0  0.1M

$\therefore$ $Ca(pam)_2$ participated = $10^{-3}\times10 = 0.01$ mole = 0.01 × 550 = 5.50 gm = 550 mg

CHAPTER

8

# Electrochemistry

## EXERCISE I (JEE MAIN)

### Electrode Potential

1. A metal rod is dipped in a solution of its ions. Its electrode potential is independent of
   (a) temperature of the solution
   (b) concentration of the solution
   (c) area of the metal exposed
   (d) nature of the metal

2. Indicator electrode is
   (a) SHE
   (b) Calomel electrode
   (c) Ag/AgCl electrode
   (d) Quinhydrone electrode

3. The position of some metals in the electrochemical series in decreasing electropositive character is given as: Mg > Al > Zn > Cu > Ag. What will happen if a copper spoon is used to stir a solution of aluminium nitrate?
   (a) The spoon will get coated with aluminium.
   (b) An alloy of copper and aluminium is formed.
   (c) The solution becomes blue.
   (d) No chemical change will take place.

4. Four colourless salt solutions are placed in separate test tubes and a strip of copper is placed in each. Which solution finally turns blue?
   (a) $AgNO_3$ (b) $Pb(NO_3)_2$
   (c) $Zn(NO_3)_2$ (d) $Cd(NO_3)_2$

5. The metal that cannot be obtained on reduction of its oxide by aluminium is
   (a) K (b) Mn
   (c) Cr (d) Fe

6. Beryllium is placed above magnesium in the group II. Beryllium dust, therefore, when added to $MgCl_2$ solution will
   (a) have no effect
   (b) precipitate Mg metal
   (c) precipitate MgO
   (d) lead to dissolution of Be metal

7. A standard hydrogen electrode has zero electrode potential because
   (a) hydrogen is easiest to oxidize
   (b) this electrode potential is assumed to be zero
   (c) hydrogen atom has only one electron
   (d) hydrogen is the lightest element

8. The standard reduction potential values of three metallic cations, X, Y and Z are +0.52, –3.03 and –1.18 V, respectively. The order of reducing power of the corresponding metals is
   (a) Y > Z > X
   (b) X > Y > Z
   (c) Z > Y > X
   (d) Z > X > Y

9. A gas 'X' at 1 atm is bubbled through a solution containing a mixture of 1 M – $Y^-$ and 1 M – $Z^-$ at 25°C. If the reduction potential of $Z^- > Y^- > X$, then,
(a) $Y^-$ will oxidize X and not $Z^-$
(b) $Y^-$ will oxidize $Z^-$ and not X
(c) $Y^-$ will oxidize both X and $Z^-$
(d) $Y^-$ will reduce both X and $Z^-$

10. Standard electrode potential data are useful for understanding the suitability of an oxidant in a redox titration. Some half-cell reactions and their standard potentials are given below:

$MnO_4^-(aq) + 8H^+(aq) + 5e^- \rightarrow Mn^{2+}(aq) + 4H_2O(l) \quad E° = 1.51$ V

$Cr_2O_7^{2-}(aq) + 14H^+(aq) + 6e^- \rightarrow 2Cr^{3+}(aq) + 7H_2O(l) \quad E° = 1.38$ V

$Fe^{3+}(aq) + e^- \rightarrow Fe^{2+}(aq) \quad E° = 0.77$ V

$Cl_2(g) + 2e^- \rightarrow 2Cl^-(aq) \quad E° = 1.40$ V

Identify the only incorrect statement regarding the quantitative estimation of aqueous $Fe(NO_3)_2$
(a) $MnO_4^-$ can be used in aqueous HCl
(b) $Cr_2O_7^{2-}$ can be used in aqueous HCl
(c) $MnO_4^-$ can be used in aqueous $H_2SO_4$
(d) $Cr_2O_7^{2-}$ can be used in aqueous $H_2SO_4$

11. The decreasing order of standard electrode potential of Mg, K, Ba and Ca, is
(a) K, Ca, Ba, Mg (b) Ba, Ca, K, Mg
(c) Ca, Mg, K, Ba (d) Mg, Ca, Ba, K

12. A metal having negative reduction potential, when dipped in the solution of its own ions, has a tendency to
(a) remain as metal atoms
(b) become electrically positive
(c) become electrically negative
(d) be deposited from the solution

13. The calomel electrode is reversible with respect to
(a) $Hg_2^{2+}$ (b) $H^+$
(c) $Hg^{2+}$ (d) $Cl^-$

14. Which one of the following does not get oxidized by bromine water?
(a) $Fe^{2+}$ to $Fe^{3+}$ (b) $Cu^+$ to $Cu^{2+}$
(c) $Mn^{2+}$ to $MnO_4^-$ (d) $Sn^{2+}$ to $Sn^{4+}$

15. In salt bridge, normally KCl is used because
(a) it is a strong electrolyte.
(b) it is good conductor of electricity.
(c) $K^+$ and $Cl^-$ ions have nearly same ionic mobility.
(d) it is an ionic compound.

16. By how much would the oxidizing power of $MnO_4^-/Mn^{2+}$ couple change if the $H^+$ ions concentration is decreased 100 times at 25°C?
(a) increases by 189 mV
(b) decreases by 189 mV
(c) will increase by 19 mV
(d) will decrease by 19 mV

17. The solution of $CuSO_4$, in which copper rod is immersed, is diluted to 10 times. The reduction electrode potential
(a) increases by 0.0295 V
(b) decreases by 0.0295 V
(c) increases by 0.059 V
(d) decreases by 0.059 V

18. The standard reduction potential of oxygen in acidic solution is +1.23 V. What is the standard reduction potential of oxygen in basic solution?
(a) +0.404 V (b) −0.404 V
(c) +2.056 V (d) −2.056 V

19. The standard reduction potentials of $Cu^{2+}$ | Cu and $Cu^{2+}$ | $Cu^+$ are 0.337 V and 0.153 V, respectively. The standard electrode potential of $Cu^+$ | Cu half-cell is
(a) 0.184 V (b) 0.827 V
(c) 0.521 V (d) 0.490 V

20. The electrode potential of hydrogen electrode in neutral solution and 298 K is
(a) −0.413 V (b) zero
(c) −0.826 V (d) +0.413 V

21. Electrode potential will be more for hydrogen electrode at pH (at the same temperature)
(a) 4 (b) 3 (c) 2 (d) 5

22. Saturated solution of $KNO_3$ is used to make 'salt bridge' because
(a) velocity of $K^+$ is greater than that of $NO_3^-$
(b) velocity of $NO_3^-$ is greater than that of $K^+$
(c) velocities of both $K^+$ and $NO_3^-$ are nearly the same
(d) $KNO_3$ is highly soluble in water

EXERCISE I

**23.** The standard reduction potentials of $Pt|Cr_2O_7^{2-}$, $Cr^{+3}$; $Pt|MnO_4^-$, $Mn^{+2}$; $Pt|Ce^{+4}$, $Ce^{+3}$ in the presence of acid are 1.33 V, 1.51 V and 1.61 V, respectively, at 25°C. The decreasing order of oxidizing power is

(a) $Cr_2O_7^{2-} > MnO_4^- > Ce^{+4}$
(b) $MnO_4^- > Cr_2O_7^{2-} > Ce^{+4}$
(c) $Ce^{+4} > MnO_4^- > Cr_2O_7^{2-}$
(d) $MnO_4^- > Ce^{+4} > Cr_2O_7^{2-}$

**24.** The standard reduction potentials at 25°C of $Li^+|Li$, $Ba^{2+}|Ba$, $Na^+|Na$ and $Mg^{2+}|Mg$ are −3.05, −2.73, −2.71 and −2.37 V, respectively. Which is the strongest reducing agent?

(a) Li (b) Ba
(c) Na (d) Mg

**25.** Some standard electrode potentials are given:

$Fe^{2+} + 2e^- \rightarrow Fe$; $E° = -0.440$ V

$Fe^{3+} + 3e^- \rightarrow Fe$; $E° = -0.036$ V

The standard electrode potential for: $Fe^{3+} + e^- \rightarrow Fe^{2+}$, is

(a) −0.476 V (b) −0.404 V
(c) +0.988 V (d) +0.772 V

**26.** The dissociation constant for $CH_3COOH$ is $1.8 \times 10^{-5}$ at 298 K. The electrode potential for the half-cell: Pt | $H_2$ (1 bar) | 0.5 M – $CH_3COOH$, at 298 K is (log 2 = 0.3; log 3 = 0.48; 2.303 $RT/F$ = 0.06)

(a) −0.3024 V (b) −0.1512 V
(c) +0.3024 V (d) +0.1512 V

**27.** At 25°C, the solubility product of CuCl is $2.0 \times 10^{-7}$ and $E^\circ_{Cl^-|CuCl|Cu}$ is 0.128 V. The value of $E^\circ_{Cu^+|Cu}$ is (log 2 = 0.3; 2.303 $RT/F$ = 0.06)

(a) +1.08 V (b) +0.53 V
(c) +1.682 V (d) +0.878 V

**28.** At 25°C, the solubility product of $Pb(OH)_2$ (s) is [Given: $E^\circ_{Pb^{2+}|Pb} = -0.13$ V; $E^\circ_{OH^-|Pb(OH)_2|Pb} = -0.55$ V and $\dfrac{2.303\,RT}{F} = 0.06$]

(a) $1.36 \times 10^{-5}$ (b) $1.0 \times 10^{-7}$
(c) $1.0 \times 10^{-14}$ (d) $1.25 \times 10^{-15}$

**29.** The standard potentials of $MnO_4^-$ | $Mn^{2+}$ and $MnO_2$ | $Mn^{2+}$ electrodes in acid solution are 1.51 and 1.23 V, respectively. Standard electrode potential for the electrode, $MnO_4^-$ | $MnO_2$ in acid solution is

(a) +1.697 V (b) +5.09 V
(c) +0.28 V (d) +1.37 V

**30.** The following reactions represent the reduction of $IO_3^-$ ion into $I^-$ ion in acidic and basic medium. Predict in which medium $IO_3^-$ ion will act as a better oxidizing agent?

$IO_3^- + 6H^+ + 6e^- \rightarrow I^- + 3H_2O$; $E° = +0.907$ V

$IO_3^- + 3H_2O + 6e^- \rightarrow I^- + 6OH^-$; $E° = +0.260$ V

(a) Acid medium (b) Basic medium
(c) Equally in both (d) Not predictable

## Galvanic Cell

**31.** The correct cell diagram for the following reaction and $E°$ for the cell is

$$2AgBr(s) + H_2(g) \rightarrow 2Ag(s) + 2H^+ + 2Br^-$$

$E^\circ_{AgBr|Ag|Br^-} = +0.10$ V.

(a) (Pt) $H_2$ | $H^+$ || $Br^-$ | AgBr | $Br_2$ (Pt); $E° = 0.10$ V
(b) (Pt) $H_2$ | $H^+$ || $Br^-$ | AgBr | $Br_2$ (Pt); $E° = -0.10$ V
(c) (Pt) $Br_2$ | AgBr | $Br^-$ || $H^+$ | $H_2$ (Pt); $E° = 0.10$ V
(d) (Pt) $Br_2$ | AgBr | $Br^-$ || $H^+$ | $H_2$ (Pt); $E° = -0.10$ V

**32.** In an experimental set-up for the measurement of EMF of a half-cell using a reference electrode and a salt bridge, when the salt bridge is removed, the voltage

(a) remains the same
(b) increases to maximum
(c) decreases half the value
(d) drops to zero

**33.** After some time, the voltage of an electrochemical cell becomes zero. This is because

(a) their electrode potential becomes zero.
(b) their reduction potential become equal but have opposite sign.
(c) their reduction potential become equal and have the same sign.
(d) the ions of the electrolyte in the salt bridge stop moving.

34. The cell reaction for the given cell is spontaneous if Pt, $Cl_2$ ($P_1$ atm) | $Cl^-$ | $Cl_2$ ($P_2$ atm), Pt
(a) $P_1 > P_2$ (b) $P_1 < P_2$
(c) $P_1 = P_2$ (d) $P_1 = 1$ atm

35. Which one of the following statements is incorrect regarding an electrochemical cell?
(a) The electrode on which oxidation takes place is called anode.
(b) Anode is the negative pole.
(c) The direction of the current is same as that of the direction of flow of electrons.
(d) The flow of current is partly due to flow of electrons and partly due to flow of ions.

36. When an electric current is drawn from a galvanic cell
(a) EMF suddenly increases.
(b) EMF gradually increases and attains a maximum value.
(c) EMF decreases and finally falls to zero.
(d) EMF must remain constant.

37. Identification of anode and cathode in an electrochemical cell is made by the use of
(a) Galvanometer (b) Salt bridge
(c) Voltmeter (d) Potentiometer

38. For the cell Zn | $Zn^{2+}$ || $Cu^{2+}$ | Cu, if the concentration of both, $Zn^{2+}$ and $Cu^{2+}$ ions are doubled, the EMF of the cell
(a) doubles (b) reduces to half
(c) remains same (d) becomes zero

39. The standard EMF of a galvanic cell can be calculated from
(a) the size of the electrode
(b) the pH of the solution
(c) the amount of metal in the anode
(d) the $E°$ values of the half-cells

40. The value of equilibrium constant for a feasible cell reaction must be
(a) <1 (b) Zero
(c) = 1 (d) >1

41. If the cell reaction is spontaneous then
(a) $\Delta G° = -ve$ (b) $E°_{red} = -ve$
(c) $E°_{red} = +ve$ (d) $\Delta G = -ve$

42. Which of the following pair of metals, when coupled, will give maximum EMF for a voltaic cell?
(a) Fe and Cu (b) Pb and Au
(c) Cu and Au (d) Ca and Cu

43. When a lead storage battery is charged
(a) $PbO_2$ dissolves.
(b) the lead electrode becomes coated with lead sulphate.
(c) sulphuric acid is regenerated.
(d) the amount of acid decreases.

44. Use of lithium metal as an electrode in high energy density batteries is due to
(a) lithium is the lightest element.
(b) lithium has the highest oxidation potential.
(c) lithium is quite reactive.
(d) lithium does not corrode readily.

45. A depolarizer used in dry cell batteries is
(a) $NH_4Cl$ (b) $MnO_2$
(c) KOH (d) $Na_3PO_4$

46. Which is correct about fuel cells?
(a) Cell continuously run as long as fuels are supplied.
(b) These are more efficient and free from pollution.
(c) These are used to provide power and drinking water to astronauts in space programme.
(d) All of these

47. When a lead storage battery is discharged
(a) $SO_2$ is evolved
(b) lead sulphate is consumed
(c) lead is formed
(d) sulphuric acid is consumed

48. For a cell reaction involving a two-electron change, the standard EMF of the cell is found to be 0.295 V at 25°C. The equilibrium constant of the reaction at 25°C will be
(a) $1 \times 10^{10}$ (b) $1 \times 10^{-10}$
(c) $29.5 \times 10^{-2}$ (d) $2 \times 10^{10}$

49. The $E_{Cell}$ for Ag(s) | AgI (satd) || $Ag^+$ (0.10 M) | Ag (s) is +0.413 V. What is the value of $K_{sp}$ of AgI?
(a) $1.0 \times 10^{-8}$ (b) $1.0 \times 10^{-7}$
(c) $1.0 \times 10^{-14}$ (d) $1.0 \times 10^{-16}$

EXERCISE I

**50.** Assuming that hydrogen behaves as an ideal gas, what is the EMF of the cell at 25°C if $P_1 = 600$ mm and $P_2 = 420$ mm: Pt | $H_2$ ($P_1$) | HCl | $H_2$ ($P_2$) | Pt? [Given: 2.303 $RT/F = 0.06$, log 7 = 0.85]

(a) −0.0045 V (b) −0.0 V
(c) +0.0045 V (d) +0.0015 V

**51.** For the electrochemical cell, M | $M^+$ || $X^-$ | X, $E^\circ_{M^+|M} = 0.44$ V and $E^\circ_{X|X^-} = 0.33$ V. From this data, one can deduce that

(a) $M + X \rightarrow M^+ + X^-$ is the spontaneous reaction
(b) $M^+ + X^- \rightarrow M + X$ is the spontaneous reaction
(c) $E_{cell} = 0.77$ V
(d) $E_{cell} = -0.77$ V

**52.** The EMF of the cell: Zn | $Zn^{2+}$ (0.01 M) || $Fe^{2+}$ (0.001 M) | Fe at 298 K is 0.2905 V, then the value of equilibrium constant for the cell reaction is

(a) $e^{\frac{0.32}{0.0295}}$ (b) $10^{\frac{0.32}{0.0295}}$
(c) $10^{\frac{0.26}{0.0295}}$ (d) $10^{\frac{0.32}{0.0591}}$

**53.** For the cell reaction: $4Br^- + O_2 + 4H^+ \rightleftharpoons 2Br_2 + 2H_2O$; $E^\circ = 0.18$ V. The value of (log $K_C$) at 298 K is [2.303 $RT/F = 0.06$]

(a) 12 (b) 6
(c) 18 (d) 3

**54.** The standard EMF for the cell reaction: Zn(s) + $Cu^{2+}$(aq) → $Zn^{2+}$(aq) + Cu(s) is 1.10 volts at 25°C. The EMF of the cell reaction when 0.1 M $Cu^{2+}$ and 0.1 M $Zn^{2+}$ solutions are used at 25°C is

(a) 1.10 V (b) 1.041 V
(c) −1.10 V (d) −1.041 V

**55.** For the cell: Ni | $Ni^{2+}$ || $Cu^{2+}$ | Cu; $E^\circ = 0.77$ V. By which of the following activity, $E_{cell}$ will increase?

(a) On decreasing [$Ni^{+2}$]
(b) On decreasing [$Cu^{+2}$]
(c) On increasing mass of Ni electrode
(d) On increasing mass of Cu electrode

**56.** The standard EMF of a Daniel cell at 298 K is $E_1$. When the concentration of $ZnSO_4$ is 1.0 M and that of $CuSO_4$ is 0.01 M, the EMF becomes $E_2$ at 298 K. The correct relationship between $E_1$ and $E_2$ is

(a) $E_1 = E_2$ (b) $E_2 = 0$
(c) $E_1 > E_2$ (d) $E_1 < E_2$

**57.** For a reaction A(s) + $2B^+$ → $A^{2+}$ + B(s); $K_C$ has been found to be $10^{12}$. The EMF of the cell is

(a) 0.354 V (b) 0.708 V
(c) 0.534 V (d) 0.453 V

**58.** An electrochemical cell is set up as follows:

Pt | $H_2$(g) (1 atm) | 0.001 M HCl || 0.1 M HA | $H_2$(g) (1 atm) | Pt

EMF of this cell is zero because (p$K_a$ of HA = 5)

(a) molar concentrations of acids are different
(b) Temperature is constant
(c) pH of two solutions are same
(d) both are standard hydrogen electrodes

**59.** The reaction: $H_2$(g) + 2AgCl(s) → $2H^+$(aq) + $2Cl^-$(aq) + 2Ag(s) occurs in the galvanic cell

(a) Ag | AgCl (s) | KCl(aq) | $AgNO_3$(aq) | Ag
(b) Pt | $H_2$(g) | HCl(aq) | $AgNO_3$(aq) | Ag
(c) Pt | $H_2$(g) | HCl(aq) | AgCl(s) | Ag
(d) Pt | $H_2$(g) | KCl(aq) | AgCl(s) | Ag

**60.** The standard reduction potentials in acidic conditions are 0.77 V and 0.53 V, respectively, for $Fe^{3+}$ | $Fe^{2+}$ and $I_3^-$ | $I^-$ couples. The equilibrium constant for the reaction: $2Fe^{3+} + 3I^- \rightleftharpoons 2Fe^{2+} + I_3^-$, is (2.303 $RT/F = 0.06$)

(a) $2 \times 10^8$ (b) $10^8$
(c) $10^4$ (d) $10^{-8}$

**61.** The following electrochemical cell has been set up: Pt(s)|$Fe^{3+}$, $Fe^{2+}$ (a = 1) || $Ce^{4+}$, $Ce^{3+}$ (a = 1) | Pt(s); $E^\circ(Fe^{3+}|Fe^{2+}) = 0.77$ V; $E^\circ(Ce^{4+}|Ce^{3+}) = 1.61$ V. If an ammeter is connected between the two platinum electrodes, predict the direction of flow of current. Will the current increase or decrease with time?

(a) Ce electrode to Fe electrode, decrease
(b) Ce electrode to Fe electrode, increase
(c) Fe electrode to Ce electrode, decrease
(d) Fe electrode to Ce electrode, increase

**62.** For the reaction: $H_2$ (1 bar) + 2AgCl(s) $\rightleftharpoons$ 2Ag(s) + $2H^+$ (0.1 M) + $2Cl^-$ (0.1 M); $\Delta G^0 = -48{,}250$ J at 25°C. The EMF of cell in which the given reaction takes place is

(a) 0.25 V (b) 0.37 V
(c) 0.13 V (d) 0.0.49 V

63. A cell contains two hydrogen electrodes. The negative electrode is in contact with a solution of $10^{-6}$ M hydrogen ions. The EMF of the cell is 0.118 V at 25°C. The concentration of hydrogen ions at the positive electrode is

(a) $10^{-6}$ M (b) $10^{-3}$ M
(c) $10^{-4}$ M (d) $10^{-5}$ M

64. The standard potentials of $OCl^-/Cl^-$ and $Cl^-/Cl_2$ are 0.94 V and −1.36 V, respectively. The $E^o$ value of $OCl^-/Cl_2$ will be

(a) 3.24 V (b) −0.42 V
(c) −2.30 V (d) 0.52 V

65. From the following $E°$ values for the half-cells:

(i) $D \rightarrow D^{2+} + 2e^-$; $E° = -1.5$ V
(ii) $B^+ + e^- \rightarrow B$; $E° = -0.5$ V
(iii) $A^{3-} \rightarrow A^{2-} + e^-$; $E° = 1.5$ V
(iv) $C^{2+} + e^- \rightarrow C^+$; $E° = +0.5$ V

Which combination of two half-cells would result in a cell with largest potential?

(a) i and iii (b) i and iv
(c) iii and iv (d) ii and iv

## Electrolysis

66. Which of the following statements does not differentiate between electrochemical cell and electrolytic cell?

(a) Spontaneous or non-spontaneous nature of the chemical process.
(b) Chemical reactions occurring at the electrodes
(c) Positive and negative nature of anode.
(d) EMF measurement.

67. The electrode through which electrons enter the electrolytic solution is

(a) cathode
(b) anode
(c) may be anode or cathode
(d) both, anode and cathode

68. Which process occurs in the electrolysis of an aqueous solution of nickel chloride at nickel anode?

(a) $Ni \rightarrow Ni^{2+} + 2e^-$ (b) $Ni^{2+} + 2e^- \rightarrow Ni$
(c) $2Cl^- \rightarrow Cl_2 + 2e^-$ (d) $2H^+ + 2e^- \rightarrow H_2$

69. If mercury is used as a cathode during the electrolysis of an aqueous NaCl solution, the ions discharged at cathode are

(a) $H^+$ (b) $Na^+$
(c) $OH^-$ (d) $Cl^-$

70. Electrochemical equivalent is more for

(a) Hydrogen (b) Silver
(c) Copper (d) Zinc

71. The electrolytic bath used in gold plating of copper articles contains

(a) Molten gold
(b) Copper sulphate(aq)
(c) $AuCl_3$(aq)
(d) $AuCl_3$ + NaCN(aq)

72. Copper can be deposited from acidified copper sulphate and alkaline cuprous cyanide. If the same current is passed for a definite time

(a) the amount of copper deposited from acidic copper sulphate will be higher
(b) the amount of copper deposited from alkaline cuprous cyanide will be higher
(c) the same amount of copper will be deposited
(d) copper will not deposit in either case

73. In the electrolytic cell, flow of electrons is from

(a) Cathode to anode in solution
(b) Cathode to anode through external supply
(c) Cathode to anode through internal supply
(d) Anode to cathode through internal supply

74. Electrolytic cell is used to convert

(a) Chemical energy to electrical energy
(b) Electrical energy to chemical energy
(c) Chemical energy to mechanical energy
(d) Electrical energy to mechanical energy

75. Faraday's law of electrolysis fails when

(a) temperature is increased
(b) inert electrodes are used
(c) a mixture of electrolytes is used
(d) in none of these cases

76. Using same quantity of current, which among Na, Mg and Al is deposited more (by mass) during electrolysis of their molten salts?

(a) Na (b) Mg
(c) Al (d) All in same

**77.** A certain current liberated 0.50 g of hydrogen in 2 h. How many grams of copper can be liberated by the same current flowing for the same time in a copper sulphate solution? (Cu = 63.5)

(a) 12.7 g (b) 15.88 g
(c) 31.75 g (d) 63.5 g

**78.** A current of 3.7 A is passed for 6 h between nickel electrodes in 0.50 L of 2 M solution of $Ni(NO_3)_2$. The molarity of $Ni^{2+}$ at the end of electrolysis is

(a) 1.172 M (b) 0.172 M
(c) 0.586 M (d) 2 M

**79.** The current efficiency of an electrodeposition of copper metal in which 9.8 g of copper is deposited by a current of 3 A for 10000 s, from aqueous copper sulphate solution, is about

(a) 60% (b) 99%
(c) 92% (d) 75%

**80.** On passing electricity through dilute $H_2SO_4$ solution, the mass of substances liberated at the cathode and anode are in the ratio of

(a) 1:8 (b) 8:1
(c) 1:32 (d) 1:16

**81.** The electrochemical equivalents of two substances are $E_1$ and $E_2$. The current that must pass to deposit the same amount at the cathodes in the same time must be in the ratio of

(a) $E_1{:}E_2$ (b) $E_2{:}E_1$
(c) $(E_1 - E_2){:}E_2$ (d) $E_1{:}(E_2 - E_1)$

**82.** The same quantity of electricity is passed through one molar solution of $H_2SO_4$ and one molar solution of HCl. The amount of hydrogen evolved from $H_2SO_4$ as compared to that from HCl is

(a) the same
(b) twice as such
(c) one half as such
(d) dependent on size of electrode

**83.** In the electrolysis of acidified $AgNO_3$ solution using Pt-electrodes, the anode reaction is

(a) $2NO_3^- \rightarrow 2NO_2 + O_2 + 2e^-$
(b) $NO_3^- \rightarrow NO + \frac{1}{2}O_2 + e^-$
(c) $2H_2O \rightarrow 4H^+ + O_2 + 4e^-$
(d) $Pt \rightarrow Pt^{3+} + 3e^-$

**84.** Two platinum electrodes were immersed in a solution of $CuSO_4$ and electric current was passed through the solution. After some time, it was found that colour of $CuSO_4$ disappeared with the evolution of gas at the electrode. The colourless solution contains

(a) platinum sulphate
(b) copper sulphate
(c) copper hydroxide
(d) sulphuric acid

**85.** A solution containing 1.0 M each of $Cu(NO_3)_2$, $Mg(NO_3)_2$, $AgNO_3$, $Hg(NO_3)_2$ is being electrolysed using inert electrodes. The values of standard electrode potential are: $Ag^+ | Ag = 0.80$ V, $Hg^{2+} | Hg = 0.79$ V, $Cu^{2+} | Cu = 0.34$ V, $Mg^{2+} | Mg = -2.37$ V. With increasing voltage, the sequence of deposit of metals on the cathode will be

(a) Ag, Hg, Cu, Mg
(b) Mg, Cu, Hg, Ag
(c) Ag, Mg, Cu
(d) Cu, Hg, Ag

**86.** During the electrolysis of an aqueous salt solution, the pH in the space near one of the electrode was increased and the other one was decreased. The salt solution was

(a) NaCl (very dilute) (b) $ZnCl_2$
(c) NaCl (Conc.) (d) $Cu(NO_3)_2$

**87.** Two electrolytic cells, one containing acidified ferrous chloride and another acidified ferric chloride, are connected in series. The mass ratio of iron deposited at cathodes in the two cells will be

(a) 3:1 (b) 2:3
(c) 1:1 (d) 3:2

**88.** A galvanic cell is set up from a zinc bar weighing 100 g and 1.0 L of 1.0 M copper sulphate solution. How long would the cell run if it is assumed to deliver a steady current of 1.0 A? (Zn = 65.4)

(a) 53.6 h (b) 26.8 h
(c) 81.97 h (d) 40.99 h

**89.** An ion is a reduced to the element when it absorbs $6 \times 10^{20}$ electrons. The number of equivalents of the ion is

(a) 0.10 (b) 0.01
(c) 0.001 (d) 0.0001

**90.** In the lead storage battery, the anode reaction is $Pb(s) + HSO_4^- + H_2O \rightarrow PbSO_4(s) + H_3O^+ + 2e^-$. How many grams of Pb will be used up to deliver 1 A for 100 h? (Pb = 208)

(a) 776 g (b) 388 g
(c) 194 g (d) 0.1 g

**91.** The copper anode of a cell containing silver nitrate solution weighs 60.0 g. After passing current for some time, it is found that 3.24 g of silver if deposited on the platinum cathode. What is the final weight of the anode? (Ag = 108, Cu = 64)

(a) 0.96 g (b) 60 g
(c) 59.04 g (d) 60.96 g

**92.** The quantity of electricity required for the reduction of 1 mole of $Fe_2O_3$ to Fe is

(a) 1F (b) 0.33F
(c) 3F (d) 6F

**93.** Three faradays of electricity is passed through molten $Al_2O_3$, aqueous solutions of $CuSO_4$ and molten NaCl. The amounts of Al, Cu and Na deposited at the cathodes will be in the molar ratio of

(a) 1:2:3 (b) 3:2:1
(c) 1:1.5:3 (d) 6:3:2

**94.** Electrolysis of a solution of $HSO_4^-$ ions produces $S_2O_8^{2-}$. Assuming 75% current efficiency, what current should be employed to achieve a production rate of 1 mole of $S_2O_8^{2-}$ per hour?

(a) 71.5 A (b) 35.7 A
(c) 53.0 A (d) 143 A

**95.** The number of Faradays required to produce 1 g-atom of Mg from $MgCl_2$ is

(a) 1 (b) 2
(c) 0.5 (d) 4

**96.** Passage of 96,500 coulomb of electricity liberates ______ L of $O_2$ at 273°C and 2 atm during electrolysis.

(a) 5.6 (b) 16.8
(c) 22.4 (d) 11.2

**97.** An electrolytic cell contains a solution of $Ag_2SO_4$ and platinum electrodes. A current is passed until 1.6 g of $O_2$ has been liberated at anode. The amount of silver deposited at cathode would be

(a) 108.0 g (b) 1.6 g
(c) 10.8 g (d) 21.6 g

**98.** When 12,000 coulombs of electricity is passed through the electrolyte, 3.0 g of a metal of atomic mass 96.5 g/mol is deposited. The electro-valency of the metal cation in the electrolyte is

(a) +4 (b) +3
(c) +2 (d) –4

**99.** How many electrons flow when a current of 5 A is passed through a solution for 200 s?

(a) $6.022 \times 10^{23}$ (b) $6.24 \times 10^{21}$
(c) $6.024 \times 10^{21}$ (d) $6.022 \times 10^{20}$

**100.** The current of 9.65 A flowing for 10 min deposits 3.0 g of a metal. The equivalent weight of the metal is

(a) 10 (b) 30
(c) 50 (d) 96.5

**101.** If a current of 1.0 A is drawn from the Daniel cell for 96.5 min, the cathode will gain in weight by (Cu = 63.5, Zn = 65.4)

(a) 1.905 g (b) 1.962 g
(c) 3.81 g (d) 3.924 g

**102.** The current required to produce oxygen at the rate of 2.8 ml (0°C, 1 atm) per second during electrolysis of acidulated water is

(a) 48.25 A/s (b) 24.12 A/s
(c) 96.5 A/s (d) 0.0048 A/s

**103.** Sodium amalgam is prepared by electrolysis of aqueous NaCl using 10 g mercury as cathode. How many Faraday of electricity is required to prepare 18.7% Na-amalgam, by weight, with a current efficiency of 50%?

(a) 0.1 F (b) 0.2 F
(c) 0.05 F (d) 0.16 F

**104.** Element A (atomic mass = 112) and element B (atomic mass = 27) form chlorides. Solutions of these chlorides are electrolysed separately and it is found that when the same quantity of electricity is passed, 5.6 g of A was deposited while only 0.9 g of B was deposited. The valency of B is 3. The valency of A is

(a) 1 (b) 2
(c) 3 (d) 4

105. The same current was passed successively through solution of zinc–ammonium sulphate and nickel–ammonium sulphate rendered alkaline with ammonia. The weights of zinc and nickel deposited in a certain time were found to be 22.89 g and 20.55 g, respectively. Given that the chemical equivalent weight of zinc is 32.7, what is the chemical equivalent weight of nickel?

(a) 58.71 (b) 29.36
(c) 14.39 (d) 36.42

## Conductance

106. Which of the following solutions have highest resistance?

(a) 1N – NaCl (b) 0.05N – NaCl
(c) 2N – NaCl (d) 0.1N – NaCl

107. Variation of molar conductance of an electrolytic solution with temperature is that it

(a) increases with increase of temperature
(b) decreases with increase of temperature
(c) first increases then decreases
(d) is not affected by temperature

108. Which pure substance will not conduct electricity?

(a) Molten NaCl (b) Molten KOH
(c) Liquefied HCl (d) Liquid Hg

109. The correct order of molar conductance at infinite dilution of LiCl, NaCl and KCl is

(a) LiCl > NaCl > KCl (b) KCl > NaCl > LiCl
(c) NaCl > KCl > LiCl (d) LiCl > KCl > NaCl

110. The molar conductance of a strong electrolyte at infinite dilution

(a) tends to a finite value, which is above that at higher concentration
(b) tends to a finite value, which is below that at higher concentration
(c) tends to zero
(d) tends to a finite value, which is equal to that at high concentration

111. The best conductor of electricity is a 0.1 M solution of

(a) Boric acid (b) Sulphuric acid
(c) Acetic acid (d) Propanoic acid

112. The specific conductance of AgCl solution in water was determined to be $1.8 \times 10^{-6}\ \Omega^{-1}\ cm^{-1}$ at 298 K. The molar conductances at infinite dilution, of $Ag^+$ and $Cl^-$ are 67.9 and 82.1 $\Omega^{-1}\ cm^2\ mol^{-1}$, respectively. What is the solubility of AgCl in water?

(a) $1.2 \times 10^{-8}$ M (b) $1.44 \times 10^{-10}$ M
(c) $1.2 \times 10^{-5}$ M (d) $1.44 \times 10^{-16}$ M

113. Equivalence conductance at infinite dilution of $NH_4Cl$, NaOH and NaCl are 129.8, 217.4 and 108.9 $\Omega^{-1}\ cm^2\ mol^{-1}$, respectively. If the equivalent conductance of 0.01 N solution of $NH_4OH$ is 9.532 $\Omega^{-1}\ cm^2\ mol^{-1}$, then the degree of dissociation of $NH_4OH$ at this temperature is

(a) 0.04% (b) 2.1%
(c) 4.0% (d) 44.7%

114. The resistance of 1 M – $CH_3COOH$ solution is 250 Ω, when measured in a cell of cell constant 125 $m^{-1}$. The molar conductivity, in $\Omega^{-1}\ m^2\ mol^{-1}$ is

(a) $5.0 \times 10^{-4}$ (b) 500
(c) $2 \times 10^{-3}$ (d) 200

115. How does the electrical conductivity of 20 ml of 0.2 M – $MgSO_4$ change when 0.5 M – $Ba(OH)_2$ solution is gradually added in it, to excess?

(a) decreases continuously
(b) increases continuously
(c) increases and then decreases
(d) decreases and then increases

116. The equivalent conductivity (in $\Omega^{-1}\ cm^2\ eq^{-1}$) of 1.0 M – $H_2SO_4$ solution of specific conductance $2.6 \times 10^{-1}\ cm^{-1}$, is

(a) $1.3 \times 10^2$ (b) $6.5 \times 10^1$
(c) $1.3 \times 10^{-1}$ (d) $2.6 \times 10^2$

117. The molar conductance of a 0.01 M solution of acetic acid was found to be 16.30 $\Omega^{-1}\ cm^{-1}\ mol^{-1}$ at 25°C. The ionic conductances of hydrogen and acetate ions at infinite dilution are 349.8 and 40.9 $\Omega^{-1}\ cm^{-1}\ mol^{-1}$, respectively, at the same temperature. What percentage of acetic acid is dissociated at this concentration?

(a) 0.04172% (b) 4.172%
(c) 41.72% (d) 0.4172%

**118.** The distance between two electrodes of a cell is 2.5 cm and area of each electrode is 5 $cm^2$. The cell constant is

(a) 0.5 $m^{-1}$ (b) 12.5 $cm^3$
(c) 2.0 cm (d) 50 $m^{-1}$

**119.** The molar conductivity of $NH_4Cl$, $OH^-$ and $Cl^-$ at infinite dilution is 150, 200 and 75 $\Omega^{-1}$ $cm^2$ $mol^{-1}$, respectively. If the molar conductivity of a 0.01 M–$NH_4OH$ solution is 22 $\Omega^{-1}$ $cm^2$ $mol^{-1}$, then its degree of dissociation is

(a) 0.146 (b) 0.063
(c) 0.080 (d) 0.293

**120.** Calculate the ionic product of water at 25°C from the following data:

Conductivity of water = $5.5 \times 10^{-6}$ mho $m^{-1}$
$\lambda^\circ_{H^+} = 0.035$ mho $m^2$ $mol^{-1}$
$\lambda^\circ_{OH^-} = 0.020$ mho $m^2$ $mol^{-1}$

(a) $2 \times 10^{-14}$ $M^2$ (b) $1 \times 10^{-7}$ M
(c) $1 \times 10^{-8}$ $M^2$ (d) $1 \times 10^{-14}$ $M^2$

**121.** Calculate $K_a$ of acetic acid if its 0.05 M solution has molar conductivity of $7.814 \times 10^{-4}$ $\Omega^{-1}$ $m^2$ $mol^{-1}$ at 25°C. Given: $\Lambda^\circ_m$ for $CH_3COOH = 3.907 \times 10^{-2}$ $\Omega^{-1}$ $m^2$ $mol^{-1}$.

(a) $2 \times 10^{-5}$ (b) $1.8 \times 10^{-5}$
(c) $4 \times 10^{-4}$ (d) 0.02

**122.** Equal volumes of 0.015 M – $CH_3COOH$ and 0.015 M – NaOH solutions are mixed together. What would be the molar conductivity of mixture if conductivity of $CH_3COONa$ is $6.3 \times 10^{-4}$ S $cm^{-1}$?

(a) 0.84 S $cm^2$ $mol^{-1}$ (b) 8.4 S $cm^2$ $mol^{-1}$
(c) 84 S $cm^2$ $mol^{-1}$ (d) 42 S $cm^2$ $mol^{-1}$

**123.** Calculate $\Lambda^\infty_m$ (in $\Omega^{-1}$ $cm^2$ $mol^{-1}$) for $SrCl_2$ at 25°C, from the following data:

| Conc. | 0.25 M | 1.0 M |
|---|---|---|
| $\Lambda_m$ (in $\Omega^{-1}$ $cm^2$ $mol^{-1}$) | 260 | 250 |

(a) 270 (b) 265
(c) 240 (d) 275

**124.** The resistance of a solution A is 50 Ω and that of solution B is 100 Ω, both solutions being taken in the same conductivity cell. If equal volumes of solution A and B are mixed, what will be the resistance of the mixture using the same cell? Assume that there is no increase in the degree of dissociation of A and B on mixing.

(a) 150 Ω (b) 75 Ω
(c) 33.33 Ω (d) 66.67 Ω

**125.** In a conductivity cell, the two platinum electrodes, each of area 10 $cm^2$ are fixed 1.5 cm apart. The cell contained 0.05 N solution of a salt. If the two electrodes are just half dipped into the solution which has a resistance of 50 Ω, the equivalent conductance of the salt solution, in $\Omega^{-1}$ $cm^2$ $eq^{-1}$, is

(a) 120 (b) 60
(c) 240 (d) 3000

# EXERCISE II (JEE ADVANCED)

## Section A (Only one Correct)

1. A student made the following observations in the laboratory:
   (I) Clean copper metal did not react with 1 M – $Pb(NO_3)_2$ solution.
   (II) Clean lead metal dissolves in 1 M – $AgNO_3$ solution and crystals of Ag metal appeared.
   (III) Clean silver metal did not react with 1 M – $Cu(NO_3)_2$ solution.
   The order of decreasing reducing character of the three metals is
   (a) Cu > Pb > Ag (b) Cu > Ag > Pb
   (c) Pb > Cu > Ag (d) Pb > Ag > Cu

2. We have an oxidation–reduction system: $[Fe(CN)_6]^{3-} + e^- \rightleftharpoons [Fe(CN)_6]^{4-}$; $E° = +0.36$ V. The ratio of concentrations of oxidized and reduced from at which the potential of the system becomes 0.24 V, is [Given: 2.303 $RT/F = 0.06$)
   (a) 2:1 (b) 1:2
   (c) 1:20 (d) 1:100

3. The standard reduction potential for the process: $[Co(H_2O)_6]^{3+} + e^- \rightarrow [Co(H_2O)_6]^{2+}$ is 1.8 V. The standard reduction potential for the process: $[Co(NH_3)_6]^{3+} + e^- \rightarrow [Co(NH_3)_6]^{2+}$ is 0.1 V. Which of the complex ion, $[Co(H_2O)_6]^{2+}$ or $[Co(NH_3)_6]^{2+}$ can be oxidized to the corresponding cobalt (III) complex, by oxygen, in basic medium, under standard condition? [Given: $E^\circ_{O_2|OH^-} = 0.4$ V]
   (a) $[Co(H_2O)_6]^{2+}$ (b) $[Co(NH_3)_6]^{2+}$
   (c) both (d) none of these

4. The standard reduction potential for the reactions: $Ag^+ + e^- \rightarrow Ag$ and $Ag(NH_3)_2^+ + e^- \rightarrow Ag + 2NH_3$ are +0.79 V and +0.37 V, respectively. From these values and the Nernst equation, what should be $K_f$ for the $Ag(NH_3)_2^+$ ion? [Given: 2.303 $RT/F = 0.06$]
   (a) $1.0 \times 10^{-7}$ (b) $1.0 \times 10^{7}$
   (c) $2.15 \times 10^{19}$ (d) $4.64 \times 10^{-20}$

5. The overall formation constant for the reaction of 6 mole of $CN^-$ with cobalt (II) is $1 \times 10^{19}$. What is the formation constant for the reaction of 6 moles of $CN^-$ with cobalt (III)? Given that
   $Co(CN)_6^{3-} + e^- \rightarrow Co(CN)_6^{4-}$; $E° = -0.83$ V
   $Co^{3+} + e^- \rightarrow Co^{2+}$; $E° = +1.81$ V
   and 2.303 $RT/F = 0.06$.
   (a) $1.0 \times 10^{63}$ (b) $1.0 \times 10^{25}$
   (c) $1.0 \times 10^{-25}$ (d) $1.0 \times 10^{-63}$

6. For the process: $Cu^{2+} + 2e^- \rightarrow Cu$; $\log[Cu^{2+}]$ vs. $E_{red}$ graph is shown in the figure, where OA = 0.34 V. The electrode potential of the half-cell of $Cu|Cu^{2+}$ (0.1 M) will be [2.303 $RT/F = 0.06$]

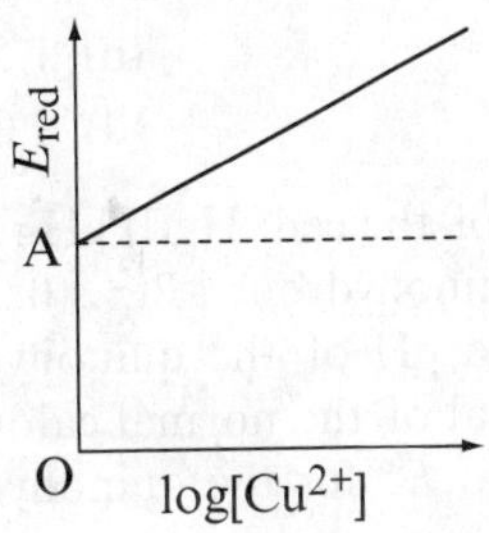


   (a) −0.31 V (b) +0.31 V
   (c) −0.37 V (d) +0.37 V

7. Tell which of the following statements accurately describes the effect of adding $CN^-$ to the cathode of a cell with a cell reaction: $Cd + 2Ag^+ \rightarrow 2Ag + Cd^{2+}$, $E° = 1.2$ V
   (a) $E°$ increases because $Cd(CN)_4^{2-}$ forms
   (b) $E°$ decreases because $Cd(CN)_4^{2-}$ forms
   (c) $E°$ increases because $Ag(CN)_2^-$ forms
   (d) $E°$ decreases because $Ag(CN)_2^-$ forms

8. For a electrochemical cell $Zn | Zn^{2+}$ ($C_1$ M) || $Cu^{2+}$ ($C_2$ M) | Cu, the decrease in free energy at a given temperature is a function of
   (a) $\ln C_1$ (b) $\ln C_2$
   (c) $\ln C_2 \cdot C_1$ (d) $\ln C_1/C_2$

9. Select the correct option if it is known that $K_{sp}$ (AgCl) > $K_{sp}$ (AgBr) > $K_{sp}$ (AgI)
   (a) $E^\circ_{I^-|AgI|Ag} > E^\circ_{Br^-|AgBr|Ag} > E^\circ_{Cl^-|AgCl|Ag}$
   (b) $E^\circ_{I^-|AgI|Ag} < E^\circ_{Br^-|AgBr|Ag} < E^\circ_{Cl^-|AgCl|Ag}$
   (c) $E^\circ_{I^-|AgI|Ag} < E^\circ_{Cl^-|AgCl|Ag} < E^\circ_{Br^-|AgBr|Ag}$
   (d) $E^\circ_{I^-|AgI|Ag} = E^\circ_{Br^-|AgBr|Ag} = E^\circ_{Cl^-|AgCl|Ag}$

10. The EMF of a galvanic cell composed of two hydrogen electrodes is 272 mV. What is the pH of the solution in which the anode is immersed if the cathode is in contact with a solution of pH = 3?

(a) 3 (b) 6.7
(c) 7.6 (d) 1.6

11. At what pH does the potential (EMF) for the disproportionation of chlorine change from a negative value to a positive value, assuming 1.0 bar pressure and 1.0 M concentration for all species except hydrogen ion? Given:

$Cl_2 + 2e^- \rightarrow 2Cl^-$ ; $E° = 1.36$ V
$2OCl^- + 4H^+ + 2e^- \rightarrow Cl_2 + 2H_2O$; $E° = 1.63$ V.
[2.303 $RT/F = 0.06$]

(a) 4.5 (b) 1.5
(c) 2.25 (d) 9.0

12. The EMF of the cell: Hg(l) | $Hg_2Cl_2$(s), KCl sol. (1.0N) | Quinohydrone | Pt, is 0.210 V at 298 K. What is the pH of the quinohydrone solution, the potential of the normal calomel electrode is 0.279 V and $E°$ for the quinohydrone electrode is 0.699 V, both at the same temperature. [2.303 $RT/F = 0.06$]

(a) 3.5 (b) 7.0
(c) 1.85 (d) −3.5

13. The potential (EMF) of a cell consisting of an anode of silver in 0.10 M – $AgNO_3$ solution and a cathode of Pt immersed in a solution of 1.6 M – $Cr_2O_7^{2-}$, 0.4 M – $Cr^{3+}$ and 0.1 M – $H^+$ is ($E°_{Ag^+|Ag} = 0.80$ V and $E°_{Cr_2O_7^{2-}|Cr^{3+}} = 1.33$ V) [2.303 $RT/F = 0.06$]

(a) 0.46 V (b) 0.60 V
(c) 0.53 V (d) −0.17 V

14. The EMF of the cell: Zn–Hg ($C_1$M) | $Zn^{2+}$(aq) | Zn–Hg ($C_2$M) at 25°C, if the concentrations of the zinc amalgams are: $C_1$ = 10 g per 100 g of mercury, $C_2$ = 1 g per 100 g mercury, is

(a) 0.059 V (b) 0.0295 V
(c) 0.59 V (d) 0.295 V

15. What is the equilibrium constant of the reaction: $2Fe^{3+} + Au^+ \rightarrow 2Fe^{2+} + Au^{3+}$? Given $E°_{Au^+|Au} = 1.68$ V, $E°_{Au^{3+}|Au} = 1.50$ V, $E°_{Fe^{3+}|Fe^{2+}} = 0.75$ V and 2.303 $RT/F = 0.06$.

(a) $1 \times 10^{22}$ (b) $1 \times 10^{-22}$
(c) $1 \times 10^{-11}$ (d) $1 \times 10^{-72}$

16. What is the EMF of the cell: Pt, $H_2$ (1 atm) | $CH_3COOH$ (0.1 M) || (0.01 M) $NH_4OH$ | $H_2$ (1 atm), Pt? Given: $K_a$ for $CH_3COOH = 1.8 \times 10^{-5}$, $K_b$ for $NH_4OH = 1.8 \times 10^{-5}$, 2.303 $RT/F$ = 0.06, log 1.8 = 0.25)

(a) 0.465 V (b) −0.465 V
(c) −0.2325 V (d) −0.93 V

17. Two electrochemical cells are assembled in which the following reactions occur:

$V^{2+} + VO^{2+} + 2H^+ \rightarrow 2V^{3+} + H_2O$; $E°_{Cell}$ = 0.616 V

$V^{3+} + Ag^+ + H_2O \rightarrow VO^{2+} + Ag(s) + 2H^+$; $E°_{Cell}$ = 0.439 V

If $E°_{Ag^+|Ag} = 0.799$ V, what is $E°_{V^{3+}|V^{2+}}$ ?

(a) −0.256 V (b) +0.256 V
(c) +1.854 V (d) −1.854 V

18. The EMF of cell: $H_2$(g) | Buffer || Normal calomel electrode, is 0.70 V at 25°C, when barometric pressure is 760 mm. What is the pH of the buffer solution? $E°_{Calomel} = 0.28$ V. [2.303 $RT/F = 0.06$]

(a) 3.5
(b) 7.0
(c) tending to zero
(d) tending to 14.0

19. What is the solubility product of a saturated solution of $Ag_2CrO_4$ in water at 298 K if the EMF of the cell: Ag | $Ag^+$ (satd. $Ag_2CrO_4$) || $Ag^+$ (0.1 M) | Ag is 0.162 V at 298 K? [2.303 $RT/F = 0.06$, log 2 = 0.3]

(a) $2.0 \times 10^{-4}$ (b) $3.2 \times 10^{-11}$
(c) $8.0 \times 10^{-12}$ (d) $4.0 \times 10^{-12}$

20. A $Tl^+$|Tl couple was prepared by saturating 0.10 M – KBr with TlBr and allowing $Tl^+$ ions form the insoluble bromide to equilibrate. This couple was observed to have a potential −0.444 V with respect to $Pb^{2+}$ | Pb couple in which $Pb^{2+}$ was 0.10 M. What is the $K_{sp}$ of TlBr. [Given: $E°_{Pb^{2+}|Pb} = -0.126$ V, $E°_{Tl^+|Tl} = -0.336$ V, log 2.5 = 0.4, 2.303 $RT/F = 0.06$]

(a) $4.0 \times 10^{-6}$ (b) $2.5 \times 10^{-4}$
(c) $4.0 \times 10^{-5}$ (d) $6.3 \times 10^{-3}$

**21.** The EMFs of the cell obtained by combining separately Zn and Cu electrodes of a Daniel cell with normal calomel electrodes are 1.083 V and $-0.018$ V, respectively, at 25° C. If the potential of normal calomel electrode is $-0.28$ V, the EMF of the Daniel cell is

(a) 1.065 V (b) 1.101 V
(c) 0.803 V (d) 0.262 V

**22.** A hydrogen electrode placed in a buffer solution of $CH_3COONa$ and $CH_3COOH$ in the ratio's $x{:}y$ and $y{:}x$ has electrode potential values $E_1$ and $E_2$ volts, respectively, at 25°C. The $pK_a$ value of acetic acid is

(a) $(E_1 + E_2)/0.118$ (b) $(E_2 - E_1)/0.118$
(c) $-(E_1 + E_2)/0.118$ (d) $(E_1 - E_2)/0.118$

**23.** The standard reduction potential at 25°C of the reaction $2H_2O + 2e^- \rightleftharpoons H_2 + 2OH^-$ is $-0.84$ V. Calculate equilibrium constant for the reaction: $2H_2O \rightleftharpoons H_3O^+ + OH^-$ at 25°C. ($2.303\ RT/F = 0.06$)

(a) $10^{-14}$ (b) $10^{14}$
(c) $10^{7}$ (d) $10^{-7}$

**24.** The voltage of the cell given below is $-0.61$ V.

Pt | $H_2$ (1 bar) | $NaHSO_3$ (0.4 M), $Na_2SO_3$ ($6.4 \times 10^{-2}$ M) || $Zn^{2+}$ (0.4 M) | Zn

If $E^\circ_{Zn^{2+}|Zn} = -0.76$ V, Calculate $K_{a2}$ of $H_2SO_4$. ($2.303\ RT/F = 0.06$)

(a) $3.2 \times 10^{-4}$ (b) $3.2 \times 10^{-2}$
(c) $3.2 \times 10^{-3}$ (d) $6.4 \times 10^{-7}$

**25.** Estimate the cell potential of a Daniel cell having 1.0 M – $Zn^{2+}$ and originally having 1.0 M – $Cu^{2+}$ after sufficient ammonia has been added to the cathode compartment to make the $NH_3$ concentration 2.0 M. Given: $E^\circ_{Zn^{2+}|Zn} = -0.76$ V, $E^\circ_{Cu^{2+}|Cu} = +0.34$ V, $K_f$ for $Cu(NH_3)_4^{2+} = 1 \times 10^{12}$ [$2.303\ RT/F = 0.06$].

(a) 1.10 V (b) 0.704 V
(c) 0.396 V (d) 1.496 V

**26.** The minimum mass of NaOH required to be added in RHS to consume all the $H^+$ present in RHS of the cell of EMF, +0.70 V at 25°C, before its use.

Zn | $Zn^{2+}$ (0.01 M) || HCl ($V$ = 500 ml) | $H_2$ (1 bar) | Pt

Given: $E^\circ_{Zn^{2+}|Zn} = -0.76$ V; $2.303\ RT/F = 0.06$

(a) 2.0 g (b) 4.0 g
(c) 0.2 g (d) 0.4 g

**27.** The EMF of the cell: Ag, AgCl in 0.1 M – KCl || satd. $NH_4NO_3$ || 0.1 M – $AgNO_3$, Ag is 0.42 V at 25°C. 0.1 M – KCl is 50% dissociated and 0.1 M – $AgNO_3$ is 40% dissociated. The solubility product of AgCl is ($2.303\ RT/F = 0.06$)

(a) $1.0 \times 10^{-10}$ (b) $2.0 \times 10^{-9}$
(c) $1.0 \times 10^{-9}$ (d) $2.0 \times 10^{-10}$

**28.** On the basis of information available from the reaction

$$4Al + 3O_2 \rightarrow 2Al_2O_3;\ \Delta G = -965 \text{ kJ/mol of } O_2$$

The minimum EMF required to carry out electrolysis of $Al_2O_3$ is

(a) 0.833 V (b) 2.5 V
(c) 5.0 V (d) 1.67 V

**29.** The theoretical efficiency of a hypothetical cell is about 84% which involves the following reaction

$$A(s) + B^{2+}(aq) \rightarrow A^{2+}(aq) + B(s);\ \Delta H = -285 \text{ kJ},$$

then, the standard EMF of the cell is

(a) 1.10 V (b) 1.24 V
(c) 2.48 V (d) 2.20 V

**30.** To a beaker containing 0.1 M – HCl, a little pure solid AgCl is added. Both a silver and a hydrogen electrode ($P_{H_2} = 1.0$ bar) are then placed in the solution. What is the approximate value of EMF measured between the electrodes at 25°C? (Given: $E^\circ_{Ag^+|Ag} = 0.80$ V and $K_{sp}$ (AgCl) = $10^{-10}$) [$2.303\ RT/F = 0.06$]

(a) 0.92 V (b) 0.32 V
(c) 1.28 V (d) 0.56 V

**31.** Although aluminium is above hydrogen in the electrochemical series, it is stable in air and water. Why?

(a) Aluminium is non-reactive metal.
(b) A layer of oxide is formed at metal surface, which prevent further reaction.
(c) The reaction is thermodynamically unfavourable, i.e., $\Delta G$ is not negative.
(d) Kinetically, the reaction has very low activation energy and hence do not occur.

32. When silver chloride is dissolved in a large excess of ammonia, practically all silver ion can be assumed to exist in form of a single ionic species $[Ag_x(NH_3)_y]^{x+}$. Compute the values of $x$ and $y$ using the two following cells:

**Cell I:** Ag | 4.0 × $10^{-4}$ M – AgCl, 1 M – $NH_3$ || 4 × $10^{-2}$ M – AgCl, 1 M – $NH_3$ | Ag; $E_{cell} = 0.118$ V at 298 K.

**Cell II:** Ag | 3 × $10^{-3}$ M – AgCl, 1 M – $NH_3$ || 3 × $10^{-3}$ M – AgCl, 0.1 M – $NH_3$ | Ag; $E_{cell} = 0.118$ V at 298 K.

(a) $x = 1, y = 2$ (b) $x = 1, y = 4$
(c) $x = 2, y = 4$ (d) $x = 1, y = 6$

33. When metallic copper is shaken with a solution of a copper salt, the reaction $Cu + Cu^{2+} \rightleftharpoons 2Cu^+$ proceeds. When equilibrium is established at 298 K, $[Cu^{2+}]/[Cu^+]^2 = 1.667 \times 10^6\ M^{-1}$. If the standard potential of the $Cu^{2+}$ | Cu half-cell is +0.3376 V, what is the standard potential of $Cu^+$| Cu half-cell? (Given: 2.303 $RT/F = 0.06$, log 2 = 0.3, log 3 = 0.48)

(a) −0.3732 V (b) 0.6752 V
(c) 0.5242 V (d) 0.151 V

34. Zinc granules are added in excess to a 500 ml of 1.0 M nickel nitrate solution at 25°C until the equilibrium is reached. If the standard reduction potential of $Zn^{2+}$|Zn and $Ni^{2+}$|Ni are −0.75 V and −0.24 V, respectively, the concentration of $Ni^{2+}$ in solution at equilibrium is (2.303 $RT/F = 0.06$)

(a) $1.0 \times 10^{-17}$ M (b) $1.0 \times 10^{17}$ M
(c) $5 \times 10^{-17}$ M (d) $2 \times 10^{-17}$ M

35. The EMF for the cell: Ag(s) | AgCl(s) | KCl(0.2 M) || KBr (0.001 M) | AgBr(s) | Ag(s) at 25°C is ($K_{sp}$(AgCl) = 2.0 × $10^{-10}$; $K_{sp}$(AgBr) = 4.0 × $10^{-13}$, 2.303 $RT/F = 0.06$, log 2 = 0.3)

(a) 0.024 V (b) −0.024 V
(c) −0.24 V (d) − 0.012 V

36. Two students use same stock solution of $ZnSO_4$ but different solutions of $CuSO_4$. The EMF of one cell is 0.03 V higher than the other. The concentration of $CuSO_4$ in the cell with higher EMF value is 0.5 M. The concentration of $CuSO_4$ in the other cell is (2.303 $RT/F = 0.06$)

(a) 0.05 M (b) 5.0 M
(c) 0.5 M (d) 0.005 M

37. Two weak acid solutions $HA_1$ and $HA_2$ each with the same concentration and having $pK_a$ values 3 and 5 are placed in contact with hydrogen electrodes (1 atm, 25°C) and are interconnected through a salt bridge. EMF of the cell is

(a) 0.0295 V (b) 0.118 V
(c) 0.0885 V (d) 0.059 V

38. Consider the reaction of extraction of gold from its ore:

$$Au(s) + 2CN^-(aq) + \frac{1}{4}O_2(g) + \frac{1}{2}H_2O(l) \rightarrow Au(CN)_2^-(aq) + OH^-(aq)$$

Use the following data to calculate $\Delta G^o$ for the above reaction.

$K_f[Au(CN)_2^-] = X$

$O_2 + 2H_2O + 4e^- \rightarrow 4OH^-;\ E° = +0.41$ V

$Au^{3+} + 3e^- \rightarrow Au;\ E° = +1.50$ V

$Au^{3+} + 2e^- \rightarrow Au^+;\ E° = +1.40$ V

(a) $-RT \ln X + 1.29$ F (b) $-RT \ln X - 1.29$ F
(c) $+RT \ln X + 2.11$ F (d) $-RT \ln X - 2.11$ F

39. $F_2$ gas cannot be obtained by the electrolysis of any aqueous fluoride salt because

(a) $F_2$ is the strongest oxidizing agent
(b) $F_2$ easily combines with water
(c) $F_2$ readily combines with the electrodes
(d) $F^-$ can never be oxidized

40. A volume of 100 ml of a buffer of 1 M – $NH_3$ and 1 M – $NH_4^+$ is placed in two half-cells connected by a salt bridge. A current of 1.5 A is passed through the cell for 20 min. If electrolysis of water takes place only and the electrode reactions are:

Right: $2H_2O + O_2 + 4e \rightarrow 4\ OH^-$
and Left: $2H_2O \rightarrow 4H^+ + O_2 + 4e$,
then, the pH of the

(a) right electrode will increase
(b) left electrode will increase
(c) both electrode will increase
(d) both electrode will decrease

41. When molten $ICl_3$ is electrolysed using platinum electrodes

(a) $I_2$ is evolved at cathode and $Cl_2$ at anode
(b) $Cl_2$ is evolved at cathode and $I_2$ at anode
(c) $I_2$ is evolved at cathode and both $I_2$ and $Cl_2$ at anode
(d) electrolysis does not take place

42. When electric current is passed through a cell having an electrolyte, positive ions move towards the cathode and negative ions towards the anode. If the cathode is pulled out of the solution, then the
(a) positive and negative ions will move towards the anode
(b) positive ions will start moving towards the anode, the negative ions will stop moving
(c) negative ions will continue to move towards the anode, the positive ions will start moving randomly
(d) positive and negative ions will start moving randomly

43. When $10^{-6}$ M – HCl is electrolysed
(a) $O_2$ is produced at the anode
(b) $H_2$ is produced at the anode
(c) $Cl_2$ is produced at the anode
(d) $Cl_2$ and $O_2$ are produced at the anode

44. A dilute aqueous solution of $Na_2SO_4$ is electrolysed using platinum electrodes. The products at the anode and cathode are, respectively,
(a) $O_2$, $H_2$ (b) $S_2O_8^{2-}$, Na
(c) $O_2$, Na (d) $S_2O_8^{2-}$, $H_2$

45. In the electrolysis of a fused salt, the mass of substance deposited on an electrode will not depend on
(a) temperature of bath
(b) current intensity
(c) time of electrolysis
(d) electrochemical equivalent of ions

46. When an electric current is passed through an aqueous solution of the following, the concentrations of cation as well as anion will not change for (assume constant volume of the electrolytic solution)
(a) CsCl (b) $KNO_3$
(c) $AgNO_3$ (d) HCl

47. Which of the following substances: Na, Hg, S, Pt and graphite can be used as electrodes in electrolytic cells having aqueous solution?
(a) Na, Pt and graphite (b) Na and Hg
(c) Hg, Pt and graphite (d) Na and S

48. An ammeter and a copper voltameter are connected in series in an electric circuit through which a constant direct current flows. The ammeter shows 0.5 A. If 0.635 g of Cu is deposited in 0.965 h, what is the percentage error of ammeter? (Cu = 63.5)
(a) 5 % (b) 10 %
(c) 9 % (d) 90 %

49. The same current is passed through acidulated water and stannous chloride solution. What volume of dry detonating gas at 0°C and 1 atm is evolved from water, when 1.20 g of tin is deposited from the other solution? (Sn = 120)
(a) 112 ml (b) 336 ml
(c) 224 ml (d) 672 ml

50. The preparation of LiOH by the electrolysis of a 35% solution of LiCl using a platinum anode led to a current efficiency of 80%. What weight of LiOH was formed by the passage of 2.5 A for 4825 s?
(a) 1.92 g (b) 2.40 g
(c) 0.96 g (d) 0.672 g

51. Assuming that copper contains only iron, silver and gold as impurities. After passage of 12.4 A for 4825 s, the mass of anode decreased by 20.00 g and the cathode increased by 19.05 g. The percentages of iron and copper in the original sample are, respectively, (Cu = 63.5, Fe = 56)
(a) 4.75%, 95.25% (b) 2.8%, 95.25%
(c) 95.25%, 4.75% (d) 95.25%, 1.91%

52. After electrolysis of an aqueous sodium chloride solution, it was found that the solution is being neutralized by 60 ml N – HCl solution. During the same period of electrolysis, 3.18 g of copper was deposited in a copper voltameter in series. What is the percentage of the theoretical yield of sodium hydroxide obtained? (Cu = 63.6)
(a) 40% (b) 60%
(c) 30% (d) 80%

53. A lead storage battery has initially 200 g of lead and 200 g of $PbO_2$ plus excess $H_2SO_4$. Theoretically, how long could this cell deliver a current of 10.0 A, without recharging, if it were possible to operate it so that the reaction goes to completion? (Pb = 208)
(a) 16083.33 s (b) 8041.67 s
(c) 44.68 s (d) 18557.7 s

**54.** A volume of 100 ml of 0.6 N–$CuSO_4$ solution is electrolysed between two platinum electrodes till the concentration in the residual liquid is 0.1 N, when a steady current of 5.0 A is used. How long should the current be passed to get the above change?

(a) 965 s (b) 96500 s
(c) 1930 s (d) 482.5 s

**55.** A constant current flowed for 2 h through a potassium iodide solution, oxidizing the iodide ion to iodine. At the end of the experiment, the iodine was titrated with 72 ml of 1.0 M–$Na_2S_2O_3$ solution. What was the average rate of current flow, in ampere?

(a) 0.965 A (b) 1.93 A
(c) 0.483 A (d) 0.0965 A

**56.** Anthracene, $C_{14}H_{10}$, can be oxidized to anthraquinone, $C_{14}H_8O_2$. What weight of anthraquinone can be produced by the passage of a current of 1 A of 40 min if the current efficiency is 96.5%?

(a) 0.862 g (b) 0.832 g
(c) 0.624 g (d) 0.738 g

**57.** Most of the copper used to make wire has been electrically refined by depositing it from copper salts solution (divalent) on to a cathode. What is the cost of electrical energy required per kg of copper if the cost of electricity is Rs 4.00 per kWh and the cell operates at 0.33 V? The electrochemical equivalent of copper is 0.00033 g/coulomb.

(a) Rs 11.11 (b) Rs 5.55
(c) Rs 2.22 (d) Rs 1.11

**58.** Electrolysis of an acetate solution produces ethane according to the reaction:

$$2CH_3COO^- \rightarrow C_2H_6(g) + 2CO_2(g) + 2e^-$$

What total volume of ethane and $CO_2$ would be produced at 0°C and 1 atm, if a current of 0.5 A is passed through the solution for 482.5 min? Assume current efficiency 80%.

(a) 1.344 L (b) 2.688 L
(c) 4.032 L (d) 1.792 L

**59.** To perform an analysis of a mixture of metal ions by electro-deposition, the second metal to be deposited must not being plating out until the concentration ratio of the second to the first is about $10^6$. What must be the minimum difference in standard potential of the two metals which form dipositive ions in order for such an analysis to be feasible?

(a) 0.177 V (b) 0.354 V
(c) 0.708 V (d) 0.088 V

**60.** During the electrolysis of 0.1 M – $CuSO_4$ solution using copper electrodes, a depletion of $Cu^{2+}$ occurs near the cathode with a corresponding excess near the anode, owing to inefficient stirring of the solution. If the local concentration of $Cu^{2+}$ near the anode and cathode are, respectively, 0.12 M and 0.08 M, the back EMF developed at 298 K is (log 1.5 = 0.18, 2.303 $RT/F$ = 0.06)

(a) 0.33 V (b) 5.4 mV
(c) 2.7 mV (d) 10.8 mV

**61.** The following galvanic cell:

Zn | $Zn(NO_3)_2$ (100 ml, 1 M) || $Cu(NO_3)_2$ (100 ml, 1 M) | Cu

was operated as an electrolytic cell as Cu as the anode and Zn as the cathode. A current of 0.4825 A was passed for 10 h and then the cell was allowed to function as galvanic cell. What would be the final EMF of the cell at 25°C? Assume that the only electrode reactions occurring were those involving Cu | $Cu^{2+}$ and Zn | $Zn^{2+}$. Given: $E^\circ_{Cu^{2+}|Cu} = +0.34$ V, $E^\circ_{Zn^{2+}|Zn} = -0.76$ V, 2.303 $RT/F$ = 0.06, log 1.9 = 0.28)

(a) 1.10 V (b) 1.0616 V
(c) 1.1084 V (d) 1.1768 V

**62.** At the Nangal fertilizer plant in Punjab, hydrogen is produced by the electrolysis of water. The hydrogen is used for the production of ammonia and nitric acid (by the oxidation of ammonia). If the average production of ammonium nitrate is 1152 kg/day, the daily consumption of electricity (in A /day) is

(a) 96,500 (b) 48,250
(c) 32,166.67 (d) 24,125

**63.** Lactic acid, $HC_3H_5O_3$, produced in 1 g sample of muscle tissue was titrated using phenolphthalein as indicator against $OH^-$ ions which were obtained by the electrolysis of water. As soon as $OH^-$ ions are produced, they react with lactic acid and at complete neutralization, immediately a pink colour is noticed. If electrolysis was made for 1158 s using 50.0 mA current to reach the end point, what was the percentage of lactic acid in muscle tissue?

(a) 5.4% (b) 2.7%
(c) 10.8% (d) 0.054%

**64.** $H_2O_2$ can be prepared by successive reactions:

$$2NH_4HSO_4 \rightarrow H_2 + (NH_4)_2S_2O_8$$

$$(NH_4)_2S_2O_8 + 2H_2O \rightarrow 2NH_4HSO_4 + H_2O_2$$

The first reaction is an electrolytic reaction and second is steam distillation. What amount of current would have to be used in first reaction to produce enough intermediate to yield 102 g pure $H_2O_2$ per hour. Assume current efficiency 50%.

(a) 643.33 A (b) 321.67 A
(c) 160.83 A (d) 1286.67 A

**65.** In an analytical determination of arsenic, a solution containing arsenious acid, $H_3AsO_3$, KI and a small amount of starch is electrolysed. The electrolysis produces free $I_2$ from $I^-$ ions and the $I_2$ immediately oxidizes the arsenious acid to hydrogen arsenate ion, $HAsO_4^{2-}$.

$$I_2 + H_3AsO_3 + H_2O \rightarrow 2I^- + HAsO_4^{2-} + 4H^+$$

When the oxidation of arsenic is complete, the free iodine combines with the starch to give a deep blue colour. If during a particular run, it takes 96.5 s for a current of 1.68 mA to give an end point (indicated by the blue colour), how many g of arsenic are present in the solution. (As = 75)

(a) $6.3 \times 10^{-4}$ g (b) $1.26 \times 10^{-4}$ g
(c) $3.15 \times 10^{-5}$ g (d) $6.3 \times 10^{-5}$ g

**66.** Electrolysis of a solution of $MnSO_4$ in aqueous sulphuric acid is a method for the preparation of $MnO_2$ as per the reaction: $Mn^{2+}(aq) + 2H_2O \rightarrow MnO_2(s) + 2H^+(aq) + H_2(g)$. Passing a current of 19.3 A for 2 h gives only 52.2 g of $MnO_2$. The current efficiency is (Mn = 55)

(a) 8.33% (b) 83.33%
(c) 41.67% (d) 100%

**67.** Copper sulphate solution (250 ml) was electrolysed using a platinum anode and a copper cathode. A constant current of 2 mA was passed for 19.3 min. It was found that after electrolysis the absorbance of the solution was reduced to 50% of its original value. Calculate the concentration of copper sulphate in the solution to begin with.

(a) $9.6 \times 10^{-5}$ M (b) $4.8 \times 10^{-5}$ M
(c) $2.4 \times 10^{-5}$ M (d) $1.2 \times 10^{-5}$ M

**68.** Calculate the quantity of electricity that would be required to reduce 12.3 g of nitrobenzene to aniline if the current efficiency for the process is 50%. If the potential drop across the cell is 3.0 V, how much energy will be consumed?

(a) 0.3 F, 86.85 kJ (b) 0.6 F, 173.7 kJ
(c) 1.2 F, 347.4 kJ (d) 0.6 F, 173.7 kJ

**69.** Perdisulphuric acid, $H_2S_2O_8$ can be prepared by electrolytic oxidation of $H_2SO_4$, oxygen and hydrogen gases are by products. In such an electrolysis, 9.08 L of $H_2$ and 2.27 L of $O_2$ were generated at STP. What is the mass of $H_2S_2O_8$ formed?

(a) 0 (b) 77.6 g
(c) 38.8 g (d) 19.4 g

**70.** During the discharge of a lead storage battery, the density of sulphuric acid fell from 1.5 to 1.1 g/ml. Sulphuric acid of density 1.5 g/ml is 40% $H_2SO_4$, by weight, and that of density 1.1 g/ml is 10% $H_2SO_4$, by weight. The battery holds 3.6 L of the acid and the volume remained practically constant during the discharge. Calculate the number of ampere-hours which the battery should have been used. The electrode reactions are:

$$Pb + SO_4^{2-} \rightarrow PbSO_4 + 2e$$

$$PbO_2 + 4H^+ + SO_4^{2-} + 2e \rightarrow PbSO_4 + 2H_2O$$

(a) 965 (b) 482.5
(c) 1930 (d) 241.25

**71.** A cell whose resistance, when filled with 0.1 M – KCl is 200 Ω, is measured to be 6400 Ω, when filled with 0.003 M – NaCl solution. What is the molar conductance of NaCl solution, in $\Omega^{-1}$ cm$^2$ mol$^{-1}$ if the molar conductance of 0.1 M – KCl is 120 $\Omega^{-1}$ cm$^2$ mol$^{-1}$?

(a) 41.67 (b) 250
(c) 125 (d) 375

**72.** At 18°C, the ionic mobilities of $NH_4^+$ and $CrO_4^{2-}$ ions are $6.6 \times 10^{-8}$ and $5.4 \times 10^{-8}$ m$^2$ volt$^{-1}$ s$^{-1}$ at infinite dilution. What is the molar conductance of ammonium chromate solution in $\Omega^{-1}$ m$^2$ mol$^{-1}$?

(a) 0.01795 (b) 0.01158
(c) $1.2 \times 10^{-8}$ (d) $1.86 \times 10^{-7}$

**73.** At 25°C, the molar conductance at infinite dilution for HCl solution is 4.25 $\Omega^{-1}$ m$^2$ mol$^{-1}$, while its specific conductance is 382.5 $\Omega^{-1}$ m$^{-1}$. If the degree of dissociation is 90%, the molarity of solution is

(a) 0.9 M (b) 1.0 M
(c) 0.1 M (d) 1.1 M

**74.** A big irregular shaped vessel contained water, conductivity of which was $2.56 \times 10^{-3}$ $S^{-1}m^{-1}$. 585 g of NaCl was then added to the water and conductivity after the addition of NaCl, was found to be $3.06 \times 10^{-3}$ $S^{-1}m^{-1}$. The molar conductivity of NaCl at this concentration is $1.5 \times 10^{-2}$ $S^{-1}m^2$ $mol^{-1}$. The capacity of vessel if it is fulfilled with water, is

(a) $3 \times 10^4$ L (b) 30 L
(c) $3 \times 10^8$ L (d) $3 \times 10^5$ L

**75.** The molar conductivity of 0.10 M solution of $MgCl_2$ is 100 mho $cm^2$ $mol^{-1}$, at 25°C. A cell with electrodes that are 1.50 $cm^2$ in surface area and 0.50 cm apart is filled with 0.10 M – $MgCl_2$ solution. How much current will flow when the potential difference between the electrodes is 5 volts?

(a) 0.03 A (b) 3.0 A
(c) 0.15 A (d) 15 A

**76.** For $Na^+$, the value of symbol $\lambda_m^\circ$ is 50.0 $\Omega^{-1}$ $cm^2$ $mol^{-1}$. The speed of $Na^+$ ion in the solution, if in the cell, electrodes are 5 cm apart and to which a potential of 19.3 volt is applied is

(a) $2 \times 10^{-3}$ cm/s (b) $1 \times 10^{-3}$ cm/s
(c) $2 \times 10^{-4}$ cm/s (d) $2 \times 10^{-2}$ cm/s

**77.** The conductivity of a saturated solution of AgCl at 298 K was found to be $3.40 \times 10^{-6}$ $\Omega^{-1}$ $cm^{-1}$; the conductivity of water used to make up the solution was $1.60 \times 10^{-6}$ $\Omega^{-1}$ $cm^{-1}$. Determine the solubility of AgCl in water in mole per litre at 298 K. The equivalent conductivity of AgCl at infinite dilution is 150.0 $\Omega^{-1}$ $cm^{-2}$ $eq^{-1}$.

(a) $1.44 \times 10^{-10}$ (b) $1.2 \times 10^{-5}$
(c) $3.33 \times 10^{-5}$ (d) $1.2 \times 10^{-8}$

**78.** Resistance of 0.2 M solution of an electrolyte is 50 Ω. The specific conductance of the solution is 1.4 S $m^{-1}$. The resistance of 0.5 M solution of the same electrolyte is 280 Ω. The molar conductivity of 0.5 M solution of the electrolyte (in S $m^2$ $mol^{-1}$) is

(a) $5 \times 10^{-3}$ (b) $5 \times 10^3$
(c) $5 \times 10^2$ (d) $5 \times 10^{-4}$

**79.** The conductivity of a saturated solution containing AgA ($K_{sp} = 3 \times 10^{-14}$) and AgB ($K_{sp} = 1 \times 10^{-14}$) is $3.75 \times 10^{-8}$ $\Omega^{-1}cm^{-1}$. If the limiting molar conductivity of $Ag^+$ and $A^-$ ion is 60 and 80 $\Omega^{-1}$ $cm^2$ $mol^{-1}$, respectively, the limiting molar conductivity of $B^-$ (in $\Omega^{-1}$ $cm^2$ $mol^{-1}$) is

(a) 135 (b) 67.5
(c) 270 (d) 190

**80.** The conductivity of saturated solution of $Ba_3(PO_4)_2$ is $1.2 \times 10^{-5}$ $\Omega^{-1}$ $cm^{-1}$. The limiting equivalent conductivities of $BaCl_2$, $K_3PO_4$ and KCl are 160, 140 and 100 $\Omega^{-1}$ $cm^2$ $eq^{-1}$, respectively. The solubility product of $Ba_3(PO_4)_2$ is

(a) $10^{-5}$ (b) $1.08 \times 10^{-23}$
(c) $1.08 \times 10^{-25}$ (d) $1.08 \times 10^{-27}$

## Section B (One or More than one Correct)

1. Which of the following statement(s) differentiate between electrochemical cell and electrolytic cell?
   (a) Spontaneous or non-spontaneous nature of the chemical process.
   (b) Chemical reactions occurring at the electrodes.
   (c) Positive and negative nature of anode.
   (d) Dependence on Faraday's law.

2. Pick up the false statement(s):
   (a) The net chemical change in a galvanic cell reaction is always redox reactions.
   (b) In a galvanic cell made of copper and cadmium electrodes, cadmium electrode may act as anode.
   (c) Standard potential increases with increasing concentration of the electrolyte.
   (d) Calomel electrode is a reference electrode having 0.00 volt potential.

3. Consider the cell: Ag(s), AgCl(s) | KCl (0.1 M) | $Hg_2Cl_2$(s), Hg(l). The cell potential
   (a) increases on increasing concentration of $Cl^-$ ions.
   (b) decreases on decreasing concentration of $Cl^-$ ions.
   (c) is independent of concentration of $Cl^-$ ions.
   (d) is independent of amounts of AgCl and $Hg_2Cl_2$.

4. The passage of electricity in the Daniel cell when Zn and Cu electrodes are connected
   (a) from Cu to Zn inside the cell
   (b) from Cu to Zn outside the cell
   (c) from Zn to Cu inside the cell
   (d) from Zn to Cu outside the cell

EXERCISE II

5. From an electrolyte, one mole of electron will deposit at cathode
(a) 63.5 g of Cu (b) 24 g of Mg
(c) 11.5 g of Na (d) 9.0 g of Al

6. Which of the following statements is/are correct?
(a) The conductivity of molten NaCl is due to movement of $Na^+$ and $Cl^-$ ions.
(b) Solid NaCl is good conductor of electricity.
(c) Molten sodium is a good conductor because of mobile electrons.
(d) Resistivity is reciprocal of molar conductivity of electrolyte.

7. Ionic conductance at infinite dilution of $Al^{3+}$ and $SO_4^{2-}$ ions are 60 and 80 $\Omega^{-1}\,cm^2\,eq^{-1}$, respectively. The correct detail(s) regarding $Al_2(SO_4)_3$ is/are
(a) the molar conductance is 140 $\Omega^{-1}\,cm^2\,mol^{-1}$.
(b) the equivalent conductance is 140$\Omega^{-1}\,cm^2\,eq^{-1}$.
(c) the molar conductance is 840 $\Omega^{-1}\,cm^2\,mol^{-1}$.
(d) the molar conductance is 23.33$\Omega^{-1}\,cm^2\,mol^{-1}$.

8. The function(s) of salt bridge in a cell is/ are
(a) It maintains standard electrode potential of cell constant which depends on several factors.
(b) It completes the electrical circuit.
(c) It departs both the solutions from each other.
(d) It maintains the electrical neutrality of both electrolytic solutions.

9. The reactions taking place in the dry cell are:
Anode: $Zn \rightarrow Zn^{2+} + 2e^-$
Cathode: $2MnO_2 + 2NH_4^+ + 2e^- \rightarrow Mn_2O_3 + 2NH_3 + H_2O$
The minimum mass of reactants, if a dry cell is to generate 0.25A for 9.65 h, are (Mn = 55, Zn = 65.4) (neglect any other chemical reactions occurring in the cell)
(a) 2.943 g Zn (b) 7.83 g $MnO_2$
(c) 1.62 g $NH_4^+$ (d) 3.915 g $MnO_2$

10. EMF of the cell: Cd(s) | $CdCl_2.5H_2O$ (sat.) | AgCl(s) | Ag(s) is +0.70 V at 0°C and +0.60 V at 50°C. If $\Delta H^o$ and $\Delta S^o$ are temperature independent, then the correct information(s) regarding the cell reaction is/are
(a) $\Delta G° = -115.8$ kJ at 50°C
(b) $\Delta G° = 135.1$ kJ at 0°C
(c) $\Delta S° = -386$ J/K
(d) $\Delta H° = -221.178$ kJ

11. Two litre solution of a buffer mixture containing 1.0 M – $NaH_2PO_4$ and 1.0 M – $Na_2HPO_4$ is placed in two compartments (one litre in each) of an electrolytic cell. The platinum electrodes are inserted in each compartment and 1.25 A current is passed for 965 min. Assuming electrolysis of only water at each compartment, what will be pH in each compartment after passage of above charge? ($pK_a$ for $H_2PO_4^- = 2.15$, log 7 = 0.85)
(a) Anode: 3.00 (b) Cathode: 3.00
(c) Anode: 1.30 (d) Cathode: 1.30

12. For the reduction of $NO_3^-$ ion in an aqueous solution, $E°$ is +0.96 V. Values of $E°$ for some metal ions are given below:
$V^{2+}(aq) + 2e^- \rightarrow V(s)$ $E° = -1.19$ V
$Fe^{3+}(aq) + 3e^- \rightarrow Fe(s)$ $E° = -0.04$ V
$Au^{3+}(aq) + 3e^- \rightarrow Au(s)$ $E° = +1.40$ V
$Hg^{2+}(aq) + 2e^- \rightarrow Hg(l)$ $E° = +0.86$ V
The pair(s) of metal that is(are) oxidized by $NO_3^-$ in aqueous solution is(are)
(a) V and Hg (b) Hg and Fe
(c) Fe and Au (d) Fe and V

13. Among the following, the intensive property is (properties are)
(a) Molar conductivity (b) Electromotive force
(c) Resistance (d) Heat capacity

14. In a galvanic cell, the salt bridge
(a) does not participate chemically in the cell reaction.
(b) stops the diffusion of ions from one electrode to another.
(c) is necessary for the occurrence of the cell reaction.
(d) ensures mixing of the two electrolytic solutions.

15. A lead storage cell is discharged which causes the $H_2SO_4$ electrolyte to change from a concentration of 40% by weight (density = 1.260 g/ml) to 28%, by weight. The original volume of solution was 1 L. Identify the correct statement(s):
(a) The overall cell reaction is: $Pb(s) + PbO_2(s) + 2H_2SO_4(aq) \rightarrow 2PbSO_4(s) + 2H_2O(l)$.
(b) A total of 2.0 moles of $H_2SO_4$ is reacted.
(c) The total charge released from anode of the cell is $1.93 \times 10^5$ coulomb.
(d) The mass of electrolytic solution has decreased.

## Section C (Comprehensions)

### Comprehension I

**Suppose that the S.H.E. was arbitrarily assigned a value of 1.00 V for $2H^+(aq) + 2e \rightarrow H_2(g)$. What would this do to the observed voltage under standard condition for each of the following: (Given: $E^\circ_{Zn^{2+}|Zn} = -0.76$ V, $E^\circ_{Cu^{2+}|Cu} = +0.34$ V when S.H.E. is 0.00 V)**

1. $E^\circ_{Zn^{2+}|Zn}$
   (a) +0.24 V (b) −0.76 V
   (c) +1.76 V (d) −0.34 V

2. $E^\circ_{Cu^{2+}|Cu}$
   (a) −0.76 V (b) +1.34 V
   (c) +1.76 V (d) +0.34 V

3. Zn–Cu Cell
   (a) +1.10 V
   (b) −1.10 V
   (c) 0.0 V
   (d) Indeterminate

### Comprehension II

**The standard reduction potential of the $Ag^+|Ag$ electrode at 298 K is 0.80 V. The solubility product of AgI is $6.4 \times 10^{-17}$ at 298 K. (2.303 $RT/F = 0.06$, log 2 = 0.3)**

4. The potential of $Ag^+ | Ag$ electrode in a saturated solution of AgI at 298 K is
   (a) −0.314 V (b) +0.314 V
   (c) −0.172 V (d) +0.172 V

5. The standard reduction potential of $I^-|AgI|Ag$ electrode at 298 K is
   (a) −0.314 V (b) +0.314 V
   (c) −0.172 V (d) +0.172 V

6. The potential of $I^-$ (0.04 M)|AgI|Ag electrode at 198 K is
   (a) −0.088 V
   (b) +0.088 V
   (c) −0.172 V
   (d) +0.172 V

### Comprehension III

**The Edison storage cell is represented as Fe(s) | FeO(s) | KOH(aq) | $Ni_2O_3$(s) | NiO(s) | Ni(s). The half-cell reactions are**

$$Ni_2O_3(s) + H_2O(l) + 2e \rightleftharpoons 2NiO(s) + 2OH^-;\ E^\circ = +0.40\ V$$

$$FeO(s) + H_2O(l) + 2e \rightleftharpoons Fe(s) + 2OH^-;\ E^\circ = -0.87\ V$$

7. What is the cell reaction?
   (a) $Ni_2O_3(s) + Fe(s) \rightarrow 2\,NiO(s) + FeO(s)$
   (b) $2NiO(s) + FeO(s) \rightarrow Ni_2O_3(s) + Fe(s)$
   (c) $Ni_2O_3(s) + FeO(s) \rightarrow 2NiO(s) + Fe(s) + O_2$
   (d) None of these

8. What is the standard cell EMF?
   (a) 1.27 V
   (b) 0.47 V
   (c) −1.27 V
   (d) −0.47 V

9. How does cell EMF depend on increasing the concentration of KOH?
   (a) increases
   (b) decreases
   (c) remains unaffected
   (d) none of these

10. What is the maximum amount of electrical energy that can be obtained from one mole of $Ni_2O_3$?
   (a) 2.54 J (b) 245.11 kJ
   (c) 122.56 kJ (d) 61.28 kJ

## Comprehension IV

**The cell potential for the unbalanced chemical reaction:**

$$Hg_2^{2+}(aq) + NO_3^-(aq) + 3H_3O^+(aq) \rightarrow 2Hg^{2+}(aq) + HNO_2(aq) + 4H_2O(l)$$

**is measured under standard conditions in the electrochemical cell shown in the accompanying diagram.**

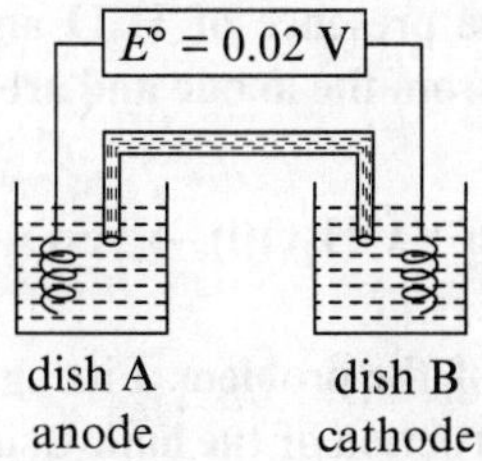


**11.** In which dish, the solution is acidic?

(a) Dish A (b) Dish B
(c) Both (d) None

**12.** How many moles of electrons pass through the circuit when 0.60 mole of $Hg^{2+}$ and 0.30 mole of $HNO_2$ are produced in the cell that contains 0.50 mole of $Hg_2^{2+}$ and 0.40 mole of $NO_3^-$ at the beginning of the reaction?

(a) 0.30 (b) 0.60
(c) 0.15 (d) 1.20

**13.** How long will it take to produce 0.10 mole of $HNO_2$ by this reaction if a current of 10 A passes through the cell?

(a) 965 s (b) 193 s
(c) 1930 s (d) 482.5 s

---

## Comprehension V

**A fuel cell is the device to convert the energy of a fuel into electrical energy without the use of heat engine, where the fuel is burnt directly. Such conversions are possible because the combustion reactions are essentially redox reactions and highly exothermic as well as highly exergonic. Electrical energy can be obtained indefinitely from a fuel cell as along as the outside supply of fuel is maintained. In hydrogen–oxygen fuel cell, the following reactions take place:**

**Anode reaction:** $2H_2(g) + 4OH^-(aq) \rightarrow 4H_2O(l) + 4e^-$

**Cathode reaction:** $O_2(g) + 2H_2O(l) + 4e^- \rightarrow 4OH^-(aq)$

**Overall reaction:** $2H_2(g) + O_2(g) \rightarrow 2H_2O(l)$

**The overall reaction has a value of $\Delta H° = -285.8$ kJ and $\Delta G° = -237.39$ kJ at 25°C per mole of $H_2O(l)$.**

**14.** What is the standard EMF of the cell?

(a) 0.615 V (b) 1.23 V
(c) 2.46 V (d) 0.74 V

**15.** What volume of gaseous $H_2$ (at STP), when combined with excess $O_2$ in the fuel cell, is needed to produce 23.739 kJ of useful work under standard conditions?

(a) 2.27 L (b) 4.54 L
(c) 1.13 L (d) 2.24 L

**16.** Suppose the concentration of hydroxide ion in the cell is doubled at 298 K. The cell voltage will be

(a) reduced by half
(b) increased by a factor of 2
(c) increased by a factor of 4
(d) unchanged

**17.** What is the approximate value of $\Delta S°$ for the fuel cell reaction at 25°C?

(a) $-0.1624$ JK$^{-1}$ (b) $-162.4$ JK$^{-1}$
(c) $+162.4$ JK$^{-1}$ (d) $+0.1624$ JK$^{-1}$

**18.** The theoretical efficiency of the fuel cell is given by

(a) 83.06% (b) 100%
(c) 67.53% (d) 97.88%

## Comprehension VI

**Billions of dollars are spent each year to replace or prevent the corrosion and subsequent destruction of government property such as ships, bridges, and metallic piping. Corrosion may result in decreased structural integrity and eventual ineffectuality. For example, a naval ship made from iron is highly susceptible to corrosion. One part of the hull acts as the anode, while another acts as the cathode, the iron hull itself connects the two parts, completing the circuit. Part of the metal hull begins oxidizing to $Fe^{2+}$ in the presence of $H_2O$ and $O_2$, the reduction reaction proceeds with an $E_{red} = 1.229$ V. It is there that $Fe^{2+}$ ions migrate from the anode and are further oxidized as**

$$2Fe^{2+}(aq) + \frac{1}{2}O_2(g) + (6+x)H_2O(l) \rightarrow Fe_2O_3{\cdot}xH_2O(s) + 4H_3O^+(aq)$$

**The $Fe_2O_3$, or rust, formed is only a small part of the problem. Pitting, or loss of solid metal, weaken the structure of the hull and may lead to serious damage and destruction of the hull. Using a sacrificial anode often successfully prevents corrosion. This metal, more easily oxidized than iron. Although the sacrificial anode eventually corrodes as well, the cost of replacing it is far less than that of replacing the iron hull.**

| Half-reaction | Potential |
|---|---|
| $O_2 + 4H^+ + 4e^- \rightarrow 2H_2O$ | $E° = 1.229$ V |
| $Fe^{3+} + e^- \rightarrow Fe^{2+}$ | $E° = 0.771$ V |
| $Cu^+ + e^- \rightarrow Cu$ | $E° = 0.521$ V |
| $O_2 + 2H_2O + 4e^- \rightarrow 4OH^-$ | $E° = 0.401$ V |
| $Cu^{2+} + 2e^- \rightarrow Cu$ | $E° = 0.342$ V |
| $Pb^{2+} + 2e^- \rightarrow Pb$ | $E° = -0.126$ V |
| $Fe^{2+} + 2e^- \rightarrow Fe$ | $E° = -0.447$ V |
| $Mn^{2+} + 2e^- \rightarrow Mn$ | $E° = -1.185$ V |
| $Mg^{2+} + 2e^- \rightarrow Mg$ | $E° = -2.372$ V |

**19.** Which of the following reactions may occur at the anode of the iron hull?

(a) $Fe^{2+}(aq) \rightarrow Fe^{3+}(aq) + e^-$
(b) $Fe(s) \rightarrow Fe^{2+}(aq) + 2e^-$
(c) $O_2(g) + 4H_3O^+(aq) \rightarrow 6H_2O(l) + 4e^-$
(d) $2H_2O(l) \rightarrow 2OH-(aq) + H_2(g) + 2e^-$

**20.** What is the standard voltage generated by the corrosion reaction on the iron hull of a ship?

(a) −1.676 V (b) −0.782 V
(c) 1.782 V (d) 1.676 V

**21.** Which of the following materials could not serve as a sacrificial anode for lead?

(a) Copper (b) Iron
(c) Magnesium (d) Manganese

**22.** An ammeter reading shows a current of 0.50 A running through the iron hull. How many grams of Fe(s) are lost in 1.0 h?

(a) 0.261 g (b) 0.522 g
(c) 1.044 g (d) 0.783 g

**23.** A ship that has sunk to the bottom of the ocean may still exhibit corrosion. One difference in the corrosion reaction between ship above water and one that has sunk of the bottom of the ocean may be that

(a) the oxidation reaction of the sunken ship does not include $H_2O$ as a reactant.
(b) the oxidation reaction of the sunken ship does not include $O_2$ as a reactant.
(c) the reduction reaction of the sunken ship does not include $H_2O$ as a reactant.
(d) the reduction reaction of the sunken ship does not include $O_2$ as a reactant.

## Comprehension VII

**A current of 15.0 A is employed to plate nickel in a $NiSO_4$ solution. Both Ni and $H_2$ are formed at the cathode. The current efficiency with respect to formation of Ni is 60%. The density of nickel = 8.9 g/ml. (Ni = 58.7)**

**24.** How much of nickel is plated on the cathode per hour?

(a) 16.43 g (b) 32.85 g
(c) 19.7 g (d) 9.85 g

**25.** What is the thickness of the plating if the cathode consists of a sheet of metal 4.0 cm$^2$, which is to be coated on both sides?

(a) 1.38 mm (b) 2.76 mm
(c) 0.69 mm (d) 23.0 mm

**26.** What volume of $H_2$ at 0°C and 1 atm is formed per hour?

(a) 6.27 L (b) 3.76 L
(c) 2.5 L (d) 5.01 L

**27.** At the end of the electrolysis, how many grams of the gaseous product appear at the anode?

(a) 4.48 g (b) 1.79 g
(c) 2.69 g (d) 7.46 g

## Comprehension VIII

**Tollen's reagent is used for the detection of aldehyde when a solution of $AgNO_3$ is added to glucose with $NH_4OH$ then gluconic acid is formed.**

$$Ag^+ + e^- \rightarrow Ag; \; E^\circ_{red} = 0.8 \text{ V}$$

$$C_6H_{12}O_6 + H_2O \rightarrow C_6H_{12}O_7 \text{ (Gluconic acid)} + 2H^+ + 2e; \; E^\circ_{oxd} = -0.05 \text{ V}$$

$$Ag(NH_3)_2^+ + e^- \rightarrow Ag(s) + 2\,NH_3; \; E^\circ_{red} = 0.337 \text{ V}$$

**[Use 2.303 $RT/F$ = 0.0592 and $F/RT$ = 38.92 at 298 K]**

**28.** $2Ag^+ + C_6H_{12}O_6 + H_2O \rightarrow 2Ag(s) + C_6H_{12}O_7 + 2H^+$. Find ln$K$ of this reaction.

(a) 66.13 (b) 58.38
(c) 28.30 (d) 46.29

**29.** When ammonia is added to the solution, pH is raised to 11. Which half-cell reaction is affected by pH and by how much?

(a) $E_{oxd}$ will increase by a factor of 0.65 from $E^\circ_{oxd}$
(b) $E_{oxd}$ will decrease by a factor of 0.65 from $E^\circ_{oxd}$
(c) $E_{red}$ will increase by a factor of 0.65 from $E^\circ_{red}$
(d) $E_{red}$ will decrease by a factor of 0.65 from $E^\circ_{red}$

**30.** Ammonia is always added in this reaction. Which of the following must be incorrect?

(a) $NH_3$ combines with $Ag^+$ to form a complex
(b) $Ag(NH_3)_2^+$ is a stronger oxidizing agent than $Ag^+$
(c) In absence of ammonia, silver salt of gluconic acid is formed
(d) $NH_3$ has affected the standard reduction potential of glucose–gluconic acid electrode

EXERCISE II

## Comprehension IX

**The standard potential of the following cell is 0.23 V at 15°C and 0.21 V at 35°C.**

$$Pt(s) \mid H_2(g) \mid HCl(aq) \mid AgCl(s) \mid Ag(s)$$

**Given: The standard reduction potential of the $Ag^+(aq)|Ag(s)$ couple is 0.80 V at 25°C.**

**$\Delta H°$ and $\Delta S°$ for the cell reaction remain unchanged in the range 15°C to 35°C.**

**31.** The overall cell reaction is

(a) $H_2(g) + 2AgCl(s) \rightarrow 2HCl(aq) + 2Ag(s)$
(b) $H_2(g) + 2Ag^+(aq) \rightarrow 2H^+(aq) + 2Ag(s)$
(c) $Ag^+(aq) + Cl^-(aq) \rightarrow AgCl(s)$
(d) $H_2(g) + Cl_2(g) \rightarrow 2HCl(aq)$

**32.** $\Delta H°$ for the cell reaction per mol of AgCl is

(a) −99,974 J (b) −49,987 J
(c) +99,974 J (d) +49,987 J

**33.** $\Delta S°$ for the cell reaction per mol of AgCl is

(a) −193 J/K (b) −96.5 J/K
(c) +96.5 J/K (d) +193 J/K

**34.** The solubility of AgCl in water at 25°C is (2.303 $RT/F = 0.058$)

(a) $10^{-10}$ M (b) $10^{-12}$ M
(c) $10^{-5}$ M (d) $10^{-6}$ M

## Comprehension X

**For the reaction: $Ag^+(aq) + Cl^-(aq) \rightleftharpoons AgCl(s)$. Given**

| Species | $\Delta G_f^\circ$ (kJ / mol) at 25°C |
|---|---|
| $Ag^+(aq)$ | +77 |
| $Cl^-(aq)$ | −129 |
| AgCl (s) | −109 |

**$Ag^+ + e^- \rightarrow Ag$; $E° = 0.80$ V**
**$Zn^{2+} + 2e^- \rightarrow Zn$; $E° = -0.76$ V**

**35.** The cell representations of the above reaction and $E^\circ_{cell}$ are

(a) $Ag(s) \mid AgCl(s) \mid Cl^-(aq) \parallel Ag^+(aq) \mid Ag(s)$; $E^\circ_{cell} = 0.59$ V
(b) $Ag(s) \mid Ag^+(aq) \parallel Cl^-(aq) \mid AgCl(s) \mid Ag(s)$; $E^\circ_{cell} = 0.59$ V
(c) $Ag(s) \mid Cl^-(aq) \parallel Ag^+(aq) \mid Ag$; $E^\circ_{cell} = 0.59$ V
(d) $Ag(s) \mid Ag^+(aq) \parallel Cl^-(aq) \mid AgCl(s) \mid Ag(s)$; $E^\circ_{cell} = 1.18$ V

**36.** The solubility product of AgCl at 298 K is

(a) $2 \times 10^{-10}$ (b) $10^{-10}$
(c) $e^{-10}$ (d) $10^{-5}$

**37.** A quantity of $6.539 \times 10^{-2}$ g of metallic zinc is added to 100 ml of saturated solution of AgCl. The value of $\log \frac{[Zn^{2+}]}{[Ag^+]^2}$ is (Zn = 65.39)

(a) 5.288 (b) 52.88
(c) 528.8 (d) 26.44

**38.** How many moles of Ag will be precipitated in the above reaction?

(a) $10^{-3}$ (b) $2 \times 10^{-3}$
(c) $10^{-6}$ (d) $10^{-5}$

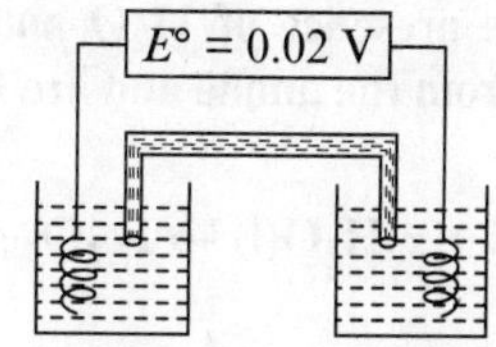


dish A anode    dish B cathode

## Comprehension XI

**A sample of water from a large swimming pool has a resistance of 10,000 Ω at 25°C, when placed in a conductivity cell. When filled with 0.02 M – KCl solution, the cell has a resistance of 100 Ω at 25°C. An amount of 585 g of NaCl was dissolved in the pool, which was thoroughly stirred. A sample of this solution gave a resistance of 8000 Ω. The molar conductivity of NaCl at this concentration is 125 $\Omega^{-1}$ $cm^2$ $mol^{-1}$ and molar conductivity of 0.02 M – KCl is 200 $\Omega^{-1}$ $cm^2$ $mol^{-1}$.**

39. Cell constant of the conductivity cell is
    (a) 4 $cm^{-1}$ (b) 0.4 $cm^{-1}$
    (c) 40 $cm^{-1}$ (d) 0.04 $cm^{-1}$

40. Conductivity of water (in $\Omega^{-1}$ $cm^{-1}$) is
    (a) 0.4 (b) 0.04
    (c) 0.0004 (d) 0.00004

41. Volume (in L) of water in the pool is
    (a) 1,25,000 (b) 12,500
    (c) 1250 (d) 125

## Comprehension XII

**Redox reactions play a pivoted role in chemistry and biology. The values of standard redox potential ($E°$) of two half-cell reactions decide which way the reaction is expected to proceed. A simple example is Daniel cell in which zinc goes into solution and copper gets deposited. Given below are a set of half-cell reactions (acidic medium) along with their $E°$ (V with respect to normal hydrogen electrode) values. Using this data, obtain the correct explanations to questions below:**

$I_2 + 2e^- \rightarrow 2I^-$ $E° = 0.54$
$Cl_2 + 2e^- \rightarrow 2Cl^-$ $E° = 1.36$
$Mn^{3+} + e^- \rightarrow Mn^{2+}$ $E° = 1.50$
$Fe^{3+} + e^- \rightarrow Fe^{2+}$ $E° = 0.77$
$O_2 + 4H^+ + 4e^- \rightarrow 2H_2O$ $E° = 1.23$

42. Among the following, identify the correct statement?
    (a) Chloride ion is oxidized by $O_2$
    (b) $Fe^{2+}$ is oxidized by iodine
    (c) Iodide ion is oxidized by chlorine
    (d) $Mn^{2+}$ is oxidized by chlorine

43. While $Fe^{3+}$ is stable, $Mn^{3+}$ is not stable in acid solution because
    (a) $O_2$ oxidizes $Mn^{2+}$ to $Mn^{3+}$
    (b) $O_2$ oxidizes both $Mn^{2+}$ to $Mn^{3+}$ and $Fe^{2+}$ to $Fe^{3+}$
    (c) $Fe^{3+}$ oxidizes $H_2O$ to $O_2$
    (d) $Mn^{3+}$ oxidizes $H_2O$ to $O_2$

## Comprehension XIII

**The concentration of potassium ions inside a biological cell is at least 20 times higher than the outside. The resulting potential difference across the cell is important in several processes such as transmission of nerve impulses and maintaining the ion balance. A simple model for such a concentration cell involving a metal 'M' is**

**M(s) | $M^+$(aq; 0.05 molar) || $M^+$(aq; 1 molar) | M(s)**

**For the above electrolytic cell, the magnitude of the cell potential, $|E_{cell}| = 70$ mV.**

44. For the above cell
    (a) $E_{cell} < 0; \Delta G > 0$ (b) $E_{cell} > 0; \Delta G < 0$
    (c) $E_{cell} < 0; \Delta G° > 0$ (d) $E_{cell} > 0; \Delta G° < 0$

45. If the 0.05 molar solution of $M^+$ is replaced by a 0.0025 molar $M^+$ solution, then the magnitude of the cell potential would be
    (a) 35 mV (b) 70 mV
    (c) 140 mV (d) 700 mV

## Comprehension XIV

**The electrochemical cell shown below is a concentration cell.**

**M(s) | $M^{2+}$ (saturated solution of a sparingly soluble salt, $MX_2$) || $M^{2+}$ (0.001 mol $dm^{-3}$) | M(s)**

**The EMF of the cell depends on the difference in concentrations of $M^{2+}$ ions at the two electrodes. The EMF of the cell at 298 K is 0.059 V.**

**46.** The value of $\Delta G$ (kJ $mol^{-1}$) for the given cell is (take 1 F = 96500 C $mol^{-1}$)

(a) −5.7  (b) 5.7
(c) 11.4  (d) −11.4

**47.** The solubility product ($K_{sp}$; $mol^3$ $dm^{-3}$) of $MX_2$ at 298 K based on the information available for the given concentration cell is (take 2.303 × $R$ × 298/F = 0.059)

(a) $1 \times 10^{-15}$  (b) $4 \times 10^{-15}$
(c) $1 \times 10^{-12}$  (d) $4 \times 10^{-12}$

## Comprehension XV

**Conductance measurements are frequently employed to find the end points of acid–base and other titrations. The principle involved is that electrical conductance of a solution depends upon the number and mobility of ions. In the conductometric titrations, the conductance of the resulting solution is measured at different stages on adding some volume of the standard solution and then a graph is plotted from the experimental observations to get the end point. Conductometric titrations have several advantages. Coloured solutions, titration of weak acid and weak base, etc. cannot be titrated by normal methods, but can be titrated successfully by this method. Further, no special care is needed in titration because the end point is determined graphically.**

**48.** Which of the following graph truly represents the titration of HCl solution against NaOH solution?

(a)

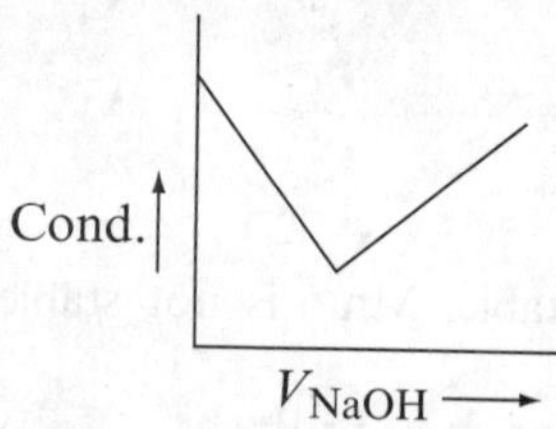


(b)

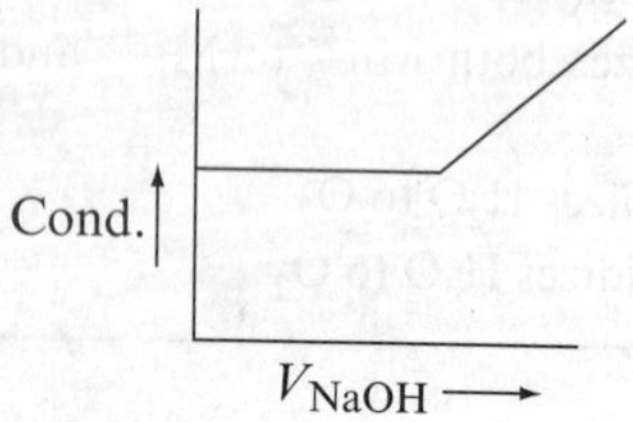


(c)

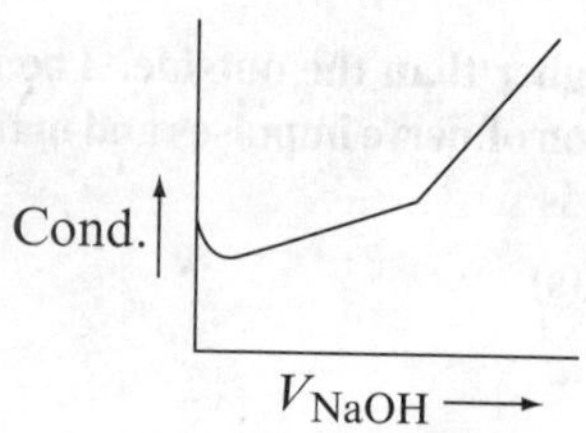


(d)

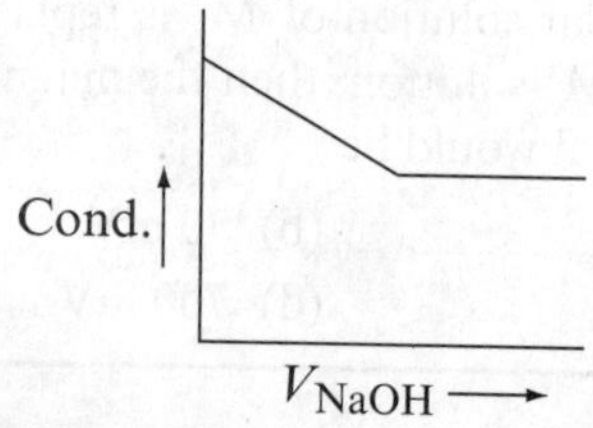


**49.** Which of the following graph truly represents the titration of $CH_3COOH$ solution against NaOH solution?

(a)

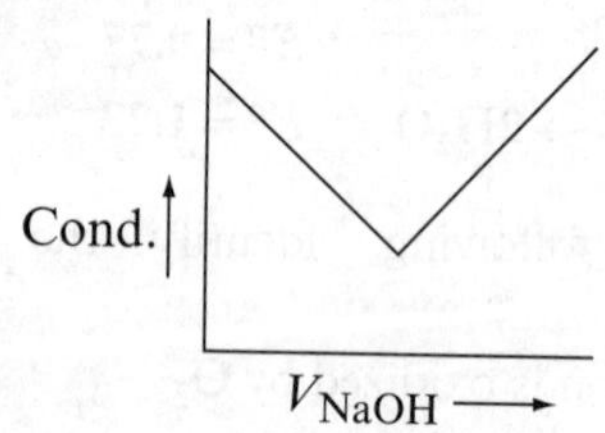


(b)

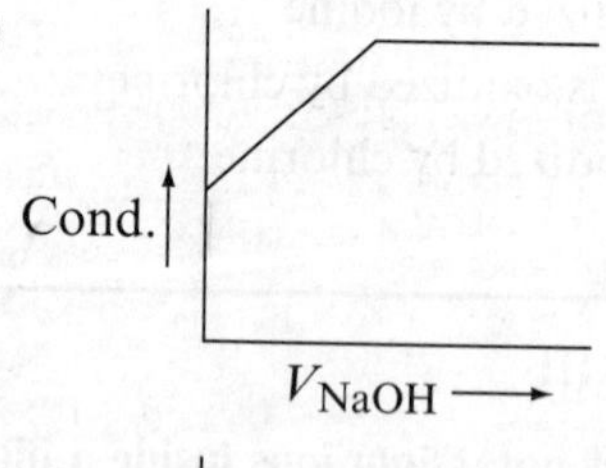


(c)

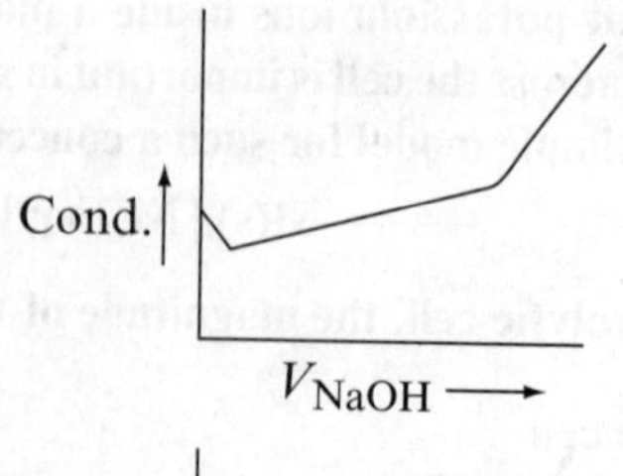


(d)

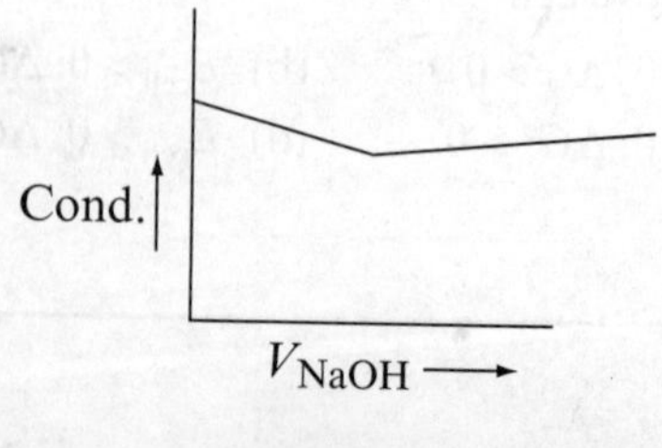

**50.** Which of the following graph truly represents the titration of HCl solution against $NH_4OH$ solution?

(a)

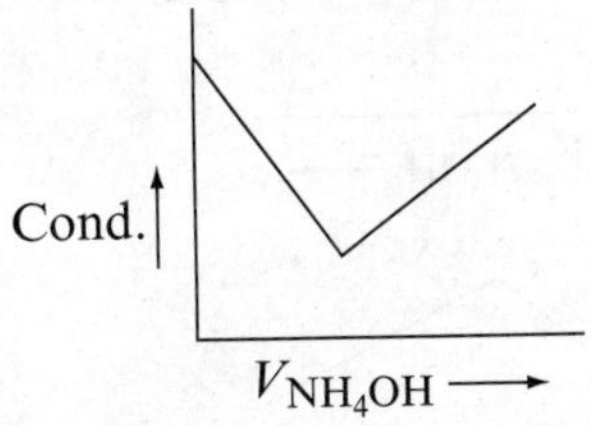


(b)

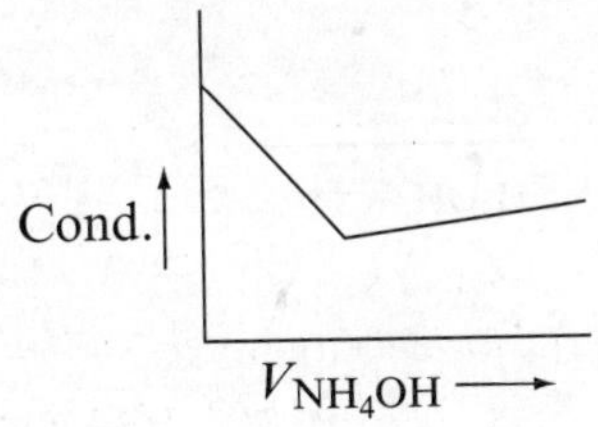


(c)

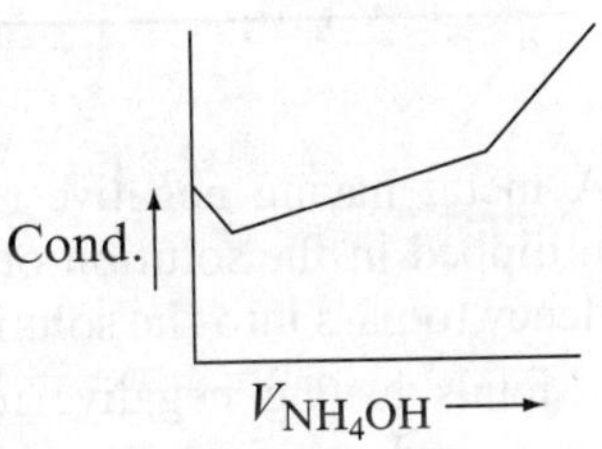


(d)

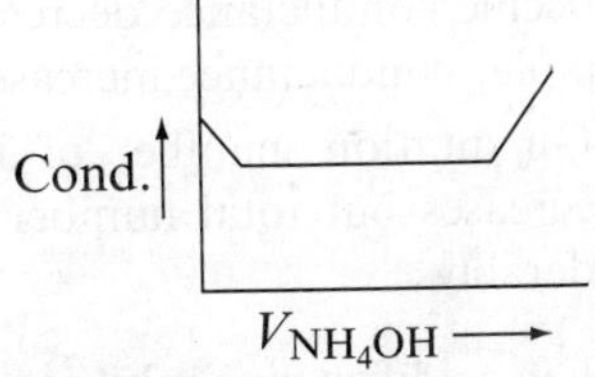


**51.** Which of the following graph truly represents the titration of a solution containing a mixture of HCl and $CH_3COOH$ against NaOH solution?

(a)

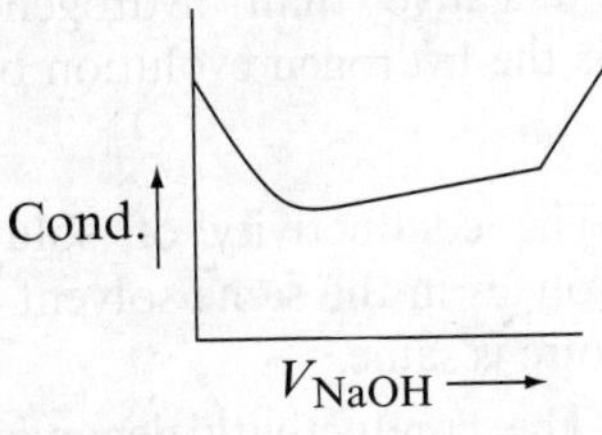


(b)

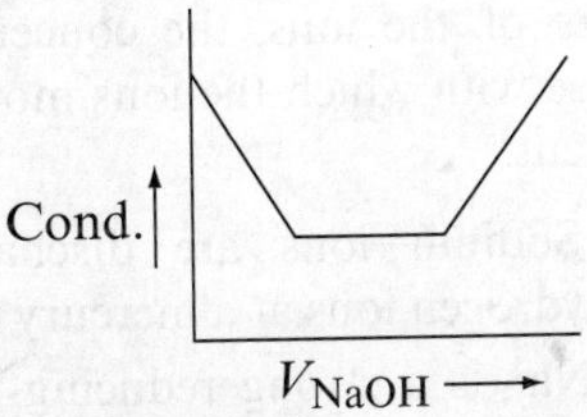


(c)

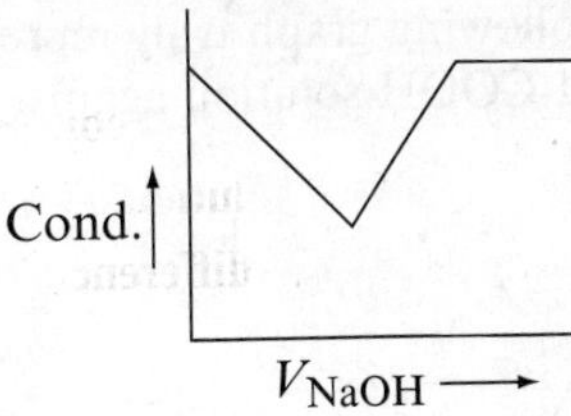


(d)

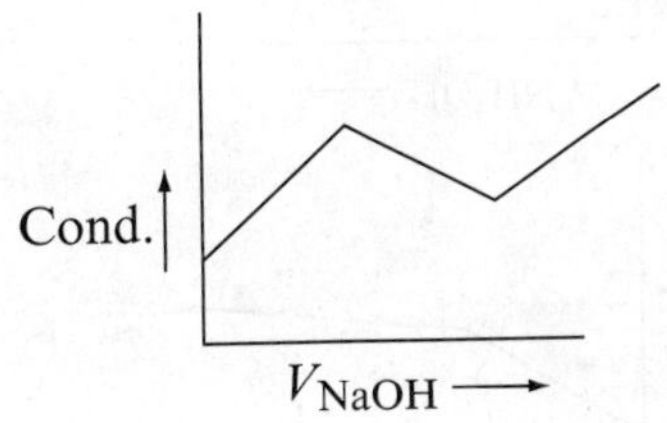


**52.** Which of the following graph truly represents the titration of $AgNO_3$ solution against KCl solution?

(a)

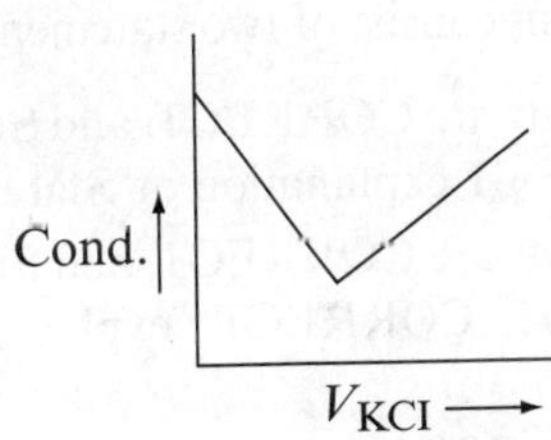


(b)

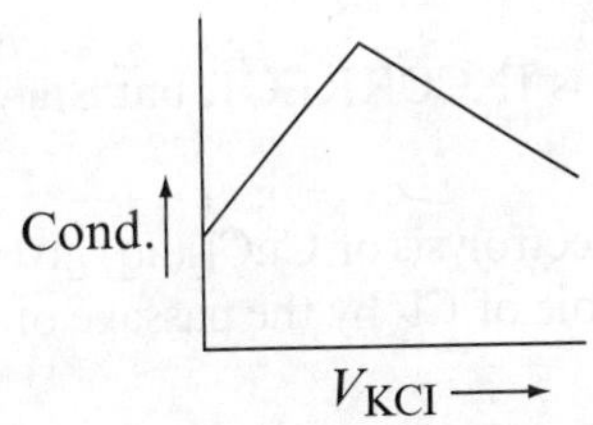


(c)

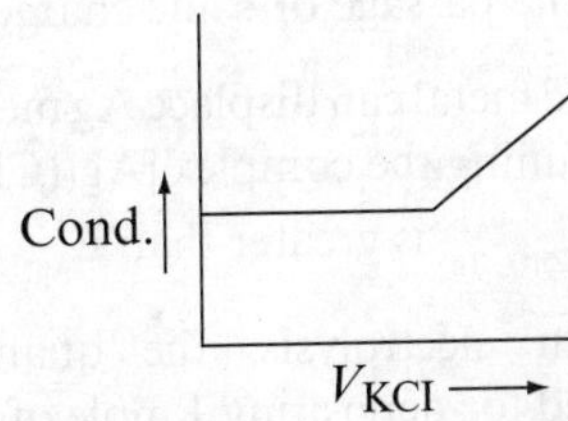


(d)

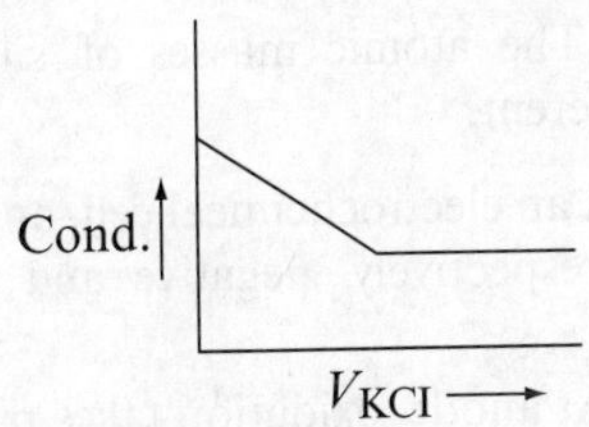

53. Which of the following graph truly represents the titration of $CH_3COOH$ solution against $NH_4OH$ solution?

(a)

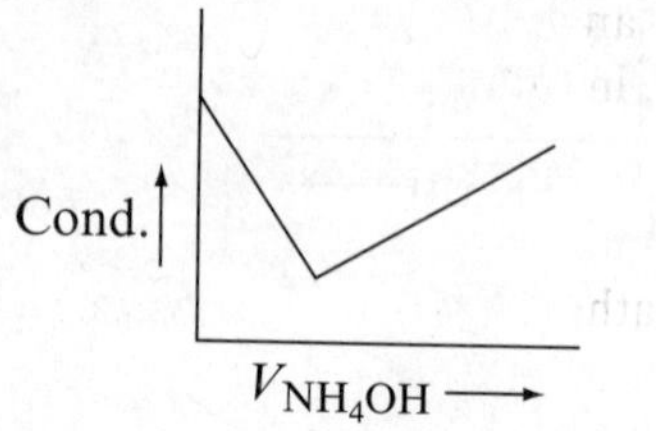


(b)

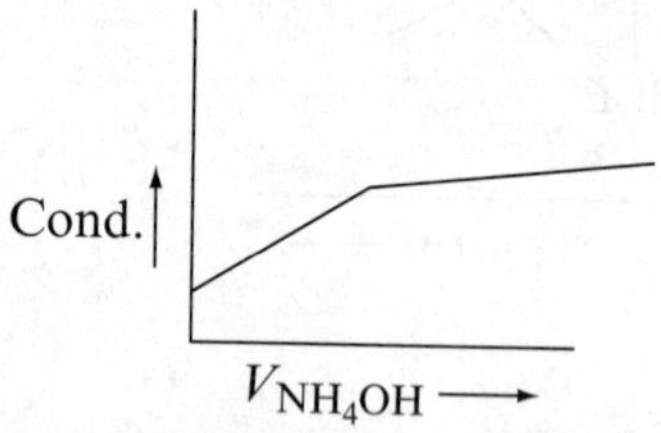


(c)

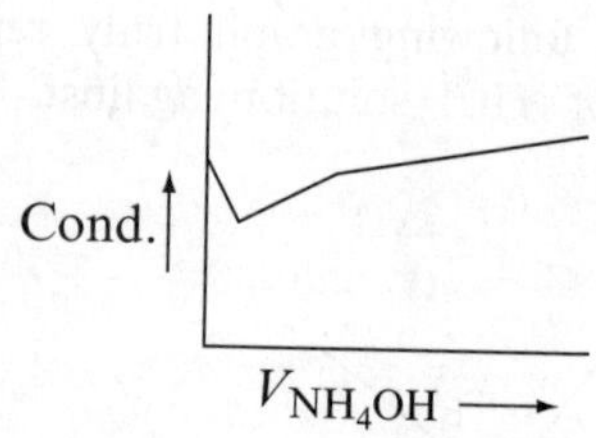


(d)

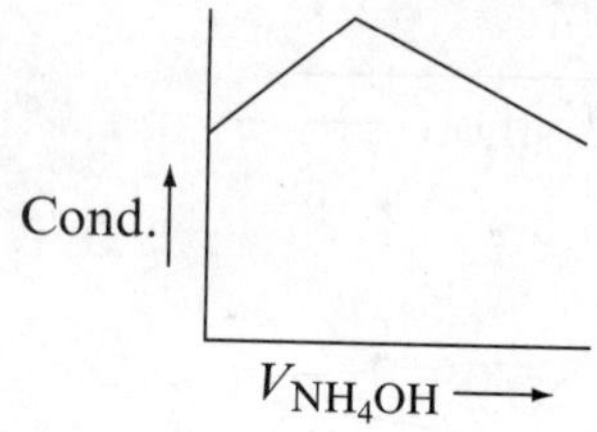


## Section D (Assertion – Reason)

The following questions consist of two statements. Mark

**(a)** If both statements are CORRECT, and **Statement II** is the CORRECT explanation of **Statement I.**

**(b)** If both statements are CORRECT, and **Statement II** is NOT the CORRECT explanation of **Statement I.**

**(c)** If **Statement I** is CORRECT, but **Statement II** is INCORRECT.

**(d)** If **Statement I** is INCORRECT, but **Statement II** is CORRECT.

1. **Statement I:** Electrolysis of $CuCl_2(aq)$ gives 1 mole of Cu and 1 mole of $Cl_2$ by the passage of suitable charge.

   **Statement II:** Equal equivalents of Cu and $Cl_2$ are formed during the passage of same charge.

2. **Statement I:** Zinc metal can displace Ag metal from a solution containing the complex $[Ag((CN)_2]^-$.

   **Statement II:** $E^\circ_{Zn^{2+}, Zn}$ is greater than $E^\circ_{Ag^+, Ag}$.

3. **Statement I:** In electrolysis, the quantity of electricity needed for depositing 1 mole of silver is different from that required for 1 mole copper.

   **Statement II:** The atomic masses of silver and copper are different.

4. **Statement I:** In an electrochemical cell, anode and cathode are, respectively, negative and positive electrode.

   **Statement II:** At anode, oxidation takes place and at cathode reduction takes place.

5. **Statement I:** A metal having negative reduction potential when dipped in the solution of its own ions has a tendency to pass into the solution.

   **Statement II:** Metals having negative reduction potential have large hydration energy.

6. **Statement I:** Specific conductance decreases with dilution while molar conductance increases.

   **Statement II:** On dilution, number of ions per unit volume decreases but total number of ions increases considerably.

7. **Statement I:** When acidified zinc sulphate solution is electrolysed between zinc electrodes, it is zinc that is deposited at the cathode and evolution of hydrogen gas does not take place.

   **Statement II:** The electrode potential of zinc becomes less negative than hydrogen as the overvoltage for the hydrogen evolution on zinc is quite large.

8. **Statement I:** The conductivity of solutions of different electrolytes in the same solvent and at a given temperature is same.

   **Statement II:** The conductivity depends on the charge and size of the ions, the concentrations of ions and ease with which the ions move under potential gradient.

9. **Statement I:** Sodium ions are discharged in preference to hydrogen ions at a mercury cathode.

   **Statement II:** $Na^+$ is a strong reducing agent in comparison to $H^+$ ion.

10. **Statement I:** The conductivity of metals decreases while that of electrolytic solution increases with increase in temperature.

    **Statement II:** Electrons in metals are very tightly held by the nucleus and are not free to move.

11. **Statement I:** An electrochemical cell can be set up only when the redox reaction is spontaneous.

    **Statement II:** A reaction is spontaneous if free energy change at constant temperature and pressure is negative.

12. **Statement I:** In the Daniel cell, if concentrations of $Cu^{2+}$ and $Zn^{2+}$ ions are doubled, the EMF of cell does not change.

    **Statement II:** If the concentration of ions in contact with the metal is doubled, the electrode potential will be doubled.

13. **Statement I:** KCl, NaCl, $NH_4Cl$, etc., cannot be used in the salt bridge of a cell containing silver ion.

    **Statement II:** A salt bridge contains concentrated solution of an inert electrolyte like KCl, $KNO_3$, $K_2SO_4$, etc., in agar-agar.

14. **Statement I:** During electrolysis of aqueous sodium acetate solution, the molar ratio of gases formed at cathode and anode is 1:3.

    **Statement II:** Acetate ion discharges at anode and $H^+$ ion, at cathode.

15. **Statement I:** Electrode potential of any electrode will change on changing any of its intensive properties.

    **Statement II:** Any intensive property of system changes on changing any of its properties, whether intensive or extensive.

## Section E (Column Match)

1. Match Column I with Column II

| Column I | Column II (Electrolysis product using inert electrodes) |
|---|---|
| (A) Dilute solution of HCl | (P) $O_2$ evolved at anode |
| (B) Dilute solution of NaCl | (Q) $H_2$ evolved at cathode |
| (C) Concentrated solution of NaCl | (R) $Cl_2$ evolved at anode |
| (D) $AgNO_3$ solution | (S) Ag deposited at cathode |

2. An aqueous solution of 'X' is added slowly to an aqueous solution of 'Y' as shown in List I. The variation in conductivity of these reactions is given in List II. Match List I with List II.

| List I (X + Y) | List II |
|---|---|
| (A) $(C_2H_5)_3N$ + $CH_3COOH$ | (P) Conductivity decreases and then increases |
| (B) KI(0.1 M) + $AgNO_3$(0.01 M) | (Q) Conductivity decreases and then does not change much |
| (C) $CH_3COOH$ + KOH | (R) Conductivity increases and then does not change much |
| (D) NaOH + HI | (S) Conductivity does not change much and then increases |

3. The standard reduction potential data at 25°C is given below:

$E°\ (Fe^{3+}, Fe^{2+}) = +0.77$ V

$E°\ (Fe^{2+}, Fe) = -0.44$ V

$E°\ (Cu^{2+}, Cu) = +0.34$ V

$E°\ (Cu^{+}, Cu) = +0.52$ V

$E°\ [O_2(g) + 4H^+ + 4e^- \rightarrow 2H_2O] = +1.23$ V

$E°\ [O_2(g) + 2H_2O + 4e^- \rightarrow 4OH^-] = +0.40$ V

$E°\ (Cr^{3+}, Cr) = -0.74$ V

$E°\ (Cr^{2+}, Cr) = -0.91$ V

Match $E°$ of the redox pair in List I with the values given in List II.

| List I | List II |
|---|---|
| (A) $E°\ (Fe^{3+}, Fe)$ | (P) −0.70 V |
| (B) $E°\ (4H_2O \rightarrow 4H^+ + 4\,OH^-)$ | (Q) −0.4 V |
| (C) $E°\ (Cu^{2+} + Cu \rightarrow 2Cu^+)$ | (R) −0.04 V |
| (D) $E°\ (Cr^{3+}, Cr^{2+})$ | (S) −0.83 V |

4. Match the following

| Column I | Column II |
|---|---|
| (A) Concentration cell | (P) $Ag \mid AgCl \mid Cl^- \parallel Ag^+ \mid Ag$ |
| (B) Spontaneous cell reaction | (Q) $Ag \mid AgCl \mid Cl^- \parallel Br^- \mid AgBr \mid Ag$ |
| (C) Non-spontaneous cell reaction | (R) $Ag \mid Ag^+ (0.1\ M) \parallel Ag^+ (1.0\ M) \mid Ag$ |
| | (S) $Ag \mid AgCl \mid Cl^- (0.1\ M) \parallel Cl^- (1.0\ M) \mid AgCl \mid Ag$ |

## Section F (Subjective)

### Single-digit Integer Type

1. Percentage of aniline hydrochloride hydrolysed in its M/40 solution at 25°C is (Given: $E_{C_6H_5NH_2.HCl|H_2} = -0.18$ V, 2.303 $RT/F = 0.06$)

2. The standard reduction potential for $Cu^{2+}|Cu$ electrode is +0.34 V. The solubility product of $Cu(OH)_2$ is $1.0 \times 10^{-19}$. The pH of solution at which the reduction potential for the above electrode becomes 0.31 V, is (2.303 $RT/F = 0.06$)

3. The reduction potential at 25°C for $Fe^{3+}|Fe^{2+}$ electrode is +0.718 V. If $E^\circ_{Fe^{2+}|Fe} = -0.44$ V and $E^\circ_{Fe^{3+}|Fe} = -0.04$ V, the ratio of molar concentrations of $Fe^{2+}$ to $Fe^{3+}$ ions in solution is (2.303 $RT/F =$ 0.06, log 5 = 0.7)

4. If $[Fe^{3+}]$ at equilibrium, when potassium iodide is added to a solution of $Fe^{3+}$ initially at 0.50 M until $[I^-] = 1.0$ M, is $x \times 10^{-5}$ M, the value of $x$ is (Given $E^\circ_{Fe^{3+}|Fe^{2+}} = 0.77$ V, $E^\circ_{I_2|I^-} = 0.53$ V, 2.303 $RT/F = 0.06$)

5. If it is desired to construct the following galvanic cell:

   Ag(s) | $Ag^+$ (saturated AgI) || $Ag^+$ (saturated AgCl, $x \times 10^{-4}$ M $Cl^-$) | Ag(s)

   to have $E_{Cell} = 0.102$ V, what should be the value of $x$ to get $[Cl^-]$, which must be present in the cathodic half-cell to achieve the desired EMF. Given $K_{sp}$ of AgCl and AgI are $1.8 \times 10^{-10}$ and $8.1 \times 10^{-17}$, respectively. (2.303 $RT/F = 0.06$, log 2 = 0.3)

6. An alloy of lead (valency = 2)-thallium (valency = 1) containing 70% Pb and 30% Tl, by weight, can be electroplated onto a cathode from a perchloric acid solution. How many hours (approx.) would be required to deposit 5.0 g of this alloy at a current of 1.10 A? (Pb = 208, Tl = 204)

7. Iridium was plated from a solution containing $IrCl_6^{\,y}$ for 2.0 h with a current of 0.075 A. The Iridium deposited on the cathode weighed 0.36 g. If the oxidation state of Ir in $IrCl_6^{\,y}$ is $x$, then the value of $(x + y)$ is (Ir = 192)

8. The electrolysis of cold sodium chloride solution produces sodium hypochlorite by reacting NaOH and $Cl_2$ thoroughly. How long (in days) will a cell operate to produce 10 L of 7.45% (by mass) solution of NaClO if the cell current is 2.5 A? Assume that the density of solution is 1.0 g/ml.

9. A volume of 500 ml of 0.1 M – $CuSO_4$ solution is electrolysed for 5 min at a current of 0.161 A. If Cu is produced at one electrode and oxygen at the other, the approximate pH of the final solution is

10. A test for complete removal of $Cu^{2+}$ ions from a solution of $Cu^{2+}$ is to add $NH_3$(aq). A blue colour signifies the formation of complex $[Cu(NH_3)_4]^{2+}$ having $K_f = 1.1 \times 10^{13}$ and thus confirms the presence of $Cu^{2+}$ in solution. 250 ml of 0.1 M – $CuSO_4$ is electrolysed by passing a current of 5 A for 1351 s. After passage of this charge, sufficient quantity of $NH_3$ is added to electrolysed solution maintaining $[NH_3] = 0.10$ M. If $[Cu(NH_3)_4]^{2+}$ is detectable up to its concentration as low as $1 \times 10^{-5}$ M, would a blue colour be shown by the electrolysed solution on addition of $NH_3$. Mark '1', if the answer is 'yes' and mark '2', if the answer is 'no'.

11. By passing a certain amount of charge through NaCl solution, 9.08 L of chlorine gas were liberated at STP. When the same amount of charge is passed through a nitrate solution of metal M, 52.8 g of the metal was deposited. If the specific heat of metal is 0.032 Cal/°C-g, the valency of metal is

12. The electrode reactions for charging of a lead storage battery are:

    $$PbSO_4 + 2e \rightarrow Pb + SO_4^{2-}$$

    $$PbSO_4 + 2H_2O \rightarrow PbO_2 + SO_4^{2-} + 4H^+ + 2e$$

    The electrolyte in the battery is an aqueous solution of sulphuric acid. Before charging, the specific gravity of the liquid was found to be 1.10 (16% $H_2SO_4$ by wt.). After charging for $\frac{965}{9}$ h, the specific gravity of the liquid was found to be 1.42 (40% $H_2SO_4$ by weight). If the battery contained 2 L of the liquid and the volume remains constant during charge, the average current (in A) used for charging the battery is

13. A dilute solution of KCl was placed between two platinum electrodes, 12 cm apart, across which a potential of 1.93 volts was applied. How far (in cm) would the $K^+$ ion move in 20 h at 25°C? Ionic conductance of $K^+$ ion at infinite dilution at 25°C is $7.5 \times 10^{-3}$ mho m$^2$ mol$^{-1}$.

14. In the refining of silver by electrolytic method, what will be the final mass (in gm) of 72.8 g silver anode (60% pure, by weight), if 9.65 A current is passed for 1 h? (Ag = 108)

15. The pH of 100 L of concentrated KCl solution after the electrolysis for 10 s using 9.65 A at 298 K is

## Four-digit Integer Type

1. The specific conductivity of a saturated solution of AgCl is $2.80 \times 10^{-4}$ mho m$^{-1}$ at 25°C. If $\lambda^\circ_{Ag^+} = 6.19 \times 10^{-3}$ mho m$^2$ mol$^{-1}$ and $\lambda^\circ_{Cl^-} = 7.81 \times 10^{-3}$ mho m$^2$ mol$^{-1}$, the solubility of silver chloride (in order of $10^{-5}$ g l$^{-1}$) at 25°C, is

2. The electrode potential (in millivolts) of $2Ag(s) + S^{2-}(aq) \rightarrow Ag_2S(s) + 2e^-$ in a solution buffered at pH = 3 and which is also saturated with 0.1 M – $H_2S$, is (Given: for $H_2S$, $K_{a1} = 10^{-8}$; $K_{a2} = 10^{-13}$, $K_{sp}$ of $Ag_2S = 4 \times 10^{-48}$, $E^\circ_{Ag^+|Ag} = 0.80$ V, log 2 = 0.3, 2.303 $RT/F = 0.06$)

3. An excess of liquid mercury is added to an acidified solution of $10^{-3}$ M – $Fe^{3+}$. It is found that 10% of $Fe^{3+}$ remains at equilibrium at 25°C. The value of $E^\circ_{Hg_2^{2+}|Hg}$ (in mV), assuming that the only reaction that occurs is: $2Hg + 2Fe^{3+} \rightarrow Hg_2^{2+} + 2Fe^{2+}$, is (Given: $E^\circ_{Fe^{3+}|Fe^{2+}} = 0.7724$ V, log2 = 0.3, log3 = 0.48, 2.303 $RT/F = 0.06$)

4. Determine potential (in mV) of the cell: Pt | $Fe^{2+}$, $Fe^{3+}$ || $Cr_2O_7^{2-}$, $Cr^{3+}$, $H^+$ | Pt in which $[Fe^{2+}] = 0.75$ M, $[Fe^{3+}] = 0.75$ M, $[Cr_2O_7^{2-}] = 2$ M, $[Cr^{3+}] = 4$ M and $[H^+] = 1$ M. Given:

   $Fe^{3+} + e^- \rightarrow Fe^{2+}$; $E^\circ = 0.77$ V

   $14H^+ + 6e + Cr_2O_7^{2-} \rightarrow 2Cr^{3+} + 7H_2O$; $E^\circ = 1.35$ V

   2.303 $RT/F = 0.06$, log 2 = 0.3, log 3 = 0.48

5. An alloy weighing 2.70 mg of Pb–Ag was dissolved in desired amount of $HNO_3$ and volume was made 250 ml. A silver electrode was dipped in solution and $E_{cell}$ of the cell:

   Pt, $H_2$(1 bar) | $H^+$(1 M) || $Ag^+$ | Ag

   was 0.50 V at 298 K. The percentage of lead in alloy is (Given: $E^\circ_{Ag^+|Ag} = 0.80$ V, Ag = 108, 2.303 $RT/F = 0.06$)

6. Given $E^\circ = 0.08$ V for $Fe^{3+}$(cyt b) | $Fe^{2+}$ (cyt b) couple and $E^\circ = 0.20$ V for $Fe^{3+}$(cyb c$_1$) | $Fe^{2+}$(cyt c$_1$) couple, where $E^\circ$ represents the standard state reduction potentials at pH = 7.0 at 25°C and cyt is an abbreviation for cytochromes. The value of $K_{eq}$ for the reaction

   $Fe^{3+}$ (cyt c$_1$) + $Fe^{2+}$ (cyt b) $\rightleftharpoons$ $Fe^{2+}$ (cyt c$_1$) + $Fe^{3+}$ (cyt b)

   is (2.303 $RT/F = 0.06$)

7. Calculate potential (in mV) of the cell: Cu | Mn(s) | $MnCl_2$(0.001 M), HCl(0.01 M) | $O_2$(0.25 bar) | Pt | Cu. Given, $E^\circ = -1.185$ V for the $Mn^{2+}$ | Mn couple and 1.229 V for the $O_2$ | $H_2O$, $H^+$ couple. (2.303 $RT/F = 0.06$, log 2 = 0.3)

8. The temperature coefficient of the cell: Zn | $ZnCl_2$ | AgCl(s) | Ag is + 0.001 V/K. The entropy change (in J/K) accompanying this reaction, at 298 K is

9. For the reaction: $4Al(s) + 3O_2(g) + 6H_2O(l) + 4OH^-(aq) \rightleftharpoons 4Al(OH)_4^-(aq)$; $E^\circ_{cell} = 2.5$ V. If $\Delta_f G^\circ$ for $H_2O(l)$ and $OH^-(aq)$ are –280.0 and –156.25 kJ/mol, respectively, the magnitude of $\Delta_f G^\circ$ for $Al(OH)_4^-(aq)$ (in kJ/mol) is

10. In a Zn–$MnO_2$ Cell, the anode is made up of Zn and cathode of carbon rod surrounded by a mixture of $MnO_2$, Carbon, $NH_4Cl$ and $ZnCl_2$ in aqueous base. If 8.7 g $MnO_2$ is present in cathodic compartment, how many days the dry cell will continue to give a current of $3.99 \times 10^{-3}$ A?

11. An amount of 19 g of molten $SnCl_2$ is electrolysed for some time. Inert electrodes are used. 1.19 g of tin is deposited at the cathode. No substance is lost during electrolysis. If the ratio of the masses of $SnCl_2$ and $SnCl_4$ after electrolysis is $x$:261, the value of $x$ is (Sn = 119)

12. An object whose surface area is 80 $cm^2$ is to be plated with an even layer of gold $8.0 \times 10^{-4}$ cm thick. The density of gold is 19.7 g/$cm^3$. The object is placed in a solution of $Au(NO_3)_3$ and a current of 2.4 A is applied. The time (in seconds) required for the electroplating to be completed, assuming that the layer of gold builds up evenly, is (Au = 197)

13. We have taken a saturated solution of AgBr. $K_{sp}$ of AgBr is $12 \times 10^{-14}$. If $10^{-7}$ moles of $AgNO_3$ is added to 1 litre of this solution, find conductivity (specific conductance) of this solution in terms of $10^{-7}$ $S{\cdot}m^{-1}$ units. Given $\lambda°(Ag^+) = 6 \times 10^{-3}$ S $m^2$ $mol^{-1}$, $\lambda°(Br^-) = 8 \times 10^{-3}$ S $m^2$ $mol^{-1}$, $\lambda°(NO_3^-) = 7 \times 10^{-3}$ $S{\cdot}m^2$ $mol^{-1}$.

14. On passing electricity through nitrobenzene solution, it is converted into azobenzene. The mass of azobenzene (in mg) produced, if the same quantity of electricity produces oxygen just sufficient to burn 96 g of fullerene ($C_{60}$), is

15. The conductivity of saturated solution of sparingly soluble salt, $Ba_3(PO_4)_2$, is $1.2 \times 10^{-5}$ $ohm^{-1}cm^{-1}$. The limiting equivalent conductances of $BaCl_2$, $K_3PO_4$ and KCl are 160, 140 and 100 $ohm^{-1}cm^2$ $eq^{-1}$, respectively. The $k_{sp}$ of $Ba_3(PO_4)_2$ (in the order of $10^{-25}$) is

## Answer Keys

**Exercise I**

### Electrode Potential

1. (c) 2. (d) 3. (d) 4. (a) 5. (a) 6. (a) 7. (b) 8. (a) 9. (a) 10. (a)
11. (d) 12. (b) 13. (a) 14. (c) 15. (c) 16. (b) 17. (b) 18. (a) 19. (c) 20. (a)
21. (c) 22. (c) 23. (c) 24. (a) 25. (d) 26. (d) 27. (c) 28. (c) 29. (a) 30. (a)

### Galvanic Cell

31. (a) 32. (d) 33. (c) 34. (b) 35. (c) 36. (c) 37. (c) 38. (c) 39. (d) 40. (d)
41. (d) 42. (d) 43. (c) 44. (b) 45. (b) 46. (d) 47. (d) 48. (a) 49. (d) 50. (c)
51. (b) 52. (b) 53. (a) 54. (a) 55. (a) 56. (c) 57. (a) 58. (c) 59. (c) 60. (b)
61. (a) 62. (b) 63. (c) 64. (d) 65. (a)

### Electrolysis

66. (b) 67. (a) 68. (a) 69. (b) 70. (b) 71. (d) 72. (b) 73. (c) 74. (b) 75. (d)
76. (a) 77. (b) 78. (d) 79. (b) 80. (a) 81. (b) 82. (a) 83. (c) 84. (d) 85. (a)
86. (a) 87. (d) 88. (a) 89. (c) 90. (b) 91. (c) 92. (d) 93. (c) 94. (a) 95. (b)
96. (a) 97. (d) 98. (a) 99. (b) 100. (c) 101. (a) 102. (a) 103. (b) 104. (b) 105. (b)

### Conductance

106. (b) 107. (a) 108. (c) 109. (b) 110. (a) 111. (b) 112. (c) 113. (c) 114. (a) 115. (d)
116. (a) 117. (b) 118. (d) 119. (c) 120. (d) 121. (a) 122. (c) 123. (a) 124. (d) 125. (a)

## Answer Keys

## Exercise II

### Section A (Only one Correct)

| | | | | | | | | | |
|---|---|---|---|---|---|---|---|---|---|
| 1. (c) | 2. (d) | 3. (b) | 4. (b) | 5. (a) | 6. (a) | 7. (d) | 8. (d) | 9. (b) | 10. (c) |
| 11. (c) | 12. (a) | 13. (a) | 14. (b) | 15. (b) | 16. (b) | 17. (a) | 18. (b) | 19. (d) | 20. (a) |
| 21. (b) | 22. (c) | 23. (a) | 24. (a) | 25. (b) | 26. (c) | 27. (d) | 28. (b) | 29. (b) | 30. (b) |
| 31. (b) | 32. (a) | 33. (c) | 34. (a) | 35. (b) | 36. (a) | 37. (d) | 38. (a) | 39. (a) | 40. (a) |
| 41. (a) | 42. (c) | 43. (a) | 44. (a) | 45. (a) | 46. (b) | 47. (c) | 48. (b) | 49. (b) | 50. (b) |
| 51. (b) | 52. (b) | 53. (a) | 54. (a) | 55. (a) | 56. (b) | 57. (d) | 58. (c) | 59. (a) | 60. (b) |
| 61. (c) | 62. (a) | 63. (a) | 64. (b) | 65. (d) | 66. (b) | 67. (a) | 68. (c) | 69. (c) | 70. (b) |
| 71. (c) | 72. (a) | 73. (c) | 74. (d) | 75. (c) | 76. (a) | 77. (b) | 78. (d) | 79. (a) | 80. (b) |

### Section B (One or More than one Correct)

| | | | |
|---|---|---|---|
| 1. (a), (c) | 2. (a), (c), (d) | 3. (c), (d) | 4. (b), (c) |
| 5. (d) | 6. (a), (c) | 7. (b), (c) | 8. (b), (c), (d) |
| 9. (a), (b), (c) | 10. (a), (c) | 11. (b), (c) | 12. (a), (b), (d) |
| 13. (a), (b) | 14. (a), (b) | 15. (a), (b), (c), (d) | |

### Section C

**Comprehension I**

1. (a) 2. (b) 3. (a)

**Comprehension II**

4. (b) 5. (c) 6. (a)

**Comprehension III**

7. (a) 8. (a) 9. (c) 10. (b)

**Comprehension IV**

11. (b) 12. (b) 13. (c)

**Comprehension V**

14. (b) 15. (a) 16. (d) 17. (b) 18. (a)

**Comprehension VI**

19. (b) 20. (d) 21. (a) 22. (b) 23. (d)

**Comprehension VII**

24. (d) 25. (a) 26. (c) 27. (a)

**Comprehension VIII**

28. (b) 29. (a) 30. (d)

**Comprehension IX**

31. (a) 32. (b) 33. (b) 34. (c)

**Comprehension X**

35. (a) 36. (b) 37. (b) 38. (c)

**Comprehension XI**

39. (b) 40. (d) 41. (a)

**Comprehension XII**

42. (c) 43. (d).

**Comprehension XIII**

44. (b) 45. (c)

**Comprehension XIV**

46. (d) 47. (b)

**Comprehension XV**

48. (a) 49. (c) 50. (b) 51. (a)
52. (c) 53. (b)

## Section D (Assertion – Reason)

1. (a) 2. (c) 3. (b) 4. (a) 5. (b) 6. (c) 7. (a) 8. (d) 9. (c) 10. (c)
11. (b) 12. (c) 13. (b) 14. (b) 15. (c)

## Section E (Column Match)

1. A → P, Q; B → P, Q; C → Q, R; D → P, S
2. A → R; B → S; C → Q; D → P
3. A → R; B → S; C → P; D → Q
4. A → R, S; B → P, R; C → Q, S

## Section F (Subjective)

### Single-digit Integer Type

1. (4) 2. (5) 3. (5) 4. (5) 5. (4) 6. (1) 7. (0) 8. (9) 9. (3) 10. (2)
11. (3) 12. (2) 13. (9) 14. (8) 15. (9)

### Four-digit Integer Type

1. (0287) 2. (0142) 3. (0815) 4. (0571) 5. (0090)
6. (0100) 7. (2633) 8. (0193) 9. (1300) 10. (0028)
11. (1520) 12. (0772) 13. (0055) 14. (0728) 15. (0108)

# HINTS AND EXPLANATIONS

## EXERCISE I (JEE MAIN)

### Electrode Potential

1. Potential of electrode is independent from the area of electrode immersed in solution.
2. Informative.
3. Copper cannot reduce $Al^{3+}$.
4. Cu will reduce $Ag^+$ and itself becomes $Cu^{2+}$.
5. K is more active metal than Al.
6. Mg is better reducing agent.
7. Informative
8. Lower the reduction potential, strong is the reducing agent.
9. $Y^- + X \longrightarrow Y^{2-} + X^+$ is possible.
10. $Mno_4^-$ Will also oxidize HCl.
11. Electrochemical series
12. $M^{n+}(aq) + ne^- \longrightarrow M(s); E = -ve, \Delta G = +ve$

    Hence, reverse reaction is spontaneous.
13. $Hg_2Cl_2(s) \rightleftharpoons Hg_2^{2+}(aq) + 2Cl^-(aq)$

    $Hg_2^{2+}(aq) + 2e^- \rightleftharpoons 2Hg(l)$
14. $E^\circ_{Mno_4^-/Mn^{2+}} > E^\circ_{Br_2/Br^-}$
15. Electrolyte used must have nearly same ionic mobility.
16. $MnO_4^- + 8H^+ + 5e^- \rightleftharpoons Mn^{2+} + 4H_2O$

    $$\Delta E = -\frac{0.059}{5} \cdot \log \frac{1}{1 \times [H^+]^8} = -\frac{0.059}{5} \cdot \log \frac{1}{(0.01)^8}$$

    $= -0.1888$ V
17. $Cu^{2+} + 2e^- \rightleftharpoons Cu$

    $$E = E^\circ - \frac{0.059}{2} \cdot \log \frac{1}{[Cu^{2+}]} = E^\circ - \frac{0.059}{2} \log \frac{1}{0.1}$$

    $= E^\circ - 0.0295$ V
18. $O_2(g) + 4H^+(aq) + 4e^- \rightleftharpoons 2H_2O(l)$

    $$E^\circ_{Basic} = E^\circ_{acidic} - \frac{0.059}{4} \cdot \log \frac{1}{(10^{-14})^4} = 0.404 \text{ V}$$
19. $$E_{Cu^+/Cu} = \frac{2 \times E_{Cu^{2+}/cu} - 1 \times E_{Cu^{2+}/Cu^+}}{2-1} = 0.521 \text{ V}$$
20. $2H^+(aq) + 2e^- \rightleftharpoons H_2(g)$

    $$E = E^\circ - \frac{0.059}{2} \cdot \log \frac{1}{(H^+)^2} = 0 - \frac{0.059}{2} \cdot \log \frac{1}{(10^{-7})^2}$$

    $= -0.0413$
21. $E = -0.059.\ P^H$
22. Ionic mobility of ions in salt bridge should be same.
23. Higher reduction potential, stronger is the oxidizing agent.
24. Lower the reduction potential, stronger is the reducing agent.
25. $$E^\circ_{Fe^{3+}/Fe^{2+}} = \frac{3 \times E^\circ_{Fe^{3+}/Fe} - 2 \times E^\circ_{Fe^{2+}/Fe}}{3-2}$$

    $= +0.772$ V
26. $H_2(g) \rightleftharpoons 2H + (aq) + 2e^-$

    $$E_{H_2/CH_3COOH} = E_{H_2/H^+} = E^\circ_{H_2/H^+} - \frac{0.06}{2} \cdot \log \frac{(H^+)^2}{P_{H_2}}$$

    $$= 0 - \frac{0.06}{2} \cdot \log \frac{\left(\sqrt{Ka \times C}\right)^2}{P_{H_2}}$$

    $$= -\frac{0.06}{2} \cdot \log \frac{1.8 \times 10^{-5} \times 0.5}{1}$$

    $= +0.1512$ V

27. $E^\circ_{Cl^-/CuCl/Cu} = E^\circ_{Cu^+/Cu} - \frac{0.06}{1}\log\frac{1}{K_{sp}}$

or, $1.28 = E^\circ_{Cu^+/Cu} - \frac{0.06}{1}\cdot\log\frac{1}{2\times10^{-7}}$

$\Rightarrow E^\circ_{Cu^+/Cu} = 1.682$ V

28. $E^\circ_{OH^-/Pb(OH)_2/Pb} = E^\circ_{Pb^{2P}/Pb} - \frac{0.06}{2}\cdot\log\frac{1}{K_{sp}}$

or, $(-0.55) = (-0.13) - \frac{0.06}{2}\cdot\log\frac{1}{K_{sp}}$

$\therefore K_{sp} = 1.0\times10^{-14}$

29. $E_{MnO_4^-/MnO_2} = \frac{5\times E_{MnO_4^-/Mn^{2+}} - 2\times E_{MnO_2/Mn^{2+}}}{5-2}$

$= \frac{5\times1.51 - 2\times1.23}{3} = 1.697$ V

30. For stronger oxidizing agent, reduction potential should be high.

## Galvanic Cell

31. Reduction occur at right electrode.

32. Circuit becomes incomplete.

33. $E_{Cell} = (ERP)_R - (ERP)_L = 0$

34. Net cell reaction is $Cl_2$ ($P_2$ atm) $\longrightarrow$ $Cl_2$ ($P_1$ atm) for spontaneous reaction: $P_2 > P_1$

35. Theoretical

36. Theoretical

37. Theoretical

38. $E_{cell} = E^\circ_{cell} - \frac{RT}{2F}\cdot\ln\frac{[Zn^{2+}]}{[Cu^{2+}]}$

39. $E^\circ_{cell}$ is independent from size or amount of electrode as well as $P^H$ of solution.

40. Theoretical

41. Theoretical

42. Ca and Cu are far away in the electrochemical series.

43. Reaction during charging of cell is 2Pb $SO_4$ (s) + $2H_2O$ (l) $\longrightarrow$ Pb(s) + $PbO_2$ (s) + $2H_2SO_4$ (aq)

44. Informative

45. Depolarizer is the term used for oxidizing agent in the cell.

46. Informative

47. Pb(s) + $PbO_2$ (s) + $2H_2SO_4$ (aq) $\longrightarrow$ $2PbSO_4$ (aq) + $2H_2O$(l)

48. $E^\circ_{cell} = \frac{0.059}{n}\cdot\log K_{eq} \Rightarrow 0.295 = \frac{0.059}{2}\cdot\log K_{eq}$

$\therefore K_{eq} = 10^{10}$

49. $L\cdot E: Ag(s) \rightleftharpoons Ag^+\left(c_1 = \sqrt{K_{SP}}\ M\right) + e^-$

$R\cdot E: Ag^+ (C_2 = 0.1\,M) + e^- \rightleftharpoons Ag(s)$

$\therefore$ Net reaction: $Ag^+(C_2) \rightleftharpoons Ag^+(C_1)$

$E_{cell} = -\frac{0.059}{1}\cdot\log\frac{C_1}{C_2}$

$\Rightarrow 0.413 = -\frac{0.059}{1}\log\frac{\sqrt{K_{sp}}}{0.1}$

$\therefore K_{sp} = 1.0\times10^{-16}$

50. Net cell reaction = $H_2(P_1) \rightleftharpoons H_2(P_2)$

$E_{cell} = -\frac{0.06}{2}\cdot\log\frac{420}{600} = +0.0045$ V

51. $E^\circ_{cell} = 0.44 - 0.33 = 0.11$ V

52. Cell reaction: $Zn + Fe^{2+} \rightleftharpoons Zn^{2+} + Fe$

$E_{cell} = E^\circ_{cell} - \frac{0.059}{2}\cdot\log\frac{[Zn^{2+}]}{[Fe^{2+}]}$

or, $0.2905 = \frac{0.059}{2}\cdot\log K_{eq} - \frac{0.059}{2}\cdot\log\frac{0.01}{0.001}$

$\therefore K_{eq} = 10^{\left(\frac{0.32}{0.0295}\right)}$

53. $E^\circ_{cell} = \frac{0.06}{n} \cdot \log K_{eq} \Rightarrow 0.18 = \frac{0.06}{4} \cdot \log K_{eq}$

$\therefore \log K_c = 12$

54. $E_{cell} = E^\circ_{cell} - \frac{RT}{2F} \cdot \ln \frac{[Zn^{2+}]}{[Cu^{2+}]}$

55. $E_{cell} = E^\circ_{cell} - \frac{RT}{2F} \cdot \ln \frac{[Ni^{2+}]}{[Cu^{2+}]}$

56. $E_2 = E_1 - \frac{0.059}{2} \cdot \log \frac{1.0}{0.01} = E_1 - 0.059$

57. $E^\circ_{cell} = \frac{0.059}{2} \cdot \log 10^{12} = 0.354$ V

58. $[H^+]_{HA} = \sqrt{K_a \cdot C} = \sqrt{10^{-5} \times 0.1} = 10^{-3}$ M

59. Theoretical

60. $E^\circ_{cell} = \frac{0.06}{n} \cdot \log K_{eq}$

$\Rightarrow (0.77 - 0.53) = \frac{0.06}{2} \cdot \log K_{eq}$

$\therefore K_{eq} = 10^8$

61. Current flow from right electrode to left electrode in external circuit.

62. $\Delta G^\circ = -nF \cdot E^\circ_{cell} \Rightarrow -48250 = -2 \times 96500 \times E^\circ_{cell}$

$\therefore E^\circ_{cell} = 0.25$ V

Now, $E_{cell} = E^\circ_{cell} - \frac{0.059}{2} \cdot \log \frac{[H^+]^2 [Cl^-]^2}{P_{H_2}}$

$= 0.25 - \frac{0.059}{2} \cdot \log \frac{(0.1)^2 \times (0.1)^2}{1} = 0.368$ V

63. Net cell reaction: $H^+(C = ?) \rightleftharpoons H^+ (10^{-6}\ M)$

$E_{cell} = -\frac{0.059}{1} \cdot \log \frac{10^{-6}}{c} = 0.118 \Rightarrow C = 10^{-4}$ M

64. $E^\circ_{OCl^-/Cl_2} = \frac{2 \times E^\circ_{OCl^-/Cl^-} - 1 \times E^\circ_{Cl_2/Cl^-}}{2-1}$

$= \frac{2 \times 0.94 - 1 \times 1.36}{1}$

$= 0.52$ V

65. For (i) & (iii); $E^\circ_{cell} = 3.0$ V

## Electrolysis

66. Oxidation occur at anode and reduction at cathode, in both.

67. Theoretical

68. Theoretical

69. Due to high over-voltage potential, reduction of $H_2O$ becomes less favorable.

70. Electrochemical equivalent, $Z = \frac{E}{F}$ and E is maximum for silver.

71. Informative

72. $W = \frac{Q}{F} \times \frac{A}{n}$

For smaller $n$, $w$ will be higher.

73. Cathode to anode in internal supply.

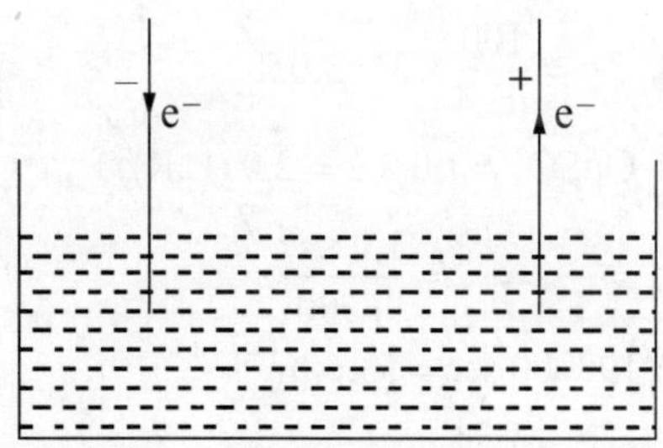


74. Informative

75. Informative

76. Equivalent weight is maximum for Na.

77. $n_{eq} H_2 = n_{eq} Cu \Rightarrow \frac{0.5}{1} = \frac{w}{63.5} \times 2$

$\Rightarrow w = 15.875$ gm

78. Ni will oxidize and $Ni^{2+}$ will reduce resulting no change in $Ni^{2+}$ Concentration.

79. $\frac{W}{E} = \frac{Q}{F} \Rightarrow \frac{9.8}{63.5} \times 2 = \frac{3 \times 10000}{96500} \times \eta \Rightarrow \eta = 0.99$

80. Cathode = $H_2$, anode = $O_2$ and $\frac{W_{H_2}}{W_{O_2}} = \frac{1}{8}$

81. $W = Z \cdot i \cdot t \Rightarrow Z \cdot i = \text{constant}$

$\therefore \quad \frac{i_1}{i_2} = \frac{Z_2}{Z_2} = \frac{E_2}{E_1}$

82. Theoretical

83. Informative

84. At cathode: $Cu^{2+}(aq) + 2e^- \longrightarrow Cu(s)$

At anode: $2H_2O(l) \longrightarrow O_2(g) + 4H^+(aq) + 4e^-$

Hence, solution becomes rich in $H_2SO_4$.

85. Reduction occur first for the species having higher reduction potential.

86. For very dilute NaCl solution,

Cathode: $2H_2O(l) + 2e^- \rightarrow H_2(g) + 2OH^-(aq)$; $P^H \uparrow$

Anode: $2H_2O(l) \rightarrow O_2(g) + 4H^+(aq) + 4e^-$; $P^H \downarrow$

87. $\frac{w_1}{w_2} = \frac{E_1}{E_2} = \frac{A/2}{A/3} = \frac{3}{2}$

88. $n_{eq}$ of Zn $= \frac{100}{65.4} \times 2 = 3.058$

$n_{eq}$ of $CuSO_4 = 1.0 \times 2 = 2.0$ (L.R.)

$\because n_{eq} = \frac{Q}{F} \Rightarrow 2 = \frac{1.0 \times t}{96500}$

$\Rightarrow t = 193000 \text{ sec} = 53.6 \text{ hr.}$

89. $n_{eq} = \frac{6 \times 10^{20}}{6 \times 10^{23}} = 10^{-3}$

90. $n_{eq} = \frac{Q}{F} \Rightarrow \frac{w}{208} \times 2 = \frac{1 \times 100 \times 3600}{96500}$

$\Rightarrow w \approx 388$ gm

91. $n_{eq}$ Ag deposited $= n_{eq}$Cu dissolved

or, $\frac{3.24}{108} \times 1 = \frac{w}{64} \times 2 \Rightarrow w = 0.96$ gm

$\therefore$ Final mass of Cu anode = 60 – 0.96 = 59.04 gm

92. $n_{eq}$ $Fe_2O_3$ reduced = $n \times n$-factor = 1 × 6

$\therefore$ Charge needed = 6 F

93. Number of moles, $n = n_{eq}/n$-factor

$\therefore n_{Al} : n_{Cu} : n_{Na} = \frac{3}{3} : \frac{3}{2} : \frac{3}{1} = 2:3:6$

94. $2HSO_4^- \rightleftharpoons S_2O_8^{2-} + 2H^+ + 2e^-$

Now, $n_{eq} = \frac{Q}{F} \Rightarrow 1 \times 2 = \frac{i \times 3600 \times 0.75}{96500}$

$\therefore i = 71.48$ A

95. Number of Faradays required $= n_{eq} = 1 \times 2 = 2$

96. $n_{eq} = n \times n\text{-factor} = \frac{Q}{F}$

or, $\frac{1 \times V}{0.0821 \times 546} \times 4 = \frac{96500}{96500} \Rightarrow V = 5.6$ L

97. $n_{eq}Ag = n_{eq}O_2 \Rightarrow \frac{w}{108} \times 1 = \frac{1.6}{32} \times 4 \Rightarrow w = 21.6$ gm

98. $n_{eq} = \frac{Q}{F} \Rightarrow \frac{3}{96.5} \times n = \frac{12000}{96500} \Rightarrow n = 4$

99. Number of electrons

$= \frac{Q}{e} = \frac{5 \times 200}{1.602 \times 10^{-19}} = 6.24 \times 10^{21}$

100. $n_{eq} = \frac{Q}{F} \Rightarrow \frac{3.0}{E} = \frac{9.65 \times 10 \times 60}{96500} \Rightarrow E = 50$

101. Cu will deposit at cathode.

$n_{eq} = \frac{Q}{F} \Rightarrow \frac{w}{63.5} \times 2 = \frac{1 \times 96.5 \times 60}{96500} \Rightarrow w = 1.905$ gm

102. $n_{eq} = n \times n\text{-factor} = \frac{Q}{F}$

or, $\frac{2.8}{22400} \times 4 = \frac{i \times 1}{96500} \Rightarrow i = 48.25$ A/sec

103. Mass of Na deposited

$= \frac{18.7}{(100 - 18.7)} \times 10 = 2.623$ gm

Now, $n_{eq} = \frac{Q}{F} \Rightarrow \frac{2.623}{23} \times 1 = \frac{Q}{F} \times 0.5 \Rightarrow Q \approx 0.2$ F

104. $n_{eq}$ A $= n_{eq}$ B

or, $\frac{5.6}{112} \times V_A = \frac{0.9}{27} \times 3$

$\therefore$ Valency of A, $V_A = 2$

105. $n_{eq}$ Zn $= n_{eq}$ Ni $\Rightarrow \frac{22.89}{32.7} = \frac{20.55}{E_{Ni}}$

$\therefore E_{Ni} = 29.36$

## Conductance

**106.** On dilution, conductance decreases and hence resistance increase.

**107.** Theoretical

**108.** HCl is molecular.

**109.** Radius of aq.ion : $K^+ < Na^+ < Li^+$

**110.** On dilution, $\wedge_m$ increases.

**111.** $H_2SO_4$ is strong electrolyte

**112.** For AgCl, $\wedge^\circ_m = \wedge_m = \lambda^\circ_{m(Ag^+)} + \lambda^\circ_m(Cl^-)$

$= 67.9 + 82.1 = 150\ ohm^{-1}\ cm^2\ mol^{-1}$

Now, $\wedge_m = \frac{\kappa}{C} \Rightarrow 150 = \frac{1.8\times10^{-6}}{s}$

$\therefore S = 1.2\times10^{-8}\ mol\ cm^{-3} = 1.2\times10^{-5}\ mol\ e^{-1}$

**113.** $\wedge^\circ_{eq(NH_4OH)} = \wedge^\circ_{eq(NH_4Cl)} + \wedge^\circ_{eq(NaOH)} - \wedge^\circ_{eq(NaCl)}$

$= 129.8 + 217.4 - 108.9 = 238.3\ ohm^{-1}\ cm^2\ mol^{-1}$

Now, $\alpha = \frac{\lambda_{eq}}{\lambda^\circ_{eq}} = \frac{9.532}{238.3} = 0.04 = 4\%$

**114.** $\wedge_m = \frac{\kappa}{C} = \frac{\frac{1}{250}\times125}{1\times10^3} = 5\times10^{-4}\ ohm^{-1}\ m^2\ mol^{-1}$

**115.** $MgSO_4(aq) + Ba(OH)_2\ (added) \longrightarrow Mg(OH)_2(s) + BaSO_4(s)$

First number of ions decreases and then increases.

**116.** $\wedge_{eq} = \frac{\kappa}{C} = \frac{2.6\times10^{-1}}{(1\times2)\times10^{-3}} = 1.3\times10^2\ ohm^{-1}\ cm^2\ eq^{-1}$

**117.** $\alpha = \frac{\wedge_m}{\wedge^\circ_m} = \frac{16.30}{(349.8+40.9)} = 0.04172$ or $4.172\%$

**118.** Cell constant, $C = \frac{l}{A} = \frac{2.5}{5} = 0.5\ cm^{-1} = 50\ m^{-1}$

**119.** $\wedge^\circ_{m(NH_4OH)} = \wedge^\circ_{m(NH_4Cl)} - \wedge^\circ_{m(Cl^-)} + \wedge^\circ_{m(OH^-)}$

$= 150 - 75 + 200 = 275\ ohm^{-1}\ cm^2\ mol^{-1}$

Now, $\alpha = \frac{\wedge_m}{\wedge^\circ_m} = \frac{22}{275} = 0.08$

**120.** $\wedge^\circ_{m(H_2O)} = \lambda^\circ_{m(H^+)} + \lambda^\circ_{m(OH^-)} = 0.035+0.020$

$= 0.055\ mho\ m^2\ mol^{-1}$

Now, $\wedge_m = \frac{\kappa}{C} \Rightarrow 0.055 = \frac{5.5\times10^{-6}}{C}$

$\therefore C = 10^{-4}\ mol\ m^{-3} = 10^{-7}\ M \Rightarrow K_w = C^2 = 10^{-14}\ M^2$

**121.** $\alpha = \frac{\wedge_m}{\wedge^\circ_m} = \frac{7.814\times10^{-4}}{3.907\times10^{-2}} = 0.02$

Now, $K_a = \frac{\alpha^2\cdot C}{1-\alpha} \simeq \alpha^2\cdot C = (0.02)^2\times0.05$

$= 2\times10^{-5}$

**122.** $n_m = \frac{\kappa}{C} = \frac{6.3\times10^{-4}}{\frac{0.015}{2}\times10^{-3}} = 84\ S\ cm^2\ mol^{-1}$

**123.** $\wedge_m = \wedge^\circ_m - A\cdot\sqrt{C}$ (for strong electrolyte)

or, $260 = \wedge^\circ_m - A\times\sqrt{0.25}$ (1)

$250 = \wedge^\circ_m - A\times\sqrt{1.00}$ (2)

$\therefore \wedge^\circ_m = 270\ ohm^{-1}\ cm^2\ mol^{-1}$

**124.** $\kappa_A = \frac{1}{50}\times G^*$ and $\kappa_B = \frac{1}{100}\times G^*$

Now, $G_{mix} = \frac{1}{G^*}\left(\frac{\kappa_A}{2} + \frac{\kappa_B}{2}\right) = \frac{3}{200}\ ohm^{-1}$

$\therefore R = \frac{1}{G} = \frac{200}{3}\ ohm$

**125.** Conductivity of solution will remain uncharged but the cell constant becomes double.

$\kappa = G\times G^* = \frac{1}{50}\times\frac{1.5}{5} = \frac{3}{5}\times10^{-2}\ ohm^{-1}cm^{-1}$

$\therefore \wedge_{eq} = \frac{K}{C} = \frac{\frac{3}{5}\times10^{-2}}{0.05\times10^{-3}} = 120\ ohm^{-1}cm^2eq^{-1}$

# EXERCISE II (JEE ADVANCED)

## Section A (Only one Correct)

1. (I) Cu cannot reduce Pb
(II) Pb can reduce Ag
(III) Ag cannot reduce Cu.
Hence, reducing power: Pb > Cu > Ag

2. $E = E° - \frac{0.06}{n} \cdot \log\frac{[R]}{[O]}$

$\Rightarrow 0.24 = 0.36 - \frac{0.06}{1}\log\frac{[R]}{[O]}$

$\therefore \frac{[O]}{[R]} = \frac{1}{100}$

3. For the complex ion to get oxidised, its reduction potential should be low.

4. $Ag^+ + 2NH_3 \rightleftharpoons Ag(NH)_3^+$;

$E° = 0.79 - 0.37 = 0.42$ V

Now, $E° - \frac{0.06}{n} \cdot \log K_{eq} \Rightarrow 0.42 = \frac{0.06}{1} \cdot \log K_f$

$\Rightarrow K_f = 10^7$

5. Given: $Co^{3+} + e^- \longrightarrow Co^{2+}$; E° = 1.81 V;

$\Delta G°_1 = -1 \times F \times 1.81$

$Co(CN)_6^{3+} + e^- \longrightarrow Co(CN)_6^{4-}$ E° = –0.83 V;

$\Delta G°_2 = -1 \times F \times (-0.83)$

$Co^{2+} + 6CN^- \longrightarrow Co(CN)_6^{4-}$; $K_f = 10^{19}$;

$\Delta G°_3 = -RT \times \ln 10^{19}$

Required $Co^{3+} + 6CN^- \longrightarrow Co(CN)_6^{3-}$ $K_f = ?$;

$\Delta G° = -RT \times \ln K_f$

Now, $\Delta G° = \Delta G°_1 - \Delta G°_2 + \Delta G°_3$

or, $-RT \cdot \ln K_f = (-1.81\,F) - 0.83\,F + (-RT\ln 10^{19})$

or, $RT \cdot \ln\frac{10^{19}}{K_f} = -2.64\,F$

$\Rightarrow \log\frac{10^{19}}{K_f} = -\frac{2.64\,F}{2.303\,RT} = -\frac{2.64}{0.06}$

$\therefore K_f = 10^{63}$

6. $E_{Cu^{2+}|Cu} = E°_{Cu^{2+}|Cu} - \frac{0.06}{2} \cdot \log\frac{1}{[Cu^{2+}]}$

$= E°_{Cu^{2+}|Cu} + 0.03\log[Cu^{2+}]$

$= 0.34 + 0.03 \times \log(0.1) = 0.31$ V

$\therefore E_{Cu/Cu^{2+}} = -0.31$ V

7. $E_{cell} = E°_{cell} - \frac{RT}{2F} \cdot \ln\frac{[Cd^{2+}]}{[Ag^+]^2}$

As CN– will form complex with Ag+ ion in the cathodic compartment, $E_{cell}$ will decrease.

8. $-\Delta G_{cell} = nF \cdot E_{cell} = nF\,[E°_{Cell} - \frac{RT}{nF} \cdot \ln\frac{[Zn^{2+}]}{[Cu^{2+}]}$

$= nFE°_{Cell} - RT \cdot \ln\frac{C_1}{C_2}$

9. $E°_{X^-|Ag\,X|Ag} = E°_{Ag^+|Ag} - \frac{RT}{F} \cdot \ln\frac{1}{K_{sp}}$

10. Cell reaction: $H^+$(cathode, $P^H = 3$) $\longrightarrow$ $H^+$ (anode, $P^H = ?$)

$E_{cell} = 0 - \frac{0.059}{1} \cdot \log\frac{[H^+]anode}{[H^+]cathode}$

$= 0.059\left[P^H_{anode} - P^H_{cathode}\right]$

or, $0.272 = 0.059\,[P^H_{anode} - 3] \Rightarrow P^H_{anode} = 7.6$

11. $Cl_2 + H_2O \rightleftharpoons Cl^- + ClO^- + 2H^+$;

$E°_{cell} = 1.36 - 1.63 = -0.27$ V

Now, $E_{cell} = E°_{cell} - \frac{0.06}{1}\log[H^+]^2$

or, $0 = -0.27 + \frac{0.06}{1} \times 2P^H = 2.25$

12. $E_{cell} = E_{quinohydrone} - E_{calomel}$

$0.210 = E_{quinohydrone} - 0.279$

$\Rightarrow E_{quinohydrone} = 0.489$ V

Quinohydrone electrode is

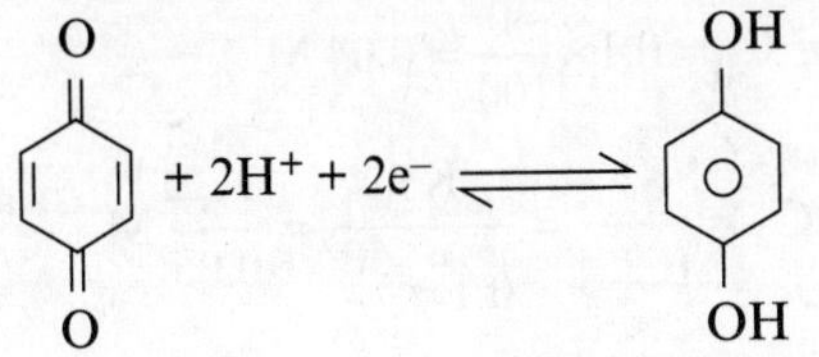

(Quinone, Q) (Hydroquinone, $H_2Q$)

$$E = E° - \frac{0.06}{2}\cdot\log\frac{1}{[H^+]} = E° - 0.06\ P^H$$

or, $0.489 = 0.699 - 0.06.P^H \Rightarrow P^H = 3.5$

**13.** Anode: $Ag(s) \longrightarrow Ag^+(aq) + e^- \ 1\times 6$

Cathode:

$$Cr_2O_7^{2-}(aq) + 14H^+(aq) + 6e^- \longrightarrow 2Cr^{3+}(aq) + 7H_2O(l)$$

Net:

$$6Ag(s) + Cr_2O_7^{2-}(aq) + 14H^+(aq) \longrightarrow 6Ag^+(aq) + 2Cr^{3+}(aq) + 7H_2O(l)$$

$$E_{cell} = E°_{cell} - \frac{0.06}{n}\cdot\log\frac{[Ag^+]^6[Cr^{3+}]^2}{[Cr_2O_7^{2-}][H^+]^{14}}$$

$$= (1.33 - 0.80) - \frac{0.06}{6}\cdot\log\frac{(0.1)^6\times(0.4)^2}{1.6\times(0.1)^{14}} = 0.46\ V$$

**14.** Net cell reaction: $Zn - Hg(C_1M) \longrightarrow Zn - Hg(C_2M)$

$$E_{cell} = 0 - \frac{0.059}{2}\log\frac{C_2}{C_1} = -\frac{0.059}{2}\log\frac{1}{10} = 0.0295\ V$$

**15.** $E°_{cell} = \frac{0.06}{2}\cdot\log K_{eq}$

$$\Rightarrow 0.75 - \frac{1.50\times 3 - 1.68\times 1}{3-1} = 0.03\log K_{eq}$$

$\therefore K_{eq} = 10^{-22}$

**16.** $[H^+]_{left} = \sqrt{Ka\cdot C} = \sqrt{1.8\times10^{-5}\times 0.1} = C_1M$

$$[H^+]_{right} = \frac{K_w}{\sqrt{Kb\cdot C}} = \frac{10^{-14}}{\sqrt{1.8\times10^{-5}\times 0.01}} = C_2M$$

Net cell reaction, assuming as concentration cell:

$$H^+(C_2M) \longrightarrow H^+(C_1M)$$

$$E_{cell} = 0 - \frac{0.06}{1}\cdot\log\frac{C_1}{C_2} = -0.465\ V$$

**17.** $E°_{V^{2+}|V^{3+}} = 1\times0.616 + 1\times0.439 - 1\times0.799$

$= 0.256\ V$

$\therefore E°_{V^{3+}|V^{2+}} = -0.256\ V$

**18.** $E_{cell} = E°_{cell} - \frac{0.06}{n}\cdot\log\frac{[H^+]^2}{P_{H_2}}$

or, $0.70 = (0.28 - 0) - \frac{0.06}{2}\cdot\log\frac{[H^+]^2}{1} \Rightarrow P^H = 7.0$

**19.** Net cell reaction:

$$Ag^+(C_1 = 0.1\,M) \longrightarrow Ag^+\left(C_2 = 2\left(\frac{K_{sp}}{4}\right)^{1/3} M\right)$$

Now, $E_{cell} = 0 - \frac{0.06}{1}\cdot\log\frac{C_2}{C_1}$

$$\Rightarrow 0.162 = -0.06\cdot\log\frac{(2K_{sp})^{1/3}}{0.1}$$

$\therefore K_{sp} = 4\times10^{-12}$

**20.** $E_{Tl^+|Tl} - E_{Pb^{2+}|Pb} = -0.444\ V$

or, $\left(E_{Tl^+|Tl} - E_{Pb^{2+}|Pb}\right) - \frac{0.06}{2}\cdot\log\frac{[Pb^{2+}]}{[Tl^+]^2} = -0.444$

or, $[(-0.336) - (-0.126)] - 0.03\cdot\log\frac{0.1}{\left(\frac{K_{sp}}{0.1}\right)^2} = -0.444$

$\therefore K_{sp} = 4\times10^{-6}$

**21.** $E_{cell} = E_{Cu} - E_{Zn} = (E_{Cu} - E_{calomel}) - (E_{Zn} - E_{calomel})$

From question $E_{calomel} - E_{Zn} = 1.083\ V$

and $E_{calomel} - E_{Cu} = -0.018\ V$

**22.** The potential of hydrogen electrode at $H_2$(1 bar) may be expressed as $E = -0.059\ P^H$

Now, $E_1 = -0.059\left[P^{Ka} + \log\frac{x}{y}\right]$

and $E_2 = -0.059\left[P^{Ka} + \log\frac{y}{x}\right]$

$$\therefore P^{Ka} = \frac{-(E_1 + E_2)}{0.118}$$

**23.** For the given reaction, $E^\circ_{cell} = E^\circ_{given} - E^\circ_{hydrogen}$

$$\therefore (-0.84) - 0 = \frac{0.06}{1} \cdot \log K_{eq} \Rightarrow K_{eq} = 10^{-14}$$

**24.** Net cell reaction may be written as

$H_2 + Zn^{2+} \rightleftharpoons 2H^+ + Zn$

$$E_{cell} = E^\circ_{cell} - \frac{0.06}{n} \log \frac{[H^+]^2}{P_{H_2} \cdot [Zn^{2+}]}$$

$$\text{or, } (-0.61) = (-0.76) - \frac{0.06}{2} \cdot \log \frac{[H^+]^2}{1 \times 0.4}$$

$\therefore [H^+] = 2 \times 10^{-3}$ M

$$\text{Now, } K_{a_2} = \frac{[H^+][SO_3^{2-}]}{[HSO_3^-]} = \frac{(2 \times 10^{-3}) \times (6.4 \times 10^{-2})}{0.4}$$

$= 3.2 \times 10^{-4}$

**25.** $Cu^{2+} + 4NH_3 \rightleftharpoons Cu(NH_3)_4^{2+}, K_f = 10^{12}$

| | | | |
|---|---|---|---|
| | 1.0 M | excess | |
| 100% | 0 | | 1.0 M |
| Equ. | $x$ | 2.0 M | 1.0 M |

$$10^{12} = \frac{1.0}{x \times (2.0)^4} \Rightarrow x = \frac{10^{-12}}{16} = 6.25 \times 10^{-14}$$

Now, cell reaction: $Zn + Cu^{2+} \rightleftharpoons Zn^{2+} + Cu$

$$E_{cell} = E^\circ_{cell} - \frac{0.06}{2} \cdot \log \frac{[Zn^{2+}]}{[Cu^{2+}]}$$

$$= [0.34 - (0.76)] - \frac{0.06}{2} \cdot \log \frac{1.0}{6.25 \times 10^{-14}} = 0.704 \text{ V}$$

**26.** Net cell reaction: $Zn + 2H^+ \rightleftharpoons Zn^{2+} + H_2$

$$E_{cell} = E^\circ_{cell} - \frac{0.06}{2} \log \frac{[Zn^{2+}] \cdot P_{H_2}}{[H^+]^2}$$

$$\text{or, } 0.70 = [0 - (-0.76)] - \frac{0.06}{2} \cdot \log \frac{0.01 \times 1}{[H^+]^2}$$

$\Rightarrow [H^+] = 0.01$ M

$$\text{Moles of HCl in RHS} = \frac{500 \times 0.01}{1000} = 5 \times 10^{-3}$$

$\therefore$ Mass of NaOH needed $= 5 \times 10^{-3} \times 40 = 0.2$ gm

**27.** On assuming concentration cell, the net cell reaction is $Ag^+$ ($C_1$M, Right) $\longrightarrow Ag^+$ ($C_2$M, left)

$$\text{Now, } C_1 = 0.1 \times \frac{40}{100} = 0.04 \text{ M}$$

$$\text{and } C_2 = \frac{K_{sp}}{[Cl^-]} = \frac{K_{sp}}{0.1 \times \frac{50}{100}} = \frac{K_{sp}}{0.05} \text{ M}$$

$$\text{Now, } E_{cell} = 0 - \frac{0.06}{1} \log \frac{C_2}{C_1}$$

$$\text{or, } 0.42 = -\frac{0.06}{1} \log \frac{K_{sp}/0.05}{0.04} \Rightarrow K_{sp} = 2 \times 10^{-10}$$

**28.** $\Delta G_{cell} = -nFE_{cell} \Rightarrow -965 \times 3 \times 10^3 = -12 \times 96500 \times E_{cell}$

$\therefore E_{cell} = 2.5$ V

**29.** $$\text{Theoretical efficiency} = \frac{-\Delta G^\circ}{-\Delta H^\circ} \Rightarrow 0.84 = \frac{-\Delta G^\circ}{285}$$

$\therefore \Delta G^\circ = -0.84 \times 285$ KJ $= -nF \cdot E^\circ_{cell}$

or, $0.84 \times 285 \times 10^3 = 2 \times 96500 \times E^\circ_{cell}$

$\Rightarrow E^\circ_{cell} = 1.24$ V

**30.** Net cell reaction: $Ag(s) + H^+ + Cl^- \rightleftharpoons AgCl(s) + \frac{1}{2} H_2(g)$ But for $E_{cell}$ calculation, reaction may be written as

$Ag(s) + H^+ \rightleftharpoons Ag^+ + \frac{1}{2} H_2$

$$\text{Now, } E_{cell} = E^\circ_{cell} - \frac{0.06}{1} \cdot \log \frac{[Ag^+] \cdot P_{H_2}^{1/2}}{[H^+]}$$

$$= [0 - 0.80] - \frac{0.06}{1} \cdot \log \frac{\left(\frac{10^{-10}}{0.1}\right) \times 1^{1/2}}{0.1} = -0.32 \text{ V}$$

It means that actual reaction is in reverse direction and $E_{cell} = 0.32$ V

**31.** Informative

**32.** $x\,Ag^+ + y\,NH_3 \rightleftharpoons Ag(NH_3)_y^{x+}$

| | | | |
|---|---|---|---|
| | aM | bM | 0 | 
| Eqn. | ? | bM | $\frac{a}{x}$M |

$b >> a$

$$K_f = \frac{a/x}{[Ag^+]^x \cdot b^y} \Rightarrow [Ag^+] = \left(\frac{a}{K_f \cdot x \cdot b^y}\right)^{1/x}$$

Cell reaction: $Ag^+$ (Right) $\longrightarrow Ag^+$ (Left)

HINTS AND EXPLANATIONS

$\therefore E_{cell} = 0 - \frac{0.059}{1} \cdot \log \frac{[Ag^+]Left}{[Ag^+]Right}$

Case-I: $0.118 = -0.059 \cdot \log\left(\frac{4\times10^{-4}}{4\times10^{-2}}\right)^{1/x} \Rightarrow x = 1$

Case-II: $0.118 = -0.059 \cdot \log\left(\frac{0.1}{1}\right)^{y/x} \Rightarrow y = 2$

**33.** $E^\circ_{cell} = \frac{0.06}{1} \cdot \log K_{eq} = 0.06 \log \frac{1}{1.667\times10^6}$

$= -0.3732$ V

Now, $E^\circ_{Cu^{2+}|Cu^+} - E^\circ_{Cu^+|Cu} = -0.3732$ V (1)

and $E^\circ_{Cu^{2+}|Cu} = \frac{1\times E^\circ_{Cu^{2+}|Cu^+} + 1\times E^\circ_{Cu^+|Cu}}{1+1}$

$= 0.3376$ V

or, $E^\circ_{Cu^{2+}|Cu^+} + E^\circ_{Cu^+|Cu} = 0.6752$ V (2)

From (1) and (2), $E^\circ_{Cu^+|Cu} = 0.5242$ V

**34.** $Zn + Ni^{2+} \rightleftharpoons Zn^{2+} + Ni$

$E^\circ_{cell} = \frac{0.06}{2} \log K_{eq}$

$\Rightarrow (-0.24) - (-0.75) = 0.03 \log K_{eq}$

$\therefore K_{eq} = 10^{17} \Rightarrow$ It means that $Ni^{2+}$ will react almost completely and $[Zn^{2+}] \approx 1.0$ M

Now, $10^{17} = \frac{1.0}{[Ni^{2+}]} \Rightarrow [Ni^{2+}] = 10^{-17}$ M

**35.** Assuming the cell as concentration cell, the cell reaction may be written as

$Ag^+\left(C_1 = \frac{4\times10^{-13}}{0.001} = 4\times10^{-10}\text{ M}\right) \longrightarrow Ag^+\left(C_2 = \frac{2\times10^{-10}}{0.2} = 10^{-9}\text{ M}\right)$

Now, $E_{cell} = 0 - \frac{0.06}{1} \cdot \log \frac{10^{-9}}{4\times10^{-10}} = -0.024$ V

**36.** $0.03 = \frac{0.06}{2} \cdot \log \frac{[Cu^{2+}]_{higher}}{[Cu^{2+}]_{lower}}$

$= 0.3 \cdot \log \frac{0.5}{[Cu^{2+}]_{lower}}$

$\therefore [Cu^{2+}]_{lower} = 0.05$ M

**37.** Cell reaction: $H^+(C_1, HA_1) \longrightarrow H^+(C_2, HA_2)$

$E_{cell} = 0 - \frac{0.059}{1} \cdot \log\left(\frac{C_2}{C_1}\right) = 0.059 \cdot \log\sqrt{\frac{Ka_2}{Ka_1}}$

$= \frac{0.059}{2}(P^{K_{a1}} - P^{K_{a2}}) = 0.059$ V

**38.** $Au^+ + 2CN^- \rightleftharpoons Au(CN)_2^-$, $\Delta G^\circ_1 = -RT \cdot \ln x$

$O_2 + 2H_2O + 4e^- \rightleftharpoons 4OH^-$; $\Delta G^\circ_2 = -4 \times F \times 0.41$

$Au^{3+} + 3e^- \rightleftharpoons Au$; $\Delta G^\circ_3 = -3 \times F \times 1.50$

$Au^{3+} + 2e^- \rightleftharpoons Au^+$; $\Delta G^\circ_4 = -2 \times F \times 1.40$

From $\Delta G^\circ_1 + \frac{1}{4}\Delta G^\circ_2 - \Delta G^\circ_3 + \Delta G^\circ_4, \Delta G^\circ_{required}$

$= -RT \ln x + 1.29$ F

**39.** Informative

**40.** $P^H$ of right electrode will increase due to formation of $OH^-$ ion.

**41.** Informative

**42.** Theoretical

**43.** At low $[Cl^-]$, $O_2$ becomes anode product

**44.** Informative

**45.** Theoretical

**46.** In the electrolysis of aq. $KNO_3$, neither $K^+$ are $NO_3^-$ participate in electrode reaction.

**47.** Na will react with water. S will not conduct electricity.

**48.** $\frac{W}{E} = \frac{Q}{F} \Rightarrow \frac{0.635}{63.5}\times2 = \frac{i\times0.965\times3600}{96500}$

$\Rightarrow i = \frac{2}{3.6}$ A

$\therefore$ % error $= \frac{\frac{2}{3.6} - 0.5}{\frac{2}{3.6}} = 10\%$

**49.** Detonating gas is mixture of $H_2$ and $O_2$

$n_{eq} S_n = n_{eq} H_2 = n_{eq} O_2 \Rightarrow \frac{1.2}{120}\times2 = n_{H_2}\times2 = n_{O_2}\times4$

$\therefore n_{H_2} = 0.01$ and $n_{O_1} = 0.005$

$\therefore$ Vol. of mixture of $H_2$ and $O_2 = 0.015 \times 22400 = 336$ ml

**50.** $n_{eq}$ of Li OH $= \frac{Q}{F} \Rightarrow \frac{w}{24}\times 1 = \frac{2.5\times 4825\times 0.8}{96500}$

**51.** $n_{eq}$ Cu deposited at cathode $= \frac{Q}{F}$

or, $\frac{w}{63.5}\times 2 = \frac{12.4\times 4825}{96500} \Rightarrow w = 19.685$ gm

But the increase in mass of cathode is only 19.05 gm It represents that 20 gm of sample contains only 19.05 gm Cu.

$\therefore$ % of Cu $= \frac{19.05}{20}\times 100 = 95.25\%$

Now, $\frac{Q}{F} = n_{eq}Cu + n_{eq}Fe$ (oxidised at anode)

or, $\frac{12.4\times 4825}{96500} = \frac{19.05}{63.5}\times 2 + \frac{w}{56}\times 2 \Rightarrow w = 0.56$ gm

$\therefore$ Percentage of Fe $= \frac{0.56}{20}\times 100 = 2.8\%$

**52.** Theoretical $n_{eq}$ of NaOH formed $= n_{eq}$Cu

$= \frac{3.18}{63.6}\times 2 = 0.1$

Actual $n_{eq}$ of NaOH formed $= \frac{60\times 1}{1000} = 0.06$

$\therefore$ Percentage yeild $= \frac{0.06}{0.1\times 100} = 60\%$

**53.** Cell reaction during discharge:

$Pb + PbO_2 + 2H_2SO_4 \longrightarrow 2PbSO_4 + 2H_2O$

$n_{Pb}$ taken $= \frac{200}{208}$ and $n_{PbO_2}$ taken $= \frac{200}{240}$

Hence, $PbO_2$ is L.R.

Now, $n_{eq}PbO_2 = \frac{Q}{F} \Rightarrow \frac{200}{240}\times 2 = \frac{10\times t}{96500}$

$\Rightarrow t = 16083.33$ sec

**55.** $\frac{Q}{F} = n_{eq}$ of $I^- = n_{eq} Na_2S_2O_3$

or, $\frac{i\times 2\times 3600}{96500} = \frac{72\times 1.0}{1000}\times 1 \Rightarrow i = 0.965$ A

**56.** $C_{14}H_{10} + 2H_2O \longrightarrow C_{14}H_8O_2 + 6H^+ + 6e^-$

$n_{eq}\ C_{14}H_8O_2 = \frac{Q}{F} \Rightarrow \frac{w}{208}\times 6 = \frac{1\times 40\times 60\times 0.965}{96500}$

$\therefore w = 0.832$ gm

**57.** Number of coulombs required $= \frac{1000}{0.00033}$ per Kg Cu

Energy required $= \frac{1000}{0.00033}\times 0.33 = 10^6$ J

$\therefore$ Cost of electricity

$= \frac{4}{10^3\times 3600}\times 10^6 = 1.11$Rupee

**58.** $n_{eq}\ CH_3Coo^-$ oxidised $= \frac{Q}{F}$

or, $n\times 1 = \frac{0.5\times 482.5\times 60\times 0.8}{96500} \Rightarrow n = 0.12$

$\therefore$ Moles of $(C_2H_6 + CO_2)$ produced

$= \frac{3}{2}\times 0.12 = 0.18$

and total volume $= 0.18 \times 22.4 = 4.032$ L

**59.** $\Delta E° = \frac{0.059}{2}\cdot \log 10^6 = 0.177$ V

**60.** Back EMF $= \frac{0.06}{2}\cdot \log\frac{0.12}{0.08} = 5.4\times 10^{-3}$ V

**61.** Equivalent of charge used

$= \frac{0.4825\times 10\times 3600}{96500} = 0.18$ F

Cell reaction during charge:

| | Cu + | $Zn^{2+}$ $\longrightarrow$ | $Cu^{2+}$ | + Zn |
|---|---|---|---|---|
| | | $\frac{100\times 1}{1000} = 0.1$ mole | $\frac{100\times 1}{1000} = 0.1$ mole | |
| | | = 0.2 eq | = 0.2 eq | |
| Final | | 0.2 – 0.1 8 | = 0.2 + 0.18 | |
| | | = 0.02 eq | = 0.38 eq | |
| | | = 0.01 mole | = 0.19 mole | |

$\therefore$ Final $[Zn^{2+}] = \frac{0.01}{100}\times 1000 = 0.1$ M

and $[Cu^{2+}] + \frac{0.19}{100}\times 1000 = 1.9$ M

Now, cell reaction as galvanic cell:

$$Zn + Cu^{2+} \longrightarrow Zn^{2+} + Cu$$

and $E_{cell} = E°_{cell} - \frac{0.06}{n} \cdot \log \frac{[Zn^{2+}]}{[Cu^{2+}]}$

$$= [0.34 - (-0.76)] - \frac{0.06}{2} \cdot \log \frac{0.1}{1.9} = 1.1084 \text{ V}$$

**62.** Reactions involved are

$$2H_2O \xrightarrow{\text{electrolysis}} 2H_2 + O_2$$

$$N_2 + 3H_2 \longrightarrow 2NH_3$$
$$NH_3 + 2O_2 \longrightarrow HNO_3 + H_2O$$
$$NH_3 + HNO_3 \longrightarrow NH_4NO_3$$

For 1 mole $NH_4NO_3$, 3 moles of $H_2O$ should be electrolyzed. Hence for 1152 Kg $NH_4NO_3$, moles of $H_2$ needed $= 3 \times \frac{1152 \times 10^3}{80} = 4.32 \times 10^4$

Now, $n_{eq} H_2 = \frac{Q}{F} \Rightarrow 4.32 \times 10^4 \times 2 = \frac{i \times 24 \times 3600}{96500}$

$\therefore i = 96500$ A/day

**63.** $n_{eq}$ $HC_3H_5O_3 = n_{OH^-} = \frac{Q}{F}$

or, $\frac{w}{90} \times 1 = \frac{50 \times 10^{-3} \times 1158}{96500} \Rightarrow w = 0.054$ gm

$\therefore$ % of lactice acid $= \frac{0.054}{1} \times 100 = 5.4\%$

**64.** $n_{H_2O_2}$ formed $= n_{(NH_4)_2S_2O_8} = n$

and $n_{eq} (NH_4)_2S_2O_8 = \frac{Q}{F}$

$\Rightarrow n \times 2 = \frac{102}{34} \times 2 = \frac{i \times 3600 \times 0.5}{96500}$

$[2SO_4^{2-} \longrightarrow S_2O_8^{2-} + 2e^-]$ $\Rightarrow i = 321.67$ A

**65.** $n_{As} = n_{H_3AsO_3} = n$ and $n_{eq}$ $H_3AsO_3 = n_{eq} I_2 = \frac{Q}{F}$

or, $\frac{w}{75} \times 2 = \frac{1.68 \times 10^{-3} \times 96.5}{96500} \Rightarrow w = 6.3 \times 10^{-5}$ gm

**66.** $\frac{w}{E} = \frac{Q}{F} \Rightarrow \frac{52.2}{87} \times 2 = \frac{19.3 \times 2 \times 3600 \times \eta}{96500}$

$\Rightarrow \eta = 0.8333$ or 83.33%

**67.** $n_{eq} Cu^{2+}$ reduced $= \frac{Q}{F} \Rightarrow n \times 2 = \frac{2 \times 10^{-3} \times 19.3 \times 60}{96500}$

$\therefore n = 1.2 \times 10^{-5}$

$\therefore [CuSO_4]_0 = \frac{(1.2 \times 10^{-5} \times 2)}{250} \times 1000 = 9.6 \times 10^{-5}$ M

**68.** $\frac{w}{E} = \frac{Q}{F} \Rightarrow \frac{12.3}{123} \times 6 = \frac{Q \times 0.5}{F} \Rightarrow Q = 1.2$ F

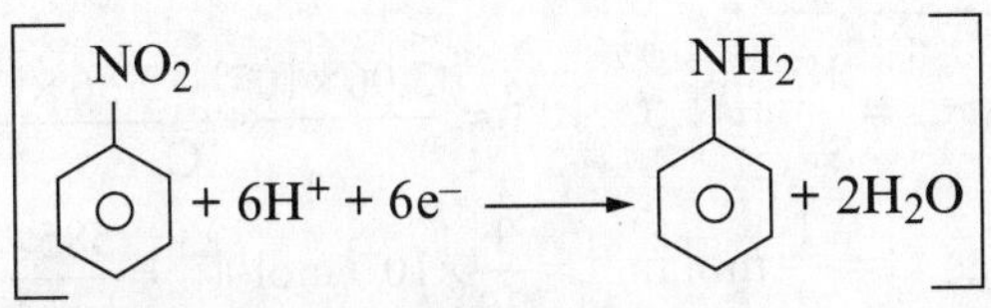


and Energy consumed = 1.2 F × 3.0 V = 347.4 KJ

**69.** $n_{eq}$ $H_2$ (at cathode) $= n_{eq}$ $O_2$ + $n_{eq}$ $H_2S_2O_8$ (at anode)

or, $\frac{9.08}{22.7} \times 2 = \frac{2.27}{22.7} \times 4 + \frac{w}{194} \times 2$ $[2SO_4^{2-} \rightarrow S_2O_8^{2-} + 2e^-]$

$\therefore w = 38.8$ gm

**70.** Initial mass of $H_2SO_4$,

$w_1 = 3600 \times 1.5 \times \frac{40}{100} = 2160$gm

Final mass of $H_2SO_4$,

$w_2 = 3600 \times 1.1 \times \frac{10}{100} = 396$ gm

$\therefore$ Moles of $H_2SO_4$ consumed $= \frac{w_1 - w_2}{98} = 18$

Now, $n_{eq}$ $H_2SO_4 = \frac{Q}{F} \Rightarrow 18 \times 1 = \frac{(\text{amp-hr}) \times 3600}{96500}$

$\therefore$ Number of ampere-hr = 482.5

**71.** $\frac{\Lambda_m(NaCl)}{\Lambda_m(KCl)} = \frac{(\kappa/C)_{NaCl}}{(\kappa/C)_{KCl}} = \frac{\frac{1}{R_{NaCl}} \cdot G^* \cdot \frac{1}{C_{NaCl}}}{\frac{1}{R_{KCl}} \cdot G^* \cdot \frac{1}{C_{KCl}}}$

$= \frac{(R \cdot C)_{KCl}}{(R \cdot C)_{NaCl}}$

or, $\frac{\Lambda_m(NaCl)}{120} = \frac{200 \times 0.1}{6400 \times 0.003}$

$\Rightarrow \Lambda_{m(NaCl)} = 125\ \Omega^{-1}\ cm^{-1}\ mol^{-1}$

**72.** $\wedge^{\circ}_{m}[(NH_4)_2CrO_4] = 2\times\lambda^{\circ}_{m}(NH_4^+) + \lambda_m(CrO_4^{2-})$
$= (2\times 6.6\times 10^{-8} + 5.4\times 10^{-8})\times 96500$
$= 0.01795\ \Omega^{-1}\ m^2\ mol^{-1}$

**73.** $\wedge_m = \alpha\cdot\wedge^{\circ}_m = \dfrac{\kappa}{C}$

or, $0.9\times 4.25\times 10^{-2} = \dfrac{382.5}{C\times 10^3} \Rightarrow C = 0.1\ M$

**74.** $\wedge_m = \dfrac{K}{C} \Rightarrow 1.5\times 10^{-2} = \dfrac{3.06\times10^{-3} - 2.56\times10^{-3}}{C}$

$\therefore C = \dfrac{1}{30}\ mol\ m^{-3} = \dfrac{1}{30}\times10^{-3}\ mol\cdot l^{-1} = \dfrac{585/58.5}{V}$

$\therefore V = 3\times 10^5\ L$

**75.** $\wedge_m = \dfrac{\kappa}{C} = \dfrac{G\cdot G^*}{C} \Rightarrow 100 = \dfrac{\frac{1}{R}\times\frac{0.5}{1.5}}{0.1\times10^{-3}}$

$\Rightarrow R = \dfrac{100}{3}\ ohm$

Now, $V = IR \Rightarrow I = \dfrac{V}{R} = \dfrac{5}{100/3} = 0.15\ A$

**76.** Ionic mobility, $\mu^{\circ} = \dfrac{\text{speed of ion}}{\text{Pot. gradient}} = \dfrac{\lambda^{\circ}_m}{F}$

or, $\dfrac{\text{speed}}{\left(\frac{19.3}{5}\right)} = \dfrac{50}{96500} \Rightarrow \text{speed} = 2\times 10^{-3}\ cm/s$

**77.** $\wedge_{eq} = \dfrac{\kappa}{C} \Rightarrow 150\times\dfrac{3.4\times10^{-6} - 1.6\times10^{-6}}{5}$

$\therefore S = 1.2\times 10^{-8}\ mol\ cm^{-3} = 1.2\times 10^{-5}\ M$

**78.** $K = G\cdot G^* \Rightarrow \dfrac{\kappa}{1.4} = \dfrac{1/280}{1/50}$

$\Rightarrow \kappa = 0.25\ s\ m^{-1}$ for 0.5 M

Now, $\wedge_m = \dfrac{\kappa}{C} = \dfrac{0.25}{0.5\times10^3} = 5\times10^{-4}\ s\ m^2 mol^{-1}$

**79.** $Ag\,A(S) \rightleftharpoons \underset{(x+y)M}{Ag^+} + \underset{xM}{A^-}$

$Ag\,B(S) \rightleftharpoons \underset{(x+y)M}{Ag^+} + \underset{yM}{B^-}$

Now, $(x+y)\cdot x = 3\times 10^{-14}$ and
$(x+y)\cdot y = 1\times 10^{-14}$

$\therefore [Ag^+] = x + y = 2\times10^{-7}\ M,$
$[A^-] = x = 1.5\times10^{-7}\ M;$
$[B^-] = y = 0.5\times10^{-7}\ M$

Now,

$\kappa_{solution} = 3.75\times10^{-8} = 2\times10^{-7}\times10^{-3}\times60 + 1.5\times10^{-7}\times10^{-3}\times80 + 0.5\times10^{-7}\times10^{-3}\ \lambda^{\circ}_B$

$\therefore \lambda^{\circ}_B = 135\ ohm^{-1}\ cm^2\ mol^{-1}$

**80.** $\wedge^{\circ}_{eq}[Be_3(Po_4)_2] = \wedge^{\circ}_{eq}[BeCl_2] + \wedge^{\circ}_{eq}[K_3Po_4] - \wedge^{\circ}_{eq}[KCl]$

$= 160 + 140 - 100 = 200\ ohm^{-1}\ cm^2\ eq^{-1}$

Now, $\eta_{eq} = \dfrac{\kappa}{C} \Rightarrow 200 = \dfrac{1.2\times10^{-5}}{C}$

$\Rightarrow C = 6\times 10^{-8}\ eq\ cm^{-3}$
$= 6\times 10^{-5}\ N$
$= 10^{-5}\ M$

Now, $K_{sp} = 108\ S^5 = 108\times(10^{-5})^5 = 1.08\times10^{-23}$

## Section B (One or More than one Correct)

**1.** Net cell reaction is spontaneous in electrochemical cell but non-spontaneous in electrolytic cell. Cathode is +ve in electrochemical cell but –ve in electrolytic cell.

**2.** For the cell: Ag(s) | Ag Cl (s) | Cl– || $Ag^+$ | Ag(s), the net cell reaction is $Ag^+ + Cl^- \rightleftharpoons AgCl(s)$.

**3.**

Left electrode: $Ag(s) + Cl^-(aq) \rightarrow AgCl(s) + e^-$ $1\times 2$

Right electrode : $Hg_2Cl_2(s) + 2e^- \rightarrow 2Hg(l) + 2Cl^-(aq)$

Net reaction: $2Ag(s) + Hg_2Cl_2(s) \rightarrow 2AgCl(s) + 2Hg(l)$

4. Informative

5. $n_{eq}\ Cu = \frac{63.5}{63.5}\times 2 = 2;\ n_{eq}\ Mg = \frac{24}{24}\times 2 = 2$

$n_{eq}\ Na = \frac{11.5}{23}\times 1 = 0.5;\ n_{eq}\ Al = \frac{9}{27}\times 3 = 1$

6. Theoretical

7. $\wedge^\circ_{eq} = 60 + 80 = 140\ ohm^{-1}\ cm^2\ eq^{-1}$

$\wedge^\circ_m = 140 \times 6 = 840\ ohm^{-1}\ cm^2\ eq^{-1}$

8. Salt bridge does not change standard potential of any electrode.

9. Moles of electron involved

$= \frac{0.25\times 9.65\times 3600}{96500} = 0.09$

$\therefore$ Mass of Zn involved $= \frac{0.09}{2}\times 65.4 = 2.943$ gm

Mass of $MnO_2$ involved $= 0.09 \times 87 = 7.83$ gm

Mass of $NH_4^+$ involved $= 0.09 \times 18 = 1.62$ gm

10. Net cell reaction is

$Cd(s) + 2AgCl(s) \rightleftharpoons 2Ag(s) + Cd^{2+}(aq) + 2Cl^-(aq)$

$\Delta G^\circ_{50^\circ C} = -nFE^\circ = -2\times 96500\times 0.6 = -115800$ J

$\Delta G^\circ_{0^\circ C} = -nFE^\circ = -2\times 96500\times 0.7 = -135100$ J

$\Delta S^\circ = nF\cdot\left(\frac{E^\circ_2 - E^\circ_1}{T_2 - T_1}\right) = 2\times 96500\times\frac{0.6-0.7}{50}$

$= -386$ J/K

$\Delta H^\circ = \Delta G^\circ + T\cdot\Delta S^\circ$

$= (-135100) + 273 \times (-386) = -240478$ J

11. Anode: $2H_2O(l) \longrightarrow O_2(g) + 4H^+(aq) + 4e^-$

Cathode: $2H_2O(l) + 2e^- \longrightarrow H_2(g) + 2OH^-(aq)$

$n_{eq}\ H^+$ produced $= n_{eq}\ OH^-$ produced $= \frac{Q}{F}$

or, $n_{H^+}\times 1 = n_{OH^-} = \frac{1.25\times 965\times 60}{96500} = 0.75$

Anode: $HPO_4^{2-} + H^+ \rightleftharpoons H_2PO_4^-$

| | | | |
|---|---|---|---|
| | 1.0 M | 0.75 M | 1.0 M |
| Final | 0.25 M | | 1.75 M |

$\therefore P^H = P^{K_a} + \log\frac{[HPO_4^{2-}]_0}{[H_2PO_4^-]} = 2.15 + \log\frac{0.25}{0.75}$

$= 1.30$

Cathode: $H_2PO_4^- + OH^- \rightleftharpoons HPO_4^{2-} + H_2O$

| | | | |
|---|---|---|---|
| | 1.0 M | 0.75 M | 1.0 M |
| Final | 0.25 M | 0 | 1.75 M |

$\therefore P^H = P^{K_a} + \log\frac{[HPO_4^{2-}]_0}{[H_2PO_4^-]}$

$= 2.15 + \log\frac{1.75}{0.25} = 3.0$

12. V, Fe and Hg will be oxidised by $NO_3^-$

13. Resistance and heat capacity depends on quantity.

14. Theoretical

15. Net cell reaction of discharge is

$Pb + PbO_2 + 2H_2SO_4 \longrightarrow 2PbSO_4 + 2H_2O$

*x* mole  *x* mole

Initial mass of $H_2SO_4$,

$w_1 = 1000 \times 1.26 \times \frac{40}{100} = 504$ gm

Final mass of $H_2SO_4$,

$w_2 = (1260 - 98x + 18x) \times \frac{28}{100}$

$= (352.8 - 22.4x)$ gm

From reaction, $504 - 98x = 352.8 - 22.4x \Rightarrow x = 2$

## Section C (Comprehensions)

### Comprehension I

1. $E^\circ_{Cell} = E^\circ_{H^+/H_2} - E^\circ_{Zn^{2+}/Zn}$ because $E^\circ_{H^+/H_2}$ was higher

or, $0.76 = 1.00 - E^\circ_{Zn^{2+}/Zn} \Rightarrow E^\circ_{Zn^{2+}/Zn} = 0.24$ V

2. $E^\circ_{Cell} = E^\circ_{Cu^{2+}/Cu} - E^\circ_{H^+/H_2}$ because $E^\circ_{Cu^{2+}/Cu}$ was higher

or, $0.34 = E^\circ_{Cu^{2+}/Cu} - 1.00 \Rightarrow E^\circ_{Cu^{2+}/Cu} = 1.34$ V

3. $E^\circ_{Cell} = E^\circ_{Cu^{2+}/Cu} - E^\circ_{Zn^{2+}/Zn} = 1.34 - 0.24 = 1.10$ V

## Comprehension II

4. $E_{Ag^+/Ag} = E^\circ_{Ag^+/Ag} - \frac{0.06}{1} \cdot \log \frac{1}{[Ag^+]}$

$= 0.80 - \frac{0.06}{1} . \log \frac{1}{\sqrt{K_{sp}}}$

$= +0.314$ V

5. $E^\circ_{I^-|AgI|Ag} = E^\circ_{Ag^+/Ag} - \frac{0.06}{1} \cdot \log K_{sp} = -0.172V$

6. $E_{I^-|AgI|Ag} = E^\circ_{I^-/AgI/Ag} - \frac{0.06}{1} \cdot \log[I^-]$

$= -0.172 - 0.06 \cdot \log 0.04 = -0.088$ V

## Comprehension III

7. As(0.40) > (–0.87), reduction of $Ni_2O_3(s)$ will occur.

8. $E^\circ_{cell} = (0.40) - (-0.87) = 1.27$ V

9. Net cell reaction is independent from $OH^-$ (aq)

10. $-\Delta G^\circ = nFE^\circ_{cell} = 2 \times 96500 \times 1.27 = 245110$ J

## Comprehension IV

11. $H_3O^+$ is only needed in balancing cathode reaction:

$NO_3^- + 3H_3O^+ + 2e^- \longrightarrow HNO_2 + 4H_2O$

12. Moles of electron needed = 2 × moles of $HNO_2$ formed

13. $n_{eq} HNO_2 = \frac{Q}{F} \Rightarrow 0.1 \times 2 = \frac{10 \times t}{96500} \Rightarrow t = 1930$ sec

## Comprehension V

14. $\Delta G^\circ = -nFE^\circ \Rightarrow -237.39 \times 10^3 = -2 \times 96500 \times E^\circ_{cell}$

$\therefore E^\circ cell = 1.23$ V

15. Moles of $H_2$ needed $= \frac{23.739}{237.39} = 0.1$

$\therefore$ Volume of $H_2$ needed $= 0.1 \times 22.7 = 2.27$ L

16. $E_{cell}$ is independent from $[OH^-]$

17. $\Delta S^\circ = \frac{\Delta H^\circ - \Delta G^\circ}{T}$

$= \frac{(-285.8 \times 10^3) - (-237.39 \times 10^3)}{298}$

$= -162.4\ J/K^{-1}$

18. $\eta = \frac{(-\Delta G^\circ)}{(-\Delta H^\circ)} = \frac{237.39}{285.8} = 0.8306$ or 83.06%

## Comprehension VI

19. Refer theory given is passage.

20. $E^\circ = E^\circ_{O_2/H_2O, H^+} - E^\circ_{Fe^{2+}/Fe} = 1.229 - (-0.447)$

$= 1.676$ V

21. $E^\circ_{Cu^{2+}/Cu} > E^\circ_{Pb^{2+}/Pb}$ and hence Cu will not oxidise easily.

22. $n_{eq}$ Fe $= \frac{Q}{F} \Rightarrow \frac{w}{56} \times 2 = \frac{0.5 \times 1.0 \times 3600}{96500}$

$\Rightarrow w = 0.522$ gm

23. Theoretical

## Comprehension VII

24. $n_{eq}$ Ni $= \frac{Q}{F} \Rightarrow \frac{w}{58.7} \times 2 = \frac{15 \times 3600 \times 0.6}{96500}$

$\Rightarrow w = 9.85$ gm

25. $V = A \times t \Rightarrow \frac{9.85}{8.9} = (4.0 \times 2) \times t \Rightarrow t = 0.138$ cm

**26.** $n_{eq}\ H_2 = \frac{Q}{F} \Rightarrow \frac{V_{H_2}}{22.4} \times 2 = \frac{15 \times 3600 \times 0.4}{96500}$

$\Rightarrow V_{H_2} = 2.5\ L$

**27.** Anode produced is only $O_2$ gas.

$n_{eq}\ O_2 = \frac{Q}{F} \Rightarrow \frac{w}{8} = \frac{15 \times 3600}{96500} \Rightarrow w = 4.477$ gm

## Comprehension VIII

**28.** $E°_{cell} = 0.8 - 0.05 = 0.75$ V

Now, $E°_{cell} = \frac{RT}{nF} \cdot \ln K \Rightarrow 0.75 = \frac{1}{2 \times 38.92} \cdot \ln K$

$\Rightarrow \ln K = 58.38$

**29.** The oxidation reaction of glucose contains $H^+$ and

$E = E° - \frac{0.0592}{2} \cdot \log[H^+]^2$

$\Rightarrow E - E° = 0.0592\ P^H = 0.6512$ V

**30.** Standard potential is independent from ammonia concentration.

## Comprehension IX

**31.** Left electron: $H_2(g) \longrightarrow 2H^+(aq) + 2e^-$

Right electrode: $2AgCl(s) + 2e^- \longrightarrow 2Ag(s) + 2Cl^-(aq)$

$\therefore$ Net reaction: $H_2(g) + 2AgCl(s) \longrightarrow 2Ag(s) + 2H^+(aq) + 2Cl^-(aq)$

**32.** $\Delta S° = nF \cdot \left(\frac{E_2 - E_1}{T_2 - T_1}\right) = 2 \times 96500 \times \frac{0.21 - 0.23}{20}$

$= -193$ J/K

$\Delta G° = -nFE° = -2 \times 96500 \times 0.23 = -44390$ J at 15°C

Now, $\Delta H° = \Delta G° + T \cdot \Delta S° = (-44390) + 288 \times (-193)$

$= -99974\ J = -49987$ J/mole AgCl

**33.** $\Delta S° = -193\ J/K = -96.5$ J/K per mole AgCl

**34.** $\Delta G°_{298} = \Delta H° - T \cdot \Delta S° = (-49987) - 298 \times (-96.5)$

or, $-1 \times 96500 \times E° = -21230$

$\Rightarrow E° = 0.22\ V = E°_{Cl^-/AgCl/Ag}$

or, $E°_{Cl^-/AgCl/Ag} = E°_{Ag^+/Ag} + \frac{0.058}{1} \log K_{sp}$

or, $0.22 = 0.80 + \frac{0.058}{1} \log K_{sp} \Rightarrow K_{sp} = 1 \times 10^{-10}$

Hence, Solubility, $S = \sqrt{K_{sp}} = 10^{-5}$ M

## Comprehension X

**35.** Add Ag(s) in both sides of given reaction to get cell reaction. Now, for $E°_{cell}$

$\Delta G° = -nFE° \Rightarrow [-109] - [77 + (-129)] \times 10^3$

$= -1 \times 96500 \times E°_{cell}$

$\therefore E°_{cell} = 0.59$ V

**36.** $E°_{cell} = \frac{0.059}{n} \log K_{eq} \Rightarrow 0.59 = \frac{0.059}{1} \cdot \log \frac{1}{K_{sp}}$

$\therefore K_{sp} = 10^{-10}$

**37.** $Zn(s) + 2Ag^+(aq) ¡ Zn^{2+}(aq) + 3Ag(s)$; $E° = 0.80 - (-0.76)$

Now, $E° = \frac{0.059}{n} \cdot \log K_{eq}$

$\Rightarrow 1.56 = \frac{0.059}{2} \cdot \log \frac{[Zn^{2+}]}{[Ag^+]^2}$

$\therefore \log \frac{[Zn^{2+}]}{[Ag^+]^2} = 52.88$

**38.** Moles of Zn added $= \frac{6.539 \times 10^{-2}}{65.39} = 10^{-3}$

Moles of $Ag^+$ present $= \frac{10^{-5}}{1000} \times 100 = 10^{-6}$ (L.R.)

$\therefore$ Moles of Ag precipitated $= 10^{-6}$

HINTS AND EXPLANATIONS

## Comprehension XI

**39.** For KCl: $\wedge_m = \frac{\kappa}{C} = \frac{G \cdot G^*}{C} \Rightarrow G^* = \wedge_m \cdot C/G$

or, $G^* = \wedge_m \cdot C \cdot R = 200 \times (0.02 \times 10^{-3}) \times 100 = 0.4$ $cm^{-1}$

**40.** $\kappa_{water} = G \cdot G^* = \frac{1}{10000} \times 0.4 = 4 \times 10^{-5}\,\Omega^{-1}cm^{-1}$

**41.** For NaCl: $\wedge_m = \frac{\kappa}{C} = \frac{G \cdot G^*}{C}$

or, $125 = \frac{\left(\frac{1}{8000} - \frac{1}{10000}\right) \times 0.4}{\left(\frac{585/58.5}{V}\right)}$

$\Rightarrow V = 1.25 \times 10^8\ cm^3$

$= 1.25 \times 10^5\ L$

## Comprehension XII

**42.** $E^\circ_{Cl_2/Cl^-} > E^\circ_{I_2/I^-}$

**43.** $E^\circ_{Mn^{3+}/Mn^{2+}} > E^\circ_{O_2/H_2O, H^+}$

## Comprehension XIII

**44.** Net reaction: $M^+(1M) \longrightarrow M^+\ (0.05\ M)$;

Higher Conc. Lower Conc.

$E_{cell} > 0, \Delta G_{cell} < 0$

**45.** $70\ mV = E^\circ - \frac{RT}{F} \cdot \ln \frac{0.05}{1}$ (1)

$E_{req} = E^\circ - \frac{RT}{F} \cdot \ln \frac{0.0025}{1} = E^\circ - \frac{RT}{F} \cdot \ln \frac{(0.05)^2}{1}$ (2)

and $E^\circ = 0$. Hence, $E_{req} = 140\ mV$

## Comprehension XIV

**46.** $\Delta G_{cell} = -nFE_{cell} = -2 \times 96500 \times 0.059 = -11387\ J$

**47.** $E_{cell} = E^\circ_{cell} - \frac{0.059}{2} \cdot \log \frac{[M^{2+}]_{Left}}{[M^{2+}]_{Right}}$

or, $0.059 = 0 - \frac{0.059}{2} \cdot \log \frac{(K_{sp}/4)^{1/3}}{0.001}$

$\Rightarrow K_{sp} = 4 \times 10^{-15}$

## Comprehension XV

**48.** $H^+ + Cl^- + \underset{\text{added}}{(NaOH)} \longrightarrow Na^+ + Cl^- + H_2O$

Conductance first decreases and $H^+$ ions are replaced by $Na^+$ ions. After equivalent point, conductance increase due to increase in number of ions.

**49.** $CH_3COOH$ + (NaOH added) $\longrightarrow CH_3COO^- + Na^+ + H_2O$

As number of ions increases, conductance increases. slight decrease initially was due to some dissociated $CH_3COOH$. After equivalence point, in place of $CH_3COO^-$ ion, number of $OH^-$ ions increases and hence slope becomes greater.

**50.** $H^+ + Cl^-$ + ($NH_4OH$ added) $\longrightarrow$ $NH_4^+ + Cl^- + H_2O$

As $H^+$ ions are replaced by $NH_4^+$ ions, conductance decreases. After equivalent point, it become almost constant as the dissociation of added NH4OH will be suppressed in presence of $NH_4^+$ ions.

**51.** HCl will neutralize first followed by $CH_3COOH$.

**52.** Ionic mobilities of $Ag^+$ and $K^+$ ions do not differ largely

**53.** First number of ions increases and then remain atmost constant.

## Section D (Assertion – Reason)

1. $CuCl_2 \longrightarrow Cu + Cl_2$
2. $E^\circ_{Zn^{2+}/Zn} < E^\circ_{Ag^+/Ag}$ (informative)
3. Reason is different charges on ions.
4. Theoretical
5. Negative reduction potential means greater tendency to get oxidised.
6. Number of ions increases considerably only for weak electrolytes.
7. Informative
8. Theoretical
9. It is due to very high over voltage potential of hydrogen at mercury cathode.
10. Theoretical
11. Theoretical
12. $E = E^\circ - \frac{RT}{nF} \cdot \log Q$
13. $Cl^-$ will combine with $Ag^+$ to precipitate AgCl.
14. At Cathode:
    $2H_2O(l) + 2e^- \longrightarrow H_2(g) + 2OH^-(aq)$
    At Anode:
    $2CH_3COO^-(aq) \longrightarrow C_2H_6(g) + 2CO_2(g) + 2e^-$
15. Theoretical

## Section E (Column Match)

1. Informative
2. Theoretical
3. $$E^\circ_{Fe^{3+}|Fe} = \frac{1 \times E^\circ_{Fe^{3+},Fe^{2+}} + 2 \times E^\circ_{Fe^{2+}|Fe}}{1+2} = -0.037\text{ V}$$
   $$E^\circ_{(4H_2O \rightarrow 4H^+ + 4OH^-)} = 0.40 - 1.23 = -0.83\text{ V}$$
   $$E^\circ_{(Cu^{2+} + Cu \rightarrow 2Cu^+)} = E^\circ_{Cu^{2+}|Cu^+} - E^\circ_{Cu^+|Cu}$$
   $$= \frac{0.34 - 0.52}{2-1} - 0.52 = -0.70\text{ V}$$
   $$E^\circ_{Cr^{3+},Cr^{2+}} = \frac{3 \times (-0.74) - 2 \times (-0.91)}{3-2} = -0.4\text{ V}$$
4. For concentration cell, both half cell must have same configuration.

P. $Ag^+ + Cl^- \longrightarrow AgCl$ sponteneous,

$$K_{eq} = \frac{1}{K_{sp}} >> 1$$

Q. $AgBr + Cl^- \longrightarrow AgCl + Br^-$;

$$K_{eq} = \frac{K_{sp}(AgBr)}{K_{sp}(AgCl)} << 1, \text{ Non-Sponteneous}$$

R. $Ag^+(1.0\text{ M}) \longrightarrow Ag^+(0.1\text{ M})$
Higher to lower concentration, spontaneous

S. $Cl^-(0.1\text{ M}) \longrightarrow Cl^-(1.0\text{ M})$
Non-spontaneous

## Section F (Subjective)

### Single-digit Integer Type

1. $$E = E^\circ - \frac{0.06}{n} \cdot \log \frac{P_{H_2}}{[H^+]^2}$$
   (assuming $2H^+ + 2e^- \rightarrow H_2$)
   $$-0.18 = 0 - \frac{0.06}{2} \cdot \log \frac{1}{[H^+]^2} \Rightarrow [H^+] = 10^{-3}\text{M}$$

$$\underset{(C-x)M}{C_6H_5NH_3^+} + H_2O \rightleftharpoons C_6H_5NH_2 + \underset{xM}{H_3O^+}$$

$$h = \frac{x}{c} = \frac{10^{-3}}{1/40} = 0.04 \text{ or } 4\%$$

2. $E = E^\circ - \frac{0.06}{2}\cdot\log\frac{1}{[Cu^+]}$

$0.31 = 0.34 - \frac{0.06}{2}\cdot\log\frac{1}{[Cu^{2+}]}$

$\Rightarrow [Cu^{2+}] = 0.1\,M$

$\therefore [OH^-] = \sqrt{\frac{K_{sp}}{Cu^{2+}}} = \sqrt{\frac{10^{-19}}{0.1}} = 10^{-9} \Rightarrow p^H = 5$

3. $E^\circ_{Fe^{3+}|Fe^{2+}} = \frac{3\times E^\circ_{Fe^{3+}|Fe} - 2\times E^\circ_{Fe^{2+}|Fe}}{3-2}$

$= \frac{3\times(-0.04) - 2\times(-0.44)}{1} = 0.76\,V$

Now, $E_{Fe^{3+}|Fe^{2+}} = E^\circ_{Fe^{3+}|Fe^{2+}} - \frac{0.06}{1}\log\frac{[Fe^{2+}]}{[Fe^{3+}]}$

or, $0.718 = 0.76 - 0.06\log\frac{[Fe^{2+}]}{[Fe^{3+}]} \Rightarrow \frac{[Fe^{2+}]}{[Fe^{3+}]} = 5$

4.

| | $2Fe^{3+}$ | + $2I^-$ → | $2Fe^{2+}$ | + $I_2$; |
|---|---|---|---|---|
| | 0.5 M | excess | | |
| 100% | 0 | 1.0 M | 0.5 M | |
| Equ. | CM | 1.0 M | 0.5 M | |

$E^\circ_{cell} = 0.77 - 0.53$
$= 0.24\,V;$

$K_{eq} = 10^8$

Now, $10^8 = \frac{(0.5)^2}{C^2\times(1.0)^2} \Rightarrow C = 5\times10^{-5}\,M$

5. Assuming the cell as concentration cell, net cell reaction is

$Ag^+\left(C_1 = \frac{K_{sp}(AgCl)}{[Cl^-]}\right) \longrightarrow Ag^+(C_2 = \sqrt{K_{sp(AgI)}})$

and $E_{cell} = O - \frac{0.06}{1}\log\frac{C_2}{C_1}$

or, 0.102

$= -0.06\cdot\log\frac{\sqrt{8.1\times10^{-17}}}{\left(\frac{1.8\times10^{-10}}{[Cl^-]}\right)} \Rightarrow [Cl^-] = 4\times10^{-4}\,M$

6. $n_{eq}\,Pb + n_{eq}\,Tl = \frac{Q}{F}$

$\Rightarrow \frac{5.0\times\frac{70}{100}}{208}\times2 + \frac{5.0\times\frac{30}{100}}{204}\times1 = \frac{1.1\times t}{96500}$

$\therefore t = 3597.4$ sec ; 1 hr

7. $n_{eq}\,I_r = \frac{Q}{F} \Rightarrow \frac{0.36}{192}\times x = \frac{0.075\times2\times3600}{96500} \Rightarrow x = 3$

Now, $x + 6(-1) = y \Rightarrow y = -3$

8. $2NaCl + 2H_2O \xrightarrow{\text{electrolysis}} 2NaOH + H_2 + Cl_2$

$2NaOH + Cl_2 \longrightarrow NaCl + NaClO + H_2O$

$n_{NaClO}$ formed = $n_{Cl_2}$ produced from electrolysis

or, $\frac{(10\times10^3\times1.0)\times\frac{7.45}{100}}{74.5}\times2 = \frac{2.5\times t}{96500}$

($n$-factor of $Cl_2$ in electrolysis)

$\therefore t = 7.72\times10^5$ sec = 8.93 days ≈ 9 days

9. Cathode: $Cu^{2+} + 2e^- \longrightarrow Cu$

Anode $2H_2O \longrightarrow O_2 + 4H^+ + 4e^-$

Moles of $e^-$ used $= \frac{Q}{F} = \frac{0.161\times5\times60}{96500} \simeq 5\times10^{-4}$

Eq. of $Cu^{2+}$ present $= \frac{500\times0.1}{1000}\times2 = 0.1$(excess)

$\therefore$ Moles of $H^+$ produced = Moles of $e^-$ = $5\times10^{-4}$

$\therefore [H^+]_{final} = \frac{5\times10^{-4}}{500}\times1000 = 10^{-3}\,M \Rightarrow P^H = 3.0$

10. cathode: $Cu^{2+} + 2e^- \longrightarrow Cu$

$\frac{250\times0.1}{1000} = 0.025$ mole  $\frac{5\times1351}{96500} = 0.07$ mole

As $Cu^+$ will not remain finally in solution, no complex formation.

11. $n_{eq}$ metal $= n_{eq}\,Cl_2 \Rightarrow \frac{52.8}{E} = \frac{9.08}{22.7}\times2 \Rightarrow E = 66$

At. wt. (approx) $= \frac{6.4}{0.032} = 200$

$\therefore$ valency $= \frac{\text{At.wt.}}{\text{Eq.wt}} = \frac{200}{66} \simeq 3$

HINTS AND EXPLANATIONS

**13.** Initial mass of $H_2SO_4$,

$$w_1 = 2000 \times 1.1 \times \frac{16}{100} = 352 \text{ gm}$$

Final mass of $H_2SO_4$,

$$w_2 = 2000 \times 1.42 \times \frac{40}{100} = 1136 \text{ gm}$$

Now, $n_{eq}$ $H_2SO_4$ produced $= \frac{Q}{F}$

$$\text{or, } \frac{(1136-352)}{98} \times 1 = \frac{i \times \left(\frac{965}{9} \times 3600\right)}{96500} \Rightarrow i = 2\text{A}$$

**13.** Ionic mobility, $\mu = \frac{\lambda°_m}{F} = \frac{\text{Speed of ion}}{\text{Potential gradient}}$

$$\text{or, } \frac{7.5 \times 10^{-3}}{96500} = \frac{\text{distance}/20 \times 3600}{1.93/0.12}$$

$\Rightarrow$ distance = 0.09 m = 9 cm

**14.** $n_{eq}$ Ag oxidised $= \frac{Q}{F} \Rightarrow \frac{w}{108} \times 1 = \frac{9.65 \times 1 \times 3600}{96500}$

$\therefore w = 38.88$ gm

$\therefore$ Mass of anode dissolved $= 38.88 \times \frac{100}{60} \times 64.8$ gm

$\therefore$ Final mass of anode = 72.8 – 64.8 = 8 gm

**15.** Cathode: $2H_2O(l) + 2e^- \longrightarrow H_2(g) + 2OH^-(aq)$

Anode: $2ce^-(aq) \longrightarrow Cl_2(g) + 2e^-$

$n_{eq}$ OH– produced $= \frac{Q}{F}$

$$\Rightarrow n_{OH^-} \times 1 = \frac{9.65 \times 10}{96500} = 10^{-3}$$

$$\therefore [OH^-]_{final} = \frac{10^{-3}}{100} = 10^{-5} \text{ M} \Rightarrow P^H = 9.0$$

## Four digit integer type

**1.** $\Lambda°_{m(AgCl)} = \lambda°_{m(AgCl)} + \lambda°_{m(Cl^-)}$

$$= 6.19 \times 10^{-3} + 7.81 \times 10^{-3}$$

$$= 14.00 \times 10^{-3} \text{ohm}^{-1}\text{m}^2\text{mol}^{-1}$$

Now, $\Lambda_m = \frac{\kappa}{C} \Rightarrow 14 \times 10^{-3} = \frac{2.8 \times 10^{-4}}{S}$

$\Rightarrow S = 0.02 \text{ mol m}^{-3}$

$= 2 \times 10^{-5}$ M

$= 287 \times 10^{-5}$ g/L

**2.** $[S^{2-}] = Ka_1 \cdot Ka_2 \cdot \frac{[H_2S]}{[H^+]^2}$

$$= 10^{-8} \times 10^{-13} \times \frac{0.1}{(10^{-3})} = 10^{-16} \text{M}$$

$$\therefore [Ag^+] = \sqrt{\frac{K_{sp}}{[S^{2-}]}} = \sqrt{\frac{4 \times 10^{-48}}{10^{-16}}} = 2 \times 10^{-16} \text{M}$$

$$\text{Now, } E_{Ag^+/Ag} = 0.80 - \frac{0.06}{1} \log \frac{1}{2 \times 10^{-16}} = -0.142 \text{ V}$$

$\therefore$ Required potential = 142 mV

**3.**

| | $2Hg(l) + 2Fe^{3+}$ | $\rightleftharpoons$ | $Hg_2^{2+}$ | $+ 2Fe^{2+}$ |
|---|---|---|---|---|
| | excess $10^{-3}$ M | | | |
| Equ. | $10^{-3} - x$ | | $\frac{x}{2}$ | $x$ |

$$= \frac{10}{100} \times 10^{-3} \text{M}$$

$\therefore x = 9 \times 10^{-4}$

$$\text{Now, } K_{eq} = \frac{\frac{x}{2} \times x^2}{(10^{-3} - x)^2} = \frac{\frac{9 \times 10^{-4}}{2} \times (9 \times 10^{-4})^2}{(1 \times 10^{-4})^2}$$

$$= \frac{9^3 \times 10^{-4}}{2}$$

$$\text{Now, } E°_{cell} = E°_{Fe^{3+}/Fe} - E°_{Hg_2^{2+}/Hg} = \frac{0.06}{2} \log K_{eq}$$

$$\text{or, } 0.7724 - E°_{Hg_2^{2+}/Hg} = \frac{0.06}{2} \cdot \log \frac{9^3 \times 10^{-4}}{2}$$

$\therefore E°_{Hg_2^{2+}/Hg} = 0.815$ V

**4.** The cell reaction is

$$6Fe^{2+} + Cr_2O_7^{2-} + 14H^+ \rightarrow 6Fe^{3+} + 2Cr^{3+} + 7H_2O$$

$$E_{cell} = E°_{cell} - \frac{0.06}{n} \cdot \log \frac{[Fe^{3+}]^6 [Cr^{3+}]^2}{[Fe^{2+}]^6 [Cr_2O_7^{2-}][H^+]^8}$$

$$= (1.35 - 0.77) - \frac{0.06}{6} \log \frac{(0.75)^6 \times (4)^2}{(0.75)^6 \times 2 \times (1)^8}$$

$= 0.571$ V

**5.** Cell reaction: $H_2(g) + 2Ag^+ \rightarrow 2H^+ + 2Ag(s)$

$$E_{cell} = E°_{cell} - \frac{0.06}{2} \log \frac{[H^+]^2}{P_{H_2} \cdot [Ag^+]^2}$$

$$\text{or, } 0.50 = 0.80 - 0.03 \log \frac{1^2}{1\times[Ag^+]^2}$$

$\Rightarrow [Ag^+] = 10^{-5}$ M

$\therefore$ Mass of Ag in alloy

$$= \left(\frac{250\times10^{-5}}{1000}\right)\times108 = 2.7\times10^{-4}\text{ gm}$$

$\therefore$ % of Pb in alloy

$$\frac{2.7\times10^{-3}-2.7\times10^{-4}}{2.7\times10^{-3}}\times100 = 90\%$$

6. $E^\circ_{cell} = \frac{0.06}{n}\log K_{eq} \Rightarrow (0.2-0.08) = \frac{0.06}{1}\log K_{eq}$

$\Rightarrow K_{eq} = 100$

7. Left electrode: $Mn(s) \rightarrow Mn^{2+} + 2e^-$

Right electrode: $2H_2O(l) \rightarrow O_2(g) + 4H^+ + 4e^-$

$\therefore$ Net cell reaction: $2Mn\ (s) + 2H_2O(l) \rightarrow 2Mn^{e+} + O_2(g) + 4H^+$

$$\text{Now, } E_{cell} = E^\circ_{cell} - \frac{0.06}{4}\log\frac{[Mn^{2+}]^2\cdot[H^+]^4\cdot P_{O_2}}{1}$$

$$= [1.229-(-1.185)] - \frac{0.06}{4}\log\frac{(0.001)^2\times(0.01)^4\times0.25}{1}$$

$= 2.633$ V

8. $\Delta S = nF\cdot\left(\frac{\partial E}{\partial T}\right)_P = 2\times96500\times0.001 = 193$ J/k

9. $\Delta_r G^\circ = -nFE^\circ = \Sigma\,\Delta_f G^\circ_{Products} - \Sigma\,\Delta_f G^\circ_{Reactants}$

$$\text{or, } \frac{-12\times9600\times2.5}{1000} = \left[4\times\Delta_f G^\circ_{Al(OH)_4^-}\right] - [4\times0+3\times0+6\times(-280)+4\times(-156.25)]$$

$\therefore \Delta_f G^\circ_{Al(OH)_4^-} = -1300$ KJ/mol

10. $n_{eq}\ MnO_2 = \frac{Q}{F} \Rightarrow \frac{8.7}{87}\times1 = \frac{3.99\times10^{-3}\times t}{96500}$

$\therefore t = 2.418\times10^6$ sec $\approx$ 28 days

11. $SnCl_2 \longrightarrow Sn^{2+} + 2Cl^-$

Cathode: $Sn^{2+} + 2e^- \longrightarrow Sn$

Anode: $2Cl^- \longrightarrow Cl_2 + 2e-$

$Cl_2 + SnCl_2 \longrightarrow SnCl_4$

Moles of $SnCl_2$ taken $= \frac{19}{190} = 0.1$

Moles of Sn produced $= \frac{1.19}{119} = 0.01$

= moles of $Cl_2$ produced

= moles of $SnCl_4$ formed

and moles of $SnCl_4$ left $= 0.1 - (0.01 + 0.01) = 0.08$

$$\therefore \frac{m_{SnCl_2}}{m_{SnCl_4}} = \frac{0.08\times190}{0.01\times261} = \frac{1520}{261}$$

12. $n_{eq}\ Au = \frac{Q}{F} \Rightarrow \frac{80\times8.0\times10^{-4}\times19.7\times3}{197} = \frac{2.4\times t}{96\,500}$

$\therefore t = 772$ sec

13. $AgBr(s) \rightleftharpoons Ag^+ + Br^-$

$(10^{-7}+x)$M $\quad x$M

Now, $(10^{-7}+x)\,x = 12\times10^{-14} \Rightarrow x = 3\times10^{-7}$

Final solution: $[Ag^+] = 4\times10^{-7}$ M, $[Br^-] = 3\times10^{-7}$ M; $[NO_3^-] = 10^{-7}$ M

Now, $\kappa_{solution} = \lambda^\circ_m(Ag^+)\times[Ag^+] + \lambda^\circ_m(Br^-)\times[Br^-] + \lambda^\circ m(NO_3^-)\times[NO_3^-]$

$= 6\times10^{-3}\times(4\times10^{-7}\times10^3) + 8\times10^{-3}\times(3\times10^{-7}\times10^3) + 7\times10^{-3}(10^{-7}\times10^{-3})$

14. $C_{60} + 60O_2 \longrightarrow 60\ CO_2$

Moles of $O_2$ needed $= 60\times n_{C_{60}} = 60\times\frac{96}{60\times12} = 8$

$n_{eq}\ O_2 = n_{eq}$ azobenzene $\Rightarrow 8\times4 = \frac{w}{182}\times8$

$\Rightarrow w = 728$ gm

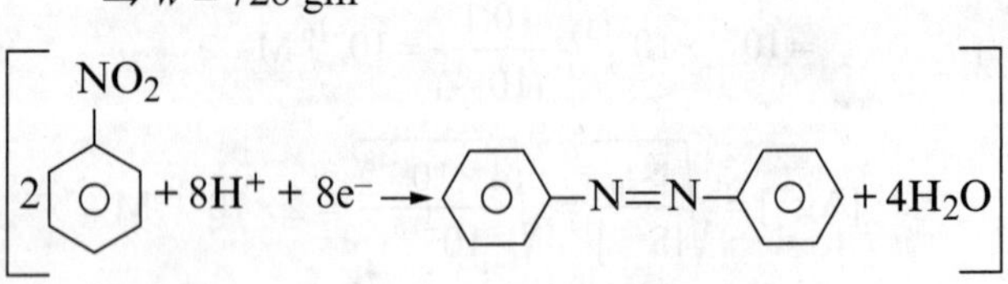

15. $\Lambda^\circ_{eq[Ba_3(PO_4)_2]} = 160 + 140 - 100$

$= 200\ Ohm^{-1}\ cm^2\ eq^{-1}$

Now, $\wedge^\circ_{eq} = \wedge_{eq} = \frac{\kappa}{C} \Rightarrow 200 = \frac{1.2\times10^{-5}}{5}$

$\therefore S = 6\times10^{-8}$ eq/cm$^3$ $= 6\times10^{-5}$ N $= 10^{-5}$ M

$\therefore K_{sp} = 108\ S^5 = 10^8\times10^{-25}\ M^5$

CHAPTER 9

# Solid State

## EXERCISE I (JEE MAIN)

### Basics

1. Most crystals show good cleavage because their atoms, ions or molecules are
   (a) weakly bonded together.
   (b) strongly bonded together.
   (c) spherically symmetrical.
   (d) arranged in planes.

2. Which of the following is the only incorrect statement regarding amorphous solids?
   (a) On heating, they may become crystalline at some temperature.
   (b) They may become crystalline when kept for a long time.
   (c) Amorphous solids can be moulded by heating.
   (d) They are anisotropic in nature.

3. A solid melt slightly above 273 K and is a poor conductor of heat and electricity. To which of the following categories does it belong?
   (a) Ionic solid (b) Covalent solid
   (c) Metallic solid (d) Molecular solid

4. A solid is soft, good conductor of electricity and has very high melting point. Its one of the allotropic forms is the hardest known substance. Hence, the solid is an example of
   (a) ionic solid (b) covalent solid
   (c) molecular solid (d) metallic solid

5. Constituent particles in quartz are bonded by
   (a) electrovalent bonds
   (b) covalent bonds
   (c) van der Waals forces
   (d) metallic bonds

6. Ionic solids are characterized by
   (a) good conductivity in solid state.
   (b) high vapour pressure.
   (c) low melting point.
   (d) solubility in polar solvents.

7. Which of the following is an example of metallic crystal solid?
   (a) C (b) Si
   (c) W (d) AgCl

8. Which of the following is an example of covalent crystal solid?
   (a) $SiO_2$ (b) Al
   (c) Ar (d) NaF

9. The particles would be stationary in a lattice only at
   (a) 0°C (b) 0 K
   (c) 298 K (d) 273 K

10. Iodine molecules are held in the crystal lattice by
    (a) London forces
    (b) dipole–dipole interactions
    (c) covalent bonds
    (d) ionic bonds

11. The unit cell of highest symmetry is
    (a) cubic (b) triclinic
    (c) hexagonal (d) monoclinic

12. The unit cell of lowest symmetry is
    (a) cubic (b) triclinic
    (c) hexagonal (d) monoclinic

13. A match box exhibits ______ geometry.
    (a) cubic (b) orthorhombic
    (c) triclinic (d) monoclinic

14. If the three inter-axial angles defining the unit cell are all equal in magnitude, the crystal cannot belong to the ____ system.
    (a) orthorhombic (b) hexagonal
    (c) tetragonal (d) cubic

15. In a crystal, the constituent particles are located at the position of
    (a) zero potential energy.
    (b) infinite potential energy.
    (c) minimum potential energy.
    (d) maximum potential energy.

16. A lattice is defined as
    (a) the amount of energy required per mole to separate the ions from their lattice positions to an infinite distance in the gas phase.
    (b) the distance separating the cations and anions.
    (c) a set of all points with identical environments within the crystal.
    (d) the arrangement of electrons in various energy levels.

17. The axial angles in triclinic crystal system are
    (a) $\alpha = \beta = \gamma = 90°$
    (b) $\alpha = \gamma = 90°, \beta \neq 90°$
    (c) $\alpha \neq \beta \neq \gamma \neq 90°$
    (d) $\alpha = \beta = \gamma \neq 90°$

18. $TiO_2$ is a well-known example of
    (a) triclinic system (b) tetragonal system
    (c) monoclinic system (d) cubic system

19. In a tetragonal crystal,
    (a) $\alpha = \beta = 90° \neq \gamma; a = b = c$
    (b) $\alpha = \beta = \gamma = 90°; a = b \neq c$
    (c) $\alpha = \beta = \gamma = 90°; a \neq b \neq c$
    (d) $\alpha = \beta = 90°; \gamma = 120°; a = b \neq c$

20. Which of the following crystalline arrangement will have at least one of the angles equal to 90° and at least two axial lengths equal?
    (a) Orthorhombic (b) Rhombohedral
    (c) Monoclinic (d) Tetragonal

## Analysis of Cubic Crystals

21. An element occurring in the body-centred cubic (BCC) structure has $1.208 \times 10^{23}$ unit cells. The total number of atoms of the element in these cells will be
    (a) $2.416 \times 10^{23}$ (b) $3.618 \times 10^{23}$
    (c) $6.04 \times 10^{22}$ (d) $1.208 \times 10^{23}$

22. The rank of a cubic unit cell is 4. The type of cell is
    (a) body-centred (b) face-centred
    (c) primitive (d) end-centred

23. An alloy of copper, silver and gold is found to have copper constituting the face-centred cubic (FCC) lattice. If silver atoms occupy the edge centres and gold is present at body centre, the alloy has a formula
    (a) $Cu_4Ag_2Au$ (b) $Cu_4Ag_4Au$
    (c) $Cu_4Ag_3Au$ (d) CuAgAu

24. A solid PQ has rock salt type structure in which Q atoms are the corners of the unit cell. If the body-centred atoms in all the unit cells are missing, then the resulting stoichiometry will be
    (a) PQ (b) $PQ_2$
    (c) $P_3Q_4$ (d) $P_4Q_3$

25. In a solid AB, having the NaCl structure, A atoms occupy the corners of the cubic unit cell. If all the face-centred atoms along one of the axes are removed, then the resulting stoichiometry of the solid is
    (a) $AB_2$ (b) $A_2B$
    (c) $A_4B_3$ (d) $A_3B_4$

EXERCISE I

26. Xenon crystallizes in FCC lattice and the edge of the unit cell is 620 pm, then the radius of xenon atom is

(a) 219.20 pm (b) 438.5 pm
(c) 265.5 pm (d) 536.94 pm

27. Metallic gold crystallizes in FCC lattice with edge-length 4.07 Å. The closest distance between gold atoms is

(a) 3.525 Å (b) 5.714 Å
(c) 2.857 Å (d) 1.428 Å

28. The distance between two nearest neighbours in BCC lattice of axial length $l$, is

(a) $l$ (b) $\frac{\sqrt{3}}{2}l$
(c) $\frac{\sqrt{2}}{2}l$ (d) $\frac{1}{2}l$

29. Sodium metal crystallizes in BCC lattice with the cell edge $a$ = 4.29 Å. What is the radius of the sodium atom?

(a) 1.86 Å (b) 2.15 Å
(c) 4.29 Å (d) 2.94 Å

30. The unit cell of a metallic element of atomic mass 108 and density 10.5 g/cm$^3$ is a cube with edge length of 409 pm. The structure of the crystal lattice is

(a) FCC (b) BCC
(c) edge-centred cubic (d) simple cubic

31. An element (atomic mass = 100) having BCC structure has unit cell edge length 400 pm. The density of this element will be ($N_A = 6 \times 10^{23}$)

(a) 5.2 g/ml (b) 10.4 g/ml
(c) 0.42 g/ml (d) 2.6 g/ml

32. Potassium has BCC structure with nearest neighbour distance ($2.5 \times \sqrt{3}$) Å. Its density will be (K = 39, $N_A = 6 \times 10^{23}$)

(a) 1.040 kg/m$^3$ (b) 104 kg/m$^3$
(c) 520 kg/m$^3$ (d) 1040 kg/m$^3$

33. Lithium borohydride ($LiBH_4$) crystallizes in an orthorhombic system with four molecules per unit cell. The unit cell dimensions are $a$ = 6.0 Å, $b$ = 4.4 Å and $c$ = 7.5 Å. The density of crystals is (Li = 7, B = 11 $N_A = 6 \times 10^{23}$)

(a) 0.74 g/cm$^3$ (b) 1.48 g/cm$^3$
(c) 0.37 g/cm$^3$ (d) 0.90 g/cm$^3$

34. FeO crystallizes in the cubic system, in which there is four formula units in each unit cell. The density of the crystal is 4.0 g/cm$^3$. The side length of each unit cell is (Fe = 56, $N_A = 6 \times 10^{23}$)

(a) 4.227 Å (b) 2.424 Å
(c) 4.932 Å (d) 2.974 Å

35. The α-form of iron exists in BCC form and γ-form of iron exists in FCC structure. Assuming that the distance between the nearest neighbours is the same in the two forms, the ratio of the density of γ-form to that of α-form is

(a) $4\sqrt{2} : 3\sqrt{3}$ (b) $4\sqrt{3} : 3\sqrt{2}$
(c) $\sqrt{3} : \sqrt{2}$ (d) 2 : 1

36. The number of atoms contained in a FCC unit cell of a diatomic molecular solid is

(a) 8 (b) 2
(c) 4 (d) 6

37. The packing fraction in simple cubic lattice is

(a) $\frac{1}{6}\pi$ (b) $\frac{\sqrt{2}}{6}\pi$
(c) $\frac{\sqrt{3}}{8}\pi$ (d) $\frac{1}{2}\pi$

38. The packing fraction in BCC lattice is

(a) $\frac{1}{6}\pi$ (b) $\frac{\sqrt{2}}{6}\pi$
(c) $\frac{\sqrt{3}}{8}\pi$ (d) $\frac{\sqrt{3}}{2}\pi$

39. The packing fraction in FCC lattice is

(a) $\frac{1}{6}\pi$ (b) $\frac{\sqrt{2}}{6}\pi$
(c) $\frac{\sqrt{3}}{8}\pi$ (d) $\frac{\sqrt{2}}{3}\pi$

40. In a BCC packing, the nearest neighbours lie along the

(a) edges of the cube.
(b) face diagonal.
(c) line joining the two opposite corners of the face.
(d) cube diagonal.

EXERCISE I

## Types of Packing

**41.** The most malleable metals (Cu, Ag, Au) have close-packing of which of the following type?

(a) Cubic close packing (CCP)
(b) Simple cubic packing
(c) BCC packing
(d) Malleability is not related to type of packing.

**42.** The number of carbon atoms per unit cell of diamond is

(a) 4 (b) 6
(c) 8 (d) 12

**43.** The number of closest neighbours of a carbon atom in diamond is

(a) 2 (b) 4
(c) 6 (d) 8

**44.** The number of closest neighbours of a carbon atom in graphite is

(a) 2 (b) 3
(c) 4 (d) 8

**45.** The low density of alkali metals is due to

(a) their BCC structure in which about 32% of the available space is unfilled.
(b) their HCP structure in which about 74% of the available space is unfilled.
(c) their cubic close packed (CCP) structure in which about 74% of the available space is unfilled.
(d) their BCC structure in which about 47% of the available space is unfilled.

**46.** The coordination number of a metal crystallizing in a HCP structure is

(a) 12 (b) 4
(c) 8 (d) 6

**47.** A metallic element crystallizes into a lattice containing a sequence of layers of ABABAB.... Any packing of spheres leaves out voids in the lattice. Approximately what percentage by volume of this lattice is empty space?

(a) 74% (b) 26%
(c) 48% (d) 32%

**48.** Which one of the following schemes of ordering close packed sheets of equal-sized spheres do not generate closest packed lattice?

(a) ABACABAC....
(b) ABCBCABC....
(c) ABCABC....
(d) ABBAA....

**49.** Which of the following has the least void space fraction in their structure?

(a) BCC (b) BCC and HCP
(c) HCP (d) FCC and HCP

**50.** The number of tetrahedral and octahedral voids in hexagonal prismatic unit cell (HCP) is

(a) 8, 4 (b) 2, 1
(c) 12, 6 (d) 6, 12

**51.** Which of the following void is smallest in close packing?

(a) Tetrahedral (b) Octahedral
(c) Cubic (d) Square

**52.** Which of the following interstitial site is formed when the three closed packed spheres of one layer is put over three closed packed spheres of the second layer, their positions being inverted with respect to each other?

(a) Tetrahedral (b) Octahedral
(c) Rhombohedral (d) Tetragonal

**53.** In an FCC arrangement of metallic atoms, what is the relative ratio of the ideal sizes of tetrahedral and octahedral voids?

(a) 0.543 (b) 0.732
(c) 0.414 (d) 0.637

**54.** Atoms of the element 'A' form HCP and atoms of element 'C' occupy only two-third of octahedral voids in it, then the general formula of the compound is

(a) CA (b) $CA_2$
(c) $C_2A_3$ (d) $C_3A_2$

**55.** A solid has three types of atoms , such as X, Y and Z. 'X' forms a FCC lattice with 'Y' atoms occupying all the tetrahedral voids and 'Z' atoms occupying half the octahedral voids. The simplest formula of solid is

(a) $X_2Y_4Z$ (b) $XY_2Z_4$
(c) $X_4Y_2Z$ (d) $X_4YZ_2$

## Radius Ratio and Packing of Crystal

**56.** The ionic radii of $Rb^+$ and $I^-$ are 1.46 and 2.16 Å, respectively. The most probable type of structure exhibited by it is

(a) CsCl type (b) NaCl type
(c) ZnS type (d) $CaF_2$ type

**57.** For an ideal ionic crystal, the simplest formula is AX and the coordination number of cation is 6. The value of radius ratio ($r_{A^+}:r_{X^-}$)

(a) must be greater than 0.732.
(b) must lie in between 0.732 and 0.414.
(c) must lie in between 0.414 and 0.225.
(d) must be less than 0.414.

**58.** In the zinc blend structure (ZnS), $S^{2-}$ adopts CCP arrangement and $Zn^{2+}$ occupies

(a) all octahedral voids.
(b) only 50% of octahedral voids.
(c) only 50% of tetrahedral voids.
(d) all tetrahedral voids.

**59.** The number of next nearest neighbours of $Cs^+$ ion in CsCl crystal is

(a) 12 (b) 8
(c) 6 (d) 4

**60.** In calcium fluoride structure, the coordination numbers of calcium and fluoride ions are, respectively,

(a) 8 and 4 (b) 6 and 8
(c) 4 and 4 (d) 4 and 8

**61.** The radius of $Ag^+$ ion is 126 pm while of $I^-$ ion is 216 pm. The coordination number of Ag in AgI is

(a) 2 (b) 4
(c) 6 (d) 8

**62.** A binary solid ($A^+ B^-$) has a rock salt structure. If the edge length is 400 pm and the radius of cation is 75 pm, then the radius of anion is

(a) 100 pm (b) 125 pm
(c) 250 pm (d) 325 pm

**63.** MgO exists in a rock salt type unit cell. Each $Mg^{2+}$ ion will be in contact with

(a) 6 $Mg^{2+}$ ions (b) 6 $O^{2-}$ ions
(c) 8 $O^{2-}$ ions (d) 4 $O^{2-}$ ions

**64.** Sodium oxide has anti-fluorite structure. The percentage of the tetrahedral voids occupied by the sodium ions is

(a) 12% (b) 25%
(c) 50% (d) 100%

**65.** CsBr has cubic structure with edge length 4.3 Å. The shortest inter ionic distance in between $Cs^+$ and $Br^-$ is

(a) 3.72 Å (b) 1.86 Å
(c) 7.44 Å (d) 4.3 Å

**66.** In $Na_2O$ structure,

(a) $O^{2-}$ ions constitute CCP and $Na^+$ ions occupy all the octahedral holes.
(b) $O^{2-}$ ions constitute CCP and $Na^+$ ions occupy all the tetrahedral holes.
(c) $O^{2-}$ ions constitute CCP and $Na^+$ ions occupy 50% of tetrahedral holes and 100% octahedral holes.
(d) $Na^+$ ions constitute CCP and $O^{2-}$ ions occupy half of octahedral holes.

**67.** NaCl crystal is

(a) FCC (b) BCC
(c) HCP (d) simple cubic

**68.** CsBr has CsCl type structure. Its density is 4.26 g/ml. The side of unit cell is (Cs = 133, Br = 80, $N_A = 6 \times 10^{23}$)

(a) 5.503 Å (b) 4.368 Å
(c) 3.225 Å (d) 2.856 Å

**69.** A solid contains $A^{n+}$ and $B^{m-}$ ions. The structure of solid is FCC for $B^{m-}$ ions and $A^{n+}$ ions are present in one-fourth of the tetrahedral voids as well as in one-fourth of octahedral voids. What is the simplest formula of solid?

(a) $A_3B_4$ (b) $A_4B_3$
(c) $AB_2$ (d) $A_2B$

**70.** An ionic solid is HCP of $Q^{2-}$ ions and $P^{x+}$ ions are in half of the tetrahedral voids. The value of $x$ should be

(a) 1 (b) 2
(c) 4 (d) 1/2

71. In a compound, oxide ions are arranged in CCP arrangement. Cations A occupy one-sixth of the tetrahedral voids and cations B occupy one-third of the octahedral voids. The formula of the compound is
(a) $AB_2O_4$ (b) $ABO_3$
(c) $ABO_2$ (d) $ABO_4$

72. Sapphire is aluminium oxide. Aluminium oxide crystallizes with aluminium ions in two-third of the octahedral voids in the closest packed array of oxide ions. What is the formula of aluminium oxide?
(a) $Al_2O_3$ (b) $AlO_2$
(c) $Al_3O_4$ (d) $Al_3O_2$

73. What is the formula of the magnetic oxide of cobalt, used in recording tapes, that crystallizes with cobalt atoms occupying one-eighth of the tetrahedral holes and one half of the octahedral holes in a closest packed array of oxide ions?
(a) $Co_2O_3$ (b) $Co_5O_8$
(c) CoO (d) $Co_3O_4$

74. Caesium chloride, on heating, changes into
(a) CsCl (g)
(b) NaCl structure
(c) antifluorite structure
(d) ZnS structure

75. BaO has a rock salt type structure. When subjected to high pressure, the ratio of the coordination number of $Ba^{2+}$ ion to $O^{2-}$ ion changes to
(a) 4 : 8 (b) 8 : 4
(c) 8 : 8 (d) 4 : 4

## Defects in Solids

76. Frenkel defect is noticed in
(a) AgBr (b) Zns
(c) AgI (d) All of these

77. Which of the following defect, if present, lowers the density of the crystal?
(a) Frenkel
(b) Schottky
(c) Metal excess defect due to excess cation.
(d) Interstitial defect.

78. Schottky defect appears in
(a) NaCl (b) CsCl
(c) AgBr (d) All of these

79. The yellow colour of ZnO and conducting nature produced in heating is due to
(a) metal excess defects due to interstitial cation.
(b) extra positive ions present in an interstitial site.
(c) trapped electrons.
(d) All of these

80. The presence of excess sodium in sodium chloride makes the crystal appearance yellow. This is due to the presence of
(a) Schottky defect (b) Frenkel defect
(c) F-centres (d) interstitial defects

81. The composition of a sample of cuprous oxide is found to be $Cu_{1.92}O_{1.00}$ due to metal deficient defect. The molar ratio of $Cu^{2+}$ and $Cu^{+}$ ions in the crystal is
(a) 4 : 5 (b) 1 : 12.5
(c) 1 : 23 (d) 1 : 24

82. NaCl is doped with $2 \times 10^{-3}$ mole % $SrCl_2$, the concentration of cation vacancies is ($N_A = 6 \times 10^{23}$)
(a) $6.0 \times 10^{18}$ mol$^{-1}$ (b) $1.20 \times 10^{19}$ mol$^{-1}$
(c) $3.0 \times 10^{18}$ mol$^{-1}$ (d) $1.20 \times 10^{21}$ mol$^{-1}$

83. Each of the following statements is correct except
(a) the lattice positions left vacant by anion and occupied by electrons are called F-centres.
(b) the presence of F-centres makes the lattice electrically neutral and diamagnetic.
(c) the presence of F-centres gives colour to the crystals.
(d) F-centres contain unpaired electrons.

84. The intrinsic or thermodynamic defect is
(a) Schottky defect
(b) metal excess defect due to excess cation.
(c) metal excess defect due to anion vacancy.
(d) metal-deficient defect.

85. The only incorrect effect on density by the given defect in solids is
(a) density must decrease by vacancy defect.
(b) density must increase by interstitial defect.
(c) density must increase by impurity defect.
(d) density does not change by dislocation defect.

## Properties of Solids

**86.** Which of the following is a ferromagnetic substance?

(a) $Fe_2O_3$  (b) $Cr_2O_3$
(c) $Fe_3O_4$  (d) $CrO_2$

**87.** Which of the following acts as a superconductor at 4 K?

(a) He  (b) Cu
(c) K  (d) Mg

**88.** Superconductors are substances which

(a) conduct electricity at low temperatures.
(b) conduct electricity at high temperature.
(c) offer very high resistance to the flow of current.
(d) offer no resistance to the flow of current.

**89.** Addition of arsenic in small amount to pure germanium will result in the formation of

(a) n-type semiconductor
(b) germanium arsenide
(c) p-type semiconductor
(d) a superconducting alloy

**90.** Which of the following arrangement shows schematic alignment of magnetic moments of anti-ferromagnetic substances?

(a) 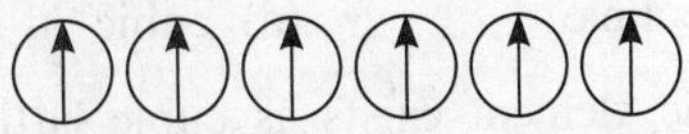

(b) 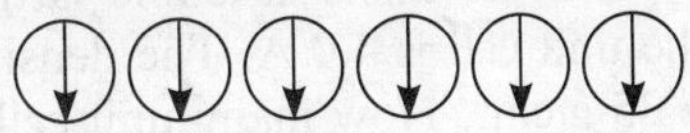

(c) 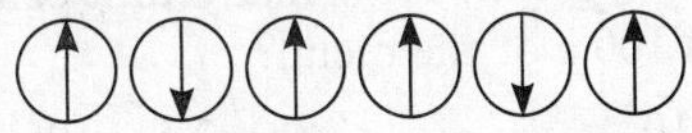

(d) 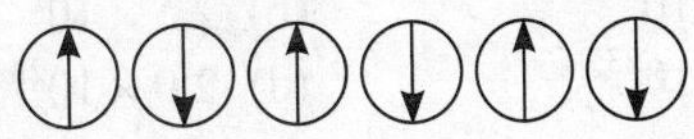

# EXERCISE II (JEE ADVANCED)

## Section A (Only one Correct)

1. For a certain crystal, the unit cell axial lengths are found to be $a$ = 5.62 Å, $b$ = 7.41 Å and $c$ = 10.13 Å. The three coordinate axes are mutually perpendicular. The crystal system to which the crystal belongs is
   (a) tetragonal (b) orthorhombic
   (c) monoclinic (d) cubic

2. A metallic element exists as cubic lattice. Each edge of the unit cell is 4.0 Å. The density of the metal is 6.25 g/cm$^3$. How many unit cells will be present in 100 g of the metal?
   (a) $1.0 \times 10^{22}$ (b) $2.5 \times 10^{23}$
   (c) $5.0 \times 10^{23}$ (d) $2.0 \times 10^{23}$

3. Aluminium crystallizes in cubic system with unit cell edge length equal to 4.0 Å. If its density is $\frac{45}{16}$ g/cm$^3$, then the atomic radius of Al-atom is ($N_A = 6 \times 10^{23}$, Al = 27)
   (a) 1.414 Å (b) 1.732 Å
   (c) 4.0 Å (d) 2.0 Å

4. Is there an expansion or contraction as iron transforms from FCC to BCC? The atomic radius of iron is 125 pm in FCC but $50\sqrt{2}$ pm in BCC.
   (a) Expansion
   (b) Contraction
   (c) Neither expansion nor contraction.
   (d) Unpredictable

5. Silver (atomic mass = 108) has an atomic radius of 144 pm and density 10.6 g/cm$^3$. To which type of cubic crystal silver belongs?
   (a) Simple (b) BCC
   (c) FCC (d) End-centred

6. Gold crystallizes with FCC lattice for which the side length of the unit cell is 5.0 Å. If the density of gold is 10.5 g/cm$^3$, then the value of Avogadro number is (Au = 198)
   (a) 6.022 × 1023 (b) 6.034 × 1023
   (c) $5.966 \times 10^{23}$ (d) $6.022 \times 10^{22}$

7. What is the void space per unit cell for metallic silver crystallizing in the FCC system, the edge length of the unit cell being 4 Å?
   (a) 47.36 Å$^3$ (b) 30.72 Å$^3$
   (c) 20.48 Å$^3$ (d) 16.64 Å$^3$

8. A metal exists as FCC crystal. If the atomic radius is $100\sqrt{2}$ pm and the density of metal is 12,500 kg/m$^3$, then the metal is (Atomic masses: Ca = 40, Co = 58.9, Sn = 119.8, Pb = 207.9 and $N_A = 6 \times 10^{23}$)
   (a) Ca (b) Co
   (c) Sn (d) Pb

9. A solid element (monoatomic) exists in cubic crystal. If its atomic radius is 1.0 Å and the ratio of packing fraction and density is 0.1 cm$^3$/g, then the atomic mass of the element is ($N_A = 6 \times 10^{23}$)
   (a) 8π (b) 16π
   (c) 80π (d) 4π

10. In FCC unit cell, what fraction of edge is not covered by atoms?
    (a) 0.134 (b) 0.293
    (c) 0.26 (d) 0.32

11. The sublimation temperature of a substance is −83°C. It exists as cubic close packed structure with unit cell edge length of 5.0 Å and density $\frac{8}{3}$ g/cm$^3$ at −100°C. What should be the density of substance at 0°C and 1 atm? ($N_A = 6 \times 10^{23}$)
    (a) 50 g/cm$^3$ (b) 2.23 g/cm$^3$
    (c) 2.23 g/L (d) 1.86 g/L

12. An alkali metal has density 4.5 g/cm$^3$. It has cubic unit cell with edge length 400 pm. The reaction of 7.68 cm$^3$ chunk of the metal with an excess of HCl solution gives a colourless gas which occupies 4.54 L at 0°C and 1 bar. The unit cell of metal is
    (a) simple cubic (b) BCC
    (c) FCC (d) end-centred

**13.** The densities of ice and water at 0°C and 1 bar are 0.96 and 0.99 g/cm$^3$, respectively. If the percentage of occupied space in ice is $x$, then the percentage of empty space in water is

(a) $\frac{32}{33}x$ (b) $\frac{33}{32}x$

(c) $100 - \frac{33}{32}x$ (d) $100 - \frac{32}{33}x$

**14.** A metal crystallizes in such a lattice in which only 70% of the total space of the crystal is occupied by the atoms. If the atomic mass of the metal is $32\pi$ g/mol and the atomic radius is 0.2 nm, then the density of the metal is

(a) 7.0 g/cm$^3$ (b) 3.5 g/cm$^3$

(c) 10.5 g/cm$^3$ (d) 14.0 g/cm$^3$

**15.** The number of octahedral voids per unit BCC cell is

(a) 1.0 (b) 2.0

(c) 1.5 (d) 0

**16.** In a hypothetical solid, 'C' atoms form CCP lattice. 'A' atoms occupy all tetrahedral voids and 'B' atoms occupy all octahedral voids, without disturbing the crystal. If a unit cell of the crystal is cut by a plane shown in the figure, then the cross section of this plane looks like

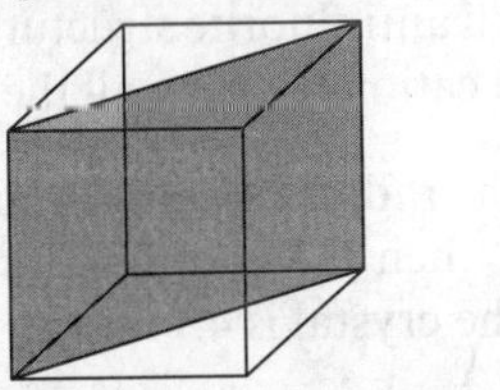

(a)

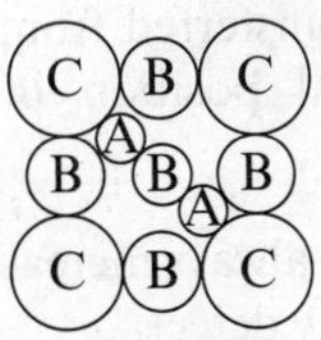


(b)

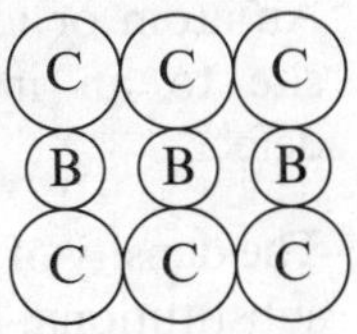


(c)

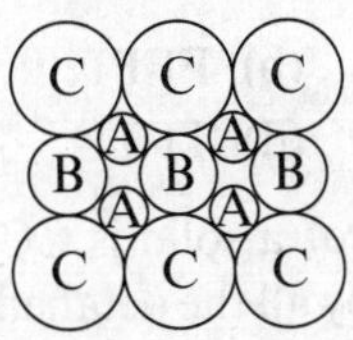


(d)

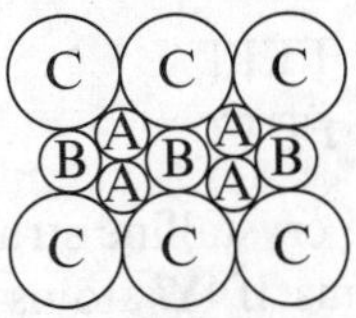


**17.** A metal having atomic mass 60.22 g/mole crystallizes in ABCABC.... type packing. The density of each metal atom if the edge length of unit cell is 10 Å is ($N_A = 6.022 \times 10^{23}$)

(a) 0.4 g/cm$^3$ (b) 40 g/cm$^3$

(c) 0.54 g/cm$^3$ (d) 54 g/cm$^3$

**18.** Packing fraction in 2D-hexagonal arrangement of identical spheres is

(a) $\frac{\pi}{3\sqrt{2}}$ (b) $\frac{\pi}{3\sqrt{3}}$

(c) $\frac{\pi}{2\sqrt{3}}$ (d) $\frac{\pi}{6}$

**19.** A close packing consists of a base of spheres, followed by a second layer where each sphere rests in the hollow at the junction of four spheres below it and the third layer then rests on these in an arrangement which corresponds exactly to that in the first layer. This packing is known as

(a) HCP (b) CCP

(c) square close packing (d) BCC packing

**20.** If the height of HCP unit cell of identical particles is $h$, then the height of octahedral voids from the base is

(a) $\frac{h}{2}$ (b) $\frac{h}{3}, \frac{2h}{3}$

(c) $\frac{h}{4}, \frac{3h}{4}$ (d) $\frac{h}{8}, \frac{7h}{8}$

**21.** If the volume occupied by a unit cell of NaCl is $4.7 \times 10^{-23}$ ml, then the volume of a crystal of NaCl weighing 1.0 g is

(a) $2.13 \times 10^{22}$ ml (b) $8.03 \times 10^{-25}$ ml

(c) 0.48 ml (d) 0.12 ml

**22.** If the unit cell length of sodium chloride crystal is 600 pm, then its density will be

(a) 2.165 g/cm$^3$ (b) 3.247 g/cm$^3$

(c) 1.79 g/cm$^3$ (d) 1.082 g/cm$^3$

**23.** The inter-metallic compound LiAg crystallizes in cubic lattice in which both Li and Ag have co-ordination number of 8. The class of crystal is

(a) simple cubic (b) BCC

(c) FCC (d) end-centred cubic

**24.** A mineral having the formula $AB_2$ crystallizes in the CCP lattice with the A atoms occupying the lattice points. What is the coordination number of the B atoms?

(a) 4 (b) 6

(c) 8 (d) 12

25. There are three cubic unit cells A, B and C. A is FCC and all of its tetrahedral voids are also occupied. B is also FCC and all of its octahedral voids are also occupied. C is simple cubic and all of its cubic voids are also occupied. If voids in all unit cells are occupied by the spheres exactly at their limiting radius, then the order of packing efficiency would be

(a) A < B < C (b) C < A < B
(c) C < B < A (d) A < C < B

26. In a cubic closest packed structure of mixed oxides, the lattice is made up of oxide ions. Tetrahedral voids are occupied by divalent $X^{2+}$ ions and octahedral voids are occupied by trivalent $Y^{3+}$ ions. If only 50% of the octahedral voids are occupied, then the percentage of tetrahedral voids occupied is

(a) 50% (b) 12.5%
(c) 25% (d) 40%

27. The distance between adjacent oppositely charged ions in rubidium chloride is 328.5 pm, in potassium chloride is 313.9 pm, in sodium bromide is 298.1 pm and in potassium bromide is 329.3 pm. The distance between adjacent oppositely charged ions in rubidium bromide is

(a) 314.7 pm (b) 338.5 pm
(c) 339.3 pm (d) 343.9 pm

28. Solid AB has a rock salt type structure. If the radius of the cation is 200 pm, then what is the maximum possible radius of the anion?

(a) 483.1 pm (b) 273.6 pm
(c) 200 pm (d) 400 pm

29. The simplest formula of a solid having CCP arrangement for 'A' atoms in which alternate face-centres are occupied by 'B' atoms and alternate edge centres are occupied by 'C' atom, is

(a) ABC (b) $A_4BC$
(c) $A_2BC$ (d) $A_4B_2C$

30. Spinel is an important class of oxides consisting of two types of metal ions with the oxide ions arranged in CCP pattern. The normal spinel has one-eighth of the tetrahedral holes occupied by one type of metal ion and one-half of the octahedral hole occupied by another type of metal ion. Such a spinel is formed by $Zn^{2+}$, $Al^{3+}$ and $O^{2-}$. The simplest formula of such spinel is

(a) $ZnAl_2O_4$ (b) $Zn_2AlO_4$
(c) $Zn_2Al_3O_4$ (d) $ZnAlO_2$

31. An ionic crystalline solid $MX_3$ has a cubic unit cell. Which of the following arrangement of the ions is consistent with the stoichiometry of the compound?

(a) $M^{3+}$ ions at the corners and $X^-$ ions at the face centres.

(b) $M^{3+}$ ions at the corners and $X^-$ ions at the body centres.

(c) $X^-$ ions at the corners and $M^{3+}$ ions at the face centres.

(d) $X^-$ ions at the corners and $M^{3+}$ ions at the body centres.

32. In a compound $XY_2O_4$, the oxide ions are arranged in CCP arrangement and cations X are present in octahedral voids. Cations Y are equally distributed between octahedral and tetrahedral voids. The fraction of the octahedral voids occupied is

(a) 1/2 (b) 1/4
(c) 1/6 (d) 1/8

33. Give the correct order of initials T (true) or F (false) for the following statements.

I. In an anti-fluorite structure, anions form FCC and cations occupy all the tetrahedral voids.

II. If the radius of cation and anion is 20 and 95 pm, then the coordination number of cation in the crystal is 4.

III. An atom or ion is transferred from a lattice site to an interstitial position in Frenkel defect.

IV. The density of crystal always increases due to substitutional impurity defect.

(a) TTTT (b) FFFF
(c) FFTT (d) TFTF

34. The crystalline structure of a solid is CCP for 'X' atoms. If 'Y' atoms occupy all the octahedral voids and 'Z' atoms occupy all the tetrahedral voids, without any distortion in the CCP arrangement of 'X' atoms, then the fraction of body diagonal not covered by the atoms is

(a) 0.26 (b) 0.24
(c) 0.76 (d) 0.81

35. In an ionic solid $AB_2O_4$, the oxide ions form CCP. 'A' and 'B' are metal ions in which one is bivalent and another is trivalent (not necessarily in given order). If all the bivalent ions occupy octahedral holes and the trivalent ions occupy tetrahedral and octahedral voids in equal numbers, then the fraction of octahedral voids unoccupied is

(a) $\frac{1}{2}$ (b) $\frac{3}{4}$

(c) $\frac{1}{4}$ (d) $\frac{7}{8}$

36. The theoretical density of ZnS is $d$ g/cm$^3$. If the crystal has 4% Frenkel defect, then the actual density of ZnS should be

(a) $d$ g/cm$^3$ (b) $0.04d$ g/cm$^3$

(c) $0.96d$ g/cm$^3$ (d) $1.04d$ g/cm$^3$

37. In a crystal at 827°C one out of $10^{10}$ lattice site is found to be vacant but at 927°C, one out of $2 \times 10^9$ lattice sites is found to be vacant. The enthalpy of vacancy formation in kJ/mol is

(a) 76.8 (b) 176.8

(c) 33.8 (d) 23.8

38. Which of the following oxide shows electrical properties like metals?

(a) $SiO_2$ (b) MgO

(c) $SO_2$(s) (d) $CrO_2$

39. Which kind of defects is introduced by doping in silicon crystal?

(a) Dislocation defect (b) Schottky defect

(c) Frenkel defect (d) Electronic defect

40. A ferromagnetic substance becomes a permanent magnet when it is placed in a magnetic field because

(a) all the domains get oriented in the direction of magnetic field.

(b) all the domains get oriented in the direction opposite to the direction of magnetic field.

(c) domains get oriented randomly.

(d) domains are not affected by magnetic field.

## Section B (One or More than one Correct)

1. Which of the following solid substance(s) will have the same refractive index when measured in different directions?

(a) Rubber (b) NaCl

(c) Plastic (d) Graphite

2. Which of the following is/are amorphous solid(s)?

(a) NaCl (b) $CaF_2$

(c) Glass (d) Plastic

3. Which of the following has face-centred Bravais lattice?

(a) Hexagonal (b) Monoclinic

(c) Cubic (d) Orthorhombic

4. If the height of HCP unit cell of identical particles is $h$, then the height of tetrahedral voids from the base is

(a) $\frac{3h}{8}, \frac{5h}{8}$ (b) $\frac{h}{3}, \frac{2h}{3}$

(c) $\frac{h}{4}, \frac{3h}{4}$ (d) $\frac{h}{8}, \frac{7h}{8}$

5. The correct information regarding a solid element (atomic radius $= r$) having ABCABC.... type packing (FCC unit cell edge length $= a$) is

(a) the distance between nearest octahedral and tetrahedral void is $\frac{\sqrt{3}a}{4}$.

(b) the distance between two nearest octahedral void is $\frac{a}{\sqrt{2}}$.

(c) the distance between two nearest tetrahedral void is $\frac{\sqrt{3}a}{2}$.

(d) the distance between successive A and B layers is $2r.\sqrt{\frac{2}{3}}$.

6. An element exists in two allotropic forms. One form is CCP and the other form is HCP arrangement of atoms of the element. If the atomic radius of element is same in both the forms, then both allotropic forms have the same

(a) density

(b) coordination number.

(c) fraction of unoccupied space.

(d) distance between two consecutive layers.

7. The correct statement(s) for the packing of identical spheres in two dimensions is/are
   (a) For square close packing, the coordination number is 4.
   (b) For HCP, the coordination number is 6.
   (c) There is only one void per sphere in both, square and HCP.
   (d) HCP is more efficiently packed than square close packing.

8. Which of the following information(s) is/are incorrect regarding the voids formed in three dimensional HCP of identical spheres?
   (a) A tetrahedral void is formed when a sphere of the second layer is present above the triangular void of the first layer.
   (b) All the triangular voids are not covered by the spheres of the second layer.
   (c) Tetrahedral voids are formed when the triangular voids in the second layer lie above the triangular voids in the first layer and the triangular shapes of these voids do not overlap.
   (d) Octahedral voids are formed when the triangular voids in the second layer exactly overlap with similar voids in the first layer.

9. Which of the following statement(s) is (are) correct?
   (a) The coordination number of each type of ions in CsCl crystal is 8.
   (b) A metal that crystallizes in BCC structure has a coordination number of 12.
   (c) A unit cell of an ionic crystal shares some of its ions with other unit cells.
   (d) The length of the unit cell in NaCl is 552 pm. ($r_{Na^+} = 95$ pm, $r_{Cl^-} = 181$ pm).

10. In which of the following crystals, alternate tetrahedral voids are occupied?
    (a) NaCl (b) ZnS
    (c) $CaF_2$ (d) $Na_2O$

11. Which of the following statement(s) is/are correct with respect to zinc blende structure?
    (a) $Zn^{2+}$ ions are present at the corners and at the centres of each face.
    (b) Only alternate tetrahedral holes are occupied by $Zn^{2+}$ ions.
    (c) The coordination number of $Zn^{2+}$ and $S^{2-}$ is 4 each.
    (d) The number of ZnS units in a unit cell is 4.

12. In a sodium chloride crystal, the unit cell edge length is *a*. The option(s) representing the correct combination is/are

| Options | Nearest neighbour of $Na^+$ ion | Ion | Distance from $Na^+$ ion | Number of ions |
|---|---|---|---|---|
| (a) | $2^{nd}$ | $Na^+$ | $0.707a$ | 12 |
| (b) | $3^{rd}$ | $Cl^-$ | $0.866a$ | 8 |
| (c) | $4^{th}$ | $Na^+$ | $a$ | 6 |
| (d) | $5^{th}$ | $Cl^-$ | $1.12a$ | 24 |

13. A quantity of 80.0 g of a salt (XY) of strong acid and weak base is dissolved in water to form 2.0 L aqueous solution. At 298 K, the pH of the solution is found to be 5.0. If XY forms CsCl type crystal and the radius of $X^+$ and $Y^-$ ions are 160 pm and 186.4 pm, respectively, then the correct information is (Given: $K_b$ of XOH $= 4\times10^{-5}$, $N_A = 6\times10^{23}$, $\sqrt{3} = 1.732$)
    (a) molar mass of the salt is 100 g/mol.
    (b) the degree of hydrolysis of salt is $2.5\times10^{-5}$.
    (c) unit cell edge length of XY crystal is 400 pm.
    (d) density of solid XY is 2.6 g/cm$^3$.

14. An ionic compound consists of only $A^+$, $B^{2+}$ and $C^{3-}$ ions. Which of the following option(s) is/are correct with respect to their possible structure?
    (a) $C^{3-}$ ions form CCP, $A^+$ ions occupy all octahedral voids and $B^{2+}$ ions occupy half of the tetrahedral voids.
    (b) $B^{2+}$ ions form HCP, $A^+$ ions occupy all octahedral voids and $C^{3-}$ ions occupy half of the tetrahedral voids.
    (c) $A^+$ ions occupy alternate corners of a simple cube, $B^{2+}$ ions occupy the other corners of the cube and $C^{3-}$ ions occupy body centre of the cube.
    (d) $B^{2+}$ ions form CCP, $C^{3-}$ ions occupy all the tetrahedral voids and $A^+$ ions occupy all the octahedral voids.

15. Potassium hexachloroplatinate (IV), $K_2[PtCl_6]$ adopts anti-fluorite crystal structure. Which of the following information is correct about ideal $K_2[PtCl_6]$ crystal ?

(a) $K^+$ ions occupy all the tetrahedral voids.

(b) The unit cell is FCC with respect to $PtCl_6^{2-}$ ions.

(c) The coordination number of $PtCl_6^{2-}$ ions is 4.

(d) The coordination number of $PtCl_6^{2-}$ ions is 8.

16. In each of the following ionic solids, the coordination number of the cation and anion is same, except

(a) MgO (b) $CaF_2$
(c) $Al_2O_3$ (d) CsCl

17. Which of the following defects in the crystals may lower the density?

(a) Interstitial defect (b) Vacancy defect
(c) Schottky defect (d) Impurity defect

18. Which of the following may have Frenkel defect?

(a) Sodium chloride (b) Zinc sulphide
(c) Silver bromide (d) Diamond

19. Which of the following statement(s) is/are incorrect regarding the defects in solids?

(a) AgBr crystal show both Schottky and Frenkel defect.

(b) Impurity defect by doping of arsenic in silicon results 'n'-type semiconductor.

(c) Doping in crystal introduces dislocation defect.

(d) Metal deficient defect can occur with extra anion present in the interstitial voids.

20. Which of the following statement is correct regarding defects in solid?

(a) Frenkel defect is usually favoured by a very small difference in the sizes of cation and anion.

(b) Frenkel defect is a dislocation defect.

(c) Trapping of an electron in the lattice leads to the formation of F-centre.

(d) Schottky defect have no effect on the physical properties of solids.

21. Which of the following is/are correct regarding point defects in solids?

(a) The density of point defect increases with the increase in rate of crystallization.

(b) In the formation of point defects, the entropy of system increases but the entropy of surrounding decreases.

(c) The density of point defect increases with the increase in temperature.

(d) All point defects results in the decreasing density of solid.

22. Which of the following is/are correct statement(s) about $Fe_3O_4$ crystal?

(a) $Fe^{2+}$ ions occupy octahedral voids only.

(b) $Fe^{3+}$ ions occupy tetrahedral voids only.

(c) $Fe^{3+}$ ions occupy octahedral as well as tetrahedral voids.

(d) $O^{2-}$ ions are present at the corner as well as the centre of alternate faces.

23. Which of the following statement(s) is/are true regarding the electrical properties of solids?

(a) (Conductivity) metals << (Conductivity) insulators < (Conductivity) semiconductors.

(b) Depending upon the temperature, $TiO_3$ can behave as insulator or conductor.

(c) $I_2(s)$ is non-conducting.

(d) n-type semiconductor will have conductivity less than pure semiconductor.

24. Which of the following solids is not an electrical conductor?

(a) Mg(s) (b) TiO(s)
(c) $I_2(s)$ (d) $H_2O(s)$

25. Which of the following oxides behave as a conductor or an insulator depending on temperature?

(a) TiO (b) $SiO_2$
(c) $TiO_3$ (d) VO

EXERCISE II

## Section C (Comprehensions)

### Comprehension I

**Potassium crystallizes in BCC lattice, with a unit cell length of $a = 5.0$ Å (K = 39, $N_A = 6 \times 10^{23}$).**

1. What is the distance between nearest neighbours?
   (a) 5.0 Å (b) 4.33 Å
   (c) 3.54 Å (d) 2.5 Å

2. What is the distance between next nearest neighbours?
   (a) 5.0 Å (b) 4.33 Å
   (c) 3.54 Å (d) 7.07 Å

3. How many nearest neighbours does each K atom have?
   (a) 8 (b) 6
   (c) 4 (d) 12

4. How many next nearest neighbours does each K atom have?
   (a) 8 (b) 6
   (c) 4 (d) 12

5. What is the theoretical density of crystalline K?
   (a) 10.4 g/cm$^3$ (b) 20.8 g/cm3
   (c) 1.04 g/cm3 (d) 2.08 g/cm$^3$

6. If the metal is melted, the density of molten metal is found to be 0.9 g/cm$^3$. What is the percentage of empty space in the molten metal?
   (a) 58.83% (b) 68.02%
   (c) 41.17% (d) 31.98%

### Comprehension II

**Copper has a FCC lattice with an unit cell edge length of $\frac{0.5}{\sqrt{2}}$ nm.**

7. What is the size of the largest atom that could fit into octahedral holes of the lattice without disturbing the lattice?
   (a) $\frac{0.207}{\sqrt{2}}$ nm (b) $\frac{0.366}{\sqrt{2}}$ nm
   (c) 0. 092 nm (d) 0.052 nm

8. What is the size of the largest atom that could fit into tetrahedral holes of the lattice without disturbing the lattice?
   (a) 0.028 nm (b) 0.052 nm
   (c) $\frac{0.1125}{\sqrt{2}}$ nm (d) $\frac{0.207}{\sqrt{2}}$ nm

### Comprehension III

**The crystalline structure of magnesium is HCP. The density of magnesium is 1.92 g/cm$^3$ [Mg = 24, $N_A = 6 \times 10^{23}$, $3\sqrt{2} = \frac{4\pi}{3} = (1.6)^3$, $\sqrt{\frac{2}{3}} = 0.8$]**

9. What is the volume of the unit cell?
   (a) $1.25 \times 10^{-22}$ cm$^3$ (b) $2.50 \times 10^{-22}$ cm$^3$
   (c) $6.25 \times 10^{-21}$ cm$^3$ (d) $1.25 \times 10^{-23}$ cm$^3$

10. What is the atomic radius of magnesium?
    (a) 5.0 Å (b) 1.56 Å
    (c) 3.125 Å (d) 2.13 Å

11. What is the height of the unit cell?
    (a) 6.24 Å (b) 2.5 Å
    (c) 5.0 Å (d) 3.12 Å

12. How many nearest neighbours does each atom have?
    (a) 6 (b) 9
    (c) 12 (d) 3

## Comprehension IV

**Titanium crystallizes in a FCC lattice. It reacts with carbon or hydrogen interstitially by allowing atoms of these elements to occupy holes in the host lattice. Hydrogen occupies tetrahedral holes but carbon occupies octahedral holes.**

**13.** Predict the formulas of titanium hydride and titanium carbide formed by saturating the titanium lattice with either 'foreign' element.

(a) TiH, TiC
(b) $TiH_2$, TiC
(c) $TiH_4$, TiC
(d) $TiH_2$, $TiC_2$

**14.** Account for the fact that hydrogen occupies tetrahedral holes while carbon occupies octahedral holes.

(a) $r_H = r_C$
(b) $r_H < r_C$
(c) $r_H > r_C$
(d) Hydrogen is more reactive than carbon.

## Comprehension V

**KCl crystallizes in the same type of lattice as does NaCl (rock salt structure). Given that $r_{Na^+}/r_{K^+} = 0.5$ and $r_{Na^+}/r_{Cl^-} = 0.7$.**

**15.** The ratio of the side of the unit cell for KCl to that for NaCl is

(a) 1.143 (b) 2.57
(c) 2.4 (d) 1.2

**16.** The ratio of density of NaCl to that of KCl is

(a) 1.49 (b) 1.17
(c) 0.853 (d) 1.143

## Comprehension VI

**$BaTiO_3$ crystallizes in the perovskite structure. This structure may be described as a cubic lattice with barium ions occupying the corners of the unit cell, oxide ions occupying the face centres and titanium ions occupying the centres of the unit cells.**

**17.** If titanium is described as occupying holes in Ba-O lattice, then what type of holes does it occupy?

(a) Tetrahedral (b) Octahedral
(c) Cubic (d) Triangular

**18.** What fraction of the holes of this type does it occupy?

(a) 0.25 (b) 0.50
(c) 1.00 (d) 0.75

## Comprehension VII

**The zinc blend structure is cubic. The unit cell may be described as a face-centred sulphide ion sub-lattice with zinc ions in the centres of alternating mini cubes made by partitioning the main cube into 8 equal parts.**

**19.** How many nearest neighbours does each $Zn^{2+}$ have?

(a) 4 (b) 6
(c) 8 (d) 2

**20.** How many nearest neighbours does each $S^{2-}$ have?

(a) 4 (b) 6
(c) 8 (d) 2

**21.** What angle is made by the lines connecting any $Zn^{2+}$ to any two of its nearest neighbours?

(a) 120° (b) 90°
(c) 109°28′ (d) 60°

**22.** What minimum $r^+/r^-$ ratio is needed to avoid anion-anion contact, if the closest cation–anion pairs are assumed to touch?

(a) 0.155 (b) 0.225
(c) 0.732 (d) 0.414

## Comprehension VIII

**A cubic unit cell contains manganese ions at the corners and fluoride ions at the centre of each edge (Mn = 55, F = 19, $N_A = 6 \times 10^{23}$).**

23. What is the empirical formula?
 (a) MnF (b) $MnF_2$
 (c) $MnF_3$ (d) $MnF_4$

24. What is the coordination number of the Mn ion?
 (a) 4 (b) 6
 (c) 8 (d) 12

25. The edge length of the unit cell, if the radius of $Mn^{3+}$ ion is 0.65 Å and that of $F^-$ ion is 1.35 Å, is
 (a) 2.00 Å (b) 4.00 Å
 (c) 3.82 Å (d) 1.30 Å

26. The density of solid is
 (a) 2.92 g/cm$^3$ (b) 5.84 g/cm$^3$
 (c) 8.75 g/cm$^3$ (d) 23.33 g/cm$^3$

## Comprehension IX

**The density of crystalline CsCl is 3.5 g/cm$^3$ [Cs = 132.5, Cl = 35.5, $N_A = 6 \times 10^{23}$, $(4.3)^3 = 80$].**

27. The volume effectively occupied by a single CsCl ion pair in the crystal is
 (a) 48 ml (b) $8 \times 10^{-23}$ ml
 (c) $3.2 \times 10^{-22}$ ml (d) $2 \times 10^{-23}$ ml

28. The smallest Cs-to-Cs internuclear distance in the crystal is
 (a) 2.0 Å (b) 4.3 Å
 (c) 3.72 Å (d) 6.08 Å

29. The smallest Cs-to-Cl internuclear distance in the crystal is
 (a) 2.0 Å (b) 4.3 Å
 (c) 3.72 Å (d) 6.08 Å

## Comprehension X

**The moles of Schottky and Frenkel defects in an ionic crystal is given as: $N = N_o \cdot e^{-E/2RT}$ and $\sqrt{N_o \cdot N_i} \cdot e^{-E/2RT}$, where $N_o$ is the number of ideal lattice points, $N_i$ is the number of available interstitial sites and $E$ is the activation energy of the formation of respective defect. For NaCl crystal, the activation energies of formation of these defects are 46 and 73.6 kcal/mol, respectively (ln 10 = 2.3).**

30. The mole fraction of Schottky defects in a NaCl crystal at 1000 K is
 (a) $1.0 \times 10^{-5}$
 (b) $1.41 \times 10^{-8}$
 (c) $2.0 \times 10^{-5}$
 (d) $5.0 \times 10^{-6}$

31. The mole fraction of Frenkel defects in a NaCl crystal at 1000 K is
 (a) $1.0 \times 10^{-5}$
 (b) $1.41 \times 10^{-8}$
 (c) $1.0 \times 10^{-8}$
 (d) $2.0 \times 10^{-8}$

## Comprehension XI

**The crystal AB (rock salt structure) has molecular mass 6.023 $y$ amu, where $y$ is an arbitrary number in amu. If the minimum distance between cation and anion is $y^{1/3}$ nm and the observed density is 20 kg/m$^3$.**

32. The theoretical density of AB(s) in kg/m$^3$ is
 (a) 20.0 kg/m$^3$ (b) 5.0 kg/m$^3$
 (c) 10.0 kg/m$^3$ (d) 2.5 kg/m$^3$

33. Which of the following is the type of defect in solids?
 (a) Frenkel defect
 (b) Schottky defect
 (c) Metal excess defect
 (d) Impurity defect

## Comprehension XII

**Silicon carbide (carborundum) (SiC) and diamond are covalent solids which crystallize in cubic structures. In SiC, carbon atoms occupy lattice points of the FCC lattice and silicon atoms occupy half of the tetrahedral voids available. In diamond, the same tetrahedral voids are occupied by carbon atoms. The densities of SiC and diamond are 3.2 and 3.6 g/cm$^3$, respectively (Si = 28, $N_A = 6 \times 10^{23}$).**

**34.** The radius of carbon atom is

(a) 0.76 Å (b) 1.12 Å
(c) 3.2 Å (d) 3.6 Å

**35.** The radius of silicon atom is

(a) 0.76 Å (b) 1.12 Å
(c) 3.2 Å (d) 3.6 Å

**36.** Which of the following will not change the density of solid SiC?

(a) Substitution of some Si-atoms by some C-atoms.
(b) Schottky defect
(c) Interchange in the position of Si and C-atoms.
(d) Decrease in the temperature of solid.

**37.** If the similar volume of SiC and diamond are considered, then which of the following is a true statement?

(a) The number of Si-atoms in SiC is 3.75 times the number of C-atoms in diamond.
(b) The number of C-atoms in diamond is 3.75 times the number of Si-atoms in SiC.
(c) The number of SiC formula units is 3.75 times the number of C-atoms in diamond.
(d) The number of C-atoms in diamond is 3.75 times the total number of atoms in SiC.

**38.** Which of the following is incorrect statement?

(a) The packing efficiency of SiC is greater than that of diamond.
(b) The coordination number of Si and C-atoms in SiC is same.
(c) The number of C–C bonds per unit cell in diamond is 16.
(d) The coordination number of C-atoms in SiC is half that of C-atoms in diamond.

## Section D (Assertion – Reason)

The following questions consist of two statements. Mark the answer as follows.

**(a)** If both statements are CORRECT, and **Statement II** is the CORRECT explanation of **Statement I**.
**(b)** If both statements are CORRECT, and **Statement II** is NOT the CORRECT explanation of **Statement I.**
**(c)** If **Statement I** is CORRECT, but **Statement II** is INCORRECT.
**(d)** If **Statement I** is INCORRECT, but **Statement II** is CORRECT.

**1. Statement I:** In closed packing of spheres, a tetrahedral void is surrounded by four spheres, whereas octahedral void is surrounded by six spheres.

**Statement II:** A tetrahedral void has a tetrahedral shape, whereas octahedral void has an octahedral shape.

**2. Statement I:** In an FCC unit cell, packing efficiency is more when all tetrahedral voids are occupied by the spheres of maximum possible size as compared with the packing efficiency when all octahedral voids are occupied in similar way.

**Statement II:** Tetrahedral voids are more in number than octahedral voids in FCC.

**3. Statement I:** Distance between the nearest lattice points in BCC is greater than the same in FCC having the same edge length.

**Statement II:** FCC has greater packing efficiency than BCC.

**4. Statement I:** Only metallic, never ionic or covalent substances, form crystals exhibiting the maximum coordination number of twelve among like-sized atoms.

**Statement II:** In metallic crystals, the forces are non-directional and hence, the arrangement of atoms frequently corresponds to the closest packing of spheres.

5. **Statement I:** In a closely packed structure, the number of octahedral voids is equal to the number of atoms present.

   **Statement II:** Octahedral voids are located midway between the two closest packed layers.

6. **Statement I:** In a closest packed structure, the number of tetrahedral voids is twice the number of atoms present.

   **Statement II:** Tetrahedral voids are located above each atom in the first layer and below each atom in the second layer.

7. **Statement I:** Atomic solids having a particular type of unit cells have fixed density irrespective of the identity of atoms.

   **Statement II:** Atomic solids crystallizing in a particular type of unit cell have a fixed value of packing fraction irrespective of the identity of atoms.

8. **Statement I:** When an element crystallizes in either CCP or HCP arrangement, the density of solid remains the same in both the arrangements.

   **Statement II:** Both CCP and HCP arrangements have the same packing fractions.

9. **Statement I:** The melting point of diamond is very high.

   **Statement II:** Diamond has very high packing efficiency.

10. **Statement I:** In sodium chloride crystal, the $Na^+$ ion occupies the octahedral void while $Cl^-$ ions occupy the vertices of octahedron.

    **Statement II:** The radius ratio ($r_{Na^+}/r_{Cl^-}$) lies between 0.414 and 0.732.

11. **Statement I:** In any ionic solid (MX) with Schottky defects, the number of positive and negative ions are same.

    **Statement II:** Equal numbers of cation and anion vacancies are present.

12. **Statement I:** Anion vacancies in alkali metal halides are produced by heating the alkali metal halide crystals with alkali metal vapour.

    **Statement II:** Electrons trapped in anion vacancies are referred to as F-centres.

13. **Statement I:** Due to Frenkel defect, there is no effect on the density of the crystalline solid.

    **Statement II:** In Frenkel defect, no cation or anion leaves the crystal.

14. **Statement I:** Frenkel defects are found in silver halides.

    **Statement II:** Frenkel defects are commonly found in ionic solids having cations sufficiently smaller than anions.

15. **Statement I:** Increasing temperature increases the density of point defects.

    **Statement II:** The process of formation of point defects in solids is endothermic and has $\Delta S > 0$.

16. **Statement I:** KCl is more likely to show Schottky defect while LiI is more likely to show Frenkel defect.

    **Statement II:** Schottky defect is more likely in ionic solids in which cations and anions are of comparable size while Frenkel defect is more likely in ionic solids in which cations and anions have large differences in their ionic sizes.

17. **Statement I:** Conductivity of silicon increases by doping it with group 15 elements.

    **Statement II:** Doping means introduction of small amount of impurities like P, As or Bi into the pure silicon crystal.

18. **Statement I:** On heating ferromagnetic or ferrimagnetic substances they become paramagnetic.

    **Statement II:** The electrons change their spin on heating.

19. **Statement I:** Anti-ferromagnetic substances possess zero magnetic moment.

    **Statement II:** MnO is an antiferromagnetic substance.

20. **Statement I:** The substances whose resistance reduces nearly to zero are called superconductors.

    **Statement II:** Superconductors are diamagnetic substances.

## Section E (Column Match)

**1.** Match Column I with Column II and Column III ($l$ = Edge length of unit cell, $r$ = radius of spherical constituent unit).

| Column I | Column II | Column III |
|---|---|---|
| (A) Simple cubic unit cell | (P) $l\sqrt{2} = 4r$ | (U) 74% occupied space |
| (B) FCC unit cell | (Q) $l = 2r$ | (V) 67.98 occupied space |
| (C) BCC unit cell | (R) $l\sqrt{3} = 4r$ | (W) 52.33% occupied space |

**2.** Match Column I with Column II and Column III.

| Column I | Column II | Column II |
|---|---|---|
| (A) NaCl | (P) FCC, anion in all tetrahedral voids. | (W) Cation (6), anion (6) |
| (B) $CaF_2$ | (Q) FCC, cation in all octahedral voids. | (X) Cation (8), anion (4) |
| (C) ZnS (zinc blende) | (R) FCC, cation in alternate tetrahedral voids. | (Y) Cation (4), anion (8) |
| (D) $Na_2O$ | (S) FCC, cation in all tetrahedral voids. | (Z) Cation (4), anion (4) |

**3.** Associate each type of solid in the left-hand column with the properties in the right-hand column. Each property may be used more than once.

| Column I | Column II |
|---|---|
| (A) Molecular solid | (P) Low-melting |
| (B) Ionic solid | (Q) High-melting |
| (C) Metallic solid | (R) Brittle |
| (D) Covalent network solid | (S) Malleable hard |
| | (T) Electrically conducting |

**4.** Match the solids in Column I with the properties in Column II.

| Column I | Column II |
|---|---|
| (A) Germanium | (P) A non-conducting solid becoming good conductor on melting. |
| (B) Potassium chloride | (Q) A high melting solid involving covalent bonded atoms. |
| (C) Sodium | (R) A solid melting far below room temperature and held together by van der Waals forces. |
| (D) Solid methane | (S) A solid having melting point about 373 K and very good conductor of electricity. |

**5.** Match the solids in Column I with the properties in Column II.

| Column I | Column II |
|---|---|
| (A) Cobalt | (P) A white solid melting at 613°C; the liquid is electrically conducting, although the solid is not. |
| (B) Iodoform | (Q) A very hard, blackish solid subliming at 2700°C. |
| (C) Lithium chloride | (R) A yellow solid with a characteristic odour having a melting point of 120°C. |
| (D) Silicon carbide | (S) A grey, lustrous solid melting at 1495°C; both the solid and liquid are electrical conductors. |

**6.** Match the columns ('$l$' is the unit cell edge length).

| Column I | Column II |
|---|---|
| (A) 0.866 $l$ | (P) Shortest distance between cation and anion in CsCl crystal. |
| (B) 0.707 $l$ | (Q) Shortest distance between two cations in $CaF_2$ crystal. |
| (C) 0.433 $l$ | (R) Shortest distance between carbon atoms in diamond. |
| | (S) Shortest distance between two cations in rock salt crystal. |

7. Match the columns.

| Column I | Column II |
|---|---|
| (A) Rock salt structure | (P) Coordination number of cation is 4. |
| (B) Zinc blend structure | (Q) $\frac{\sqrt{3}\,a}{4} = r^+ + r^-$ |
| (C) Fluorite structure | (R) Coordination number of cation and anion is same. |
| | (S) Distance between two nearest anion is $\frac{a}{\sqrt{2}}$. |

8. Match the crystal system/unit cells mentioned in Column I with their characteristic features mentioned in Column II.

| Column I | Column II |
|---|---|
| (A) Simple cubic and FCC | (P) have these cell parameters $a = b = c$ and $\alpha = \beta = \gamma$. |
| (B) Cubic and rhombohedral | (Q) are two crystal systems. |
| (C) Cubic and tetragonal | (R) have only two crystallographic angles of 90°. |
| (D) Hexagonal and monoclinic | (S) belongs to same crystal system. |

9. Match Column I having different characteristics with Column II having different type of unit cell.

| Column I | Column II |
|---|---|
| (A) At least two interfacial angles be 90°. | (P) Unit cell of monoclinic sulphur. |
| (B) At least two crystallographic axis length be same. | (Q) Unit cell of diamond. |
| (C) None of the axial lengths to be same. | (R) Orthorhombic crystal. |
| (D) None of the interfacial angles to be same. | (S) Tetragonal crystal. |
| | (T) Triclinic crystal. |

10. Match the columns (assume the unit cell edge length equal to $a$).

| Column I | Column II |
|---|---|
| (A) $a$ | (P) Two times the smallest distance between two similarly charged ions in fluorite structure. |
| (B) $\frac{a}{\sqrt{2}}$ | (Q) Distance between the next nearest ions in NaCl structure |
| (C) $\sqrt{3}\,a$ | (R) Distance between the next nearest ions in CsCl structure. |
| (D) $\sqrt{2}\,a$ | (S) Four times the smallest distance of tetrahedral voids from the corner of FCC unit cell. |
| | (T) Smallest distance between the nearest similarly charged ions in antifluorite structure. |

11. Match the columns (assume the unit cell edge length equal to $a$).

| Column I | Column II |
|---|---|
| (A) $a$ | (P) Distance between the nearest particle in simple cubic crystal. |
| (B) $\frac{a}{\sqrt{2}}$ | (Q) Distance between the nearest particle in BCC crystal. |
| (C) $\frac{\sqrt{3}}{2}a$ | (R) Distance between the nearest particle in FCC crystal. |
| (D) $\sqrt{2}\,a$ | (S) Distance between the next nearest particle in simple cubic crystal. |
| | (T) Distance between the next nearest particle in BCC crystal. |

## Section F (Subjective)

### Single-digit Integer Type

1. The density of crystalline sodium is 2.167 g/cm$^3$. What would be the side length (in cm) of cube containing one mole of NaCl?

2. Iron(II) oxide crystal has a cubic structure and each edge of the unit cell is 0.493 nm. If the density of the oxide as 4.0 g/cm$^3$, then the number of iron (II) oxide formula units in each unit cell is (Fe = 56)

3. Methane crystallizes in a cubic unit cell with edge length 0.6 nm. The density of solid $CH_4$ is only slightly greater than the density of liquid $CH_4$. If the density of liquid $CH_4$ is 0.5 g/cm$^3$, then the number of $CH_4$ molecules per unit cell is

4. Ice crystallizes in a hexagonal lattice. At low temperatures at which the structure was determined, the lattice constants were $a$ = 4.53 Å and $b$ = 7.41 Å as shown in the figure. How many $H_2O$ molecules are contained in an unit cell? (Given that density of ice = 0.92 g/cm$^3$)

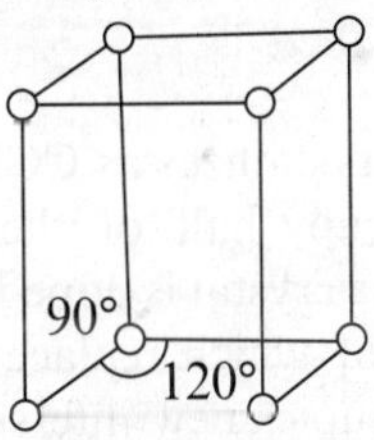


5. The density of graphite is 2.4 g/ml and the spacing between the layers is found by X-ray diffraction to be $2\sqrt{3}$ Å. If the carbon–carbon distance in the molecular layers is $x$ Å, then the value of $1.08 \times x^2$ is

6. X-ray diffraction analysis of crystalline $C_{60}$ (Buckminsterfullerene) shows that the crystal structure at 300 K can be regarded as FCC with a one-molecule basis and unit cell edge length of $\sqrt{2}$ nm. The molecules are orientationally disordered due to rotation. The centre-to-centre distance (in nm) between the nearest neighbour $C_{60}$ molecules in the solid at 300 K is

7. KF crystallizes in the NaCl type structure. If the radius of $K^+$ ion is $\frac{186}{\sqrt{2}}$ pm and $F^-$ ion is $\frac{214}{\sqrt{2}}$ pm, then what is the shortest distance between $K^+ - K^+$ ions (in Å)?

8. X-ray diffraction shows that the unit cell edge length of NaCl is 500 pm. If the measured density of NaCl is 2.964 g/ml, what percent of $Na^+$ and $Cl^-$ ions must be missing? ($N_A = 6 \times 10^{23}$)

9. The following are some statements related with solid state of matter:
   (i) All crystalline solids are isotropic.
   (ii) Amorphous solids are super cooled liquids with high viscosity.
   (iii) In crystals, short range order exists.
   (iv) Unit cell is the smallest building unit in the crystals.
   (v) Every object possesses an identity element.
   (vi) Each lattice point in a crystal has the same environment.
   (vii) The basic unit must have the same stoichiometric composition as the entire crystal.
   (viii) A crystal spacc lattice cannot have a positive ion at one lattice point and a negative ion at another lattice point.

   The number of true statements is

10. A metal 'M' (atomic mass = 31.25) crystallizes in CCP but it has some vacancy defect. If the edge length of the unit cell is 500 pm and the density of the metal is 1.6075 g/cm$^3$, then the number of moles of metal atoms missing per litre of the crystal is (1 amu = $1.67 \times 10^{-24}$ g)

11. The total number of ions per unit cell in sphalerite (zinc blend) structure if there is a Schottky defect of one anion per unit cell, is

12. The number of hexagonal faces that are present in a truncated octahedron is

13. A metal (atomic mass = 75) exist in BCC structure. If the uncovered distance between the atoms along the edge is equal to 67 pm, then the density of crystal (in g/cm$^3$) is ($N_A = 6 \times 10^{23}$, $\sqrt{3} = 1.732$)

**14.** A strong current of trivalent gaseous boron passed through a silicon crystal decreases the density of the crystal due to part replacement of silicon by boron and due to interstitial vacancies created by missing Si atoms. In one such experiment, one gram of silicon is taken and the boron atoms are found to be 1000 ppm by mass when the density of the Si crystal decreases by 12%. The percentage of missing vacancies due to silicon, which are filled up by boron atoms is (Atomic masses: Si = 30, B = 11)

**15.** The number of square faces in a truncated octahedron is '$x$' and the number of carbon atoms per unit cell in diamond is '$y$', then the value of $(y - x)$ is

## Four-digit Integer Type

**1.** Gold has a closely packed structure which can be viewed as spheres occupying 0.74 of the total volume. If the density of gold is 19.7 g/ml, then the atomic radius of gold (in pm) is [Au = 197, $(1.43)^3 \times 4\pi = 37$, $N_A = 6 \times 10^{23}$]

**2.** The approximate percentage of vacant space in a silicon cubic cell having crystal structure similar to diamond is

**3.** Insulin forms crystals of orthorhombic type with unit cell dimensions of $12.5 \times 8.0 \times 3.0$ nm$^3$. If the density of the crystal is $1.5 \times 10^3$ kg/m$^3$ and there are six insulin molecules per unit cell, the molar mass of insulin (in kg/mol) is ($N_A = 6 \times 10^{23}$)

**4.** The olivine series of minerals consists of crystals in which Fe and Mg ions may substitute for each other. The density of forsterite ($Mg_2SiO_4$) is 3.3 g/ml and that of fayalite ($Fe_2SiO_4$) is 4.2 g/ml. What is the percentage of fayalite in an olivine with a density of 3.9 g/ml?

**5.** The unit cell of $TlAl(SO_4)_2{\cdot}xH_2O$ is FCC with $a = 1.26$ nm. If density = 2.14 g/ml, then the value of '$x$' is [Tl = 207, Al = 27, $(1.26)^3 = 2$, $N_A = 6 \times 10^{23}$]

**6.** The density of diamond is $2\sqrt{3}$ g/cm$^3$ at 25°C. The carbon-carbon bond distance (in pm) in diamond is [$(1.55)^3 = 3.75$, $N_A = 6 \times 10^{23}$]

**7.** You are given a marble of diameter 10 mm. They are to be placed such that their centres are lying in a square bond by four lines each of length 40 mm. The maximum number of marbles which may be placed inside this area is

**8.** The edge length of the unit of LiCl (NaCl-like structure) is 520 pm. Assuming that the lithium ion is small enough that the chloride ions are in contact, the ionic radius of chloride ion (in pm) is $\left(\frac{1}{\sqrt{2}} = 0.7\right)$

**9.** Rutile, a mineral that contains only titanium and oxygen atoms has structure which can be described as a closely packed array of oxide ions with titanium ions in one-half of the octahedral voids (Ti = 48).

$ab$ = the mass percentage of titanium in rutile.

$c$ = magnitude of oxidation state of titanium in rutile.

$d = 1$, if the oxidation state of Ti is positive and 2 if the oxidation state of Ti is negative.

The value of '$abcd$' is

**10.** A solid 'AB' crystallizes as CCP for 'A' atoms and 'B' atoms occupy half of the tetrahedral voids. One litre of the crystal is doped with 1 mole atoms of 'C', some of which replace the 'B' atoms and remaining occupies new interstitial voids without affecting the dimensions of the crystal. If the density of the crystal before doping is 4.8 g/cm$^3$ and the density after doping is 4.795 g/cm$^3$, then the percentage of 'C' atoms which replace 'B' atoms is (Atomic masses of A, B and C are 40, 30 and 15, respectively)

**11.** An ideal crystal AB has rock salt structure in which $A^+$ is occupying octahedral voids. The crystal is doped with atoms of 'C' which occupy all the tetrahedral voids without distorting the lattice. If the crystal shows defect such that the body-centred atom is missing, then the percentage of body diagonal covered by ions/atoms in the defective crystal is

**12.** A non-stoichiometric compound $Fe_7S_8$ consists of iron in both $Fe^{2+}$ and $Fe^{3+}$ forms and sulphur is present as sulphide ions. If the percentage of cation vacancies relative to $Fe^{2+}$ ions present initially (as ideal crystal of FeS) is '$x$', then the value of '**10$x$**' is

**13.** The density of solid argon is $4/\pi$ g/cm$^3$ at $-233°C$. If the argon atoms are assumed to be spheres of radius $1.5 \times 10^{-8}$ cm, the approximate percentage of empty space in solid argon is (Ar = 40)

**14.** An element (atomic mass = 125) crystallizes in simple cubic structure. If the diameter of the largest sphere which can be placed in the crystal without disturbing the crystal is 366 pm and the density of crystal is '$d$' g/cm$^3$, then the value of '300 $d$' is ($N_A = 6 \times 10^{23}$)

**15.** The structure of crystalline macromolecules may be determined by X–ray diffraction techniques using method similar to those for smaller molecules. Fully crystalline polyethylene has its chain aligned in an orthorhombic unit cell of dimensions 667 pm × 500 pm × 280 pm. There are two repeating $CH_2CH_2$ units per unit cell. The theoretical density of fully crystalline polyethylene (in kg/m$^3$) is ($N_A = 6 \times 10^{23}$)

## Answer Keys

**Exercise I**

### Basics

1. (d) 2. (d) 3. (d) 4. (b) 5. (b) 6. (d) 7. (c) 8. (a) 9. (b) 10. (a)
11. (a) 12. (b) 13. (b) 14. (b) 15. (c) 16. (c) 17. (c) 18. (b) 19. (b) 20. (d)

### Analysis of Cubic Crystals

21. (a) 22. (b) 23. (c) 24. (c) 25. (d) 26. (a) 27. (c) 28. (b) 29. (a) 30. (a)
31. (a) 32. (d) 33. (a) 34. (c) 35. (a) 36. (a) 37. (a) 38. (c) 39. (b) 40. (d)

### Types of Packing

41. (a) 42. (c) 43. (b) 44. (b) 45. (a) 46. (a) 47. (b) 48. (d) 49. (d) 50. (c)
51. (a) 52. (b) 53. (a) 54. (c) 55. (a)

### Radius Ratio and Packing of Crystal

56. (b) 57. (b) 58. (c) 59. (c) 60. (a) 61. (c) 62. (b) 63. (b) 64. (d) 65. (a)
66. (b) 67. (a) 68. (b) 69. (a) 70. (b) 71. (b) 72. (a) 73. (d) 74. (b) 75. (c)

### Defects in Solids

76. (d) 77. (b) 78. (d) 79. (c) 80. (c) 81. (c) 82. (b) 83. (b) 84. (a) 85. (c)

### Properties of Solids

86. (d) 87. (a) 88. (d) 89. (a) 90. (d)

## Answer Keys

**Exercise II**

### Section A (Only one Correct)

1. (b) 2. (b) 3. (a) 4. (b) 5. (c) 6. (b) 7. (d) 8. (c) 9. (a) 10. (b)
11. (c) 12. (b) 13. (c) 14. (b) 15. (d) 16. (c) 17. (c) 18. (b) 19. (b) 20. (c)
21. (d) 22. (c) 23. (a) 24. (a) 25. (b) 26. (b) 27. (d) 28. (a) 29. (b) 30. (a)
31. (a) 32. (a) 33. (d) 34. (b) 35. (a) 36. (a) 37. (b) 38. (d) 39. (d) 40. (a)

### Section B (One or More than one Correct)

1. (a), (c) 2. (c), (d) 3. (c), (d) 4. (a), (d)
5. (a), (b), (d) 6. (a), (b), (c), (d) 7. (a), (b), (d) 8. (c), (d)
9. (a), (c), (d) 10. (b) 11. (b), (c), (d) 12. (a), (b), (c), (d)
13. (a), (b), (c), (d) 14. (a), (b) 15. (a), (b), (d) 16. (b), (c)
17. (b), (c), (d) 18. (b), (c) 19. (c) 20. (b), (c)
21. (a), (b), (c) 22. (a), (c) 23. (b), (c) 24. (c), (d)
25. (c), (d)

### Section C

**Comprehension I**

1. (b) 2. (a) 3. (a) 4. (b) 5. (c)
6. (c)

**Comprehension II**

7. (d) 8. (a)

**Comprehension III**

9. (a) 10. (b) 11. (c) 12. (c)

**Comprehension IV**

13. (b) 14. (b)

**Comprehension V**

15. (a) 16. (b)

**Comprehension VI**

17. (b) 18. (a)

**Comprehension VII**

19. (a) 20. (a) 21. (c) 22. (b)

**Comprehension VIII**

23. (c) 24. (b) 25. (b) 26. (a)

**Comprehension IX**

27. (b) 28. (b) 29. (c)

**Comprehension X**

30. (a) 31. (b)

**Comprehension XI**

32. (b) 33. (d)

**Comprehension XII**

34. (a) 35. (b) 36. (c) 37. (d) 38. (d)

### Section D (Assertion – Reason)

1. (c) 2. (d) 3. (a) 4. (a) 5. (a) 6. (a) 7. (d) 8. (a) 9. (c) 10. (a)
11. (a) 12. (b) 13. (a) 14. (b) 15. (a) 16. (a) 17. (b) 18. (a) 19. (b) 20. (b)

## Section E (Column Match)

1. $A \rightarrow Q, W; B \rightarrow P, U; C \rightarrow R, V$
2. $A \rightarrow Q, W; B \rightarrow P, X; C \rightarrow R, Z; D \rightarrow S, Y$
3. $A \rightarrow P; B \rightarrow Q, R; C \rightarrow Q, S, T; D \rightarrow Q, R$
4. $A \rightarrow Q; B \rightarrow P; C \rightarrow S; D \rightarrow R$
5. $A \rightarrow S; B \rightarrow R; C \rightarrow P; D \rightarrow Q$
6. $A \rightarrow P; B \rightarrow Q, S; C \rightarrow R$
7. $A \rightarrow R, S; B \rightarrow P, Q, R, S; C \rightarrow Q$
8. $A \rightarrow P, S; B \rightarrow P, Q; C \rightarrow Q; D \rightarrow Q, R$
9. $A \rightarrow P, Q, R, S; B \rightarrow Q, S; C \rightarrow P, R, T; D \rightarrow T$
10. $A \rightarrow P, Q, R; B \rightarrow T; C \rightarrow S; D \rightarrow P$
11. $A \rightarrow P, T; B \rightarrow R; C \rightarrow Q; D \rightarrow S$

## Section F (Subjective)

**Single-digit Integer Type**

| | | | | | | | | | |
|---|---|---|---|---|---|---|---|---|---|
| 1. (3) | 2. (4) | 3. (4) | 4. (4) | 5. (2) | 6. (1) | 7. (4) | 8. (5) | 9. (5) | 10. (2) |
| 11. (6) | 12. (8) | 13. (2) | 14. (2) | 15. (2) | | | | | |

**Four-digit Integer Type**

| | | | | |
|---|---|---|---|---|
| 1. (0143) | 2. (0066) | 3. (0045) | 4. (0072) | 5. (0012) |
| 6. (0155) | 7. (0018) | 8. (0182) | 9. (6041) | 10. (0067) |
| 11. (0059) | 12. (0125) | 13. (0073) | 14. (4000) | 15. (1000) |

# HINTS AND EXPLANATIONS

## EXERCISE I (JEE MAIN)

### Basics

1. Theory based
2. Amorphous solids are isotropic.
3. Melting point is very low.
4. Graphite is covalent solid.
5. Quartz is a covalent solid having network of —O—Si—O— (with bonds above and below Si)
6. Ionic compounds are normally soluble in polar solvents.
7. Tungston, W
8. $SiO_2$ has a network of —O—Si—O— (with bonds above and below Si)
9. At absolute zero, the particles have no thermal energy.
10. $I_2$ is a non-polar molecule and hence, London force.
11. Cubic
12. Triclinic
13. Orthorhombic ($a \neq b \neq c, \alpha = \beta = \gamma = 90°$)
14. Hexagonal ($\alpha = \beta = 90°, \gamma = 120°$)
15. Theory based
16. Informative
17. Informative
18. Informative
19. Informative
20. Informative

### Analysis of Cubic Crystals

21. For BCC : $Z = 2$

$\therefore$ Number of atoms $= 2 \times 1.208 \times 10^{23} = 2.416 \times 10^{23}$

22. $Z = 4 \Rightarrow$ FCC

23. $Z_{Cu} = 8 \times \frac{1}{8} + 6 \times \frac{1}{2} = 4$

$Z_{Ag} = 12 \times \frac{1}{4} = 3$

$Z_{Au} = 1$

$\therefore$ Formula: $Cu_4Ag_3Au$

24. Q form FCC and P is at body center (missing) and edge center. Now,

$Z_Q = 8 \times \frac{1}{8} + 6 \times \frac{1}{2} = 4$

$Z_P = 12 \times \frac{1}{4} = 3$

$\therefore$ Formula: $P_3Q_4$

25. 'A' atoms are at corners and hence, at face centers. Along any one axis, there are two faces and hence two 'A' atoms are missing.

$Z_A = 8 \times \frac{1}{8} + 4 \times \frac{1}{2} = 3$

$Z_B = 1 + 12 \times \frac{1}{4} = 4$

$\therefore$ Formula: $A_3B_4$

26. $\sqrt{2}a = 4r \Rightarrow r = \frac{\sqrt{2} \times 620}{4} = 219.20\,\text{pm}$

27. Closest distance between two atoms in FCC

$= \frac{a}{\sqrt{2}} = \frac{4.07}{\sqrt{2}} = 2.857\,\text{Å}$

28. Closest distance between two atoms in BCC

$= \frac{\sqrt{3}\,l}{2}$

29. $\sqrt{3}\,a = 4r \Rightarrow r = \frac{\sqrt{3} \times 4.29}{4} = 1.86\,\text{Å}$

**30.** $d = \dfrac{Z \cdot M}{N_A \cdot V}$

$\Rightarrow 10.5 = \dfrac{Z \times 108}{(6.022 \times 10^{23}) \times (409 \times 10^{-10})^3}$

$\therefore Z = 4 \Rightarrow \text{FCC}$

**31.** $d = \dfrac{Z \cdot M}{N_A \cdot V} = \dfrac{2 \times 100}{(6 \times 10^{23}) \times (400 \times 10^{-10})^3}$

$= 5.2 \text{ gm/cm}^3$

**32.** $\dfrac{\sqrt{3}\,a}{2} = 2.5 \times \sqrt{3}\text{ Å} \Rightarrow a = 5\text{ Å}$

Now, $d = \dfrac{Z \cdot M}{N_A \cdot V} = \dfrac{2 \times 39}{(6 \times 10^{23}) \times (5 \times 10^{-8})^3}$

$= 1.04 \dfrac{\text{gm}}{\text{cm}^3} = 1040 \text{ kg/m}^3$

**33.** $d = \dfrac{Z \cdot M}{N_A \cdot V}$

$= \dfrac{4 \times (7 + 11 + 4 \times 1)}{(6 \times 10^{23}) \times (6 \times 10^{-8} \times 4.4 \times 10^{-8} \times 7.5 \times 10^{-8})}$

$= 0.74 \text{ g/cm}^3$

**34.** $d = \dfrac{Z.M}{N_A.V}$

$\Rightarrow 4.0 = \dfrac{4 \times 72}{(6 \times 10^{23}) \times a^3}$

$\Rightarrow a = 4.932 \times 10^{-8} \text{ cm}$

**35.** $\dfrac{d_\gamma}{d_\alpha} = \dfrac{(Z.M/N_A \cdot V)_\gamma}{(Z.M/N_A \cdot V)_\alpha} = \dfrac{4/(4r/\sqrt{2})^3}{2/(4r/\sqrt{3})^3} = \dfrac{4\sqrt{2}}{3\sqrt{3}}$

**36.** Number of atoms $= Z \times 2 = 4 \times 2 = 8$

**37.** $\text{P.F.} = \dfrac{V_{\text{particles}}}{V_{\text{unit cell}}} = \dfrac{1 \times \frac{4}{3}\pi r^3}{(2r)^3} = \dfrac{\pi}{6}$

**38.** $\text{P.F.} = \dfrac{V_{\text{particles}}}{V_{\text{unit cell}}} = \dfrac{2 \times \frac{4}{3}\pi r^3}{\left(\frac{4r}{\sqrt{3}}\right)^3} = \dfrac{\sqrt{3}\pi}{8}$

**39.** $\text{P.F.} = \dfrac{V_{\text{particles}}}{V_{\text{unit cell}}} = \dfrac{4 \times \frac{4}{3}\pi r^3}{\left(\frac{4r}{\sqrt{2}}\right)^3} = \dfrac{\pi}{3\sqrt{2}}$

**40.** Theory based

## Types of Packing

**41.** Malleability increases with increase in C.N. in metallic solids.

**42.** Carbon atoms form FCC and also occupy alternate tetrahedral voids.

$Z = 8 \times \dfrac{1}{8} + 6 \times \dfrac{1}{2} + 4 = 8$

**43.** C.N. of C-atom in diamond = 4

Each atom is tetrahedrally bonded with 4 C-atoms.

**44.** C.N. of C-atom in graphite = 3

**45.** Informative

**46.** Informative

**47.** HCP $\Rightarrow$ Void space $\simeq 26\%$

**48.** For closest packing, each successive layer should be different.

**49.** FCC and HCP (P.E. = 74%)

**50.** $Z = 6 \Rightarrow \text{T.V.} = 2 \times 6 = 12$ and $\text{O.V.} = 1 \times 6 = 6$

**51.** Informative

**52.** 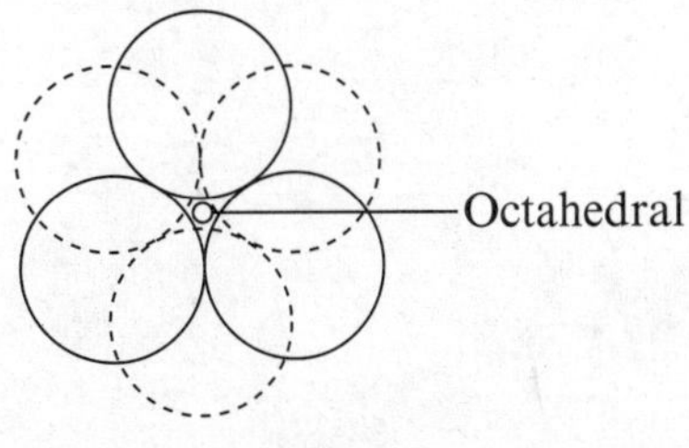


**53.** $\dfrac{r_{\text{tetrahedral}}}{r_{\text{octahedral}}} = \dfrac{0.225\text{ R}}{0.414\text{ R}} = 0.543$

54. $Z_A = 6$ and $Z_C = 6\times\frac{2}{3} = 4$

$\therefore$ Formula = $A_6C_4 = A_3C_2$

55. $Z_X = 4$; $Z_Y = 2\times 4 = 8$; $Z_Z = \frac{1}{2}\times 4 = 2$

$\therefore$ Formula = $X_6Y_4Z_2 = X_2Y_4Z$

## Radius Ratio and Packing of Crystal

56. $\frac{r_{Rb^+}}{r_{I^-}} = \frac{1.46}{2.16} = 0.676 < 0.732$

$\Rightarrow$ NaCl structure

57. Cation is in octahedral void and hence, $r_{A^+} : r_{X^-}$ lies between 0.414 to 0.732.

58. Informative

59. The next nearest neighbour is $Cs^+$ ion at '$a$' distance and hence, its number is 6.

60. Informative

61. $\frac{r_{Ag^+}}{r_{I^-}} = \frac{126}{216} = 0.583$

$\Rightarrow$ Ochahedral void (0.414 – 0.732)

$\therefore$ C.N. of $Ag^+ = 6$

62. For rock salt structure, $a = 2\,(r^+ + r^-)$

or, $400 = 2\left(75 + r^-\right)$

$\Rightarrow r^- = 125$ pm

63. C.N. of $Mg^{2+}$ ion = 6

64. $Na^+$ ions occupy all the tetrahedral voids.

65. $\sqrt{3}a = 2\left(r_{Cs^+} + r_{Br^-}\right)$

$\therefore$ Shortest distance between ions, $r_{Cs^+} + r_{Cl^-} = \frac{\sqrt{3}a}{2}$

$= \frac{\sqrt{3}\times 4.3}{2} = 3.72\,\text{Å}$

66. Informative

67. Informative

68. $d = \frac{Z\cdot M}{N_A\cdot V}$

$\Rightarrow 4.26 = \frac{1\times(133+80)}{\left(6\times 10^{23}\right)\times a^3}$

$\Rightarrow a = 4.368\times 10^{-8}$ cm

69. $Z_{B^{m-}} = 8\times\frac{1}{8} + 6\times\frac{1}{2} = 4$

$Z_{A^{n+}} = \frac{1}{4}\times 8 + \frac{1}{4}\times 4 = 3$

$\therefore$ Formula = $A_3B_4$

70. $Z_{Q^{2-}} = 6$

$Z_{P^{x+}} = \frac{1}{2}\times 12 = 6$

As compound should be neutral,

$6\times(-2) + 6\times(+x) = 0$

$\therefore x = 2$

71. $Z_O = 4$

$Z_A = \frac{1}{6}\times 8 = \frac{4}{3}$

$Z_B = \frac{1}{3}\times 4 = \frac{4}{3}$

$\therefore$ Formula = $A_{\frac{4}{3}}B_{\frac{4}{3}}O_4 \equiv ABO_3$

72. $Z_O = 1$ (say), then $Z_{Al} = \frac{2}{3}\times 1 = \frac{2}{3}$

$\therefore$ Formula = $Al_{\frac{2}{3}}O_1 \equiv Al_2O_3$

73. $Z_O = 1$ (say), then $Z_{Co} = \frac{1}{8}\times 2 + \frac{1}{2}\times 1 = \frac{3}{4}$

$\therefore$ Formula = $Co_{\frac{3}{4}}O_1 \equiv Co_3O_4$

74. Expansion may result decrease in C.N.

75. Compression may result increase in C.N.

## Defects in Solids

76. Informative

77. Informative

78. Informative

79. Informative

80. Informative

81. Let $x$ atoms are in +2 state.

$$x(+2)+(1.92-x)(+1)+1.00(-2)=0$$

$$\Rightarrow x=0.08$$

$$\therefore \frac{N_{cu^{2+}}}{N_{cu^{+}}}=\frac{0.08}{1.92-0.08}=\frac{1}{23}$$

82. One $Sr^{2+}$ ion will result removal of two $Na^+$ ion from crystal for neutrality of crystal. One position of $Na^+$ ion will be occupied by $Sr^{2+}$, but other position will remain vacant.

Number of cation vacancies = Number of $Sr^{2+}$ ions

$$=\frac{2\times10^{-3}}{100}\times6\times10^{23}=1.2\times10^{19}\text{ per mol}$$

83. Informative

84. Informative

85. In impurity defect, the density may increase or decrease or even may remain unchanged depending on impurity.

## Properties of Solids

86. Informative

87. Informative

88. Informative

89. Arsenic belongs to group 15.

90. Informative

# EXERCISE II (JEE ADVANCE)

## Section A (Only one Correct)

1. $a\neq b\neq c, \alpha=\beta=\gamma=90°$

$\Rightarrow$ Orthorhombic

2. Volume of metal taken $=\frac{m}{d}=\frac{100}{6.25}=16\text{ cm}^3$

Volume of each unit cell $=\left(4\times10^{-8}\text{ cm}\right)^3$

$=64\times10^{-24}\text{ cm}^3$

$\therefore$ Number of unit cells $=\frac{16}{64\times10^{-24}}=2.5\times10^{23}$

3. $d=\frac{Z\cdot M}{N_A\cdot V}$

$$\Rightarrow \frac{45}{16}=\frac{Z\times27}{\left(6\times10^{23}\right)\times\left(4\times10^{-8}\right)^3}$$

$$\Rightarrow Z=4$$

Hence, Al crystal is FCC.

For FCC: $\sqrt{2}a=4r$

$$\Rightarrow r=\frac{\sqrt{2}\times4.0}{4}=1.414\text{ Å}$$

4. $$d_{FCC}=\frac{Z\cdot M}{N_A\cdot V}=\frac{4\times56}{\left(6\times10^{23}\right)\left(\frac{4\times125\times10^{-10}}{\sqrt{2}}\right)^3}$$

$$=8.45\text{ gm/cm}^3$$

$$d_{BCC}=\frac{Z\cdot M}{N_A\cdot V}=\frac{2\times56}{\left(6\times10^{23}\right)\left(\frac{4\times50\sqrt{2}\times10^{-10}}{\sqrt{3}}\right)^3}$$

$$=42.87\text{ gm/cm}^3$$

As the density of iron is increased, there is contraction.

5. Packing fraction $= \dfrac{\left(6\times10^{23}\right)\times\frac{4}{3}\pi\times\left(144\times10^{-10}\right)^3}{\left(\frac{108}{10.6}\right)}$

$= 0.736$

Hence, the crystal should be FCC.

6. $d = \dfrac{Z\cdot M}{N_A\cdot V}$

$\Rightarrow 10.5 = \dfrac{4\times198}{N_A\times\left(5\times10^{-8}\right)^3}$

$\Rightarrow N_A = 6.034\times10^{23}$

7. Void space per unit cell $= 0.26\times V_{\text{unit cell}}$

$= 0.26\times\left(4\,\text{Å}\right)^3$

$= 16.64\,\text{Å}^3$

8. $d = \dfrac{Z\cdot M}{N_A\cdot V}$

$\Rightarrow 12.5\dfrac{\text{gm}}{\text{cm}^3} = \dfrac{4\times M\text{ gm}}{\left(6\times10^{23}\right)\times\left(\frac{4\times100\sqrt{2}\times10^{-10}}{\sqrt{2}}\right)^3\text{cm}^3}$

$\therefore M = 120 \Rightarrow$ Metal is Sn.

9. $\dfrac{\text{P.F.}}{d} = \dfrac{V_{\text{particle}}/V_{\text{unit cell}}}{m_{\text{particle}}/V_{\text{unit cell}}} = \dfrac{V_{\text{particle}}}{m_{\text{particle}}}$

or, $0.1 = \dfrac{Z\times\frac{4}{3}\pi\left(1.0\times10^{-8}\right)^3}{Z\times M/6\times10^{23}}$

$\Rightarrow M = 8\pi$

10. Fraction of edge covered by atoms

$= \dfrac{2r}{a} = \dfrac{2r}{4r/\sqrt{2}} = 0.707$

Hence, fraction of edge not covered = 0.293

11. $d = \dfrac{Z\cdot M}{N_A\cdot V}$

$\Rightarrow \dfrac{8}{3} = \dfrac{4\times M}{\left(6\times10^{23}\right)\times\left(5\times10^{-8}\right)^3}$

$\Rightarrow M = 50$ gm/mol

At 0°C, the substance will exist as gas and its density

$= \dfrac{50\text{ gm}}{22.4\text{ L}} = 2.23$ g/L

12. $\text{M} + \text{HCl} \rightarrow \text{MCl} + \frac{1}{2}\text{H}_2$

1 mole $\qquad \frac{1}{2}$ mole

= A gm $\qquad = \frac{1}{2}\times 22.7$ L at 0°C and 1 bar

$\therefore$ (7.68 × 4.5) gm $\qquad \dfrac{11.35}{\text{A}}\times(7.68\times4.5) = 4.54$

$\therefore \text{A} = 86.4$

Now, $d = \dfrac{Z\cdot M}{N_A\cdot V}$

$\Rightarrow 4.5 = \dfrac{Z\times86.4}{\left(6\times10^{23}\right)\times\left(400\times10^{-10}\right)^3}$

$\therefore Z = 2$

$\Rightarrow$ Unit cell is BCC.

13. Percentage of occupied space in water

$= x\times\dfrac{0.99}{0.96} = \dfrac{33x}{32}$

$\therefore$ Percentage of empty space in water is $100 - \dfrac{33x}{32}$.

14. (Volume of crystal containing one mole metal) $\times\dfrac{70}{100} = \left(6\times10^{23}\right)\times\frac{4}{3}\pi\times\left(0.2\times10^{-7}\text{ cm}\right)^3$

$\therefore V_{\text{crystal}} = \dfrac{64}{7}\pi\text{ cm}^3$

$\therefore$ Density $= \dfrac{32\pi}{\left(\frac{64}{7}\right)\pi} = 3.5\text{ gm/cm}^3$

15. There is no octahedral voids in BCC. However, all the face centres are distorted octahedral voids, which are not considered because they are not regular voids.

16. None of the tetrahedral as well as octahedral voids will be in contact with other tetrahedral and octahedral voids, respectively.

**17.** Packing is FCC for which $\sqrt{2}a = 4r$.

$$\therefore\ r = \frac{\sqrt{2}\times 10\,\text{Å}}{4} = 2.5\sqrt{2}\,\text{Å}$$

Now, density of metal atom $= \dfrac{m_{\text{atom}}}{V_{\text{atom}}}$

$$= \frac{60.22}{\left(6.022\times 10^{23}\right)\times \frac{4}{3}\pi\times\left(2.5\sqrt{2}\times 10^{-8}\right)^3}$$

$= 0.54\ \text{gm/cm}^3$

**18.** $$\text{P.F.} = \frac{V_{\text{particle}}}{V_{\text{unit cell}}} = \frac{3\times\frac{4}{3}\pi r^3}{\left\{6\times\frac{\sqrt{3}}{4}\times(2r)^2\right\}\times 2r} = \frac{\pi}{3\sqrt{3}}$$

**19.**

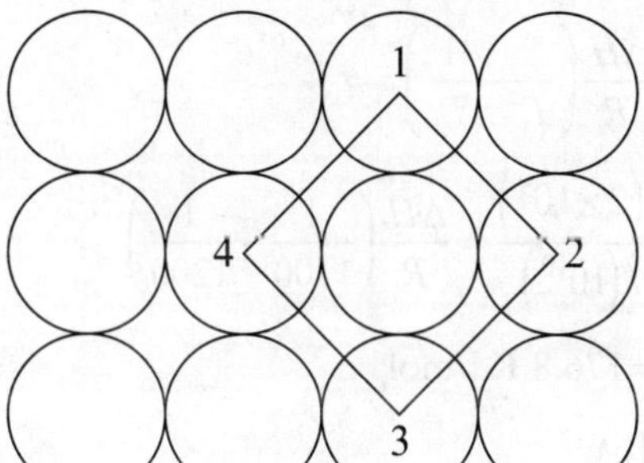


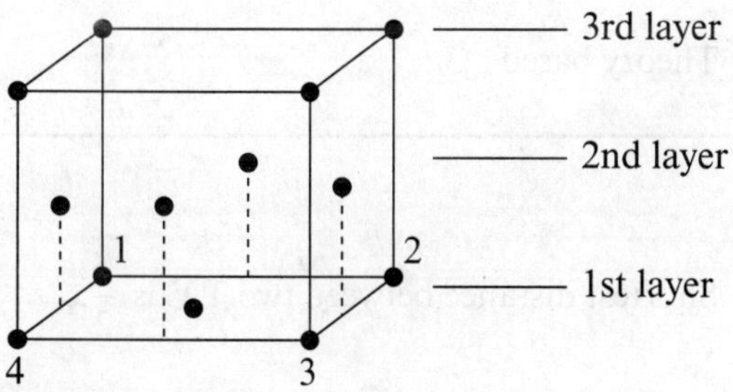


**20.**

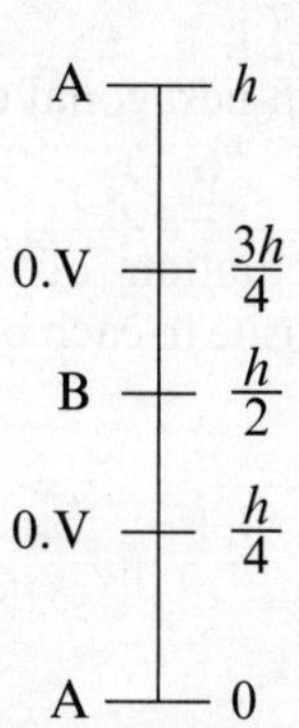


**21.** Number of NaCl formula units in 1 gm

$$= \frac{1}{58.5}\times\left(6\times 10^{23}\right)$$

Each unit cell contains 4 NaCl formula units and hence, volume of crystal

$$= \frac{6\times 10^{23}}{58.5\times 4}\times 4.7\times 10^{-23} = 0.12\,\text{ml}$$

**22.** $$d = \frac{Z\cdot M}{N_{\text{A}}\cdot V} = \frac{4\times 58.5}{\left(6\times 10^{23}\right)\times\left(600\times 10^{-10}\right)^3}$$

$$= 1.8\ \text{gm/cm}^3$$

**23.** The structure is simple cubic for both metals.

**24.** 'B' should occupy all the tetrahedral voids and hence, its C.N. = 4.

**25.** $$(\text{P.E.})_{\text{A}} = \frac{4\times\frac{4}{3}\pi r^3 + 8\times\frac{4}{3}\pi(0.225r)^3}{\left(\frac{4r}{\sqrt{2}}\right)^3} = 0.76$$

$$(\text{P.E.})_{\text{B}} = \frac{4\times\frac{4}{3}\pi r^3 + 4\times\frac{4}{3}\pi(0.414r)^3}{\left(\frac{4r}{\sqrt{2}}\right)^3} = 0.79$$

$$(\text{P.E.})_{\text{C}} = \frac{1\times\frac{4}{3}\pi r^3 + 1\times\frac{4}{3}\pi(0.732r)^3}{(2r)^3} = 0.72$$

**26.** $Z_{\text{O}^{2-}} = 4;\ Z_{\text{X}^{2+}} = 8\times\dfrac{x}{100};\ Z_{\text{Y}^{3+}} = 4\times\dfrac{50}{100} = 2$

For neutrality of crystal,

$$4\times(-2) + \frac{8x}{100}\times(+2) + 2\times(+3) = 0$$

$$\Rightarrow x = 12.5$$

**27.** $r_{\text{Rb}^+} + r_{\text{Cl}^-} = 328.5\,\text{pm} = x$

$r_{\text{K}^+} + r_{\text{Cl}^-} = 313.9\,\text{pm} = y$

$r_{\text{Na}^+} + r_{\text{Br}^-} = 298.1\,\text{pm} = z$

$r_{\text{K}^+} + r_{\text{Br}^-} = 329.3\,\text{pm} = w$

$\therefore\ r_{\text{Rb}^+} + r_{\text{Br}^-} = x + w - y = 343.9\,\text{pm}$

**28.** $\dfrac{r^+}{r^-} = 0.414$

$\Rightarrow r^- = \dfrac{200}{0.414} = 483.1\text{pm}$

**29.** $Z_A = 4;\quad Z_B = 2\times\dfrac{1}{2} = 1;\quad Z_C = 4\times\dfrac{1}{4} = 1$

$\therefore$ Formula = $A_4BC$

**30.** $Z_O = 4;$

$Z_{\text{Metal I(M)}} = \dfrac{1}{8}\times 8 = 1$

$Z_{\text{Metal II(N)}} = \dfrac{1}{2}\times 4 = 2$

$\therefore$ Formula = $MN_2O_4$

For neutrality, M = $Zn^{2+}$ and N = $Al^{3+}$

**31.** $Z_{M^{3+}} = 8\times\dfrac{1}{8} = 1$ (If $M^{3+}$ are at corners)

$Z_{X^-} = 6\times\dfrac{1}{2} = 3$ (If $F^-$ are at face centres)

**32.** Octahedral voids in FCC = 4, but only one is occupied by '*x*' and one by '*y*'.

**33.** (II) $\dfrac{r^+}{r^-} = \dfrac{20}{95} = 0.21\ (0.155-0.225) \Rightarrow \text{C.N.} = 3$

**34.**

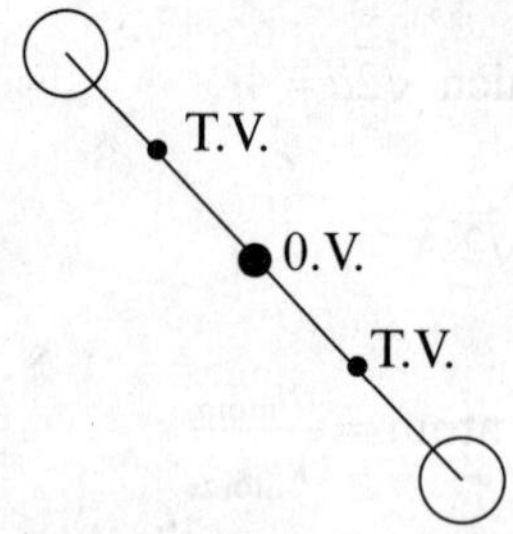


Fraction of body diagonal covered

$$= \frac{2r + 2\times 0.414r + 4\times 0.225r}{\sqrt{3}\times\left(\dfrac{4r}{\sqrt{2}}\right)} = 0.76$$

$\therefore$ Fraction, not covered = 1 – 0.76 = 0.24

**35.** For electrical neutrality, A = bivalent and B = trivalent. Now, one octahedral void is occupied by 'A' and one by 'B'.

**36.** Frenkel defect does not change the density.

**37.** $\ln\dfrac{f_2}{f_1} = \dfrac{\Delta H}{R}\left(\dfrac{1}{T_1} - \dfrac{1}{T_2}\right)$

or, $\ln\dfrac{1/(2\times 10^9)}{1/(10^{10})} = \dfrac{\Delta H}{R}\left(\dfrac{1}{1100} - \dfrac{1}{1200}\right)$

$\Rightarrow \Delta H = 176.8$ KJ/mol

**38.** Informative

**39.** Informative

**40.** Theory based

## Section B (One or More than one Correct)

1. Amorphous
2. Informative
3. Informative
4. 

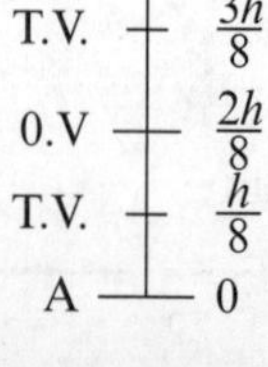


5. Shortest distance between two T.V. is $\dfrac{a}{2}$.
6. Both have same packing efficiency and C.N.
7. There are the voids per sphere in hexagonal close packing.
8. For octahedral void, the orientation of both tetrahedral voids should be opposite to each other.
9. $a = 2\left(r_{Na^+} + r_{Cl^-}\right)$
10. Informative
11. Informative
12. Informative

**13.** (a) $P^{H} = 7 + \frac{1}{2}\left(P^{K_b} + \log c\right)$

$\Rightarrow 5.0 = 7 - \frac{1}{2}(4.4 + \log c)$

$\therefore c = 0.4 = \frac{80/M}{2} \Rightarrow M = 100$

(b) $h = \sqrt{\frac{K_h}{K_b \times c}} = \sqrt{\frac{10^{-14}}{4\times10^{-5}\times0.4}} = 2.5\times10^{-5}$

(c) $\sqrt{3}\,a = 2\left(r_{x^+} + r_{y^-}\right)$

$\Rightarrow a = \frac{2\times(160+186.4)}{\sqrt{3}} = 400\text{ pm}$

(d) $d = \frac{ZM}{N_A \cdot V} = \frac{1\times100}{6\times10^{23}\times\left(400\times10^{-10}\right)^3}$

$= 2.6\text{ gm/cm}^3$

**14.** (a) $Z_{C^{3-}} = 4,\ Z_{A^+} = 4,\ Z_{B^{2+}} = \frac{1}{2}\times8 = 4$

Crystal is electrically neutral and hence, it is possible.

(b) $Z_{B^{2+}} = 6,\ Z_{A^+} = 6,\ Z_{C^{3-}} = \frac{1}{2}\times12 = 6$

Crystal is electrically neutral and hence, it is possible.

(c) $Z_{A^+} = 4\times\frac{1}{8} = \frac{1}{2},\ Z_{B^{2+}} = 4\times\frac{1}{8} = \frac{1}{2},\ Z_{C^{3-}} = 1$

Crystal is negatively changed and hence, it is not possible.

(d) $Z_{B^{2+}} = 4,\ Z_{C^{3-}} = 8,\ Z_{A^+} = 4$

Crystal is negatively changed and hence, it is not possible.

**15.** Informative

**16.** $CaF_2\ (8:4),\ Al_2O_3\ (6:4)$

**17.** Informative

**18.** Informative

**19.** Metal deficient defect can occur with extra anion present in the interstitial voids, but it is very rare.

**20.** Informative

**21.** $\Delta H$ = +ve and hence, the surroundings must lose heat.

**22.** $O^{2-} = CCP$

$Fe^{2+} = \frac{1}{4}\times\text{O.V.} = 1$

$Fe^{3+} = 1$ in T.V. and 1 in O.V.

**23.** Informative

**24.** Informative

**25.** Informative

## Section C (Comprehensions)

### Comprehension I

**1.** $\sqrt{3}a = 4r \Rightarrow 2r = \frac{\sqrt{3}\times5.0}{2} = 4.33\ \text{Å}$

**2.** Next nearest neighbours are at '$a$' distance.

**3.** C.N. = 8

**4.** Number of next nearest neighbours = 6

**5.** $d = \frac{Z\cdot M}{N_A\cdot V} = \frac{2\times39}{\left(6\times10^{23}\right)\times\left(5\times10^{-8}\right)^3} = 1.04\text{ gm/cm}^3$

**6.** Fractional space occupied by atoms

$= \frac{V_{\text{atoms}}}{V_{\text{liquid}}} = \frac{V_{\text{atoms/mol}}}{V_{\text{liquid/mol}}}$

$= \frac{\left(6\times10^{23}\right)\times\frac{4}{3}\pi\times\left(\frac{\sqrt{3}\times5\times10^{-8}}{4}\right)^3}{(39/0.9)} = 0.5883$

$\therefore$ Percentage of empty space = (1 – 0.5883) × 100 = 41.17%

HINTS AND EXPLANATIONS

## Comprehension II

7. $\sqrt{2}a = 4r \Rightarrow r = \dfrac{\sqrt{2}\times\dfrac{0.5}{\sqrt{2}}}{4} = 0.125\,\text{nm}$

Now, size of octahedral void

$= 0.414 \times 0.0125 = 0.052$ nm

8. Size of tetrahedral void $= 0.225 \times 0.0125$

$= 0.028$ nm

## Comprehension III

9. $\text{Volume} = \dfrac{\text{Mass}}{\text{Density}} = \dfrac{6\times 24}{(6\times 10^{23})\times 1.92}$

$= 1.25\times 10^{-22}\,\text{cm}^3$

or, $6\times\dfrac{4}{3}\pi r^3 = \dfrac{\pi}{3\sqrt{2}}\times 1.25\times 10^{-22}$

$\Rightarrow r = 1.5625\times 10^{-8}\,\text{cm}$

10. Volume occupied by particles $= \dfrac{\pi}{3\sqrt{2}}\times V_{\text{unit all}}$

11. Height of unit cell $= 4r\cdot\sqrt{\dfrac{2}{3}} = 5\,\text{Å}$

12. Number of nearest neighbours = 12

## Comprehension IV

13. Number of T.V. per particle = 2

Number of O.V. per particle = 1

14. T.V. are smaller than O.V.

## Comprehension V

15. $\dfrac{a_{\text{KCl}}}{a_{\text{NaCl}}} = \dfrac{2(r_{\text{K}^+} + r_{\text{Cl}^-})}{2(r_{\text{Na}^+} + r_{\text{Cl}^-})} = \dfrac{\dfrac{r_{\text{K}^+}}{r_{\text{Na}^+}} + \dfrac{r_{\text{Cl}^-}}{r_{\text{Na}^+}}}{1 + \dfrac{r_{\text{Cl}^-}}{r_{\text{Na}^+}}}$

$= \dfrac{\dfrac{1}{0.7} + \dfrac{1}{0.5}}{1 + \dfrac{1}{0.5}} = 1.143$

16. $\dfrac{d_{\text{NaCl}}}{d_{\text{KCl}}} = \dfrac{58.5/a^3_{\text{NaCl}}}{74.5/a^3_{\text{KCl}}} = \dfrac{58.5}{74.5}\times(1.143)^3 = 1.172$

## Comprehension VI

17. Centre is octahedral void.

18. Number of O.V. = 4, but only one is occupied.

## Comprehension VII

19. C.N. of $Zn^{2+} = 4$

20. C.N. of $S^{2-} = 4$

21. $Zn^{2+}$ is tetrahedrally linked with 4 $S^{2-}$ ions.

22. $\frac{r^+}{r^-} > 0.225$

## Comprehension VIII

23. $Z_{Mn} = 8 \times \frac{1}{8} = 1$

$Z_F = 12 \times \frac{1}{4} = 3$

$\therefore$ Formula = $MnF_3$

24. C.N. of Mn = 6

25. $a = 2\left(r_{Mn^{3+}} + r_{F^-}\right) = 2(0.65 + 1.35) = 4.00$ Å

26. $d = \frac{1 \times (55 + 3 \times 19)}{\left(6 \times 10^{23}\right) \times \left(4 \times 10^{-8}\right)^3} = 2.92 \text{ gm/cm}^3$

## Comprehension IX

27. $V_{\text{formula unit}} = \frac{m}{d} = \frac{(132.5 + 35.5)}{6 \times 10^{23} \times 3.5} = 8 \times 10^{-23} \text{ cm}^3$

28. $d = \frac{Z.M}{N_A.V}$

$\Rightarrow 3.5 = \frac{1 \times 168}{\left(6 \times 10^{23}\right) \times a^3}$

$\Rightarrow a = 4.3 \times 10^{-8}$ cm

Hence, nearest Cs – Cs distance = $a$ = 4.3 Å

29. Nearest Cs – Cl distance $= \frac{\sqrt{3}a}{2} = 3.72$ Å

## Comprehension X

30. $\frac{N}{N_O} = e^{-E/2RT} = e^{-\frac{46 \times 10^3}{2 \times 2 \times 1000}} = 1.0 \times 10^{-5}$

31. In NaCl, all O.V. are occupied. Hence, the available voids are only tetrahedral.

$\therefore N_i = 2 \times N_o$

Now, $N = \sqrt{N_o \times 2N_o} \cdot e^{-E/2RT}$

$\therefore \frac{N}{N_o} = \sqrt{2} \times e^{-E/2RT} = \sqrt{2} \times e^{-\frac{73.6 \times 1000}{2 \times 2 \times 1000}}$

$= 1.41 \times 10^{-8}$

## Comprehension XI

32. $d = \frac{Z \cdot M}{N_A \cdot V}$

$= \frac{4 \times 6.023Y}{\left(6.023 \times 10^{23}\right) \times \left(2 \cdot Y^{1/3} \times 10^{-7}\right)^3}$

$= \frac{1}{200} \text{ gm/cm}^3 = 5 \text{ kg/m}^3$

33. $d_{\text{observed}} >> d_{\text{theoretical}}$

Such large difference is possible due to impurity defect.

## Comprehension XII

34. For diamond crystal, $\sqrt{3}a = 8r$ and $Z = 8$

Now, $d = \frac{Z \cdot M}{N_A \cdot V}$

$\therefore 3.6 = \frac{8 \times 12}{(6 \times 10^{23}) \times \left(\frac{8r}{\sqrt{3}}\right)^3}$

$\therefore r = 0.76 \times 20^{-8}$ cm

35. For SiC, $\sqrt{3}a = 4(r_c + r_{si})$ and $Z = 4$

Now, $d = \frac{Z \cdot M}{N_A \cdot V} \Rightarrow 3.2$

$= \frac{4 \times 40}{(6 \times 10^{23}) \times \left(\frac{4(r_c + r_{si})}{\sqrt{3}}\right)^3}$

$\therefore r_{si} = 1.12 \times 10^{-8}$ cm

36. Interchange between C and si atoms will neither change '$Z$' nor '$a$'.

37. For the same volume, $\frac{m_{dia}}{3.6} = \frac{m_{sic}}{3.2}$

or, $\frac{n_c \times 12}{3.6} = \frac{n_{sic} \times 40}{3.2} \Rightarrow \frac{n_{sic}}{n_c} = \frac{1}{3.75}$

38. Packing efficiency of SiC is greater due to unequal size of particles.

## Section D (Assertion – Reason)

1. Voids are named according to the orientation of spheres constituting the voids.
2. Relative increase in packing efficiency is high when larger voids are occupied.
3. BCC : $2r = \frac{\sqrt{3}a}{2}$

   FCC : $2r = \frac{\sqrt{2}a}{2}$
4. Informative
5. Informative
6. Informative
7. Density depends on mass and size of particles but not the packing efficiency.
8. Theoretical
9. Packing efficiency of diamond is only 0.34.
10. Informative
11. Theoretical
12. Informative
13. Informative
14. Informative
15. $\Delta H = +\text{ve}, \Delta S = +\text{ve}$

    Hence for $-\text{ve}\ \Delta G$, the temperature should be high.
16. Informative
17. Theoretical
18. Theoretical
19. Informative
20. Informative

## Section E (Column Match)

1. Theoretical (Cubic crystals)
2. Informative (Ionic solids)
3. Theoretical (Classification of solids)
4. Theoretical (Classification of solids)

5. Theoretical (Classification of solids)

6. $CaCl : \sqrt{3}a = 2\left(r^+ + r^-\right)$

$CaF_2 : \sqrt{3}a = 4\left(r^+ + r^-\right)$ and $\sqrt{2}a = 4.r^+$

Diamond : $\sqrt{3}a = 8r$

Nacl:

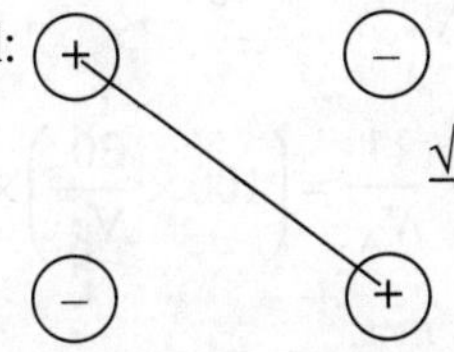


$$\frac{\sqrt{2}a}{2} = \frac{a}{\sqrt{2}}$$

7. Theoretical (Ionic solids)

8. Informative (Basic crystal system)

9. Informative (Basic crystal system)

10. $CaF_2 : d_{F^- - F^-} = \frac{a}{2}$ and $d_{Ca^{2+} - Ca^{2-}} = \frac{a}{\sqrt{2}}$

$NaCl : d_{Na^+ - Na^-} = a$

$CsCl : d_{Cs^+ - Cs^+} = a$

$d_{T.V.}$ from corner $= \frac{\sqrt{3}a}{4}$

$Na_2O : d_{Na^+ - Na^+} = \frac{a}{2}$ and $d_{O^{2-}} - d_{O^{2-}} = \frac{a}{\sqrt{2}}$

11. SC : Nearest = $a$, Next nearest = $\sqrt{2}a$

BCC : $= \frac{\sqrt{3}a}{2}$ $= a$

FCC : $= \frac{\sqrt{2}a}{2}$ $= a$

## Section F (Subjective)

### Single-digit Integer Type

1. Volume, $l^3 = \frac{m}{d} = \frac{58.5}{2.167} = 27\,\text{cm}^3 \Rightarrow l = 3\,\text{cm}$

2. $d = \frac{Z \cdot M}{N_A \cdot V} \Rightarrow 4 = \frac{Z \times 72}{6\times10^{23}\times\left(0.493\times10^{-7}\right)^3}$

$\Rightarrow Z = 4$

3. $d_{CH_4(l)} = d_{CH_4(s)} = \frac{Z \cdot M}{N_A \cdot V}$

or, $0.5 = \frac{Z\times16}{6\times10^{23}\times\left(0.6\times10^{-7}\right)^3} \Rightarrow Z = 4$

4. $d = \frac{Z \cdot M}{N_A \cdot V}$

$$0.92 = \frac{Z\times18}{6\times10^{23}\times\left\{2\times\frac{\sqrt{3}}{4}\times\left(4.53\times10^{-8}\right)^2\times 7.41\times10^{-8}\right\}}$$

$\Rightarrow$

$\therefore Z = 4$

5. $d = \frac{Z \cdot M}{N_A \cdot V}$

$$2.4 = \frac{Z\times12}{6\times10^{23}\times\left\{6\times\frac{\sqrt{3}}{4}\cdot\left(x\times10^{-8}\right)^2\times 2\sqrt{3}\times10^{-8}\right\}}$$

$\Rightarrow$

$\therefore x^2 = \frac{200}{108}$

6. $\sqrt{2}\,a = 4r$

$\Rightarrow 2r = \frac{\sqrt{2}\,a}{2} = 1\,\text{nm}$

7.

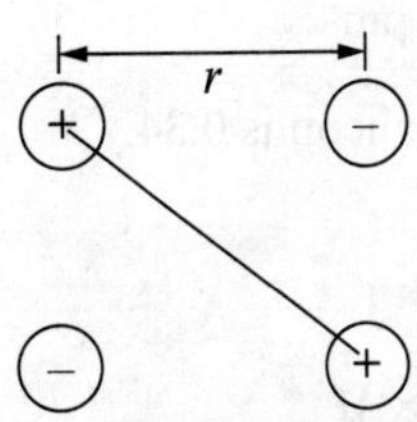


$$\sqrt{2}r = \sqrt{2}\left(\frac{186}{\sqrt{2}} + \frac{214}{\sqrt{2}}\right)\text{pm} = 400\,\text{pm}$$

8. $d = \dfrac{Z \cdot M}{N_A \cdot V}$

$$= \frac{4 \times 58.5}{6 \times 10^{23} \times (500 \times 10^{-10})^3} = 3.12 \text{ gm/cm}^3$$

$\therefore$ Percentage vacancy $= \dfrac{3.12 - 2.964}{3.12} \times 100 = 5\%$

9. (ii), (iv), (v), (vi), (vii) are true statements.

10. $d_{\text{theo}} = \dfrac{4 \times 31.25 \times 1.67 \times 10^{-24}}{(500 \times 10^{-10})^3} = 1.67 \text{ gm/cm}^3$

Now, $m_{\text{theo}} = 1670$ gm per litre

$m_{\text{actual}} = 1607.5$ gm per litre

$\therefore$ Moles of metal missing per litre

$$= \frac{1670 - 1607.5}{31.25} = 2$$

11. $Z_{\text{ZnS}} = 4$ but due to defect, $Z_{\text{ZnS}} = 3$.

12. Octahedron has eight triangular faces. Hence, truncated octahedron will have eight hexagonal faces.

13. BCC: Fraction of edge covered by atom $= \dfrac{2r}{a}$.

$\therefore$ Fraction of edge uncovered

$$= 1 - \frac{2r}{a} = 1 - 2 \times \frac{\sqrt{3}}{4} = 0.134$$

From question: 0.134 a = 67 pm $\Rightarrow$ a = 500 pm

Now, $d = \dfrac{Z \cdot M}{N_A \cdot V} = \dfrac{2 \times 75}{(6 \times 10^{23}) \times (500 \times 10^{-10})^3}$

$= 2 \text{ gm/cm}^3$

14. Let the ideal crystal was having 100 Si-atom, After doping, $x$ Si-atom are missing and $y$ B-atom are doped. Now,

$$(100 - x) \times \frac{30}{N_A} + y \times \frac{11}{N_A} = \left(100 \times \frac{30}{N_A}\right) \times \frac{88}{100}$$

or, $30x - 11y = 360$ (1)

and $\dfrac{\dfrac{(100 - x)}{N_A} \times 30}{\dfrac{y}{N_A} \times 11} = \dfrac{(1 - 0.001)}{0.001}$

$\Rightarrow \dfrac{3000 - 30x}{11y} = 999$ (2)

From (1) and (2), $\dfrac{y}{x} \times 100 = 2\%$

15. In truncated octahedron, all corner of octahedron become square faces and hence, its number = 6. And, the number of c-atoms per unit cell in diamond = 8.

---

## Four-digit Integer Type

1. $\dfrac{m}{d} \times 0.74 = N_A \times \dfrac{4}{3} \pi r^3$

or, $\dfrac{197}{19.7} \times 0.74 = 6 \times 10^{23} \times \dfrac{4}{3} \pi \times r^3$

$\Rightarrow r = 1.43 \times 10^{-8}$ cm = 143 pm

2. P.E. of diamond and here silicon is 0.34.

3. $d = \dfrac{Z \cdot M}{N_A \cdot V}$

$$1.5 \times 10^3 = \frac{6 \times M}{(6 \times 10^{23}) \times \begin{bmatrix} 12.5 \times 8.0 \times 3.0 \\ \times (10^{-9})^3 \end{bmatrix}}$$

$\Rightarrow M = 45$ Kg/mol

4. Let the percentage of fayalite be $x$.

$V_{\text{olivine}} = V_{\text{fayalite}} + V_{\text{fosterite}}$

or, $\dfrac{100}{3.9} = \dfrac{x}{4.2} + \dfrac{100 - x}{3.3}$

$\Rightarrow x = 71.79$

5. $d = \dfrac{Z \cdot M}{N_A \cdot V}$

$$\Rightarrow 2.14 = \frac{4 \times (426 + 18x)}{(6 \times 10^{23}) \times (1.26 \times 10^{-7})^3}$$

$\Rightarrow x = 12$

HINTS AND EXPLANATIONS

6. For diamond: $\sqrt{3}a = 4 \times d_{C-C}$

Now, $d = \dfrac{Z \cdot M}{N_A \cdot V}$

$$\Rightarrow 2\sqrt{3} = \frac{8 \times 12}{(6 \times 10^{23}) \times \left(\frac{4 \times d_{C-C}}{\sqrt{3}}\right)^3}$$

$\therefore d_{C-C} = 1.55 \times 10^{-8}$ cm $= 155$ pm

7. The maximum backing efficiency of identical spheres in 2D is 0.90.

Hence, $(40)^2 \times 0.90 = N \times \pi \times \left(\frac{10}{2}\right)^2$

$\Rightarrow N = 18$

8. $4.r_{Cl^-} = \sqrt{2} \times 520$

$\Rightarrow r_{Cl^-} = 182$ pm

9. $Z_{Ti} = \frac{1}{2} \times 1 = \frac{1}{2}$

$Z_O = 1$ (assume)

$\therefore$ Formula $= Ti_{1/2}O_1 \cong T_iO_2$

Now, $Ti = \dfrac{48}{48+32} \times 100 = 60\%$

and oxidation state of $Ti = +4$

10. For one litre crystal,

$m_{\text{initial}} - m_{\text{B removed}} + m_{\text{C added}} = m_{\text{final}}$

or, $4800 - x \times 30 + 1 \times 15 = 4795$

$\Rightarrow x = \dfrac{2}{3}$

$\therefore$ Percentage of C-atoms which replaced B-atoms

$= \dfrac{x}{1} \times 100 = 67\%$

11. Percentage of body diagonal covered

$$= \frac{2r_{B^-} + 4r_C}{\sqrt{3}\left[2\left(r_{A^+} + r_{B^-}\right)\right]} \times 100\%$$

$$= \frac{2r_{B^-} + 4 \times 0.225 r_{B^-}}{2\sqrt{3}\left(0.414 r_{B^-} + r_{B^-}\right)} \times 100\% = 59.2\%$$

12. Out of 8 $Fe^{2+}$ in original crystal, 1 is missing.

$\therefore$ Percentage of cation vacancy $= \frac{1}{8} \times 100 = 12.5\%$

13. Percentage occupied space

$$= \frac{V_{\text{particles}}}{V_{\text{solid}}} \times 100\%$$

$$= \frac{6 \times 10^{23} \times \frac{4}{3}\pi\left(1.54 \times 10^{-8}\right)^3}{\left(\frac{40}{4/\pi}\right)} \times 100\% = 27\%$$

$\therefore$ Empty space = 73%

14. For cubic void, $\dfrac{r}{a} = 0.732$

$\Rightarrow a = \dfrac{366/2}{0.732} = 250$ pm

Now, $d = \dfrac{Z \cdot M}{N_A \cdot V} = \dfrac{1 \times 125}{(6 \times 10^{23}) \times (250 \times 10^{-10})^3}$

$= \dfrac{100}{7.5}$ gm/cm$^3$

15. $d = \dfrac{Z \cdot M}{N_A \cdot V} = \dfrac{2 \times 28}{(6 \times 10^{23}) \times (667 \times 500 \times 280 \times 10^{-30})}$

$= 1$ gm/cm$^3$

$= 1000$ kg/cm$^3$

CHAPTER

# 10 Liquid Solution

## EXERCISE I (JEE MAIN)

### Solution of Gas in Liquid

1. Low concentration of oxygen in blood and tissues of people living at high altitude is due to
   (a) low temperature.
   (b) low atmospheric pressure.
   (c) high atmospheric pressure.
   (d) both low temperature and high pressure.

2. Which of the following statement is correct regarding the solubility of gas in water?
   (a) Solubility increases with increase in temperature.
   (b) A more polar gas will be less soluble.
   (c) Solubility increases with increase in pressure.
   (d) Solubility is always an endothermic process.

3. The only incorrect statement regarding Henry's law for a gas dissolving in water is
   (a) The value of $K_H$ (Henry's constant) is greater for $O_2$ than $H_2$ at same temperature.
   (b) The value of $K_H$ increases with increase in temperature.
   (c) Henry's law is not applicable for HCl(g).
   (d) Henry's law is applicable when the solubility of gas is very low.

4. Henry's law constant for $CO_2$ in water is $1.6 \times 10^8$ Pa at 298 K. The quantity of $CO_2$ in 500 g of soda water when packed under 3.2 bar pressure at 298 K is
   (a) 2.44 g (b) 24.4 g
   (c) 0.244 g (d) 0.61 g

5. $H_2S$, a toxic gas with rotten egg like smell, is used for the quantitative analysis. If the solubility of $H_2S$ in water at STP is 0.2 m, then Henry's law constant for $H_2S$ in water at 273 K is
   (a) $3.6 \times 10^8$ Pa (b) $5.0 \times 10^8$ Pa
   (c) $5.0 \times 10^5$ Pa (d) $2.78 \times 10^7$ Pa

### Vapour Pressure

6. On increasing the altitude at constant temperature, the vapour pressure of a liquid
   (a) increases
   (b) decreases
   (c) remains the same.
   (d) depends upon climate.

7. The boiling points of $C_6H_6$, $CH_3OH$, $C_6H_5NH_2$ and $C_6H_5NO_2$ are 80°C, 65°C, 184°C and 212°C, respectively. Which of the following will have the highest vapour pressure at room temperature?
   (a) $C_6H_6$ (b) $CH_3OH$
   (c) $C_6H_5NH_2$ (d) $C_6H_5NO_2$

8. Which of the following is not a characteristic property of the polar liquids?
(a) They have high boiling points.
(b) They have high heat of vaporization.
(c) They have low viscosity.
(d) They have low vapour pressure.

9. Vapour pressure of the liquid
(a) increases with increase in temperature.
(b) decreases with increase in temperature.
(c) is independent of temperature.
(d) either increases or decreases with increase in temperature, depending on the nature of liquid.

10. The vapour pressure of water at 300 K in a closed container is 0.4 atm. If the volume of the container is doubled, then its vapour pressure at 300 K will be
(a) 0·8 atm (b) 0·2 atm
(c) 0·4 atm (d) 0·6 atm

## Solution of Liquid in Liquid

11. Which of the following behaviour is true about the ideal binary liquid solution of liquids 'A' and 'B', if $P_A^\circ < P_B^\circ$?
(a) Plot of $P_{total}$ vs $X_A$ is non-linear.
(b) Plot of $P_{total}$ vs $X_B$ is linear with +ve slope.
(c) Plot of $P_{total}$ vs $X_B$ is linear with slope = 0.
(d) Plot of $P_{total}$ vs $X_B$ is linear with –ve slope.

12. For an ideal solution of A and B, $Y_A$ is the mole fraction of A in the vapour phase at equilibrium. Which of the following plot should be linear?
(a) $P_{total}$ vs $Y_A$ (b) Ptotal vs YB
(c) $\frac{1}{P_{total}}$ vs YA (d) $\frac{1}{P_{total}}$ vs $\frac{1}{Y_A}$

13. At 323 K, the vapour pressure (in mm Hg) of a methanol–ethanol solution is represented as
$$P = 120X + 140,$$
where $X$ is the mole fraction of methanol in liquid solution at equilibrium. The value of $P^\circ_{EtOH}$ is
(a) 120 mm (b) 140 mm
(c) 260 mm (d) 20 mm

14. The vapour pressure of pure benzene at 88°C is 960 mm and that of toluene at the same temperature is 380 mm of benzene. At what mole fraction of benzene, the mixture will boil at 88°C?
(a) 0.655 (b) 0.345
(c) 0.05 (d) 0.25

15. Heptane and octane form ideal solution. At 373 K, the vapour pressures of the pure liquids are 106 kPa and 46 kPa, respectively. What will be the vapour pressure, in bar, of a mixture of 30.0 g of heptane and 34.2 g of octane?
(a) 76 bar (b) 152 bar
(c) 1.52 bar (d) 0.76 bar

16. Benzene and toluene form an ideal solution. The vapour pressures of benzene and toluene are 75 mm and 25 mm, respectively, at 20°C. If the mole fractions of benzene and toluene in vapour are 0.75 and 0.25, respectively, then the vapour pressure of the ideal solution is
(a) 62.5 mm (b) 50 mm
(c) 30 mm (d) 40 mm

17. For which of the following pair, the heat of mixing, $\Delta H_{mix}$, is approximately zero?
(a) $CH_3COOCH_3 + CHCl_3$
(b) $CH_3COOH + H_2O$
(c) $C_2H_5OH + CH_3OH$
(d) $CH_3COCH_3 + C_6H_6$

18. Each of the following pair shows a positive deviation from the Raoult's law except
(a) $(C_2H_5)O + HCl$
(b) $CCl_4 + CH_3OH$
(c) $CHCl_3 + C_2H_5OH$
(d) $C_2H_5OH + H_2O$

19. The vapour pressure of a solution of two liquids, A ($P^\circ$ = 80 mm, $X$ = 0.4) and B ($P^\circ$ = 120 mm, $X$ = 0.6) is found to be 100 mm. It shows that the solution exhibits
(a) negative deviation from ideal behaviour.
(b) positive deviation from ideal behaviour.
(c) ideal behaviour.
(d) positive deviation at lower concentration

20. When 25 ml of $CCl_4$ and 25 ml of toluene is mixed, the total volume of the solution will be
(a) 50 ml (b) > 50 ml
(c) < 50 ml (d) indefinite

EXERCISE I

21. The immiscible liquid system containing aniline-water boils at 98°C under a pressure of 760 mm. At this temperature, the vapour pressure of water is 700 mm. If aniline is distilled in steam at 98°C, then what percent of total weight of the distillate will be aniline?

(a) 7.89 (b) 8.57
(c) 30.7 (d) 44.3

22. Azeotropic mixture of liquids can only be separated by

(a) simple distillation.
(b) fractional distillation.
(c) distillation under reduced pressure.
(d) chemical means.

23. The boiling point of an azeotropic mixture of water and ethanol is less than that of water and ethanol, separately. The mixture shows

(a) no deviation from Raoult's law.
(b) positive deviation from Raoult's law.
(c) negative deviation from Raoult's law.
(d) that the solution is unsaturated.

24. Which of the following can be separated into its pure components by fractional distillation?

(a) $C_6H_6 + C_7H_8$ (b) $H_2O + HCl$
(c) $H_2O + HNO_3$ (d) $H_2O + C_2H_5OH$

25. Pure water boils at 373 K and nitric acid at 359 K. The azeotropic mixture of water and nitric acid boils at 393.5 K. On distillation of the azeotropic mixture,

(a) pure nitric acid will distil over first.
(b) pure water will distil over first.
(c) one of them will distil over with small amount of the other.
(d) both of them will distil over in the same composition as they are in the mixture.

## Lowering of Vapour Pressure

26. The ratio between lowering of vapour pressure of solution and mole fraction of solute is equal to

(a) relative lowering of vapour pressure.
(b) vapour pressure of pure solvent.
(c) vapour pressure of solution.
(d) molar mass of solvent.

27. The vapour pressure of a dilute solution of a non-volatile solute is not influenced by the

(a) temperature of the solution
(b) melting point of the solute
(c) mole fraction of the solute
(d) degree of dissociation of solute

28. Addition of 4.0 g of a non-volatile solute to 54.0 g water results in a lowering of vapour pressure by 1.0%. What is the molar mass of the non-volatile solute?

(a) 132 (b) 66
(c) 133.33 (d) 333

29. An aqueous solution of sucrose is 0.5 molal. What is the vapour pressure of water above this solution? The vapour pressure of pure water is 25.0 mm Hg at this temperature.

(a) 24.8 mm Hg (b) 0.45 mm Hg
(c) 2.22 mm Hg (d) 20.3 mm Hg

30. When mass fraction of solute in a very dilute solution is doubled, its mole fraction is

(a) doubled (b) halved
(c) increased (d) remains constant

31. The vapour pressure of a solution of non-volatile solute is

(a) less than that of solvent.
(b) equal to that of solvent.
(c) more than that of solvent.
(d) equal to or more than that of solvent.

32. Addition of solute to a given solvent in a closed container

(a) decreases the rate of condensation.
(b) decreases the rate of evaporation.
(c) increases the rate of evaporation.
(d) decreases both the rate of evaporation and condensation.

33. The mass of a non-volatile solute (molecular mass = 40) which should be dissolved in 114 g octane to reduce its vapour pressure to 80% is

(a) 8 g (b) 12 g
(c) 4 g (d) 10 g

34. The vapour pressure of water at room temperature is lowered by 5% on dissolving a non-volatile solute in it. The molality of the solution is

(a) 2.0 (b) 1.4
(c) 2.78 (d) 2.92

35. How many grams of glucose ($C_6H_{12}O_6$) should be dissolved in 0.5 kg of water at 25°C to reduce the vapour pressure of the water by 1.0%?

(a) 50.5 g (b) 50.0 g
(c) 18.0 g (d) 18.2 g

36. An ideal solution is obtained by dissolving $n$ moles of non-volatile, non-electrolyte solute in $N$ moles of solvent. If the vapour pressure of solution is $P$ and the vapour pressure of pure solvent is '$P^o$', then

(a) $\frac{P^o - P}{P} = \frac{n}{N}$ (b) $\frac{P^o - P}{P^o} = \frac{n}{N}$

(c) $\frac{P^o - P}{P^o} = \frac{N}{n}$ (d) $\frac{P^o - P}{P} = \frac{N}{n}$

37. Dry air was passed successively through a solution of 5 g of a solute in 80 g of water and then through pure water. The loss in mass of solution was 2.5 g and that of pure solvent was 0.04 g. What is the molecular mass of the solute?

(a) 70.3 (b) 71.43
(c) 14.28 (d) 14.06

38. An ideal solution was obtained by mixing methanol and ethanol. If the partial vapour pressures of methanol and ethanol are 2.8 and 4.2 kPa respectively, the mole fraction of methanol in the vapour at equilibrium is

(a) 0.67 (b) 0.4
(c) 0.6 (d) 0.33

39. For an ideal solution containing two liquid components A and B, the Gibb's free energy of mixing is minimum, when the molar ratio of the liquids is

(a) 1 : 1 (b) 1 : 2
(c) 1 : 10 (d) 1 : 1000

40. Liquids A and B form an ideal solution. A certain solution of A and B contains 25 mole percent of A, whereas the vapours in the equilibrium with the solution at 298 K contains 50 mole percent of A. The ratio of vapour pressures of pure A to that of pure B at 298 K, is

(a) 1 : 1 (b) 3 : 1
(c) 1 : 3 (d) 2 : 1

## Osmotic Pressure

41. Which of the following serves best as a semipermeable membrane?

(a) Copper ferrocyanide
(b) Vegetable membrane
(c) Animal membrane
(d) Cellophane

42. Two aqueous solutions $S_2$ and $S_2$ are separated by a semipermeable membrane. $S_1$ has lower vapour pressure than $S_2$. Which of the following statement is correct?

(a) More solvent will flow from $S_2$ to $S_2$.
(b) More solute will flow from $S_2$ to $S_1$.
(c) Solvent from $S_1$ to $S_2$ and $S_2$ to $S_1$ will flow at equal rates.
(d) No flow of solvent will take place.

43. After removing the hard shell of an egg by dissolving in dilute HCl, a semipermeable membrane can be visible. If such an egg is kept in a saturated solution of common salt, the size of egg will

(a) shrink
(b) grow
(c) remain unchanged.
(d) first shrink, then grow.

44. The process of getting fresh water from sea water is known as

(a) osmosis (b) filtration
(c) pressure distillation (d) reverse osmosis

45. Blood is isotonic with

(a) 0.9% (w/v) – NaCl solution
(b) 0.9 M – NaCl solution
(c) 0.9 M – NaCl solution
(d) 9.0% (w/v) – NaCl solution

46. The solution containing 4.0 g of PVC in 1 L of dioxane was found to have osmotic pressure of 0.006 atm at 300 K. The molecular mass of the polymer PVC is

(a) 16,420 (b) 1642
(c) 1,64,200 (d) 4105

47. A semipermeable membrane separates a solution which is 0.012 M in glucose from one that is 0.250 M in glucose. On which of these solutions must pressure be applied to prevent a net flow of water through the membrane?
(a) On the 0.012 M solution.
(b) On the 0.250 M solution.
(c) Equal pressure on both the solutions.
(d) The pressure on 0.012 M solution should be double the pressure on 0.250 M solution.

48. If 0.1 molar solution of glucose is separated from 0.1 molar solution of cane sugar by a semipermeable membrane, then which one of the following statements is correct?
(a) Water will flow from glucose solution into cane sugar solution.
(b) Cane sugar will flow across the membrane into glucose solution.
(c) Glucose will flow across the membrane into cane sugar solution.
(d) There will be no net movement across the semipermeable membrane.

49. Five percent solution of a solute (X) is isotonic with 0.855% solution of sucrose (molecular weight = 342). What is the molecular mass of solute (X)?
(a) 200 (b) 58.482
(c) 400 (d) 2000

50. What will be the osmotic pressure of decimolar solution of glucose at 30°C?
(a) 2.48 atm (b) 8.24 atm
(c) 3.48 atm (d) 0.241 atm

51. The solution having higher osmotic pressure than the reference solution is called
(a) hypertonic solution (b) isotonic solution
(c) hypotonic solution (d) ideal solution

52. Due to osmosis, the volume of concentrated solution
(a) decreases slowly (b) increases slowly
(c) increases rapidly (d) decreases rapidly

53. The osmotic pressure of a solution may be increased by
(a) decreasing the temperature of solution.
(b) diluting the solution.
(c) increasing the moles of solute.
(d) increasing the moles of solvent.

54. The relationship between osmotic pressure at 273 K when 10 g glucose ($\pi_1$), 10 g urea ($\pi_2$) and 10 g sucrose ($\pi_3$) are dissolved in 250 ml of water, is
(a) $\pi_1 > \pi_2 > \pi_3$ (b) $\pi_3 > \pi_1 > \pi_2$
(c) $\pi_2 > \pi_1 > \pi_3$ (d) $\pi_2 > \pi_3 > \pi_1$

55. A solution having 54 g of glucose per litre has an osmotic pressure of 4.56 bar. If the osmotic pressure of a urea solution is 1.52 bar at the same temperature, then what would be its concentration?
(a) 1.0 M (b) 0.5 M
(c) 0.3 M (d) 0.1 M

## Elevation in Boiling Point

56. Which of the following will cook food more easily?
(a) Saline water (b) Pure water
(c) Alcoholic water (d) All in the same rate

57. The elevation in boiling point method is used for the determination of molecular masses of
(a) non-volatile and soluble solute.
(b) non-volatile and insoluble solute.
(c) volatile and soluble solute.
(d) volatile and insoluble solute.

58. The ebullioscopic constant of a liquid solvent is the elevation of boiling point of
(a) one molar solution of non-volatile and non-electrolyte solute in it.
(b) one normal solution of non-volatile and non-electrolyte solute in it.
(c) one formal solution of non-volatile and non-electrolyte solute in it.
(d) one molal solution of non-volatile and non-electrolyte solute in it.

59. A solution containing 2.60 g of a non-volatile and non-electrolyte solute in 200 g of water boils at 100.130°C at 1 atm. What is the molar mass of the solute? [$K_b$ ($H_2O$) = 0.52 K-kg/mol]
(a) 52.0 g $mol^{-1}$ (b) 152.0 g $mol^{-1}$
(c) 104 g $mol^{-1}$ (d) 204 g $mol^{-1}$

60. Water should boil at 90°C at pressure
(a) 1 atm (b) > 1 atm
(c) < 1 atm (d) Any of these

61. The latent heat of vaporization of a liquid of molar mass, 80 g/mol and boiling point, 127°C is 8 kcal/mol. The ebullioscopic constant of the liquid is
(a) 3.2 K-kg/mol (b) 0.04 K-kg/mol
(c) 0.32 K-kg/mol (d) 0.52 K-kg/mol

62. Which of the following aqueous solution will have the highest boiling point?
(a) 34.2%(w/w) Sugar (b) 18%(w/w) Glucose
(c) 6.00%(w/w) Urea (d) All the same

63. The molal boiling point elevation constant of water is 0.513°C kg mol$^{-1}$. When 0.1 mole of sugar is dissolved in 200 g of water, the solution boils under a pressure of 1 atm at
(a) 100.513°C (b) 102.565°C
(c) 100.256°C (d) 101.025°C

64. The rise in the boiling point of a solution containing 1.8 g of glucose in 100 g of a solvent is 0.1°C. The molal elevation constant of the solvent is
(a) 1 K/m (b) 10 K/m
(c) 0.01 K/m (d) 2731 K/m

65. An aqueous solution of a non-volatile and non-electrolyte solute (molecular mass = 150) boils at 373.26 K. The composition of solution, in terms of mass percent of the solute, is ($K_b$ of water = 0.52)
(a) 50% (b) 7.5%
(c) 6.98% (d) 75%

## Depression in Freezing Point

66. An aqueous solution of 10% NaCl (consider ideal behaviour of the solution) is cooled. It will allow some
(a) NaCl to crystallize.
(b) water to freeze.
(c) water to solidify along with some NaCl.
(d) precipitation of NaCl.

67. If 1 mole of a non-volatile and non-electrolyte solute in 1000 g of water depresses the freezing point by 1.86°C, then what will be the freezing point of a solution of 1 mole of the solute in 500 g of water?
(a) −0.93°C (b) −1.86°C
(c) 3.72°C (d) −3.72 °C

68. What is the molecular mass of a non-ionizing solid if 10 g of this solid, when dissolved in 100 g of water forms a solution which freezes at −1.24°C? $K_f(H_2O) = 1.86$°C kg mol$^{-1}$.
(a) 250 (b) 150
(c) 120 (d) 75

69. It is more convenient to obtain the molecular mass of an unknown solute by measuring the freezing point depression than by measuring the boiling point elevation because
(a) freezing point depression is a colligative property, whereas boiling point elevation is not.
(b) freezing point depressions are larger than boiling point elevations for the same solution.
(c) freezing point depressions are smaller than boiling point elevations for the same solution.
(d) freezing point depressions depends more on the amount of solute than boiling point elevation.

70. When the depression in freezing point is carried out, the equilibrium exist between
(a) liquid solvent and solid solvent.
(b) liquid solute and solid solvent.
(c) liquid solute and solid solute.
(d) liquid solvent and solid solute.

71. The molal depression constant for four liquids P, Q, R and S, respectively, are 1.84, 2.20, 3.15 and 3.92. If 0.1 m urea solution is made in all the solvents, then the solution in which solvent will show maximum depression in freezing point is
(a) P (b) Q
(c) R (d) S

72. Among the colligative properties of solution, which one is the best method for the determination of molecular masses of proteins and polymers?
(a) Osmotic pressure
(b) Lowering in vapour pressure
(c) Lowering in freezing point
(d) Elevation in boiling point

**73.** 5 g of urea is dissolved in one kg of water. Up to what temperature, the solution may be cooled before ice starts crystallizing out? ($K_f$ of water = 1.86)

(a) –0.310°C (b) –0.240°C
(c) –0.195°C (d) –0.155°C

**74.** In the above problem, if the solution is cooled to –0.200°C, then how many grams of ice would separate?

(a) 200 (b) 225
(c) 325 (d) 175

**75.** Of the following measurements, the one most suitable for the determination of the molecular mass of oxyhaemoglobin, a molecule with a molecular mass of many thousands, is

(a) the elevation of the boiling point.
(b) the depression of the freezing point.
(c) the osmotic pressure.
(d) any of the previous three, as they are all equally good.

## Abnormal Colligative Properties

**76.** Phenol dimerizes in benzene. If the observed molecular mass of phenol in solution is 120, then its degree of dimerization is

(a) 0.600 (b) 0.433
(c) 0.277 (d) 0.866

**77.** The van't Hoff factor for a dilute solution of $K_3[Fe(CN)_6]$ is

(a) 4.0 (b) 0.25
(c) 5.0 (d) 3.0

**78.** The molecular mass of NaCl as determined by osmotic pressure measurement is

(a) 58.5 (b) less than 58.5
(c) more than 58.5 (d) unpredictable

**79.** The limiting value of van't Hoff factor for $Na_2SO_4$ is

(a) 2 (b) 3
(c) 4 (d) 5

**80.** The ratio of the elevation in boiling points of NaCl solution to that of glucose solution of same molality is nearly

(a) 1 : 2 (b) 1 : 1
(c) 2 : 1 (d) 58.5 : 180

**81.** The mass of KCl required to depress the freezing point of 500 g water by 2 K is ($K_f$ = 1.86, K = 39)

(a) 10.01 g (b) 40.05 g
(c) 7.45 g (d) 20.03 g

**82.** For each of the following dilute solutions, van't Hoff factor is equal of 3, except

(a) $Na_2SO_4$ (b) $CaF_2$
(c) $K_3PO_4$ (d) $(NH_4)_2CO_3$

**83.** Which of the following aqueous solution will have the lowest freezing point?

(a) 0.10 m-sucrose (b) 0.10 m-$NiCl_2$
(c) 0.10 m-$CuSO_4$ (d) 0.10 m-$NH_4NO_3$

**84.** At the same temperature, each of the following solution has the same osmotic pressure except

(a) 0.140 M-sucrose (b) 0.07 M-KCl
(c) 0.070 M-$Ca(NO_2)_2$ (d) 0.140 M-urea

**85.** The 0.1 m aqueous solutions, each of urea, common salt and sodium sulphate, are taken. The ratio of their elevation in boiling points is

(a) 1 : 1 : 1 (b) 1 : 2 : 1
(c) 1 : 2 : 3 (d) 3 : 2 : 1

**86.** $FeCl_3$ solution (side X) on reaction with $K_4[Fe(CN)_6]$ (side Y) in aqueous solution gives blue colour. These are separated by a semipermeable membrane as shown. Due to osmosis, there is

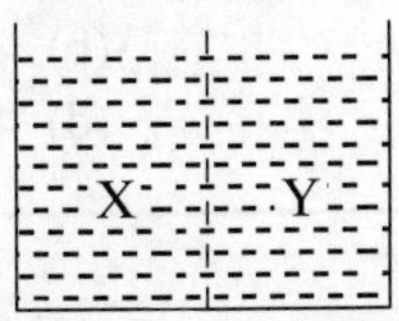


(a) blue colour formation in the side X.
(b) blue colour formation in the side Y.
(c) blue colour formation in both sides of X and Y.
(d) no blue colour formation.

**87.** Which one of the following pairs of solution can we expect to be isotonic at the same temperature?

(a) 0.1 M-urea and 0.1 M-NaCl
(b) 0.1 M-urea and 0.1 M-$MgCl_2$
(c) 0.1 M-$Na_2SO_4$ and 0.1 M-NaCl
(d) 0.1 M – $Na_2SO_4$ and 0.1 M – $Ca(NO_3)_2$

88. Aqueous solutions of 0.004 M-$Na_2SO_4$ and 0.01 M-Glucose are isotonic. The percentage dissociation of $Na_2SO_4$ is
(a) 25% (b) 60%
(c) 75% (d) 40%

89. Under the condition of similar temperature, which of the following solution will have minimum vapour pressure?
(a) 0.1 M-sugar (b) 0.1 M-NaCl
(c) 0.1 M-$BaCl_2$ (d) 0.1 M-$Al_2(SO_4)_3$

90. Maximum osmotic pressure will be shown by which of the following solution (all 90% ionized)?
(a) 0.1 M-$AlCl_3$ (b) 0.1 M-$BaCl_2$
(c) 0.1 M-NaCl (d) 0.1 M-sugar

91. If 0.1 molal aqueous solution of sodium bromide freezes at –0.3348°C at one atmospheric pressure, then the percent dissociation of salt in the solution is ($K_f$ = 1.86)
(a) 90 (b) 80
(c) 60 (d) 20

92. An aqueous solution containing 0.25 moles of a non-volatile but strong electrolyte solute 'X' in 500 g water freezes at –2.79°C. The number of ions furnished in water per formula unit of 'X' is ($K_f$ = 1.86)
(a) 1 (b) 2
(c) 3 (d) 4

93. An aqueous solution (0.85%) of $NaNO_3$ is apparently 90% dissociated at 27°C. The osmotic pressure of solution is
(a) 2.463 atm (b) 4.68 atm
(c) 24.63 atm (d) 46.8 atm

94. The van't Hoff factor for $Ba(NO_3)_2$ solution is 2.74. The degree of dissociation of salt is
(a) 0.913 (b) 0.87
(c) 1.00 (d) 0.74

95. The elevation in boiling point, when 13.44 g of freshly prepared $CuCl_2$ is added to one kg of water, is (Some useful data: $K_b$ of water = 0.52 K – kg/mol and molecular weight of $CuCl_2$ = 134.4)
(a) 0.052 (b) 0.104
(c) 0.156 (d) 0.208

96. When 20 g of naphtholic acid ($C_{11}H_8O_2$) is dissolved in 50 g of benzene ($K_f$ = 1.72 K kg $mol^{-1}$), a freezing point depression of 2 K is observed. The van't Hoff factor (i) is
(a) 0.5 (b) 1
(c) 2 (d) 3

97. The degree of dissociation ($\alpha$) of a weak electrolyte, $A_xB_y$, is related to van't Hoff factor (i) by the expression
(a) $\alpha = \frac{i-1}{x+y-1}$ (b) $\alpha = \frac{i-1}{x+y+1}$
(c) $\alpha = \frac{x+y-1}{i-1}$ (d) $\alpha = \frac{x+y+1}{i-1}$

98. The amino acid alanine has two isomers, such as $\alpha$-alanine and $\beta$-alanine. When equal masses of these two compounds are dissolved in equal mass of a solvent, the solution of $\alpha$-alanine freezes at relatively lower temperature. Which forms of $\alpha$-alanine or $\beta$-alanine, has the larger equilibrium constant for ionization?
(a) $\alpha$-alanine (b) $\beta$-alanine
(c) Same for both (d) Unpredictable

EXERCISE I

# EXERCISE II (JEE ADVANCED)

## Section A (Only one Correct)

1. The Henry's law constant for the solubility of $N_2$ gas in water at 298 K is $1.0 \times 10^5$ atm. The mole fraction of $N_2$ in air is 0.8. The number of moles of $N_2$ from air dissolved in 10 moles of water at 298 K and 5 atm pressure is
   (a) $4.0 \times 10^{-4}$ (b) $4.0 \times 10^{-5}$
   (c) $5.0 \times 10^{-4}$ (d) $5.0 \times 10^{-5}$

2. At a particular temperature and at a pressure of 1 atm, the mass and volume of $O_3$ that can be dissolved in 1 L of $H_2O$ is 'm' gram and 'V' ml, respectively. Assuming that the solution obeys Henry's law, the mass and volume of $O_3$ dissolved in 2 L of water at the same temperature and at 5 atm will be
   (a) 5 $m$ gram, 5$V$ ml
   (b) 10 $m$ gram, 10$V$ ml
   (c) 10 $m$ gram, 2$V$ ml
   (d) 10 $m$ gram, 5$V$ ml

3. The vapour pressure of a substance of low volatility can be measured by passing an unreactive gas over a sample of the substance and analysing the composition of the resulting gas mixture. When nitrogen was passed over mercury at 23°C, the mixture was analysed to 0.8 mg of Hg, 50.4 g of nitrogen at a total pressure of 720 torr. The vapour pressure of mercury at this temperature is (Hg = 200)
   (a) $8 \times 10^{-4}$ torr (b) $1.6 \times 10^{-3}$ torr
   (c) $3.2 \times 10^{-3}$ torr (d) $1.6 \times 10^{-4}$ torr

4. A beaker (A) contains 20 g sugar in 100 g water. Another beaker (B) contains 10 g of sugar in 100 g water. Both the beakers are placed under a bell jar and allowed to stand until equilibrium is reached. How much water will be transferred from one beaker to the other? Neglect the mass of water vapour present at equilibrium.
   (a) 0
   (b) 33.33 g water will transfer from beaker–A to beaker–B.
   (c) 33.33 g water will transfer from beaker–B to beaker–A.
   (d) 50.00 g water will transfer from beaker–B to beaker–A.

5. Liquids A and B form an ideal solution. The plot of $\frac{1}{X_A}$ ($Y$-axis) versus $\frac{1}{Y_A}$ ($X$-axis) (where $X_A$ and $Y_A$ are the mole fractions of A in liquid and vapour phases at equilibrium, respectively) is linear whose slope and intercept, respectively, are given as
   (a) $\frac{P_A^o}{P_B^o}, \frac{(P_A^o - P_B^o)}{P_B^o}$ (b) $\frac{P_A^o}{P_B^o}, \frac{(P_B^o - P_A^o)}{P_B^o}$
   (c) $\frac{P_B^o}{P_A^o}, \frac{(P_A^o - P_B^o)}{P_B^o}$ (d) $\frac{P_B^o}{P_A^o}, \frac{(P_B^o - P_A^o)}{P_B^o}$

6. Liquids P and Q form an ideal solution. The vapour pressures of pure liquids P and Q at 80°C are 300 mm and 100 mm of Hg, respectively. Suppose that the vapour above a solution composed of 1.0 mole of P and 1.0 mole of Q at 80°C is collected and condensed. This condensate is then heated to 80°C and vapour is again condensed to form a liquid R. What is the mole fraction of P in R?
   (a) 0.5 (b) 0.75
   (c) 0.90 (d) 0.10

7. A mixture contains 1 mole of volatile liquid A ($P_A{}^o$ = 100 mm Hg) and 3 moles of volatile liquid B ($P_B{}^o$ = 80 mm Hg). If the solution behaves ideally, the total vapour pressure of the distillate is
   (a) 85 mm Hg (b) 85.88 mm Hg
   (c) 90 mm Hg (d) 92 mm Hg

8. Pressure over an ideal binary liquid solution containing 10 moles each of liquid A and B is gradually decreased isothermally. At what pressure, half of the total amount of liquid will get converted into vapour? ($P_A^o$ = 200 torr, $P_B^o$ = 100 torr)
   (a) 150 torr (b) 166.5 torr
   (c) 133.3 torr (d) 141.4 torr

9. A liquid mixture of 'A' and 'B' (assume ideal solution) is placed in a cylinder and piston arrangement. The piston is slowly pulled out isothermally so that the volume of liquid decreases and that of the vapour increases. At the instant when the quantity of the liquid still remaining is negligibly small, the mole fraction of 'A' in the vapour is 0.4. If $P_A° = 0.4$ atm and $P_B° = 1.2$ atm at this temperature, then the total pressure at which the liquid has almost evaporated is
(a) 0.667 atm (b) 1.5 atm
(c) 0.8 atm (d) 0.545 atm

10. A liquid solution is formed by mixing 10 moles of aniline and 20 moles of phenol at a temperature where the vapour pressures of pure liquid aniline and phenol are 90 and 87 mm Hg, respectively. The possible vapour pressure of solution at that temperature is
(a) 82 mm Hg (b) 88 mm Hg
(c) 90 mm Hg (d) 94 mm Hg

11. At 35°C, the vapour pressure of $CS_2$ is 512 mm Hg and that of acetone $CH_3COCH_3$ is 344 mm Hg. A solution of $CS_2$ and acetone, in which the mole fraction of $CS_2$ is 0.25 has a total vapour pressure of 600 mmHg. Which the following statement about the solution of acetone and $CS_2$ is true?
(a) A mixture of 100 ml of acetone and 100 ml of $CS_2$ has a total volume of 200 ml.
(b) When acetone and $CS_2$ are mixed at 35°C, heat is released.
(c) When acetone and $CS_2$ are mixed at 35°C, heat is absorbed.
(d) Raoult's law is obeyed by both $CS_2$ and acetone for the solution in which the mole fraction of $CS_2$ is 0.25.

12. Mixing of the components in which of the following pair rises the temperature?
(a) Trichloromethane and ethanol.
(b) Tetrachloroethane and methanol.
(c) Trichloromethane and ethoxy ethane.
(d) Trichloroethane and ethoxy ethane.

13. Water and chlorobenzene are immiscible liquids. Their mixture boils at 90°C under a reduced pressure of $9.031 \times 10^4$ Pa. The vapour pressure of pure water at 90°C is $7.031 \times 10^4$ Pa and the molecular mass of chlorobenzene is 112.5. On mass percent basis, chlorobenzene in the distillate is equal to
(a) 50 (b) 22.2
(c) 64 (d) 36

14. A liquid solvent is in equilibrium with its vapour. When a non-volatile solute is added to this liquid, the instant effect is the rate at which the solvent molecules leaves the
(a) vapour phase and decreases.
(b) vapour phase and increases.
(c) solution and increases.
(d) solution and decreases.

15. The vapour pressure of a solution of a non-volatile and non-electrolyte solute in a solvent is 95% of the vapour pressure of the pure solvent at the same temperature. If the molecular mass of the solvent is 0.3 times that of solute, then the mass ratio of solvent and solute is
(a) 3 : 20 (b) 57 : 10
(c) 1 : 5 (d) 4 : 1

16. The vapour pressure of a solvent is decreased by 10 mm of Hg when a non-volatile solute was added to the solvent. The mole fraction of solute in the solution is 0.2. What would be the mole fraction of solvent if decrease in vapour pressure is 20 mm of Hg?
(a) 0.8 (b) 0.6
(c) 0.4 (d) 0.7

17. Vapour pressure of a solution of 5 g of non-electrolyte in 100 g of water at a particular temperature is 2985 $N/m^2$. If the vapour pressure of pure water at this temperature is 3000 $N/m^2$, then the molecular mass of the solute is
(a) 60 (b) 120
(c) 180 (d) 360

18. The vapour pressure of pure benzene at a certain temperature is 0.85 bar. A non-volatile, non-electrolyte solid weighing 0.5 g is added to 39.0 g of benzene. The vapour pressure of the solution then is 0.845 bar. Molar mass of the solid substance is
(a) 169 g/mol (b) 170 g/mol
(c) 85 g/mol (d) 39 g/mol

19. The vapour pressure of water is 12.3 kPa at 300 K. What is the vapour pressure of 1 molal aqueous solution of a non-volatile solute at 300 K?
(a) 1.208 kPa (b) 12.08 kPa
(c) 0.22 kPa (d) 12.13 kPa

**20.** A solution containing 30 g of a non-volatile solute exactly in 90 g water has a vapour pressure of 2.8 kPa at 298 K. On adding 18 g of water in the solution, the new vapour pressure becomes 2.9 kPa at 298 K. The molecular mass of the solute is

(a) 23 (b) 69
(c) 34 (d) 43

**21.** Lowering of vapour pressure in 1 molal aqueous solution at 100°C is

(a) 13.44 mm Hg (b) 14.12 mm Hg
(c) 31.2 mm Hg (d) 35.2 mm Hg

**22.** When 2 g non-volatile hydrocarbon containing 94.4% carbon by mass is dissolved in 100 g of benzene, the vapour pressure of benzene at 30°C is lowered from 89.78 mm to 89.0 mm. The molecular formula of the hydrocarbon is

(a) $C_{12}H_{34}$ (b) $C_{13}H_{22}$
(c) $C_{14}H_{12}$ (d) $C_{14}H_{10}$

**23.** Air was drawn through a solution containing 40 g of solute (non-volatile and non-electrolyte) in 100 g of water and then through water. The loss of mass of water was 0.05 g and the total mass of water absorbed in sulphuric acid tube was 2.05 g. The molecular mass of the solute is

(a) 267.86 (b) 288.0
(c) 295.2 (d) 302.4

**24.** A quantity of 10 g of solute 'A' and 20 g of solute 'B' is dissolved in 500 ml water. The solution is isotonic with the solution obtained by dissolving 6.67 g of 'A' and 30 g of 'B' in 500 ml water at the same temperature. The ratio of molar masses, $M_A : M_B$, is

(a) 1 : 1 (b) 3 : 1
(c) 1 : 3 (d) 2 : 3

**25.** Equal volumes of $M/20$ glucose solution at 300 K and $M/20$ sucrose solution at 300 K are mixed without change in temperature. If the osmotic pressure of glucose solution, sucrose solution and the mixture of two solutions are $\pi_1$, $\pi_2$ and $\pi_3$, respectively, then

(a) $\pi_1 = \pi_2 = \pi_3$ (b) $\pi_1 > \pi_2 > \pi_3$
(c) $\pi_1 < \pi_2 < \pi_3$ (d) $\pi_1 = \pi_2 < \pi_3$

**26.** A glucose solution is isopiestic with blood. What is the osmolarity of glucose solution?

(a) 0.31 M (b) 5.5 M
(c) 3.1 M (d) 0.154 M

**27.** Based upon the technique of reverse osmosis, the approximate minimum pressure required to desalinate sea water containing 2.5% (w/v) NaCl at 27°C should be

(a) 10.5 atm (b) 21 atm
(c) 2.1 atm (d) 1.05 atm

**28.** A liquid is in equilibrium with its vapour at its boiling point. On average, the molecules in the two phases have equal

(a) potential energy
(b) intermolecular force
(c) kinetic energy
(d) total energy

**29.** The boiling point of a 2% (w/w) aqueous solution of a non-volatile and non-electrolyte solute is 0.102°C higher than that of pure water. If $K_b$ for water is 0.52 K-kg/mol, the molecular mass of the solute is

(a) 180 (b) 102
(c) 40 (d) 104

**30.** Which one of the following statement is incorrect about a liquid solution containing a non-volatile solute?

(a) The fraction of total molecules in solution that has sufficient kinetic energy to overcome the binding forces of the liquid decreases as more solute molecules are added.
(b) The boiling point of the solution increases as we add more solute.
(c) The higher the concentration of the solute, the lower the vapour pressure of the solvent.
(d) Addition of a non-volatile solute increases the vapour pressure of the solvent.

**31.** The molar latent heat of vaporization of a certain liquid is 4.9 kcal/mol. The normal boiling point of the liquid is 77°C. The change in boiling point per atm increase in pressure at the normal boiling point of the liquid is

(a) 50 deg/atm (b) 0.02 deg/atm
(c) 5.0 deg/atm (d) 0.05 deg/atm

**32.** 72 g of glucose is dissolved in 1 kg of water in a sauce pan. At what temperature will the solution boil at 1.013 bar pressure? The value of $K_b$ for water is 0.52 K-kg mol$^{-1}$.

(a) 337.208 K (b) 373.208°C
(c) 373.208 K (d) 375.08 K

33. The boiling point of pure benzene is 353.23 K. When 1.80 g of a non-volatile solute was dissolved in 90 g of benzene, the boiling point is raised to 354.11 K. Calculate the molar mass of the solute. The value of $K_b$ for benzene is 2.53 K kg $mol^{-1}$.

(a) 85 g $mol^{-1}$ (b) 57.5 g $mol^{-1}$
(c) 23 g $mol^{-1}$ (d) 38.4 g $mol^{-1}$

34. At 100°C, the vapour pressure of a solution of 4.0 g of solute in 100 g of water is 750 mm. The boiling point of the solution is ($k_b$ of water = 0.52 K-kg/mol)

(a) 100°C (b) 100.04°C
(c) 100.4°C (d) 104.0°C

35. A quantity of 6.4 g sulphur dissolved in 200 g of $CS_2$ raises the boiling point of the solvent by 0.32°C. Boiling point of $CS_2$ = 47°C and latent heat of vaporization of $CS_2$ = 80.0 cal/g. The molecular formula of sulphur in this liquid is

(a) $S_8$ (b) $S_4$
(c) $S_2$ (d) S

36. What is the boiling point of a solution of 1.00 g of naphthalene dissolved in 94.0 g of toluene? The normal boiling point of toluene is 110.75°C and $K_{x,b}$ = 32.0 K for toluene.

(a) 110.95°C (b) 111.0°C
(c) 113.41°C (d) 110.75°C

37. The vapour pressure of an aqueous solution is found to be 750 torr at a temperature *T*, and the same solution shows an elevation in boiling point equal to 1.04 K. If *T* is the boiling point of pure water, then the atmospheric pressure should be ($K_b$ of water = 0.52 K-kg/mol)

(a) 760 torr (b) 777 torr
(c) 746 torr (d) 750 torr

38. When glycerin is added to a litre of water, then which of the following is observed?

(a) Water evaporates more easily.
(b) The temperature of water increases.
(c) The freezing point of water is lowered.
(d) The viscosity of water is lowered.

39. The molar mass of substance forming 7.0% (by mass) solution in water which freezes at −0.93°C, is (The cryoscopic constant of water is 1.86°C kg $mol^{-1}$).

(a) 140 g $mol^{-1}$ (b) 150.5 g $mol^{-1}$
(c) 160 g $mol^{-1}$ (d) 155 g $mol^{-1}$

40. When 36.0 g of a solute having the empirical formula $CH_2O$ is dissolved in 1.20 kg of water, the solution freezes at −0.93°C. What is the molecular formula of the solute? ($K_f$ = 1.86°C kg $mol^{-1}$)

(a) $C_6H_{12}O_6$ (b) $C_3H_6O_3$
(c) $CH_2O$ (d) $C_2H_4O_2$

41. The approximate molality of ethylene glycol ($C_2H_6O_2$) in an aqueous solution which freezes at a temperature not higher than −15°C is ($k_f$ of water = 1.86°C kg $mol^{-1}$)

(a) 8.06 m (b) 0.806 m
(c) 0.145 m (d) 1.5 m

42. An aqueous solution of non-volatile and non-electrolyte solute boils at 100.78°C. It should freeze at (for water, $K_f$ and $K_b$ are 1.86 and 0.52 K-kg/mol, respectively)

(a) 2.79°C (b) 270.21°C
(c) −2.79°C (d) −270.21°C

43. How many grams of sucrose (molecular weight = 342) should be dissolved in 100 g water in order to produce a solution with a 104.76°C difference between the freezing point and the boiling point temperature? ($K_f$ = 1.86, $K_b$ = 0.52)

(a) 34.2 g (b) 68.4 g
(c) 684 g (d) 313.06 g

44. Molal depression constant for Pb is 8.5°C kg/mol. How many grams of Sn must be dissolved in 425 g of Pb to produce an alloy with melting point of 322°C? The melting point of pure Pb is 327°C. (Sn = 120)

(a) 30.0 g (b) 15.0 g
(c) 86.7 g (d) 41.3 g

45. The boiling point and freezing point of a solvent 'A' are 90.0°C and 3.5°C, respectively. The $K_f$ and $K_b$ values of the solvent are 17.5 and 5.0 K-kg/mol, respectively. What is the boiling point of a solution of 'B' (non-volatile, non-electrolyte solute) in 'A', if the solution freezes at 2.8°C?

(a) 90.0°C (b) 89.8°C
(c) 90.2°C (d) 90.7°C

46. A very small amount of non-volatile solute (that does not dissociate) is dissolved in 56.8 $cm^3$ of benzene (density 0.889 g $cm^{-3}$). At room temperature, vapour pressure of this solution is 100 mm Hg while that of benzene is 102 mm Hg. If the freezing temperature of this solution is 1.3 degree lower than that of benzene, then what is the value of molal freezing point depression constant of benzene?

(a) 5.07 deg/molal (b) 1.3 deg/molal
(c) 3.9 deg/molal (d) 4.97 deg/molal

47. A liquid freezes at 7°C and boils at 77°C. If the $K_f$ and $K_b$ values for the liquid are 5.6 and 2.5 deg/molal, respectively, then the ratio of molar latent heat of fusion to molar latent heat of vaporization is

(a) 1 : 1 (b) 1 : 3
(c) 2 : 7 (d) 7 : 3

48. The normal freezing point of nitrobenzene is 278.82 K. A non-volatile solute is dissolved in it and a solution of molality 0.25 m is prepared. If the observed freezing point of the solution is 276.82 K, the value for $K_f$ of nitrobenzene is

(a) 6 K-kg $mol^{-1}$ (b) 8 K-kg $mol^{-1}$
(c) 7 K-kg $mol^{-1}$ (d) 5 K-kg $mol^{-1}$

49. A 5% (by mass) solution of cane sugar in water has freezing point of 272.85 K. Calculate the freezing point of 5% (by mass) solution of glucose in water, if the freezing point of pure water is 273.15 K.

(a) 0.57 K (b) 272.58 K
(c) 272.85 K (d) 273.72 K

50. Two elements A and B form compounds having molecular formula $AB_2$ and $AB_4$. When dissolved in 20 g of $C_6H_6$, 1 g of $AB_2$ lowers the freezing point by 2.55 K, whereas 1.0 g of $AB_4$ lowers it by 1.7 K. The molar depression constant for benzene is 5.1 K-kg $mol^{-1}$. The atomic masses of A and B are

(a) 50, 25 (b) 50, 50
(c) 25, 50 (d) 75, 25

51. Ice begins to separate at –0.744°C from an aqueous solution containing 2.4 g of a non-volatile, non-electrolyte in 100 g water. The osmotic pressure of the solution at 27°C is ($K_f$ for water = 1.86, assume molality = molarity)

(a) 0.4 atm (b) 9.852 atm
(c) 997.68 atm (d) 240 atm

52. The osmotic coefficient of a non-electrolyte is related to the freezing point depression by the expression, $\phi = \Delta T_f/(m \cdot K_f)$. The depression in freezing point of 0.4 molal aqueous solution of sucrose is 0.93°C. The osmotic coefficient is ($K_f$ of water = 1.86 K-kg/mol)

(a) 0.8 (b) 1.0
(c) 1.25 (d) 0.125

53. A solution of a non-volatile solute in water freezes at –1.0°C. The vapour pressure of pure water at 298 K is 24.24 mm Hg and $K_f$ for water is 1.80 K-kg $mol^{-1}$. The vapour pressure of this solution at 298 K is

(a) 24.00 mm Hg (b) 22.04 mm Hg
(c) 24.12 mm Hg (d) 23.76 mm Hg

54. A solution of 'x' mole of sucrose in 100 g of water freezes at – 0.2°C. As ice separates out, the freezing point goes down to – 0.25°C. How many grams of ice would have separated?

(a) 18 g (b) 20 g
(c) 80 g (d) 25 g

55. In KI solution, mercuric iodide is added. The osmotic pressure of resultant solution will

(a) increase
(b) decrease
(c) remains unchanged.
(d) increase or decrease, depending on amount.

56. If 0.1 M solution of NaCl has same osmotic pressure as 0.1 M solution of $CaCl_2$, then the degree of dissociation of $CaCl_2$ will be

(a) 0.5 (b) 0.2
(c) 0.4 (d) 0.33

57. The osmolarity of 0.2 M-$Na_2SO_4$ is

(a) 0.6 M (b) 0.4 M
(c) 0.2 M (d) 0.8 M

58. $PtCl_4 \cdot 6H_2O$ can exist as a hydrated complex. 1.0 molal aqueous solution has depression in freezing point of 3.72°C. Assume 100% ionization and $K_f$ of water = 1.86°C $mol^{-1}$ kg. The complex is

(a) $[Pt(H_2O)_6]Cl_4$
(b) $[Pt(H_2O)_4Cl_2]Cl_2 \cdot 2H_2O$
(c) $[Pt(H_2O)_3Cl_3]Cl \cdot 3H_2O$
(d) $[Pt(H_2O)_2Cl_4] \cdot 4H_2O$

59. One molal solution of benzoic acid in benzene boils at 81.53°C. The normal boiling point of benzene is 80.10°C. Assuming that the solute is 90% dimerized, the value of $K_b$ for benzene is
(a) 3.5 deg/molal (b) 5.2 deg/molal
(c) 2.6 deg/molal (d) 0.75 deg/molal

60. A quantity of 2 g of $C_6H_5COOH$ dissolved in 25 g of benzene shows a depression in freezing point equal to 1.96 K. Molar depression constant for benzene is 4.9 K-kg $mol^{-1}$. What is the percentage association of acid if it forms double molecules (dimer) in solution?
(a) 39% (b) 78%
(c) 61% (d) 19.5%

61. The pH of a 0.1 M solution of a monobasic acid is 2.0. Its osmotic pressure at a given temperature $T$ K is
(a) $0.1RT$ (b) $0.11RT$
(c) $0.09RT$ (d) $0.01RT$

62. To one litre of 0.1 M-HCl solution, 0.025 mole of solid $NH_4Cl$ is added. Assuming complete dissociation of solutes, the freezing point of solution is ($K_f$ of water = 1.86 K-kg $mol^{-1}$)
(a) −0.465°C (b) −0.93°C
(c) −0.372°C (d) −0.279°C

63. 'x' moles of KCl and 'y' mole of $BaCl_2$ are dissolved in 1 kg of water. If $(x + y) = 0.1$, then the observed range of $\Delta T_f$ is ($K_f$ of water = 1.86 K-kg $mol^{-1}$)
(a) 0.372°C to 0.558°C
(b) 0.186°C to 0.93°C
(c) 0.558°C to 0.93°C
(d) 0.372°C to 0.93°C

64. A quantity of 3.125 g of a mixture of KCl and NaCl dissolved in 1 kg of water produces a depression of 0.186°C in freezing point. The molar ratio of KCl to NaCl in the solution (assuming complete dissociation of the salts) is ($K_f$ = 1.86 deg/ molal)
(a) 1:3 (b) 2:3
(c) 1:1 (d) 3:1

65. A dilute solution contains $n$ mole of solute 'A' in 1 kg of a solvent with molal elevation constant $K_b$. The solute A undergoes dimerization as $2A \rightleftharpoons A_2$. If $\Delta T_b$ is the elevation in boiling point of a given solution and molality = molarity, the equilibrium constant $K_C$ for dimer formation is
(a) $\frac{K_b(n.K_b - \Delta T_b)}{(2.\Delta T_b - n.K_b)}$ (b) $\frac{K_b(n.K_b - \Delta T_b)}{(2.\Delta T_b - n.K_b)^2}$
(c) $\frac{K_b.n.\Delta T_b}{(2.\Delta T_b - n.K_b)^2}$ (d) $\frac{(n.K_b - \Delta T_b)}{(n.\Delta T_b - 2.K_b)^2}$

66. Two solvents 'A' and 'B' have $K_f$ values 1.86 and 2.79 K $mol^{-1}$ kg, respectively. A given amount of a substance when dissolved in 500 g of 'A' completely dimerizes and when the same amount of the substance is dissolved in 500 g of 'B', the solute undergoes trimerization. The ratio of observed lowering of freezing point in two cases is
(a) 2 : 3 (b) 4 : 9
(c) 1 : 1 (d) 1 : 2

67. Which of the following solution will show minimum osmotic effect?
(a) KCl solution
(b) Colloidal gold sol
(c) $CaCl_2$ solution
(d) $Na_3PO_4$ solution

68. A 0.001 molal solution of a complex $MA_8$ in water has the freezing point of −0.0054°C. Assuming 100% ionization of the complex in water, which of the following is the correct representation of the complex? ($K_f$ of water = 1.86 $Km^{-1}$)
(a) $[MA_8]$ (b) $[MA_7]A$
(c) $[MA_6]A_2$ (d) $[MA_5]A_3$

69. The vapour pressure of a saturated solution of sparingly soluble salt ($XCl_3$) was 17.20 mm Hg at 27°C. If the vapour pressure of pure water is 17.25 mm Hg at 27°C, then what is the solubility of the sparingly soluble salt $XCl_3$ in mole per litre?
(a) $4.04 \times 10^{-2}$ (b) $8.08 \times 10^{-2}$
(c) $2.02 \times 10^{-2}$ (d) $4.04 \times 10^{-3}$

**70.** A non-volatile solute 'X' completely dimerizes in water, if the temperature is below −3.72°C and the solute completely dissociates as $X \rightarrow Y + Z$, if the temperature is above 100.26°C. In between these two temperatures (including both temperatures), the solute is neither dissociated nor associated. One mole of 'X' is dissolved in 1.0 kg water ($K_b = 0.52$ K-kg/mol, $K_f = 1.86$ K-kg/ mol). Identify the incorrect information related with the solution.

(a) The freezing point of solution is −1.86°C.
(b) The boiling point of solution is 101.04°C.
(c) When the solution is cooled to −7.44°C, 75% of water present initially will separate as ice.
(d) When the solution is heated to 102.08°C, 50% of water present initially will escape out as vapour.

## Section B (One or More than one Correct)

**1.** A cylinder fitted with a movable piston contains liquid water in equilibrium with water vapour at 25°C. Which of the following operation(s) results in a decrease in the equilibrium vapour pressure at 25°C?

(a) Moving the piston downward for a short distance.
(b) Removing small amount of vapour.
(c) Removing a small amount of liquid water.
(d) Dissolving some salt in the water.

**2.** Which is a colligative property?

(a) Osmotic pressure (b) Vapour pressure
(c) Freezing point (d) Boiling point

**3.** The colligative properties of a solution may depend on

(a) Temperature
(b) Nature of solute
(c) Nature of solvent
(d) Amount of solution

**4.** The vapour pressures of pure liquids A, B and C are 75, 22 and 10 torr, respectively. Which of the following is/are possible value(s) of vapour pressure of binary or ternary solutions having equimolar amounts of these liquids? Assume ideal behaviour for all possible solutions.

(a) 53.5 torr (b) 35.67 torr
(c) 48.5 torr (d) 16 torr

**5.** Which of the following statement(s) is/are correct?

(a) Minimum boiling azeotropic mixture boils at temperature lower than either of the two pure components.
(b) Maximum boiling azeotropic mixture boils at temperature higher than either of the two pure components.
(c) Minimum boiling azeotropic mixture shows (+) ve deviation.
(d) Maximum boiling azeotropic mixture shows (−) ve deviation.

**6.** Which of the following liquid pairs do not show positive deviation from Raoult's law?

(a) Acetone–Chloroform
(b) Benzene–Methanol
(c) Water–Nitric acid
(d) Water–Hydrochloric acid

**7.** A mixture of liquids 'A' and 'B' in the molar ratio 1 : 2 forms a maximum boiling azeotrope. Identify the incorrect statement, if 'A' is more volatile. (Molar masses: A = 100, B = 50).

(a) A liquid solution of 'A' and 'B' having mass % of A = 50 will have vapours having mass % of A = 50.
(b) A liquid solution of 'A' and 'B' having mass % of A > 50 will have vapours having mass % of A > 50.
(c) A mixture of 'A' and 'B' in the molar ratio 1 : 3 can be separated into azeotropic mixture and pure 'A'.
(d) A mixture of 'A' and 'B' in the molar ratio 2 : 3 can be separated into azeotropic mixture and pure 'A'.

**8.** Which of the following statement(s) is/are correct?

(a) Two sucrose solution of same molality prepared in different solvent will have the same freezing point depression.
(b) Osmotic pressure ($\pi$) of a solution is given by $\pi = MRT$, where $M$ is the molarity of the solution.
(c) At same temperature, the correct order of osmotic pressure for 0.01 M aqueous solution of each compound is $BaCl_2 > KCl > CH_3COOH >$ Sucrose.
(d) Raoult's law states that the vapour pressure of a component over a solution is proportional to its mole fraction.

9. Consider 0.1 m aqueous solutions of two solutes X and Y. The solute X behaves as a univalent electrolyte, while the solute Y dimerizes in solution. Which of the following statement(s) is/are correct regarding these solutions?
   (a) The boiling point of the solution of X will be higher than that of Y.
   (b) The osmotic pressure of the solution of X will be higher than that of Y.
   (c) The freezing point of the solution of X will be higher than that of Y.
   (d) The relative lowering of vapour pressure of both the solution will be the same.

10. In which of the following liquid pair, there is contraction in volume on mixing?
    (a) $C_6H_{14} + C_6H_6$ (b) $H_2O + HCl$
    (c) $H_2O + HNO_3$ (d) $H_2O + C_2H_5OH$

11. In a binary electrolyte (AB type), the observed lowering of vapour pressure as compared to the theoretically calculated one for non-electrolyte can never be
    (a) more than double.
    (b) exactly equal to double.
    (c) less than double.
    (d) more than one.

12. Benzene and naphthalene forms an ideal solution at room temperature. For this process, the true statement(s) is(are)
    (a) $\Delta G = +ve$ (b) $\Delta S_{system} = +ve$
    (c) $\Delta S_{surrounding} = 0$ (d) $\Delta H = 0$

13. An amount of 1 mole of a non-volatile solid is dissolved in 200 moles of water. The solution is cooled to a temperature 'T' K (lower than the freezing point of the solution) to cause ice formation. After the removal of ice, the remaining solution is heated to 373 K, where the vapour pressure of solution is observed to be 740 mm Hg. Identify the correct information(s) from the following ($K_f$ of water = 2.0 K-kg/mol).
    (a) 163 moles of ice is formed at $T$ K.
    (b) $T\text{K} = \left(273 - \frac{2000}{37\times18}\right)$ K
    (c) Freezing point of the original solution is $-\frac{10}{18}$ °C.
    (d) Relative lowering of vapour pressure of the final solution is $\frac{1}{38}$.

14. Blood cells in the human body have semipermeable membrane and depending upon the concentration of solution inside blood cells and outside (in the blood), **'Lysis'** (expansion of blood cells) and **'Crenation'** (contraction of blood cells) may occur. Kidneys are responsible for keeping the solution inside blood cell and blood at the same concentration. Identify the correct information(s).
    (a) Lysis will occur when blood cells are kept in a solution which is isotonic with blood.
    (b) Crenation will occur when blood cells are kept in a solution which is hypertonic with blood.
    (c) Blood cells will have normal shape when placed in an isotonic solution with blood.
    (d) Lysis will occur when blood cells are kept in a solution which is hypotonic with blood.

15. Two liquids 'A' and 'B' form a solution which shows significant positive deviation from Raoult's law. The normal boiling point of pure liquid 'A' is higher than that of 'B'. Which of the following information(s) related with the solution must be incorrect?
    (a) Normal boiling point of the solution of any composition will always be less than that of 'A'.
    (b) Azeotropic mixture of 'A' and 'B' will have normal boiling point lower than that of 'B'.
    (c) If a small amount of liquid 'B' is added in the liquid solution, the normal boiling point of solution will always decrease.
    (d) On distillation of the solution at constant pressure of 1 atm, the distillate will have pure liquid 'Y' and the residual liquid will be pure liquid 'X'.

## Section C (Comprehensions)

### Comprehension I

**The composition of vapour over a binary ideal solution is determined by the composition of the liquid. $X_A$ and $Y_A$ are the mole fractions of 'A' in the liquid and vapour at equilibrium, respectively.**

1. The value of $X_A$ for which $(Y_A - X_A)$ is maximum is

(a) $\frac{\sqrt{P_A^o \cdot P_B^o} - P_A^o}{P_A^o - P_B^o}$ (b) $\frac{\sqrt{P_A^o \cdot P_B^o} - P_B^o}{P_A^o - P_B^o}$

(c) $\frac{\sqrt{P_A^o \cdot P_B^o} - P_B^o}{P_A^o}$ (d) $\frac{\sqrt{P_A^o \cdot P_B^o} - P_A^o}{P_B^o}$

2. The vapour pressure of solution at this composition is

(a) $\sqrt{P_A^o \cdot P_B^o}$ (b) $(P_A^o - P_B^o)$

(c) $(P_A^o + P_B^o)$ (d) $0.5\,(P_A^o + P_B^o)$

### Comprehension II

**Liquids 'A' and 'B' forms an ideal solution. At $T$ K, $P_A^o = 0.4$ bar and $P_B^o = 0.6$ bar. In a cylinder piston arrangement, 2.0 moles of vapours of liquid 'A' and 3.0 moles of vapours of liquid 'B' are taken at $T$ K and 0.3 bar.**

3. Predict whether the vapours will condense or not, at the given pressure.

(a) Yes, but only a single drop of liquid will form.
(b) Yes, but only partial condensation will occur.
(c) Yes, complete condensation will occur.
(d) No

4. If the vapours are compressed slowly and isothermally, then at what pressure, the first drop of liquid will appear?

(a) 0.4 bar (b) 0.5 bar
(c) 0.52 bar (d) 0.6 bar

5. If the initial volume of vapours was 10.0 dm$^3$, then at what volume, the first drop of liquid will form?

(a) 6.0 dm$^3$ (b) 4.0 dm$^3$
(c) 10.0 dm$^3$ (d) 1.0 dm$^3$

6. What is the composition of first drop of liquid formed?

(a) $X_A = 0.4$ (b) $X_A = 0.5$
(c) $X_A = 0.6$ (d) $X_A = 1.0$

7. If the vapours are compressed further, then at what pressure, almost complete condensation of vapours will occur?

(a) 0.4 bar (b) 0.5 bar
(c) 0.52 bar (d) 0.6 bar

8. What is the composition of last traces of vapours remained?

(a) $Y_A = \frac{2}{5}$ (b) $Y_A = \frac{1}{2}$

(c) $Y_A = \frac{9}{13}$ (d) $Y_A = \frac{4}{13}$

9. Which of the following is the only incorrect information regarding the composition of the system at 0.51 bar pressure?

(a) $X_A = 0.45$

(b) $Y_A = \frac{6}{17}$

(c) $n_{A(\text{liquid})} = \frac{12}{11}$

(d) $n_{A(\text{vapour})} = \frac{12}{11}$

10. At what pressure, 20% of the total moles of vapours will liquefy? ($\sqrt{145} = 12.04$)

(a) 0.5 bar (b) 0.504 bar
(c) 0.48 bar (d) 0.508 bar

## Comprehension III

**One mole of liquid 'A' and two moles of liquid 'B' are mixed at 27°C to form an ideal binary solution. (ln 2 = 0.7, ln 3 = 1.1)**

**11.** The value of $\Delta H_{mix}$ is

(a) 0 (b) + ve
(c) − ve (d) any of these

**12.** The value of $\Delta G_{mix}$ is

(a) 0 (b) + 1140 cal/mol
(c) −1140 cal/mol (d) −380 cal/mol

**13.** The value of $\Delta S_{mix}$ is

(a) 0
(b) + 3.8 cal/K
(c) −3.8 cal/K
(d) + 3.8 cal/K-mol

## Comprehension IV

**Benzene and toluene form an ideal solution. A liquid solution is formed by mixing 10 moles each of benzene and toluene at 300 K and 200 mm Hg pressure. At 300 K, the vapour pressures of pure benzene and toluene are 100 and 40 mm Hg, respectively.**

**14.** If the pressure over the mixture at 300 K is reduced, then at what pressure does the first vapour form?

(a) 40 mm Hg (b) 70 mm Hg
(c) 100 mm Hg (d) 199 mm Hg

**15.** What is the composition (mole fraction of benzene) of the first traces of vapour formed?

(a) 0.5 (b) 0.857
(c) 0.714 (d) 0.286

**16.** If the pressure is reduced further, then at what pressure does the last trace of liquid disappear?

(a) 57.14 mm Hg (b) 40 mm Hg
(c) 100 mm Hg (d) 66.67 mm Hg

**17.** What is the composition (mole fraction of benzene) of the last trace of liquid?

(a) 0.5 (b) 0.857
(c) 0.714 (d) 0.286

**18.** At what pressure, half of the total moles of liquid solution will vaporize?

(a) 57.14 mm Hg (b) 63.25 mm Hg
(c) 70 mm Hg (d) 66.67 mm Hg

## Comprehension V

**The two liquids 'A' and 'B' have the same molecular mass and form ideal solution. The solution has the vapour pressure 700 mm at 80°C. The above solution is distilled without reflux. Three-fourth of the solution is collected as condensate. The mole fraction of 'A' in the condensate is 0.75 and that in the residue is 0.3. The vapour pressure of the residue at 80°C is 600 mm.**

**19.** The mass ratio of 'B' and 'A' in the initial solution at equilibrium at 80°C was

(a) 1 : 1 (b) 29 : 51
(c) 51 : 80 (d) 29 : 80

**20.** The vapour pressure of pure liquid 'A' at 80°C is

(a) 807.4 mm (b) 511.1 mm
(c) 755.6 mm (d) 533.3 mm

**21.** The vapour pressure of pure liquid 'B' at 80°C is

(a) 807.4 mm (b) 511.1 mm
(c) 755.6 mm (d) 533.3 mm

## Comprehension VI

**Liquids A (molar mass = 100 g/mol and density = 1.25 g/ml) and B (molar mass = 50 g/mol and density = 1.0 g/ml) form an ideal solution. The molal volume of liquid 'A' increases by a factor of 3800 as it vaporizes at 27°C and that of liquid 'B' increases by a factor of 7600 at 27°C. A solution of liquids 'A' and 'B' at 27°C has a vapour pressure of 54.0 torr. (R = 0.08 L-atm/K-mol).**

22. The vapour pressure of pure liquid 'A' at 27°C is
    (a) 60 torr (b) 48 torr
    (c) 49.26 torr (d) 61.58 torr

23. The vapour pressure of pure liquid 'B' at 27°C is
    (a) 60 torr (b) 48 torr
    (c) 49.26 torr (d) 61.58 torr

24. The mole fraction of 'A' in the vapour above the solution in equilibrium at 27°C is
    (a) 0.5 (b) 0.44
    (c) 0.56 (d) 0.75

## Comprehension VII

**Two solutions of non-volatile solutes 'A' and 'B' are prepared. The molar mass ratio is $M_A : M_B = 1 : 3$. Both are prepared a 5% solution by mass in water.**

25. The ratio of the freezing point depressions $\Delta T_f$ (A):$\Delta T_f$ (B) of the solution is
    (a) 1:3 (b) 3:1
    (c) 1:1 (d) 5:3

26. If the above solutions are mixed to prepare two solutions $S_1$ and $S_2$, in the ratio 2 : 3 and 3 : 2, by volume, respectively, then what would be the ratio $\Delta T_f (S_1) : \Delta T_f (S_2)$?
    (a) 1 : 1 (b) 4 : 9
    (c) 9 : 11 (d) 2 : 3

## Comprehension VIII

**Properties such as boiling point, freezing point and vapour pressure of a pure solvent change when solute molecules are added to get homogeneous solution. These are called colligative properties. Applications of colligative properties are very useful in day-to-day life. One of its examples is the use of ethylene glycol and water mixture as anti-freezing liquid in the radiator of automobiles.**

**A solution M is prepared by mixing ethanol and water. The mole fraction of ethanol in the mixture is 0.9.**

**Freezing point depression constant of water ($K_f^{water}$) = 1.86 K kg mol$^{-1}$**
**Freezing point depression constant of ethanol ($K_f^{ethanol}$) = 2.0 K kg mol$^{-1}$**
**Boiling point elevation constant of water ($K_b^{water}$) = 0.52 K kg mol$^{-1}$**
**Boiling point elevation constant of ethanol ($K_b^{ethanol}$) = 1.2 K kg mol$^{-1}$**
**Standard freezing point of water = 273 K**
**Standard freezing point of ethanol = 155.7 K**
**Standard boiling point of water = 373 K**
**Standard boiling point of ethanol = 351.5 K**
**Vapour pressure of pure water = 32.8 mm Hg**
**Vapour pressure of pure ethanol = 40 mm Hg**
**Molecular weight of water = 18 g mol$^{-1}$**
**Molecular weight of ethanol = 46 g mol$^{-1}$**

**In answering the following questions, consider the solutions to be ideal dilute solutions and solutes to be non-volatile and non-dissociative.**

27. The freezing point of the solution **M** is

(a) 268.7 K (b) 268.5 K
(c) 234.2 K (d) 150.9 K

28. The vapour pressure of the solution **M** is

(a) 39.3 mm Hg (b) 36.0 mm Hg
(c) 29.5 mm Hg (d) 28.8 mm Hg

29. Water is added to the solution **M** such that the mole fraction of water in the solution becomes 0.9. The boiling point of this solution is

(a) 380.4 K (b) 376.2 K
(c) 375.5 K (d) 354.7 K

## Comprehension IX

**The lowering of vapour pressure on adding a non-volatile solute in a pure liquid solvent may be measured by Ostwald–Walker method. In this method, dry air is first passed through a series of vessels having the solution, then through a series of vessels having pure solvent and finally through a vessel (normally U-tube) having the absorbent of the solvent. The masses of solution and pure solvent decreases due to removal of vapours of solvent in the flow of air and the mass of absorbent increases due to absorption of the vapour of solvent in order to make the air dry. By measuring the changes in mass, we may determine the lowering of vapour pressure and hence, the composition of solution using Raoult's law.**

30. If the mass of absorbent is increased by 0.24 g and the mass of pure solvent (water) is decreased by 0.02 g, then the mass percent of solute (glucose) in its aqueous solution is

(a) $\frac{1000}{21}\%$ (b) $\frac{100}{12}\%$

(c) $\frac{1000}{22}\%$ (d) $\frac{100}{11}\%$

31. If the experiment is performed with aqueous $AlCl_3$ solution ($a = 0.8$) prepared by dissolving 1 mole of $AlCl_3$ in 17 mole of water and the decrease in the mass of solution in the experiment is found to be 0.18 g, then the increase in the mass of absorbent should be

(a) 0.216 g (b) 0.225 g
(c) 0.191 g (d) 1.08 g

32. If the arrangement of vessels is changed in the experiment as

$$\text{Dry air} \xrightarrow{\text{Pure Solvent}} \xrightarrow{\text{Solution}} \xrightarrow{\text{Absorbent}} \text{Dry air}$$

then, which of the following must be incorrect?

(a) The mass of pure solvent will decrease.
(b) The mass of solution will decrease.
(c) The mass of absorbent will increase.
(d) The experiment will obey the law of conservation of mass.

## Section D (Assertion – Reason)

The following questions consist of two statements. Mark the answer as follows.

**(a)** If both statements are CORRECT, and **Statement II** is the CORRECT explanation of **Statement I**.

**(b)** If both statements are CORRECT, and **Statement II** is NOT the CORRECT explanation of **Statement I**.

**(c)** If **Statement I** is CORRECT, but **Statement II** is INCORRECT.

**(d)** If **Statement I** is INCORRECT, but **Statement II** is CORRECT.

1. **Statement I:** The solubility of the gas in a liquid increases with increase in pressure.

**Statement II:** The solubility of a gas in a liquid, expressed by mole fraction of gas in the liquid solution at equilibrium is directly proportional to the pressure of the gas.

2. **Statement I:** At the same temperature, water has higher vapour pressure than acetic acid.

**Statement II:** Hydrogen bonding in water is weaker than in acetic acid.

3. **Statement I:** 0.1 m aqueous solution of glucose has higher depression in freezing point than 0.1 m aqueous solution of urea.

**Statement II:** $k_f$ has same value in both.

EXERCISE II

4. **Statement I:** The boiling point of 0.1 m of urea solution is less than that of 0.1 m of KCl solution.
**Statement II:** Elevation of boiling point is directly proportional to the number of species present in the solution.

5. **Statement I:** Addition of ethylene glycol (non-volatile) to water lowers the freezing point of water.
**Statement II:** Addition of any substance to water lowers the freezing point of water.

6. **Statement I:** Addition of a non-volatile solute to a volatile solvent increases the boiling point.
**Statement II:** Addition of non-volatile solute to a volatile solvent results decrease in the vapour pressure.

7. **Statement I:** Addition of a non-volatile solute to a volatile solvent decreases the freezing point.
**Statement II:** Addition of non-volatile solute to a volatile liquid solvent in equilibrium with the solid solvent results in the fusion of some solid solvent. In order to re-achieve the equilibrium, the temperature of the system must be decreased.

8. **Statement I:** When an ideal binary solution of liquids A and B is distilled, the boiling point of the distillate becomes lower than that of the parental liquid solution.
**Statement II:** When an ideal binary solution of liquids A and B is distilled, the distillate contains relatively larger amount of the more volatile liquid.

9. **Statement I:** Except osmotic pressure, all other colligative properties depend on the nature of solvent.
**Statement II:** Colligative properties are intensive properties.

10. **Statement I:** The vapours above a liquid solution of two volatile liquids is always more rich in more volatile component.
**Statement II:** More volatile component has lower boiling point.

## Section E (Column Match)

1. Match the solution in Column I with its nature in Column II.

| Column I | Column II |
|---|---|
| (A) Benzene + Toluene | (P) Non-ideal solution |
| (B) Ethanol + Water | (Q) Ideal solution |
| (C) Benzene + Ethanol | (R) $\Delta H_{mix} > 0$ |
| (D) Acetone + Chloroform | (S) $\Delta H_{mix} < 0$ |

2. Match the solution in Column I with its nature in Column II.

| Column I | Column II |
|---|---|
| (A) *n*-hexane + *n*-heptane | (P) Can be perfectly separated by distillation. |
| (B) Acetone + Chloroform | (Q) Maximum boiling azeotrope. |
| (C) Acetone + Aniline | (R) Cannot be perfectly separated by distillation. |
| (D) Ethanol + Water | (S) Minimum boiling azeotrope. |

3. Match the solutions in Column I with their osmotic properties in Column II.

| Column I | Column II |
|---|---|
| (A) $S_1$ : 0.1 M-glucose; $S_2$ : 0.1 M-urea | (P) $S_1$ and $S_2$ are isotonic. |
| (B) $S_1$ : 0.1 M-NaCl; $S_2$ : 0.1 M-$Na_2SO_4$ | (Q) $S_1$ is hypertonic to $S_2$. |
| (C) $S_1$ : 0.1 M-NaCl; $S_2$ : 0.1 M-KCl | (R) $S_1$ is hypotonic to $S_2$. |
| (D) $S_1$ : 0.1 M-$CuSO_4$ $S_2$ : 0.1 M-sucrose | |

4. Match the following if the molecular masses of X, Y and Z are same.

| Column I (Solvent and its boiling point) | Column II (Molal elevation constant, $K_b$) |
|---|---|
| (A) X (100°C) | (P) 0.68 |
| (B) Y (27°C) | (Q) 0.53 |
| (C) Z (253°C) | (R) 0.98 |
| (D) W (182°C) | (S) 0.79 |

EXERCISE II

5. Match the columns.

| Column I (Solute in water) | Column II (Degree of dissociation to make van't Hoff factor equal to 2) |
|---|---|
| (A) KCl | (P) 0.33 |
| (B) $Na_2SO_4$ | (Q) 1.00 |
| (C) $K_4[Fe(CN)_6]$ | (R) 0.50 |
| (D) $K_3[Fe(CN)_6]$ | (S) 0.25 |

6. Match the columns.

| Column I (Solute in water) | Column II (van't Hoff factor considering complete dissociation) |
|---|---|
| (A) Glucose | (P) 2.0 |
| (B) NaCl | (Q) 1.0 |
| (C) $MgCl_2$ | (R) 3.0 |
| (D) $AlCl_3$ | (S) 4.0 |

7. Match the columns.

| Column I (Colligative properties) | Column II (Aqueous solution; molality = molarity) |
|---|---|
| (A) $\Delta T_f = 0.3K_f$ | (P) 0.1 M-$Ca(NO_3)_2$ |
| (B) $\Delta T_b = 0.28K_b$ | (Q) 0.14 M-NaBr |
| (C) $\pi = 0.19RT$ | (R) 0.1 M-$MgCl_2$ ($\alpha = 0.9$) |
| (D) $\frac{P^\circ - P}{P^\circ} = \frac{\left(\frac{\Delta T_f}{K_f}\right)}{\frac{1000}{18} + \left(\frac{\Delta T_f}{K_f}\right)}$ | (S) 0.28 M-urea |
| | (T) 0.1 M-HA (monobasic acid, $K_a = 0.81$) |

## Section F (Subjective)

### Single Digit Integer Type

1. A quantity of 4 ml of a gas at 1 atm and 300 K is dissolved in 1 L of water. The volume (in ml) of gas that will dissolve in 0.5 L of water at 4 atm and 300 K is

2. A '100 proof' solution of ethanol in water consists of 50.00 ml of $C_2H_5OH(l)$ and 50.00 ml of $H_2O(l)$ mixed at 15.56°C. The density of the solution is 0.9344 g/ml, that of pure $H_2O$ is 1.0000 g/ml and that of pure $C_2H_5OH$ is 0.7939 g/ml. Is the solution ideal? Answer '1', if the solution is ideal and answer '2', if the solution is non-ideal.

3. A mixture of ideal gases is cooled up to liquid helium temperature (4.22 K) to form an ideal solution. Is this statement true or false? Answer '1' for true and '2' for false.

4. The vapour pressure of water over an aqueous solution of $NH_3$ at 70°C is 186.8 torr. Given that the mole fraction of $NH_3$ in the solution at equilibrium is 0.20 and the vapour pressure of pure $H_2O(l)$ at 70°C is 233.5 torr, determine whether the solution is ideal or not. Answer '1', if the solution is ideal and answer '2', if the solution is non-ideal.

5. Solutions A and B have osmotic pressures of 2.4 atm and 4.2 atm, respectively, at a certain temperature. The osmotic pressure (in atm) of a solution prepared by mixing the solutions in 2 : 1 volume ratio, respectively, at the same temperature is

6. At 27°C, a solution containing 0.2 g of polyisobutylene in 100 ml of benzene developed a rise of 2.463 mm in osmotic equilibrium. The molar mass of polyisobutylene (in $10^5$ g/mol) is (Given: The density of final solution is 1.013 g/ml, $g = 10$ m/s$^2$, $R = 0.0821$ L-atm/K-mol)

7. Consider the following arrangement, in which a solution containing 20 g of haemoglobin in 1 $dm^3$ of the solution is placed in right compartment and pure water is placed in left compartment, separated by SPM. At equilibrium, the height of liquid in the right compartment is 74.5 mm in excess of that in the left compartment. The temperature of the system is maintained at 298 K. The number of millimoles in 320 g of haemoglobin is (Given: The density of final solution is 1.013 g/ ml, g = 10 m/$s^2$, $R$ = 0.08 L-atm/ K-mol)

8. At 10°C, the osmotic pressure of urea solution is 500 mm. The solution is diluted and the temperature is raised to 25°C, when the osmotic pressure is found to be 105.3 mm. The extent of dilution (ratio of final to initial volume of solution) is

9. A quantity of 1.04 g of ($CoCl_3 \cdot 6NH_3$) (molecular weight = 267) was dissolved in 100 g of $H_2O$. The freezing point of the solution was −0.29°C. How many moles of solute particle exist in solution for each mole of solute introduced? The value of $K_f$ for water = 1.86° C $m^{-1}$.

10. The freezing point of an aqueous solution of KCN containing 0.2 mole/kg water was −0.80°C. On adding 0.1 mole of $Hg(CN)_2$ in the solution containing 1 kg of water, the freezing point of the solution was −0.6°C. Assuming that the complex is formed according to the following equation

$$Hg(CN)_2 + m\,CN^- \rightarrow Hg\,(CN)_{m+2}^{m-}$$

and Hg $(CN)_2$ is the limiting reactant, the value of $m$ is

## Four Digit Integer Type

1. Vapour pressure of solution containing 6 g of a non-volatile solute in 180 g water is 20 torr. If 1 mole of water is further added, then vapour pressure increases by 0.02 torr. The molar mass (in g/mol) of the non-volatile solute is

2. A solution containing compound 'X' in water and a solution containing glucose in water were put in a closed system. By doing this, some water vapour was removed from one solution and got condensed in the other. It is found that when both the solutions were at equilibrium vapour pressure, one solution contains 10% (w/w) of 'X' and the other 5% (w/w) glucose. The molar mass (in g/mol) of 'X' is

3. At 293 K, the vapour pressure of water is 2400 Pa and the vapour pressure of an aqueous urea solution is 2300 Pa. The osmotic pressure (in atm) of solution at 300 K, if the density of solution at this temperature is 1185 kg/$m^3$, is ($R$ = 0.08 L-atm/ K-mol)

4. How many grams of ethylene glycol should be mixed in 795 g water to form an antifreeze solution that will not start to freeze until the temperature reaches −30°C? ($K_f$ for water = 1.86)

5. The mass of ice that will separate out on cooling a solution containing 50 g of ethylene glycol in 375 g water to −6.0°C is ($K_f$ for water = 1.86 K $mol^{-1}$ kg)

6. A solution of 38.4 g of naphthalene in 185 g camphor depresses the freezing point of the latter by 13.5 K while 11.6 g of another substance when dissolved in 185 g of camphor gives depression of 9.0 K in freezing point. The molecular mass of the second substance is

7. Equal masses of two aqueous solutions, one of 3.6 g glucose per 100 g of water and the other of 3.6 g urea per 100 g of water are mixed. If the freezing point of resulting solution is $-x$°C, then the value of '$1000x$' is ($K_f$ of water = 1.86)

8. Safrole is contained in oil of sassafras and was once used to flavour root beer. A 2.4 mg sample of safrole was dissolved in 100.0 mg of diphenyl ether. The solution had a freezing of 25.64°C. The freezing point of pure diphenyl ether is 26.84°C and the freezing-point-depression constant $K_f$, is 8.00°C/m. The molecular mass of safrole is

9. A quantity of 2.0 g of benzoic acid dissolved in 26.0 g of benzene shows a depression in freezing point equal to 1.60 K. Molal depression constant of benzene is 4.88 K-kg $mol^{-1}$. The percentage association of the acid is

10. The value of '$x$' g of a non-electrolytic compound ($M$ = 200) is dissolved in 1 L of 0.05 M NaCl aqueous solution. The osmotic pressure of this solution is found to be 4.92 atm at 27°C. Assume complete dissociation of NaCl and ideal behaviour for solution. The value of '$x$' is ($R$ = 0.082 L-atm $mol^{-1}K^{-1}$)

11. At 20°C, the vapour pressure of 0.1 molal aqueous solution of urea is 0.03 mm less than that of water and the vapour pressure of 0.1 molal solution of KCl is 0.0594 mm less than that of water. The apparent percentage dissociation of KCl in water at the given temperature is (Neglect the moles of solute particles in comparison to the moles of water in both solutions.)

12. A current of dry air was passed through a series of bulbs containing 1.25 g of a solute $A_2B$ (molar mass = 90 g/mol) in 49 g of water and then through pure water. The loss in weight of the former series of bulbs was 0.98 g and in the later series 0.01 g. The percentage dissociation of solute is

13. A quantity of 75.2 g of phenol is dissolved in 1 kg of solvent of $k_f = 14$. If the depression in freezing point is 7 K, then the percentage of phenol that dimerizes is

14. An aqueous solution contains 12% (w/w) $MgSO_4$ and 9.5% (w/w) $MgCl_2$. If the sulphate dissociates up to 80% and chloride dissociates up to 60%, the boiling point of solution (in Kelvin) is ($K_b$ of water = 0.785 K-kg/mol)

15. When cells of the skeletal vacuole of a frog were placed in a series of NaCl solutions of different concentrations at 6°C, it was observed microscopically that they remained unchanged in $x$% NaCl solution, it shrank in more concentrated solutions and swells in more dilute solutions. Water freezes from the $x$% salt solution at –0.40°C. If the osmotic pressure of the cell cytoplasm at 6°C is '$y$' × 0.0821 atm, then the value of '$y$' is ($K_f = 1.86$ K mol$^{-1}$ kg)

## Answer Keys

**Exercise I**

### Solution of Gas in Liquid

1. (b) 2. (c) 3. (a) 4. (a) 5. (d)

### Vapour Pressure

6. (c) 7. (b) 8. (c) 9. (a) 10. (c)

### Solution of Liquid in Liquid

11. (b) 12. (c) 13. (b) 14. (a) 15. (d) 16. (b) 17. (c) 18. (a) 19. (a) 20. (b)
21. (c) 22. (d) 23. (b) 24. (a) 25. (d)

### Lowering of Vapour Pressure

26. (b) 27. (b) 28. (a) 29. (a) 30. (a) 31. (a) 32. (b) 33. (d) 34. (d) 35. (a)
36. (a) 37. (a) 38. (b) 39. (a) 40. (b)

### Osmotic Pressure

41. (a) 42. (b) 43. (a) 44. (d) 45. (a) 46. (a) 47. (b) 48. (d) 49. (d) 50. (a)
51. (a) 52. (b) 53. (c) 54. (c) 55. (d)

### Elevation in Boiling Point

56. (a) 57. (a) 58. (d) 59. (a) 60. (c) 61. (a) 62. (a) 63. (c) 64. (a) 65. (c)

### Depression in Freezing Point

66. (b) 67. (d) 68. (b) 69. (b) 70. (a) 71. (d) 72. (a) 73. (d) 74. (b) 75. (c)

### Abnormal Colligative Properties

76. (b) 77. (a) 78. (b) 79. (b) 80. (c) 81. (d) 82. (c) 83. (b) 84. (c) 85. (c)
86. (d) 87. (d) 88. (c) 89. (d) 90. (a) 91. (b) 92. (c) 93. (b) 94. (b) 95. (c)
96. (a) 97. (a) 98. (a)

## Answer Keys — Exercise II

### Section A (Only one Correct)

1. (a) 2. (c) 3. (b) 4. (c) 5. (b) 6. (c) 7. (b) 8. (d) 9. (a) 10. (a)
11. (c) 12. (c) 13. (c) 14. (d) 15. (b) 16. (b) 17. (c) 18. (a) 19. (b) 20. (a)
21. (a) 22. (d) 23. (b) 24. (c) 25. (a) 26. (a) 27. (b) 28. (c) 29. (d) 30. (d)
31. (a) 32. (c) 33. (b) 34. (c) 35. (a) 36. (b) 37. (b) 38. (c) 39. (b) 40. (d)
41. (a) 42. (c) 43. (b) 44. (a) 45. (c) 46. (a) 47. (c) 48. (b) 49. (b) 50. (a)
51. (b) 52. (c) 53. (a) 54. (b) 55. (b) 56. (a) 57. (a) 58. (c) 59. (c) 60. (b)
61. (b) 62. (a) 63. (a) 64. (a) 65. (b) 66. (c) 67. (b) 68. (c) 69. (a) 70. (c)

### Section B (One or More than one Correct)

1. (d) 2. (a) 3. (a), (c) 4. (b), (c), (d)
5. (a), (b), (c), (d) 6. (a), (c), (d) 7. (c) 8. (b), (c), (d)
9. (a), (b) 10. (b), (c) 11. (a) 12. (b), (c), (d)
13. (a), (b), (c), (d) 14. (b), (c), (d) 15. (c), (d)

### Section C

**Comprehension I**

1. (b) 2. (a)

**Comprehension II**

3. (d) 4. (b) 5. (a) 6. (b)
7. (c) 8. (d) 9. (d) 10. (b)

**Comprehension III**

11. (a) 12. (d) 13. (b)

**Comprehension IV**

14. (b) 15. (c) 16. (a) 17. (d) 18. (b)

**Comprehension V**

19. (b) 20. (a) 21. (b)

**Comprehension VI**

22. (a) 23. (b) 24. (c)

**Comprehension VII**

25. (b) 26. (c)

**Comprehension VIII**

27. (d) 28. (b) 29. (b)

**Comprehension IX**

30. (a) 31. (a) 32. (b)

## Section D (Assertion – Reason)

1. (a) 2. (a) 3. (d) 4. (a) 5. (c) 6. (a) 7. (a) 8. (a) 9. (d) 10. (d)

## Section E (Column Match)

1. A → Q, B → P, R, C → P, R; D → P, S
2. A → P, B → Q, R, C → R, S; D → R, S
3. A → P, B → R, C → P; D → Q
4. A → P; B → Q; C → R; D → S
5. A → Q; B → R; C → S; D → P
6. A → Q; B → P; C → R; D → S
7. A → P; B → Q, R, S; C → T; D → P, Q, R, S, T

## Section F (Subjective)

### Single-digit Integer Type

1. (2) 2. (2) 3. (2) 4. (1) 5. (3) 6. (2) 7. (5) 8. (5) 9. (4) 10. (2)

### Four-digit Integer Type

1. (0054) 2. (0380) 3. (0060) 4. (0795) 5. (0125)
6. (0058) 7. (0744) 8. (0160) 9. (0096) 10. (0020)
11. (0098) 12. (0050) 13. (0075) 14. (0377) 15. (0060)

# HINTS AND EXPLANATIONS

## EXERCISE I (JEE MAIN)

### Solution of Gas in Liquid

1. Low pressure unfavours the solubility of gas in liquid.
2. Solubility increases on increasing pressure.
3. $O_2$ is less soluble in water than $H_2$.
4. $P = K_H \cdot X \Rightarrow 3.2\times10^5 = 1.6\times10^8 \times \dfrac{n_{CO_2}}{n_{CO_2} + \dfrac{500}{18}}$

$\therefore n_{CO_2} \approx 5.5\times10^{-2} \Rightarrow n_{CO_2} \approx 2.44$ gm

5. $P = K_H \cdot X \Rightarrow 1 \text{ bar} = K_H \times \dfrac{\dfrac{0.2}{1000}}{0.2 + \dfrac{1000}{18}}$

$\therefore K_H = 277.78 \text{ bar} \simeq 2.78\times10^7$ Pa

### Vapour Pressure

6. Vapour pressure is a function of temperature only.
7. Low boiling point means weak intermolecular forces and hence, higher vapour pressure.
8. Viscosity should be high.
9. Theory based
10. Vapour pressure is a function of temperature only.

### Solution of Liquid in Liquid

11. $P_{total} = X_A \cdot P_A^\circ + X_B \cdot P_B^\circ = (1 - X_B)\cdot P_A^\circ + X_B \cdot P_B^\circ$

$= P_A^\circ + X_B(P_B^\circ - P_A^\circ)$

Graph is a straight line with slope $(P_B^\circ - P_A^\circ) = +$ve

12. $\dfrac{1}{P_{total}} = \dfrac{Y_A}{P_A^\circ} + \dfrac{Y_B}{P_B^\circ} = \dfrac{Y_A}{P_A^\circ} + \dfrac{1 - Y_A}{P_A^\circ}$

$= \dfrac{1}{P_A^\circ} + Y_A\left(\dfrac{1}{P_A^\circ} - \dfrac{1}{P_B^\circ}\right)$

Hence, the graph of $\dfrac{1}{P_{total}}$ vs $Y_A$ is straight line.

13. When $X = 0 \Rightarrow P = P_{EtOH}^\circ = 140$ mm Hg

14. $P_{total} = X_{Benzene} \cdot P_{Benzene}^\circ + X_{toluene} \cdot P_{toluene}^\circ$

or, $760 = x \times 960 + (1 - x) \times 380 \Rightarrow x = 0.655$

15. $n_{neptane}, C_7H_{16} = \dfrac{30}{100} = 0.3 = n_1$

$n_{octane}, C_8H_{18} = \dfrac{34.2}{114} = 0.3 = n_2$

Now, $P = X_1 \cdot P_1^\circ + X_2 \cdot P_2^\circ = \dfrac{0.3}{0.6}\times106 + \dfrac{0.3}{0.6}\times46$

$= 76$ KPa $= 0.76$ bar

16. $\dfrac{1}{p} = \dfrac{Y_{Benzene}}{P_{Benzene}^\circ} + \dfrac{Y_{Toluene}}{P_{Toluene}^\circ} = \dfrac{0.75}{75} + \dfrac{0.25}{25}$

$\therefore P = 50$ mm

17. Solution of methanol and ethanol is almost ideal.
18. Solution of ether and HCl shows negative deviation.
19. $P_{ideal} = 0.4\times80 + 0.6\times120 = 104 \text{ mm} > 100$ mm

$\Rightarrow$ Negative deviation

20. $> 50$ ml, as the solution will show positive deviation.
21. $\dfrac{n_{aniline}}{n_{water}} = \dfrac{P_{aniline}^\circ}{P_{water}^\circ} \Rightarrow \dfrac{m_a/93}{m_w/18} = \dfrac{760 - 700}{700}$

$\Rightarrow \dfrac{m_a}{m_w} = 0.44$

$\therefore$ Mass percent of aniline in distillate

$= \dfrac{m_a}{m_a + m_w}\times100 = 30.7\%$

22. Azeotropic mixture cannot be separated by any kind of distillation.

23. Theory based.

24. A solution of benzene and aniline is almost ideal.

25. The composition of distillate and solution remains same in azeotrope.

## Lowering of Vapour Pressure

26. $\frac{P^\circ - P}{X_1} = P^\circ$

27. Nature of solute does not affect any colligative property.

28. $\frac{P^\circ - P}{P} = \frac{n_1}{n_2} \Rightarrow \frac{1}{99} = \frac{4/M}{54/18} \Rightarrow M = 132$

29. $P = X_2 \cdot P^\circ = \frac{\frac{1000}{18}}{0.5 + \frac{1000}{18}} \times 25 = 24.78 \text{ mm Kg}$

30. $n_{\text{Solution}} = n_{\text{Solvent}}$

31. Theory based

32. Rate of evaporation decreases as the available surface area for solvent molecules decreases.

33. $\frac{P^\circ - P}{P} = \frac{n_1}{n_2} \Rightarrow \frac{20}{80} = \frac{w/40}{114/114} \Rightarrow w = 10 \text{ gm}$

34. $\frac{P^\circ - P}{P} = \frac{n_1}{n_2} \Rightarrow \frac{5}{95} = \frac{n_1}{n_2}$

$\therefore$ Molality of solution $= \frac{n_1}{n_2 \times 18} \times 1000 = 2.92$

35. $\frac{P^\circ - P}{P} = \frac{n_1}{n_2} \Rightarrow \frac{1}{99} = \frac{w/180}{500/18} \Rightarrow w = 50.5 \text{ gm}$

36. $\frac{P^\circ - P}{P} = \frac{n_{\text{Solute}}}{n_{\text{Solvent}}} = \frac{n}{N}$

37. Loss in mass of solution $\alpha\, P$

and loss in mass of water $\alpha\, (P^\circ - P)$.

$\therefore \frac{\text{Loss in mass of water}}{\text{Loss in mass of solution}} = \frac{P^\circ - P}{P} = \frac{n_1}{n_2}$

or $\frac{0.04}{2.5} = \frac{5/M}{80/18} \Rightarrow M = 70.3$

38. $Y_{\text{methanol}} = \frac{P_{\text{methanol}}}{P_{\text{total}}} = \frac{2.8}{2.8 + 4.2} = 0.4$

39. Theory based

## Osmotic Pressure

40. $\frac{Y_A}{Y_B} = \frac{X_A}{X_B} \cdot \frac{P_A^\circ}{P_B^\circ} \Rightarrow \frac{0.5}{0.5} = \frac{0.25}{0.75} \times \frac{P_A^\circ}{P_B^\circ} \Rightarrow \frac{P_A^\circ}{P_B^\circ} = \frac{3}{1}$

41. Information

42. Solvent will move from higher vapour pressure to lower vapour pressure.

43. Water will come out as the concentration of salt solution will be very high.

44. Informative

45. Informative

46. $\pi = CRT \Rightarrow 0.006 = \frac{4}{M} \times 0.0821 \times 300$

$\Rightarrow M = 16420$

47. Solution of higher concentration as osmosis occurs from lower to higher concentration.

48. Theory based

49. $C_1 = C_2 \Rightarrow \frac{5}{M} = \frac{0.855}{342} \Rightarrow M = 2000$

50. $\pi = CRT = 0.1 \times 0.0821 \times 303 = 2.487 \text{ atm}$

51. Informative

52. Osmosis is not a very fast process.

53. $\pi\, \alpha\, C = \frac{n_{\text{Solute}}}{V_{\text{Solution}}}$

54. $\pi_1 = \frac{10}{180} \cdot RT;\ \pi_2 = \frac{10}{60} \cdot RT;\ \pi_1 = \frac{10}{342} \cdot RT$

55. $4.56 = \frac{54}{180} \cdot RT$ and $1.52 = C.RT \Rightarrow C = 0.1\text{ M}$

56. Saline water, as it increases boiling points.

## Elevation in Boiling Point

**57.** Theory based

**58.** $m = 1 \Rightarrow \Delta T_b = K_b$

**59.** $\Delta T_b = K_b \cdot m \Rightarrow 0.13 = 0.52 \times \frac{2.6/M}{200/1000} \Rightarrow M = 52$

**60.** Boiling point decreases on decreasing the external pressure.

**61.** $K_b = \frac{RT^2 \cdot M}{1000 \cdot \Delta H_{vap}} = \frac{2 \times (400)^2 \times 80}{1000 \times 8 \times 10^3}$

$= 3.2\ K.Kg/mol$

**62.** $m_{Sugar} = \frac{34.2/342}{65.8/1000} m;\ m_{glucose} = \frac{18/180}{82/1000} m$

and $m_{urea} = \frac{6/60}{94/1000} m$

**63.** $\Delta T_b = K_b \cdot m = 0.513 \times \frac{0.1}{200/1000} = 0.2565°C$

$\therefore$ B.P. of solution = 100.2565°C

**64.** $\Delta T_b = K_b \cdot m$

$\Rightarrow 0.1 = K_b \times \frac{1.8/180}{100/1000} \Rightarrow K_b = 1.0\ K/m$

**65.** $\Delta T_b = K_b \cdot m \Rightarrow 0.26 = 0.52 \times \frac{w_1/150}{w_2/1000} \Rightarrow \frac{w_1}{w_2} = \frac{3}{40}$

$\therefore$ Mass percent of solute $= \frac{w_1}{w_1 + w_2} \times 100\% = 6.98\%$

## Depression in Freezing Point

**66.** Only solvent freezes.

**67.** $\Delta T_f = K_f \cdot m$

$1.86 = K_f \cdot \frac{1}{1}$ (1)

$\Delta T_f = K_f \cdot \frac{1}{0.5}$ (2)

$\therefore$ Required $\Delta T_f = 3.72°\ C \Rightarrow$ Freezing point = – 3.72°C

**68.** $\Delta T_f = K_f \cdot m \Rightarrow 1.29 = 1.86 \times \frac{10/M}{100/1000}$

$\Rightarrow M = 150$

**69.** For any solvent, $K_f > K_b$ and hence, $\Delta T_f > \Delta T_b$.

**70.** Theory based

**71.** Greater $K_f$, greater $\Delta T_f$.

**72.** Osmotic pressure is easily measurable even at very concentration of solute.

**73.** $\Delta T_f = K_f \cdot m = 1.86 \times \frac{5/60}{1} = 0.155°C$

$\therefore$ F.P. = – 0.155° C

**74.** $\Delta T_f = K_f \cdot m \Rightarrow 0.2 = 1.86 \times \frac{5/60}{m_{Solvent}}$

$\Rightarrow M_{Solvent} = 0.775\ Kg$

$\therefore$ Mass of water separate as ice = 1 – 0.775 = 0.225 Kg

**75.** Osmotic pressure is easily measurable.

## Abnormal Colligative Properties

**76.** $\alpha = \frac{M_O - M}{(n-1) \cdot M} = \frac{94 - 120}{\left(\frac{1}{2} - 1\right) \times 120} = 0.433$

[For $2A \rightleftharpoons A_2, n = \frac{1}{2}$]

**77.** $K_3[Fe(CN)_6] \rightleftharpoons 3K^+ + Fe(CN)_6^{3-} \Rightarrow i = 4$

**78.** Dissociation results in the decrease of molar mass.

**79.** $Na_2SO_4 \rightleftharpoons 2Na^+ + SO_4^{2-} \Rightarrow i = 3$

**80.** Effective molality of NaCl = 2 × molality of glucose

**81.** $\Delta T_f = K_f \cdot m \Rightarrow 2 = 1.86 \times \frac{(w/74.5) \times 2}{500/1000}$

$\Rightarrow w = 20.03$ gm

**82.** For $K_3PO_4$, $i = 4$

**83.** For greater number of solute particles, $\Delta T_f$ is maximum and hence freezing point will be lowest.

84. Effective molarity of $Ca(NO_3)_2 = 0.07 \times 3 = 0.21$ M

85. Effective molality of urea = 0.1 m
Effective molality of NaCl = $0.1 \times 2 = 0.2$ m
Effective molality of $Na_2SO_4 = 0.1 \times 3 = 0.3$ m

86. As there is no movement of solute particles through SPM, the colour will not change in any side.

87. The effective molarity of solutions should be same.

88. $Na_2SO_4 \rightleftharpoons 2Na^+ + SO_4^{2-}$

$0.004(1-\alpha)$M $\quad 0.004\times2\alpha \quad 0.004\times\alpha$

Effective molarity = $0.004(1-\alpha) + 0.008\alpha + 0.004\alpha$
$= 0.004 + 0.008\alpha$

For solution to be isotonic,
$0.004 + 0.008\alpha = 0.01 \Rightarrow \alpha = 0.75$

89. Greater the number of particles, lower will be vapour pressure.

90. Greater the number of particles, higher will be osmotic pressure.

91. $\Delta T_f = K_f \cdot m \Rightarrow 0.3348 = 1.86 \times 0.1(1+\alpha)$
$\Rightarrow \alpha = 0.8$

92. $\Delta T_{f(\text{theo})} = K_f \cdot m = 1.86 \times \dfrac{0.25}{500/1000} = 0.93\text{K}$

Now, $i = \dfrac{\Delta T_{f(\text{exp})}}{\Delta T_{f(\text{theo})}} = \dfrac{2.79}{0.93} = 3$

As, $i = 1 + \alpha(n-1) \Rightarrow 3 = 1 + 1(n-1) \Rightarrow n = 3$

93. Molarity of solution = $\dfrac{0.85/85}{100/1000} = 0.1$ M

$\pi_{\text{theo}} = \text{CRT} = 0.1 \times 0.0821 \times 300 = 2.463$ atm
$i = 1 + \alpha(n-1) = 1 + 0.9(2-1) = 1.9$
$\therefore \pi_{\text{exp}} = i \cdot \pi_{\text{theo}} = 1.9 \times 2.463 = 4.68$ atm

94. $i = 1 + \alpha(n-1) \Rightarrow 2.74 = 1 + \alpha(3-1) \Rightarrow \alpha = 0.87$

95. $\Delta T_{b(\text{theo})} = K_b \cdot m = 0.52 \times \dfrac{13.44}{134.4} = 0.052$ K

$i = 1 + \alpha(n-1) = 1 + 1(3-1) = 3$

$\therefore \Delta T_{b(\text{exp})} = i \cdot \Delta T_{b(\text{theo})} = 3 \times 0.52 = 0.156$ K

96. $\Delta^T_{f(\text{theo})} = K_f \cdot m = 1.72 \times \dfrac{20/172}{50/1000} = 4\text{K}$

Now, $i = \dfrac{\Delta^T_f(\text{exp})}{\Delta^T_f(\text{theo})} = \dfrac{2}{4} = 0.5$

97. $i = 1 + \alpha(n-1) = 1 + \alpha(x+y-1)$

$\therefore \alpha = \dfrac{i-1}{x+y-1}$

98. Lower freezing point means greater number of particles and hence, the extent of dissociation of $\alpha$-alanine is greater.

---

## EXERCISE II (JEE ADVANCED)

### Section A (Only one Correct)

1. $P_{N_2} = K_H \cdot X_{N_2(\text{Solution})}$

or $5 \times 0.8 = (1.0 \times 10^5) \times \dfrac{n_{N_2}}{n_{N_2} + 10} \approx 10^5 \times \dfrac{n_{N_2}}{10}$

$\therefore n_{N_2} = 4 \times 10^{-4}$

2. $p = K_H \cdot X \approx K_H \cdot \dfrac{n_{\text{gas}}}{n_{\text{liq}}} \Rightarrow m_{\text{gas}} \,\alpha\, P \cdot V_{\text{liq}}$

$\therefore \dfrac{m_2}{m_1} = \dfrac{P_2 \cdot V_2}{P_1 \cdot V_1} \Rightarrow \dfrac{m_2}{m} = \dfrac{5\times2}{1\times1} \Rightarrow m_2 = 10$ m

Now, $P = K_H \cdot \dfrac{n_{\text{gas}}}{n_{\text{liq}}} = \dfrac{K_H}{n_{\text{liq}}} \cdot \dfrac{P \cdot V}{RT}$

$\therefore$ Volume of gas dissolved, $V = \dfrac{RT}{K_H} \cdot n_{\text{liq}}$

(Volume of gas dissolved is independent of pressure of gas)

$\therefore \dfrac{V_2}{V_1} = \dfrac{V_2, \text{liquid}}{V_1, \text{liquid}} \Rightarrow \dfrac{V_2}{V \text{ ml}} = \dfrac{2}{1} \Rightarrow V_2 = 2\text{V ml}$

3. $P_{Hg} = X_{Hg} \cdot P_{total} = \dfrac{\dfrac{0.8\times10^{-3}}{200}}{\dfrac{0.8\times10^{-3}}{200}+\dfrac{50.4}{28}}\times720$

$=1.6\times10^{-3}$ torr

4. Final vapour pressure and hence, the composition of both solutions must be same. As solution in beaker (A) has higher concentration, its vapour pressure is low. Hence, water from (B) will transfer in (A) as vapour.

$\therefore \quad \dfrac{20}{200+x}=\dfrac{10}{100-x} \Rightarrow x = 33.33$

5. $Y_A = \dfrac{P_A}{P_{total}} = \dfrac{X_A \cdot P_A{}^{\circ}}{P_{total}}$

$\dfrac{1}{X_A} = \dfrac{1}{Y_A}\cdot P_A{}^{\circ}\cdot\left(\dfrac{Y_A}{P_A{}^{\circ}}+\dfrac{(1-Y_A)}{P_B{}^{\circ}}\right)$

$= \dfrac{1}{Y_A}\cdot\dfrac{P_A{}^{\circ}}{P_B{}^{\circ}}+\dfrac{P_B{}^{\circ}-P_A{}^{\circ}}{P_B{}^{\circ}}$

6. $\left(\dfrac{n_P}{n_Q}\right)_{2^{nd}\text{ condense}} = \left(\dfrac{n_P}{n_Q}\right)_{\text{initial}}\cdot\left(\dfrac{P_P{}^{\circ}}{P_Q{}^{\circ}}\right)^n$

Where $n$ = number of condensation steps.

Now, $\left(\dfrac{n_P}{n_Q}\right)_{\text{final}} = \dfrac{1}{1}\times\left(\dfrac{300}{100}\right)^2 = \dfrac{9}{1}$

$\therefore X_P = \dfrac{9}{9+1} = 0.90$

7. The mole fraction of A in distillate,

$X'_A = Y_A = \dfrac{X_A \cdot P_A{}^{\circ}}{P_{total}} = \dfrac{\dfrac{1}{4}\times100}{\dfrac{1}{4}\times100+\dfrac{3}{4}\times80} = \dfrac{5}{17}$

Now, V.P. of distillate, $P = X'_A \cdot P_A{}^{\circ} + X'_B \cdot P_B{}^{\circ}$

$= \dfrac{5}{17}\times100+\dfrac{12}{17}\times80 = 85.88$ mm Kg

8. Let the final composition:
liquid (10 mole): A = $x$ mole, B = $(10 - x)$ mole
Vapour (10 mole): A = $(10 - x)$ mole, B = $x$ mole

Now, $\dfrac{Y_A}{Y_B} = \dfrac{X_A}{X_B}\cdot\dfrac{P_A{}^{\circ}}{P_B{}^{\circ}} \Rightarrow \dfrac{10-x}{x} = \dfrac{x}{10-x}\cdot\dfrac{200}{100}$

$\therefore x = 4.14$

Now, $p = X_A \cdot P_A{}^{\circ} + X_B \cdot P_B{}^{\circ}$

$= \dfrac{x}{10}\times200+\dfrac{10-x}{10}\times100$

$= 141.4$ torr

9. $\dfrac{1}{P_{total}} = \dfrac{Y_A}{P_A{}^{\circ}}+\dfrac{Y_B}{P_B{}^{\circ}} = \dfrac{0.4}{0.4}+\dfrac{0.6}{1.2} = \dfrac{3}{2}$

$\Rightarrow P_{total} = \dfrac{2}{3}$ atm

10. If the solution were ideal,

$P_{total} = \dfrac{10}{30}\times90+\dfrac{20}{30}\times87 = 88$ mm kg

As the solution of phenol and aniline shows negative deviation, the V.P. must be less than 88 mm kg.

11. For ideal behavior,

$P_{total} = 0.25 \times 512 + 0.725 \times 344$

$= 386$ mm Hg $< 600$ mm Hg

Hence, the solution shows positive deviation

$\Rightarrow \Delta H_{mix}$ = positive

12. For increase in temperature, the solution shows negative deviation ($\Delta H$ = negative).

13. $\dfrac{n_{\text{Chlorobenzene}}}{n_{\text{water}}} = \dfrac{P^{\circ}_{\text{Chlorobenzene}}}{P^{\circ}_{\text{water}}}$

or, $\dfrac{x/112.5}{(100-x)/18} = \dfrac{9.031\times10^4 - 7.031\times10^4}{7.031\times10^4}$

$\Rightarrow x = 64$

14. As the available surface area for solvent molecules decreases, the rate of vaporization decreases.

15. $\dfrac{P^{\circ}-P}{P} = \dfrac{n_1}{n_2} \Rightarrow \dfrac{5}{95} = \dfrac{m_1/\text{M}}{m_2/0.3\,\text{M}} \Rightarrow \dfrac{m_2}{m_1} = \dfrac{57}{10}$

16. $P^{\circ} - P = X_1 \cdot P^{\circ}$

$10 = 0.2 \times P^{\circ}$ (1)

$20 = X_1 \times P^{\circ}$ (2)

From (1) and (2): $X_1 = 0.4 \Rightarrow X_2 = X_{\text{solvent}} = 0.6$

17. $\dfrac{P^{\circ}-P}{P} = \dfrac{n_1}{n_2} \Rightarrow \dfrac{3000-2985}{2985} = \dfrac{5/\text{M}}{100/18}$

$\Rightarrow \text{M} = 179.1$

18. $\dfrac{P^{\circ}-P}{P} = \dfrac{n_1}{n_2} \Rightarrow \dfrac{0.85-0.845}{0.845} = \dfrac{0.5/\text{M}}{39/78} \Rightarrow \text{M} = 169$

**19.** $P = X_2 \cdot P° = \dfrac{\frac{1000}{18}}{1+\frac{1000}{18}} \times 12.3 = 12.08 \text{ K Pa}$

**20.** $P = X_2 \cdot P°$

$$2.8 = \dfrac{\frac{90}{18}}{\frac{30}{M}+\frac{90}{18}} \times P° \text{ and } 2.9 = \dfrac{\frac{108}{18}}{\frac{30}{M}+\frac{108}{18}} \cdot P°$$

$\therefore M = 23$

**21.** $P° - P = X_1 \cdot P° = \dfrac{1}{1+\frac{1000}{18}} \times 760 = 13.44 \text{ mm kg}$

**22.** $\dfrac{P° - P}{P} = \dfrac{n_1}{n_2} \Rightarrow \dfrac{89.78-89}{89} = \dfrac{2/M}{100/78} \Rightarrow M = 178$

Now, the number of C-atoms in each molecule

$$= \dfrac{178 \times \frac{94.4}{100}}{12} = 14$$

and the number of H-atoms in each molecule

$$= \dfrac{178 \times \frac{5.6}{100}}{1} = 10$$

$\therefore$ Hydrocarbon is $C_{14}H_{10}$.

**23.** Loss is mass of solvent, $w_1 \propto (P° - P)$

and gain is mass of absorbent, $w_2 \propto P°$

$$\therefore \dfrac{w_1}{w_2} = \dfrac{P° - P}{P°} = X_1 \Rightarrow \dfrac{0.05}{2.05} = \dfrac{\frac{40}{M}}{\frac{40}{M}+\frac{100}{18}} \Rightarrow M = 288$$

**24.** $C_1 = C_2 \Rightarrow \dfrac{10}{M_A} + \dfrac{20}{M_B} = \dfrac{6.67}{M_A} + \dfrac{30}{M_B} \Rightarrow \dfrac{M_A}{M_B} = \dfrac{1}{3}$

**25.** As solution have same concentration, mixing will not change the total molar concentration.

**26.** Isopiestic refers to the same pressure. Blood is isotonic with 0.9% (*w*/*v*) NaCl solution.

$\therefore$ Osmolarity $= \dfrac{0.9/58.5}{100/1000} \times 2 \approx 0.31 \text{ M}$

**27.** $\pi = CRT = \left(\dfrac{2.5/58.5}{100/1000} \times 2\right) \times 0.0821 \times 300$

$= 21.05$ atm

**28.** Theory based

**29.** $\Delta T_b = K_b \cdot m \Rightarrow 0.104 = 0.52 \times \dfrac{2/M}{98/1000}$

$\Rightarrow M = 104$

**30.** Theory based

**31.** Clausius–Clapeyron equation:

$$\dfrac{dT}{dP} = \dfrac{RT^2}{\Delta H.P} = \dfrac{2 \times (350)^2}{4.9 \times 10^3} = 50 \text{ K/atm}$$

**32.** $\Delta T_b = K_b \cdot m = 0.52 \times \dfrac{72}{180} = 0.208 \text{ K}$

$\therefore$ B.P. of solution = 373.208 K

**33.** $\Delta T_b = K_b \cdot m$

$\Rightarrow (354.11 - 353.23) = 2.53 \times \dfrac{1.8/M}{90} \times 1000$

$\therefore M = 57.5$

**34.** If molality of solution is '*m*', then

$$\dfrac{P° - P}{P} = \dfrac{n_1}{n_2} \Rightarrow \dfrac{760-750}{750} = \dfrac{m}{1000/18} \Rightarrow m = \dfrac{20}{27}$$

Now, $\Delta T_b = K_b \cdot m = 0.52 \times \dfrac{20}{27} = 0.385 \text{ K}$

$\therefore$ B.P. of solution = 100.385° C

**35.** $K_b = \dfrac{0.002\, T_o^2}{\Delta H_{vap}(\text{Cal/gm})} = \dfrac{0.002 \times (320)^2}{80}$

$= 2.56$ K/m

Now, $\Delta T_b = K_b \cdot m$

$\Rightarrow 0.32 = 2.56 \times \dfrac{6.4/32x}{200/1000} \Rightarrow x = 8$

$\therefore$ Molecular formula of sulphur $= S_x = S_8$

**36.** $\Delta T_b = K_{x,b} \cdot X_{\text{solute}} = 32 \times \dfrac{1/128}{\frac{1}{128}+\frac{94}{94}} \approx 0.25°\text{C}$

$\therefore$ B.P. of solution = 110.75 + 0.25 = 111° C

**37.** $\Delta T_b = K_b \cdot m \Rightarrow 1.04 = 0.52 \times m \Rightarrow m = 2$

Now, $\dfrac{P° - P}{P} = \dfrac{n_1}{n_2} \Rightarrow \dfrac{P° - 750}{750} = \dfrac{2}{1000/18}$

$\Rightarrow P° = 777$ torr

**38.** Theory based

**39.** $\Delta T_f = K_f \cdot m \Rightarrow 0.93 = 1.86 \times \dfrac{7/M}{93/1000}$

$\Rightarrow M = 150.5$

**40.** $\Delta T_f = K_f \cdot m \Rightarrow 0.93 = 1.86 \times \frac{36/M}{1.2} \Rightarrow M = 60$

If the molecular formula is $C_xH_{2x}O_x$, then

$12x + 2x + 16x = 60 \Rightarrow x = 2$

$\therefore$ Molecular formula = $C_2H_4O_2$

**41.** $\Delta T_f = K_f \cdot m \Rightarrow 15 = 1.86 \times m \Rightarrow m = 8.06$

**42.** $\frac{\Delta T_f}{\Delta T_b} = \frac{K_f}{K_b} \Rightarrow \frac{\Delta T_f}{0.78} = \frac{1.86}{0.52} \Rightarrow \Delta T_f = 2.79°C$

$\therefore$ F.P. of solution = –2.79° C

**43.** $\Delta T_f + \Delta T_b = (K_f + K_b) \cdot m$

or $4.76 = (1.86 + 0.52) \times \frac{w/342}{100/1000} \Rightarrow w = 68.4$ gm

**44.** $\Delta T_f = K_f \cdot m \Rightarrow 5 = 8.5 \times \frac{w/120}{425/1000} \Rightarrow w = 30$ gm

**45.** $\frac{\Delta T_f}{\Delta T_b} = \frac{K_f}{K_b} \Rightarrow \frac{\Delta T_b}{0.7} = \frac{5}{17.5} \Rightarrow \Delta T_b = 0.2°C$

$\therefore$ B.P. of solution = 90.2°C

**46.** If the molarity of solution is *m*, then

$$\frac{P° - P}{P} = \frac{n_1}{n_2} \Rightarrow \frac{2}{100} = \frac{m}{1000/78} \Rightarrow m = \frac{10}{39}$$

Now, $\Delta T_f = K_f \cdot m \Rightarrow 1.3 = K_f \times \frac{10}{39}$

$\Rightarrow K_f = 5.07$ K/m

**47.** $\frac{K_f}{K_b} = \frac{T_f°^2/\Delta H_{fus}}{T_b°^2/\Delta H_{vap}} \Rightarrow \frac{\Delta H_{fus}}{\Delta H_{vap}} = \left(\frac{T_f°}{T_b°}\right)^2 \times \frac{K_b}{K_f}$

$= \left(\frac{280}{350}\right)^2 \times \frac{2.5}{5.6} = \frac{2}{7}$

**48.** $\Delta T_f = K_f \cdot m \Rightarrow 2.0 = K_f \times 0.25 \Rightarrow K_f = 8$ K/m

**49.** $\Delta T_f = K_f \cdot m$

For cane sugar solution:

$(273.15 - 272.85) = K_f \cdot \frac{5/342}{95/1000}$ (1)

For glucose solution:

$(273.15 - T_f) = K_f \cdot \frac{5/180}{95/1000}$ (2)

From (1) and (2), $T_f = 272.58$ K

**50.** $\Delta T_f = K_f \cdot m$

For $AB_2$ solution: $2.55 = 5.1 \times \frac{1/(x+2y)}{20/1000}$ (1)

For $AB_4$ solution: $1.7 = 5.1 \times \frac{1/(x+4y)}{20/1000}$ (2)

$\therefore$ Atomic mass of A = $x$ = 50

Atomic mass of B = $y$ = 25

**51.** $\Delta T_f = K_f \cdot m \Rightarrow 0.744 = 1.86 \times m \Rightarrow m = 0.4$

Now, $\pi = CRT = 0.4 \times 0.0821 \times 300 = 9.852$ atm

**52.** $\phi = \frac{\Delta T_f}{K_f \cdot m} = \frac{0.93}{1.86 \times 0.4} = 1.25$

**53.** $\Delta T_f = K_f \cdot m \Rightarrow 1.0 = 1.80 \times m \Rightarrow m = \frac{1}{1.8}$

$$P = X_2 \cdot P° = \frac{\frac{1000}{18}}{\frac{1}{1.8} + \frac{1000}{18}} \times 24.24 = 24 \text{ mm Hg}$$

**54.** $\Delta T_f = K_f \cdot m$

$\therefore 0.2 = K_f \times \frac{x}{100/1000}$ and $0.25 = K_f \times \frac{x}{y/1000}$

$\therefore$ Mass of ice separated out = 100 – y = 20 gm

**55.** $2\,KI(aq) + HgI_2 \longrightarrow K_2HgI_4(aq)$

(4 particles) (added) (3 particles)

As the number of ions in solution decreases, osmotic pressure decreases.

**56.** Osmolarity of both solution should be equal

$\therefore 0.1 \times 2 = 0.1 \times (1 + 2\alpha) \Rightarrow \alpha = 0.5$

**57.** Osmolarity = $0.2 \times 3 = 0.6$ M

**58.** $\Delta T_f = K_f \cdot m \Rightarrow 3.72 = 1.86 \times 1.0 \times n \Rightarrow n = 2$

$\therefore$ From each particle, two ions should form.

**59.** $\Delta T_b = K_b \cdot m \Rightarrow 1.43 = K_b \times 1\left[1 + 0.9\left(\frac{1}{2} - 1\right)\right]$

$\therefore K_b = 2.6$ K/m

**60.** $\Delta T_f = K_f \cdot m\left[1 + \alpha\left(\frac{1}{n} - 1\right)\right]$

$\therefore 1.96 = 4.9 \times \frac{2/122}{25/1000}\left[1 + \alpha\left(\frac{1}{2} - 1\right)\right] \Rightarrow \alpha = 0.78$

**61.** $HA \rightleftharpoons H^+ + A^-$

$(0.1-x)M \quad xM \quad xM$
$=0.01M$

$\therefore$ Osmolarity $= (0.1 - x) + x + x = 0.11$ M

Now, $\pi = CRT = 0.11$ RT

**62.** $\Delta T_f = K_f \cdot m = 1.86 \times [0.1 \times 2 + 0.025 \times 2]$

$= 0.465°C$

$\therefore$ F.P. of solution = – 0.465° C

**63.** The effective molality will be in the range of 0.2 to 0.3 and $\Delta T_f = K_b \cdot m$ will be in the range of $1.86 \times 0.2 = 0.372°$ C to $1.86 \times 0.3 = 0.558°$ C.

**64.** Let the mixture contain $x$ mole KCl and $y$ mole NaCl. Then

$x \times 74.5 + y \times 58.5 = 3.125$ (1)

and $(x+y)\times 2 = \dfrac{\Delta T_f}{K_f} = \dfrac{0.186}{1.86} = 0.1$ (2)

$\therefore \dfrac{x}{y} = \dfrac{1}{3}$

**65.**

| | $2A \rightleftharpoons$ | $A_2$ |
|---|---|---|
| Initial conc. | nM | 0 |
| Equilibrium conc. | $(n-x)$M | $\frac{x}{2}$M |

Now, $\Delta T_b = K_b \cdot m = K_b \cdot \left[n - x + \dfrac{x}{2}\right]$

$\therefore x = 2n - \dfrac{2\Delta T_b}{K_b}$

Now, $K_c = \dfrac{[A_2]}{[A]^2} = \dfrac{\left(\frac{x}{2}\right)}{(n-x)^2} = \dfrac{n - \frac{\Delta T_b}{K_b}}{\left(\frac{2\Delta T_b}{K_b} - n\right)^2}$

$= \dfrac{K_b(n \cdot K_b - \Delta T_b)}{(2\Delta T_b - n \cdot K_b)^2}$

**66.** $\dfrac{\Delta T_f(A)}{\Delta T_f(B)} = \dfrac{K_f, A \cdot m_1}{K_f, B \cdot m_2} = \dfrac{1.86 \times m/2}{2.79 \times m/3} = \dfrac{1}{1}$

**67.** Colloidal solutions have low value of any colligative property than the true solution of same composition.

**68.** If the complex dissociates into $n$ ions, then

$\Delta T_f = K_f \cdot m \Rightarrow 0.0054 = 1.86 \times [0.001 \times n] \Rightarrow n \approx 3$

**69.** $XCl_3(s) \rightleftharpoons X^{3+} + 3Cl^-$

$\qquad\qquad\qquad$ SM $\quad$ 3SM

$\dfrac{P° - P}{P} = \dfrac{n_1}{n_2} \Rightarrow \dfrac{17.25 - 17.20}{17.20} = \dfrac{4S}{1000/18}$

$\Rightarrow S = 4.04 \times 10^{-2}$ M

**70.** (a) $\Delta T_f = K_f \cdot m = 1.86 \times 1 = 1.86°C$

$\therefore$ F.P. of solution = – 1.86° C

(b) $\Delta T_b = K_b \cdot m = 0.52 \times 1 = 0.52°C$

$\therefore$ B.P. of solution should be 100.52° C. As the solute dissociates completely above 100.26° C, its actual $\Delta T_b = 0.52 \times 2 = 1.04°C$ and hence, B.P. = 101.04° C.

(c) $\Delta T_f = K_f \cdot m \Rightarrow 7.44 = 1.86 \times \dfrac{1/2}{m_{solvent}}$

(as solute dimerizes)

$\therefore m_{Solvent}$ left = 0.125 kg

$\therefore$ Percentage of water separated as ice = $(1 - 0.125) \times 100 = 87.5\%$

(d) $\Delta T_b = K_b \cdot m \Rightarrow 2.08 = 0.52 \times \dfrac{2}{m_{solvent}}$

(Complete dissociation)

$\therefore m_{Solvent}$ left = 0.5 kg

Percentage of water evaporated = $(1 - 0.5) \times 100 = 50\%$

## Section B (One or More than one Correct)

1. At constant temperature, the vapour pressure may be changed by changing the composition.
2. Theory based
3. Theory based
4. $P_{AB} = \dfrac{75 + 22}{2} = 48.5$ torr

$P_{BC} = \dfrac{22 + 10}{2} = 16$ torr

$P_{AC} = \dfrac{75 + 10}{2} = 42.5$ torr

$P_{ABC} = \dfrac{75 + 22 + 10}{2} = 35.67$ torr

5. Theory based
6. A, C, D shows negative deviation.
7. (a) Mass percent of A = 50 $\Rightarrow n_A : n_B = 1 : 2$ $\Rightarrow$ Azeotrope

Hence, vapour will have the same composition of liquid.

(b) Mass percent of A > 50 $\Rightarrow n_A : n_B = 1 : 2$

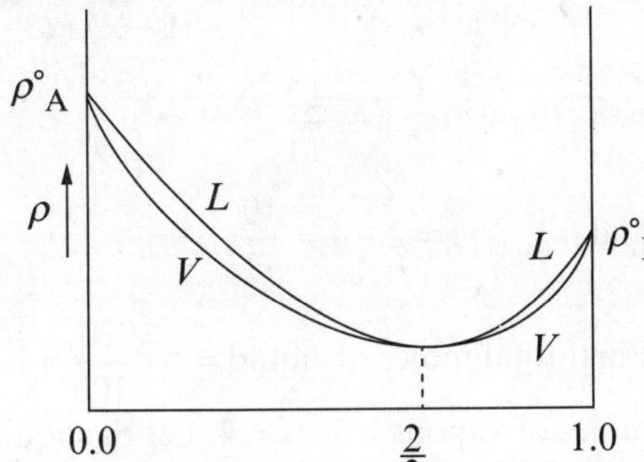


In this case, the vapour must be more rich in A than liquid.

(c) $X_B = \frac{3}{4} = 0.75 > \frac{2}{3}$ $\Rightarrow$ Pure $A$ cannot be obtained.

(d) $X_B = \frac{3}{5} = 0.60 < \frac{2}{3} \Rightarrow$ Pure $A$ cannot be obtained in traces.

8. (a) On changing the solvent, $K_f$ will change.
9. F. P. and V. P. will become lower for $X$.
10. Informative
11. Theory based
12. Theory based
13. (a) $\frac{P^\circ - P}{P} = \frac{n_1}{n_2} \Rightarrow \frac{760-740}{740} = \frac{1}{n_2} \Rightarrow n_2 = 37$

$\therefore$ Moles of water separated as ice $= 200 - 37 = 163$

(b) $\Delta T_f = K_f \cdot m = 2.0 \times \frac{1}{(37\times18)/1000}$

$= \frac{2000}{37\times18}$ K

$\therefore T = \left(273 - \frac{2000}{37\times18}\right) K$

(c) For original solution:

$\Delta T_f = K_f \cdot m = 2 \times \frac{1}{(200\times18)/1000}$

$\therefore$ F. P. $= 0 - \Delta T_f = -\frac{10^\circ}{18} C$

(d) For final solution: $\frac{P^\circ - P}{P^\circ} = X_1 = \frac{1}{1+37} = \frac{1}{38}$

14. Theory based
15.

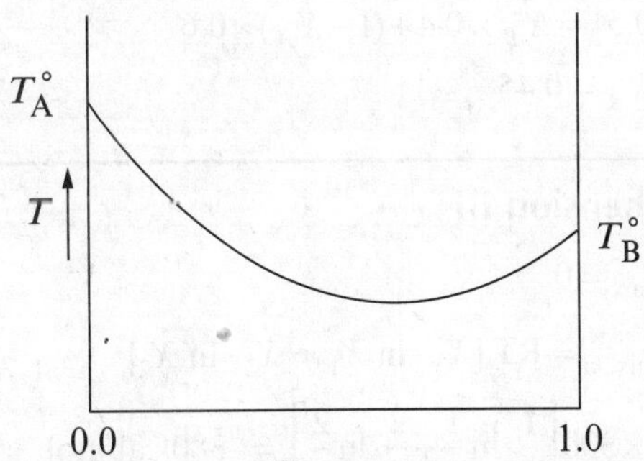


## Section C (Comprehensions)

### Comprehension I

1. $Y_A - X_A = \frac{X_A \cdot P_A{}^\circ}{P_B{}^\circ + X_A \cdot (P_A{}^\circ - P_B{}^\circ)} - X_A$

$= \frac{X_A(P_A{}^\circ - P_B{}^\circ) - X_A^2(P_B{}^\circ - P_A{}^\circ)}{P_B{}^\circ + X_A \cdot (P_A{}^\circ - P_B{}^\circ)} = f(X_A)$

For maximum $(Y_A - X_A)$, $\frac{d(Y_A - X_A)}{dX_A} = 0$

$\therefore X_A = \frac{\sqrt{P_A{}^\circ \cdot P_B{}^\circ} - P_B{}^\circ}{P_A{}^\circ - P_B{}^\circ}$

2. $P_{\text{total}} = P_B{}^\circ + X_A \cdot (P_A{}^\circ - P_B{}^\circ) = \sqrt{P_A{}^\circ \cdot P_B{}^\circ}$

## Comprehension II

3. $\dfrac{1}{P_{\text{total}}}=\dfrac{Y_A}{P_A{^\circ}}+\dfrac{Y_B}{P_B{^\circ}}=\dfrac{2/5}{0.4}+\dfrac{3/5}{0.6}$

$\Rightarrow P_{\text{total}}=0.5>0.3$ bar

As the applied pressure is less than equilibrium pressure, the system must be 100% vapour.

4. First drop of liquid will form at 0.5 bar.

5. $P_1V_1=P_2V_2 \Rightarrow 0.3\times 10=0.5\times V_2 \Rightarrow V_2=6.0\ \text{dm}^3$

6. $P_{\text{total}}=X_A\cdot P_A{^\circ}+X_B\cdot P_B{^\circ}$

$\Rightarrow 0.5=X_A\times 0.4+(1-X_A)\times 0.6$

$\therefore X_A=0.5$

7. Liquid composition: $A\approx 2$ mole, $B\approx 3$ mole

$\therefore\ P_{\text{total}}=\dfrac{2}{5}\times 0.4+\dfrac{3}{5}\times 0.6=0.52$ bar

8. $Y_A=\dfrac{X_A\cdot P_A{^\circ}}{P_{\text{total}}}=\dfrac{\frac{2}{5}\times 0.4}{0.52}=\dfrac{4}{13}$

9. $P_{\text{total}}=X_A\cdot P_A{^\circ}+X_B\cdot P_B{^\circ}$

$\Rightarrow 0.51=X_B\times 0.4+(1-X_A)\times 0.6$

$\therefore X_A=0.45$

Now, $Y_A=\dfrac{X_A\cdot P_A{^\circ}}{P_{\text{total}}}=\dfrac{0.45\times 0.4}{0.51}=\dfrac{6}{17}$

Let moles of $A$ and $B$ in liquid form is $x$ and $y$, respectively.

$X_A=\dfrac{x}{x+y}=0.45$ and $Y_A=\dfrac{2-x}{(2-x)+(3-y)}=\dfrac{6}{17}$

$\therefore n_A\,(\text{liquid})=x=\dfrac{12}{11}$

and $n_{A(\text{vapour})}=2-x=\dfrac{10}{11}$

10. Final total moles of liquid $=5\times\dfrac{20}{100}=1$ and total moles of vapour $=5-1=4$. Let the liquid contain $x$ mole $A$.

$P_{\text{total}}=\dfrac{x}{1}\times 0.4+\dfrac{1-x}{1}\times 0.6=0.6-0.2x$ (1)

and $Y_A=\dfrac{X_A\cdot P_A{^\circ}}{P_{\text{total}}}\Rightarrow\dfrac{2-x}{4}=\dfrac{\frac{x}{1}\times 0.4}{0.6-0.2x}$ (2)

From (2): $x=0.48$

$\therefore$ From (1): $P_{\text{total}}=0.504$ bar

## Comprehension III

11. $\Delta H_{\text{mix}}=0$

12. $\Delta G_{\text{mix},\,m}=\text{RT}\,[X_1\cdot\ln X_1+X_2\cdot\ln X_2]$

$=2\times 300\left[\dfrac{1}{3}\cdot\ln\dfrac{1}{3}+\dfrac{2}{3}\cdot\ln\dfrac{2}{3}\right]=-380$ cal/mol

13. $\Delta S_{\text{mix},m}=-\dfrac{\Delta G_{\text{mix},m}}{T}=-\dfrac{-380}{300}=\dfrac{3.8}{3}$ cal / K-mol

and $\Delta S_{\text{mix}}=3\times\dfrac{3.8}{3}=3.8$ cal / K

## Comprehension IV

14. $P_{\text{total}}=X_B\cdot P_B{^\circ}+X_T\cdot P_T{^\circ}$

$=\dfrac{10}{20}\times 100+\dfrac{10}{20}\times 40=70\ \text{mm Hg}$

15. $Y_A=\dfrac{X_B\cdot P_B{^\circ}}{P_{\text{total}}}=\dfrac{0.5\times 100}{70}=\dfrac{5}{7}=0.714$

16. The vapour will contain almost 10 moles of both

$\dfrac{1}{P_{\text{total}}}=\dfrac{Y_B}{P_B{^\circ}}+\dfrac{Y_T}{P_T{^\circ}}=\dfrac{0.5}{100}+\dfrac{0.5}{40}$

$\Rightarrow P_{\text{total}}=57.14$ mm kg

17. $X_B=\dfrac{Y_B\cdot P_{\text{total}}}{P_B{^\circ}}=\dfrac{0.5\times 57.14}{100}=0.286$

18. Final system contains 10 moles of liquid and 10 moles of vapour. Let the moles of benzene in liquid be $x$.

$P_{\text{total}}=X_B\cdot P_B{^\circ}+X_T\cdot P_T{^\circ}=\dfrac{x}{10}\times 100+\dfrac{10-x}{10}\times 40$

or, $P_{\text{total}}=40+6x$ (1)

$Y_B=\dfrac{X_B\cdot P_B{^\circ}}{P_{\text{total}}}\Rightarrow\dfrac{10-x}{10}=\dfrac{\frac{x}{10}\times 100}{40+6x}\Rightarrow x=3.87$

From Equation (1): $P_{\text{total}}=63.25$ mm kg

## Comprehension V

A + B

($n_A + n_B$) moles

Residual solution $= \frac{1}{4}(n_A + n_B)$ moles

A = $x$ mole

B = $y$ mole

Condensate $= \frac{3}{4}(n_A + n_B)$

A = $(n_A - x)$ mole

B = $(n_B - y)$ mole

From question:

$$700 = \frac{n_A}{n_A + n_B} \times P_A^\circ + \frac{n_B}{n_A + n_B} \times P_B^\circ \quad (1)$$

$$600 = \frac{x}{x+y} \times P_A^\circ + \frac{y}{x+y} \times P_B^\circ \quad (2)$$

$$x + y = \frac{1}{4}(n_A + n_B) \quad (3)$$

$$\frac{x}{x+y} = 0.3 \quad (4)$$

$$\frac{n_A - x}{\frac{3}{4}(n_A + n_B)} = 0.75 \quad (5)$$

19. $n_B : n_A = 29.51$

20. $P_A^\circ = 807.4$ mm

21. $P_B^\circ = 511.1$ mm

## Comprehension VI

22. $P_A^\circ = \frac{RT}{V_m} = \frac{0.08 \times 300}{\frac{100}{1.25} \times \frac{3800}{1000}} \times 760 = 60$ torr

23. $P_B^\circ = \frac{RT}{V_m} = \frac{0.08 \times 300}{\frac{50}{1.00} \times \frac{7600}{1000}} \times 760 = 48$ torr

24. $\frac{1}{P_{\text{total}}} = \frac{Y_A}{P_A^\circ} + \frac{Y_B}{P_B^\circ} \Rightarrow \frac{1}{54} = \frac{Y_A}{60} + \frac{1 - Y_A}{48} \Rightarrow Y_A = \frac{5}{9}$

## Comprehension VII

25. $\Delta T_f = K_f \cdot m = K_f \cdot \frac{50}{\text{M}}$

$\therefore \Delta T_f(A) : \Delta T_f(B) = \text{M}_B : \text{M}_B = 3:1$

26. Average molar mass of solute in $S_1$

$$\text{M}(S_1) = \frac{2 \times \text{M}_A + 3 \times \text{M}_B}{2+3} = \frac{11}{5}\text{M}_A$$

and average molar mass of solute in $S_2$.

$$\text{M}(S_2) = \frac{3\text{M}_A + 2 \times M_B}{2+3} = \frac{9}{5}\text{M}_A$$

$\therefore \Delta T_f(S_1) : \Delta T_f(S_2) = M(S_2) : M(S_1) = 9:11$

## Comprehension VIII

27. $\Delta T_f = K_f \cdot m = 2.0 \times \frac{0.1}{0.9 \times 46} \times 1000 = 4.8$ K

$\therefore$ Freezing point of solution = 155.7 – 4.8
= 150.9 K

28. $P = X_2 \cdot P^\circ = 0.9 \times 40 = 36$ mm kg

29. $\Delta T_b = K_b \cdot m = 0.52 \times \frac{0.1}{0.9 \times 18} \times 1000 = 3.2$ K

$\therefore$ B.P. of solution = 373 + 3.2 = 376.2 K

## Comprehension IX

30. Increase in mass of absorber $\alpha\, P^\circ$ and decrease in mass of pure solvent $\alpha\, (P^\circ - P)$.

$$\therefore \frac{P^\circ - P}{P^\circ} = \frac{0.02}{0.24} = X_1 = \frac{\frac{x}{180}}{\frac{x}{180} + \frac{100 - x}{18}}$$

$\therefore$ Mass percent of glucose, $x = \frac{1000}{21}\%$

31. $AlCl_3 \rightleftharpoons Al^{3+} + 3Cl^-$

$\underset{=0.2}{1-0.8} \quad 0.8 \quad \underset{=2.4}{3\times0.8}$

Total effective mole of solute = 0.2 + 0.8 + 2.4
= 3.4

Now, decrease in mass of solution $\alpha P$ and increase in mass of absorber $\alpha P°$.

$\therefore \frac{P}{P°} = X_2 = \frac{17}{17+3.4} = \frac{5}{6} = \frac{0.18}{|\Delta m|_{\text{absorber}}}$

$\therefore$ Increase in mass of absorber = 0.216 gm

32. As $P° > P$, some vapour above the solution from saturated moist air coming will condense and hence, the mass of solution will increase.

## Section D (Assertion – Reason)

1. Henry's law
2. Theory based
3. Both have same $\Delta T_f$
4. KCl will dissociate.
5. Theory based
6. Theory based
7. Theory based
8. Theory based
9. Relative lowering of V.P. is also independent of solvent.
10. Deviation may occur in non-ideal solution.

## Section E (Column Match)

1. Informative
2. Informative
3. (A) Some concentrations
   (B) Osmolarity : NaCl = 0.2 M, $Na_2SO_4$ = 0.3 M
   (C) Osmolarity : NaCl = KCl = 0.2 M
   (D) Osmolarity : $CuSO_4$ = 0.2 M,
   Sucrose = 0.1 M
4. Higher the B.P. of solvent, normally higher is its $K_b$ value.
5. (A) $2 = 1 + \alpha(2-1) \Rightarrow \alpha = 1.00$
   (B) $2 = 1 + \alpha(3-1) \Rightarrow \alpha = 0.50$
   (C) $2 = 1 + \alpha(5-1) \Rightarrow \alpha = 0.25$
   (D) $2 = 1 + \alpha(4-1) \Rightarrow \alpha = 0.33$
6. (A) $i = 1$
   (B) $i = 1 + 1(2-1) = 2$
   (C) $i = 1 + 1(3-1) = 3$
   (D) $i = 1 + 1(4-1) = 4$
7. (P) Effective conc. = $0.1 \times 3 = 0.3$ M = 0.3 $m$
   (Q) Effective conc. = $0.14 \times 2 = 0.28$ M = 0.28 $m$
   (R) Effective conc. = 0.1 [1+0.9(3 – 1) = 0.28 M = 0.28 $m$
   (S) Effective conc. = 0.28 M = 0.28 $m$
   (T) $\underset{(0.1-x)M}{HA} \rightleftharpoons \underset{xM}{H^+} + \underset{xM}{A^-}$

   $K_a = 0.81 = \frac{x \cdot x}{0.1-x} \Rightarrow x = 0.09$

   $\therefore$ Effective conc. = $(0.1 + x) = 0.19$ M = 0.19 $m$

## Section F (Subjective)

### Single Digit Integer Type

1. $V_{gas}\ \alpha\ n_{Solvent}$ but independent of pressure.

$\therefore \left(\frac{V_2}{V_1}\right)_{gas} = \left(\frac{V_2}{V_1}\right)_{Solvent} \Rightarrow \frac{V_2}{4\text{ ml}} = \frac{0.5}{1} \Rightarrow V_2 = 2\text{ ml}$

2. $m_{solution} = m_{water} + m_{ethanol}$

or, $V \times 0.9344 = 50 \times 1.000 + 50 \times 0.7939$

$\therefore V \approx 96\text{ ml} < 100\text{ ml}$

Solution is non-ideal with negative deviation.

HINTS AND EXPLANATIONS

3. Ideal gas can never be liquefied.

4. $P = X_2 \cdot P^\circ = 0.8 \times 233.5 = 186.8$ torr $= P_{exp}$

$\therefore$ Solution is ideal.

5. $\pi = \dfrac{\pi_1 V_1 + \pi_2 V_2}{(V_1 + V_2)} = \dfrac{2.4 \times 2V + 4.2 \times V}{2V + V} = 3$ atm

6. $\pi = CRT = \rho g h$

or $\dfrac{0.2/\text{M}}{100/1000} \times 0.0821 \times 300 = \dfrac{1.013 \times 1000 \times 0.2463}{1.013 \times 10^6}$

$\therefore \text{M} = 2 \times 10^5$

7. $\pi = CRT = \rho g h$

or $\dfrac{n}{1} \times 0.08 \times 298 = \dfrac{1.013 \times 1000 \times 7.45}{1.013 \times 10^6}$

$\therefore n = \dfrac{25 \times 10^{-3}}{80}$

$\therefore$ Millimoles in 320 gm $= \dfrac{25}{80} \times \dfrac{320}{20} = 5$

8. $\dfrac{V_2}{V_1} = \dfrac{C_1}{C_2} = \dfrac{\pi_1 / T_1}{\pi_2 / T_2} = \dfrac{500/283}{105.3/298} \approx 5$

9. $\Delta T_f = K_f \cdot m \Rightarrow 0.29 = 1.86 \times \dfrac{1.04/267}{100/1000} \times n$

$\Rightarrow n \approx 4$

10. $\Delta T_f = K_f \cdot m$

For KCN solution: $0.80 = K_f \times 0.2 \times 2$ (1)

$$\begin{array}{lccc} & \text{Hg(CN)}_2 + & \text{mCN}^- & \rightleftharpoons \text{Hg(CN)}_{m+2}^{m-} \\ & 0.1\ \text{mole} & 0.2\ \text{mole} & 0 \\ \text{Final} & 0 & (0.2-0.1\text{m})\ \text{mole} & 0.1\ \text{mole} \end{array}$$

Final effective molality = (0.2 – 0.1 m) + 0.1 + 0.2
= 0.5 – 0.1 m

Now, $0.60 = K_f \times (0.5 - 0.1\ \text{m})$ (2)

From (1) and (2): $m = 2$

## Four Digit Integer Type

1. $P = X_2 \cdot P^\circ$

$20 = \dfrac{180/18}{\dfrac{6}{\text{M}} + \dfrac{180}{18}} \times P^\circ$ (1)

and $20.02 = \dfrac{11}{\dfrac{6}{\text{M}} + 11} \times P^\circ$ (2)

$\therefore \text{M} = 54$

2. Mole fraction of solvent is same in both.

$\therefore \dfrac{\dfrac{90}{18}}{\dfrac{10}{\text{M}} + \dfrac{90}{18}} = \dfrac{\dfrac{95}{18}}{\dfrac{5}{180} + \dfrac{95}{18}} \Rightarrow \text{M}_X = 380$

3. $\dfrac{P^\circ - P}{P^\circ} = X_1 = \dfrac{2400 - 2300}{2400} = \dfrac{1}{24}$

1 mole solution $\begin{cases} \text{Urea} = \dfrac{1}{24}\ \text{mole} = \dfrac{1}{24} \times 60 = 2.5\ \text{gm} \\ \text{Water} = \dfrac{23}{24}\ \text{mole} = \dfrac{23}{24} \times 18 = 17.25\ \text{gm} \end{cases}$

$\therefore$ Volume of 1 mole solution $= \dfrac{2.5 + 17.25}{1.185} = \dfrac{50}{3}$ ml

Now, $\pi = CRT = \left( \dfrac{1/24}{50/3} \times 1000 \right) \times 0.08 \times 300$

$= 60$ atm

4. $\Delta T_f = K_f \cdot m$

or, $30 = 1.86 \times \dfrac{w/62}{795/1000} \Rightarrow w = 795$ gm

5. $\Delta T_f = K_f \cdot m$

or, $6 = 1.86 \times \dfrac{50/62}{w/1000} \Rightarrow w_{\text{water (final)}} = 250$ gm

$\therefore$ Mass of water separated as ice = 375 – 250
= 125 gm

6. $\Delta T_f = K_f \cdot m$

Naphthalene solution: $13.5 = K_f \times \dfrac{38.4/128}{185/1000}$ (1)

Unknown substance solution:

$9.0 = K_f \times \dfrac{11.6/\text{M}}{185/1000}$ (2)

$\therefore \text{M} = 58$

7. $\Delta T_f = K_f \cdot m = 1.86 \times \dfrac{\left( \dfrac{3.6}{180} + \dfrac{3.6}{60} \right)}{200} \times 1000$

$= 0.744°\text{C}$

$\therefore$ F. P. of solution = – 0.744°C

**8.** $\Delta T_f = K_f \cdot m$

$$(26.84 - 25.64) = 8 \times \frac{2.4 \times 10^{-3} / \text{M}}{100 \times 10^{-3} / 1000} \Rightarrow \text{M} = 160$$

**9.** $\Delta T_f = K_f \cdot m$

$$\Rightarrow 1.60 = 4.88 \times \frac{2/122}{26/1000} \times \left[1 + \alpha\left(\frac{1}{2} - 1\right)\right]$$

$\therefore \alpha = 0.96$ or 96%

**10.** $\pi = CRT \Rightarrow 4.92 = \left(\frac{x}{200} + 0.05 \times 2\right) \times 0.08 \times 300$

$\therefore x = 20$

**11.** $P° - P = X_1 \cdot P° = \frac{n_1}{n_1 + n_2}, P° \approx \frac{n_1}{n_2} \cdot P°$

Urea solution: $0.03 = \frac{0.1}{\left(\frac{1000}{18}\right)} \times P°$

KCl solution: $0.0594 = \frac{0.1[1 + \alpha(2 - 1)]}{\left(\frac{1000}{18}\right)} \times P°$

$\therefore \alpha = 0.98$ or 98%

**12.** Loss in weight of solution $\alpha\, P$

Loss in weight of water $\alpha(P° - P)$

Now, $\frac{P° - P}{P} = \frac{n_1}{n_2} \Rightarrow \frac{0.01}{0.98} = \frac{\frac{1.25}{90}[1 + \alpha(3 - 1)]}{\frac{49}{18}}$

$\therefore \alpha = 0.50$ or 50%

**13.** $\Delta T_f = K_f \cdot m \Rightarrow 7 = 14 \times \frac{75.2}{94} \times \left[1 + \alpha\left(\frac{1}{2} - 1\right)\right]$

$\therefore \alpha = 0.75$ or 75%

**14.**

100 gm solution (Say)
- → $MgSO_4$ = 12 gm = $\frac{12}{120}$ = 0.1 mole
- → $MgCl_2$ = 9.5 gm = $\frac{9.5}{9.5}$ = 0.1 mole
- → Water = 100 – (12 + 9.5) = 78.5 gm

Effective moles of solute

= 0 · 1 [1 + 0.8(2 – 1)] + 0.1 [1 + 0.6 (3– 1)] = 0.4

Now, $\Delta T_b = K_b \cdot m = 0.785 \times \frac{0.4}{78.5/1000} = 4$ K

∴ B. P. of solution = 373 + 4 = 377 K

**15.** $\Delta T_f = K_f \cdot m \Rightarrow 0.40 = 1.86 \times m$

∴ Effective molality of NaCl solution = $\frac{20}{93}$ m

Now, $\pi = CRT = \frac{20}{93} \times 0.0821 \times 279$

$= 60 \times 0.0821$ atm

CHAPTER 11

# Chemical Kinetics

## EXERCISE I (JEE MAIN)

### Rate of Reaction, Rate Law, Order and Molecularity of a Reaction

1. Azomethane ($CH_3NNCH_3$) decomposes according to the following equation.

$$CH_3–N \equiv N–CH_3(g) \rightarrow C_2H_6(g) + N_2(g)$$

The initial concentration of azomethane was $1.50 \times 10^{-2}$ M. After 10 minutes, the concentration was decreased to $1.29 \times 10^{-2}$ M. The average rate of reaction during this time interval is

(a) $3.5 \times 10^{-6}$ $Ms^{-1}$ (b) $2.1 \times 10^{-4}$ $Ms^{-1}$
(c) $3.5 \times 10^{-6}$ M $h^{-1}$ (d) $2.1 \times 10^{-3}$ $Mmin^{-1}$

2. For the reaction $2HI \rightarrow H_2 + I_2$, the expression $-\frac{1}{2}\frac{d[HI]}{dt}$ represents

(a) the rate of formation of HI.
(b) the rate of disappearance of HI.
(c) the instantaneous rate of the reaction.
(d) the average rate of the reaction.

3. For the reaction 4A + B → 2C + 2D, the only incorrect statement is

(a) the rate of disappearance of B is one-fourth the rate of disappearance of A.
(b) the rate of appearance of C is half the rate of disappearance of B.
(c) the rate of formation of D is half the rate of consumption of A.
(d) the rates of formation of C and D are equal.

4. For the reaction $2N_2O_5(g) \rightarrow 4NO_2(g) + O_2(g)$, the concentration of $NO_2$ increases by $2.4 \times 10^{-2}$ M in 6 s. What will be the average rate of appearance of $NO_2$ and the average rate of disappearance of $N_2O_5$?

(a) $2 \times 10^{-3}$ $Ms^{-1}$, $4 \times 10^{-3}$ $Ms^{-1}$
(b) $2 \times 10^{-3}$ $Ms^{-1}$, $1 \times 10^{-3}$ $Ms^{-1}$
(c) $2 \times 10$ $Ms^{-1}$, $2 \times 10^{-3}$ $Ms^{-1}$
(d) $4 \times 10^{-3}$ $Ms^{-1}$, $2 \times 10^{-3}$ $Ms^{-1}$

5. Which of the following statement is wrong about the rate of reaction?

(a) Rate of a reaction cannot be negative.
(b) Rate of a reaction is the change in concentration of the reactant per unit time per unit stoichiometric coefficient of that component.
(c) Average rate and instantaneous rate are always different.
(d) Rate may depend upon the surface area of the reactants.

6. For a reaction $2NH_3 \rightarrow N_2 + 3H_2$, it is observed that

$$-\frac{d[NH_3]}{dt} = K_1[NH_3]; \; +\frac{d[N_2]}{dt} = K_2[NH_3] \text{ and } +\frac{d[H_2]}{dt} = K_3[NH_3]$$

The correct relation between $K_1$, $K_2$ and $K_3$ is

(a) $K_1 = K_2 = K_3$ (b) $2K_1 = 3K_2 = 6K_3$
(c) $3K_1 = 6K_2 = 2K_3$ (d) $6K_1 = 3K_2 = 2K_3$

7. The rate constant of a reaction depends upon
(a) concentration (b) pressure
(c) temperature (d) All of these

8. Sometimes, reaction rates can be estimated by knowing
(a) the atmospheric pressure.
(b) the number of bond changes.
(c) the number of apparatus used in the experiment.
(d) the net ionic equation.

9. The rate expression for the reaction A(g) + B(g) → C(g) is rate = $K{\cdot}C_A^{\,2}{\cdot}C_B^{\,1/2}$. What changes in the initial concentration of A and B will cause the rate of reaction increase by a factor of eight?
(a) $C_A \times 2, C_B \times 2$ (b) $C_A \times 2, C_B \times 4$
(c) $C_A \times 1, C_B \times 4$ (d) $C_A \times 4, C_B \times 1$

10. The rate constant of a second-order reaction is $10^{-2}$ lit mole$^{-1}$ s$^{-1}$. The rate constant when expressed in ml molecule$^{-1}$ min$^{-1}$ is ($N_A = 6 \times 10^{23}$)
(a) 0.01 (b) $1.0 \times 10^{-21}$
(c) $1.0 \times 10^{21}$ (d) $3.6 \times 10^{20}$

11. The rate of a reaction A + 2B → P is $2 \times 10^{-2}$ M/min, when concentrations of each A and B are 1.0 M. If the rate of reaction $r = K[A]^2[B]$, the rate of reaction when half of the B has reacted should be
(a) $5.625 \times 10^{-3}$ M/min (b) $3.75 \times 10^{-3}$ M/min
(c) 9.375 M/min (d) $2.5 \times 10^{-3}$ M/min

12. For the reaction $A_2(g) \rightarrow B(g) + \frac{1}{2}C(g)$, pressure of the system increases from 100 to 120 mm in 5 minutes. The average rate of disappearance of $A_2$ (in mm/min) in this time interval is
(a) 4 (b) 8
(c) 2 (d) 16

13. The condition at which average rate can be equal to instantaneous rate of the reaction is
(a) $\Delta n = 0$
(b) $\Delta t \rightarrow 0$
(c) reaction is elementary.
(d) reaction is complex.

14. The rate constant for the reaction $N_2O_5 \rightarrow 4NO_2 + O_2$ is $3.0 \times 10^{-5}$ s$^{-1}$. If the rate of reaction is $2.4 \times 10^{-5}$ mol litre$^{-1}$ s$^{-1}$, then the concentration of $N_2O_5$ (in mole litre$^{-1}$) is
(a) 1.4 (b) 1.2
(c) 0.04 (d) 0.8

15. Consider the chemical reaction $N_2(g) + 3H_2(g) \rightarrow 2NH_3(g)$. The rate of this reaction can be expressed in terms of time derivative of concentration of $N_2(g)$, $H_2(g)$ or $NH_3(g)$. Identify the correct relationship amongst the rate expressions.

(a) $\text{Rate} = -\dfrac{d[N_2]}{dt} = -\dfrac{1}{3}\dfrac{d[H_2]}{dt} = +\dfrac{1}{2}\dfrac{d[NH_3]}{dt}$

(b) $\text{Rate} = -\dfrac{d[N_2]}{dt} = -3\dfrac{d[H_2]}{dt} = +2\dfrac{d[NH_3]}{dt}$

(c) $\text{Rate} = -\dfrac{d[N_2]}{dt} = \dfrac{1}{3}\dfrac{d[H_2]}{dt} = +\dfrac{1}{2}\dfrac{d[NH_3]}{dt}$

(d) $\text{Rate} = -\dfrac{d[N_2]}{dt} = -\dfrac{d[H_2]}{dt} = +\dfrac{d[NH_3]}{dt}$

16. For the reaction $N_2(g) + 3H_2(g) \rightarrow 2NH_3(g)$ under certain conditions of temperature and partial pressure of the reactants, the rate of formation of $NH_3$ is $10^{-3}$ kg h$^{-1}$. The rate of consumption of $H_2$ under same condition is
(a) $1.5 \times 10^{-3}$ kg h$^{-1}$ (b) $1.76 \times 10^{-4}$ kg h$^{-1}$
(c) $6.67 \times 10^{-4}$ kg h$^{-1}$ (d) $3 \times 10^{-3}$ kg h$^{-1}$

17. The reaction $2FeCl_3 + SnCl_2 \rightarrow 2FeCl_2 + SnCl_4$ is a reaction of
(a) zero-order (b) first-order
(c) second-order (d) third-order

18. Which one of the following statement is wrong regarding molecularity of a reaction?
(a) It may be 1, 2 or 3.
(b) It may be either a whole number or fraction.
(c) It is the number of molecules of the reactants taking part in a single-step chemical reaction.
(d) It depends on the rate determining step reaction in the reaction.

**19.** Which of the following statement is incorrect?

(a) Rate law is an experimental fact, whereas law of mass action is a theoretical proposal.

(b) Rate law is always different from the expression of law of mass action.

(c) Rate law is more informative than law of mass action for the development of mechanism.

(d) Order of a reaction is equal to the sum of powers of concentration terms in the rate law.

**20.** The rate constant is numerically the same for first, second and third-order reactions, respectively. Which one is true for rate of three reactions, if concentration of reactant is lesser than 1 M?

(a) $r_1 = r_2 = r_3$ (b) $r_1 > r_2 > r_3$
(c) $r_1 < r_2 < r_3$ (d) None of these

**21.** For the elementary reaction $2NO + O_2 \rightarrow 2NO_2$, the volume of a container is tripled. The rate becomes

(a) 3 times (b) 27 times
(c) $\frac{1}{27}$ times (d) as initial

**22.** A zero-order reaction is one

(a) in which reactants do not react.

(b) in which one of the reactants is in large excess.

(c) whose rate does not change with time.

(d) whose rate increases with time.

**23.** Which of the following statement is incorrect

(a) Stoichiometry of a reaction tells about the order of the elementary reactions.

(b) For a zero-order reaction, rate and the rate constant are identical.

(c) A zero-order reaction is controlled by factors other than concentration of reactants.

(d) A zero-order reaction is always an elementary reaction.

**24.** If the rate of a gaseous reaction is independent of partial pressure of reactant, then the order of reaction is

(a) 0 (b) 1
(c) 2 (d) 3

**25.** When the concentration of reactant in the reaction $A \rightarrow B$ is increased by 8 times, the rate increases only by 2 times. The order of reaction is

(a) 3 (b) $\frac{1}{3}$
(c) 2 (d) $\frac{1}{2}$

**26.** The rate constant ($K$) for the reaction $2A + B \rightarrow$ Products was found to be $2.58 \times 10^{-5}$ L mol$^{-1}$ s$^{-1}$ after 15 s, $2.59 \times 10^{-5}$ L mol$^{-1}$ s$^{-1}$ after 30 s and $2.57 \times 10^{-1}$ L mol$^{-1}$ s$^{-1}$ after 50 s. The overall order of the reaction is

(a) 0 (b) 1
(c) 2 (d) 3

**27.** Which of the following statements is incorrect?

(a) The order of reaction is the sum of powers of all the concentration terms in the rate equation.

(b) The order of reaction with respect to one reactant is the ratio of the change of logarithm of the rate of the reaction to the change in the logarithm of the concentration of the particular reactant, keeping the concentrations of all other reactants constant.

(c) Orders of reactions can be whole numbers or fractional numbers.

(d) The order of a reaction can only be determined from the stoichiometric equation for the reaction.

**28.** For a chemical reaction $X \rightarrow Y$, the rate of reaction increases by a factor of 1.837 when the concentration of X is increased by 1.5 times. The order of the reaction with respect to X is

(a) 1 (b) 1.5
(c) 2 (d) −1

**29.** The order of a reaction is zero. It will be definitely

(a) exothermic (b) endothermic
(c) elementary (d) complex

**30.** Which one of the following statements for order of a reaction is not correct?

(a) Order can be determined experimentally.

(b) Order of reaction is equal to sum of the powers of concentration terms in differential rate law.

(c) It is not affected with the stoichiometric coefficient of the reactants.

(d) Order cannot be fractional.

**31.** Select the rate law that corresponds to the data shown for the following reaction

| Exp. no. | $[A_0]$ (M) | $[B_0]$ (M) | Initial rate (M s$^{-1}$) |
|---|---|---|---|
| 1 | 0.012 | 0.035 | 0.1 |
| 2 | 0.024 | 0.070 | 0.8 |
| 3 | 0.024 | 0.035 | 0.1 |
| 4 | 0.012 | 0.070 | 0.8 |

(a) $r = K[B]^3$ (b) $r = K[B]^4$
(c) $r = K[A][B]^3$ (d) $r = K[A]^2[B]^3$

**32.** The rate law for a reaction between the substances A and B is given by rate = $K[A]^n [B]^m$. On doubling the concentration of A and halving the concentration of B, the ratio of the new rate to the earlier rate of the reaction will be as

(a) $1/2^{m+n}$ (b) $(m+n)$
(c) $(n-m)$ (d) $2^{(n-m)}$

**33.** Which of the following reactions is of fractional order?

(a) Decomposition of $N_2O_5$.
(b) Isomerization: $CH_3-NC \rightarrow CH_3-CN$
(c) Thermal decomposition of $CH_3-CHO$
(d) Isomerization: $\triangle \rightarrow CH_3-CH=CH_2$

**34.** For a reaction $2A+B+3C \rightarrow D+3E$, the following data is obtained.

| Exp. No. | Concentration in mole per litre A | B | C | Initial rate of formation of D (torr s$^{-1}$) |
|---|---|---|---|---|
| 1 | 0.01 | 0.01 | 0.01 | $2.5 \times 10^{-4}$ |
| 2 | 0.02 | 0.01 | 0.01 | $1.0 \times 10^{-3}$ |
| 3 | 0.01 | 0.02 | 0.01 | $2.5 \times 10^{-4}$ |
| 4 | 0.01 | 0.02 | 0.02 | $5.0 \times 10^{-4}$ |

The order with respect to A, B and C are, respectively,

(a) 0, 1, 2 (b) 2, 0, 1
(c) 1, 0, 2 (d) 2, 1, 1

**35.** In the gas phase, two butadiene molecules can dimerize to give larger molecules according to the reaction $2C_4H_6(g) \rightarrow C_8H_{12}(g)$. The rate law for this reaction is $r = K[C_4H_6]^2$ with $K = 6.1 \times 10^{-2}$ L mol$^{-1}$ s$^{-1}$ at the temperature of reaction. The rate of formation of $C_8H_{12}$ when the concentration of $C_4H_6$ is 0.02 M is

(a) $2.44 \times 10^{-5}$ Ms$^{-1}$ (b) $1.22 \times 10^{-5}$ Ms$^{-1}$
(c) $1.22 \times 10^{-3}$ Ms$^{-1}$ (d) $2.44 \times 10^{-6}$ Ms$^{-1}$

**36.** For the reactions of first, second and third orders, $K_1 = K_2 = K_3$, where concentrations are expressed in 'M'. The correct relation in $K_1$, $K_2$ and $K_3$, if concentrations are expressed is mol ml$^{-1}$ is

(a) $K_1 = K_2 = K_3$
(b) $K_1 \times 10^{-3} = K_2 = K_3 \times 10^3$
(c) $K_1 = K_2 \times 10^{-3} = K_3 \times 10^{-6}$
(d) $K_1 \times 10^3 = K_2 \times 10^{-3} = K_3$

**37.** The rate of the reaction $3A + 2B \rightarrow$ Products is given by the rate expression rate = $K[A][B]^2$. If A is taken in large excess, then the reaction start obeying

(a) third-order kinetics
(b) second-order kinetics
(c) first-order kinetics
(d) zero-order kinetics

**38.** The hydrolysis of an ester was carried out separately with 0.05 M HCl and 0.05 M $H_2SO_4$. Which of the following is true for the first order rate constant of the reaction?

(a) $K_{HCl} > K_{H_2SO_4}$ (b) $K_{HCl} < K_{H_2SO_4}$
(c) $K_{HCl} = K_{H_2SO_4}$ (d) $K_{H_2SO_4} = (K_{HCl})^2$

**39.** The hydrolysis of methyl formate in acid solution has the rate expression rate = $K[HCOOCH_3][H^+]$. The balanced equation is $HCOOCH_3 + H_2O \rightarrow HCOOH + CH_3OH$. The rate law contains $[H^+]$ though the balanced equation does not contain $H^+$, because

(a) more for convenience to express the rate law.
(b) $H^+$ ion is a catalyst of the reaction.
(c) $H^+$ ion is an important constituent of any reaction.
(d) all acids contain $H^+$ ions.

**40.** In the acid hydrolysis reaction $A + H_2O + H^+ \rightarrow$ Products, where $[H^+] = 0.1$ M and $H_2O$ is present in large excess, the apparent rate constant is $1.5 \times 10^{-5}$ s$^{-1}$.

$$r = K[A][H_2O][H^+] = K_{app}[A]$$

The true rate constant (third-order rate constant) of reaction is

(a) $2.7 \times 10^{-6}$ M$^{-1}$ s$^{-1}$ (b) $1.5 \times 10^{-4}$ M$^{-2}$ s$^{-1}$
(c) $2.7 \times 10^{-6}$ M$^{-2}$ s$^{-1}$ (d) $2.7 \times 10^{-6}$ M$^2$ s$^{-1}$

## Mechanism of Reaction

**41.** A complex reaction $2X + Y \rightarrow Z$, takes place in the following two steps.

$$X + Y \xrightarrow{K_1} 2W,$$

$$X + 2W \xrightarrow{K_2} Z$$

If $K_1 << K_2$, then the order of reaction will be

(a) 1 (b) 2
(c) 3 (d) 0

**42.** For the reaction $A \xrightarrow{K_1} B \xrightarrow{K_2} C \xrightarrow{K_3} D$, where $K_3 > K_2 > K_1$; the rate of formation of D is determined by

(a) the rate of reaction: $A \rightarrow B$
(b) the rate of reaction: $B \rightarrow C$
(c) the rate of reaction: $C \rightarrow D$
(d) the rate of reaction: $A \rightarrow B$ with rate constant $K_1 . K_2 . K_3$

**43.** For hypothetical reaction $A \rightarrow B$ takes place according to

$$A \underset{}{\overset{K_1}{\rightleftharpoons}} C \text{ (fast)}$$

$$A + C \xrightarrow{K_2} B \text{ (slow)}$$

the rate of reaction is ($K_1$ is equilibrium constant)

(a) $K_2[B][C]$ (b) $K_1K_2[A]$
(c) $K_1K_2[A]^2$ (d) $K_1[B][C]$

**44.** The overall rate $+\frac{d[P]}{dt}$ for the reaction $2A + C \rightarrow P$ from the mechanism

$$2A \overset{K}{\rightleftharpoons} B \text{ (fast)}$$

$$B + C \xrightarrow{K_f} P \text{ (slow)}$$

where, $K$ = equilibrium constant and $K_f$ = forward rate constant, is given by

(a) $\frac{d[P]}{dt} = K K_f [A]^2[C]$

(b) $\frac{d[P]}{dt} = K [A][B]$

(c) $\frac{d[P]}{dt} = K_f [B][C]$

(d) $\frac{d[P]}{dt} = K K_f [A]^2[B][C]$

**45.** The suggested mechanism for the reaction $CHCl_3(g) + Cl_2(g) \rightarrow CCl_4(g) + HCl(g)$ is

$$Cl_2 \underset{K_2}{\overset{K_1}{\rightleftharpoons}} 2\dot{Cl} \text{ (fast)}$$

$$CHCl_3 + \dot{Cl} \xrightarrow{K_3} HCl + \dot{C}Cl_3 \text{ (slow)}$$

$$\dot{C}Cl_3 + \dot{Cl} \xrightarrow{K_4} CCl_4 \text{ (fast)}$$

The experimental rate law consistent with the mechanism is

(a) rate $= K_3[CHCl_3][Cl_2]$
(b) rate $= K_4[CCl_3][Cl]$
(c) rate $= K_{eq}[CHCl_3][Cl_2]$
(d) rate $= K_3 K_{eq}^{1/2}[CHCl_3][Cl_2]^{1/2}$

## Integrated Rate Law

**46.** For a zero-order reaction,

(a) the reaction rate is doubled when the initial concentration is doubled.
(b) the time for half change is half the time taken for completion of the reaction.
(c) the time for half change is independent of the initial concentration.
(d) the time for completion of the reaction is independent of the initial concentration.

**47.** The time taken in 75% completion of a zero-order reaction is 10 h. In what time, the reaction will be 90% completed?

(a) 12.0 h (b) 16.6 h
(c) 10.0 h (d) 20.0 h

**48.** A zero-order reaction $A \rightarrow B$. At the end of 1 h, A is 75% reacted. How much of it will be left unreacted at the end of 2 h?

(a) 12.5% (b) 6.25%
(c) 3.12% (d) 0%

**49.** Which of the following represents the expression for 3/4th the life of a first-order reaction?

(a) $\frac{k}{2.303} \log \frac{4}{3}$ (b) $\frac{2.303}{k} \log \frac{4}{3}$

(c) $\frac{2.303}{k} \log 4$ (d) $\frac{2.303}{k} \log 3$

**50.** The decomposition of $H_2O_2$ can be followed by titration with $KMnO_4$ and is found to be a first-order reaction. The rate constant is $4.5 \times 10^{-2}$. In an experiment, the initial titre value was 25 ml. The titre will be 5 ml after

(a) $4.5 \times 10^{-2} \times 5$ min (b) $\frac{\ln 5}{4.5 \times 10^{-2}}$ min

(c) $\frac{\ln(5/4)}{4.5 \times 10^{-2}}$ min (d) $4.5 \times 1.25$ min

**51.** For the first-order reaction $t_{99\%} = x \times t_{90\%}$. The value of '$x$' will be

(a) 10 (b) 6

(c) 3 (d) 2

**52.** The half-life of a first-order reaction is 10 minutes. In what time, the rate of reaction will decrease from $6.0 \times 10^{21}$ molecules $ml^{-1}$ $s^{-1}$ to $4.5 \times 10^{25}$ molecules $litre^{-1}$ $min^{-1}$? ($N_A = 6.0 \times 10^{23}$)

(a) 10 min (b) 20 min

(c) 30 min (d) 40 min

**53.** For the second-order reaction $2A \rightarrow B$, time taken for the [A] to fall to one-fourth value is how many times the time it takes for [A] to fall to half of its initial value?

(a) Two (b) Three

(c) Four (d) Seven

**54.** The half-life for a given reaction is doubled if the initial concentration of the reactant is doubled. The order of the reaction is

(a) 0 (b) 1

(c) 2 (d) 3

**55.** At the point of intersection of the two curves shown, the concentration of B is given as ($A \rightarrow nB$)

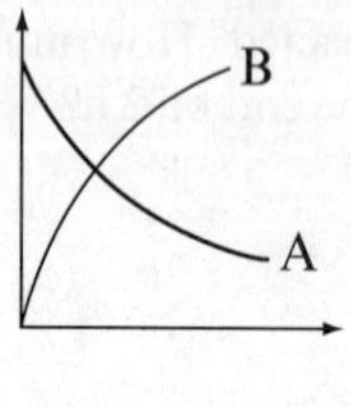


(a) $\frac{A_0}{n}$ (b) $\frac{A_0}{n-1}$

(c) $\frac{nA_0}{n+1}$ (d) $\frac{n-1}{n+1}A_0$

**56.** Which of the following graphs is correct representation for the reaction of type $A + 2B \rightarrow C$?

(a)

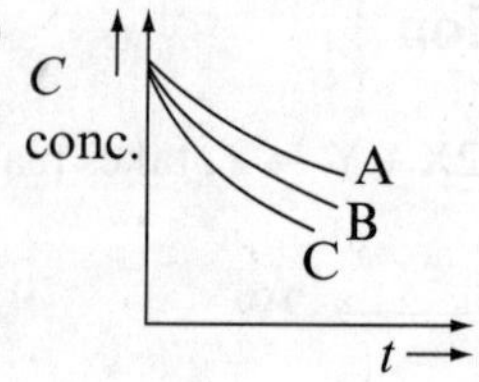


(b)

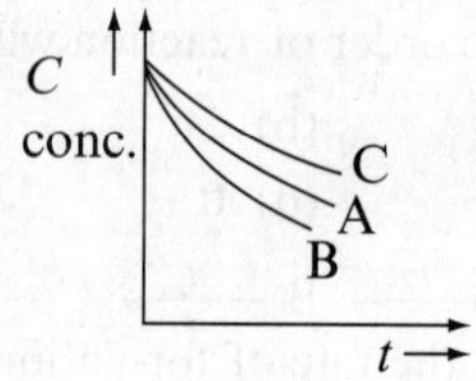


(c)

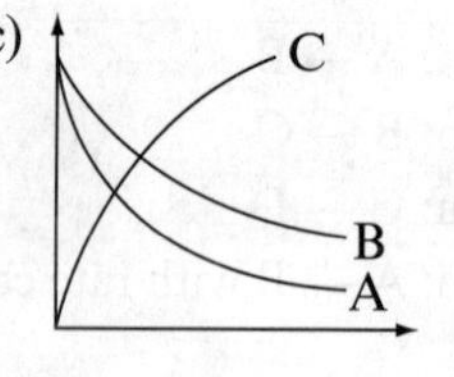


(d)

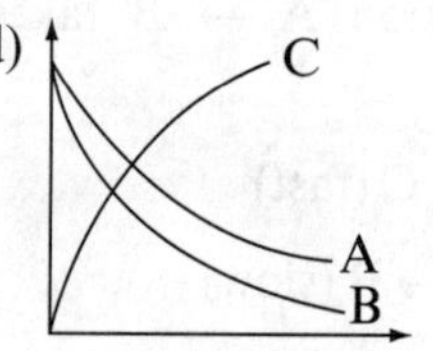


**57.** For a given reaction, the concentration of the reactant plotted against time gave a straight line with negative slope. The order of the reaction is

(a) 3 (b) 2

(c) 1 (d) 0

**58.** For a certain reaction involving a single reactant, it is found that $C_0\sqrt{T}$ is constant, where $C_0$ is the initial concentration of the reactant and $T$ is the half-life. What is the order of the reaction?

(a) 1 (b) 1.5

(c) 2 (d) 3

**59.** For a certain reaction of order '$n$', the time for half change $t_{1/2}$ is given by $t_{1/2} = \frac{[2-\sqrt{2}]}{k} \times C_0^{1/2}$, where $k$ is the rate constant and $C_0$ is the initial concentration. The value of $n$ is

(a) 1 (b) 2

(c) 1.5 (d) 0.5

60. The reaction $H_3C–CH_2–NO_2 + OH^- \rightarrow H_3C–CH^-–NO_2 + H_2O$ obeys the rate law for pseudo first-order kinetics in the presence of a large excess of hydroxide ion. If 1% of nitroethane undergoes reaction in half a minute when the reactant concentration is 0.002 M, what is the pseudo first-order rate constant?
(a) $2 \times 10^{-2}$ min$^{-1}$ (b) $6 \times 10^{-3}$ min$^{-1}$
(c) $4 \times 10^{-2}$ min$^{-1}$ (d) $1 \times 10^{-2}$ min$^{-1}$

61. Which of the following is an incorrect statement about first-order reaction?
(a) The time for 93.75% completion of reaction is two times the time for 75% completion of reaction.
(b) Half-life is the half of the time of complete reaction.
(c) About 63% reaction completes in time, equal to $1/K$.
(d) The time for 90% reaction will be equal to $2.303/K$, irrespective of the initial concentration of the reactant.

62. For the first-order reaction $A(g) \rightarrow 2B(g) + C(g)$, the total pressure after time $t$ from the start of reaction with A is P and after infinite time, it is $P_\infty$. Then the rate constant of the reaction is
(a) $\frac{1}{t}\ln\frac{P_\infty}{P}$ (b) $\frac{1}{t}\cdot\ln\frac{2P_\infty}{3(P_\infty - P)}$
(c) $\frac{1}{t}\cdot\ln\frac{2P_\infty}{3P_\infty - P}$ (d) $\frac{1}{t}\ln\frac{2P_\infty}{P_\infty - 3P}$

63. As the initial concentration increases from 0.75 to 1.55 M in a reaction, $t_{1/2}$ decreases from 60 to 29 s. The order of the reaction is
(a) zero (b) first
(c) second (d) third

64. Which of the following is correct about first-order reaction?
(a) $t_{0.5} = 50$ s, $t_{0.75} = 100$ s
(b) $t_{0.5} = 50$ s, $t_{0.75} = 75$ s
(c) $t_{0.5} = 50$ s, $t_{0.75} = 50$ s
(d) $t_{0.5} = 50$ s, $t_{0.25} = 25$ s

65. $t_{1/2}$ of a reaction: A → Products $\left(\text{order} = \frac{3}{2}\right)$ is represented by $t_{1/2} \propto \frac{1}{[A_0]^m}$. The value of $m$ is
(a) 0.5 (b) −0.5
(c) 1.5 (d) −1.5

66. If $t_{1/2}$ of a second-order reaction is 1.0 h. After what time, the amount will be 25% of the initial amount?
(a) 1.5 h (b) 2 h
(c) 2.5 h (d) 3 h

67. In a first-order reaction, the activity of reactant drops from 800 mol/dm$^3$ to 50 mol/dm$^3$ in $2 \times 10^4$ s. The rate constant of the reaction in s$^{-1}$ is
(a) $1.386 \times 10^{-4}$ (b) $1.386 \times 10^{-3}$
(c) $1.386 \times 10^{-5}$ (d) $5.0 \times 10^3$

68. The decomposition of $N_2O_5$ according to the equation $2N_2O_5(g) \rightarrow 4NO_2(g) + O_2(g)$ is a first-order reaction. After 30 minutes, from the start of the decomposition in a closed vessel, the total pressure developed is found to be 300 mm of Hg and on complete decomposition, the total pressure is 600 mm of Hg. The rate constant of the reaction is (ln 1.2 = 0.18)
(a) $6.0 \times 10^{-3}$ min$^{-1}$ (b) 0.18 min$^{-1}$
(c) $6.0 \times 10^{-2}$ min$^{-1}$ (d) $3.0 \times 10^{-3}$ min$^{-1}$

69. For a first-order reaction, the ratio of time for 99.9% of the reaction to complete and half of the reaction to complete is
(a) 1 (b) 2 (c) 4 (d) 10

70. For the second-order reaction A + B → Products, the rate constant $k$, is given as

$$k = \frac{2.303}{(a-b)t}\log\frac{b(a-x)}{a(b-x)},$$

where, $a$ and $b$ are the initial concentrations of 'A' and 'B' and $x$ is the change in concentration after time $t$. If $b >> a$, the reaction reduces to
(a) first-order with respect to 'A'.
(b) zero-order with respect to 'A'.
(c) first-order with respect to 'B'.
(d) overall zero-order.

71. For a first-order reaction, $t_{0.75} = 1386$ s. Its specific reaction rate is
(a) $10^{-3}$ s$^{-1}$ (b) $10^{-2}$ s$^{-1}$
(c) $10^{-4}$ s$^{-1}$ (d) $10^{-5}$ s$^{-1}$

72. It takes 40 minutes for the completion of 50% of the reaction A → Products. For the completion of next 50% of the reaction, the time taken is 80 minutes. The order of the reaction is
(a) 2 (b) 0
(c) 3 (d) 1

**73.** At 325°C, 1, 3-butadiene dimerizes according to the equation: $2C_4H_6(g) \rightarrow C_8H_{12}(g)$. It is found that the partial pressure $P$, of the reactant obeys the law: $\frac{1}{P} = kt + \frac{1}{P_0}$. The order of the reaction is

(a) 1 (b) 0.5
(c) 2 (d) 1.5

**74.** The half-life for the first-order reaction $H_2O_2(aq) \rightarrow H_2O(l) + \frac{1}{2}O_2(g)$ is 30 minutes. If the volume of $O_2(g)$ collected is 100 ml after a long time, then the volume of $O_2(g)$ collected (at the same pressure and temperature) after 60 minutes from the start of reaction is

(a) 25 ml (b) 12.5 ml
(c) 75 ml (d) 50 ml

**75.** The decomposition of hydrogen peroxide in aqueous solution is a first-order reaction

| Time in minutes | 0 | 10 |
|---|---|---|
| Volume ($V$, in ml) | 25.0 | 20.0 |

Here, $V$ is the number of ml of potassium permanganate required to decompose a definite volume of hydrogen peroxide solution at the given time. From the following data, the rate constant (in $min^{-1}$) of reaction is (ln 5 = 1.6, ln 2 = 0.7)

(a) 0.09 (b) 0.02
(c) 0.2 (d) 0.16

## Parallel, Sequential and Reversible Reactions

**76.** For irreversible elementary reactions in parallel $A \xrightarrow{K_1} R$ and $A \xrightarrow{K_2} S$, the rate of disappearance of reactant 'A' is

(a) $(k_1 - k_2)\,C_A$ (b) $(k_1 + k_2)\,C_A$
(c) $1/2\,(k_1 + k_2)\,C_A$ (d) $k_1 C_A$

**77.** For the consecutive unimolecular-type first-order reaction $A \xrightarrow{k_1} R \xrightarrow{k_2} S$, the concentration of component 'R', $C_R$, at any time, '$t$' is given by

$$C_R = C_A^{\circ} . K_1 \left[ \frac{e^{-k_1 t}}{(k_2 - k_1)} + \frac{e^{-k_2 t}}{(k_1 - k_2)} \right]$$

If $C_A = C_A^{\circ}$, $C_R = C_S = 0$ at $t = 0$, the time at which the maximum concentration of 'R' occurs is

(a) $t_{max} = \frac{k_2 - k_1}{\ln(k_2 / k_1)}$ (b) $t_{max} = \frac{\ln(k_2 / k_1)}{k_2 - k_1}$

(c) $t_{max} = \frac{e^{k_2 / k_1}}{k_2 - k_1}$ (d) $t_{max} = \frac{e^{k_2 - k_1}}{k_2 - k_1}$

**78.** Consider the following consecutive first-order reaction.

$$A \xrightarrow{K_1} B \xrightarrow{K_2} C$$

If $K_1 = 0.01\ min^{-1}$ and $K_1 : K_2 = 1:2$, after what time from the start of reaction, the concentration of 'B' will be maximum? (ln 2 = 0.7)

(a) 70 min (b) 140 min
(c) 35 min (d) 700 min

**79.** For the consecutive first-order reactions:

$$A \xrightarrow{K_1} B \xrightarrow{K_2} C,$$

in what condition, $[C] = [A_0].\ (1 - e^{-K_1 t})$?

(a) $K_1 < K_2$ (b) $K_1 << K_2$
(c) $K_2 < K_1$ (d) $K_2 << K_1$

**80.** When excess toluene-α-d ($C_6H_5CH_2D$) was photochemically monochlorinated at 80°C with 0.1 mole of chlorine, there was obtained 0.0212 mole DCl and 0.0848 mole HCl. What is the value of the isotope effect $K^H/K^D$?

(a) $\frac{1}{4}$ (b) $\frac{4}{1}$

(c) $\frac{5}{1}$ (d) $\frac{1}{5}$

## Dependence of Reaction Rate on Temperature and Catalyst

**81.** The activation energy for a chemical reaction is primarily dependent on

(a) temperature
(b) nature of the reacting species.
(c) collision frequency.
(d) concentration of reactants.

**82.** The activation energy of a reaction is zero. The rate constant of the reaction

(a) increases with increase in temperature.
(b) decreases with increase in temperature.
(c) increases with decrease in temperature.
(d) is nearly independent in temperature.

**83.** The activation energy of a reaction can be lowered by
(a) increasing temperature.
(b) lowering temperature.
(c) adding a catalyst.
(d) removing one or more products.

**84.** At a certain temperature, the reaction between NO and $O_2$ to form $NO_2$ is fast, while that between CO and $O_2$ is slow. It may be concluded that
(a) NO is more reactive than CO.
(b) CO is smaller in size than NO.
(c) activation energy for the reaction $2NO + O_2 \rightarrow 2NO_2$ is less.
(d) activation energy for the reaction $2NO + O_2 \rightarrow 2NO_2$ is high.

**85.** Rate of which type of elementary reaction increases with increase in temperature?
(a) Thermal (b) Exothermic
(c) Endothermic (d) All of these

**86.** In gaseous reactions important for the understanding of the upper atmosphere, $H_2O$ and O react bimolecularly to form two OH radicals. $\Delta H$ for this reaction is 72 kJ at 500 K and $E_a$ is 77 kJ/mol, then $E_a$ for the bimolecular recombination of two OH radicals to form $H_2O$ and O is
(a) 149 kJ/mol (b) 72 kJ/mol
(c) 5 kJ/mol (d) 7 kJ/mol

**87.** The rate constant is given by the equation $K = P \cdot A \cdot e^{-E_a/RT}$. Which factor should register a decrease for the reaction to proceed more rapidly?
(a) $T$ (b) $A$
(c) $E_a$ (d) $P$

**88.** According to the collisions theory, the rate of reaction increases with temperature due to
(a) increase in number of collisions between reactant molecules.
(b) increase in speed of reacting molecules.
(c) increase in the number of molecules having sufficient energy for reaction.
(d) decrease in activation energy of reaction.

**89.** Select the incorrect statement from the following.
(a) It is possible to change the rate constant for a reaction by changing the temperature.
(b) The rate constant for a reaction is independent of reactant concentrations.
(c) The rate of a catalysed reaction is always independent from the concentration of the catalyst.
(d) In multistep reactions, the rate determining step is the slowest one.

**90.** On increasing the temperature, the half-life of reactions
(a) remain unchanged
(b) increases
(c) decreases
(d) increases or decreases

**91.** According to the Arrhenius equation, straight line is to be obtained by plotting the logarithm of the rate constant of a chemical reaction (log $K$) against
(a) $T$ (b) $\log T$
(c) $\frac{1}{T}$ (d) $e^T$

**92.** The rate of a reaction is approximately doubled for every 10°C rise in temperature. If the temperature is raised by 50°C, then the reaction rate increases by about
(a) 10 times (b) 16 times
(c) 32 times (d) 64 times

**93.** In general, the rate of a reaction can be increased by all the factors except
(a) increasing the temperature.
(b) increasing the concentration of reactants.
(c) increasing the activation energy.
(d) using a catalyst.

**94.** The rate of chemical reactions is independent of
(a) molecularity
(b) nature of reactants.
(c) concentration of reactants.
(d) temperature of reaction.

**95.** The activation energies of two reactions are $E_1$ and $E_2$ with $E_1 > E_2$. If the temperature of the reacting system is increased from $T_1$ (rate constants are $K_1$ and $K_2$) to $T_2$ (rate constants are $K_1'$ and $K_2'$), then predict which of the following alternative is correct?
(a) $\frac{K_1'}{K_1} = \frac{K_2'}{K_2}$ (b) $\frac{K_1'}{K_1} > \frac{K_2'}{K_2}$
(c) $\frac{K_1'}{K_1} < \frac{K_2'}{K_2}$ (d) $K_1' < K_2'$

96. Which of the following statements is not correct?
(a) The efficiency of a solid catalyst depends upon its surface area.
(b) Catalyst operates by providing alternate path for the reaction that involves lower activation energy.
(c) Catalyst lowers the energy of activation of the forward direction without affecting the energy of activation of the backward direction.
(d) Catalyst does not affect the overall enthalpy change of the reaction.

97. Collision theory of chemical reactions explains
(a) rate of reactions.
(b) order of reactions.
(c) molecularity of reactions.
(d) All of these

98. The values of enthalpies of reactants and products are $x$ and $y$ J/mol, respectively. If the activation energy for the backward reaction is $z$ J/mol, then the activation energy for forward reaction will be (in J/mol)
(a) $x - y - z$ (b) $x - y + z$
(c) $y - x - z$ (d) $y - x + z$

99. The rate constant, the activation energy and the Arrhenius parameter of a chemical reaction at 25°C are $3.0 \times 10^{-4}$ s$^{-1}$, 104.4 kJ/mol and $6.0 \times 10^{14}$ s$^{-1}$, respectively. The value of the rate constant as $T \to \infty$, is
(a) $2.0 \times 10^{18}$ s$^{-1}$ (b) $6.0 \times 10^{14}$ s$^{-1}$
(c) infinity (d) $3.0 \times 10^{-4}$ s$^{-1}$

100. If $I$ is the intensity of absorbed light and $C$ is the concentration of AB for the photochemical process $AB + h\nu \rightarrow AB^*$, the rate of formation of AB* is directly proportional to
(a) $C$ (b) $I$
(c) $I^2$ (d) $C \cdot I$

# EXERCISE II (JEE ADVANCED)

## Section A (Only one Correct)

1. The rate of a reaction may be expressed as $+\frac{1}{2}\frac{d[C]}{dt} = -\frac{1}{3}\frac{d[D]}{dt} = +\frac{1}{4}\frac{d[A]}{dt} = -\frac{d[B]}{dt}$. The reaction is

(a) $4A + B \rightarrow 2C + 3D$ (b) $B + 3D \rightarrow 4A + 2C$
(c) $4A + 2C \rightarrow B + 3D$ (d) $2A + 3B \rightarrow 4C + D$

2. For a reversible reaction $2NO_2 \underset{K_2}{\overset{K_1}{\rightleftharpoons}} N_2O_4$, the rate of disappearance of $NO_2$ is equal to

(a) $\frac{2k_1}{k_2}[NO_2]^2$
(b) $2K_1[NO_2]^2 - 2K_2[N_2O_4]$
(c) $2K_1[NO_2]^2 - K_2[N_2O_4]$
(d) $(2K_1 - K_2)[NO_2]$

3. For a gaseous reaction A(g) → Products (order = $n$), the rate may be expressed as

(i) $-\frac{dP_A}{dt} = K_1 \cdot P_A{}^n$

(ii) $-\frac{1}{V}\frac{dn_A}{dt} = K_2 \cdot C_A{}^n$

The rate constants, $K_1$ and $K_2$ are related as ($P_A$ and $C_A$ are the partial pressures and molar concentration of A at time '$t$', respectively)

(a) $K_1 = K_2$ (b) $K_2 = K_1 \cdot (RT)^n$
(c) $K_2 = K_1 \cdot (RT)^{1-n}$ (d) $K_2 = K_1 \cdot (RT)^{n-1}$

4. For the reaction $H_2 + I_2 \underset{K_2}{\overset{K_1}{\rightleftharpoons}} 2HI$, the rate law expression is

(a) $-\frac{1}{2}\frac{d[HI]}{dt} = K_1[H_2][I_2]$

(b) $-\frac{1}{2}\frac{d[HI]}{dt} = \frac{K_1[HI]^2}{K_2[H_2][I_2]}$

(c) $-\frac{1}{2}\frac{d[HI]}{dt} = K_1[H_2][I_2] - K_2[HI]^2$

(d) $-\frac{1}{2}\frac{d[HI]}{dt} = K_2[HI]^2 - K_1[H_2][I_2]$

5. For the reaction aA + bB → P, $r = K[A]^a \cdot [B]^b$, if the concentration of A is doubled, then the rate is doubled. If the concentration of B is doubled, then the rate becomes four times. The correct relation is

(a) $-\frac{d[A]}{dt} = -\frac{d[B]}{dt}$ (b) $-\frac{d[A]}{dt} = -2.\frac{d[B]}{dt}$

(c) $-2.\frac{d[A]}{dt} = -\frac{d[B]}{dt}$ (d) $-\frac{d[A]}{dt} = +\frac{d[B]}{dt}$

6. The reaction: A(g) + 2B(g) → C(g) + D(g) is an elementary process. In an experiment, the initial partial pressure of A and B are $P_A = 0.60$ atm and $P_B = 0.80$ atm. When $P_B = 0.20$ atm, the rate of reaction, relative to the initial rate is

(a) $\frac{1}{16}$ (b) $\frac{1}{24}$

(c) $\frac{1}{32}$ (d) $\frac{1}{48}$

7. The following reactions are occurring simultaneously in a vessel such that the temperature of the system is not changing.

$A + 3B \rightarrow P$: $\Delta H = -2x$ kJ
$M \rightarrow 2Q + R$: $\Delta H = +x$ kJ

If the rate of disappearance of 'B' is $y$ M $s^{-1}$, the rate of appearance of 'Q' (in M $s^{-1}$) is

(a) $y$ M $s^{-1}$ (b) 1.33 $y$ M $s^{-1}$
(c) 0.75 $y$ M $s^{-1}$ (d) $2y$ M $s^{-1}$

8. What is the order of a chemical reaction A + 2B → C, if the rate formation of C increases by a factor of 2.82 on doubling the concentration of A and increases by a factor of 9 on tripling the concentration of B?

(a) $\frac{7}{2}$ (b) $\frac{7}{4}$

(c) $\frac{5}{2}$ (d) $\frac{5}{4}$

9. For the chemical reaction between mercuric chloride and potassium oxalate, the mass of $Hg_2Cl_2$ precipitated in different time intervals at 100°C is given below.

| Exp. no. | $[HgCl_2]_0$ (M) | $[K_2C_2O_4]_0$ (M) | Time (minutes) | $Hg_2Cl_2$ formed (g) |
|---|---|---|---|---|
| 1 | 0.0836 | 0.404 | 65 | 0.0068 |
| 2 | 0.0836 | 0.202 | 120 | 0.0031 |
| 3 | 0.0418 | 0.404 | 60 | 0.0032 |

What is the rate law expression of this reaction?

(a) $r = K[HgCl_2][K_2C_2O_4]$  (b) $r = K[HgCl_2]^2[K_2C_2O_4]$

(c) $r = K[HgCl_2]^2[K_2C_2O_4]^2$  (d) $r = K[HgCl_2][K_2C_2O_4]^2$

10. Iodide ion is oxidized to hypoiodite ion ($IO^-$) by hypochlorite ion ($ClO^-$) in basic solution as:

$$I^-(aq) + ClO^-(aq) \xrightarrow{OH^-} IO^-(aq) + Cl^-(aq)$$

The following initial-rate experiments were run and, for each, the initial rate of formation of $IO^-$ was determined.

| Exp. no. | Initial concentration (mol/L) | | | Initial rate (mol $L^{-1}s^{-1}$) |
|---|---|---|---|---|
| | $I^-$ | $ClO^-$ | $OH^-$ | |
| 1 | 0.010 | 0.020 | 0.010 | $12.2 \times 10^{-2}$ |
| 2 | 0.020 | 0.010 | 0.010 | $12.2 \times 10^{-2}$ |
| 3 | 0.010 | 0.010 | 0.010 | $6.1 \times 10^{-2}$ |
| 4 | 0.010 | 0.010 | 0.020 | $3.0 \times 10^{-2}$ |

The correct rate law for the reaction is

(a) $r = K[I^-][ClO^-][OH^-]^0$  (b) $r = K[I^-]^2[ClO^-]^2[OH^-]^0$

(c) $r = K[I^-][ClO^-][OH^-]$  (d) $r = K[I^-][ClO^-][OH^-]^{-1}$

11. For the chemical reaction A → products, the rate of disappearance of A is a given by

$$r_A = -\frac{dC_A}{dt} = \frac{K_1.C_A}{1+K_2.C_A}$$

At low concentration of A, expressed as $C_A$, the reaction is of the first-order with the rate constant

(a) $\frac{K_1}{K_2}$  (b) $K_1$

(c) $\frac{K_2}{K_1}$  (d) $\frac{K_1}{K_1+K_2}$

12. For a two-step reaction:

$$A \underset{K_2}{\overset{K_1}{\rightleftharpoons}} R + B$$

$$R + C \xrightarrow{K_3} \text{Products}$$

(where R is a reactive intermediate whose concentration is maintained at some low steady-state throughout the reaction), the rate law expression will be

(a) $\frac{dx}{dt} = \frac{k_1[A]}{1+\frac{k_2[B]}{k_3[C]}}$

(b) $\frac{dx}{dt} = k_1[A]$

(c) $\frac{dx}{dt} = k_1[A]^2$

(d) $\frac{dx}{dt} = k_1[A][B][R]$

13. The chemical reaction $2O_3 \rightarrow 3O_2$ proceeds as

$$O_3 \underset{K_2}{\overset{K_1}{\rightleftharpoons}} O_2 + O \text{ (fast)},$$

$$O + O_3 \xrightarrow{K_3} 2O_2 \text{ (slow)}$$

The rate law expression will be

(a) rate $= K_1[O][O_3]$  (b) rate $= K_1[O_3]$

(c) rate $= \frac{K_1.K_3}{K_2}\frac{[O_3]^2}{[O_2]}$  (d) rate $= K_3\frac{[O_3]^2}{[O_2]}$

**14.** The reaction X(g) → Y(g) follows first-order kinetics. The correct graph representing the rate of formation (R) of Y(g) with time ($t$) is

(a)

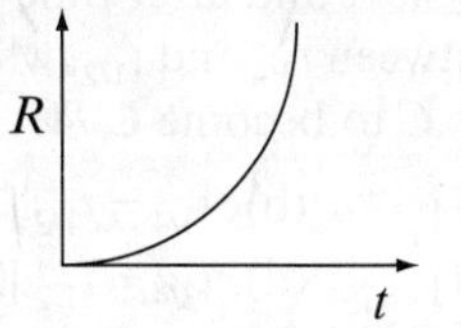

(b)

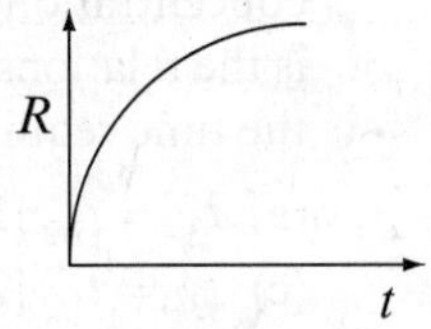

(c)

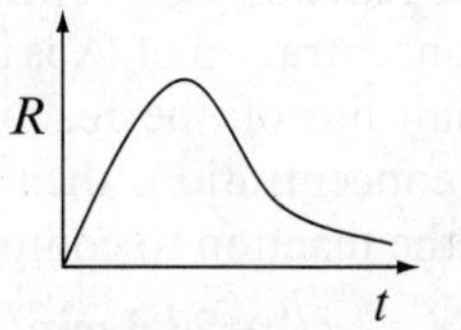

(d)

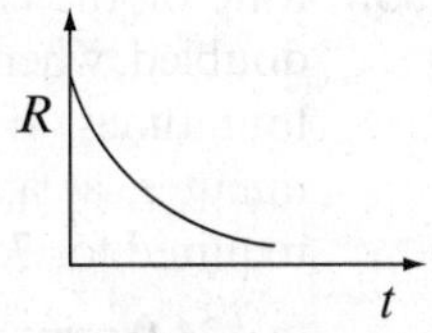

**15.** The initial rate of a zero-order reaction A(g) → 2B(g) is 0.01 M min$^{-1}$. If the initial concentration of 'A' is 0.1 M, then the concentration of 'B' after 60 s is
(a) 0.09 M (b) 0.01 M
(c) 0.02 M (d) 1.2 M

**16.** For a zero-order reaction: $2NH_3(g) \rightarrow N_2(g) + 3H_2(g)$, the rate of reaction is 0.1 atm/s. Initially, only $NH_3(g)$ was present at 3 atm and the reaction is performed at constant volume and temperature. The total pressure of gases after 10 s from the start of reaction will be
(a) 4 atm (b) 5 atm
(c) 3.5 atm (d) 4.5 atm

**17.** For the reaction 3A → 2B, the following graph is obtained experimentally. The rate of reaction when [A] = 0.2 M, is

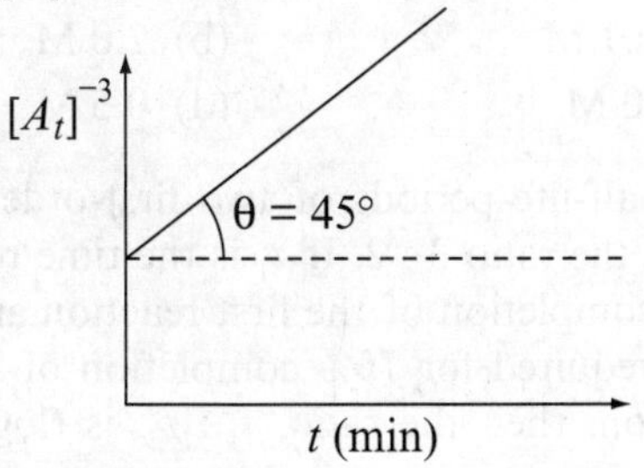

(a) $4 \times 10^{-4}$ M min$^{-1}$ (b) $\frac{16}{9} \times 10^{-4}$ M min$^{-1}$
(c) $1.6 \times 10^{-3}$ M min$^{-1}$ (d) $\frac{16}{3} \times 10^{-3}$ M min$^{-1}$

**18.** A substance 'A' decomposes in solution following first-order kinetics. Flask 1 contains 1 L of 1 M solution of 'A' and flask 2 contains 100 ml of 0.6 M solution of 'A'. After 8.0 h, the concentration of 'A' in flask 1 becomes 0.25 M. In what time, the concentration of 'A' in flask 2 becomes 0.3 M?
(a) 8.0 h (b) 3.2 h
(c) 4.0 h (d) 9.6 h

**19.** A kinetic study of the reaction A → products provides the data $t$ = 0 s, [A] = 2.00 M; 500 s, 1.00 M; 1500 s, 0.50 M; 3500 s, 0.25 M. In the simplest possible way determine whether this reaction is of
(a) zero-order (b) first-order
(c) second-order (d) third-order

**20.** After 20% completion, the rate of reaction A → products is 10 unit and after 80% completion, the rate is 0.625 unit. The order of the reaction is
(a) zero (b) first
(c) second (d) third

**21.** At a certain temperature and fixed volume, the rate constant for the decomposition of $C_2H_6(g)$ into $C_2H_4(g)$ and $H_2(g)$ is 0.0015 Pa$^{-1}$ h$^{-1}$. If the initial pressure of $C_2H_6(g)$ is 3 bar, the time for the pressure of system (reaction mixture) to become 5 bar is
(a) 16 s (b) 0.01 min
(c) 0.06 min (d) 36 s

**22.** For the irreversible reaction A → products in a batch reactor, 80% reactant ([$A_0$] = 1 M) is converted in a 8 minute run and conversion is 90% after 18 minute. The order of this reaction is
(a) 1 (b) 2
(c) 1/2 (d) 3/2

**23.** The inversion of cane sugar proceeds with half-life of 500 minute at pH = 5 for any concentration of sugar. However, if pH = 6, the half-life changes to 50 minute. The rate law expression for the sugar inversion can be written as
(a) $r = K[\text{sugar}]^2[H^+]^0$ (b) $r = K[\text{sugar}][H^+]^2$
(c) $r = K[\text{sugar}][H^+]$ (d) $r = K[\text{sugar}][H^+]^{-1}$

**24.** Hydrolysis of ethyl acetate is catalysed by HCl. The rate is proportional to the concentration of both the ester and HCl. The rate constant is 0.1 M$^{-1}$h$^{-1}$. What is the half-life, if the initial concentrations are 0.02 M for ester and 0.01 M for the catalysing acid?
(a) 347 h (b) 519 h
(c) 836 h (d) 693 h

**25.** The order of the reaction A → products is 0.5. The time for complete reaction $T$ and time for 50% reaction, $t_{1/2}$ will be related as
(a) $T = 2 \times t_{1/2}$ (b) $T = 4 \times t_{1/2}$
(c) $t_{1/2} = 0.3T$ (d) $t_{1/2} = 0.7T$

26. Two substances 'A' and 'B' are initially present as $[A_0]=8[B_0]$ and $t_{1/2}$ for the first-order decomposition of 'A' and 'B' are 10 and 20 minutes, respectively. If they start decomposing at the same time, after how much time, the concentration of both of them would be same?
(a) 20 min (b) 40 min
(c) 60 min (d) 200 min

27. For a first-order reaction A → Product, the initial concentration of A is 0.1 M and after 40 minutes of time, it becomes 0.025 M. What is the rate of reaction at reactant concentration 0.01 M?
(a) $3.465 \times 10^{-4}$ mol lit$^{-1}$ min$^{-1}$
(b) $3.465 \times 10^{-5}$ mol lit$^{-1}$ min$^{-1}$
(c) $6.93 \times 10^{-4}$ mol lit$^{-1}$ min$^{-1}$
(d) $1.7325 \times 10^{-4}$ mol lit$^{-1}$ min$^{-1}$

28. The rate of change of concentration of 'B' for the reaction A → B is given as $+\frac{d[B]}{dt} = K[A]^{1/3}$. The half-life period of the reaction will be
(a) $\frac{3[A_0]^{2/3}(2^{2/3}-1)}{2^{5/3}.K}$ (b) $\frac{3[A_0]^{2/3}(2^{2/3}-1)}{2^{2/3}.K}$
(c) $\frac{3[A_0]^{2/3}(2^{2/3}-1)^2}{2^{5/3}.K}$ (d) $\frac{2[A_0]^{2/3}(2^{2/3}-1)}{3\times 2^{2/3}\times K}$

29. A solution of $N_2O_5$ in $CCl_4$ yields by decomposition at 45°C, 4.8 ml of $O_2$, 20 minutes after the start of the experiment and 9.6 ml of $O_2$ after a very long time. The decomposition obeys first-order kinetics. What volume of $O_2$ would have evolved, 40 minutes after the start?
(a) 7.2 ml (b) 2.4 ml
(c) 9.6 ml (d) 6.0 ml

30. The decomposition of $NH_3$ on tungsten surface follows zero-order kinetics. The half-life is 315 s for an initial pressure of 70 mm of $NH_3$. If the initial pressure had been 150 mm, then what would be the half-life?
(a) 315 s (b) 472.5 s
(c) 675 s (d) 630 s

31. A first-order reaction A(g) → $n$B(g) is started with 'A'. The reaction takes place at constant temperature and volume. If the initial pressure was $P_0$ and the rate constant of reaction is '$K$', then at any time $t$, the total pressure of the reaction system will be
(a) $P_0[n+(1-n)e^{-kt}]$ (b) $P_0(1-n)e^{-kt}$
(c) $P_0 \cdot n \cdot e^{-kt}$ (d) $P_0[n-(1-n)e^{-kt}$

32. For a reaction of order $n$, the integrated form of the rate equation is $(n-1)\cdot K\cdot t = (C_0)^{1-n} - (C)^{1-n}$, where $C_0$ and $C$ are the values of the reactant concentration at the start and after time '$t$'. What is the relationship between $t_{3/4}$ and $t_{1/2}$, where $t_{3/4}$ is the time required for $C$ to become $C_0/4$.
(a) $t_{3/4} = t_{1/2}\cdot[2^{n-1}+1]$ (b) $t_{3/4} = t_{1/2}\cdot[2^{n-1}-1]$
(c) $t_{3/4} = t_{1/2}\cdot[2^{n+1}-1]$ (d) $t_{3/4} = t_{1/2}\cdot[2^{n+1}+1]$

33. Rate of the chemical reaction A → products, is doubled, when the concentration of 'A' is increased four times. If the half-life of the reaction is 16 minutes, at a given concentration, then the time required for 75% of the reaction to complete is
(a) 24.0 min (b) 27.3 min
(c) 48 min (d) 49.4 min

34. $SO_2Cl_2 \rightarrow SO_2 + Cl_2$ is a first-order gaseous reaction with $K = 2.5 \times 10^{-5}$ s$^{-1}$ at 320°C. The percentage of $SO_2Cl_2$ decomposed on heating for 100 minutes is (ln 1.16 = 0.15)
(a) 86.2 (b) 15.0
(c) 85.0 (d) 13.8

35. When the concentration of 'A' is 0.1 M, it decomposes to give 'X' by a first-order process with a rate constant of $6.93 \times 10^{-2}$ min$^{-1}$. The reactant 'A' in the presence of catalyst gives 'Y' by a second-order mechanism with the rate constant of 0.2 min$^{-1}$ M$^{-1}$. In order to make half-life of both the processes same, one should start the second-order reaction with an initial concentration of 'A' equal to
(a) 0.01 M (b) 2.0 M
(c) 1.0 M (d) 0.5 M

36. The half-life periods of two first-order reactions are in the ratio 3 : 2. If $t_1$ is the time required for 25% completion of the first reaction and $t_2$ is the time required for 75% completion of the second reaction, then the ratio, $t_1 : t_2$, is (log 3 = 0.48, log 2 = 0.3)
(a) 3 : 10 (b) 12 : 25
(c) 3 : 5 (d) 3 : 2

37. If 1% of a substance undergoing decomposition is consumed in 1 minute, when the concentration is 0.02 M and in 15 s, when the concentration is 0.04 M, the order of the reaction is
(a) 3 (b) 2
(c) 1 (d) zero

**38.** In the biological processes, the time taken by certain virus or bacteria to double its population is called generation time. In milk, at 37°C, *Lactobacillus acidophilus* has a generation time of about 75 minutes. The population relative to the initial value at 60 minutes is (ln 2 = 0.7)

(a) 0.8 (b) $e^{0.56}$
(c) $e^{0.8}$ (d) $e^{0.875}$

**39.** A study of the conversion of ammonium cyanate to urea in the solution $NH_4CNO \rightarrow CO(NH_2)_2$ gave the following results.

| Initial concentration (moles/litre) | 0.05 | 0.10 | 0.20 |
|---|---|---|---|
| Time for half conversion (h) | 37.82 | 18.95 | 9.45 |

The value of $t_{1/2}$ of reaction for an initial concentration of 0.15 M is

(a) ≈56.77 h (b) ≈12.6 h
(c) ≈14.2 h (d) ≈13.7 h

**40.** Thermal decomposition of dibromosuccinic acid (DBSA) taking place according to the following equation, obeys first-order kinetics.

| CH (Br) COOH |  | CHCOOH |  |
|---|---|---|---|
| \| | ⟶ | ‖ | + HBr |
| CH (Br) COOH |  | CBrCOOH |  |

The progress of reaction may be followed by means of alkali titration of the solution (definite volume of reaction mixture) at various time intervals. If $T_0$ and $T_t$ be the ml of alkali solution at zero time and at any time $t$, respectively and $a$ and $(a - x)$ be the concentrations of DBSA at zero time and at any time $t$, respectively, then the value of $\frac{a}{a-x}$ is

(a) $\frac{T_0}{3T_0 - 2T_t}$ (b) $\frac{T_0}{T_t}$

(c) $\frac{T_0}{T_0 - T_t}$ (d) $\frac{T_0}{2T_0 - T_t}$

**41.** In a certain first-order reaction, $B^{n+}$ is getting converted to $B^{(n+4)+}$ in solution. The rate constant of this reaction is measured by titrating a volume of the solution with a reducing agent which reacts only with $B^{n+}$ and $B^{(n+4)+}$. In the process, it converts $B^{n+}$ to $B^{(n-2)+}$ and $B^{(n+4)+}$ to $B^{(n-1)+}$. At $t = 0$, the volume of reagent consumed is 25 ml and at $t$ = 10 minute, the volume used is 32.5 ml. The rate constant for the conversion of $B^{n+}$ of $B^{(n+4)+}$ is (ln 2 = 0.7, ln 5 = 1.6)

(a) 0.2 $min^{-1}$ (b) 0.8 $min^{-1}$
(c) 0.02 $min^{-1}$ (d) 0.08 $min^{-1}$

**42.** The reaction A(g) → 2B(g) + C(g) follows first-order kinetics. The reaction is started with pure 'A' in a rigid closed vessel maintained at constant temperature. After 10 s, a pin hole is developed in the vessel. If the molar ratio of gases 'A' and 'B' coming out initially is 1 : 2, the rate constant of reaction is (Molar masses of A, B and C are 16, 4 and 8 g/mol, respectively, ln 2 = 0.7, ln 3 = 1.1)

(a) 0.04 $s^{-1}$ (b) 0.4 $s^{-1}$
(c) 0.11 $s^{-1}$ (d) 0.07 $s^{-1}$

**43.** For the reaction A + B → products, the rate law is $r = K[A]^2[B]$, where $K = 0.5\ M^{-2} min^{-1}$. The time in which the concentration of 'A' becomes half of its initial concentration if the initial concentrations of 'A' and 'B' are 0.002 M and 2.0 M, respectively, is

(a) 50 min (b) 500 min
(c) 1.0 min (d) 100 min

**44.** The initial rate of hydrolysis of methyl acetate (1.0 M) by a weak acid (HA, 1.0 M) is 1/100 th of that of a strong acid (HX, 1.0 M) at 25°C. The $K_a$ of HA is

(a) $1 \times 10^{-4}$ (b) $1 \times 10^{-5}$
(c) $1 \times 10^{-6}$ (d) $1 \times 10^{-3}$

**45.** Reaction A + B → C + D follows the rate law: $r = (2.31 \times 10^{-3}\ s^{-1})[A]^{1/2}[B]^{1/2}$. The reaction is started with 1.0 mole each of 'A' and 'B'. In what time, the moles of 'A' becomes 0.25?

(a) 150 s (b) 300 s
(c) 450 s (d) 600 s

**46.** The rate expression for a reaction is $\frac{-dC}{dt} = \frac{\alpha.C}{1+\beta C}$, where α, β are constants and $C$ is the concentration of reactant at time $t$. The half-life for this reaction is

(a) $\frac{1}{\alpha}\ln 2 + \frac{\beta.C_0}{2\alpha}$ (b) $\frac{1}{\beta}\ln 2 + \frac{\beta.C_0}{2\alpha}$

(c) $\frac{\beta.\ln 2}{\alpha}$ (d) $\frac{\alpha}{\beta}$

**47.** The acid catalysed reaction of acetic acid with ethanol:

$$CH_3COOH + C_2H_5OH \rightarrow CH_3COOC_2H_5 + H_2O$$

follows the rate law $-\frac{d[CH_3COOH]}{dt} = K[H^+][CH_3COOH][C_2H_5OH] = K'[CH_3COOH][C_2H_5OH]$. When $[CH_3COOH]_0 = [C_2H_5OH]_0 = 0.2$ M and pH = 3, the half-life for the reaction is 50 minutes. The value of true rate constant, $K$, of the reaction is

(a) $1.386 \times 10^{-2}$ min$^{-1}$ (b) 0.1 M$^{-1}$ min$^{-1}$
(c) 100 M$^{-2}$ min$^{-1}$ (d) 13.86 min$^{-1}$

**48.** The reaction A + B → C + D, has been studied under the following initial conditions.

| Case I | Case II |
|---|---|
| $[A]_0 = 2.0$ M | $[A]_0 = 0.02$ M |
| $[B]_0 = 0.01$ M | $[B]_0 = 1.0$ M |

The variation of concentration with time was noted as follows.

| Case I | Time (minute) | 0 | 10 | 20 | 30 |
|---|---|---|---|---|---|
| | [B] | 0.0100 M | 0.0080 M | 0.0064 M | 0.00512 M |
| Case II | Time (minute) | 0 | 10 | 20 | 30 |
| | [A] | 0.0200 M | 0.0180 M | 0.0162 M | 0.0146 M |

The rate constant of the reaction is (ln 3 = 1.1, ln 5 = 1.6, ln 10 = 2.3)

(a) 0.02 min$^{-1}$ (b) 0.01 min$^{-1}$
(c) 0.01 M$^{-1}$min$^{-1}$ (d) 0.02 M$^{-1}$min$^{-1}$

**49.** In Lindemann theory of unimolecular reactions, it is shown that the apparent rate constant for such a reaction is $k_{app} = \frac{k_1C}{1+\alpha C}$, where $C$ is the concentration of the reactant, $k_1$ and $\alpha$ are constants. The value of $C$ for which $k_{app}$ has 90% of its limiting value at $C$ tending to infinitely large is ($\alpha = 9 \times 10^5$)

(a) $10^{-6}$ mole/litre (b) $10^{-4}$ mole/litre
(c) $10^{-5}$ mole/litre (d) $5 \times 10^{-5}$ mole/litre

**50.** Decomposition of a non-volatile solute 'A' into another non-volatile solute 'B' and 'C' in aqueous solution follows first-order kinetics as follows:

$$A \rightarrow 2B + C$$

When one mole of 'A' is dissolved in 180 g water and left for decomposition, the vapour pressure of solution was found to be 20 mm Hg after 12 h. What is the vapour pressure of solution after 24 h? Assume constant temperature of 25°C throughout. The vapour pressure of water at 25°C is 24 mm Hg.

(a) 18 mm Hg
(b) 19.2 mm Hg
(c) 10 mm Hg
(d) 16 mm Hg

**51.** A substance undergoes first-order decomposition. The decomposition follows two parallel first-order reaction with $K_1 = 1.26 \times 10^{-4}$ s$^{-1}$ for the formation of 'B' and $K_2 = 3.15 \times 10^{-5}$ s$^{-1}$ for the formation of 'C'. The percentage distribution of 'B' and 'C' are

(a) 80% B, 20% C (b) 75% B, 25% C
(c) 90% B, 10% C (d) 60% B, 40% C

**52.** The rate equation for an autocatalytic reaction

$$A + R \xrightarrow{k} R + R \text{ is } r_A = -\frac{dC_A}{dt} = kC_AC_R$$

The rate of disappearance of reactant A is maximum when

(a) $C_A = 2C_R$ (b) $C_A = C_R$
(c) $C_A = C_R/2$ (d) $C_A = (C_R)^{1/2}$

**53.** Surface-catalysed reactions that are incorporated by the product, obey the differential rate expression, $\frac{dy}{dt} = \frac{k[C_0 - y]}{1+by}$, where $C_0$ = initial concentration and $k$ and $b$ are constants. The half-life of reaction is

(a) $[(1 + C_0b)\ln 2 - \frac{C_0b}{2}]\frac{1}{K}$

(b) $[(1 - C_0b)\ln 2 + \frac{C_0b}{2}]\frac{1}{K}$

(c) $[(1 - C_0b)\ln 2 - \frac{C_0b}{2}]\frac{1}{K}$

(d) $[(1 + C_0b)\ln 2 + \frac{C_0b}{2}]\frac{1}{K}$

**54.** For the following first-order competing reaction,

$$A + \text{Reagent} \rightarrow \text{Product}$$
$$B + \text{Reagent} \rightarrow \text{Product}$$

the ratio of $K_1/K_2$, if only 50% of 'B' will have been reacted when 94% of 'A' has been reacted is (log 2 = 0.3, log 3 = 0.48)

(a) 4.06 (b) 0.246
(c) 8.33 (d) 0.12

**55.** In 80% ethanol at 55°C, isopropyl bromide reacts with hydroxide ion according to the following kinetics.

$$-\frac{d[RX]}{dt} = (4.8 \times 10^{-5}\ M^{-1}\ s^{-1})[RX][OH^-] + 2.4 \times 10^{-6}\ s^{-1}\ [RX]$$

What percentage of isopropyl bromide reacts by the $S_{N_2}$ mechanism when $[OH^-] = 0.01$ M?

(a) 16.67% (b) 83.33%
(c) 66.67% (d) 33.33%

**56.** For the series of competitive reactions.

$$H + HO_2 \xrightarrow{K_1} H_2 + O_2$$
$$H + HO_2 \xrightarrow{K_2} 2\ OH$$
$$H + HO_2 \xrightarrow{K_3} H_2O + O$$

It has been found that $K_1{:}K_2{:}K_3 = 0.60{:}0.30{:}0.10$. The molar ratio of the products, $H_2$, $O_2$, OH, $H_2O$ and O, at time is

(a) 6 : 6 : 6 : 1 : 1 (b) 6 : 6 : 3 : 1 : 1
(c) 3 : 3 : 6 : 0.5 : 0.5 (d) 3 : 3 : 3 : 1 : 1

**57.** In the parallel reactions A $\xrightarrow{K_1 = \ln 3\ \text{min}^{-1}}$ B and A $\xrightarrow{K_2 = \ln 3\ \text{min}^{-1}}$ C, the time when the concentrations of A, B and C becomes equal is

(a) 0.5 s (b) 30 s
(c) 45 s (d) 60 s

**58.** For the sequential reactions A $\xrightarrow{K_1 = 0.02\ \text{min}^{-1}}$ B $\xrightarrow{K_2 = 0.02\ \text{min}^{-1}}$ C, the initial concentration of 'A' was 0.2 M and initially 'B' and 'C' were absent. The time at which the concentration of 'B' becomes maximum and the maximum concentration of 'B' are, respectively,

(a) 50 min, $\left(\frac{0.2}{e}\right)$ M (b) 50 min, 0.2 M
(c) infinite, 0.2 M (d) 25 min, $\left(\frac{0.2}{e}\right)$ M

**59.** In the given first-order sequential reactions A $\xrightarrow{K_1}$ B $\xrightarrow{K_2}$ C $\xrightarrow{K_3}$ D, what is the ratio of number of atoms of A to the number of atoms of B after long time interval starting with pure A?
$\left(K_1 = \frac{\ln 2}{1200} \text{ and } K_2 = \frac{\ln 2}{30}\right)$

(a) 0.67 (b) 10
(c) 20 (d) 40

**60.** An organic compound A decomposes the following two parallel first-order reactions A $\xrightarrow{K_1}$ B and A $\xrightarrow{K_2}$ C. If $K_1$ is $1.25 \times 10^{-5}\ s^{-1}$ and $\frac{K_1}{K_2} = \frac{1}{9}$, then the value of $\frac{[C]}{[A]}$ after one hour of start of reaction by taking only A, is (ln 1.568 = 0.45)

(a) $\frac{1}{9}$ (b) 0.5112
(c) 1.4112 (d) $\frac{9}{20}$

**61.** For the consecutive first-order reactions: A $\xrightarrow{K_1}$ B $\xrightarrow{K_2}$ C, the concentrations of A and B are 0.2 M and 0.01 M, respectively, at steady state. If $K_1$ is $2.5 \times 10^{-4}\ \text{min}^{-1}$, then what is the value of $K_2$?

(a) $5.0 \times 10^{-3}\ \text{min}^{-1}$ (b) $2.5 \times 10^{-4}\ \text{min}^{-1}$
(c) $1.25 \times 10^{-5}\ \text{min}^{-1}$ (d) $5.0 \times 10^{-4}\ \text{min}^{-1}$

**62.** For the consecutive first-order reactions A $\xrightarrow{K_1}$ B $\xrightarrow{K_2}$ C, B is in steady state. What will be the concentration of C at this stage? At $t = 0$, $[A] = [A]_0$ and $[B] = [C] = 0$.

(a) 0 (b) $[A]_0$
(c) $[A]_0.(1- e^{-K_1t})$ (d) $[A]_0.\ e^{-K_1t}$

63. For the first-order gaseous reaction $A(g) \xrightarrow{K_1} 2B(g)$, $A(g) \xrightarrow{K_2} C(g)$, the initial pressure in a container of fixed volume, V litre, is 1 atm. Pressure of the system is 1.4 atm at $t = 10$ minute and the pressure is 1.5 atm after a very long time. The only correct information about the reactions is (ln 2 = 0.7, ln 10 = 2.3)
(a) $2K_1 = K_2 = 0.08\ min^{-1}$
(b) $K_1 = K_2 = 0.08\ min^{-1}$
(c) $K_1 = 2K_2 = 0.08\ min^{-1}$
(d) $K_1 = K_2 = 0.16\ min^{-1}$

64. For a bimolecular gaseous reaction of type 2A → Products, the average speed of reactant molecules is $2 \times 10^4$ cm/s, the molecular diameter is 4 Å and the number of reactant molecules per $cm^3$ is $2 \times 10^{19}$. The maximum rate of reaction should be
(a) $4.72 \times 10^7\ mol\ l^{-1}\ s^{-1}$
(b) $1.18 \times 10^7\ mol\ l^{-1}\ s^{-1}$
(c) $9.44 \times 10^7\ mol\ l^{-1}\ s^{-1}$
(d) $2.36 \times 10^7\ mol\ l^{-1}\ s^{-1}$

65. For certain reaction it is observed that $\ln K = \alpha + \beta \ln T - \frac{\gamma}{T}$, where $K$ is the rate constant, $T$ is the temperature (Kelvin) and α, β and γ are constants. What is the Arrhenius activation energy?
(a) $\beta RT + \gamma R$ (b) $\gamma R$
(c) $\beta + \gamma R$ (d) $\beta RT - \gamma R$

66. For two parallel first-order reactions, what is the overall activation energy of reaction? The yields of B and C in products are 40% and 60%, respectively.

$$A \xrightarrow{Ea = 20\ kcal/mol} B \quad A \xrightarrow{Ea = 40\ kcal/mol} C$$

(a) 60 kcal/mol (b) 32 kcal/mol
(c) 28 kcal/mol (d) 20 kcal/mol

67. Rate of an uncatalysed first-order reaction at $T$ K is half of the rate of catalysed reaction at '$0.5T$' K. If the catalyst lowers the threshold energy by 20 kcal, what is the activation energy of uncatalysed reaction? ($T$ = 300 K, ln 2 = 0.7)
(a) 39.58 kcal/mol (b) 19.58 kcal/mol
(c) 40.42 kcal/mol (d) 20.42 kcal/mol

68. For the first-order parallel reactions $A \xrightarrow{K_1} B$ and $A \xrightarrow{K_2} C$, $K_1 = 8\ min^{-1}$ and $K_2 = 2\ min^{-1}$ at 300 K. If the activation energies for the formation of B and C are 20 and 28.314 kJ/mol, respectively, then find the temperature at which B and C will be obtained in 2 : 1 mole ratio. (ln 2 = 0.7)
(a) 400 K (b) 385.5 K
(c) 379.75 K (d) 412.25 K

69. The two reactions (I) A → Products, and (II) B → Products follow first-order kinetics. The rate of reaction-I is doubled when temperature is raised from 300 to 310 K. The half-life for this reaction at 310 K is 30 minutes. At the same temperature, B decomposes twice as fast as A. If the energy of activation for the reaction-II is half that of reaction-I, the rate constant of reaction-II at 300 K is
(a) $0.0233\ min^{-1}$ (b) $0.0327\ min^{-1}$
(c) $0.0164\ min^{-1}$ (d) $0.0654\ min^{-1}$

70. A first-order reaction A → B, activation energy is 9.6 kcal/mol. When a 20% solution of 'A' was kept at 27°C for 21.6 minutes, 75% decomposition took place. What will be the percent decomposition in 8.0 minutes in a 30% solution maintained at 47°C? Assume that activation energy remains constant in this range of temperature ($e = 2.7$).
(a) 25% (b) 50%
(c) 75% (d) 87.5%

71. Arrhenius equation gives the change in rate constant (and hence rate of reaction) with temperature. If the activation energy of the reaction is found to be equal to $RT$, then
(a) the rate of reaction does not depend upon initial concentration.
(b) the rate constant becomes about 37% of the Arrhenius constant $A$.
(c) the rate constant becomes equal to 73% of the Arrhenius constant $A$.
(d) the rate of the reaction becomes infinite or zero.

72. The rate of the reaction gets doubled when the temperature changes from 7°C to 17°C. By what factor will it change for the temperature change from 17°C to 27°C? [Antilog(0.30) = 2.0, Antilog(0.280) = 1.91, Antilog(0.322) = 2.1]
(a) 1.81 (b) 1.71
(c) 1.91 (d) 2.1

EXERCISE II

**73.** For the combustion of carbon, $\Delta H = -$ve and $\Delta S$ = +ve and hence, thermodynamically the process is spontaneous at all temperatures. But coal stored in coal depots does not burn automatically due to

(a) very high threshold energy barrier.
(b) thermodynamic stability of coal.
(c) lower energy of activation needed for burning.
(d) low temperature in coal depots.

**74.** Which of the following statements is incorrect?

(a) For endothermic reactions, energy of activation is always greater than the heat of reaction.
(b) For exothermic reactions, the energy of activation is always smaller than heat of reaction.
(c) Rate of elementary reaction always increases with increase in temperature.
(d) Temperature coefficient of a reaction is 1.0, when $T \to \infty$.

**75.** The mechanism of the reaction $A + 2B \to D + E$; $\Delta H = -$ve is

Step I: $A + 2B \to C$ (slow)
Step II: $C \to D + E$ (fast)

The energy profile of the reaction should be

(a)

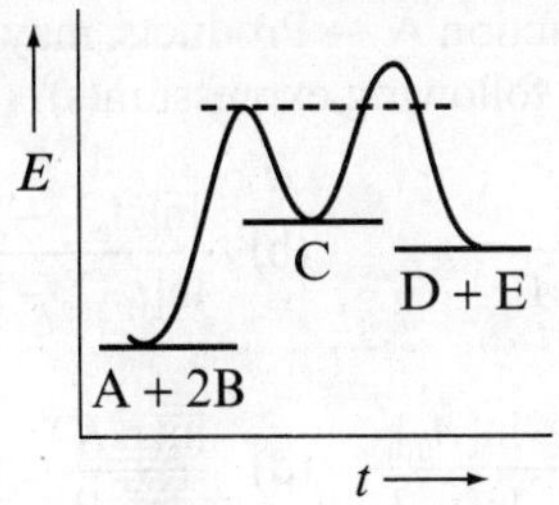


(b)

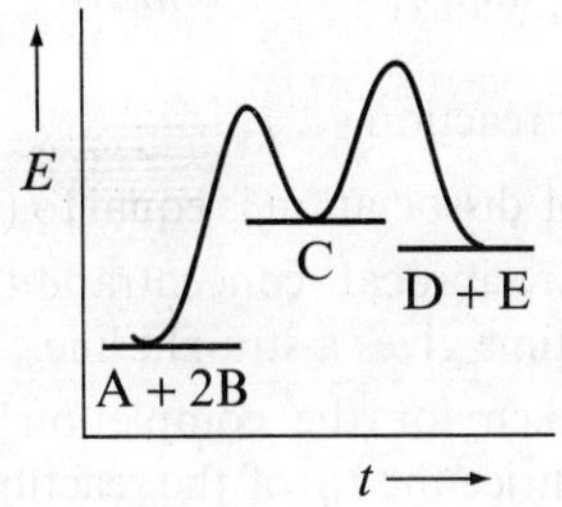


(c)

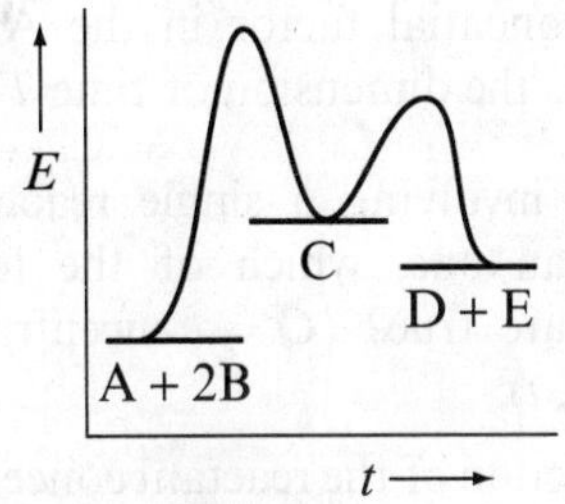


(d)

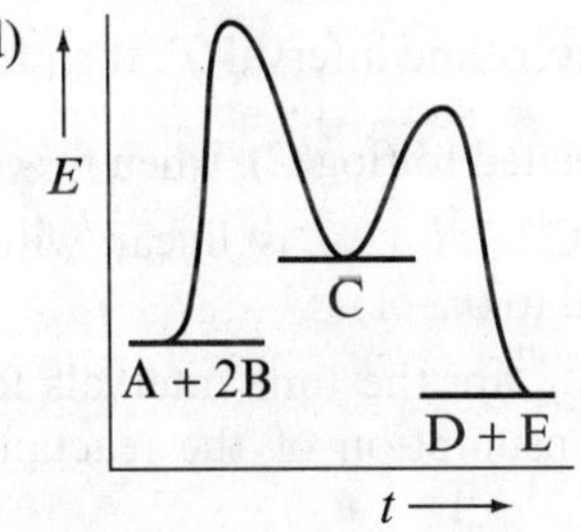


## Section B (One or More than one Correct)

**1.** For a complex (multistep) reaction,

(a) the overall rate of reaction is the rate of slowest step.
(b) the overall molecularity has no significance.
(c) molecularity and order may or may not be same.
(d) the number of rate determining steps may be more than one.

**2.** Which of the following statements is true?

(a) The rate law for a reaction must depend on the concentrations of all reactants that appear in the stoichiometric equation.
(b) The rate of a catalysed reaction must be independent of the concentration of catalyst.
(c) The rate constant for the reaction is independent of the concentrations of the reacting species.
(d) There is a single rate determining step in any reaction mechanism.

**3.** Which of the following is pseudo first-order reaction?

(a) The basic hydrolysis of ethyl acetate.
(b) The inversion of sucrose in the presence of an acid.
(c) The acidic hydrolysis of ethyl acetate.
(d) The decomposition of ammonium nitrite in aqueous solution.

4. The order of reaction A $\rightarrow$ Products, may be given by which of the following expression(s)?

(a) $\dfrac{\ln r_2 - \ln r_1}{\ln[A]_2 - \ln[A]_1}$ (b) $\dfrac{\ln[A_0]_2 - \ln[A_0]_1}{\ln[t_{1/2}]_2 - \ln[t_{1/2}]_1}$

(c) $1 + \dfrac{\ln[A_0]_2 - \ln[A_0]_1}{\ln[t_{1/2}]_2 - \ln[t_{1/2}]_1}$ (d) $\dfrac{\ln(r/K)}{\ln[A]}$

5. For a first-order reaction,
   (a) the degree of dissociation is equal to $(1 - e^{-kt})$.
   (b) a plot of reciprocal concentration of the reactant vs time gives a straight line.
   (c) the time taken for the completion of 75% reaction is thrice the $t_{1/2}$ of the reaction.
   (d) the pre-exponential factor in the Arrhenius equation has the dimension of time $T^{-1}$.

6. For a reaction involving a single reactant and order other than one, which of the following statement(s) is/are true? ($C$ = concentration of reactant at time, $t$)
   (a) If $f$ is the fraction of the reactant concentration consumed over time interval '$t$', then $\log\left(\dfrac{f}{t}\right)$ is linearly related to (log C), when f << 1.
   (b) A plot of $C^{(n-1)}$ vs $t$ is linear with slope proportional to $(n - 1)$.
   (c) If $t_{1/2}$ and $t_{3/4}$ are the time intervals for 50% and 75% consumption of the reactant, then $t_{3/4} : t_{1/2} = 2^{n-1} + 1$
   (d) A plot of log C vs log $t$ is linear with slope = 1.

7. Initial concentration of reactant for $n$th order reaction is $C_0$. Which of the following relation is/are not correct about $t_{1/2}$ of the reaction?
   (a) $\ln t_{1/2} = \text{Constant} - (n - 1) \ln C_0$
   (b) $\ln t_{1/2} = \ln n + \text{Constant} - \ln C_0$
   (c) $t_{1/2} \cdot \ln n = \text{Constant} + \ln C_0$
   (d) $\ln t_{1/2} = n \ln C_0$

8. Consider the following statements. The law for the acid-catalysed hydrolysis of an ester being given as rate = $K[H^+]$[ester] = $K'$[ester]. If the acid (strong) concentration is doubled at constant ester concentration,
   1. the second-order rate constant, $K$, is doubled.
   2. the pseudo first-order rate constant, $K'$ is doubled.
   3. the rate of reaction is doubled.

   Which of the above statement is correct?
   (a) 1 and 2 (b) 2 and 3
   (c) 1 and 3 (d) 1, 2 and 3

9. The rate law for the reaction

$$\text{RCI} + \text{NaOH(aq)} \rightarrow \text{ROH} + \text{NaCl}$$

is given by, rate = $K_1$[RCl]. The rate of the reaction will be
   1. doubled on doubling the concentration of NaOH.
   2. halved on reducing the concentration of alkyl halide to one half.
   3. increased on increasing the temperature of the reaction.
   4. unaffected by increasing the temperature of the reaction.

   The correct answer is
   (a) 1, 2, 3 (b) 3, 4
   (c) 1, 3 (d) 2, 3

10. $SO_3$ gas is entering the environment at a constant rate of $6.93 \times 10^{-6}$ g/L/day due to emission of polluting gases from thermal power plant at Kota but at the same time it is decomposing and following first-order kinetics with half-life of 100 days. Based on these details, select the correct statement(s) from the following.
   (a) Concentration of $SO_3$ in Kota is $1.25 \times 10^{-5}$ M (assume $SO_3$ present in air reaches steady state).
   (b) If $10^3$ L of air is passed through 1 L pure water (assuming all $SO_3$ is dissolved in it) and the resulting solution is titrated against 1 N NaOH solution, 15 ml of NaOH solution is required to reach the end point.
   (c) An industry is manufacturing $H_2SO_4$ at the rate of 980 kg per day with the use of $SO_3$ in air and it should use $8 \times 10^5$ L air per day.
   (d) If $SO_3$ emission is stopped, then after 1000 days, its concentration will reduce to $1.22 \times 10^{-8}$ M.

**11.** The decomposition reaction $3A(g) \rightarrow 2B(g) + 2C(s)$ follows first-order kinetics. Starting with pure 'A' (at 6 atm), the pressures of system after 20 minutes and after a very long time are 5.05 atm and 4.05 atm, respectively. Identify the correct statement(s) related with the reaction.

(a) Time for 75% completion of reaction is slightly more than 40 minutes.
(b) Time for 87.5% completion of reaction is slightly less than 60 minutes.
(c) Time for 93.75% completion of reaction is exactly 80 minutes.
(d) The pressure of system after 40 minutes will be 4.55 atm.

**12.** To very good approximations, the cooling of a hot body to room temperature follows first-order kinetics (in this case, however, the unit that is changing is temperature (in kelvin), not molarity). If the rate constant for a body is $0.04\ s^{-1}$, then [ln 2 = 0.7, ln(323/25) = 2.6]

(a) the time taken for that body to go from 323°C to 25°C is 17.5 s.
(b) the time taken for that body to go from 1192 K to 298 K is 35 s.
(c) the time taken for that body to go from 323°C to 25°C is 65 s.
(d) the time taken for that body to go from 1192 K to 298 K is 130 s.

**13.** For a gaseous reaction $A + B \rightarrow 2C$, the rate law is $r = k[A]^{-1}[B]^2$. Which of the following statement is incorrect?

(a) If the initial concentration of 'A' is double of 'B', then the half-life of reaction will remain same throughout the reaction.
(b) If initially 'A' is taken in very large amount, then the half-life of reaction will keep on increasing as reaction proceeds.
(c) If initially 'B' is taken in very large amount, then the half-life of reaction will remain same throughout the reaction.
(d) If 'A' and 'B' are taken in equal concentration, then both will be reduced to half of the original amount in the same time.

**14.** For two reactions $A \xrightarrow{K_1} P$ and $B \xrightarrow{K_2} Q$ of first and second order, respectively, if the initial concentrations of A and B are same (1 M), then the time taken by A and B to reach at 0.5 M concentration is same. Which of the following is/are correct statement(s) regarding the reactions?

(a) The initial rate of reaction of A is greater.
(b) The initial rate of reaction of B is greater.
(c) The magnitude of $K_1$ is greater than that of $K_2$.
(d) The magnitude of $K_2$ is greater than that of $K_1$.

**15.** Consider the following first-order decomposition reaction.

$$A_4(g) \rightarrow 4A(g)$$

Which of the following statement(s) is/are correct regarding the reaction? (log 2 = 0.3)

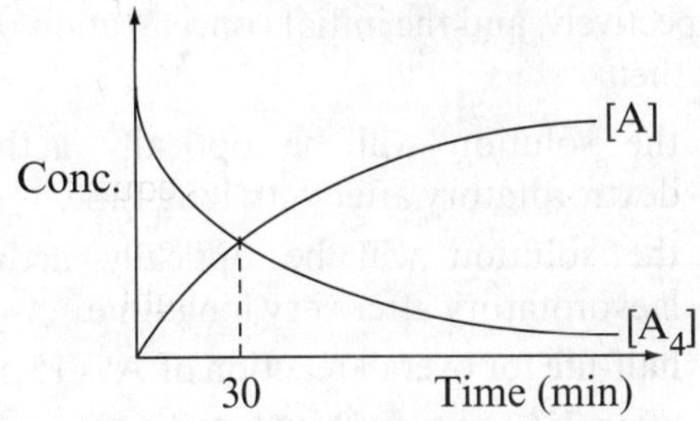


(a) At 30 min, only 20% reaction is complete.
(b) $t_{1/2}$ of reaction is 90 min.
(c) Rate of reaction decreases linearly with time.
(d) The time for intersection of two curves is independent of initial concentration of $A_4$.

**16.** α-maltose ($C_{12}H_{22}O_{11}$) can be hydrolysed to glucose ($C_6H_{12}O_6$) as follows.

$$C_{12}H_{22}O_{11}(aq) + H_2O(l) \rightarrow 2C_6H_{12}O_6(aq)$$

On the basis of following data, identify the correct statement(s) related with the reaction.

$$\Delta_f H^\circ_{C_{12}H_{22}O_{11}(aq)} = -2238 \text{ kJ/mol},$$

$$\Delta_f H^\circ_{C_6H_{12}O_6(aq)} = -1263 \text{ kJ/mol},$$

$$\Delta_f H^\circ_{H_2O(l)} = -285 \text{ kJ/mol}$$

| Time (minute) | 0 | 50 | 100 |
|---|---|---|---|
| Conc. of α-maltose (M) | 0.4 | 0.1 | 0.025 |

(a) The hydrolysis of α-maltose is an exothermic process.
(b) Heat liberated in the combustion of 1.0 mole of α-maltose must be greater than the heat liberated in combustion of 2.0 moles of glucose.
(c) On increasing the temperature, the extent of hydrolysis of α-maltose will decrease.
(d) The hydrolysis of α-maltose follows first-order kinetics.

**17.** For the reaction A → Products (order = $n$), the theoretical time of completion of reaction is

(a) Infinite for all values of $n$.
(b) Infinite for $n \geq 1$.
(c) $\dfrac{[A_0]^{1-n}}{K(1-n)}$ for $n \neq 1$
(d) $\dfrac{[A_0]^{1-n}}{K(1-n)}$ for $n < 1$

**18.** Consider the following reactions.

$$A \xrightarrow{K_1 = 6.93\times10^{-2}\ \text{min}^{-1}} B \text{ and}$$

$$A \xrightarrow{K_2 = 13.86\times10^{-2}\ \text{min}^{-1}} C$$

Here, A, B and C all are optically active compounds. If the optical rotations of A, B and C per unit concentration are 60°, –72° and 42°, respectively, and the initial concentration of A is 2 M, then

(a) the solution will be optically active and dextrorotatory after very long time.

(b) the solution will be optically active and laevorotatory after very long time.

(c) half-life for overall reaction of A is 15 minutes.

(d) after 75% conversion of A into B and C, the optical rotation of solution will be 36°.

**19.** For the first-order parallel reactions, starting with 1 M of A, which of the following statement(s) is/are correct?

$$A \xrightarrow{3K} B$$

$$A \xrightarrow{2K} 2C$$

$$A \xrightarrow{K} 3D$$

(a) $[B]_t : [C]_t : [D]_t$ is independent of time.

(b) $[C]_t > [B]_t = [D]_t$

(c) $[C]_\infty = 0.67$ M

(d) If $K = 0.5\ \text{min}^{-1}$, then [B] = 1 M after 1.386 min.

**20.** When secondary alkyl halide undergoes $S_N$ reaction, it follows $S_{N_1}$ (first-order) and $S_{N_2}$ (second-order) pathway simultaneously. The dominant pathway depends on large number of factors. An alkyl halide undergoes both pathway with rate constants $3\times10^{-4}\ s^{-1}$ and $5\times10^{-4}\ M^{-1}\ s^{-1}$, respectively. Which of the following is/are correct information for the reaction?

(a) If the concentration of nucleophile is 0.1 M, then the dominant pathway of reaction is $S_{N_1}$.

(b) If the concentration of nucleophile is 1.0 M, then the dominant pathway of reaction is $S_{N_2}$.

(c) The percentage of $S_{N_1}$ product will be 50, when the nucleophile concentration is 0.6 M.

(d) The percentage of $S_{N_1}$ product will be 40, when the nucleophile concentration is 0.9 M.

**21.** For the reversible reaction $A \rightleftharpoons B$, the time at which the rate of change in concentration of 'A' is equal to the rate of change in concentration of 'B' is/are

(a) $t = 0$ (b) $t = t_{1/2}$

(c) $t = t_{\text{equilibrium}}$ (d) $t =$ infinite

**22.** For the parallel reactions $A \xrightarrow{K_1} B$ and $A \xrightarrow{K_2} C$, the initial concentration of 'A' is $C_{A_0}$ and initially 'B' and 'C' are absent. Concentrations of A, B and C at any time '$t$' is $C_A$, $C_B$ and $C_C$, respectively. The correct relation(s) is/are

(a) $C_A + C_B + C_C = C_{A_0}$

(b) $\dfrac{dC_A}{dt} + \dfrac{dC_B}{dt} + \dfrac{dC_C}{dt} = 0$

(c) $\dfrac{C_B}{C_C} = \dfrac{K_1}{K_2}$

(d) $\dfrac{C_B}{C_{A_0} - C_A} = \dfrac{K_1}{K_1 + K_2}$

**23.** Which of the following statement(s) is/are correct for a photochemical reaction?

(a) Photochemical reactions are complex (multistep) reactions.

(b) The overall order of such reaction is always zero.

(c) Only the first step of such reaction follows zero-order kinetics.

(d) The rate of such reaction depends on the intensity of radiations absorbed.

**24.** Which of the following statement is/are incorrect?

(a) When $\Delta t$ is infinitesimally small, the average rate equals the instantaneous rate.

(b) Activation energy for the forward reaction equals activation energy for the reverse reaction in a catalysed reaction.

(c) For a reversible reaction, an increase in temperature, increase the rate for both the forward and backward reaction.

(d) Larger the initial reactant concentration for a zero-order reaction, shorter is the half-life.

**25.** In a consecutive reaction system A $\xrightarrow{E_1}$ B $\xrightarrow{E_2}$ C when $E_1$ is much greater than $E_2$, the yield of B increases with

(a) increase in temperature.
(b) decreases in temperature.
(c) increase in initial concentration of A.
(d) decrease in initial concentration of A.

**26.** When the temperature of a reaction system is increased by 10°C (from 25°C to 35°C) at constant volume, the reaction rate doubles. It is due to

(a) the total number of collisions between reacting molecules increases nearly by 100%.
(b) the number of collisions between the molecules possessing sufficiently high energy increases nearly by 100%.
(c) the total number of collisions between reacting molecules increases nearly by 1 to 2%.
(d) the average translational kinetic energy of molecules becomes nearly double.

**27.** Which of the following quantity/quantities is/are temperature independent?

(a) A (Pre-exponential factor)
(b) $E_a$ (Activation energy)
(c) R (Universal gas constant)
(d) $K$ (Rate constant)

**28.** Decomposition of $A_2$(g) and $B_3$(g) follows first-order kinetics as follows.

$$A_2(g) \rightarrow 2A(g);\ K_1\ (/h^{-1}) = e^{-\frac{14000(/J)}{RT}+5}$$

$$B_3(g) \rightarrow 3B(g);\ K_2\ (/h^{-1}) = e^{-\frac{20000(/J)}{RT}+10}$$

Here, $K_1$ and $K_2$ are the rate constants with respect to disappearance of $A_2$ and $B_3$, respectively. One mole, each of $A_2$(g) and $B_3$(g), is taken in a 100 L evacuated flask and at some temperature at which they start decomposing at the same rate. The incorrect information regarding the reactions is

(a) The temperature at which the reactions are performed is $\frac{1200}{8.314}$ K.
(b) At any instant, $\left(\frac{P_{A_2}}{P_{B_3}}\right)$ will be constant and equal to 1.0, where $P_{A_2}$ and $P_{B_3}$ are the partial pressures of gases at that time.
(c) At some instant, the total pressure of all gases in the flask may be less than 0.2 atm.
(d) At any instant, $\left(\frac{P_A}{P_B}\right)$ will be constant and equal to 1.0, where $P_A$ and $P_B$ are the partial pressures of gases at that time.

**29.** In a study of effect of temperature on reaction rate, the value of $\frac{1}{K}\cdot\frac{dK}{dT}$ is found to be $\frac{1.25\times10^6}{T^3}$ $K^{-1}$. Identify the correct statement(s).

(a) The activation energy for the reaction at 250 K is 10 kcal/mol.
(b) The activation energy for the reaction at 2000 K is 1.25 kcal/mol.
(c) The rate of increase of rate constant with the increase in temperature is higher at lower temperature than at higher temperature.
(d) The value of $\frac{d(\ln K)}{dT}$ is 0.625 at 1000 K.

**30.** For the reversible reaction A $\rightleftharpoons$ B; ΔH = –2 kcal, the pre-exponential factors are same for the forward and backward reactions. If the activation energy of backward reaction is 8 kcal/mol, then which of the following statement is correct about the reaction?

(a) The activation energy of forward reaction is 6 kcal/mol.
(b) At 500 K, the fraction of 'A' molecules crossing the energy barrier for forward reaction is $e^{-6}$.
(c) At 500 K, the fraction of 'B' molecules crossing the energy barrier for forward reaction is $e^{-8}$.
(d) Equilibrium constant for the reaction is $e^{-2}$ at 500 K.

## Section C (Comprehensions)

### Comprehension I

**The thermal decomposition of $N_2O_5$ occurs as $2N_2O_5 \rightarrow 4NO_2 + O_2$. Experimental studies suggest that the rate of decomposition of $N_2O_5$, rate of formation of $NO_2$ or rate of formation of $O_2$, all becomes double if concentration of $N_2O_5$ is doubled.**

1. The correct mechanism for the decomposition of $N_2O_5$ may be

(a) Step-I: $N_2O_5 \xrightarrow{\text{slow}} NO_2 + NO_3$
Step-II: $N_2O_5 + NO_3 \xrightarrow{\text{fast}} 3NO_2 + O_2$

(b) Step-I: $N_2O_5 \xrightarrow{\text{fast}} NO_2 + NO_3$
Step-II: $N_2O_5 + NO_3 \xrightarrow{\text{slow}} 3NO_2 + O_2$

(c) $N_2O_5 \xrightarrow{\text{fast}} 2NO_2 + \frac{1}{2}O_2$

(d) Step-I: $N_2O_5 \xrightarrow{\text{fast}} NO + NO_2 + O_2$
Step-II: $N_2O_5 + NO_2 \xrightarrow{\text{slow}} 3NO_2 + \frac{1}{2}O_2$

2. If the rate constants for decomposition of $N_2O_5$, formation of $NO_2$ and formation of $O_2$ are $K_1$, $K_2$ and $K_3$, respectively, then

(a) $K_1 = K_2 = K_3$ (b) $2K_1 = K_2 = 4K_3$
(c) $K_1 = 2K_2 = K_3$ (d) $2K_1 = 4K_2 = K_3$

3. If the rate of formation of $O_2$ is 16 g/h, then the rate of decomposition of $N_2O_5$ and the rate of formation of $NO_2$ are

(a) 108 g/h, 92 g/h (b) 54 g/h, 46 g/h
(c) 32 g/h, 64 g/hr (d) 16 g/h, 16 g/h

### Comprehension II

**The following Rice–Herzfeld mechanism has been suggested for the formation of phosgene $COCl_2$.**

**(1)** $Cl_2 \longrightarrow 2Cl$ $[K_1 = A_1 \cdot e^{-E_{a_1}/RT}]$
**(2)** $2Cl \longrightarrow Cl_2$ $[K_2 = A_2 \cdot e^{-E_{a_2}/RT}]$
**(3)** $Cl + CO \longrightarrow COCl$ $[K_3 = A_3 \cdot e^{-E_{a_3}/RT}]$
**(4)** $COCl \longrightarrow Cl + CO$ $[K_4 = A_4 \cdot e^{-E_{a_4}/RT}]$
**(5)** $COCl + Cl_2 \longrightarrow COCl_2 + Cl$ $[K_5 = A_5 \cdot e^{-E_{a_5}/RT}]$

4. Assuming steady-state approximation for [Cl] and [COCl], the rate law for the formation of $COCl_2$ is

(a) $\dfrac{K_5K_2K_1^{1/2}[Cl_2]^{3/2}[CO]}{K_3^{1/2}(K_4 + K_5[Cl_2])}$

(b) $\dfrac{K_5K_3K_1^{1/2}[Cl_2][CO]^{3/2}}{K_2^{1/2}(K_4 + K_5[Cl_2])}$

(c) $\dfrac{K_5K_3K_1^{1/2}[Cl_2]^{3/2}[CO]}{K_2^{1/2}(K_4 + K_5[Cl_2])}$

(d) $\dfrac{K_5K_2K_1^{1/2}[Cl_2][CO]^{3/2}}{K_3^{1/2}(K_4 + K_5[Cl_2])}$

5. The rate law will be if we simplify the rate law obtained in previous question assuming that the reactions (1) – (4) are much faster than reaction (5)

(a) $\dfrac{K_4K_3K_2^{1/2}[Cl_2]^{3/2}[CO]}{K_4K_2}$

(b) $\dfrac{K_4K_3K_2^{1/2}[Cl_2][CO]^{3/2}}{K_2K_3^{1/2}}$

(c) $\dfrac{K_5K_3K_1^{1/2}[Cl_2][CO]^{3/2}}{K_4^{1/2}K_2}$

(d) $\dfrac{K_5K_3K_1^{1/2}[Cl_2]^{3/2}[CO]}{K_4K_2^{1/2}}$

6. What is the overall pre-exponential factor of reaction, if steps (1) to (4) are much faster than (5)?

(a) $\dfrac{A_5A_3A_1^{1/2}}{A_4A_2^{1/2}}$ (b) $\dfrac{A_5A_3^{1/2}A_1}{A_2A_4^{1/2}}$

(c) $\dfrac{A_5A_4A_1^{1/2}}{A_4A_2^{1/2}}$ (d) $\dfrac{A_4A_3A_2^{1/2}}{A_2A_4^{1/2}}$

EXERCISE II

**7.** What is the overall activation energy of reaction, if steps (1) to (4) are much faster than (5)?

(a) $\frac{Ea_5.Ea_3.Ea_1^{1/2}}{Ea_4.Ea_2^{1/2}}$

(b) $Ea_5 + Ea_3 + \frac{1}{2}Ea_1 - \frac{1}{2}Ea_2 - Ea_4$

(c) $Ea_1 + Ea_3 + Ea_5 - Ea_2 - Ea_4$

(d) $-Ea_5 - Ea_3 - \frac{1}{2}Ea_1 + Ea_2 + \frac{1}{2}Ea_4$

## Comprehension III

**Chain reactions are a specific class of reactions in which highly reactive species are produced as intermediates which carry the reaction at a rapid rate for a long time. These reactions were first studied in 1934 by Frank O. Rice and Karl F. Herzfeld and are referred to as Rice–Herzfeld chain reactions. The steady-state approximation has proved highly useful in accounting for their rate laws. A typical chain reaction is that between $H_2$ and $Br_2$ to form HBr: $H_2(g) + Br_2(g) \rightarrow 2HBr(g)$, for which a five step mechanism has been proposed.**

**Step I:** $Br_2 \xrightarrow{K_1} 2Br$ **(Chain initiation)**

**Step II:** $Br + H_2 \xrightarrow{K_2} HBr + H$ **(Chain propagation)**

**Step III:** $H + Br_2 \xrightarrow{K_3} HBr + Br$ **(Chain propagation)**

**Step IV:** $H + HBr \xrightarrow{K_4} H_2 + Br$ **(Chain retardation)**

**Step V:** $2Br \xrightarrow{K_5} Br_2$ **(Chain termination)**

**8.** The concentration of bromine radical is

(a) $K_1[Br_2]^2$

(b) $\frac{K_1}{K_5}[Br_2]^2$

(c) $\frac{K_1}{K_5}[Br_2]$

(d) $\left(\frac{K_1}{K_5}[Br_2]\right)^{1/2}$

**9.** The concentration of hydrogen radical is

(a) $\frac{K_2K_1[Br_2][H_2]}{K_5(K_3[Br_2]+K_4[HBr])}$

(b) $\frac{K_2K_1[Br_2]^{1/2}[H_2]}{K_5(K_3[Br_2]+K_4[HBr])}$

(c) $\frac{K_2K_1^{1/2}[Br_2]^{1/2}[H_2]}{K_5^{1/2}(K_3[Br_2]+K_4[HBr])}$

(d) $\frac{K_2K_1^{1/2}[Br_2]^{1/2}[H_2]^2}{K_5^{1/2}(K_3[Br_2]^{1/2}+K_4[HBr])}$

**10.** The rate of formation of HBr, $+\frac{d[HBr]}{dt}$ is

(a) $\frac{2K_3K_2K_1^{1/2}[Br_2]^{3/2}[H_2]}{K_5^{1/2}(K_3[Br_2]+K_4[HBr])}$

(b) $\frac{2K_2K_1^{1/2}[Br_2]^{3/2}[H_2]}{K_5^{1/2}(K_3[Br_2]+K_4[HBr])}$

(c) $\frac{K_2K_1^{1/2}[Br_2]^{1/2}[H_2]}{K_5^{1/2}(K_3[Br_2]+K_4[HBr])}$

(d) $\frac{2K_2K_1^{1/2}[Br_2]^{1/2}[H_2]}{K_5^{1/2}}$

**11.** What is the initial rate of formation of HBr, $+\frac{d[HBr]}{dt}$?

(a) $\frac{2K_2K_1^{1/2}[Br_2]^{1/2}[H_2]}{K_5^{1/2}}$

(b) $\frac{2K_2K_1^{1/2}[Br_2]^{1/2}[H_2]}{K_5^{1/2}K_3}$

(c) $\frac{K_2K_1^2[Br_2]^{1/2}[H_2]}{K_5^2K_3}$

(d) Zero

## Comprehension IV

**A definite volume of $H_2O_2$ undergoing decomposition required 25.6 ml of standard $KMnO_4$ solution for titration. After 10 and 20 minutes, the volumes of permanganate required were 16.0 and 10.0 ml, respectively ($\log 2 = 0.3$, $(1.6)^5 = (3.2)^2$).**

12. The order of reaction is
(a) 1 (b) 2
(c) 3 (d) 0

13. The time (in minutes) required for the decomposition to be half completed is
(a) 1.5 (b) 7.5
(c) 15.0 (d) 13.3

14. The fraction of $H_2O_2$ decomposed after 25 minutes is
(a) $\frac{5}{16}$ (b) 0.6
(c) $\frac{1}{3.2}$ (d) $\frac{11}{16}$

15. The reaction $2H_2O_2(aq) \rightarrow 2H_2O(l) + O_2(g)$ is an example of
(a) pseudo unimolecular
(b) pseudo bimolecular
(c) elementary reaction
(d) photochemical reaction

## Comprehension V

**The first-order decomposition of di-tert-butyl peroxide (DTBP) to acetone is given by the following equation.**

$$C_8H_{18}O_2(g) \rightarrow 2CH_3COCH_3(g) + C_2H_6(g)$$

**The reaction has a half-life of 80 minutes at 147°C. Starting with pure DTBP in a flask of constant volume at a pressure of 800 torr,**

16. In what time (in min), the partial pressure of DTBP is 100 torr?
(a) 60 (b) 120
(c) 180 (d) 240

17. In what time (in min), the partial pressure of acetone is 1200 torr?
(a) 80 (b) 160
(c) 120 (d) 240

18. What will be the total gas pressure (in torr), when the partial pressure of DTBP is 700 torr?
(a) 1000 (b) 700
(c) 900 (d) 1100

## Comprehension VI

**The reaction of the reactive fragment OH with $H_2S$: $4OH(g) + H_2S(g) \rightarrow SO_2(g) + 2H_2O(g)$ was studied at a particular temperature.**

| [OH] mol/litre | $[H_2S]$ mol/litre | Rate of disappearaces of $H_2S$ mol $l^{-1}$ $s^{-1}$ |
|---|---|---|
| $1.3 \times 10^{-8}$ | $2.1 \times 10^{-8}$ | $1.4 \times 10^{-6}$ |
| $3.9 \times 10^{-8}$ | $2.1 \times 10^{-8}$ | $4.2 \times 10^{-6}$ |
| $3.9 \times 10^{-8}$ | $4.2 \times 10^{-8}$ | $8.4 \times 10^{-6}$ |

19. What is the rate law expression for the reaction?
(a) $r = K[OH]^2[H_2S]$ (b) $r = K[OH][H_2S]^2$
(c) $r = K[OH]^{-1}[H_2S]$ (d) $r = K[OH][H_2S]$

20. What is the rate constant of the reaction at the given temperature?
(a) $5.1 \times 10^9$ $M^{-1}s^{-1}$ (b) $2.8 \times 10^{15}$ $M^{-1}s^{-1}$
(c) $1.5 \times 10^{10}$ $M^{-1}s^{-1}$ (d) $3.9 \times 10^{17}$ $M^{-2}s^{-1}$

**21.** What is the rate, mol $l^{-1}s^{-1}$, at the instant when $[OH] = 1.7 \times 10^{-8}$ M and $[H_2S] = 1.0 \times 10^{-8}$ M?

(a) $8.7 \times 10^{-6}$ (b) $8.7 \times 10^{-8}$

(c) $8.7 \times 10^{-7}$ (d) $1.15 \times 10^{6}$

**22.** What is the rate, mol $s^{-1}$, at the instant when $[OH] = 1.7 \times 10^{-8}$ M and $[H_2S] = 1.0 \times 10^{-8}$ M and the volume of the reacting system is 0.1 litre?

(a) $8.7 \times 10^{-6}$ (b) $8.7 \times 10^{-8}$

(c) $8.7 \times 10^{-7}$ (d) $1.15 \times 10^{6}$

## Comprehension VII

**At constant temperature and volume, X decomposes as $2X(g) \rightarrow 3Y(g) + 2Z(g)$. $P_x$ is the partial pressure of X.**

| Observation no. | Time (min) | $P_x$ (mm Hg) |
|---|---|---|
| 1 | 0 | 800 |
| 2 | 100 | 400 |
| 3 | 200 | 200 |

**23.** The order of reaction with respect to 'X' is

(a) 0 (b) 1

(c) 2 (d) 0.5

**24.** The rate constant of reaction is

(a) $6.93 \times 10^{-3}$ min$^{-1}$

(b) $1.386 \times 10^{-2}$ min$^{-1}$

(c) $2 \times 10^{-4}$ M$^{-1}$min$^{-1}$

(d) $3.465 \times 10^{-3}$ min$^{-1}$

**25.** The time for 87.5% completion of the reaction is

(a) 350 min (b) 300 min

(c) 400 min (d) 600 min

**26.** The total pressure, when pressure of 'X' is 700 mm of Hg, is

(a) 1000 mm Hg (b) 900 mm Hg

(c) 950 mm Hg (d) 250 mm Hg

## Comprehension VIII

**Consider the following reaction occurring at constant volume.**

| | A(g) | + | 2B(g) | → | C(g) | + | D(g) |
|---|---|---|---|---|---|---|---|
| $t = 0$ | a M | | b M | | 0 | | 0 |

**The rate of reaction is $r = k \cdot C_B$, where $C_B$ = concentration of 'B' at any time '$t$' and $k$ = specific reaction rate.**

**27.** The concentration of 'A' at time '$t$' is

(a) $C_A = a - kt$

(b) $C_A = a \cdot e^{-kt}$

(c) $C_A = a - \frac{b}{2}(1 - e^{-2kt})$

(d) $C_A = a - b(1 - e^{-kt})$

**28.** The time at which the concentration of 'A' reduces half of its initial value is

(a) $\frac{a}{2k}$ (b) $\frac{a}{k}$

(c) $\frac{1}{2k} \ln \frac{b}{b-a}$ (d) $\frac{\ln 2}{k}$

**29.** The condition for $(t_{1/2})_A = (t_{1/2})_B$ is

(a) $b = 2a$ (b) $b = a$

(c) $a = 2b$ (d) none of these

## Comprehension IX

**The gaseous reaction $n_1A\,(g) \rightarrow n_2B\,(g)$ is first order with respect to 'A'. The rate constant of reaction is '$K$'. The reaction is studied at a constant pressure and temperature. Initially, the moles of 'A' was '$a$' and 'B' was not present.**

**30.** How many moles of 'A' are present at time $t$?

(a) $a \cdot e^{-kt}$ (b) $a \cdot e^{-n_1kt}$

(c) $a \cdot e^{-n_2kt}$ (d) $a\left(1-e^{-n_1kt}\right)$

**31.** If the initial volume of system was $V_0$, then the volume of system after time $t$, will be

(a) $\dfrac{n_1V_0}{n_2}$

(b) $\dfrac{n_2V_0}{n_1}$

(c) $V_0\left[\dfrac{n_2}{n_1}+\left(1-\dfrac{n_2}{n_1}\right)\cdot e^{-n_1kt}\right]$

(d) $V_0\left[\dfrac{n_1}{n_2}-\left(\dfrac{n_2}{n_1}-1\right)\cdot e^{-n_1kt}\right]$

**32.** What will be the concentration of 'A' at time $t$, if $n_1 = 1$ and $n_2 = 2$?

(a) $[A_0]\cdot e^{-kt}$ (b) $[A_0]\left(\dfrac{e^{-kt}}{2-e^{-kt}}\right)$

(c) $[A_0]\left(\dfrac{e^{-kt}}{1-e^{-kt}}\right)$ (d) $[A_0](1-2\cdot e^{-kt})$

## Comprehension X

**The oxidation of metals is generally a slow electrochemical reaction involving many steps. These steps involve electron transfer reactions. A particular type of oxidation involve overall first-order kinetics with respect to fraction of unoxidized metal surface thickness $(1-f)$ relative to the maximum thickness $(T)$ of oxidized surface, when metal surface is exposed to air for a considerable period of time.**

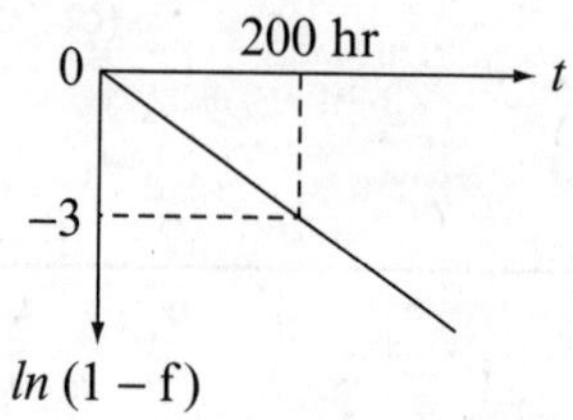


**Rate law:** $\dfrac{df}{dt} = K(1-f)$,

**where,** $f = x/T$,

**$x$ = thickness of the oxide film at time '$t$'**

**and $T$ = thickness of the oxide film at $t = \infty$**

**For the oxidation of a particular metal, a graph is shown in the figure.**

**33.** The time taken for thickness to grow 50% of $T$ is

(a) 23.1 h (b) 46.2 h

(c) 100 h (d) 92.4 h

**34.** The exponential variation of $f$ with $t$ (h) is

(a) $f = 1 - e^{-3t/200}$ (b) $f = e^{-3t/200} - 1$

(c) $f = e^{-3t/200}$ (d) $f = e^{3t/200}$

## Comprehension XI

***Trans*-1,2-dideuterocyclopropane (A) undergoes first-order decomposition. The observed rate constant at a certain temperature, measured in terms of the disappearance of A was $1.5 \times 10^{-4}\ s^{-1}$. Analysis of the reaction products showed that the reaction follows two parallel paths, one leading to dideuteropropene (B) and the other to *cis*-1,2-dideuterocyclopropane (C). The compound 'B' was found to constitute 10% of the reaction product, independent of the extent of reaction.**

**35.** What is the order of the reaction for each of the paths?

(a) 0 (b) 0.5

(c) 1 (d) 2

**36.** The rate constant for the formation of 'B' is

(a) $1.5 \times 10^{-4}\ s^{-1}$ (b) $1.35 \times 10^{-5}\ s^{-1}$

(c) $1.5 \times 10^{-5}\ s^{-1}$ (d) $1.5 \times 10^{-3}\ s^{-1}$

## Comprehension XII

**Consider the following first-order consecutive reactions $A \xrightarrow{K_1} B \xrightarrow{K_2} C$. The initial concentration of A is 1.0 M and $K_1 : K_2 = 2.0 : 3.0$. Given that $K_1 = 0.04\ min^{-1}$. ($e = 2.72$, $e^{-1} = 0.368$, $e^{1.5} = 4.48$, $e^{-1.5} = 0.223$, $e^{-2} = 0.135$, $e^{0.7} = 2$, $e^{1.1} = 3$, $e^{-0.8} = 0.45$)**

**37.** What is the concentration of 'A' after 25 minutes?

(a) 0.5 M (b) 0.368 M
(c) 0.135 M (d) 0.223 M

**38.** What is the concentration of 'B' after 25 minutes?

(a) 0.223 M (b) 0.29 M
(c) 0.27 M (d) 0.145 M

**39.** What is the concentration of 'C' after 25 minutes?

(a) 0.0 M (b) 0.342 M
(c) 0.29 M (d) 0.145 M

**40.** In what time, the concentration of 'B' will become maximum?

(a) 10 minutes (b) 20 minutes
(c) 25 minutes (d) 40 minutes

**41.** What is the maximum concentration of 'B'?

(a) 0.290 M (b) 0.296 M
(c) 0.50 M (d) 0.3 M

**42.** In what time, the concentrations of A, B and C become equal?

(a) 20 min (b) 27.5 min
(c) 22.5 min (d) Never

**43.** What is the concentration of 'A' when $-\frac{d[A]}{dt} = +\frac{d[C]}{dt}$?

(a) 0.704 M (b) 0.45 M
(c) 0.33 M (d) 0.5 M

**44.** What is the maximum rate of formation of 'C'?

(a) 0.342 M/min
(b) $1.776 \times 10^{-2}$ M/min
(c) $2.052 \times 10^{-2}$ M/min
(d) $1.8 \times 10^{-2}$ M/min

## Comprehension XIII

**For the first-order reaction opposed by first-order reaction $A \underset{K_b}{\overset{K_f}{\rightleftharpoons}} B$, $[A]_0 = 0.15$ M, $[A]_{eq} = 0.10$ M, $[A]_{10} = 0.125$, where times are given in minutes.**

**45.** What is the rate constant of forward reaction?

(a) $2.31 \times 10^{-2}\ min^{-1}$ (b) $4.62 \times 10^{-2}\ min^{-1}$
(c) $2.31 \times 10^{-3}\ min^{-1}$ (d) $1.165 \times 10^{-2}\ min^{-1}$

**46.** What is the rate constant of backward reaction?

(a) $2.31 \times 10^{-2}\ min^{-1}$ (b) $4.62 \times 10^{-2}\ min^{-1}$
(c) $2.31 \times 10^{-3}\ min^{-1}$ (d) $1.165 \times 10^{-2}\ min^{-1}$

**47.** What is the equilibrium constant of the reaction?

(a) 1.5 (b) 0.67
(c) 0.5 (d) 2.0

**48.** Half-life of a reversible reaction may be defined as the time in which half of the amount of reactant which should react up to equilibrium has been reacted. The half-life of the reaction is

(a) 10 minutes (b) 15 minutes
(c) 5 minutes (d) 20 minutes

## Comprehension XIV

**A substance 'X' dissociates simultaneously into 'Y' and 'Z' as**

$$X(g) \xrightarrow{A_1 = 2\times10^{14}\ s^{-1},\ Ea_1 = 12\ kcal/mol} Y(g)$$

$$X(g) \xrightarrow{A_1 = 2\times10^{14} \times e^{-2}\ s^{-1},\ Ea_1 = 9\ kcal/mol} Z(g)$$

**49.** Temperature at which the net activation energy for the dissociation of 'X' will be 10.5 kcal/mol is

(a) 200 K (b) 750 K
(c) 900 K (d) 625 K

EXERCISE II

50. Which of the following statement is correct regarding the major product of the reactions?
(a) 'Y' will be the major product at 300 K.
(b) 'Y' will be the major product at 700 K.
(c) 'Y' will be the major product at 900 K.
(d) 'Z' will be the major product at 900 K.

51. If the temperature is increased, then
(a) the rate of formation of 'Y' will increase but the rate of formation of 'Z' will decrease.
(b) the rate of formation of both, 'Y' and 'Z' will increase in such a way that the ratio of rate constants of their formation remains constant.
(c) the rate of formation of both, 'Y' and 'Z' will increase in such a way that the ratio of rate constants of their formation decreases.
(d) the rate of formation of both, 'Y' and 'Z' will increase in such a way that the ratio of rate constants of their formation increases.

## Comprehension XV

**For the reaction sequence $A \underset{K_{-1}}{\overset{K_1}{\rightleftharpoons}} B \underset{K_{-2}}{\overset{K_2}{\rightleftharpoons}} C \underset{K_{-3}}{\overset{K_3}{\rightleftharpoons}} D$, D is more stable than A by 42.8 kcal/mol, B is more stable than C by 13.2 kcal/mol and A is more stable than C by 20.8 kcal/mol. The transition state going from A to B is more stable than the transition state going from B to C by 10.2 kcal/mol, but it is less stable than the transition state going from C to D by 2.4 kcal/mol. The energy of activation of C for transition state C to D is 4.3 kcal/mol.**

52. Which of the following is the faster step in the forward direction?
(a) A ⟶ B
(b) B ⟶ C
(c) C ⟶ D
(d) Cannot be predicted

53. Which of the following is the faster step in the backward direction?
(a) D ⟶ C
(b) C ⟶ B
(c) B ⟶ A
(d) Cannot be predicted

54. Which of the following is fastest step among these six steps?
(a) C ⟶ D (b) C ⟶ B
(c) B ⟶ A (d) B ⟶ C

55. Which of the following is therate-determining step in forward direction?
(a) A ⟶ B (b) B ⟶ C
(c) C ⟶ D (d) Cannot be predicted

56. Which of the following is the rate-determining step in backward direction?
(a) D ⟶ C (b) C ⟶ B
(c) B ⟶ A (d) Cannot be predicted

## Section D (Assertion – Reason)

The following questions consist of two statements. Mark the answer as follows.
(a) If both statements are CORRECT, and **Statement II** is the CORRECT explanation of **Statement I**.
(b) If both statements are CORRECT, and **Statement II** is NOT the CORRECT explanation of **Statement I**.
(c) If **Statement I** is CORRECT, but **Statement II** is INCORRECT.
(d) If **Statement I** is INCORRECT, but **Statement II** is CORRECT.

1. **Statement I:** A fractional-order reaction must be a complex reaction.
**Statement II:** Fractional order of rate-determining step equals to overall order of a complex reaction.

2. **Statement I:** The time of completion of reactions of type A → products (order $< 1$) may be determined.
**Statement II:** Reactions with order $\geq 1$ are either too slow or too fast and hence, the time of completion cannot be determined.

3. **Statement I:** The temperature coefficient of a single-step reaction may be negative.
**Statement II:** The rate of reaction having negative order with respect to a reactant decreases with increase in concentration of that reactant.

EXERCISE II

4. **Statement I:** The overall rate of a reversible reaction may decrease with increase in temperature.

   **Statement II:** When the activation energy of forward reaction is less than that of backward reaction, then the increase in the rate of backward reaction is more than that of forward reaction on increasing the temperature.

5. **Statement I:** In a reversible endothermic reaction, the activation energy of forward reaction is higher than that of backward reaction.

   **Statement II:** The threshold energy of forward reaction is more than that of backward reaction.

6. **Statement I:** A catalyst provides an alternative path to the reaction in which conversion of reactants into products takes place quickly.

   **Statement II:** The catalyst forms an activated complex of lower potential energy with the reactants by which more number of molecules are able to cross the barrier per unit time.

7. **Statement I:** In zero-order reactions, half-life is half of the completion time of reaction.

   **Statement II:** Half-life is half of the time for completion of reaction of any order.

8. **Statement I:** $2A + B \rightarrow 2C + D$ must be a third-order reaction.

   **Statement II:** The rate constant for a third-order reaction has unit $L^2\ mol^{-2}\ s^{-1}$.

9. **Statement I:** For zero-order reaction ($A \rightarrow B$), successive half-life of reaction decreases with the progress of reaction.

   **Statement II:** Rate of zero-order reaction remains unchanged with the progress of reaction.

10. **Statement I:** The concentration of pure solid and pure liquid is taken as constant at constant temperature.

    **Statement II:** The concentration of pure solid and liquids depends on the density and molecular mass.

## Section E (Column Match)

1. Match the columns.

| Column I (Reactions) | Column II (Unit of rate constant, K) |
|---|---|
| (A) $SO_2Cl_2 \rightarrow SO_2 + Cl_2$ | (P) $s^{-1}$ |
| (B) $CH_3COOC_2H_5 + NaOH \rightarrow CH_3COONa + C_2H_5OH$ | (Q) $min^{-1}$ |
| (C) $2H_2O_2 \rightarrow 2H_2O + O_2$ | (R) $L\ mol^{-1}\ min^{-1}$ |
| (D) $H_2O_2 + 2I^- + 2H^+ \rightarrow 2H_2O + I_2$ | (S) $L\ mol^{-1}\ s^{-1}$ |

2. Match the columns.

| Column I (Reactions) | Column II (Increase in rate when concentration of reactant is doubled) |
|---|---|
| (A) $CH_3CHO \rightarrow CH_4 + CO$ | (P) 2 times |
| (B) $SO_2Cl_2 \rightarrow SO_2 + Cl_2$ | (Q) $2^{1.5}$ times |
| (C) $2\ H_2O_2 \rightarrow 2H_2O + O_2$ | (R) 2 times in acid medium |
| (D) $CH_3COOC_2H_5 \xrightarrow{HOH} CH_3COOH + C_2H_5OH$ | (S) $2^2$ times in basic medium |

3. Match the columns.

| Column I | Column II |
|---|---|
| (A) Rate $= k \times$ Intensity of light | (P) Second-order |
| (B) Rate $= k[A]^1[B]^1$ | (Q) Zero-order |
| (C) Rate $= k[A]^{3/2}[B]^{1/2}$ | (R) First-order when A is excess. |
| (D) Rate $= k[A]^2[B]^1$ | (S) Second-order when B is excess. |

4. Match the kinetic equation of Column I with the unit of its rate constant in Column II.

| Column I (Kinetic equation) | Column II (Unit of rate constant) |
|---|---|
| (A) $x = kt$ | (P) $s^{-1}$ |
| (B) $k = \frac{2.303}{t}\log_{10}\left(\frac{a}{a-x}\right)$ | (Q) $L\ mol^{-1}\ s^{-1}$ |
| (C) $k = \frac{1}{t}\left[\frac{1}{(a-x)} - \frac{1}{a}\right]$ | (R) $mol\ L^{-1}\ s^{-1}$ |
| (D) $k = \frac{2.303}{t(a-b)}\log\left[\frac{b(a-x)}{a(b-x)}\right]$ | (S) $atm^{-1}\ s^{-1}$ |

5. Match the half-life in Column I with the order in Column II for the reaction: A $\rightarrow$ Products.

| Column I | Column II |
|---|---|
| (A) $t_{1/2}$ = Constant | (P) First-order |
| (B) $t_{1/2} \propto$ a | (Q) Third-order |
| (C) $t_{1/2} \propto$ 1/a | (R) Second-order |
| (D) $t_{1/2} \propto$ 1/a$^2$ | (S) Zero-order |

Here, $a$ = Initial concentration of the reactant.

6. Match Column I with Column II and Column III for the reaction A $\rightarrow$ Products.

| Column I | Column II | Column III |
|---|---|---|
| (A) $x$ is plotted against $t$ in zero-order reaction. | (P) Slope = $-k/2.303$ | (U) Intercept = $a$ |
| (B) log $(a - x)$ is plotted against '$t$' in first-order reaction. | (Q) Slope = $2k$ | (V) Intercept = $1/a$ |
| (C) $(a - x)^{-1}$ is plotted against '$t$' in second order reaction. | (R) Slope = $-k$ | (W) Intercept = $1/a^2$ |
| (D) $(a - x)^{-2}$ is plotted against '$t$' in second order. | (S) Slope = $k$ | (X) Intercept = log $a$ reaction |

Here, 'a' = Initial concentration reactant, '$x$' = Extent of reaction in time '$t$' s.

7. Match Column I with Column II for the reaction A $\rightarrow$ Products, with initial concentration of 'A' equal to 'a' M.

| Column I | Column II |
|---|---|
| (A) $\frac{0.693}{k}$ | (P) Half-life of zero-order reaction. |
| (B) $\frac{a}{2k}$ | (Q) Half-life of first-order reaction. |
| (C) $\frac{1}{k}$ | (R) Average life of zero-order reaction. |
| (D) $\frac{1}{a \cdot k}$ | (S) Average life of first-order reaction. |
| | (T) Half-life of second-order reaction. |

8. Match the columns.

| Column I (Graphs for reaction: A $\rightarrow$ Products) | Column II (Coordinates, $y$-axis vs. $x$-axis) |
|---|---|
| (A) 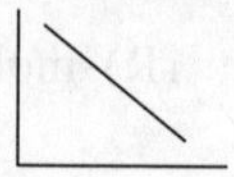 | (P) ln[A] vs. $t$ (order = 1) |
| (B) 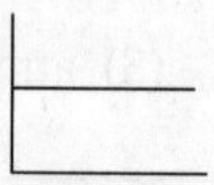 | (Q) $t_{1/2}$ vs. $[A_0]$ (order = 1) |
| (C) 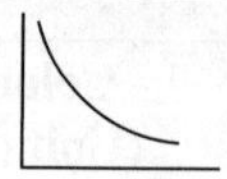 | (R) $r$ vs. $t$ (Order > 0) |
| (D) 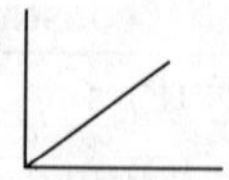 | (S) $r$ vs. $t$ (Order = 0) |
| | (T) $t_{1/2}$ vs. $[A_0]$ (Order > 1) |
| | (U) $r$ vs. $[A]$ (Order = 1) |

9. For the reaction: A $\rightarrow$ Products, $t_{x/y}$ represents the time in which $x/y$ fraction of reactant is converted into products.

| Column I | Column II |
|---|---|
| (A) $t_{5/9}$ | (P) Equal to 54 s, if $t_{1/3}$ is 18 s in case of first-order reaction. |
| (B) $t_{19/27}$ | (Q) Equal to 32 s, if $t_{1/4}$ is 16 s in case of first-order reaction. |
| (C) $t_{7/8}$ | (R) Equal to 56 s, if $t_{1/3}$ is 4 s in case of second-order reaction. |
| (D) $t_{7/16}$ | (S) Equal to 30 s, if $t_{1/3}$ is 18 s in case of zero-order reaction. |
| | (T) Equal to 28 s, if $t_{1/2}$ is 16 s in case of zero-order reaction. |

10. For the first-order consecutive reactions $A \xrightarrow{K_1} B \xrightarrow{K_2} C$, the reaction is started with some amount of only 'A'. Match the Column I (observations) with Column II (time) related with reactions.

| Column I | Column II |
|---|---|
| (A) Rate of formation of 'C' is maximum. | (P) $t = t_{1/2}$ of A, if $K_2 = 2K_1$ |
| (B) $[B]_t = [C]_t$ | (Q) $t = \frac{1}{K_2 - K_1} . \ln\left(\frac{K_2}{K_1}\right)$ |
| (C) $[A]_t = [B]_t$ | (R) $t$ = time at which the rate of formation of 'B' is maximum. |
| | (S) $t = \frac{1}{K_1 - K_2} \cdot \ln \left(\frac{2K_1 - K_2}{K_1}\right)$ |

## Section F (Subjective)

### Single Digit Integer Type

1. The number of reactions among the following, as written, could describe elementary reaction, is

(i) $Cl_2 + CO \rightarrow COCl_2$; $r = K[Cl_2]^{3/2}[CO]$

(ii) $PCl_3 + Cl_2 \rightarrow PCl_5$; $r = K[PCl_3][Cl_2]$

(iii) $2NO + H_2 \rightarrow N_2O + H_2O$; $r = K[NO][H_2]$

(iv) $2NO + O_2 \rightarrow 2NO_2$; $r = K[NO]^2[O_2]$

(v) $NO + O_3 \rightarrow NO_2 + O_2$; $r = K[NO][O_3]$

2. The following mechanisms are proposed for the reaction $CO + NO_2 \rightarrow CO_2 + NO$, at low temperature.

Mechanism 1:

$2NO_2 \rightleftharpoons N_2O_4$ (fast)

$N_2O_4 + 2CO \rightarrow 2CO_2 + 3NO$ (slow)

Mechanism 2:

$2NO_2 \rightarrow NO_3 + NO$ (slow)

$NO_3 + CO \rightarrow NO_2 + CO_2$ (fast)

Which of the above mechanism is consistent with the observed rate law: $+\frac{d[CO_2]}{dt} = K[NO_2]^2$?

3. For the reaction $3BrO^- \rightarrow BrO_3^- + 2Br^-$ in alkaline aqueous solution, the value of the second-order rate constant (in $BrO^-$) in rate law for $-\frac{d[BrO^-]}{dt}$ was found to be 0.06 $M^{-1}s^{-1}$. The rate constant (in $M^{-1}s^{-1}$), when the rate law is written as $+\frac{d[BrO_3^-]}{dt}$ is $a$ $M^{-1}s^{-1}$ and as $+\frac{d[Br^-]}{dt}$ is $b$ $M^{-1}s^{-1}$, then the value of $(a + b) \times 100$ is

4. In a certain polluted atmosphere containing $O_3$ at a steady-state concentration of $2 \times 10^{-8}$ M, the hourly production of $O_3$ by all sources was estimated as $7.2 \times 10^{-15}$ M. If only mechanism for the destruction of $O_3$ is second-order reaction, then the rate constant (in ml $mol^{-1}$ $s^{-1}$) for the destruction reaction is

5. For the reaction $2NO(g) + H_2(g) \rightarrow N_2O(g) + H_2O(g)$, the value of $-\frac{dP}{dt}$ was found to be 1.5 Pa $s^{-1}$ for a pressure of 372 Pa for NO and 0.25 Pa $s^{-1}$ for a pressure of 152 Pa for NO, the pressure of $H_2$ being constant. If pressure of NO was kept constant, the value of $-\frac{dP}{dt}$ was found to be 1.60 Pa $s^{-1}$ for a pressure of 289 Pa for $H_2$ and 0.79 Pa $s^{-1}$ for a pressure of 144 Pa for $H_2$. If the order of reaction with respect to NO and $H_2$ are $a$ and $b$, respectively, then the value of $(a + b)$ is

6. For a chemical reaction A + B → Products, the order is one with respect to each A and B. The sum of $x$ and $y$ from the following data is as follows.

| Rate ($mol\ l^{-1}s^{-1}$) | [A] ($mol\ l^{-1}$) | [B] ($mol\ l^{-1}$) |
|---|---|---|
| 0.10 | 0.20 | 0.05 |
| 0.40 | $x$ | 0.05 |
| 0.80 | 0.40 | $y$ |

7. In the presence of excess thiocyanate ion ($SCN^-$) the following reaction is first-order in chromium (III) ion, $Cr^{3+}$; the rate constant is $9.0 \times 10^{-5}\ s^{-1}$.

$$Cr^{3+}(aq) + SCN^-(aq) \rightarrow Cr(SCN)^{2+}(aq)$$

If 80.0% reaction is required to obtain a noticeable colour from the formation of the $Cr(SCN)^{2+}$ ion, then how many hours are required? (ln 5 = 1.62)

8. The reaction of thioacetamide with water is shown by the equation below.

$$CH_3C(S)NH_2(aq) + H_2O \rightarrow H_2S(aq) + CH_3C(O)NH_2(aq)$$

The rate of reaction is given by the rate law Rate = $k[H_3O^+][CH_3C(S)NH_2]$. Consider 1 L of solution that is 0.20 M in $CH_3C(S)NH_2$ and 0.15 M in HCl at 25°C. For each of the changes listed below, the number of changes that decreases the rate of reaction is

(i) a 4.0 g sample of NaOH is added to the solution.

(ii) 500 ml of water is added to the solution.

(iii) the 0.15 M HCl solution is replaced by 0.15 M acetic acid solution.

(iv) the reaction is carried out at 35°C instead of 25°C.

9. If 0.01% of a substance undergoing decomposition is consumed in 1 ms when the concentration is 0.02 M and in 0.25 ms when the concentration is 0.04 M. The order of reaction is

10. Ethylene is produced as $C_4H_8 \rightarrow 2C_2H_4$. The rate constant is $\frac{25}{18} \times 10^{-5}\ s^{-1}$. In what time (in hours) will the molar ratio of the ethylene to cyclobutane in the reaction mixture attain the value 1.0? (ln 2 = 0.7, ln 3 = 1.10)

11. A viral preparation was inactivated in a chemical bath. The inactivation process was found to be first order in virus concentration and at the beginning of the experiment, 2.0% of the virus was found to be inactivated per minute. If the rate constant for the inactivation process is $K\ min^{-1}$, then the value of 400 $K$ is

12. Hydrogen peroxide in aqueous solution decomposes by a first-order reaction to water and oxygen. The rate constant for this decomposition is $7.50 \times 10^{-4}\ s^{-1}$. What quantity of heat (in J) is initially liberated per second from 2.0 L of solution that is 0.02 M of $H_2O_2$? Given $\Delta_f H$ of $H_2O_2(l)$ = –187.0 kJ/mol and $\Delta_f H$ $H_2O(l)$ = –287.0 kJ/mol.

13. An optically active drug has one chiral centre and only dextrorotatory isomer is effective. Moreover, it becomes ineffective when its optical activity is reduced to 40% of the original. It was found that mutarotation of this drug was first-order reaction with rate constant $\frac{4.5}{3.1536} \times 10^{-8}\ s^{-1}$. The expiration time of the drug (in years) is (ln 2 = 0.7, ln 3 = 1.1, ln 5 = 1.6)

14. For the reaction A(g) → $n$B(g), the rate constant is $6.93 \times 10^{-4}\ s^{-1}$. The reaction is performed at constant volume and temperature, starting with pure A(g). If after 1000 s from the start of reaction, the pressure of system becomes 3 times the initial pressure, the value of $n$ is

15. The acidic hydrolysis of ethyl acetate in aqueous solution is first order with respect to ethyl acetate. Upon varying the pH of solution, the first-order rate constant varies as follows.

| pH | 3 | 2 | 1 |
|---|---|---|---|
| $K_1$ ($\times 10^{-3}\ s^{-1}$) | 1.0 | 10 | 100 |

If the order of reaction with respect to $H^+$ is $x$ and the value of the rate constant (in $M^{-1}s^{-1}$) is $y$, then the value of ($x + y$) is

16. The complex $[Co(NH_3)_5F]^{2+}$ reacts with water as $[Co(NH_3)_5F]^{2+} + H_2O \rightarrow [Co(NH_3)_5(H_2O)]^{3+} + F^-$. The rate of reaction may be given as $r = K[\text{complex}]^a[H^+]^b$. The reaction is acid catalysed and hence, $[H^+]$ does not change during the reaction. Thus, the rate may be given as $r = K'[\text{complex}]^a$ where $K' = K[H^+]^b$. The value of ($a + b$) is

| [Complex] | $[H^+]$ | $t_{1/2}$ (h) | $t_{3/4}$ (h) |
|---|---|---|---|
| 0.1 M | 0.01 M | 1.0 | 2.0 |
| 0.2 M | 0.02 M | 0.5 | 1.0 |

**17.** If for a reaction A → products, the concentration of A are $C_o$, $a \cdot C_o$, $a^2 \cdot C_o$, $a^3 \cdot C_o$, ....... after time interval 0, $t$, $2t$, $3t$ ........, where $a$ is a constant and $0 < a < 1$. The order of reaction is

**18.** Dissociation of salt $A_2B_3$(aq) in water follows first-order kinetics $A_2B_3(aq) \rightarrow 2A^{3+}(aq) + 3B^{2-}(aq)$. Starting from $t = 0$, a definite volume of solution is taken at different instant and its osmotic rise is measured. If the osmotic rise was 2 mm at $t = 0$ and 6 mm at $t = 10$ min, then the value of $\frac{x}{y}$ is, where $x$ = osmotic pressure of solution (in Pascal) at $t = 20$ minutes and $y = t_{3/4}$ of the dissociation reaction (in min) [Given: Density of final solution in each case = 1.0 g/ml, $g = 10$ m/s$^2$]

**19.** A substance A undergoes a reaction such that its concentration at any time $t$ (in seconds) can be represented by the equation $[A_t]^4 = \frac{1}{(t+1)}$. The rate of disappearance of 'A' (in $10^{-5}$ M s$^{-1}$) at the concentration, [A] = 0.2 M is

**20.** In order to determine the order of reaction A(g) → 2B(g), the vapour density of the system is determined at different stages of reaction at constant temperature. The reaction is started with pure A(g). From the following data, the order of reaction is as follows.

| **Time (min)** | 0 | 10 | 20 |
|---|---|---|---|
| **Vapour density** | 42 | 35 | 30 |

**21.** In the reaction A → Products, the rate is doubled when the concentration of 'A' is quadrupled. If 50% of the reaction occurs in $8\sqrt{2}$ h, then how long (in hours) would it take for the completion of next 50% reaction?

**22.** A gaseous compound 'A' decomposes simultaneously into gaseous products 'B', 'C' and 'D' with the rate constants $2 \times 10^{-3}$ s$^{-1}$, $3 \times 10^{-3}$ s$^{-1}$ and $1.93 \times 10^{-3}$ s$^{-1}$, respectively. If only A(g) was present initially at a pressure of 13.86 atm and the reaction is performed at constant volume and temperature, the partial pressure (in atm) of B(g) in the reaction mixture after 100 s from the start of reaction is

**23.** Consider two first-order reactions I and II. The frequency factor of I is 100 times that of II, the activation energy of I is 4.606 kcal higher than that of II. If the ratio of rate constants for the reactions I and II is $x : 1$ at 227°C, then the value of $x$ is

**24.** If the activation energy of a reaction is 83.14 kJ/mole, the fraction of molecules at 500 K which have enough energy to form products is $x \times 10^{y}$. The value of $(y - x)$ is ($\ln 2 \times 10^{-9} = -20$, $\ln 4 \times 10^{-5} = -10$, $\ln 4 \times 10^{-18} = -40$)

**25.** A given sample of milk turns sour at room temperature (27°C) in 16 h. In a refrigerator at 7°C, milk can be stored three times as long before it sours. How long (in hours) it takes for milk to sour at 57°C? (ln 2 = 0.7, ln 3 = 1.1)

## Four Digit Integer Type

**1.** The rate constant for the decomposition of ozone following second-order kinetics is $5 \times 10^{-4}$ L mol$^{-1}$ s$^{-1}$ at a certain temperature. The rate of reaction in terms of number of molecules decomposed per ml per minute, when concentration of ozone is $2 \times 10^{-8}$ M, is ($N_A = 6 \times 10^{23}$)

**2.** In thermal decomposition of $C_2H_5Br$(g) to $C_2H_4$(g) and HBr(g), the initial pressure changes from 200 to 390 mm of Hg at the end of reaction. Percentage of $C_2H_5Br$(g) undecomposed, when the pressure of mixture was 342.5 mm of Hg, is

**3.** The generation time of any substance is the time in which its concentration doubles. The growth of a certain bacteria follows first-order kinetics. If the population of bacteria triples in 96 h, then what is the generation time (in h) for it? (log 2 = 0.3, log 3 = 0.48)

**4.** Nitric oxide reacts with oxygen to give nitrogen dioxide.

$$2NO(g) + O_2(g) \rightarrow 2NO_2(g)$$

The rate law is $-\Delta[NO]/\Delta t = k[NO]^2[O_2]$, where the rate constant is $1.6 \times 10^5$ M$^{-2}$ s$^{-1}$ at 327°C. A vessel contains NO and $O_2$ at 327°C. The initial partial pressures of NO and $O_2$ are 190 and 288 mm Hg, respectively. The initial rate of decrease of partial pressure of NO (in mm Hg per second) is ($R = 0.08$ L-atm/K-mol)

5. A drop (0.05 ml) of solution contains $3 \times 10^{-7}$ moles of $H^+$. If the rate constant of disappearance of $H^+$ is $1.0 \times 10^{-7}$ mol $l^{-1}s^{-1}$, then how long (in min) would it take for $H^+$ in drop to disappear?

6. A substance is reduced to one third of its original concentration in 100 minutes by first-order kinetics. The time (in min) in which it will be reduced to one ninth of its original value is

7. A drug is known to be ineffective after it has decomposed 30%. The original concentration of a sample was 500 units/ml. When analysed 20 months later, the concentration was found to be 420 units/ml. Assuming that the decomposition is first order, what will be the expiration time (in month) of the drug? (ln 2 = 0.7, ln 5 = 1.6, ln 7 = 1.9, ln 3 = 1.1)

8. The decomposition of benzene diazonium chloride in aqueous solution is a reaction of first order which proceeds as $C_6H_5N_2Cl \rightarrow C_6H_5Cl + N_2(g)$. A certain solution of benzene diazonium chloride contains initially an amount of this compound which gives 80 $cm^3$ of nitrogen on complete decomposition. It is found that at 30°C, 40 $cm^3$ of nitrogen are evolved in 40 minutes. How long (in min) after the start of the decomposition, will 70 $cm^3$ of nitrogen have been evolved?

9. The decomposition of a compound 'P' at temperature $T$, according to the following equation

$$2P(g) \rightarrow 4Q(g) + R(g) + S(l)$$

is a first-order reaction. After 30 minutes, from the start of the decomposition in a closed vessel, the total pressure developed is found to be 445 mm and after a long time, the total pressure is 625 mm. The vapour pressure of S($l$) at this temperature is 25 mm. The total pressure after 60 minutes is

10. The half-life for the first-order decomposition of nitramide

$$NH_4NO_2(aq) \rightarrow N_2O(g) + H_2O(l)$$

is 123 minutes at 27°C. If 200 ml of 0.02 M $NH_4NO_2$ solution is allowed to decompose, then how long (in min) must the reaction proceed to produce 49.26 ml of 'wet' $N_2O(g)$ measured at 27°C and 785 mm Hg? The vapour pressure of water at 27°C is 25 mm Hg.

11. The reaction 2A + B → C + D goes to completion and follows the rate law $-\frac{d[A]}{dt} = K[A]^2[B]$. The sum of values of $x$ and $y$ in the following data are

| Set | $[A_0] \times 10^6$ M | $[B_0] \times 10^6$ M | Half-life (s) |
|---|---|---|---|
| 1 | 300 | 4 | 62.5 |
| 2 | 300 | 6 | $x$ |
| 3 | 5 | 300 | 625 |
| 4 | 10 | 300 | $y$ |

12. The gas phase decomposition of dimethyl ether follows first-order kinetics.

$$CH_3–O–CH_3 \rightarrow CH_4(g) + H_2(g) + CO(g)$$

The reaction is carried out in a constant volume container at 500°C and has a half-life of 14.5 minutes. Initially, only dimethyl ether is present at a pressure of 4.0 atmosphere. What is the total pressure of the system (in atm) after 43.5 minutes? Assume ideal gas behaviour.

13. The rate of a first-order reaction is 0.04 mol $litre^{-1}s^{-1}$ at 10 minutes and 0.03 mol $litre^{-1}s^{-1}$ at 22 minutes after initiation. The half-life of the reaction (in seconds) is (ln 2 = 0.7, ln 3 = 1.1)

14. An acid-catalysed hydrolysis of ester is carried out in the presence of HCl. Small but same volume of reaction mixture were taken out at various instant and were titrated with standard NaOH solution. The following data was obtained.

| Time (in min) | 0 | 120 | ∞ |
|---|---|---|---|
| Volume (ml) of NaOH solution needed for titration | 20 | 55 | 60 |

If the initial concentration of ester was 6.0 M, then calculate the value of ***abcd***, where

***ab*** = half-life for the hydrolysis of ester (in minutes)

***cd*** = molar concentration of HCl used in the experiment.

15. For the reaction A(g) → 2B(g) + C(g), the rate constant is $1.386 \times 10^{-3}$ $s^{-1}$. The reaction is performed at constant pressure and temperature starting with only 'A'. The initial concentration of 'A' was 0.1 M. The concentration of 'A' (in millimole per litre) after 500 s from the start of reaction is

**16.** An optically active substance decomposes into optically active substances 'B' and 'C' as follows.

$$A \xrightarrow{K = 0.001\ \text{min}^{-1}} 2B + C$$

The specific rotations of A, B and C are +40°, +10° and −30° per mole, respectively. If initially A and C were present in 4 : 3 mole ratio, the time (in min), after which the sample becomes optically inactive is (ln 2 = 0.7, ln 5 = 1.6, ln 7 = 2.0, ln 13 = 2.5)

**17.** For a reaction $2A + B \rightarrow C + D$, the following data is collected by experiments.

| Set | $[A_0] \times 10^{-2}$ M | $[B_0] \times 10^{-2}$ M | $T_{1/2}$ (min) | $T_{7/8}$ (min) |
|---|---|---|---|---|
| 1 | 500 | 2 | 30 | 90 |
| 2 | 3 | 400 | 10 | 70 |
| 3 | 6 | 800 | a | b |
| 4 | 250 | 1 | c | d |

The value of $(a + b + c + d)$ is

**18.** The gas phase decomposition of acetic acid at 1189 K proceeds by way of two parallel reactions.

$$CH_3COOH \longrightarrow CH_4 + CO_2,\quad K_1 = 3.2\ s^{-1}$$

$$CH_3COOH \longrightarrow H_2C = C = O + H_2O,\ K_2 = 4.8\ s^{-1}$$

The maximum percentage yield of the ketene ($CH_2CO$) obtainable at this temperature is

**19.** For overall reaction $A + 2B + 3C \rightleftharpoons D$, the reaction rate is given as follows.

$$r\ (\text{M/min}) = 2 \times 10^{-6}\,[A]^2 - \frac{1.4\times10^{-6}\,[D]^2}{[B]\,[C]}$$

If initially each of the reactant has concentrate 1.0 M, calculate the rate of reaction when the concentration of C becomes 0.7 M assuming no D is present at the start of the reaction. If the rate is '$r$' M/min, then the value of $(r \times 10^9)$ is

**20.** A mixture of two substances A and B gives the same product C by two parallel first-order reactions.

$$A \xrightarrow{T_A} C \text{ and } B \xrightarrow{T_B} C$$

If the initial concentrations of A and B are 1.0 and 0.8 M, respectively and their half-lives, $T_A$ and $T_B$ are 10 minutes and 15 minutes, respectively, the concentration of C (in millimole per litre) after 30 minutes is?

**21.** For the reversible reaction $A(aq) \rightleftharpoons B(aq)$, which is first-order in both directions, $K_f = \frac{1.38}{300}\ \text{min}^{-1}$. The variation in concentrations with time is shown in the figure.

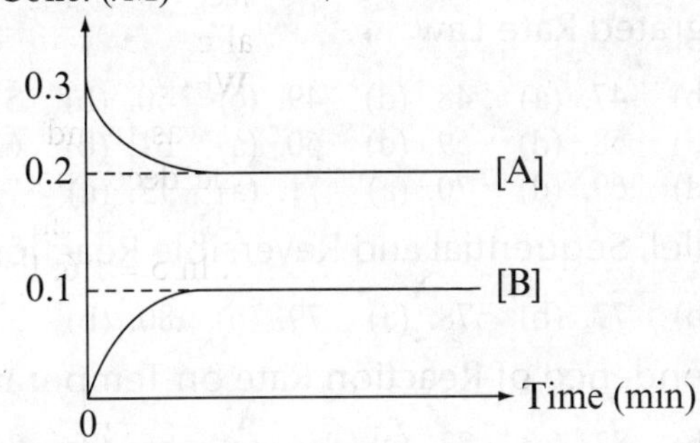


The time (in min) in which 25% of 'A' would be converted into 'B', is (ln 2 = 0.69)

**22.** At 25°C, a reaction is complete in 4 h. Approximately what temperature (in °C) must be used to cause the reaction to be completed in 30 minutes? The temperature coefficient of reaction is 2.

**23.** The rate of decomposition for $CH_3NO_2$ and $C_2H_5NO_2$ can be given in terms of rate constant (in $s^{-1}$) $K_1$ and $K_2$, respectively. The energy of activation for these reactions is 152.30 and 171.39 kJ/mol and the frequency factors are $10^{13}$ and $10^{14}$ $s^{-1}$, respectively. The temperature (in °C) at which rate constant will be same for both decomposition reactions, is ($R$ = 8.3 J/K-mol, ln 10 = 2.3)

**24.** The activation energy of a reaction is 149.4 kJ/mol and its pre-exponential factor is $5 \times 10^{13}\ s^{-1}$. At what temperature (in K) will the reaction have half-life of 1 minute? (ln 2 = 0.7, ln 3 = 1.1, ln 7 = 2.0, ln $10^{-16}$ = −36.9, $R$ = 8.3 J/K-mol)

**25.** A hydrogenation reaction is carried out at 500 K. If the same reaction is carried out in the presence of a catalyst at the same rate, the temperature required is 400 K. The activation energy of reaction (in kJ/mol), if the catalyst lowers the activation energy by 20 kJ/mol, is

## Answer Keys

### Exercise I

### Rate of Reaction, Rate Law, Order and Molecularity of a Reaction

1. (a) 2. (c) 3. (b) 4. (d) 5. (c) 6. (c) 7. (c) 8. (b) 9. (b) 10. (b) 11. (a)
12. (b) 13. (b) 14. (d) 15. (a) 16. (b) 17. (d) 18. (b) 19. (b) 20. (b)
21. (c) 22. (c) 23. (d) 24. (a) 25. (b) 26. (c) 27. (d) 28. (b) 29. (d) 30. (d) 31. (a)
32. (d) 33. (c) 34. (b) 35. (a) 36. (b) 37. (b) 38. (b) 39. (b) 40. (c)

### Mechanism of Reaction

41. (b) 42. (a) 43. (c) 44. (a) 45. (d)

### Integrated Rate Law

46. (b) 47. (a) 48. (d) 49. (c) 50. (b) 51. (d) 52. (c) 53. (b) 54. (a) 55. (c) 56. (d)
57. (d) 58. (d) 59. (d) 60. (a) 61. (b) 62. (b) 63. (c) 64. (a) 65. (a) 66. (a) 67. (a)
68. (d) 69. (d) 70. (a) 71. (a) 72. (a) 73. (c) 74. (c) 75. (b)

### Parallel, Sequential and Reversible Reactions

76. (b) 77. (b) 78. (a) 79. (b) 80. (b)

### Dependence of Reaction Rate on Temperature and Catalyst

81. (b) 82. (a) 83. (c) 84. (c) 85. (d) 86. (c) 87. (c) 88. (c) 89. (c) 90. (c) 91. (c)
92. (c) 93. (c) 94. (a) 95. (b) 96. (c) 97. (a) 98. (d) 99. (b) 100. (b)

---

## Answer Keys

### Exercise II

### Section A (Only one Correct)

1. (b) 2. (b) 3. (d) 4. (d) 5. (c) 6. (c) 7. (b) 8. (a) 9. (d) 10. (d)
11. (b) 12. (a) 13. (c) 14. (d) 15. (c) 16. (b) 17. (b) 18. (c) 19. (c) 20. (c)
21. (a) 22. (b) 23. (d) 24. (d) 25. (c) 26. (c) 27. (a) 28. (a) 29. (a) 30. (c)
31. (a) 32. (a) 33. (b) 34. (d) 35. (d) 36. (a) 37. (a) 38. (b) 39. (b) 40. (a)
41. (c) 42. (a) 43. (b) 44. (a) 45. (d) 46. (a) 47. (c) 48. (c) 49. (c) 50. (b)
51. (a) 52. (b) 53. (a) 54. (a) 55. (a) 56. (a) 57. (b) 58. (a) 59. (d) 60. (b)
61. (a) 62. (c) 63. (b) 64. (a) 65. (a) 66. (b) 67. (a) 68. (c) 69. (b) 70. (c)
71. (b) 72. (c) 73. (a) 74. (b) 75. (d)

### Section B (One or More than one Correct)

1. (a), (b), (c), (d) 2. (c) 3. (b), (c) 4. (a), (c), (d)
5. (a), (d) 6. (a), (c) 7. (b), (c), (d) 8. (b)
9. (d) 10. (a), (d) 11. (c), (d) 12. (a), (b)
13. (a), (c) 14. (b), (d) 15. (a),(b), (d) 16. (a), (c), (d)
17. (b), (d) 18. (a), (d) 19. (a), (b), (c) 20. (a), (b), (c), (d)
21. (a), (b), (c), (d) 22. (a), (b), (c), (d) 23. (a), (c), (d) 24. (b), (d)
25. (a), (c) 26. (b) 27. (c) 28. (c), (d)
29. (a), (b), (c) 30. (a), (b), (c)

### Section C

**Comprehension I**

1. (a) 2. (b) 3. (a)

**Comprehension II**

4. (c) 5. (d) 6. (a) 7. (b)

**Comprehension III**

8. (d) 9. (c) 10. (a) 11. (a)

**Comprehension IV**

12. (a) 13. (c) 14. (d) 15. (a)

**Comprehension V**

16. (d) 17. (b) 18. (a)

**Comprehension VI**

19. (d) 20. (a) 21. (c) 22. (b)

**Comprehension VII**

23. (b) 24. (d) 25. (b) 26. (c)

**Comprehension VIII**

27. (c) 28. (c) 29. (a)

**Comprehension IX**

30. (b) 31. (c) 32. (b)

**Comprehension X**

33. (b) 34. (a)

**Comprehension XI**

35. (c) 36. (c)

**Comprehension XII**

37. (b) 38. (b) 39. (b) 40. (b)
41. (d) 42. (b) 43. (b) 44. (d)

**Comprehension XIII**

45. (a) 46. (b) 47. (c) 48. (a)

**Comprehension XIV**

49. (b) 50. (c) 51. (d)

**Comprehension XV**

52. (c) 53. (b) 54. (a) 55. (b) 56. (a)

## Section D (Assertion – Reason)

1. (c) 2. (c) 3. (d) 4. (a) 5. (c) 6. (a) 7. (c) 8. (d) 9. (a) 10. (a)

## Section E (Column Match)

1. A → P, Q; B → R, S; C → P, Q; D → R, S
2. A → Q; B → P; C → P; D → R, S
3. A → Q; B → P, R; C → P; D → R, S
4. A → R; B → P; C → Q, S; D → Q, S
5. A → P; B → S; C → R; D → Q
6. A → R, U; B → P, X; C → S, V; D → Q, W
7. A → Q; B → P, R; C → S; D → T
8. A → P; B → Q, S; C → R, T; D → U
9. A → S; B → P; C → R, T; D → Q
10. A → P, Q; B → P, R; C → S

## Section F (Subjective)

**Single-digit Integer Type**

1. (3) 2. (2) 3. (6) 4. (5) 5. (3) 6. (1) 7. (5) 8. (3) 9. (3) 10. (2)
11. (8) 12. (3) 13. (2) 14. (5) 15. (2) 16. (2) 17. (1) 18. (4) 19. (8) 20. (0)
21. (8) 22. (2) 23. (1) 24. (7) 25. (4)

**Four-digit Integer Type**

1. (7200) 2. (0025) 3. (0060) 4. (1250) 5. (1000)
6. (0200) 7. (0040) 8. (0120) 9. (0535) 10. (0123)
11. (0375) 12. (0011) 13. (1680) 14. (4003) 15. (0025)
16. (0500) 17. (0260) 18. (0060) 19. (1595) 20. (1475)
21. (0100) 22. (0055) 23. (0727) 24. (0500) 25. (0100)

# HINTS AND EXPLANATIONS

## EXERCISE I (JEE MAIN)

### Rate of Reaction, Rate Law, Order and Molecularity of a Reaction

1. $r_{av} = -\frac{\Delta C}{\Delta t} = \frac{-(1.29\times10^{-2} - 1.50\times10^{-2})}{10}$

$= 2.1\times10^{-4}$ M min$^{-1}$

$= \frac{2.1\times10^{-4}}{60} = 3.5\times10^{-6}$ Ms$^{-1}$

2. $r_{inst} = -\frac{1}{2}\cdot\frac{d[HI]}{dt}$

3. $r = \frac{r_A}{4} = \frac{r_B}{1} = \frac{r_C}{2} = \frac{r_D}{2}$

4. $r_{av,NO_2} = \frac{2.4\times10^{-2}}{6} = 4\times10^{-3}$ Ms$^{-1}$

And, $\frac{r_{N_2O_5}}{2} = \frac{r_{NO_2}}{4}$

$\Rightarrow r_{N_2O_2} = 2\times10^{-3}$ Ms$^{-1}$

5. $r_{av}$ and $r_{inst}$ may to same.

6. $r_{rxn} = -\frac{1}{2}\cdot\frac{d[NH_3]}{dt} = +\frac{d[N_2]}{dt} = +\frac{1}{3}\cdot\frac{d[H_2]}{dt}$

or, $\frac{1}{2}\cdot K_1[NH_3] = K_2[NH_3] = \frac{1}{3}\cdot K_3[NH_3]$

$\therefore 3K_1 = 6K_2 = 2K_3$

7. Rate constant depends on temperature but it is independent from concentration or pressure.

8. Normally, the rate of reaction decreases with increase in number of bonds involved.

9. Check options

10. $K = 10^{-2}$ L mol$^{-1}$ s$^{-1}$ = $10^{-2}\times(1000$ ml$)$

$\times \frac{1}{6\times10^{23}\text{molecule}} \times \frac{60}{\text{min}}$

$= 1.0\times10^{-21}$ ml molecule$^{-1}$ min$^{-1}$

11. A + 2B $\longrightarrow$ P

| | A | 2B |
|---|---|---|
| $t_1$ | 1.0 M | 1.0 M |
| $t_2$ | 1.0 – 0.25 = 0.75 M | 1.0 – 0.5 = 0.5 M |

Now, $\frac{r_2}{r_1} = \frac{K[A_2]^2[B_2]}{K[A_1]^2[B_1]} \Rightarrow \frac{r_2}{2\times10^{-2}} = \frac{(0.75)^2\times0.5}{(1.0)^2\times1.0}$

$\therefore r_2 = 5.625\times10^{-3}$ M/min

12. $A_2 \longrightarrow B + \frac{1}{2}C$

| | $A_2$ | B | $\frac{1}{2}$C |
|---|---|---|---|
| $t = 0$ | 100 mm | 0 | 0 |
| $t = 5$ min | $(100 - x)$ mm | $x$ mm | $\frac{x}{2}$ mm |

Now, $(100 - x) + x + \frac{x}{2} = 120 \Rightarrow x = 40$

$\therefore r_{av,A_2} = -\frac{\Delta P}{\Delta t} = \frac{x}{t} = \frac{40}{5} = 8$ mm/ min

13. Theoretical

14. $r = K[N_2O_5]$

$\Rightarrow [N_2O_5] = \frac{r}{K} = \frac{2.4\times10^{-5}\text{M s}^{-1}}{3.0\times10^{-5}\text{s}^{-1}} = 0.8$ M

15. $r = -\frac{d[N_2]}{dt} = -\frac{1}{3}\cdot\frac{d[H_2]}{dt} = +\frac{1}{2}\cdot\frac{d[NH_3]}{dt}$

16. $\frac{r_{H_2}}{3} = \frac{r_{NH_3}}{2}$

$\Rightarrow r_{H_2} = \frac{3}{2}\times\frac{10^{-3}}{17}\times10^3$ mol hr$^{-1}$

$= \frac{3}{34}\times\frac{2}{10^3}$ Kg hr$^{-1}$ $= 1.76\times10^{-4}$ Kg hr$^{-1}$

17. Informative

18. Molecularity can even be fractional.

19. Rate law and law of mass action may be identical.

HINTS AND EXPLANATIONS

**20.** $r_1 = K[A], r_2 = K[A]^2, r_3 = K[A]^3$ and $[A] < 1$ M

$\therefore r_1 > r_2 > r_3$

**21.** $r = K[NO]^2[O_2]$

On tripling the volume, the concentration becomes $\frac{1}{3}$ and hence, rate becomes $\frac{1}{27}$ times.

**22.** Theoretical

**23.** Molecularity can never be zero.

**24.** $r = K \cdot P^0$

**25.** $r = K \cdot [A]^n$

$$\therefore \frac{r_2}{r_1} = \left(\frac{[A_2]}{[A_1]}\right)^n$$

$$\Rightarrow 2 = (8)^n \Rightarrow n = \frac{1}{3}$$

**26.** From the unit of rate constant, order = 2.

**27.** Order is independent from stoichiometry.

**28.** $r = K \cdot [A]^n$

$$\therefore \frac{r_2}{r_1} = \left(\frac{[A_2]}{[A_1]}\right)^n \Rightarrow 1.837 = (1.5)^n \Rightarrow n = 1.5$$

**29.** Order is not related with ΔH.

**30.** Order may be fractional.

**31.** $r = K[A]^x[B]^y$

$$\text{Exp (1)} \div \text{Exp (3)}: \frac{0.1}{0.1} = \frac{K(0.012)^x(0.035)^y}{K(0.024)^x(0.035)^y}$$

$$\Rightarrow x = 0$$

$$\text{Exp (2)} \div \text{Exp (3)}: \frac{0.8}{0.1} = \frac{K(0.024)^x(0.070)^y}{K(0.024)^x(0.035)y}$$

$$\Rightarrow y = 3$$

**32.** $$\frac{r_2}{r_1} = \frac{K(2[A])^n \cdot ([B]/2)^m}{K[A]^n \cdot [B]^m} = 2^{n-m}$$

**33.** Informative

**34.** $r = K[A]^x[B]^y[C]^z$

$$\text{Exp (1)} \div \text{Exp (2)}: \frac{2.5\times10^{-4}}{1.0\times10^{-3}} = \frac{K(0.01)^x(0.01)^y(0.01)^z}{K(0.02)^x(0.01)^y(0.01)^z}$$

$$\Rightarrow x = 2$$

$$\text{Exp (1)} \div \text{Exp (3)}: \frac{2.5\times10^{-4}}{2.5\times10^{-4}} = \frac{K(0.01)^x(0.01)^y(0.01)^z}{K(0.01)^x(0.02)^y(0.01)^z}$$

$$\Rightarrow y = 0$$

$$\text{Exp (3)} \div \text{Exp (4)}: \frac{2.5\times10^{-4}}{5.0\times10^{-4}} = \frac{K(0.01)^x(0.02)^y(0.01)^z}{K(0.01)^x(0.02)^y(0.02)^z}$$

$$\Rightarrow z = 1$$

**35.** $r = r_{C_8H_{12}} = K \cdot [C_4H_6]^2$

$= (6.1\times10^{-2}\ M^{-1}s^{-1})\times(0.02\ M)^2$

$= 2.44 \times 10^{-5}\ M\ s^{-1}$

**36.** $K_1(Ms^{-1}) = K_2(s^{-1}) = K_3(M^{-1}s^{-1})$

or, $K_1 \times 10^{-3}$ (mol ml$^{-1}$ s$^{-1}$) = $K_2$ (s$^{-1}$) = $K_3 \times 10^{-3}$ (mol ml$^{-1}$ s$^{-1}$)

**37.** $r = K[A][B]^2 = K'[B]^2$

**38.** $r = K[\text{ester}][H^+]$

**39.** Informative

**40.** $$K = \frac{K_{app}}{[H_2O][H^+]} = \frac{1.5\times10^{-5}s^{-1}}{\left(\frac{1000}{18}M\right)(0.1\,M)}$$

$$= 2.7\times10^{-6}\ M^{-2}s^{-1}$$

## Mechanism of Reaction

**41.** 1st step is RDS and hence, $r = K_1[X][Y]$

**42.** $A \xrightarrow{K_1} B$ is RDS

**43.** $r = K_2[A][C]$ and $K_1 = \frac{[C]}{[A]}$

$\therefore r = K_1K_2[A]^2$

**44.** $r = K_f[B][C]$ and $K = \frac{[B]}{[A]^2}$

$\therefore r = K \cdot K_f \cdot [A]^2[C]$

**45.** $r = K_3[CHCl_3][\dot{C}l][Cl]$ and $\frac{K_1}{K_2} = \frac{[\dot{C}l]^2}{[Cl_2]}$

$$\therefore r = \sqrt{\frac{K_1}{K_2}} \cdot K_3[CHCl_3][Cl_2]^{1/2}$$

## Integrated Rate Law

**46.** $t_{1/2} = \frac{[A_0]}{2K}$ and $t_{100\%} = \frac{[A_0]}{K}$

**47.** $K = \frac{[A_0]-[A]}{t}$ = Constant

or, $\frac{a-0.25a}{10} = \frac{a-0.10a}{t} \Rightarrow t = 12$ hours

**48.** Reaction will complete before 2 hours

**49.** when $t = t_{3/4}$, $[A] = [A_0] - \frac{3}{4}[A_0] = \frac{[A_0]}{4}$

Now, $t_{3/4} = \frac{1}{K}\cdot \ln \frac{[A_0]}{\left(\frac{[A_0]}{4}\right)} = \frac{\ln 4}{K} = \frac{2.303}{K}\cdot \log 4$

**50.** $Kt = \ln \frac{[H_2O_2]_0}{[H_2O_2]_t} = \ln \frac{V_0}{V_t}$

$\therefore\ t = \frac{1}{4.5\times 10^{-2}\ \text{min}}\cdot \ln \frac{25}{5} = \frac{\ln 5}{4.5\times 10^{-2}}$ min

**51.** $t_{99\%} = x \times t_{90\%}$

$\Rightarrow \frac{1}{K}\cdot \ln \frac{100}{1} = x\cdot \frac{1}{K}\cdot \ln \frac{100}{10}$

$\Rightarrow x = 2$

**52.** $r_0 = 6.0\times 10^{21}$ molecules $ml^{-1}\ s^{-1}$

$= 6\times 10^{21}\times 10^3\times 60$

$= 36\times 10^{25}$ molecules $L^{-1}\ min^{-1}$

Now, $t = \frac{t_{1/2}}{\log 2}\cdot \log \frac{r_0}{r} = \frac{10\,\text{min}}{\log 2}\cdot \log \frac{36\times 10^{25}}{4.5\times 10^{25}}$

$= 30$ min

**53.** For 2nd order reaction, $t_{1/2} = \frac{1}{K\cdot [A_0]}$

Hence, second $t_{1/2}$ is double of first $t_{1/2}$.

$[A_0] \xrightarrow{x\ \text{min}} \frac{[A_0]}{2} \xrightarrow{2x\ \text{min}} \frac{[A_0]}{4}$

**54.** $t_{1/2}\ \alpha\ [A_0]^{1-n}$

**55.**

| | $A$ | $\longrightarrow$ | $nB$ |
|---|---|---|---|
| $t=0$ | $[A_0]$ | | 0 |
| $t=t$ | $[A_0]-x$ | | $nx$ |

At intersection, $[A] = [B] \Rightarrow [A_0] - x = nx$

$\therefore\ [B] = nx = n.\frac{[A_0]}{n+1}$

**56.** At any instant, $[B] < [A]$ if $[B_0] = [A_0]$

**57.** For zero order: $[A] = [A_0] - K\cdot t$

**58.** $t_{1/2}\ \alpha\ [A_0]^{1-n}$

Given: $C_0\cdot \sqrt{T}$ = constant $\Rightarrow T$ = constant $\times C_0^{-2}$

As $1-n=-2 \Rightarrow n=3$

**59.** $1-n = \frac{1}{2} \Rightarrow n = \frac{1}{2}$

**60.** $K = \frac{1}{t}\cdot \ln \frac{[A_0]}{[A]} = \frac{1}{0.5\ \text{min}}\cdot \ln \frac{100}{99} = 0.02\ \text{min}^{-1}$

**61.** The first-order reaction never completes 100%.

**62.**

| | A | $\longrightarrow$ | 2B | + | C |
|---|---|---|---|---|---|
| $t=0$ | $P_0$ | | 0 | | 0 |
| $t=t$ | $P_0-x$ | | $2x$ | | $x$ |
| $t=\infty$ | $\approx 0$ | | $2P_0$ | | $P_0$ |

From question: $P_\infty = 2P_0 + P_0$

$\Rightarrow P_0 = \frac{P_\infty}{3}$

and $P = (P_0 - x) + 2x + x$

$\Rightarrow x = \frac{P-P_0}{2} = \frac{3P-P_\infty}{6}$

Now, $K = \frac{1}{t}\cdot \ln \frac{P_A{}^\circ}{P_A} = \frac{1}{t}\cdot \ln \frac{P_0}{P_0-x}$

$= \frac{1}{t}\cdot \ln \frac{P_\infty/3}{\frac{P_\infty}{3} - \frac{3P-P_\infty}{6}}$

$= \frac{1}{t}\cdot \ln \frac{2P_\infty}{3(P_\infty - P)}$

**63.** $t_{1/2}\alpha[A_0]^{1-n}$

$\Rightarrow \frac{29}{60} = \left(\frac{1.55}{0.75}\right)^{1-n}$

$\Rightarrow n \approx 2$

64. $t_{0.75} = 2 \times t_{0.50}$

65. $t_{1/2} \; \alpha \; [A_0]^{1-n} \quad \Rightarrow 1 - \frac{3}{2} = -\text{m} \quad \Rightarrow m = 0.5$

66. $100 \xrightarrow{1.0\,\text{hr}} 50 \xrightarrow{2.0\,\text{hr}} 25$ as $t_{1/2} = \frac{1}{K[A_0]}$

67. $K = \frac{1}{t} \cdot \ln \frac{[A_0]}{[A]} = \frac{1}{2 \times 10^4\,\text{s}} \cdot \ln \frac{800}{50} = 1.386 \times 10^{-4}\,\text{s}^{-1}$

68.

| | $2N_2O_5 \longrightarrow$ | $4NO_2$ + | $O_2$ |
|---|---|---|---|
| $t = 0$ | $P_0$ | 0 | 0 |
| $t = 30$ min | $P_0 - x$ | $2x$ | $x/2$ |
| $t = \infty$ | $\approx 0$ | $2P_0$ | $P_0/2$ |

From question: $2P_0 + \frac{P_0}{2} = 600$

$\Rightarrow P_0 = 240$ mm

and $(P_0 - x) + 2x + \frac{x}{2} = 300 \Rightarrow x = 40$ mm

Now, $K_{N_2O_5} = \frac{1}{t} \cdot \ln \frac{P_{N_2O_5}{}^{\circ}}{P_{N_2O_5}} = \frac{1}{30\text{ min}} \cdot \ln \frac{240}{240 - 40}$

$= 6 \times 10^{-3}\ \text{min}^{-1}$

$\therefore K_{rxn} = \frac{K_{N_2O_5}}{2} = 3 \times 10^{-3}\ \text{min}^{-1}$

69. $t_{99.9\%} = \frac{t_{1/2}}{\log 2} . \log \frac{100}{0.1} \quad \Rightarrow \frac{t_{99.9\%}}{t_{1/2}} \approx 10$

70. $(b - a) \simeq b$ and $(b - x) \simeq b$

$\therefore Kt = \frac{2.303}{a-b} \cdot \ln \frac{b(a-x)}{a(b-x)} \simeq \frac{2.303}{(-b)} \cdot \ln \frac{b \cdot (a-x)}{a \cdot b}$

or, $(b \cdot K) \cdot t = 2.303 \cdot \ln \frac{a}{a-x}$

$\therefore$ Reaction becomes first-order with respect to $A$.

71. $t_{0.75} = 2 \times t_{0.50} \quad \Rightarrow 1386\ \text{sec} = 2 \times \frac{0.693}{K}$

$\therefore K = 10^{-3}\ \text{s}^{-1}$

72. Next $t_{1/2}$ is double than initial and hence, reaction is second-order.

73. Formula for second-order reaction.

74. $t = \frac{t_{1/2}}{\log 2} \cdot \log \frac{V_\infty}{V_\infty - V_t}$

$\Rightarrow 60 = \frac{30}{\log 2} \cdot \log \frac{100}{100 - V_t}$

$\therefore V_t = 75$ min

75. $K = \frac{1}{t} \ln \frac{V_0}{V_t} = \frac{1}{10\text{ min}} \cdot \ln \frac{25}{20} - 0.02\ \text{min}^{-1}$

## Parallel, Sequential and Reversible Reactions

76. $r_A = \left(-\frac{d[A]}{dt}\right)_1 + \left(-\frac{d[A]}{dt}\right)_2 = K_1\ [A] + K_2\ [A] = (K_1 + K_2) \cdot [A]$

77. For $C_R$ max, $\frac{dC_R}{dt} = 0$

$\Rightarrow t_{\max} = \frac{\ln(K_2 / K_1)}{K_2 - K_1}$

78. $t_{\max} = \frac{\ln\left(\frac{K_2}{K_1}\right)}{K_2 - K_1} = \frac{\ln\left(\frac{2}{1}\right)}{0.02 - 0.01} = 70$ min

79. On derivation,

$[C] = [A_0]\left[1 - \frac{K_2 \cdot e^{-K_1 t} - K_1 \cdot e^{-K_2 t}}{K_2 - K_1}\right]$ (when $K_1 \neq K_2$)

If $K_1 << K_2$, [C] becomes $[A_0].(1 - e^{-K_1 t})$

80. $\frac{K^H}{K^D} = \frac{0.0848}{0.0212} = \frac{4}{1}$

## Dependence of Reaction Rate on Temperature and Catalyst

**81.** Theoretical

**82.** $K = A \cdot e^{-E_a/RT} = A \,\alpha \sqrt{T}$ (if $E_a = 0$)

**83.** Theoretical

**84.** Lower $E_a$, faster reaction.

**85.** The rate of every elementary reaction increases with increase in temperature.

**86.** $\Delta H = E_{af} - E_{ab}$

$\Rightarrow 72 = 77 - E_{ab} \Rightarrow E_{ab} = 5$ kJ/mol

**87.** $E_a \downarrow K \uparrow$

**88.** Theoretical

**89.** The ratio of catalytic reaction may depend on catalyst concentration.

**90.** $t_{1/2} \,\alpha\, \frac{1}{K} \quad \Rightarrow T\uparrow \quad K\uparrow \quad t_{1/2}\downarrow$

**91.** $K = A \cdot e^{-E_a/RT} = \log K = \log A - \frac{E_a}{2.303} \cdot \frac{1}{T}$

**92.** $\frac{r_2}{r_1} \approx (2)^5 = 32$

**93.** $E_a \uparrow \quad r \downarrow$

**94.** Theoretical

**95.** Reaction with higher $E_a$ are more sensitive towards temperature changes.

**96.** Catalyst lowers the activation energies of forward as well as backward reactions by the same extent.

**97.** Theoretical

**98.** $\Delta H = E_{af} - E_{ab} \Rightarrow y - x = E_{af} - z \Rightarrow E_{af} = y - x + z$

**99.** $K = A \cdot e^{-E_a/RT} = A$ (when $T \to \infty$)

**100.** The step of absorption of light depends only on the intensity of absorbed light and follows zero-order kinetics (independent from reactant concentration).

# EXERCISE II (JEE ADVANCED)

## Section A (Only one Correct)

**1.** Negative sign is for reactants and positive for products.

**2.** $r_{rxn} = -\frac{1}{2} \cdot \frac{d[NO_2]}{dt} = K_1[NO_2]^2 - K_2[N_2O_4]$

$\therefore$ Rate of disappearance of $NO_2$ is given by,

$$-\frac{d[NO_2]}{dt} = 2K_1[NO_2]^2 - 2K_2[N_2O_4]$$

**3.** $P_A = \frac{n_A RT}{V} \Rightarrow \frac{dP_A}{dt} = \frac{RT}{V} \cdot \frac{dn_A}{dt}$

or, $(-K_1 \cdot P_A^n) = RT(-K_2 \cdot C_A^n)$

$$\therefore K_2 = \frac{K_1}{RT} \cdot \left(\frac{P_A}{C_A}\right)^n = \frac{K_1}{RT} \cdot (RT)^n = (RT)^{n-1}$$

**4.** $-\frac{1}{2} \cdot \frac{d[HI]}{dt} = K_2[HI]^2 - K_1[H_2][I_2]$

**5.** Order = 1 $\longleftarrow$ with respect to $A \Rightarrow a = 1$

Order = 2 $\longleftarrow$ with respect to $B \Rightarrow b = 2$

Hence, reaction is A + 2B $\longrightarrow$ P.

$$r = -\frac{d[A]}{dt} = -\frac{1}{2}\frac{d[B]}{dt}$$

**6.**

| | A | + | 2B | $\longrightarrow$ C + D |
|---|---|---|---|---|
| $t = 0$ | 0.6 atm | | 0.8 atm | |
| $t = t$ | $0.6 - x$ | | $0.8 - 2x$ | |
| | = 0.3 atm | | = 0.2 atm | $\Rightarrow x = 0.3$ |

$$\therefore \frac{r_t}{r_0} = \frac{K \times 0.3 \times (0.2)^2}{K \times 0.6 \times (0.8)^2} = \frac{1}{32}$$

7. For no change in temperature, $\Delta H_{net} = 0$ and hence, for 3 moles of $B$ reacted, 4 moles of $Q$ should form.

8. $2.82 = (2)^x \Rightarrow x = \frac{3}{2}$

$9 = (3)^y \Rightarrow y = 2$

$\therefore$ overall order $= x + y = \frac{7}{2}$

9. $r_1 = 0.0068/65 = 1.046 \times 10^{-4}$ gm/min

$r_2 = 0.0031/120 = 2.583 \times 10^{-5}$ gm/min

$r_3 = 0.0032/60 = 5.333 \times 10^{-5}$ gm/min

From (1) and (2) : order with respect to $K_2C_2O_4 = 2$

From (1) and (3) : order with respect to $HgCl_2 = 1$

10. From (2) and (3) : order with respect to $I^- = 1$

From (1) and (3) : order with respect to $ClO^- = 1$

From (3) and (4) : order with respect to $OH^- = 1$

11. $(1 + K_2 \cdot C_A) \approx 1$

12. For steady state, $+\frac{d[R]}{dt} = 0$

or, $K_1[A] - K_2[R][B] - K_3[R][C] = 0$

$\therefore [R] = \frac{K_1[A]}{K_2[B] + K_3[C]}$

Now, $\frac{dx}{dt} = K_3[R][C] = \frac{K_3K_1[A][C]}{K_2[B] + K_3[C]}$

13. $r = K_3[O][O_3]$

For 1st step, $\frac{K_1}{K_2} = \frac{[O_2][O]}{[O_3]}$

$\therefore r = K_3[O][O_3] = \frac{K_3K_1[O_3]^2}{K_2[O_2]}$

14. $r = -\frac{d[X]}{dt} = +\frac{d[Y]}{dt}$ and rate decreases with time.

15.

| | A | $\longrightarrow$ | 2B |
|---|---|---|---|
| $t = 0$ | 0.1 M | | 0 |
| $t = 1$ min | $0.1 - x$ | | $2x$ |

For zero order reaction: $[A_0] - [A] = Kt$

or, $0.1 - (0.1 - x) = 0.01 \times 1 \Rightarrow x = 0.01$

$\therefore [B] = 2x = 0.02$ M

16. $2NH_3 \longrightarrow N_2 + 3H_2$

$r_{rxn} = \frac{r_{NH_3}}{2} = \frac{r_{N_2}}{1} = \frac{r_{H_2}}{3} = 0.1$ atm/s = Constant

Hence, after 10 seconds:

$P_{NH_3} = 3 - 2 \times 0.1 \times 10 = 1$ atm

$P_{N_2} = 0.1 \times 10 = 1$ atm

$P_{H_2} = 3 \times 0.1 \times 10 = 3$ atm

$\therefore P_{total} = 1 + 1 + 3 = 5$ atm

17. For $-\frac{d[A]}{dt} = K[A]^n$ and $n \neq 1$

$[A]^{1-n} = [A_0]^{1-n} - K(1-n) \cdot t$

For given graph, $1 - n = -3 \Rightarrow n = 4$

and $-K(1 - n) = \tan 45° \Rightarrow K(4 - 1) = 1$

$\therefore K = \frac{1}{3} M^{-3} min^{-1}$

Now, $r_{rxn} = -\frac{1}{3} \cdot \frac{d[A]}{dt} = \frac{1}{3} \cdot K \cdot [A]^4$

$= \frac{1}{3} \times \frac{1}{3} \times (0.2)^4 = \frac{16}{9} \times 10^{-4}$ M min$^{-1}$

18. $1M \xrightarrow[8.0 \text{ hr}]{t = 2 \times t_{1/2}} 0.25M \quad \Rightarrow t_{1/2} = 4.0$ hr

$0.6M \xrightarrow[4.0 \text{ hr}]{t = t_{1/2}} 0.3M$

19. The successive $t_{1/2}$ are double of previous one and hence, order = 2.

20. $r = K[A]^n$

$10 = K(0.8)^n$ (1)

$0.625 = K(0.2)^n$ (2)

$\therefore n = 2$

21.

| | $C_2H_6 \longrightarrow$ | $C_2H_4$ | + $H_2$ |
|---|---|---|---|
| $t = 0$ | 3 bar | 0 | 0 |
| $t = ?$ | $(3 - x)$ bar | $x$ bar | $x$ bar |

From question: $(3 - x) + x + x = 5 \quad \Rightarrow x = 2$

From the unit of rate constant, the order of reaction is 2, hence,

$t = \frac{1}{K}\left(\frac{1}{P_{C_2H_6}} - \frac{1}{P^\circ_{C_2H_6}}\right) = \frac{1}{0.0015}\left(\frac{1}{3 - x} - \frac{1}{3}\right) \times \frac{1}{10^5}$

$= 4.44 \times 10^{-3}$ hr = 16 seconds

22. Let the reaction be first-order.

$K_1 = \frac{1}{8} \cdot \ln\frac{100}{20} = 0.201$

$K_2 = \frac{1}{18} \cdot \ln\frac{100}{10} = 0.128$

As $K_1 \neq K_2$, the reaction is not first-order. Let the reaction be second-order.

$$K_1 = \frac{1}{8}\left(\frac{1}{0.2} - \frac{1}{1}\right) = 0.5$$

$$K_2 = \frac{1}{18}\left(\frac{1}{0.1} - \frac{1}{1}\right) = 0.5$$

As $K_1 = K_2$, order = 2

**23.** Here, $t_{1/2}$ is independent from sugar concentration and hence, the order with respect to sugar is 1.

Now, $r = K\,[\text{Sugar}][\text{H}^+]^n = K' \cdot [\text{Sugar}]$

$$t_{1/2} = \frac{\ln 2}{K'} = \frac{\ln 2}{K[\text{H}^+]^n}$$

$$500 = \frac{\ln 2}{K \cdot (10^{-5})^n} \text{ and } 50 = \frac{\ln 2}{K \cdot (10^{-6})^n}$$

$\therefore n = -1$

**24.** $r = K\,[\text{ester}]\,[H^+] = K' \cdot [\text{ester}]$

$$\therefore t_{1/2} = \frac{\ln 2}{K'} = \frac{\ln 2}{K[\text{H}^+]} = \frac{0.693}{0.1 \times 0.01} = 693 \text{ hr}$$

**25.** For $n$th order reaction ($n \neq 1$)

$$Kt = \frac{[A_0]^{1-n} - [A]^{1-n}}{1-n}$$

For $n = 0.5$, $Kt = \dfrac{[A_0]^{1/2} - [A]^{1/2}}{\frac{1}{2}}$

Now, $t_{100\%} = T = \dfrac{2([A_0]^{1/2} - 0^{1/2})}{K} = \dfrac{2[A_0]^{1/2}}{K}$

and $t_{50\%} = t_{1/2} = \dfrac{2\left([A_0]^{1/2} - \left(\dfrac{[A_0]}{2}\right)^{1/2}\right)}{K}$

$$= \frac{2[A_0]^{1/2}\left(1 - \frac{1}{\sqrt{2}}\right)}{K}$$

$$\therefore \frac{T}{t_{1/2}} = \frac{1}{1 - \frac{1}{\sqrt{2}}} = \frac{1}{0.3}$$

**26.** For $A$: $t = \dfrac{t_{1/2}}{\log 2} \cdot \log \dfrac{[A_0]}{[A]} = \dfrac{10}{\log 2} \cdot \log \dfrac{8[B_0]}{[A]}$

For $B$: $t = \dfrac{t_{1/2}}{\log 2} \cdot \log \dfrac{[B_0]}{[B]} = \dfrac{20}{\log 2} \cdot \log \dfrac{[B_0]}{[A]}$

From question, $\dfrac{10}{\log 2} \cdot \log \dfrac{8[B_0]}{[A]} = \dfrac{20}{\log 2} \cdot \log \dfrac{[B_0]}{[A]}$

$$\therefore \frac{[B_0]}{[A]} = 8 \Rightarrow t = \frac{20}{\log 2} \cdot \log \frac{[B_0]}{[A]} = 60 \text{ min}$$

Alternate method:

$$A: 8[B_0] \xrightarrow{10} 4[B_0] \xrightarrow{10} 2[B_0] \xrightarrow{10}$$

$$[B_0] \xrightarrow{10} \frac{[B_0]}{2} \xrightarrow{10} \frac{[B_0]}{4} \xrightarrow{10} \frac{[B_0]}{8}$$

$$B: [B_0] \xrightarrow{20} \frac{[B_0]}{2} \xrightarrow{20} \frac{[B_0]}{4} \xrightarrow{20} \frac{[B_0]}{8}$$

**27.** $0.1\text{ M} \xrightarrow[40\text{ min}]{t = 2t_{1/2}} 0.025\text{ M}$

$\Rightarrow t_{1/2} = 20$ min

Now, $r = K[A] = \dfrac{\ln 2}{t_{1/2}}[A] = \dfrac{0.693}{20 \text{ min}} \times 0.01 \text{ M}$

$= 3.465 \times 10^{-4} \text{ M min}^{-1}$

**28.** $+\dfrac{d[B]}{dt} = -\dfrac{d[A]}{dt} = K[A]^{1/3}$

or, $-\displaystyle\int_{[A_0]}^{[A_0]/2} \frac{d[A]}{[B]^{1/3}} = K \cdot \int_0^{t_{1/2}} dt$

$$t_{1/2} = \frac{3[A_0]^{2/3}(2^{2/3} - 1)}{2^{5/3} \cdot K}$$

**29.**

| | $N_2O_5 \longrightarrow$ | $2NO_2 + \frac{1}{2}O_2$ |
|---|---|---|
| $t = 0$ | $a$ mole | 0 |
| $t = t$ | $(a - x)$ mole | $\frac{x}{2}$ mole $\alpha\, V_t$ |
| $t = \infty$ | $\approx 0$ | $\frac{a}{2}$ mole $\alpha\, V_\infty$ |

$$K = \frac{1}{t} \cdot \ln \frac{[N_2O_5]_0}{[N_2O_5]} = \frac{1}{t} \cdot \ln \frac{a}{a - x} = \frac{1}{t} \cdot \ln \frac{V_\infty}{V_\infty - V_t}$$

Now, $\dfrac{1}{20} \cdot \ln \dfrac{9.6}{9.6 - 4.8} = \dfrac{1}{40} \cdot \ln \dfrac{9.6}{9.6 - V_t} \Rightarrow V_t = 7.2$ ml

**30.** For zero order reaction, $t_{1/2}\ \alpha\ P^{\circ}_{NH_3}$

$$\therefore \frac{315}{t_{1/2}} = \frac{70}{150} \Rightarrow t_{1/2} = 675 \text{ sec}$$

**31.**

| | A $\longrightarrow$ | $n$B |
|---|---|---|
| $t = 0$ | $P_0$ | 0 |
| $t = t$ | $P_0 - x$ | $n.x$ |

Now, $(P_0 - x) = P_0 \cdot e^{-Kt} \Rightarrow x = P_0(1 - e^{-Kt})$

Now, $P_{total} = (P_0 - x) + nx$
$= P_0 \cdot e^{-Kt} + n \cdot P_0 (1 - e^{-Kt})$
$= P_0[n + (1 - n) e^{-Kt}]$

**32.** $$t_{1/2} = \frac{(C_0)^{1-n} - \left(\frac{C_0}{2}\right)^{1-n}}{K(1-n)} = \frac{(C_0)^{1-n}[1-2^{n-1}]}{K(1-n)}$$

and $$t_{3/4} = \frac{(C_0)^{1-n} - \left(\frac{C_0}{4}\right)^{1-n}}{K(1-n)} = \frac{(C_0)^{1-n}[1-2^{2(n-1)}]}{K(1-n)}$$

$$\therefore \frac{t_{3/4}}{t_{1/2}} = \frac{1-2^{2(n-1)}}{1-2^{(n-1)}} = 1+2^{(n-1)}$$

**33.** $$\frac{r_2}{r_1} = \left(\frac{[A_2]}{[A_1]}\right)^n \Rightarrow 2 = (4)^n \quad \Rightarrow n = \frac{1}{2}$$

Now, $t_{1/2} \,\alpha\, [A_0]^{1-n} \quad \Rightarrow t_{1/2} \,\alpha\, [A_0]^{1/2}$

$$100 \xrightarrow{t=16 \text{ min}} 50 \xrightarrow{t=16/\sqrt{2} \text{ min}} 25$$

$\therefore$ Time for 75% reaction $= 16 + \frac{16}{\sqrt{2}} = 27.3$ min

**34.** $$Kt = \ln\frac{a}{a-x} = \ln\frac{1}{\left(1-\frac{x}{a}\right)}$$

or, $$2.5\times10^{-5}\times(100\times60) = \ln\frac{1}{\left(1-\frac{x}{a}\right)}$$

$$\Rightarrow \qquad \frac{x}{a} = 0.138$$

$\therefore$ Percentage decomposition $= \frac{x}{a}\times100 = 13.8\%$

**35.** $$(t_{1/2})_1 = (t_{1/2})_2 \Rightarrow \frac{0.693}{K_1} = \frac{1}{K_2[A_0]}$$

$$\therefore [A_0] = \frac{K_1}{0.693\,K_2} = \frac{6.93\times10^{-2}}{0.693\times0.2} = 0.5 \text{ M}$$

**36.** $$\frac{t_1}{t_2} = \frac{\frac{(t_{1/2})_1}{\log 2}\cdot\log\frac{100}{75}}{\frac{(t_{1/2})_2}{\log 2}\cdot\log\frac{100}{25}} = \frac{(t_{1/2})_1}{(t_{1/2})_2}\cdot\frac{\log\left(\frac{4}{3}\right)}{\log 4}$$

$$= \frac{3}{2}\times\frac{0.6-0.48}{0.6} = \frac{3}{10}$$

**37.** $r = K[A]^n$

$$\frac{\frac{1}{100}\times0.02}{60} = K(0.02)^n \qquad (1)$$

$$\frac{\frac{1}{100}\times0.04}{15} = K(0.04)^n \qquad (2)$$

$\therefore n = 3$

**38.** $$t = \frac{T_{gen}}{\ln 2}\cdot\ln\frac{[A]}{[A_0]} \Rightarrow 60 = \frac{75}{\ln 2}\cdot\ln\frac{[A]}{[A_0]}$$

$$\therefore \frac{[A]}{[A_0]} = e^{0.56}$$

**39.** $$\frac{(t_{1/2})_1}{(t_{1/2})_2} = \left(\frac{[A_0]_1}{[A_0]_2}\right)^{1-n} \Rightarrow \frac{37.82}{18.95} = \left(\frac{0.05}{0.10}\right)^{1-n}$$

$\Rightarrow n = 2$

Now, $$\frac{(t_{1/2})}{37.82} = \left(\frac{0.15}{0.05}\right)^{-1} \Rightarrow t_{1/2} = 12.6 \text{ hr}$$

**40.**

$$\begin{matrix}\text{CH (Br) COOH}\\ |\\ \text{CH (Br) COOH}\end{matrix} \rightarrow \begin{matrix}\text{CHCOOH}\\ \|\\ \text{C Br COOH}\end{matrix} + \text{H Br}$$

| | | | |
|---|---|---|---|
| $t = 0$ | $a$ mole | 0 | 0 |
| $t = t$ | $(a - x)$ mole | $x$ mole | $x$ mole |

Now, $$\frac{T_0\cdot N}{1000} = a\times2 \qquad (1)$$

and $$\frac{T_t\cdot N}{1000} = (a-x)\times2 + x\times2 + x\times1 \qquad (2)$$

$$\therefore \frac{a}{a-x} = \frac{T_0}{3T_0 - 2T_t}$$

**41.**

| | $B^{n+}$ $\longrightarrow$ | $B^{(n+4)+}$ |
|---|---|---|
| $t = 0$ | $a$ mole | 0 |
| $t = 10$ min | $(a - x)$ mol | $x$ mol |

Now, $$\frac{25\times N}{1000} = a\times2 \qquad (1)$$

and, $$\frac{32.5\times N}{1000} = (a-x)\times2 + x\times5 \qquad (2)$$

Now, $$K = \frac{1}{t}\cdot\ln\frac{a}{a-x} = \frac{1}{10 \text{ min}}\cdot\ln\frac{12.5}{12.5-2.5}$$

$= 0.02 \text{ min}^{-1}$

**42.** $$A \longrightarrow 2B + C$$

| | A | 2B | C |
|---|---|---|---|
| $t=0$ | $a$ mol | 0 | 0 |
| $t=10$ sec | $(a-x)$ mol | $2x$ mol | $x$ mol |

Now, $\frac{r_A}{r_B}=\frac{P_A}{P_B}\cdot\sqrt{\frac{M_B}{M_A}} \Rightarrow \frac{1}{2}=\frac{a-x}{2x}\cdot\sqrt{\frac{4}{16}}$

$\Rightarrow x=\frac{a}{3}$

Now, $K=\frac{1}{t}\cdot\ln\frac{a}{a-x}=\frac{1}{10\text{ sec}}\cdot\ln\frac{a}{a-\frac{2a}{3}}$

$=0.04\text{ sec}^{-1}$

**43.** $r=K[A]^2[B]=K'\cdot[A]^2$ as $[B_0]>>[A_0]$

$\therefore t_{1/2}=\frac{1}{K'\cdot[A_0]}=\frac{1}{K[B_0][A_0]}=\frac{1}{0.5\times0.002\times2.0}$

$=500$ min

**44.** $r=K[\text{ester}][H^+]$

$\therefore \frac{r_{HA}}{r_{HX}}=\frac{1}{100}=\frac{[H^+]_{HA}}{1.0} \Rightarrow [H^+]_{HA}=0.01$ M

Now, $Ka_{(HA)}=\frac{[H^+][A^-]}{[HA]}=\frac{0.01\times0.01}{(1-0.01)}\approx10^{-4}$

**45.** As $[A_0]=[B_0]$ and the stoichiometric coefficients of both A and B are 1, at any time $[A]=[B]$. Hence, $r=K[A]^{1/2}[B]^{1/2}=K[A]$.

Required time $=2\times t_{1/2}=2\times\frac{0.693}{2.31\times10^{-3}}=600$ sec

**46.** $-\frac{dC}{dt}=\frac{\alpha C}{1+\beta C} \Rightarrow -\int_{Co}^{Co/2}\left(\frac{1}{C}+\beta\right)\cdot dC=\alpha\cdot\int_0^{t_{1/2}}dt$

$\therefore t_{1/2}=\frac{1}{\alpha}\left(\ln2+\frac{\beta\cdot C_0}{2}\right)$

**47.** $r=K'[CH_3COOH][C_2H_5OH]=K'\cdot[CH_3COOH]^2$

$\therefore t_{1/2}=\frac{1}{K'\cdot[CH_3COOH]_0}$

$\Rightarrow 50=\frac{1}{(K\times10^{-3})\times0.2}$

$\therefore$ K = 100 $M^{-2}$ $min^{-1}$

**48.** $r=K[A]^x[B]^y$

**For case-I :** $r=K[A]^x[B]^y=K'\cdot[B]^y$ where $K'=K[A_0]^x$

In equal time interval, the concentrations of $B$ are in G.P. and hence, $y=1$ and

$K'=\frac{1}{10}\cdot\ln\frac{0.01}{0.008}=0.02\text{ min}^{-1}$

**For case-II:** $r=K[A]^x[B]^y=K''[A]^x$ where $K''=K[B_0]^y$

In equal time interval, the concentration of $A$ are in G.P. and hence, $x=1$ and

$K''=\frac{1}{10}\cdot\ln\frac{0.02}{0.018}=0.01\text{ min}^{-1}$

Now, $r=K[A][B]$

$\therefore K=\frac{K'}{[A_0]}$ or, $\frac{K''}{[B_0]}=\frac{0.02}{2.0}$ or $\frac{0.01}{1.0}$

$=0.01\text{ M}^{-1}\text{ min}^{-1}$

**49.** $K_{app}=\frac{K_1\cdot C}{1+\alpha\cdot C}=\frac{K_1}{\frac{1}{C}+\alpha}$

$\lim_{C\to\infty}K_{app}=\frac{K_1}{\alpha}$

From question, $\frac{K_1\cdot C}{1+\alpha\cdot C}=\frac{90}{100}\times\frac{K_1}{\alpha}$

or, $\frac{C}{1+9\times10^5C}=\frac{90}{100}\times\frac{1}{9\times10^5} \Rightarrow C=10^{-5}$ M

**50.** $$A \longrightarrow 2B + C$$

| | A | 2B | C |
|---|---|---|---|
| $t=0$ | 1 | 0 | 0 |
| $t=12$ hr | $1-x$ | $2x$ | $x$ |
| $t=24$ hr | $1-y$ | $2y$ | $y$ |

V.P. of solution, $P=X_2\cdot P^o$

or, $20=\frac{180/18}{\frac{180}{18}+(1+2x)}\times24 \Rightarrow x=0.5$

$\therefore t=12\text{ hr}=t_{1/2}$

Now, $t=24\text{ hr}=2\times t_{1/2} \Rightarrow y=0.75$

Now, V.P. of solution, $P=X_2\cdot P^o$

$=\frac{10}{10+(1+2y)}\times24$

$=19.2$ mm Hg

**51.** $\frac{[B]}{[C]}=\frac{K_1}{K_2}=\frac{1.26\times10^{-4}}{3.15\times10^{-5}}=\frac{4}{1}$

$\therefore$ Percentage of $B=\frac{4}{5}\times100=80\%$

**52.**

$$\begin{array}{lccc} & A & \longrightarrow & R \\ t=t & a-x & & x \end{array}$$

$\therefore r = K(a-x)\cdot x$

For maximum rate, $\frac{dr}{dx} = 0 \Rightarrow x = \frac{a}{2} \Rightarrow C_A = C_R$

**53.** $K\cdot\int_0^{t_{1/2}} dt = \int_0^{Co/2} \frac{1+by}{Co-y}\cdot dy$

$\Rightarrow t_{1/2} = \frac{1}{K}\left[(1+Co\cdot b)\cdot \ln 2 - \frac{Co\cdot b}{2}\right]$

**54.** $t_{50\%, B} = t_{94\%, A}$

or, $\frac{1}{K_2}\cdot\ln\frac{100}{50} = \frac{1}{K_1}\cdot\ln\frac{100}{6} \Rightarrow \frac{K_1}{K_2} = \frac{4.067}{1}$

**55.** Percentage product by $S_{N^2}$ mechanism

$= \frac{(4.8\times10^{-5})[RX](0.01)}{(4.8\times10^{-5})[RX](0.01)+2.4\times10^{-6}[RX]}\times100$

$= 16.67\%$

**56.** $[H_2]:[O_2]:[OH]:[H_2O]:[O]$

$= K_1 : K_1 : 2K_2 : K_3 : K_3$

$= 0.60 : 0.60 : 2\times0.30 : 0.10 : 0.10 = 6:6:6:1:1$

**57.** $[A]+[B]+[C]=[A_0]$

when $[A]=[B]=[C]$, $[A] = \frac{[A_0]}{3} = [A_0]\cdot e^{-Kt}$

or, $\frac{1}{3} = e^{-(\ln3+\ln3)\cdot t} \Rightarrow t = 0.5$ min

**58.** As $K_1 = K_2 = K$ (Say),

$t_{max} = \frac{1}{K} = \frac{1}{0.02} = 50$ min

and $[B]_{max} = \frac{[A_0]}{e} = \frac{0.2}{e}$ M

**59.** After long time, $r_A = r_B \Rightarrow K_1[A] = K_2[B]$

$\therefore \frac{[A]}{[B]} = \frac{K_2}{K_1} = 40$

**60.** $\frac{[C]}{[A]} = \frac{\frac{K_2[A_0]}{K_1+K_2}(1-e^{-(K_1+K_2)t})}{[A_0]\cdot e^{-(K_1+K_2)\cdot t}}$

$= \frac{K_2}{K_1+K_2}(e^{(K_1+K_2)\cdot t}-1)$

$= \frac{9K_1}{10K_1}(e^{10K_1\cdot t}-1) = \frac{9}{10}[e^{10\times1.25\times10^{-5}\times3600}-1]$

$= 0.5112$

**61.** At steady state, $K_1[A] = K_2[B]$

$\therefore K_2 = \frac{K_1[A]}{[B]} = \frac{2.5\times10^{-4}\times0.2}{0.01} = 5\times10^{-3}\ \text{min}^{-1}$

**62.** Reaction may be considered as $A \xrightarrow{K_1} C$

$\therefore [C] = [A_0]\ (1-e^{-K_1t})$

**63.**

$$\begin{array}{lcccc} & A \xrightarrow{K_1} & 2B & A \xrightarrow{K_2} & C \\ t=0 & 1\text{ atm} & 0 & 1\text{ atm} & 0 \\ t=10\text{ min} & (1-x-y) & 2x & 1-x-y & y \\ t=\infty & (1-a-b) & 2a & (1-a-b) & b \\ & \approx 0 & & \approx 0 & \end{array}$$

From question, $a+b=1$ and $2a+b=1.5$

$\therefore a = b = 0.5$

Now, $\frac{P_B}{P_C} = \frac{2K_1}{K_2} = \frac{2a}{b} = \frac{2x}{y} \Rightarrow \frac{K_1}{K_2} = 1 = \frac{x}{y}$

Now, $P_{10\text{ min}} = (1-x-y)+2x+y = 1.4$

$\Rightarrow x = y = 0.4$

$\therefore P_A = 1-x-y = 0.2$ atm at $t = 10$ min

Now, $K_1 + K_2 = \frac{1}{t}\cdot\ln\frac{P_A°}{P_A} = \frac{1}{10}\cdot\ln\frac{1}{0.2}$

$= 0.16\ \text{min}^{-1}$

$\therefore K_1 = K_2 = 0.08\ \text{min}^{-1}$

**64.** $r_{max} = z_{11} = \frac{1}{\sqrt{2}}\pi\sigma^2\cdot u_{av}\cdot N^{*2}$

$= \frac{1}{\sqrt{2}}\pi\times(4\times10^{-8}\text{cm})^2\times(2\times10^4\text{cm s}^{-1})$

$\times(2\times10^{19}\ \text{cm}^{-3})^2$

$= 2.842\times10^{28}\ \text{cm}^{-3}\ \text{s}^{-1} = 4.74\times10^7\ \text{mol l}^{-1}\ \text{s}^{-1}$

**65.** $\frac{d(\ln K)}{dT} = \frac{E_a}{RT^2}$

or, $0+\frac{\beta}{T}+\frac{\gamma}{T^2} = \frac{E_a}{RT^2} \Rightarrow E_a = (\beta T+\gamma)R$

**66.** $\frac{K_1}{K_2} = \frac{[B]}{[C]} = \frac{40}{60} = \frac{2}{3}$

Now, $E_{a(\text{overall})} = \frac{K_1\cdot E_{a1}+K_2\cdot E_{a2}}{K_1+K_2} = 32$ kcal/mol

**67.** $r_{\text{uncat}} = \frac{1}{2} \times r_{\text{cat}} \Rightarrow K_{\text{uncat}} = \frac{1}{2} \times K_{\text{cat}}$

or, $A \cdot e^{-E_{a(\text{uncat})}/RT} = \frac{1}{2} \times A \cdot e^{-E_{a(\text{cat})}/RT \times 0.5T}$

or, $\ln 2 - \frac{E_{a(\text{uncat})}}{RT} = -\frac{E_{a(\text{uncat})} - 20}{0.5RT}$

$\therefore E_{a(\text{uncat})} = 38.58$ kcal/mol

**68.** For A $\longrightarrow$ B; $K_1 = 8\ \text{min}^{-1}$ at $T = 300$ K

$K_1' = ?$ at $T = ?$

$$\ln \frac{K_1'}{8} = \frac{20\text{KJ}}{R}\left(\frac{1}{300} - \frac{1}{T}\right) \quad (1)$$

For A $\longrightarrow$ C; $K_2 = 2\ \text{min}^{-1}$ at $T = 300$ K

$K_2' = ?$ at $T = ?$

$$\ln \frac{K_2'}{2} = \frac{28.314\ \text{KJ}}{R}\left(\frac{1}{300} - \frac{1}{T}\right) \quad (2)$$

From (1) and (2), $\ln \frac{K_2'/2}{K_1'/8} = \frac{8.314 \times 10^3}{8.314}\left(\frac{1}{300} - \frac{1}{T}\right)$

or, $\ln\left(\frac{1}{2} \times \frac{8}{2}\right) = \left(\frac{1}{300} - \frac{1}{T}\right) \times 10^3 \Rightarrow T = 379.75$ K

**69.** Given: $\frac{K_{1(310)}}{K_{1(300)}} = 2$, $K_{1(310)} = \frac{\ln 2}{30\ \text{min}}$

$\frac{K_{1(310)}}{K_{1(310)}} = 2$ and $E_{a_2} = \frac{1}{2} E_{a_1}$

For reaction 1: $\ln\left[\frac{K_{1(310)}}{K_{1(300)}}\right] = \frac{E_{a_1}}{R}\left(\frac{1}{300} - \frac{1}{310}\right)$ (1)

For reaction 2: $\ln\left[\frac{K_{2(310)}}{K_{2(300)}}\right] = \frac{E_{a_2}}{R}\left(\frac{1}{300} - \frac{1}{310}\right)$ (2)

From (1) ÷ (2) : $\frac{\ln 2}{\ln\left[\frac{K_{2(310)}}{K_{2(300)}}\right]} = 2$

or, $\left[\frac{K_{2(310)}}{K_{2(300)}}\right] = \sqrt{2}$

$$\Rightarrow K_{2(300)} = \frac{K_{2(310)}}{\sqrt{2}} = \frac{2 \times \frac{\ln 2}{30}}{\sqrt{2}}$$

$= 0.0327\ \text{min}^{-1}$

**70.** At 27°C, $K_1 = \frac{1}{21.6} \cdot \ln \frac{100}{25} = \frac{\ln 2}{10.8}\ \text{min}^{-1}$

Now, $\ln \frac{K_2}{K_1} = \frac{E_a}{R}\left(\frac{1}{T_1} - \frac{1}{T_2}\right)$

$= \frac{9.6 \times 10^3}{2}\left(\frac{1}{300} - \frac{1}{320}\right) = 1.0$

$\therefore \frac{K_2}{K_1} = e = 2.7$

$\Rightarrow K_2 = 2.7 \times \frac{\ln 2}{10.8} = \frac{\ln 2}{4}$

$\Rightarrow (t_{1/2})_2 = 4$ min

$\therefore$ Percentage decomposition in 8.0 min = 75%

**71.** $K = A \cdot e^{-E_a/Rt} = A \cdot e^{-RT/RT} = \frac{A}{e} \simeq 0.37\ A$

**72.** $\ln 2 = \frac{E_a}{R}\left(\frac{1}{280} - \frac{1}{290}\right)$ (1)

and $\ln x = \frac{E_a}{R}\left(\frac{1}{290} - \frac{1}{300}\right)$ (2)

From (2) ÷ (1), $\frac{\ln x}{\ln 2} = \frac{280}{300} \Rightarrow x = 1.91$

**73.** Informative

**74.** Theory based

**75.** First energy barrier is high as Step-I is slow.

## Section B (One or More than one Correct)

1. Theory based
2. Theory based
3. Informative

**4.** $r = K \cdot [A]^n \quad \Rightarrow n = \frac{\ln(r/K)}{\ln [A]}$

Now, $\frac{r_2}{r_1} = \left(\frac{[A_2]}{[A_1]}\right)^n \Rightarrow n = \frac{\ln r_2 - \ln r_1}{\ln[A_2] - \ln[A_1]}$

And, $t_{1/2} \propto [A_0]^{1-n} \Rightarrow \frac{(t_{1/2})_2}{(t_{1/2})_1} = \left(\frac{[A_0]_2}{[A_0]_1}\right)^{1-n}$

$\therefore\ n = 1 - \frac{\ln[A_0]_2 - \ln[A_0]_1}{\ln(t_{1/2})_2 - \ln(t_{1/2})_1}$

5. $[A] = [A_0](1-\alpha) = [A_0]\cdot e^{-Kt} \Rightarrow \alpha = 1 - e^{-Kt}$

6. (a) $-\frac{d[A]}{dt} = K\cdot[A]^n \quad f = \frac{[A_1]-[A_2]}{[A_1]} = \frac{-d[A]}{[A]}$

From question, $f = \frac{-d[A]}{[A]}$

$\therefore\ \frac{f[A]}{t} = K[A]^n \Rightarrow \frac{f}{t} = K\cdot[A]^{n-1}$

or, $\log\left(\frac{f}{t}\right) = \log K + (n-1)\cdot\log[A]$

(b) $\frac{[A_0]^{1-n} - [A]^{1-n}}{1-n} = Kt$

$\Rightarrow [A]^{1-n} = [A_0]^{1-n} + (n-1)\cdot Kt$

(c) $\frac{t_{3/4}}{t_{1/2}} = \frac{[A_0]^{1-n} - \left(\frac{[A_0]}{4}\right)^{1-n}}{[A_0]^{1-n} - \left(\frac{[A_0]}{2}\right)^{1-n}} = \frac{1-(2^2)^{n-1}}{1-2^{n-1}}$

$= 1 + 2^{n-1}$

7. $t_{1/2} = C\cdot(C_0)^{1-n} \Rightarrow \ln t_{1/2} = \ln C + (1-n)\cdot\ln C_0$

8. $K' = K\cdot[H^+]$

On doubling $[H^+]$, $K'$ will double but $K$ will remain unchanged.

9. Theory based

10. (a) For steady state, $\frac{6.93\times10^{-6}}{80} = \frac{0.693}{100}\times[SO_3]$

$\therefore\ [SO_3] = 1.25\times10^{-5}$ M

(b) $n_{eq}\ SO_3 = n_{eq}\ NaOH \Rightarrow 1.25\times10^{-5}\times10^3\times2 = V_{NaOH}\times1$

$\therefore\ V_{NaOH} = 2.5\times10^{-2}$ L $= 25$ ml

(c) Mole of $SO_3$ needed $= \frac{980\times10^3}{98} = 10^4$

$\therefore$ Air needed $= \frac{10^4}{1.25\times10^{-5}} = 8\times10^8$ L

(d) 1000 days $= 10\ t_{1/2}$

$\therefore\ [SO_3] = \frac{1.25\times10^{-5}}{2^{10}} \approx 1.25\times10^{-8}$ M

11.

| | $3A(g) \longrightarrow$ | $2B(g)$ + | $2C(s)$ |
|---|---|---|---|
| $t = 0$ | 6 atm | 0 | – |
| $t = 20$ min | $(6-x)$ atm | $\frac{2}{3}x$ atm | 0.05 atm |
| $t = \infty$ | $\approx 0$ | 4 atm | 0.05 atm |

But from question, $P_\infty = 4.05$ atm and hence, $(4.05-4) = 0.05$ atm is the vapour pressure of C(s).

Now, $P_{20} = (6-x) + \frac{2}{3}x + 0.05 = 5.05 \Rightarrow x = 2$

$\therefore\ t = 20$ min $= t_{1/2}$

12. $-\frac{d\theta}{dt} = K\cdot\theta \Rightarrow Kt = \ln\frac{\theta_0}{\theta}$

(a) $t = \frac{1}{K}\cdot\ln\frac{\theta_0}{\theta} = \frac{1}{0.04}\cdot\ln\frac{596}{298} = 17.5$ sec

(b) $t = \frac{1}{0.04}\cdot\ln\frac{1192}{298} = 35$ sec

13. (a)

| | A + | B $\longrightarrow$ | 2C |
|---|---|---|---|
| $t = 0$ | $2a$ | $a$ | |
| $t = t$ | $2a - x$ | $a - x$ | |

As [A] ≠ [B] throughout, the overall reaction is not first-order.

(b) $r = K[A]^{-1}[B]^2 = K'\cdot[B]^2 \Rightarrow t_{1/2} = \frac{1}{K'[B_0]}$

(c) $r = K[A]^{-1}[B]^2 = K''[A]^{-1}$

(d) As [A] – [B] – stoichiometric ratio, then the mole ratio will remain constant throughout.

14. $A \xrightarrow{K_1} P\ ;\ t_{1/2} = \frac{0.693}{K_1}$

$B \xrightarrow{K_2} Q\ ;\ t_{1/2} = \frac{1}{K_2[B_0]} = \frac{1}{K_2}$

From question, $\frac{0.693}{K_1} = \frac{1}{K_2} \Rightarrow K_2 > K_1$

15.

| | $A_4 \longrightarrow$ | 4A |
|---|---|---|
| $t = 0$ | $a$M | 0 |
| $t = 30$ min | $(a-x)$ M | $4x$M |

As $a - x = 4x \Rightarrow x = \frac{a}{5}$

$\therefore$ Percentage reaction at $t = 30$ min

$= \frac{x}{a}\times100 = 20\%$

Now, $30 = \frac{t_{1/2}}{\log 2}\cdot\log\frac{a}{a-x} \Rightarrow t_{1/2} = 90$ min

**16.** (a) $\Delta_r H = \Sigma \Delta_f H_{\text{Products}} - \Sigma \Delta_f H_{\text{Reactants}}$
$= 2 \times (-1263) - [(-2238) + (-285)] = -3$ KJ/mol

(b) Can not confirm because in aqueous medium, there is no combustion.

(d) Concentration in G.P. in equal time interval.

**17.** $n = 1 \Rightarrow t_{100\%} = \frac{1}{K} \cdot \ln \frac{[A_0]}{0} = \text{Infinite}$

$$n \neq 1 \Rightarrow t_{100\%} = \frac{[A_0]^{1-n} - (0)^{1-n}}{K(1-n)} = \frac{[A_0]^{1-n}}{K(1-n)} \text{ if } n < 1$$
$$= \text{Infinite if } n > 1$$

**18.** $\frac{[B]}{[C]} = \frac{K_1}{K_2} = \frac{1}{2} \Rightarrow [C] > [B]$

Hence, after long time, the solution will be dextrorotatory.

Now, $K = K_1 + K_2 = 6.93 \times 10^{-2} + 13.86 \times 10^{-2}$
$= 3 \times 6.93 \times 10^{-2}\ \text{min}^{-1}$

$\therefore t_{1/2} = \frac{\ln 2}{K} = \frac{0.693}{3 \times 6.93 \times 10^{-2}} = \frac{10}{3}$ min

| | A ⟶ B | | A ⟶ C | |
|---|---|---|---|---|
| $t = 0$ | 2M | 0 | 2M | 0 |
| $t = t$ | $2 - (x + y)$M | $x$M | $2 - (x + y)$M | $y$M |

From question, $x + y = 1.5$ and $\frac{x}{y} = \frac{1}{2}$

$\therefore x = 0.5, y = 1.0$

Hence, total rotation $= 0.5 \times 60° + 0.5 \times (-72°) + 1.0 \times 42° = 36°$

**19.** $[B] : [C] : [D] = 1 \times 3K : 2 \times 2K : 3 \times K = 3 : 4 : 3$

| | A ⟶ B | | A ⟶ 2C | | A ⟶ 3D | |
|---|---|---|---|---|---|---|
| $t = 0$ | 1M | 0 | 1M | 0 | 1M | 0 |
| $t = t$ | $1 - (x + y + z)$ | $x$M | $1 - (x + y + z)$ | $2y$M | $1 - (x + y + z)$ | $3z$M |
| $t = \infty$ | $1 - (a + b + c)$ | $a$M | $1 - (a + b + c)$ | $2b$M | $1 - (a + b + c)$ | $3c$M |

As $a : 2b : 3c = 3 : 4 : 3$ and $a + b + c = 1$

$[C] = 2b = 0.67$ M

As $[A_0] = 1$ M, $[B] \neq 1$M

**20.** For $S_N1$ path : $r_1 = (3 \times 10^{-4}\ \text{s}^{-1})$ [RX]

For $S_N2$ path : $r_2 = (5 \times 10^{-4}\ \text{M}^{-1}\ \text{s}^{-1})$ [RX][$\bar{\bar{N}}u$]

(a) [$\bar{\bar{N}}u$] = 0.1 M, then $r_1 > r_2$

(b) [$\bar{\bar{N}}u$] = 1.0 M, then $r_1 < r_2$

(c) [$\bar{\bar{N}}u$] = 0.6 M, then $r_1 = r_2$

(d) [$\bar{\bar{N}}u$] = 0.4 M, then $\frac{r_1}{r_2} = \frac{2}{3}$

$\therefore$ Percentage product by $S_N1 = \frac{2}{2+3} \times 100 = 40\%$

**21.** $-\frac{d[A]}{dt} = +\frac{d[B]}{dt}$ always

**22.** As mole is not changing, $C_A + C_B + C_C = C_{A_0}$

Now, $\frac{C_B}{C_{A_0} - C_A} = \frac{C_B}{C_B + C_C} = \frac{K_1}{K_1 + K_2}$

**23.** Informative

**24.** Theoretical

**25.** Increase in temperature will result in greater increase in the rate of reaction A → B than B → C.

**26.** Informative

**27.** Informative

**28.** $K_1 = K_2 \Rightarrow -\frac{14000}{RT} + 5 = -\frac{20000}{RT} + 10$

$\Rightarrow T = \frac{1200}{8.314} K$

Now, $\frac{P_{A_2}}{P_{B_3}} = \frac{1 \times e^{-K_1 t}}{1 \times e^{-K_2 t}} = \frac{1}{1}$

Now, initial pressure

$$P_0 = \frac{(1+1) \times 0.0821 \times \frac{1200}{8.314}}{100} = 0.237 \text{ atm}$$

As number of moles will increase on reaction, the total pressure can never be less than 0.2 atm

Now, $\frac{P_A}{P_B} = \frac{2K_1}{3K_2} = \frac{2}{3}$

**29.** $\frac{1}{K} \cdot \frac{dK}{dT} = \frac{d(\ln K)}{dT} = \frac{1.25 \times 10^6}{T^3} = \frac{E_a}{RT^2}$

$\therefore E_a = \frac{1.25 \times 10^6 R}{T} = \frac{1.25 \times 10^6 \times 2}{250} = 10^4$ cal/mol

**30.** $\Delta H = E_{a_f} - E_{a_b}$

$\Rightarrow -2 = E_{a_f} - 8$

$\Rightarrow E_{a_f} = 6$ kcal/mol

Now, the fraction of molecules crossing energy barrier $= e^{-E_a/RT}$

and $K_{eq} = e^{-\Delta H/Rt}$

## Section C (Comprehensions)

### Comprehension I

1. The overall reaction is first-order.

2. $\frac{K_1}{2}=\frac{K_2}{4}=\frac{K_3}{1} \Rightarrow 2K_1 = K_2 = 4K_3$

3.

| $2N_2O_5$ | $\longrightarrow$ | $4NO_2$ | + | $O_2$ |
|---|---|---|---|---|
| $2\times 108$ gm | | $4\times 46$ gm | | 32 gm |
| 108 gm | | 92 gm | | 16 gm |

### Comprehension II

4. $CO(g) + Cl_2(g) \longrightarrow COCl_2(g)$

$$r_{COCl_2} = +\frac{d[COCl_2]}{dt} = K_5\cdot[COCl][Cl_2] \quad (1)$$

Now, for steady state of COCl, $+\frac{d[COCl]}{dt}=0$

or $K_3[Cl][CO] - K_4[COCl] - K_5[COCl][Cl_2] = 0$

$$\therefore [COCl] = \frac{K_3[Cl][CO]}{K_4 + K_5[Cl_2]} \quad (2)$$

$\therefore$ For steady state of Cl, $\frac{d[Cl]}{dt}=0$

or $2K_1[Cl_2] - 2K_2[Cl]^2 - K_3[Cl][CO] + K_4[COCl] + K_5[COCl][Cl_2] = 0$

$$\therefore [Cl]=\left(\frac{K_1[Cl_2]}{K_2}\right)^{1/2} \quad (3)$$

From (1), (2), (3), $r_{COCl_2} = \frac{K_1^{1/2}K_5K_3[CO][Cl_2]^{3/2}}{K_2^{1/2}(K_4+K_5[Cl_2])}$

5. $r_4 >> r_5$

or $K_4[COCl] >> K_5[COCl][Cl_2]$

or $K_4 >> K_5[Cl_2]$

$$\therefore r_{COCl_2} = \frac{K_1^{1/2}K_3K_5[CO][Cl_2]^{3/2}}{K_2^{1/2}(K_4+K_5[Cl_2])} \approx \frac{K_1^{1/2}K_3K_5[CO][Cl_2]^{3/2}}{K_2^{1/2}K_4}$$

6. $A_{overall} = \frac{A_1^{1/2}\cdot A_3\cdot A_5}{A_2^{1/2}\cdot A_4}$

7. $E_{a_{overall}} = \frac{1}{2}E_{a_1} + E_{a_3} + E_{a_5} - \frac{1}{2}E_{a_2} - E_{a_4}$

### Comprehension III

For steady state of Br, $+\frac{d[Br]}{dt}=0$

or, $2K_1[Br_2] - K_2[Br][H_2] + K_3[H][Br_2] + K_4[H][HBr] - 2K_5[Br]^2 = 0 \quad (1)$

For steady state of H, $+\frac{d[H]}{dt}=0$

or, $K_2[Br][H_2] - K_3[H][Br_2] - K_4[H][HBr] = 0 \quad (2)$

8. From (1) and (2), $[Br]=\left(\frac{K_1[Br_2]}{K_5}\right)^{1/2}$

9. $$[H] = \frac{K_2[Br][H_2]}{K_3[Br_2]+K_4[HBr])} = \frac{K_2\cdot K_1^{1/2}\cdot[Br_2]^{1/2}\cdot[H_2]}{K_5^{1/2}(K_3[Br_2]+K_4[HBr])}$$

10. $$+\frac{d[HBr]}{dt} = K_2[Br][H_2] + K_3[H][Br_2] - K_4[H][HBr] = 2K_3[H][Br_2] = \frac{2K_3K_2K_1^{1/2}[Br_2]^{3/2}[H_2]}{K_5^{1/2}(K_3[Br_2]+K_4[HBr])}$$

11. At $t = 0$, [HBr] = 0 and hence, initial rate is given by,

$$r_0 = \frac{2K_2K_1^{1/2}[Br_2]^{1/2}[H_2]}{K_5^{1/2}}$$

## Comprehension IV

12. $K = \frac{1}{t} \cdot \ln \frac{[H_2O_2]_0}{[H_2O_2]_t} = \frac{1}{t} \cdot \ln \frac{V_0}{V_t}$

For $t = 10$ min, $K_1 = \frac{1}{10} \cdot \ln \frac{25.6}{16} = \frac{\ln 1.6}{10}$ min$^{-1}$

For $t = 20$ min, $K_2 = \frac{1}{20} \cdot \ln \frac{25.6}{10} = \frac{\ln 1.6}{10}$ min$^{-1}$

As $K_1 = K_2$, order of reaction = 1

13. $t_{1/2} = \frac{\ln 2}{K} = \frac{\log 2}{\left(\frac{\log 1.6}{10}\right)} = 15\,\text{min}$

14. $Kt = \ln \frac{[H_2O_2]_0}{[H_2O_2]_t} = \ln \frac{a}{a-x} = \ln \frac{1}{1-\frac{x}{a}}$

or, $\frac{\ln 1.6}{10} \times 25 = \ln \frac{1}{1-\frac{x}{a}}$

$\Rightarrow \frac{x}{a} = \frac{11}{16}$

15. Order = 1, but molecularity = 2 (as per reaction).

## Comprehension V

$$\begin{array}{llll} & C_8H_{18}O_2(g) \rightarrow & 2CH_3COCH_3(g) & + C_2H_6(g) \\ t=0 & 800\text{ torr} & 0 & 0 \\ t=t & (800-x)\text{ torr} & 2x\text{ torr} & x\text{ torr} \end{array}$$

16. $P_{C_8H_{18}O_2} = 800\text{ torr} \xrightarrow{t=3\times t_{1/2}} 100\text{ torr}$

$\therefore t = 3 \times 80 = 240$ min

17. $P_{acetone} = 2x = 1200 \Rightarrow x = 600$

$\therefore P_{C_8H_{18}O_2} = 800\text{ torr} \xrightarrow{t=2\times t_{1/2}} 200\text{ torr}$

$\therefore t = 2 \times 80 = 160$ min

18. $800 - x = 700 \Rightarrow x = 100$

$\therefore P_{total} = (800 - x) + 2x + x = 1000$ torr

## Comprehension VI

19. From (1) and (2) data : order w.r.t OH = 1

From (2) and (3) data : order w.r.t $H_2S$ = 1

$\therefore r = K[H_2S][OH]$

20. $K = \frac{r}{[H_2S][OH]} = \frac{1.4\times10^{-6}\text{ M s}^{-1}}{(2.1\times10^{-8}\text{ M})\times(1.3\times10^{-8}\text{ M})}$

$= 5.1 \times 10^9\text{ M}^{-1}\text{ s}^{-1}$

21. $r = K[H_2S][OH] = 5.1 \times 10^9 \times (1.0 \times 10^{-8}) \times (1.7 \times 10^{-8})$

$= 8.67 \times 10^{-7}\text{ M s}^{-1}$

22. $r = 8.67 \times 10^{-7} \times 0.1 = 8.67 \times 10^{-8}\text{ mol s}^{-1}$

## Comprehension VII

23. $K = \frac{1}{t} \cdot \ln \frac{P_x{}^\circ}{P_x}$

For $t = 100$ min, $K_1 = \frac{1}{100} \cdot \ln \frac{800}{400} = \frac{\ln 2}{100}$ min$^{-1}$

For $t = 200$ min, $K_2 = \frac{1}{200} \cdot \ln \frac{800}{200} = \frac{\ln 2}{100}$ min$^{-1}$

As $K_1 = K_2$, order of reaction = 1

24. $K = \frac{\ln 2}{100} = 6.93\times10^{-3}\text{ min}^{-1}$

$\therefore K_{rxn} = \frac{K}{2} = 3.465\times10^{-3}\text{ min}^{-1}$

25. Time for 87.5% reaction $= 3\times t_{1/2} = 3\times \frac{\ln 2}{6.93\times10^{-3}}$

$= 300$ min

26.

$$\begin{array}{llll} & 2X(g) \longrightarrow & 3Y(g) & + \; 2Z(g) \\ t=0 & 800 & 0 & 0 \\ t=t & 800-x & \frac{3}{2}x & x \\ & = 700 & & \end{array}$$

$\therefore P_{total} = 800 + \frac{3}{2}x = 950$ torr

## Comprehension VIII

**27.** $$\begin{array}{lccccc} & A & + & 2B & \longrightarrow C & + \ D \\ t=0 & a\,M & & b\,M & 0 & 0 \\ t=t & (a-x)\,M & & (b-2x)\,M & & \end{array}$$

Now, $r = K \cdot C_B \Rightarrow -\dfrac{d[A]}{dt} = K(b-2x)$

$$\Rightarrow \frac{dx}{dt} = K(b-2x)$$

or, $\int_0^x \dfrac{dx}{b-2x} = K \cdot \int_0^t dt \Rightarrow x = \dfrac{b}{2}(1-e^{-2Kt})$

$$\therefore C_A = a - x = a - \frac{b}{2}(1-e^{-2Kt})$$

**28.** For $C_A = \dfrac{a}{2}$,

$$\frac{a}{2} = a - \frac{b}{2}(1-e^{-2Kt})$$

$$\therefore (t_{1/2})_A = \frac{1}{2K} \cdot \ln\frac{b}{b-a}$$

**29.** For $(t_{1/2})_A = (t_{1/2})_B$, $\dfrac{[A]}{[B]} = \dfrac{a}{b} = \dfrac{1}{2}$

## Comprehension IX

**30.** $r_{rxn} = -\dfrac{1}{n_1} \cdot \dfrac{dn_A}{dt} = K \cdot n_A \Rightarrow n_A = n_A{}^{\circ} \cdot e^{-n_1 kt}$

**31.** $$\begin{array}{lccc} & n_1 A & \longrightarrow & n_2 A \\ t=0 & a \text{ mole} & & 0 \\ t=t & (a-x)\text{mole} & & \dfrac{n_2}{n_1} \cdot x \text{ mole} \\ & = a \cdot e^{-n_1 kt} & & \end{array}$$

$$\therefore x = a(1-e^{-n_1 kt})$$

Now, $\dfrac{V_2}{V_1} = \dfrac{n_{\text{final}}}{n_{\text{initial}}} = \dfrac{(a-x) + \dfrac{n_2}{n_1} \cdot x}{a}$

or, $\dfrac{V_2}{V_0} = \dfrac{a + x\left(\dfrac{n_2}{n_1} - 1\right)}{a} = 1 + \left(\dfrac{n_2}{n_1} - 1\right) \cdot (1-e^{-n_1 kt})$

$$\therefore V_2 = V_0\left[\frac{n_2}{n_1} + \left(1 - \frac{n_2}{n_1}\right)e^{-n_1 kt}\right]$$

**32.** If $n_1 = 1$, $n_2 = 2$, then $V_2 = V_0(2 - e^{-kt})$

Now, $[A] = \dfrac{n_A}{V_2} = \dfrac{n_A{}^{\circ} \cdot e^{-kt}}{V_0 \cdot (2-e^{-kt})} = [A_0] \cdot \left(\dfrac{e^{-kt}}{2-e^{-kt}}\right)$

## Comprehension X

**33.** $\int_0^f \dfrac{df}{1-f} = K \cdot \int_0^t dt$

$$\Rightarrow t = -\frac{\ln(1-f)}{K}$$

Now, $K = \dfrac{-(-3)}{200} = \dfrac{3}{200}\,\text{hr}^{-1}$

$$\therefore t_{1/2} = \frac{\ln 2}{K} = \frac{0.693}{\left(\dfrac{3}{200}\right)} = 46.2 \text{ hr}$$

**34.** $t = -\dfrac{\ln(1-f)}{K} \Rightarrow f = 1 - e^{-Kt} = 1 - e^{-3t/200}$

## Comprehension XI

**35.** Unit of $K = s^{-1} \Rightarrow$ order = 1

**36.** $K_B = \dfrac{10}{100} \times K = \dfrac{10}{100} \times 1.5\times10^{-4} = 1.5\times10^{-5}\ s^{-1}$

## Comprehension XII

**37.** $[A]=[A_0]\cdot e^{-K_1t}=(1.0\,\text{M})\times e^{-0.04\times25}=0.368\text{ M}$

**38.** $[B]=\dfrac{K_1[A_0]}{K_2-K_1}(e^{-K_1t}-e^{-K_2t})$

$$=\frac{0.04\times(1.0\,\text{M})}{0.06-0.04}(e^{-0.04\times25}-e^{-0.06\times25})$$

$$=0.29\text{ M}$$

**39.** $[C]=[A_0]-[A]-[B]=1.0-0.368-0.29=0.342\text{ M}$

**40.** $t_{\text{max}}=\dfrac{\ln\left(\dfrac{K_2}{K_1}\right)}{K_2-K_1}=\dfrac{\ln\left(\dfrac{3}{2}\right)}{0.06-0.04}=20\text{ min}$

**41.** $[B]_{\text{max}}=[A_0]\left(\dfrac{K_2}{K_1}\right)^{\frac{-K_2}{K_2-K_1}}=(1.0\text{ M})\times\left(\dfrac{3}{2}\right)^{\frac{-0.06}{0.06-0.04}}$

$$=0.3\text{ M}$$

**42.** $[A]=[B]=[C]=\dfrac{[A_0]}{3}$

Now, $t=\dfrac{1}{K_1}\cdot\ln\dfrac{[A_0]}{[A]}=\dfrac{\ln 3}{K_1}=\dfrac{1.1}{0.04}=27.5\text{ min}$

**43.** $-\dfrac{d[A]}{dt}=+\dfrac{d[C]}{dt}\Rightarrow K_1[A]=K_2[B]\Rightarrow t=20\text{ min}$

$\therefore\ [A]=[A_0]\cdot e^{-K_1t}=(1.0\text{ M})\cdot e^{-0.04\times20}=0.45\,\text{M}$

**44.** $(r_C)_{\text{max}}=K_2[B]_{\text{max}}=0.06\times0.3=1.8\times10^{-2}\text{ M/ min}$

## Comprehension XIII

| | A | $\rightleftharpoons$ | B |
|---|---|---|---|
| $t=0$ | 0.15 M | | 0 |
| $t=10$ | $(0.15-x)$ M | | $x$ M |
| | $=0.125$ M | | $=0.025$ M |
| $t=t_{\text{eq}}$ | $(0.15-x_{\text{eq}})$ M | | $x_{\text{eq}}$ M |
| | $=0.10$ M | | $=0.05$ M |

Now, $K_{eq}=\dfrac{K_f}{K_b}=\dfrac{0.05}{0.10}=\dfrac{1}{2}$ (1)

and $t_{1/2}=\dfrac{\ln 2}{K_f+K_b}\Rightarrow 10\text{ min}=\dfrac{0.693}{K_f+K_b}$ (2)

**45.** From (1) and (2), $K_f=2.31\times10^{-2}\text{ min}^{-1}$

**46.** From (1), $K_b=4.62\times10^{-2}\text{ min}^{-1}$

**47.** $K_{eq}=0.5$

**48.** $t_{1/2}=10\text{ min}$

## Comprehension XIV

**49.** $E_{a(\text{overall})}=\dfrac{K_1\cdot E_{a_1}+K_2\cdot E_{a_1}}{K_1+K_2}$

or, $10.5=\dfrac{K_1\times12+K_2\times9}{K_1+K_2}\Rightarrow K_1=K_2$

or, $A_1\cdot e^{-E_{a_1}/RT}=A_2\cdot e^{-E_{a_2}/RT}$

or, $\dfrac{E_{a_1}-E_{a_2}}{RT}=\ln\dfrac{A_1}{A_2}$

$\Rightarrow\dfrac{(12-9)\times10^3}{2\times T}=\ln\dfrac{2\times10^{14}}{2\times10^{14}\times e^{-2}}$

$\therefore\ T=750\ K$

**50.** Above 750 K, $Y$ will be the major product and below 750 K, $Z$ will be the major product as $E_{a_1}>E_{a_2}$.

**51.** Reactions with higher $E_a$ are more sensitive towards temperature change.

### Comprehension XV

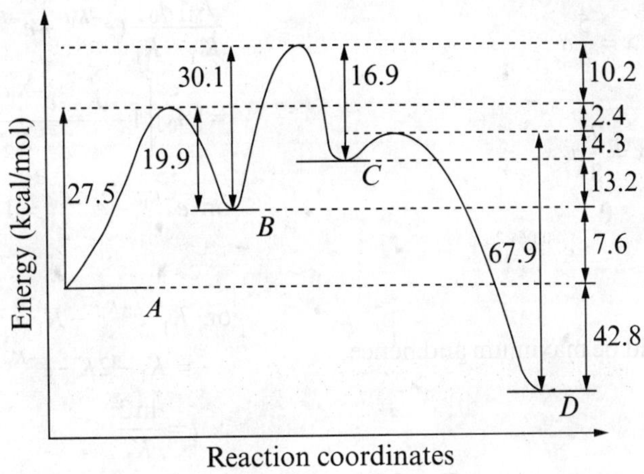


**52.** C → D $[E_a : A \xrightarrow{27.5} B \xrightarrow{30.1} C \xrightarrow{4.3} D]$

**53.** C → B $[E_a : D \xrightarrow{67.9} C \xrightarrow{16.9} B \xrightarrow{19.9} A]$

**54.** C → D [Lowest $E_a$]

**55.** B → C [Highest $E_a$]

**56.** D → C [Highest $E_a$]

### Section D (Assertion – Reason)

1. Molecularity can never be fractional.
2. $t_{100\%} = \dfrac{[A_0]^{1-n}}{K(1-n)}$ when $n < 1$

   = Infinite when $n \geq 1$
3. For a particular step, rates always increase with increase in temperature.
4. Relative increase in rate constant with increase in temperature is higher for the reaction with higher activation energy.
5. $\Delta H = E_{a_f} - E_{a_b}$
6. Theoretical
7. For zero order reaction : $t_{1/2} = \dfrac{[A_0]}{2K}, t_{100\%} = \dfrac{[A_0]}{K}$
8. Order is in dependent from stoichiometry of reaction.
9. $t_{1/2} = \dfrac{[A_0]}{2K}$
10. Theoretical

### Section E (Column Match)

1. Informative
2. Informative
3. Theoretical
4. Theoretical
5. Theoretical
6. Theoretical
7. Theoretical
8. Theoretical
9. (P) $a \xrightarrow[= 18 \text{ sec}]{t_{1/3}} \dfrac{2a}{3} \xrightarrow[= 18 \text{ sec}]{t_{1/3}} \dfrac{4a}{9} \xrightarrow[= 18 \text{ sec}]{t_{1/3}} \dfrac{8a}{27}$

   $t_{19/27} = 54$ sec

   (Q) $a \xrightarrow[= 16 \text{ sec}]{t_{1/4}} \dfrac{3a}{4} \xrightarrow[= 16 \text{ sec}]{t_{1/4}} \dfrac{9a}{16}$

   $t_{7/16} = 32$ sec

   (R) $K = \dfrac{1}{4}\left[\dfrac{1}{2a/3} - \dfrac{1}{a}\right] = \dfrac{1}{56}\left[\dfrac{1}{a-x} - \dfrac{1}{a}\right] \Rightarrow x = \dfrac{7}{8}a$

(S) $K = \dfrac{a - \frac{2a}{3}}{18} = \dfrac{x}{30} \Rightarrow x = \dfrac{5}{9}a$

(T) $K = \dfrac{a - \frac{a}{2}}{16} = \dfrac{x}{28} \Rightarrow x = \dfrac{7}{8}a$

**10.** (A) $\dfrac{d[C]}{dt} = K_2[B]$

For $\left(\dfrac{d[C]}{dt}\right)_{max}$, $[B]$ should be maximum and hence

$$t = \frac{\ln\left(\frac{K_2}{K_1}\right)}{K_2 - K_1} = \frac{\ln 2}{K_1} \quad \text{(when } K_2 = 2K_1\text{)}$$

Now, $(t_{1/2})_A = \dfrac{\ln 2}{K_1}$

(B) Rate of formation of $B$ is maximum at $t = 0$, at which $[B] = [C] = 0$

Now, $[B] = [C]$

$$\frac{K_1[A_0]}{K_2 - K_1}(e^{-K_1t} - e^{-K_2t})$$

$$= [A_0]\left[1 - \frac{K_2 \cdot e^{-K_1t} - K_1 \cdot e^{-K_2t}}{K_2 - K_1}\right]$$

or, $e^{-K_1t} - e^{-2K_1t} = 1 - \dfrac{2K_1 \cdot e^{-K_1t} - K_1 \cdot e^{-2K_1t}}{K_1}$

(when $K_2 = 2K_1$)

or, $K_1 \cdot e^{-K_1t} - K_1 \cdot e^{-2K_1t}$

$= K_1 - 2K_1 \cdot e^{-K_1t} - K_1 \cdot e^{-2K_1t}$

$\therefore\ t = \dfrac{\ln 2}{K_1}$

(C) [A] = [B]

$$[A_0] \cdot e^{-K_1t} = \frac{K_1[A_0]}{K_2 - K_1}(e^{-K_1t} - e^{-K_2t})$$

$$\frac{K_2 - K_1}{K_1} = 1 - e^{(K_1 - K_2)t}$$

$$\therefore\ t = \frac{1}{K_1 - K_2} \cdot \ln \frac{2K_1 - K_2}{K_1}$$

## Section F (Subjective)

### Single Digit Integer Type

**1.** ii, iv, v

**2.** Theoretical

**3.** $\dfrac{K_{BrO^-}}{3} = \dfrac{K_{BrO_3^-}}{1} = \dfrac{K_{Br^-}}{2}$

$\therefore\ K_{BrO_3^-} = a = \dfrac{0.06}{3} = 0.02\ M^{-1}s^{-1}$

and $K_{Br^-} = b = \dfrac{2}{3} \times 0.06 = 0.04\ M^{-1}s^{-1}$

**4.** $\dfrac{7.2 \times 10^{-15}}{3600}\ M\ s^{-1} = K \times (2 \times 10^{-8}\ M)^2$

$K = \dfrac{1}{200}\ M^{-1}\ s^{-1} = 5\ ml\ mol^{-1}\ s^{-1}$

**5.** $-\dfrac{dP}{dt} = K \cdot P_{NO}^a \cdot P_{H_2}^b$

$\dfrac{1.5}{0.25} = \left(\dfrac{372}{152}\right)^a \Rightarrow a = 2$

and $\dfrac{1.60}{0.79} = \left(\dfrac{289}{144}\right)^b \Rightarrow b = 1$

**6.** $\dfrac{0.1}{0.4} = \dfrac{0.20}{x} \Rightarrow x = 0.8$

$\dfrac{0.1}{0.8} = \dfrac{0.2 \times 0.05}{0.4 \times y} \Rightarrow y = 0.2$

**7.** $t = \dfrac{1}{K} \cdot \ln \dfrac{[Cr^{3+}]_0}{[Cr^{3+}]}$

$= \dfrac{1}{9 \times 10^{-5}\ s^{-1}} \cdot \ln \dfrac{100}{100 - 80}$

$= 1.8 \times 10^4$ sec $= 5$ hrs

**8.** (i) Addition of NaOH will decrease $[H_3O^+]$.

(ii) Addition of water will decrease the concentration of both.

(iii) Acetic acid is a weak acid and hence, $[H_3O^+]$ will decrease.

(iv) Increase in temperature increases the reaction rate.

**9.** Time for certain progress of reaction, $t \alpha [A_0]^{1-n}$

$$\frac{1\times10^{-3}}{0.25\times10^{-3}} = \left(\frac{0.02}{0.04}\right)^{1-n} \Rightarrow n = 3$$

**10.**

| | $C_4H_8$ | $\longrightarrow$ | $2C_2H_4$ |
|---|---|---|---|
| $t = 0$ | $a$ mole | | 0 |
| $t = t$ | $(a - x)$ mole | | $2x$ mole |

As, $a - x = 2x \Rightarrow x = \frac{a}{3}$

Now,

$$t = \frac{t}{K}\cdot\ln\frac{a}{a-x} = \frac{1}{\frac{25}{18}\times10^{-5}\text{ s}^{-1}}\times\ln\frac{a}{a-\frac{a}{3}} = 2\text{hrs}$$

**11.** For 2% reaction, we may assume that rate is almost constant.

$$r = K[A] \Rightarrow \frac{2}{100}\times[A]\text{ min}^{-1} = K[A]$$
$$\Rightarrow K = 0.02\text{ min}^{-1}$$

**12.** $H_2O_2\,(aq) \longrightarrow H_2O(l) + \frac{1}{2}\,O_2(g)$

$\Delta H = (-287) - (-187) = -100$ KJ/mol

Moles of $H_2O_2$ reacted per sec $= 7.5\times10^{-4}\times0.02\times2$
$= 3\times10^{-5}$

$\therefore$ Heat produced per sec $= 3\times10^{-5}\times(100\times10^3)$
$= 3$ J

**13.** $$t = \frac{1}{K}\cdot\ln\frac{a}{a-x} = \frac{1}{\left(\frac{4.5}{3.1536}\times10^{-8}\text{ s}^{-1}\right)}\cdot\ln\frac{100}{40}$$

$$= \frac{0.9\times3.1536\times10^8}{4.5\times3.1536\times10^7}\text{Year} = 2\text{ Years}$$

**14.** $$t_{1/2} = \frac{0.693}{6.93\times10^{-4}} = 1000\text{ sec}$$

| | A | $\longrightarrow$ | $n$B |
|---|---|---|---|
| $t = 0$ | $a$ mole | | 0 |
| $t = 1000$ sec | $\frac{a}{2}$ mole | | $\frac{n\cdot a}{2}$ mole |

Now, $$\frac{\frac{a}{2}+\frac{n\cdot a}{2}}{a} = 3 \Rightarrow n = 5$$

**15.** $r = K[\text{ester}][H^+]^x = k_1[\text{ester}]$

$K_1 = K\cdot[H^+]^x$

$$\frac{1.0}{10} = \left(\frac{10^{-3}}{10^{-2}}\right)^x \Rightarrow x = 1$$

and $$K_1 = \frac{K_1}{[H^+]} = \frac{1.0\times10^{-3}}{10^{-3}} = 1 = y$$

**16.** $t_{3/4} = 2\times t_{1/2}$ and hence, $a = 1$.

Now, $$t_{1/2} = \frac{\ln 2}{K} = \frac{\ln 2}{K[H^+]^b}$$

$$\frac{1.0}{0.5} = \left(\frac{0.02}{0.01}\right)^b \Rightarrow b = 1$$

**17.** Concentrations are in G.P. and hence, order = 1.

**18.**

| | $A_2B_3(aq)$ | $\longrightarrow$ | $2A^{3+}\,(aq)$ + | $3\,B^{2-}\,(aq)$ |
|---|---|---|---|---|
| $t = 0$ | $a$ mole | | 0 | 0 |
| $t = 10$ min | $a - x$ | | $2x$ | $3x$ |

Now, $\pi = CRT = \rho gh \Rightarrow$ total mole $\alpha\, h$

$$\therefore \frac{a}{a+4x} = \frac{2}{6} \Rightarrow t = 10\text{ min} = t_{1/2}$$

Now, at $t = t_{3/4} = 2\times t_{1/2} = 20$ min, $x = \frac{3a}{4}$

$$\therefore \frac{a}{a+\left(4\times\frac{3a}{4}\right)} = \frac{2}{h} \Rightarrow h = 8\text{ mm}$$

$$\pi = x = \rho gh = \left(1.0\frac{\text{gm}}{\text{cm}^3}\right)\times\left(1000\frac{\text{cm}}{\text{s}^2}\right)\times(0.8\text{ cm})$$
$$= 800\frac{\text{dyne}}{\text{cm}^2} = 80\text{ pascal}$$

Now, $$\frac{x}{y} = \frac{80}{20} = 4$$

**19.** $$[A]^4 = \frac{1}{t+1} \Rightarrow 4[A]^3\cdot\frac{d[A]}{dt} = -\frac{1}{(t+1)^2} = -[A]^8$$

$$\therefore -\frac{d[A]}{dt} = \frac{[A]^5}{4} = \frac{(0.2)^5}{4} = 8\times10^{-5}\text{ M s}^{-1}$$

**20.**

| | A | $\longrightarrow$ | 2B |
|---|---|---|---|
| $t = 0$ | $a$ mole | | 0 |
| $t = t$ | $(a - x)$ mole | | $2x$ mole |

From mass conservation, $a\times M_0 = (a + x)\times M_t$

$$\therefore x = \frac{a(M_0 - M_t)}{M_t}$$

If the reaction is zero order, then

$$K = \frac{a-(a-x)}{t} = \frac{x}{t} = \frac{a(M_0 - M_t)}{t\cdot M_t}$$

For $t = 10$ min, $K = \frac{a(42-35)}{10\times 35} = \frac{a}{50}$

For $t = 20$ min, $K = \frac{a(42-30)}{20\times 30} = \frac{a}{50}$

As $K$ values are same, the reaction is of zero-order.

**21.** $r = K[A]^n$ and $2r = K(4[A])^n \Rightarrow n = \frac{1}{2}$

$\therefore t_{1/2} \alpha [A_0]^{1-n} = [A_0]^{1/2}$

Next $t_{1/2}$ will be $\frac{1}{\sqrt{2}}$ times of previous one and

hence, $t = \frac{8\sqrt{2}}{\sqrt{2}} = 8$ hr.

**22.** $P_B = \frac{K_B \cdot P_A{}^{\circ}}{K_A + K_B + K_C}\cdot[1 - e^{-(K_A+K_B+K_C)\cdot t}]$

$= \frac{2\times 10^{-3}\times 13.86}{6.93\times 10^{-3}}[1 - e^{-6.93\times 10^{-3}\times 100}] = 2$ atm

**23.** $\frac{K_{\text{I}}}{K_{\text{II}}} = \frac{A_{\text{I}}\cdot e^{-E_{a_{\text{I}}}/RT}}{A_{\text{II}}\cdot e^{-E_{a_{\text{II}}}/RT}} = \frac{A_{\text{I}}}{A_{\text{II}}}\cdot e^{-(E_{a_{\text{I}}} - E_{a_{\text{II}}})/RT}$

$= = 100\times e^{-4.606\times 10^3/2\times 500} = 1$

**24.** Fraction of molecules having sufficient energy

$= e^{-E_a/RT} = e^{-83.14\times 10^3/8.314\times 500} = 2\times 10^{-9}$

**25.** $\ln\frac{K_2}{K_1} = \ln\frac{t_1}{t_2} = \frac{E_a}{R}\left(\frac{1}{T_1} - \frac{1}{T_2}\right)$

or, $\ln\frac{1}{3} = \frac{E_a}{R}\left(\frac{1}{300} - \frac{1}{280}\right)$ (1)

and $\ln\frac{16}{t} = \frac{E_a}{R}\left(\frac{1}{300} - \frac{1}{330}\right)$ (2)

From (1) and (2), $t = 4$ hrs

## Four Digit Integer Type

**1.** $r = K[O_3]^2 = 5\times 10^{-4}\times(2\times 10^{-8})^2$

$= 2\times 10^{-19}$ mol l$^{-1}$ s$^{-1}$

$= 2\times 10^{-19}\times 6\times 10^{23}\times 10^{-3}\times 60$

$= 7200$ molecules ml$^{-1}$ min$^{-1}$

**2.** At $t = \infty$, $P_{\text{total}}$ should be 400 mm, but as it is only 390 mm, some unreactive gas should also be present in the vessel. Let $P^{\circ}_{C_2H_5Br} = P_0$ mm then $P_{\text{unreactive}}$ gas $= (200 - P_0)$ mm.

| | $C_2H_5Br(g)$ | $\longrightarrow$ $C_2H_4(g)$ | + $HBr(g)$ |
|---|---|---|---|
| $t = 0$ | $P_0$ | 0 | 0 |
| $t = t$ | $P_0 - x$ | $x$ | $x$ |
| $t = \infty$ | 0 | $P_0$ | $P_0$ |

From question, $P_0 + P_0 + (200 - P_0) = 390$

$\Rightarrow P_0 = 190$

and $(P_0 - x) + x + x + (200 - P_0) = 342.5$

$\Rightarrow x = 142.5$

$\therefore$ Percentage $C_2H_5$ Br undecomposed

$= \frac{P_0 - x}{P_0} = 25\%$

**3.** $t = \frac{t_{\text{gen}}}{\log 2}\cdot\log\frac{[A]}{[A_0]} \Rightarrow 96 = \frac{t_{\text{gen}}}{0.30}\cdot\log 3$

$\Rightarrow t_{\text{gen}} = 60$ hrs

**4.** $r = -\frac{dP_{\text{NO}}}{dt} = K'\cdot P_{\text{NO}}^2\cdot P_{O_2}$ and

$K' = \frac{1.6\times 10^5}{(0.08\times 600)^2}$ atm$^{-2}$s$^{-1}$

$\therefore r = \frac{1.6\times 10^5}{48\times 48}\times\left(\frac{190}{760}\right)^2\times\left(\frac{288}{760}\right)$

$= \frac{1250}{760}$ atm s$^{-1}$ $= 1250$ mm s$^{-1}$

**5.** From the unit of rate constant, the process is zero order.

$\therefore t_{100\%} = \frac{[H^+]_0}{K}$

$= \frac{\left(\frac{3\times 10^{-7}}{0.05}\times 1000\right)}{1.0\times 10^{-7}} = 6\times 10^4$ sec $= 1000$ min

**6.** $K = \frac{1}{t}\ln\frac{[A_0]}{[A]} = \text{Constant}$

$\therefore \frac{1}{100}\cdot\ln\frac{[A_0]}{[A_0]/3} = \frac{1}{t}\cdot\ln\frac{[A_0]}{[A_0]/9} \Rightarrow t = 200$ min

**7.** $K = \frac{1}{t}\cdot\ln\frac{[A_0]}{[A]} = \text{Constant}$

$\therefore \frac{1}{20} \cdot \ln \frac{500}{420} = \frac{1}{t} \cdot \ln \frac{100}{70} \Rightarrow t = 40 \text{ min}$

**8.** $K = \frac{1}{t} \cdot \ln \frac{V_\infty}{V_\infty - V_t} = \text{Constant}$

$\therefore \frac{1}{40} \cdot \ln \frac{80}{80-40} = \frac{1}{t} \cdot \ln \frac{80}{80-70} \Rightarrow t = 120 \text{ min}$

**9.**

| | 2P $\longrightarrow$ | 4Q + | R + S(l) | |
|---|---|---|---|---|
| $t = 0$ | $P_0$ | 0 | 0 | |
| $t = 30$ min | $P_0 - x$ | $2x$ | $\frac{x}{2}$ | V.P. = 25 |
| $t = 60$ min | $P_0 - y$ | $2y$ | $\frac{y}{2}$ | V.P. = 25 |
| $t = \infty$ | 0 | $2P_0$ | $\frac{P_0}{2}$ | V.P. = 25 |

From question, $2P_0 + \frac{P_0}{2} + 25 = 625 \Rightarrow P_0 = 240$

and $(P_0 - x) + 2x + \frac{x}{2} + 25 = 445 \Rightarrow x = 120$

Now, $\frac{1}{30} \cdot \ln \frac{P_0}{P_0 - x} = \frac{1}{60} \cdot \ln \frac{P_0}{P_0 - y} \Rightarrow y = 180$

$\therefore P_{60} = (P_0 - y) + 2y + \frac{y}{2} + 25 = 535 \text{ mm}$

**10.** Initial moles of $NH_4NO_2 = \frac{200 \times 0.02}{1000} = 0.004$

and moles of $N_2O$ formed

$= \frac{\frac{(785-25)}{760} \times \frac{49.26}{1000}}{0.0821 \times 300} = 0.002$

$\therefore t_{req} = t_{1/2} = 123 \text{ min}$

**11.** For set 1 and 2, $r = K'[B]$ as $[A_0] >> [B_0]$

and $t_{1/2} = \frac{\ln 2}{K'} = \frac{\ln 2}{K[A_0]^2} \Rightarrow x = 62.5$

For set 3 and 4, $r = K''[A]^2$ as $[B_0] >> [A_0]$

and $t_{1/2} = \frac{1}{K''[A_0]} = \frac{1}{K[B_0][A_0]}$

$\Rightarrow y = \frac{625}{2} = 312.5$

$\therefore x + y = 62.5 + 312.5 = 375$

**12.** $t = 43.5 \text{ min} = 3t_{1/2}$

Hence, $P_{ether} = \frac{4}{2^3} = 0.5 \text{ atm} \Rightarrow \Delta P_{ether} = 3.5 \text{ atm}$

$\therefore P_{final} = 0.5 + 3.5 \times 3 = 11 \text{ atm}$

**13.** $\Delta t = \frac{t_{1/2}}{\ln 2} \cdot \ln \frac{r_1}{r_2} \Rightarrow 12 = \frac{t_{1/2}}{\ln 2} \cdot \ln\left(\frac{0.04}{0.03}\right)$

$\therefore t_{1/2} = 28 \text{ min} = 1680 \text{ sec}$

**14.** $t = \frac{t_{1/2}}{\ln 2} \cdot \ln \frac{V_\infty - V_0}{V_\infty - V_t} \Rightarrow 120 = \frac{t_{1/2}}{\ln 2} \cdot \ln \frac{60-20}{60-55}$

$\therefore t_{1/2} = 40 \text{ min}$ $\quad ab = 40$

Now, $\frac{[\text{HCl}]}{[\text{ester}]} = \frac{V_0}{V_\infty - V_0} \Rightarrow \frac{[\text{HCl}]}{6.0} = \frac{20}{60-20}$

$\Rightarrow [\text{HCl}] = 3.0 \text{ M}$

$\therefore cd = 03$

**15.** $t_{1/2} = \frac{0.693}{1.386 \times 10^{-3}} = 500 \text{ sec}$

Let the initial moles of $A = x$, then after 500 sec,

| A $\longrightarrow$ | 2B | + C |
|---|---|---|
| $x - \frac{x}{2}$ | $2 \times \frac{x}{2}$ | $\frac{x}{2}$ |
| $= \frac{x}{2}$ | $= x$ | $= \frac{x}{2}$ |

Total moles becomes $\left(\frac{x}{2} + x + \frac{x}{2}\right) = 2x$. As moles becomes double, volume becomes double and hence,

$[A]_{req} = \frac{0.1}{2 \times 2} = 0.025 \text{ M} = 25$ millimole per litre

**16.**

| | A $\longrightarrow$ | 2B | + C |
|---|---|---|---|
| $t = 0$ | $4a$ | 0 | $3a$ |
| $t = t$ | $4a - x$ | $2x$ | $3a + x$ |

From question $(4a - x)(40°) + 2x(10°) + (3a + x)(-30°) = 0°$

$\therefore x = \frac{7}{5}a$

Now, $t = \frac{1}{K} \cdot \ln \frac{4a}{4a - \frac{7}{5}a} = \frac{1}{0.001} \cdot \ln \frac{20}{13} = 500 \text{ min}$

**17.** Exp (1): $r = K'[B]^y$ as $[A_0] >> [B_0]$

$\because t_{7/8} = 3\times t_{1/2} \Rightarrow y = 1$

Exp (2): $r = K''[A]^x$ as $[A_0] << [B_0]$

$\because t_{7/8} = 7\times t_{1/2} \Rightarrow x = 2$

Now, for exp (2) and (3), $t_{1/2} = \dfrac{1}{K''[A_0]} = \dfrac{1}{K[B_0][A_0]}$

$\therefore\ a = \dfrac{10}{2\times 2} = 2.5$ and $b = 7\times 2.5 = 17.5$

And for exp (1) and (4), $t_{1/2} = \dfrac{\ln 2}{K'} = \dfrac{\ln 2}{K[A_0]}$

$\therefore c = 30\times 2 = 60$ and $d = 3\times 60 = 180$

**18.** Percentage yield

$$= \frac{K_2}{K_1+K_2}\times 100 = \frac{4.8}{3.2+4.8}\times 100 = 60\%$$

**19.**

| | A + | 2B + | 3C $\rightleftharpoons$ | D |
|---|---|---|---|---|
| $t=0$ | 1.0 M | 1.0 M | 1.0 M | 0 |
| $t=t$ | $1-x$ $=0.9$ | $1-2x$ $=0.8$ | $1-3x$ $=0.7$ (given) | $x$ $=0.1$ |

$$\therefore r = 2\times 10^{-6}\times(0.9)^2 - \frac{1.4\times 10^{-6}\times(0.1)^2}{0.8\times 0.7}$$

$$= 1.595\times 10^{-6}$$

**20.** $[C] = 0.875 + 0.6 = 1.475$ M

**21.** $K_{eq} = \dfrac{K_f}{K_b} = \dfrac{[B]}{[A]} \Rightarrow \dfrac{1.38/300}{K_b} = \dfrac{0.1}{0.2}$

$\Rightarrow K_b = \dfrac{1.38}{150}\ \text{min}^{-1}$

Now, $t = \dfrac{1}{K_f + K_b}\cdot\ln\dfrac{x_{e,B}}{x_{e,B} - x_B}$

$$= \frac{1}{\frac{1.38}{300}+\frac{2.76}{300}}\cdot\ln\frac{0.1}{0.1 - 0.3\times\frac{25}{100}}$$

$$= \frac{300}{6\times\ln 2}\cdot\ln 4 = 100 \text{ min}$$

**22.** For completion in 30 min, the rate should be increased by $\dfrac{4\times 60}{30} = 8$ times. Assuming temperature coefficient constant, the approximate temperature is $25 + \left(10\times\dfrac{8}{2}\right) = 55°C$.

**23.** $K_1 = K_2 \Rightarrow A_1\cdot e^{-E_{a_1}/RT} = A_2\cdot e^{-E_{a_2}/RT}$

$\therefore \ln\dfrac{A_2}{A_1} = \dfrac{E_{a_2} - E_{a_1}}{RT}$

$\ln\dfrac{10^{14}}{10^{13}} = \dfrac{(171.39 - 152.30)\times 10^3}{8.3\times T}$

or, $T = 1000$ K $= 727°$ C

**24.** $t_{1/2} = \dfrac{\ln 2}{K} = \dfrac{\ln 2}{A\cdot e^{-E_a/RT}}$

or, $1\times 60 = \dfrac{0.7}{5\times 10^{13}\times e^{-149.4\times 10^3/8.3\times T}}$

$\therefore T = 500$ K

**25.** $K_{cat} = K_{uncat}$

or, $A\cdot e^{-E_{a_{cat}}/RT_1} = A\cdot e^{-E_{a_{uncat}}/RT_2}$

or, $\dfrac{E_{a_{cat}}}{T_1} = \dfrac{E_{a_{uncat}}}{T_2} \Rightarrow \dfrac{E_{a_{uncat}} - 20}{400} = \dfrac{E_{a_{uncat}}}{500}$

$\therefore E_{a_{uncat}} = 100$ kJ/mol

CHAPTER 12

# Surface Chemistry

## EXERCISE I (JEE MAIN)

### Adsorption

1. Which of the following interface can never be obtained?
   (a) Liquid–Liquid (b) Solid–Solid
   (c) Liquid–Gas (d) Gas–Gas

2. When a chalk stick is dipped in ink, then
   (a) the coloured pigments in the ink gets absorbed in the chalk.
   (b) the solvent gets adsorbed at the surface of chalk.
   (c) the inside of the chalk will remain as white while the surface will have coloured pigments.
   (d) the inside of the chalk will have coloured pigments while the surface will remain white.

3. Physical adsorption is appreciable at
   (a) higher temperature
   (b) lower temperature
   (c) at room temperature
   (d) 100°C

4. The rate of chemisorption
   (a) decreases with increase of pressure.
   (b) is independent of pressure.
   (c) is maximum at one atmospheric pressure.
   (d) increases with increase of pressure.

5. Chromatography is a technique based on
   (a) solubilities of solute.
   (b) adsorption of solute.
   (c) chemical adsorption followed by dispersion.
   (d) differential adsorption of different constituents of a mixture.

6. Which of the following is not a characteristic of chemisorption?
   (a) Adsorption is irreversible.
   (b) Adsorption decreases with increase in temperature.
   (c) Adsorption is specific.
   (d) Adsorption increases with increase in surface area.

7. Which one of the following is not a correct statement?
   (a) Physical adsorption is reversible in nature.
   (b) Physical adsorption involves van der Waals forces.
   (c) Rate of physical adsorption increases with increase of pressure on the adsorbate.
   (d) High activation energy is involved in physical adsorption.

8. Which gas will be adsorbed on a solid to greater extent?
   (a) A gas having non-polar molecules.
   (b) A gas having higher critical temperature.
   (c) A gas having lowest critical temperature.
   (d) A gas having higher critical pressure.

9. The nature of bonding forces in adsorption is
(a) purely physical like van der Waals force.
(b) purely chemical.
(c) both, physical and chemical always.
(d) none of these

10. Which one of the following is not applicable to chemisorption?
(a) High magnitude of $\Delta H$.
(b) Occurs at higher temperature.
(c) It is reversible.
(d) It forms mono layer.

11. Which characteristic of adsorption is wrong?
(a) Physical adsorption decreases with increase in temperature.
(b) Physical adsorption increases with increase in temperature.
(c) Physical adsorption is a reversible process.
(d) Adsorption is limited only to the surface.

12. Sorption is the term used when
(a) only adsorption takes place.
(b) only absorption takes place.
(c) adsorption and absorption occur simultaneously.
(d) desorption takes place.

13. The volumes of gases $H_2$, $CH_4$, $CO_2$ and $NH_3$ adsorbed by 1 g of charcoal at 288 K are in the order
(a) $H_2 > CH_4 > CO_2 > NH_3$
(b) $CH_4 > CO_2 > NH_3 > H_2$
(c) $CO_2 > NH_3 > H_2 > CH_4$
(d) $NH_3 > CO_2 > CH_4 > H_2$

14. Which plot is the adsorption isobar for chemisorption?

(a)
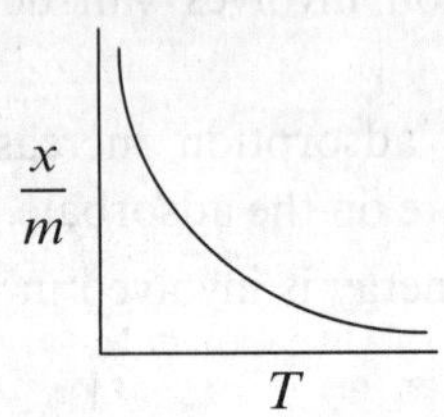


(b)
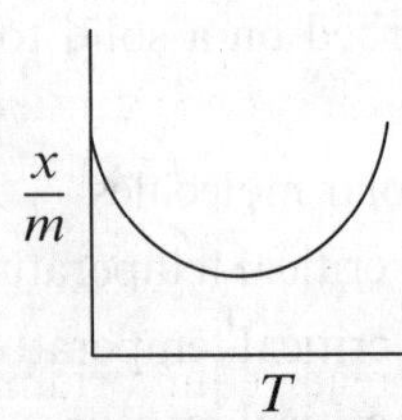


(c)
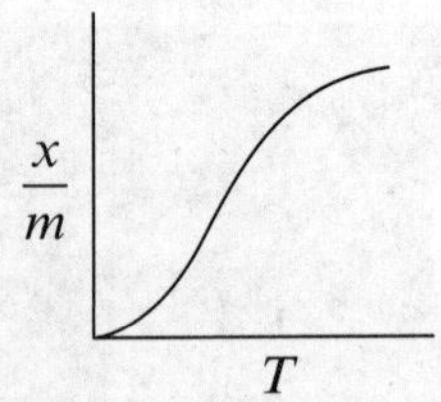


(d)
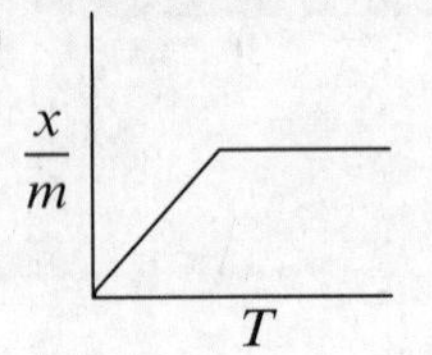


15. In the Freundlich adsorption isotherm, the slope of the straight line graph between log ($x/m$) and log P for the adsorption of a gas on solid is
(a) $k$ (b) log $k$
(c) $n$ (d) $1/n$

16. Which one of the following characteristics is not correct for physical adsorption of a gas on solid?
(a) Adsorption is reversible.
(b) Adsorption increases with increase in temperature.
(c) Absorption is spontaneous.
(d) Both enthalpy and entropy of adsorption are negative.

17. The enthalpy of adsorption in physisorption lies in the range of
(a) (40 – 400) kJ/mol
(b) (80 – 240) kJ/mol
(c) (20 – 40) kJ/mol
(d) (1 – 10) kJ/mol

18. In physical adsorption, the forces associated are
(a) ionic bond (b) covalent bond
(c) van der Waals forces (d) all of these

19. How many layers are adsorbed in chemical adsorption?
(a) 1 (b) 2
(c) 3 (d) 1 or more

20. Rate of physisorption increases with
(a) decrease in temperature.
(b) increase in temperature.
(c) decrease in pressure.
(d) decrease in surface area.

## Catalysis

21. Which one of the following is not an example of homogeneous catalysis?
(a) Formation of $SO_3$ in the chamber process.
(b) Formation of $SO_3$ in the contact process.
(c) Hydrolysis of an ester in the presence of acid.
(d) Decomposition of $KClO_3$ in the presence of $MnO_2$.

22. The decomposition of hydrogen peroxide can be slowed down by the addition of a small amount of acetamide. The latter acts as
(a) inhibitor (b) promoter
(c) moderator (d) poison

23. Efficiency of the catalyst depends on its
(a) molecular mass
(b) number of free valencies
(c) physical state
(d) amount used

24. Which of the following types of metals make the most efficient catalysts?
(a) Transition metals
(b) Alkali metals
(c) Alkaline earth metals
(d) Radioactive metals

25. In the reaction $KMnO_4 + H_2SO_4 + H_2C_2O_4 \rightarrow$ products, $Mn^{++}$ ions act as
(a) positive catalyst (b) negative catalyst
(c) auto catalyst (d) enzyme catalyst

26. In the Haber process of synthesis of $NH_3$,
(a) Mo acts as a catalyst and Fe as a promoter.
(b) Fe acts as a catalyst and Mo as a promoter.
(c) Fe acts as inhibitor and Mo as a catalyst.
(d) Fe acts as promoter and Mo as auto-catalyst.

27. TEL minimizes the knocking effect when mixed with petrol. It acts as
(a) positive catalyst (b) negative catalyst
(c) auto-catalyst (d) induced catalyst

28. Platinized asbestos is used as a catalyst in the manufacture of $H_2SO_4$. It is an example of
(a) homogeneous catalyst
(b) heterogeneous catalyst
(c) auto-catalyst
(d) induced catalyst

29. The catalyst used in the manufacture of hydrogen by Bosch's process is
(a) $Fe_2O_3$ (b) $Cr_2O_3$
(c) $Fe_2O_3 + Cr_2O_3$ (d) Cu

30. In the Ostwald's process for the manufacture of $HNO_3$, the catalyst used is
(a) Fe (b) Pt
(c) $V_2O_5$ (d) Mo

31. In a reversible reaction, a catalyst
(a) increases the rate of forward reaction only.
(b) increases the rate of forward reaction to a greater extent than that of the backward reaction.
(c) increases the rate of forward reaction and decreases than that of the backward reaction.
(d) increases the rate of forward and backward reaction by the same factor.

32. The decomposition of $H_2O_2$ may be checked by adding a small quantity of phosphoric acid. It is an example of
(a) neutralization
(b) negative catalysis
(c) positive catalysis
(d) catalytic poisoning

33. In the manufacture of $H_2SO_4$ by contact process, the presence of $As_2O_3$ acts as
(a) catalytic promoter (b) catalytic poison
(c) induced catalyst (d) auto catalyst

34. Air can oxidize sodium sulphate in aqueous solution but cannot do so in the case of sodium arsenite. However, if, air is passed through a solution containing both sodium sulphite and sodium arsenite then both are oxidized. This is an example of
(a) positive catalysis (b) negative catalysis
(c) induced catalysis (d) auto catalysis

35. A catalyst is a substance which
(a) increases the equilibrium concentration of the product.
(b) changes the equilibrium constant of the reaction.
(c) shortens the time to reach equilibrium.
(d) supplies energy to the reaction.

36. A catalyst
   (a) increases the free energy change in the reaction.
   (b) decreases the free energy change in the reaction.
   (c) does not increase and decreases the free energy change in the reaction.
   (d) can either decrease or increase the free energy change depending on what catalyst we use.

37. The work of enzymes in living system is
   (a) oxygen transfer.
   (b) to provide immunity.
   (c) to catalyse biochemical reactions.
   (d) to provide energy.

38. Which is not correct for heterogeneous catalysis?
   (a) The catalyst decreases the energy of activation.
   (b) The surface of catalyst plays an important role.
   (c) The catalyst forms an intermediate with the reactants at the surface.
   (d) There is no change in the energy of activation.

39. In the case of autocatalysis
   (a) reactant catalyses.
   (b) heat produced in the reaction catalyses.
   (c) product catalyses.
   (d) solvent catalyses.

40. Zeolites are used as a catalyst in
   (a) petrochemical industries during cracking.
   (b) in the preparation of $H_2SO_4$.
   (c) in the hydrolysis of ester.
   (d) All the above.

## Colloids

41. Butter is a colloid formed when
   (a) liquid fat is dispersed in liquid fat.
   (b) liquid fat is dispersed in water.
   (c) water is dispersed in liquid fat.
   (d) solid fat is dispersed in water.

42. Lyophobic colloids are
   (a) reversible (b) irreversible
   (c) water loving (d) solvent loving

43. Sulphur sol contains
   (a) discrete sulphur atoms.
   (b) discrete sulphur molecules.
   (c) water dispersed in solid sulphur.
   (d) large aggregates of sulphur molecules.

44. When freshly precipitated $Fe(OH)_3$ is boiled with water in the presence of few drops of dil. HCl, a hydrated ferric oxide sol is obtained. The method is termed as
   (a) electro dispersion (b) peptization
   (c) ultra filtration (d) dialysis

45. Bleeding is stopped by the application of ferric chloride. This is because
   (a) the blood starts flowing in opposite direction.
   (b) the ferric chloride seals the blood vessel.
   (c) the blood reacts and forms a solid which seals the blood vessel.
   (d) the blood is coagulated and thus, the blood vessel is sealed.

46. $Fe^{3+}$ ions coagulate blood. This shows that blood contains colloidal particles bearing
   (a) negative charge.
   (b) positive charge.
   (c) no charge.
   (d) either positive or negative charge.

47. Gold number is a measure of
   (a) the amount of gold present in the colloidal solution.
   (b) the amount of gold required to break the colloid.
   (c) the amount of gold required to protect the colloid.
   (d) the protective power of lyophilic colloid.

48. On addition of one ml solution of 10% NaCl to 10 ml red gold sol in the presence of 0.0025 g of starch, the coagulation is just prevented. The gold number of starch is
   (a) 25 (b) 2.5
   (c) 0.25 (d) 0.025

**49.** All colloidal solution shows
(a) very high osmotic pressure
(b) high osmotic pressure
(c) low osmotic pressure
(d) no osmotic pressure

**50.** The gold numbers of A, B, C and D are 0.04, 0.002, 10 and 25, respectively. The protective powers of A, B, C and D, respectively, are in the order
(a) A > B > C > D (b) B > A > C > D
(c) D > C > B > A (d) C > A > B > D

**51.** The charge of $As_2S_3$ sol is due to the absorbed
(a) $H^+$ (b) $OH^-$
(c) $O^{2-}$ (d) $S^{2-}$

**52.** The movement of dispersion medium in an electric field when the dispersed particles are prevented from moving is called
(a) cataphoresis
(b) electrophoresis
(c) electro-osmosis
(d) brownian movement

**53.** To coagulate gelatin sol, which of the following is most effective?
(a) NaCl (b) $Na_3PO_4$
(c) $AlCl_3$ (d) Alcohol

**54.** The potential difference between the fixed charged layer and the diffused layer having opposite charge is called
(a) colloidal potential
(b) zeta potential
(c) electrostatic potential
(d) electrode potential

**55.** An examples of micelle is
(a) $As_2O_3$ solution
(b) ruby glass
(c) $Na_2CO_3$ solution
(d) concentrated sodium stearate solution

**56.** A freshly prepared $Fe(OH)_3$ precipitate is peptized by adding $FeCl_3$ solution. The charge on the colloidal particle is due to preferential absorption of
(a) $Cl^-$ ions (b) $Fe^{+++}$ ions
(c) $OH^-$ ions (d) $H^+$ ions

**57.** Hardy–Schulze rule state that
(a) non-electrolytes have better coagulating action on colloids than electrolytes.
(b) sols are coagulated by effective ions whose charge is opposite to that of sol and the ions of higher charge are much more effective than the ions of lower charge.
(c) charge of the ions has no effect on the coagulation of a sol.
(d) sols are coagulated only by those ions whose charge is similar to that of the sol.

**58.** Among the following, which is the correct statement about milk?
(a) Milk is an emulsion of liquid fat in water.
(b) Milk in the emulsion of protein in water.
(c) Milk is stabilized by protein.
(d) Milk is stabilized by fat.

**59.** Colloidal particles carry charge. This can be shown by which of the following effect?
(a) Tyndall effect (b) Electrophoresis
(c) Brownian movement (d) Dialysis

**60.** Which of the following is most effective in causing the coagulation of ferric hydroxide sol?
(a) KCl (b) $KNO_3$
(c) $K_2SO_4$ (d) $K_3Fe(CN)_6$

**61.** On adding $AgNO_3$ solution into KI solution, a negatively charged colloidal sol is obtained when they are in
(a) 100 ml of 0.1 M $AgNO_3$ + 100 ml of 0.1 M KI
(b) 100 ml of 0.1 M $AgNO_3$ + 50 ml of 0.2 M KI
(c) 100 ml of 0.1 M $AgNO_3$ + 50 ml of 0.1 M KI
(d) 100 ml of 0.1 M $AgNO_3$ + 100 ml of 0.15 M KI

**62.** Substances whose solutions can readily diffuse through animal membranes are called
(a) colloids (b) crystalloids
(c) electrolytes (d) non-electrolytes

**63.** The size of the colloidal particles is in between
(a) $10^{-7}$ and $10^{-9}$ cm (b) $10^{-9}$ and $10^{-11}$ cm
(c) $10^{-4}$ and $10^{-7}$ cm (d) $10^{-2}$ and $10^{-3}$ cm

**64.** If dispersed phase is liquid and the dispersion medium is solid, the colloid is known as
(a) sol (b) gel
(c) emulsion (d) foam

65. An emulsion is a colloidal solution consisting of
(a) two solids.
(b) two liquids.
(c) two gases.
(d) one solid and one liquid.

66. Peptization is a process of
(a) precipitating the colloidal particles.
(b) purifying the colloidal sol.
(c) dispersing the precipitate into colloidal sol.
(d) movement of colloidal particles towards the opposite charged electrodes.

67. Colloids are purified by
(a) Brownian motion (b) precipitation
(c) dialysis (d) filtration

68. Which of the following substance gives a positively charged sol?
(a) Gold
(b) Arsenious sulphide
(c) Starch
(d) Ferric hydroxide

69. Blood may be purified by
(a) dialysis (b) electro-osmosis
(c) coagulation (d) filtration

70. Tanning of leather is
(a) colouring of leather by chemicals.
(b) drying process to make the leather hard.
(c) polishing of leather to make it look attractive.
(d) coagulative hardening of the leather by chemicals.

71. Surface tension of lyophilic sol is
(a) lower than that of dispersion medium.
(b) more than that of dispersion medium.
(c) equal than that of dispersion medium.
(d) any of these, depending on the dispersion medium.

72. Which of the following forms a colloidal solution in water?
(a) NaCl (b) Glucose
(c) Starch (d) Barium nitrate

73. When dilute aqueous solution of $AgNO_3$ (excess) is added to KI solution, positively charged sol particles of AgI are formed due to the adsorption of which of the following ion?
(a) $K^+$ (b) $Ag^+$
(c) $I^-$ (d) $NO_3^-$

74. Which one of the following substances is not used for preparing lyophilic sols?
(a) Metal sulphide (b) Gum
(c) Gelatin (d) Starch

75. Which of the following is not a colloidal system?
(a) Sugar in water (b) Muddy water
(c) Concrete (d) Bread

76. Tyndall effect in colloidal solution is due to
(a) absorption of light.
(b) scattering of light.
(c) reflection of light.
(d) presence of electrically charged particles.

77. Which of the following substances will have least CMC value?
(a) $C_6H_{13}COOK$ (b) $C_9H_{19}SO_3Na$
(c) $C_{11}H_{23}NH_4Cl$ (d) $C_8H_{17}COONa$

78. In the styrene foam, the dispersed phase and dispersion medium are, respectively,
(a) gas, gas (b) gas, solid
(c) solid, solid (d) liquid, solid

79. What happens when a lyophilic sol is added to a lyophobic sol?
(a) Lyophilic sol gets protected by the lyophobic sol.
(b) Lyophobic sol is coagulated.
(c) Lyophilic sol is coagulated.
(d) Lyophobic sol is protected by the lyophobic sol.

80. Lyophilic sols are more stable than lyophobic sols because
(a) the colloidal particles have positive charge.
(b) the colloidal particles have no charge.
(c) the colloidal particles are highly solvated.
(d) there are strong electrostatic repulsions between the charged colloidal particles.

# EXERCISE II (JEE ADVANCED)

## Section A (Only one Correct)

1. Which among the following statements is false?
   (a) Increase in pressure increases the amount of adsorption.
   (b) Increase in temperature may decrease the amount of adsorption.
   (c) The adsorption may be monolayered or multilayered.
   (d) Particle size of the adsorbent will not affect the amount of adsorption.

2. Adsorption of gases on solid surface is exothermic because
   (a) chemical bonds are always formed.
   (b) adsorbate decreases the surface energy of adsorbent.
   (c) entropy increases.
   (d) free energy increases.

3. Graph between log $\left(\frac{x}{m}\right)$ and log $P$ is a straight line at an angle 45° with intercept on $y$-axis, 0.3010. The amount (in g) of the gas absorbed per g of the adsorbent when pressure is 0.2 atm is (assume that the adsorption obey Freundlich isotherm)
   (a) 0.4 (b) 0.6
   (c) 0.8 (d) 0.2

4. Desorption of a gas from metal surface follows first-order kinetics. The rate constant of desorption can be given by Arrhenius equation. If the desorption of hydrogen on manganese is found to increases 10 times on increasing the temperature from 600 to1000 K, the activation energy of desorption is (ln 10 = 2.3)
   (a) 6.0 kcal/mol (b) 6.9 kcal/mol
   (c) 3.0 kcal/mol (d) 57.4 kcal/mol

5. According to Freundlich adsorption isotherm, which of the following is correct?
   (a) $\frac{x}{m} \propto P^{1/n}$
   (b) $\frac{x}{m} \propto P^{0}$
   (c) $\frac{x}{m} \propto P^{1}$
   (d) All the above are correct for different ranges of pressure.

6. At 70 K, the adsorption of $N_2(g)$ at iron surface obeys Freundlich adsorption isotherm. The following data is collected experimentally.

| $P$ (/bar) | 4 | 25 | 64 |
|---|---|---|---|
| $\frac{x}{m}$ | 0.2 | 0.5 | 0.8 |

   Here, $\frac{x}{m}$ is the mass (in g) of $N_2(g)$ adsorbed per g of iron at $P$ bar pressure. The moles of $N_2(g)$ adsorbed per g iron at 36 bar and 70 K is
   (a) $\frac{6}{10}$ (b) $\frac{3}{140}$
   (c) $\frac{3}{70}$ (d) $\frac{3}{280}$

7. The following information is available for the reaction R → P.

$\text{R} \longrightarrow \text{P}$: rate = $r_1$
$\text{R} \xrightarrow{\text{A}} \text{P}$: rate = $r_2$
$\text{R} \xrightarrow{\text{B}} \text{P}$: rate = $r_3$
$\text{R} \xrightarrow{\text{C}} \text{P}$: rate = $r_4$
$\text{R} \xrightarrow{\text{D}} \text{P}$: rate = $r_5$
$\text{R} \xrightarrow{\text{A+C}} \text{P}$: rate = $r_6$
$\text{R} \xrightarrow{\text{A+D}} \text{P}$: rate = $r_7$

   If $r_3 < r_1 = r_4 = r_5 < r_7 < r_2 < r_6$, then the only incorrect statement is
   (a) B is an inhibitor.
   (b) C is a catalytic promoter of catalyst A.
   (c) D is a catalytic poison of catalyst A.
   (d) A is a catalytic promoter of catalyst C.

8. Which of the following statements is incorrect about catalyst?
   (a) A catalyst catalyses a spontaneous chemical reaction.
   (b) A catalyst catalyses all chemical reaction, whether spontaneous or non-spontaneous.
   (c) A catalyst provides alternate path of chemical reaction.
   (d) A catalyst does not involve in a chemical reaction.

9. Which of the following statements is incorrect?
(a) Enzymes exist in colloidal state.
(b) Enzymes are catalysts.
(c) Enzymes can catalyse any reaction.
(d) Urease is an enzyme.

10. Which of the following statements is false for catalyst?
(a) A catalyst initiates the reaction.
(b) It does not alter the position of equilibrium in a reversible reaction.
(c) A catalyst remains unchanged in quantity and composition at the end of reaction.
(d) Catalysts are sometimes very specific with respect to a reaction.

11. A catalytic poison renders the catalyst ineffective because
(a) it is preferentially adsorbed on the catalyst.
(b) it adsorbs the molecules of the reactants.
(c) it always combines chemically with the catalyst.
(d) it combines with one of the reactant.

12. Regarding criteria of catalyst, which one of the following statements is not true?
(a) The catalyst is unchanged chemically after the reaction.
(b) A small quantity of catalyst is often sufficient to bring about a considerable amount of the reaction.
(c) In reversible reaction, the catalyst alters the equilibrium position.
(d) The catalyst alters the rate of reaction.

13. Enzyme catalysis is an example of
(a) auto-catalysis
(b) heterogeneous catalysis
(c) homogenous catalysis
(d) induced catalysis

14. Which of the following is wrong in case of enzyme catalysis?
(a) Enzymes work best at an optimum temperature.
(b) Enzymes work best at an optimum pH.
(c) Enzymes are highly specific for substrates.
(d) An enzyme raises activation energy.

15. Which of the following statements about a catalyst is true?
(a) A catalyst accelerates the reaction by bringing down the free energy of activation.
(b) A catalyst does not take part in the reaction mechanism.
(c) A catalyst makes the reaction more feasible by making the $\Delta G^\circ$ more negative.
(d) A catalyst makes the equilibrium constant of the reaction more favourable for the forward reaction.

16. Which of the following is not true with respect to a catalyst?
(a) A catalyst increases the rate of a reaction while inhibitor decreases the reaction rate.
(b) A promoter cannot catalyse the reaction on its own without the presence of a catalyst.
(c) The activity of enzyme catalyst increases in the presence of metal ions, like $Mn^{2+}$, $Co^{2+}$, etc.
(d) When a solid catalyst is used for gaseous reaction, energy released in physisorption increases the rate of a reaction.

17. Which of the following is not a property of hydrophilic sol?
(a) High concentration of dispersed phase can be easily attained.
(b) Coagulation is reversible.
(c) Viscosity and surface tension are nearly as that of water.
(d) The charge on the particles may depend on the pH value and it may be positive, negative or even zero.

18. Which metal sol (dispersion medium being water) cannot be prepared by Bredig's arc method?
(a) K (b) Cu
(c) Au (d) Pt

19. The formed colloid in following reaction is $SnO_2$ + HCl (Excess) →
(a) $SnCl_4/Cl^-$ (b) $SnCl_4/O^{2-}$
(c) $SnCl_4/H^+$ (d) $SnCl_4/Sn^{4+}$

20. When some special substances such as protein particles, blood corpuscles, etc., are separated by a permeable membrane. The process is called
(a) dialysis (b) diffusion
(c) exosmosis (d) endosmosis

21. The simplest way to check whether a system is colloidal is by which of the following method?
(a) Tyndall effect
(b) Brownian movement
(c) Electrodialysis
(d) Finding out particle size

22. Which one of the following colloidal solution is positive sol?
(a) Blood
(b) Clay sol
(c) Smoke
(d) Gelatin in strongly acidic solution

23. When a lyophobic colloidal solution is observed in Tyndall effect, we can see
(a) the size of colloidal particles.
(b) the light scattered by colloidal particles.
(c) the shape of colloidal particles.
(d) the size of particles of dispersion medium.

24. Identify the statement which is correct with respect to surface phenomenon.
(a) Osmotic pressure of rubber sol will be same as that of sucrose solution having same mass mixed in the same mass of water.
(b) A gas may show physisorption at low temperature and chemisorption at high temperature.
(c) Soap sol of sodium palmitate will coagulate near cathode in electrophoresis.
(d) Gold sol on mixing with starch sol causes stabilization of starch sol.

25. Which of the following reaction is not used in the preparation of colloidal solution?
(a) $2H_2S + SO_2 \rightarrow 3S + 2H_2O$
(b) $2Mg + CO_2 \rightarrow 2MgO + C$
(c) $FeCl_3 + 3H_2O \rightarrow Fe(OH)_3 + 3HCl$
(d) $2AuCl_3 + 3SnCl_2 \rightarrow 2Au + 3SnCl_4$

26. A quantity of $1.9 \times 10^{-4}$ g of the metal having density 19 g/ml is dispersed in 1 L of water to give a sol having spherical metal particles of radius 10 nm. The approximate number of metal sol particles per $cm^3$ of the sol is
(a) $2.39 \times 10^9$ (b) $4 \times 10^{10}$
(c) $1.9 \times 10^9$ (d) $2.8 \times 10^{-12}$

27. Among the following, the surfactant that will form micelles in aqueous solution at the lowest molar concentration at ambient conditions is
(a) $CH_3(CH_2)_{15}N^+(CH_3)_3Br^-$
(b) $CH_3(CH_2)_{11}OSO_3^-Na^+$
(c) $CH_3(CH_2)_6COO^-Na^+$
(d) $CH_3(CH_2)_{11}N^+(CH_3)_3Br^-$

28. Among the electrolytes $Na_2SO_4$, $CaCl_2$, $Al_2(SO_4)_3$ and $NH_4Cl$, the most effective coagulating agent for $Sb_2S_3$ is
(a) $Na_2SO_4$ (b) $CaCl_2$
(c) $Al_2(SO_4)_3$ (d) $NH_4Cl$

29. Methylene blue, from its aqueous solution, is adsorbed on activated charcoal at 25°C. For this process, which of the following statement is correct?
(a) The adsorption requires activation at 25°C.
(b) The adsorption is accompanied by a decrease in enthalpy.
(c) The adsorption increases with increase in temperature.
(d) The adsorption is irreversible.

30. Match the columns.

| Column I | Column II |
|---|---|
| (1) Coagulation | (P) Scattering of light |
| (2) Peptization | (Q) Purification of colloidal solution |
| (3) Tyndall effect | (R) Addition of an electrolyte |
| (4) Dialysis | (S) Precipitation of colloidal solution |

| | 1 | 2 | 3 | 4 |
|---|---|---|---|---|
| (a) | P | Q | R | S |
| (b) | S | R | P | Q |
| (c) | R | S | Q | P |
| (d) | Q | R | P | S |

## Section B (One or More than one Correct)

1. Which of the following statement(s) is/are correct regarding adsorption?
   (a) Activated adsorption is highly specific.
   (b) van der Waals adsorption is reversible.
   (c) Adsorption is always exothermic.
   (d) Adsorption is an iso-entropic process.

2. Adsorption is accompanied by
   (a) decrease in entropy of the system.
   (b) decrease in enthalpy of the system.
   (c) decrease in free energy of the system.
   (d) decrease in surface energy of adsorbent.

3. On increasing temperature at constant pressure, the amount of gas adsorbed at the surface per unit mass of adsorbent
   (a) decreases in case of physisorption.
   (b) decreases in case of chemisorption.
   (c) increases in case of chemisorption.
   (d) increases in case of physisorption.

4. Which of the following statement(s) regarding adsorption is/are not correct?
   (a) Surface particles of adsorbent are not in the same environment as the particles inside the bulk.
   (b) During adsorption, there is always decrease in the residual forces on the surface.
   (c) At equilibrium in adsorption, the values of $\Delta H$ and $\Delta S$ become equal.
   (d) The extent of adsorption increases with increase in surface area per unit mass of adsorbent.

5. Which of the following statement(s) is/are incorrect regarding activated adsorption?
   (a) Its extent increases with increase in pressure and it may change to multilayer adsorption at high pressure.
   (b) Its extent increases with decrease in temperature and it may change to multilayer adsorption at low temperature.
   (c) Its extent increases with increase in surface area of adsorbent.
   (d) When oxygen is adsorbed at graphite surface, none of the gas leaves the surface on strong heating.

6. Which of the following statement is correct with regard to adsorption of a gas on solid surface?
   (a) Adsorption is always exothermic.
   (b) Physisorption may transform into chemisorption at high temperature.
   (c) Physisorption increases with increasing temperature but chemisorption decreases with increasing temperature.
   (d) Chemisorption is more exothermic than physisorption however, it is relatively slow due to higher energy of activation.

7. Zeolites
   (a) are microporous aluminosilicates.
   (b) have the general formula $M_{x/n}\cdot(AlO_2)_x\cdot SiO_2\cdot mH_2O$
   (c) have pore sizes between 260 pm to 740 pm.
   (d) acts as shape-selective catalyst.

8. Which of the following acts as negative catalyst?
   (a) Tetraethyl lead as antiknock compound.
   (b) Glycerol in decomposition of $H_2O_2$.
   (c) Ethanol in the oxidation of chloroform.
   (d) Lindlar's catalyst.

9. According to adsorption theory of catalysis, the reaction rate increases, because
   (a) adsorption produces heat which increases the rate of reaction.
   (b) in the process of adsorption, the kinetic energy of the molecules increases.
   (c) the concentration of reactants at the active centres becomes high due to adsorption.
   (d) the activation energy of the reaction becomes high due to adsorption.

10. Identify the incorrect statement(s) from the following.
   (a) On changing the catalyst in a heterogeneous catalysis, the product of reaction may change.
   (b) Enzyme catalysed reactions are elementary (single step) reactions.
   (c) Lactobacilli enzyme is responsible for the conversion of milk into curd.
   (d) A catalytic promoter may increase the reaction rate even in the absence of catalyst.

EXERCISE II

**11.** Which of the following option(s) is/are correct regarding catalytic activity?

(a) Conversion of proteins into amino acids by hydrolysis in intestine is catalysed by the enzyme pancreatic trypsin.

(b) In case of fever, the activity of biochemical catalyst gets adversely affected.

(c) In heterogeneous catalysis, the reactants must get adsorbed reasonably strong on the catalyst surface but not so strongly that desorption of product becomes impossible.

(d) In the decomposition of methyl urea, enzyme urease acts as a catalyst.

**12.** Enzymes are highly efficient catalyst. An enzyme 'catalase' increases the rate of decomposition of $H_2O_2(aq)$ into $H_2O(l)$ and $O_2(g)$ $e^{20}$ times at 300 K. The activation energy of enzyme catalysed reaction is 2.0 kcal/mol. Which of the following information(s) is/are correct assuming that the value of pre-exponential factor is unaffected by temperature change or use of catalyst and the activation energy is independent of temperature.

(a) The activation energy of uncatalysed reaction is 14.0 kcal/mol.

(b) Uncatalysed reaction is an elementary (single step) reaction.

(c) Catalysed reaction is a complex (multi step) reaction.

(d) On increasing the temperature, the ratio of rates of catalysed and uncatalysed reactions becomes less than $e^{20}$

**13.** During electro-osmosis of $Fe(OH)_3$ sol,

(a) sol particles move towards anode.

(b) sol particles move towards cathode.

(c) the dispersion medium move towards anode.

(d) the sol particles do not move in either direction.

**14.** Colloidal particles in a sol can be coagulated by

(a) heating.

(b) addition of electrolyte.

(c) addition of oppositely charged sol.

(d) addition of dispersion medium.

**15.** The nature of electric charge on colloidal particles can be experimentally determined by which of the following process?

(a) Brownian movement (b) Electrophoresis

(c) Electro-osmosis (d) Ultramicroscope

**16.** Identify the statement(s) which is/are not correct with respect to the surface phenomenon.

(a) If on adding electrolyte in an emulsion, the conductivity decreases, then it will be oil in water type emulsion.

(b) Tyndall effect is observed when the refractive indices of the dispersed phase and dispersion medium differ largely.

(c) Macromolecular colloids are generally lyophobic in nature.

(d) Gases which can react with the adsorbents generally show chemisorption.

**17.** Select the correct statement(s) from the following.

(a) If the critical micelle concentration (CMC) of a soap is $10^{-3}$ M, then $10^{-4}$ M solution of this soap will have colloidal nature.

(b) Charge on the colloidal particles may be determined with the help of dialysis.

(c) Peptization is the process of conversion of a colloidal sol into precipitate by adding some electrolyte.

(d) When some oil-soluble dye is added in the water in oil emulsion, the background becomes coloured.

**18.** For the coagulation of a particular colloidal solution, the order of coagulation power of some electrolytes is in the order $Na_3PO_4 > BaSO_4 > AlCl_3$. Which of the following information(s) may be correct for such colloid?

(a) In electro-osmosis, the dispersion medium move towards the anode.

(b) Diffused layer around colloidal particles in the electric double layer mostly contains negatively charged particles.

(c) It may be basic dye such as methylene blue.

(d) It may be metal sol.

19. The following experimental data is obtained for the osmotic pressure of the solution by adding different moles of $C_{17}H_{35}COONa$ in sufficient water to get 1.0 L solution in each case.

| Moles of $C_{17}H_{35}COONa$ | 0 | 0.01 | 0.02 | 0.03 | 0.04 |
|---|---|---|---|---|---|
| $\frac{\pi}{RT}$ (moles/litre) | 0 | 0.02 | 0.04 | 0.058 | 0.076 |

Select the correct information(s) regarding the nature of solution.

(a) At 0.02 M concentration, the solution behaves as true solution.

(b) At 0.04 M concentration, the solution behaves as colloidal solution.

(c) The critical micelle concentration (CMC) of $C_{17}H_{35}COONa$ is in between 0.02 M and 0.03 M.

(d) At 0.05 M concentration, the mixture of $C_{17}H_{35}COONa$ and water will be homogeneous.

20. Select the incorrect statement related to colloids.

(a) Muddy water can be coagulated more effectively by alums as compared to NaCl.

(b) Surface tension of lyophilic colloids is less than that of dispersion medium.

(c) When excess of $AgNO_3$ is added in aqueous KI solution and the sol is subjected to electrophoresis, coagulation occurs near the anode.

(d) Cloud bursting in atmosphere occurs due to large amount of water present in the cloud.

21. Which of the following statement is incorrect?

(a) The conductivity of a soap solution decreases sharply at CMC.

(b) Tyndall effect is more effective in gold sol in comparison to the rubber sol.

(c) The elevation in boiling point of an alcoholic solution of sulphur is less than that of its sol in water if mass of sulphur present per unit volume of mixture is same in both cases.

(d) CMC value of $CH_3(CH_2)_9NH_3Cl$ will be less than that of $CH_3(CH_2)_6COONa$.

22. Which of the following statement is correct?

(a) Potassium ferrocyanide can cause greater coagulation in a basic dye as compared to $Na_2HPO_3$.

(b) A starch aquasol can act as protective colloid for $Fe(OH)_3$ sol.

(c) The slope of the Freundlich adsorption isotherm ($\log\frac{x}{m}$ vs. $\log P$) keeps on changing for a long range of pressure and is constant over a limited range of pressure.

(d) Higher the zeta potential, greater is the stability of colloid.

23. Which of the following will show Tyndall effect?

(a) Aqueous solution of sodium stearate above critical micelle concentration (CMC).

(b) Aqueous solution of sodium stearate below critical micelle concentration (CMC).

(c) Aqueous solution of sodium stearate above Krafft temperature.

(d) Aqueous solution of sugar.

24. Choose the correct reason(s) for the stability of the lyophobic colloidal particles.

(a) Preferential adsorption of ions on their surface from the solution.

(b) Preferential adsorption of solvent on their surface from the solution.

(c) Attraction between different particles having opposite charges on their surface.

(d) Potential difference between the fixed layer and the diffused layer of opposite charges around the colloidal particles.

25. The dispersed phase in colloidal iron(III) hydroxide and colloidal gold is positively and negatively charged, respectively. Which of the following statement(s) is/are not correct?

(a) Magnesium chloride solution coagulates the gold sol more readily than the iron(III) hydroxide sol.

(b) Sodium sulphate solution causes coagulation in both sols.

(c) Mixing of the sols has no effect.

(d) Coagulation in both sols can be brought about by electrophoresis.

## Section C (Comprehensions)

### Comprehension I

**Whenever a mixture of gases is allowed to come in contact with a particular adsorbent under the same conditions, the relatively stronger adsorbate is adsorbed to greater extent irrespective of its amount present. For example, $H_2O$ is adsorbed on silica gel to greater extent than $N_2$ or $O_2$. This shows that some adsorbates are preferentially adsorbed. It is also observed that the preferentially adsorbable adsorbents can displace a weakly adsorbable adsorbate from the surface of an adsorbent.**

1. Which of the following gas is adsorbable to the maximum extent?
   (a) He (b) Ne
   (c) Ar (d) Xe
2. Which of the following gas can displace all the remaining gases?
   (a) $O_2$ (b) $N_2$
   (c) CO (d) $H_2$
3. When the temperature is increased,
   (a) the extent of adsorption increases.
   (b) the extent of adsorption decreases.
   (c) the extent of adsorption remains unchanged.
   (d) the extent of adsorption first decreases and then increases.

### Comprehension II

**A chemist was studying the phenomenon of adsorption by placing blood charcoal in KCl solution. He/She observed the difference in behaviour with dilute KCl solution and with concentrated KCl solution. He/She also made a detailed study of the adsorption of gases on solid adsorbents. He/She observed that at the same temperature, different amounts of gases, like $NH_3$, $SO_2$, $CO_2$, HCl were adsorbed by the same amount of the adsorbent. He/She further studied the effect of temperature on adsorption and observed that in some cases, adsorption showed a regular trend while in some other cases, the trend was not regular.**

4. Which of the following result, the chemist must have observed about his studies with KCl solution?
   (a) Dilute KCl solution shows no adsorption, whereas concentrated KCl solution shows adsorption.
   (b) Concentrated KCl solution shows no adsorption, whereas dilute KCl solution shows adsorption.
   (c) Dilute KCl solution shows positive adsorption, whereas concentrated KCl shows negative adsorption.
   (d) Concentrated KCl shows positive adsorption, whereas dilute KCl shows negative adsorption.
5. The correct order of adsorption of gases studied will be
   (a) $NH_3 > SO_2 > CO_2 > HCl$
   (b) $CO_2 > SO_2 > NH_3 > HCl$
   (c) $SO_2 > NH_3 > HCl > CO_2$
   (d) $HCl > SO_2 > NH_3 > CO_2$
6. Which of the following is correct?
   (a) Adsorption is always exothermic.
   (b) Adsorption is always endothermic.
   (c) Physical adsorption is endothermic, whereas chemical adsorption is exothermic.
   (d) Chemical adsorption is endothermic, whereas physical adsorption is exothermic.

### Comprehension III

**Specific surface area of a solid adsorbent is the surface area of the unit mass of the adsorbent. For any adsorbent, the specific surface area may be increased by taking the adsorbent in powdered form or by increasing the pores in the solid.**

7. If the unit mass of a solid taken as a cube of volume 8 $cm^3$ is powdered into identical $10^{12}$ cubes, then the specific surface area of the solid increases by
   (a) $10^{12}$ times (b) $10^4$ times
   (c) $10^{-8}$ times (d) $10^3$ times

8. A volume of 112 $cm^3$ hydrogen gas is adsorbed uniformly at the surface of 5 g palladium at 273°C and 2 atm. If the effective surface area of each hydrogen molecule is 0.4 $nm^2$, then the specific surface area of palladium is ($N_A = 6 \times 10^{23}$)

(a) $1.2 \times 10^7$ $cm^2/g$
(b) $6.0 \times 10^7$ $cm^3/g$
(c) $2.4 \times 10^6$ $cm^2/g$
(d) $6 \times 10^5$ $cm^2/g$

## Comprehension IV

**The adsorption of a gas at a metal surface is called occlusion. The extent of gas adsorbed at metal surface at a certain temperature depends on the pressure as**

$$\frac{x}{m} = K \cdot P^{1/n}$$

**where $K$ and $n$ are constants for a particular combination of gas and metal at fixed temperature and $\frac{x}{m}$ is the mass of gas adsorbed per unit mass of metal. For hydrogen gas at platinum surface, the following graph is obtained.**

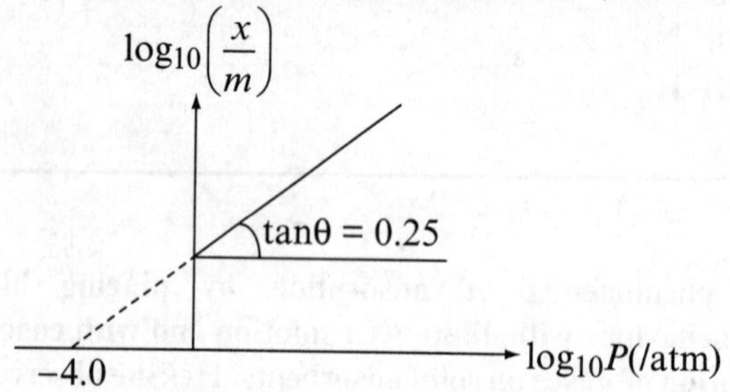


9. The correct values of 'K' and '$n$' are, respectively,

(a) 1.0, 0.25
(b) 10.0, 4.0
(c) 10.0, 0.25
(d) 0.10, 4.0

10. The mass of hydrogen gas adsorbed on the surface of 10 g platinum at 16 atm is

(a) 200 g
(b) 40 g
(c) 20 g
(d) 4 g

11. The pressure at which 810 g of hydrogen gas will adsorb on the surface of 1.0 g platinum is

(a) 2 atm
(b) 9 atm
(c) 3 atm
(d) 4 atm

## Comprehension V

**The clouds consist of charged particles of water dispersed in air. Some of them are +vely charged and some are –vely charged. When +vely charged clouds come closer, they cause lightening and thundering, whereas when +vely and –vely charged clouds come closer, they cause heavy rain by aggregation of minute particles. It is possible to cause artificial rain by throwing electrified sand or silver iodide from an aeroplane and thus coagulating the mist hanging in air.**

12. When excess of $AgNO_3$ is treated with KI solution, AgI forms

(a) positively charged sol
(b) negatively charged sol
(c) neutral sol
(d) true solution

13. AgI helps in artificial rain because

(a) it helps in condensation process
(b) it helps in dispersion process
(c) it helps in coagulation
(d) all of these

14. Smoke screens consist of

(a) fine particles of $TiO_2$ dispersed in air by aeroplane.
(b) fine particles of AgI dispersed in air by aeroplane.
(c) fine particles of $Al_2O_3$ dispersed in air by aeroplane.
(d) fine particles of carbon dispersed in air by aeroplane.

## Comprehension VI

**The charge on colloidal particle is due to the selective adsorption of ions present in the dispersion medium. The adsorbed ion on the colloidal particles is responsible for the development of electric charge. For example, the selective adsorption of $H^+$ ions lead to the development of positive charge on the colloidal particles and the selective adsorption of OH- ions lead to the development of negative charge on the colloidal particles.**

**15.** A colloidal solution of ferric hydroxide sol is prepared by the hydrolysis of $FeCl_3$. The colloidal particles of ferric hydroxide is
(a) positively charged
(b) negatively charged
(c) neutral
(d) none of these

**16.** The colloidal solution of $AgNO_3$ in AgI can be represented as
(a) AgI, $Ag^+$, $I^-$ (b) AgI, $Ag^+$, $NO_3$–
(c) $Ag^+$, $NO_3^-$, AgI (d) $AgNO_3$, $Ag^+$, $I^-$

**17.** Which of the following ion forms primary electrical double layer on the colloidal particle of $As_2S_3$?
(a) $As^{3+}$ (b) $S^{2-}$
(c) $H^+$ (d) $OH^-$

## Comprehension VII

**The protective power of the lyophilic colloids is expressed in terms of gold number, a term introduced by Zsigmondy. Gold number is the number of milligrams of the protective colloid which prevents the coagulation of 10 ml of red gold sol, when 1 ml of a 10% solution of sodium chloride is added to it. Thus, smaller the gold number of the lyophilic colloid, the greater is its protective power.**

**18.** On addition of 1 ml of 10% solution of NaCl to 10 ml red gold sol in the presence of 0.025 g of the starch, the coagulation is just prevented. The gold number of starch is
(a) 0.025 (b) 0.25
(c) 2.5 (d) 25

**19.** Gold number gives an indication of
(a) protective nature of colloid.
(b) purity of gold in suspension.
(c) charge on a colloidal solution of gold.
(d) mole of gold per litre of colloidal solution.

**20.** Which of the following statements is true?
(a) Lower the gold number, more will be its protective power.
(b) Higher the gold number, more will be its protective power.
(c) Higher the coagulation value, more will be its coagulation power.
(d) Higher the coagulation value, less will be its coagulating power.

## Section D (Assertion – Reason)

The following questions consist of two statements. Mark the answer as follows.

**(a)** If both statements are CORRECT, and **Statement II** is the CORRECT explanation of **Statement I.**

**(b)** If both statements are CORRECT, and **Statement II** is NOT the CORRECT explanation of **Statement I.**

**(c)** If **Statement I** is CORRECT, but **Statement II** is INCORRECT.

**(d)** If **Statement I** is INCORRECT, but **Statement II** is CORRECT.

**1.** **Statement I:** When a finely divided active carbon or clay is stirred into a dilute solution of dye, the intensity of colour in the solution is decreased.

**Statement II:** The dye is adsorbed on the solid surface.

2. **Statement I:** The particles at the surface of solid have higher average energy than the particles in the bulk of solid.
   **Statement II:** Due to adsorption, the surface energy of solid decreases.

3. **Statement I:** Adsorption always decreases with increase in temperature.
   **Statement II:** Adsorption is temperature dependent.

4. **Statement I:** For adsorption, $\Delta G$, $\Delta S$ and $\Delta H$, all have negative values.
   **Statement II:** Adsorption is a spontaneous exothermic process in which randomness decreases due to force of attraction between adsorbent and adsorbate.

5. **Statement I:** A gas with higher critical temperature gets adsorbed to more extent than a gas with lower critical temperature.
   **Statement II:** The easily liquefiable gases get adsorbed to greater extent.

6. **Statement I:** A catalyst provides an alternative path to the reaction in which conversion of reactants into products takes place quickly.
   **Statement II:** The catalyst forms an activated complex of lower potential energy than the uncatalysed reaction, due to which more number of molecules become able to cross the energy barrier per unit time.

7. **Statement I:** Lyophilic colloids are known as reversible sols.
   **Statement II:** Lyophilic sols are liquid loving.

8. **Statement I:** Addition of small amount of $AgNO_3$ to excess of KI solution gives negative sol, whereas addition of small amount of KI to excess of $AgNO_3$ solution gives positive sol of AgI.
   **Statement II:** The sol particles adsorb the common ions present in solution and acquire their charge.

9. **Statement I:** Sol particles show Tyndall effect.
   **Statement II:** The scattering of light occurs due to suitable size of sol particles.

10. **Statement I:** The micelle formed by sodium stearate in water has $-COO^-$ groups at the surface.
    **Statement II:** Surface tension of water is reduced by the addition of stearate.

11. **Statement I:** A colloidal solution of cellulose nitrate in water cannot be prepared directly.
    **Statement II:** Lyophilic sols can be prepared directly by mixing the substance with the dispersion medium.

12. **Statement I:** Colloidal sols scatter light while true solutions do not.
    **Statement II:** The particles in the colloidal sol move much slower than that of the true solution.

13. **Statement I:** Colloidal solution is electrically neutral.
    **Statement II:** Due to similar nature of the charge carried by the particles of dispersed phase in lyophobic colloids, they repel each other and do not combine to form bigger particles.

14. **Statement I:** Isoelectric point is pH at which colloids can move towards either of electrode.
    **Statement II:** At isoelectric point, colloidal particles become electrically neutral.

15. **Statement I:** When $AgNO_3$ is added in excess of KI, colloidal particles gets attracted towards anode in electrophoresis.
    **Statement II:** Colloidal particles adsorb common ions (excess) and thus become charged.

16. **Statement I:** A colloid gets coagulated by addition of an electrolyte.
    **Statement II:** The rate of coagulation depends on the magnitude and sign of the charge of the coagulant ion.

17. **Statement I:** $Fe^{3+}$ can be used for coagulation of $As_2S_3$ sol.
    **Statement II:** $Fe^{3+}$ reacts with $As_2S_3$ to give $Fe_2S_3$.

18. **Statement I:** The conversion of fresh precipitate to colloidal state is called peptization.
    **Statement II:** It is caused by addition of common ions, which neutralizes the charge on colloidal particles.

19. **Statement I:** Proteins, starch and rubber are lyophilic colloids.
    **Statement II:** They have strong interaction with the dispersion medium.

20. **Statement I:** Micelles are formed by surfactant molecules above the critical micellar concentration (CMC).
    **Statement II:** The conductivity of a solution having surfactant molecules decreases sharply at the CMC.

## Section E (Column Match)

1. Match the columns.

| Column I (Colloidal solution) | Column II (Dispersed phase) | Column III (Dispersion medium) |
|---|---|---|
| (A) Colloidion | (P) Water | (U) Ethanol |
| (B) Fog | (Q) Cellulose | (V) Oil |
| (C) Butter | (R) Fat | (W) Air |
| (D) Milk | | (X) Water |

2. Match the columns.

| Column I | Column II |
|---|---|
| (A) Cottrell precipitation | (P) Purification of blood. |
| (B) Electrophoresis | (Q) Precipitation of colloidal particles by addition of electrolytes. |
| (C) Hemodialysis | (R) Removal of pollutants from industrial waste gases. |
| (D) Coagulation | (S) Movement of charged colloidal particles towards oppositely charged electrode. |

3. Match the columns.

| Column I | Column II |
|---|---|
| (A) Removal of water by silica gel. | (P) Absorption |
| (B) Removal of water by anhydrous $CaCl_2$. | (Q) Adsorption |
| (C) Surface phenomena | (R) Evaporation of liquid. |
| (D) Bulk phenomena | (S) Boiling of liquid. |

4. Match the columns.

| Column I (Reaction) | Column II (Suitable enzyme catalyst) |
|---|---|
| (A) Starch → Maltose | (P) Urease |
| (B) Sucrose → Glucose + Fructose | (Q) Diastase |
| (C) Urea → Ammonia + $CO_2$ | (R) Zymase |
| (D) Glucose → Ethanol + $CO_2$ | (S) Invertase |

## Section F (Subjective)

### Single-digit Integer Type

1. Molecular formula of starch can be represented as $(C_6H_{10}O_5)_n$. If the gold number of one such starch is 6.48 and 0.01 millimoles of this starch is required to be added in 10 ml of red gold sol to just prevent coagulation by adding 1 ml of 10% sodium chloride solution, then the value of '$n$' is

2. Adsorption of hydrogen gas over palladium surface obey Freundlich's adsorption isotherm as shown in the graph. The mass (in g) of hydrogen gas adsorbed on the surface of 1.0 g palladium at a pressure of $8.1 \times 10^{-3}$ atm is

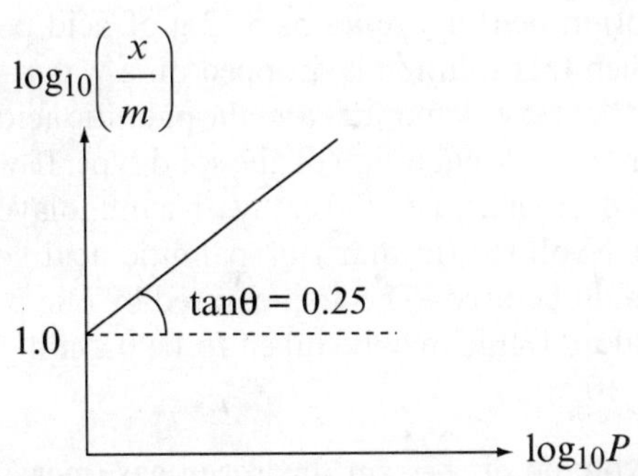


3. The desorption of gas molecules from the adsorbent surface obeys Arrhenius equation. The average time (in sec) up to which a $N_2$ molecule may remain adsorbed at Pt-surface at 400 K is (Arrhenius parameters for desorption are: $A = 1.25 \times 10^8\ s^{-1}$, $E_a = 16$ kcal/mol; $e^{20} = 5 \times 10^8$)

4. Among the following colloids, the number of colloids having negatively charged colloidal particles is:
Soap solution of sodium palmitate, sol of AgCl obtained on dissolving excess $AgNO_3$ into KCl solution, gold sol, $Fe(OH)_3$ sol, silicic acid sol, basic dye, acidic dye, metal sulphide sol, sol of AgCl obtained by dissolving excess KCl in $AgNO_3$ solution.

5. Nitrogen gas adsorbed on charcoal to the extent of 0.387 $cm^3/g$ at a pressure of 1.6 atm and at temperature of 200 K, but at 250 K the same amount of adsorption was achieved only when the pressure was increased to 32 atm. The magnitude of molar enthalpy of adsorption (in kcal/mol) of nitrogen on charcoal is (ln 20 = 3.0)

6. A quantity of 1.0 g of charcoal adsorbs 100 ml 0.5 M $CH_3COOH$ to form a monolayer, and thereby the molarity of $CH_3COOH$ reduces to 0.49 M. The surface area (in $10^{-19}$ $m^2$) of the charcoal adsorbed by each molecule of acetic acid is (Surface area of charcoal $= 3.0 \times 10^2$ $m^2/g$) ($N_A = 6 \times 10^{23}$).

7. Twenty per cent of surface sites are occupied by $N_2$ molecules. The density of surface site is 6.023 $\times$ $10^{14}$ $cm^{-2}$ and the total surface area is 1000 $cm^2$. The catalyst is heated to 300 K while $N_2$ is completely desorbed into a pressure of 0.001 atm and volume of 2.46 $cm^3$. The number of active sites occupied by each $N_2$ molecule is

8. A monolayer of $N_2$ molecules (effective area 0.15 $nm^2$) is absorbed on the surface of 1.00 g of Fe/$Al_2O_3$ catalyst at 77 K, the boiling point of liquid nitrogen. Upon warming, the nitrogen occupies 2.24 $cm^3$ at 0°C and 760 torr. The surface area of catalyst (in $m^2$) is (Avogadro's number $N_A$ = $6 \times 10^{23}$)

9. The minimum number of phases in a colloidal system is

10. The chemisorption of hydrogen on manganese is activated, but only weakly, so careful measurements have shown that it proceeds 20 times faster at 1000 K than at 600 K. The activation energy for chemisorption (in kcal/mol) is ($e^3 = 20$)

## Four-digit Integer Type

1. A volume of 585 ml of 1%(w/w) NaCl solution of density 1.2 g/ml is required for complete coagulation of 200 ml of a gold sol in two hours. The coagulation value of NaCl (in millimole/L) is

2. A solution of palmitic acid (Molar mass = 256 g/mol) in benzene contains 5.12 g of acid per $dm^3$. When this solution is dropped on a water surface the benzene evaporates and the palmitic acid forms a monomolecular film of the solid type. If we wish to cover an area of 480 $cm^2$ with a monolayer, then what volume (in $mm^3$) of palmitic acid solution should be used? The area covered by one palmitic acid molecule may be taken to be 0.2 $nm^2$. ($N_A$ = $6 \times 10^{23}$)

3. A volume of $\frac{224}{\pi}$ $cm^3$ hydrogen gas (measured at 0°C and 1 atm) forms a uniform monolayer at 25 g copper. The density of liquid hydrogen is $\frac{0.16}{\pi}$ g/$cm^3$. The specific surface area of copper (in $m^2/g$) is ($N_A = 6 \times 10^{23}$)

4. The average time for which an oxygen atom remains adsorbed to a tungsten surface is 0.36 s at 2500 K and 0.72 s at 2000 K. The activation energy for desorption (in kcal/mol) is (ln 2 = 0.7)

5. The activation energy for the desorption of hydrogen gas from iron and charcoal surfaces are 8.0 and 20.0 kcal/mol, respectively. Assume that the pre-exponential factor have the same value of $10^{12}$ $s^{-1}$ in both and the ratio of half-lives of hydrogen molecules at the iron and charcoal surfaces is 1:$e^x$ at 327°C, the value of $x$ is

6. It was found that 0.10 mg of an adsorbate covered 0.06 $m^2$ of a solid surface. If the molar mass of the adsorbate is 0.25 kg/mol, then the effective area of each adsorbate molecule (in $10^{-20}$ $m^2$) is ($N_A = 6.0 \times 10^{23}$).

7. A sample of charcoal weighing 6.00 g was brought into contact with a gas contained in a vessel of 1.52 litre capacity at 27°C. The pressure of the gas was found to fall from 700 to 400 mm of Hg. The volume of the gas (in ml) reduced to 0°C and 1 atm that is adsorbed per gram of adsorbent under the experimental condition is (neglect the volume of the solid)

**8.** At 0°C and 1 atm pressure, the volume of nitrogen gas required to cover a sample of silica gel, assuming Langmuir monolayer adsorption is found to be 168 $cm^3/g$ of the gel. The surface area (in $m^2$) per gram of silica gel is (the area occupied by each nitrogen molecule is 0.16 $nm^3$) ($N_A = 6 \times 10^{23}$)

**9.** The $\Delta H$ of adsorption at constant amount of vapour adsorbed is called the isosteric enthalpy of adsorption $\Delta_\Theta H$. The adsorption of $N_2$ on charcoal amounted to 0.894 $cm^3$ (STP)/g at (0.4 atm and 200 K) and (at 59.2 atm and $T$ K). If $\Delta_\Theta H = -16.628$ kJ/mol, then the value of $T$ is (ln 148 = 5)

**10.** The mechanism of enzyme catalysed reaction is given by Michaelis and Menten as follows.

Step I: $E + S \underset{K_{-1}}{\overset{K_1}{\rightleftharpoons}} ES$ (Fast)

Step II: $ES \xrightarrow{K_2} P + E$ (Slow)

The rate of product formation may be given as $+\frac{dP}{dt} = \frac{K_1 K_2 [E]_0 [S]}{K_{-1} + K_2 + K_1 [S]}$, where $[E]_0$ is the total enzyme concentration. For an enzyme-substrate system obeying simple Michaelis and Menten mechanism, the rate of product formation when the substrate concentration is very large has the limiting value 0.02 $Ms^{-1}$. At a substrate concentration of 250 $mg/dm^3$, the rate is half this value. The value of $K_1/K_{-1}$ (in $dm^3/kg$) assuming that $K_2 << K_{-1}$ is

## Answer Keys

Exercise I

### Adsorption

1. (d) 2. (c) 3. (b) 4. (d) 5. (d) 6. (b) 7. (d) 8. (b) 9. (d) 10. (c)
11. (b) 12. (c) 13. (d) 14. (c) 15. (d) 16. (b) 17. (c) 18. (c) 19. (a) 20. (b)

### Catalysis

21. (b) 22. (a) 23. (b) 24. (a) 25. (c) 26. (b) 27. (b) 28. (b) 29. (c) 30. (b)
31. (d) 32. (b) 33. (b) 34. (c) 35. (c) 36. (c) 37. (c) 38. (d) 39. (c) 40. (a)

### Colloids

41. (c) 42. (b) 43. (d) 44. (b) 45. (d) 46. (a) 47. (d) 48. (b) 49. (c) 50. (b)
51. (d) 52. (c) 53. (c) 54. (b) 55. (d) 56. (b) 57. (b) 58. (a) 59. (b) 60. (d)
61. (d) 62. (b) 63. (c) 64. (b) 65. (b) 66. (c) 67. (c) 68. (d) 69. (a) 70. (d)
71. (a) 72. (c) 73. (b) 74. (a) 75. (a) 76. (b) 77. (c) 78. (b) 79. (d) 80. (c)

## Answer Keys

**Exercise II**

### Section A (Only one Correct)

1. (d) 2. (b) 3. (a) 4. (b) 5. (a) 6. (b) 7. (d) 8. (d) 9. (c) 10. (a)
11. (a) 12. (c) 13. (c) 14. (d) 15. (a) 16. (d) 17. (c) 18. (a) 19. (a) 20. (a)
21. (a) 22. (d) 23. (b) 24. (b) 25. (b) 26. (a) 27. (a) 28. (c) 29. (b) 30. (b)

### Section B (One or More than one Correct)

1. (a), (b), (c) 2. (a), (b), (c), (d) 3. (a), (c) 4. (c)
5. (a), (d) 6. (a), (b), (d) 7. (a), (b), (c), (d) 8. (a), (b), (c)
9. (a), (c) 10. (b), (d) 11. (a), (b), (c) 12. (a), (c), (d)
13. (c), (d) 14. (a), (b), (c), (d) 15. (b), (c) 16. (a), (c)
17. (d) 18. (a), (b), (c) 19. (a), (b), (c) 20. (c), (d)
21. (c) 22. (a), (b), (d) 23. (a), (c) 24. (a), (d)
25. (c)

### Section C

**Comprehension I**

1. (d) 2. (c) 3. (b)

**Comprehension II**

4. (d) 5. (c) 6. (a)

**Comprehension III**

7. (b) 8. (c)

**Comprehension IV**

9. (b) 10. (a) 11. (c)

**Comprehension V**

12. (a) 13. (c) 14. (a)

**Comprehension VI**

15. (a) 16. (b) 17. (b)

**Comprehension VII**

18. (d) 19. (a) 20. (a)

### Section D (Assertion – Reason)

1. (a) 2. (b) 3. (d) 4. (a) 5. (a) 6. (a) 7. (b) 8. (a) 9. (a) 10. (b)
11. (d) 12. (b) 13. (b) 14. (d) 15. (a) 16. (a) 17. (c) 18. (c) 19. (a) 20. (b)

### Section E (Column Match)

1. A → Q, U; B → P, W; C → P, V; D → R, X
2. A → R; B → S; C → P; D → Q
3. A → Q; B → P; C → Q, R; D → P, S
4. A → Q; B → S; C → P; D → R

### Section F (Subjective)

**Single-digit Integer Type**

1. (4) 2. (3) 3. (4) 4. (6) 5. (6)
6. (5) 7. (2) 8. (9) 9. (2) 10. (9)

**Four-digit Integer Type**

1. (0600) 2. (0020) 3. (0015) 4. (0014) 5. (0010)
6. (0025) 7. (0091) 8. (0720) 9. (0400) 10. (4000)

# HINTS AND EXPLANATIONS

## EXERCISE I (JEE MAIN)

### Adsorption

1. Gaseous mixtures are homogeneous and hence, there is no interface.
2. Solute of ink gets adsorbed but solvent gets absorbed in chalk.
3. Physisorption is favourable at low temperature.
4. $M(s)+X(g) \rightarrow M-X(s)$

   Rate of forward reaction: $r \propto P_{X(g)}$
5. Theory based
6. Chemisorption increases with increase in temperature.
7. For physisorption, activation energy is not needed.
8. Gases having higher critical temperature have greater intermolecular forces.
9. Either van der Waal type forces or chemical bonds.
10. Chemisorption is irreversible.
11. Physisorption decreases on increasing temperature.
12. Theory based
13. Greater the polarity or polarizability, greater will be the extent of adsorption. Hence, $H_2 < CH_4 < CO_2 < NH_3$.
14. Chemisorption increases with increase in temperature, but not linearly.
15. $\frac{x}{m} = k \cdot p^{1/n}$

    or, $\log\frac{x}{m} = \log K + \frac{1}{n} \cdot \log P$

    $\Rightarrow$ Slope $= \frac{1}{n}$
16. Physisorption decreases or increasing temp.
17. $\Delta H_{\text{Physisorption}} = -(20-40)\text{KJ/mol}$
18. van der Waal type weak intermolecular forces only.
19. Chemisorption is monolayer.
20. Rate increases with increase in temperature. but the extent of adsorption decreases.

### Catalysis

21. $2SO_2(g)+O_2(g) \underset{}{\overset{Pt(s)}{\rightleftharpoons}} 2SO_3(g)$ (Contact process)
22. Negative catalyst = Inhibitor
23. Greater the free valency, greater will be the involvement of catalyst.
24. Transition metals. For better catalysis, the extent of adsorption should be high but the bonds formed between adsorbate and adsorbent should not be enough stronger.
25. Auto catalyst
26. Fe is the catalyst and Mo is the promotor.
27. Informative
28. Informative
29. Informative
30. Informative
31. Theory based
32. Theory based
33. Informative

34. Theory based

35. Catalyst has no effect on equilibrium constant and hence, equilibrium composition.

36. Catalyst does not change the reactants and products and hence, no change in $\Delta_rH$, $\Delta_rS$, $\Delta_rG$, etc.

37. Informative

38. Catalyst decreases the activation energy.

39. Theory based

40. Informative

## Colloids

41. Butter is a colloidal system in which water is dispersed in liquid or sold fat.

42. Theory based

43. Informative

44. Theory based

45. $FeCl_3$ coagulates blood.

46. Blood is negatively charged colloid. Haemoglobin is positively charged

47. Smaller the gold number, greater is the protective power of lyophilic colloid.

48. Gold number = $0.0025 \times 1000 = 2.5$

49. Colloidal solutions show lower value of any of the colligative property.

50. Smaller the gold number, greater is the protective power of lyophilic colloid

51. Informative

52. Definition of electro-osmosis.

53. Natural colloids are mostly negatively charged.

54. Definition of zeta potential.

55. Informative

56. $Fe^{3+}$ is common ion.

57. Theory based

58. Informative

59. Charge on colloidal particles may be determined experimentally by electrophoresis or electro-osmosis.

60. Ferric hydroxide sol is positively charged and anion should carry higher charge.

61. For negative charge on colloidal particles. $[Ag^+] < [I^-]$

62. Definition of crystalloids.

63. 1 to 1000 nm

64. Definition of gel

65. Definition of emulsion

66. Definition of peptisation

67. Informative

68. Metal oxide and hydroxides are positively charged colloid.

69. Informative

70. Informative

71. Theory based

72. NaCl, Glucose and $Ba(NO_3)_2$ forms true solution in water

73. As $Ag^+$ ions are in excess, they get adsorbed more preferentially.

74. Metal sulphides are water insoluble and cannot be lyophilic.

75. Sugar in water is true solution.

76. Theory based

77. Larger the size of molecule, smaller is the CMC value.

78. Informative

79. Protective action

80. Theory based

# EXERCISE II (JEE ADVANCED)

## Section A (Only one Correct)

1. Greater the specific surface area of adsorbent, greater will be the extent of adsorption.

2. Adsorption decreases the surface energy.

3. $\frac{x}{m} = K.P^{\frac{1}{n}} \Rightarrow \log\frac{x}{m} = \log K + \frac{1}{n}\cdot\log P$

   From question, $\log K = 0.3010 = \log_2 \Rightarrow K = 2$

   And $\frac{1}{n} = \tan 45° = 1 \Rightarrow n = 1$

   $\therefore \frac{x}{m} = 2 \times P = 2 \times 0.2 = 0.4$

4. $K = A.e^{-E_a/RT}$

   $\therefore \ln\frac{K_2}{K_1} = \frac{E_a}{R}\left(\frac{1}{T_1} - \frac{1}{T_2}\right)$

   or, $\ln 10 = \frac{E_a}{R}\left(\frac{1}{600} - \frac{1}{1000}\right) \Rightarrow E_a = 6900$ cal/mol

5. For a particular combination of adsorbent, adsorbate and temperature, only one value of '$n$' is permissible.

6. $\frac{x}{m} = K \cdot P^{\frac{1}{n}}$

   $\Rightarrow 0.2 = K \times (4)^{\frac{1}{n}}$ (1)

   $0.5 = K \times (25)^{\frac{1}{n}}$ (2)

   $0.8 = K \times (64)^{\frac{1}{n}}$ (3)

   From (1), (2) and (3), $K = \frac{1}{10}$ and $n = 2$

   $\therefore \left(\frac{x}{m}\right)_{\text{required}} = K \times (36)^{\frac{1}{n}} = 0.6$

   Hence, moles of $N_2$ adsorbed per gm of iron $= \frac{0.6}{28} = \frac{3}{140}$

7. $r_1 < r_2 \Rightarrow$ A is the catalyst.

   $r_3 < r_1 \Rightarrow$ B is the catalyst.

   $r_1 = r_4 = r_5 \Rightarrow$ C and D are not catalysts.

   $r_1 < r_7 < r_2 \Rightarrow$ D is catalytic poison.

   $r_1 < r_2 < r_6 \Rightarrow$ C is catalytic promotor.

8. A catalyst always involve in the reaction.

9. Enzymes are specific.

10. A catalyst does not initiate the reaction.

11. Theory based

12. Catalyst does not alter the equilibrium position.

13. Homogeneous catalysis, because the physical states of both reactant and catalyst is aqueous (liquid).

14. Activation energy is decreased.

15. Catalyst lowers the activation energy.

16. Catalysis occurs through chemisorption.

17. Viscosity is higher and surface tension is smaller than water.

18. $K$, as it reacts vigorously in water.

19. $SnCl_4$ formed by reaction will adsorb some common $Cl^-$ ions.

20. Informative

21. True solution or suspension does not show Tyndall effect.

22. Blood, clay and smoke are negative sol. In strong acidic solution, gelatin adsorbs some $H^+$ ions and become positive.

23. Theory

24. Informative

25. Informative

**26.** Volume of metal used $= \dfrac{1.9\times10^{-4}}{19} = 10^{-5}\,\text{cm}^3$

$\therefore\ N\times\dfrac{4}{3}\times(10\times10^{-7}\,\text{cm})^3 = 10^{-5}\,\text{cm}^3$

$\Rightarrow N = 2.39\times10^{12}$

Hence, number of particles per cm$^3$

$= \dfrac{2.39\times10^{12}}{1000} = 2.39\times10^{9}$

**27.** Larger the carbon chain, normally smaller is CMC.

**28.** Sulphide sol have negative charge on colloidal particles

**29.** Adsorption is physisorption.

**30.** Theory based

## Section B (One or More than one Correct)

**1.** Entropy decreased in adsorption.

**2.** Theory based

**3.** Theory based

**4.** ΔH can never be equal to ΔS.

**5.** (a) Chemisorption does not change into physisorption at higher pressure.

(b) CO or $CO_2$ gases leave the surfaces.

**6.** Theory based

**7.** Informative

**8.** In Lindlar's catalyst, catalytic poison is used

**9.** Theory based

**10.** All catalytic reaction is multistep reaction

**11.** Informative

**12.** (a) $\dfrac{K_{\text{cat}}}{K_{\text{uncat}}} = e^{20} = \dfrac{A.e^{E_{a'}/RT}}{A.e^{E_{a'}/RT}} = e^{(E_a - E_{a'})/RT}$

$\therefore\ 20 = \dfrac{E_a - E_{a'}}{RT} = \dfrac{E_a - 2\ \text{Kcal}}{\dfrac{2}{1000}\times300}$

$\Rightarrow E_a = 14$ Kcal/mol

(b), (c) Reaction: $2H_2O_2(aq) \rightarrow 2H_2O(l)+O_2(g)$ is first order.

(d) Rate of uncatalysed reaction increases to greater extent on increasing temperature because its activation energy is high.

**13.** Sol particles are restricted to move. Solvent particles move in opposite direction to the expected movement of sol particles. $Fe(OH)_3$ sol is positively charged.

**14.** Informative

**15.** Electrophoresis and electro-osmosis are the experimental methods to determine charge on colloidal particles.

**16.** Informative

**17.** Below CMC, the solution is true solution.

**18.** $PO_4^{3-} > SO_4^{2-} > Cl^-$

$\Rightarrow$ Sol particles are positively charged.

**19.** $RCOONa \rightleftharpoons RCOO^- + Na^+$

As true solution, one mole of RCOONa will become two moles in solution. But, as micelle formation starts, the total number of particles start decreasing due to association.

**20.** Due to excess Ag+, sol particles will be positively charged.

**21.** (a) It is due to sharp decrease in number of ions.

(b) Tyndall effect is better shown by lyophobic colloid.

(c) Colloidal solutions have lower value of colligative properties.

(d) Larger the carbon chain, normally lower CMC value.

**22.** (a) Basic dye is positively charged and hence, $Fe(CN)_6^{4-} > HPO_3^{2-}$

(c) Slope should not change in Freundlich's isotherm.

**23.** Tyndall effect is shown by colloids.

**24.** Theory based

**25.** Charge : $Mg^{2+}$ (2 unit) > $Cl^-$ (1 unit)

Hence, better coagulation for negatively charged gold sol.

## Section C (Comprehensions)

### Comprehension I

1. Polarizability is maximum in Xe.
2. CO is polar and hence, more preferential adsorption.
3. Adsorption decreases on increasing temperature.

### Comprehension II

4. In case of concentrated KCl, KCl adsorbs on blood charcoal surface, but in case of dilute KCl, blood charcoal dissolves in KCl solution.
5. Greater critical temperature, greater the extent of adsorption.
6. Adsorption is always exothermic.

### Comprehension III

7. Initial surface area, $A_1 = 6\times(2\text{ cm})^2 = 24\text{ cm}^2$

Final volume of each cube $= \dfrac{8\text{ cm}^3}{10^{12}}$

$\therefore$ Final side length of each cube $= \left(8\times10^{12}\text{ cm}^3\right)^{1/3}$

$= 2\times10^{-4}$ cm

Hence, final surface area of each cube,

$A_2 = 6\times(2\times10^{-4}\text{ cm})^2 = 24\times10^{-8}\text{ cm}^2$

$\therefore \dfrac{\text{Final total surface area}}{\text{Initial surface area}} = \dfrac{24\times10^{-8}\times10^{12}}{24} = 10^4$

8. Number of $H_2$ molecules

$= \dfrac{2\times0.112}{0.0821\times546}\times6\times10^{23} = 3\times10^{21}$

$\therefore$ Specific surface area

$= \dfrac{3\times10^{21}\times0.4\times(10^{-7}\text{ cm})^2}{5\text{gm}}$

$= 2.4\times10^6\text{ cm}^2/\text{gm}$

### Comprehension IV

9. $\log\left(\dfrac{x}{m}\right) = \log K + \dfrac{1}{n}\cdot\log P$

Slope $= \dfrac{1}{n} = 0.25 \Rightarrow n = 4$

and for $x$-intercept, $\log\left(\dfrac{x}{m}\right) = 0$

$\Rightarrow \log K = -\dfrac{1}{n}\cdot\log P$

$= -\dfrac{1}{4}\times(-4) = 1.0$

$\therefore K = 10$

10. $\dfrac{x}{m} = 10\times(16)^{1/4} = 20$

$\Rightarrow x = 20\times10 = 200$ gm

11. $\dfrac{810}{1.0} = 10\times(P)^{1/4}$

$\Rightarrow P = 3$ atm

### Comprehension V

**12.** Positively charged due to adsorption of $Ag^+$ ions.

**13.** Theory based

**14.** Informative

### Comprehension VI

**15.** Positively charged due to adsorption of $Fe^{3+}$ ions.

**16.** $AgI, Ag^+, NO_3^-$

$\downarrow$ fixed layer $\downarrow$ diffused layer

**16.** $S^{2-}$ ions get adsorbed.

### Comprehension VII

**18.** Gold number $= 0.025 \times 1000 = 25$

**19.** Theory based

**20.** Theory based

## Section D (Assertion – Reason)

**1.** Colour become less intense due to adsorption.

**2.** Surface particles have higher energy due to unbalanced forces.

**3.** Word 'always' is not suitable because chemisorption increased with increase in temperature.

**4.** Theory based

**5.** Theory based

**6.** Theory based

**7.** Theory based

**8.** Theory based

**9.** Theory based

**10.** Theory based

**11.** Cellulose nitrate sol is lyophilic.

**12.** Scattering is not related to speed of particles.

**13.** Theory based

**14.** Theory based

**15.** Colloidal particles are negatively charged due to adsorption of $I^-$ ions and hence, it moves towards anode.

**16.** Theory based

**17.** Informative

**18.** Peptization occurs due to adsorption of common ion.

**19.** Natural colloids are normally lyophilic.

**20.** Theory based

## Section E (Column Match)

**1.** Informative

**2.** Informative

**3.** Informative

**4.** Informative

## Section F (Subjective)

### Single-digit Integer Type

1. $6.48 = (0.01\times10^{-3}\times162n)\times10^3$

$\Rightarrow n = 4$

2. $\log\frac{x}{m} = \log K + \frac{1}{n}\cdot\log P$

From the given graph, $\log K = 1.0 \Rightarrow K = 10$

and $\frac{1}{n} = 0.25 \Rightarrow n = 4$

Now, $\frac{x}{m} = K.P^{\frac{1}{n}}$

$\Rightarrow \frac{x}{1.0} = 10\times(8.1\times10^{-3})^{1/4}$

$\Rightarrow x = 3$ gm

3. $t_{av} = \frac{1}{k} = \frac{1}{A.e^{-E_a/RT}}$

$= \frac{1}{(1.25\times10^8 s^{-1})\times e^{-16\times10^3/2\times400}} = 4\,\text{sec}$

4. Soap solution of sodium palmitate, gold sol, silicic acid sol, acidic dye, metal sulphide sol, sol of AgCl by excess KCl in $AgNO_3$.

5. $\ln\frac{P_1}{P_2} = \frac{\Delta H_{ads}}{R}\left(\frac{1}{T_1} - \frac{1}{T_2}\right)$

or, $\ln\frac{1.6}{32} = \frac{\Delta H_{ads}}{R}\left(\frac{1}{200} - \frac{1}{250}\right)$

$\Rightarrow \Delta H_{ads} = 6000$ cal/mol

6. Number of $CH_3COOH$ molecules adsorbed

$= \frac{100\times(0.5-0.49)}{1000}\times6\times10^{23} = 6\times10^{20}$

$\therefore$ Surface area of each molecule

$= \frac{3\times10^2}{6\times10^{20}} = 5\times10^{-19}\,\text{m}^2$

7. Number of $N_2$ molecules

$= \frac{0.001\times2.46\times10^{-3}}{0.082\times300}\times6.023\times10^{23} = 6.023\times10^{16}$

$\therefore$ Number of active sites per molecule

$= \frac{1000\times6.023\times10^{14}\times\frac{20}{100}}{6.023\times10^{16}} = 2$

8. Number of $N_2$ molecules absorbed

$= \frac{2.24\times10^{-3}}{22.4}\times6\times10^{23} = 6\times10^{19}$

$\therefore$ Specific surface area

$= 6\times10^{19}\times0.15\times(10^{-9})^2 = 9$

9. Colloid is a heterogeneous system $\Rightarrow$ min = 2 phases

10. $\ln 20 = \frac{E_a}{R}\left(\frac{1}{600} - \frac{1}{1000}\right)$

$\Rightarrow E_a = 9000$ cal/mol.

### Four-digit Integer Type

1. Mass of NaCl used

$= (585\times1.2)\times1/100 = 5.85\times1.2$ gm

Moles of NaCl used $= \frac{5.85\times1.2}{58.5} = 0.12$

$\therefore$ Coagulation value

$= \frac{0.12\times10^3}{200/1000} = 600$ millimole/litre

2. Number of palmitic acid molecules needed

$= \frac{480\ \text{cm}^2}{0.2\times(10^{-7}\,\text{cm})^2} = 2.4\times10^{17}$

Moles of palamitic acid molecules

$= \frac{2.4\times10^{17}}{6\times10^{23}} = 4\times10^{-7}$

$\therefore$ Volume of solution needed

$= \frac{1\ \text{dm}^3}{5.12/256}\times4\times10^{-7} = 2\times10^{-5}\ \text{dm}^3 = 20\ \text{mm}^3$

3. The radius of hydrogen molecule $= \left(\frac{m}{d} \times \frac{3}{4\pi}\right)^{1/3}$

$$= \left(\frac{2}{6\times10^{23}\times\frac{0.16}{\pi}} \times \frac{3}{4\pi}\right)^{1/3} = 2.5\times10^{-8}\text{cm}$$

Number of hydrogen molecules at the surface per gm Cu

$$= \frac{224/\pi}{22400} \times \frac{6\times10^{23}}{25} = \frac{2.4\times10^{20}}{\pi}$$

∴ Specific surface area of Cu

$$= \frac{2.4\times10^{20}}{\pi} \times \pi \times (2.5\times10^{-8})^2$$

$$= 150000 \text{ cm}^2/\text{gm} = 15 \text{ m}^2/\text{gm}$$

4. $t_{av} = \frac{1}{k}$

Now, $\ln\frac{k_2}{k_1} = \ln\frac{t_1}{t_2} = \frac{E_a}{R} = \left(\frac{1}{T_1} - \frac{1}{T_2}\right)$

or, $\ln\frac{0.36}{0.72} = \frac{E_a}{R}\left(\frac{1}{2500} - \frac{1}{2000}\right)$

$\Rightarrow E_a = 14000$ cal/mol

5. $$\frac{(t_{1/2})_{\text{Fe}}}{(t_{1/2})_{\text{charcoal}}} = \frac{K_{\text{charcoal}}}{K_{Fe}} = \frac{A.e^{-20\times10^3/2\times600}}{A.e^{-8\times10^3/2\times600}} = \frac{1}{e^{10}}$$

6. Number of adsorbate molecules

$$= \frac{0.10\times10^{-3}}{0.25\times10^3} \times 6\times10^{23} = 2.4\times10^7$$

∴ Effective surface area $= \frac{0.06}{2.4\times10^{17}} = 25\times10^{-20}\text{ m}^2$

7. Moles of gas adsorbed per gm of charcoal

$$= \frac{(700-400)\times1.52}{760\times R\times300\times6}$$

Volume of gas adsorbed per gm of charcoal

(at 0°C and 1 atm) $= \frac{300\times1.52}{760\times R\times300\times6} \times \frac{R\times273}{1}$

$= 0.091$ litre

8. Specific surface area of silica gel

$$= \frac{168}{22400} \times 6\times10^{23} \times 0.16 \times (10^{-9})^2 = 720\text{m}^2/\text{gm}$$

9. $\ln\frac{P_1}{P_2} = \frac{\Delta_\theta H}{R}\left(\frac{1}{T_1} - \frac{1}{T_1}\right)$

or, $\ln\frac{0.4}{59.2} = \frac{-16.628\times10^3}{8.314}\left(\frac{1}{200} - \frac{1}{T}\right)$

$\Rightarrow T = 400$ K

10. $r = \frac{K_1K_2[E_0][S]}{K_{-1}+K_2+K_1[S]} \simeq \frac{K_1K_2[E_0][S]}{K_{-1}+K_1[S]}$

For $r_{\max}$, $K_1[S] \gg K_{-1}$

$\therefore r_{\max} = \frac{K_1K_2[E_0][S]}{K_1[S]} = K_2[E_0] = 0.02$ M

From question,

$$\frac{K_2[E_0]}{2} = \frac{K_1K_2[E_0][S]}{K_{-1}+K_1[S]}$$

$\Rightarrow K_{-1} + K_1[S] = 2K_1[S]$

$$\therefore \frac{K_1}{K_{-1}} = \frac{1}{[S]} = \frac{1}{250\frac{\text{mg}}{\text{dm}^3}} = \frac{\text{dm}^3}{250\times10^{-6}\text{kg}} = 4000\frac{\text{dm}^3}{\text{kg}}$$

CHAPTER 13

# Atomic Structure

## EXERCISE I (JEE MAIN)

### Fundamental Particles

1. Gases are bad conductors of electricity. Their conductivity may be increased by
   (a) increasing the pressure as well as potential difference between the electrodes.
   (b) decreasing the pressure as well as potential difference between the electrodes.
   (c) decreasing the pressure and/or increasing the potential difference between the electrodes.
   (d) increasing the pressure and/or decreasing the potential difference between the electrodes.

2. Which of the following is true for cathode ray?
   (a) It is not deflected by magnetic field.
   (b) It is an electromagnetic wave.
   (c) It emits X-ray, when strikes a metal.
   (d) It consist of all the negative particles present in the atoms.

3. The specific charge of cathode rays
   (a) depends on the nature of the gas.
   (b) depends on the material of the discharge tube.
   (c) depends on the potential difference between cathode and anode.
   (d) is a universal constant.

4. Which of the following is not a fundamental particle?
   (a) Electron (b) Proton
   (c) Neutron (d) X-rays

5. The presence of charge particles in the atoms was first confirmed by
   (a) Rutherford (b) Thomson
   (c) Faraday (d) Goldstein

6. From the discharge tube experiment, it is concluded that
   (a) mass of proton is fractional.
   (b) matter contains electrons.
   (c) matter contains nucleus.
   (d) positive rays are heavier than protons.

7. The cathode rays experiment demonstrated that
   (a) α-particles are the nuclei of He atoms.
   (b) the *e*/*m* ratio for the particles of the cathode rays varies gas to gas.
   (c) cathode rays are streams of negatively charged particles.
   (d) the mass of an atom is essentially all contained in its very small nucleus.

8. Which of the following is not the possible path of cathode rays ejecting from the surface of cathode?

(a)
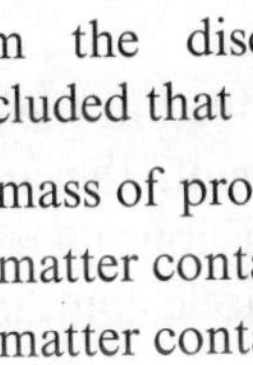


(b)
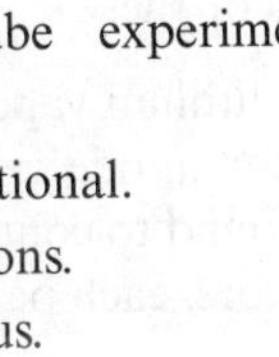


(c)
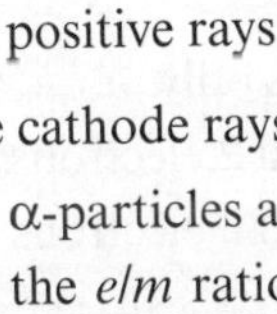


(d)
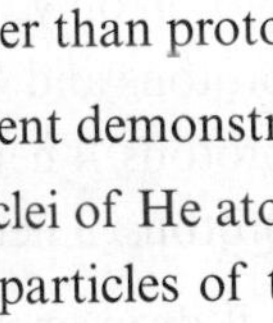

9. Cathode rays are made up of electrons. Anode rays are made up of
   (a) only protons.
   (b) only nucleus of atoms.
   (c) positive residue of atoms.
   (d) only from all the positive particles present in the atoms.

10. Which of the following statement is incorrect
   (a) Cathode rays are emitted out from the surface of cathode.
   (b) Cathode rays travel in straight line.
   (c) Anode rays are heavier than cathode rays.
   (d) Anode rays are emitted out from the surface of anode.

11. The *e*/*m* ratio of anode rays produced in the discharge tube, depends on the
   (a) nature of the gas filled in the tube.
   (b) nature of anode material.
   (c) nature of cathode material.
   (d) all of these

12. When lithium vapours were filled in the discharge tube for anode rays experiment, the anode rays were found to contain only $Li^+$ ions ($A = 7, Z = 3$). Therefore, each particle of anode contains
   (a) 1 proton only.
   (b) 3 protons and 4 neutrons only.
   (c) 3 protons, 4 neutrons and 2 electrons.
   (d) 3 protons, 3 neutrons and 3 electrons.

13. In an oil drop experiment, the following charges (in arbitrary units) were found on a series of oil droplets $4.5 \times 10^{-18}$, $3.0 \times 10^{-18}$, $6.0 \times 10^{-18}$, $7.5 \times 10^{-18}$, $9.0 \times 10^{-18}$. The charge on electron (in the same unit) should be
   (a) $3.0 \times 10^{-18}$ (b) $9.0 \times 10^{-18}$
   (c) $1.5 \times 10^{-18}$ (d) $1.6 \times 10^{-19}$

14. In Wilson cloud chamber experiment, two particles were found to show equal deviations but in opposite directions. The names positron and negatron were given to these particles by Anderson. Hence, Negatron is
   (a) neutron (b) neutrino
   (c) proton (d) electron

15. Which of the following particle is not deflected in the magnetic field?
   (a) Electron (b) Proton
   (c) Neutron (d) Deuteron

16. Which of the following particle have non-zero *e*/*m* ratio?
   (a) Neutron (b) Neutrino
   (c) Positron (d) Neutral meson

17. The *e*/*m* ratio is maximum for
   (a) $Na^+$ (b) $Al^{3+}$
   (c) $H^+$ (d) $Mg^{2+}$

18. The potential difference between cathode and anode in a cathode ray tube is $V$. The speed acquired by the electrons is proportional to
   (a) $V$ (b) $\sqrt{V}$
   (c) $V^2$ (d) $1/\sqrt{V}$

19. The ratio of specific charges of α–particle and deuteron is
   (a) 1 : 2 (b) 2 : 1
   (c) 1 : 1 (d) 4 : 1

20. The *e*/*m* ratio of a particle of charge 2 unit and mass 4 amu is
   (a) $4.8 \times 10^7$ C/kg
   (b) 0.5 C/kg
   (c) $4.8 \times 10^4$ C/kg
   (d) $8 \times 10^{-20}$ C/kg

## Rutherford's Atomic Models

21. Atoms have void spaces. It was first suggested by
   (a) Rutherford (b) Thomson
   (c) Lenard (d) Dalton

22. Rutherford's experiment, which established the nuclear model of the atom, used a beam of
   (a) β-particles, which impinged on a metal foil and got absorbed.
   (b) γ-rays, which impinged on a metal foil and ejected electrons.
   (c) helium atoms, which impinged on a metal foil and got scattered.
   (d) helium nuclei, which impinged on a metal foil and got scattered.

23. Which of the following is not a conclusion of Rutherford's atomic model?
   (a) Most of the part inside an atom is empty.
   (b) Almost all mass of an atom is concentrated in the nucleus.
   (c) The size of nucleus is very small in comparison to the size of atom.
   (d) Electron revolves around the nucleus in definite orbits.

24. Which of the following is not a correct statement according to Rutherford's atomic model?
   (a) 99% of mass of an atom is centred in the nucleus.
   (b) Most of the part inside the atom is empty.
   (c) The size of nucleus is very small in comparison to the atoms.
   (d) Electrons revolve round the nucleus.

25. When β-particles are sent through a tin metal foil, most of them go straight through the foil as
   (a) β-particles are much heavier than electron.
   (b) most part of the atom is empty space.
   (c) β-particles are positively charged.
   (d) β-particles move with high velocity.

26. A proton and a deuteron are projected towards the stationary gold nucleus in different experiments with the same speed. The distance of closest approach will be
   (a) same for both.
   (b) greater for proton.
   (c) greater for deuteron.
   (d) depends on speed.

27. Two particles A and B having same *e*/*m* ratio are projected towards silver nucleus in different experiments with the same speed. The distance of closest approach will be
   (a) same for both.
   (b) greater for A.
   (c) greater for B.
   (d) depends on speed.

28. α-particles are projected towards the nucleus of following metals with the same kinetic energy. Towards which metal, the distance of closest approach will be minimum?
   (a) Cu ($Z = 29$)   (b) Ag ($Z = 47$)
   (c) Au ($Z = 79$)   (d) Ca ($Z = 20$)

29. In different experiments, α-particles, proton, deuteron and neutron are projected towards gold nucleus with the same kinetic energy. The distance of closest approach will be minimum for
   (a) α-particle   (b) proton
   (c) deuteron   (d) neutron

30. The following charged particles accelerated from rest through the same potential difference are projected towards gold nucleus in different experiments. The distance of closest approach will be maximum for
   (a) α–particle   (b) proton
   (c) deuteron   (d) same for all

31. In the Rutherford scattering experiment, the number of alpha particles scattered at an angle $\theta = 60°$ is 36 per minute. The number of alpha particles per minute scattered at angles $\theta = 90°$ is (Assume all other conditions to be identical)
   (a) 144   (b) 9
   (c) 36   (d) 16

32. If nucleus and atom are considered as perfect spheres with diameters $4 \times 10^{-15}$ m and $2 \times 10^{-10}$ m, respectively, then the ratio of the volumes of nucleus and atom should be
   (a) $2 \times 10^{-5}:1$   (b) $8 \times 10^{-15}:1$
   (c) $1.25 \times 10^{14}:1$   (d) $8 \times 10^{15}:1$

33. With what velocity should an α–particle travel towards the nucleus of copper atoms so as to arrive at a distance $10^{-12}$ m from the nucleus of the copper atom? ($4.8 \times \sqrt{29 \times 60} = 200$, $N_A = 6 \times 10^{23}$, $e = 1.6 \times 10^{-19}$ C)
   (a) $2 \times 10^3$ ms$^{-1}$   (b) $2 \times 10^6$ ms$^{-1}$
   (c) $2 \times 10^5$ ms$^{-1}$   (d) $2 \times 10^7$ ms$^{-1}$

34. An α-particle accelerated through *V* volt is fired towards a nucleus. The distance of closest approach is *r*. If a proton accelerated through the same potential is fired towards the same nucleus, the distance of closest approach of the proton will be
   (a) *r*   (b) 2*r*
   (c) *r*/2   (d) *r*/4

35. The distance of closest approach of an α-particle fired towards a nucleus with momentum 'P' is *r*. What will be the distance of closest approach when the momentum of the α-particle is 2P?
   (a) 2*r*   (b) 4*r*
   (c) *r*/2   (d) *r*/4

## Planck's Quantum Theory, Photoelectric Effect and Moseley's Experiment

36. Small packets of light is called
(a) proton (b) quanta
(c) photon (d) spectrum

37. A radio station emits radiations of 400 kHz. The metre band of the station is
(a) 400 (b) 750
(c) 1333.33 (d) 7.5

38. Which of the following electromagnetic radiation have greater frequency?
(a) X–rays (b) Ultraviolet rays
(c) Radio waves (d) Visible rays

39. As its closest approach, the distance between the Mars and the Earth is found to be 60 million km. When the planets are at this closest distance, how long would it take to send a radio message from a space probe sent to Mars from Earth?
(a) 5 s (b) 200 s
(c) 0.2 s (d) 20 s

40. Two electromagnetic radiations have wave numbers in the ratio 2 : 3. Their energies per quanta will be in the ratio
(a) 3 : 2 (b) 9 : 4
(c) 4 : 9 (d) 2 : 3

41. A radio station is emitting the radiations of frequency $2 \times 10^4$ Hz. If its frequency is doubled,
(a) wavelength will be doubled.
(b) energy per quanta will be doubled.
(c) wave number will be halved.
(d) all of these

42. The eyes of a certain member of the reptile family pass a single visual single to the brain when the visual receptors are stuck by photons of wavelength 662.6 nm. If a total energy of $3.0 \times 10^{-14}$ J is required to trap the signal, what is the minimum number of photons that must strike the receptor?
(a) $1.0 \times 10^5$ (b) $1.0 \times 10^6$
(c) 1000 (d) 1

43. A photon of 400 nm is absorbed by a gas molecule and then the molecule re-emits two photons. One re-emitted photon has wavelength 500 nm. Assuming that there is no change in the energy of molecule, the wavelength of second re-emitted photon is
(a) 100 nm (b) 2000 nm
(c) –100 nm (d) 900 nm

44. A green bulb and a red bulb are emitting the radiations with equal power. The correct relation between numbers of photons emitted by the bulbs per second is
(a) $n_g = n_r$ (b) $n_g < n_r$
(c) $n_g > n_r$ (d) unpredictable

45. A dye emits 50% of the absorbed energy as fluorescence. If the number of quanta absorbed and emitted out is in the ratio 1 : 2 and it absorbs the radiation of wavelength '$x$' Å, then the wavelength of the emitted radiation will be
(a) $x$ Å (b) $0.5x$ Å
(c) $4x$ Å (d) $0.25x$ Å

46. Wavelength of photon which have energy equal to average of energy of photons with $\lambda_1 = 4000$ Å and $\lambda_2 = 6000$ Å will be
(a) 5000 Å (b) 4800 Å
(c) 9600 Å (d) 2400 Å

47. Bond dissociation on energy of $Br_2$ is 200 kJ/mole. The longest wavelength of photon that can break this bond would be ($N_A \times hc = 0.12$ J)
(a) $6.0 \times 10^{-5}$ m (b) $1.2 \times 10^{-5}$ m
(c) $6.0 \times 10^{-7}$ m (d) $1.2 \times 10^{-7}$ m

48. Wavelength of photon having energy 1 eV would be
(a) $1.24 \times 10^{-4}$ m (b) $1.24 \times 10^{-6}$ m
(c) $1.24 \times 10^{-5}$ m (d) $1.24 \times 10^{4}$ m

49. In the emission of photoelectrons, the number of photoelectrons emitted per unit time depends upon
(a) energy of the incident radiation.
(b) intensity of the incident radiation.
(c) frequency of the incident radiation.
(d) wavelength of the incident radiation.

50. Radiations of frequency, ($\nu$) are incident on a photosensitive metal. The maximum kinetic energy of photoelectrons is $E$. When the frequency of the incident radiations is doubled, then what is the maximum kinetic energy of the photoelectrons?
(a) $2E$ (b) $E/2$
(c) $E + h\nu$ (d) $E - h\nu$

51. A photo sensitive surface is receiving light of wavelength 5000 Å at the rate of $10^{-7}$ J/s. The number of photons received per second is

(a) $2.5 \times 10^{11}$ (b) $3.0 \times 10^{32}$
(c) $2.5 \times 10^{18}$ (d) $2.5 \times 10^{9}$

52. In order to increase the kinetic energy of ejected photoelectrons, there should be an increase in

(a) intensity of radiation.
(b) wavelength of radiation.
(c) frequency of radiation.
(d) both wavelength and intensity of radiation.

53. The threshold wavelength for ejection of electrons from a metal is 330 nm. The work function for the photoelectric emission from the metal is ($h = 6.6 \times 10^{-34}$ J-s)

(a) $1.2 \times 10^{-18}$J (b) $6.0 \times 10^{-19}$ J
(c) $1.2 \times 10^{-20}$ J (d) $6.0 \times 10^{-12}$ J

54. The ratio of wavelengths of $K_\alpha$-characteristic X-rays produced when iron ($Z = 26$) and scandium ($Z = 21$) are used as anticathode is

(a) 26 : 21 (b) 4 : 5
(c) 16 : 25 (d) 25 : 16

55. The wavelength of the $K_\alpha$ line for an element of atomic number 57 is λ. What is the wavelength of $k_\alpha$ line for the element of atomic number 29?

(a) λ (b) 2λ
(c) 4λ (d) λ/4

## Bohr's Atomic Model

56. Bohr's model may be applied to

(a) $Na^{10+}$ ion (b) He atom
(c) $Be^{2+}$ ion (d) $C^{6+}$ ion

57. If the radius of first orbit of H–atom is $x$ Å, then the radius of the second orbit of $Li^{2+}$ ion will be

(a) $x$ Å (b) $\frac{4x}{3}$ Å
(c) $\frac{9x}{2}$ Å (d) $4x$ Å

58. According to Bohr's model, the radius of $Ne^{9+}$ ion in ground state should be

(a) 0.529 Å (b) 0.0529 Å
(c) 5.29 Å (d) 52.9 Å

59. The ratio of spacing between the third and fourth orbit to the spacing between sixth and seventh orbit of H–atom is

(a) 7 : 13 (b) 13 : 7
(c) 16 : 49 (d) 1 : 1

60. What would be the approximate quantum number ($n$) for a circular orbit of hydrogen, $1 \times 10^{-5}$ cm in diameter?

(a) 31 (b) 43
(c) 40 (d) 39

61. If the mass of electron is doubled, the radius of first orbit of H-atom becomes approximately

(a) 0.529 Å (b) 0.265 Å
(c) 1.058 Å (d) 0.32 Å

62. The ratio of circumference of third and second orbits of $He^+$ ion is

(a) 3 : 2 (b) 2 : 3
(c) 9 : 4 (d) 4 : 9

63. If the mass of electron is doubled, the speed of electron revolving around $Li^{2+}$ nucleus will

(a) remain same.
(b) be doubled.
(c) be halved.
(d) be quadrupled.

64. What is the orbit number of the $He^+$ ion in which electron have speed $\frac{1}{205.67}$ times the speed of light?

(a) 1 (b) 2
(c) 3 (d) 4

65. The speed of electron revolving in the fourth orbit of a hydrogen-like atom or ion is 1094 km/s. The atom or ion is

(a) H (b) $He^+$
(c) $Li^{2+}$ (d) $Be^{3+}$

66. The escape velocity for earth is 11.2 km/s. The orbit number for H-atom in which speed of electron is about 19.54 times the escape velocity is

(a) 4 (b) 8
(c) 10 (d) infinite

67. The ratio of the speed of the electron in the ground state of hydrogen atom to the speed of light in vacuum is

(a) 1 : 1 (b) 1 : 100
(c) 1 : 137 (d) 2 : 3

68. An electron revolves round $Li^{2+}$ nucleus at a distance of 1.587 Å. The speed of electron should be

(a) $2.188 \times 10^6$ m/s (b) $6.564 \times 10^6$ m/s
(c) $7.293 \times 10^5$ m/s (d) $7.293 \times 10^6$ m/s

69. How much distance an electron revolving in 3rd orbit of $He^+$ ion will travel in one second

(a) $1.458 \times 10^6$ m (b) $3.28 \times 10^6$ m
(c) $4.862 \times 10^5$ m (d) $2.917 \times 10^6$ m

70. The ratio of time taken by electron in revolution around the H-nucleus in the second and third orbits is

(a) 2 : 3 (b) 4 : 8
(c) 8 : 27 (d) 27 : 8

71. For hydrogen atom, the number of revolutions of the electron per second in the orbit of quantum number, $n$, is proportional to

(a) $n^3$ (b) $\sqrt{n}$
(c) $n^{-3}$ (d) $n^{-1}$

72. Which of the following is not a permissible value of angular momentum of electron in H-atom?

(a) $1.5\,\frac{h}{\pi}$ (b) $0.5\,\frac{h}{\pi}$
(c) $1.25\,\frac{h}{\pi}$ (d) All of these

73. If an electron in H-atom jumps from one orbit to other, its angular momentum doubles. The distance of electron from nucleus becomes ____ times the initial distance.

(a) 2 (b) 4
(c) $\frac{1}{2}$ (d) $\frac{1}{4}$

74. The angular momentum of electron revolving in the second orbit of H-atom is '$x$' J·s. The angular momentum of electron in the second orbit of $He^+$ ion should be

(a) $x$ J·s (b) $2x$ J·s
(c) $0.5x$ J·s (d) $4x$ J·s

75. The angular momentum of electron revolving around the nucleus of H–atom is directly proportional to

(a) $r$ (b) $r^{1/2}$
(c) $r^{-1/2}$ (d) $r^{-1}$

76. What is the angular speed of an electron revolving in the third orbit of $He^+$ ion?

(a) $6.12 \times 10^{15}$ s$^{-1}$ (b) $1.63 \times 10^{-16}$ s
(c) $1.92 \times 10^{16}$ s$^{-1}$ (d) $1.95 \times 10^{15}$s$^{-1}$

77. The force of attraction on electron by the nucleus is directly proportional to

(a) $\frac{n^3}{Z^4}$ (b) $\frac{Z^3}{n^4}$
(c) $\frac{n^4}{Z^2}$ (d) $\frac{Z^2}{n^4}$

78. The K.E. of electron in $He^+$ will be maximum in

(a) third orbit (b) first orbit
(c) seventh orbit (d) infinite orbit

79. As the orbit number increases, the K.E. and P.E. for an electron

(a) both increases.
(b) both decreases.
(c) K.E. increases but P.E. decreases.
(d) P.E. increases but K.E. decreases.

80. The ratio of energies of first excited state of $He^+$ ion and ground state of H–atom is

(a) 1 : 1 (b) 4 : 1
(c) 1 : 4 (d) 16 : 1

81. For which atom or ion, the energy level of the second excited state is –13.6 eV?

(a) H (b) $He^+$
(c) $Li^{2+}$ (d) Li

82. The orbit from which when electron will jump in other orbit, the energy may be absorbed but not emitted out, will be

(a) first orbit (b) second orbit
(c) seventh orbit (d) infinite orbit

83. In Bohr's model of the hydrogen atom, let $r$, $v$ and $E$ represent the orbit radius, speed of an electron and the total energy of the electron, respectively. Which of following relation is proportional to the orbit number $n$?

(a) $v \cdot r$ (b) $r/E$
(c) $r/v$ (d) $r \cdot E$

84. The ratio of potential energy of electron in the third orbit of $Li^{2+}$ ion to the kinetic energy of electron in the fourth orbit of $He^+$ ion should be

(a) 8 : 1 (b) –8 : 1
(c) –16 : 1 (d) 1 : 1

85. Which of the following quantity for an electron revolving around the H-nucleus is independent to the mass of electron?

(a) Distance from nucleus
(b) Kinetic energy
(c) Potential energy
(d) Speed

86. The potential energy of electron revolving in the ground state of H atom is

(a) –13.6 eV (b) –6.8 eV
(c) –27.2 eV (d) Zero

87. An electron is revolving around the nucleus of $He^+$ ion with a speed of $2.188 \times 10^6$ m/s. The potential energy of the electron is

(a) –13.6 eV (b) –6.8 eV
(c) –27.2 eV (d) Zero

88. As the orbit number increases, the difference in two consecutive energy levels

(a) remain constant (b) increases
(c) decreases (d) is unpredictable

89. The amount of energy released when an electron jumps from the seventh excited state to the first excited state in $He^+$ ion is

(a) 13.32 eV (b) 53.28 eV
(c) 12.75 eV (d) 26.08 eV

90. The energy difference will be minimum for which of the following energy levels of H-atom?

(a) $n = 2$ and $n = 3$ (b) $n = 3$ and $n = 4$
(c) $n = 1$ and $n = 2$ (d) $n = 1$ and $n = 4$

91. For which transition in H-atom, the amount of energy released will be maximum?

(a) $n = 4$ to $n = 2$ (b) $n = 5$ to $n = 2$
(c) $n = 2$ to $n = 1$ (d) $n = 7$ to $n = 2$

92. How much energy is needed for an electron revolving in the second orbit of $He^+$ ion in order to double its angular momentum?

(a) 40.8 eV (b) 2.55 eV
(c) 10.2 eV (d) 12.09 eV

93. The ionization energy of a hypothetical atom is 50 eV. If this atom obeys Bohr's atomic model, the energy of electron in its fifth orbit will be

(a) –1250 eV (b) + 2 eV
(c) –2 eV (d) +1250 eV

94. An electron revolving around H-nucleus in the ground state absorbs 10.2 eV energy. Its angular momentum increases by

(a) $\frac{h}{2\pi}$ (b) $\frac{h}{\pi}$
(c) $\frac{2h}{\pi}$ (d) $\frac{h}{4\pi}$

95. The ionization energy of $He^+$ ion is $x$ eV. The ionization energy of $Be^{3+}$ ion should be

(a) $4x$ eV (b) $2x$ eV
(c) $\frac{x}{4}$ eV (d) $\frac{x}{2}$ eV

96. The excitation energy of an electron from second orbit to third orbit of a hydrogen-like atom or ion with +Ze nuclear charge is 47.2 eV. If the energy of H-atom in the lowest energy state is –13.6 eV, then the value of $Z$ is

(a) 4 (b) 5
(c) 6 (d) 7

97. Electromagnetic radiations of wavelength 240 nm are just sufficient to ionize sodium atom. The ionization energy of sodium (in kJ/mol) is

(a) 5.167 (b) 498.58
(c) 118.83 (d) 51.67

98. The ionization energy of He-atom in the ground state may be

(a) 13.6 eV (b) 54.4 eV
(c) 108.8 eV (d) 27.0 eV

**99.** The binding energy for the third electron in the ground state of Li-atom should be

(a) 108.8 eV (b) 122.4 eV
(c) 30.6 eV (d) 27.2 eV

**100.** Suppose that means were available for stripping 29 electrons from $_{30}Zn$ in vapours of this metal. The ionization energy for the last electron is

(a) 11.5 keV (b) 12.24 keV
(c) 13.6 eV (d) 408 eV

## Spectrum

**101.** Suppose the mass of electron is decreased by 25%. How will it affect the Rydberg constant?

(a) It remains unchanged.
(b) It becomes one-fourth.
(c) It reduces to 75% of its original value.
(d) It is doubled.

**102.** The charge on the electron and proton is reduced to half. Let the present value of the Rydberg constant be *R*. What will be the new value of the Rydberg constant?

(a) $\frac{R}{2}$ (b) $\frac{R}{4}$
(c) $\frac{R}{8}$ (d) $\frac{R}{16}$

**103.** Rydberg is

(a) also called Rydberg constant and it is the universal constant.
(b) the unit of wavelength and one Rydberg equals $1.09 \times 10^7$ m.
(c) the unit of wave number and one Rydberg equals $1.09 \times 10^7\ m^{-1}$.
(d) the unit of energy and one Rydberg equals 13.6 eV.

**104.** For the same electronic transition in the following atom or ion, the frequency of the emitted radiation will be maximum for

(a) H-atom (b) D-atom
(c) $He^+$ ion (d) $Li^{2+}$ ion

**105.** An electron jumps from the fourth orbit to the first orbit in a H-atom. The number of photons liberated out will be

(a) 1 (b) 2
(c) 3 (d) 6

**106.** The wavelength of radiation emitted out in the transition $n = 4$ to $n = 1$ in $Li^{2+}$ ion is

(a) $\frac{135R}{16}$ (b) $\frac{16}{135R}$
(c) $\frac{16R}{135}$ (d) $\frac{135}{16R}$

**107.** What is the frequency of the second line of the Paschen series in the spectrum of $He^+$ ion?

(a) $\frac{64\ R.C}{225}$ (b) $\frac{64\ R}{225}$
(c) $\frac{225}{64\ R}$ (d) $\frac{225\ C}{64\ R}$

**108.** What is the wave number of the radiation of lowest frequency in the Lyman series of the spectrum of $Li^{2+}$ ion?

(a) $\frac{4}{27R}$ (b) $\frac{27R}{4}$
(c) $\frac{27RC}{4}$ (d) $\frac{4C}{27R}$

**109.** The wavelength of a spectral line obtained by an electronic transition is inversely proportional to

(a) the number of transit electrons.
(b) the nuclear charge of the atom.
(c) the energy difference of the related energy levels.
(d) Speed of the transit electron.

**110.** In H-atom, the wave number ratio is 108 : 7 is for

(a) first Lyman and first Balmer transition.
(b) first Lyman and first Brackett transition.
(c) first Lyman and first Paschen transition.
(d) first Lyman and second Balmer transition.

**111.** Wave number of the first line in the Balmer series of $Be^{3+}$ is $2.5 \times 10^5\ cm^{-1}$. Wave number of the second line of the Paschen series of $Li^{2+}$ is

(a) $7.2 \times 10^4\ cm^{-1}$ (b) $7.2 \times 10^5\ cm^{-1}$
(c) $7.2 \times 10^{-4}\ cm^{-1}$ (d) $1.8 \times 10^4\ cm^{-1}$

**112.** When an electron jumps from $n$th orbit to 1st orbit in an imaginary atom obeying Bohr's model, it emits two radiations of wavelengths 400 nm and 300 nm. The frequency of radiation emitted out in the transition $n = n$ to $n = 1$ will be

(a) $7.5 \times 10^{14}$ Hz  (b) $1.0 \times 10^{15}$ Hz
(c) $8.75 \times 10^{14}$ Hz  (d) $1.75 \times 10^{15}$ Hz

**113.** The given diagram indicates the energy levels of a certain atom. When the system moves from $2E$ level to $E$ level, a photon of wavelength $\lambda$ is emitted. The wavelength of the photon emitted during its transition from $4E/3$ level to $E$ level is.

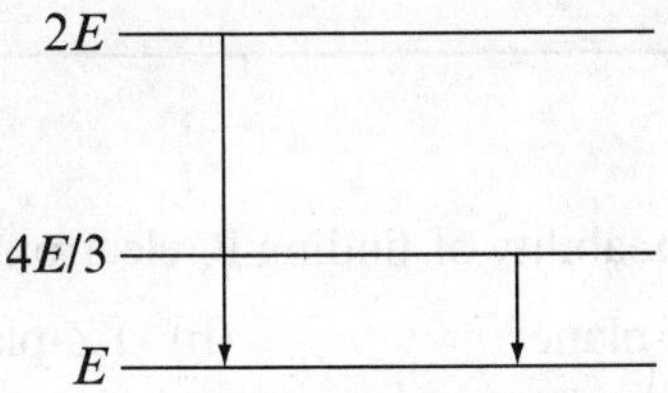


(a) $\lambda/3$  (b) $3\lambda/4$
(c) $4\lambda/3$  (d) $3\lambda$

**114.** What transition in the hydrogen spectrum would have the same wavelength as the Balmer transition $n = 4$ to $n = 2$ of $He^+$ spectrum?

(a) $n = 4$ to $n = 2$  (b) $n = 4$ to $n = 1$
(c) $n = 2$ to $n = 1$  (d) $n = 3$ to $n = 2$

**115.** The number of possible spectral lines in the bracket series in hydrogen spectrum, when electrons present in the ninth excited state return to the ground state, is

(a) 36  (b) 45
(c) 5  (d) 6

## Heisenberg's Uncertainty Principle

**116.** The uncertainty in measuring the speed of a particle is zero. Uncertainty in measuring its position will be

(a) zero  (b) $\dfrac{h}{4\pi}$
(c) $\dfrac{h}{4\pi m}$  (d) infinite

**117.** Uncertainty in measuring the speed of a particle is numerically equal to the uncertainty in measuring its position. The value of these uncertainties will be

(a) equal to $\sqrt{\dfrac{h}{4\pi m}}$.

(b) less than $\sqrt{\dfrac{h}{4\pi m}}$.

(c) greater than $\sqrt{\dfrac{h}{4\pi m}}$.

(d) (a) or (c)

**118.** If uncertainty in position and momentum of a particle is numerically equal, then the minimum uncertainty in speed of the particle should be

(a) $\sqrt{\dfrac{h}{2\pi}}$  (b) $\dfrac{1}{2m}\sqrt{\dfrac{h}{\pi}}$
(c) $\sqrt{\dfrac{h}{\pi}}$  (d) $\dfrac{1}{m}\sqrt{\dfrac{h}{\pi}}$

**119.** The mass of a particle is $10^{-10}$ g and its diameter is $10^{-4}$ cm. If its speed is $10^{-6}$ cm/s with 0.0001% uncertainty in measurement, the minimum uncertainty in its position is

(a) $5.28 \times 10^{-8}$ m  (b) $5.28 \times 10^{-7}$ m
(c) $5.28 \times 10^{-6}$ m  (d) $5.28 \times 10^{-9}$ m

**120.** Uncertainty in the position of an electron (mass $= 9.1 \times 10^{-31}$ kg) moving with a velocity 300 m/s accurate up to 0.001% will be

(a) $5.76 \times 10^{-3}$ m  (b) $1.92 \times 10^{-2}$ m
(c) $3.84 \times 10^{-3}$ m  (d) $19.2 \times 10^{-4}$ m

## De Broglie's Equation

**121.** The ratio of de Broglie wavelength of electron and proton moving with the same speed is about

(a) 1836 : 1  (b) 1 : 1836
(c) 1 : 1  (d) 1 : 2

122. An electron makes five crests during one revolution around H-nucleus. The electron belongs from the
(a) first orbit (b) fourth orbit
(c) fifth orbit (d) sixth orbit

123. The circumference of the third orbit of $He^+$ ion is $x$ m. The de Broglie wavelength of electron revolving in this orbit will be
(a) $\frac{x}{3}$ m (b) $3x$ m
(c) $\frac{x}{9}$ m (d) $9x$ m

124. The momentum of a photon of wavelength 6626 nm will be
(a) $10^{-28}$ kg ms$^{-1}$ (b) $10^{-25}$ kg ms$^{-1}$
(c) $10^{-31}$ kg m$^{-1}$ (d) zero

125. If λ be the de Broglie wavelength of a thermal neutron at 27°C, then the wavelength of the same neutron at 927°C is
(a) λ (b) 0.5λ
(c) 2λ (d) 0.25λ

## Quantum Numbers

126. The energy of different orbitals in an atom or ion having only one electron depends on
(a) $n$ only (b) $n$ and $l$ only
(c) $n$, $l$ and $m$ only (d) $n$, $l$, $m$ and $s$

127. The size of an orbital is given by
(a) principal quantum number.
(b) azimuthal quantum number.
(c) magnetic quantum number.
(d) spin quantum number.

128. The types and number of orbitals belonging from the fifth orbit are, respectively,
(a) 5, 25 (b) 25, 5
(c) 4, 16 (d) 5, 5

129. The electron in the same orbital may be identified with the quantum number
(a) $n$ (b) $l$
(c) $m$ (d) $s$

130. The orbital angular momentum of an electron is 2$s$ orbital is
(a) $+\frac{1}{2}.\frac{h}{2\pi}$ (b) 0
(c) $\frac{h}{2\pi}$ (d) $\sqrt{2}\frac{h}{2\pi}$

131. The orbital angular momentum of a 4$p$ electron will be
(a) $4.\frac{h}{2\pi}$ (b) $\sqrt{2}.\frac{h}{2\pi}$
(c) $\sqrt{6}.\frac{h}{4\pi}$ (d) $\sqrt{2}.\frac{h}{4\pi}$

132. The probability of finding $P_y$ electron is zero in
(a) $XY$-plane (b) $YZ$-plane
(c) $XZ$-plane (d) $Y$-axis

133. The quantum number which determines the shape of the orbital is
(a) magnetic quantum number.
(b) azimuthal quantum number.
(c) principal quantum number.
(d) spin quantum number.

134. Orbital with maximum symmetry is
(a) $p$-orbital (b) $s$-orbital
(c) $d_{xy}$-orbital (d) $d_{z^2}$-orbital

135. In the presence of external magnetic field, $p$-orbital is
(a) 3-fold degenerate (b) 5-fold degenerate
(c) 7-fold degenerate (d) non-degenerate

136. The number of orbitals of g-type
(a) 5 (b) 7
(c) 9 (d) 11

137. Which of the following orbital does not exist according to quantum theory?
(a) 5g (b) 4f
(c) 5h (d) 6h

138. Which of the following set of quantum numbers is permissible?
(a) 4, 1, +2, +1/2 (b) 4, 2, –1, +1/2
(c) 4, 0, 0, 1 (d) 4, 4, +2, –1/2

EXERCISE I

**139.** Number of orbitals represented by $n = 3$, $l = 2$ and $m = +2$ is

(a) 1 (b) 2
(c) 3 (d) 4

**140.** The quantum numbers +1/2 and –1/2 for the electron spin represent

(a) rotation of the electron in clockwise and anticlockwise direction, respectively.
(b) rotation of the electron in anticlockwise and clockwise direction, respectively.
(c) magnetic moment of the electron pointing up and down, respectively.
(d) two quantum mechanical spin states which have no classical analogue.

## Schrodinger's Equation

**141.** The number of nodal planes in $2p_x$ orbital is

(a) zero (b) 1
(c) 2 (d) infinite

**142.** Which of the following orbital is represented by the complete wave function $\psi_{410}$?

(a) 4*s* (b) 3*p*
(c) 4*p* (d) 4*d*

**143.** Number of nodal surface in 5*s* orbital is

(a) 5 (b) 4
(c) 3 (d) 0

**144.** The orbital having two nodal surfaces is

(a) 1*s* (b) 2*s*
(c) 3*s* (d) 2*p*

**145.** The number of radial nodes of 3*s*, 3*p* and 3*d* electrons are, respectively,

(a) 0, 1, 2 (b) 2, 1, 0
(c) 2, 2, 2 (d) 1, 3, 5

## Electronic Configuration

**146.** The process of successive addition of protons to the nucleus followed by an addition of the same number of electrons to the available orbitals in the sequence of increasing energy to obtain the electronic configuration of many electronic configuration of many electron atom is known as

(a) Pauli's exclusion principle
(b) Hund's rule
(c) Heisenberg's uncertainty principle
(d) Aufbau principle

**147.** When the value of azimuthal quantum number is 3, the maximum and minimum values of spin multiplicity are

(a) 1, 8 (b) 8, 1
(c) 6, 1 (d) 7, 0

**148.** A completely filled *d*-orbital ($d^{10}$) is of

(a) spherical symmetry
(b) octahedral symmetry
(c) tetrahedral symmetry
(d) unsymmetry

**149.** An atom has $d^8$ configuration. The maximum number of electrons in the same spin is

(a) 5 (b) 3
(c) 8 (d) 2

**150.** The number of orbitals having $(n + l) < 5$ is

(a) 9 (b) 8
(c) 4 (d) 10

**151.** The total number of orbitals for $(n + l) = 4$ is

(a) 4 (b) 16
(c) 32 (d) 9

**152.** Which of the following configuration is violating Pauli's exclusion principle?

(a) 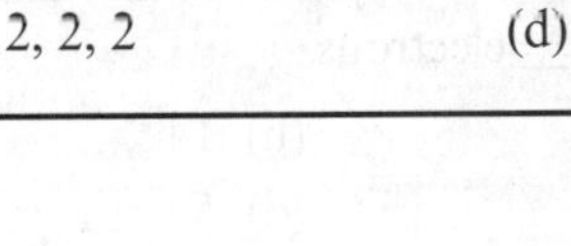


(b) 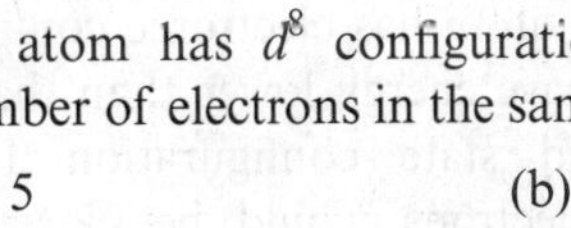


EXERCISE I

(c) 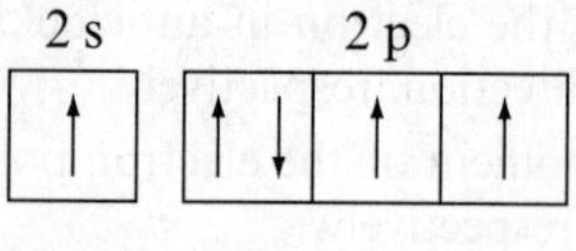


(d) (b) and (c)

**153.** If there are three possible values (–1/2, 0, +1/2) for the spin quantum number, then the maximum capacity of second orbit will become

(a) 8 electrons (b) 6 electrons
(c) 12 electrons (d) 27 electrons

**154.** The electrons identified by quantum numbers $n$ and $l$,

(i) $n = 4, l = 1$ (ii) $n = 4, l = 0$
(iii) $n = 3, l = 2$ (iv) $n = 3, l = 1$

can be placed in the order of increasing energy from the lowest to highest, as

(a) iv < ii < iii < i (b) ii < iv < i < iii
(c) i < iii < ii < iv (d) iii < i < iv < ii

**155.** If the numbers of orbitals of a particular type were $(3l + 1)$, but spin quantum numbers were only +1/2 and –1/2, then $d$-type orbitals will contain a maximum of ___ electrons.

(a) 10 (b) 14
(c) 7 (d) 5

**156.** If the nitrogen atom has electronic configuration $1s^7$, it would have energy lower than that of the normal ground state configuration $1s^2 2s^2 2p^3$, because the electrons would be closer to the nucleus. Yet $1s^7$ is not observed because it violates

(a) Heisenberg's uncertainty principle
(b) Hund's rule
(c) Pauli's exclusion principle
(d) Bohr postulate of stationary orbits

**157.** Which quantum number differs for the two electrons present in K-shell of an atom?

(a) Principal quantum number
(b) Azimuthal quantum number
(c) Magnetic quantum number
(d) Spin quantum number

**158.** The correct set of four quantum numbers for the unpaired electron of chloride atom is

(a) 3, 2, 0, +1/2 (b) 3, 1, 0, +1/2
(c) 3, 1, +1, 0 (d) 3, 0, –1, +1/2

**159.** The correct set of four quantum numbers for the valence electron of rubidium ($Z = 37$) is

(a) 5, 0, 0, +1/2 (b) 5, 1, 0, +1/2
(c) 5, 1, 1, +1/2 (d) 6, 0, 0, +1/2

**160.** The correct set of quantum numbers defining the highest energy electron in scandium (I) ion is

(a) $n = 3, l = 1, m = 0, s = -1/2$
(b) $n = 3, l = 0, m = 0, s = -1/2$
(c) $n = 4, l = 0, m = 0, s = +1/2$
(d) $n = 3, l = 2, m = 2, s = +1/2$

**161.** How many unpaired electrons are present in the ground state of chromium ($Z = 24$)?

(a) 1 (b) 5
(c) 6 (d) 0

**162.** The K and L shell of an element are completely filled and there are 16 electrons in M-shell and 2 electrons in N-shell. The atomic number of the element is

(a) 18 (b) 28
(c) 22 (d) 26

**163.** The penultimate and outermost orbit of an element contains 10 and 2 electrons, respectively. If the outermost orbit is fourth orbit, then the atomic number of the element should be

(a) 12 (b) 22
(c) 32 (d) 40

**164.** The number of unpaired electron in G. S., first E.S. and second E.S. of S ($Z = 16$) are, respectively,

(a) 0, 2 and 4 (b) 2, 4 and 6
(c) 0, 4 and 6 (d) 2, 4 and 4

**165.** The electronic structure of zinc ($Z = 30$) is 2, 8, 18, 2. The electronic structure of gallium ($Z = 31$) will be

(a) 2, 8, 18, 2, 1 (b) 2, 8, 19, 2
(c) 2, 8, 18, 3 (d) 2, 8, 19, 3

**166.** Which of the following ion have the same number of unpaired electrons as in $Fe^{2+}$ ($Z = 26$)?

(a) $Fe^{3+}$ ($Z = 26$) (b) $Ni^{2+}$ ($Z = 28$)
(c) $Co^{3+}$ ($Z = 27$) (d) $Cr^{+}$ ($Z = 24$)

**167.** Which of the following will have magnetic moment about 4.9 BM?

(a) $Cr^{+}$ ($Z = 24$) (b) $Ti^{4+}$ ($Z = 22$)
(c) $Fe^{2+}$ ($Z = 26$) (d) $Cu^{2+}$ ($Z = 29$)

168. Which of the following ion is diamagnetic?

(a) $Sc^{3+}$ ($Z = 21$) (b) $Ti^{2+}$ ($Z = 22$)
(c) $V^{3+}$ ($Z = 23$) (d) $Fe^{2+}$ ($Z = 26$)

169. Which of the following ion will have maximum magnetic moment?

(a) $Fe^{3+}$ ($Z = 26$) (b) $Cr^{3+}$ ($Z = 24$)
(c) $Ti^{4+}$ ($Z = 22$) (d) $Co^{3+}$ ($Z = 27$)

170. For which of the following element, all of its existing ion $M^{x+}$ will be diamagnetic?

(a) Cu (b) Fe
(c) Cr (d) Na

171. The magnetic moment of $Ni^{x+}$ ion ($Z = 28$) is about 2.82 BM. The value of $x$ is

(a) 2 (b) 4
(c) 1 (d) 3

172. A compound of vanadium has a magnetic moment of 1.73 BM. The electronic configuration of vanadium ion in the compound is

(a) $[Ar]3d^2$ (b) $[Ar]3d^1$
(c) $[Ar]3d^3$ (d) $[Ar]4s^1$

173. Which of the following is paramagnetic?

(a) $Zn^{2+}$ ($Z = 30$) (b) $Ni^{2+}$ ($Z = 28$)
(c) $Sc^{3+}$ ($Z = 21$) (d) $O^{2-}$ ($Z = 8$)

174. Which of the following ion is expected to be coloured?

(a) $Zn^{2+}$ ($Z = 30$) (b) $Ca^{2+}$ ($Z = 20$)
(c) $Sn^{2+}$ ($Z = 50$) (d) $V^{2+}$ ($Z = 23$)

175. Which of the following ion is expected to be colourless?

(a) $Ni^{2+}$ ($Z = 28$) (b) $Mn^{2+}$ ($Z = 25$)
(c) $Zn^{2+}$ ($Z = 30$) (d) $Cu^{2+}$ ($Z = 29$)

## EXERCISE II (JEE ADVANCED)

### Section A (Only one Correct)

1. The *e*/*m* ratio of cathode rays is $x$ unit, when hydrogen is filled in the discharge tube. What will be its value, when deuterium ($D_2$) is filled in it?
   (a) $x$ unit (b) $x/2$ unit
   (c) $2x$ unit (d) $x/4$ unit

2. The specific charges of two particles A and B are in the ratio 2 : 3. If the mass ratio $m_A : m_B$ is 2 : 3, then the ratio of their charges $e_A : e_B$, is
   (a) 1 : 1 (b) 4 : 9
   (c) 9 : 4 (d) 2 : 3

3. An electron at rest is accelerated through a potential difference of 200 V. If the specific charge of electron is $1.764 \times 10^{11}$ C/kg, then the speed acquired by the electron is about
   (a) $8.4 \times 10^6$ cm/s (b) $8.4 \times 10^6$ m/s
   (c) $4.2 \times 10^6$ m/s (d) $4.2 \times 10^6$ cm/s

4. The mass of a negative meson is 208 time the mass of electron. If the *e*/*m* ratio of α-particle is $x$ unit, then the *e*/*m* ratio of this meson will be
   (a) $\frac{x}{208}$ unit (b) $\frac{x}{52}$ unit
   (c) 17.65 $x$ unit (d) 104 $x$ unit

5. An electron and a proton are accelerated through a potential $V$. If $P_e$ and $P_p$ are their momentum, then $P_P : P_e$ ratio is approximately equal to
   (a) 1 : 1836 (b) 1 : 1
   (c) 1836 : 1 (d) 43 : 1

6. The frequency of an electromagnetic radiation which makes $2 \times 10^6$ waves per 50 cm is
   (a) $1.2 \times 10^{15}$ Hz (b) 150 Hz
   (c) $6 \times 10^{14}$ Hz (d) $1.2 \times 10^{13}$ Hz

7. A certain laser transition emits $6.0 \times 10^{15}$ quanta per second per square metre of λ = 662.6 nm. What is the power output in joule per second per square metre?
   (a) $1.8 \times 10^{-3}$ (b) $6.626 \times 10^{-4}$
   (c) $1.8 \times 10^3$ (d) $6.626 \times 10^{-12}$

8. A bulb emits light of wavelength $\frac{1987.8}{7}$ nm. The bulb is rated as 200 W and 14% of the energy is emitted as light. How many photons are emitted by the bulb per second?
   (a) $1.2 \times 10^{21}$ (b) $4 \times 10^{10}$
   (c) $1.33 \times 10^{11}$ (d) $4 \times 10^{19}$

9. An amount of $1.75 \times 10^{-4}$ mole of HI decomposes by the absorption of photons of wavelength 2500 Å. If one molecule is decomposed per absorbed photon, the total energy absorbed is ($N_A \times hc = 0.12$ J)
   (a) 42.0 J (b) 4.2 J
   (c) 8.4 J (d) 84 J

10. The dye acriflavine when dissolved in water has its maximum light absorption at 4530 Å and has maximum florescence emission at 5080 Å. The number of fluorescence quanta is about 53% of the number of quanta absorbed. What percentage of absorbed light energy is emitted as fluorescence?
    (a) 41% (b) 47%
    (c) 74% (d) 63%

11. The vapours of Hg absorb some electron accelerated by a potential difference of 5.0 V as a result of which light is emitted. If the full energy of single incident electron is supposed to be converted into light emitted by single Hg-atom, then the wavelength of the emitted light is
    (a) 2480 nm (b) 248 nm
    (c) 6200 nm (d) 620 nm

12. At one time the meter was defined as 1650763.73 wavelength of the orange light emitted by a light source containing $Kr^{86}$ atoms. What is the corresponding photon energy of this radiation?
    (a) $3.28 \times 10^{-19}$ J/quanta
    (b) $1.2 \times 10^{-31}$ J/quanta
    (c) $1.09 \times 10^{-27}$ J/quanta
    (d) 2.048 J/quanta

**13.** A ruby laser produces radiations of wavelength 662.6 nm in pulses whose duration are $1.0 \times 10^{-9}$ s. If the laser produces 0.36 J of energy per pulse, then how many photons are produced in each pulse?

(a) $1.2 \times 10^9$ (b) $1.2 \times 10^{27}$
(c) $1.2 \times 10^{18}$ (d) $1.2 \times 10^{15}$

**14.** $O_2$ undergoes photochemical dissociation into one normal oxygen atom and one oxygen atom, 1.2 eV more energetic than normal. The dissociation of $O_2$ into two normal atoms of oxygen requires 482.5 kJ/mol. The maximum wavelength effective for photochemical dissociation of $O_2$ is (1 eV = 96.5 kJ/mol)

(a) 248 nm (b) 1033.3 nm
(c) 1236.2 nm (d) 200 nm

**15.** Photodissociation of water $H_2O(l) + h\nu \rightarrow H_2(g) + \frac{1}{2}O_2(g)$ has been suggested as a source of hydrogen. The heat absorbed in this reaction is 289.5 kJ/mole of water decomposed. The maximum wavelength that would provide the necessary energy assuming that one photon causes the dissociation of one water molecule is (1 eV = 96.5 kJ/mol)

(a) 413.33 nm (b) 826.67 nm
(c) 206.67 nm (d) 4.3 nm

**16.** The dissociation energy of $H_2$ is 429.0 kJ/mol. If $H_2$ is dissociated by illumination with radiation of wavelength 270.0 nm, then what percentage of radiant energy will be converted into kinetic energy? ($h = 6.6 \times 10^{-34}$ J·s, $N_A = 6 \times 10^{23}$)

(a) 1.25% (b) 2.5%
(c) 5.0% (d) 7.5%

**17.** In a measurement of the quantum efficiency of photosynthesis in green plants, it was found that 9 quanta of red light at 6900 Å were needed to evolve 1 molecule of $O_2$. The average energy storage in the photosynthesis process is 111.6 kcal/mol of $O_2$ evolved. What is the energy conversion efficiency in this experiment? ($\frac{h \cdot c}{e} = 1.24 \times 10^{-6}$ nm.eV, 1 eV = 23 kcal/mol)

(a) 70% (b) 50%
(c) 40% (d) 30%

**18.** For a photochemical reaction $A \rightarrow B$, $1 \times 10^{-5}$ moles of 'B' were formed on absorption of $6.626 \times 10^7$ erg at 360 nm. The quantum efficiency (molecules of 'B' formed per photon) is ($N_A = 6 \times 10^{23}$)

(a) 1.0 (b) 0.25
(c) 0.5 (d) 2.0

**19.** Light of wavelength $\lambda$, falls on a metal having work function $hc/\lambda_0$. Photoelectric effect will take place only if

(a) $\lambda \geq \lambda_0$ (b) $\lambda \geq 2\lambda_0$
(c) $\lambda \leq \lambda_0$ (d) $\lambda \leq \lambda_0/2$

**20.** Light of wavelength ($\lambda$) strikes a metal surface with intensity $X$ and the metal emits $Y$ electrons per second of maximum kinetic energy $Z$. What will happen to $Y$ and $Z$ if $X$ is halved?

(a) $Y$ will be halved and $Z$ will be doubled.
(b) $Y$ will be doubled and $Z$ will be halved.
(c) $Y$ will be halved and $Z$ will remain the same.
(d) $Y$ will remain the same and $Z$ will be halved.

**21.** Photoelectric emission is observed from a metal surface for frequencies $\nu_1$ and $\nu_2$ of the incident radiation ($\nu_1 > \nu_2$). If maximum kinetic energies of the photoelectrons in the two cases are in the ratio $1 : K$, then the threshold frequency for the metal is given by

(a) $\frac{\nu_2 - \nu_1}{K-1}$ (b) $\frac{K\nu_2 - \nu_1}{K-1}$
(c) $\frac{K\nu_1 - \nu_2}{K}$ (d) $\frac{K\nu_1 - \nu_2}{K-1}$

**22.** Photons of frequency ($\nu$) fall on metal surface for which the threshold of frequency is $\nu_0$. Which of the following statement is correct?

(a) All ejected electrons have the same kinetic energy $h(\nu - \nu_0)$.
(b) The ejected electrons have a distribution of kinetic energy from zero to $h(\nu - \nu_0)$.
(c) The most energetic electron has kinetic energy $h\nu$.
(d) The average kinetic energy of ejected electrons is $h(\nu - \nu_0)$.

**23.** If $\lambda_0$ is the threshold wavelength for photoelectric emission from a metal surface, $\lambda$ is the wavelength of light falling on the surface of metal and $m$ is the mass of electron, then the maximum speed of ejected electrons is given by

(a) $\left[\frac{2h}{m}(\lambda_0 - \lambda)\right]^{1/2}$ (b) $\left[\frac{2hc}{m}(\lambda_0 - \lambda)\right]^{1/2}$
(c) $\left[\frac{2hc}{m}\left(\frac{\lambda_0 - \lambda}{\lambda_0\lambda}\right)\right]^{1/2}$ (d) $\left[\frac{2h}{m}\left(\frac{1}{\lambda_0} - \frac{1}{\lambda}\right)\right]^{1/2}$

**24.** Which of the following graphs is correct for the photoelectric effect?

(a)

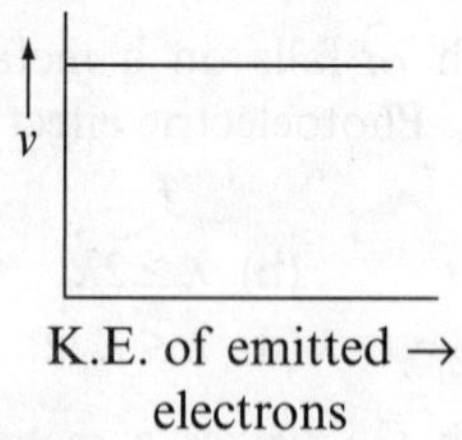


(b)

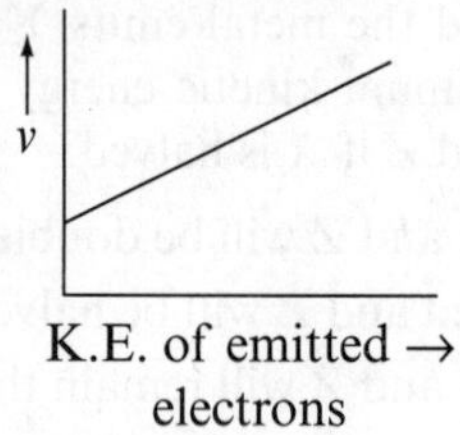


(c)

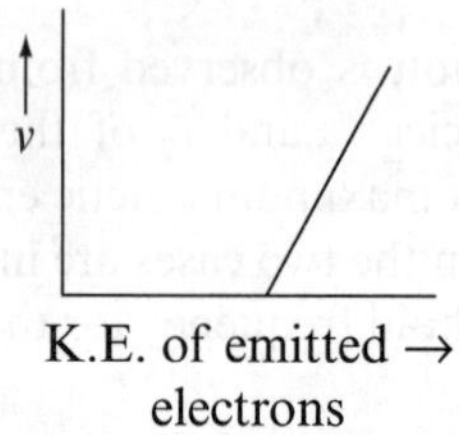


(d)

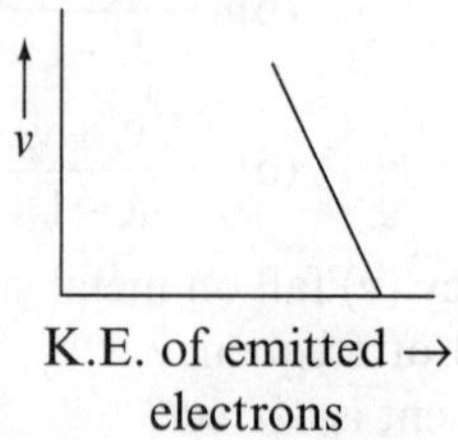


**25.** The wavelength of $K_\alpha$-characteristic X-rays produced is $\lambda$, when cathode rays strike on a metal of atomic number $Z$. What should be the atomic number of metal such that it can produce the $K_\alpha$-characteristic X-rays of wavelength $4\lambda$?

(a) $\frac{Z}{16}$ (b) $\frac{Z}{2}$
(c) $\frac{Z+1}{2}$ (d) $2Z-1$

**26.** Two carbon discs, 1.0 g each are 1.0 cm apart have equal and opposite charges. If the force of attraction between them is $10^{-5}$ N, then the ratio of excess electrons to the total atoms on the negatively charged disc is ($N_A = 6 \times 10^{23}, e = 1.6 \times 10^{-19}$ C)

(a) $2.4 \times 10^{-12} : 1$ (b) $10^{-14} : 2.4$
(c) $10^{12} : 2.4$ (d) $2.4 : 10^{12}$

**27.** The radius of the hydrogen atom in its ground state is $5.3 \times 10^{-11}$ m. After collision with an electron it is found to have a radius of $21.2 \times 10^{-11}$ m. The principal quantum number of final state of the atom is

(a) 2 (b) 3
(c) 4 (d) 16

**28.** For which orbit in $He^+$ ion, the circumference is 26.5 Å?

(a) 2 (b) 3
(c) 4 (d) 16

**29.** The radius of the second orbit of H-atom is equal to the radius of

(a) second orbit of $He^+$ ion.
(b) third orbit of $Li^{2+}$ ion.
(c) fourth orbit of $He^+$ ion.
(d) fourth orbit of $Be^{3+}$ ion.

**30.** As the orbit number increases, the distance between two consecutive orbits ($r_1$ = radius of first orbit)

(a) increases by $2r_1$.
(b) increases by $(2n - 1)r_1$, where $n$ is the lower orbit number.
(c) increases by $(2n - 1)r_1$, where $n$ is the higher orbit number.
(d) remains constant.

**31.** The ratio of the areas within the electron orbits for the first excited state to the ground state for the hydrogen atom is

(a) 2 : 1 (b) 4 : 1
(c) 8 : 1 (d) 16 : 1

**32.** When an electron jumps from the second orbit to fourth orbit, its distance from nucleus increases by 2.116 Å. The atom or ion should be

(a) H atom (b) $He^+$ ion
(c) $Li^{2+}$ ion (d) $Be^{3+}$ ion

**33.** What is the distance travelled by an electron revolving in the second orbit of $Be^{3+}$ ion in 100 revolutions?

(a) $3.32 \times 10^{-8}$ m (b) $5.29 \times 10^{-8}$ m
(c) $6.64 \times 10^{-8}$ m (d) $1.16 \times 10^{-8}$ m

**34.** Which of the following cannot be circumference of an orbit in H-atom? ($r_0$ = radius of the first orbit)

(a) $2\pi r_0$ (b) $4\pi r_0$
(c) $8\pi r_0$ (d) $18\pi r_0$

35. The speed of electron revolving around H-nucleus is $0.547 \times 10^6$ m/s. The distance of electron from the nucleus is

(a) 2.116 Å (b) 4.761 Å
(c) 8.464 Å (d) 0.529 Å

36. The speed of electron in H-atom is directly proportional to

(a) $r$ (b) $\sqrt{r}$
(c) $1/r$ (d) $1/\sqrt{r}$

37. The ratio of the speed of the electron in the first Bohr orbit of hydrogen and the speed of light is equal to (where $e$, $h$ and $c$ have their usual meanings)

(a) $2\pi hc/e^2$ (b) $e^2c/2\pi h$
(c) $e^2h/2\pi c$ (d) $e^2/2\epsilon_0 hc$

38. In the Bohr's atomic model, the electrostatic force of attraction between nuclear charge ($Ze$) and electron of charge $e$ is balanced by the centripetal force acting towards the centre of atom. If $\epsilon_0$ be the permittivity of vacuum and $r$ be the radius of orbit in which electron is revolving, the speed of electron is

(a) $\sqrt{\dfrac{Ze^2}{(4\pi\epsilon_0)mr}}$ (b) $\sqrt{\dfrac{(4\pi\epsilon_0)mr}{Ze^2}}$
(c) $\sqrt{(4\pi\epsilon_0)mrZe^2}$ (d) $\dfrac{e}{\sqrt{(4\pi\epsilon_0)mr}}$

39. If an electron is revolving around the nucleus of $He^+$ ion at a distance of 4.0 Å, the magnitude of centripetal force on electron by the nucleus is ($e = 1.6 \times 10^{-19}$ C)

(a) $2.88 \times 10^{-9}$ N (b) $2.88 \times 10^{-7}$ N
(c) $1.152 \times 10^{-18}$ N (d) $1.44 \times 10^{-9}$ N

40. The time period of revolution in the third orbit of $Li^{2+}$ ion is $x$ second. The time period of revolution in the second orbit of $He^+$ ion should be

(a) $x$ s (b) $\frac{3}{2}x$ s
(c) $\frac{2}{3}x$ s (d) $\frac{8}{27}x$ s

41. In two hydrogen atoms A and B, the electrons are revolving around the nucleus in circular orbits of radius $r$ and $4r$, respectively. The ratio of times taken by them to complete one revolution is

(a) 1 : 2 (b) 1 : 4
(c) 1 : 8 (d) 1 : 64

42. The time period of revolution of electron in H-atom is directly proportional to

(a) $r$ (b) $r^{1/2}$
(c) $r^{3/2}$ (d) $r^2$

43. The average lifetime of an electron in an excited state of hydrogen atom is about $10^{-8}$ s. How many revolutions does an electron in the $n = 2$ state make before dropping to the $n = 1$ state?

(a) $10^8$ (b) $8.33 \times 10^6$
(c) $6.67 \times 10^7$ (d) $1.04 \times 10^6$

44. According to Maxwell's theory of electrodynamics, an electron going in a circle should emit radiation of frequency equal to its frequency of revolution. What should be the wavelength of the radiation emitted by a hydrogen atom in the ground state if this rule is followed?

(a) 4500 nm (b) 450 nm
(c) 45 nm (d) 4.5 nm

45. The angular momentum of electron in Bohr's orbit is J. What will be the K.E. of electron in that Bohr's orbit?

(a) $\frac{1}{2}\frac{Jv}{r}$ (b) $\frac{Jv}{r}$
(c) $\frac{J^2}{2m}$ (d) $\frac{J^2}{2r}$

46. The kinetic energy of an electron in the second Bohr orbit of a hydrogen atom is ($a_o$ is the Bohr radius)

(a) $\dfrac{h^2}{4\pi^2 ma_o^2}$ (b) $\dfrac{h^2}{16\pi^2 ma_o^2}$
(c) $\dfrac{h^2}{32\pi^2 ma_o^2}$ (d) $\dfrac{h^2}{64\pi^2 ma_o^2}$

47. The ionization energy of a hydrogen-like atom is 14.4 eV. The amount of energy released when electron jumps from the fourth orbit to the first orbit in this atom is

(a) 13.5 eV (b) 10.8 eV
(c) 0.9 eV (d) 12.75 eV

48. The radius of first orbit of H-atoms is 0.529 Å. The radius of first orbit of D-atoms should be

(a) exactly 0.529 Å.
(b) slightly less than 0.529 Å.
(c) slightly greater than 0.529 Å.
(d) 1.058 Å.

**49.** The ionization energy of H-atoms is 13.6 eV. The ionization energy of deuterium atom should be

(a) exactly 13.6 eV.
(b) slightly less than 13.6 eV.
(c) slightly greater than 13.6 eV.
(d) 27.2 eV.

**50.** An antiproton has the mass of a proton but a charge of $-e$. If a proton and an antiproton orbited each other, then how far apart would they be in ground state of such a system? Mass of a proton is 1836 times the mass of an electron.

(a) 0.058 pm (b) 0.029 pm
(c) 0.014 pm (d) 194.25 nm

**51.** A lithium atom has three electrons. Assume the following simple picture of the atom. Two electrons move close to the nucleus making up a spherical cloud around it and the third moves outside this cloud in a circular orbit. Bohr's model can be used for the motion of this third electron but $n = 1$ state is not available to it. The ionization energy of lithium in ground state, using the above picture, is

(a) 13.6 eV (b) 10.2 eV
(c) 3.4 eV (d) 1.51 eV

**52.** An electron in a hydrogen atom in its ground state absorbs 1.5 times as much energy as the minimum required for it to escape from the atom. What is the speed of the emitted electron?

(a) $1.55 \times 10^6$ m/s (b) $2.68 \times 10^6$ m/s
(c) $2.19 \times 10^6$ m/s (d) $1.02 \times 10^6$ m/s

**53.** Sodium atoms emit a spectral line with a wavelength in the yellow, 589.6 nm. What is the approximate difference in energy between the two energy levels involved in the emission of this spectral line?

(a) $3.37 \times 10^{-19}$ J (b) 2.1 eV
(c) 48.35 kcal/mol (d) All of these

**54.** A certain molecule has an energy level diagram for its vibrational energy in which two levels are 0.0141 eV apart. The wavelength of the emitted line for the molecule as it falls from one of these levels to the other is about

(a) 88 μm (b) 88 mm
(c) 174.84 m (d) 88 nm

**55.** In a discharge tube, there are only two hydrogen atoms. If the electrons in both atoms are de-exciting from 4th orbit, the minimum and maximum number of spectral lines should, respectively, be

(a) 1, 4 (b) 4, 1
(c) 3, 4 (d) 1, 6

**56.** Electrons are de-exciting from the fifth orbit in hydrogen atoms but the first orbit is not available for them. The maximum number of spectral lines should be

(a) 10 (b) 6
(c) 15 (d) 3

**57.** From a hydrogen discharge tube, only three photons are picked up. The energies of these three photons were 10.2, 12.1 and 1.9 eV. These photons are coming from

(a) only one atoms. (b) two atoms.
(c) three atoms. (d) two or more atoms.

**58.** When electron jumps from the fourth orbit to the second orbit in $He^+$ ion, the radiation emitted out will fall in

(a) ultraviolet region (b) visible region
(c) infrared region (d) radio wave region

**59.** When electrons are de-exciting to the ground state from $n$th orbit of hydrogen atoms, 15 spectral lines are formed. The shortest wavelength among these will be

(a) $\frac{11}{900}R$ (b) $\frac{900}{11R}$

(c) $\frac{35}{36}R$ (d) $\frac{36}{35R}$

**60.** Rydberg given the equation for all visible radiation in the hydrogen spectrum as $\lambda = \frac{kn^2}{n^2 - 4}$. The value of $k$ in terms of Rydberg constant is

(a) $4R$ (b) $\frac{R}{4}$

(c) $\frac{4}{R}$ (d) $R$

**61.** The wavelengths of the first Lyman lines of hydrogen, $He^+$ and $Li^{2+}$ ions are $\lambda_1$, $\lambda_2$, $\lambda_3$. The ratio of these wavelengths is

(a) 1 : 4 : 9 (b) 9 : 4 : 1
(c) 36 : 9 : 4 (d) 6 : 3 : 2

62. An excited hydrogen atom emits a photon of wavelength $\lambda$ in returning to the ground state. If $R$ is the Rydberg constant, then the quantum number $n$ of the excited state is

(a) $\sqrt{\lambda R}$ (b) $\sqrt{\lambda R - 1}$

(c) $\sqrt{\frac{\lambda R}{\lambda R - 1}}$ (d) $\sqrt{\lambda R(\lambda R - 1)}$

63. Suppose that in any Bohr atom or ion, orbits are only in even numbers like 2, 4, 6, …. The maximum wavelength of radiation emitted in the visible region of H-spectrum should be

(a) $\frac{4}{R}$ (b) $\frac{R}{4}$

(c) $\frac{36}{5R}$ (d) $\frac{16}{3R}$

64. The wavelength of first line of Lyman series of H-atom is 1216 Å. What will be the wavelength of first line of Lyman series in 10 time ionized sodium atom ($Z = 11$)?

(a) 1216 Å (b) 12.16 Å
(c) 10 Å (d) 110 Å

65. In a sample of hydrogen atoms, all the atoms are in a particular excited state. If the emission spectrum of this sample has only 4 spectral lines in the visible region, then the total number of spectral lines in IR region possible from that state is

(a) 6 (b) 5
(c) 7 (d) 15

66. Imagine an atom made up of a stationary proton and a hypothetical particle of double the mass of electron but having the same charge as the electron. Apply Bohr's atomic model and consider all possible transitions of this hypothetical particle directly to the first excited state. The longest wavelength photon that will be emitted has wavelength (given in terms of Rydberg constant $R$ for the hydrogen atom) equal to

(a) $\frac{9}{5R}$ (b) $\frac{36}{5R}$

(c) $\frac{18}{5R}$ (d) $\frac{4}{R}$

67. Which of the following expression represent the wave number of spectral lines in Balmer series (if $n$ is the principal quantum number of higher energy level)?

(a) $\frac{R(n^2 - 4)}{n^2}$ (b) $\frac{R(n-2)(n+2)}{n^2}$

(c) $\frac{R(n-2)(n+2)}{4n^2}$ (d) $\frac{R(n-1)(n+1)}{4n^2}$

68. The energy emitted when electron of 1.0 g atom of hydrogen undergoes transition giving the spectral lines of lowest energy in the visible region of its atomic spectra is

(a) 301.22 kJ (b) 328 kJ
(c) 984 kJ (d) 182.22 kJ

69. A series of lines in the spectrum of atomic hydrogen lines at wavelength 656.46, 486.27, 434.17, 410.29 nm. What is the wavelength of next line in this series?

(a) 397.12 nm (b) 407.83 nm
(c) 389.01 nm (d) 360.54 nm

70. To what series does the spectral line of atomic hydrogen belong if its wave number is equal to the difference between the wave numbers of the following two lines of the Balmer series: 486.1 and 410.2 nm?

(a) Lyman series (b) Balmer series
(c) Paschen series (d) Brackett series

71. The value of Rydberg constant $R$, if $He^+$ ions are known to have the wavelength difference between the first (of the longest wavelength) lines of the Balmer and Lyman series equal to 132 nm, is

(a) $2.07 \times 10^{16}$ m$^{-1}$ (b) $1.11 \times 10^7$ m$^{-1}$
(c) $9 \times 10^{-8}$ m$^{-1}$ (d) $1.936 \times 10^7$ m$^{-1}$

72. The wavelength of the first line of the $He^+$ ion spectral series whose interval between the extreme lines is $2.725 \times 10^6$ m$^{-1}$ is ($R = 1.09 \times 10^7$ m$^{-1}$)

(a) 471.82 nm (b) 4718.2 nm
(c) 1019.37 nm (d) 165.14 nm

73. The binding energy of an electron in the ground state of hydrogen-like ions in whose spectrum, the third line of the Balmer series is equal to 108.5 nm, is

(a) 13.6 eV (b) 54.4 eV
(c) 122.4 eV (d) 14.4 eV

**74.** A stationary $He^+$ ion emitted a photon corresponding to the first line of the Lyman series. That photon liberated a photoelectron from a stationary hydrogen atom in the ground state. The velocity of the photoelectron is

(a) $3.1 \times 10^6$ m/s (b) $3.1 \times 10^5$ m/s
(c) $9.56 \times 10^{12}$ m/s (d) $9.56 \times 10^6$ m/s

**75.** A single electron species in energy level (orbit number $n$) with energy $X$ was provided with excess of energy so that it jumps to higher energy level with energy $Y$. If it can emit radiations of six different wavelengths on de-excitement between these two energy levels, then the correct relation is

(a) $\frac{X}{Y} = (n-1)^2$ (b) $\frac{X}{Y} = 1 + \frac{3}{n}$
(c) $\sqrt{\frac{X}{Y}} = 1 + \frac{3}{n}$ (d) $\frac{X}{Y} = 1 + \frac{n}{3}$

**76.** When an electron de-excites from higher orbit in H-atom, two radiations are emitted out in Paschen and Lyman series. The wavelength of radiation emitted out in Lyman series is

(a) $\frac{8R}{9}$ (b) $\frac{3R}{4}$
(c) $\frac{4}{3R}$ (d) $\frac{9}{8R}$

**77.** If the radius of first Bohr orbit is $x$ unit, then de Broglie wavelength of electron in the third orbit is

(a) $2\pi x$ unit (b) $6\pi x$ unit
(c) $9x$ unit (d) $18\pi x$ unit

**78.** If $E_1$, $E_2$ and $E_3$ are the kinetic energies of an electron, an α-particle and a proton with the same de Broglie wavelength, then

(a) $E_1 > E_3 > E_2$ (b) $E_2 > E_3 > E_1$
(c) $E_1 > E_2 > E_3$ (d) $E_1 = E_2 = E_3$

**79.** A proton and an α-particle are accelerated through the same potential difference. The ratio of their de Broglie wavelengths is

(a) 1 : 1 (b) 2 : 1
(c) $\sqrt{2}$ : 1 (d) $2\sqrt{2}$ : 1

**80.** The de Broglie wavelength of a vehicle moving with velocity $v$ is λ. Its load is changed so that the velocity as well as kinetic energy is doubled. What will be the new de Broglie wavelength?

(a) λ (b) 2λ
(c) 4λ (d) λ/2

**81.** When accelerated electrons are directed against an anticathode in an X-ray tube, the radiation obtained has a continuous spectrum with a wavelength minimum, $\lambda_{min} = \frac{1.24 \times 10^{-6}}{V}$ m, where V is the voltage used for accelerating the electrons. $\lambda_{min}$ for $V = 5 \times 10^4$ V is

(a) 0.124 nm (b) 24.8 nm
(c) 2.48 nm (d) 1.24 nm

**82.** The dynamic mass (in kg) of the photon with a wavelength corresponding to the series limit of the Balmer transitions of the $He^+$ ion is

(a) $4.22 \times 10^{-36}$ (b) $2.24 \times 10^{-34}$
(c) $2.42 \times 10^{-35}$ (d) $4.22 \times 10^{-35}$

**83.** An electron is continuously accelerated in a vacuum tube by the application of a potential difference. If its de Broglie wavelength decreases by 1% over a path length of $l$ cm, then its kinetic energy

(a) increases by 1% (b) increases by 2%
(c) decreases by 2% (d) increases by 0.5%

**84.** Assume that the uncertainty in the position of a particle is equal to its de Broglie wavelength. The minimum uncertainly in its velocity is equal to

(a) 0.25 times its velocity.
(b) $\frac{\pi}{4}$ times its velocity.
(c) $\frac{1}{4\pi}$ times its velocity.
(d) $\frac{4}{\pi}$ times its velocity.

**85.** What should be the increase in kinetic energy of electron in order to decrease its de Broglie wavelength from 100 nm to 50 nm?

(a) 0.451 keV (b) $4.51 \times 10^{-4}$ eV
(c) $4.51 \times 10^{-3}$ eV (d) 0.0451 eV

**86.** An α–particle is accelerated from rest through a potential difference of 6.0V. Its de Broglie wavelength is

(a) 5 Å (b) 4.15 pm
(c) 414.6 Å (d) 5 nm

**87.** The de Broglie wavelength of electron of $He^+$ ion is 3.32 Å. If the photon emitted upon de-excitation of this $He^+$ ion is made to hit H-atom in its ground state so as to liberate electron from it, then what will be the de Broglie wavelength of photoelectron?

(a) 2.348 Å (b) 1.917 Å
(c) 3.329 Å (d) 1.66 Å

**88.** Photoelectrons are liberated by ultraviolet light of wavelength 3000 Å from a metallic surface for which the photoelectric threshold is 4000 Å. The de Broglie wavelength of electrons emitted with maximum kinetic energy is

(a) 1000 Å (b) 42.43 Å
(c) 12.05 Å (d) 3.54 Å

**89.** The minimum uncertainty in de Broglie wavelength of an electron accelerated from rest by a potential difference of 6.0V, if the uncertainty in measuring the position is $\frac{1}{\pi}$ nm, is

(a) 6.25 Å (b) 6.0 Å
(c) 0.625 Å (d) 0.3125 Å

**90.** A photon of 2.55 eV is emitted out by an electronic transition in hydrogen atom. The change in de Broglie wavelength of the electron is

(a) 3.32 Å (b) 4.98 Å
(c) 6.64 Å (d) 9.96 Å

**91.** The orbital angular momentum of 2*p* and 3*p*-orbitals

(a) are same.
(b) are different and more for 2*p*-orbital.
(c) are different and more for 3*p*-orbital.
(d) depends on the type of atom or ion.

**92.** Which of the following energy level may bring absorption of photon but never emission of photon?

(a) 3*d* (b) 2*p*
(c) 1*s* (d) 2*s*

**93.** An electron that has the quantum numbers $n = 3$ and $m = 2$

(a) must have spin quantum number value +1/2.
(b) must have $l = 2$, 3 or 4.
(c) must have $l = 0$, 1 or 2.
(d) must have $l = 2$.

**94.** If an electron has spin quantum number of + ½ and magnetic quantum number of –1, then it cannot be present in

(a) *s*-orbital (b) *p*-orbital
(c) *d*-orbital (d) *f*-orbital

**95.** In which of the following orbital, electron will be closer to the nucleus?

(a) 6*s* (b) 4*f*
(c) 5*d* (d) 6*p*

**96.** In the absence of external magnetic field, *d*-orbital is

(a) 3-fold degenerate (b) 5-fold degenerate
(c) 7-fold degenerate (d) non-degenerate

**97.** The following electronic transitions occur when Lithium atoms are sprayed into a hot flame;

$$2s \xrightarrow{\text{I}} 2p \xrightarrow{\text{II}} 3d \xrightarrow{\text{III}} 3p \xrightarrow{\text{IV}} 4s \xrightarrow{\text{V}} 3p,$$

which of these transition would result in the emission of light?

(a) I, II and IV (b) III and V
(c) III, IV and V (d) all of these steps

**98.** The possible set of quantum numbers for which $n = 4$, $l = 3$ and $s = +\frac{1}{2}$ is

(a) 14 (b) 7
(c) 5 (d) 10

**99.** Which of the following graph represents the radial probability function of 3*d* electron?

(a)

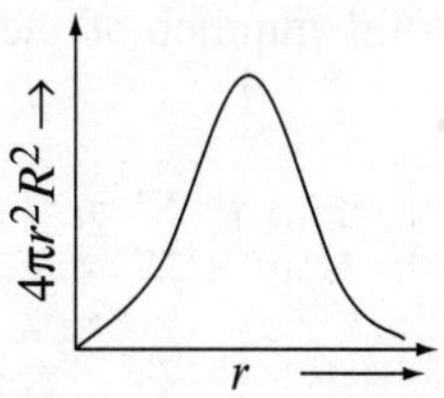


(b)

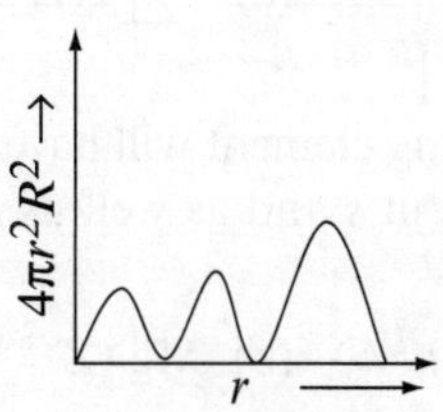

(c)

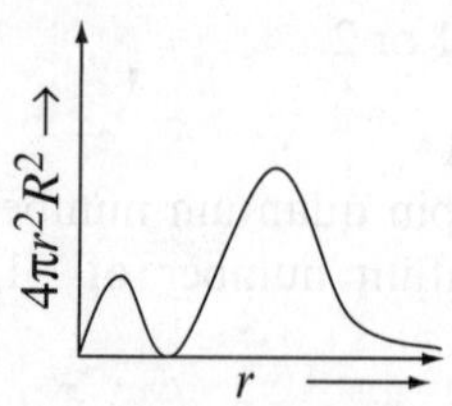


(d)

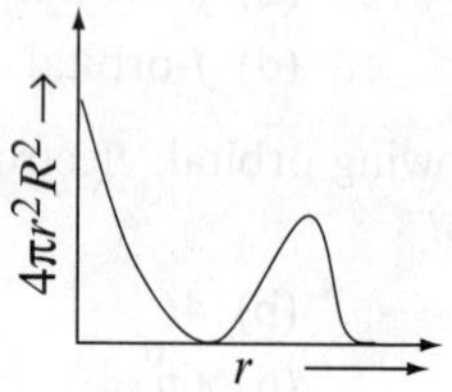


**100.** The wave function of 1*s* orbital of H-atom is $\psi = \frac{1}{\sqrt{\pi}}\left(\frac{1}{a_0}\right)^{3/2} e^{-r/a_0}$, where $a_o$ = Bohr's radius. The probability of finding the electrons at a distance '*r*' from the nucleus is given by

(a) $\psi = \psi^2\, dr$ (b) $\int \psi^2 4\pi r^2 dr$

(c) $\psi^2 4\pi r^2 dr$ (d) $\psi \cdot dV$

**101.** For an electron in a hydrogen atom, the wave function is given by $\psi_{1s} = (\pi/\sqrt{2})\, e^{-r/a_0}$ where $a_o$ is the radius of first Bohr's orbit and *r* is the distance from the nucleus with which the probability of finding electron varies. What will be the ratio of probabilities of finding electrons at the nucleus to first Bohr's orbit $a_o$?

(a) 0 (b) *e*

(c) $e^2$ (d) $\frac{1}{e^2}$

**102.** If *n* and *l* are, respectively, the principal and azimuthal quantum numbers, then the expression for calculating the total number of electrons in any energy level is

(a) $\sum_{l=1}^{l=n} 2(2l+1)$ (b) $\sum_{l=1}^{l=n-1} 2(2l+1)$

(c) $\sum_{l=0}^{l=n+1} 2(2l+1)$ (d) $\sum_{l=0}^{l=n-1} 2(2l+1)$

**103.** Which of the following element will have the same number of electrons in *s* and as well as *p*-type of orbitals?

(a) Fe ($Z = 26$) (b) Mg ($Z = 12$)

(c) Ne ($Z = 10$) (d) Ar ($Z = 18$)

**104.** Electronic configuration of an element is $1s^2, 2s^1, 2p^2$. It is

(a) ground state configuration of B ($Z = 5$).
(b) excited state configuration of B.
(c) ground state configuration of $C^+$ ion ($Z = 6$).
(d) impossible configuration.

**105.** The number of electrons having $m = 0$ for sodium atom is

(a) 2 (b) 5

(c) 7 (d) 3

**106.** The magnitude of the orbital angular momentum of an electron is given by $L = \sqrt{5}\, h/\pi$. How many orbitals of this kind are possible, belonging to an orbit?

(a) 4 (b) 5

(c) 11 (d) 9

**107.** What is the most probable distance of a 1*s* electron in a $He^+$ ion. The wave function for 1*s* orbital is given by $\psi = \sqrt{\left(\frac{Z^3}{\pi a_o^3}\right)} e^{-Zr/a_o}$, where $a_o$ = radius of first Bohr's orbit in H-atom = 52.9 pm.

(a) 52.9 pm (b) 13.25 pm

(c) 6.61 pm (d) 26.45 pm

**108.** The average and the most probable distance from the nucleus for 1*s* electron in hydrogen atom are, respectively ($a_o$ is the first Bohr radius.),

(a) $a_o$, $a_o$ (b) $a_o$, $1.5a_o$

(c) $1.5a_o$, $a_o$ (d) $1.5a_o$, $1.5a_o$

**109.** For an atom or ion having single electron, compare the energies of the following orbitals.

$S_1$ = A spherically symmetrical orbital having two spherical nodes.

$S_2$ = An orbital which is double dumb-bell and has no radial node.

$S_3$ = An orbital with orbital angular momentum zero and three radial nodes.

$S_4$ = An orbital having one planar and one radial node.

(a) $S_1 = S_2 = S_3 = S_4$ (b) $S_1 = S_2 = S_4 < S_3$

(c) $S_1 > S_2 > S_3 > S_4$ (d) $S_1 < S_4 < S_3 < S_2$

**110.** The orbital angular momentum of an electron is $\sqrt{3}\,\frac{h}{\pi}$. Which of the following may be the permissible value of angular momentum of this electron revolving in unknown Bohr's orbit?

(a) $\frac{h}{\pi}$ (b) $\frac{h}{2\pi}$

(c) $\frac{3h}{2\pi}$ (d) $\frac{2h}{\pi}$

## Section B (One or More than one Correct)

1. An α-particle having kinetic energy 4.0 MeV is projected towards tin nucleus ($Z = 50$). Select the correct information(s) regarding the α-particle.
   (a) Its distance of closest approach towards the nucleus is $3.6 \times 10^{-14}$ m.
   (b) Its potential energy at a distance of $9.0 \times 10^{-14}$ m from the nucleus is 1.6 MeV.
   (c) Its kinetic energy at a distance of $4.5 \times 10^{-14}$ m from the nucleus is 0.8 MeV.
   (d) At a moment, the distance between α-particle and the nucleus becomes $2.0 \times 10^{-16}$ m.

2. The energy of an electron in the first Bohr's orbit of H-atom is 13.6 eV. The possible energy value(s) of the excited state(s) for electrons in Bohr's orbits to hydrogen is
   (a) –3.4 eV (b) –10.2 eV
   (c) –1.51 eV (d) –0.85 eV

3. According to Bohr's atomic theory, which of the following relations is/are correct?
   (a) Kinetic energy of electron $\alpha\ \frac{z^2}{n^2}$
   (b) The product of speed of electron and the principal quantum number $\alpha\ z^2$
   (c) Frequency of revolution of the electron in an orbit $\alpha\ \frac{z^2}{n^3}$
   (d) Coloumbic force of attraction on the electron $\alpha\ \frac{z^3}{n^4}$

4. As an electron jumps from the fourth orbit to the second orbit in $Be^{3+}$ ion, its
   (a) kinetic energy increases.
   (b) speed increases.
   (c) frequency of revolution increases.
   (d) potential energy decreases.

5. Which of the following statement(s) is/are correct about the Bohr's model of hydrogen atom?
   (a) The acceleration of the electron in the $n = 2$ orbit is more than that in the $n = 1$ orbit.
   (b) The angular momentum of the electron in then $n = 2$ orbit is more than that in the $n = 1$ orbit.
   (c) The kinetic energy of electron in the $n = 2$ orbit is less than that in the $n = 1$ orbit.
   (d) The centripetal force of electron in the $n = 2$ orbit is more than that in the $n = 1$ orbit.

6. The frequency of certain line of the Lyman series ($n_2 = 4$ to $n_1 = 1$) of the atomic spectrum of hydrogen can satisfy the following conditions.
   (a) It is the sum of the frequencies of a Lyman line and a Balmer line.
   (b) It is the sum of the frequencies of a certain Lyman line, a Balmer line and a Brackett line.
   (c) It is the sum of the frequencies of a Lyman line, a Balmer line and a Paschen line.
   (d) It is the sum of the frequencies of a Lyman and a Paschen line.

7. A sample of hydrogen atoms in ground state is exposed to electromagnetic radiations of 1028 Å. The wavelengths of the induced radiation(s) is/are
   (a) 1028 Å (b) 1218.4 Å
   (c) 6579.2 Å (d) 190.4 Å

8. Some hydrogen-like atoms in ground state absorbs '*n*' photons having same the energy and on de-excitement, it emits exactly '*n*' photons. The energy of absorbed photon may be
   (a) 91.8 eV (b) 40.8 eV
   (c) 48.4 eV (d) 54.4 eV

**9.** Which of the following statement(s) is/are correct?

(a) The ratio of the radii of the first three Bohr orbits of hydrogen atom is 1 : 8 : 27.

(b) The ratio of magnitude of total energy : kinetic energy : potential energy for electron in any orbit of hydrogen atom is 1 : 1 : 2.

(c) The frequency of a green light is $6 \times 10^{14}$ Hz, then its wavelength is 500 nm.

(d) The ratio of de Broglie wavelength of a H-atom, He-atom and $CH_4$-molecule moving with equal kinetic energy is 4 : 2 : 1.

**10.** When photons of energy 4.25 eV strikes the surface of a metal 'A', the ejected photoelectrons have maximum kinetic energy $T_A$ (in eV) and de Broglie wavelength $\lambda_A$. The maximum kinetic energy of photoelectrons liberated from another metal 'B' by photons of energy 4.20 eV is $T_B$ (= $T_A$ – 1.50 eV). If the de Broglie wave length of these photoelectrons is $\lambda_B$ (= $2\lambda_A$), then

(a) the work function of 'A' is 2.25 eV.

(b) the work function of 'B' is 3.70 eV.

(c) $T_A$ = 2.00 eV

(d) $T_B$ = 2.75 eV

**11.** Which of the following suggested that de Broglie wavelengths is not possible for the electron in a Bohr's orbit of the hydrogen atom?

(a) 3.20 Å (b) 4.98 Å

(c) 9.96 Å (d) 6.64 Å

**12.** The magnitude of spin angular momentum of an electron is given by

(a) $\left\{\sqrt{s(s+1)}\right\}\dfrac{h}{2\pi}$ (b) $\dfrac{sh}{2\pi}$

(c) $\dfrac{\sqrt{3}}{2}\dfrac{h}{2\pi}$ (d) $\dfrac{h}{4\pi}$

**13.** Which of the following information is true?

(a) 3*s* orbital is spherically symmetrical with two nodes.

(b) $d_{x^2-y^2}$ orbitals has lobes of electron density in *XY*-plane along *X* and *Y*-axis.

(c) The radial probability curve of 1*s*, 3*p* and 5*d* have one, two and three regions of maximum probability.

(d) $3d_{z^2}$ has zero electron density in *XY*-plane.

**14.** The correct statement(s) regarding $3p_y$ orbital is/are

(a) angular part of wave function is independent of angles $\theta$ and $\phi$.

(b) number of maxima in $4\pi r^2 R^2(r)$ vs. *r* curve is 2.

(c) *XZ* plane is the nodal plane.

(d) magnetic quantum number must be −1.

**15.** The angular part of the wave function depends on the quantum numbers

(a) *n* (b) $\ell$

(c) *m* (d) *s*

**16.** The electronic configuration of carbon atom in the excited state is $1s^2\ 2s^1\ 2p^3$. Which of the following is/are incorrect statement(s) about it?

(a) The number of unpaired electron is 4.

(b) There are five electrons in the same spin.

(c) There are only two unpaired electrons.

(d) The spin of all *p*-electrons are similar.

**17.** Which of the following statement(s) is (are) correct?

(a) The electronic configuration of Cr is [Ar] $3d^5 4s^1$ (Atomic number of Cr = 24).

(b) The magnetic quantum number may have a negative value.

(c) In silver atom, 23 electrons have a spin of one type and 24 of the opposite type.

(d) Azimuthal quantum number may have a negative value.

**18.** Which is/are correct for sodium atom in the ground state?

(a) There is only one unpaired electrons.

(b) There are five pairs of electrons.

(c) 6 electrons are in one spin and other 5 in opposite spin.

(d) There are ten electrons in the same spin.

**19.** Select the correct statement(s) among the following.

(a) Outside any orbital, the probability of finding the electron is zero.

(b) For single electronic atom or ion, the most probable distance of electron in an orbital having no radial node is $\frac{n^2 a_o}{Z}$ from the nucleus, where $a_o$ is the first Bohr's radius.

(c) The average distance of electron (belonging from the same orbit) from the nucleus decreases with increase in the value of angular momentum quantum number for the orbital.

(d) The angular wave function of any s-orbital is independent from $\theta$ and $\phi$.

**20.** Among the following, select the correct information(s).

(a) The opposite lobes in any *d*-orbital have the same sign of wave function.

(b) 1*s* orbital is the only orbital for which the sign of wave function does not have radial as well as angular dependency.

(c) The number of radial nodes is always greater than that of angular nodes.

(d) All the orbitals belonging from an orbit have the same number of total nodes.

## Section C (Comprehensions)

### Comprehension I

**Three laser guns labelled as I, II and III have power 2, 3 and 5 W (not necessary in the same order) are used to produce photocurrent from metal plate. Number of photons emitted by laser guns are $4 \times 10^{18}$, $5 \times 10^{18}$ and $9 \times 10^{18}$ per second. Metal plate have threshold energy $4.5 \times 10^{-19}$ J. Neither the power nor the number of photons emitted by a particular laser gun is known and it is known that all capable photons emit a photoelectron. ($N_A = 6 \times 10^{23}$, $e = 1.6 \times 10^{-19}$ C)**

**1.** Minimum possible wavelength of emitted photoelectron is

(a) $\sqrt{680}$ Å (b) $\sqrt{30}$ Å

(c) $\sqrt{480}$ Å (d) $\sqrt{120}$ Å

**2.** Minimum photocurrent which must be passed through the circuit is

(a) 2.88 A (b) 1.44 A

(c) 2.08 A (d) 0.64 A

**3.** Ratio of maximum to minimum photocurrent which can be passed through the circuit is

(a) 5 : 4 (b) 9 : 5

(c) 9 : 4 (d) 9 : 2

### Comprehension II

**Let us assume a different atomic model in which electron revolves around the nucleus (proton) at a separation *r* under the action of force which is different from electrostatic force of attraction. The potential energy between an electron and the proton due to this force is given by $U = -klr^4$, where *k* is a constant. This hypothetical atom is obeying Bohr's quantization condition.**

**4.** The radius of *n*th Bohr's orbit is

(a) $r = \frac{\pi}{nh}\sqrt{km}$ (b) $r = \frac{2\pi}{nh}\sqrt{km}$

(c) $r = \frac{4\pi}{nh}\sqrt{km}$ (d) $r = \frac{8\pi}{nh}\sqrt{km}$

**5.** The speed of electron in the *n*th orbit is

(a) $V = \frac{nh}{8\pi^2 m\sqrt{km}}$ (b) $V = \frac{n^2 h}{8\pi^2 m\sqrt{km}}$

(c) $V = \frac{nh^2}{4\pi^2 m\sqrt{km}}$ (d) $V = \frac{n^2 h^2}{8\pi^2 m\sqrt{km}}$

**6.** The total energy of the electron in the *n*th orbit is

(a) $\text{T.E.} = \frac{-n^4 h^4}{128\pi^4 m^2 k}$ (b) $\text{T.E.} = \frac{n^4 h^4}{128\pi^4 m^2 k}$

(c) $\text{T.E.} = \frac{n^4 h^4}{256 k\pi^4 m^2}$ (d) $\text{T.E.} = \frac{-n^4 h^4}{256\pi^4 m^2}$

## Comprehension III

**A hydrogen-like atom (atomic number *Z*) is in a higher excited state of quantum number *n*. This excited atom can make a transition to the first excited state by successively emitting two photons of energies 10.20 and 17.00 eV, respectively. Alternatively, the atom from the same excited state can make a transition to the second excited state by successively emitting two photons of energies 4.25 and 5.95 eV, respectively.**

7. The value of '$n$' is
   (a) 4 (b) 5
   (c) 6 (d) 7

8. The value of $Z$ is
   (a) 2 (b) 3
   (c) 4 (d) 5

9. How much energy will be emitted when an electron in this atom moves from $(n + 1)$ to ground state?
   (a) 114.75 eV (b) 117.5 eV
   (c) 119.9 eV (d) 122.74 eV

## Comprehension IV

**A certain gas of identical hydrogen-like atoms has all its atoms in a particular upper energy level. The atoms make transition to a higher level when a monochromatic radiation having wavelength 1654 Å is incident on it. Subsequently, the atoms emit radiation of only three different photon energies.**

10. The initial energy level of atoms was
    (a) $n = 1$ (b) $n = 2$
    (c) $n = 3$ (d) $n = 4$

11. The final energy level of atoms is
    (a) $n = 1$ (b) $n = 2$
    (c) $n = 3$ (d) $n = 4$

12. The atom/ion is
    (a) H (b) D
    (c) $He^+$ (d) $Li^{+2}$

13. The energy of photon required to remove electron from higher energy level is
    (a) 6.04 eV (b) 13.6 eV
    (c) 27.2 eV (d) 36.8 eV

## Comprehension V

**A gas of identical H-like atom has some atoms in the lowest (ground) energy level 'A' and some atoms in a particular upper (excited) energy level 'B' and there are no atoms in any other energy level. The atoms of the gas make transition to a higher energy level by absorbing monochromatic light of photon energy 2.7 eV. Subsequently, the atoms emit radiation of only six different photons energies. Some of the emitted photons have energy 2.7 eV. Some have more and some have less than 2.7 eV.**

14. The principal quantum number of initially excited level 'B' is
    (a) 1 (b) 2
    (c) 3 (d) 4

15. The ionization energy for gas atoms is
    (a) 3.4 eV (b) 12.8 eV
    (c) 14.4 eV (d) 13.6 eV

16. The minimum energy of emitted photon is
    (a) 0.2 eV (b) 13.5 eV
    (c) 6.7 eV (d) 0.7 eV

## Comprehension VI

**A muon is an unstable elementary particle whose mass is $207m_e$ and whose charge is either $+e$ or $-e$. A negative muon ($\mu^-$) can be captured by a proton to form a muonic atom. This atom follows Bohr's quantization condition. Answer the following, neglecting reduced mass effect.**

**17.** What is the radius of the first Bohr's orbit of this atom?

(a) 52.9 pm (b) 0.256 pm
(c) 0.256 nm (d) 10.9 nm

**18.** What is the ionization energy of the atom?

(a) 13.6 eV (b) 0.066 eV
(c) 0.583 MeV (d) 2.84 keV

**19.** What is the wavelength of the photon emitted when muon drops from second orbit to the ground state in this atom?

(a) $2.53 \times 10^{-5}$ m
(b) $1.22 \times 10^{-7}$ m
(c) $5.91 \times 10^{-10}$ m
(d) $2.85 \times 10^{-12}$ m

## Comprehension VII

**A sample of H-atoms contains all atoms in the ground state. If the atoms are irradiated by photons of $x$ Å, the atoms get excited to a particular energy level. When these atoms de-excites, they emit radiations of six different photon energies.**

**20.** What is the value of $x$?

(a) 978.6 (b) 0.098
(c) 1032 (d) 1223

**21.** What is the orbit number for the excited state?

(a) 6 (b) 3
(c) 4 (d) 2

**22.** What is the maximum wavelength among emitted radiations?

(a) 978.6 Å (b) 1223 Å
(c) 18,872.87 nm (d) 18,872.87 Å

**23.** What is the maximum frequency among emitted radiations?

(a) $2.453 \times 10^{15}$ Hz (b) $3.066 \times 10^{15}$ Hz
(c) $5.912 \times 10^{16}$ Hz (d) $1.081 \times 10^{18}$ Hz

**24.** What are the wavelengths of all infrared radiations coming out?

(a) 1887.3 nm, 4077.5 nm
(b) 1887.3 nm
(c) 122.3 nm, 103.2 nm, 97.9 nm
(d) 122.3 nm, 103.2 nm

**25.** What are the wavelengths of all visible radiations coming out?

(a) 660.5 nm, 489.3 nm
(b) 660.5 nm, 489.3 nm, 436.9 nm
(c) 660.5 nm, 489.3 nm, 436.9 nm, 412.8 nm
(d) 660.5 nm, 436.9 nm

## Comprehension VIII

**When a sample of hydrogen atoms is irradiated by electromagnetic radiations of suitable wavelength, all the electrons jump from the ground state to the fifth orbit. As the electrons cannot remain permanently in the fifth orbit, they de-excites to the ground state making one or more than one transitions. Each transition results in the emission of an electro-magnetic radiation of a particular wavelength. When these radiations pass through a prism, they deviate with different angles resulting in the spectral lines.**

**26.** What is the maximum number of spectral lines, if the sample is containing only one atom?

(a) 5 (b) 4
(c) 6 (d) 10

**27.** What is the maximum number of spectral lines, if the sample is containing only two atoms?

(a) 5 (b) 6
(c) 7 (d) 10

**28.** What is the maximum number of spectral lines, if the sample is containing only three atoms?

(a) 5 (b) 6
(c) 7 (d) 10

29. What is the maximum number of spectral lines, if the sample contains infinite number of atoms?
(a) 5 (b) 6
(c) 7 (d) 10

30. What should be the minimum number of hydrogen atoms in the sample to get a maximum of 10 spectral lines?
(a) 1 (b) 6
(c) 8 (d) 10

## Comprehension IX

**A hydrogen-like atom of atomic number $Z$ is in an excited state of quantum number $2n$. It can emit a maximum energy photon of 204 eV. If it makes a transition to quantum state $n$, a photon of energy 40.8 eV is emitted.**

31. What is the initial excited state of the atom?
(a) 4 (b) 8
(c) 16 (d) 2

32. What is the value of $Z$?
(a) 2 (b) 4
(c) 8 (d) 16

33. What is the energy of ground state of this atom?
(a) −13.6 eV (b) −54.4 eV
(c) −217.6 eV (d) −870.4 eV

34. What is the minimum energy of the photons emitted by this atom during de-excitation?
(a) 204 eV (b) 4.16 eV
(c) 10.6 eV (d) 13.6 eV

## Comprehension X

**In stars, the Pickering series is found in the $He^+$ spectrum. It is emitted when the electron in $He^+$ jumps from higher levels to the level with $n = 4$.**

35. The wavelengths of the lines in this series are given by $\lambda = \frac{Cn^2}{n^2 - 16}$, in which $n$ = 5, 6, 7... and $C$ is a constant. What is the value of $C$ in nm?
(a) 366.97 (b) 91.74
(c) 22.93 (d) 1467.89

36. What is the wavelength of the series limit?
(a) 22.93 nm (b) 366.97 nm
(c) 1019.36 nm (d) 63.71 nm

37. In what region(s) of the spectrum does this series occur? Visible region is from 360 to 780 nm.
(a) Visible only.
(b) Infrared only.
(c) Visible and infrared.
(c) Visible and ultraviolet.

## Comprehension XI

**The hydrogen-like species $Li^{2+}$ is in a spherically symmetric state $S_1$ with one radial node. Upon absorbing light, the ion undergoes transition to a state $S_2$. The state $S_2$ has one radial node and its energy is equal to the ground state energy of the hydrogen atom.**

38. The state $S_1$ is
(a) 1*s* (b) 2*s*
(c) 2*p* (d) 3*s*

39. Energy of the state $S_1$ in units of the hydrogen atom ground state energy is
(a) 0.75 (b) 1.50
(c) 2.25 (d) 4.50

40. The orbital angular momentum quantum number of the state $S_2$ is
(a) 0 (b) 1
(c) 2 (d) 3

EXERCISE II

## Comprehension XII

**Suppose a particle has four quantum numbers such that the permitted values are**

**$n = 1, 2, 3......$**

**$l = (n - 1), (n - 3), (n - 5), ...,$ but no negative value**

**$j = \left(l + \frac{1}{2}\right)$ or $\left(l - \frac{1}{2}\right)$, the latter is not negative**

**$m = -j$ in integer step to $+j$.**

**41.** The other permitted values for $n = 2$ is/are

(a) $l = 1, j = \frac{3}{2}, m = -\frac{3}{2}$

(b) $l = 0, j = \frac{1}{2}; m = -\frac{1}{2}$

(c) $l = 1, j = \frac{1}{2}; m = -\frac{3}{2}$

(d) All of these

**42.** For $n = 3$, the possible sets of $(n, l, j, m)$ is

(a) 10 (b) 8

(c) 12 (d) 6

## Comprehension XIII

**The wave function for an atomic orbital of single electron atom or ion is**

$$\psi(r, \theta, \phi) = \frac{2}{3}\left(\frac{Z}{3a_o}\right)^{3/2}(1-\sigma)(12 - 8\sigma + \sigma^2)\cdot\sigma\cdot e^{-\sigma/2}\cdot\cos\theta$$

**where $\sigma = \frac{2Zr}{na_o}$ and $a_o = 0.529$ Å. All other parameters have their usual meaning.**

**43.** The number of radial and angular nodes for the orbital is, respectively,

(a) 3, 1 (b) 2, 1

(c) 3, 2 (d) 2, 2

**44.** The atomic orbital should be

(a) $4p$ (b) $5p$

(c) $5d$ (d) $5f$

**45.** If $\theta$ is the angle measured from $Z$-axis, then the orbital should be

(a) $p_x$ (b) $p_y$

(c) $p_z$ (d) $d_{z^2}$

**46.** The maximum distance of radial node from the nucleus is

(a) $\frac{a_o}{Z}$ (b) $\frac{3a_o}{Z}$

(c) $\frac{6a_o}{Z}$ (d) $\frac{15a_o}{Z}$

## Section D (Assertion – Reason)

The following questions consist of two statements. Mark the answer as follows.

**(a)** If both statements are CORRECT, and **Statement II** is the CORRECT explanation of **Statement I.**

**(b)** If both statements are CORRECT, and **Statement II** is NOT the CORRECT explanation of **Statement I.**

**(c)** If **Statement I** is CORRECT, but **Statement II** is INCORRECT.

**(d)** If **Statement I** is INCORRECT, but **Statement II** is CORRECT.

**1.** **Statement I:** The orbital angular momentum of an electron in any $s$-atomic orbital is zero.

**Statement II:** For any $s$-electron, $l = 0$.

**2.** **Statement I:** The kinetic energy of the photoelectron ejected increases with increase in frequency of incident light.

**Statement II:** Increase in intensity of incident light increases the photoelectric current.

EXERCISE II

3. **Statement I:** Threshold frequency is the maximum frequency required for the ejection of electron from the metal surface.

   **Statement II:** Threshold frequency is the characteristic of a metal.

4. **Statement I:** Spin quantum number can have two values $+\frac{1}{2}$ and $-\frac{1}{2}$.

   **Statement II:** +ve and –ve signs signify the positive and negative wave functions.

5. **Statement I:** For $He^+$ ion, the energy of electron in 3*p*-orbital is greater than that in 3*s*-orbital.

   **Statement II:** Energy of an electron in single electron system depends only on principal quantum number.

6. **Statement I:** Helium and beryllium have similar outer electronic configuration $ns^2$.

   **Statement II:** Both are chemically inert.

7. **Statement I:** Sum of the radial and angular nodes of all the occupied orbitals in ground state of oxygen atom is 4.

   **Statement II:** The electronic configuration of oxygen atom is $1s^2\ 2s^2\ 2p^4$. The number of radial and angular nodes for an orbital is $(n - l - 1)$ and $l$, respectively.

8. **Statement I:** The *XY*-plane is the nodal plane for $d_{z^2}$ orbital.

   **Statement II:** For a *d*-orbital, the number of angular nodes is 2.

9. **Statement I:** The radial wave functions for $3p_x$ and $3p_y$ orbitals are different.

   **Statement II:** The radial function depends on *n* and *l* while the angular function depends on *l* and *m*.

10. **Statement I:** In multi-electron system (atom or ion), 4*s* orbital is filled first then 3*d* orbital is filled.

    **Statement II:** 4*s* orbital is closer to the nucleus than 3*d* orbital.

## Section E (Column Match)

1. Match the columns.

| Column I (Orbitals) | Column II (Nodal properties) |
|---|---|
| (A) 2*s* | (P) Angular node = 1 |
| (B) 1*s* | (Q) Radial node = 0 |
| (C) 2*p* | (R) Radial node = 1 |
| (D) 3*p* | (S) Angular node = 0 |

2. Match Column I with Column II in hydrogen atom spectrum.

| Column I | Column II |
|---|---|
| (A) Lyman series | (P) Visible region |
| (B) Balmer series | (Q) Infrared region |
| (C) Paschen series | (R) Absorption spectrum |
| (D) Brackett series | (S) Ultraviolet region |

3. According to Bohr's theory, $E_n$ = Total energy, $K_n$ = Kinetic energy, $V_n$ = Potential energy, $r_n$ = Radius of *n*th orbit. Match the columns.

| Column I | Column II |
|---|---|
| (A) $V_n/K_n = ?$ | (P) 0 |
| (B) If radius of *n*th orbit $\propto E_n^x$; $x = ?$ | (Q) –1 |
| (C) Orbital angular momentum in lowest energy | (R) –2 |
| (D) $\frac{1}{r^n} \propto Z^y$; $y = ?$ | (S) 1 |

4. Match list I with list II.

| List I | List II |
|---|---|
| (A) Number of values of *l* for an energy level. | (P) 0, 1, 2, ……, $(n-1)$ |
| (B) Actual values of *l* for an energy level. | (Q) $+l$, ……, 0,……, $-l$ |
| (C) Number of *m* values for a particular type of orbital. | (R) $2l + 1$ |
| (D) Actual value of '*m*' for a particular type of orbital. | (S) *n* |

**5.** Match the columns.

| Column I (Orbital) | Column II (*R* vs. *r* Graph) |
|---|---|
| (A) 3*s* | (P) 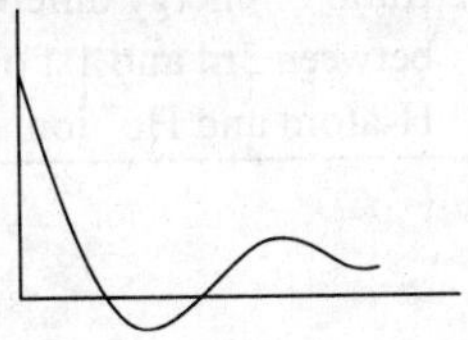 |
| (B) 4*s* | (Q) 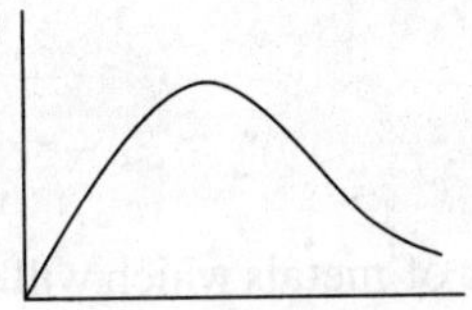 |
| (C) 2*p* | (R) 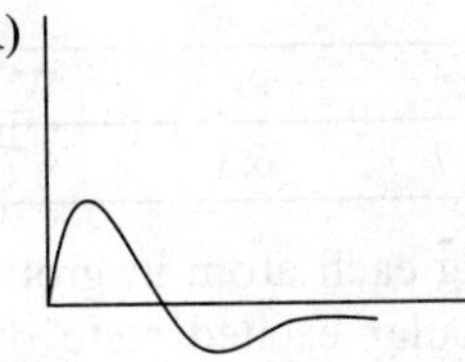 |
| (D) 3*p* | (S) 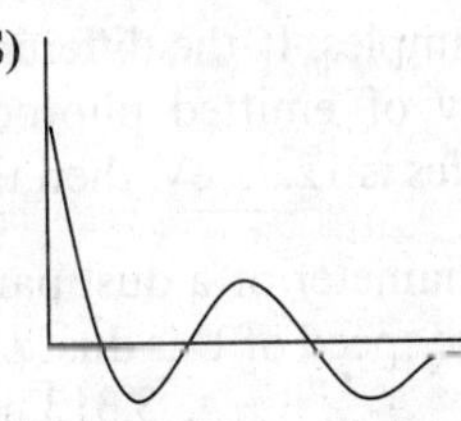 |

**6.** Match the columns.

| Column I (Parameters for H-like atoms) | Column II (Dependence on *n* and *Z*) |
|---|---|
| (A) Radius of orbit. | (P) $n^2$ |
| (B) Speed of electron. | (Q) $1/n$ |
| (C) Centripetal force between electron and nucleus. | (R) $1/n^3$ |
| (D) Frequency of revolution of electron. | (S) $1/n^4$ |
| | (T) Z |
| | (U) 1/Z |
| | (V) $Z^2$ |
| | (W) $Z^3$ |

**7.** Match the columns.

| Column I | Column II |
|---|---|
| (A) 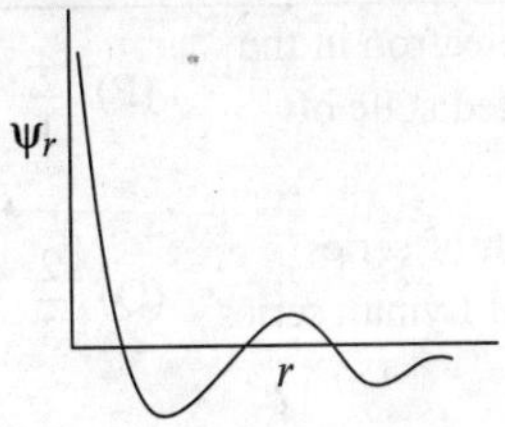  | (P) 4*s* |
| (B) 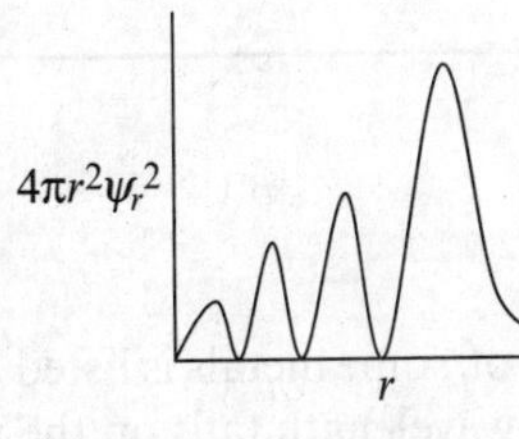  | (Q) Any of the 5*p* orbital. |
| (C) Angular wave function independent from θ and φ. | (R) 3*s* |
| (D) At least one angular node. | (S) any of the 6*d* orbital. |

**8.** Match the entries in Column I with the correctly related quantum number(s) in Column II.

| Column I | Column II |
|---|---|
| (A) Orbital angular momentum of the electron in a hydrogen-like atomic orbital. | (P) Principle quantum number |
| (B) A hydrogen-like one-electron wave function obeying Pauli's principle. | (Q) Azimuthal quantum number |
| (C) Shape, size and orientation of hydrogen-like atomic orbital. | (R) Magnetic quantum number |
| (D) Probability density of electron at the nucleus in hydrogen-like atom. | (S) Electron spin quantum number |

**9.** Match the columns.

| Column I | Column II |
|---|---|
| (A) Orbital with equal number of radial and angular nodes. | (P) $3d_{x^2-y^2}$ |
| (B) Orbitals with number of radial nodes less than the number of angular nodes. | (Q) $2p_z$ |
| (C) Orbitals with zero radial nodes but two angular nodes. | (R) $3p_x$ |
| | (S) $5d_{xy}$ |

EXERCISE II

10. Match the columns.

| Column I | Column II |
|---|---|
| (A) Ratio of speed of electron in the fifth and third excited state of H-atom. | (P) $\frac{4}{1}$ |
| (B) Ratio of wavelength of series limit of Balmer and Lyman series of H-spectrum. | (Q) $\frac{2}{3}$ |
| (C) Ratio of wavelength of photon corresponding to β-line of Lyman series and γ-line of Paschen series of H-spectrum. | (R) $\frac{1}{4}$ |
| (D) Ratio of energy difference between 3rd and 1st orbits of H-atom and $He^+$ ion. | (S) $\frac{3}{32}$ |

## Section F (Subjective)

### Single-digit Integer Type

1. The work function (ϕ) of some metals is listed below. The number of metals which will show photoelectric effect when light of 300 nm wavelength falls on the metal is

| Metal | Li | Na | K | Mg | Cu | Ag | Fe | Pt | W |
|---|---|---|---|---|---|---|---|---|---|
| ϕ (eV) | 2.4 | 2.3 | 2.2 | 3.7 | 4.8 | 4.3 | 4.7 | 6.3 | 4.75 |

2. The average life time for then $n = 3$ excited state of a hydrogen-like atom is $4.8 \times 10^{-8}$ s and that for the $n = 2$ state is $1.28 \times 10^{-7}$ s. Ratio of the average number of revolutions made in the $n = 3$ state to the average number of revolutions made in the $n = 2$ state before any transitions can take place from these states is $1 : x$. The value of $x$ is

3. The quantum number $n$ corresponding to the excited state of $He^+$ ion if on transition to the ground state that ion emits two photons in succession with wavelengths 108.5 and 30.4 nm is

4. The atomic number of hydrogen-like ion has the wavelength difference between the first line of Balmer and Lyman series equal to 59.3 nm.

5. A sample of hydrogen atoms containing all the atoms in a particular excited state absorb radiations of a particular wavelength by which the atoms get excited to another excited state. When the atoms finally de-excite to the ground state, they emit the radiations of 10 different wavelengths. Out of these 10 radiations, 7 have wavelengths shorter than the absorbed radiation and 2 have wavelength longer than the absorbed radiation. The orbit number for the initial excited state of atoms is

6. Photons of same energy were allowed to strike on two different samples of hydrogen atoms, one having each atom in ground state and other in a particular excited state of orbit number $n$. The photonic beams ionize the hydrogen atoms in both the samples. If the difference in maximum kinetic energy of emitted photoelectron from both the samples is 12.75 eV, then the value of $n$ is

7. The diameter of a dust particle of mass $10^{-3}$ g is 2 Å. The speed of this dust particle is measured with the uncertainty of $\frac{3.313}{\pi} \times 10^{-3}$ m/s. The minimum uncertainty in measuring the position of the duct particle (in order of $10^{-26}$ m) is

8. The uncertainty in position and velocity of a particle are $10^{-11}$ m and $5.27 \times 10^{-24}$ m·s$^{-1}$, respectively. The minimum mass of the particle (in kg) is

9. The circumference of the second orbit of an atom or ion having single electron is 4 nm. The de Broglie wavelength of electron (in nm) revolving in this orbit is

10. The Schrodinger wave equation for hydrogen atom is $\Psi_{2s} = \frac{1}{4\sqrt{2\pi}}\left(\frac{1}{a_0}\right)^{3/2}\left(2 - \frac{r_0}{a_0}\right)e^{-\frac{r_0}{a_0}}$, where $a_0$ is Bohr's radius. If the radial node is $2s$ be at $r_0$, then the value of $\frac{r_o}{a_o}$ is

EXERCISE II

## Four-digit Integer Type

1. An α-particle of momentum $3.2 \times 10^{-20}$ kg ms$^{-1}$ is projected towards the nucleus of an atom of an element. If the distance of closest approach of α-particle is $1.5 \times 10^{-13}$ m, then the atomic number of element is (Mass of α-particle = 4 amu, charge on electron = $1.6 \times 10^{-19}$ coulomb, $N_A = 6 \times 10^{23}$)

2. Suppose a satellite is in geostationary orbit at an average distance of 12600 km from the earth surface. This satellite is telecasting a cricket match between Indian and Pakistan. The radiowaves from the stadium are first received by the satellite and then from satellite to our TV sets. If a batsman hits a ball, then after what time (approximate, in milliseconds), we will see this action in our TV sets?

3. A cobalt target is bombarded with electrons and the wavelength of its characteristic spectrum is measured. A second fainter characteristic spectrum is also found due to an impurity in the target. The wavelengths of the $K_\alpha$ lines are 180.0 pm (cobalt, $Z = 27$) and 144.0 pm (impurity). The atomic number of impurity is

4. How many moles of photon would contain sufficient energy to raise the temperature of 245 g of water from 19.5°C to 99.5°C? The specific heat of water is 4.2 J/°C-g and frequency of light radiation used is $2.45 \times 10^{10}$ per second. ($6.626 \times 5.04 = 33.6$, $N_A = 6 \times 10^{23}$)

5. A quantity of 1.0 g hydrogen atoms are excited to radiations. The study of spectra indicates that 45% of the atoms are in third energy level and 40% of atoms in second energy level and the rest in ground state. The total energy (in kJ) evolved, when all the atoms return to ground state is (Ionization energy of hydrogen atom is 1310 kJ/mol)

6. A proton and an electron, both at rest initially, combines to form a hydrogen atom in the ground state. A single photon is emitted in this process. The wavelength (in nm) of emitted photon is

7. The energy (in kJ) required to excite 1 L of hydrogen gas at 1 atm and 300 K to the first excited state of atomic hydrogen is (The energy for the dissociation of H–H bond is 436 kJ mol$^{-1}$ and the ionization energy of hydrogen is 1312 kJ mol$^{-1}$). $R = 0.08$ L-atm/K-mol

8. In the hydrogen spectrum, the longest wavelength in the Lyman series is 120 nm and the shortest wavelength in the Balmer series is 360 nm. From this data, the longest wavelength (in nm) of light that could ionize hydrogen atom is

9. An electron beam can undergo diffraction by crystals. Through what potential (in Volt) should a beam of electron be accelerated so that its wavelength becomes equal to 0.25 nm?

10. The atomic masses of He and Ne are 4 and 20 amu, respectively. The value of the de Broglie wavelength of He gas at −73°C is $M$ times that of the de Broglie wavelength of Ne at 727°C. The value of $M$ is

## Answer Keys

**Exercise I**

### Fundamental Particles

1. (c) 2. (c) 3. (d) 4. (d) 5. (c) 6. (b) 7. (c) 8. (d) 9. (c) 10. (d)
11. (a) 12. (c) 13. (c) 14. (d) 15. (c) 16. (c) 17. (c) 18. (b) 19. (c) 20. (a)

### Rutherford's Atomic Models

21. (c) 22. (d) 23. (d) 24. (a) 25. (b) 26. (b) 27. (a) 28. (d) 29. (d) 30. (d)
31. (b) 32. (b) 33. (b) 34. (a) 35. (d)

### Planck's Quantum Theory, Photoelectric Effect and Moseley's Experiment

36. (c) 37. (b) 38. (a) 39. (b) 40. (d) 41. (b) 42. (a) 43. (b) 44. (b) 45. (c)
46. (b) 47. (c) 48. (b) 49. (b) 50. (c) 51. (a) 52. (c) 53. (b) 54. (c) 55. (c)

## Bohr's Atomic Mode

56. (a) 57. (b) 58. (b) 59. (a) 60. (a) 61. (b) 62. (c) 63. (a) 64. (c) 65. (b)
66. (c) 67. (c) 68. (a) 69. (a) 70. (c) 71. (c) 72. (c) 73. (b) 74. (a) 75. (b)
76. (a) 77. (b) 78. (b) 79. (d) 80. (a) 81. (c) 82. (a) 83. (a) 84. (b) 85. (d)
86. (c) 87. (c) 88. (c) 89. (c) 90. (b) 91. (c) 92. (c) 93. (c) 94. (a) 95. (a)
96. (b) 97. (b) 98. (d) 99. (b) 100. (b)

## Spectrum

101. (c) 102. (d) 103. (d) 104. (d) 105. (a) 106. (b) 107. (a) 108. (b) 109. (c) 110. (c)
111. (a) 112. (d) 113. (d) 114. (c) 115. (d)

## Heisenberg's Uncertainty Principle

116. (d) 117. (d) 118. (b) 119. (a) 120. (b)

## De Broglie's Equation

121. (a) 122. (c) 123. (a) 124. (a) 125. (b)

## Quantum Numbers

126. (a) 127. (a) 128. (a) 129. (d) 130. (b) 131. (b) 132. (c) 133. (b) 134. (b) 135. (d)
136. (c) 137. (c) 138. (b) 139. (a) 140. (d)

## Schrodinger's Equation

141. (b) 142. (c) 143. (b) 144. (c) 145. (b)

## Electronic Configuration

146. (d) 147. (b) 148. (a) 149. (a) 150. (d) 151. (a) 152. (b) 153. (c) 154. (a) 155. (b)
156. (c) 157. (d) 158. (b) 159. (a) 160. (c) 161. (c) 162. (b) 163. (b) 164. (b) 165. (c)
166. (c) 167. (c) 168. (a) 169. (a) 170. (d) 171. (a) 172. (b) 173. (b) 174. (d) 175. (c)

# Answer Keys

## Exercise II

## Section A (Only one Correct)

1. (a) 2. (b) 3. (b) 4. (c) 5. (d) 6. (a) 7. (a) 8. (d) 9. (d) 10. (b)
11. (b) 12. (a) 13. (c) 14. (d) 15. (a) 16. (b) 17. (d) 18. (c) 19. (c) 20. (c)
21. (d) 22. (b) 23. (c) 24. (b) 25. (c) 26. (b) 27. (a) 28. (c) 29. (d) 30. (c)
31. (d) 32. (c) 33. (a) 34. (b) 35. (c) 36. (d) 37. (d) 38. (a) 39. (a) 40. (c)
41. (c) 42. (c) 43. (b) 44. (c) 45. (a) 46. (c) 47. (a) 48. (b) 49. (c) 50. (a)
51. (c) 52. (a) 53. (d) 54. (a) 55. (a) 56. (b) 57. (d) 58. (a) 59. (d) 60. (c)
61. (c) 62. (c) 63. (d) 64. (b) 65. (a) 66. (c) 67. (c) 68. (d) 69. (a) 70. (d)
71. (b) 72. (a) 73. (b) 74. (a) 75. (c) 76. (d) 77. (b) 78. (a) 79. (d) 80. (a)
81. (b) 82. (c) 83. (b) 84. (c) 85. (b) 86. (b) 87. (a) 88. (c) 89. (c) 90. (c)
91. (a) 92. (c) 93. (d) 94. (a) 95. (b) 96. (b) 97. (b) 98. (b) 99. (a) 100. (a)
101. (a) 102. (d) 103. (b) 104. (b) 105. (c) 106. (d) 107. (d) 108. (c) 109. (b) 110. (d)

## Section B (One or More than one Correct)

1. (a), (b), (c)
2. (a), (c), (d)
3. (a), (c), (d)
4. (a), (b), (c), (d)
5. (b), (c)
6. (a), (c), (d)
7. (a), (b), (c)
8. (a), (b)
9. (b), (c), (d)
10. (a), (b), (c)
11. (a), (b)
12. (a), (c)
13. (a), (b), (c)
14. (b), (c)
15. (b), (c)
16. (c)
17. (a), (b), (c)
18. (a), (b), (c)
19. (b), (c), (d)
20. (a), (b), (d)

## Section C

**Comprehension I**

1. (b) 2. (d) 3. (c)

**Comprehension II**

4. (c) 5. (d) 6. (c)

**Comprehension III**

7. (c) 8. (b) 9. (c)

**Comprehension IV**

10. (b) 11. (c) 12. (c) 13. (a)

**Comprehension V**

14. (b) 15. (c) 16. (d)

**Comprehension VI**

17. (b) 18. (d) 19. (c)

**Comprehension VII**

20. (a) 21. (c) 22. (d) 23. (a)
24. (b) 25. (a)

**Comprehension VIII**

26. (b) 27. (b) 28. (c)
29. (d) 30 (b)

**Comprehension IX**

31. (a) 32. (b) 33. (c) 34. (c)

**Comprehension X**

35. (a) 36. (b) 37. (c)

**Comprehension XI**

38. (b) 39. (c) 40. (b)

**Comprehension XII**

41. (a) 42. (c)

**Comprehension XIII**

43. (a) 44. (b) 45. (c) 46. (d)

## Section D (Assertion – Reason)

1. (a) 2. (b) 3. (d) 4. (c) 5. (d) 6. (c) 7. (a) 8. (d) 9. (d) 10. (c)

## Section E (Column Match)

1. $A \rightarrow R, S; B \rightarrow Q, S; C \rightarrow P, Q; D \rightarrow P, R$
2. $A \rightarrow R, S; B \rightarrow P, S; C \rightarrow Q; D \rightarrow Q$
3. $A \rightarrow R; B \rightarrow Q; C \rightarrow P; D \rightarrow S$
4. $A \rightarrow S; B \rightarrow P; C \rightarrow R; D \rightarrow Q$
5. $A \rightarrow P; B \rightarrow S; C \rightarrow Q; D \rightarrow R$
6. $A \rightarrow P, U; B \rightarrow Q, T; C \rightarrow S, W; D \rightarrow R, V$
7. $A \rightarrow P; B \rightarrow P, Q, S; C \rightarrow P, R; D \rightarrow Q, S$
8. $A \rightarrow Q, R; B \rightarrow P, Q, R, S; C \rightarrow P, Q, R; D \rightarrow P, Q$
9. $A \rightarrow R, S; B \rightarrow Q, P; C \rightarrow P$
10. $A \rightarrow Q; B \rightarrow P; C \rightarrow S; D \rightarrow R$

## Section F (Subjective)

**Single-digit Integer Type**

1. (4) 2. (6) 3. (5) 4. (3) 5. (3)
6. (4) 7. (5) 8. (1) 9. (2) 10. (2)

**Four-digit Integer Type**

1. (0025) 2. (0084) 3. (0030) 4. (8400) 5. (0917)
6. (0091) 7. (0100) 8. (0090) 9. (0024) 10. (0005)

# HINTS AND EXPLANATIONS

## EXERCISE I (JEE MAIN)

### Fundamental Particles

1. Theory based
2. Theory based
3. Specific charge is $\left(\frac{e}{m}\right)$ ratio
4. x-rays = electromagnetic radiation
5. Faraday, by his experiments of electrolysis.
6. Informative
7. Theoretical
8. Cathode rays comes out normal to the surface of cathode.
9. $X(g) \rightleftharpoons X^+(g) + e^-$
   anode rays
10. Anode rays do not come out from the surface of anode.
11. Theoretical
12. Number of protons = $Z = 3$
    Number of electrons = $3 - 1 = 2$
    Number of neutrons = $A - Z = 4$
13. Charge on electron = HCF of charges in oil droplets = $1.5 \times 10^{-18}$ unit
14. $\left(\frac{e}{m}\right)$ ratio of positron and electron are same.
15. Neutron is electrically neutral.
16. Except positron, all are chargeless.
17. $H^+ > Al^{3+} > Mg^{2+} > Na^+ \left(\equiv 1 > \frac{1}{9} > \frac{1}{12} > \frac{1}{23}\right)$
18. $e \cdot V = \frac{1}{2} m v^2 \Rightarrow v = \sqrt{2 \times \left(\frac{e}{m}\right) \times V} \alpha \sqrt{V}$
19. $\frac{(e/m)_\alpha}{(e/m)_\alpha} \approx \frac{2/4}{1/2} = \frac{1}{1}$
20. $\frac{e}{m} = \frac{2 \times 1.602 \times 10^{-19}\ C}{4 \times 1.67 \times 10^{-27}\ Kg} \simeq 4.8 \times 10^7\ C/Kg$

### Rutherford's Atomic Models

21. Informative.
22. $\alpha$-particles are $He^{2+}$ ions or Helium nucleus.
23. The concept of definite paths was not the part of Rutherford's model.
24. Almost all (not 99%) mass of an atom is concentrated in the nucleus.
25. Theoretical
26. Distance of closest approach, $r = \frac{K \cdot q_1 q_2}{\left(\frac{1}{2} m v^2\right)}$

    From question: $r \alpha \frac{1}{m}$

    $\therefore \frac{r_p}{r_d} = \frac{m_d}{m_p} = \frac{2}{1}$
27. $r = \frac{K \cdot q_1 q_2}{\left(\frac{1}{2} m v^2\right)}$

    As $\left(\frac{q_1}{m}\right)$, $q_2$ and $v$, all are same, $r$ is same.
28. For $r_{min}$, $q_2$ = nuclear charge and hence, $z$ should be minimum.
29. Neutron is chargeless.
30. $r = \frac{K \cdot q_1 q_2}{q_1 \cdot V} = \frac{K \cdot q_2}{V}$ = Constant

31. $N_\alpha \alpha \dfrac{1}{\sin^4\left(\dfrac{\theta}{2}\right)}$

$\therefore \dfrac{N_1}{N_2} = \left[\dfrac{\sin\left(\dfrac{90}{2}\right)}{\sin\left(\dfrac{60}{2}\right)}\right]^4 \Rightarrow \dfrac{36}{N_2} = \dfrac{4}{1} \Rightarrow N_2 = 9$

32. $\dfrac{V_{\text{nucleus}}}{V_{\text{atom}}} = \left(\dfrac{4\times10^{-15}}{2\times10^{-10}}\right) = \dfrac{8\times10^{-15}}{1}$

33. $\dfrac{1}{2}\text{m v}^2 = K\cdot\dfrac{q_1q_2}{r}$

or, $\dfrac{1}{2}\times\left(\dfrac{4\times10^{-3}}{6\times10^{23}}\right)\times V^2$

$= 9\times10^9\times\dfrac{(2\times1.6\times10^{-19})\times(29\times1.6\times10^{-19})}{10^{-12}}$

$\therefore v = 2\times10^6$ m/s

34. $r = K\cdot\dfrac{q_1q_2}{q_1\cdot V} = \text{Constant}$

35. $r = K\cdot\dfrac{q_1q_2}{(P^2/2\,\text{m})} \Rightarrow r\ \alpha \dfrac{1}{P^2}$

## Planck's Quantum Theory, Photoelectric Effect and Moseley's Experiment

36. Informative

37. $\lambda = \dfrac{c}{\upsilon} = \dfrac{3\times10^8}{400\times10^3} = 750$ m

38. Informative

39. $\text{Time} = \dfrac{\text{Distance}}{\text{Speed}}$

$= \dfrac{60\times10^6\times10^3}{3\times10^8} = 200$ sec

40. $E = n\cdot h\overline{\upsilon} \Rightarrow E_1 : E_2 = \overline{\upsilon_1} : \overline{\upsilon_2} = 2:3$

41. $E = n\cdot h\upsilon$

42. $E = n\cdot\dfrac{hc}{\lambda}$

$\Rightarrow n = \dfrac{E\cdot\lambda}{hc} = \dfrac{(3\times10^{-14})\times(662.6\times10^{-9})}{(6.626\times10^{-34})\times(3\times10^8)} = 10^5$

43. $E = E_1 + E_2 \Rightarrow \dfrac{hc}{\lambda} = \dfrac{hc}{\lambda_1} + \dfrac{hc}{\lambda_2}$

or, $\dfrac{1}{400} = \dfrac{1}{500} + \dfrac{1}{\lambda_2} \Rightarrow \lambda_2 = 2000$ nm

44. $E_g = E_r \Rightarrow n_g\cdot\dfrac{h_c}{\lambda_g} = n_r\cdot\dfrac{hc}{\lambda_r} \Rightarrow \dfrac{n_g}{n_r} = \dfrac{\lambda_g}{\lambda_r} < 1$

45. $E_{\text{emitted}} = \dfrac{50}{100}\times E_{\text{absorbed}}$

or, $n_1\cdot\dfrac{hc}{\lambda_1} = \dfrac{1}{2}\times n_2\times\dfrac{hc}{\lambda_2}$

or $\dfrac{\lambda_1}{\lambda_2} = \dfrac{2}{1}\times\dfrac{n_1}{n_2} = \dfrac{2}{1}\times\dfrac{2}{1} \Rightarrow \dfrac{\lambda_1}{x\text{Å}} = \dfrac{4}{1} \Rightarrow \lambda_1 = 4x\text{Å}$

46. $E = \dfrac{1}{2}(E_1 + E_2) \Rightarrow \dfrac{1}{\lambda} = \dfrac{1}{2}\left(\dfrac{1}{4000} + \dfrac{1}{6000}\right)$

$\therefore \lambda = 4800$ Å

47. $E = n\cdot\dfrac{hc}{\lambda} \Rightarrow 200\times10^3 = \dfrac{N_A\times hc}{\lambda} = \dfrac{0.12}{\lambda}$

$\therefore \lambda = 6\times10^{-7}$ m

48. $\lambda = \dfrac{12400}{E}\text{Å} - \text{eV} = \dfrac{12400}{1} = 12400\text{Å}$

$= 12400\times10^{-10} = 1.24\times10^{-6}$ m

49. Theoretical

50. $(\text{K.E.})_{\max} : E = h\upsilon - \phi$

and $E' = h\cdot 2\upsilon - \phi$

$\therefore E' = E + h\upsilon$

51. $E = n\cdot\dfrac{hc}{\lambda}$

$\Rightarrow \dfrac{E\cdot\lambda}{hc} = \dfrac{10^{-7}\times5000\times10^{-10}}{6.626\times10^{-34}\times3\times10^8} = 2.5\times10^{11}$

51. Theoretical

53. $\phi = \dfrac{hc}{\lambda} = \dfrac{6.6\times10^{-34}\times3\times10^8}{330\times10^{-9}} = 6\times10^{-19}$ J

54. $\sqrt{\frac{c}{\lambda}} = a(z-b)$

$\therefore \frac{\lambda_{iron}}{\lambda_{scandium}} = \frac{(21-1)^2}{(26-1)^2} = \frac{16}{25}$

55. $\frac{\lambda_2}{\lambda_1} = \left(\frac{z_1-b}{z_2-b}\right)^2 \Rightarrow \frac{\lambda_2}{\lambda} = \left(\frac{57-1}{29-1}\right)^2 \Rightarrow \lambda_2 = 4\lambda$

## Bohr's Atomic Model

56. $Na^{10+}$ ion has single electron

57. $r_{n,z} = 0.529 \times \frac{n^2}{z}$ Å

$\therefore \frac{r_{2,Li^{2+}}}{r_{1,H}} = \frac{0.529 \times \frac{2^2}{3}}{0.529 \times \frac{1^2}{1}} \Rightarrow \frac{r_{2,Li^{2+}}}{x\text{Å}} = \frac{4}{3}$

$\Rightarrow r_{2,Li^{2+}} = \frac{4}{3}x\text{Å}$

58. $r_{1,Ne^{9+}} = 0.529 \times \frac{1^2}{10} = 0.0529\text{Å}$

59. $\frac{r_4 - r_3}{r_7 - r_6} = \frac{r_1 \times 4^2 - r_1 \times 3^2}{r_1 \times 7^2 - r_1 \times 6^2} = \frac{7}{13}$

60. $r_{n,z} = 0.529 \times \frac{n^2}{z}\text{Å} \Rightarrow \frac{10^{5}}{2} = 0.529 \times \frac{n^2}{1} \times 10^{-8}$

$\Rightarrow n \approx 31$

61. $r_{n,z} = \frac{(4\pi\varepsilon_0)n^2h^2}{4\pi^2 mze^2} \Rightarrow r_{n,z} \alpha \frac{1}{m}$

$\therefore$ Radius of 1st orbit of H-atom $= \frac{0.529}{2} = 0.2645$ Å

62. $\frac{(2\pi r)_3}{(2\pi r)_2} = \frac{3^2}{2^2} = \frac{9}{4}$

63. Speed is independent of mass.

64. $V_{n,z} = 2.188 \times 10^6 \frac{z}{n}$ m/s

$\frac{1}{205.67} \times 3 \times 10^8 = 2.188 \times 10^6 \times \frac{2}{n} \Rightarrow n = 3$

65. $V_{n,z} = 2.188 \times 10^6 \frac{z}{n}$ m/s

or, $1094 \times 10^3 = 2.188 \times 10^6 \times \frac{z}{4} \Rightarrow z = 2$

$\Rightarrow He^+$ ion

66. $V_{n,z} = 2.188 \times 10^6 \frac{z}{n}$ m/s

$19.54 \times 11.2 \times 10^3 = 2.188 \times 10^6 \times \frac{1}{n} \Rightarrow n \approx 10$

67. $\frac{V}{c} = \frac{2.188 \times 10^6}{3 \times 10^8} \approx \frac{1}{137}$

68. $r_{n,z} = 0.529 \times \frac{n^2}{z} \Rightarrow 1.587 = 0.529 \times \frac{n^2}{3} \Rightarrow n = 3$

$\therefore V_{n,z} = 2.188 \times 10^6 \times \frac{z}{n} = 2.188 \times 10^6 \times \frac{3}{3}$

$= 2.188 \times 10^6$ m/s

69. $V_{n,z} = 2.188 \times 10^6 \times \frac{z}{n} = 2.188 \times 10^6 \times \frac{2}{3}$

$= 1.459 \times 10^6$ m/s

$\therefore$ Distance travelled by electron in 1 second $= 1.459 \times 10^6$ m

70. $T_{n,z} = 1.5 \times 10^{-16} \times \frac{n^3}{z^2}$ sec

$\therefore \frac{T_{2,M}}{T_{3,M}} = \frac{2^3}{3^3} = \frac{8}{27}$

71. $f_{n,z} = \frac{1}{T_{n,z}} \alpha \frac{1}{n^3}$

72. Angular momentum $= n \cdot \frac{h}{2\pi}$ and $n$ must be an integer.

73. $n_2 = 2 \times n_1$

$\therefore \frac{r_2}{r_1} = \frac{n_2^2}{n_1^2} = \frac{4}{1}$

74. Angular momentum is independent from atomic number.

75. Angular momentum $\alpha\, n\, \alpha \sqrt{r}$

76. $w = \frac{V}{r} = \frac{2.188 \times 10^6 \times \frac{2}{3}}{0.529 \times 10^{-10} \times \frac{3^2}{2}} = 6.13 \times 10^{15}\ s^{-1}$

**77.** $F = \frac{mv^2}{r} \alpha \frac{\left(\frac{z}{n}\right)^2}{\left(\frac{n^2}{z}\right)} = \frac{z^3}{n^4}$

**78.** $\text{K.E.} = \frac{1}{2}\text{mv}^2 = \frac{1}{2}m\left(2.188\times10^6\times\frac{z}{n}\right)^2 \alpha \frac{1}{n^2}$

$\therefore$ For maximum K.E., $n = 1$ (minimum)

**79.** $\text{K.E.} = \frac{1}{2}\text{mv}^2$, $\text{P.E.} = -\text{mv}^2$ and $V \alpha \frac{1}{n}$

Therefore, with increase in orbit number, K.E. decreased but P.E. increases.

**80.** $E_{n,z} = -13.6\times\frac{z^2}{n^2}$ eV

$\therefore \frac{E_{2,He^+}}{E_{1,H}} = \frac{-13.6\times\frac{2^2}{2^2}}{-13.6\times\frac{1^2}{1^2}} = \frac{1}{1}$

**81.** $E_{n,z} = -13.6\frac{z^2}{n^2}$ eV $\Rightarrow -13.6 = -13.6\times\frac{z^2}{n^2} \Rightarrow z = 3$

**82.** First orbit has lowest energy.

**83.** (a) $v\cdot r \alpha \frac{z}{n}\times\frac{n^2}{z} = n$

**84.** $\frac{(\text{P.E.})_{3,Li^{2+}}}{(\text{K.E.})_{4,He^+}} = \frac{-27.2\times\frac{3^2}{3^2}}{13.6\times\frac{2^2}{4^2}} = -\frac{8}{1}$

**85.** Speed, $V_{n,z} = \frac{2\pi ze^2}{(4\pi\varepsilon_0)nh}$

**86.** $\text{P.E.} = -27.2\times\frac{z^2}{n^2} = -27.2$ eV

**87.** $V_{n,z} = 2.188\times10^6\frac{z}{n}$ m/s

$\Rightarrow 2.188\times10^6 = 2.188\times10^6\times\frac{2}{n} \Rightarrow n = 2$

Now, $\text{P.E.} = -27.2\times\frac{z^2}{n^2} = -27.2\times\frac{2^2}{2^2} = -27.2$ eV

**88.** Theoretical

**89.** $\Delta E = 13.6z^2\left(\frac{1}{n_1^2}-\frac{1}{n_2^2}\right)\text{eV} = 13.6\times2^2\left(\frac{1}{2^2}-\frac{1}{8^2}\right)$

$= 12.75$ eV

**90.** Energy difference decreases on increasing the orbit number.

**91.** $\Delta E_{2\to1}$ is even greater that $\Delta E_{\infty\to2}$.

**92.** The required transition is $n = 2$ to $n = 4$.

**93.** $E_1 = -\text{I. E.} = -50$ eV

Now, $E_n = \frac{E_1}{n^2} \Rightarrow E_5 = \frac{-50}{5^2} = -2$ eV

**94.** $\Delta E = 13.6z^2\left(\frac{1}{n_1^2}-\frac{1}{n_2^2}\right)\text{eV}$

$\therefore 10.2 = 13.6\times1^2\left(\frac{1}{1^2}-\frac{1}{n_2^2}\right)$

$\Rightarrow n_2 = 2$

Now, change in angular momentum $= \frac{2\times h}{2\pi}-\frac{1\times h}{2\pi}$

$= \frac{h}{2\pi}$

**95.** I. E. $\alpha\, z^2$

$\therefore \frac{(\text{I.E.})_{Be^{3+}}}{(\text{I.E.})_{He^+}} = \left(\frac{z_{Be^{2+}}}{z_{He^+}}\right)^2 \Rightarrow \frac{(\text{I.E.})_{Be^{3+}}}{x} = \left(\frac{9}{2}\right)^2$

$\therefore (\text{I.E.})_{Be^{3+}} = 4x$ eV

**96.** $\Delta E = 13.6z^2\left(\frac{1}{n_1^2}-\frac{1}{n_2^2}\right)\text{eV}$

or $47.2 = 13.6\times z^2\left(\frac{1}{2^2}-\frac{1}{3^2}\right) \Rightarrow z \approx 5$

**97.** $\text{I.E.} = \frac{hc}{\lambda} = \frac{6.626\times10^{-34}\times3\times10^8}{240\times10^{-9}}\times6.022\times10^{23}$

$= 4.98\times10^5$ J/mol

**98.** Second I.E. of He is $13.6\times2^2 = 54.4$ eV. Hence, I.E. of He-atom must be greater than 13.6 eV but less than 54.4 eV.

**99.** Required B.E. $= 13.6\times3^2 = 122.4$ eV

**100.** Required I.E. $= 13.6\times30^2 = 12240$ eV

## Spectrum

**101.** $R = \dfrac{2\pi^2 me^4}{(4\pi\varepsilon_0)^2 h^3 c}$

$\Rightarrow R \alpha m$

Here, $m$ becomes $\left(m - \dfrac{m}{4} = \dfrac{3m}{4}\right)$, then $R$ becomes $\dfrac{3}{4}R$.

**102.** $R \alpha e^4$

$\therefore \dfrac{R_2}{R_1} = \left(\dfrac{e_2}{e_1}\right)^4 \Rightarrow \dfrac{R_2}{R} = \left(\dfrac{e/2}{e}\right)^4 \Rightarrow R_2 = \dfrac{R}{16}$

**103.** Informative

**104.** For maximum $\upsilon$, $z$ should be maximum.

**105.** Only one photon is involved in one electronic transition.

**106.** $\dfrac{1}{\lambda} = Rz^2\left(\dfrac{1}{n_1^2} - \dfrac{1}{n_2^2}\right)$

$= R \times 3^2\left(\dfrac{1}{1^2} - \dfrac{1}{4^2}\right) = \dfrac{135\,R}{16}$

$\therefore \lambda = \dfrac{16}{135\,R}$

**107.** Second line of Panchen series is $5 \rightarrow 3$.

Now, $\upsilon = \dfrac{c}{\lambda} = CRz^2\left(\dfrac{1}{n_1^2} - \dfrac{1}{n_2^2}\right)$

$= CR \times 2^2\left(\dfrac{1}{3^2} - \dfrac{1}{5^2}\right)$

$= \dfrac{64CR}{225}$

**108.** For lowest frequency in Lyman series, the transition is $2 \rightarrow 1$.

Now, $\bar{\upsilon} = Rz^2\left(\dfrac{1}{n_1^2} - \dfrac{1}{n_2^2}\right) = R \times 3^2\left(\dfrac{1}{1^2} - \dfrac{1}{2^2}\right)$

$= \dfrac{27R}{4}$

**109.** $\Delta E = \dfrac{hc}{\lambda}$

**110.** $\dfrac{\lambda_2}{\lambda_1} = \dfrac{\left(\dfrac{1}{n_1^2} - \dfrac{1}{n_2^2}\right)_1}{\left(\dfrac{1}{n_1^2} - \dfrac{1}{n_2^2}\right)_2} = \dfrac{\left(\dfrac{1}{1^2} - \dfrac{1}{2^2}\right)}{\left(\dfrac{1}{3^2} - \dfrac{1}{4^2}\right)} = \dfrac{108}{7}$

**111.** $\dfrac{\overline{\upsilon_2}}{\overline{\upsilon_1}} = \dfrac{z_2^2\left(\dfrac{1}{n_1^2} - \dfrac{1}{n_2^2}\right)_1}{z_1^2\left(\dfrac{1}{n_1^2} - \dfrac{1}{n_2^2}\right)_2}$

$\Rightarrow \dfrac{\overline{\upsilon_2}}{2.5 \times 10^5} = \dfrac{3^2\left(\dfrac{1}{3^2} - \dfrac{1}{5^2}\right)}{4^2\left(\dfrac{1}{2^2} - \dfrac{1}{3^2}\right)}$

$\therefore \overline{\upsilon_2} = 7.2 \times 10^4\ \text{cm}^{-1}$

**112.** $\upsilon = \dfrac{c}{\lambda} = c\left(\dfrac{1}{\lambda_1} + \dfrac{1}{\lambda_2}\right)$

$= 3 \times 10^8\left(\dfrac{1}{400 \times 10^{-9}} + \dfrac{1}{300 \times 10^{-9}}\right)$

$= 1.75 \times 10^{15}$ Hz

**113.** $\Delta E = \dfrac{hc}{\lambda} \Rightarrow \Delta E \alpha \dfrac{1}{\lambda}$

Now, $\dfrac{\lambda_2}{\lambda 1} = \dfrac{\Delta E_1}{\Delta E_2} \Rightarrow \dfrac{\lambda_2}{\lambda} = \dfrac{2E - E}{\dfrac{4E}{3} - E} \Rightarrow \lambda_2 = 3\lambda$

**114.** $\bar{\upsilon}_H = \bar{\upsilon}_{He^+}$

or, $R \times 1^2\left(\dfrac{1}{n_1^2} - \dfrac{1}{n_2^2}\right) = R \times 2^2\left(\dfrac{1}{2^2} - \dfrac{1}{4^2}\right)$

$\therefore n_1 = 1$ and $n_2 = 2$

**115.** $10 \rightarrow 4, 9 \rightarrow 4, 8 \rightarrow 4, 7 \rightarrow 4, 6 \rightarrow 4, 5 \rightarrow 4$

## Heisenberg's Uncertainty Principle

**116.** $\Delta x = \dfrac{h}{4\pi m \cdot \Delta V_{\min}}$

**117.** $\Delta x \cdot m\Delta v \geq \dfrac{h}{4\pi}$

If the numerical values of $\Delta x$ and $\Delta v$ is $x$, then

$x \geq \sqrt{\dfrac{h}{4\pi m}}$

**118.** If the numerical values of $\Delta x$ and $\Delta p$ is $x$, then

$$x \geq \sqrt{\frac{h}{4\pi}}$$

$$\therefore \Delta v_{\min} = \frac{1}{m} \cdot \sqrt{\frac{h}{4\pi}}$$

**119.** $\Delta x_{\min} = \dfrac{h}{4\pi m \cdot \Delta v} = \dfrac{6.626 \times 10^{-34}}{4\pi \times 10^{-13} \times \dfrac{0.001}{100} \times 10^{-8}}$

$= 5.28 \times 10^{-8}$ m

**120.** $\Delta x_{\min} = \dfrac{h}{4\pi m \cdot \Delta v}$

$$= \frac{6.626 \times 10^{-34}}{4\pi \times (9.1 \times 10^{-31}) \times \left(300 \times \dfrac{0.0001}{100}\right)}$$

$= 1.93 \times 10^{-2}$ m

## De-Broglie's Equation

**121.** $\lambda = \dfrac{h}{mv} \Rightarrow \dfrac{\lambda_e}{\lambda_p} = \dfrac{m_p}{m_e} = \dfrac{1836}{1}$

**122.** Number of orbit = Number of waves.

**123.** $2\pi r = n\lambda \Rightarrow x = 3 \times \lambda \Rightarrow \lambda = \dfrac{x}{3}$ m

**124.** $p = \dfrac{h}{\lambda} = \dfrac{6.626 \times 10^{-34}}{6626 \times 10^{-9}} = 10^{-28}$ Kg m s$^{-1}$

**125.** $\lambda \alpha \dfrac{1}{v} \alpha \dfrac{1}{\sqrt{T}} \Rightarrow \dfrac{\lambda_2}{\lambda_1} = \sqrt{\dfrac{T_1}{T_2}} \Rightarrow \dfrac{\lambda_2}{\lambda_1} = \sqrt{\dfrac{300}{1200}}$

$$\therefore \lambda_2 = \frac{\lambda}{2}$$

## Quantum Numbers

**126.** Theoretical

**127.** Theoretical

**128.** Type of orbitals = $n$

Number of orbitals = $n^2$

**129.** Theoretical

**130.** Orbital angular momentum = $\sqrt{l(l+1)} \cdot \dfrac{h}{2\pi}$ and for $s$-orbital, $l = 0$.

**131.** For $p$-orbital, $l = 1$

$\therefore$ Orbital angular momentum = $\sqrt{l(l+1)} \cdot \dfrac{h}{2\pi}$

$$= \sqrt{1(1+1)} \cdot \frac{h}{2\pi} = \sqrt{2} \cdot \frac{h}{2\pi}$$

**132.** $xz$ plane is the nodal plane for $p_y$-orbital.

**133.** The shape is given by $l$.

**134.** The $s$-orbital is symmetrical in all directions.

**135.** Theoretical

**136.** Number of orbitals of a particular type = $2l + 1$ and for $g$-orbital, $l = 4$

**137.** For 5$h$, $n = 5$ and $l = 5$, which is not permissible.

**138.** Theoretical

**139.** A set of $n$, $l$, $m$ represent a single orbital.

**140.** Informative

## Schrodinger's Equation

**141.** The $p$-orbital has only one nodal plane.

**142.** $n = 4$, $l = 1$, $m = 0 \Rightarrow$ orbital = 4$p$

**143.** Number of nodal surface of $s$-orbital = $n - 1$

**144.** Theoretical

**145.** Number of radial nodes = $n - l - 1$

## Electronic Configuration

**146.** Informative

**147.** Spin multiplicity = 2*s* + 1, where *s* = magnitude of total spin quantum number.

$l = 3 \Rightarrow$ Number of orbitals = 2 × 3 + 1 = 7

$\therefore$ Maximum spin multiplicity $= 2\times\frac{7}{2}+1=8$

(When each orbital has single electron in same spin)

**148.** Informative

**149.** *d*-orbitals are five in number.

**150.** Orbitals are 1*s*, 2*s*, 2*p*, 3*s*, 3*p*, 4*s*.

No. of orbitals = 1 + 1 + 3 + 1 + 3 + 1 = 10

**151.** Orbitals are 3*p*, 4*s*. Number = 3 + 1 = 4.

**152.** Two electrons in the same orbital cannot have same spin.

**153.** $3n^2 = 3 \times 2^2 = 12$

**154.** $3p < 4s < 3d < 4p$

**155.** (3 × 2 + 1) × 2 = 14

**156.** Pauli's exclusion principle limits the maximum capacity of electrons in an orbital equal to 2.

**157.** *K*-shell $\Rightarrow n = 1$

**158.** 3*p*-orbital $\Rightarrow n = 3, l = 1$

**159.** Valence electron $\Rightarrow n = 3, l = 1$

**160.** Orbital is 4*s*.

**161.** $1s^2\ 2s^2\ 2p^6\ 3s^2\ 3p^6\ \underbrace{3d^5\ 4s^1}_{\text{6 unpaired electrons}}$

**162.** 2 + 8 + 16 + 2 = 28

**163.** 2 + 8 + 10 + 2 = 22

**164.**

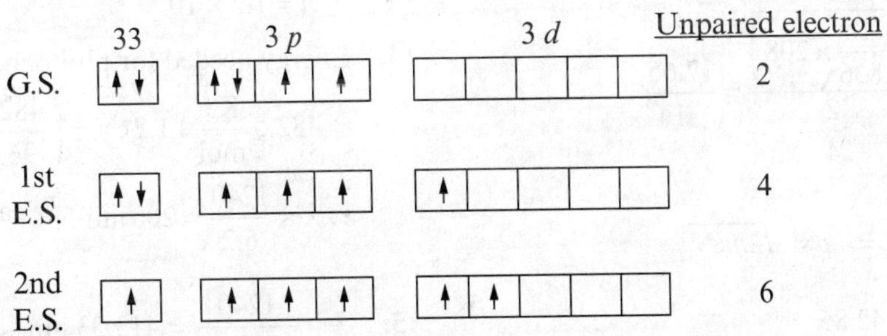


**165.** Outermost shell can have a maximum of 8 electrons.

**166.** Number of unpaired electron in $Fe^{2+}$ is 4.

**167.** Number of unpaired electron should be 4.

**168.** Number of unpaired electron should be zero.

**169.** Number of unpaired electron should be maximum.

**170.** $Na^+$ is diamagnetic.

**171.** Number of unpaired electron is 2. Hence, the ion is $Ni^{2+}$ ($d^8$ configuration).

**172.** Number of unpaired electron is 1.

**173.** $Ni^{2+}$ has unpaired electron is 1.

**174.** Ions having $d^{1-9}$ configuration are expected to be coloured.

**175.** Ions having $d^0$ or $d^{10}$ configuration are expected to be colourless.

## EXERCISE II (JEE ADVANCE)

### Section A (Only one Correct)

1. $\frac{e}{m}$ ratio of cathode rays is independent to the nature of gas.

2. $$\frac{\left(\frac{e}{m}\right)_A}{\left(\frac{e}{m}\right)_B} = \frac{e_A}{e_B} \times \frac{m_B}{mA}$$

$$\Rightarrow \frac{2}{3} = \frac{e_A}{e_B} \times \frac{3}{2}$$

$$\Rightarrow \frac{eA}{e_B} = \frac{4}{9}$$

3. $$eV = \frac{1}{2}mv^2 \Rightarrow v = \sqrt{2 \times \frac{e}{m} \times V}$$

$$= \sqrt{2 \times 1.764 \times 10^{11} \times 200} = 8.2 \times 10^6 \text{ m/s}$$

4. $$\frac{\left(\frac{e}{m}\right)_{\text{mesor}}}{\left(\frac{e}{m}\right)_{\alpha\text{-particle}}} = \frac{\frac{1}{\left(\frac{1}{1836} \times 208\right)}}{\frac{2}{4}} = \frac{17.65}{1}$$

5. $$eV = \frac{1}{2}mv^2 = \frac{p^2}{2m} \Rightarrow p = \sqrt{2meV}$$

$$\therefore \frac{p_p}{p_e} = \sqrt{\frac{1836}{1}} = \frac{42.85}{1}$$

6. $$v = c\bar{v} = \left(3 \times 10^{10} \frac{\text{cm}}{s}\right) \times \left(\frac{2 \times 10^6}{50} \text{cm}^{-1}\right)$$

$$= 1.2 \times 10^{15} \text{H}_z$$

7. $$p = \frac{E}{t} = \frac{n \cdot hc}{t \cdot \lambda} = \frac{6 \times 10^{15} \times 6.626 \times 10^{-34} \times 3 \times 10^8}{1 \times 662.6 \times 10^{-9}}$$

$$= 1.8 \times 10^{-3} \text{ J/s} - \text{m}^2$$

8. $$p = \frac{E}{t} = \frac{n \cdot hc}{t \cdot \lambda}$$

$$\Rightarrow \frac{14}{100} \times 200 = \frac{n \times 6.626 \times 10^{-34} \times 3 \times 10^8}{1 \times \frac{1987.8}{7} \times 10^{-9}}$$

$$\therefore n = 4 \times 10^{19} \text{ s}^{-1}$$

9. $$E = n \cdot \frac{hc}{\lambda} = (1.75 \times 10^{-4} \times N_A) \times \frac{hc}{2500 \times 10^{-10}}$$

$$= 84 \text{ J}$$

10. $$E_{abs} \times \frac{x}{100} = E_{\text{emit}} \Rightarrow n_1 \cdot \frac{hc}{\lambda_1} \cdot \frac{x}{100} = n_2 \cdot \frac{hc}{\lambda_2}$$

$$\therefore x = \frac{n_2}{n_1} \times \frac{\lambda_1}{\lambda_2} = \frac{53}{100} \times \frac{4530}{5080} = 47.3$$

11. $$\lambda = \frac{1240}{5} = 248 \text{ nm}$$

12. $$E = n \cdot hc\bar{v} = 1 \times 6.626 \times 10^{-34} \times 3 \times 10^8 \times 1650763.73$$

$$= 3.28 \times 10^{-19} \text{ J/quanta.}$$

13. $$E = \frac{nhc}{\lambda} \Rightarrow 0.36 = \frac{n \times 6.626 \times 10^{-34} \times 3 \times 10^8}{662.6 \times 10^{-9}}$$

$$\therefore n = 1.2 \times 10^{18}$$

14. Energy needed for photochemical dissociation

$$= 482.5 \frac{\text{KJ}}{\text{mol}} + 1.2 \text{eV} = \left(\frac{482.5}{96.5} + 1.2\right) \text{eV} = 6.2 \text{ eV}$$

$$\therefore \lambda \approx \frac{1240}{6.2} = 200 \text{ nm}$$

15. $$\lambda \approx \frac{1240}{\left(\frac{289.5}{96.5}\right)} = 413.33 \text{ nm}$$

16. Energy absorbed per mole of $H_2$

$$= 6 \times 10^{23} \times \frac{6.6 \times 10^{-34} \times 3 \times 10^8}{270 \times 10^{-9}} \times 10^{-3} = 440 \text{ KJ}$$

$\therefore$ Percentage of absorbed energy corrected into

$$\text{K.E.} = \frac{440 - 429}{440} \times 100$$

$$= 2.5\ \%$$

17. $$E = 9 \times \frac{12400}{6900} \times 23 = 372 \text{ kcal/mole}$$

$\therefore$ Energy conversion efficiency

$$= \frac{111.6}{372} \times 100 = 30\%$$

18. $E = n\cdot\frac{hc}{\lambda} \Rightarrow 6.626 = n\times\frac{6.626\times10^{-34}\times3\times10^{8}}{360\times10^{-9}}$

$\therefore$ Mole of photons absorbed

$= \frac{n}{\sim_A} = \frac{1.2\times10^{19}}{6\times10^{23}} = 2\times10^{-5}$

$\therefore$ Quantum efficiency $= \frac{1\times10^{-5}}{2\times10^{-5}} 0.5$

19. Theoretical

20. Theoretical

21. $h\upsilon_1 = h\upsilon_0 + E$ and $h\upsilon_2 = h\upsilon_0 + \mathrm{E\cdot K}$

$\therefore \upsilon_0 = \frac{K\upsilon_1 - \upsilon_2}{K-1}$

22. Theoretical

23. $\frac{1}{2}mv_{\max}^2 = \frac{hc}{\lambda} - \frac{hc}{\lambda_0} \Rightarrow v_{\max} = \left[\frac{2hc}{m}\left(\frac{\lambda_0-\lambda}{\lambda_0\lambda}\right)\right]^{\frac{1}{2}}$

24. $h\upsilon = \mathrm{K.E.} + h\upsilon_0 \Rightarrow \upsilon = \frac{1}{h}\cdot(\mathrm{K.E.}) + \upsilon_0$

25. $\sqrt{\frac{c}{\lambda}} = a(z-1)$ and $\sqrt{\frac{c}{4\lambda}} = a(z'-1)$

$\therefore z' = \frac{z+1}{z}$

26. Number of atoms in the disc

$= \frac{1}{12}\times6\times10^{23} = 5\times10^{22}$

Now, $F = K\cdot\frac{q_1q_2}{r^2} \Rightarrow 10^{-5} = 9\times10^9\times\frac{q^2}{(10^{-2})^2}$

$\Rightarrow q = \frac{10^{-10}}{3}\mathrm{C}$

$\therefore$ Number of excess electron on negatively charged disc $= \frac{10^{-10}/3}{1.6\times10^{-19}} = \frac{10^9}{4.8}$

Hence, $\frac{\text{Number of excess electron}}{\text{Number of atoms}} = \frac{10^9/4.8}{5\times10^{22}}$

$= \frac{10^{-14}}{2.4}$

27. $r_n = r_1\times n^2 \Rightarrow 21.2\times10^{-11} = 5.3\times10^{-11}\times n^2$
$\Rightarrow n = 2$

28. $2\pi r_n = 26.5\ \text{Å} \Rightarrow 2\pi\times0.529\times\frac{n^2}{2} = 26.5 \Rightarrow n = 4$

29. $r_n = 0.529\times\frac{n^2}{z}$ Å

30. $r_n = r_1\times n^2$

$\therefore r_n - r_{n-1} = r_1\times n^2 - r_1\times(n-1)^2 = (2n-1)\cdot r_1$

Where $n$ is the higher orbit.

31. $\frac{A_2}{A_1} = \frac{(\pi r^2)_2}{(\pi r^2)_1} = \left(\frac{r_2}{r_1}\right)^2 = \left(\frac{r_1\times2^2}{r_1}\right)^2 = \frac{16}{1}$

32. $r_4 - r_2 = 2.116\ \text{Å} \Rightarrow 0.529\times\frac{4^2}{z} - 0.529\times\frac{2^2}{z}$
$= 2.116$

$\therefore z = 3 \Rightarrow Li^{2+}$ ion

33. $d = 2\pi r\times100 = 2\pi\times\left(0.529\times\frac{2^2}{4}\times10^{-10}\ \mathrm{m}\right)\times100$
$= 3.32\times10^{-8}$ m

34. Circumference $= Z\pi r = 2\pi\cdot r_0\times\frac{n^2}{1} = 2\pi r_0 n^2$ and $n = 1, 2, 3,\ldots.$

35. $V_n = 2.188\times10^6\ \frac{z}{n}$ m/s

$\Rightarrow 0.547\times10^6 = 2.188\times10^6\times\frac{1}{n}$

$\therefore n = 4$

Now, $r_n = 0.529\times\frac{n^2}{z} = 0.529\times\frac{4^2}{1} = 8.464$ Å

36. $\frac{mv^2}{r} = \frac{1}{4\pi\varepsilon_0}.\frac{ze^2}{r^2} \Rightarrow v = \sqrt{\frac{ze^2}{(4\pi\varepsilon_0)mr}}$

37. $\frac{v}{c} = \frac{2\pi ze^2}{(4\pi\varepsilon_0)nh\cdot c}$

38. Solution of Q.36

39. $F = \frac{1}{4\pi\varepsilon_0}\frac{ze^2}{r^2} = 9\times10^9\times\frac{2\times(1.6\times10^{-19})^2}{(4\times10^{-10})^2}$
$= 2.88\times10^{-9}$ N

40. $T_{n,z} = 1.5\times10^{-16}\frac{n^3}{z^2}$ sec

$\frac{T_{2,He^+}}{T_{3,Li^{2+}}} = \frac{2^3/2^2}{3^3/3^2} \Rightarrow T_{2,He^+} = \frac{2}{3}x$ sec

41. $\frac{r_1}{r_2} = \frac{n_1^2}{n_2^2} \Rightarrow \frac{r}{4r} = \frac{n_2^1}{n_2^2} \Rightarrow \frac{n_1}{n_2} = \frac{1}{2}$

$\therefore \frac{T_1}{T_2} = \frac{n_1^3}{n_2^3} = \frac{1}{8}$

42. $T_n \alpha n^3$ and $n \alpha \sqrt{r_n} \Rightarrow T_n \alpha r_n^{3/2}$

43. $N = \frac{10^{-8\text{sec}}}{T_{n,z}} = \frac{10^{-8}}{1.5 \times 10^{-16} \times \frac{2^3}{1^2}} = 8.33 \times 10^6$

44. $\lambda = \frac{c}{v} = c \cdot T_{n,z} = 3 \times 10^8 \times 1.5 \times 10^{-16} \times \frac{1^3}{1^2}$

$= 4.5 \times 10^{-8}$ m

45. $\text{K.E.} = \frac{1}{2} mv^2 = \frac{1}{2} \cdot \frac{(mvr) \cdot v}{r} = \frac{\text{J} \cdot \text{V}}{2r}$

46. $\text{K.E.} = \frac{1}{2} mv^2 = \frac{1}{2} m \left( \frac{nh}{2\pi \cdot mr} \right)^2 = \frac{n^2 h^2}{8\pi^2 mr^2}$

$= \frac{n^2 h^2}{8\pi^2 m \cdot (a_0^2 \cdot n^4)} = \frac{h^2}{8\pi^2 ma_0^2 \cdot n^2}$

47. $\Delta E = (\text{I.E.}) \left( \frac{1}{n_1^2} - \frac{1}{n_2^2} \right) = 14.4 \times \left( \frac{1}{1^2} - \frac{1}{4^2} \right) = 13.5 \text{ eV}$

48. Reduced mass effect: $r' = r \cdot \left( 1 + \frac{m_e}{m_n} \right)$

On increasing the nuclear mass, radius decreases.

49. Reduced mass effect:

$(\text{I.E.})' = (\text{I.E.}) \cdot \frac{1}{\left( 1 + \frac{m_e}{m_n} \right)}$

On increasing the nuclear mass, ionisation energy increases.

50. $u = \frac{m_1 m_2}{m_1 + m_2} = \frac{m_p \cdot m_p}{m_p + m_p} = \frac{m_p}{2} = \frac{1836 \times m_e}{2}$

$\therefore r = \frac{0.529}{918} \text{Å} = 0.058 pm$

51. $n = 2$ but $z = 3 - 2 = 1$

$\therefore E_2 = -13.6 \times \frac{1^2}{2^2} = -3.4 \text{ eV}$

and I.E. = 3.4 eV

52. K.E. of emitted electron = $0.5 \times 13.6$ eV

Now, $\text{K.E.} = \frac{1}{2} \text{mV}^2$

or, $6.8 \times 1.6 \times 10^{-19} = \frac{1}{2} \times 9.1 \times 10^{-31} \times v^2$

$\Rightarrow v = 1.55 \times 10^6$ m/s

53. $\Delta E = \frac{1240}{589.6} = 2.1 \text{ eV} = 3.37 \times 10^{-19} \text{J}$

$= 48.5$ kcal/mol

54. $\lambda = \frac{1240}{0.0141} = 8.8 \times 10^4 \text{ nm} = 8.8 \times 10^{-5} \text{ m} = 88 \text{ nm}$

55. Minimum is 1 ($4 \rightarrow 1$ transition in both atoms) and maximum is 4 ($4 \rightarrow 3 \rightarrow 2 \rightarrow 1$ in one atom and any other transition in other atom).

56. Number of available orbits is only 4. Hence, maximum number of spectral lines = $^4C_2 = 6$.

57. 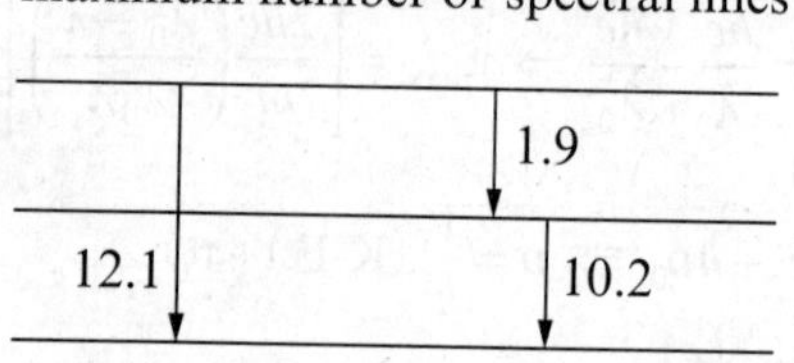


At least two atoms are needed for these three transitions.

58. $\frac{1}{\lambda} = Rz^2 \left( \frac{1}{n_1^2} - \frac{1}{n_2^2} \right) = R \times 2^2 \left( \frac{1}{2^2} - \frac{1}{4^2} \right)$

$\Rightarrow \lambda = 1223$ Å

$\therefore$ UV region.

59. $\frac{n(n-1)}{2} = 15 \Rightarrow n = 6$

Now, for shortest wavelength, required transition is $6 \rightarrow 1$.

$\therefore \frac{1}{\lambda} = R \times 1^2 \left( \frac{1}{1^2} - \frac{1}{6^2} \right) = \frac{35}{36} R \Rightarrow \lambda = \frac{36}{35R}$

60. $\frac{1}{\lambda} = R \times 1^2 \left( \frac{1}{2^2} - \frac{1}{n^2} \right) = \frac{R \cdot (n^2 - 4)}{4n^2}$

$\therefore \lambda = \frac{4n^2}{R(n^2 - 4)} = \frac{K \cdot n^2}{n^2 - 4} \Rightarrow K = \frac{4}{R}$

61. $\lambda \alpha \frac{1}{z^2} \Rightarrow \lambda_{\text{H}} : \lambda_{\text{He}^+} : \lambda_{\text{Li}^{2+}} = \frac{1}{1^2} : \frac{1}{2^2} : \frac{1}{3^2} = 36 : 9 : 4$

62. $\frac{1}{\lambda} = R \times 1^2 \times \left( \frac{1}{1^2} - \frac{1}{n^2} \right) \Rightarrow n = \sqrt{\frac{\lambda R}{\lambda R - 1}}$

**63.** Required transition is $4 \rightarrow 2$

$$\frac{1}{\lambda} = R \times 1^2\left(\frac{1}{2^2} - \frac{1}{4^2}\right) = \frac{3R}{16} \Rightarrow \lambda = \frac{16}{3R}$$

**64.** $\lambda \alpha \frac{1}{z^2} \Rightarrow \frac{\lambda_{Na^{10+}}}{\lambda_H} = \frac{1^2}{10^2} \Rightarrow \lambda_{Na^{10+}} = 12.16$ Å

**65.** Excited state is $n = 6$ [$3 \rightarrow 2, 4 \rightarrow 2, 5 \rightarrow 2, 6 \rightarrow 2$]

$\therefore$ Number of spectral lines in 1*R* region = 6.

**66.** Required transition is $3 \rightarrow 2$.

Modified Rydberg constant is given by,

$$R' = R \times 2 \text{ as } R = \frac{2\pi^2 me^4}{(4\pi\varepsilon_0)^2 h^3 c}$$

$$\text{Now, } \frac{1}{\lambda} = R' \times 1^2\left(\frac{1}{2^2} - \frac{1}{3^2}\right) = 2R \cdot \frac{5}{36}$$

$$\therefore \lambda = \frac{18}{5R}$$

**67.** $\bar{\upsilon} = R \cdot \left(\frac{1}{2^2} - \frac{1}{n^2}\right) = \frac{R(n^2 - 4)}{4n^2}$

**68.** $\Delta E = 1312 \times \left(\frac{1}{2^2} - \frac{1}{3^2}\right) = 182.22$ KJ

**69.** All are visible radiations. Next line is from transition $7 \rightarrow 2$.

**70.**

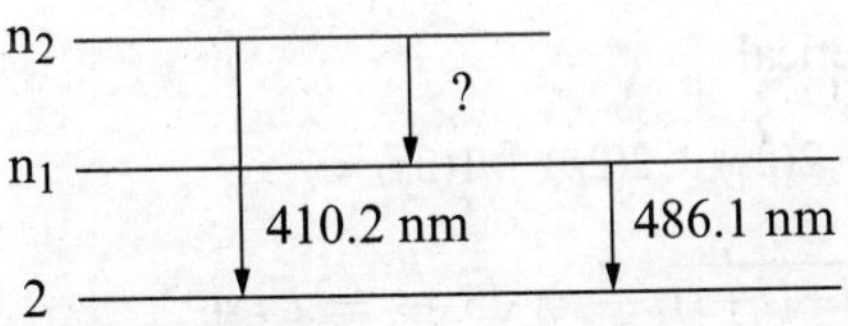


Both are visible radiations. For required series, we get only $n_1$.

$$\text{Now, } \frac{1}{486.1 \times 10^{-9}} = 1.09 \times 10^7 \times 1^2\left(\frac{1}{2^2} - \frac{1}{n_1^2}\right)$$

$\Rightarrow n_1 = 4$.

Hence, the series is Brackett series.

**71.** $\frac{1}{\lambda_1} = R \times 2^2\left(\frac{1}{2^2} - \frac{1}{3^2}\right) \Rightarrow \lambda_1 = \frac{9}{5R}$

$$\frac{1}{\lambda_2} = R \times 2^2\left(\frac{1}{1^2} - \frac{1}{2^2}\right) \Rightarrow \lambda_2 = \frac{1}{5R}$$

From question, $\lambda_1 - \lambda_2 = 132$ nm

or, $\frac{9}{5R} - \frac{1}{3R} = 132 \times 10^{-9}$ m $\Rightarrow$ R = $1.11 \times 10^9$ m$^{-1}$

**72.**

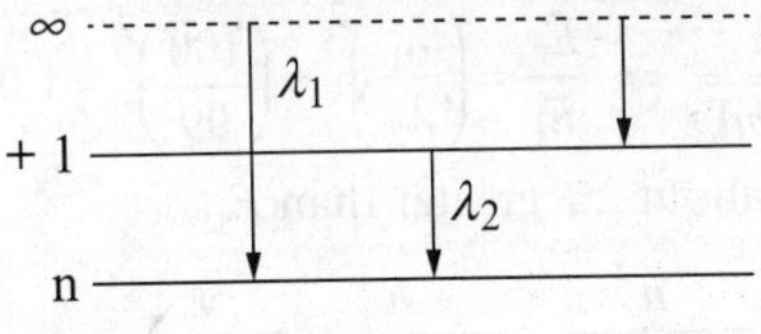


$$\bar{\upsilon} = 2.725 \times 10^6 = 1.09 \times 10^7 \times 1^2\left(\frac{1}{(n+1)^2} - \frac{1}{\infty^2}\right)$$

$\therefore n = 3$

$$\text{Now, } \frac{1}{\lambda_{req}} = 1.09 \times 10^7 \times 2^2\left(\frac{1}{3^2} - \frac{1}{4^2}\right)$$

$\Rightarrow \lambda_{req} = 471.8$ nm

**73.** $\frac{1240}{108.5} = \text{B.E.}\left(\frac{1}{2^2} - \frac{1}{5^2}\right) \Rightarrow$ B.E. = 54.4 eV

**74.** K.E. of electron

$$= 13.6 \times 2^2\left(\frac{1}{1^2} - \frac{1}{2^2}\right) - 13.6 = 27.2 \text{ eV}$$

$$\text{Now, } 27.2 \times 1.6 \times 10^{-19} = \frac{1}{2} \times 9.1 \times 10^{-31} \times v^2$$

$\therefore$ v $\approx 3.1 \times 10^6$ m/s

**75.** $X = \frac{E}{n^2}$ and $Y = \frac{E}{(n+3)^2}$

$$\therefore \sqrt{\frac{X}{Y}} = 1 + \frac{3}{n}$$

**76.** $\frac{1}{\lambda} = R \times 1^2 \times \left(\frac{1}{1^2} - \frac{1}{3^2}\right) \Rightarrow \lambda = \frac{9}{8R}$

**77.** $n\lambda = 2\pi r \Rightarrow \frac{2\pi \times x \times 3^2}{3} = 6\pi x$

**78.** $\lambda = \frac{h}{\sqrt{2mE}} \Rightarrow E \alpha \frac{1}{m}$ (for same $\lambda$)

**79.** $\lambda = \frac{h}{\sqrt{2mE}} \Rightarrow \frac{\lambda_p}{\lambda_\alpha} = \sqrt{\frac{4 \times 2}{1 \times 1}} = \frac{2\sqrt{2}}{1}$

**80.** $p = mv = \frac{\frac{1}{2}mv^2}{\frac{1}{2}v} = \text{const} \Rightarrow \lambda = \text{Constant}$

**81.** $\lambda_{min} = \frac{1.24 \times 10^{-6}}{5 \times 10^4} = 2.48 \times 10^{-11}$ m

**82.** $m = \frac{h}{c \cdot \lambda} = \frac{h}{c} \times R \times 2^2\left(\frac{1}{2^2} - \frac{1}{\infty^2}\right) = 2.4 \times 10^{-35}$ kg

83. $\lambda = \frac{h}{\sqrt{2mE}} \Rightarrow \frac{E_2}{E_1} = \left(\frac{\lambda_1}{\lambda_2}\right)^2 = \left(\frac{100}{99}\right)^2 \approx 1.02$

$\therefore E_2$ is about 2% greater than $E_1$.

84. $\Delta v_{min} = \frac{h}{4\pi m \cdot \Delta x} = \frac{h}{4\pi m \cdot \frac{h}{mv}} = \frac{v}{4\pi}$

85. $\lambda = \frac{h}{\sqrt{2mE}} \Rightarrow \Delta E = \frac{h^2}{2m}\left(\frac{1}{\lambda_2^2} - \frac{1}{\lambda_1^2}\right)$

$= \frac{(6.626\times10^{-34})^2}{2\times9.1\times10^{-31}}\left(\frac{1}{(50\times10^{-9})^2} - \frac{1}{(100\times10^{-9})^2}\right)$

$= 7.24 \times 10^{-23}$ J $= 4.5 \times 10^{-4}$ eV

86. $\lambda = \frac{h}{\sqrt{2mE}} = \frac{h}{\sqrt{2m2V}}$

$= \frac{6.626\times10^{-34}}{\sqrt{2\times4\times1.66\times10^{-27}\times2\times1.6\times10^{-19}\times6}}$

$= 4.15 \times 10^{-12}$ m

87. $\lambda = 3.32 \cdot \frac{n}{z}$ Å $\Rightarrow 3.32 = 3.32 \times \frac{n}{2} \Rightarrow n = 2$

Energy of photon liberated in $2 \rightarrow 1$ transition,

$\Delta E = 13.6 \times 2^2 \times \left(\frac{1}{1^2} - \frac{1}{2^2}\right) = 40.8$ eV

$\therefore$ K.E. of emitted electron from H-atom $= 40.8 - 13.6 = 27.2$ eV

Hence, its de Broglie wavelength is given by,

$\lambda = \sqrt{\frac{150}{27.2}} = 2.348$Å

88. K.E. of electrons $= \frac{12400}{3000} - \frac{12400}{4000} = 1.03$ eV

$\therefore \lambda = \sqrt{\frac{150}{1.03}} = 12.05$ Å

89. $\Delta x \cdot \Delta\lambda \geq \frac{\lambda^2}{4\pi}$ and $\lambda = \sqrt{\frac{150}{6}} = 5$Å

$\therefore \Delta\lambda_{min} = \frac{\lambda^2}{4\pi \cdot \Delta x} = \frac{(5\times10^{-10})^2}{4\pi\times\left(\frac{1}{\pi}\times10^{-9}\right)}$

$= 6.25\times10^{-11}$m

90. $\Delta E = 2.55 \text{ eV} = 13.6\times1^2\left(\frac{1}{n_1^2} - \frac{1}{n_2^2}\right)$eV

$\therefore n_1 = 2$ and $n_2 = 4$

Now, $\Delta\lambda = 3.32\times\frac{4}{1} - 3.32\times\frac{2}{1} = 6.64$ Å

91. Orbital angular momentum $= \sqrt{l(l+1)} \cdot \frac{h}{2\pi}$

92. Electron of 1*s* level can never emit photon.

93. Maximum permissible value of $l = (n - 1)$

94. $m = -1 \Rightarrow l \geq 1 \Rightarrow$ can not be *s*-orbital.

95. Theoretical

96. Theoretical

97. Energy $2s < 2p < 3s < 3p < 4s < 3d$

98. $m = -3, -2, -1, 0, +1, +2, +3$

99. Number of radial nodes $= n - l - 1 = 3 - 2 - 1 = 0$

100. Theoretical

101. Probability of finding electron at the nucleus = 0

102. Theoretical

103. Mg($z$ = 12) $1s^2\,2s^2\,2p^6\,3s^2$

104. Theoretical

105. 2(1*s*) + 2(2*s*) + 2(2*p*) + 1(3*s*) = 7

106. $L = \sqrt{l+(l+1)} \cdot \frac{h}{2\pi} = \sqrt{5} \cdot \frac{h}{\pi} \Rightarrow l = 4$

Number of orbitals $= 2l + 1 = 9$

107. $r_{mp} = \frac{a_0}{z} = 26.45$ pm

108. For $n - l - 1 = 0$, $r_{mp} = \frac{n^2 a_0}{z}$

and for all orbitals, $r_{av} = \frac{n^2 a_0}{z}\left[1 + \frac{1}{2}\left(1 - \frac{l(l+1)}{n^2}\right)\right]$

109. $S_1 = 3s$; $S_2 = 3d$; $S_3 = 4s$; $S_4 = 3p$

For single electron system $3s = 3p = 3d < 4s$

110. $\sqrt{l(l+1)} \cdot \frac{h}{2\pi} = \sqrt{3} \cdot \frac{h}{\pi} \Rightarrow l = 3 \Rightarrow n \geq 4$

## Section B (One or More than one Correct)

1. (a) $(\text{K.E.})_{\text{Initial}} = (\text{P.E.})_{\text{at distance of closest approach}}$

or $4.0 \text{ MeV} = \text{K.} \dfrac{q_1 q_2}{r}$

or, $4 \times 10^6 \times 1.6 \times 10^{-19} = 9 \times 10^9 \times$

$$\frac{(2\times1.6\times10^{-19})\times(50\times1.6\times10^{-19})}{r}$$

$\therefore$ Distance of closest approach, $r = 3.6 \times 10^{-14}$ m

(b) $\text{P.E.} = \text{K}\cdot\dfrac{q_1 q_2}{r} = 9 \times 10^9 \times$

$$\frac{(2\times1.6\times10^{-19})\times(50\times1.6\times10^{-19})}{9\times10^{-14}}$$

$$= 10^{25} \times (1.6 \times 10^{-19})^2 \text{ J} = \frac{10^{25}\times(1.6\times10^{-19})^2}{1.6\times10^{-19}} \text{e V}$$

$= 1.6$ MeV

(c) $\text{P.E.} = K\cdot\dfrac{q_1 q_2}{r}$

$$= \frac{9\times10^9\times(2\times1.6\times10^{-19})\times(50\times1.6\times10^{19})}{4.5\times10^{-14}\times(1.6\times10^{-19}\times10^6)}$$

$= 3.2$ MeV

$\therefore$ K.E. of $\alpha$-particle at this distance $= 4.0 - 3.2 = 0.8$ MeV

2. $\varepsilon_n = \dfrac{\varepsilon_1}{n^2}$

3. Theoretical

4. Theoretical

5. Theoretical

6.

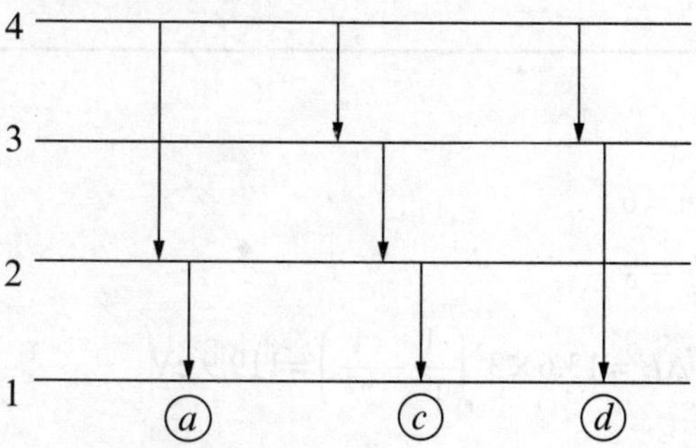


7. $\dfrac{1}{\lambda} = RZ^2\left(\dfrac{1}{n_1^2} - \dfrac{1}{n_2^2}\right)$

$$\Rightarrow \frac{1}{1028\times10^{-10}} = 1.09\times10^7\times1^2\left(\frac{1}{1^2}-\frac{1}{n^2}\right)$$

$\therefore n = 3$

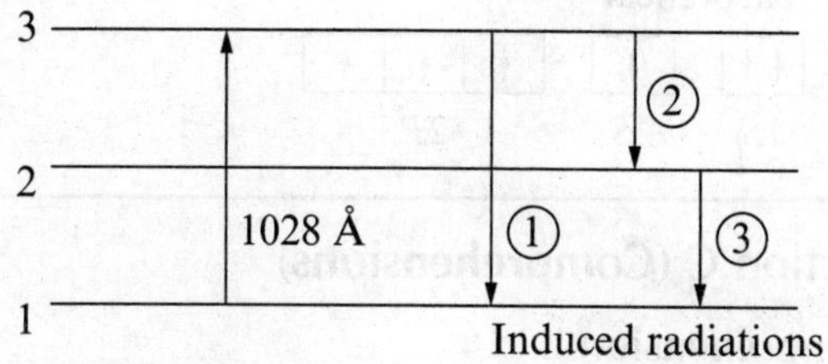


$\lambda_1 = 1028$ Å

$$\frac{\lambda_2}{\lambda_1} = \frac{\left(\frac{1}{1^2}-\frac{1}{3^2}\right)}{\left(\frac{1}{2^2}-\frac{1}{3^2}\right)} \Rightarrow \lambda_2 = 6579.2 \text{ Å}$$

$$\frac{\lambda 3}{\lambda_1} = \frac{\left(\frac{1}{1^2}-\frac{1}{3^2}\right)}{\left(\frac{1}{1^2}-\frac{1}{2^2}\right)} \Rightarrow \lambda_3 = 1218.4 \text{ Å}$$

8. Per atom only one photon is emitted out and hence, the concerned transition is $2 \rightarrow 1$.

$$\Delta E = 13.6\, Z^2\left(\frac{1}{1^2}-\frac{1}{2^2}\right) = 10.2\, Z^2 \text{ eV}$$

9. (a) $r_1 : r_2 : r_3 = 1^2 : 2^2 : 3^2 = 1 : 4 : 9$

(c) $\lambda = \dfrac{c}{v} = \dfrac{3\times10^8}{6\times10^{14}} = 5\times10^{-7}\text{m} = 500$ nm

(d) $\lambda = \dfrac{h}{\sqrt{2mE}} \Rightarrow \lambda\, \alpha \dfrac{1}{\sqrt{m}}$

$$\therefore \lambda_{\text{H}} : \lambda_{\text{H}_e} : \lambda_{cn_4} = \frac{1}{\sqrt{1}} : \frac{1}{\sqrt{4}} : \frac{1}{\sqrt{16}} = 4:2:1$$

10. $hv = \phi + (\text{K.E.})_{\max}$

For A: $4.25 = \phi_{\text{A}} + T_{\text{A}}$ and $\lambda_{\text{A}} = \dfrac{h}{\sqrt{2mT_{\text{A}}}}$

For B: $4.20 = \phi_{\text{B}} + T_{\text{B}}$ and $\lambda_{\text{B}} = \dfrac{h}{\sqrt{2mT_{\text{B}}}}$

As $T_{\text{B}} = T_{\text{A}} - 1.50$ and $\lambda_{\text{B}} = 2\,\lambda_{\text{A}}$

$\phi_{\text{A}} = 2.25$ eV; $\phi_{\text{B}} = 3.70$ eV; $T_{\text{A}} = 2.0$ eV; $T_{\text{B}} = 0.5$ eV

11. $l = 3.32\, \dfrac{n}{z}$ Å

12. Theoretical

13. Theoretical

14. Theoretical

15. Theoretical

16. [↑↓] [↑] [↑ | ↑ | ↑]
1S 2S 2P

17. Theoretical

18. Na (11) $1s^2\ 2s^2\ 2p^6\ 3s^1$

19. Theoretical

20. Total nodes = $n - 1$

## Section C (Comprehensions)

### Comprehension I

1. For minimum $l$, K.E. of photoelectron should be maximum. For it, the power should be maximum and number of photons is minimum.

$$E_{max} \text{ for photon} = \frac{5}{4\times10^{18}} = 1.25\times10^{-18}\text{ J}$$

$\therefore$ (K.E.)$_{max}$ of photoelectron = $1.25 \times 10^{-18} - 4.5 \times 10^{-19}$

$= 8.0 \times 10^{-19}$ J = 5 eV

$$\therefore \lambda_{min} = \sqrt{\frac{150}{5}} = \sqrt{30}\ \text{Å}$$

2. $i_{min} = 4 \times 10^{18} \times 1.6 \times 10^{-19} = 0.64$ A

3. $$\frac{i_{max}}{i_{min}} = \frac{9\times10^{18}\times1.6\times10^{-19}}{4\times10^{18}\times1.6\times10^{-19}} = \frac{9}{4}$$

### Comprehension II

4. $$F = -\frac{du}{dr} = \frac{4K}{r^5} = \frac{MV^2}{r} \Rightarrow V^2 = \frac{4K}{mr^4} \quad (1)$$

From Bohr's quantization, $$V^2 = \frac{n^2h^2}{4\pi^2m^2r^2} \quad (2)$$

$$\therefore \frac{4K}{mr^4} = \frac{n^2h^2}{4\pi^2m^2r^2} \Rightarrow r = \sqrt{\frac{16\pi^2mK}{n^2h^2}} = \frac{4\pi}{nh}.\sqrt{mK}$$

5. $$V = \frac{nh}{2\pi mr} = \frac{nh}{2\pi m.\dfrac{4\pi}{nh}\sqrt{mK}} = \frac{n^2h^2}{8\pi^2m\sqrt{mK}}$$

6. $$\text{E} = \text{K.E.} + \text{P.E.} = \frac{1}{2}mv^2 + \left(-\frac{K}{r^4}\right)$$

$$= \frac{1}{2}m\cdot\frac{4K}{mr^4} - \frac{K}{r^4} = \frac{K}{\left(\dfrac{4\pi}{nh}.\sqrt{mK}\right)^4}$$

$$\therefore E = \frac{n^4h^4}{256\pi^4m^2K}$$

### Comprehension III

$$\Delta E = 13.6z^2\left(\frac{1}{n_1^2} - \frac{1}{n_2^2}\right)\text{eV}$$

$$10.2 + 17.0 = 13.6\ z^2\left(\frac{1}{2^2} - \frac{1}{n^2}\right) \quad (1)$$

$$4.25 + 5.95 = 13.6\ z^2\left(\frac{1}{3^2} - \frac{1}{n^2}\right) \quad (2)$$

7. $n = 6$

8. $z = 3$

9. $$\Delta E = 13.6\times3^2\left(\frac{1}{1^2} - \frac{1}{7^2}\right) = 119.9\text{ eV}$$

## Comprehension IV

**10.** After excitation, $n = 3$. Hence, initial excited state is $n = 2$.

**11.** $n = 3$

**12.** $\frac{1}{\lambda} = RZ^2\left(\frac{1}{n_1^2} - \frac{1}{n_2^2}\right) \Rightarrow \frac{1}{1654 \times 10^{-10}}$

$= 1.09 \times 10^7 \times z^2\left(\frac{1}{2^2} - \frac{1}{3^2}\right)$

$\therefore z = 2$ Þ $He^+$ ion

**13.** $\Delta E = 13.6\, z^2\left(\frac{1}{n_1^2} - \frac{1}{n_2^2}\right) = 13.6 \times 2^2\left(\frac{1}{3^2} - \frac{1}{\infty^2}\right)$

$= 6.04$ eV

## Comprehension V

**14.** Final excited state, after absorption of 2.7 eV, is 4. On de-excitation, the sample emit radiations equal to less than or more than 2.7 eV and hence, the initial excited state must be 2.

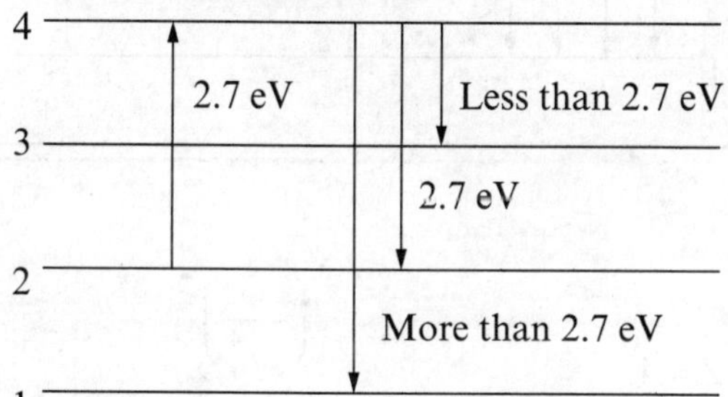


**15.** $\Delta E = I.E.\left(\frac{1}{n_1^2} - \frac{1}{n_2^2}\right) \Rightarrow 2.7 = I.E.\left(\frac{1}{2^2} - \frac{1}{4^2}\right)$

$\therefore I.E. = 14.4$ eV

**16.** $\Delta E_{min} = I.E.\left(\frac{1}{n_1^2} - \frac{1}{n_2^2}\right) = 14.4 \times \left(\frac{1}{3^2} - \frac{1}{4^2}\right)$

$= 0.7$ eV

## Comprehension VI

**17.** $r = \frac{(4\pi\varepsilon_0)n^2h^2}{4\pi^2 mze^2} = 0.529 \times \frac{n^2}{z}$ Å (for H-like atom)

For this system,

$r = \frac{0.529}{1 \times 207} = 2.56 \times 10^{-3}$ Å $= 0.256$ pm

**18.** $I.E. = \frac{2\pi^2 mz^2e^4}{(4\pi\varepsilon_0)^2 n^2h^2} = 13.6 \times \frac{z^2}{n^2}$ eV (for H-like atom)

For this system, $I.E. = 13.6 \times 207 = 2835.9$ eV

**19.** Rydberg constant for this system $= 1.09 \times 10^7 \times 207\ m^{-1}$

$\therefore \frac{1}{\lambda} = (1.09 \times 10^7 \times 207)\left(\frac{1}{1^2} - \frac{1}{2^2}\right)$

$\Rightarrow \lambda = 5.91 \times 10^{-10}$ m

## Comprehension VII

**20.** $\frac{n(n-1)}{2} = 6 \Rightarrow n = 4$

Now, $\frac{1}{\lambda} = RZ^2\left(\frac{1}{n_1^2} - \frac{1}{n_2^2}\right)$

$\Rightarrow \frac{1}{x \times 10^{-10}} = 1.09 \times 10^7 \times 1^2\left(\frac{1}{1^2} - \frac{1}{4^2}\right)$

$\therefore x = 978.6$

**21.** $n = 4$

**22.** For max $\lambda$, transition : $n = 4$ to $n = 3$

$\therefore \frac{1}{\lambda_{max}} = 1.09 \times 10^7 \times 1^2\left(\frac{1}{3^2} - \frac{1}{4^2}\right)$

$\Rightarrow \lambda_{max} = 1.887 \times 10^{-6}$ m

**23.** For max $\nu$, transition : $n = 4$ to $n = 1$

$$\therefore \quad \nu_{\max} = \frac{c}{\lambda} = \frac{3\times10^{8}}{978.6\times10^{-10}} = 3.066\times10^{15}\,\text{Hz}$$

**24.** 1R radiations involve transition : $n = 4$ to $n = 3$ only.

**25.** Visible radiation involve transitions : $n = 4$ to $n = 2$ (489.3 nm) and $n = 3$ to $n = 2$ (660.5 nm).

---

### Comprehension VIII

**26.** $5 \xrightarrow{1} 4 \xrightarrow{2} 3 \xrightarrow{3} 2 \xrightarrow{4} 1$

**27.** $5 \longrightarrow 3 \longrightarrow 2 \longrightarrow 1$ and

$5 \longrightarrow 4 \longrightarrow 3 \longrightarrow 1$

**28.** 6 + 1 (any possibility after than Q.27)

**29.** $\dfrac{5(5-1)}{2} = 10$

**30.**

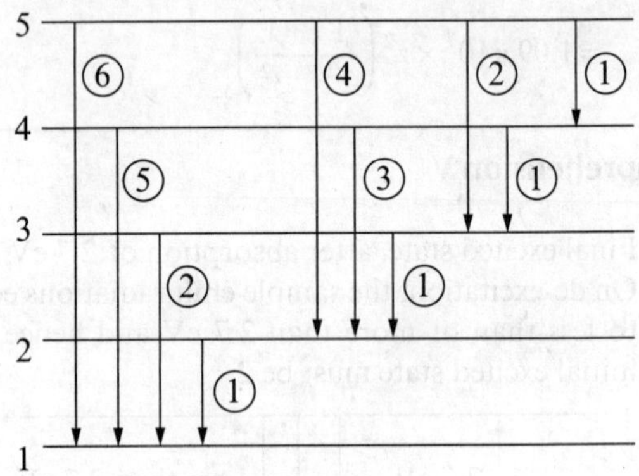


---

### Comprehension IX

$$\Delta E = 13.6Z^2\left(\frac{1}{n_1^2} - \frac{1}{n_2^2}\right)eV$$

$$204 = 13.6Z^2\left(\frac{1}{1^2} - \frac{1}{(2n)^2}\right) \qquad (1)$$

$$40.8 = 13.6Z^2\left(\frac{1}{n^2} - \frac{1}{(2n)^2}\right) \qquad (2)$$

**31.** $n = 2 \quad \Rightarrow \quad 2n = 4$

**32.** $Z = 4$

**33.** $E_1 = -13.6\times\dfrac{4^2}{1^2} = -217.6$ eV

**34.** $\Delta E_{\min} = 13.6\times4^2\left(\dfrac{1}{3^2} - \dfrac{1}{4^2}\right) = 10.58$ eV

---

### Comprehension X

**35.** $$\frac{1}{\lambda} = RZ^2\left(\frac{1}{n_1^2} - \frac{1}{n_2^2}\right) = R\times2^2\left(\frac{1}{4^2} - \frac{1}{n^2}\right)$$

$$= \frac{4R(n^2-16)}{16n^2}$$

$$\therefore \lambda = \frac{4n^2}{R(n^2-16)} = \frac{cn^2}{n^2-16} \Rightarrow C = \frac{4}{R} = 366.97 \text{ nm}$$

**36.** For series limit, $n = \infty$

$\therefore \lambda = C = 366.97$ nm

**37.** For $n = 5$, $\quad \lambda = 1019.36$ nm

For $n = \infty$, $\quad \lambda = 366.97$ nm

---

### Comprehension XI

**38.** $S_1 = 2s$

**39.** $E_{S_1} = -13.6\times\dfrac{3^2}{2^2} = E_H\times2.25$

**40.** $S_2 = 3p \Rightarrow l = 1$

## Comprehension XII

**41.** $n = 2$

$l = 1 \qquad j = \frac{3}{2}$ or $\frac{1}{2}$

$m = = -\frac{3}{2}, -\frac{1}{2}, +-\frac{1}{2}, +-\frac{3}{2}$ for $j = \frac{3}{2}$

$= -\frac{1}{2}, +\frac{1}{2}$ for $j = \frac{1}{2}$

**42.** $n = 3$

$l = 0 \quad \Rightarrow j = \frac{1}{2} \quad \Rightarrow m = -\frac{1}{2}, +\frac{1}{2}$

$= 2 \quad \Rightarrow j = \frac{3}{2}$

$\Rightarrow m = -\frac{3}{2}, -\frac{1}{2}, +\frac{1}{2}, +\frac{3}{2}$

$= \frac{5}{2} \Rightarrow m = -\frac{5}{2}, -\frac{3}{2}, -\frac{1}{2}, +\frac{1}{2}, +\frac{3}{2}, +\frac{5}{2}$

## Comprehension XIII

**43.** Radial nodes $= n - l - 1 = 3$

Angular nodes = 1

**44.** $n = 5, l = 1 \Rightarrow$ Orbital = $5p$

**45.** $p_z$

**46.** For radial nodes, $\sigma = 1, 2, 6 = \frac{2zr}{5a_0}$

$\therefore r_{max} = \frac{15\, a_0}{Z}$

## Section D (Assertion – Reason)

1. Orbital angular momentum $= \sqrt{l(l+1)} \cdot \frac{h}{2\pi}$
2. Theory based
3. Theory based
4. Spin quantum number is independent from wave function.
5. Theory based
6. Be is the reactive element.
7. $2p \Rightarrow 2p_x + 2p_y + 2p_z \Rightarrow$ Total 3 angular nodes.
8. $dz^2$ has two conical nodes.
9. $3p_x$ and $3p_y$ differs in angular function.
10. 4s energy level is lower than $3d$.

## Section E (Column Match)

1. Radial nodes $= n - l - 1$

   Angular nodes = $l$
2. Theory based
3. (A) $\frac{V_n}{K_n} = \frac{P.E.}{K.E.} = \frac{-mV^2}{\frac{1}{2}mV^2} = -2$

   (B) $\varepsilon_n \propto (-\frac{1}{r_n})$

   (C) Lowest energy level is 1$s$.

   (D) $r_n \propto \frac{1}{z}$
4. Theory based
5. Graph of $s$-orbital status with some value but for other orbitals, it starts from zero.

   Radial nodes: $3s = 2$, $4s = 3$, $2p = 0$, $3p = 1$
6. (A) $r \propto \frac{n^2}{z}$

   (B) $V \propto \frac{z}{n}$

   (C) $F = \frac{mv^2}{r} \propto \frac{z^3}{n^4}$

   (D) $f = \frac{v}{2\pi r} \propto \frac{z^2}{n^3}$

7. (A) 3 radial nodes $\Rightarrow$ 4*s*, 5*p*, 6*d* but graph does not start from origin and hence, only 4*s*.
(B) 3 radial nodes $\Rightarrow$ 4*s*, 5*p*, 6*d*
(C) Only *s*-orbital
(D) $l \geq 1$

8. Theory based

9.

| Orbital | Radial nodes | Angular nodes |
|---|---|---|
| 3*d* | 0 | 2 |
| 2*p* | 0 | 1 |
| 3*p* | 1 | 1 |
| 5*d* | 2 | 2 |

10. (A) $\frac{V_6}{V_4} = \frac{4}{6} = \frac{2}{3}$ $\left(V \propto \frac{1}{n}\right)$

(B) $\frac{\lambda 3}{\lambda 2} = \frac{\left(\frac{1}{1^2} - \frac{1}{\infty^2}\right)}{\left(\frac{1}{2^2} - \frac{1}{\infty^2}\right)} = \frac{4}{1}$

(C) $\frac{\lambda_c}{\lambda_p} = \frac{\left(\frac{1}{3^2} - \frac{1}{6^2}\right)}{\left(\frac{1}{1^2} - \frac{1}{3^2}\right)} = \frac{3}{3^2}$

(D) $\frac{\Delta E_n}{\Delta E_{H_e^+}} = \frac{1^2}{2^2} = \frac{1}{4}$

## Section F (Subjective)

### Single-digit Integer Type

1. $\varepsilon = \frac{1240}{300} = 4.13$ eV

For photoelectric effect, $\varepsilon \geq \phi \Rightarrow N_0 = 4$

2. Frequency of reduction $\propto \frac{z^2}{n^3}$

$\therefore \frac{T_3}{T_2} = \frac{\frac{1^2}{3^3} \times 4.8 \times 10^{-8}}{\frac{1^2}{2^2} \times 1.28 \times 10^{-7}} = \frac{1}{6}$

3. $\frac{1}{\lambda} = RZ^2\left(\frac{1}{n_1^2} - \frac{1}{n_2^2}\right)$

$\Rightarrow \frac{1}{108.5 \times 10^{-7}} + \frac{1}{30.4 \times 10^{-7}}$

$= 1.09 \times 10^7 \times 2^2 \times \left(\frac{1}{1^2} - \frac{1}{n^2}\right)$

$\therefore n = 5$

4. $\lambda_3 - \lambda_2 = 59.3$ nm

or, $\frac{1}{RZ^2\left(\frac{1}{2^2} - \frac{1}{3^2}\right)} - \frac{1}{RZ^2\left(\frac{1}{1^2} - \frac{1}{2^2}\right)} = 59.3$ nm

$\Rightarrow z = 3$

5. Final excited state = 5th orbit
As only the wavelengths are longer than absorbed radiation initial excited state = 3rd orbit

6. $\Delta E = 12.75 = 13.6 \times 1^2\left(\frac{4nm}{2} - \frac{1}{n^2}\right) \Rightarrow n = 4$

7. $\Delta x_{\min} = \frac{h}{4\pi m \Delta V} = \frac{6.626 \times 10^{-34}}{4\pi \times 10^{-6} \times \frac{3.313}{\pi} \times 10^{-3}}$

$= 5 \times 10^{-26}$ m

8. $m_{\min} = \frac{h}{4\pi \cdot \Delta x \cdot \Delta v}$

$= \frac{6.626 \times 10^{-34}}{4\pi \times 10^{-11} \times 5.27 \times 10^{-24}} = 1$ kg

9. $2\pi r = nl \quad \Rightarrow l = \frac{4 \text{ nm}}{2} = 2$ nm

10. For radial node, $\psi_{23} = 0 \quad \Rightarrow r_0 = 2\,a_0$

### Four-digit Integer Type

1. Initial K. E. = P. E. at distance of closest approach

or, $\frac{p^2}{2\text{m}} = \frac{1}{4\pi\varepsilon_0} \cdot \frac{q_1 q_0}{r}$.

or

$\frac{(3.2 \times 10^{-20})^2}{2 \times 4 \times \frac{10^{-3}}{6 \times 10^{23}}} = 9 \times 10^9 \times \frac{2 \times 1.6 \times 10^{-19} \times z \times 1.6 \times 10^{-19}}{1.5 \times 10^{-13}}$

$\therefore z = 25$

**2.** $t = \dfrac{\text{Distance}}{\text{Speed}} = \dfrac{2\times 12600\times 10^3}{3\times 10^8} = 0.084 \text{ sec}$

**3.** $\sqrt{\dfrac{c}{\lambda}} = a(z-6)$

$\sqrt{\dfrac{c}{180}} = a(27-1)$

$\sqrt{\dfrac{c}{144}} = a(z-1) \Rightarrow z = 30$

**4.** $nh\nu = \text{ms}\cdot\Delta T$

or, $n \times 6.626 \times 10^{-34} \times 2.45 \times 10^{10} = 245 \times 4.2 \times (99.5 - 19.5)$

$\therefore$ Number of photons $= 5.04 \times 10^{27}$

$\therefore$ Moles of photon $= \dfrac{5.04\times 10^{27}}{6\times 10^{23}} = 8400$

**5.** $\Delta E = \Delta E_1 + \Delta E_2$

$= 1310\left(\dfrac{1}{1^2} - \dfrac{1}{3^2}\right)\times\dfrac{45}{100} + 1310\times\left(\dfrac{1}{1^2} - \dfrac{1}{2^2}\right)\times\dfrac{40}{100} = 917 \text{ kJ}$

**6.** $\lambda = \dfrac{1240}{13.6} = 91.17 \text{ nm}$

**7.** Moles of $H_2 = \dfrac{PV}{RT} = \dfrac{1\times 1}{0.08\times 300} = x$

$\Delta E = 436\times x + 1312\left(\dfrac{1}{1^2} - \dfrac{1}{2^2}\right)\times 2x = 100.16 \text{ kJ}$

**8.**

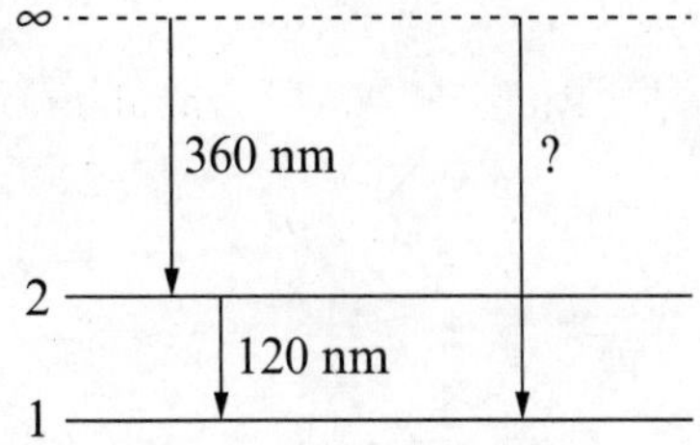


$\dfrac{1}{\lambda} = \dfrac{1}{120} + \dfrac{1}{360} \Rightarrow \lambda = 90 \text{ nm}$

**9.** $\lambda = \sqrt{\dfrac{150}{\text{V}}}\text{-}\ V^{\frac{1}{2}} \Rightarrow 2.5 = \sqrt{\dfrac{150}{\text{V}}} \Rightarrow V = 24$

**10.** $\lambda = \dfrac{h}{\sqrt{2\text{mE}}} - \dfrac{h}{\sqrt{2m\cdot\dfrac{3}{2}KT}} \Rightarrow \lambda\alpha\dfrac{1}{\sqrt{m\cdot T}}$

$\therefore \dfrac{\lambda_{\text{H}_\text{e}}}{\lambda_{\text{N}_\text{e}}} = \sqrt{\dfrac{20\times 1000}{4\times 200}} = 5$

CHAPTER

14

# Nuclear Chemistry

## EXERCISE (JEE ADVANCED)

### Section A (Only one Correct)

### Nuclear Stability

1. Let $F_{pp}$, $F_{pn}$ and $F_{nn}$ denote the magnitudes of net force by a proton on a proton, by a proton on a neutron and by a neutron on a neutron, respectively. Neglect gravitational force when the separation is 1 fm,
(a) $F_{pp} > F_{pn} = F_{nn}$ (b) $F_{pp} = F_{pn} = F_{nn}$
(c) $F_{pp} > F_{pn} > F_{nn}$ (d) $F_{pp} < F_{pn} = F_{nn}$

2. Which of the following element has a relatively more stable nucleus?
(a) U (b) H
(c) Fe (d) Ra

3. Which of the following is least likely to be stable?
(a) $Ca^{40}$ (b) $Al^{30}$
(c) $Sn^{119}$ (d) $Mn^{55}$

4. The energy required to separate the nucleons from a nucleus is called
(a) nuclear energy (b) ionization energy
(c) binding energy (d) lattice energy

5. The value of packing fraction of carbon-12 is
(a) positive (b) negative
(c) zero (d) infinite

6. Of the following isotopes, which one is likely to be the most stable?
(a) $Zn^{63}$ (b) $Zn^{67}$
(c) $Zn^{71}$ (d) $Zn^{64}$

7. Which stable nucleus has half of the radius of nucleus of $Fe^{56}$?
(a) $Cd^{112}$ (b) $N^{14}$
(c) $Si^{28}$ (d) $Li^{7}$

8. The largest stable nucleus is
(a) $U^{238}$ (b) $Bi^{209}$
(c) $U^{235}$ (d) $Pb^{206}$

9. Among the isotopes of all the elements (radioactive as well as non-radioactive), the *n*/*p* ratio is minimum for
(a) ${}_{1}H^{1}$ (b) ${}_{2}He^{4}$
(c) ${}_{83}Bi^{209}$ (d) ${}_{26}Fe^{56}$

10. Among the isotopes of all the elements (radioactive as well as non-radioactive), the *n*/*p* ratio is maximum for
(a) ${}_{1}H^{1}$ (b) ${}_{1}H^{3}$
(c) ${}_{83}Bi^{209}$ (d) ${}_{2}He^{4}$

11. Among the isotopes of all the elements (only non-radioactive), the *n*/*p* ratio is maximum for
(a) ${}_{1}H^{1}$ (b) ${}_{1}H^{3}$
(c) ${}_{83}Bi^{209}$ (d) ${}_{2}He^{4}$

12. Among the isotopes of all the elements (radioactive as well as non-radioactive), the packing fraction is maximum for
(a) ${}_{1}H^{1}$ (b) ${}_{6}C^{12}$
(c) ${}_{1}H^{3}$ (d) ${}_{26}Fe^{56}$

## Radioactivity

13. α-particle is considered identical to He-nucleus because
    (a) He-nucleus is present in the nuclei of all α-emitters.
    (b) He-nucleus has two protons and two neutrons.
    (c) any sealed vessel containing some α-emitter is found to contain He gas after some time.
    (d) He-nucleus is the most stable nucleus.

14. Emission of one α-particle from a nucleus results in the loss of two protons and two neutrons from the nucleus. These four particles (two protons and two neutrons) comes out from the nucleus
    (a) all at a time.
    (b) one by one, both protons followed by both neutrons.
    (c) one by one, both neutrons followed by both protons.
    (d) one by one, protons and neutrons alternatively.

15. In β-decay, an electron comes out from an atom. The electron comes out due to nuclear change, not from the orbit of the atom. It may be explained by the fact that on β-decay,
    (a) the atomic number decreases by one unit.
    (b) the mass number remains unchanged.
    (c) the atomic species gets changed.
    (d) the atomic species remains unchanged.

16. The isotope of $_{90}Ra^{231}$ can be converted to $_{90}Th^{227}$ by
    (a) One alpha emission.
    (b) Four beta emission.
    (c) Two alpha and two beta emissions.
    (d) One alpha and two beta emissions.

17. When a nucleus reverts from an excited state to the ground state, the energy difference between the two states is emitted as
    (a) α-particle (b) β-particle
    (c) γ-rays (d) Neutrino

18. A free neutron decays to a proton but a free proton does not decay to a neutron. This is because
    (a) neutron is a composite particle made of a proton and an electron, whereas proton is a fundamental particle.
    (b) neutron is a uncharged particle, whereas proton is a charged particle.
    (c) neutron has larger rest mass than a proton.
    (d) weak forces can operate in a neutron but not in a proton.

19. During the transformation of $_aX^b$ to $_cY^d$, the number of β-particles emitted are
    (a) $\frac{b-d}{4}$
    (b) $(a-c)+\frac{b-d}{2}$
    (c) $(c-a)+\frac{b-d}{2}$
    (d) $(b-d)+2(c-a)$

20. The only stable nucleus of fluorine is $F^{19}$. What type of decay is expected from $F^{18}$?
    (a) α-decay (b) β-decay
    (c) Positron decay (d) K-capture

21. A positron is emitted from $_{11}Na^{23}$. The ratio of the mass number and atomic number in the resulting nuclide is
    (a) $\frac{22}{10}$ (b) $\frac{22}{11}$
    (c) $\frac{23}{10}$ (d) $\frac{23}{12}$

22. $Na^{23}$ is the more stable isotope of Na. Find out the process by which $Na^{24}$ can undergo radioactive decay.
    (a) β-emission (b) α-emission
    (c) $\beta^+$ emission (d) Electron capture

23. $C^{14}$ is a beta active nucleus. A sample of $C^{14}H_4$ gas kept in a closed vessel shows increase in pressure with time. This is due to
    (a) the formation of $N^{14}H_3$ and $H_2$.
    (b) the formation of $B^{11}H_3$ and $H_2$.
    (c) the formation of $C^{14}{}_2H_4$ and $H_2$.
    (d) the formation of $C^{12}H_3$, $N^{14}H_2$ and $H_2$.

24. Which of the following is not emitted by radioactive substance?
    (a) α-rays (b) β-rays
    (c) Positron (d) Proton

EXERCISE

25. Consider the beta decay, $Au^{198} \rightarrow Hg^{198*}$, where $Hg^{198*}$ represents a mercury nucleus in an excited state at energy 1.063 MeV above the ground state. What can be the maximum kinetic energy of the electron emitted? The atomic masses of $Au^{198}$ and $Hg^{198}$ are 197.968 u and 197.966 u, respectively. (1 u = 931.5 MeV)

(a) 0.8 MeV (b) 1.863 MeV
(c) 1.063 MeV (d) 1.0 MeV

## Rate Law

26. If the amount of radioactive substance increases three times, the number of disintegration per unit time will be

(a) doubled (b) one-third
(c) triple (d) uncharged

27. The radioactive decay follows

(a) zero order (b) first order
(c) second order (d) third order

28. If a radioactive element is placed in an evacuated container, its rate of disintegration

(a) will be increased.
(b) will be decreased.
(c) will change very slightly.
(d) will remain unchanged.

29. A radioactive element has atomic mass 90 amu and a half-life of 28 years. The number of disintegrations per second per gram is

(a) $5.24\times10^{10}$ (b) $5.24\times10^{8}$
(c) $5.24\times10^{-10}$ (d) $5.24\times10^{12}$

30. Four vessels 1, 2, 3 and 4 respectively contains, 10 g-atom ($t_{1/2}$ = 10 hr), 1 g-atom ($t_{1/2}$ = 5 hr), 5 g-atom ($t_{1/2}$ = 2 hr) and 2 g-atom ($t_{1/2}$ = 1 hr) of different radioactive nuclides. In the beginning, the maximum radioactivity would be exhibited by the vessel

(a) 4 (b) 3
(c) 2 (d) 1

31. If 8 g of a radioactive isotope has a half-life of 10 hr. The half-life of 2 g of the same substance is

(a) 2.5 hr (b) 5 hr
(c) 10 hr (d) 40 hr

32. If $N_o$ is the initial number of nuclei, the number of nuclei remaining undecayed at the end of $n$th half-life is

(a) $2^n \cdot N_o$ (b) $2^{-n} \cdot N_o$
(c) $n^2 \cdot N_o$ (d) $n^{-2} \cdot N_o$

33. A radioactive isotope having a half-life of 3 days was received after 12 days. It was found that there were 3 g of the isotope in the container. The initial mass of the isotopes when packed was

(a) 12 g (b) 24 g
(c) 36 g (d) 48 g

34. 1 g-atom of an α-emitting nuclide $_ZX^A$ ($t_{1/2}$ = 10 hr) was placed in a sealed container. The time required for the accumulation of $4.5\times10^{23}$ helium atoms in the container is ($N_A = 6\times10^{23}$)

(a) 4.52 hr (b) 9.40 hr
(c) 10.0 hr (d) 20.0 hr

35. The radioactivity of a sample is $R_1$ at a time $T_1$ and $R_2$ at a time $T_2$. If the half-life of the specimen is $T$, then the number of atoms that have disintegrated in the time $(T_2 - T_1)$ is equal to

(a) $(R_1T_1 - R_2T_2)$ (b) $(R_1 - R_2)$
(c) $\frac{R_1 - R_2}{T}$ (d) $\frac{(R_1 - R_2)\cdot T}{0.693}$

36. A freshly prepared radioactive source of half-life 2 hr emits radiations of intensity which is 64 times the permissible safe level. The minimum time after which it would be possible to work safely with this source is

(a) 6 hr (b) 12 hr
(c) 24 hr (d) 128 hr

37. Two isotopes 'P' and 'Q' of atomic masses 10 and 20, respectively, are mixed in equal amount by mass. After 20 days, their mass ratio is found to be 1 : 4. Isotope 'P' has a half-life of 10 days. The half-life of isotope 'Q' is

(a) Zero (b) 5 day
(c) 20 day (d) Infinite

38. A radioactive sample has an initial activity of 28 dpm. Half an hour later, the activity is 14 dpm. How many atoms of the radioactive nuclide were there originally? (ln 2 = 0.7)

(a) 1200 (b) 200
(c) 600 (d) 300

**39.** Tritium has a half-life of 12.26 years. A 5.0 ml sample of triturated water has an activity of 2.4 $\times 10^9$ cpm. How many years will it take for the activity to fall to $3.0 \times 10^8$ cpm?

(a) 6.13 (b) 24.52
(c) 36.78 (d) 49.04

**40.** The half-life of $Tc^{99}$ is 6.0 hr. The delivery of a sample of $Tc^{99}$ from the reactor to the nuclear medicine lab of a certain hospital takes 3.0 hr. What is the minimum amount of $Tc^{99}$ that must be shipped in order for the lab to receive 10.0 mg?

(a) 20.0 mg (b) 15.0 mg
(c) 14.1 mg (d) 12.5 mg

**41.** $K^{40}$ consists 0.012% of the potassium in nature. The human body comprises 0.35% potassium by weight. Calculate the total radioactivity resulting from $K^{40}$ decay in a 75 kg human. Half-life from $K^{40}$ is $1.3 \times 10^9$ years.

(a) 8017.64 dps (b) 4008.82 dps
(c) 16035.28 dps (d) 345.24 dpm

**42.** The activity of a certain preparation decreases 2.5 times after 7.0 days. Find its half-life. (log 2 = 0.3)

(a) 10.58 days (b) 2.65 days
(c) 5.25 days (d) 4.2 days

**43.** A mixture of $Pu^{239}$ and $Pu^{240}$ has a specific activity of $6 \times 10^9$ dps per gram sample. The half-lives of the isotopes are $2.40 \times 10^4$ years and $7.17 \times 10^3$ years, respectively. The mass percent of $Pu^{239}$ in the sample is

(a) 31.12% (b) 68.88%
(c) 10.37% (d) 89.63%

**44.** An ore of uranium is found to contain ${}^{238}_{92}U$ and ${}^{206}_{82}Pb$ in the mass ratio of 1 : 0.1. The half-life period of ${}^{238}_{92}U$ is $4.5 \times 10^9$ years. Age of the ore is (log 2 = 0.3, $\log \frac{114.9}{103} = 0.048$)

(a) $7.2 \times 10^8$ years (b) $7.2 \times 10^7$ years
(c) $7.2 \times 10^9$ years (d) $2.16 \times 10^9$ years

**45.** In nature a decay chain starts with $Th^{232}$ and finally terminates at $Pb^{208}$. A thorium ore sample was found to contain $6.72 \times 10^{-5}$ ml of He (at 273K and 1 atm) and $4.64 \times 10^{-7}$ g of $Th^{232}$. Find the age of the sample assuming that source of He to be only due to decay of $Th^{232}$. Also assume complete retention of He within the ore ($t_{1/2}$ of $Th^{232} = 1.38 \times 10^{10}$ years, log 2 = 0.3).

(a) $2.3 \times 10^{10}$ years
(b) $2.3 \times 10^9$ years
(c) $4.6 \times 10^9$ years
(d) $9.2 \times 10^9$ years

## Parallel and Sequential Decay

**46.** The radioactive series to which ${}_{88}Ra^{224}$ belongs is

(a) Actinium series (b) Thorium series
(c) Uranium series (d) Neptunium series

**47.** Actinium series starts with A and ends at Z. Here, A and Z are

(a) ${}_{90}Th^{232}$, ${}_{82}Pb^{206}$ (b) ${}_{90}Th^{235}$, ${}_{82}Pb^{207}$
(c) ${}_{92}U^{235}$, ${}_{82}Pb^{207}$ (d) ${}_{90}Ac^{227}$, ${}_{82}Bi^{209}$

**48.** Bismuth is the end product of radioactive disintegration series known as

(a) $4n$ (b) $4n + 1$
(c) $4n + 2$ (d) $4n + 3$

**49.** The end product of ($4n$ +2) disintegration series is

(a) ${}_{82}Pb^{204}$ (b) ${}_{82}Pb^{208}$
(c) ${}_{82}Pb^{209}$ (d) ${}_{82}Pb^{206}$

**50.** Consider the following process of decay,

$${}_{92}U^{234} \rightarrow {}_{90}Th^{230} + {}_{2}He^{4};\ t_{1/2} = 2{,}50{,}000 \text{ years}$$

$${}_{90}Th^{230} \rightarrow {}_{88}Ra^{226} + {}_{2}He^{4};\ t_{1/2} = 80{,}000 \text{ years}$$

$${}_{88}Ra^{226} \rightarrow {}_{86}Rn^{222} + {}_{2}He^{4};\ t_{1/2} = 1600 \text{ years}$$

After the above process has occurred for a long time, a state is reached where for every two thorium atoms formed from ${}_{92}U^{234}$, one decomposes to form ${}_{88}Ra^{226}$ and for every two ${}_{88}Ra^{226}$ formed, one decomposes. The ratio of ${}_{90}Th^{230}$ to ${}_{88}Ra^{226}$ will be

(a) 25/8 (b) 100/1
(c) 50/1 (d) 25/4

EXERCISE

**51.** $Ac^{227}$ has a half-life of 22 years with respect to radioactive decay. The decay follows two parallel paths, one leading to $Th^{227}$ and the other leading to $Fr^{223}$. The percentage yields of these two daughter nuclides are 2 % and 98 %, respectively. Which of the following is the only incorrect information related to decay?

(a) The decay constant with respect to radioactive decay of $Ac^{227}$ is $3.15 \times 10^{-2}\ yr^{-1}$.

(b) The decay constant for the decay of $Ac^{227}$ into $Th^{227}$ is $6.3 \times 10^{-4}\ yr^{-1}$.

(c) The decay constant for the decay of $Ac^{227}$ into $Fr^{223}$ is $3.087 \times 10^{-2}\ yr^{-1}$.

(d) The mass ratio of $Th^{227}$ and $Fr^{223}$ will be 1 : 49, after 22 years.

**52.** A radioactive isotope is being produced at a constant rate $dN/dt = R$ in an experiment. The isotope has a half-life $t_{1/2}$. After a time $t >> t_{1/2}$, the number of active nuclei will become constant. The value of this constant is

(a) $R$  (b) 1
(c) $R/\lambda$  (d) $\lambda/R$

**53.** The $t_{1/2}$ of $Pb^{212}$ is 8.0 hr. It undergoes decay to its daughter (unstable) element $Bi^{212}$ of half-life 60.0 minute. The time at which daughter element will have maximum activity is

(a) 205.7 min  (b) 3.429 min
(c) 60.0 min  (c) 67.5 min

**54.** A radionuclide 'A' decays simultaneously into 'B' and 'C' by α- and β-emission respectively. The half-lives for the decay are 20 and 60 min, respectively. The time in which 87.5% of 'A' will decay is

(a) 15 min  (b) 30 min
(c) 45 min  (d) 60 min

**55.** For the parallel radioactive decay,

$$A \xrightarrow{\lambda_1 = 0.05\ min^{-1}} B + 40\ MeV$$

$$A \xrightarrow{\lambda_2 = 0.15\ min^{-1}} C + 80\ MeV$$

the average energy released per atom decay of 'A' is

(a) 60 MeV  (b) 70 MeV
(c) 66.67 MeV  (d) 75 MeV

## Nuclear Reactions

**56.** The number of neutrons accompanying the formation of $_{54}Xe^{139}$ and $_{38}Sr^{94}$ from the absorption of slow neutron by $_{92}U^{235}$ by nuclear fission is

(a) 0  (b) 2
(c) 1  (d) 3

**57.** Complete the following nuclear reaction: $_{25}Mn^{55}$ $(n, \gamma)$ ______

(a) $_{25}Mn^{55}$  (b) $_{24}Cr^{56}$
(c) $_{24}Cr^{54}$  (d) $_{25}Mn^{56}$

**58.** In the reaction, $_4Be^9 + X \rightarrow {}_5B^{10} + \gamma$, X is

(a) proton  (b) deuteron
(c) α-particle  (d) neutron

**59.** Which one of the following particles is used to bombard $_{13}Al^{27}$ to give $_{15}P^{30}$ and a neutron?

(a) $_1H^2$  (b) γ
(c) α  (d) β

**60.** Which of the following nuclear reaction occurs in the nature of the formation of tritium?

(a) $_3Li^6 + {}_0n^1 \rightarrow {}_2He^4 + {}_1H^3$

(b) $_5B^{10} + {}_0n^1 \rightarrow 2\,{}_2He^4 + {}_1H^3$

(c) $_7N^{14} + {}_0n^1 \rightarrow {}_6C^{12} + {}_1H^3$

(d) $_4Be^9 + {}_1D^2 \rightarrow 2\,{}_2He^4 + {}_1H^3$

**61.** Positron annihilated during collision with electron produces

(a) γ-rays  (b) neutron
(c) proton  (d) α-particle

**62.** The mass defect of nuclear reaction, $_4Be^{10} \rightarrow {}_5B^{10} + e^-$ is

(a) $\Delta m$ = Atomic mass of $_4Be^{10}$ – Atomic mass of $_5B^{10}$

(b) $\Delta m$ = Atomic mass of $_4Be^{10}$ – Atomic mass of $_5B^{10}$ + mass of one electron

(c) $\Delta m$ = Atomic mass of $_4Be^{10}$ – Atomic mass of $_5B^{10}$ – mass of one electron

(d) $\Delta m$ = Atomic mass of $_4Be^{10}$ – Atomic mass of $_5B^{10}$ + Mass of two electrons

**63.** The mass defect of the nuclear reaction: $_5B^8 \rightarrow {}_4Be^8 + e^+$ is

(a) $\Delta m$ = Atomic mass of $_5B^8$ – Atomic mass of $_4Be^8$

(b) $\Delta m$ = Atomic mass of $_5B^8$ – Atomic mass of $_4Be^8$ + mass of one electron

(c) $\Delta m$ = Atomic mass of $_5B^8$ – Atomic mass of $_4Be^8$ – mass of one electron

(d) $\Delta m$ = Atomic mass of $_5B^8$ – Atomic mass of $_4Be^8$ + mass of two electrons

64. The transformations of protons into neutrons and vice versa are ___ -order reactions.

(a) zero (b) first
(c) second (d) half

65. Bombardment of aluminium be α-particle leads to its artificial disintegration in two way (i) and (ii) as shown. Products X, Y and Z, respectively, are

$^{27}_{13}Al \xrightarrow{(i)} {}^{30}_{14}Si + X$

$^{27}_{13}Al \xrightarrow{(ii)} {}^{30}_{15}P + Y$

$^{30}_{15}P \longrightarrow {}^{30}_{14}Si + Z$

(a) proton, neutron, positron
(b) neutron, positron, proton
(c) proton, positron, neutron
(d) positron, proton, neutron

## Section B (One or More than one Correct)

1. Decrease in atomic number is observed during

(a) alpha decay (b) beta decay
(c) positron decay (d) electron capture

2. Consider a sample of a pure beta-active material and select the correct information(s).

(a) Beta particles emitted may have different energy.
(b) The beta particles originally exist inside the nucleus and are ejected at the time of beta decay.
(c) The antineutrino emitted in beta decay has zero mass and hence zero momentum.
(d) The active nucleus changes to one of its isobars after the beta decay.

3. Assuming that only particles emitted during natural radioactive decay are alpha and beta particles, which of the following atoms could not possibly result from the natural decay of $_{92}U^{235}$ atoms?

(a) $_{90}Th^{231}$ (b) $_{89}Ac^{227}$
(c) $_{89}Ac^{235}$ (d) $_{82}Pb^{207}$

4. The activity of radioactive isotopes do not change with change in

(a) temperature
(b) pressure
(c) chemical environment
(d) amount

5. Half-life period of a radio-isotope is independent of

(a) temperature
(b) state of chemical combination
(c) amount of radio-isotope
(d) pressure

6. $_{84}Rn^{219}$ is a member of actinium series. Another member of the same series is

(a) $_{92}U^{235}$ (b) $_{90}Th^{232}$
(c) $_{89}Rn^{225}$ (d) $_{15}P^{35}$

7. Which of the following is not a fissionable material?

(a) $U^{238}$ (b) $U^{235}$
(c) $U^{233}$ (d) $Pu^{239}$

8. Which one of the following notations shows the product correctly?

(a) $_{96}Cm^{242}$ (α, 2n) $_{97}Bk^{243}$
(b) $_5B^{10}$ (α, n) $_7N^{13}$
(c) $_7N^{14}$ (n, p) $_6C^{14}$
(d) $_{14}Si^{28}$ (d, n) $_{15}P^{29}$

9. Which of the following is (n, p) type reaction?

(a) $_5C^{13} + {}_1H^1 \rightarrow {}_6C^{14}$
(b) $_7N^{14} + {}_1H^1 \rightarrow {}_8O^{15}$
(c) $_{13}Al^{27} + {}_0n^1 \rightarrow {}_{12}Mg^{27} + {}_1H^1$
(d) $_{92}U^{235} + {}_0n^1 \rightarrow {}_{54}Xe^{140} + {}_{88}Sr^{94}$

10. Which of the following is/are correct statements?

(a) All the nuclei having *n*/*p* ratio in between 1.0 and 1.52 have stable nucleus.
(b) All the nuclei having binding energy per nucleon greater than that of $Bi^{209}$ have unstable nucleus.
(c) All the nuclei having even number of protons as well as neutrons have stable nucleus.
(d) There are very few stable nuclei having odd number of protons as well as neutrons.

11. Nuclear reaction(s) accompanied with emission of neutron(s) are
(a) ${}^{27}_{13}Al + {}^{4}_{2}He \rightarrow {}^{30}_{15}P$
(b) ${}^{12}_{6}C + {}^{1}_{1}H \rightarrow {}^{13}_{7}N$
(c) ${}^{30}_{15}P \rightarrow {}^{30}_{14}Si$
(d) ${}^{241}_{96}Am + {}^{4}_{2}He \rightarrow {}^{244}_{97}Bk$

12. In the nuclear transmutation, ${}^{9}_{4}Be + X \rightarrow {}^{8}_{4}Be + Y$, (X, Y) is (are)
(a) ($\gamma$, $n$)
(b) ($p$, $D$)
(c) ($n$, $D$)
(d) ($\gamma$, $p$)

## Section C (Comprehensions)

### Comprehension I

**The isotopic mass of any atom is always less than its theoretical mass. The difference between the expected (theoretical) mass (sum of the masses of protons, neutrons and electrons present in the atom) and the actual mass of an isotope is called mass defect for that isotope of the element. It is denoted by $\Delta m$. The reason behind the mass defect is conversion of some mass into energy which is released in the formation of that atom in the combination or binding the fundamental particles together according to Einstein's equation $E = \Delta m \cdot C^2$. This energy released in binding the particles together is called binding energy. If $\Delta$m is 1 amu, then $E = 931.5$ MeV.**

1. The binding energy per nucleon in $O^{16}$ nucleus is (mass of proton = 1.0072 u, mass of neutron = 1.0086 u)
(a) 14.72 MeV
(b) 7.36 MeV
(c) 117.7 MeV
(d) 4.8 MeV

2. What is the energy required for separation of an $O^{16}$ nucleus into four identical particles? (Nuclear masses: $O^{16} = 15.9944$ u, $He^4 = 4.0026$ u)
(a) 14.904 MeV
(b) 0.9315 MeV
(c) 1.863 MeV
(d) 698.15 MeV

3. Find the energy required for separation of a $Ne^{20}$ nucleus into two alpha-particles and a $C^{12}$ nucleus if it is known that the binding energies per on nucleon in $Ne^{20}$, $He^4$ and $C^{12}$ nuclei are equal to 8.03, 7.07 and 7.68 MeV, respectively.
(a) 6.72 MeV
(b) 40.16 MeV
(c) 11.88 MeV
(d) 5.8 MeV

### Comprehension II

**In any nuclear reaction, the total mass of products is found to be less than the total mass of reactants. The difference in mass is due to the conversion of some mass into energy and hence, energy is produced in the nuclear reactions. ($E = 931.5$ MeV/amu.)**

4. The nuclide $Sc^{50}$, mass 49.9516 amu is neutron rich. It decays to form $Ti^{50}$, mass 49.94479 amu. If the emitted $\beta$-particle has a kinetic energy of 0.80 MeV, then what is the kinetic energy of the antineutrino emitted simultaneously?
(a) 2.87 MeV
(b) 5.54 MeV
(c) 9.36 MeV
(d) 6.34 MeV

5. The alpha particles emitted by radium have energies of 4.795 and 4.611 MeV. What is the wavelength of the gamma rays accompanying the decay? The difference in energies of alpha particle emitted out is equal to the energy of gamma rays.
(a) 6.74 pm
(b) 9.87 pm
(c) 3.37 pm
(d) 4.58 pm

6. $Th^{228}$ emits an alpha particle to reduce to $Ra^{224}$. Calculate the kinetic energy of the alpha particle emitted in the following decay: $Th^{228} \rightarrow Ra^{224*} + \alpha$; $Ra^{224*} \rightarrow Ra^{224} + \gamma$ (217 KeV) Atomic masses of $Th^{228}$, $Ra^{224}$ and $He^4$ are 228.028726 u, 224.020196 u and 4.00260 u, respectively.
(a) 5.21 MeV
(b) 3.24 MeV
(c) 4.46 MeV
(d) 5.31 MeV

## Comprehension III

**Let a radionuclide 'A' decays into another nuclide 'B' and 'B' is also radioactive decaying into 'C'.**

$$A \xrightarrow{\lambda_A} B \xrightarrow{\lambda_B} C$$

**The number of radio-nuclides present at time, *t*:**

$$N_A = N_{A_0} \cdot e^{-\lambda_A t}$$

$$N_B = \frac{N_{A_0} \cdot \lambda_A}{\lambda_B - \lambda_A}\left(e^{-\lambda_A t} - e^{-\lambda_B t}\right)$$

$$N_C = N_{A_0} - (N_A + N_B)$$

**7.** $Co^{57}$ decays to $Fe^{57}$ by $\beta^+$ emission. The resulting $Fe^{57}$ is in its excited state and comes to the ground state by emitting $\gamma$-rays. The half-life of $\beta^+$ decay is 270 days and that of the $\gamma$-emission is $10^{-8}$ s. A sample of $Co^{57}$ gives $5.0 \times 10^9$ gamma rays per second. How much time will elapse before the emission rate of gamma rays drops to $2.5 \times 10^9$ per second?

(a) 135 days
(b) 270 days
(c) $10^{-8}$ s
(d) $5 \times 10^{-9}$ s

**8.** Consider $Th^{228} \rightarrow Ra^{224} \rightarrow Rn^{220}$, where $t_{1/2}(Th^{228}) = 1.913$ years, $t_{1/2}(Ra^{224}) = 3.64$ days. Determine the N(Th)/N(Ra) ratio at transient equilibrium in which $t_{1/2}$(parent) > $t_{1/2}$(daughter) has been established.

(a) 191 : 1
(b) 380 : 1
(c) 95 : 1
(d) 5727 : 1

**9.** Consider $Ra^{224} \rightarrow Rn^{220} \rightarrow Po^{216}$, where $t_{1/2}(Ra^{224}) = 3.64$ days, $t_{1/2}(Rn^{220}) = 55$ s. Determine the N(Ra)/N(Rn) ratio at secular equilibrium in which $t_{1/2}$(parent) >> $t_{1/2}$(daughter) has been established.

(a) 5734 : 1
(b) 1 : 5734
(c) 11468 : 1
(d) 1 : 11468

## Comprehension IV

**The age of earth may be calculated by the determination of amounts of any radioactive substance and its decay product in any sample of rock. It is assumed that the decay product was not present at the time of origin of earth and it is also assumed that the rock is present from the time of origin of earth.**

**10.** A sample of pitch blend is found to contain 59.5% Uranium and 12.875% Lead. Of this lead, only 80% was $Pb^{206}$ isotope but uranium was almost $U^{238}$ isotope. If the disintegration constant of $U^{238}$ is $1.5 \times 10^{-10}$ $yr^{-1}$, then how old could be the pitch blend deposit? (ln 2 = 0.7, ln 3 = 1.1, ln 10 = 2.3)

(a) $1.33 \times 10^9$ years
(b) $1.33 \times 10^8$ years
(c) $4 \times 10^9$ years
(d) $4 \times 10^8$ years

**11.** Analysis of potassium and argon atoms in a moon rock sample by a mass spectrometer shows that the ratio of the number of (stable) $Ar^{40}$ atoms present to the number of (radioactive) $K^{40}$ atoms is 10.3 : 1. Assume that all the argon atoms were produced by the decay of potassium atoms with a half-life of $1.25 \times 10^9$ years. How old is the rock? (log 2 = 0.3, log 10.3 = 1.02, log 11.3 = 1.05)

(a) $4.375 \times 10^9$ years
(b) $1.25 \times 10^8$ years
(c) $4.375 \times 10^{10}$ years
(d) $1.25 \times 10^9$ years

**12.** A rock recovered from far underground is found to contain 0.86 mg of $U^{238}$, 0.15 mg of $Pb^{206}$ and 1.6 mg of $Ar^{40}$. How much $K^{40}$ will it likely contain? Half-lives of $U^{238}$ and $K^{40}$ are $4.47 \times 10^9$ years and $1.25 \times 10^9$ years, respectively.

(a) 1.7 mg
(b) 3.4 mg
(c) 5.1 mg
(d) 0.85 mg

EXERCISE

## Comprehension V

**Carbon-14 is used to determine the age of organic material. The procedure is based on the formation of $C^{14}$ by neutron capture in the upper atmosphere.**

$${}_7N^{14} + {}_0n^1 \rightarrow {}_6C^{14} + {}_1H^1$$

**$C^{14}$ is absorbed by living organism during photosynthesis. The $C^{14}$ content is constant in living organism. Once the plant or animal dies, the uptake of carbon dioxide by it ceases and the level of $C^{14}$ in the dead being falls due to beta decay which $C^{14}$ undergoes.**

$${}_6C^{14} \rightarrow {}_7N^{14} + \beta$$

**The half-life period of $C^{14}$ is 5770 years. The decay constant ($\lambda$) can be calculated by using the following formula $\lambda = 0.693/t_{1/2}$.**

**The comparison of the β-activity of the dead matter with that of the carbon still in the circulation enables measurement of the period of the isolation of the material from the living cycle. However, the method ceases to be accurate over periods longer than 30,000 years. The proportion of $C^{14}$ to $C^{12}$ in living matter is 1 : 1012.**

13. Which of the following option is correct?
    (a) In a living organism, the circulation of $C^{14}$ from atmosphere is high so that the carbon content is constant in organism.
    (b) Carbon dating can be used to find out the age of earth crust and rocks.
    (c) Radioactive absorption due to cosmic radiation is equal to the rate of radioactive decay and hence, the carbon content remains constant in living organism.
    (d) Carbon dating can be used to determine the concentration of $C^{14}$ in dead beings.

14. What should be the age of fossil for meaningful determination of its age?
    (a) 6 years
    (b) 6000 years
    (c) 60000 years
    (d) It can be used to calculate any age.

15. A nuclear explosion has taken place leading to increase in concentration of $C^{14}$ in nearby areas. $C^{14}$ concentration is $C_1$ in nearby areas and $C_2$ in areas far away. If the age of the fossil is determined to be $T_1$ and $T_2$ at the places respectively, then
    (a) the age of the fossil will increase at the place where explosion has taken and $T_1 - T_2 = \frac{1}{\lambda}\ln\frac{C_1}{C_2}$.
    (b) the age of the fossil will decrease at the place where explosion has taken and $T_1 - T_2 = \frac{1}{\lambda}\ln\frac{C_1}{C_2}$.
    (c) the age of fossil will be determined to be same.
    (d) $\frac{T_1}{T_2} = \frac{C_1}{C_2}$

## Comprehension VI

**A small amount of solution containing $Na^{24}$ radionuclide with activity 1260 dps was administrated into blood of patient in a hospital. After 5 hr, a sample of the blood drawn out from the patient showed an activity of 15 dpm per ml. $t_{1/2}$ for $Na^{24}$ radionuclide is 15 hr.**

16. The volume of blood in the patient is $[(1.26)^3 = 2]$
    (a) 2.0 L (b) 3.0 L
    (c) 4.0 L (d) 5.0 L

17. Activity of blood sample (in dpm per ml) drawn after a further time of 5 hr is
    (a) 9.45 (b) 630
    (c) 10.5 (d) 11.9

## Comprehension VII

**Study the following transition series.**

$$A \xrightarrow{-\alpha} B \xrightarrow{-\alpha} C \xrightarrow{-\beta} D \xrightarrow{-\beta} E \xrightarrow{-\alpha} F \xrightarrow{-\beta} G \xrightarrow{-\alpha} H$$

**18.** Which of the following is not an isotopic pair?

(a) A and D (b) B and E
(c) C and F (d) D and G

**19.** If the mass number of 'A' is 230, then what is the mass number of 'H'?

(a) 230 (b) 214
(c) 217 (d) 216

**20.** If the atomic number of 'H' is 88, then what is the atomic number of 'A'?

(a) 91 (b) 92
(c) 93 (d) 94

## Comprehension VIII

**Nuclei of a radioactive element A are being produced at a constant rate, $\alpha$. The element has a decay constant $\lambda$. At time, $t = 0$, there are $N_o$ nuclei of the element.**

**21.** The number of nuclei of 'A' at time t is

(a) $\frac{\alpha}{\lambda}(1-e^{-\lambda t})$

(b) $N_o \cdot e^{-\lambda t}$

(c) $\frac{1}{\lambda}\,[\alpha-(\alpha-\lambda N_o)e^{-\lambda t}]$

(d) $\frac{N_o \cdot \alpha}{\lambda}\left[1-\left(1-\frac{\lambda}{\alpha}\right)e^{-\lambda t}\right]$

**22.** If $\alpha = 2N_o\lambda$, the number of nuclei of 'A' after one half-life of 'A' is

(a) zero (b) $0.5N_o$
(c) $1.5N_o$ (d) $2.0N_o$

**23.** If $\alpha = 2N_o\lambda$, the limiting value of $N$ as $t \to \infty$ is

(a) zero (b) $0.5N_o$
(c) $1.5N_o$ (d) $2.0N_o$

## Comprehension IX

**A human body excretes certain material through sweating by law similar to radioactive decay. If $Tc^{99}$ is injected in some form in human body, the body excretes half of the amount in 24 hours. Simultaneously, $Tc^{99}$ undergoes radioactive decay with half-life of 8 hr. The activity of $Tc^{99}$, just after the injection, was 32 μCi.**

**24.** How much time will elapse before the activity of patient falls to 16 μCi?

(a) 3.8 hr
(b) 4.8 hr
(c) 6.0 hr
(d) 8.0 hr

**25.** What will be the activity after 48 hr of the overall excreted material till that time?

(a) 0.125 μCi
(b) 0.375 μCi
(c) 0.5 μCi
(d) 1.0 μCi

## Section D (Assertion – Reason)

The following questions consist of two statements. Mark the answer as follows.

(a) If both statements are CORRECT, and **Statement II** is the CORRECT explanation of **Statement I.**

(b) If both statements are CORRECT, and **Statement II** is NOT the CORRECT explanation of **Statement I.**

(c) If **Statement I** is CORRECT, but **Statement II** is INCORRECT.

(d) If **Statement I** is INCORRECT, but **Statement II** is CORRECT.

1. **Statement I:** As the number of protons increases, the number of neutrons increases relatively faster.

 **Statement II:** The stability of nucleus increases continuously as *n*/*p* ratio increases.

2. **Statement I:** The binding energy of nucleus increases with increase in atomic number.

 **Statement II:** Heavier elements have greater number of non-radioactive isotopes than radioactive isotopes.

3. **Statement I:** The isotopes of lighter elements are more likely to undergo nuclear fusion reactions while the isotopes of higher elements are more likely to undergo nuclear fission reactions.

 **Statement II:** Nuclear binding energy per nucleon first increases to a maximum and then decreases with increase in atomic number.

4. **Statement I:** As the atomic number increases, the *n*/*p* ratio required for stability of nucleus increases.

 **Statement II:** As the number of protons increases in the small volume of nucleus, more and more neutrons are required to compensate the repulsion between the protons.

5. **Statement I:** The radioisotope $C^{14}$ decays spontaneously by β-emission.

 **Statement II:** β-emission leads to conversion of a neutron to proton, resulting in the decrease in *n*/*p* ratio.

6. **Statement I:** Most of the heavy nuclei ($Z > 83$) decay through a number of steps producing $_{82}Pb^{206}$.

 **Statement II:** $_{82}Pb^{206}$ has both the number of protons and neutrons, magic numbers.

7. **Statement I:** If a nucleus may undergo positron decay as well as K-capture, the amount of energy released in positron decay is more than that in K-capture.

 **Statement II:** The loss in mass in K-capture is equal to the difference in atomic masses of parent nucleus and the daughter nucleus but in positron decay, additional mass loss equal to the mass of two electrons also occur.

8. **Statement I:** If the *n*/*p* ratio of a radioactive nucleus is greater than that of stable nucleus, the possible way of decay is beta decay.

 **Statement II:** Beta decay is energetically more favourable than all other kinds of decay.

9. **Statement I:** The activity of a sample of $U^{238}$ is equal to the activity of a sample of equal mass of $U^{238}F_6$.

 **Statement II:** The specific activity of a particular radionuclide is independent from all physical as well as chemical factors.

10. **Statement I:** The ratio of mean life and half-life of all the radionuclide is same.

 **Statement II:** The half-life of any radionuclide is 0.693 times its mean life.

11. **Statement I:** Nucleons are held together due to their fast exchange into one-another through mesons.

 **Statement II:** Mesons are massless and chargeless particles.

12. **Statement I:** The neutrons are better initiators of nuclear reactions, than the protons, deuterons or α-particles of the same energy.

 **Statement II:** Neutrons are uncharged particles and hence, they are not repelled by positively charged nucleus.

13. **Statement I:** Nuclide $^{30}_{13}Al$ is less stable than $^{40}_{20}Ca$.

 **Statement II:** Nuclides having odd numbers of protons and neutrons are generally unstable.

14. **Statement I:** The plot of atomic number (*y*-axis) versus number of neutrons (*x*-axis) for stable nuclei shows a curvature towards *x*-axis from the line of 45° slope as the atomic number is increased.

 **Statement II:** Proton-proton electrostatic repulsions begin to overcome attractive forces involving protons and neutrons in heavier nuclides.

15. **Statement I:** All the radio nuclides decay α, β as well as γ-rays.

 **Statement II:** Normally, a particular radionuclide either emits α or β-particle.

## Section E (Column Match)

1. Match the columns.

| Column I | Column II |
|---|---|
| (A) Isotopes | (P) Have same mass numbers. |
| (B) Isotones | (Q) Have same number of neutrons but different number of protons. |
| (C) Isobars | (R) Have same number of protons but different number of neutrons. |
| (D) Isodiapheres | (S) Have same number of excess neutrons than protons. |

2. Match the columns.

| Column I (Mode of decay) | Column II (Effect on *n*/*p* ratio) |
|---|---|
| (A) α-decay | (P) Increases |
| (B) β-decay | (Q) Decreases |
| (C) γ-decay | (R) Remains unchanged |
| (D) Positron decay | |
| (E) K-capture | |

3. Match the columns.

| Column I (Mode of decay) | Column II (Effect on atomic number and mass number) |
|---|---|
| (A) α-decay | (P) Atomic number increases by 1 unit. |
| (B) β-decay | (Q) Atomic number decreases by 2 unit. |
| (C) γ-decay | (R) Atomic number remains unchanged. |
| (D) Positron decay | (S) Mass number decreases by 4 units. |
| (E) K-capture | (T) Mass number remains unchanged. |

4. Match the columns.

| Column I (Isotope) | Column II (Characteristic) |
|---|---|
| (A) $^{40}_{20}Ca$ | (P) Unstable, alpha emitter |
| (B) $^{133}_{53}I$ | (Q) Unstable, beta emitter |
| (C) $^{121}_{53}I$ | (R) Unstable, positron emitter |
| (D) $^{232}_{90}Th$ | (S) Stable |

5. Match the columns.

| Column I (Parent nucleus of the radioactive series) | Column II (Number of α- and β-particles emitted out to get the end product of that series) |
|---|---|
| (A) $^{235}_{92}U$ | (P) 6α and 4β |
| (B) $^{238}_{92}U$ | (Q) 8α and 5β |
| (C) $^{241}_{94}Pu$ | (R) 8α and 6β |
| (D) $^{232}_{90}Th$ | (S) 7α and 4β |

## Section F (Subjective)

### Single-digit Integer Type

1. One of the hazards of nuclear explosion is the generation of $^{90}Sr$ and its subsequent incorporation in bones. This nuclide has a half-life of 2.81 years. Suppose 2.048 mg was absorbed by a new-born child, then how much $^{90}Sr$ (in microgram) will remain in his bones after 28.1 years?

2. A sample of sodium phosphate ($Na_3PO_4$) weighing 86.4 mg contains radioactive $P^{32}$. If 0.164% of the phosphorus atoms in the compound is $P^{32}$, activity of the sample (in order of $10^{11}$ dps) is ($P^{32}$ has a half-life of 14.0 days, $N_A = 6 \times 10^{23}$, ln 2 = 0.7)

3. A radioactive isotope $_ZA^m$ ($t_{1/2}$ = 10 days) decays to give $_{Z-6}B^{m-12}$ stable atom. If 'm' g of 'A' is taken and kept in a sealed tube and 5.6 × 'x' litre of He gas at 1 atm and 0°C is accumulated in 20 days, then the value of 'x' is

4. There is a steam of neutrons with a kinetic energy of 0.335 eV. If the half-life of neutron is 700 s, then what percent will decay before they travel a distance of 80 km. Mass of neutron is $1.675 \times 10^{-27}$ kg (ln 1.01 = 0.01, ln 2 = 0.7).

5. The isotopes $U^{238}$ and $U^{235}$ occur in nature in the atomic ratio 140 : 1. The half-life periods of $U^{238}$ and $U^{235}$ are $4.5 \times 10^9$ and $7.2 \times 10^8$ years, respectively. Assuming that at the time of earth formation, they were present in equal numbers, the age of earth (in order of $10^9$ years) is (log 2 = 0.3, log 140 = 2.1)

## Four-digit Integer Type

1. An experiment requires minimum beta activity produced at the rate of $5.0 \times 10^{15}$ beta particles per minute. The half-life period of ${}^{99}_{42}Mo$, which is a beta emitter is 69.3 hr. The minimum moles (in the order $10^{-7}$) of ${}^{99}_{42}Mo$ required to carry out the experiment in 6.93 hr is [$(1.07)^{10} = 2$, $\log 2 = 0.3$, $N_A = 6 \times 10^{23}$]

2. The nucleidic ratio of ${}^{3}_{1}H$ to ${}^{1}_{1}H$ in a sample of water is $8.0 \times 10^{-18}$:1. Tritium undergoes decay with a half-life period of 12.3 years. How many tritium atoms would 0.9 mg of such a sample contain 36.9 years after the original sample is collected? ($N_A = 6 \times 10^{23}$)

3. To investigate the beta-decay of $Mg^{23}$ radionuclide, a counter was activated at the moment $t = 0$. It registered $N_1$ beta-particles by a moment $t_1 = 2.0$ s, and by a moment $t_2 = 3t_1$ the number of registered beta-particles was $\frac{8}{3}$ times greater. The mean life-time of the given nuclei (in seconds) is ($\sqrt{\frac{23}{3}} = 2.768$, $\ln 0.884 = 0.125$)

4. A man weighing 80 kg was injected with 5 ml of water containing tritium giving $9 \times 10^9$ counts per minute. After some time, the titrated water had equilibrium with the body water of the patient. A 1.0 ml sample of plasma water then showed an activity of $2.25 \times 10^5$ counts per minute. The mass percent of water in the body is

5. ${}^{64}Cu$ (half-life = 12.8 hr) decays by $\beta^-$ emission (32%), $\beta^+$ emission (19%) and electron capture (49%). The partial half-life (in hours) for $\beta^-$ emission is

6. The mean lives of a radioactive substance are 1620 years and 405 years for α-emission and β-emission, respectively. The time (in years) during which three fourth of a sample will decay if it is decaying both by α and β-emission simultaneously is

7. $Y^{90}$ has a half-life of 64 hr and $Sr^{90}$, 32 years. $Sr^{90}$ decays to $Y^{90}$ by β-emission. What will be the amount (in mg) of $Y^{90}$ in equilibrium with 730 g of $Sr^{90}$?

8. The binding energy per nucleon for the nuclei of mass numbers 240 and 120 are 7.2 and 8.1 MeV, respectively. If a nucleus of mass number 240 breaks into two exactly identical nuclei, the amount of energy released (in MeV) is

9. It is proposed to use the nuclear fusion reaction $H^2 + H^2 \rightarrow He^4$ in a nuclear reactor of 200 MW rating. If the energy from the above reaction is used with 25% efficiency in the reactor, then how many moles of deuterium atoms will be needed per day? The masses of $H^2$ and $He^4$ are 2.0021 and 4.0026 amu, respectively (1 amu = $1.5 \times 10^{-10}$ J).

10. A sample of ${}^{131}_{53}I$, as iodide ion, was administered to a patient in a carrier consisting of $10\sqrt{2}$ mg of iodide ion. After 4.0 days, 10% of the initial radioactivity was detected in the thyroid gland of the patient. What mass (in mg) of the iodide ion had migrated to the thyroid gland? The half-life of ${}^{131}_{53}I$ is 8.0 days.

## Answer Keys

### Section A (Only one Correct)

**Nuclear Stability**

1. (d) 2. (c) 3. (b) 4. (c) 5. (c) 6. (d) 7. (d) 8. (b) 9. (a) 10. (b)
11. (c) 12. (a)

**Radioactivity**

13. (c) 14. (a) 15. (c) 16. (d) 17. (c) 18. (c) 19. (c) 20. (c) 21. (c) 22. (a)
23. (a) 24. (d) 25. (a)

**Rate Law**

26. (c) 27. (b) 28. (d) 29. (d) 30. (b) 31. (c) 32. (b) 33. (d) 34. (d) 35. (d)
36. (b) 37. (d) 38. (a) 39. (c) 40. (c) 41. (a) 42. (c) 43. (a) 44. (a) 45. (c)

**Parallel and Sequential Decay**

46. (b) 47. (c) 48. (b) 49. (d) 50. (b) 51. (d) 52. (c) 53. (a) 54. (c) 55. (b)

**Nuclear Reactions**

56. (d) 57. (d) 58. (a) 59. (c) 60. (c) 61. (a) 62. (a) 63. (d) 64. (b) 65. (a)

## Section B (One or More than one Correct)

1. (a), (c), (d) 2. (a), (d) 3. (c) 4. (a), (b), (c)
5. (a), (b), (c), (d) 6. (a) 7. (a) 8. (b), (c), (d)
9. (c) 10. (d) 11. (a) 12. (a), (b)

## Section C

**Comprehension I**

1. (b) 2. (a) 3. (c)

**Comprehension II**

4. (b) 5. (a) 6. (a)

**Comprehension III**

7. (b) 8. (a) 9. (a)

**Comprehension IV**

10. (a) 11. (a) 12. (a)

**Comprehension V**

13. (c) 14. (a) 15. (a)

**Comprehension VI**

16. (c) 17. (d)

**Comprehension VII**

18. (a) 19. (b) 20. (c)

**Comprehension VIII**

21. (c) 22. (c) 23. (d)

**Comprehension IX**

24. (c) 25. (b)

## Section D (Assertion – Reason)

1. (c) 2. (c) 3. (a) 4. (a) 5. (a) 6. (d) 7. (a) 8. (c) 9. (d) 10. (a)
11. (c) 12. (a) 13. (b) 14. (a) 15. (d)

## Section E (Column Match)

1. A → R; B → Q; C → P; D → S
2. A → P; B → Q; C → R; D → P; E → P
3. A → Q, S; B → P, T; C → R, T; D → T; E → T
4. A → S; B → Q; C → R; D → P
5. A → S; B → R; C → Q; D → P

## Section F (Subjective)

**Single-digit Integer Type**

1. (2) 2. (3) 3. (9) 4. (1) 5. (6)

**Four-digit Integer Type**

1. (0535) 2. (0060) 3. (0016) 4. (0050) 5. (0040)
6. (0449) 7. (0167) 8. (0216) 9. (0960) 10. (0002)

# HINTS AND EXPLANATIONS

## EXERCISE (JEE ADVANCED)

### Section A (Only one Correct)

### Nuclear Stability

1. Nuclear forces are same in between any two nucleon and it is attractive at 1 fm but repulsive forces are also there between protons
2. Informative (B.E./nucleon is maximum for Fe)
3. For lighter nuclei, $\frac{n}{p} > 1$ may make the nucleus unstable
4. Theory based
5. Informative
6. Number of $n$ and $p$, both is even in ${}_{30}Zn^{64}$.
7. $r_N \propto A^{1/3}$

$\Rightarrow r_1 = \frac{1}{2} \times r_2$

$\Rightarrow (A_1)^{1/3} = \frac{1}{2} \times (56)^{1/3}$

$\Rightarrow A_1 = 7$

8. Informative
9. For ${}_1H^1$, $\frac{n}{p} = \frac{0}{1} = 0$
10. For ${}_1H^3$,

$$\frac{0.693}{32 \times 365 \times 24} \times \left(\frac{730}{90} \times N_A\right) = \frac{0.693}{64} \times \left(\frac{w}{90} \times N_A\right)$$

(Some isotopes having $\frac{n}{p}$ ratio greater that, ${}_1H^3$ are also know, like ${}_2He^8$)

11. Informative
12. Informative

### Radioactivity

13. Experimental reason behind considering $\alpha$-particle as He- nucleus.
14. Experimental fact
15. Experimental reason behind considering b-emission as nuclear charge.
16. Isotope formation
17. Reason of $g$ -emission
18. For an increase in mass, large amount of energy is needed and hence, it is non-spontaneous.
19. $b - N_\alpha \times 4 = \alpha$ and $a - 2 \times N_\alpha + 1 \times N_\beta = c$

$\therefore N_\alpha = \frac{b-\alpha}{4}$ and $N_\beta = (c-a) + 2 \times \frac{b-d}{4}$

20. $\left(\frac{n}{p}\right)_{F^{18}} < \left(\frac{n}{p}\right)_{F^{19}}$,

Hence, $F^{18}$ should undergo $a$-decay on $b^+$- decay on $k$-capture. Normally, $a$-decay and $k$-capture is not found in lighter nuclei.

21. ${}_{11}Na^{23} \xrightarrow{F^+} {}_{10}Ne^{23}$
22. $\left(\frac{n}{p}\right)_{Na^{24}} > \left(\frac{n}{p}\right)_{Na^{23}}$

Hence, $Na^{24}$ should undergo $\beta$-decay.

23. $c^{14} \xrightarrow{-\beta} N^{14}$
24. Informative
25. $\Delta m = m_{Au^{198}} - m_{Hg^{198}} = 197.968 - 197.966 = 0.002u$

$\therefore Q - \text{value} = 0.002 \times 931.5 = 1.8630$ MeV

But $Hg^{198}$ is having energy 1.063 MeV greater than $Hg^{198}$ and hence, maximum K.E. of emitted $b$–particle = 1.863 – 1.063 = 0.8MeV.

**Rate Law**

**26.** $r \propto N$

**27.** $r \propto N'$

**28.** Rate is independent from all external factors.

**29.** $r = \lambda N = \dfrac{0.693}{28\times 3.15\times 10^7}\times\left(\dfrac{1}{90}\times 6\times 10^{23}\right)$

$= 5.24\times 10^{12}$ dpspg

**30.** $r_1 = \dfrac{0.693}{10}\times(10\times N_A);$

$r_2 = \dfrac{0.693}{5}\times(1\times N_A)$

$r_3 = \dfrac{0.693}{2}\times(5\times N_A);$

$r_4 = \dfrac{0.693}{1}\times(2\times N_A)$

**31.** $t$ ½ is independent from amount.

**32.** $N = N_0\times\left(\dfrac{1}{2}\right)^n$

**33.** $w = w_o\times\left(\dfrac{1}{2}\right)^n$

$\Rightarrow 3g = w_o\times\left(\dfrac{1}{2}\right)^{12/3}$

$\Rightarrow w_o = 48$ gm

**34.** Moles of He formed $= \dfrac{4.5\times 10^{23}}{6\times 10^{23}} = 0.75 =$ Moles of decayed

$\therefore t = 2\times t_{1/2} = 20$ hrs

**35.** Number of atoms present at time $T_1$, $N_1 = \dfrac{R_1}{0.693/T}$

Number of atoms present at time $T_2$, $N_2 = \dfrac{R_2}{0.693/T}$

$\therefore$ Number of atoms decayed $= \dfrac{(R_1 - R_2)T}{0.693}$

**36.** Rate should decrease $\dfrac{1}{64} = \dfrac{1}{2^6}$ times and hence,

$t = 6\times t_{1/2} = 12$ hrs

**37.** $P: \dfrac{w}{10} \xrightarrow{10\text{ day}} \dfrac{w}{20} \xrightarrow{10\text{ day}} \dfrac{w}{40}$

$Q: \dfrac{w}{20} \xrightarrow{20\text{ day}} \dfrac{w}{20}$

(As final mass ratio is 1 : 4)

Hence, $Q$ is non-radioactive.

**38.** $t$ ½ = 30 min

Now, $r = \lambda N$

$\Rightarrow 28 = \dfrac{0.7}{30}\times N \Rightarrow N = 1200$

**39.** $\dfrac{r}{r_o} = \dfrac{3\times 10^8}{3\times 10^8} = \dfrac{1}{8} = \dfrac{1}{2^3}$

$\Rightarrow t = 3\times t_{1/2} = 3\times 12.26 = 36.78$ yrs

**40.** $w = w_o\times\left(\dfrac{1}{2}\right)^n$

$\Rightarrow 10\,\text{mg} = w_o\left(\dfrac{1}{2}\right)^{3/6}$

$\Rightarrow w_o = 14.14$ mg

**41.** $r = \lambda N = \dfrac{0.693}{1.3\times 10^9\times 365\times 24\times 3600}\times\left(\dfrac{75\times 10^3\times\dfrac{0.35}{100}\times\dfrac{0.012}{100}}{40}\times 6.022\times 10^{23}\right)$

$= 017.64$ dps

**42.** $t = \dfrac{t_{1/2}}{\log 2}\cdot\log\dfrac{r_o}{r} \Rightarrow > \dfrac{t_{1/2}}{\log 2}\cdot\log 2.5$

$\therefore t_{1/2} = 5.25$ days

**43.** Let the sample contains $x$ gm $Pu^{239}$.

Now, $r = (\lambda N)_{Pu^{239}} + (\lambda N)_{Pu^{240}}$

or $6\times 10^9 = \dfrac{0.693}{2.4\times 10^4\times 365\times 24\times 3600}\times\left(\dfrac{x}{239}\times 6.022\times 10^{23}\right)$

$+\dfrac{0.693}{7.17\times 10^3\times 365\times 24\times 3600}\times\left(\dfrac{1-x}{240}\times 6.022\times 10^{23}\right)$

$\Rightarrow x = 0.3112$

Hence, mass percent of $Pu^{239} = \dfrac{x}{1}\times 100 = 31.12\%$

**44.** $$U^{238} \longrightarrow Pb^{206}$$

| | | |
|---|---|---|
| Initial | $a$ | 0 |
| Present | $a-x$ | $x$ |

From question, $\dfrac{(a-x)\times 238}{x\times 206} = \dfrac{1}{0.1}$

$$\Rightarrow x = \frac{238}{2298}a$$

Now, age of ore, $t = \dfrac{t_{1/2}}{\log 2}\cdot \log\dfrac{a}{a-x}$

$$= \frac{4.5\times10^9}{0.3}\cdot\log\frac{1}{1-\dfrac{238}{2298}} = 7.2\times10^8 \text{ years}$$

**45.** $$Th^{232} \longrightarrow Pb^{208} + 6He^4$$

| | | |
|---|---|---|
| Initial | $a$ mole | 0 |
| Present | $(a-x)$ mole | $6x$ mole |

$$= \frac{4.64\times10^{-7}}{232} = 2\times10^{-9} = \frac{6.72\times10^{-5}}{22400} \Rightarrow 5\times10^{-10}$$

$\therefore a = 2.5\times10^{-9}$

Now, age of sample, $t = \dfrac{t_{1/2}}{\log 2}\cdot\log\dfrac{a}{a-x}$

$$= \frac{1.38\times10^{10}}{0.3}\log\frac{2.5\times10^{-9}}{2.5\times10^{-9}-2.5\times10^{-9}} = 4.6\times10^9 \text{ years}$$

## Parallel and Sequential Decay

**46.** 224 is an integer multiple of 4 and hence, $Ra^{224}$ belongs to 4n series, which is thorium series.

**47.** Informative

**48.** Informative

**49.** Informative

**50.** $r_{Th} = r_{Ra}$

$$\Rightarrow \frac{N_{Th}}{(t_{1/2})Th} = \frac{N_{Th}}{(t_{1/2})Ra}$$

$$\Rightarrow \frac{N_{Th}}{N_{Ra}} = \frac{80000}{1600}$$

**51.** (a) $\lambda_{AC^{227}} = \dfrac{0.693}{22} = 3.15\times10^{-2}\,\text{Yr}^{-1}$

(b) $l$ for the formation of $Th^{229}$

$$= \frac{2}{100}\times3.15\times10^{-2} = 6.3\times10^{-4}\,\text{Yr}$$

(c) $l$ for the formation of $Fr^{223}$

$$= \frac{98}{100}\times3.15\times10^{-2} = 3.087\times10^{-2}\,\text{Yr}^{-1}$$

(d) $\dfrac{N_{Th^{227}}}{N_{Fr^{223}}} = \dfrac{2}{98} \Rightarrow \dfrac{m_{Th^{227}}}{m_{Fe^{223}}} = \dfrac{2\times227}{98\times223} \neq \dfrac{1}{49}$

**52.** The net rate of formation of radioisotope,

$$+\frac{dn}{dt} = R - \lambda.N$$

After very long time, steady state will be achieved, at which $+\dfrac{dn}{dt} = 0$. Hence, $N = \dfrac{R}{\lambda}$.

**53.** $Pb^{212} \xrightarrow{t_{1/2}=8\text{hr}} Bi^{212} \xrightarrow{t_{1/2}=1\text{hr}}$

Time for maximum nuclei and hence, maximum activity of Bi212, $t_{max} = \dfrac{1}{\lambda_2-\lambda_1}\cdot\ln\dfrac{\lambda_2}{\lambda_1}$

$$= \frac{1}{\dfrac{\ln2}{1} - \dfrac{\ln2}{8}}\cdot\ln\frac{1/1}{1/8} = 3.429\text{hr}$$

$= 205.7$ min

**54.** $\lambda_{overall} = \lambda_\alpha + \lambda_\beta \Rightarrow \dfrac{0.693}{t_{1/2}} = \dfrac{0.693}{20} + \dfrac{0.693}{60}$

$\therefore t_{1/2} = 15$ min

$\therefore$ For 87.5% decay, $t = 3\times t_{1/2} = 45$ min

**55.** Average energy released

$$= \frac{0.05\times40 + 0.15\times80}{0.05+0.15} = 70 \text{ MeV}$$

## Nuclear Reactions

56. $_{92}U^{235} + \overset{1}{\underset{0}{n}} \longrightarrow {}_{54}Xe^{139} + {}_{38}Sr^{94} + 3\overset{1}{\underset{0}{n}}$

57. $_{25}Mn^{55} + \overset{1}{\underset{0}{n}} \longrightarrow {}_{25}Mn^{56} + \gamma$

58. $_4Be^9 + \underset{\text{(proton)}}{{}_1H^1} \longrightarrow {}_5B^{10} + \gamma$

59. $_{13}Al^{27} + \underset{(\alpha\text{-particle})}{{}_2He^4} \longrightarrow {}_{15}P^{30} + {}_0n^1$

60. Informative

61. Informative

62. Theory based

63. Theory based

64. Informative

65. $_{13}Al^{27} + {}_2He^4 \longrightarrow {}_{14}Si^{30} + {}_1H^1 (= X)$

$_{13}Al^{27} + {}_2He^4 \longrightarrow {}_{15}P^{30} + {}_0n^1 (= Y)$

$_{15}P^{30} \longrightarrow {}_{14}Si^{30} + {}_{+1}e^1 (= Z)$

## Section B (One or More than one Correct)

1. Theory based
2. $Q$-value is distributed between $\beta$-particle and anti-neutrino.
3. $_{92}U^{235} \xrightarrow{-\alpha} {}_{90}Th^{23} \xrightarrow{-\alpha} {}_{88}Ra^{227} \xrightarrow{-\beta} {}_{89}Ac^{227}$

$\downarrow -\beta$

$_{93}Np^{235} \xrightarrow{-\alpha} {}_{91}Pa^{231}$

$_{89}AC^{235}$ is not possible.

4. Activity is independent from all external factors.
5. Half-life of a radio isotope is its characteristic property, independent from all factors.
6. Actinum series: $_{92}U^{235} \rightarrow {}_{82}Pb^{207}$
7. Informative
8. $_{96}Cm^{242} + {}_2He^4 \longrightarrow {}_{97}Bk^{293} + 2\,{}_0n^1$ (Incorrect)

$_5B^{10} + {}_2He^4 \longrightarrow {}_7N^{13} + {}_0n^1$ (Correct)

$_7N^{19} + {}_0n^1 \longrightarrow {}_6C^{14} + {}_1H^1$ (Correct)

$_{19}Si^{28} + {}_1H^2 \longrightarrow {}_{15}P^{29} + {}_0n^1$ (Correct)

9. Neutron is projectile and proton is emitted particles.
10. Informative
11. $_{13}Al^{27} + {}_2He^4 \longrightarrow {}_{15}P^{30} + {}_0n^1$

$_6C^{12} + {}_1He^1 \longrightarrow {}_7N^{13} + \gamma$

$_{15}P^{30} \longrightarrow {}_{14}Si^{30} + {}_{+1}e^0$

$_{96}Au^{241} + {}_2He^4 \longrightarrow {}_{97}BK^{244} + {}_1H^1$

12. $_4Be^9 + \gamma \longrightarrow {}_4Be^8 + {}_0n^1$

$_4Be^9 + {}_1H^1 \longrightarrow {}_4Be^8 + {}_1H^2$

## Section C (Comprehensions)

### Comprehension I

1. $\Delta m = (8\times1.0072 + 8\times1.0086) - 16 = 0.1264u$

$\therefore$ B.E. per nucleon $= \dfrac{0.1264\times931.5}{16} = 7.36\,\text{MeV}$

2. $_8O^{16} \longrightarrow 4\,{}_2He^4$

$\Delta m = 15.9944 - 4\times4.0026 = -0.016u$

$\therefore$ Energy required in separation $= 0.016\times931.5 = 14.904\,\text{MeV}$

3. $_{10}Ne^{20} \longrightarrow {}_6C^{12} + 2\,{}_2He^4$

Energy required

$= (20\times8.03) - (12\times7.68 + 2\times4\times7.07)$

$= 11.88\,\text{MeV}$

## Comprehension II

**4.** $SC^{50} \longrightarrow \tau i^{50} + \beta + \bar{v}$

$$Q-\text{value} = (49.9516 - 49.94479) \times 931.5$$
$$= 6.34 \text{ MeV}$$

$\therefore$ K.E. of $\bar{v} = 6.34 - 0.80 = 5.54$ MeV

**5.** $$\lambda = \frac{LC}{\Delta E} = \frac{6.626\times10^{-39}\times3\times10^{8}}{(4.795-4.611)\times10^{6}\times1.6\times10^{-19}}$$
$$= 6.75\times10^{-12}$$

**6.** $Th^{228} \xrightarrow{-\alpha} Ra^{224} \xrightarrow{-\gamma} Ra^{224}$

$$Q\text{-value} = \left[(228.028726) - (224.020196 + 4.0026 + 4.0026)\right]$$
$$\times 931.6 - 217\times10^{-3}$$
$$= 5.307 \text{ MeV}$$

$$\therefore \text{K.E. of } \alpha\text{-particle} = \frac{224}{228}\times5.307$$
$$= 5.214 \text{ MeV}$$

## Comprehension III

**7.** Overall rate is the rate of slowest step and hence,

$T_{\text{required}} = 270$ days

**8.** At transient equilibrium, $N_B = \dfrac{\lambda_A}{\lambda_B - \lambda_A}\cdot N_A$

$$\text{or, } \frac{N_A}{N_B} = \frac{N_{Th}}{N_{Ra}} = \frac{\lambda_{Ra} - \lambda_{Th}}{\lambda_{Th}} = \frac{\dfrac{\ln 2}{3.64} - \dfrac{\ln 2}{1.913\times365}}{\dfrac{\ln 2}{1.913\times365}}$$
$$= \frac{190.8}{1}$$

**9.** At secular equilibrium, $N_B = \dfrac{\lambda_A}{\lambda_B}\cdot N_A$

$$\text{or, } \frac{N_A}{N_B} = \frac{N_{Ra}}{N_{Rn}} = \frac{\lambda_{Rn}}{\lambda_{Ra}} = \frac{\dfrac{\ln 2}{55}}{\dfrac{\ln 2}{3.65\times24\times3600}} = 5733.8$$

## Comprehension IV

**10.**

| | $U^{238} \longrightarrow$ | $Pb^{206}$ |
|---|---|---|
| Initial | $a$ mole | 0 |
| Present | $(a-x)$ mole | $x$ mole |

$$= \frac{59.5}{238} = \frac{12.875}{206}\times\frac{80}{100}$$

$\therefore a = 0.30$

Now,

$$t = \frac{1}{\lambda}\cdot\ln\frac{a}{a-x} = \frac{1}{1.52\times10^{-10}}\cdot\ln\frac{0.3}{0.25}$$
$$= 1.33\times10^{9} \text{ Yrs}$$

**11.**

| | $K^{40} \longrightarrow$ | $Ar^{40}$ |
|---|---|---|
| Initial | $a$ mole | 0 |
| Present | $(a-x)$ mole | $x$ mole |
| | $= 1$ | $= 10.3$ |

$\therefore a = 11.3$

$$\text{Now, } t = \frac{t_{1/2}}{\log 2}\cdot\log\frac{a}{a-x} = \frac{1.25\times10^{9}}{0.3}\cdot\log\frac{11.3}{1}$$
$$= 4.375\times10^{9} \text{ years}$$

**12.**

| | $U^{238} \longrightarrow$ | $Pb^{206}$ | $K^{40} \longrightarrow$ | $Ar^{40}$ |
|---|---|---|---|---|
| Initial | mole | 0 | $b$ mole | 0 |
| Present | $(a-x)$ mole | $x$ mole | $(b-y)$ mole | $y$ mole |

$$= \frac{0.86\times10^{-3}}{238} = \frac{0.15\times10^{-3}}{206} = \frac{10\times10^{-3}}{40} = \frac{1.6\times10^{-3}}{40}$$

$$t = \frac{(t_{1/2})_{U^{238}}}{\log 2}\cdot\log\frac{a}{a-x} = \frac{(t_{1/2})_{K^{40}}}{\log 2}\cdot\log\frac{b}{b-y}$$

$\therefore w = 1.7$ mg

## Comprehension V

**13.** Given in paragraph

**14.** For radioactive tracing, time should be comparable to $t_{1/2}$.

**15.** $T_1 = \frac{1}{\lambda} \cdot \ln \frac{c_1}{c}$ and $T_2 = \frac{1}{\lambda} \cdot \ln \frac{c_2}{c}$

As $c_1 > c_2, T_1 > T_2$ and $T_1 - T_2 = \frac{1}{\lambda} \cdot \ln \frac{c_1}{c_2}$

## Comprehension VI

**16.** Let the volume of blood be $V$ ml.

$$t_1 = 2 = T_{av} \cdot \ln \frac{N_0}{N_0 - N_1}$$

$$\Rightarrow 5 = \frac{15}{\log 2} \cdot \log \frac{1260}{15V/60}$$

$\therefore V = 4000$

**17.** $r_0 = \frac{1260 \times 60}{4000} = 18.9$ dpm per ml

Now, $\frac{r_0}{r_5} = \frac{r_5}{r_{10}}$

$$\Rightarrow \frac{18.9}{15} = \frac{15}{r_{10}}$$

$\Rightarrow r_{10} = 11.9$ dpm per ml

## Comprehension VII

**18.** Isotopes are B and E, C and F, D and G.

**19.** Mass number of H $= 230 - 4 \times 4 = 214$

**20.** $Z_A - 4 \times 2 + 3 \times 1 = 88$

$\Rightarrow Z_A = 93$

## Comprehension VIII

**21.** Net rate of formation $+\frac{dn}{dt} = \alpha - \lambda N$

$$\text{or, } \int_{N_0}^{N} \frac{dN}{\alpha - \lambda N} = \int_0^t dt$$

$$\Rightarrow N = \frac{1}{\lambda}\left[\alpha - (\alpha - \lambda N_0).e^{-\lambda t}\right]$$

**22.** $t = t_{1/2} = \frac{\ln 2}{\lambda}$ and $\alpha = 2N_0\lambda$

$\therefore N = 1.5\, N_0$

**23.** $t \to \infty$, then $N = \frac{\alpha}{\lambda} = 2N_0$

## Comprehension IX

**24.** $\lambda = \lambda_1 + \lambda_2$

$$\Rightarrow \frac{\ln 2}{t_{1/2}} = \frac{\ln 2}{24} + \frac{\ln 2}{8}$$

$\Rightarrow t_{1/2} = 6$ hours.

**25.** Activity of excreted material in 48 hours $\frac{N_1}{N_0} = 0.116$ and $T_{av} = 16$ sec. But as $T_c$ is simultaneously decaying with $t_{1/2}$ = 8 hrs. Final activity after 48 hrs $= \frac{24}{2^6} = 0.375 \mu ci.$

## Section D (Assertion – Reason)

**1.** $\frac{n}{p}$ ratio does not increases continuously.

**2.** Binding energy increases but the binding energy per nucleon first increases and then decreases.

**3.**

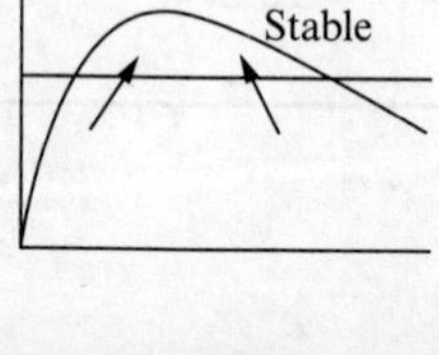

4. Informative
5. Theory based
6. All heavy nuclei should not produce ${}_{82}Pb^{206}$.
7. Theory based
8. $\beta$-decay occurs to decrease $\frac{n}{p}$ ratio.
9. Same mass of $U^{238}$ and $U^{238}F_6$ have different numbers of $U^{238}$ nuclei.
10. $\frac{t_{1/2}}{T_{av}} = \frac{0.693/\lambda}{1/\lambda} = 0.693 =$ Same for all
11. Mesons have mass 200 to 300 times mass of electrons.
12. Theory based
13. ${}_{13}Ae^{30}$ have high $\frac{n}{p}$ ratio than its stable nucleus ${}_{13}Ae^{27}$.
14. Theory based
15. Theory based

## Section E (Column Match)

1. Informative
2. Theory based
3. Theory based
4. ${}_{53}I^{127}$ is stable and hence, $I^{333}$ is beta emitter and $I^{121}$ is positron emitter.
5. (a) ${}_{92}U^{235} \longrightarrow {}_{82}Pb^{207}$

$$N_\alpha = \frac{235-207}{4} = 7$$
$$N_\beta = 82-(92-2\times 7) = 4$$

(b) ${}_{92}U^{238} \longrightarrow {}_{82}Pb^{206}$

$$N_\alpha = \frac{238-206}{4} = 8$$
$$N_\beta = 82-(92-2\times 8) = 6$$

(c) ${}_{94}Pu^{241} \longrightarrow {}_{83}Bi^{209}$

$$N_\alpha = \frac{241-209}{4} = 8$$
$$N_\beta = 83-(94-2\times 8) = 5$$

(d) ${}_{90}Th^{232} \longrightarrow {}_{82}Pb^{208}$

$$N_\alpha = \frac{232-208}{4} = 6$$
$$N_\beta = 82-(90-2\times 6) = 4$$

## Section F (Subjective)

### Single-digit Integer Type

1. Number of half-lifes $\frac{28.1}{2.81} = 10$

$\therefore$ Mass of $Sr^{90}$ remained

$$= \frac{2.048\times 10 \text{ gm}}{2^{10}} = 2\times 10^{-6} \text{ gm}$$

2. $r = \lambda N = \frac{0.7}{14\times 24\times 3600}$

$$\times\left(\frac{86.4\times 10^{-3}}{164}\times\frac{0.164}{100}\times 6\times 10^{23}\right)$$
$$= 3\times 10^{11} \text{ dps}$$

3. ${}_ZA^m \longrightarrow {}_{Z-6}B^{m-12} + 3\,{}_2He^4$

| | | |
|---|---|---|
| $t = 0$ | 1 mole | 0 |
| $t = 20$ days | $1-\frac{3}{4}$ | $3\times\frac{3}{4}$ mole |

$= \frac{1}{4}$ mole $\quad \therefore\ V = \frac{9}{4}\times 22.4$

$= 9\times 5.6\ L$ at 0°C and 1 atom

4. $\frac{1}{2}mV^2 = 0.335$ eV

$\therefore\ V=\sqrt{\dfrac{2\times0.335\times1.6\times10^{-19}}{1.675\times10^{-27}}}=8000\,\text{m/s}$

Hence, time for travelling 80 km,

$t=\dfrac{d}{v}=\dfrac{80\times10^{3}}{8000}=10\,\text{sec}$

Now, $t=\dfrac{t_{1/2}}{\ln 2}\cdot\ln\dfrac{N_0}{N}$

or, $10=\dfrac{700}{\ln 2}\cdot\dfrac{100}{100-x}$

$\Rightarrow x=0.99\approx 1$

5. $t=\left(\dfrac{t_{1/2}}{\log 2}\cdot\log\dfrac{N_0}{N}\right)_{U^{238}}=\left(\dfrac{t_{1/2}}{\log 2}\cdot\log\dfrac{N_0}{N}\right)_{U_{235}}$

or, $\dfrac{4.5\times10^{9}}{\log 2}\cdot\log\dfrac{N_0}{140x}=\dfrac{7.2\times10^{8}}{\log 2}\log\dfrac{N_0}{x}$

$\therefore\ \log\dfrac{N_0}{x}=2.5$

$\therefore$ Age of earth, $t=\dfrac{7.2\times10^{8}}{\log 2}\cdot\log\dfrac{N_0}{x}=6\times10^{9}$ years

## Four-digit Integer Type

1. $t=\dfrac{t_{1/2}}{\log 2}\cdot\log\dfrac{r_0}{r}$

or, $6.93=\dfrac{6.93}{\log 2}\cdot\log\dfrac{r_0}{5\times10^{15}}$

$\Rightarrow r_0=5.35\times10^{15}$ dpm

Now, $r_0=\lambda.N_0$

or, $5.35\times10^{15}=\dfrac{0.693}{69.3\times60}\times(n\times6\times10^{23})$

$\therefore\ n=5.35\times20^{-5}$

2. Initial number of $H^3$ atoms =

$\dfrac{0.93\times10^{-3}}{18}\times6\times10^{23}\times2\times8\times10^{-18}$

$=4.8\times10^{2}$

Number of half lives $=\dfrac{36.9}{12.3}=3$

$\therefore$ Final number of $H^3$ atoms

3. $t_1=2=T_{av}\cdot\ln\dfrac{N_0}{N_0-N_1}$

and $3t_1=6=T_{av}\cdot\ln\dfrac{N_0}{N_0-\frac{8}{3}N_1}$

$\therefore\ \dfrac{N_1}{N_0}=0.116$ and $T_{av}=16$ sec

4. Let the mass of water present in body = $w$ gm. Now, $9\times10^{9}=2.25\times10^{5}\times w\Rightarrow w=4\times10^{4}$ gm $=40$ kg

$\therefore$ Mass per cent of water in body $=\dfrac{40}{80}\times100=50\%$

5. $\lambda_{\beta-}=\dfrac{32}{100}\times\lambda_{\text{overall}}$

$\therefore\ (t_{1/2})_{\beta-}=\dfrac{100}{32}\times(t_{1/2})_{\text{overall}}=\dfrac{100}{32}\times12.8=40$ hr

6. $\lambda=\dfrac{1}{1620}+\dfrac{1}{405}$

$\Rightarrow \lambda=\dfrac{1}{324}Yr^{-1}$

$\therefore\ t_{\text{required}}=2\times t_{1/2}=2\times0.693\times324$

$=449.064$ years

7. $Sr^{90}\longrightarrow Y^{90}\longrightarrow$ other format

For radioactive equilibrium,

$(\lambda.N)_{Sr^{90}}=(\lambda.N)_{Y^{90}}$

or, $\dfrac{0.693}{32\times365\times24}\times\left(\dfrac{730}{90}\times N_A\right)$

$=\dfrac{0.693}{64}\times\left(\dfrac{w}{90}\times N_A\right)$

$\therefore\ w=0.1667$ gm

8. Energy released $=2\times120\times8.1-240\times7.2$
$=216$ MeV

**9.** $\Delta m = 2 \times 2.0021 - 4.0026 = 0.0016$ amu

Now, let $n$ moles of $H^2$ be required.

$$\frac{n}{2} \times 6 \times 10^{23} \times 0.0016 \times 1.5 \times 10^{-10} \times \frac{25}{100}$$

$$= 200 \times 10^6 \times 3600 \times 24$$

$\therefore n = 960$

**10.** After $\frac{2}{8} = \frac{2}{2}$ half-life, detectable activity $= \frac{100}{\sqrt{2}}\%$

But actual detected activity is 10%. Hence, mass of iodine migrated in thyroid gland

$$= 10\sqrt{2} \times \frac{10}{100\sqrt{2}} = 2 \text{ mg}$$